Cohomological Induction

and

Unitary Representations

Anthony W. Knapp

and

David A. Vogan, Jr.

Princeton University Press
Princeton, New Jersey
1995

Library of Congress Cataloging-in-Publication Data

Knapp, Anthony W., 1941–
    Cohomological induction and unitary representations / Anthony W.
Knapp and David A. Vogan, Jr.
        p.   cm. — (Princeton mathematical series ; 45)
    Includes bibliographical references (p.   –   ) and indexes.
    ISBN 0-691-03756-6 (acid-free)
    1. Semisimple Lie groups. 2. Representations of groups.
3. Homology theory. 4. Harmonic analysis. I. Vogan, David A.,
1954–   . II. Title. III. Series.
QA387.K565   1995
512′.55—dc20   94–48602 CIP

ISBN 0-691-03756-6

This book has been composed in Adobe Times-Roman, using caps-and-small-caps
from the Computer Modern group, MathTime Fonts created by
the TeXplorators Corporation, and mathematics symbol fonts and
Euler Fraktur fonts distributed by the American Mathematical Society

The publisher would like to acknowledge Anthony W. Knapp
and David A. Vogan, Jr., for providing the camera-ready copy
from which this book was printed

1 3 5 7 9 10 8 6 4 2

Cohomological Induction

and

Unitary Representations

Princeton Mathematical Series

EDITORS: LUIS A. CAFFARELLI, JOHN N. MATHER, *and* ELIAS M. STEIN

To Susan and Lois

CONTENTS

PREFACE

In the kind of analysis accomplished by representation theory, unitary representations play a particularly important role because they are the most convenient to decompose. However, only in rare cases does one use a classification to identify candidates for the irreducible constituents. More often one or more particular constructions will suffice to produce enough candidates.

In 1978 lectures, Gregg Zuckerman introduced a new construction, now called cohomological induction, of representations of semisimple Lie groups that were expected often to be irreducible unitary. Philosophically, cohomological induction is based on complex analysis in the same sense that George Mackey's construction of induced representations is based on real analysis. Zuckerman's construction thus serves as a natural complement to Mackey's, and it was immediately clear that the new method might go a long way toward explaining the most mysterious features of unitary representation theory. Zuckerman used his construction to produce algebraic models of the Bott-Borel-Weil Theorem, of Harish-Chandra's discrete series representations, of some special representations arising in mathematical physics, and of many more representations that are neither induced nor familiar.

Although Zuckerman's construction is based on complex analysis, it is in fact completely algebraic. In the complex-analysis setting the representation space is supposed to be a space of Dolbeault-cohomology sections over a noncompact complex homogeneous space of the group in question. But this setting turned out to be difficult to study in detail, and Zuckerman created an algebraic analog by abstracting the notion of passing to Taylor coefficients. If K is a maximal compact subgroup of the semisimple group G, the representation of G of interest is to be replaced by its subspace of vectors that transform in finite-dimensional spaces under K. This subspace, known as a $(\mathfrak{g}, K)$ module, is compatibly a representation space for K and the complexified Lie algebra $\mathfrak{g}$ of G.

Whether the construction is complex analytic or algebraic, the goal is to produce irreducible unitary representations. Zuckerman's representations, however, carry no obvious inner products, and construction of a candidate for the inner product is a serious project. By contrast, in Mackey's real-analysis theory of induced representations, the space of a representation is always the Hilbert space of square-integrable functions on some measure space, and the inner product is immediately at hand.

This book is an exposition of five fundamental theorems about co-homological induction, all related directly or indirectly to such inner products. We call them the Duality Theorem, the Irreducibility Theorem, the Signature Theorem, the Unitarizability Theorem, and the Transfer Theorem. The Introduction explains these theorems in the context of their history and motivation.

A chapter-by-chapter list of prerequisites for the book appears on p. xv. Roughly speaking, it is assumed for the first three chapters that the reader knows about elementary Lie theory, universal enveloping algebras, the abstract representation theory of compact groups, distributions on manifolds as in Appendix B, and elementary homological algebra as in Appendix C. Later chapters assume also the Cartan-Weyl theory for semisimple Lie algebras and compact connected Lie groups, some basic facts about real forms and parabolic subalgebras, and spectral sequences as in Appendix D.

Zuckerman introduced the Duality Theorem (§III.2 below) as a con-jecture, showing how it could be used to construct (possibly indefinite) Hermitian forms on cohomologically induced representations. With P. Trauber, he proposed several ideas toward proofs. Among other things, Zuckerman and Trauber showed how to write down the pairing in the Duality Theorem; what was not obvious was that the pairing was invariant under the representation. Enright-Wallach [1980]* gave a proof of this invariance, and therefore of the Duality Theorem.

Zuckerman's algebraic construction of $(\mathfrak{g}, K)$ modules via Taylor co-efficients uses a functor Γ that defines away the question of conver-gence. The functor Γ is not exact, only left exact, and the degrees of its derived functors play the role of the degrees of the Dolbeault cohomology classes. The initial definition of an invariant Hermitian form on a cohomologically induced representation involved a mixture of the derived functors of Γ and another algebraic construction ("ind" in Chapter 6 of Vogan [1981a]) that did not fit well. Zuckerman recognized that this combination was incongruous and searched for a right-exact functor Π to replace Γ. His search was unsuccessful, and a first version of Π was not announced until Bernstein [1983]. The use of a properly defined Π is critical to the approach taken in this book, and our definition is in terms of a change of rings.

We begin, following a 1970s idea of Flath and Deligne that was developed in Knapp-Vogan [1986], by introducing a "Hecke algebra" $R(\mathfrak{g}, K)$, which may be regarded as the set of bi-K finite distributions on

---

*A name followed by a bracketed year is an allusion to the list of References at the end of the book.

the underlying group G with support in the compact subgroup K. The set $R(\mathfrak{g}, K)$ is a complex associative algebra with an approximate identity, and $(\mathfrak{g}, K)$ modules coincide with "approximately unital" modules for $R(\mathfrak{g}, K)$. From this point of view the Zuckerman functor Γ becomes a Hom type change-of-ring functor of the kind studied in Cartan-Eilenberg [1956]. This fact immediately suggests using the corresponding tensor product change-of-ring functor as Π.

In fact, from the Hecke algebra point of view, the functors "ind" and "pro" in Chapter 6 of Vogan [1981a] are also change-of-ring functors constructed from $\otimes$ and Hom, respectively, and the same thing is true of the functors "coinvariants" and "invariants," whose derived functors give Lie algebra homology and cohomology. Thus there are really just two master functors in the theory, having to do with changes of rings by $\otimes$ and Hom. These functors are called P and I in this book because of their effect on projectives and injectives. Many fundamental results (including versions of Frobenius reciprocity) are consequences of standard associativity formulas for $\otimes$ and Hom.

With these general results in hand, Chapter V takes up the definition and first properties of cohomological induction. The functors $\mathcal{R}$ considered in Vogan [1981a] are built from Γ and pro, thus from the master functor I mentioned above. To construct Hermitian forms, it is essential to use instead $\mathcal{L}$, constructed analogously from Π and ind, thus from the master functor P.

Once invariant Hermitian forms have been constructed with the aid of the Duality Theorem, the question arises whether the forms are definite and hence are inner products. The Signature Theorem, proved in Chapter VI, addresses this question on that part of the cohomologically induced representation that is most easily related to the inducing representation. (The subspace in question is the "bottom layer" first considered in Speh-Vogan [1980].) The theorem says that cohomological induction always preserves a part of the signature of a Hermitian form. More precisely it identifies subspaces of the inducing and cohomologically induced representations and says that the Hermitian forms on these two subspaces have the same signature. An important feature of the Signature Theorem is that it makes no positivity assumption on the parameters of the inducing representation.

By contrast, the remaining three main theorems do include some positivity hypothesis. The Irreducibility Theorem (Chapter VIII) gives conditions under which cohomological induction carries irreducible representations to irreducible representations, and the Unitarizability Theorem (Chapter IX) gives conditions under which cohomological induction carries unitary representations to unitary representations. Zuckerman visualized the Irreducibility Theorem as a consequence of the Duality Theorem and gave a number of the ideas needed for a proof; all of the ideas are in place in Vogan [1981a]. The Unitarizability Theorem is newer and was first proved in Vogan [1984]. Together the Irreducibility Theorem and Unitarizability Theorem finally give confirmation that cohomological induction is actually a construction of irreducible unitary representations.

Once cohomological induction has constructed irreducible unitary representations, the question is what these representations are and how they can be related to each other. This topic is addressed in Chapter XI. A key tool in the investigation is the last of the five main theorems, the Transfer Theorem, which permits analysis of the effect of a "change of polarization" in constructions like cohomological induction. Consequently one can compare cohomological induction with Mackey induction and locate many cohomologically induced representations in the Langlands classification.

A few words are in order about the origins of this book. David Vogan sketched a proof of the Signature Theorem as early as 1984. Anthony Knapp began to study this sketch in 1985, in order to be able to use the result in some joint work with M. W. Baldoni-Silva. This study revealed various gaps and difficulties in the proof and in the literature on which it was based, and the first fruit of the study was Knapp-Vogan [1986]. Among other things, this preprint gave a rigorous development of the functor Π. The expected publication of the Signature Theorem was delayed because of other developments in the theory, and the authors eventually decided on a more complete treatment of cohomological induction. The present work may be regarded as a revision and extension of Knapp [1988] and Vogan [1981a].

For the most part, attributions of theorems appear in the end Notes. That section also mentions related papers and tells of some further results beyond those in the text.

The authors are grateful to Renée Burton for reading and criticizing extensive portions of the manuscript. The typesetting was by $\mathcal{A}_{\mathcal{M}}\mathcal{S}$-TEX. Knapp received financial support from a visiting position at the Massachusetts Institute of Technology and from National Science Foundation Grants DMS 85-01793 and DMS 91-00367, and Vogan received financial support from National Science Foundation Grants DMS 85-04029 and DMS 90-11483.

April 1994

PREREQUISITES BY CHAPTER

This book assumes knowledge of first-year graduate linear algebra, abstract algebra, and real analysis. Additional prerequisites are listed here by chapter. Most mathematics assumed for a particular chapter is generally needed for later chapters, as well.

CHAPTER I. Elementary Lie groups and Lie algebras. Universal enveloping algebra: universal mapping property, Poincaré-Birkhoff-Witt Theorem, symmetrization mapping. Haar measure on Lie groups. Abstract representation theory of compact groups: Schur orthogonality, Frobenius reciprocity, characters. Miscellaneous algebra as in Appendix A, distributions on manifolds as in Appendix B.

CHAPTER II. Elementary homological algebra as in Appendix C, §§1–2. Starting in §6: derived functors as in Appendix C, §§3–4.

CHAPTER III. No additional prerequisites.

CHAPTER IV. Roots and weights: semisimple Lie algebras over $\mathbb{C}$, roots, Weyl group, Theorem of the Highest Weight, highest weights in tensor products, complete reducibility of finite-dimensional representations, structure of Lie algebra of a compact group, normalizers and centralizers. Starting in §3: real semisimple Lie groups and Cartan involutions. Some acquaintance with Cartan subalgebras and parabolic subalgebras would be helpful. Starting in §7: Peter-Weyl Theorem, linearity of compact Lie groups. Starting in §8: Nullstellensatz.

CHAPTER V. Starting in §5: Euler-Poincaré Principle as in Appendix C, §5. Starting in §8: spectral sequences as in Appendix D.

CHAPTER VI. No additional prerequisites.

CHAPTER VII. Starting in §6: generic and regular elements in a complex Lie algebra, general Cartan subalgebras.

CHAPTERS VIII–X. No additional prerequisites.

CHAPTER XI. Structure theory of real reductive groups. Some acquaintance with parabolic induction and discrete series would be helpful.

CHAPTER XII. No additional prerequisites.

STANDARD NOTATION

Item	Meaning
$\#S$ or $|S|$	number of elements in S
$\emptyset$	empty set
E^c	complement of set, contragredient module
n positive	$n > 0$
$\mathbb{Z}, \mathbb{Q}, \mathbb{R}, \mathbb{C}$	integers, rationals, reals, complex numbers
Re z, Im z	real and imaginary parts of z
$\bar{z}$	complex conjugate of z
1	multiplicative identity
1 or I	identity matrix
dim V	dimension of vector space
V^*	dual of vector space
$V_{\mathbb{C}}$	complexification of vector space
Tr A	trace of A
A^t	transpose of A
$[A : B]$	index or multiplicity of B in A
$\bigoplus V_i$	direct sum of the V_i
$\langle S \rangle$	linear span of S
A^*	conjugate transpose of A
G_0	identity component of group G
GL, SL	general linear, special linear
O, SO	orthogonal, special orthogonal
U, SU	unitary, special unitary
Sp	symplectic
C^∞	infinitely differentiable
$\mathrm{Hom}_R(A, B)$	R linear maps of A into B
$\mathrm{End}_R(A)$	R linear maps of A into itself
$\mathrm{Aut}(V)$	automorphism group of V
Ad	adjoint representation of group
ad	adjoint representation of Lie algebra
$\ltimes$	semidirect product
$U(\mathfrak{g})$	universal enveloping algebra
$U_n(\mathfrak{g})$	n^{th} member of filtration of $U(\mathfrak{g})$
$S(\mathfrak{g})$	symmetric algebra
$S^n(\mathfrak{g})$	n^{th} homogeneous summand of $S(\mathfrak{g})$
σ	symmetrization from $S(\mathfrak{g})$ to $U(\mathfrak{g})$

Cohomological Induction

and

Unitary Representations

INTRODUCTION

This Introduction provides historical background and motivation for cohomological induction and gives an overview of the five main theorems. The section of Notes at the end of the book points to expositions where more detail can be found, and it gives references for the results that are cited. The Introduction is not logically necessary for the remainder of the book, and it occasionally uses mathematics that is not otherwise a prerequisite for the book.

The first part of the Introduction tells the sense in which representation theory of a semisimple Lie group G with finite center reduces to the study of "$(\mathfrak{g}, K)$ modules," and it describes the early constructions of infinite-dimensional group representations. One of the constructions is from complex analysis and produces representations in spaces of Dolbeault cohomology sections over a complex homogeneous space of G. This construction is expected to lead often to irreducible unitary representations. Passage to Taylor coefficients leads to an algebraic analog of this construction and to the definition of the left-exact Zuckerman functor Γ.

Cohomological induction involves more than the Zuckerman functor and its derived functors; it involves also the passage from a parabolic subalgebra of $\mathfrak{g}$ to $\mathfrak{g}$ itself. The original construction of Zuckerman's does not lend itself naturally to the introduction of invariant Hermitian forms, and for this reason a right-exact version of the Zuckerman functor, known as the Bernstein functor Π, is introduced. The definition of Π depends on introduction of a "Hecke algebra" $R(\mathfrak{g}, K)$, and Π is then given as a change of rings.

1. Origins of Algebraic Representation Theory

Harish-Chandra's first work in representation theory used particular representations of specific noncompact groups to address problems in mathematical physics. In the late 1940s, long after Élie Cartan and Hermann Weyl had completed their development of the representation theory of compact connected Lie groups, Harish-Chandra turned his attention to compact groups and reworked the Cartan-Weyl theory in his own way. Introducing what are now known as Verma modules, he gave a uniform, completely algebraic construction of the irreducible representations of such groups. (Chevalley independently gave a different such algebraic construction. See the Notes for details.)

Motivated by a question in Mautner [1950] of whether connected semisimple Lie groups are of "type I," and perhaps emboldened by the success with finite-dimensional representations, Harish-Chandra began an algebraic treatment of the infinite-dimensional representations of noncompact groups, concentrating largely on connected real semisimple groups. Although he initially allowed arbitrary connected semisimple groups, he eventually imposed the hypothesis that the groups have finite center, and we shall concentrate on that case. Let G be such a group.

Harish-Chandra worked with representations of G on a complex Banach space V, with the continuity property that the action $G \times V \to V$ is continuous. His early goal was to strip away any need for real or complex analysis with such representations and to handle them purely in terms of algebra.

If π is a continuous representation of G on the Banach space V, then the norms of the operators $\pi(x)$ are uniformly bounded for x in compact neighborhoods of 1, and we can average π by L^1 functions of compact support:

$$\pi(f)v = \int_G f(x)\pi(x)v \, dx \qquad \text{for } v \in V \text{ and } f \text{ compactly supported in } L^1(G).$$

Here dx is a Haar measure on G and is two-sided invariant because G is semisimple. The integral may be interpreted either as a vector-valued "Bochner integral" or as an ordinary Lebesgue integral for the value of every continuous linear functional $\langle \cdot, l \rangle$ on $\pi(f)v$, namely $\langle \pi(f)v, l \rangle = \int_G f(x)\langle \pi(x)v, l \rangle \, dx$.

We say that $v \in V$ is a C^∞ **vector** if $x \mapsto \pi(x)v$ is a C^∞ function from G into V. Gårding observed that the subspace $C^\infty(V)$ of C^∞ vectors is dense in V. In fact, if v is in V and if $f_n \geq 0$ is a sequence of compactly supported C^∞ functions on G of integral 1 and with support shrinking to $\{1\}$, then $\pi(f_n)v$ is a C^∞ vector for each n and $\pi(f_n)v \to v$.

Let $\mathfrak{g}_0$ be the Lie algebra of G. We can make $\mathfrak{g}_0$ act on the space of C^∞ vectors by the definition

$$\pi(X)v = \frac{d}{dt}\pi(\exp tX)v|_{t=0} \qquad \text{for } X \in \mathfrak{g}_0, \ v \in C^\infty(V),$$

and one can check that π becomes a representation of $\mathfrak{g}_0$:

$$\pi([X, Y]) = \pi(X)\pi(Y) - \pi(Y)\pi(X) \qquad \text{on } C^\infty(V).$$

As a consequence if $\mathfrak{g} = \mathfrak{g}_0 \otimes_\mathbb{R} \mathbb{C}$ denotes the complexification of $\mathfrak{g}_0$ and if $U(\mathfrak{g})$ denotes the universal enveloping algebra of $\mathfrak{g}$, then π extends uniquely from a linear map $\pi : \mathfrak{g}_0 \to \text{End}_\mathbb{C}(C^\infty(V))$ to a complex-linear algebra homomorphism $\pi : U(\mathfrak{g}) \to \text{End}_\mathbb{C}(C^\infty(V))$ sending 1 to 1. In this way $C^\infty(V)$ becomes a $U(\mathfrak{g})$ module.

Under this construction a group representation leads to a $U(\mathfrak{g})$ module in such a way that closed G invariant subspaces W yield $U(\mathfrak{g})$ submodules

$C^\infty(W)$. This correspondence, however, is inadequate as a reduction to algebra of the analytic aspects of representation theory. For one thing, simple examples show that the closure of a $U(\mathfrak{g})$ submodule of $C^\infty(V)$ need not be G invariant. For another, $U(\mathfrak{g})$ has countable vector-space dimension while $C^\infty(V)$ typically has uncountable dimension; thus $C^\infty(V)$ is usually not close to being irreducible (i.e., simple as a $U(\mathfrak{g})$ module) even if V is irreducible.

Harish-Chandra rectified the first problem by using the $U(\mathfrak{g})$ invariant subspace $C^\omega(V) \subseteq C^\infty(V)$ of **analytic vectors**, the subspace of v's for which $x \mapsto \pi(x)v$ is real analytic on G. It is not hard to check that the closure in V of a $U(\mathfrak{g})$ invariant subspace of $C^\omega(V)$ is G invariant. It is still true that $C^\omega(V)$ is dense in V, but this fact is much more difficult to prove than its C^∞ analog and we state it as a theorem.

Theorem 0.1 (Harish-Chandra). If π is a continuous representation of G on a Banach space V, then the subspace $C^\omega(V)$ of analytic vectors is dense in V.

To deal with the problem that $C^\infty(V)$ and even $C^\omega(V)$ are too large, Harish-Chandra made use of a maximal compact subgroup K of the semisimple group G. We say that $v \in V$ is K **finite** if $\pi(K)v$ spans a finite-dimensional space. The subspace of K finite vectors breaks into a (possibly infinite) direct sum of finite-dimensional subspaces on which K operates irreducibly. The subspace V_K of K finite vectors is dense in V, by an averaging argument similar to the proof that $C^\infty(V)$ is dense. Actually even the subspace $C^\omega(V)_K$ of K finite vectors in $C^\omega(V)$ is dense and is a direct sum of finite-dimensional subspaces on which K operates irreducibly.

It is a simple matter to show that $C^\omega(V)_K$ and $C^\infty(V)_K$ are $U(\mathfrak{g})$ sub-modules of $C^\infty(V)$. Thus $C^\infty(V)_K$ and $C^\omega(V)_K$ are both $U(\mathfrak{g})$ modules and representation spaces for K, and the $U(\mathfrak{g})$ and K structures evidently satisfy certain compatibility conditions. Following terminology introduced by Lepowsky [1973], we call $C^\infty(V)_K$ with its $U(\mathfrak{g})$ and K structures the **underlying** $(\mathfrak{g}, K)$ **module** of V. (See Chapter I for the precise definition of $(\mathfrak{g}, K)$ module.) For consistency of terminology, we often refer to the representation of π on V as "the representation V."

Even the correspondence $V \to C^\omega(V)_K$ is not one-one. For example, V might be the closure in a suitable norm of a G invariant space of functions on a homogeneous space of G. If the L^2 norm is used, one version of V results, while if an L^2 norm on the function and its first partial derivatives is used, another version of V results. Especially because of Theorem 0.6 below, it is customary to define away this problem. We say that V_1 and V_2 are **infinitesimally equivalent** if $C^\infty(V_1)_K$ and $C^\infty(V_2)_K$ are equivalent algebraically—i.e., if there is a $\mathbb{C}$ linear isomorphism of $C^\infty(V_1)_K$ onto $C^\infty(V_2)_K$ respecting the $U(\mathfrak{g})$ and K actions.

The reduction to algebra works best for representations that are irreducible or almost irreducible. Theorems 0.2 and 0.3 below prepare the setting. We say that V or its underlying $(\mathfrak{g}, K)$ module is **quasisimple** if the center $Z(\mathfrak{g})$ of $U(\mathfrak{g})$ operates as scalars in the $(\mathfrak{g}, K)$ module. Theorem 0.2 should be regarded as a version of Schur's Lemma.

Theorem 0.2 (Segal, Mautner). If V is an irreducible unitary representation of G on a Hilbert space, then V is quasisimple.

If τ is an irreducible finite-dimensional representation of K, let V_τ be the sum of all K invariant subspaces of V_K for which the K action under π is equivalent with τ. We say that V is **admissible** if each V_τ is finite-dimensional. From the denseness of $C^\omega(V)_K$ in V given in Theorem 0.1, it follows that $C^\omega(V)_\tau$ is dense in V_τ. Consequently if V_τ is finite-dimensional, then $C^\omega(V)_\tau = C^\infty(V)_\tau = V_\tau$.

Theorem 0.3 (Harish-Chandra). If V is an irreducible quasisimple representation of G on a Banach space, then V is admissible.

Thus admissibility of a $(\mathfrak{g}, K)$ module is a reasonable way to make precise the idea of being almost irreducible. Theorem 0.1 has the following easy but important consequence.

Theorem 0.4 (Harish-Chandra). If V is an admissible representation of G on a Banach space, then the closed G invariant subspaces W of V stand in one-one correspondence with the $U(\mathfrak{g})$ invariant subspaces S of $V_K = C^\infty(V)_K$, the correspondence $W \leftrightarrow S$ being

$$S = W_K \qquad \text{and} \qquad W = \overline{S}.$$

For many purposes it is the unitary representations that are of primary interest. In the case of a unitary representation π on a Hilbert space V with Hermitian inner product $\langle \cdot, \cdot \rangle$, we see immediately that the underlying $(\mathfrak{g}, K)$ module $C^\infty(V)_K$ has the properties that

(0.5)
$$\langle \pi(X)v_1, v_2 \rangle = -\langle v_1, \pi(X)v_2 \rangle \quad \text{for } X \in \mathfrak{g}_0 \text{ and } v_1, v_2 \in V$$
$$\langle \pi(k)v_1, \pi(k)v_2 \rangle = \langle v_1, v_2 \rangle \qquad \text{for } k \in K \text{ and } v_1, v_2 \in V.$$

In the reverse direction, we say that a Hermitian form $\langle \cdot, \cdot \rangle$ on a $(\mathfrak{g}, K)$ module is **invariant** if (0.5) holds. The $(\mathfrak{g}, K)$ module is **infinitesimally unitary** if it admits a positive definite invariant Hermitian form.

Classifying irreducible unitary representations is the same as classifying irreducible admissible infinitesimally unitary $(\mathfrak{g}, K)$ modules, as a result of the following theorem. The theorem is due to Harish-Chandra for G linear. For general G, extra steps due independently to Lepowsky and Rader are needed for the proof.

Theorem 0.6.

(a) Any irreducible admissible infinitesimally unitary $(\mathfrak{g}, K)$ module is the underlying $(\mathfrak{g}, K)$ module of an irreducible unitary representation of G on a Hilbert space.

(b) Two irreducible unitary representations of G on Hilbert spaces are unitarily equivalent if and only if they are infinitesimally equivalent.

2. Early Constructions of Representations

One of the fundamental problems in the representation theory of semisimple groups is to classify and categorize the irreducible unitary representations. Bargmann classified the irreducible unitary representations of

$$G = SL(2, \mathbb{R}) = \left\{ \begin{pmatrix} a & b \\ c & d \end{pmatrix} \middle| ad - bc = 1 \right\}$$

by classifying the candidates for underlying $(\mathfrak{g}, K)$ modules and then exhibiting unitary representations corresponding to each. About the time of Bargmann's work, Gelfand and Naimark classified the irreducible unitary representations of $SL(2, \mathbb{C})$ by using global methods. The representations for these two groups were later taken by other people as models for constructions in other semisimple groups.

Let us describe two of the series of representations obtained by Bargmann. The first of these, now known as the **principal series**, consists of one representation of $G = SL(2, \mathbb{R})$ for each parameter $(\pm, iv)$, where $\pm$ is a sign and iv is a purely imaginary complex number. The space in each case is $L^2(\mathbb{R})$, and the action for the representation with parameter $(\pm, iv)$ is

$$\left(\pi \begin{pmatrix} a & b \\ c & d \end{pmatrix} f \right)(x) = \begin{cases} |-bx + d|^{-1-iv} f(\frac{ax-c}{-bx+d}) & \text{if } + \\ \text{sgn}(-bx + d)|-bx + d|^{-1-iv} f(\frac{ax-c}{-bx+d}) & \text{if } - . \end{cases}$$

For the second series, now known as the **discrete series**, it is more convenient to work with the isomorphic group

$$G = SU(1, 1) = \left\{ \begin{pmatrix} \alpha & \beta \\ \bar{\beta} & \bar{\alpha} \end{pmatrix} \middle| |\alpha|^2 - |\beta|^2 = 1 \right\}.$$

This series consists of one representation of G for each parameter $(\pm, n)$, where $\pm$ is a sign and n is an integer ≥ 2. For the series with the sign $-$, the Hilbert space is the set of analytic f in the unit disc for which

$$\|f\|^2 = \int_{|z|<1} |f(z)|^2 (1 - |z|^2)^{n-2} \, dx \, dy < \infty,$$

and the action is

$$\left(\pi \begin{pmatrix} \alpha & \beta \\ \bar{\beta} & \bar{\alpha} \end{pmatrix} f \right)(z) = (-\bar{\beta}z + \alpha)^{-n} f\left(\frac{\bar{\alpha}z - \beta}{-\bar{\beta}z + \alpha} \right).$$

The series with the sign $+$ is obtained by taking the complex conjugate of $\begin{pmatrix} \alpha & \beta \\ \bar{\beta} & \bar{\alpha} \end{pmatrix}$ before applying π as above.

Meanwhile Mackey was developing a theory of induced representations for locally compact groups, and he was apparently the first to realize that the principal-series representations of $SL(2, \mathbb{R})$ were of this form. (See the Notes.) More particularly the principal series were a case of what we shall call "parabolic induction." An account of the general case appears in Chapter XI below. In the special case of $SL(2, \mathbb{R})$, define subgroups of $G = SL(2, \mathbb{R})$ by

$$K = \{k_\theta\} = \left\{ \begin{pmatrix} \cos\theta & \sin\theta \\ -\sin\theta & \cos\theta \end{pmatrix} \right\},$$

$$M = \left\{ \pm \begin{pmatrix} 1 & 0 \\ 0 & 1 \end{pmatrix} \right\}, \quad A = \left\{ \begin{pmatrix} e^t & 0 \\ 0 & e^{-t} \end{pmatrix} \right\}, \quad N = \left\{ \begin{pmatrix} 1 & x \\ 0 & 1 \end{pmatrix} \right\}.$$

To the parameter $(\pm, iv)$, we associate a one-dimensional character of M and the differential of a one-dimensional character of A by

$$\sigma \begin{pmatrix} \varepsilon & 0 \\ 0 & \varepsilon \end{pmatrix} = \begin{cases} \varepsilon & \text{if } \pm \text{ is } - \\ 1 & \text{if } \pm \text{ is } + \end{cases}$$

and

$$v \begin{pmatrix} t & 0 \\ 0 & -t \end{pmatrix} = ivt.$$

Then $man \mapsto e^{v \log a} \sigma(m)$ is a representation of the upper triangular subgroup MAN, and we shall define the corresponding classical induced representation $\tilde{\pi}$ of G. The definition of $\tilde{\pi}$ involves a shift in parameter by

$$\rho \begin{pmatrix} t & 0 \\ 0 & -t \end{pmatrix} = t.$$

in order to get the representation $\tilde{\pi}$ to be unitary. A dense subspace of the Hilbert space is

$$\{F : G \to \mathbb{C} \text{ of class } C^\infty \mid F(xman) = e^{-(\nu+\rho)\log a}\sigma(m)^{-1}F(x)\}$$

with

$$\|F\|^2 = \frac{1}{2\pi}\int_0^{2\pi} |F(k_\theta)|^2\, d\theta,$$

and the action is simply

$$(\tilde{\pi}(g)F)(x) = F(g^{-1}x).$$

The actual Hilbert space is the completion of the above dense subspace, with action by the continuous extension of each $\tilde{\pi}(g)$. The correspondence $F \mapsto f$ with the more classical realization is given by $f(y) = F\begin{pmatrix} 1 & 0 \\ y & 1 \end{pmatrix}$, except that a constant factor needs to be introduced if the correspondence is to be unitary.

Generalizing the principal series to semisimple groups is then just a matter of a little structure theory. The upper triangular subgroup gets replaced by a parabolic subgroup of G, this parabolic subgroup decomposes suitably as MAN, σ gets replaced by a suitable kind of irreducible unitary representation of M, and ν gets replaced by an imaginary-valued linear functional on the Lie algebra of A. The result is **parabolic induction**. When ρ is correctly generalized, parabolic induction carries unitary representations to unitary representations.

It was Harish-Chandra who found how to generalize the discrete series. While the generalization of the principal series used real analysis, the generalization of the discrete series required complex analysis. For the group $G = SU(1, 1)$, the analytic group of matrices with Lie algebra $\mathfrak{g}$ is $G_{\mathbb{C}} = SL(2, \mathbb{C})$. Within $G_{\mathbb{C}}$, let

$$B = \left\{\begin{pmatrix} a & 0 \\ c & a^{-1} \end{pmatrix}\right\}.$$

Then we readily check that

(a) every element of the subset $GB \subseteq G_{\mathbb{C}}$ has a unique decomposition as a product

(0.7) $\begin{pmatrix} 1 & z \\ 0 & 1 \end{pmatrix}\begin{pmatrix} \gamma^{-1} & 0 \\ 0 & \gamma \end{pmatrix}\begin{pmatrix} 1 & 0 \\ \zeta & 1 \end{pmatrix}$ with $\zeta \in \mathbb{C}$, $\gamma \in \mathbb{C}^\times$, $|z| < 1$,

and every matrix (0.7) is in GB.

(b) GB is an open subset of $SL(2, \mathbb{C})$, and its product complex structure obtained from (0.7) is the same as what it inherits from $G_{\mathbb{C}}$. In particular, left translation by any member of G is a holomorphic automorphism of GB.

To construct the generalizable version of the discrete series with parameter $(-, n)$, let ξ_n be the one-dimensional holomorphic representation of B given by

$$\xi_n \begin{pmatrix} a & 0 \\ c & a^{-1} \end{pmatrix} = a^{-n}.$$

The Hilbert space is taken as

(0.8)
$$\left\{ F : GB \to \mathbb{C} \;\middle|\; \begin{array}{l} \text{(i)} \;\; F \text{ is holomorphic} \\ \text{(ii)} \;\; F(xb) = \xi_n(b)^{-1}F(x) \text{ for } x \in GB, \, b \in B \\ \text{(iii)} \;\; \|F\|^2 = \int_G |F(x)|^2 \, dx < \infty \end{array} \right\},$$

and the action is

$$(\tilde{\pi}(g)F)(x) = F(g^{-1}x) \qquad \text{for } F \text{ as in (0.8)}, \, g \in G, \, x \in GB.$$

Except for a constant depending on the normalization of dx, the correspondence $F \mapsto f$ with the more classical realization is given by $f(z) = F(z, 1, 0)$ relative to the coordinates (0.7), and the inverse is $f \mapsto F$ with $F(z, \gamma, \zeta) = \gamma^{-n} f(z)$.

The generalization of this construction to other semisimple groups involves some special assumption on the group, and the resulting representations (when nonzero) are called the "holomorphic discrete series." If G is noncompact simple, the special assumption is that G/K is a complex manifold on which G operates holomorphically, and then G/K arises from the generalization of (0.7) as the z's allowed in the matrices $\begin{pmatrix} 1 & z \\ 0 & 1 \end{pmatrix}$. But Harish-Chandra phrased the condition in terms of roots. He assumed that a maximal torus of K is maximal abelian in G, and he hypothesized that in some ordering on the roots "every noncompact positive root is totally positive." With the terminology of roots stripped away, this condition says that the centralizer in G of the identity component of the center of K is K itself.

At any rate the condition is satisfied if G is compact connected, and Harish-Chandra's construction therefore gives global realizations of the irreducible representations of compact connected groups. Other authors came upon the same realizations of representations of compact connected groups independently at about the same time, starting from the point of view of algebraic geometry, and the result has come to be known as the Borel-Weil Theorem. Let us give a precise statement. The irreducible representations of a compact connected G are given by the Theorem of the Highest Weight, and we shall describe a global realization in terms of the highest weight as parameter. Regard G as a matrix group, let $G_\mathbb{C}$ be its complexification, introduce a maximal torus T of G, and fix a system of positive roots. Let B be the analytic subgroup of $G_\mathbb{C}$ whose Lie algebra contains the complexified Lie algebra $\mathfrak{t}$ of T, as well as the root spaces for all the negative roots. It turns out that GB is open and closed in $G_\mathbb{C}$ and hence $G_\mathbb{C} = GB$.

Theorem 0.9 (Borel-Weil Theorem). For the compact connected Lie group G, if $\lambda \in \mathfrak{t}^*$ is dominant and analytically integral and if ξ_λ denotes the corresponding holomorphic one-dimensional representation of B, then a realization of an irreducible representation of G with highest weight λ is in the space

(0.10)
$$\left\{ F : GB \to \mathbb{C} \;\middle|\; \begin{array}{l} \text{(i)} \ \ F \text{ is holomorphic} \\ \text{(ii)} \ \ F(xb) = \xi_\lambda(b)^{-1} F(x) \text{ for } x \in GB, \ b \in B \\ \text{(iii)} \ \ \|F\|^2 = \int_G |F(x)|^2 \, dx < \infty \end{array} \right\}$$

with G acting by

$$(\tilde{\pi}(g)F)(x) = F(g^{-1}x) \qquad \text{for } F \text{ as in (0.10), } g \in G, \ x \in GB.$$

Condition (iii) is automatic in the presence of (i), and we can drop it. Also we can replace GB by $G_\mathbb{C}$, but we prefer to emphasize the parallel with the construction for G noncompact by leaving GB in place.

From a geometric point of view the setting underlying the Borel-Weil Theorem is a bundle that we can view two ways

$$
\begin{array}{ccc}
GB & \longrightarrow & G_\mathbb{C} \\
\downarrow & & \downarrow \\
G/T & \longrightarrow & G_\mathbb{C}/B
\end{array}
$$

The left column is the one of interest for representations, and the map is the quotient by B. The horizontal maps are inclusions with image open since $G \cap B = T$, and they are onto since G is compact. The map on the right is the quotient map by the closed complex subgroup B. At the bottom right the quotient $G_\mathbb{C}/B$ is a complex manifold, and G/T therefore acquires an invariant complex structure. In a way that will be described in the next section, the functions F of (0.10) may be identified with the holomorphic sections of the holomorphic line bundle over G/T

associated to the character ξ_λ, and G acts on the space of sections in the natural way. In short, the irreducible representation with highest weight λ is realized as the space of global holomorphic sections of a certain holomorphic line bundle.

Suppose now that λ is analytically integral but no longer dominant. The space (0.10) still makes sense, but it is now zero, i.e., the holomorphic line bundle has no nontrivial sections. In order to find an interesting representation, we need an additional idea.

In complex geometry an operator $\bar\partial$ allows the introduction of a cohomology theory in such a way that the 0th-degree cohomology is just the space of global holomorphic sections. The operator $\bar\partial$ has a formula like that of the deRham d, except that $\dfrac{\partial}{\partial x_j}$ and dx_j get replaced by $\dfrac{\partial}{\partial \bar z_j}$ and $d\bar z_j$. We shall describe $\bar\partial$ more precisely in the next section.

In any event, in the holomorphic line bundle associated to the character ξ_λ, it is possible to speak of smooth $(0, k)$ cochain sections, and the image of one level of $\bar\partial$ is contained in the kernel of the next level. Let the representation space of the one-dimensional representation ξ_λ be denoted $\mathbb{C}_\lambda$, and let the spaces of cocycles and coboundaries be called

$$Z^{0,k}(G/T, \mathbb{C}_\lambda) \qquad \text{and} \qquad B^{0,k}(G/T, \mathbb{C}_\lambda),$$

respectively. The group G acts on these, and the quotient

$$H^{0,k}(G/T, \mathbb{C}_\lambda) = Z^{0,k}(G/T, \mathbb{C}_\lambda)/B^{0,k}(G/T, \mathbb{C}_\lambda)$$

is called the $(0, k)^{\text{th}}$ space of Dolbeault cohomology sections. From the point of view of representation theory, it is desirable to have a topology on these spaces such that the topology on $H^{0,k}(G/T, \mathbb{C}_\lambda)$ is obtained as the quotient topology from the other two. The Hilbert-space topology on $Z^{0,k}(G/T, \mathbb{C}_\lambda)$ from square integrability on G is not convenient, because use of the completion makes it necessary to address the meaning of $\bar\partial$ on nonsmooth cochain sections. But we shall be able to give $Z^{0,k}(G/T, \mathbb{C}_\lambda)$ a satisfactory C^∞ type topology below. Since $H^{0,k}(G/T, \mathbb{C}_\lambda)$ will be Hausdorff if and only if $B^{0,k}(G/T, \mathbb{C}_\lambda)$ is a closed subspace, it is important to know whether $B^{0,k}(G/T, \mathbb{C}_\lambda)$ is closed in $Z^{0,k}(G/T, \mathbb{C}_\lambda)$. For G compact it is indeed closed, but it is not a trivial matter to prove this fact. The Bott-Borel-Weil Theorem identifies the space $H^{0,k}(G/T, \mathbb{C}_\lambda)$. The notation is

(0.11)
$$\begin{aligned}
&\Delta = \{\text{roots of } (\mathfrak{g}, \mathfrak{t})\} \\
&\Delta^+ = \text{a positive system for } \Delta \\
&\delta = \tfrac{1}{2}\sum_{\alpha \in \Delta^+} \alpha \\
&W = \text{Weyl group of } \Delta \\
&B = \text{Borel subgroup built from } negative \text{ roots} \\
&G/T\text{'s complex structure from } G_{\mathbb{C}}/B.
\end{aligned}$$

Theorem 0.12 (Bott-Borel-Weil Theorem, first form).   With G compact and with notation as in (0.11), and let $\lambda \in \mathfrak{t}^*$ be integral.

(a) If $\langle \lambda + \delta, \alpha \rangle = 0$ for some $\alpha \in \Delta$, then $H^{0,k}(G/T, \mathbb{C}_\lambda) = 0$ for all k.

(b) If $\langle \lambda + \delta, \alpha \rangle \neq 0$ for all $\alpha \in \Delta$, let

(0.13) $$q = \#\{\alpha \in \Delta^+ \mid \langle \lambda + \delta, \alpha \rangle < 0\}.$$

Choose $w \in W$ with $w(\lambda + \delta)$ dominant, and put $\mu = w(\lambda + \delta) - \delta$. Then

$$H^{0,k}(G/T, \mathbb{C}_\lambda) = \begin{cases} 0 & \text{if } k \neq q \\ F^\mu & \text{if } k = q, \end{cases}$$

where F^μ is a finite-dimensional irreducible representation of G with highest weight μ.

Before taking up the detailed discussion of representations in spaces of Dolbeault cohomology sections, let us generalize our definition of representation suitably. Let V be a locally convex, complete, complex linear topological (Hausdorff) space. A (continuous) **representation** of the Lie group G on V is a homomorphism $\pi : G \to \text{Aut}\, V$ such that the map $G \times V \to V$ is continuous. With no change in the formalism, we can define the subspace $C^\infty(V)$ of C^∞ **vectors**. The assumption that V is locally convex and complete allows us to define the integral of a continuous function from a compact Hausdorff space X into V, with respect to a Borel measure on X. Taking X to be a compact neighborhood of 1 in G, we can apply Gårding's argument given above to see that $C^\infty(V)$ is dense in V.

For G semisimple with maximal compact subgroup K, we can again speak of the subspace V_K of K finite vectors. If τ is an irreducible finite-dimensional representation of K and if Θ_{τ^*} and d_τ are the character and degree of the contragredient of τ, define

$$\pi(\chi_\tau)v = \int_K d_\tau \Theta_{\tau^*}(k)\pi(k)v\, dk \qquad \text{for } v \in V.$$

Then $\pi(\chi_\tau)$ is a continuous projection whose image is V_τ and whose kernel contains all $V_{\tau'}$ for τ' inequivalent with τ. With this definition in place, we can argue that V_K is dense in V and that $C^\infty(V)_\tau$ is dense in V_τ.

A representation π on V as above is said to be **smooth** if $C^\infty(V) = V$. When a representation is given to us on a Banach space, the subspace of C^∞ vectors becomes a smooth representation if $C^\infty(V)$ is retopologized using the family of seminorms $\| \cdot \|_u$ parametrized by $u \in U(\mathfrak{g})$ and defined by $\|v\|_u = \|\pi(u)v\|$. Any smooth representation becomes a $U(\mathfrak{g})$ module under the definition $u \cdot v = \pi(u)v$, and its subspace of K finite vectors is a $(\mathfrak{g}, K)$ module called the **underlying** $(\mathfrak{g}, K)$ **module** of V.

3. Sections of Homogeneous Vector Bundles

This section describes a representation-theoretic construction by complex analysis that generalizes what happens for the holomorphic discrete series and the Bott-Borel-Weil Theorem. The expectation is that many of the resulting representations will be irreducible unitary and that we will therefore have a complex-analysis construction to complement the real-analysis construction given by parabolic induction. It is assumed that the reader is acquainted with some elementary structure theory of semisimple groups; discussion of this topic may be found in Chapter IV below. We shall make use of vector bundles in the construction. Although a full analytic theory requires understanding vector bundles with infinite-dimensional fiber, we shall restrict to the finite-dimensional case.

Throughout this section we work with the following setting, sometimes limiting ourselves to special cases: G is a connected linear reductive Lie group with complexification $G_{\mathbb{C}}$, K is a fixed maximal compact subgroup, T is a compact connected abelian subgroup of K (hence a torus), and $L = Z_G(T)$ is the centralizer of T in G. From Lemma 5.10 below, it is known that L is connected. Therefore the complexification $L_{\mathbb{C}}$ is meaningful as a subgroup of $G_{\mathbb{C}}$, namely the analytic subgroup of $G_{\mathbb{C}}$ with Lie algebra the complex subalgebra generated by the Lie algebra of L. Let Q be a parabolic subgroup of $G_{\mathbb{C}}$ with Levi factor $L_{\mathbb{C}}$.

We denote Lie algebras of Lie groups A, B, etc., by $\mathfrak{a}_0$, $\mathfrak{b}_0$, etc., and we denote their complexifications by $\mathfrak{a}$, $\mathfrak{b}$, etc. The complex Lie algebras of complex Lie groups $G_{\mathbb{C}}$, $L_{\mathbb{C}}$, Q are denoted $\mathfrak{g}$, $\mathfrak{l}$, $\mathfrak{q}$. We use an overbar to denote the conjugation of $\mathfrak{g}$ with respect to $\mathfrak{g}_0$.

We can decompose the Lie algebra $\mathfrak{q}$ of Q as a vector-space direct sum $\mathfrak{q} = \mathfrak{l} \oplus \mathfrak{u}$, where $\mathfrak{u}$ is the nilradical. Then $\mathfrak{u}$ and $\bar{\mathfrak{u}}$ are both nilpotent complex Lie algebras, and we have $[\mathfrak{l}, \mathfrak{u}] \subseteq \mathfrak{u}$ and $[\mathfrak{l}, \bar{\mathfrak{u}}] \subseteq \bar{\mathfrak{u}}$.

We assume that $\mathfrak{q}$ is a θ **stable parabolic**; this condition means that

$$(0.14a) \qquad\qquad \mathfrak{g}_0 \cap \mathfrak{q} = \mathfrak{l}_0.$$

It is equivalent to assume a vector-space direct-sum decomposition

$$(0.14b) \qquad\qquad \mathfrak{g} = \bar{\mathfrak{u}} \oplus \mathfrak{l} \oplus \mathfrak{u}.$$

Under the condition (0.14), the natural mapping $G/L \to G_{\mathbb{C}}/Q$ is an inclusion, and the image is an open set. Thus the choice of Q has made G/L into a complex manifold with G operating holomorphically.

A noncompact example to keep in mind is the group $G = U(m,n)$ of complex matrices that preserve an indefinite Hermitian form. Here $G_{\mathbb{C}} = GL(m+n, \mathbb{C})$. If we take T to be any closed connected subgroup of the diagonal of the form

$$T = \operatorname{diag}(e^{i\theta_1}, \ldots, e^{i\theta_1}, e^{i\theta_2}, \ldots, e^{i\theta_2}, \cdots, e^{i\theta_r}, \ldots, e^{i\theta_r}),$$

then L will be a block-diagonal subgroup within G with r blocks, and L will necessarily be connected. We can choose $\mathfrak{u}$ to be the complex Lie algebra of corresponding block-upper-triangular matrices and $\bar{\mathfrak{u}}$ to consist of the corresponding block-lower-triangular matrices.

We take as known that

$$(0.15) \qquad\qquad p : G \to G/L$$

is a C^∞ principal fiber bundle with structure group L. Let V be a finite-dimensional real or complex vector space, let $GL(V)$ be its general linear group, and let $\rho : L \to GL(V)$ be a C^∞ homomorphism. The **associated vector bundle**

$$(0.16\mathrm{a}) \qquad\qquad p_V : G \times_L V \to G/L$$

is a vector bundle with structure group $GL(V)$ whose bundle space is given by

$$(0.16\mathrm{b}) \qquad G \times_L V = \{(g,v)/\sim\} \qquad \text{with} \qquad (gl,v) \sim (g, \rho(l)v)$$

for $g \in G$, $l \in L$, and $v \in V$. Let $[(g,v)]$ denote the class of (g,v). We omit a description of the bundle structure.

The space of C^∞ sections of (0.16) is denoted $\mathcal{E}(G \times_L V)$. The group G acts on $G \times_L V$ by left translation: $g_0[(g,v)] = [(g_0 g, v)]$ in the notation of (0.16b). This action induces a well-defined action of G on $\mathcal{E}(G \times_L V)$ by $(g_0 \gamma)(gL) = g_0(\gamma(g_0^{-1} gL))$ for $\gamma \in \mathcal{E}(G \times_L V)$. When V is complex, this construction yields a representation of G (understood to be on a complex vector space). This representation is continuous in the sense that $(g_0, \gamma) \mapsto g_0 \gamma$ is continuous from $G \times \mathcal{E}(G \times_L V)$ to $\mathcal{E}(G \times_L V)$ if $\mathcal{E}(G \times_L V)$ is given its usual C^∞ topology. It is a smooth representation in the sense of §2.

Similarly

$$(0.17) \qquad\qquad p : G_{\mathbb{C}} \to G_{\mathbb{C}}/Q$$

is a holomorphic principal fiber bundle with structure group Q. In the above situation if V is complex and if ρ extends to a holomorphic homomorphism $\rho : Q \to GL(V)$, then we can construct an associated vector bundle

(0.18a) $$p_V : G_{\mathbb{C}} \times_{\mathbb{C}} V \to G_{\mathbb{C}}/Q$$

with bundle space given by

(0.18b) $G_{\mathbb{C}} \times_Q V = \{(g_{\mathbb{C}}, v)/\sim\}$ with $(g_{\mathbb{C}}q, v) \sim (g_{\mathbb{C}}, \rho(q)v)$.

The bundle (0.18) is a holomorphic vector bundle.

The inclusion $G/L \hookrightarrow G_{\mathbb{C}}/Q$ induces via pullback from (0.18a) a bundle map

(0.19) $$G \times_L V \hookrightarrow G_{\mathbb{C}} \times_Q V.$$

In terms of (0.16b) and (0.18b), this map is given simply by $(g, v) \mapsto (g, v)$. The result is that the C^∞ complex vector bundle $G \times_L V$ acquires the structure of a holomorphic vector bundle. We can regard the space of holomorphic sections $\mathcal{O}(G \times_L V)$ of $G \times_L V$ as a vector subspace of $\mathcal{E}(G \times_L V)$. (Actually less is needed about ρ than extendibility to Q in order to get the homomorphic structure on $G \times_L V$. See the Notes for details.)

To any section γ of $G \times_L V$ we can associate a function $\varphi_\gamma : G \to V$ by the definition

(0.20a) $$\gamma(gL) = [(g, \varphi_\gamma(g))] \in G \times_L V.$$

Under this correspondence, C^∞ sections γ go to C^∞ functions φ_γ, and we obtain an isomorphism

(0.20b)
$$\mathcal{E}(G \times_L V) \cong \left\{ \varphi : G \to V \;\middle|\; \begin{array}{l} \varphi \text{ of class } C^\infty, \\ \varphi(gl) = \rho(l)^{-1}\varphi(g) \text{ for } l \in L, g \in G \end{array} \right\}.$$

The group G acts on the right member of (0.20b) by the left regular action, and the isomorphism respects the actions by G. The usual C^∞ topology on $\mathcal{E}(G \times_L V)$ corresponds to the C^∞ topology on the space of φ's. It is under this correspondence that we can identify the functions F in (0.8) and (0.10) with sections of holomorphic line bundles.

The correspondence $\gamma \leftrightarrow \varphi_\gamma$ works locally as well, with sections over an open set $U \subseteq G/L$ corresponding to functions φ on the open subset $p^{-1}(U)$ of G transforming as in (0.20b). Again γ of class C^∞ corresponds to φ_γ of class C^∞. Let $\mathcal{E}(U)$ be the space of C^∞ sections over U.

In the special case that $G \times_L V$ admits the structure of a holomorphic vector bundle because of (0.19) and (0.18), we can speak of the space of holomorphic sections $\mathcal{O}(U)$ over an open set $U \subseteq G/L$. The proposition below tells how to use φ_γ to decide whether γ is holomorphic.

Proposition 0.21. Suppose that ρ extends to a holomorphic homomorphism $\rho : Q \to GL(V)$ and thereby makes $G \times_L V$ into a holomorphic vector bundle. Let $U \subseteq G/L$ be open, let γ be in $\mathcal{E}(U)$, and let φ_γ be the corresponding function from $p^{-1}(U)$ to V given by (0.20). Then γ is holomorphic if and only if

$$(0.22a) \qquad (Z\varphi_\gamma)(g) = -\rho(Z)(\varphi_\gamma(g))$$

for all $g \in p^{-1}(U)$ and $Z \in \mathfrak{q}$, with Z acting on φ_γ as a complex left-invariant vector field.

In typical applications to representation theory, ρ in the proposition is given on L and extends holomorphically to $L_{\mathbb{C}}$. The extension to Q is taken to be trivial on the unipotent radical of Q. Equation (0.22a) holds for $Z \in \mathfrak{l}_0$ for any C^∞ section, and it extends to $Z \in \mathfrak{l}$ by complex linearity. Thus (0.22a) may be replaced in this situation by the condition

$$(0.22b) \qquad Z\varphi_\gamma = 0 \qquad \text{for all } Z \in \mathfrak{u}.$$

The special case $\rho = 1$ shows how to recognize holomorphic functions on open subsets of G/L.

Let M be a complex manifold, and let p be in M. We denote by $T_p(M)$ the tangent space of M (considered as a C^∞ manifold) at p, consisting of derivations of the algebra of smooth germs at p, and we let $T(M)$ be the tangent bundle. Also we denote by $T_{\mathbb{C},p}(M)$ the complex vector space of derivations of the algebra of holomorphic germs at p, and we let $T_{\mathbb{C}}(M)$ be the corresponding bundle. There is a canonical $\mathbb{R}$ isomorphism

$$(0.23a) \qquad T_p(M) \xrightarrow{\sim} T_{\mathbb{C},p}(M)$$

given by

$$(0.23b) \qquad \xi \mapsto \zeta, \quad \text{where } \zeta(u + iv) = \xi(u) + i\xi(v).$$

Let J_p be the member of $GL(T_p(M))$ that corresponds under (0.23) to multiplication by i in $T_{\mathbb{C},p}(M)$. Then $J = \{J_p\}$ is a bundle map from $T(M)$ to itself whose square is -1.

The following proposition allows us to relate these considerations to associated vector bundles.

Proposition 0.24. There are canonical bundle isomorphisms

$$(0.25a) \qquad T(G/L) \cong G \times_L (\mathfrak{g}_0/\mathfrak{l}_0)$$

and

$$(0.25b) \qquad T_{\mathbb{C}}(G_{\mathbb{C}}/Q) \cong G_{\mathbb{C}} \times_Q (\mathfrak{g}/\mathfrak{q})$$

with L and Q acting on $\mathfrak{g}_0/\mathfrak{l}_0$ and $\mathfrak{g}/\mathfrak{q}$, respectively, by Ad.

The inclusion $G/L \subseteq G_{\mathbb{C}}/Q$ allows us to regard

$$(0.26) \qquad\qquad T_{\mathbb{C}}(G/L) \cong GQ \times_Q (\mathfrak{g}/\mathfrak{q}).$$

At any point $p = gL$ of G/L, the left sides of (0.25a) and (0.26), namely $T(G/L)$ and $T_{\mathbb{C}}(G/L)$, are $\mathbb{R}$ isomorphic via (0.23). It is easy to check that the corresponding isomorphism of the right sides of (0.25a) and (0.26) at p is given by

$$(g, X + \mathfrak{l}_0) \mapsto (g, X + \mathfrak{q}) \qquad \text{for } g \in G, \ X \in \mathfrak{g}_0.$$

This result allows us to compute the effect of J.

Complexifying (0.25a), we have

$$T(G/L)_{\mathbb{C}} \cong G \times_L (\mathfrak{g}_0/\mathfrak{l}_0)_{\mathbb{C}},$$

and J acts in the fiber at each point. We let $T(G/L)^{1,0}$ and $T(G/L)^{0,1}$ be the subbundles of $T(G/L)_{\mathbb{C}}$ corresponding to the respective eigenvalues i and $-i$ of J, so that

$$(0.27a) \qquad\qquad T(G/L)_{\mathbb{C}} \cong T(G/L)^{1,0} \oplus T(G/L)^{0,1}.$$

We have

$$(0.27b) \qquad\qquad (\mathfrak{g}_0/\mathfrak{l}_0)_{\mathbb{C}} \cong \mathfrak{g}/\mathfrak{l} \cong \bar{\mathfrak{u}} \oplus \mathfrak{u}$$

as L modules, and a little calculation shows that (0.27b) gives the decomposition of the fibers under J corresponding to (0.27a). In other words

$$
(0.27c) \qquad
\begin{aligned}
T(G/L)^{1,0} &\cong G \times_L \bar{\mathfrak{u}} \\
T(G/L)^{0,1} &\cong G \times_L \mathfrak{u}.
\end{aligned}
$$

Taking duals in (0.27a) and forming alternating tensors, we have

$$(0.28) \qquad \wedge^{p,q} T^*(G/L)_{\mathbb{C}} \cong G \times_L ((\wedge^p \bar{\mathfrak{u}})^* \otimes (\wedge^q \mathfrak{u})^*).$$

Via (0.28), members of $\mathcal{E}(\wedge^{p,q} T^*(G/L)_{\mathbb{C}})$ correspond to functions from G to $(\wedge^p \bar{\mathfrak{u}})^* \otimes (\wedge^q \mathfrak{u})^*$ transforming on the right under L by $\mathrm{Ad}^* \otimes \mathrm{Ad}^*$.

The scalar $\bar{\partial}$ operator for a complex manifold M is an operator

$$\bar{\partial} : \mathcal{E}(\wedge^{p,q} T^*(M)_{\mathbb{C}}) \to \mathcal{E}(\wedge^{p,q+1} T^*(M)_{\mathbb{C}}),$$

and it has $\bar{\partial}^2 = 0$. For the case that $M = G/L$, we can interpret $\bar{\partial}$ in terms of (0.28).

We can construct also a vector-valued version of $\bar{\partial}$. Namely let $G \times_L V$ be a holomorphic vector bundle as above. We introduce $\bar{\partial}_V = \bar{\partial} \otimes 1$ as an operator

$$\bar{\partial}_V : \mathcal{E}(\wedge^{p,q} T^*(G/L)_{\mathbb{C}} \otimes (G \times_L V)) \to \mathcal{E}(\wedge^{p,q+1} T^*(G/L)_{\mathbb{C}} \otimes (G \times_L V));$$

$\bar{\partial}_V$ is well defined because the transition functions for $G \times_L V$ are holomorphic. Also $\bar{\partial}_V^2 = 0$. Using (0.28) and dropping the subscript "V" on $\bar{\partial}_V$, we can interpret $\bar{\partial}_V$ as an operator

$$\bar{\partial} : \mathcal{E}(G \times_L ((\wedge^p \bar{\mathfrak{u}})^* \otimes (\wedge^q \mathfrak{u})^* \otimes V)) \to \mathcal{E}(G \times_L ((\wedge^p \bar{\mathfrak{u}})^* \otimes (\wedge^{q+1} \mathfrak{u})^* \otimes V)).$$

In representation theory one works with the case $p = 0$. We define

$$C^{0,q}(G/L, V) = \mathcal{E}(G \times_L ((\wedge^q \mathfrak{u})^* \otimes V)).$$

As always, this is the representation space for a continuous representation of G. The operator $\bar{\partial}$ is continuous and the kernel is closed. H.-W. Wong has shown (under the standing hypothesis of finite-dimensional V) that the image of $\bar{\partial}$ is closed and therefore that the quotient is Hausdorff. Thus we can define the **Dolbeault cohomology space** $H^{0,q}(G/L, V)$ as

$$(0.29) \qquad H^{0,q}(G/L, V) = \ker(\bar{\partial}|_{C^{0,q}(G/L,V)})/\mathrm{image}(\bar{\partial}|_{C^{0,q-1}(G/L,V)}).$$

Since $\bar{\partial}$ commutes with G, the topological vector space $H^{0,q}(G/L, V)$ carries a continuous representation of G.

The Bott-Borel-Weil Theorem identifies the spaces $H^{0,q}(G/L, V)$ of (0.29) in the case that G is compact. In this situation it has long been known that $\bar{\partial}$ has closed image. If one introduces a Hermitian inner product on V, then the formal adjoint $\bar{\partial}^*$ of $\bar{\partial}$ is meaningful, and it has long been known also that (0.29) can be computed alternatively as the representation on $\ker \bar{\partial} \cap \ker \bar{\partial}^*$ in $C^{0,q}(G/L, V)$. Members of $\ker \bar{\partial} \cap \ker \bar{\partial}^*$ are called **strongly harmonic**; this alternate approach shows that each cohomology class has exactly one strongly harmonic representative.

We have already stated the Bott-Borel-Weil Theorem in the special case that $L = T$ and $Q = B$. For general G/L with G compact and L the centralizer of a torus, the notation is

$$
\begin{aligned}
&G = \text{compact connected Lie group} \\
&T = \text{a torus in } G \\
&L = Z_G(T) \\
&T \text{ extended to a maximal torus } \tilde{T} \text{ in } L \\
&\Delta = \{\text{roots of } (\mathfrak{g}, \tilde{\mathfrak{t}})\} \\
(0.30)\quad &\Delta(\mathfrak{l}) = \{\text{roots of } (\mathfrak{l}, \tilde{\mathfrak{t}})\} \subseteq \Delta \\
&\Delta^+ \text{ chosen with } \Delta(\mathfrak{l}) \text{ generated by simple roots} \\
&\delta = \tfrac{1}{2} \sum_{\alpha \in \Delta^+} \alpha \\
&W = \text{Weyl group} \\
&Q = \text{built from } \mathfrak{l} \text{ and } \textit{negative} \text{ roots} \\
&G/L\text{'s complex structure from } G_{\mathbb{C}}/Q.
\end{aligned}
$$

Theorem 0.31 (Bott-Borel-Weil Theorem, second form). With G compact and with notation as in (0.30), let V^λ be irreducible for L with highest weight λ.

(a) If $\langle \lambda + \delta, \alpha \rangle = 0$ for some $\alpha \in \Delta$, then $H^{0,j}(G/L, V^\lambda) = 0$ for all j.

(b) If $\langle \lambda + \delta, \alpha \rangle \neq 0$ for all $\alpha \in \Delta$, define q as in (0.13), choose $w \in W$ so that $w(\lambda + \delta)$ is dominant, and put $\mu = w(\lambda + \delta) - \delta$. Then

$$
H^{0,j}(G/L, V^\lambda) = \begin{cases} 0 & \text{if } j \neq q \\ F^\mu & \text{if } j = q, \end{cases}
$$

where F^μ is a finite-dimensional irreducible representation of G with highest weight μ.

Historically the next cases of our construction to be considered were those for discrete-series representations. For a unimodular group G, an irreducible unitary representation π is in the **discrete series** if it is a direct summand of the right regular representation on $L^2(G)$, or equivalently if some (or equivalently every) nonzero matrix coefficient $(\pi(g)v_1, v_2)$ is in $L^2(G)$. Holomorphic discrete series for $SU(1, 1)$ as in (0.7) and (0.8) provide examples.

Let G be linear connected semisimple, and let K be a maximal compact subgroup. For G compact (so that $K = G$), every irreducible unitary representation is in the discrete series. For G noncompact, the discrete-series representations were parametrized by Harish-Chandra. We shall not recite the parametrization now, but it has features in common with the Theorem of the Highest Weight and the Weyl character formula. At this time we need to know only that discrete-series representations exist for G if and only if a maximal torus T of K is maximal abelian in G.

Langlands conjectured that all of Harish-Chandra's discrete series could be realized globally in a fashion similar to that in the Bott-Borel-Weil Theorem (with base space G/T). In making this conjecture, Langlands imposed square integrability on his allowable cocycles and coboundaries. The virtue of this choice is that it makes the conjecture correct (as was later shown by Schmid); the difficulty is that parallel square-integrability restrictions are not available in the general setting of (0.29) when L is noncompact.

The problem with allowing arbitrary cocycles and coboundaries can already be seen in $SU(1, 1)$. Since the unit disc is a Stein manifold, we can get nonzero cohomology only in degree 0 (by H. Cartan's Theorem B). Thus if we fix the one-dimensional holomorphic representation ξ_n of B, the interest is the space of functions $F : GB \to \mathbb{C}$ satisfying (i) and (ii) in (0.8). A feature of the theory of holomorphic discrete series is that all nonzero K finite F's satisfying (i) and (ii) also satisfy (iii), or else none do. When $n > 1$, (iii) holds and we get a unitary representation. But when $n \leq 1$, (iii) fails. For example, when $n = -1$, the space of functions F has a two-dimensional invariant subspace equivalent with the standard representation of $SU(1, 1)$, which is not unitary.

It would be nice to have a setting where the L^2 cohomology and the Dolbeault cohomology are compatible, and Schmid discovered such a setting. His idea was to adapt Δ^+ (and hence the complex structure) to the parameter, making the parameter dominant. Then the degree of interest for cohomology is $S = \dim_{\mathbb{C}}(K/T) = \dim_{\mathbb{C}}(\mathfrak{u} \cap \mathfrak{k})$. Under some hypotheses Schmid proved that the natural map from L^2 cohomology in degree S into Dolbeault cohomology is one-one.

If we rephrase the Bott-Borel-Weil Theorem with this idea in place, the notation is as follows: We let G, T, and Δ be as in (0.11). Let $\lambda_0 \in \mathfrak{t}^*$ be a given nonsingular parameter (λ_0 corresponds to $\lambda + \delta$ in Theorem 0.31), and suppose that $\lambda_0 - \delta_0$ is analytically integral for the half sum δ_0 of positive roots in some (or equivalently each) positive system. Define

$$\begin{aligned}
&\Delta^+ = \{\alpha \in \Delta \mid \langle \lambda_0, \alpha \rangle > 0\} \\
&\delta = \tfrac{1}{2} \sum_{\alpha \in \Delta^+} \alpha \\
&\lambda = \lambda_0 - \delta \\
\text{(0.32)} \quad &V^\lambda = \text{irreducible finite-dimensional representation} \\
&\qquad\quad \text{of } L \text{ with highest weight } \lambda \\
&Q \text{ built from } \mathfrak{l} \text{ and } \Delta^+ \text{ instead of } -\Delta^+ \\
&G/L\text{'s complex structure from } G_{\mathbb{C}}/Q.
\end{aligned}$$

Theorem 0.33 (Bott-Borel-Weil Theorem, third form). Let G be compact connected, with notation as in (0.32). Then

$$H^{0,j}(G/L, \; V^\lambda \otimes_{\mathbb{C}} \textstyle\bigwedge^{\mathrm{top}} \mathfrak{u}) = \begin{cases} 0 & \text{if } j \neq \dim_{\mathbb{C}}(G/L) \\ F^\lambda & \text{if } j = \dim_{\mathbb{C}}(G/L). \end{cases}$$

Schmid proved an analogous theorem about realizing discrete series. For Schmid's setting, G is a noncompact semisimple group, K is a maximal compact subgroup, and T is a maximal torus of K that is also maximal abelian in G. In this setting under the assumption that the parameter is dominant and very nonsingular, Schmid proved that $\bar{\partial}$ has closed image, that nonzero Dolbeault cohomology occurs only in degree $S = \dim_{\mathbb{C}}(K/T) = \dim_{\mathbb{C}}(\mathfrak{u} \cap \mathfrak{k})$, and that the smooth representation in degree S is infinitesimally equivalent with the expected discrete-series representation. Aguilar-Rodriguez extended Schmid's theorem to handle all discrete series.

Handling further cases of $H^{0,j}(G/L, V)$ presents formidable problems. One difficulty is in proving that $\bar{\partial}$ has closed image; this step was carried out for general G and finite-dimensional V by H.-W. Wong. Another difficulty is that $H^{0,j}(G/L, V)$ carries no obvious inner product. In parabolic induction, the inner products arise by integration, with the norm given by that for a vector-valued $L^2(K)$. However, $H^{0,j}(G/L, v)$ is a space of Dolbeault cohomology classes on a noncompact complex manifold. To construct an inner product analytically, one must show that the K finite cohomology classes have strongly harmonic representatives, face the fact that the L invariant Hermitian form on each fiber $(\bigwedge^j \mathfrak{u})^* \otimes_{\mathbb{C}} V$ may not be positive definite if L is noncompact, and prove that the strongly harmonic representatives of the K finite cohomology classes are square integrable on G/L. Except in isolated special cases chiefly in mathematical physics, the first progress in this direction was due to Rawnsley, Schmid, and Wolf, and came under various complex-analysis assumptions on G/L. Barchini, Knapp, and Zierau showed how to obtain strongly harmonic representatives with a mild real-analysis restriction on G/L, and Barchini was able to drop this restriction (retaining only the assumption of finite-dimensional fiber). Zierau has shown how in some cases square integrability on G/L may be deduced for the strongly harmonic representatives.

In any event, direct progress with the analytic setting has been slow in coming. Zuckerman's contribution, introduced in the next section, was to create an algebraic analog of this complex-analysis setting, thereby bypassing many of the analytic difficulties.

4. Zuckerman Functors

Zuckerman functors provide an algebraic analog of the complex-analysis construction in §3. They were introduced by Zuckerman in a series of lectures in 1978 and were developed further by Vogan [1981a]. For this section we use the following notation:

(0.34)
$$
\begin{aligned}
G &= \text{linear connected reductive Lie group} \\
K &= \text{a maximal compact subgroup} \\
T &= \text{a torus in } G \\
L &= Z_G(T) \\
Q &= \text{parabolic subgroup in } G_{\mathbb{C}} \text{ as in §1} \\
\mathfrak{q} &= \mathfrak{l} \oplus \mathfrak{u} \\
(\sigma, V) &= \text{smooth representation of } L.
\end{aligned}
$$

The space V can be infinite dimensional, but we shall treat it as finite dimensional for the current purposes of motivation. The representation (σ, V) gives us a representation of $\mathfrak{l}$, and we extend this to a representation of $\mathfrak{q}$ by making $\mathfrak{u}$ act as 0. It will be helpful for purposes of motivation to think of the representation of $\mathfrak{q}$ on V as coming from a holomorphic representation of Q on V, but this assumption can be avoided.

In the analytic setting, $\bar{\partial}$ is an operator

$$
(0.35) \qquad \bar{\partial} : \mathcal{E}(G \times_L ((\wedge^j \mathfrak{u})^* \otimes V)) \to \mathcal{E}(G \times_L ((\wedge^{j+1} \mathfrak{u})^* \otimes V)).
$$

Using the isomorphism (0.20), we regard $\bar{\partial}$ as an operator with domain equal to the space of smooth functions φ from G into $(\wedge^j \mathfrak{u})^* \otimes V$ satisfying

$$
(0.36) \qquad \varphi(gl) = (\mathrm{Ad}(l)^{-1} \otimes \sigma(l)^{-1}) \varphi(g) \qquad \text{for } g \in G,\ l \in L
$$

and with range equal to the corresponding space of functions into $(\wedge^{j+1} \mathfrak{u})^* \otimes V$.

In the algebraic analog we try to construct only the K finite vectors of $H^{0,j}$, thus obtaining a $(\mathfrak{g}, K)$ module. Let $\mathcal{C}(\mathfrak{g}, K)$ be the category of all $(\mathfrak{g}, K)$ modules.

The idea is to work with the Taylor coefficients at $g = 1$ of the function φ in (0.36), regarding each coefficient as attached to a left-invariant complex derivative (of some order) of φ at $g = 1$. Thus the idea of passing to Taylor coefficients gives us a linear map

$$
\varphi \mapsto \varphi^{\#} \in \mathrm{Hom}_{\mathbb{C}}(U(\mathfrak{g}), (\wedge^j \mathfrak{u})^* \otimes V).
$$

The transformation law (0.36) forces

(0.37) $\varphi^\# \in \mathrm{Hom}_\mathfrak{l}(U(\mathfrak{g}), (\wedge^j \mathfrak{u})^* \otimes V)$,

where $\mathfrak{l}$ acts on $U(\mathfrak{g})$ on the right. If we assume that φ is K finite, then the action of $L \cap K$ on the left of φ gives an action of $L \cap K$ on $\varphi^\#$ by $\mathrm{Hom}(\mathrm{Ad}, \mathrm{Ad}^* \otimes \sigma)$, and $\varphi^\#$ will be $L \cap K$ finite. Thus $\varphi^\#$ lies in a subspace that we denote

(0.38) $\mathrm{Hom}_\mathfrak{l}(U(\mathfrak{g}), (\wedge^j \mathfrak{u})^* \otimes V)_{L \cap K}$

to indicate the $L \cap K$ finiteness. On (0.38) we have a representation of $\mathfrak{g}$ (via the action of $U(\mathfrak{g})$ on the left) and the representation of $L \cap K$, and (0.38) is a $(\mathfrak{g}, L \cap K)$ module.

The passage from the space of φ's as in (0.36) to the space of $\varphi^\#$'s in (0.38) loses information because

  (a) φ need not be analytic, and hence $\varphi \mapsto \varphi^\#$ is not one-one
  (b) formal power series do not have to converge and convergent power series do not have to globalize, and hence $\varphi \mapsto \varphi^\#$ is not onto.

We can get around the difficulties in (a) and (b) by defining away the problem. Let $\Gamma = \Gamma^{\mathfrak{g},K}_{\mathfrak{g}, L \cap K}$ be the functor

$$\Gamma : \mathcal{C}(\mathfrak{g}, L \cap K) \to \mathcal{C}(\mathfrak{g}, K)$$

given by

$\Gamma(V) = $ sum of all finite-dimensional $\mathfrak{k}$ invariant subspaces
            of V for which the action of $\mathfrak{k}$ globalizes to K,

$\Gamma(\psi) = \psi|_{\Gamma(V)}$ if $\psi \in \mathrm{Hom}(V, W)$.

The functor Γ is covariant and left exact and is called the **Zuckerman functor**.

IDEA. Impose $\bar{\partial}$ between spaces

(0.39) $\Gamma(\mathrm{Hom}_\mathfrak{l}(U(\mathfrak{g}), (\wedge^j \mathfrak{u})^* \otimes V)_{L \cap K})$,

and take the kernel/image as a $(\mathfrak{g}, K)$ module analog of $H^{0,j}(G/L, V)$.

Let us bring in homological algebra, temporarily assuming that $L \subseteq K$. Then we make the following observations:

1) For the case $j = 0$ at least when V is finite dimensional, the condition that $\bar{\partial}\varphi^{\#} = 0$ is that $Z\varphi = 0$ for all $Z \in \mathfrak{u}$, in view of (0.22b). Thus the kernel/image space for $j = 0$ should be regarded as

$$(0.40) \qquad \Gamma(\mathrm{Hom}_{\mathfrak{q}}(U(\mathfrak{g}), V)_{L \cap K}).$$

2) Identification of (0.40) as the space of interest for $j = 0$ suggests looking at the sequence

(0.41)
$$0 \longrightarrow \mathrm{Hom}_{\mathfrak{q}}(U(\mathfrak{g}), V)_{L \cap K} \longrightarrow \mathrm{Hom}_{\mathfrak{l}}(U(\mathfrak{g}), (\wedge^0 \mathfrak{u})^* \otimes V)_{L \cap K}$$
$$\longrightarrow \mathrm{Hom}_{\mathfrak{l}}(U(\mathfrak{g}), (\wedge^1 \mathfrak{u})^* \otimes V)_{L \cap K} \longrightarrow \cdots$$

in the category $\mathcal{C}(\mathfrak{g}, L \cap K)$. In fact, it can be proved that (0.41) is an injective resolution of $\mathrm{Hom}_{\mathfrak{q}}(U(\mathfrak{g}), V)_{L \cap K}$ in the category $\mathcal{C}(\mathfrak{g}, L \cap K)$.

3) The category $\mathcal{C}(\mathfrak{g}, L \cap K)$ has enough injectives. Combining (2) and the idea above about (0.39), we see that the jth space of interest, namely the jth kernel/image of (0.39), is

$$(0.42) \qquad \Gamma^j(\mathrm{Hom}_{\mathfrak{q}}(U(\mathfrak{g}), V)_{L \cap K}),$$

where Γ^j is the jth right derived functor of Γ. (In fact, (0.42) is defined as the jth cohomology of the complex obtained by applying Γ to (0.41), since (0.41) is an injective resolution.)

4) The space (0.42) gives the underlying $(\mathfrak{g}, K)$ module of K finite vectors of $H^{0,j}(G/L, V)$ for the cases of compact groups and the discrete series. These results are due essentially to Zuckerman and are proved in Vogan [1981a].

These observations lead us to the second crucial idea.

IDEA. Even when L is not compact, *define* the jth space of interest to be (0.42).

In short, the Zuckerman construction is to pass from V in $\mathcal{C}(\mathfrak{l}, L \cap K)$ first to $\mathrm{Hom}_{\mathfrak{q}}(U(\mathfrak{g}), V)_{L \cap K}$ in $\mathcal{C}(\mathfrak{g}, L \cap K)$ and then to $\Gamma^j(\mathrm{Hom}_{\mathfrak{q}}(U(\mathfrak{g}), V)_{L \cap K})$ in $\mathcal{C}(\mathfrak{g}, K)$.

5. Cohomological Induction

Let G be connected semisimple with finite center. In keeping with the ideas of §4, we call

$$\mathcal{R}^j(Z) = \Gamma^j(\text{pro}_{\mathfrak{q},L\cap K}^{\mathfrak{g},L\cap K}(Z^{\#}))$$

a **cohomological induction** functor. Here

$$Z^{\#} = Z \otimes_{\mathbb{C}} \bigwedge^{\text{top}} \mathfrak{u} \quad \text{and} \quad \text{pro}_{\mathfrak{q},L\cap K}^{\mathfrak{g},L\cap K}(V) = \text{Hom}_{\mathfrak{q}}(U(\mathfrak{g}), V)_{L\cap K}.$$

The passage $Z \mapsto Z^{\#}$ is a normalization included to be consistent with the notation in the third form of the Bott-Borel-Weil Theorem (Theorem 0.33), and the compositions $\mathcal{R}^j$ carry $\mathcal{C}(\mathfrak{l}, L \cap K)$ to $\mathcal{C}(\mathfrak{g}, K)$.

What §4 shows is that the functors $\mathcal{R}^j$ provide a reasonable algebraic analog of the Dolbeault cohomology functors $H^{0,j}(G/L, Z^{\#})$. In order to discuss unitarity, we need to see how these functors affect Hermitian forms. Here we find an unpleasant surprise: $\mathcal{R}^j$ cannot be applied naturally to Hermitian forms. Roughly speaking, the problem is that

$$\text{pro}_{\mathfrak{q},L\cap K}^{\mathfrak{g},L\cap K}(Z^{\#}) = \text{Hom}_{\mathfrak{q}}(U(\mathfrak{g}), Z^{\#})_{L\cap K} \cong \text{Hom}_{\mathbb{C}}(U(\bar{\mathfrak{u}}), Z^{\#})_{L\cap K}$$

is simply too large to carry such a form. (Actually the imposed $L \cap K$ finiteness allows one to find invariant Hermitian forms on $\text{pro}_{\mathfrak{q},L\cap K}^{\mathfrak{g},L\cap K}(Z^{\#})$, but not in any natural way.) The only consolation is that the Dolbeault cohomology has a parallel problem: For Z finite-dimensional, it follows from the work of Wong that $H^{0,j}(G/L, Z^{\#})$ can carry an invariant Hermitian form only when the cohomology is finite dimensional. The forms arising in Schmid's construction of the discrete series, for example, are defined only on certain dense subspaces of cohomology.

So we start over. Suppose again that Z is an $(\mathfrak{l}, L \cap K)$ module. The first step is to regard $Z^{\#}$ as a $(\bar{\mathfrak{q}}, L \cap K)$ module on which $\bar{\mathfrak{u}}$ acts by 0. The second step is to apply an "algebraic induction" functor to form a $(\mathfrak{g}, L \cap K)$ module

$$\text{ind}_{\bar{\mathfrak{q}},L\cap K}^{\mathfrak{g},L\cap K}(Z^{\#}) = U(\mathfrak{g}) \otimes_{\bar{\mathfrak{q}}} Z^{\#} \cong U(\mathfrak{u}) \otimes_{\mathbb{C}} Z^{\#}.$$

The third step is to apply some projective version $\Pi_j = (\Pi_{\mathfrak{g},L\cap K}^{\mathfrak{g},K})_j$ of Γ^j to get a $(\mathfrak{g}, K)$ module:

$$\mathcal{L}_j(Z) = \Pi_j(\text{ind}_{\bar{\mathfrak{q}},L\cap K}^{\mathfrak{g},L\cap K}(Z^{\#})).$$

We refer to $\mathcal{L}_j$ also as a **cohomological induction** functor.

The geometric setting of §3 does not suggest what Π should be. Instead we look for a direct algebraic definition, aiming to have many maps associated with Π go in the opposite direction of maps for Γ and to have Π be right exact rather than left exact. With this goal in mind, Π should be related to "largest K finite quotients" in the same way that Γ is related to "largest K finite subspaces." But largest K finite quotients do not always exist, and the definition of Π takes a little care. The first rigorous definition of Π was given by Bernstein in 1983. Let us postpone discussion of what is necessary to the next section. Historically the original attacks on unitarizability took Theorem 0.44a below as a definition of Π and its derived functors, and we can use this somewhat unsatisfactory approach as an interim measure.

An invariant sesquilinear form on a module V arises from a map of V into its Hermitian dual V^h. (See §VI.2 below for the definition of V^h.) To carry an invariant Hermitian form from Z to $\mathcal{L}_j(Z)$, we need a procedure for passing from a map $Z \to Z^h$ to a map $\mathcal{L}_j(Z) \to [\mathcal{L}_j(Z)]^h$. We give this procedure one step at a time. In our three-step construction, the map $Z \to Z^h$ easily gives a map from $Z^\#$ to $[Z^\#]^h$ and then a map from the $(\bar{\mathfrak{q}}, L \cap K)$ module Z to the $(\mathfrak{q}, L \cap K)$ module $[Z^\#]^h$. The second and third steps are handled by the proposition and theorem that follow.

We say that the $(\mathfrak{l}, L \cap K)$ module Z has **finite length** if Z has a (finite) composition series whose quotients are irreducible. In this case, any irreducible representation of the compact group $L \cap K$ occurs in Z with only finite multiplicity (see Theorem 10.1 below).

Proposition 0.43. Suppose Z is an $(\mathfrak{l}, L \cap K)$ module. Then

(a) there is a natural $(\mathfrak{g}, L \cap K)$ map

$$\mathrm{ind}_{\bar{\mathfrak{q}},L\cap K}^{\mathfrak{g},L\cap K}(Z^\#) \to \mathrm{pro}_{\mathfrak{q},L\cap K}^{\mathfrak{g},L\cap K}(Z^\#)$$

that is nonzero if Z is nonzero,

(b) there is a natural isomorphism

$$[\mathrm{ind}_{\bar{\mathfrak{q}},L\cap K}^{\mathfrak{g},L\cap K}(Z^\#)]^h \cong \mathrm{pro}_{\mathfrak{q},L\cap K}^{\mathfrak{g},L\cap K}([Z^\#]^h),$$

(c) any nonzero invariant Hermitian form $\langle \cdot, \cdot \rangle_L$ on Z induces a nonzero invariant Hermitian form $\langle \cdot, \cdot \rangle_\mathfrak{g}$ on $\mathrm{ind}_{\bar{\mathfrak{q}},L\cap K}^{\mathfrak{g},L\cap K}(Z^\#)$.

This proposition is elementary and is addressed at the beginning of §VI.4 below. In particular, the composition of the map in (a), followed by pro of the map $Z^\# \to [Z^\#]^h$ and then the inverse of the map in (b), carries $\mathrm{ind}_{\bar{\mathfrak{q}},L\cap K}^{\mathfrak{g},L\cap K}(Z^\#)$ to its Hermitian dual and defines the form $\langle \cdot, \cdot \rangle_\mathfrak{g}$ in (c).

Theorem 0.44. Let $S = \dim \mathfrak{u} \cap \mathfrak{k}$. Then

(a) there is a natural isomorphism of functors

$$(\Pi_{\mathfrak{g},L\cap K}^{\mathfrak{g},K})_j \cong (\Gamma_{\mathfrak{g},L\cap K}^{\mathfrak{g},K})^{2S-j} \qquad \text{for } 0 \leq j \leq 2S,$$

(b) for $W \in \mathcal{C}(\mathfrak{g}, L \cap K)$, there is a natural isomorphism

$$[\Pi_j(W)]^h \cong \Gamma^j(W^h) \qquad \text{for } 0 \leq j \leq 2S,$$

(c) any invariant Hermitian form $\langle \cdot, \cdot \rangle_{\mathfrak{g}}$ on a $(\mathfrak{g}, L \cap K)$ module W induces an invariant Hermitian form $\langle \cdot, \cdot \rangle_G$ on $\Pi_S(W)$.

In the approach that we shall take in this book, parts (a) and (b) are substantially the Duality Theorem, the first main theorem of the book, which is proved in Chapter III below. Part (a) is most of Hard Duality, and part (b) is an instance of Easy Duality. Part (c) is then a formal consequence. If, as an interim measure as suggested above, (a) is taken as a definition of Π and its derived functors, then (b) is substantially the Duality Theorem in its original form as stated by Zuckerman and Enright-Wallach [1980].

Corollary 0.45. If Z is an $(\mathfrak{l}, L \cap K)$ module of finite length, then an invariant Hermitian form $\langle \cdot, \cdot \rangle_L$ on Z induces an invariant Hermitian form $\langle \cdot, \cdot \rangle_G$ on $\mathcal{L}_S(Z)$.

Recall that we have been seeking a complex-analysis construction (or an algebraic analog of one) yielding irreducible unitary representations and complementing the real-analysis construction of parabolic induction. We intend for cohomological induction with $\mathcal{L}_S$ to be that construction. Before considering how close we are to the desired goal, we mention one more theorem as background.

Theorem 0.46. If Z is an $(\mathfrak{l}, L \cap K)$ module of finite length, then

(a) $\text{ind}_{\bar{\mathfrak{q}},L\cap K}^{\mathfrak{g},L\cap K}(Z^\#)$ and $\text{pro}_{\mathfrak{q},L\cap K}^{\mathfrak{g},L\cap K}(Z^\#)$ have finite length, and they have the same irreducible composition factors and multiplicities

(b) all the $(\mathfrak{g}, K)$ modules $\mathcal{L}_j(Z)$ and $\mathcal{R}^j(Z)$ have finite length, and

$$\sum_j (-1)^j (\mathcal{L}_j(Z)) = \sum_j (-1)^j (\mathcal{R}^j(Z))$$

in the Grothendieck group of finite-length $(\mathfrak{g}, K)$ modules.

Part (a) is proved in §V.2 and §V.7 under a positivity hypothesis on Z, and the general case may be deduced from the special case by an argument with tensor products. The conclusion of finite length in (b) is proved in §V.2 and §V.4, and the identity in the Grothendieck group follows from part (a), Theorem 0.44a, and the long exact sequences for the derived functors of Π and Γ.

The discussion before Theorem 0.33 suggests aiming for interesting $(\mathfrak{g}, K)$ modules to occur as $\mathcal{R}^j(Z)$ with $j = S$, and Corollary 0.45 suggests that the $(\mathfrak{g}, K)$ module to consider for unitarity is $\mathcal{L}_j(Z)$ with $j = S$. Referring to Theorem 0.46, we see a way for $\mathcal{L}_S(Z)$ to match $\mathcal{R}^S(Z)$, namely that they be irreducible and that $\mathcal{L}_j(Z) = \mathcal{R}^j(Z) = 0$ for $j \neq S$. We are thus led to consider the following two problems.

PROBLEM A. Under what conditions can we conclude that $\mathcal{L}_j(Z)$ and $\mathcal{R}^j(Z)$ are 0 for $j \neq S$ and that $\mathcal{L}_S(Z)$ is irreducible?

PROBLEM B. When is $\mathcal{L}_S(Z)$ infinitesimally unitary?

For the most part, we shall need to assume some positivity condition on Z in order to make much progress. But there is one thing that can be said without assuming any positivity condition. Starting from the $(\mathfrak{l}, L \cap K)$ module Z, we can forget part of the action and regard Z as an $(\mathfrak{l} \cap \mathfrak{k}, L \cap K)$ module. The cohomological induction functor $\mathcal{L}_S$ for G has an analog for K given by

$$\mathcal{L}_S^K(Z) = (\Pi_{\mathfrak{k}, L \cap K}^{\mathfrak{k}, K})_S(\text{ind}_{\bar{\mathfrak{q}} \cap \mathfrak{k}, L \cap K}^{\mathfrak{k}, L \cap K}(Z^{\#})),$$

and this operates summand by summand on the irreducible constituents of the fully reducible $(\mathfrak{l} \cap \mathfrak{k}, L \cap K)$ module Z. A version of the third form of the Bott-Borel-Weil Theorem (Theorem 0.33) for $\mathcal{L}_S^K$ shows that an $L \cap K$ irreducible constituent of Z maps to an irreducible K representation or 0 depending on whether a certain translate of its highest weight is dominant for K. A K type (i.e., an equivalence class of irreducible representations of K) is said to be in the **bottom layer** if it occurs in $\mathcal{L}_S^K(Z)$. In §V.6 below it is shown that the **bottom-layer map**

$$\mathcal{B} : \mathcal{L}_S^K(Z) \rightarrow \mathcal{L}_S(Z) \qquad \text{given by} \qquad \Pi_S \circ \text{inclusion}$$

is one-one onto the full K isotypic subspaces for the K types of the bottom layer in $\mathcal{L}_S(Z)$.

The second main theorem of the book is the Signature Theorem. A special case of it says that if $\langle \cdot, \cdot \rangle_L$ is positive definite on the $L \cap K$

types of Z for which $\mathcal{L}_S^K$ is nonzero, then $\langle \cdot, \cdot \rangle_G$ is positive definite on the K types of the bottom layer in $\mathcal{L}_S(Z)$. More generally it says that an invariant notion of signature is preserved in passing from these $L \cap K$ types of Z to the K types of the bottom layer in $\mathcal{L}_S(Z)$.

6. Hecke Algebra and the Definition of Π

The beginnings of a definition of Π date back to Zuckerman's 1978 lectures and to ideas proposed at the time by Trauber and Borel. In connection with a possible proof of Hard Duality for Γ, Trauber and Borel suggested introducing the complex convolution algebra $R(\mathfrak{g}, K)$ of all left and right K finite distributions on G with support in K. In the same way that $\mathfrak{g}$ modules amount to the same thing as left $U(\mathfrak{g})$ modules in which 1 acts as 1, $(\mathfrak{g}, K)$ modules are identified with certain $R(\mathfrak{g}, K)$ modules. The algebra $R(\mathfrak{g}, K)$ usually does not have an identity, only an "approximate identity," and the condition that 1 act as 1 should be replaced by the condition "approximately unital," i.e., that, on each element of the module, members far out in the approximate identity act as 1. With this definition, $(\mathfrak{g}, K)$ modules amount to the same thing as left $R(\mathfrak{g}, K)$ modules that are approximately unital. (See §I.4 below.)

We call $R(\mathfrak{g}, K)$ the **Hecke algebra** for $(\mathfrak{g}, K)$. As is shown in Proposition 2.70 below, the functor $\Gamma = \Gamma_{\mathfrak{g}, L \cap K}^{\mathfrak{g}, K}$ is then given by

$$\Gamma(W) = \mathrm{Hom}_{R(\mathfrak{g}, L \cap K)}(R(\mathfrak{g}, K), V)_K.$$

In the language of homological algebra of rings and modules, Γ is a Hom type change-of-rings functor (except for the condition of K finiteness carried in the subscript K). The theory of change-of-rings functors suggests looking also at the corresponding tensor-product type functor, and this we may take as Π:

$$\Pi(W) = R(\mathfrak{g}, K) \otimes_{R(\mathfrak{g}, L \cap K)} W.$$

This is the definition that was used in Knapp-Vogan [1986]. Bernstein [1983] had earlier introduced an equivalent definition of Π in the course of investigating the correspondence between two different classifications of irreducible $(\mathfrak{g}, K)$ modules, and consequently we call Π the **Bernstein functor**.

The definitions of ind and pro as

$$\mathrm{ind}_{\bar{\mathfrak{q}}, L \cap K}^{\mathfrak{g}, K}(V) = U(\mathfrak{g}) \otimes_{U(\bar{\mathfrak{q}})} V$$

and

$$\mathrm{pro}_{\mathfrak{q}, L \cap K}^{\mathfrak{g}, K}(V) = \mathrm{Hom}_{U(\mathfrak{q})}(U(\mathfrak{g}), V)_{L \cap K}$$

appear to be further changes of rings, at first glance from $U(\mathfrak{q})$ or $U(\bar{\mathfrak{q}})$ to $U(\mathfrak{g})$. But use of $U(\mathfrak{q})$, $U(\bar{\mathfrak{q}})$, and $U(\mathfrak{g})$ as the rings ignores the operation of $L \cap K$. The changes of rings should be from the rings appropriate to $(\mathfrak{q}, L \cap K)$ or $(\bar{\mathfrak{q}}, L \cap K)$ modules to the algebra $R(\mathfrak{g}, L \cap K)$, which is appropriate for $(\mathfrak{g}, L \cap K)$ modules.

Here we encounter a complication. The definition of $R(\mathfrak{g}, K)$ in terms of distributions assumed that $\mathfrak{g}$ is the complexification of the Lie algebra $\mathfrak{g}_0$ of a group G in which K is a subgroup, and $(\mathfrak{q}, L \cap K)$ and $(\bar{\mathfrak{q}}, L \cap K)$ do not fit this description. Thus we cannot immediately define $R(\mathfrak{q}, L \cap K)$ and $R(\bar{\mathfrak{q}}, L \cap K)$ in terms of distributions. Of course, we could attempt a definition of $R(\mathfrak{q}, L \cap K)$ and $R(\bar{\mathfrak{q}}, L \cap K)$ as subalgebras of $R(\mathfrak{g}, L \cap K)$, but fixing a total $\mathfrak{g}$ in which to operate would surely result in trouble eventually.

Thus what is needed is an algebraic construction of $R(\mathfrak{g}, K)$. Early joint work of Knapp and Vogan on such a construction appears in Knapp [1988]. By separating the parallel and transverse parts of the distributions that appear in $R(\mathfrak{g}, K)$, we show in §I.4 that

$$(0.47a) \qquad R(\mathfrak{g}, K) \cong R(K) \otimes_{U(\mathfrak{k})} U(\mathfrak{g}),$$

where $R(K)$ denotes the algebra of left and right K finite distributions on K (which are simply the K finite functions times Haar measure). The trouble with the isomorphism (0.47a) is that the multiplication law is lost. By separating parts in the reverse order, however, we obtain a second isomorphism

$$(0.47b) \qquad R(\mathfrak{g}, K) \cong U(\mathfrak{g}) \otimes_{U(\mathfrak{k})} R(K).$$

Understanding the relationship between (0.47a) and (0.47b) leads to the multiplication rule, which can then be used to define an abstract version of $R(\mathfrak{g}, K)$.

Chapter I below gives a version of this algebraic construction that improves on what is in Knapp [1988]. With the construction in place, Chapter II takes up the question of change of rings. In an expression

$$\Pi^{\mathfrak{g},K}_{\mathfrak{g},L\cap K}(\mathrm{ind}^{\mathfrak{g},L\cap K}_{\bar{\mathfrak{q}},L\cap K}(V)) \qquad \text{with } V \in \mathcal{C}(\bar{\mathfrak{q}}, L \cap K),$$

both operations Π and ind are changes of rings, first from $R(\bar{\mathfrak{q}}, L \cap K)$ to $R(\mathfrak{g}, L \cap K)$ and then from $R(\mathfrak{g}, L \cap K)$ to $R(\mathfrak{g}, K)$. They can therefore be telescoped into a single change from $R(\bar{\mathfrak{q}}, L \cap K)$ to $R(\mathfrak{g}, K)$:

$$P^{\mathfrak{g},K}_{\bar{\mathfrak{q}},L\cap K}(V) = R(\mathfrak{g}, K) \otimes_{R(\bar{\mathfrak{q}},L\cap K)} V.$$

Similar remarks apply to Γ and pro. The one-step Hom type change-of-rings functor is

$$I^{\mathfrak{g},K}_{\mathfrak{q},L\cap K}(V) = \text{Hom}_{R(\bar{\mathfrak{q}},L\cap K)}(R(\mathfrak{g}, K), V)_K.$$

More generally we see that $P^{\mathfrak{g},K}_{\mathfrak{h},B}$ and $I^{\mathfrak{g},K}_{\mathfrak{h},B}$ make sense whenever $\mathfrak{h} \subseteq \mathfrak{g}$ and $B \subseteq K$ compatibly. In fact, the inclusions can be replaced by maps $i_{\text{alg}} : \mathfrak{h} \to \mathfrak{g}$ and $i_{\text{gp}} : B \to K$ with suitable compatibility properties. The extended definitions of the functors P and I for this situation are

$$P^{\mathfrak{g},K}_{\mathfrak{h},B}(V) = R(\mathfrak{g}, K) \otimes_{R(\mathfrak{h},B)} V$$

and
$$I^{\mathfrak{g},K}_{\mathfrak{h},B}(V) = \text{Hom}_{R(\mathfrak{h},B)}(R(\mathfrak{g}, K), V)_K.$$

Chapter II develops the theory in this generality. The functors P and I will have as special cases Π and Γ, ind and pro, $\Pi \circ \text{ind}$ and $\Gamma \circ \text{pro}$, and coinvariants and invariants. The derived functors of P and I will have as special cases Π_j and Γ^j, $(\Pi \circ \text{ind})_j \cong \Pi_j \circ \text{ind}$ and $(\Gamma \circ \text{pro})^j \cong \Gamma^j \circ \text{pro}$, and Lie algebra homology and cohomology. Thus P and I, along with their derived functors, are pervasive in the theory.

7. Positivity and the Good Range

Let us return to Problems A and B in §5. Again G is a connected semisimple Lie group with finite center, K is a maximal compact subgroup, and $\mathfrak{g}$ is the complexified Lie algebra of G. As mentioned, we need to assume some positivity condition on the $(\mathfrak{l}, L \cap K)$ module Z in order to make much progress on the two problems.

At the same time that Harish-Chandra was introducing Verma modules (see §1), he investigated the center $Z(\mathfrak{g})$ of the universal enveloping algebra $U(\mathfrak{g})$. Let $\mathfrak{h}$ be any Cartan subalgebra of $\mathfrak{g}$. Harish-Chandra introduced a map $\gamma_{\mathfrak{g}}$ from $Z(\mathfrak{g})$ into the symmetric algebra $S(\mathfrak{h})$ and showed that $\gamma_{\mathfrak{g}}$ is an algebra isomorphism of $Z(\mathfrak{g})$ onto the algebra of Weyl-group invariants in $S(\mathfrak{h})$. (See §IV.7 below.) In terms of this map he proved that every homomorphism $\chi : Z(\mathfrak{g}) \to \mathbb{C}$ is of the form $\chi = \chi_\lambda$ for some $\lambda \in \mathfrak{h}^*$, where

$$\chi_\lambda(z) = \lambda(\gamma_{\mathfrak{g}}(z)).$$

Moreover, $\chi_\lambda = \chi_{\lambda'}$ if and only if λ and λ' are in the same orbit of the Weyl group. (See §IV.8 below.)

We say that a $U(\mathfrak{g})$ module V has **infinitesimal character** λ if $Z(\mathfrak{g})$ operates by scalars in V and if the homomorphism $\chi : Z(\mathfrak{g}) \to \mathbb{C}$ defined by those scalars is $\chi = \chi_\lambda$. For any irreducible $U(\mathfrak{g})$ module V, a version of Schur's Lemma due to Dixmier (Proposition 4.87 below) says that V has an infinitesimal character.

For our situation with L and G, let $\mathfrak{h}$ be a Cartan subalgebra of $\mathfrak{l}$. Then $\mathfrak{h}$ is also a Cartan subalgebra of $\mathfrak{g}$, and infinitesimal characters for $\mathfrak{l}$ and $\mathfrak{g}$ can both be given as members of $\mathfrak{h}^*$. We shall assume from now on that our $(\mathfrak{l}, L \cap K)$ module Z has an infinitesimal character, as well as finite length.

Let us pause for some examples. Let $\Delta(\mathfrak{g}, \mathfrak{h})$ be the set of roots of $\mathfrak{g}$, and let $\Delta(\mathfrak{u})$ and $\Delta(\mathfrak{l}, \mathfrak{h})$ denote the subsets of roots whose root vectors lie in $\mathfrak{u}$ and $\mathfrak{l}$, respectively. If we introduce a positive system $\Delta^+(\mathfrak{l}, \mathfrak{h})$ for $\mathfrak{l}$, then we can take $\Delta(\mathfrak{u}) \cup \Delta^+(\mathfrak{l}, \mathfrak{h})$ as a positive system $\Delta^+(\mathfrak{g}, \mathfrak{h})$ for $\mathfrak{g}$. Let δ_L, $\delta(\mathfrak{u})$, and δ be half the sum of the members of $\Delta^+(\mathfrak{l}, \mathfrak{h})$, $\Delta(\mathfrak{u})$, and $\Delta^+(\mathfrak{g}, \mathfrak{h})$, respectively. Note that $\delta = \delta_L + \delta(\mathfrak{u})$. If Z is an irreducible finite-dimensional $\mathfrak{l}$ module with highest weight μ, then Z has infinitesimal character $\mu + \delta_L$. The unique weight of $\bigwedge^{\text{top}}\mathfrak{u}$ is $2\delta(\mathfrak{u})$, and thus, in this case, $Z^\#$ has highest weight $\mu + 2\delta(\mathfrak{u})$ and infinitesimal character $\mu + \delta + \delta(\mathfrak{u})$. The following proposition is proved below in §V.2.

Proposition 0.48. If the $(\mathfrak{l}, L \cap K)$ module Z has infinitesimal character λ, then $Z^\#$ has infinitesimal character $\lambda + 2\delta(\mathfrak{u})$, while

$$\text{ind}_{\bar{\mathfrak{q}}, L \cap K}^{\mathfrak{g}, L \cap K}(Z^\#), \quad \text{pro}_{\mathfrak{q}, L \cap K}^{\mathfrak{g}, L \cap K}(Z^\#), \quad \mathcal{L}_j(Z), \quad \text{and} \quad \mathcal{R}^j(Z)$$

all have infinitesimal character $\lambda + \delta(\mathfrak{u})$.

In order to formulate positivity conditions, let $\langle \cdot, \cdot \rangle$ denote the Killing form on $\mathfrak{g}_0$. More generally, if $\mathfrak{g}_0 = \mathfrak{k}_0 \oplus \mathfrak{p}_0$ is the Cartan decomposition of $\mathfrak{g}_0$ relative to $\mathfrak{k}_0$, we can use as $\langle \cdot, \cdot \rangle$ any $\text{Ad}(G)$ invariant nondegenerate symmetric bilinear form on $\mathfrak{g}_0$ that is negative definite on $\mathfrak{k}_0$, is positive definite on $\mathfrak{p}_0$, and has $\mathfrak{k}_0$ orthogonal to $\mathfrak{p}_0$. This form extends by complexification to all of $\mathfrak{g}$, by restriction to nondegenerate forms on both $\mathfrak{l}$ and $\mathfrak{h}$, and by dualization to $\mathfrak{h}^*$. The form is positive definite on the real span of the roots in $\mathfrak{h}^*$. For purposes of this Introduction, we make the following definition.

DEFINITION 0.49. With $\langle \cdot, \cdot \rangle$ as above, suppose that the $(\mathfrak{l}, L \cap K)$ module Z has an infinitesimal character λ. We say that Z or λ is in the **good range** or that Z is **good** (relative to $\mathfrak{q}$ and $\mathfrak{g}$) if

$$\text{Re}\langle \lambda + \delta(\mathfrak{u}), \alpha \rangle > 0 \qquad \text{for all } \alpha \in \Delta(\mathfrak{u}).$$

We say that Z or λ is **weakly good** if

$$\mathrm{Re}\langle \lambda + \delta(\mathfrak{u}), \alpha \rangle \geq 0 \qquad \text{for all } \alpha \in \Delta(\mathfrak{u}).$$

These definitions are independent of the choice of the form $\langle \cdot, \cdot \rangle$ on $\mathfrak{g}_0$, of the choice of Cartan subalgebra $\mathfrak{h}$ of $\mathfrak{l}$, and of the choice of λ from within its orbit under the Weyl group of $\mathfrak{l}$. A first answer to Problem A in §5 is as follows.

Theorem 0.50. Let Z be an $(\mathfrak{l}, L \cap K)$ module of finite length with infinitesimal character λ, and suppose that Z is weakly good. Then

  (a) $\mathcal{L}_j(Z) = \mathcal{R}^j(Z) = 0$ for $j \neq S$
  (b) $\mathcal{L}_S(Z) \cong \mathcal{R}^S(Z)$
  (c) Z irreducible implies $\mathcal{L}_S(Z)$ is irreducible or zero.

If Z is assumed actually to be good, then (c) can be strengthened to

  (c′) Z irreducible implies $\mathcal{L}_S(Z)$ is irreducible.

Parts (a) and (b) are given as a vanishing theorem in §V.7. Parts (c) and (c′) are essentially the Irreducibility Theorem (Theorem 8.2 below), the third main theorem of the book.

The need for the hypotheses in Theorem 0.50 can be understood already in the compact case. When G is compact and Z is not good, $\mathcal{L}_S(Z)$ is zero. If Z is not good and no root is orthogonal to $\lambda + \delta(\mathfrak{u})$, the vanishing result in (a) will fail.

A first answer to Problem B in §5 is as follows.

Theorem 0.51. Let Z be an $(\mathfrak{l}, L \cap K)$ module of finite length with infinitesimal character λ, let $\langle \cdot, \cdot \rangle_L$ be a nonzero invariant Hermitian form on Z, and let $\langle \cdot, \cdot \rangle_G$ be the corresponding invariant Hermitian form on $\mathcal{L}_S(Z)$. If Z is weakly good, then

  (a) $\langle \cdot, \cdot \rangle_L$ nondegenerate implies $\langle \cdot, \cdot \rangle_G$ nondegenerate
  (b) $\langle \cdot, \cdot \rangle_L$ positive definite implies $\langle \cdot, \cdot \rangle_G$ positive definite.

In particular, if Z is weakly good and Z is infinitesimally unitary, then $\mathcal{L}_S(Z)$ is infinitesimally unitary.

Part (a) is an observation in §VI.4 below. Part (b) is the Unitarizability Theorem (Theorem 9.1 below), the fourth main theorem of the book.

8. One-Dimensional Z and the Fair Range

We continue with the notation of §7. For special kinds of $(\mathfrak{l}, L \cap K)$ modules Z, some improvement is possible in Theorems 0.50 and 0.51. In this section we examine especially the case of one-dimensional Z. If λ' is the unique weight of Z, we write $Z = \mathbb{C}_{\lambda'}$. The infinitesimal character of $\mathbb{C}_{\lambda'}$ is $\lambda = \lambda' + \delta_L$, and the good range is given by $\langle \lambda + \delta(\mathfrak{u}), \alpha \rangle > 0$ for $\alpha \in \Delta(\mathfrak{u})$.

Let $\mathfrak{z}$ be the center of $\mathfrak{l}$. This is automatically a subspace of the Cartan subalgebra $\mathfrak{h}$.

DEFINITION 0.52. With $\langle \cdot, \cdot \rangle$ as above, suppose that the $(\mathfrak{l}, L \cap K)$ module Z has an infinitesimal character λ. We say that Z or λ is in the **fair range** or that Z is **fair** (relative to $\mathfrak{q}$ and $\mathfrak{g}$) if

$$\text{Re}\langle \lambda + \delta(\mathfrak{u}), \alpha|_{\mathfrak{z}} \rangle > 0 \qquad \text{for all } \alpha \in \Delta(\mathfrak{u}).$$

We say that Z or λ is **weakly fair** if

$$\text{Re}\langle \lambda + \delta(\mathfrak{u}), \alpha|_{\mathfrak{z}} \rangle \geq 0 \qquad \text{for all } \alpha \in \Delta(\mathfrak{u}).$$

When $Z = \mathbb{C}_{\lambda'}$ and $\lambda = \lambda' + \delta_L$, we have

$$\langle \lambda + \delta(\mathfrak{u}), \alpha|_{\mathfrak{z}} \rangle = \langle \lambda' + \delta(\mathfrak{u}), \alpha \rangle.$$

Thus the conditions "fair" and "weakly fair" say for all $\alpha \in \Delta(\mathfrak{u})$ that $\langle \lambda' + \delta(\mathfrak{u}), \alpha \rangle$ is > 0 or ≥ 0, respectively.

Whether or not Z is one-dimensional, it is not hard to see that if Z is in the good range, then Z is in the fair range. Also if Z is in the weakly good range, then Z is in the weakly fair range.

Unfortunately the "fair" hypothesis does not imply analogs of Theorems 0.50 and 0.51 of §7 in general. But here is a first hint that there are positive results to be found.

Theorem 0.53. If Z is a weakly fair one-dimensional $(\mathfrak{l}, L \cap K)$ module with infinitesimal character λ, then

(a) $\mathcal{L}_j(Z) = \mathcal{R}^j(Z) = 0$ for $j \neq S$
(b) $\mathcal{L}_S(Z) \cong \mathcal{R}^S(Z)$
(c) the action of $U(\mathfrak{g})$ on $\mathcal{L}_S(Z)$ extends naturally to an algebra $D(G_{\mathbb{C}}/Q)_{(\lambda + \delta(\mathfrak{u}))|_{\mathfrak{z}}}$ of "twisted differential operators," and $\mathcal{L}_S(Z)$, as a $D(G_{\mathbb{C}}/Q)_{(\lambda + \delta(\mathfrak{u}))|_{\mathfrak{z}}}$ module, is irreducible or zero.

Parts (a) and (b) are in §V.7 below, along with parts (a) and (b) of Theorem 0.50. Discussion of (c) begins in §VIII.5 below and continues in Chapter XII. The definition of $D(G_{\mathbb{C}}/Q)_{(\lambda+\delta(\mathfrak{u}))|_{\mathfrak{z}}}$ will not concern us at this time. The point is that $U(\mathfrak{g})$ can be enlarged to a naturally defined algebra that always acts irreducibly on $\mathcal{L}_S(Z)$ if $\mathcal{L}_S(Z) \neq 0$. However, $U(\mathfrak{g})$ itself sometimes acts irreducibly and sometimes acts reducibly. Strengthening the hypothesis "weakly fair" to "fair" in Theorem 0.53 does not yield a conclusion (c) that is closer to (c) or (c′) of Theorem 0.50; for example, one cannot guarantee that $\mathcal{L}_S(Z)$ is nonzero, or that it is irreducible or zero as a $(\mathfrak{g}, K)$ module.

There is again a parallel result for unitarity.

Theorem 0.54. If Z is a weakly fair one-dimensional infinitesimally unitary $(\mathfrak{l}, L \cap K)$ module with invariant form $\langle \cdot, \cdot \rangle_L$, then the corresponding form $\langle \cdot, \cdot \rangle_G$ on $\mathcal{L}_S(Z)$ is positive definite, and consequently $\mathcal{L}_S(Z)$ is infinitesimally unitary.

It is natural to try to understand what it is about one-dimensional representations that makes Theorems 0.53 and 0.54 work. Doing so involves looking in detail at the proofs, and we postpone this project to Chapter XII. Examination of the proof of Theorem 0.54 leads to the definition of "weakly unipotent" $(\mathfrak{l}, L \cap K)$ modules. When such a module Z is weakly fair, we obtain the same conclusion as in Theorem 0.54, that Z infinitesimally unitary implies $\mathcal{L}_S(Z)$ infinitesimally unitary. The situation with generalizing Theorem 0.53 is more complicated. The algebra $D(G_{\mathbb{C}}/Q)_{(\lambda+\delta(\mathfrak{u}))|_{\mathfrak{z}}}$ leads to "Dixmier algebras," and irreducibility is expressed in terms of them. The algebra $U(\mathfrak{g})$ maps into a Dixmier algebra, with very large image, and a conclusion of irreducibility of $\mathcal{L}_S(Z)$ is valid when $U(\mathfrak{g})$ maps onto the Dixmier algebra. These matters are discussed in Chapter XII below.

9. Transfer Theorem

Now that we have a construction that often yields irreducible unitary representations, we want to be able to use them. In practice, being able to use these representations requires understanding how they fit into a classification and understanding how they can be constructed in other ways.

The first ingredient in this analysis is to realize parabolic induction on the level of $(\mathfrak{g}, K)$ modules. In parabolic induction we induce a smooth or

Hilbert-space representation from a parabolic subgroup of G to G itself. The usual convention is to start with an irreducible representation of M, a one-dimensional representation of A, and the trivial representation of N, and then to proceed as in §2. Translating the data by a certain nonunitary one-dimensional representation e^ρ on A ensures that unitary representations of MA lead to unitary representations of G. Let $\mathfrak{l} = \mathfrak{m} \oplus \mathfrak{a}$ be the complexified Lie algebra of MA, let $\mathfrak{n}$ be the complexified Lie algebra of N, and put $\mathfrak{q} = \mathfrak{l} \oplus \mathfrak{n}$. If we write Z for the underlying $(\mathfrak{l}, L \cap K)$ module of the representation of MA and denote by $Z^\natural$ the effect of putting ρ in place, then the underlying $(\mathfrak{g}, K)$ module turns out to be

$$\Gamma_{\mathfrak{g},L \cap K}^{\mathfrak{g},K}(\mathrm{pro}_{\mathfrak{q},L \cap K}^{\mathfrak{g},L \cap K}(Z^\natural)).$$

This conclusion remains valid if Z is replaced by any $(\mathfrak{l}, L \cap K)$ module of finite length. Moreover

$$(\Gamma_{\mathfrak{g},L \cap K}^{\mathfrak{g},K})^j(\mathrm{pro}_{\mathfrak{q},L \cap K}^{\mathfrak{g},L \cap K})(Z^\natural)) = 0 \qquad \text{for } j > 0.$$

Thus the $(\mathfrak{g}, K)$ analog of parabolic induction is notationally similar to cohomological induction except on two points:

(a) the normalization $Z \mapsto Z^\natural$ is different from the earlier normalization $Z \mapsto Z^\#$.

(b) the representation of interest occurs in cohomology of degree 0 rather than degree S.

Let us drop the normalizations for the remainder of this Introduction. (In practice, we eventually want some normalization back in place in order to make unitary representations go to unitary representations.) Suppose $\mathfrak{q}$ is any parabolic subalgebra of $\mathfrak{g}$ and $\mathfrak{q}^-$ is the opposite parabolic. Let us suppose that $\mathfrak{q} \cap \mathfrak{q}^- = \mathfrak{l}$ is the complexification of a real Lie subalgebra $\mathfrak{l}_0$ of $\mathfrak{g}_0$. We write $\mathfrak{u}$ for the nilpotent radical of $\mathfrak{q}$, so that $\mathfrak{q} = \mathfrak{l} \oplus \mathfrak{u}$. By $L \cap K$ we mean any closed subgroup of K whose Lie algebra is $\mathfrak{l}_0 \cap \mathfrak{k}_0$ such that $\mathrm{Ad}(L \cap K)\mathfrak{u} \subseteq \mathfrak{u}$. Then we can form

(0.55) $(\Gamma_{\mathfrak{g},L \cap K}^{\mathfrak{g},K})^j(\mathrm{pro}_{\mathfrak{q},L \cap K}^{\mathfrak{g},L \cap K})(Z))$ and $(\Pi_{\mathfrak{g},L \cap K}^{\mathfrak{g},K})_j(\mathrm{ind}_{\mathfrak{q},L \cap K}^{\mathfrak{g},L \cap K})(Z))$.

The problem is to understand the $(\mathfrak{g}, K)$ modules (0.55), relating them to each other as $\mathfrak{u}$ and j vary. There are two tools for doing so, and they can then be iterated:

(a) the Transfer Theorem addresses a one-step change in $\mathfrak{u}$ in the special case that $\mathfrak{l}$ reduces to a Cartan subalgebra. Under a condition on Z, the theorem matches the module in degree j

for one choice of u with the module in degree $j + 1$ for another choice of u.

(b) the double-induction spectral sequence addresses what happens when two Γ type constructions or two Π type constructions are composed.

These results are made precise and proven in Chapter XI. The Transfer Theorem is the fifth main theorem of the book.

The Transfer Theorem leads to striking relationships among $(\mathfrak{g}, K)$ modules (0.55) when L is a Cartan subgroup of G. Under some restrictions on Z, such a $(\mathfrak{g}, K)$ module is called **standard**. Various classification theorems are formulated in terms of quotients or submodules of standard modules. One such is the Langlands classification, which realizes irreducible representations as the result of a three-step process consisting of

(i) construction of discrete series and "limits of discrete series"

(ii) passage to a standard representation by Mackey induction

(iii) extraction of an irreducible quotient or subrepresentation (depending on the particular version of the Langlands classification, and depending on the use of Π or Γ in the classification).

Step (i) is given by cohomological induction, and step (ii) is parabolic induction. The Transfer Theorem implies that the same $(\mathfrak{g}, K)$ modules result if one goes through a three-step process consisting of

(i) construction of a "principal-series" representation of a group "split modulo center," using Mackey parabolic induction

(ii) passage to a standard representation by cohomological induction

(iii) extraction of an irreducible subrepresentation or quotient (depending on the use of Π or Γ in the classification).

One of the uses of these results is to place cohomologically induced modules in the Langlands classification, at least in the weakly good range. Using the results, one can transfer the Signature Theorem to a theorem cast solely in terms of the Langlands classification. The transferred theorem is a powerful tool for exhibiting Langlands parameters that do not correspond to unitary representations.

CHAPTER I

HECKE ALGEBRAS

The goal of this chapter is to attach an associative algebra $R(\mathfrak{g}, K)$ to any pair $(\mathfrak{g}, K)$ in which $\mathfrak{g}$ is a complex Lie algebra and K is a compact Lie group with certain compatibility properties relative to $\mathfrak{g}$. The algebra $R(\mathfrak{g}, K)$ usually has no identity, but it does have an approximate identity. Representation theory suggests a natural definition of $(\mathfrak{g}, K)$ modules, and the category of $(\mathfrak{g}, K)$ modules gets identified with the category of left $R(\mathfrak{g}, K)$ modules. Familiar homological constructions from module theory may then be applied to representation theory.

Special cases of $R(\mathfrak{g}, K)$ are treated first: When $K = \{1\}$, $R(\mathfrak{g}, K)$ reduces to the universal enveloping algebra of $\mathfrak{g}$. When $\mathfrak{g}$ is the complexified Lie algebra of K, $R(\mathfrak{g}, K)$ reduces to the convolution algebra of distributions on K that transform in finite-dimensional spaces under K. More generally when K is a subgroup of a Lie group G whose complexified Lie algebra is $\mathfrak{g}$, $R(\mathfrak{g}, K)$ reduces to the distributions on G that are supported in K and transform in finite-dimensional spaces under K. In the completely general case, $R(\mathfrak{g}, K)$ is constructed from a convolution algebra of distributions on K and the universal enveloping algebra of $\mathfrak{g}$ by an operation called smash product, followed by passage to a quotient.

1. Distributions on Lie Groups

Our initial work with Hecke algebras will involve distributions of compact support on Lie groups. The necessary analytic background concerning distributions on manifolds appears in Appendix B. Here we simply recall some definitions and notation. For a manifold X, we let $C^\infty(X)$ be the space of complex-valued smooth functions on X, and we let $C^\infty_{\text{com}}(X)$ be the subspace of such functions with compact support. The space $C^\infty(X)$ has a natural topology given by a family of seminorms, and the members of the continuous dual $\mathcal{E}'(X)$ are called **distributions of compact support**. The operation of a distribution T on a function f will be denoted

$$(1.1) \qquad \langle T, f \rangle \qquad \text{or} \qquad \int_X f(x) \, dT(x).$$

The latter notation will be especially helpful when using Fubini's Theorem, which is given as Theorem B.20.

Turning to Lie groups, we begin with some general notation. Suppose G is a real Lie group. We write G_0 for the identity component of G, and $\mathfrak{g}_0 = \mathrm{Lie}(G)$ for the Lie algebra. The complexification of $\mathfrak{g}_0$ is called $\mathfrak{g}$, and the universal enveloping algebra of $\mathfrak{g}$ is called $U(\mathfrak{g})$.

Sometimes we start with a real Lie algebra $\mathfrak{g}_0$, without any group G, and we still use the notation $\mathfrak{g}$ and $U(\mathfrak{g})$. If we start with a complex Lie algebra $\mathfrak{g}$, we still use the notation $U(\mathfrak{g})$.

Write l and r for the left and right regular representations on $C^\infty(G)$:

$$(1.2) \qquad [l(g)f](x) = f(g^{-1}x) \qquad \text{and} \qquad [r(g)f](x) = f(xg).$$

These are smooth representations, and we use the same letters to denote the differentials, which are representations of $\mathfrak{g}_0$ or (by natural extension) of $U(\mathfrak{g})$. For example,

$$[r(Y)f](x) = \frac{d}{dt}(f(x \exp tY))|_{t=0}$$

for Y in $\mathfrak{g}_0$.

The representations l and r of G and $U(\mathfrak{g})$ preserve compactly supported functions and therefore give rise to regular representations on $\mathcal{E}'(G)$. Explicitly

$$(1.3) \quad \begin{aligned} \langle l(g)T, f \rangle &= \langle T, l(g^{-1})f \rangle & \text{and} & \quad \langle l(u)T, f \rangle = \langle T, l(u^t)f \rangle \\ \langle r(g)T, f \rangle &= \langle T, r(g^{-1})f \rangle & \text{and} & \quad \langle r(u)T, f \rangle = \langle T, r(u^t)f \rangle \end{aligned}$$

for $T \in \mathcal{E}'(G)$, $f \in C^\infty(G)$, $g \in G$, and $u \in U(\mathfrak{g})$. Here u^t refers to the antiautomorphism **transpose** of $U(\mathfrak{g})$ characterized by

$$(1.4) \qquad\qquad X^t = -X \qquad \text{and} \qquad (uv)^t = v^t u^t$$

for $X \in \mathfrak{g}$ and $u, v \in U(\mathfrak{g})$.

It is often convenient to identify $U(\mathfrak{g})$ with the space of distributions on G supported at the identity. When confusion is possible, we may write $\partial(u)$ for the distribution given by evaluation at 1 of the left-invariant differential operator corresponding to u:

$$(1.5) \qquad\qquad \langle \partial(u), f \rangle = (r(u)f)(1).$$

If we define $\partial(u)^t = \partial(u^t)$, we have defined an operation transpose on the subspace of $\mathcal{E}'(G)$ supported at the identity. We can extend this definition to all distributions as follows: For smooth f, let f^t be defined by $f^t(g) = f(g^{-1})$. Then **transpose** extends from $U(\mathfrak{g})$ to $\mathcal{E}'(G)$ by the formula

(1.6a) $$\langle T^t, f \rangle = \langle T, f^t \rangle.$$

Transpose has an interpretation in terms of push-forward as defined in (B.26). If $i : G \to G$ is the inversion map $i(x) = x^{-1}$, then

(1.6b) $$T^t = i_*(T).$$

In calculating with the distribution integrals of (1.1), it is convenient to insert a purely notational step at the end of (1.6a), replacing x by x^{-1} to obtain

(1.6c) $$\int_G f(x)\, dT^t(x) = \int_G f(x^{-1})\, dT(x) = \int_G f(x)\, dT(x^{-1}).$$

The third expression is just new notation for the second expression, but it makes a number of calculations easier.

Now we introduce convolution. If S and T are in $\mathcal{E}'(G)$, Theorem B.20 associates a distribution $S \times T$ in $\mathcal{E}'(G \times G)$. Let $m : G \times G \to G$ denote multiplication. Then the push-forward m_* of m is a mapping $m_* : \mathcal{E}'(G \times G) \to \mathcal{E}'(G)$. We define **convolution** as the composition of these operations:

(1.7a) $$S * T = m_*(S \times T).$$

More concretely

(1.7b) $$\langle S * T, f \rangle = \int_{G \times G} f(xy)\, dS(x)\, dT(y)$$

for $f \in C^\infty(G)$.

Proposition 1.8. With convolution as multiplication, $\mathcal{E}'(G)$ is an associative algebra with the **Dirac distribution** δ_1 (evaluation-at-1) as identity. The transpose mapping is an antiautomorphism of order 2.

PROOF. The only algebra property that requires checking is associativity of convolution. Let $m_1 : G \times G \times G \to G$ be given by

$$m_1(a, b, c) = m \circ (1, m)(a, b, c) = a(bc),$$

so that (B.27) implies

$$(m_1)_* = m_* \circ (1, m)_* = m_* \circ (1_*, m_*).$$

Then (1.7a) gives

$$R * (S * T) = m_*(R \times (S * T))$$
$$= m_*(R \times m_*(S \times T))$$
$$= m_* \circ (1_*, m_*)(R \times S \times T) = (m_1)_*(R \times S \times T).$$

Similarly if $m_2(a, b, c) = (ab)c$, then

$$(R * S) * T = (m_2)_*(R \times S \times T).$$

By associativity in G, $m_1 = m_2$. Thus $R * (S * T) = (R * S) * T$.

Certainly transpose has order 2. To see its effect on convolution, we let $i : G \to G$ be inversion and $s : G \times G \to G \times G$ be interchange of coordinates. The two maps $m_1 = i \circ m$ and $m_2 = m \circ (i, i) \circ s$ of $G \times G$ into G are equal, sending (a, b) to $(ab)^{-1}$ and $b^{-1}a^{-1}$, respectively. Thus $(m_1)_* = (m_2)_*$. By (1.6b) and (1.7a), we have

$$(m_1)_*(S \times T) = i_* \circ m_*(S \times T) = i_*(S * T) = (S * T)^t$$

$$(m_2)_*(S \times T) = m_* \circ (i, i)_* \circ s_*(S \times T) = m_*(T^t \times S^t) = T^t * S^t.$$

Hence $(S * T)^t = T^t * S^t$ as required. This completes the proof.

Under the definition (1.7), it is easy to check that multiplication in $U(\mathfrak{g})$ corresponds to convolution of distributions:

$$(1.9) \qquad\qquad \partial(uv) = \partial(u) * \partial(v).$$

In fact, we have

$$\langle \partial(u) * \partial(v), f \rangle = \int_{G \times G} f(xy) \, d\partial(u)(x) \, d\partial(v)(y)$$

$$= \langle \partial(u)_x, \langle \partial(v), f(x \cdot) \rangle \rangle \qquad \text{by Fubini's Theorem}$$

$$= \langle \partial(u)_x, r(v)l(x^{-1})f \rangle$$

$$= \langle \partial(u), r(v)f \rangle \qquad \text{since } r(v)l(x^{-1}) = l(x^{-1})r(v)$$

$$= r(u)r(v)f$$

$$= r(uv)f$$

$$= \langle \partial(uv), f \rangle.$$

Other useful special cases of convolution involve the Dirac distributions and are

$$(1.10) \qquad\qquad \begin{aligned} \delta_g * T &= l(g)T \\ T * \delta_{g^{-1}} &= r(g)T. \end{aligned}$$

These follow immediately from (1.7b) and Fubini's Theorem. Combining them, we obtain, for $u \in U(\mathfrak{g})$,

$$(1.11) \qquad \delta_g * \partial(u) * \delta_{g^{-1}} = l(g)r(g)\partial(u) = \partial(\mathrm{Ad}(g)u),$$

the second equality following by arguing first for $u \in \mathfrak{g}_0$ and then extending the result to $U(\mathfrak{g})$.

The algebra $\mathcal{E}'(G)$ together with the antiautomorphism transpose is for many purposes an excellent substitute for the group algebra in the case of Lie groups. Formula (1.9) shows that it contains $U(\mathfrak{g})$; it also contains the group G itself (as the Dirac distributions δ_x).

Once we fix a right Haar measure $d_r x$ on G, we can convert members h of $C^\infty_{\mathrm{com}}(G)$ into distributions $h\, d_r x$ in $\mathcal{E}'(G)$. The action of the distribution $h\, d_r x$ is given by integration:

$$(1.12) \qquad \langle h\, d_r x,\, f \rangle = \int_G f(x)h(x)\, d_r x.$$

We shall examine the effect of convolution on this special class of distributions. We have

$$
\begin{aligned}
\langle (h\, d_r x) * T,\, f \rangle &= \int_{G \times G} f(xy)h(x)\, d_r x\, dT(y) \\
&= \int_G \left[\int_G h(xy^{-1})\, dT(y) \right] f(x)\, d_r x \\
&= \int_G \left[\int_G (r(x^{-1})h^t)(y)\, dT(y) \right] f(x)\, d_r x.
\end{aligned}
$$

In other words,
$$(h\, d_r x) * T = \langle T, r(x^{-1})h^t \rangle\, d_r x.$$

It is customary to simplify this formula by defining $h * T$ directly:

$$(1.13a) \qquad (h * T)(x) = \langle T, r(x^{-1})h^t \rangle.$$

The right side of (1.13a) may be written also as $\langle T, h^t(\cdot x^{-1}) \rangle$, and Lemma B.23 says that this is a C^∞ function of x; in fact, the lemma says that $h * T$ makes sense as a C^∞ function even if h is only in $C^\infty(G)$, not in $C^\infty_{\mathrm{com}}(G)$.

In order to work similarly with convolution in the reverse order, we introduce the **modular function** $\Delta : G \to \mathbb{R}^+$:

$$d_r(t \cdot) = \Delta(t)\, d_r(\cdot).$$

The function Δ is a smooth homomorphism of G into $\mathbb{R}^+$. We have

$$\langle T * (h\, d_r x),\ f \rangle = \int_{G \times G} f(xy) h(y)\, dT(x)\, d_r y$$

$$= \int_G \left[\int_G \Delta(x^{-1}) h(x^{-1}y)\, dT(x) \right] f(y)\, d_r y$$

$$= \int_G \left[\int_G (\Delta^{-1} \cdot l(y) h')(x)\, dT(x) \right] f(y)\, d_r y.$$

In other words,

$$T * (h\, d_r x) = \langle T,\ \Delta^{-1} \cdot l(x) h' \rangle\, d_r x.$$

We can define $T * h$ directly as a function by the formula

(1.13b) $(T * h)(x) = \langle T,\ \Delta^{-1} \cdot l(x) h' \rangle.$

Lemma B.23 shows that $T * h$ is a C^∞ function, even if h does not have compact support.

A special case of (1.13) worth noting is that

(1.14) $\langle T, f \rangle = (f' * T)(1) = (T * \Delta^{-1} f')(1).$

Other special cases of (1.13) allow us to write the representations of $U(\mathfrak{g})$ on $C^\infty(G)$ as

(1.15) $\Delta^{-1} l(u)(\Delta f) = \partial(u) * f$ and $r(u) f = f * \partial(u').$

To prove (1.15), we use the identities

(1.16) $r(g) f' = (l(g) f)'$ and $r(u) f' = (l(u) f)',$

which are immediate from (1.2). The identity for $r(u) f$ in (1.15) follows from the computation

$$(f * \partial(u'))(x) = \langle \partial(u'), r(x^{-1}) f' \rangle = \langle \partial(u)', r(x^{-1}) f' \rangle$$

$$= \langle \partial(u), (r(x^{-1}) f')' \rangle = \langle \partial(u), l(x^{-1}) f \rangle$$

$$= r(u) l(x^{-1}) f(1) = l(x^{-1}) r(u) f(1) = r(u) f(x),$$

and the identity for $l(u) f$ follows from the computation

$$(\partial(u) * f)(x) = \langle \partial(u),\ \Delta^{-1} \cdot l(x) f' \rangle = r(u)(\Delta^{-1} \cdot l(x) f')(1)$$

$$= (l(u)(\Delta \cdot r(x) f))'(1) = l(u)(\Delta \cdot r(x) f)(1)$$

$$= (l(u) \Delta(x)^{-1} r(x)(\Delta f))(1) = \Delta(x)^{-1} l(u)(\Delta f)(x).$$

Unless G is unimodular ($\Delta \equiv 1$), the imbedding of $C_{\text{com}}^{\infty}(G)$ into $\mathcal{E}'(G)$ does not commute with transpose. Thus we cannot expect $(h * T)^t$ to equal $T^t * h^t$, for example. The correct identity for functions is a consequence of the formula $d_r(x^{-1}) = \Delta(x)^{-1} d_r x$ and is

$$(h \, d_r x)^t = \Delta^{-1} h^t \, d_r x.$$

Formulas (1.13a) and (1.13b) can be specialized further to the case where T is given by a function: $T = k \, d_r x$. Then (1.13a) becomes

$$h * (k \, d_r x) = \int_G h(xy^{-1})k(y) \, d_r y,$$

which is the traditional formula for a convolution of functions $h * k(x)$, and (1.13b) becomes

$$(k \, d_r x) * h = \int_G k(xy^{-1})h(y) \, d_r y,$$

which is also the traditional formula for convolution of functions.

What all this interaction between distributions and functions proves is that $C_{\text{com}}^{\infty}(G)$ is a module (on either side) for the associative algebra $\mathcal{E}'(G)$, with convolution as the product of a member of $\mathcal{E}'(G)$ and a member of $C_{\text{com}}^{\infty}(G)$.

2. Hecke Algebras for Compact Lie Groups

Throughout this section, let K be a compact Lie group. Suppose (π, V) is a representation of K in the algebraic sense; we make no topological assumptions yet. A vector $v \in V$ is called **formally K finite** if the subspace $\langle \pi(K)v \rangle$ generated by v under the action of K is finite dimensional. In this case the subspace carries a natural topology; if the representation of K on $\langle \pi(K)v \rangle$ is continuous (or, equivalently, smooth), we say that v is K **finite**. Set

(1.17) $$V_K = \left\{ v \in V \,\middle|\, \begin{array}{l} \dim\langle \pi(K)v \rangle < \infty, \text{ and } K \text{ acts} \\ \text{continuously on } \langle \pi(K)v \rangle \end{array} \right\},$$

the subspace of K **finite vectors** in V.

We say that (π, V) is **locally K finite** if $V_K = V$. It is clear that locally K finite representations are closed under passage to subrepresentations, quotients, and arbitrary direct sums. For any algebraic representation (π, V), the pair (π, V_K) is a representation of K, and it is locally K finite, i.e., $(V_K)_K = V_K$.

Let $\widehat{K}$ be the set of equivalence classes of finite-dimensional irreducible smooth representations of K. A member of $\widehat{K}$ is called a K **type**, and a finite-dimensional irreducible smooth representation of K is said to be of **type** γ if its equivalence class (in $\widehat{K}$) is γ.

Proposition 1.18. Let (π, V) be a locally K finite representation.

(a) The space V is the (possibly infinite) algebraic direct sum of finite-dimensional K invariant subspaces on which π acts irreducibly. In fact, if V_γ is the sum of all irreducible subspaces of V of type γ, then V_γ is isomorphic to a (possibly infinite) algebraic direct sum of irreducible representations of type γ, and

(1.19)
$$V = \bigoplus_{\gamma \in \widehat{K}} V_\gamma.$$

(b) If W is any K invariant subspace, then there exists a K invariant subspace W' such that $V = W \oplus W'$.

REMARKS. When V_γ is written as a direct sum of irreducible representations of type γ, the number of summands is called the **multiplicity** of γ in (π, V). The subspace V_γ is called the K **isotypic** component of type γ.

PROOF. Let $\mathbb{C}[K]$ be the abstract **group algebra** of K over $\mathbb{C}$, which is an associative algebra over $\mathbb{C}$ with a basis $\{v_k\}$ parametrized by the elements k of K and with a multiplication law $v_k v_{k'} = v_{kk'}$. It is immediate that the unital left $\mathbb{C}[K]$ modules are the arbitrary representations of K, and the locally K finite representations of K form a good category $\mathcal{C}$ in the sense of §A.1. We apply the theory of §A.3 to this $\mathcal{C}$. For $v \in V$, $\langle \pi(K)v \rangle$ is a finite-dimensional continuous representation of K and is completely reducible, by the representation theory of compact groups. By Proposition A.8 every locally K finite representation of K is completely reducible. The rest of (a) follows from Corollary A.16a, and (b) follows from Proposition A.9b.

Distributions on K enter representation theory through the following proposition and corollary. The proofs will use the extensions of members of $\mathcal{E}'(K)$ to vector-valued functions as in (B.17) of Appendix B. So as to be consistent with (1.1), we shall write $\langle T, f \rangle_V$ for the action of $T \in \mathcal{E}'(K)$ on $f \in C^\infty(K, V)$, rather than $T(f)$ as in Appendix B. We retain the notation $\langle T, f \rangle$ for the action on scalar-valued functions. Let dk refer to a fixed normalization of (two-sided) Haar measure on K.

Proposition 1.20. Let (π, V) be a locally K finite representation of K. For $T \in \mathcal{E}'(K)$, either of the formulas

(1.21a) $$\pi(T)v = \langle T, \pi(\cdot)v \rangle_V \qquad \text{for } v \in V$$

or

(1.21b) $$\langle \pi(T)v, v^* \rangle = \langle T, \langle \pi(\cdot)v, v^* \rangle \rangle \qquad \text{for } v \in V, \ v^* \in V^*$$

defines $\pi(T) \in \mathrm{End}_{\mathbb{C}} V$ in such a way that

    (a) $\pi(\delta_{k_0}) = \pi(k_0)$ for $k_0 \in K$
    (b) $\pi(\partial(u)) = \pi(u)$ for $u \in U(\mathfrak{g})$
    (c) $\pi(f \, dk)v = \int_K f(k)\pi(k)v \, dk$ for $f \in C^\infty(K)$ and $v \in V$
    (d) $\pi(T_1 * T_2) = \pi(T_1)\pi(T_2)$
    (e) $\pi(l(k_0)T) = \pi(k_0)\pi(T)$ for $k_0 \in K$
    (f) $\pi(r(k_0)T) = \pi(T)\pi(k_0)^{-1}$ for $k_0 \in K$.

Moreover, if $V_0 \subseteq V$ is an invariant subspace and π_0 is the restriction of π to V_0, then

(1.21c) $$\pi_0(T) = \pi(T)|_{V_0} \qquad \text{for all } T \in \mathcal{E}'(K).$$

In addition, if (π, V) and (π', V') are two such representations and if $E : V \to V'$ is K equivariant, then

(1.21d) $$\pi'(T)E = E\pi(T) \qquad \text{for all } T \in \mathcal{E}'(K).$$

Finally if V is finite-dimensional, then $\pi(T)$ is given also by

(1.21e) $$\pi(T) = \langle T, \pi \rangle_{\mathrm{End}_{\mathbb{C}} V}$$

and conclusion (c) may be strengthened to

    (c') $\pi(f \, dk) = \int_K f(k)\pi(k) \, dk$ for $f \in C^\infty(K)$.

REMARK. This result has an extension to smooth representations of general Lie groups on complete locally convex Hausdorff linear topological spaces. See Warner [1972], vol. I, 253. We shall use only the special case stated in the proposition.

PROOF. By (B.17), (1.21a) defines $\pi(T)$. Formula (1.21b) results from applying (B.17c) to the linear map $A : V \to \mathbb{C}$ given by $A(v) = \langle v, v^* \rangle$. Moreover, (1.21b) uniquely determines $\pi(T)$ and hence may be used as an alternate definition of $\pi(T)$ in place of (1.21a). The consistency condition (1.21c) for invariant subspaces $V_0 \subseteq V$ is immediate from (1.21b).

Conclusion (c) is a special case of (1.21a). If V is finite dimensional, then (1.21e) makes sense. It determines (1.21a) as a result of (B.17c) applied to the linear map $A : \mathrm{End}_{\mathbb{C}} V \to V$ given by $A(E) = Ev$. Conversely (1.21b) determines (1.21e) as a result of the definition in (B.17b). Conclusion (c') is a special case of (1.21e).

Conclusion (a) is immediate from (1.21a). Conclusion (e) follows by applying (B.17c) with $A : V \to V$ given by $A = \pi(k_0)$, while conclusion (f) follows by applying (1.21a) with v replaced by $\pi(k_0)^{-1}v$.

For (d), let v be in V. Then

$$\pi(T_1)\pi(T_2)v = \langle T_1, \pi(\cdot)\pi(T_2)v \rangle_V \qquad \text{by (1.21a)}$$

$$= \int_K \pi(k_1)\pi(T_2)v \, dT_1(k_1)$$

$$= \int_K \pi(k_1) \left[\int_K \pi(k_2)v \, dT_2(k_2) \right] dT_1(k_1) \qquad \text{by (1.21a)}$$

$$= \int_K \left[\int_K \pi(k_1 k_2)v \, dT_2(k_2) \right] dT_1(k_1) \qquad \text{since } \pi(k_1) \text{ is linear}$$

$$= \int_{K \times K} \pi(k_1 k_2)v \, dT_1(k_1) \, dT_2(k_2) \qquad \text{by Fubini's Theorem}$$

$$= \langle T_1 * T_2, \pi(\cdot)v \rangle \qquad \text{by (1.7b)}$$

$$= \pi(T_1 * T_2)v \qquad \text{by (1.21a).}$$

For (b), conclusion (d) and formula (1.9) show that it is enough to prove that $\pi(\partial(X)) = \pi(X)$ for $X \in \mathfrak{g}_0$. We have

$$\langle \pi(\partial(X))v, v^* \rangle = \langle \partial(X), \langle \pi(\cdot)v, v^* \rangle \rangle$$

$$= (r(X)\langle \pi(\cdot)v, v^* \rangle)(1)$$

$$= \frac{d}{dt}\langle \pi(\exp tX)v, v^* \rangle|_{t=0}$$

$$= \langle \pi(X)v, v^* \rangle,$$

and (b) follows. Finally to prove (1.21d), we use (1.21a), equivariance, and (B.17c) to calculate

$$E\pi(T)v = E\langle T, \pi(\cdot)v \rangle = \langle T, E\pi(\cdot)v \rangle$$

$$= \langle T, \pi(\cdot)Ev \rangle = \pi(T)Ev,$$

and the proof is complete.

We shall apply the preceding general considerations to some specific representations. To do so, we need a way of identifying the subspaces V_γ of (1.19). We use the following notation: If (π, V) is a representation of K, then (π^*, V^*) is the **contragredient** representation with $\pi^*(k) = \pi(k^{-1})^t$. The **character** Θ_γ of $\gamma \in \widehat{K}$ is the function $\Theta_\gamma : K \to \mathbb{C}$ given by $\Theta_\gamma(k) = \operatorname{Tr} \pi(k)$ for any irreducible representation π of type γ.

Any finite-dimensional smooth representation (π, V) of K can be made unitary by introducing a suitable Hermitian inner product in V. In such an inner product $(\cdot, \cdot)$, the Schur orthogonality relations for a single irreducible π read

$$(1.22a) \qquad \int_K (\pi(k)v, w)\overline{(\pi(k)v', w')}\, dk = \frac{\operatorname{vol} K}{\dim \pi}(v, v')\overline{(w, w')}.$$

This formula can be rewritten as

$$(1.22b) \qquad \int_K \langle \pi(k)v, w \rangle \langle \pi(k)^{-1}w', v' \rangle\, dk = \frac{\operatorname{vol} K}{\dim \pi} \langle v, v' \rangle \langle w', w \rangle,$$

where v and w' are in V, w and v' are in V^*, and $\langle \cdot, \cdot \rangle$ is the usual pairing of V and V^*; in this notation, unitarity of π plays no role.

As a consequence of the fact that any finite-dimensional smooth representation (π, V) of K can be made unitary, we have the formula

$$\Theta_{\gamma^*}(k) = \overline{\Theta_\gamma(k)} \qquad \text{for } k \in K, \, \gamma \in \widehat{K}.$$

We define χ_γ to be the member of $\mathcal{E}'(K)$ given by

$$(1.23) \qquad\qquad \chi_\gamma = \frac{(\dim \gamma)\Theta_{\gamma^*}\, dk}{\operatorname{vol} K}.$$

Proposition 1.24. If (π, V) is a locally K finite representation of K and if $V = \bigoplus_{\gamma \in \widehat{K}} V_\gamma$ is the canonical direct-sum decomposition into isotypic components, then $\pi(\chi_\gamma)$ is the projection operator of V on V_γ.

REMARK. The operator in question is constructed in Proposition 1.20.

PROOF. Let $v_1, \ldots, v_n$ be a basis for an irreducible invariant subspace V_0 of type γ'. We may assume this basis is orthonormal with respect to an invariant Hermitian inner product on V_0. We write $\pi(k)v_j = \sum_i \gamma'_{ij}(k)v_i$. Then Proposition 1.20c gives

$$\left\langle \pi\left(\frac{(\dim \gamma)\Theta_{\gamma^*}\, dk}{\operatorname{vol} K}\right)v_j, v \right\rangle = \frac{\dim \gamma}{\operatorname{vol} K}\int_K \left\langle \sum_l \overline{\gamma_{ll}(k)}\pi(k)v_j, v \right\rangle dk$$

$$= \frac{\dim \gamma}{\operatorname{vol} K}\int_K \sum_{l,i}\langle \overline{\gamma_{ll}(k)}\gamma'_{ij}(k)v_i, v \rangle\, dk.$$

Schur orthogonality says this is 0 if γ' and γ are inequivalent, and it is $\sum_{l,i}\delta_{li}\delta_{jl}\langle v_i, v\rangle = \langle v_j, v\rangle$ if $\gamma' = \gamma$. Thus the operator is the identity on V_γ and is 0 on the other $V_{\gamma'}$'s.

The specific representations that we shall study are the regular representations of K on functions and distributions. The one of chief interest is the one on distributions, since distributions map covariantly with points and functions map contravariantly. Curiously the K finite distributions will all be smooth functions, and thus we can identify the K finite distributions with the K finite functions once we fix a normalization of Haar measure. This identification is useful for computational purposes, but for conceptual purposes the reader is advised to distinguish the distributions from the functions.

We begin by treating functions. Let (π, F_γ) be a finite-dimensional representation of K of type $\gamma \in \widehat{K}$, and let (π^*, F_γ^*) be its contragredient. If $v \in F_\gamma$ and $v^* \in F_{\gamma^*}$, the function

$$(1.25a) \qquad\qquad f_{v^*,v}(k) = \langle \pi(k)v, v^* \rangle$$

is called a **matrix coefficient** of π.

If we fix $v \in F_\gamma$, then $v^* \mapsto f_{v^*,v}$ is equivariant from (π^*, F_γ^*) into $(l, C^\infty(K))$ because

$$f_{\pi^*(k)v^*,v}(\cdot) = \langle \pi(\cdot)v, \pi^*(k)v^* \rangle = \langle \pi(k^{-1} \cdot)v, v^* \rangle = l(k)f_{v^*,v}.$$

Hence $f_{v^*,v}$ is of type γ^* under l. Similarly if we fix $v^* \in F_\gamma^*$, then $v \mapsto f_{v^*,v}$ is equivariant from (π, F_γ) into $(r, C^\infty(K))$. Hence $f_{v^*,v}$ is of type γ under r. The bilinear map

$$(1.25b) \qquad\qquad (v^*, v) \mapsto f_{v^*,v} = \langle \pi(\cdot)v, v^* \rangle$$

extends to a linear map

(1.25c)
$$\mathcal{M} : F_\gamma^* \otimes_{\mathbb{C}} F_\gamma \to C^\infty(K) \quad \text{given by} \quad \mathcal{M}(v^* \otimes v) = f_{v^*,v} = \langle \pi(\cdot)v, v^* \rangle;$$

$\mathcal{M}$ is one-one as a consequence of Schur's Lemma, since $F_\gamma^* \otimes_{\mathbb{C}} F_\gamma$ may be regarded as an irreducible representation of $K \times K$.

Proposition 1.26. Let $C^\infty(K)_{K,l}$ and $C^\infty(K)_{K,r}$ be the subspaces of K finite vectors under the regular representations of l and r of K on $C^\infty(K)$. If γ is in $\widehat{K}$, then

$$C^\infty(K)_{K,l,\gamma^*} = C^\infty(K)_{K,r,\gamma} = \mathcal{M}(F_\gamma^* \otimes_{\mathbb{C}} F_\gamma)$$

under the map (1.25). Consequently $C^\infty(K)_{K,l} = C^\infty(K)_{K,r}$.

PROOF.   Let f be a member of $C^\infty(K)_{K,r,\gamma}$ not in $\mathcal{M}(F_\gamma^* \otimes_{\mathbb{C}} F_\gamma)$. Without loss of generality we may assume that f is orthogonal to all members of the finite-dimensional space $\mathcal{M}(F_\gamma^* \otimes_{\mathbb{C}} F_\gamma)$. In particular, f is orthogonal to every translate of

$$\Theta_\gamma(\cdot) = \sum_i \langle \pi(\cdot)v_i, v_i^* \rangle,$$

where the sum is taken over a basis $\{v_i\}$ of F_γ and its dual basis $\{v_i^*\}$ of F_γ^*. But then Proposition 1.24 gives

$$f(\cdot) = r\left(\frac{(\dim \gamma)\Theta_{\gamma^*}\, dk}{\mathrm{vol}\, K}\right) f(\cdot) = \frac{\dim \gamma}{\mathrm{vol}\, K} \int_K \Theta_{\gamma^*}(k) r(k) f(\cdot)\, dk$$

$$= \frac{\dim \gamma}{\mathrm{vol}\, K} \int_K \overline{\Theta_\gamma(k)} f(\cdot k)\, dk$$

$$= \frac{\dim \gamma}{\mathrm{vol}\, K} \int_K \overline{\Theta_\gamma(\cdot^{-1}k)} f(k)\, dk$$

$$= 0.$$

This proves that $C^\infty(K)_{K,r,\gamma} = \mathcal{M}(F_\gamma^* \otimes_{\mathbb{C}} F_\gamma)$, and the equality with $C^\infty(K)_{K,l,\gamma^*}$ is proved similarly.

Taking advantage of the conclusion of Proposition 1.26 that left and right K finite functions are the same, we can define

(1.27a) $C(K) = C^\infty(K)_K = \{K \text{ finite smooth functions on } K\}$

without specifying l or r. Let

(1.27b) $C(K)_\gamma = \mathcal{M}(F_\gamma^* \otimes_{\mathbb{C}} F_\gamma)$

with $\mathcal{M}$ as in (1.25). Proposition 1.26 gives us the direct sum decomposition

(1.28) $C(K) = \bigoplus_{\gamma \in \widehat{K}} C(K)_\gamma = \bigoplus_{\gamma \in \widehat{K}} \mathcal{M}(F_\gamma^* \otimes_{\mathbb{C}} F_\gamma).$

This formula is the decomposition (1.19) for the representation $(r, C(K))$, with the action of K on $\mathcal{M}(F_\gamma^* \otimes_{\mathbb{C}} F_\gamma)$ given by γ on F_γ and given by the trivial representation on F_γ^*. For the representation $(l, C(K))$, the γth term of (1.28) corresponds to the $\gamma^{*\text{th}}$ term of (1.19), with the action of K on $\mathcal{M}(F_\gamma^* \otimes_{\mathbb{C}} F_\gamma)$ given by γ^* on F_γ^* and given by the trivial representation on F_γ.

Proposition 1.29. Under the regular representations l and r of K on $C^\infty(K)_K$, the action of $\mathcal{E}'(K)$ is given as follows:

$$l(T) = \text{left convolution by } T$$
$$r(T) = \text{right convolution by } T^t.$$

PROOF. We apply Proposition 1.20 with $V = C(K)$. Formula (1.21a) with $\pi = l$ gives

$$l(T)f = \int_K l(k) f(\cdot) \, dT(k) = \int_K f(k^{-1} \cdot) \, dT(k).$$

Applying (B.17c) to evaluate both sides at k_0, we obtain

$$l(T)f(k_0) = \int_K f(k^{-1}k_0) \, dT(k) = \int_K f^t(k_0^{-1}k) \, dT(k) = \langle T, l(k_0)f^t \rangle,$$

and this is $T * f(k_0)$ by (1.13b). Hence $l(T)f = T * f$. A similar computation using (1.13a) gives $r(T)f = f * T^t$.

Now we treat distributions. With no assumption on K finiteness, we have the following result.

Proposition 1.30. The distributions χ_γ satisfy

$$\chi_\gamma * T = T * \chi_\gamma$$

for all $T \in \mathcal{E}'(K)$.

PROOF. Let f be in $C^\infty(K)$. It is enough to observe that

$$
\begin{aligned}
\langle \chi_\gamma * T, \ f \rangle &= \frac{\dim \gamma}{\text{vol } K} \int_K \left[\int_K f(xy) \Theta_{\gamma^*}(x) \, dx \right] dT(y) \\
&= \frac{\dim \gamma}{\text{vol } K} \int_K \left[\int_K f(x) \Theta_{\gamma^*}(xy^{-1}) \, dx \right] dT(y) \\
&= \frac{\dim \gamma}{\text{vol } K} \int_K \left[\int_K f(x) \Theta_{\gamma^*}(y^{-1}x) \, dx \right] dT(y) \\
&= \frac{\dim \gamma}{\text{vol } K} \int_K \left[\int_K f(yx) \Theta_{\gamma^*}(x) \, dx \right] dT(y) \\
&= \langle T * \chi_\gamma, \ f \rangle.
\end{aligned}
$$

Once we impose K finiteness on distributions, we obtain a result parallel to Proposition 1.29.

Proposition 1.31. Let $\mathcal{E}'(K)_{K,l}$ and $\mathcal{E}'(K)_{K,r}$ be the subspaces of K finite vectors under the regular representations l and r of K on $\mathcal{E}'(K)$. Then the action of $\mathcal{E}'(K)$ on $\mathcal{E}'(K)_{K,l}$ and $\mathcal{E}'(K)_{K,r}$ is given as follows:

$$\text{on } \mathcal{E}'(K)_{K,l}: \qquad l(T) = \text{left convolution by } T$$
$$\text{on } \mathcal{E}'(K)_{K,r}: \qquad r(T) = \text{right convolution by } T^t.$$

Every member of either $\mathcal{E}'(K)_{K,l}$ or $\mathcal{E}'(K)_{K,r}$ is a smooth function, i.e., is of the form $f\,dk$ with f a C^∞ function K finite under l or r respectively. Consequently $\mathcal{E}'(K)_{K,l} = \mathcal{E}'(K)_{K,r}$.

PROOF. We apply Proposition 1.20 with $V = \mathcal{E}'(K)_{K,l}$ or $\mathcal{E}'(K)_{K,r}$ as appropriate. Members h of $C^\infty(K)$ certainly yield members of V^* to which we can apply (1.21b). Thus we have

$$\begin{aligned}
\langle l(T)S, h\rangle &= \langle T,\, \langle l(\cdot)S, h\rangle\rangle \\
&= \langle T,\, \langle S, l(\cdot)^{-1}h\rangle\rangle \\
&= \int_K \Big[\int_K h(\cdot k)\,dS(k)\Big]\,dT(\cdot) \\
&= \langle T * S, h\rangle.
\end{aligned}$$

Since h is arbitrary, $l(T)S = T * S$. Similarly $r(T)S = S * T^t$. By Proposition 1.24 the members of $\mathcal{E}'(K)_{K,l,\gamma}$ are the left K finite distributions T such that

$$(1.32) \qquad l\Big(\frac{(\dim \gamma)\Theta_{\gamma^*}\,dk}{\text{vol } K}\Big)T = T.$$

We have just seen that the left side is given by convolution. Since one of the factors is a C^∞ function (times dk), so is the left side. Thus (1.32) shows that T is given by a C^∞ function (times dk). The action of l on T is compatible with the action of l on this function, and hence this function is left K finite. By Proposition 1.26, it is also right K finite. The result follows.

By Proposition 1.31 we can define

$$(1.33) \qquad R(K) = \mathcal{E}'(K)_K = \{K \text{ finite distributions on } K\}$$

without specifying l or r. $R(K)$ is a two-sided ideal of $\mathcal{E}'(K)$, as a consequence of Proposition 1.31 and Proposition 1.20 (applied with $\pi = l$ and $\pi = r$. We call it the **Hecke algebra** of K. Its importance lies in converting the study of locally K finite representations into module theory; a quantitative result in this direction will be given as Theorem 1.57.

For a K type γ, the map $f \in C^\infty(K) \mapsto f\,dk \in \mathcal{E}'(K)$ gives us K equivariant isomorphisms

$$(1.34) \qquad \begin{aligned} C^\infty(K)_{K,l,\gamma^*} &\xrightarrow{\sim} \mathcal{E}'(K)_{K,l,\gamma^*} \\ C^\infty(K)_{K,r,\gamma} &\xrightarrow{\sim} \mathcal{E}'(K)_{K,r,\gamma}, \end{aligned}$$

by Proposition 1.31. Since the two spaces on the left are equal (by Proposition 1.26), we see that

$$(1.35) \qquad \mathcal{E}'(K)_{K,l,\gamma^*} = \mathcal{E}'(K)_{K,r,\gamma}.$$

Define

$$(1.36) \qquad R(K)_\gamma = \mathcal{E}'(K)_{K,l,\gamma},$$

so that

$$(1.37) \qquad R(K) = \bigoplus_{\gamma \in \widehat{K}} R(K)_\gamma = \bigoplus_{\gamma \in \widehat{K}} \mathcal{E}'(K)_{K,l,\gamma}.$$

This formula is the decomposition (1.19) for the representation $(l, R(K))$. For the representation $(r, R(K))$, the γth term of (1.37) corresponds to the $\gamma^{*\text{th}}$ term of (1.19). Notice that the roles of l and r relative to γ and γ^* have been reversed in passing from $C(K)$ in (1.28) to $R(K)$ in (1.37). Fixing a normalization of dk gives us an isomorphism

$$R(K)_\gamma \cong C(K)_{\gamma^*},$$

as a result of (1.34). We shall see that $R(K)_\gamma$ and $C(K)_\gamma$ are canonically dual to each other (independently of any normalization of dk).

To get at this duality, we shall examine $R(K)_\gamma$ more closely. In connection with (1.25), we used a representation (π, F_γ) of K of type γ. In order to simplify the notation, we shall abuse notation slightly and refer to this representation as (γ, F_γ). Formulas (1.38) below extend identities we know for representations of K to identities about $\mathcal{E}'(K)$, and they are therefore immediate consequences of (1.21a): With $f_{v^*,v}$ as in (1.25a) and T in $\mathcal{E}'(K)$,

$$(1.38) \qquad l(T)f_{v^*,v} = f_{\gamma^*(T)v^*,v} \qquad \text{and} \qquad r(T)f_{v^*,v} = f_{v^*,\gamma(T)v}.$$

Proposition 1.39. Fix $\gamma \in \widehat{K}$. Then

(a) $R(K)_\gamma$ is a two-sided ideal in $R(K)$.
(b) the map $T \mapsto \gamma(T)$ exhibits an algebra isomorphism $R(K)_\gamma \cong \text{End}_{\mathbb{C}} F_\gamma$.
(c) $R(K)_\gamma$ is canonically dual to $C(K)_\gamma$ under the pairing

$$\langle T, f_{v^*,v} \rangle = \langle \gamma(T)v, v^* \rangle.$$

PROOF.

(a) The definition $R(K)_\gamma = \mathcal{E}'(K)_{K,l,\gamma}$ in (1.36) makes it clear that $R(K)_\gamma$ is spanned by the images of all linear maps $E : F_\gamma \to R(K)$ that are K equivariant for γ and l. If v is in F_γ and S is in $R(K)$ (or even $\mathcal{E}'(K)$), application of Proposition 1.31 and (1.21d) yields

$$S * Ev = l(S)(Ev) = E(\gamma(S)v).$$

Hence $S * R(K)_\gamma \subseteq R(K)_\gamma$. For multiplication on the other side, we argue similarly from the characterization of $R(K)_\gamma$ as $\mathcal{E}'(K)_{K,r,\gamma^*}$ given in (1.35).

(b) Proposition 1.20d shows that $T \mapsto \gamma(T)$ is an algebra homomorphism. We have $R(K)_\gamma = \mathcal{E}'(K)_{K,l,\gamma} \cong C^\infty(K)_{K,l,\gamma} = \mathcal{M}(F_\gamma \otimes_{\mathbb{C}} F_\gamma^*)$. Since $\mathcal{M}$ is one-one, $\dim R(K)_\gamma = (\dim F_\gamma)^2$. But $\dim(\mathrm{End}_{\mathbb{C}} F_\gamma) = (\dim F_\gamma)^2$. Thus $T \mapsto \gamma(T)$ will be an isomorphism on $R(K)_\gamma$ if we show it is one-one. Suppose $T \in R(K)_\gamma$ has $\gamma(T) = 0$. Formulas (1.38) show that $r(T)f_{v^*,v} = 0$ for all v and v^*. Hence (1.14) gives

$$0 = r(T)f_{v^*,v}(1) = f_{v^*,v} * T'(1) = \langle T', f_{v^*,v}^t \rangle = \langle T, f_{v^*,v} \rangle.$$

Since T is in $R(K)_\gamma$, we can write $T = f\,dk$ with $f \in C(K)_{\gamma^*}$. If $f = \sum_i c_i f_{v_i,v_i^*}$, then we have

$$0 = \langle T, f_{v^*,v} \rangle = \sum_i c_i \int_K f_{v_i,v_i^*}(k) f_{v^*,v}\, dk$$

$$= \sum_i c_i \int_K \langle v_i, \gamma^*(k)v_i^* \rangle \langle \gamma(k)v, v^* \rangle\, dk$$

$$= \frac{\mathrm{vol}\, K}{\dim \gamma} \sum_i c_i \langle v_i, v^* \rangle \langle v, v_i^* \rangle$$

for all $v \in F_\gamma$ and $v^* \in F_\gamma^*$, by (1.22b). It follows that all c_i are 0 and that $T = 0$.

(c) The two sides in the formula both equal $\langle T, \langle \gamma(\cdot)v, v^* \rangle \rangle$ and hence equal each other. For fixed T the formula defines a bilinear map $F_{\gamma^*} \times F_\gamma \to \mathbb{C}$ and hence yields a linear map $F_\gamma^* \otimes_{\mathbb{C}} F_\gamma \to \mathbb{C}$. Identifying $F_\gamma^* \otimes_{\mathbb{C}} F_\gamma$ with $C(K)_\gamma$ via $\mathcal{M}$, we thus obtain a map $R(K)_\gamma \to C(K)_\gamma^*$ that is evidently linear. Part (b) shows that this map is an isomorphism. This completes the proof.

We have canonical isomorphisms

$$C(K)_\gamma \cong F_\gamma^* \otimes_{\mathbb{C}} F_\gamma \qquad \text{and} \qquad R(K)_\gamma \cong \mathrm{End}_{\mathbb{C}} F_\gamma.$$

The duality in Proposition 1.39c can be explained in a different way by using the canonical isomorphism

$$F_\gamma \otimes_\mathbb{C} F_\gamma^* \xrightarrow{\sim} \mathrm{End}_\mathbb{C} F_\gamma$$

given by sending $v \otimes v^*$ to the endomorphism $w \mapsto v^*(w)v$. The duality amounts to nothing more than dualizing $F_\gamma^* \otimes_\mathbb{C} F_\gamma$ factor by factor. This operation of dualizing factor by factor is consistent with the left and right actions of K on $C(K)_\gamma$ and $R(K)_\gamma$.

Another property of $R(K)_\gamma$ is given in the following lemma, which will be used in Chapter II.

Lemma 1.40. Let (π, V) be a locally K finite representation of K, fix $\gamma \in \widehat{K}$, let

$$M = \mathrm{Hom}_K(R(K)_\gamma, V)$$

$$= \left\{ \varphi \in \mathrm{Hom}_\mathbb{C}(R(K)_\gamma, V) \left|\begin{array}{l} \varphi(l(k)T) = \pi(k)(\varphi(T)) \\ \text{for } k \in K,\ T \in R(K)_\gamma \end{array}\right.\right\},$$

and define a representation $(\tilde{\pi}, M)$ of K by $(\tilde{\pi}(k)\varphi)(T) = \varphi(T * \delta_k)$. Then the map $\varphi \mapsto x_\varphi$ of M into V given by $x_\varphi = \varphi(\chi_\gamma)$ is an isomorphism of $(\tilde{\pi}, M)$ onto (π, V_γ).

PROOF. The element x_φ is in V_γ since

$$\pi(\chi_\gamma)(x_\varphi) = \pi(\chi_\gamma)(\varphi(\chi_\gamma)) = \varphi(\chi_\gamma * \chi_\gamma) = \varphi(\chi_\gamma) = x_\varphi,$$

and thus $\varphi \mapsto x_\varphi$ is $\mathbb{C}$ linear from $\mathrm{Hom}_K(R(K)_\gamma, V)$ into V_γ. It respects $\tilde{\pi}$ and π since

$$\begin{aligned}
x_{\tilde{\pi}(k)\varphi} &= (\tilde{\pi}(k)\varphi)(\chi_\gamma) \\
&= \varphi(\chi_\gamma * \delta_k) \\
&= \varphi(\delta_k * \chi_\gamma) && \text{by Proposition 1.30} \\
&= \varphi(l(k)\chi_\gamma) && \text{by Proposition 1.31} \\
&= \pi(k)\varphi(\chi_\gamma) \\
&= \pi(k)x_\varphi.
\end{aligned}$$

Also the map $\varphi \mapsto x_\varphi$ into V_γ is invertible with inverse associating to any $x \in V_\gamma$ the map $\varphi_x \in M$ defined by

$$\varphi_x(T) = \pi(T)(x);$$

this is a K map since

$$\varphi_x(l(k)T) = \pi(l(k)T)(x) = \pi(k)\pi(T)x = \pi(k)(\varphi_x(T))$$

by Proposition 1.20e. Finally we check that $x \mapsto \varphi_x$ inverts $\varphi \mapsto x_\varphi$. First,

$$\varphi_{x_\varphi}(T) = \pi(T)x_\varphi = \pi(T)\varphi(\chi_\gamma) = \varphi(l(T)\chi_\gamma) = \varphi(T * \chi_\gamma) = \varphi(T)$$

since T is in $R(K)_\gamma$. Second,

$$x_{\varphi_x} = \varphi_x(\chi_\gamma) = \pi(\chi_\gamma)x = x$$

since x is in V_γ. This proves the lemma.

A number of calculations with $R(K)$ simplify greatly by using vector-valued distributions and taking advantage of the isomorphism (B.18):

$$(1.41) \qquad \mathcal{E}'(K) \otimes_{\mathbb{C}} V \cong \mathcal{E}'(K) \otimes_{C^\infty(K)} C^\infty(K, V)_f.$$

The map from left to right here is the obvious inclusion; the point is that the inclusion is onto. We shall make use of a version of this isomorphism in which $R(K)$ replaces $\mathcal{E}'(K)$.

We begin with a version for $C(K)$ of the isomorphism

$$(1.42) \qquad C^\infty(K) \otimes_{\mathbb{C}} V \cong C^\infty(K, V)_f$$

of (B.17a). Here the map from left to right is $h(k) \otimes v \mapsto h(k)v$. Suppose V carries a locally K finite representation π of K. Then $C^\infty(K, V)_f$ admits representations L and R of K defined by

$$L(k_0)F(k) = \pi(k_0)(F(k_0^{-1}k))$$
$$R(k_0)F(k) = \pi(k_0)(F(kk_0)),$$

and (1.42) transforms $l \otimes \pi$ and $r \otimes \pi$ to L and R, respectively.

Lemma 1.43. If (π, V) is a locally K finite representation of K, then, under either of the representations $l \otimes \pi$ or $r \otimes \pi$ of K,

$$(C^\infty(K) \otimes_{\mathbb{C}} V)_K = C(K) \otimes_{\mathbb{C}} V.$$

PROOF. Since the tensor product of locally K finite representations of K is locally K finite, the right side is contained in the left side. In the reverse direction, let $v_1, \ldots, v_n$ be linearly independent in V and let $\sum f_j \otimes v_j$ in $C^\infty(K) \otimes_{\mathbb{C}} V$ be K finite under $l \otimes \pi$ or $r \otimes \pi$. We are to show that each f_j is in $C(K)$. Without loss of generality we may assume that V is finite dimensional.

Form the contragredient V^*, which is locally K finite because it is finite dimensional. The map of $V \otimes_{\mathbb{C}} V^*$ into $\mathbb{C}$ given by $v \otimes v^* \mapsto \langle v, v^* \rangle$ is easily checked to be K equivariant, and hence the induced map

$$(C^\infty(K) \otimes_{\mathbb{C}} V) \otimes_{\mathbb{C}} V^* \to C^\infty(K)$$

is K equivariant, as well. Therefore

$$(C^\infty(K) \otimes_{\mathbb{C}} V)_K \otimes_{\mathbb{C}} V^*$$

maps into $C(K)$. Applying this fact to $\sum f_j \otimes v_j$ and a member of V^* that picks off the coefficient of v_i, we see that f_i is in $C(K)$ for each i. This proves the lemma.

Let us continue with V as a locally K finite representation of K. Define

(1.44) $C(K, V) = (C^\infty(K, V)_f)_K.$

Taking into account (1.42) and Lemma 1.43, we obtain the isomorphism

(1.45) $C(K) \otimes_{\mathbb{C}} V \cong C(K, V).$

Before giving the version of (1.41) with $R(K)$ in place of $\mathcal{E}'(K)$, we need one more lemma. Recall from (B.11) that if T is in $\mathcal{E}'(K)$ and p is in $C^\infty(K)$, then pT is defined by $f \mapsto T(pf)$ for $f \in C^\infty(K)$.

Lemma 1.46. If T is in $R(K)$ and p is in $C(K)$, then pT is in $R(K)$.

PROOF. Fix a normalization of Haar measure dk. By Proposition 1.31, we can write $T = h\,dk$ for some $h \in C(K)$. The distribution in question is $hp\,dk$. Since h and p are sums of matrix coefficients, so is hp (from the tensor product). Thus hp is in $C(K)$, and $hp\,dk$ is in $R(K)$.

Proposition 1.47. If V is a locally K finite representation of K, then

$$R(K) \otimes_{\mathbb{C}} V \cong R(K) \otimes_{C(K)} C(K, V)$$

as vector spaces, the map from left to right being $r \otimes v \mapsto r \otimes (k \mapsto v)$.

REMARKS. Because $C(K)$ is a commutative algebra, the left $C(K)$ module structure on $R(K)$ defined by Lemma 1.46 may be regarded also as a right $C(K)$ module structure. In addition, $C(K, V)$ is a left $C(K)$ module under pointwise multiplication. Thus the tensor product on the right is well defined.

PROOF. Since $C(K)$ is a ring with identity, the left side is

$$\cong (R(K) \otimes_{C(K)} C(K)) \otimes_{\mathbb{C}} V \cong R(K) \otimes_{C(K)} (C(K) \otimes_{\mathbb{C}} V).$$

By (1.45) the right side is isomorphic to $R(K) \otimes_{C(K)} C(K, V)$, as we wished to show.

We shall need to know how to invert this isomorphism explicitly. Let $v(\cdot)$ be in $C(K, V)$, let $\{v_i\}$ be a basis for a finite-dimensional subspace of V containing the image of $v(\cdot)$, and let $\{v_i^*\}$ be the dual basis for the dual space, so that

$$v(k) = \sum_i \langle v(k), v_i^* \rangle v_i.$$

Each function $\langle v(\cdot), v_i^* \rangle$ is in $C(K)$ by the proof of (1.45). Suppose that T is in $R(K)$, so that $T \otimes v(\cdot)$ is a typical generating element of $R(K) \otimes_{C(K)} C(K, V)$. Lemma 1.46 shows that $\langle v(\cdot), v_i^* \rangle T$ is in $R(K)$. Hence

$$(1.48) \qquad T \otimes v(\cdot) = \sum_i T \otimes \langle v(\cdot), v_i^* \rangle v_i = \sum_i \langle v(\cdot), v_i^* \rangle T \otimes v_i.$$

Thus $T \otimes v(\cdot)$ is the image of the element $\sum_i \langle v(\cdot), v) i^* \rangle T \otimes v_i$ of $R(K) \otimes_{\mathbb{C}} V$.

EXAMPLE. Let (π, V) be a locally K finite representation of K, let x be in V, and let T be in $R(K)$. Then $T \otimes \pi(\cdot)x$ is a member of $R(K) \otimes_{C(K)} C(K, V)$. Proposition 1.47 shows us how to regard $T \otimes \pi(\cdot)x$ as a member of $R(K) \otimes_{\mathbb{C}} V$, namely

$$(1.49) \qquad T \otimes \pi(\cdot)x = \sum_i \langle \pi(\cdot)x, x_i^* \rangle T \otimes x_i.$$

We shall return to this example in §3.

We conclude this section by reviewing **transpose** in the context of $R(K)$. This operation was defined on all of $\mathcal{E}'(K)$ in (1.6a), and its definition leads to the identity

$$l(k)T^t = (r(k)T)^t \qquad \text{for } k \in K.$$

It follows that if T is right K finite, then T^t is left K finite. Hence transpose carries $R(K)$ into itself.

3. Approximate Identities

By way of introduction, let K again be a compact Lie group. If γ is a K type, we recall from (1.23) that $\chi_\gamma \in \mathcal{E}'(K)$ is given by

$$\chi_\gamma = \frac{(\dim \gamma)\Theta_{\gamma^*}\, dk}{\text{vol } K}.$$

Proposition 1.24 says that $\pi(\chi_\gamma)$ is the projection operator on V_γ whenever (π, V) is a locally K finite representation of K. We apply this fact with $(\pi, V) = (l, R(K))$, taking $r \in R(K)_{\gamma'}$ and using (1.36) and Proposition 1.31. The result is

$$(1.50) \qquad\qquad \chi_\gamma * r = \begin{cases} r & \text{if } \gamma = \gamma' \\ 0 & \text{if } \gamma \neq \gamma'. \end{cases}$$

Suppose now that A is a finite subset of $\widehat{K}$. Let

$$(1.51) \qquad\qquad \chi_A = \sum_{\gamma \in A} \chi_\gamma.$$

Summing the conclusion of Proposition 1.30 over $\gamma \in A$, we obtain

$$(1.52) \qquad\qquad \chi_A * T = T * \chi_A$$

for all $T \in \mathcal{E}'(K)$. If A and B are finite subsets of $\widehat{K}$, then (1.50) implies

$$(1.53a) \qquad\qquad \chi_A * \chi_B = \chi_{A \cap B}.$$

In particular, χ_A is idempotent. If r belongs to $R(K)_{\gamma'}$, then

$$(1.53b) \qquad\qquad \chi_A * r = \begin{cases} r & \text{if } \gamma' \in A \\ 0 & \text{if } \gamma' \notin A. \end{cases}$$

These properties suggest the following definition.

DEFINITION 1.54. Suppose R is a ring and S is a directed set. An **approximate identity** for R indexed by S is a collection $\{\chi_s \mid s \in S\}$ of elements of R having the following properties: If $s \leq s'$, then $\chi_s \chi_{s'} = \chi_{s'} \chi_s = \chi_s$; and if $r \in R$, there is an $s \in S$ such that $\chi_s r = r \chi_s = r$. A left R module M is called **approximately unital** if for any $m \in M$ there is an $s \in S$ such that $\chi_s m = m$.

Approximately unital modules are discussed briefly in §A.1. We shall take the discussion of that section as known at this point.

In the notation of Definition 1.54, any left R module M has a largest approximately unital submodule

$$M_u = \{m \in M \mid \chi_s m = m \text{ for some } s \in S\}.$$

Proposition 1.55. The functor $M \to M_u$ is covariant and exact in the category of left R modules.

PROOF. Covariance is clear. Let

$$A \xrightarrow{\psi} B \xrightarrow{\varphi} C$$

be exact, and let

$$A_u \xrightarrow{\psi|_{A_u}} B_u \xrightarrow{\varphi|_{B_u}} C_u$$

be the complex obtained by applying the functor. We know that $\text{image}(\psi|_{A_u}) \subseteq \ker(\varphi|_{B_u})$. If b is in $\ker(\varphi|_{B_u})$, then $b = \psi(a)$ for some $a \in A$ by exactness of the given sequence, and $b = \chi_s b$ for some s in S since b is in B_u. The element $a' = \chi_s a$ is in A_u since $\chi_s a' = \chi_s^2 a = \chi_s a$, and $\psi(a') = \psi(\chi_s a) = \chi_s \psi(a) = \chi_s b = b$. Thus $\text{image}(\psi|_{A_u}) = \ker(\varphi|_{B_u})$.

The system $\{\chi_A\}$ defined in (1.51) is an approximate identity for $R(K)$ indexed by the set of finite subsets of $\widehat{K}$. In this case it is convenient to write M_K instead of M_u for the largest approximately unital submodule of M; as a consequence of Theorem 1.57 below, this will not conflict with the notation (1.17) when both are defined.

In Definition 1.54, any genuine identity in R must belong to $\{\chi_s\}$; if an identity exists, any approximately unital R module must be unital. In the case of $R(K)$, the identity element ought to be the Dirac distribution at 1 in K. This distribution is *not* K finite unless K is a finite group, since its translates under the regular representation constitute the (linearly independent) set of all Dirac distributions at points of K.

Let (π, V) be a locally K finite representation of K. Proposition 1.20 allows us to regard V as a left $R(K)$ module under the definition

$$(1.56) \qquad Tv = \pi(T)v \qquad \text{for } T \in R(K) \text{ and } v \in V.$$

This left $R(K)$ module is approximately unital. In fact, let v be in $V_{\gamma_1} \oplus \cdots \oplus V_{\gamma_n}$, and put $A = \{\gamma_1, \ldots, \gamma_n\}$. Then $\pi(\chi_A)$ is the identity on $V_{\gamma_1} \oplus \cdots \oplus V_{\gamma_n}$, and hence $\chi_A v = \pi(\chi_A)v$. This proves the direct part of the following theorem, which identifies locally K finite representations of K with approximately unital left $R(K)$ modules.

Theorem 1.57. If (π, V) is a locally K finite representation of K, then the definition (1.56) makes V into an approximately unital left $R(K)$ module. Conversely suppose V is an approximately unital left $R(K)$ module. Then V is a complex vector space, and the map $R(K) \times V \to V$ given by $(T, v) \mapsto Tv$ is $\mathbb{C}$ bilinear. If $T \in \mathcal{E}'(K)$ and $v \in V$, choose a finite subset $A \subseteq \widehat{K}$ such that $\chi_A v = v$, and define

(1.58) $\pi(T)v = (T * \chi_A)v.$

Then $\pi(T)$ is well defined on V, and the system $\{\pi(T) \mid T \in \mathcal{E}'(K)\}$ is the extension to $\mathcal{E}'(K)$ of a locally K finite representation (π, V) of K. These two constructions are inverse to one another. Under this identification of representations with modules, the K equivariant maps between such representations coincide with the $R(K)$ homomorphisms between such modules.

PROOF. Suppose V is an approximately unital left $R(K)$ module. Then V is a complex vector space by Proposition A.4.

To see that $\pi(T)v$ is well defined, let $B \subseteq \widehat{K}$ be another finite set. Without loss of generality let $B \supseteq A$, so that $\chi_B * \chi_A = \chi_A$. Then we have

$$(T * \chi_B)v = (T * \chi_B)(\chi_A v) = ((T * \chi_B) * \chi_A)v$$
$$= (T * (\chi_B * \chi_A))v = (T * \chi_A)v.$$

So $\pi(T)$ is well defined on v. Since we can always find, for given v and v', a sufficiently large B so that $\chi_B v = v$ and $\chi_B v' = v'$, it follows also that $\pi(T)v$ is linear in v. Clearly it is linear in T. Also we have $\pi(1) = 1$.

Let S and T and v be given, and choose $B \subseteq \widehat{K}$ large enough so that $\chi_B v = v$ and $\chi_B(\pi(T)v) = \pi(T)v$. Then

$$\pi(S)\pi(T)v = (S * \chi_B)(\pi(T)v)$$
$$= (S * \chi_B)((T * \chi_B)v)$$
$$= ((S * \chi_B) * (T * \chi_B))v$$
$$= (S * T * \chi_B * \chi_B)v \qquad \text{by Proposition 1.30}$$
$$= ((S * T) * \chi_B)v$$
$$= \pi(S * T)v.$$

Since $\pi(\delta_1) = 1$, we conclude that $\pi : \mathcal{E}'(K) \to \mathrm{End}_{\mathbb{C}} V$ is an algebra homomorphism.

Now that $\pi(\delta_k)$ is defined for every Dirac distribution δ_k, we can set

(1.59) $\pi(k) = \pi(\delta_k),$

and π will be a homomorphism of K into $\mathrm{Aut}_{\mathbb{C}} V$.

If $v \in V$ is given and if $\chi_A v = v$, then

$$\pi(\mathcal{E}'(K))v = (\mathcal{E}'(K) * \chi_A)v = \sum_{\gamma \in A} R(K)_\gamma v,$$

and the right side is finite dimensional by Proposition 1.39b. Hence each v is formally K finite.

To prove that (π, V) is a locally K finite representation of K, let V_0 be a finite-dimensional K invariant subspace of V. It is enough to check that $k \mapsto \pi(k)|_{V_0}$ is continuous. Choose a finite set $A \subseteq \widehat{K}$ so that $\chi_A v = v$ for all $v \in V_0$. The map $k \mapsto l(k)\chi_A$ is continuous into the finite-dimensional subspace

$$R' = \sum_{\gamma \in A} R(K)_\gamma$$

of $R(K)$, the map $T \in R' \mapsto Tv \in V_0$ is linear between finite-dimensional spaces and hence is continuous, and the map $k \mapsto \pi(k)v$ is the composition, since

$$\pi(k)v = \pi(\delta_k)v = (\delta_k * \chi_A)v = (l(k)\chi_A)v.$$

Thus $k \mapsto \pi(k)v$ is continuous for each $v \in V_0$, and $k \mapsto \pi(k)|_{V_0}$ is continuous.

Before showing that the two constructions are inverse to one another, let us check the following: When we start with an approximately unital left $R(K)$ module, the constructed action by $\mathcal{E}'(K)$ is the extension to distributions of the restriction to K of the original module action. More specifically if we start with an approximately unital left $R(K)$ module and define successively

  (i) $\pi(T)$, for $T \in \mathcal{E}'(K)$, as in (1.58)
  (ii) $\pi(k)$, for $k \in K$, as in (1.58) and (1.59)
  (iii) $\pi_{\mathrm{new}}(T)$, for $T \in \mathcal{E}'(K)$, as in Proposition 1.20,

then $\pi_{\mathrm{new}}(T) = \pi(T)$ for all $T \in \mathcal{E}'(K)$.

To prove this statement, let $v \in V$ be given, and choose a finite subset $A \subseteq \widehat{K}$ with $\chi_A v = v$. Then every $T \in \mathcal{E}'(K)$ satisfies

$$\begin{aligned}
\pi_{\mathrm{new}}(T)v &= \langle T, \pi(\cdot)v \rangle && \text{by Proposition 1.20} \\
&= \langle T, (\delta_\cdot * \chi_A)v \rangle && \text{by (1.58) and (1.59)} \\
&= \langle T, (l(\cdot)\chi_A)v \rangle && \text{by (1.10)} \\
&= (l(T)\chi_A)v && \text{by Proposition 1.20} \\
&= (T * \chi_A)v && \text{by Proposition 1.31} \\
&= \pi(T)v && \text{by (1.58).}
\end{aligned}$$

Hence $\pi_{\text{new}}(T) = \pi(T)$, as asserted.

Now the verification that the two constructions are inverse to each other is formal. There are two things to check. One is that if we start with an approximately unital left $R(K)$ module V, make it into a locally K finite representation of K, extend it to distributions, and restrict to $R(K)$, we get back what we started with. To check this, let $T \in R(K)$, $v \in V$, and $\chi_A v = v$. Then the above consistency property shows that the extension of the action to T is $(T * \chi_A)v$, which is $T(\chi_A v) = Tv$ by the module property of V. The other thing to check is that if we start with a locally K finite representation (π, V) of K, extend it to distributions, restrict to $R(K)$, and then define a representation π_{new} as in (1.58), then $\pi_{\text{new}} = \pi$. To do so, let $k \in K$, $v \in V$, and $\pi(\chi_A)v = v$. Then

$$
\begin{aligned}
\pi_{\text{new}}(k)v &= (\delta_k * \chi_A)v && \text{by (1.58)} \\
&= \pi(\delta_k * \chi_A)v && \text{by the construction of the} \\
& && R(K) \text{ module} \\
&= \pi(l(k)\chi_A)v && \text{by Proposition 1.31} \\
&= \pi(k)\pi(\chi_A)v && \text{by Proposition 1.20e} \\
&= \pi(k)v.
\end{aligned}
$$

Finally if (π, V) and (π', V') are locally K finite representations of K, we show that

$$(1.60) \qquad\qquad \text{Hom}_K(V, V') = \text{Hom}_{R(K)}(V, V').$$

The left side is contained in the right side, according to (1.21d). For the reverse inclusion, let $E \in \text{Hom}_{R(K)}(V, V')$, $v \in V$, $\chi_B v = v$, and $\chi_B(Ev) = Ev$. Then

$$E\pi(k)v = E(\delta_k * \chi_B)v = (\delta_k * \chi_B)Ev = \pi'(k)Ev,$$

and E is in $\text{Hom}_K(V, V')$.

The use of approximate identities makes it possible to use integral notation in dealing with vector-valued distributions in situations as in Proposition 1.47 and the example after it. We continue with that example.

EXAMPLE, continued. We have already expanded $T \otimes \pi(\cdot)x$ by means of (1.49). Let A be a finite subset of $\widehat{K}$ large enough so that

$$(1.61) \qquad\qquad \langle \pi(\cdot)x, x_i^* \rangle T * \chi_A = \langle \pi(\cdot)x, x_i^* \rangle T$$

for all i and for all x in a finite-dimensional space of interest. By Proposition 1.31 the left side of (1.61) is

$$= l(\langle \pi(\cdot)x, x_i^* \rangle T)(\chi_A).$$

In turn we can use (1.21a) to rewrite this expression as

$$= \int_K l(k)\chi_A \, d(\langle \pi(\cdot)x, x_i^* \rangle T)(k) = \int_K \langle \pi(k)x, x_i^* \rangle l(k)\chi_A \, dT(k).$$

If we multiply by x_i and sum on i, then this expression becomes

$$\int_K (l(k)\chi_A \otimes \sum_i \langle \pi(k)x, x_i^* \rangle x_i) \, dT(k).$$

Therefore (1.49) gives

$$(1.62) \qquad T \otimes \pi(\cdot)x = \int_K (l(k)\chi_A \otimes \pi(k)x) \, dT(k).$$

The right side of (1.62) is simple to manipulate, yet it does not present any ambiguity in handling convolutions.

A locally K finite representation (π, V) of K is said to be **admissible** if each K type γ has finite multiplicity in (π, V). The rest of this section fits logically into the discussion at this point but is not needed until much later in the book. The reader may wish to skip this material at this time and return to it when it is invoked later.

In the terminology of Appendix A, the locally K finite representations of K form a good category of left $R(K)$ modules by Theorem 1.57. With the same kind of convention as in §A.4 for finite-dimensional vector spaces, the subcategory $\mathcal{C}$ of admissible locally K finite representations is a small good category of left $R(K)$ modules, and therefore its Grothendieck group $K(\mathcal{C})$ is well defined.

Proposition 1.63. Let $\mathcal{C}$ be the small good category of admissible locally K finite representations (considered as left $R(K)$ modules), and let $K(\mathcal{C})$ be its Grothendieck group.

(a) For each $\gamma \in \widehat{K}$ and any admissible locally K finite representation (π, V), let $n_\gamma(V)$ be the multiplicity of γ in V. Then the map $V \to n_\gamma(V)$ of the objects of $\mathcal{C}$ into $\mathbb{Z}$ extends to a well-defined homomorphism $n_\gamma(\cdot)$ of $K(\mathcal{C})$ into $\mathbb{Z}$.

(b) The map $V \to \{n_\gamma(V)\}_{\gamma \in \widehat{K}}$ of the objects of $\mathcal{C}$ into $\mathbb{Z}^{\widehat{K}}$ extends to an isomorphism $K(\mathcal{C}) \cong \mathbb{Z}^{\widehat{K}}$.

PROOF.

(a) The map extends to a well-defined homomorphism of the free abelian group $\mathcal{F}(\mathcal{C})$ into $\mathbb{Z}$, as a consequence of admissibility. An exact sequence

$$0 \longrightarrow A \longrightarrow B \longrightarrow C \longrightarrow 0$$

in $\mathcal{C}$ defines a relation $(B) = (A) + (C)$ in $\mathcal{F}(\mathcal{C})$, and we are to see that n_γ maps this relation to 0. Proposition 1.18b implies that $B \cong A \oplus C$. In the notation of (1.19), we must have $B_\gamma \cong A_\gamma \oplus C_\gamma$. Hence $n_\gamma(B) = n_\gamma(A) + n_\gamma(C)$, as required.

(b) We obtain a homomorphism $K(\mathcal{C}) \to \mathbb{Z}^{\widehat{K}}$ as a result of (a). Let us show it is onto. First, if $\gamma \in \widehat{K}$ is given, let (π_γ, F_γ) be a finite-dimensional representation of K of type γ. If $m \geq 1$ is an integer, let $V_{\gamma,m}$ be the space of functions from $\{1, \ldots, m\}$ to F_γ, and let $\pi_{\gamma,m}$ be the representation

$$\pi_{\gamma,m}(k)\{f(j)\}_{j=1}^m = \{\pi_\gamma(k)f(j)\}_{j=1}^m.$$

Then $n_\gamma(V_{\gamma,m}) = m$ and $n_{\gamma'}(V_{\gamma,m}) = 0$ if $\gamma' \neq \gamma$.

Now let $\{m_{\gamma'}\}_{\gamma' \in \widehat{K}}$ be a member of $\mathbb{Z}^{\widehat{K}}$ with $m_{\gamma'} \geq 0$ for all γ'. Then $\bigoplus_{\gamma' \in \widehat{K}} V_{\gamma',m_{\gamma'}}$ is the representation space of an admissible locally K finite representation with $n_\gamma\left(\bigoplus_{\gamma' \in \widehat{K}} V_{\gamma',m_{\gamma'}}\right) = m_\gamma$ for all $\gamma \in \widehat{K}$. Hence $\{m_{\gamma'}\}_{\gamma' \in \widehat{K}}$ is in the image of our homomorphism $K(\mathcal{C}) \to \mathbb{Z}^{\widehat{K}}$. Since the image is a subgroup of $\mathbb{Z}^{\widehat{K}}$, the image is all of $\mathbb{Z}^{\widehat{K}}$.

Finally we prove that $K(\mathcal{C}) \to \mathbb{Z}^{\widehat{K}}$ is one-one. We are to prove that any member of $\mathcal{F}(\mathcal{C})$ that maps to 0 in $\mathbb{Z}^{\widehat{K}}$ is in the relation subgroup. Let

$$\sum_j r_j(V_j) - \sum_k s_k(V_k')$$

be such a member of $\mathcal{F}(\mathcal{C})$ with all $r_j \geq 1$ and all $s_k \geq 1$. As in the proof of Proposition A.31, this member of $\mathcal{F}(\mathcal{C})$ is congruent modulo the relation subgroup to some $(V) - (V')$, where V and V' are admissible locally K finite representations. The assumption is that

$$n_\gamma((V) - (V')) = 0 \qquad \text{for all } \gamma \in \widehat{K},$$

i.e., that
$$n_\gamma(V) = n_\gamma(V') \qquad \text{for all } \gamma \in \widehat{K}.$$

By Proposition 1.18a we must have

$$V \cong \bigoplus_{\gamma \in \widehat{K}} V_{\gamma,n_\gamma(V)} \cong V'.$$

It follows that $(V) - (V')$ is in the relation subgroup, and $K(\mathcal{C}) \to \mathbb{Z}^{\widehat{K}}$ is one-one.

4. Hecke Algebras in the Group Case

Let $\mathfrak{g}$ be a finite-dimensional complex Lie algebra, and let K be a compact Lie group, possibly disconnected. We call $(\mathfrak{g}, K)$ a **pair** if the following compatibility conditions are satisfied:

(1.64)

    (i) the complexified Lie algebra $\mathfrak{k}$ of K is a subalgebra of $\mathfrak{g}$;

    (ii) K acts on $\mathfrak{g}$ by automorphisms $\mathrm{Ad}(k)$ extending the adjoint action on $\mathfrak{k}$; and

    (iii) the differential of $\mathrm{Ad}(K)$ is $\mathrm{ad}(\mathfrak{k}) \subseteq \mathrm{ad}(\mathfrak{g})$.

Since K is compact, we can choose a subspace $\mathfrak{p} \subseteq \mathfrak{g}$ stable under $\mathrm{Ad}(K)$ such that $\mathfrak{g} = \mathfrak{k} \oplus \mathfrak{p}$.

We shall attach to the pair $(\mathfrak{g}, K)$ an associative algebra $R(\mathfrak{g}, K)$ called the Hecke algebra of the pair. Under the special assumption (1.65) below for $(\mathfrak{g}, K)$, the definition of $R(\mathfrak{g}, K)$ is concrete and natural, and we give it shortly. This special assumption is often satisfied in the applications to group-representation theory. In the general case the definition of $R(\mathfrak{g}, K)$ is rather abstract, consisting of an algebraic construction that mimics what happens in the special case. We postpone treatment of the general case to §§5–6.

The special assumption is as follows:

(1.65)

    There is a real Lie group G with complexified Lie algebra $\mathfrak{g}$ such that K is a compact subgroup of G for which $\mathfrak{k}$ and Ad are compatible with the definitions imposed by G.

To see that (1.65) is not automatically satisfied by a pair $(\mathfrak{g}, K)$, one can take $\mathfrak{g}$ to be semisimple and K to be a nontrivial covering of a compact torus in a simply connected group corresponding to a real form of $\mathfrak{g}$. Alternatively one can put $K = \{1\}$ and choose $\mathfrak{g}$ to be any complex Lie algebra that is not the complexification of a real form $\mathfrak{g}_0$; such Lie algebras $\mathfrak{g}$ exist already in dimension 3.

In any event, suppose (1.65) is satisfied for the pair $(\mathfrak{g}, K)$. To keep the notation simple, we shall assume that G is **unimodular**, i.e., that $\Delta = 1$. Then the **Hecke algebra** $R(\mathfrak{g}, K)$ is the algebra of left K finite distributions on G with support in K, with convolution as multiplication. This is a subalgebra of $\mathcal{E}'(G)$ because

$$\text{support}(S * T) \subseteq (\text{support } S)(\text{support } T).$$

Proposition 1.83 will show that left K finite actually implies two-sided K finite. Corollary 1.71 and Proposition 1.80 will show that the choice of G in (1.65) does not affect $R(\mathfrak{g}, K)$, apart from isomorphisms.

EXAMPLES.

1) If $K = \{1\}$, $R(\mathfrak{g}, \{1\})$ may be identified with the universal enveloping algebra $U(\mathfrak{g})$ of $\mathfrak{g}$. In fact, by definition $R(\mathfrak{g}, \{1\})$ is the convolution algebra of distributions on G supported at $\{1\}$. By Corollary B.33 (with $X = \{1\}$, $Y = G$, and $p = 1$), the distributions in question are all derivatives of all orders at the point 1, and these are the elements of $U(\mathfrak{g})$. Multiplication in $R(\mathfrak{g}, \{1\})$ corresponds to multiplication in $U(\mathfrak{g})$ as a consequence of (1.9).

2) If $\mathfrak{g}$ is equal to $\mathfrak{k}$, then we can always arrange for (1.65) to be satisfied by taking $G = K$. In this case, $R(\mathfrak{k}, K)$ reduces to the Hecke algebra $R(K)$ discussed in §2. This identification is a consequence of the definitions and of Proposition 1.31, which implies that left K finite and two-sided K finite amount to the same thing in this situation.

Returning to a general pair $(\mathfrak{g}, K)$ related to a unimodular group G, we shall describe some elements in the algebra $R(\mathfrak{g}, K)$. Temporarily let $i : K \to G$ be the inclusion. In the notation of (B.26), $i_*(T)$, for $T \in R(K)$, is a distribution on G supported on K. It is left K finite because

(1.66) $l(i(k))(i_*(T)) = i_*(l(k)T).$

(This identity follows from the computation

$$\langle l(i(k))i_*(T), f\rangle = \langle i_*(T), l(i(k))^{-1}f\rangle = \langle T, l(i(k))^{-1}f \circ i\rangle$$
$$= \langle T, l(k)^{-1}(f \circ i)\rangle = \langle l(k)T, f \circ i\rangle = \langle i_*(l(k)T), f\rangle,$$

in which the third equality uses the fact that i is a homomorphism.) Consequently $i_*(T)$ is in $R(K)$. The mappings i and i_* are cumbersome to carry along, and we shall henceforth suppress them from the notation.

More generally suppose T is in $R(K)$ and u is in $U(\mathfrak{g})$. Then $T * \partial(u)$ is a distribution on G supported on K, and it is left K finite because

$$l(k)(T * \partial(u)) = \delta_k * (T * \partial(u)) = (\delta_k * T) * \partial(u).$$

Thus we have a bilinear map

(1.67a) $(T, u) \in R(K) \times U(\mathfrak{g}) \mapsto T * \partial(u) \in R(\mathfrak{g}, K)$

and hence a corresponding linear map

(1.67b) $R(K) \otimes_{\mathbb{C}} U(\mathfrak{g}) \to R(\mathfrak{g}, K)$.

If v is in $U(\mathfrak{k})$, then

$$\langle r(v^t)T, u \rangle = \langle r(\partial(v)^t)T, u \rangle = \langle T * \partial(v), u \rangle \mapsto T * \partial(v) * \partial(u)$$

and $\langle T, l(v)u \rangle = \langle T, vu \rangle \mapsto T * \partial(vu) = T * \partial(v) * \partial(u)$.

Hence the linear map (1.67b) descends to a linear map

(1.67c) $R(K) \otimes_{U(\mathfrak{k})} U(\mathfrak{g}) \to R(\mathfrak{g}, K)$.

We shall show in Corollary 1.71 that the linear map (1.67c) is a vector-space isomorphism. But first we identify a slightly more general class of distributions.

Proposition 1.68. Suppose G is a real unimodular Lie group and H is a closed subgroup. Write $\mathcal{E}'(G, H)$ for the algebra of compactly supported distributions on G with support contained in H. Then

$$\mathcal{E}'(G, H) \cong \mathcal{E}'(H) \otimes_{U(\mathfrak{h})} U(\mathfrak{g}),$$

the isomorphism from right to left being given by $T \otimes u \mapsto T * \partial(u)$. More explicitly, for $f \in C^{\infty}(G)$, the element of $\mathcal{E}'(G, H)$ acts by

$$\langle T * \partial(u), f \rangle = \langle T, (r(u)f)|_H \rangle.$$

The isomorphism from left to right sends the left action of $\mathcal{E}'(H)$ on $\mathcal{E}'(G, H)$ to the left action on $\mathcal{E}'(H)$, and it sends the right action of $U(\mathfrak{g})$ on $\mathcal{E}'(G, H)$ to the right action on $U(\mathfrak{g})$.

PROOF. The identity for the action of $T * \partial(u)$ follows from the computation

$$\langle T * \partial(u), f \rangle = \langle r(u^t)T, f \rangle = \langle T, r(u)f \rangle = \langle T, (r(u)f)|_H \rangle,$$

which uses successively Proposition 1.31, (1.3), and the definition of the suppressed mapping i_* of (B.26).

Sending $T \otimes u$ to $T * \partial(u)$ obviously defines a map from $\mathcal{E}'(H) \otimes_{\mathbb{C}} U(\mathfrak{g})$ to $\mathcal{E}'(G, H)$. That it descends to the tensor product over $U(\mathfrak{h})$ is clear from the same computation as in (1.67).

Next we prove that $\mathcal{E}'(H) \otimes_{U(\mathfrak{h})} U(\mathfrak{g}) \to \mathcal{E}'(G, H)$ is one-one. Let $\mathfrak{p}$ be a vector-space complement to $\mathfrak{h}$ in $\mathfrak{g}$, let $\{X_1, \ldots, X_l\}$ be a basis of $\mathfrak{p}$, and suppose

(1.69a) $\displaystyle\sum_n T_n \otimes u_n$ maps to 0.

That is, suppose

(1.69b) $\displaystyle\sum_n \langle T_n, (r(u_n)f)|_H \rangle = 0$ for all $f \in C^\infty(G)$.

Using the tensor product relation over $U(\mathfrak{h})$ and the Poincaré-Birkhoff-Witt Theorem, we may assume in (1.69a) and hence also in (1.69b) that the u_n's are distinct monomials $X_1^{j_1} \ldots X_l^{j_l}$. Canonical coordinates of the second kind near the identity in G show that the map of $H \times \mathbb{R}^l$ to G given by

(1.70) $(h, x_1, \ldots, x_l) \mapsto h \exp x_1 X_1 \cdots \exp x_l X_l$

is a local diffeomorphism about $(1, 0, \ldots, 0)$, and we are thus allowed to take f above to be any function that in these coordinates is of the form $f_1(k) f_2(x_1, \ldots, x_l)$, where f_1 is in $C^\infty(H)$ with support near the identity and f_2 is in $C^\infty(\mathbb{R}^l)$ with support near the origin. Our monomial $X_1^{j_1} \cdots X_l^{j_l}$ operates just on the f_2 part and gives

$$\frac{\partial^{j_1 + \cdots + j_l} f_2}{\partial x_1^{j_1} \ldots \partial x_l^{j_l}}(0, \ldots, 0).$$

Choosing f_2 suitably, we can arrange that (1.69b) reduces to

           $\langle T_n, f_1 \rangle = 0$ for all $f_1 \in C^\infty(H)$ supported near 1,

for any single n that we please. In other words, 1 is not in the support of T_n. Left translating (1.69a) by $h_0 \in H$ and repeating the argument, we see that h_0 is not in the support of T_n. By Corollary B.14, $T_n = 0$. Thus the map is one-one.

Let us prove that the map is onto. Let $T \in \mathcal{E}'(G, H)$ be given. Using a partition of unity, we can reduce to the case that T is supported in a small G neighborhood of a point h_0 in H. On such a neighborhood we can use the variant

              $(h, x_1, \ldots, x_l) \mapsto h_0 h \exp x_1 X_1 \ldots \exp x_l X_l$

of (1.70) to introduce the structure of the product of two manifolds. Application of Corollary B.33 then shows that T has the required form $\sum T_n \times \partial(u_n)$ in the product coordinates, hence the form $\sum T_n * \partial(u_n)$ on G.

Finally the assertion about left actions by $\mathcal{E}'(H)$ follows from the same computation as for (1.66), and the assertion about right actions by $U(\mathfrak{g})$ follows from (1.9) and the associativity of convolution.

Corollary 1.71. Suppose G is a real unimodular Lie group and K is a compact subgroup. Then the algebra $R(\mathfrak{g}, K)$ of left K finite distributions on G supported on K is isomorphic as a vector space to

$$R(K) \otimes_{U(\mathfrak{k})} U(\mathfrak{g})$$

by the map $\qquad\qquad T \otimes u \mapsto T * \partial(u).$

This isomorphism sends the left action of $\mathcal{E}'(K)$ on $R(\mathfrak{g}, K)$ to the left action on $R(K)$, and it sends the right action of $U(\mathfrak{g})$ on $R(\mathfrak{g}, K)$ to the right action on $U(\mathfrak{g})$.

PROOF. In view of Proposition 1.68, we need to prove only that the map $T \otimes u \mapsto T * \partial(u)$ is onto $R(\mathfrak{g}, K)$. By Proposition 1.68, any element T of $R(\mathfrak{g}, K)$ may be written as $\sum T_i * \partial(u_i)$, with T_i a distribution on K. Suppose T is left K finite. Application to $(l, R(K))$ of the direct part of Theorem 1.57 shows that there is a finite subset A of $\widehat{K}$ such that $\chi_A * T = T$. Then

$$T = \sum (\chi_A * T_i) * \partial(u_i).$$

Since $R(K)$ is an ideal in $\mathcal{E}'(K)$, each $\chi_A * T_i$ belongs to $R(K)$.

REMARKS. $R(K)$ is a subalgebra of $R(\mathfrak{g}, K)$, but $U(\mathfrak{g})$ is usually not, despite its presence in the notation. In fact, if K is infinite, $R(\mathfrak{g}, K)$ does not have an identity.

We shall need an explicit formula for multiplication in $R(\mathfrak{g}, K)$ when $R(\mathfrak{g}, K)$ is realized as $R(K) \otimes_{U(\mathfrak{k})} U(\mathfrak{g})$. In forming a product

$$(S * \partial(u)) * (T * \partial(v)),$$

we want somehow to rewrite $\partial(u) * T$ with the factors in the other order.

Let us consider this problem without being completely rigorous yet. The value of $T' * \partial(u')$ on f according to Proposition 1.68 is $\int_K (r(u')f)(k) \, dT'(k)$, and we can write this relationship symbolically as

(1.72) $$\langle T' * \partial(u'), f \rangle = \int_K f(ku') \, dT'(k).$$

Our interest is in

(1.73a) $$\langle \partial(u) * T, f \rangle = \langle l(u)T, f \rangle = \langle T, l(u^t)f \rangle,$$

which in the same symbolic notation is

$$(1.73b) \qquad = \int_K f(uk)\,dT(k) = \int_K f(k(\mathrm{Ad}(k)^{-1}u))\,dT(k).$$

(We shall justify this formal manipulation in a moment.) The difficulty in comparing the right side of (1.73b) with the right side of (1.72) is that in place of the single element u', we have the $U(\mathfrak{g})$ valued function $k \mapsto \mathrm{Ad}(k)^{-1}u$. To go further, let $\{u_i\}$ be a basis for the finite-dimensional span $\langle \mathrm{Ad}(K)u \rangle$, and let $\{u_i^*\}$ be the dual basis for $\langle \mathrm{Ad}(K)u \rangle^*$. Define members p_i of $C(K)$ by

$$(1.74a) \qquad p_i(k) = \langle \mathrm{Ad}(k)^{-1}u, u_i^* \rangle,$$

so that

$$(1.74b) \qquad \mathrm{Ad}(k)^{-1}u = \sum_i p_i(k)u_i.$$

Then (1.73b) is

$$(1.75) \qquad = \sum_i \int_K f(ku_i)p_i(k)\,dT(k).$$

Now write $T_i = p_i T$ for the distribution defined by $f \mapsto T(p_i f)$, as in (B.11). By Lemma 1.46, T_i is in $R(K)$. Combining (1.73) and (1.75), we obtain

$$(1.76a) \qquad \langle \partial(u) * T, f \rangle = \sum_i \langle T_i, f(\cdot u_i) \rangle,$$

which we can rewrite using (1.72) as

$$(1.76b) \qquad \langle \partial(u) * T, f \rangle = \sum_i \langle T_i * \partial(u_i), f \rangle.$$

Since f was arbitrary, this means that

$$(1.76c) \qquad \partial(u) * T = \sum_i T_i * \partial(u_i).$$

The first step in this argument, the equality in (1.73b), leans a little too heavily on symbolic notation, using $f(uk)$ in place of $(l(u')f)(k)$. To pass from (1.73) to (1.76) more carefully, we note first that

$$(1.77) \qquad (l(u')f)(k) = (r(\mathrm{Ad}(k)^{-1}u)f)(k)$$

by verifying this identity for $u \in \mathfrak{g}_0$ and then extending it to $u \in U(\mathfrak{g})$. Substituting from (1.74b), we have

$$(1.78) \qquad l(u^t)f = \sum_i p_i r(u_i)f,$$

and then (1.76b) results from the computation

$$
\begin{aligned}
\langle \partial(u) * T, f \rangle &= \langle T, l(u^t)f \rangle & \text{by (1.73a)} \\
&= \sum_i \langle T, p_i r(u_i)f \rangle & \text{by (1.78)} \\
&= \sum_i \langle T_i, r(u_i)f \rangle \\
&= \sum_i \langle r(u_i^t)T_i, f \rangle \\
(1.79) \qquad &= \sum_i \langle T_i * \partial(u_i), f \rangle.
\end{aligned}
$$

We can summarize this result as follows.

Proposition 1.80. Let $S * \partial(u)$ and $T * \partial(v)$ be members of

$$R(\mathfrak{g}, K) \cong R(K) \otimes_{U(\mathfrak{k})} U(\mathfrak{g}).$$

Let $\{u_i\}$ be a basis of $\langle \operatorname{Ad}(K)u \rangle$, and let $\{u_i^*\}$ be the dual basis of $\langle \operatorname{Ad}(K)u \rangle^*$. Define distributions T_i by means of (B.11) as

$$T_i = \langle \operatorname{Ad}(\cdot)^{-1}u, u_i^* \rangle T.$$

Then the product formula in $R(\mathfrak{g}, K)$ is

$$(S * \partial(u)) * (T * \partial(v)) = \sum_i (S * T_i) * \partial(u_i v).$$

PROOF. Lemma 1.46 shows that T_i is in $R(K)$, and the computation (1.79) shows that $\partial(u) * T = \sum_i T_i * \partial(u_i)$. Convolving this identity on the left side by S and on the right side by $\partial(v)$, we obtain the desired product formula.

REMARKS. Taking advantage of vector-valued distributions as in (1.49) and Proposition 1.47, we can abbreviate the result of the computation (1.79) as

$$(1.81) \qquad \partial(u) * T = T(\cdot) * \partial(\mathrm{Ad}(\cdot)^{-1}u).$$

(We shall use somewhat different notation in later sections.) In the notation of (1.81), the product formula becomes

$$(1.82) \qquad (S * \partial(u)) * (T * \partial(v)) = (S * T(\cdot)) * \partial((\mathrm{Ad}(\cdot)^{-1}u)v).$$

WARNING. Although the notation (1.82) for the product formula is more elegant than that in Proposition 1.80, it suffers from the weakness that it is hard to tell which of the function multiplication and the convolution occurs first. This distinction is important because the two operations do not commute in general. The notation in Proposition 1.80 is clearer in this regard. When multiplications and convolutions are both present, it helps to make use of the expansion in (1.49). Better yet is the integral notation in (1.62), which keeps multiplications and convolutions straight without any need to choose a basis.

Proposition 1.83. Every member of $R(\mathfrak{g}, K)$ is two-sided K finite.

PROOF. Let $T * \partial(u)$ be in $R(\mathfrak{g}, K) \subseteq \mathcal{E}'(G, K)$; $T * \partial(u)$ is assumed left K finite. Then

$$
\begin{aligned}
r(k)(T * \partial(u)) &= T * \partial(u) * \delta_{k^{-1}} & \text{by (1.10)} \\
&= (T * \delta_{k^{-1}}) * (\delta_k * \partial(u) * \delta_{k^{-1}}) \\
&= r(k)T * \partial(\mathrm{Ad}(k)u) & \text{by (1.10) and (1.11).}
\end{aligned}
$$

As $k \in K$ varies, the right side remains in the image under convolution of the finite-dimensional space $\langle r(K)T \rangle \otimes \langle \partial(\mathrm{Ad}(K)u) \rangle$. Hence $T * \partial(u)$ is right K finite.

Thus we have an isomorphism

$$R(\mathfrak{g}, K) \cong U(\mathfrak{g}) \otimes_{U(\mathfrak{k})} R(K)$$

that respects the left action of $U(\mathfrak{g})$ and the right action of $R(K)$. The formulas corresponding to (1.81) and (1.82) are

$$(1.84) \qquad S * \partial(v) = \partial(\mathrm{Ad}(\cdot)v) * S(\cdot)$$

and

$$(1.85) \qquad (\partial(u) * S) * (\partial(v) * T) = \partial(u(\mathrm{Ad}(\cdot)v)) * (S(\cdot) * T).$$

The members T of $R(K)$ embed as $T * \partial(1)$ and $\partial(1) * T$ in the two realizations of $R(\mathfrak{g}, K)$; this fact is consistent with (1.81) and (1.84).

The operation **transpose** carries $\mathcal{E}'(G)$ into itself, sending (support T) into (support $T)^{-1}$. Hence it carries $\mathcal{E}'(G, K)$ into itself. It follows from Proposition 1.83 that it carries $R(\mathfrak{g}, K)$ into itself.

The algebra $R(\mathfrak{g}, K)$ has an approximate identity in the sense of Definition 1.54. Namely the approximate identity $\{\chi_A\}$ of $R(K)$ embeds as an approximate identity of $R(\mathfrak{g}, K)$ under the inclusion $R(K) \subseteq R(\mathfrak{g}, K)$. The property $\chi_A * \chi_B = \chi_A$ when $A \subseteq B$ was proved in (1.53a). The fact that each $r \in R(\mathfrak{g}, K)$ has some A with $\chi_A * r = r * \chi_A = r$ follows from Proposition 1.83 and Theorem 1.57: The proposition says that $(l, R(K))$ and $(r, R(K))$ are locally K finite representations of K, and the theorem says that the correponding $R(K)$ modules are approximately unital.

For any pair $(\mathfrak{g}, K)$, a $(\mathfrak{g}, K)$ **module** is a complex vector space V carrying representations of $\mathfrak{g}$ and K such that

(1.86)

    (i) the K representation is locally K finite;

    (ii) the differentiated version of the K action is the restriction to $\mathfrak{k}$ of the $\mathfrak{g}$ action; and

    (iii) $(\mathrm{Ad}(k)u)x = k(u(k^{-1}x))$ for $k \in K$, $u \in U(\mathfrak{g})$, and $x \in V$.

If V and W are two $(\mathfrak{g}, K)$ modules, a $(\mathfrak{g}, K)$ **map** from V to W is a linear map that respects the $\mathfrak{g}$ representations and also the K representations. We denote by $\mathrm{Hom}_{\mathfrak{g}, K}(V, W)$ the vector space of all $(\mathfrak{g}, K)$ maps from V to W.

Under the assumption (1.65) that G exists, the algebra $R(\mathfrak{g}, K)$ plays a role for $(\mathfrak{g}, K)$ modules that extends the role of $R(K)$ in Theorem 1.57. We state the result now without proof. In §6 we shall restate the result for pairs $(\mathfrak{g}, K)$ that need not satisfy (1.65), and we shall supply a proof at that time.

Theorem. If $(\mathfrak{g}, K)$ is a pair and V is a $(\mathfrak{g}, K)$ module, then V is an approximately unital left $R(\mathfrak{g}, K)$ module in a natural way. Conversely every approximately unital left $R(\mathfrak{g}, K)$ module comes from a $(\mathfrak{g}, K)$ module by an inverse construction. Under this correspondence of representations and modules,

$$\mathrm{Hom}_{\mathfrak{g}, K}(V, W) = \mathrm{Hom}_{R(\mathfrak{g}, K)}(V, W).$$

5. Abstract Construction

In this section we begin the abstract construction of the Hecke algebra associated to a pair $(\mathfrak{g}, K)$ as defined in (1.64). The goal is to produce an algebra $R(\mathfrak{g}, K)$ with the same properties as in §4 (specifically Corollary 1.71 through formula (1.85)) such that $(\mathfrak{g}, K)$ modules are the same as approximately unital left $R(\mathfrak{g}, K)$ modules, but we no longer assume that $(\mathfrak{g}, K)$ arises from a group G.

We shall carry out the construction in two steps. The first step, to be done in this section, is to produce an algebra that is isomorphic to $R(K) \otimes_{\mathbb{C}} U(\mathfrak{g})$ as a vector space and that converts a weak version of $(\mathfrak{g}, K)$ modules into approximately unital left modules for the algebra. This algebra is called the "smash product" of $R(K)$ and $U(\mathfrak{g})$ and will be denoted $R(K) \# U(\mathfrak{g})$. In §6 we define $R(\mathfrak{g}, K)$ as a quotient of this smash product, and ordinary $(\mathfrak{g}, K)$ modules will be the same as approximately unital left $R(\mathfrak{g}, K)$ modules.

The first step involves forgetting that $U(\mathfrak{k})$ acts on both $R(K)$ and $U(\mathfrak{g})$ and that these actions get reflected in property (ii) of $(\mathfrak{g}, K)$ modules in (1.86). We shall introduce terminology that encourages us to forget these actions of $U(\mathfrak{k})$.

Suppose that $\mathcal{A}$ is a complex associative algebra with an identity and that K is a compact Lie group. We say that $(\mathcal{A}, K)$ is a **weak pair** if we are given a locally K finite representation

$$(1.87) \qquad\qquad \mathrm{Ad} : K \to \mathrm{Aut}(\mathcal{A})$$

of K on $\mathcal{A}$ by automorphisms of the associative-algebra structure. A **weakly compatible module** for the weak pair $(\mathcal{A}, K)$ is a complex vector space M having the structure of a left $\mathcal{A}$ module and a locally K finite representation of K, subject to the compatibility condition

$$(1.88) \qquad k(am) = (\mathrm{Ad}(k)a)(km) \qquad \text{for } k \in K, \, a \in \mathcal{A}, \, m \in M.$$

When no confusion is possible, we may refer to M as an $(\mathcal{A}, K)$ module.

If M and N are two weakly compatible $(\mathcal{A}, K)$ modules, we denote by $\mathrm{Hom}_{\mathcal{A}, K}(M, N)$ the vector space of linear maps from M into N that respect the $\mathcal{A}$ module structures and also the K representations.

EXAMPLES.

1) If $(\mathfrak{g}, K)$ is a pair in the sense of (1.64), then $(U(\mathfrak{g}), K)$ is a weak pair. Any $(\mathfrak{g}, K)$ module in the sense of (1.86) is a weakly compatible module for $(U(\mathfrak{g}), K)$.

2) If $(\mathfrak{g}, K)$ is a pair, then $(S(\mathfrak{g}/\mathfrak{k}), K)$ is a weak pair, where S refers to a symmetric algebra. Weakly compatible modules for this pair arise as associated graded modules for $(\mathfrak{g}, K)$ modules. They make an appearance in Chapter X, but we shall not need the theory of the present section at that time.

We propose to construct an algebra $R(K) \# \mathcal{A}$ with the property that weakly compatible $(\mathcal{A}, K)$ modules are the same thing as approximately unital left $R(K) \# \mathcal{A}$ modules. This algebra will be obtained by imposing a multiplication on $R(K) \otimes_{\mathbb{C}} \mathcal{A}$. To motivate the definition, suppose M is a weakly compatible $(\mathcal{A}, K)$ module. Convert the representation of K on M into a left $R(K)$ module structure by (1.56), and define maps

$$(1.89a) \qquad \begin{aligned} \mu_1 &: (R(K) \otimes_{\mathbb{C}} \mathcal{A}) \times M \to M \\ \mu_2 &: (\mathcal{A} \otimes_{\mathbb{C}} R(K)) \times M \to M \end{aligned}$$

out of the $\mathcal{A}$ module structure and the $R(K)$ module structure by

$$(1.89b) \qquad \begin{aligned} \mu_1(T \otimes a)(m) &= T(am) \\ \mu_2(a \otimes T)(m) &= a(Tm) \end{aligned} \qquad \text{for } T \in R(K),\, a \in \mathcal{A},\, m \in M.$$

Ultimately we want these maps to represent ring actions by $R(K) \# \mathcal{A}$, with elements taken either from $R(K) \otimes_{\mathbb{C}} \mathcal{A}$ or from $\mathcal{A} \otimes_{\mathbb{C}} R(K)$. We need formulas analogous to (1.81) and (1.84) relating these two maps. For this purpose we introduce a vector space isomorphism

$$(1.90a) \qquad \tau : R(K) \otimes \mathcal{A} \to \mathcal{A} \otimes R(K),$$

analogous to identity (1.84). In the vector notation of (1.49) and Proposition 1.47, the map τ is given by

$$(1.90b) \qquad \tau(T \otimes a) = \mathrm{Ad}(\cdot)a \otimes T$$

for $T \in R(K)$ and $a \in \mathcal{A}$.

The mapping τ is a vector-space isomorphism, with τ^{-1} in vector notation given by

$$(1.91) \qquad \tau^{-1}(b(\cdot) \otimes S) = S \otimes \mathrm{Ad}(\cdot)^{-1}b(\cdot).$$

In fact, $\tau^{-1}\tau = 1$ since

$$\tau^{-1}(\tau(T \otimes a(\cdot))) = \tau^{-1}(\mathrm{Ad}(\cdot)a(\cdot) \otimes T)$$
$$= T \otimes \mathrm{Ad}(\cdot)^{-1}\mathrm{Ad}(\cdot)a(\cdot) = T \otimes a(\cdot);$$

similarly $\tau\tau^{-1} = 1$.

Let us write out τ and τ^{-1} in the scalar notation of the right side of (1.49). If we choose a basis $\{a_i\}$ for an $\mathrm{Ad}(K)$ invariant finite-dimensional space containing the span $\langle \mathrm{Ad}(K)a \rangle$ and if we let $\{a_i^*\}$ be the dual basis, then

$$(1.92a) \qquad\qquad \mathrm{Ad}(k)a = \sum_i \langle \mathrm{Ad}(k)a, a_i^* \rangle a_i,$$

with each coefficient $\langle \mathrm{Ad}(\cdot)a, a_i^* \rangle$ in $C(K)$. In terms of this expansion, T is given by

$$(1.92b) \qquad\qquad \tau(T \otimes a) = \sum_i a_i \otimes \langle \mathrm{Ad}(\cdot)a, a_i^* \rangle T.$$

Similarly the formula for τ^{-1} in scalar notation is

$$(1.93) \qquad\qquad \tau^{-1}(b \otimes S) = \sum_j \langle \mathrm{Ad}(\cdot)^{-1}b, b_j^* \rangle S \otimes b_j.$$

DEFINITION 1.94. Suppose $(\mathcal{A}, K)$ is a weak pair. The **smash product** $R(K) \# \mathcal{A}$ is the algebra with underlying vector space $R(K) \otimes_{\mathbb{C}} \mathcal{A}$ and with multiplication defined as follows. Suppose S and T are in $R(K)$ and a and b are in $\mathcal{A}$. Write

$$\tau^{-1}(a \otimes T) = T \otimes \mathrm{Ad}(\cdot)^{-1}a = \sum_j \langle \mathrm{Ad}(\cdot)^{-1}a, a_j^* \rangle T \otimes a_j$$

and $\qquad\qquad\qquad T_j = \langle \mathrm{Ad}(\cdot)^{-1}a, a_j^* \rangle T$

in accordance with (1.91) and (1.93). Then

$$(S \otimes a)(T \otimes b) = \sum_j (ST_j \otimes a_j b).$$

Here the multiplications on the right are the (convolution) multiplication in $R(K)$ and the algebra multiplication in $\mathcal{A}$.

Multiplication is well defined, independently of $\{a_j\}$ and its span, since the product rule amounts to applying τ^{-1} to $a \otimes T$, left multiplying by S, and right multiplying by b. The distribution T_j is in $R(K)$ by Lemma 1.46.

WARNING. It is tempting to think that multiplication in $R(K) \# \mathcal{A}$ has a simple expression in vector notation using τ. Certainly the result should be left-by-S and right-by-b on $\tau^{-1}(a \otimes T)$. But left-by-S does not consist simply in putting S at the left of the $R(K)$ factor in $R(K) \otimes_{C(K)} C(K, \mathcal{A})$. Formula (1.102) below will show that the correct operation involving S is

$$(\text{left-by-}S)(T \otimes c(\cdot)) = (l \otimes l)(S)(T \otimes c(\cdot))$$

instead of $(l \otimes 1)(S)(T \otimes c(\cdot))$ on the right side. (The latter is not even well defined over $C(K)$.) When $c(\cdot) \in C(K, V)$ is a constant function, the two expressions amount to the same thing, and a simple formula results. To make manageable computations with multiplication, we must therefore either switch to scalar notation and use the formulas in Definition 1.94 or else use integral notation for distributions.

Lemma 1.95. Suppose $(\mathcal{A}, K)$ is a weak pair, M is a weakly compatible $(\mathcal{A}, K)$ module, and α is in $R(K) \otimes_{\mathbb{C}} \mathcal{A}$. If m is in M, then the maps μ_1 and μ_2 in (1.89b) are related by

$$\mu_1(\alpha)(m) = \mu_2(\tau(\alpha))(m).$$

PROOF. Without loss of generality, let $\alpha = T \otimes a$. With a_i as in (1.92a), put $T_i = \langle \text{Ad}(\cdot)a, a_i^* \rangle T$. We are to prove that

$$(1.96) \qquad T(am) = \sum_i a_i(T_i m).$$

From (1.88) we have

$$(1.97) \qquad k(am) = (\text{Ad}(k)a)(km) = \sum_i \langle \text{Ad}(k)a, a_i^* \rangle a_i(km).$$

Thus

$$
\begin{aligned}
T(am) &= \int_K k(am)\, dT(k) && \text{by (1.21a)} \\
&= \sum_i \int_K \langle \text{Ad}(k)a, a_i^* \rangle a_i(km)\, dT(k) && \text{by (1.97)} \\
&= \sum_i \int_K a_i(km)\, dT_i(k) && \\
&= \sum_i a_i(T_i m) && \text{by (1.21a).}
\end{aligned}
$$

This proves (1.96) and the lemma.

Lemma 1.98. Suppose $(\mathcal{A}, K)$ is a weak pair. Then the definitions

(1.99)
$$k(\tau^{-1}(b \otimes T)) = \tau^{-1}(\mathrm{Ad}(k)b \otimes l(k)T)$$
$$a(\tau^{-1}(b \otimes T)) = \tau^{-1}(ab \otimes T)$$

for $a, b \in \mathcal{A}$, $T \in R(K)$, and $k \in K$ make $M = R(K) \otimes_{\mathbb{C}} \mathcal{A} = R(K) \# \mathcal{A}$ into a weakly compatible module for the weak pair $(\mathcal{A}, K)$ in such a way that the map $\mu_1 : (R(K) \otimes_{\mathbb{C}} \mathcal{A}) \times M \to M$ of (1.89) coincides with the multiplication mapping within $R(K) \# \mathcal{A}$.

PROOF. Using Proposition 1.47 to write

$$M = R(K) \otimes_{\mathbb{C}} \mathcal{A} \cong R(K) \otimes_{C(K)} C(K, \mathcal{A}),$$

let us observe that the definitions of actions by k and a become

(1.100)
$$k(\tau^{-1}(b(\cdot) \otimes T)) = \tau^{-1}(\mathrm{Ad}(k)(b(k^{-1} \cdot)) \otimes l(k)T)$$
$$a(\tau^{-1}(b(\cdot) \otimes T)) = \tau^{-1}(ab(\cdot) \otimes T)$$

in $R(K) \otimes_{C(K)} C(K, \mathcal{A})$. In fact, it is enough to verify that the actions by k and a in (1.100) are well defined in the tensor product over $C(K)$, since they clearly reduce to the given definitions when $b(\cdot)$ is constant. In other words we are to check that the effects of k and a on a relation

$$b(\cdot)c(\cdot) \otimes T - b(\cdot) \otimes cT,$$

namely τ^{-1} of

(1.101)
$$\mathrm{Ad}(k)(b(k^{-1} \cdot)c(k^{-1} \cdot)) \otimes l(k)T - \mathrm{Ad}(k)(b(k^{-1} \cdot)) \otimes l(k)(cT)$$
$$ab(\cdot)c(\cdot) \otimes T - ab(\cdot) \otimes cT,$$

are relations. When we apply τ^{-1} to the second expression in (1.101), we obtain

$$T \otimes \mathrm{Ad}(\cdot)^{-1}(ab(\cdot)c(\cdot)) - cT \otimes \mathrm{Ad}(\cdot)^{-1}(ab(\cdot)),$$

which is a relation since $c(\cdot)$ is scalar valued and is acted upon trivially by $\mathrm{Ad}(\cdot)^{-1}$. For the first expression in (1.101), we have

$$\mathrm{Ad}(k)(b(k^{-1} \cdot)c(k^{-1} \cdot)) = \mathrm{Ad}(k)(b(k^{-1} \cdot))l(k)c$$

and $l(k)(cT) = (l(k)c)(l(k)T)$. Thus τ^{-1} of the first expression equals

$$l(k)T \otimes \mathrm{Ad}(\,\cdot\,)^{-1}\mathrm{Ad}(k)(b(k^{-1}\cdot))l(k)c$$
$$- (l(k)c)(l(k)T) \otimes \mathrm{Ad}(\,\cdot\,)^{-1}\mathrm{Ad}(k)(b(k^{-1}\cdot)),$$

and this is a relation.

Since τ is a vector-space isomorphism, km and am are well defined for all $m \in M$. It is clear that the definitions (1.100) make K act on M by a locally K finite representation and make M into a left $\mathcal{A}$ module. Using (1.100) repeatedly, we obtain

$$\begin{aligned}
k(a(\tau^{-1}(b(\,\cdot\,) \otimes T))) &= \tau^{-1}(\mathrm{Ad}(k)(ab(k^{-1}\cdot)) \otimes l(k)T) \\
&= \tau^{-1}((\mathrm{Ad}(k)a)\mathrm{Ad}(k)(b(k^{-1}\cdot)) \otimes l(k)T) \\
&= (\mathrm{Ad}(k)a)\tau^{-1}(\mathrm{Ad}(k)(b(k^{-1}\cdot)) \otimes l(k)T) \\
&= (\mathrm{Ad}(k)a)(k(\tau^{-1}(b(\,\cdot\,) \otimes T))),
\end{aligned}$$

and therefore M is a weakly compatible module for the weak pair $(\mathcal{A}, K)$.

The main step is to compute the action of K and $\mathcal{A}$ in (1.100) without the τ^{-1}. Let us verify that

(1.102)
$$k(T \otimes b(\,\cdot\,)) = l(k)T \otimes l(k)b(\,\cdot\,)$$
$$a(T \otimes b(\,\cdot\,)) = T \otimes (\mathrm{Ad}(\,\cdot\,)^{-1}a)b(\,\cdot\,).$$

For the first one, we have

$$\begin{aligned}
k(T \otimes b(\,\cdot\,)) &= k(\tau^{-1}(\tau(T \otimes b(\,\cdot\,)))) \\
&= k(\tau^{-1}(\mathrm{Ad}(\,\cdot\,)(b(\,\cdot\,)) \otimes T)) \\
&= \tau^{-1}(\mathrm{Ad}(k)(\mathrm{Ad}(k^{-1}\cdot)b(k^{-1}\cdot)) \otimes l(k)T) \quad \text{by (1.100)} \\
&= l(k)T \otimes \mathrm{Ad}(\,\cdot\,)^{-1}(\mathrm{Ad}(\,\cdot\,)(b(k^{-1}\cdot))) \quad \text{by (1.91)} \\
&= l(k)T \otimes l(k)b.
\end{aligned}$$

For the second one,

$$\begin{aligned}
a(T \otimes b(\,\cdot\,)) &= a(\tau^{-1}(\tau(T \otimes b(\,\cdot\,)))) \\
&= a(\tau^{-1}(\mathrm{Ad}(\,\cdot\,)(b(\,\cdot\,)) \otimes T)) \quad \text{by (1.90b)} \\
&= \tau^{-1}(a\mathrm{Ad}(\,\cdot\,)(b(\,\cdot\,)) \otimes T) \quad \text{by (1.100)} \\
&= T \otimes \mathrm{Ad}(\,\cdot\,)^{-1}(a\mathrm{Ad}(\,\cdot\,)(b(\,\cdot\,))) \quad \text{by (1.91)} \\
&= T \otimes (\mathrm{Ad}(\,\cdot\,)^{-1}a)b(\,\cdot\,).
\end{aligned}$$

Now we can check that (1.89) indeed implements multiplication in M. To handle multiplication we switch to scalar notation. Let b be in $\mathcal{A}$, as opposed to $C(K, \mathcal{A})$. In view of (1.102), the action of K yields

$$(1.103) \qquad\qquad S(T \otimes b) = ST \otimes b,$$

by Propositions 1.20 and 1.31. (Again ST refers to convolution.) The module action is

$$
\begin{aligned}
\mu_1(S \otimes a)(T \otimes b) &= S(a(T \otimes b)) && \text{by the first formula in (1.89b)} \\
&= S \sum_j T_j \otimes a_j b && \text{by (1.102) for } a(T \otimes b) \\
&= \sum_j ST_j \otimes a_j b && \text{by (1.103),}
\end{aligned}
$$

and this coincides with the product $(S \otimes a)(T \otimes b)$ in Definition 1.94.

Proposition 1.104. Suppose $(\mathcal{A}, K)$ is a weak pair.

(a) The smash product $R(K) \# \mathcal{A}$ is an associative algebra, and the formula for μ_1 in (1.89b) defines a module structure under this algebra for any weakly compatible $(\mathcal{A}, K)$ module.

(b) If the isomorphism τ of (1.92b) is used to identify $R(K) \otimes_C \mathcal{A}$ with $\mathcal{A} \otimes_C R(K)$, then, in this realization of the underlying vector space, the algebra structure on $R(K) \# \mathcal{A}$ is

$$(a \otimes S)(b \otimes T) = \sum_i (ab_i) \otimes (S_i T),$$

provided $\tau(S \otimes b) = \sum_i b_i \otimes S_i$.

PROOF.

(a) Suppose M is a weakly compatible $(\mathcal{A}, K)$ module. If α and β belong to $R(K) \# \mathcal{A}$ and m is an element of M, then the claim is that

$$(1.105) \qquad\qquad \mu_1(\alpha)(\mu_1(\beta)m) = \mu_1(\alpha\beta)(m).$$

To prove this, we may suppose $\alpha = S \otimes a$ and $\beta = T \otimes b$ as in Definition 1.94. Then

$$
\begin{aligned}
\mu_1(\alpha)(\mu_1(\beta)m) &= S(a(T(bm))) && \text{by (1.89)} \\
&= S(\mu_2(a \otimes T)(bm)) && \text{by (1.89)} \\
&= S(\mu_1(\sum T_j \otimes a_j)(bm)) && \text{by Lemma 1.95} \\
&= \sum S(T_j(a_j(bm))) && \text{by (1.89).}
\end{aligned}
$$

Since M is a module for $\mathcal{A}$ and for $R(K)$, the above expression is

$$
\begin{aligned}
&= \sum (ST_j)((a_j b)m)\\
&= \mu_1\Big(\sum (ST_j) \otimes (a_j b)\Big)m \qquad \text{by (1.89).}
\end{aligned}
$$

Now Definition 1.94 shows that this is exactly the right side of (1.105).

Lemma 1.98 shows that $R(K) \# \mathcal{A}$ itself is a weakly compatible $(\mathcal{A}, K)$ module and that (1.89) yields the product formula. Thus (1.105) establishes that multiplication in $R(K) \# \mathcal{A}$ is associative. Consequently $R(K) \# \mathcal{A}$ is an associative algebra. Reference to (1.105) shows that (1.89) makes M into a left $R(K) \# \mathcal{A}$ module.

(b) It is understood that $\mathrm{Ad}(\cdot)b = \sum_i \langle \mathrm{Ad}(\cdot)b,\, b_i^* \rangle b_i$ and that $S_i = \langle \mathrm{Ad}(\cdot)b,\, b_i^* \rangle S$. Then (1.21a) gives

$$
\begin{aligned}
S(\tau^{-1}(b \otimes T)) &= \int_K k(\tau^{-1}(b \otimes T))\, dS(k)\\
&= \int_K \tau^{-1}(\mathrm{Ad}(k)b \otimes l(k)T)\, dS(k) && \text{by (1.99)}\\
&= \sum_i \int_K \tau^{-1}(b_i \otimes l(k)T)\langle \mathrm{Ad}(\cdot)b,\, b_i^* \rangle\, dS(k)\\
&= \sum_i \langle S_i,\, \tau^{-1}(b_i \otimes l(\cdot)T)\rangle\\
&= \sum_i \tau^{-1}(b_i \otimes l(S_i)T) && \text{by (1.21a)}\\
(1.106)\qquad &= \sum_i \tau^{-1}(b_i \otimes S_i T) && \text{by Proposition 1.31}
\end{aligned}
$$

if we again abbreviate the convolution as $S_i T$. Consequently the product in question is

$$
\begin{aligned}
\tau^{-1}(a \otimes S)\tau^{-1}(b \otimes T) &= \mu_1(\tau^{-1}(a \otimes S))(\tau^{-1}(b \otimes T)) && \text{by Lemma 1.98}\\
&= a(S(\tau^{-1}(b \otimes T))) && \text{by Lemma 1.95}\\
&= a(\tau^{-1}(\sum_i b_i \otimes S_i T)) && \text{by (1.106)}\\
&= \tau^{-1}(\sum_i a b_i \otimes S_i T) && \text{by (1.99).}
\end{aligned}
$$

The identification we are to make in (b) amounts to dropping τ^{-1} from the notation, and the desired alternate product formula is the result. This completes the proof.

From now on, we shall drop μ_1, μ_2, and τ from the notation. Every element of $R(K) \# A$ can be written in two ways, one in $R(K) \otimes_{\mathbb{C}} A$ and one in $A \otimes_{\mathbb{C}} R(K)$; these are related by the now-suppressed isomorphism τ.

From (1.90) we have $1 \otimes T = T \otimes 1$ if T is in $R(K)$. Thus Definition 1.94 gives $(S \otimes 1)(T \otimes 1) = ST \otimes 1$, with the product ST again referring to convolution. Consequently the one-one mapping of $R(K)$ into $R(K) \# A$ given by $T \mapsto T \otimes 1$ is an algebra homomorphism.

Let $\{\chi_A\}$ be the approximate identity (1.51) of $R(K)$. Then the image $\chi_A \otimes 1$ is an approximate identity in $R(K) \# A$. In fact, any element r of $R(K) \# A$ is a finite sum $\sum S_l \otimes a_l$. We can find a finite subset $A \subseteq \widehat{K}$ with $\chi_A S_l = S_l$ for all l, and then $(\chi_A \otimes 1)r = r$. The same element r is also a finite sum $\sum b_m \otimes T_m$. We can find B with $T_m \chi_B = T_m$ for all m, and then $r(1 \otimes \chi_B) = r$. Hence

$$(\chi_{A \cup B} \otimes 1)r = r = r(1 \otimes \chi_{A \cup B}),$$

and $\{\chi_A \otimes 1\}$ is indeed an approximate identity.

Proposition 1.107. Suppose (A, K) is a weak pair. If M is a weakly compatible module for (A, K), then the definition (1.89) makes M into an approximately unital left $R(K) \# A$ module. Conversely every approximately unital left $R(K) \# A$ module arises from one and only one weakly compatible module for (A, K) in this way, and the two constructions are inverse to one ·another. Under this correspondence of representations and modules,

$$\mathrm{Hom}_{A,K}(M, N) = \mathrm{Hom}_{R(K)\#A}(M, N).$$

PROOF. If M is a weakly compatible module for (A, K), then the first part of Proposition 1.104 says that M becomes a left $R(K) \# A$ module. At the same time, Theorem 1.57 says that M, being the space of a locally K finite representation of K, becomes an approximately unital $R(K)$ module. The two actions of $R(K)$ are compatible (under the identification $T \in R(K) \mapsto T \otimes 1 \in R(K) \# A$), and therefore the $R(K) \# A$ module structure is approximately unital.

Conversely let M be an approximately unital left $R(K) \# A$ module. It is then an approximately unital left $R(K)$ module, and Theorem 1.57 uniquely determines a locally K finite representation of K yielding the $R(K)$ module structure. For uniqueness of the A action, we apply Lemma 1.95 and Proposition 1.104 to see that we must have

$$a(Tm) = (a \otimes T)m \qquad \text{for } a \in A, \ T \in R(K), \ m \in M.$$

Since M is approximately unital, $\chi_A m = m$ for A a sufficiently large finite subset of $\widehat{K}$, and hence we must have

(1.108) $am = (a \otimes \chi_A)m$ for all $a \in \mathcal{A}$.

This proves uniqueness.

For existence of the weakly compatible module structure, we define the action of $\mathcal{A}$ by (1.108), where again A is chosen so that $\chi_A m = m$. This choice is possible since $R(K) \# \mathcal{A}$ is approximately unital. To prove that M is a left $\mathcal{A}$ module, we need to verify associativity. Fix A and a as above, and choose B large enough so that

$$\chi_B(a \otimes \chi_A) = a \otimes \chi_A.$$

This is possible since M is approximately unital. For such B,

$$\begin{aligned}
b(am) &= (b \otimes \chi_B)((a \otimes \chi_A)m) && \text{by (1.108)} \\
&= ((b \otimes \chi_B)(a \otimes \chi_A))m && \text{since } M \text{ is an } R(K) \# \mathcal{A} \text{ module} \\
&= (b(\chi_B(a \otimes \chi_A)))m && \text{by Lemma 1.98} \\
&= (b(a \otimes \chi_A))m && \\
&= ((ba) \otimes \chi_A)m && \text{by Lemma 1.98} \\
&= (ba)m && \text{by (1.108).}
\end{aligned}$$

So (1.108) makes M a left $\mathcal{A}$ module. To check the weak compatibility, fix A as in (1.108). Then the definition in Theorem 1.57 gives for all sufficiently large B

$$\begin{aligned}
k(am) &= (\delta_k \chi_B)(am) && \text{by definition of } K \text{ action on } M \\
&= (\delta_k \chi_B)((a \otimes \chi_A)m) && \text{by (1.108)} \\
&= ((\delta_k \chi_B)(a \otimes \chi_A))m && \text{since } M \text{ is an } R(K) \# \mathcal{A} \text{ module} \\
&= (k(a \otimes \chi_A))m && \text{from the } K \text{ action on } R(K) \# \mathcal{A} \\
&&& \text{for } B \text{ sufficiently large} \\
&= (\mathrm{Ad}(k)a \otimes l(k)\chi_A)m && \text{by Lemma 1.98} \\
&= (\mathrm{Ad}(k)a \otimes \delta_k \chi_A)m && \\
&= (\mathrm{Ad}(k)a)((\delta_k \chi_A)m) && \text{by Lemma 1.95} \\
&= (\mathrm{Ad}(k)a)(km) && \text{by definition of } K \text{ action on } M.
\end{aligned}$$

Thus the module is weakly compatible. Since we have both existence and uniqueness for the K representation structure and the $\mathcal{A}$ module structure, it follows that the two constructions are inverse to one another.

To check equality of the Hom's, let $E : M \rightarrow N$ be given. First suppose E is in $\mathrm{Hom}_{\mathcal{A}, K}(M, N)$, i.e., $E(km) = k(Em)$ and $E(am) = a(EM)$ for $k \in K$, $a \in \mathcal{A}$, $m \in M$. Formula (1.21d) gives $E(Tm) = T(Em)$ for $T \in R(K)$. Two applications of Proposition 1.104 then give

$$(T \otimes a)(Em) = T(a(Em)) = T(E(am)) = E(T(am)) = E((T \otimes a)m).$$

So E is in $\mathrm{Hom}_{R(K) \# \mathcal{A}}(M, N)$.

Conversely if E is in $\mathrm{Hom}_{R(K) \# \mathcal{A}}(M, N)$, then $(T \otimes a)(Em) = E((T \otimes a)m)$. So Proposition 1.104 gives

$$(1.109) \qquad T(a(Em)) = (T \otimes a)(Em) = E((T \otimes a)m) = E(T(am)).$$

Since M and N are approximately unital, we can choose A large enough so that $\chi_A(a(Em)) = a(Em)$ and $\chi_A(am) = am$. Taking $T = \chi_A$ in (1.109) gives $a(Em) = E(am)$. Taking $a = 1$ in (1.109) gives $T(Em) = E(Tm)$. Hence E is in $\mathrm{Hom}_{\mathcal{A}, K}(M, N)$.

A particular example of an approximately unital left $R(K) \# \mathcal{A}$ module is $R(K) \# \mathcal{A}$ itself, with the algebra acting on the left. The fact that $\{\chi_A \otimes 1\}$ is an approximate identity in the algebra implies that the module is approximately unital. The proposition says that K and $\mathcal{A}$ act on $R(K) \# \mathcal{A}$. By Lemma 1.98 these actions are

$$(1.110) \qquad \begin{aligned} k(b \otimes T) &= \mathrm{Ad}(k)b \otimes l(k)T \\ a(b \otimes T) &= ab \otimes T \end{aligned}$$

for $k \in K$, $a \in \mathcal{A}$, $b \in \mathcal{A}$, $T \in R(K)$.

There is an obvious definition of approximately unital right $R(K) \# \mathcal{A}$ module. A **weakly compatible right module** for the weak pair $(\mathcal{A}, K)$ is a complex vector space M having the structure of a right $\mathcal{A}$ module and a locally K finite linear action by K on the right, subject to the compatibility condition

$$(ma)k = (mk)(\mathrm{Ad}(k^{-1})a) \qquad \text{for } k \in K, \ a \in \mathcal{A}, \ m \in M.$$

Proposition 1.107 has an analog for this situation that we can state as follows.

Proposition 1.107′. Suppose $(\mathcal{A}, K)$ is a weak pair. If M is a weakly compatible right module for $(\mathcal{A}, K)$, then M becomes an approximately unital right $R(K) \# \mathcal{A}$ module. Conversely every approximately unital right $R(K) \# \mathcal{A}$ module arises from one and only one weakly compatible right module for $(\mathcal{A}, K)$ in this way, and the two constructions are inverse to one another. Under this correspondence of representations and modules,

$$\mathrm{Hom}_{\mathcal{A}, K}(M, N) = \mathrm{Hom}_{R(K) \# \mathcal{A}}(M, N).$$

An instance of Proposition 1.107′ arises if we consider $M = R(K) \# \mathcal{A}$ itself as an approximately unital right $R(K) \# \mathcal{A}$ module. The action of K and $\mathcal{A}$ for this case are

(1.111)
$$(T \otimes b)k = r(k)^{-1} T \otimes \mathrm{Ad}(k)^{-1} b \qquad \text{for } k \in K$$
$$(T \otimes b)a = T \otimes ba \qquad \text{for } a \in \mathcal{A}.$$

6. Hecke Algebras for Pairs $(\mathfrak{g}, K)$

In this section we complete the abstract construction of $R(\mathfrak{g}, K)$ for a pair $(\mathfrak{g}, K)$ in the sense of (1.64). The discussion in §5 concerned weak pairs $(\mathcal{A}, K)$, of which $(U(\mathfrak{g}), K)$ is an example. The action of K by Ad on $U(\mathfrak{g})$, hence of $\mathfrak{k}$ by ad, was incorporated into the definitions of weak pairs and weakly compatible modules. But the action of $\mathfrak{k}$ on $U(\mathfrak{g})$ by left or right translation was not incorporated into the definition of a weak pair. The new step, which will allow us to complete the definition of $R(\mathfrak{g}, K)$, will be to take this second action of $\mathfrak{k}$ into account. The result will be that $R(\mathfrak{g}, K)$ is a quotient algebra of the smash product $R(K) \# \mathcal{A}$.

We should mention that it is possible to abstract the additional condition on a weak pair $(\mathcal{A}, K)$ relative to $\mathfrak{k}$ so that the theory of this section will apply. A weak pair satisfying the additional condition will be called a "generalized pair." Generalized pairs play a role only in the advanced parts of the theory, starting with Chapter XII, and thus we prefer to work just with pairs $(\mathfrak{g}, K)$ for the time being. The theory of this section, however, does carry over to generalized pairs. A precise definition of "generalized pair," together with a tantalizing example and discussion of the sense in which the theory does carry over, appears at the end of this section. Some readers may want to read the material at the end of this section now, in order to be able to verify that the proofs of this section really do generalize.

Without further reference, we shall let a and b be members of $U(\mathfrak{g})$, X be a member of $\mathfrak{k}$, and u and v be members of $U(\mathfrak{k})$. If M is a $(\mathfrak{g}, K)$

module and m is in M, then Xm has two interpretations — one as the result of applying the $\mathfrak{g}$ action and another as the result $\partial(X)m$ of applying the differential of the K action. These two interpretations yield the same result, according to axiom (ii) of the definition (1.86) of a $(\mathfrak{g}, K)$ module. Thus the identities

(1.112)
$$(au)m = a(um)$$
$$(ua)m = u(am)$$

simply reflect the associativity property for a $U(\mathfrak{g})$ module if um or $u(am)$ is regarded as computed from the action of $U(\mathfrak{g})$, but they become nontrivial properties when we reinterpret um and $u(am)$ as computed from the differentiated action of K. We shall use (1.112) in proving (c) of the next lemma.

Lemma 1.113. Let $(\mathfrak{g}, K)$ be a pair. Then

(a) the linear span of all $b \otimes \partial(v)T - bv \otimes T$ is a left ideal in $R(K) \# U(\mathfrak{g})$,

(b) the linear span of all $S \otimes ua - S\partial(u) \otimes a$ is a right ideal in $R(K) \# U(\mathfrak{g})$,

(c) the elements in (a) and (b) act as 0 in any $(\mathfrak{g}, K)$ module M when M is reinterpreted as an approximately unital left $R(K) \# U(\mathfrak{g})$ module.

(d) the two linear spans in (a) and (b) are equal and therefore form a two-sided ideal $I_{\mathfrak{k}}$ in $R(K) \# U(\mathfrak{g})$,

PROOF. For (a), the left ideals are the invariant subspaces of $R(K) \# \mathcal{A}$ for the weakly compatible $(U(\mathfrak{g}), K)$ action of (1.110) on the left. The images under left-by-k and left-by-a of the element in (a) are

$$\mathrm{Ad}(k)b \otimes l(k)(\partial(v)T) - (\mathrm{Ad}(k)b)(\mathrm{Ad}(k)v) \otimes l(k)T$$

and
$$ab \otimes \partial(v)T - abv \otimes T.$$

The second of these is obviously in the linear span, while the first is in the linear span since

$$l(k)(\partial(v)T) \quad \text{means} \quad \delta_k * \partial(v) * T \qquad\qquad \text{by (1.10)}$$
$$= (\delta_k * \partial(v) * \delta_{k^{-1}}) * (\delta_k * T)$$
$$= \partial(\mathrm{Ad}(k)v) * l(k)T \qquad\qquad \text{by (1.10) and (1.11)}$$

and since the right side is $(\partial(\mathrm{Ad}(k)v))l(k)T$ in our notation with convolution signs suppressed. Therefore the linear span in (a) is a left ideal.

The proof of (b) is similar, using right modules and (1.111) in place of left modules and (1.110).

For (c), Lemmas 1.95 and 1.98 give

$$(b \otimes \partial(v)T)(m) = b((\partial(v)T)(m)) = b(\partial(v)(Tm)) = (bv)(Tm),$$

the last equality holding by (1.112). A second application of Lemmas 1.95 and 1.98 shows that the right side equals $(bv \otimes T)(m)$. Hence the elements in (a) act as 0. Similarly the elements in (b) act as 0 because of the computation

$$(S\partial(u) \otimes a) = (S\partial(u))(a(m)) = S(\partial(u)(a(m)))$$
$$= S(ua(m)) = (S \otimes ua)(m).$$

This proves (c).

For (d), we make $\overline{M} = U(\mathfrak{g}) \otimes_{U(\mathfrak{k})} R(K)$ into a $(\mathfrak{g}, K)$ module by having $U(\mathfrak{g})$ act by left multiplication on the first factor alone and by having K act by $\mathrm{Ad} \otimes l$. We need to check that the action of K is well defined over $U(\mathfrak{k})$, but this was proved in the course of proving (a). Also we need to check that the differentiated K action is consistent with the $U(\mathfrak{g})$ action, i.e.,

$$\mathrm{ad}\, X \otimes 1 + 1 \otimes l(X) = l(X) \otimes 1.$$

In fact, we have

$$(\mathrm{ad}\, X \otimes 1 + 1 \otimes l(X))(b \otimes T) = (Xb - bX) \otimes T + b \otimes \partial(X)T$$
$$\equiv (Xb - bX) \otimes T + bX \otimes T$$
$$= Xb \otimes T = (l(X) \otimes 1)(b \otimes T).$$

By Proposition 1.107, $\overline{M}$ becomes an approximately unital left $R(K) \# U(\mathfrak{g})$ module. Write M for the weakly compatible $(U(\mathfrak{g}), K)$ module

$$M = R(K) \# U(\mathfrak{g}) = U(\mathfrak{g}) \otimes_{\mathbb{C}} R(K).$$

The kernel of the homomorphism $M \to \overline{M}$ is by definition the linear span $I_{(a)}$ of the elements of (a), and this homomorphism respects the K and $U(\mathfrak{g})$ actions. Thus we have an exact sequence of $R(K) \# U(\mathfrak{g})$ modules

(1.114) $$0 \longrightarrow I_{(a)} \longrightarrow M \longrightarrow \overline{M} \longrightarrow 0.$$

Let

$$J = \{\alpha \in R(K) \# U(\mathfrak{g}) \mid \alpha \bar{m} = 0 \text{ for all } \bar{m} \in \overline{M}\}.$$

Then J is a two-sided ideal, and (c) tells us that $I_{(a)} \subseteq J$.

We shall prove that $I_{(a)} = J$. Thus let α be in J. Choose a finite subset $A \subseteq \widehat{K}$ large enough so that $\alpha(1 \otimes \chi_A) = \alpha$ in $R(K) \# U(\mathfrak{g})$, i.e., in M. Passing from M to $\overline{M}$ and remembering that (1.114) respects $R(K) \# U(\mathfrak{g})$ actions, we obtain

$$\alpha\overline{(1 \otimes \chi_A)} = \bar{\alpha},$$

where overbars denote images in $\overline{M}$. Since α is in J, the left side is 0 and hence also $\bar{\alpha} = 0$. By the exactness of (1.114), α is in $I_{(a)}$. Thus $I_{(a)} = J$.

If $I_{(b)}$ denotes the linear span of the elements of (b), then (c) tells us that $I_{(b)} \subseteq J$. Hence $I_{(b)} \subseteq I_{(a)}$. The reverse inclusion is established by repeating the proof with right modules in place of left. (See Proposition 1.107′ at the end of §5.) This completes the proof.

DEFINITION 1.115. Suppose $(\mathfrak{g}, K)$ is a pair. The **Hecke algebra** of $(\mathfrak{g}, K)$ is the quotient $R(\mathfrak{g}, K)$ of the smash product $R(K) \# U(\mathfrak{g})$ by the two-sided ideal $I_\mathfrak{k}$ of Lemma 1.113c.

The Hecke algebra of $(\mathfrak{g}, K)$ is thus an associative algebra. As a vector space, it satisfies

$$(1.116) \qquad R(\mathfrak{g}, K) \cong R(K) \otimes_{U(\mathfrak{k})} U(\mathfrak{g}) \cong U(\mathfrak{g}) \otimes_{U(\mathfrak{k})} R(K).$$

Following (1.116), we shall write elements in $R(\mathfrak{g}, K)$ as sums of tensors $S \otimes a$ or as sums of tensors $b \otimes T$. We often write just T for an element $T \otimes 1 = 1 \otimes T$ of the subalgebra $R(K)$. The elements $\chi_A = \chi_A \otimes 1 = 1 \otimes \chi_A$ form an approximate identity as a consequence of the corresponding property of $R(K) \# U(\mathfrak{g})$.

Theorem 1.117. Suppose $(\mathfrak{g}, K)$ is a pair with $R(\mathfrak{g}, K)$ as its Hecke algebra. Then

(a) the second isomorphism in (1.116) is implemented by the formulas (1.92c) and (1.93) for τ and τ^{-1}.

(b) formulas (1.89) make any $(\mathfrak{g}, K)$ module into an approximately unital left $R(\mathfrak{g}, K)$ module.

(c) each approximately unital left $R(\mathfrak{g}, K)$ module arises via (1.89) from one and only one $(\mathfrak{g}, K)$ module structure.

(d) under the correspondence of representations and modules in (b) and (c),

$$\mathrm{Hom}_{\mathfrak{g}, K}(M, N) = \mathrm{Hom}_{R(\mathfrak{g}, K)}(M, N).$$

PROOF. Part (a) is a consequence of the calculations with τ and τ^{-1} in §5 and of Lemma 1.113c. For (b), let M be a $(\mathfrak{g}, K)$ module. Proposition 1.104 says that we can regard M as an approximately unital left $R(K) \# U(\mathfrak{g})$ module. By Lemma 1.113, M becomes a left $R(\mathfrak{g}, K)$ module. M is approximately unital for $R(\mathfrak{g}, K)$ since the approximate identity in $R(\mathfrak{g}, K)$ is the image (under passage to the quotient modulo $I_{\mathfrak{k}}$) of the approximate identity in $R(K) \# U(\mathfrak{g})$.

For (c), let M be an approximately unital left $R(\mathfrak{g}, K)$ module. Lifting the action, we can regard M as an approximately unital left $R(K) \# U(\mathfrak{g})$ module. By Proposition 1.107, M is a weakly compatible module for $(\mathfrak{g}, K)$. We have to prove (1.112). After our lifting, the elements of the ideal $I_{\mathfrak{k}}$ act on M as 0. Let m be in M. Then

$$(bv \otimes T)m = (b \otimes \partial(v)T)m.$$

The action here is given by (1.89), and we have

$$(bv)(Tm) = b((\partial(v)T)m) = b(\partial(v)(Tm)).$$

If $A \subseteq \widehat{K}$ is a finite subset chosen so large that $(1 \otimes \chi_A)m = m$, we can take $T = 1 \otimes \chi_A$ and conclude that

$$(bv)m = b(\partial(v)m).$$

By Proposition 1.20b, this equality yields (1.112).

Finally (d) follows immediately from the corresponding statement in Proposition 1.107. This completes the proof of the theorem.

The Hecke algebra $R(\mathfrak{g}, K)$ itself is a left $R(\mathfrak{g}, K)$ module, and the existence of the approximate identity makes this module approximately unital. Theorem 1.117c says that $R(\mathfrak{g}, K)$ is a $(\mathfrak{g}, K)$ module. Here $U(\mathfrak{g})$ and K act on the left as follows:

(1.118a)
$$\begin{aligned} a(b \otimes T) &= ab \otimes T \\ a(T \otimes b) &= T \otimes (\mathrm{Ad}(\cdot)^{-1}a)b \end{aligned} \qquad \text{for } a \in U(\mathfrak{g})$$

(1.118b)
$$\begin{aligned} k(b \otimes T) &= \mathrm{Ad}(k)b \otimes l(k)T \\ k(T \otimes b) &= l(k)T \otimes b = \delta_k T \otimes b \end{aligned} \qquad \text{for } k \in K.$$

The action of $\mathcal{E}'(K)$ or $R(K)$ that corresponds to (1.118b) is

(1.118c)
$$S(T \otimes b) = l(S)T \otimes b = ST \otimes b,$$

with ST referring to convolution.

With obvious changes our theory for left modules of the kinds we have considered is applicable to right modules. If $(\mathfrak{g}, K)$ is a pair, a **right $(\mathfrak{g}, K)$ module** is a complex vector space V that is a right $U(\mathfrak{g})$ module and has a linear action by K on the right such that

(1.86′)
- (i) the associated representation π of K given by $\pi(k)x = xk^{-1}$ is locally K finite;
- (ii) the differentiated version of the K action is the restriction to $\mathfrak{k}$ of the $\mathfrak{g}$ action; and
- (iii) $x(\mathrm{Ad}(k)u) = ((xk)u)k^{-1}$ for $k \in K$, $u \in U(\mathfrak{g})$, and $x \in V$.

Theorem 1.117 has an analog for this situation that we can state as follows.

Theorem 1.117′. Suppose $(\mathfrak{g}, K)$ is a pair with $R(\mathfrak{g}, K)$ as its Hecke algebra. Any right $(\mathfrak{g}, K)$ module becomes an approximately unital right $R(\mathfrak{g}, K)$ module via the definitions

$$x(T \otimes b) = (xT)b = \left(\int_K xk \, dT(k) \right) b$$

and
$$x(b \otimes T) = (xb)T = \int_K (xb)k \, dT(k),$$

and each approximately unital right $R(\mathfrak{g}, K)$ module arises via these formulas from one and only one right $(\mathfrak{g}, K)$ module structure. Under this correspondence of representations and modules,

$$\mathrm{Hom}_{\mathfrak{g}, K}(M, N) = \mathrm{Hom}_{R(\mathfrak{g}, K)}(M, N).$$

The Hecke algebra $R(\mathfrak{g}, K)$ itself is a right $R(\mathfrak{g}, K)$ module, and the existence of the approximate identity makes this module approximately unital. Theorem 1.117′ says that $R(\mathfrak{g}, K)$ is a right $(\mathfrak{g}, K)$ module. Here $U(\mathfrak{g})$ and K act on the right as follows:

(1.118′a)
$$(b \otimes T)a = b(\mathrm{Ad}(\cdot)a) \otimes T$$
$$(T \otimes b)a = T \otimes ba$$
for $a \in U(\mathfrak{g})$

(1.118′b)
$$(b \otimes T)k = b \otimes r(k)^{-1}T = b \otimes T\delta_k$$
$$(T \otimes b)k = r(k)^{-1}T \otimes \mathrm{Ad}(k)^{-1}b$$
for $k \in K$.

The action of $\mathcal{E}'(K)$ or $R(K)$ that corresponds to (1.118′b) is

(1.118′c)
$$(b \otimes T)S = b \otimes r(S^t)T = b \otimes TS,$$

with TS referring to convolution.

The identities with Hom in Theorems 1.117 and 1.117′ have the following analog for tensor products.

Theorem 1.119. Suppose $(\mathfrak{g}, K)$ is a pair with $R(\mathfrak{g}, K)$ as Hecke algebra. Let V be an approximately unital right $R(\mathfrak{g}, K)$ module, and let W be an approximately unital left $R(\mathfrak{g}, K)$ module. Regard V as a right $(\mathfrak{g}, K)$ module and W as a $(\mathfrak{g}, K)$ module. Then $V \otimes_{R(\mathfrak{g}, K)} W$ is the quotient of $V \otimes_{\mathbb{C}} W$ by the linear span

(1.120a)
$$\langle xX \otimes y - x \otimes Xy, \, xk \otimes y - x \otimes ky \mid X \in \mathfrak{g}, \, k \in K, \, x \in V, \, y \in W \rangle.$$

REMARKS. We are allowed to regard V as a right $(\mathfrak{g}, K)$ module and W as a $(\mathfrak{g}, K)$ module by Theorems 1.117′ and 1.117. In obvious notation the conclusion of the theorem is that

$$V \otimes_{R(\mathfrak{g}, K)} W = V \otimes_{\mathfrak{g}, K} W.$$

PROOF. The theorem will follow if we show that (1.120a) coincides with the linear span

(1.120b) $$\langle xr \otimes y - x \otimes ry \mid r \in R(\mathfrak{g}, K), \, x \in V, \, y \in W \rangle.$$

Let $xX \otimes y - x \otimes Xy$ be given with $X \in \mathfrak{g}$. If $A \subseteq \widehat{K}$ is a finite subset large enough so that $(xX)\chi_A = xX$ and $\chi_A y = y$, then the element $r = X \otimes \chi_A$ satisfies

$$xr \otimes y - x \otimes ry = (xX)\chi_A \otimes y - x \otimes X(\chi_A y) = xX \otimes y - x \otimes Xy.$$

Thus the first kind of generator of (1.120a) is in (1.120b).

Let $xk \otimes y - x \otimes ky$ be given with $k \in K$. Let $A \subseteq \widehat{K}$ be a finite subset large enough so that $x\chi_A = x$ and $\chi_A y = y$. Proposition 1.30 gives $\delta_k \chi_A = \chi_A \delta_k$. Then the element $r = \delta_k \chi_A = \chi_A \delta_k$ satisfies

$$xr \otimes y - x \otimes ry = (x\chi_A)\delta_k \otimes y - x \otimes \delta_k(\chi_A y)$$
$$= x\delta_k \otimes y - x \otimes \delta_k y$$
$$= xk \otimes y - x \otimes ky \qquad \text{by Proposition 1.20.}$$

Thus the second kind of generator of (1.120a) is in (1.120b). Therefore (1.120a) is contained in (1.120b).

For the reverse inclusion, first let $r = T \in R(K)$ be given. In terms of the effect of distributions on vector-valued functions as in (B.17), we have

$$xr \otimes y - x \otimes ry = xT \otimes y - x \otimes Ty = \int_K (xk \otimes y - x \otimes ky) \, dT(k).$$

From (B.17b) we see that the vector value of the right side is a finite linear combination of the vector values of the integrand. Thus it is of the form

$$\sum_j c_j(xk_j \otimes y - x \otimes k_j y)$$

and is in (1.120a).

To handle a general $u \otimes T$ with $u \in U(\mathfrak{g})$ and $T \in R(K)$, we can argue inductively on the order of u. The inductive step is to show that if $r = u \otimes T$ has all $xr \otimes y - x \otimes ry$ in (1.120a), then so does $r' = Xu \otimes T$ for $X \in \mathfrak{g}$. We have

$$
\begin{aligned}
xr' \otimes y &- x \otimes r'y \\
&= xXuT \otimes y - x \otimes XuTy \\
&= ((xX)uT \otimes y - xX \otimes uTy) + (xX \otimes uTy - x \otimes XuTy).
\end{aligned}
$$

The first term is in (1.120a) by inductive hypothesis, and the second term is one of the generators of (1.120a). Hence (1.120b) is contained in (1.120a).

The algebra $U(\mathfrak{g})$ has a special property not shared by general associative algebras $\mathcal{A}$ that can be combined with K to form weak pairs $(\mathcal{A}, K)$, namely the existence of the transpose mapping. This map, defined in (1.4), exists whether or not $\mathfrak{g}$ is related to a group G or real Lie algebra $\mathfrak{g}_0$. If $(\mathfrak{g}, K)$ is a pair, we can use this map to define an antiautomorphism **transpose** first of $R(K) \# U(\mathfrak{g})$ and then of $R(\mathfrak{g}, K)$ by

$$
(1.121) \qquad
\begin{aligned}
(T \otimes a)^t &= a^t \otimes T^t \\
(a \otimes T)^t &= T^t \otimes a^t.
\end{aligned}
$$

Here $T^t \in R(K)$ is defined by (1.6a) and is in $R(K)$ by Proposition 1.83.

Proposition 1.122. If $(\mathfrak{g}, K)$ is a pair, then the transpose map is a well-defined antiautomorphism of $R(K) \# U(\mathfrak{g})$ and descends to a well-defined antiautomorphism of $R(\mathfrak{g}, K)$.

PROOF. It is well defined as a linear map of $R(K) \# U(\mathfrak{g})$ to itself. Also

$$((S \otimes a)(T \otimes b))^t = \left(\sum_i S(\langle \mathrm{Ad}(\cdot)^{-1}a, a_i^* \rangle T) \otimes a_i b \right)^t$$

$$= \left(\sum_i ST_i \otimes a_i b \right)^t$$

(1.123)
$$= \sum_i b^t a_i^t \otimes T_i^t S^t.$$

Here

(1.124)
$$\langle T_i^t, f \rangle = \langle T_i, f^t \rangle = \langle T, \langle \mathrm{Ad}(\cdot)^{-1}a, a_i^* \rangle f(\cdot^{-1}) \rangle = \langle T^t, \langle \mathrm{Ad}(\cdot)a, a_i^* \rangle f \rangle.$$

In these computations, $\{a_i\}$ is a basis of an $\mathrm{Ad}(K)$ invariant subspace containing a. For $Y \in \mathfrak{g}$, $\mathrm{Ad}(k)Y^t = -\mathrm{Ad}(k)Y = (\mathrm{Ad}(k)Y)^t$. It follows that

(1.125)
$$\mathrm{Ad}(k)a^t = (\mathrm{Ad}(k)a)^t$$

and then that $\{a_i^t\}$ is a basis of an $\mathrm{Ad}(K)$ invariant subspace containing a^t. Moreover, applying transpose to

$$\mathrm{Ad}(k)a = \sum_i \langle \mathrm{Ad}(k)a, a_i^* \rangle a_i$$

and using (1.125) gives

$$\mathrm{Ad}(k)a^t = \sum_i \langle \mathrm{Ad}(k)a, a_i^* \rangle a_i^t,$$

from which it follows that

(1.126)
$$\langle \mathrm{Ad}(k)a^t, (a_i^t)^* \rangle = \langle \mathrm{Ad}(k)a, a_i^* \rangle.$$

Therefore

$$(T \otimes b)^t (S \otimes a)^t = (b^t \otimes T^t)(a^t \otimes S^t)$$

$$= \sum_i b^t a_i^t \otimes (\langle \mathrm{Ad}(\cdot)a^t, (a_i^t)^* \rangle T^t) S^t$$

$$= \sum_i b^t a_i^t \otimes (\langle \mathrm{Ad}(\cdot)a, a_i^* \rangle T^t) S^t \qquad \text{by (1.126)}$$

$$= \sum_i b^t a_i^t \otimes T_i^t S^t \qquad \text{by (1.124),}$$

and the right side agrees with (1.123). Thus transpose is an antiautomorphism of $R(K) \# U(\mathfrak{g})$.

For transpose to descend to $R(\mathfrak{g}, K)$, it must map $I_{\mathfrak{k}}$ to itself. We have

$$(S \otimes ua - S\partial(u) \otimes a)^t = a^t u^t \otimes S^t - a \otimes \partial(u)^t S^t$$
$$= a^t u^t \otimes S^t - a \otimes \partial(u^t) S^t$$

by definition of $\partial(u)^t$. Since u^t is in $U(\mathfrak{k})$, Lemma 1.113c shows that the
result is in $I_{\mathfrak{k}}$. Hence transpose descends to $R(\mathfrak{g}, K)$.

Now we shall take up the matter of "generalized pairs" that was introduced at the beginning of this section. The reader who is not interested
in preparation for the advanced applications in Chapter XII may skip
the remainder of this section.

Suppose that $(\mathcal{A}, K)$ is a weak pair. We say that $(\mathcal{A}, K)$ is a **generalized
pair** if a homomorphism $\iota : U(\mathfrak{k}) \to \mathcal{A}$ with $\iota(1) = 1$ is specified that is
compatible with Ad on $\mathcal{A}$ and the usual adjoint action of K on $U(\mathfrak{k})$:

$$(1.127) \qquad \mathrm{Ad}(k)(\iota(u)) = \iota(\mathrm{Ad}(k)(u)) \qquad \text{for } k \in K, \, u \in U(\mathfrak{k}).$$

A **compatible** $(\mathcal{A}, K)$ **module** is a weakly compatible $(\mathcal{A}, K)$ module M
satisfying

$$(1.128) \qquad (\iota(u))m = um \qquad \text{for } u \in U(\mathfrak{k}), \, m \in M.$$

Here the action of u on M on the right side is by the differential of the
action of K.

To simplify the notation, it will be convenient to regard $\mathcal{A}$ as a two-
sided $U(\mathfrak{k})$ module using ι:

$$(1.129) \qquad ua = \iota(u)a \quad \text{and} \quad au = a\iota(u) \qquad \text{for } a \in \mathcal{A}, \, u \in U(\mathfrak{k}).$$

Then (1.128) implies the following identities that more visibly generalize
(1.112):

$$(1.130) \qquad \begin{aligned} (au)m &= a(um) \\ (ua)m &= u(am). \end{aligned}$$

for $a \in \mathcal{A}$, $u \in U(\mathfrak{k})$, and $m \in M$. In fact, if we apply a to both sides of
(1.128), we obtain

$$a(um) = a((\iota(u))m) = (a\iota(u))m$$

since M is an $\mathcal{A}$ module, and this is the first identity of (1.130). The second identity follows similarly by using (1.128) with am in place of m.

With these definitions, Lemma 1.113 extends to generalized pairs $(\mathcal{A}, K)$ with only notational changes. We can therefore extend Definition 1.115 by defining the **Hecke algebra** of the generalized pair $(\mathcal{A}, K)$ to be

$$(1.131) \qquad R(\mathcal{A}, K) = (R(K) \# U(\mathfrak{g}))/I_{\mathfrak{k}}.$$

This is an associative algebra. As a vector space, it satisfies

$$R(\mathcal{A}, K) \cong R(K) \otimes_{U(\mathfrak{k})} \mathcal{A} \cong \mathcal{A} \otimes_{U(\mathfrak{k})} R(K).$$

The elements $\chi_A \otimes 1 = 1 \otimes \chi_A$ form an approximate identity.

If $(\mathcal{A}, K)$ is a generalized pair, then the expected generalization of Theorem 1.117 is valid, and the given proof needs only notational changes: Compatible $(\mathcal{A}, K)$ modules correspond to approximately unital left $R(\mathcal{A}, K)$ modules. $R(\mathcal{A}, K)$ may be viewed also as a two-sided $\mathcal{A}$ module and a two-sided K module, via (1.118).

For an arbitrary generalized pair $(\mathcal{A}, K)$, the algebra $\mathcal{A}$ need not have as part of its definition a distinguished antiautomorphism that we can use to define a transpose mapping for $R(\mathcal{A}, K)$. Instead we make a separate definition. Suppose $\mathcal{A}$ is endowed with a linear antiautomorphism $a \mapsto a^t$ satisfying

$$(1.132) \qquad \begin{array}{ll} \text{(i) } \mathrm{Ad}(k)a^t = (\mathrm{Ad}(k)a)^t & \text{for } k \in K, \, a \in \mathcal{A}, \text{ and} \\ \text{(ii) } (au)^t = u^t a^t & \text{for } u \in U(\mathfrak{k}), \, a \in \mathcal{A}. \end{array}$$

In this case the antiautomorphism **transpose** of $R(K) \# \mathcal{A}$ and then of $R(\mathcal{A}, K)$ is defined by

$$(1.133) \qquad \begin{array}{l} (T \otimes a)^t = a^t \otimes T^t \\ (a \otimes T)^t = T^t \otimes a^t. \end{array}$$

The proof that transpose is an antiautomorphism and descends to $R(\mathcal{A}, K)$ is the same as for Proposition 1.122. Condition (i) serves in place of (1.125), and condition (ii) is what is needed to carry through the argument that $I_{\mathfrak{k}}$ is sent into itself.

EXAMPLE. Let $\mathcal{A}$ be the **Weyl algebra** in one variable, i.e.,

$$\mathcal{A} = \left\{ \sum_{n \geq 0} P_n(x) \left(\frac{d}{dx}\right)^n \;\middle|\; P_n \text{ is a polynomial, and the sum is finite} \right\}.$$

This associative algebra is written also as

$$\mathcal{A} = \mathbb{C}\left[x, \frac{d}{dx}\right] \quad \text{with the relation} \quad \frac{d}{dx} x = x \frac{d}{dx} + 1.$$

The vector space

$$V = \{P(x)e^{-x^2/2} \mid P \text{ is a polynomial}\}$$

is a left $\mathcal{A}$ module in a natural way. Shortly we shall introduce a compact group $\tilde{K}$ such that $(\mathcal{A}, \tilde{K})$ is a generalized pair and V is a compatible $(\mathcal{A}, \tilde{K})$ module.

Observe that V is a simple $\mathcal{A}$ module. In fact, let V_1 be a nonzero $\mathcal{A}$ submodule, and choose $P(x)e^{-x^2/2} \neq 0$ in V_1 with $\deg P$ as small as possible. The $\mathcal{A}$ element $\frac{d}{dx} + x$ carries this to $P'(x)e^{-x^2/2}$, and the minimality of $\deg P$ means $P' = 0$. So $e^{-x^2/2}$ is in V_1. Applying the elements $P(x)$ of $\mathcal{A}$ to $e^{-x^2/2}$, we see that $V_1 = V$.

Let $G = SL(2, \mathbb{R})$, let K be the rotation subgroup, let $\tilde{G}$ be a double cover of G, and let $\tilde{K} \subseteq \tilde{G}$ be the corresponding double cover of K. The complexified Lie algebra of G and $\tilde{G}$ is $\mathfrak{g} = \mathfrak{sl}(2, \mathbb{C})$, and $\mathfrak{k} = \mathbb{C}\left(\begin{smallmatrix} 0 & 1 \\ -1 & 0 \end{smallmatrix}\right)$. We introduce a nonobvious algebra homomorphism $\varphi : U(\mathfrak{g}) \to \mathcal{A}$ given as follows: For $\left(\begin{smallmatrix} a & b \\ c & -a \end{smallmatrix}\right) \in \mathfrak{g}$,

$$\varphi \begin{pmatrix} a & b \\ c & -a \end{pmatrix} = \tfrac{1}{2} \begin{pmatrix} -ix & d/dx \end{pmatrix} \begin{pmatrix} a & b \\ c & -a \end{pmatrix} \begin{pmatrix} i\, d/dx \\ -x \end{pmatrix}.$$

In terms of the usual basis

$$h = \begin{pmatrix} 1 & 0 \\ 0 & -1 \end{pmatrix}, \qquad e = \begin{pmatrix} 0 & 1 \\ 0 & 0 \end{pmatrix}, \qquad f = \begin{pmatrix} 0 & 0 \\ 1 & 0 \end{pmatrix},$$

φ is given by

$$\varphi(h) = \tfrac{1}{2}\left(x \tfrac{d}{dx} + \tfrac{d}{dx} x\right)$$

$$\varphi(e) = \tfrac{1}{2} i x^2$$

$$\varphi(f) = \tfrac{1}{2} i \left(\tfrac{d}{dx}\right)^2.$$

Using this basis, we readily check that

$$\varphi([Y_1, Y_2]) = \varphi(Y_1)\varphi(Y_2) - \varphi(Y_2)\varphi(Y_1) \qquad \text{for } Y_1, Y_2 \in \mathfrak{g}.$$

Therefore φ extends to an algebra homomorphism of $U(\mathfrak{g})$ into $\mathcal{A}$ with $\varphi(1) = 1$. The homomorphism φ has a big kernel and is onto the set of all even elements in $\mathcal{A}$ (i.e., elements having all terms of even total degree in x and d/dx).

Composing φ with the $\mathcal{A}$ module structure on V makes V into a representation space for $\mathfrak{g}$. Let us see the extent to which the action of $\mathfrak{k}$ globalizes. We have

$$\varphi\begin{pmatrix} 0 & 1 \\ -1 & 0 \end{pmatrix} = \tfrac{i}{2}(x^2 - (\tfrac{d}{dx})^2)$$

and

$$\tfrac{i}{2}(x^2 - (\tfrac{d}{dx})^2)(P(x)e^{-x^2/2}) = \tfrac{i}{2}e^{-x^2/2}(P''(x) - 2xP'(x) - P(x)).$$

We look for eigensolutions. The equation

$$P''(x) - 2xP'(x) - P(x) = \lambda P(x)$$

can be solved by power series; it has a unique monic polynomial solution P_n if $\lambda \in \{-(2n + 1)\}_{n=0}^{\infty}$, and P_n has degree n. (Essentially P_n is the nth **Hermite polynomial**.) Since there is one such polynomial of each degree, $\{P_n\}$ is a basis of the space of all polynomials. The operator $\tfrac{i}{2}(x^2 - (\tfrac{d}{dx})^2)$ is diagonal in this basis, with

$$\tfrac{i}{2}(x^2 - (\tfrac{d}{dx})^2)(P_n(x)e^{-x^2/2}) = -\tfrac{i}{2}(2n + 1)P_n(x)e^{-x^2/2}, \quad n \geq 0.$$

According to this equation, $\exp t \begin{pmatrix} 0 & 1 \\ -1 & 0 \end{pmatrix}$ is trying to act on $P_n(x)e^{-x^2/2}$ by the scalar

$$\exp(t(-\tfrac{i}{2})(2n + 1)).$$

For $t = 2\pi$, we never get 1. Thus the action of $\mathfrak{k}$ does not globalize to K. However, for $t = 4\pi$, we do get 1 for all n. Thus the action of $\mathfrak{k}$ does globalize to $\tilde{K}$. It follows that V is a $(\mathfrak{g}, \tilde{K})$ module.

The fact that V is a simple $\mathcal{A}$ module, in combination with knowledge of the image of $U(\mathfrak{g})$ in $\mathcal{A}$, leads us to an analysis of reducibility. The operator $(\tfrac{d}{dx} + x)^2$ is in the image of $U(\mathfrak{g})$ and carries $P(x)e^{-x^2/2}$ to $P''(x)e^{-x^2/2}$. Consequently any nonzero invariant subspace contains some nonzero $P(x)e^{-x^2/2}$ with $P'' = 0$. We readily verify that

$$V^+ = \{P(x)e^{-x^2/2} \mid P \text{ is even}\} \quad \text{and} \quad V^- = \{P(x)e^{-x^2/2} \mid P \text{ is odd}\}$$

are invariant subspaces with $V = V^+ \oplus V^-$. Each of V^+ and V^- has a cyclic vector ($e^{-x^2/2}$ or $xe^{-x^2/2}$), and the above analysis proves irreducibility of each. Hence $V = V^+ \oplus V^-$ is the full story about reducibility. The representation of $(\mathfrak{g}, \tilde{K})$ on V is called the **metaplectic representation**.

We can make $(\mathcal{A}, \tilde{K})$ into a generalized pair and V into a compatible $(\mathcal{A}, \tilde{K})$ module. The homomorphism of $U(\mathfrak{k})$ into $\mathcal{A}$ is just $\varphi|_{U(\mathfrak{k})}$. To get the homomorphism $\mathrm{Ad} : \tilde{K} \to \mathrm{Aut}(\mathcal{A})$, we differentiate (1.88) to obtain the necessary condition

$$((\mathrm{ad}\, X)a)m = X(am) - a(Xm)$$

for $X = \begin{pmatrix} 0 & 1 \\ -1 & 0 \end{pmatrix}$. This condition will be satisfied if

$$(\mathrm{ad}\, X)a = Xa - aX.$$

In turn, $\mathcal{A}$ is generated by x and $\frac{d}{dx}$. A little calculation gives

$$(\mathrm{ad}\, X)x = -i\frac{d}{dx}$$

and
$$(\mathrm{ad}\, X)\frac{d}{dx} = -ix.$$

In other words, $\mathrm{ad}\, X$ is given in the basis $\{x, \frac{d}{dx}\}$ by the matrix $\begin{bmatrix} 0 & -i \\ -i & 0 \end{bmatrix}$. Therefore $\mathrm{Ad}(\exp t X)$ is given in the same basis by the matrix

$$\mathrm{Ad}(\exp t X) \leftrightarrow \begin{bmatrix} \cos t & -i\sin t \\ -i\sin t & \cos t \end{bmatrix}.$$

With this definition, it is easy to check that $(\mathcal{A}, \tilde{K})$ is indeed a generalized pair and that V is a compatible $(\mathcal{A}, \tilde{K})$ module. This example will be discussed further in Chapter XII.

CHAPTER II

THE CATEGORY $\mathcal{C}(\mathfrak{g}, K)$

Once all $(\mathfrak{g}, K)$ modules are reinterpreted as a good category $\mathcal{C}(\mathfrak{g}, K)$ of left $R(\mathfrak{g}, K)$ modules, it is possible to introduce change-of-ring functors P and I, with P defined by $\otimes$ and I defined by Hom. These are the master functors on which the whole theory of cohomological induction is built. Each is adjoint to an exact forgetful-like functor $\mathcal{V}^\vee$ or $\mathcal{F}$, and this adjointness is the first key to developing the properties of P and I. Special cases include ind and pro (Lie algebra induction and production), the basic functors Π and Γ, and coinvariants and invariants.

Within $\mathcal{C}(\mathfrak{g}, K)$, one can construct the tensor product and K finite Hom of two $(\mathfrak{g}, K)$ modules. If one of the two modules is held fixed, the construction is an exact functor in the other module. The Mackey isomorphisms make explicit the facts that P commutes with tensor product and I commutes with Hom.

The derived functors of P and I are of paramount interest. Special cases include Lie algebra homology and cohomology, relative Lie algebra homology and cohomology, the relative Ext functors, and the derived functors of the Bernstein and Zuckerman functors.

In $\mathcal{C}(\mathfrak{g}, K)$ the trivial one-dimensional $(\mathfrak{g}, K)$ module has a concrete projective resolution known as the Koszul resolution. From the Koszul resolution one can form a standard projective resolution and a standard injective resolution for each module in $\mathcal{C}(\mathfrak{g}, K)$. The standard resolutions enable one to write specific complexes whose homology or cohomology gives the derived functors of P or I on a given module.

1. Functors P and I

Let $(\mathfrak{g}, K)$ be a pair in the sense of (1.64), and let $\mathcal{C}(\mathfrak{g}, K)$ be the category of all $(\mathfrak{g}, K)$ modules and $(\mathfrak{g}, K)$ maps. Now that we have introduced the Hecke algebra $R(\mathfrak{g}, K)$, Theorem 1.117 allows us to regard $\mathcal{C}(\mathfrak{g}, K)$ as the category of all left $R(\mathfrak{g}, K)$ modules that are approximately unital relative to the approximate identity $\{\chi_A\}$. This category is "good" in the sense of (A.1). Theorem 1.117 is so fundamental that we shall often use it without specific reference. We write $\mathcal{C}(\mathfrak{g})$ for $\mathcal{C}(\mathfrak{g}, \{1\})$.

We denote by $\tilde{\mathcal{C}}(\mathfrak{g}, K)$ the good category of all left $R(\mathfrak{g}, K)$ modules. If M is a module in $\tilde{\mathcal{C}}(\mathfrak{g}, K)$, Definition 1.54 allows us to define

$$(2.1) \qquad M_K = \{m \in M \mid \chi_A m = m \text{ for some finite } A \subseteq \widehat{K}\}.$$

Then M_K is in $\mathcal{C}(\mathfrak{g}, K)$, and $M \mapsto M_K$ is a functor from $\tilde{\mathcal{C}}(\mathfrak{g}, K)$ to $\mathcal{C}(\mathfrak{g}, K)$ called the K **finite part**, the effect on maps being given by restriction. The K finite-part functor has the property that

(2.2) $(\cdot)_K$ is covariant and exact,

according to Proposition 1.55.

In the special case of $\mathcal{C}(\mathfrak{k}, K)$, where $\mathfrak{k}$ is the complexified Lie algebra of K, we have

(2.3) $R(\mathfrak{k}, K) \cong R(K)$

as algebras, since, under the isomorphism of vector spaces

$$R(\mathfrak{k}, K) \cong R(K) \otimes_{U(\mathfrak{k})} U(\mathfrak{k}) \cong R(K),$$

the left K actions are identified by $l \leftrightarrow l \otimes 1 \leftrightarrow l$ and the right K actions are identified by $r \leftrightarrow r \otimes \mathrm{Ad} \leftrightarrow r$. The category $\mathcal{C}(\mathfrak{k}, K)$ has the following special feature.

Lemma 2.4. In the category $\mathcal{C}(\mathfrak{k}, K)$, every module is projective and injective.

PROOF. Let P be a module in $\mathcal{C}(\mathfrak{k}, K)$, and let the first diagram in (C.5) be given. By Proposition 1.18b, write $C = \ker \psi \oplus C'$ for some K invariant subspace C'. Then $\psi|_{C'}$ is an isomorphism onto B. If we let $\psi_1 = (\psi|_{C'})^{-1}$, then we can take $\sigma = \psi_1 \circ \tau$. Hence P is projective. A similar argument in the context of (C.7) shows that every module is injective.

In this chapter we shall make extensive use of the results of Appendix C. Through §5 we shall use §§C.1–C.2, and §§C.3–C.4 will come into play starting in §II.6.

We shall work throughout this chapter with one or more pairs $(\mathfrak{g}, K)$. Even though we concentrate on left $R(\mathfrak{g}, K)$ modules, we shall sometimes encounter right $R(\mathfrak{g}, K)$ modules. Thus both Theorems 1.117 and 1.117′ are relevant to our study. The results of §§1–2, except for the examples, extend with the obvious interpretations to generalized pairs. If one works with generalized pairs, it is occasionally necessary to convert a left $R(A, K)$ module into a right module, and we need to assume the existence of a transpose map in order to make this conversion. In the case of pairs $(\mathfrak{g}, K)$, the existence of the transpose map is automatic, and this question does not arise.

The subject of this section will be the passage between the categories of left modules associated to two pairs. Let $(\mathfrak{h}, L)$ and $(\mathfrak{g}, K)$ be two pairs. The definition of pair means that $\mathfrak{l}$ embeds in $\mathfrak{h}$ and $\mathfrak{k}$ embeds in $\mathfrak{g}$ with certain compatibility conditions. For the time being, let us give names to these Lie algebra inclusions:

(2.5a) $\qquad\qquad\qquad \iota_L : \mathfrak{l} \to \mathfrak{h} \qquad$ and $\qquad \iota_K : \mathfrak{k} \to \mathfrak{g};$

until §3, it will not matter whether ι_L and ι_K are one-one. (For the theory of generalized pairs, one replaces ι_L and ι_K with homomorphisms between the associative algebras that are not necessarily one-one.) The compatibility condition in the definition of pair forces

(2.5b) $\qquad \begin{aligned} &\mathrm{Ad}_L(l) \circ \iota_L = \iota_L \circ \mathrm{Ad}_L(l) \qquad &&\text{for } l \in L \\ &\mathrm{Ad}_K(k) \circ \iota_K = \iota_K \circ \mathrm{Ad}_K(k) \qquad &&\text{for } k \in K. \end{aligned}$

A **map of pairs**

(2.6a) $\qquad\qquad\qquad\qquad i : (\mathfrak{h}, L) \to (\mathfrak{g}, K)$

consists of two maps

(2.6b) $\qquad \begin{aligned} &i_{\mathrm{alg}} : \mathfrak{h} \to \mathfrak{g}, \qquad &&\text{a Lie algebra homomorphism} \\ &i_{\mathrm{gp}} : L \to K, \qquad &&\text{a Lie group homomorphism} \end{aligned}$

satisfying the compatibility conditions

(2.6c) $\qquad$ (i) $i_{\mathrm{alg}} \circ \iota_L = \iota_K \circ di_{\mathrm{gp}}$, where di_{gp} is the differential of i_{gp};

$\qquad\qquad$ (ii) $i_{\mathrm{alg}} \circ \mathrm{Ad}_L(l) = \mathrm{Ad}_K(i_{\mathrm{gp}}(l)) \circ i_{\mathrm{alg}}$ for $l \in L$.

(For the theory with generalized pairs, one will want to replace i_{alg} with a homomorphism between the associative algebras.) When there is no possibility of confusion, we shall refer to all of i_{alg}, i_{gp}, and di_{gp} simply as i.

    EXAMPLES.

    1) An **inclusion of pairs**. Everything is to take place in $(\mathfrak{g}, K)$. That is, i_{alg} is an inclusion $\mathfrak{h} \hookrightarrow \mathfrak{g}$ and i_{gp} is an inclusion $L \hookrightarrow K$. Condition (i) is that the composite inclusions $\mathfrak{l} \subseteq \mathfrak{k} \subseteq \mathfrak{g}$ and $\mathfrak{l} \subseteq \mathfrak{h} \subseteq \mathfrak{g}$ are the same, and condition (ii) is that the two possible interpretations of $\mathrm{Ad}(l)$ on $\mathfrak{g}$ are consistent. This example leads to constructions for passing from a representation of a subgroup to a representation of the whole group.

This is the primary example of interest. In Knapp [1988] the discussion of P and I was limited to this example. We often write $(\mathfrak{h}, L) \hookrightarrow (\mathfrak{g}, K)$ for an inclusion of pairs.

2) The **trivial map** $i : (\mathfrak{g}, K) \to (0, \{1\})$. That is, $i_{\mathrm{alg}}(\mathfrak{g}) = 0$ and $i_{\mathrm{gp}}(K) = \{1\}$. The compatibility conditions (2.6c) are automatically satisfied. This example leads to relative Lie algebra homology and cohomology.

3) A **special semidirect product map**. Let $(\mathfrak{q}, L)$ be a pair in which $\mathfrak{q} = \mathfrak{l} \oplus \mathfrak{u}$ with $\mathfrak{u}$ an ideal and with $\mathfrak{u}$ and $\mathfrak{l}$ both stable under $\mathrm{Ad}(L)$. If we let $i_{\mathrm{alg}} : \mathfrak{q} \to \mathfrak{l}$ be the projection and $i_{\mathrm{gp}} : L \to L$ be the identity, then $i : (\mathfrak{q}, L) \to (\mathfrak{l}, L)$ is a map of pairs. This example leads to $\mathfrak{u}$ homology and cohomology with an action of L in place. It will be generalized significantly in Chapter III. A concrete instance of this situation arises with $G = U(n)$ equal to the unitary group of n-by-n matrices and $\mathfrak{g}$ equal to its complexified Lie algebra. Let T be the diagonal subgroup of G, $\mathfrak{b}$ be the full upper-triangular subalgebra of $\mathfrak{g}$, and $\mathfrak{n}$ be the strictly upper-triangular subalgebra of $\mathfrak{g}$. Then $\mathfrak{b} = \mathfrak{t} \oplus \mathfrak{n}$ as vector spaces, with $\mathfrak{t}$ a Lie subalgebra and $\mathfrak{n}$ an ideal. The special semidirect product map of pairs is $i : (\mathfrak{b}, T) \to (\mathfrak{t}, T)$, with $i_{\mathrm{alg}} : \mathfrak{b} \to \mathfrak{t}$ equal to the quotient map modulo $\mathfrak{n}$ and with i_{gp} equal to the identity. This semidirect product map of pairs leads to $\mathfrak{n}$ homology and cohomology with an action of T in place. In Chapter IV, we shall see that an example of a semidirect product map arises in the corresponding situation for any compact connected Lie group; these examples will be the subject of Kostant's Theorem.

4) A **covering map** of pairs $i : (\mathfrak{g}, \tilde{K}) \to (\mathfrak{g}, K)$. Here i_{alg} is the identity and i_{gp} is onto with finite kernel.

Suppose $(\mathfrak{h}, L)$ and $(\mathfrak{g}, K)$ are two pairs related by a map of pairs as in (2.6). Then $R(\mathfrak{g}, K)$ in a natural way is a left and right $(\mathfrak{h}, L)$ module. The left actions are

(2.7)
$$b(a \otimes T) = i_{\mathrm{alg}}(b)a \otimes T \qquad \text{for } b \in U(\mathfrak{h}), \ a \in U(\mathfrak{g}), \ T \in R(K)$$
$$l(a \otimes T) = \mathrm{Ad}_K(i_{\mathrm{gp}}(l))a \otimes \delta_l * T \quad \text{for } l \in L, \ a \in U(\mathfrak{g}), \ T \in R(K),$$

and the right actions are defined analogously on elements $T \otimes a$. By Theorems 1.117 and 1.117′, $R(\mathfrak{g}, K)$ becomes an approximately unital $R(\mathfrak{h}, L)$ module on each side.

If V is in $\mathcal{C}(\mathfrak{h}, L)$, we define

(2.8)
$$P(V) = P^{\mathfrak{g}, K}_{\mathfrak{h}, L}(V) = R(\mathfrak{g}, K) \otimes_{R(\mathfrak{h}, L)} V$$

as a left $R(\mathfrak{g}, K)$ module under $r(s \otimes v) = (rs) \otimes v$. The module $P(V)$ is approximately unital as a left $R(\mathfrak{g}, K)$ module since for fixed $s \in R(\mathfrak{g}, K)$ we can choose a finite subset $A \subseteq \widehat{K}$ so that $\chi_A s = s$ and then

$$\chi_A(s \otimes v) = (\chi_A s) \otimes v = s \otimes v.$$

By (iii) of (C.13), $R(\mathfrak{g}, K) \otimes_{R(\mathfrak{h}, L)} (\cdot)$ is covariant and right exact. Thus we have the following result.

Proposition 2.9. The functor P is a right-exact covariant functor from $\mathcal{C}(\mathfrak{h}, L)$ to $\mathcal{C}(\mathfrak{g}, K)$.

Similarly we can make $\mathrm{Hom}_{R(\mathfrak{h}, L)}(R(\mathfrak{g}, K), V)$ into a left $R(\mathfrak{g}, K)$ module by letting $(r\varphi)(s) = \varphi(sr)$. This $R(\mathfrak{g}, K)$ module may not be approximately unital, but its K finite part is. Thus we define

$$(2.10) \qquad I(V) = I_{\mathfrak{h}, L}^{\mathfrak{g}, K}(V) = \mathrm{Hom}_{R(\mathfrak{h}, L)}(R(\mathfrak{g}, K), V)_K.$$

Proposition 2.11. The functor I is a left-exact covariant functor from $\mathcal{C}(\mathfrak{h}, L)$ to $\mathcal{C}(\mathfrak{g}, K)$.

PROOF. By (ii) of (C.13), $\mathrm{Hom}_{R(\mathfrak{h}, L)}(R(\mathfrak{g}, K), \cdot)$ is covariant and left exact from $\mathcal{C}(\mathfrak{h}, L)$ to $\tilde{\mathcal{C}}(\mathfrak{g}, K)$. The K finite-part functor is covariant exact from $\tilde{\mathcal{C}}(\mathfrak{g}, K)$ to $\mathcal{C}(\mathfrak{g}, K)$, and I is the composition of these two functors.

The functors P and I are the "master functors" mentioned in the introduction. The notation is chosen because P sends projectives to projectives and because the derived functors of P are defined by projective resolutions, while I sends injectives to injectives and its derived functors are defined by injective resolutions.

EXAMPLES.

1) Inclusion of pairs.

1a) Suppose $K = L$, so that the pairs are $(\mathfrak{h}, L)$ and $(\mathfrak{g}, L)$. In Proposition 2.57 we shall see that P and I reduce to functors ind and pro that are called **Lie algebra induction** and **Lie algebra production** and that are defined by the right-hand equalities below:

$$(2.12) \qquad \begin{aligned} P_{\mathfrak{h}, L}^{\mathfrak{g}, L}(V) &\cong \mathrm{ind}_{\mathfrak{h}, L}^{\mathfrak{g}, L}(V) = U(\mathfrak{g}) \otimes_{U(\mathfrak{h})} V \\ I_{\mathfrak{h}, L}^{\mathfrak{g}, L}(V) &\cong \mathrm{pro}_{\mathfrak{h}, L}^{\mathfrak{g}, L}(V) = \mathrm{Hom}_{U(\mathfrak{h})}(U(\mathfrak{g}), V)_L. \end{aligned}$$

Here L acts on $U(\mathfrak{g}) \otimes_{U(\mathfrak{h})} V$ by

$$(2.13a) \qquad\qquad l(u \otimes x) = \text{Ad}(l)u \otimes lx,$$

and it acts on $\text{Hom}_{U(\mathfrak{h})}(U(\mathfrak{g}), V)_L$ by

$$(2.13b) \qquad\qquad (l\varphi)(u) = l(\varphi(\text{Ad}(l)^{-1}u)).$$

Unfortunately the notational correspondences $P \leftrightarrow \text{ind}$ and $I \leftrightarrow \text{pro}$ interchange the first letters, but we are stuck with the traditional definitions of ind and pro. The mnemonic is that ind and pro are the *only* backward aspects of the notation. Actually for the definitions and isomorphisms (2.12) amd (2.13), it is not necessary to assume that i_{alg} be one-one, only that i_{gp} be one-one onto; see the remarks with Proposition 2.57.

1b) Suppose $\mathfrak{g} = \mathfrak{h}$, so that the pairs are $(\mathfrak{g}, L)$ and $(\mathfrak{g}, K)$. In these cases P and I are given the special names Π and Γ. The functor Γ is called the **Zuckerman functor** after the discoverer of the preliminary version of Γ given in Example 1c below; the notation Γ is supposed to suggest sections, in keeping with the discussion of §§3–4 of the Introduction. The functor Π is sometimes called the **Bernstein functor** after the discoverer of a preliminary version of Π, and sometimes Π is called the **dual Zuckerman functor** because of Theorem 3.1 below. The notation for Π is supposed to suggest a dual geometric object (cycles or periods), even though Π does not actually have a rigorous analytic counterpart. In Proposition 2.69 we shall see that

$$
(2.14) \quad
\begin{aligned}
P^{\mathfrak{g},K}_{\mathfrak{g},L}(V) &= \Pi^{\mathfrak{g},K}_{\mathfrak{g},L}(V) = \Pi(V) \cong R(K) \otimes_{R(\mathfrak{k},H)} V \\
I^{\mathfrak{g},K}_{\mathfrak{g},L}(V) &= \Gamma^{\mathfrak{g},K}_{\mathfrak{g},L}(V) = \Gamma(V) \cong \text{Hom}_{R(\mathfrak{k},H)}(R(K), V)_K
\end{aligned}
$$

as natural isomorphisms of locally K finite representations of K.

1c) Suppose $\mathfrak{g} = \mathfrak{h}$, so that the pairs are $(\mathfrak{g}, L)$ and $(\mathfrak{g}, K)$, and suppose K is connected. In Proposition 2.70 we shall see that the I functor $\Gamma^{\mathfrak{g},K}_{\mathfrak{g},L}(V)$ is the "largest K finite subspace" of V, i.e., the subspace spanned by all vectors x such that $U(\mathfrak{k})x$ is finite-dimensional and the action of $\mathfrak{k}$ on $U(\mathfrak{k})x$ is the differential of some action of K. (See pp. 325–327 of Vogan [1981a] for a formulation of Γ in these terms when K is not necessarily connected.) Unfortunately there is no simple analog of this relationship for the P functor $\Pi^{\mathfrak{g},K}_{\mathfrak{g},L}(V)$: There need not be a "largest K finite quotient" of V. For example, let $\mathfrak{g} = \mathfrak{k}$, $H = \{1\}$, and $V = U(\mathfrak{k})$. Then every finite-dimensional K invariant subspace of $L^2(K)$ is a quotient of $U(\mathfrak{k})$, but every K finite quotient of $U(\mathfrak{k})$ is finite dimensional; hence there can be no largest K finite quotient.

1d) Suppose $\mathfrak{g} = \mathfrak{h} = 0$, so that the pairs are $(0, L)$ and $(0, K)$ with $L \subseteq K$ and both groups finite. In Proposition 2.73 we shall see that $I_{0,L}^{0,K}$ coincides with the classical notion of induction of representations in the sense of Frobenius and Schur.

1e) Suppose that $\mathfrak{g} = \mathfrak{h}$ and that L has finite index in K. This case has features in common with both (1b) and (1d) but is closer to (1d). We shall introduce a variant induced$_{\mathfrak{g},L}^{\mathfrak{g},K}$ of the "classical induction" functor and show in Proposition 2.77 that P and I are both naturally isomorphic to this functor.

2) Pairs $(\mathfrak{g}, K)$ and $(0, \{1\})$ with the trivial map of pairs. In Proposition 2.88 we shall see that P and I reduce to functors that are called **coinvariants** $(\,\cdot\,)_{\mathfrak{g},K}$ and **invariants** $(\,\cdot\,)^{\mathfrak{g},K}$, namely

(2.15a) $\qquad P_{\mathfrak{g},K}^{0,\{1\}}(V) \cong V_{\mathfrak{g},K} \qquad$ and $\qquad I_{\mathfrak{g},K}^{0,\{1\}}(V) \cong V^{\mathfrak{g},K},$

where

$$V^{\mathfrak{g},K} = \{v \in V \mid Xv = 0 \text{ and } kv = v \text{ for all } X \in \mathfrak{g},\, k \in K\}$$

(2.15b)

$$V_{\mathfrak{g},K} = V/\langle Xv, (1-k)v \mid X \in \mathfrak{g},\, k \in K,\, v \in V\rangle \cong (V/\mathfrak{g}V)^K.$$

(The superscript $(\,\cdot\,)^K$ in the isomorphic expression for $V_{\mathfrak{g},K}$ refers to invariants.) When $K = \{1\}$, the K is normally dropped from the notation and one writes

(2.16) $\qquad\qquad V_{\mathfrak{g}} = V_{\mathfrak{g},\{1\}} \quad$ and $\quad V^{\mathfrak{g}} = V^{\mathfrak{g},\{1\}}.$

3) A special semidirect product map of pairs $i : (\mathfrak{q}, L) \to (\mathfrak{l}, L)$ with $\mathfrak{q} = \mathfrak{l} \oplus \mathfrak{u}$. There is a second relevant map of pairs in this situation, namely the inclusion $j : (\mathfrak{u}, \{1\}) \hookrightarrow (\mathfrak{q}, L)$. This is called the **associated inclusion** to i. A $(\mathfrak{q}, L)$ module V becomes a $(\mathfrak{u}, \{1\})$ module $\mathcal{F}_{\mathfrak{q},L}^{\mathfrak{u},\{1\}}(V)$ by forgetting the unnecessary structure. In Proposition 2.89 we shall see that $P_{\mathfrak{q},L}^{\mathfrak{l},L}$ and $I_{\mathfrak{q},L}^{\mathfrak{l},L}$ lead to $\mathfrak{u}$ coinvariants and invariants with an action by L in place. Namely

$$P_{\mathfrak{q},L}^{\mathfrak{l},L}(V) = \mathcal{F}_{\mathfrak{q},L}^{\mathfrak{u},\{1\}}(V)_{\mathfrak{u}} \qquad \text{with an } L \text{ action}$$

and $\qquad\qquad I_{\mathfrak{q},L}^{\mathfrak{l},L}(V) = \mathcal{F}_{\mathfrak{q},L}^{\mathfrak{u},\{1\}}(V)^{\mathfrak{u}} \qquad \text{with an } L \text{ action}.$

4) A covering map of pairs $i : (\mathfrak{g}, \tilde{K}) \to (\mathfrak{g}, K)$, where i_{alg} is the identity and i_{gp} is onto with finite kernel. We shall introduce an averaging functor $A_{\mathfrak{g},\tilde{K}}^{\mathfrak{g},K}$ that averages the $\tilde{K}$ action to obtain a K action, and we shall show in Proposition 2.92 that P and I are both naturally isomorphic to this functor.

2. Properties of P and I

In this section we shall assemble the elementary properties of the P and I functors. Unless otherwise stated, $(\mathfrak{g}, K)$ will denote a pair. When two pairs are present, it will be understood that $i : (\mathfrak{h}, L) \to (\mathfrak{g}, K)$ is a map of pairs. We begin with a lemma that says that $P_{\mathfrak{g},K}^{\mathfrak{g},K}$ and $I_{\mathfrak{g},K}^{\mathfrak{g},K}$ are naturally isomorphic to the identity functor.

Lemma 2.17. If V is in $\mathcal{C}(\mathfrak{g}, K)$, then

$$(2.18a) \qquad R(\mathfrak{g}, K) \otimes_{R(\mathfrak{g},K)} V \cong V$$

and

$$(2.18b) \qquad \operatorname{Hom}_{R(\mathfrak{g},K)}(R(\mathfrak{g}, K), V)_K \cong V$$

as natural isomorphisms within $\mathcal{C}(\mathfrak{g}, K)$.

PROOF. For (2.18a), the map $r \otimes v \mapsto rv$ is an $R(\mathfrak{g}, K)$ map of the left side of (2.18a) into the right side. It is onto because of the presence of approximate identities. Let us prove it is one-one. If $\sum (r_i \otimes v_i)$ maps to 0, then $\sum r_i v_i = 0$. Choose a finite subset $A \subseteq \widehat{K}$ with $\chi_A r_i = r_i$ for all i. Then

$$\begin{aligned} \sum (r_i \otimes v_i) &= \sum (\chi_A r_i \otimes v_i) \\ &= \sum (\chi_A r_i \otimes v_i) - \sum (\chi_A \otimes r_i v_i) \\ &= \sum (\chi_A r_i \otimes v_i - \chi_A \otimes r_i v_i) \\ &= 0, \end{aligned}$$

and the map is one-one. Clearly the isomorphism is natural in V.

For (2.18b), let $F(V)$ be the left side of (2.18b), and let X be arbitrary in $\mathcal{C}(\mathfrak{g}, K)$. Then we have natural isomorphisms (in V and X)

$$\begin{aligned} \operatorname{Hom}_{R(\mathfrak{g},K)}&(X, F(V)) \\ &= \operatorname{Hom}_{R(\mathfrak{g},K)}(X, \operatorname{Hom}_{R(\mathfrak{g},K)}(R(\mathfrak{g}, K), V)) \\ &\cong \operatorname{Hom}_{R(\mathfrak{g},K)}(R(\mathfrak{g}, K) \otimes_{R(\mathfrak{g},K)} X, V) \qquad \text{by (C.20)} \\ &\cong \operatorname{Hom}_{R(\mathfrak{g},K)}(X, V) \qquad\qquad\qquad\quad \text{by (2.18a).} \end{aligned}$$

(The first equality holds since the image of any element of X is automatically K finite.) The result follows by applying (C.22).

Proposition 2.19. Let $i : (\mathfrak{h}, L) \to (\mathfrak{g}, K)$ and $j : (\mathfrak{j}, M) \to (\mathfrak{h}, L)$ be maps of pairs, so that $j \circ i : (\mathfrak{j}, M) \to (\mathfrak{g}, K)$ is a third map of pairs. For V in $\mathcal{C}(\mathfrak{j}, M)$, there are natural isomorphisms

$$P_{\mathfrak{h},L}^{\mathfrak{g},K} \circ P_{\mathfrak{j},M}^{\mathfrak{h},L}(V) \cong P_{\mathfrak{j},M}^{\mathfrak{g},K}(V)$$

and

$$I_{\mathfrak{h},L}^{\mathfrak{g},K} \circ I_{\mathfrak{j},M}^{\mathfrak{h},L}(V) \cong I_{\mathfrak{j},M}^{\mathfrak{g},K}(V).$$

PROOF. Formulas (C.21) and (2.18a) give

$$P_{\mathfrak{h},L}^{\mathfrak{g},K} \circ P_{\mathfrak{j},M}^{\mathfrak{h},L}(V) = R(\mathfrak{g}, K) \otimes_{R(\mathfrak{h},L)} R(\mathfrak{h}, L) \otimes_{R(\mathfrak{j},M)} V$$

$$\cong R(\mathfrak{g}, K) \otimes_{R(\mathfrak{j},M)} V = P_{\mathfrak{j},M}^{\mathfrak{g},K}(V),$$

while (C.20) and (2.18a) give

$$I_{\mathfrak{h},L}^{\mathfrak{g},K} \circ I_{\mathfrak{j},M}^{\mathfrak{h},L}(V) = \mathrm{Hom}_{R(\mathfrak{h},L)}(R(\mathfrak{g}, K), \mathrm{Hom}_{R(\mathfrak{j},M)}(R(\mathfrak{h}, L), V)_L)_K$$

$$= \mathrm{Hom}_{R(\mathfrak{h},L)}(R(\mathfrak{g}, K), \mathrm{Hom}_{R(\mathfrak{j},M)}(R(\mathfrak{h}, L), V))_K$$

$$\cong \mathrm{Hom}_{R(\mathfrak{j},M)}(R(\mathfrak{g}, K) \otimes_{R(\mathfrak{h},L)} R(\mathfrak{h}, L), V)_K$$

$$\cong \mathrm{Hom}_{R(\mathfrak{j},M)}(R(\mathfrak{g}, K), V)_K = I_{\mathfrak{j},M}^{\mathfrak{g},K}(V).$$

REMARK. The importance of naturality of the isomorphisms is that it yields isomorphisms of the derived functors, by (C.26).

Along with P and I we shall need "forgetful functors" going from $\mathcal{C}(\mathfrak{g}, K)$ to $\mathcal{C}(\mathfrak{h}, L)$. Suppose X is a $(\mathfrak{g}, K)$ module. We write $\mathcal{F}(X)$ or $\mathcal{F}_{\mathfrak{g},K}^{\mathfrak{h},L}(X)$ for the $(\mathfrak{h}, L)$ module with the same underlying vector space as X and with actions

(2.20a) $\qquad bx = i_{\mathrm{alg}}(b)x \qquad$ and $\qquad lv = i_{\mathrm{gp}}(l)x.$

This definition makes it clear that $\mathcal{F}$ is a covariant exact functor. Using the left $R(\mathfrak{h}, L)$ module structure on $R(\mathfrak{g}, K)$ obtained from (2.7), we may identify $\mathcal{F}(X)$ naturally as

(2.20b) $\qquad\qquad \mathcal{F}(X) \cong R(\mathfrak{g}, K) \otimes_{R(\mathfrak{g},K)} X.$

The map from right to left sends $r \otimes x$ to rx, just as in (2.18a), and is one-one onto by the proof of Lemma 2.17.

In the theory of finite groups, when $L \subseteq K$ are given, the multiplicity of an irreducible representation of K in an induced representation of K equals the multiplicity of the inducing representation of L in the restriction to L of the irreducible representation of K. This is classical **Frobenius reciprocity**. Example 1d of §1 noted that the classical notion of induction of representations for finite groups is an instance of the I functor. As the next result shows, the reciprocity result works in all settings of the I functor.

Proposition 2.21 (Frobenius reciprocity). The functor I is right adjoint to $\mathcal{F}$. That is, there is a natural isomorphism

$$\mathrm{Hom}_{R(\mathfrak{g},K)}(X, I(V)) \cong \mathrm{Hom}_{R(\mathfrak{h},L)}(\mathcal{F}(X), V)$$

for V in $C(\mathfrak{h}, L)$ and X in $C(\mathfrak{g}, K)$.

PROOF.

$$\mathrm{Hom}_{R(\mathfrak{g},K)}(X, I(V))$$
$$= \mathrm{Hom}_{R(\mathfrak{g},K)}(X, \mathrm{Hom}_{R(\mathfrak{h},L)}(R(\mathfrak{g}, K), V)_K)$$
$$= \mathrm{Hom}_{R(\mathfrak{g},K)}(X, \mathrm{Hom}_{R(\mathfrak{h},L)}(R(\mathfrak{g}, K), V))$$
$$\cong \mathrm{Hom}_{R(\mathfrak{h},L)}(R(\mathfrak{g}, K) \otimes_{R(\mathfrak{g},K)} X, V) \qquad \text{by (C.20)}$$
$$\cong \mathrm{Hom}_{R(\mathfrak{h},L)}(\mathcal{F}(X), V) \qquad \text{by (2.20b)}.$$

REMARK. If Φ and Φ' are corresponding members of the two Hom's, inspection of the proof shows that they are related by

$$(2.22) \qquad \Phi(x)(r) = \Phi'(rx) \qquad \text{for } r \in R(\mathfrak{g}, K), \, x \in X.$$

Corollary 2.23. I carries injectives to injectives.

PROOF. If V is injective in $C(\mathfrak{h}, L)$, then Proposition 2.21 says that

$$(2.24) \qquad X \mapsto \mathrm{Hom}_{R(\mathfrak{g},K)}(X, I(V))$$

is naturally isomorphic with

$$X \mapsto \mathrm{Hom}_{R(\mathfrak{h},L)}(\mathcal{F}(X), V),$$

which is exact by (C.14), being the composition of the exact functors $\mathcal{F}$ and $\mathrm{Hom}_{R(\mathfrak{h},L)}(\cdot, V)$. By (C.16), (2.24) is exact. By (C.14) again, $I(V)$ is injective.

Corollary 2.25. If V is in $C(\mathfrak{k}, K)$, then $I_{\mathfrak{k},K}^{\mathfrak{g},K}(V)$ is injective in $C(\mathfrak{g}, K)$.

PROOF. We combine Lemma 2.4 and Corollary 2.23.

REMARK. An injective in $C(\mathfrak{g}, K)$ of the form $I_{\mathfrak{k},K}^{\mathfrak{g},K}(V)$ is called a **standard injective**.

Corollary 2.26.

(a) If X is in $\mathcal{C}(\mathfrak{g}, K)$, then X is a submodule of a standard injective. Consequently $\mathcal{C}(\mathfrak{g}, K)$ has enough injectives.

(b) If X is injective in $\mathcal{C}(\mathfrak{g}, K)$, then X is a direct summand of a standard injective.

PROOF. (a) Let I be the standard injective $I = I_{\ell,K}^{\mathfrak{g},K}(\mathcal{F}_{\mathfrak{g},K}^{\ell,K}(X))$. By Proposition 2.21 we have

$$\mathrm{Hom}_{\mathfrak{g},K}(X, I) \cong \mathrm{Hom}_{\ell,K}(\mathcal{F}_{\mathfrak{g},K}^{\ell,K}(X), \mathcal{F}_{\mathfrak{g},K}^{\ell,K}(X)).$$

If φ is the member of the left side that corresponds to 1 on the right side, then (2.22) says that φ is defined by

$$\varphi(x)(r) = rx \qquad \text{for } r \in R(\mathfrak{g}, K), \ x \in X.$$

To prove (a), we have only to check that φ is one-one. If $\varphi(x) = 0$, then $rx = 0$ for all r. Since X is approximately unital, $x = 0$.

(b) With I as in (a), we obtain an exact sequence

$$0 \longrightarrow X \longrightarrow I \longrightarrow I/X \longrightarrow 0.$$

Since X is injective, (C.33) says this exact sequence splits. Hence X is a direct summand of I. This completes the proof.

To obtain the corresponding assertions for the functor P is a little more delicate. Suppose X is a $(\mathfrak{g}, K)$ module. Guided by (2.20b), we define

$$(2.27) \qquad \mathcal{F}^{\vee}(X) = (\mathcal{F}^{\vee})_{\mathfrak{g},K}^{\mathfrak{h},L}(X) = \mathrm{Hom}_{R(\mathfrak{g},K)}(R(\mathfrak{g}, K), X)_L.$$

Here we are using the right $R(\mathfrak{h}, L)$ module structure on $R(\mathfrak{g}, K)$, defining $(r\varphi)(s) = \varphi(sr)$ to make the Hom into a left $R(\mathfrak{h}, L)$ module, and taking the L finite part.

Lemma 2.28. If X in $\mathcal{C}(\mathfrak{g}, K)$, when regarded as an $R(K)$ module, decomposes according to (1.19) as $X = \bigoplus_{\gamma \in \widehat{K}} X_{\gamma}$, then

$$\mathrm{Hom}_{R(\mathfrak{g},K)}(R(\mathfrak{g}, K), X) \cong \prod_{\gamma \in \widehat{K}} X_{\gamma}$$

as an $R(K)$ module, and this isomorphism is natural in X. If a map of pairs $i : (\mathfrak{h}, L) \to (\mathfrak{g}, K)$ is given, then this isomorphism respects the actions of $R(L)$: on the left as in (2.27) and on the right by the product of the action on each X_{γ}.

REMARK. $\prod_{\gamma \in \widehat{K}} X_\gamma$ is meant to allow arbitrary tuples $\{x_\gamma\}_{\gamma \in \widehat{K}}$, not just those with finitely many nonzero terms.

PROOF. The decomposition

$$R(K) \cong \bigoplus_{\gamma \in \widehat{K}} R(K)_\gamma$$

of (1.37) respects multiplication, by Proposition 1.39a, and hence it respects the K action on both left and right, as well as the L action on the right if a map of pairs is given. Tensoring with $U(\mathfrak{g})$ over $U(\mathfrak{k})$, we have

$$(2.29) \qquad R(\mathfrak{g}, K) \cong U(\mathfrak{g}) \otimes_{U(\mathfrak{k})} R(K) \cong \bigoplus_{\gamma \in \widehat{K}} U(\mathfrak{g}) \otimes_{U(\mathfrak{k})} R(K)_\gamma.$$

This isomorphism still respects the actions by K and L on the right. If a left action by K is defined on each factor by $\mathrm{Ad} \otimes l$, the isomorphism respects the action by K on the left, and it is clear that it respects left multiplication by $U(\mathfrak{g})$. Applying $\mathrm{Hom}_{R(\mathfrak{g},K)}(\,\cdot\,, X)$ to both sides of (2.29), we therefore have

(2.30)
$$\mathrm{Hom}_{R(\mathfrak{g},K)}(R(\mathfrak{g}, K), X) \cong \prod_{\gamma \in \widehat{K}} \mathrm{Hom}_{R(\mathfrak{g},K)}(U(\mathfrak{g}) \otimes_{U(\mathfrak{k})} R(K)_\gamma, X)$$

as left $R(K)$ and $R(L)$ modules, and this isomorphism is natural in X.

Fix $\gamma \in \widehat{K}$. If Φ is in $\mathrm{Hom}_{R(\mathfrak{g},K)}(U(\mathfrak{g}) \otimes_{U(\mathfrak{k})} R(K)_\gamma, X)$, we define φ in $\mathrm{Hom}_{R(K)}(R(K)_\gamma, X)$ by

$$\varphi(T) = \Phi(1 \otimes T).$$

The map $\Phi \mapsto \varphi$ is invertible with inverse

$$\Phi(a \otimes T) = a(\varphi(T)),$$

and thus we have an isomorphism of left $R(K)$ and $R(L)$ modules

$$(2.31) \qquad \mathrm{Hom}_{R(\mathfrak{g},K)}(U(\mathfrak{g}) \otimes_{U(\mathfrak{k})} R(K)_\gamma, X) \cong \mathrm{Hom}_{R(K)}(R(K)_\gamma, X),$$

natural in X.

By Theorem 1.57 and Lemma 1.40, we have isomorphisms of left $R(K)$ and $R(L)$ modules

$$(2.32) \qquad \mathrm{Hom}_{R(K)}(R(K)_\gamma, X) \cong \mathrm{Hom}_K(R(K)_\gamma, X) \cong X_\gamma$$

natural in X. Combining (2.30), (2.31), and (2.32), we obtain the isomorphism of the lemma.

Proposition 2.33. The functor $\mathcal{F}^\vee$ is an exact covariant functor from $\mathcal{C}(\mathfrak{g}, K)$ to $\mathcal{C}(\mathfrak{h}, L)$. For X in $\mathcal{C}(\mathfrak{g}, K)$ the map

$$j : \mathcal{F}(X) \to \mathcal{F}^\vee(X) \qquad \text{with} \qquad (j(x))(r) = rx$$

is one-one and is natural in X. Moreover, j is onto if and only if every $\tau \in \widehat{L}$ has the following property: There are at most finitely many $\gamma \in \widehat{K}$ such that $X_\gamma \neq 0$ and τ occurs in the representation $\gamma \circ i_{\mathrm{gp}}$ of L. In particular, j is an isomorphism if either

(i) $i_{\mathrm{gp}}(L)$ has finite index in K or
(ii) $X_\gamma \neq 0$ for only finitely many $\gamma \in \widehat{K}$.

REMARKS. The main case of (i) is that $L = K$, and the main case of (ii) is that X is finite dimensional.

PROOF. Certainly $\mathcal{F}^\vee$ is covariant from $\mathcal{C}(\mathfrak{g}, K)$ to $\mathcal{C}(\mathfrak{h}, L)$. Put $X^\vee = \mathrm{Hom}_{R(\mathfrak{g},K)}(R(\mathfrak{g}, K), X)$, as a member of $\tilde{\mathcal{C}}(\mathfrak{h}, L)$. The functor $\mathcal{F}^\vee$ is the composition of $X \mapsto X^\vee$ from $\mathcal{C}(\mathfrak{g}, K)$ to $\tilde{\mathcal{C}}(\mathfrak{h}, L)$ and the L finite part $(\cdot)_L$. The second of these functors is exact by Proposition 1.55, and $\mathcal{F}^\vee$ will be exact if $X \mapsto X^\vee$ is exact. Lemma 2.28 says that $X^\vee = \prod_{\gamma \in \widehat{K}} X_\gamma$ as an $R(L)$ module. Hence it is enough for the exactness of $\mathcal{F}^\vee$ to prove that

$$X = \bigoplus_{\gamma \in \widehat{K}} X_\gamma \mapsto \prod_{\gamma \in \widehat{K}} X_\gamma = X^\vee$$

is exact from $\mathcal{C}(\mathfrak{k}, K)$ to $\tilde{\mathcal{C}}(\mathfrak{l}, L)$. By (2.3) and Lemma 2.28, this functor is naturally isomorphic with the functor $F = \mathrm{Hom}_{R(K)}(R(K), \cdot)$, which is left exact by (C.13). Hence if

$$0 \longrightarrow Y \longrightarrow X \xrightarrow{\rho} X/Y \longrightarrow 0.$$

is an exact sequence in $\mathcal{C}(\mathfrak{k}, K)$, the complex

$$0 \longrightarrow F(Y) \longrightarrow F(X) \xrightarrow{F(\rho)} F(X/Y).$$

is exact. We shall show $F(\rho)$ is onto. The given exact sequence splits in $\mathcal{C}(\mathfrak{k}, K)$, by Proposition 1.18b. If $\rho' : X/Y \to X$ is a $(\mathfrak{k}, K)$ map with $\rho\rho' = 1_{X/Y}$, then $F(\rho)F(\rho') = 1_{F(X/Y)}$ in $\mathcal{C}(\mathfrak{l}, L)$, and it follows that $F(\rho)$ is onto. By (C.11), $\mathcal{F}^\vee$ is exact.

The definition of j shows that j is an $R(\mathfrak{h}, L)$ map, and it is clearly natural in X. If $j(x) = 0$, then $rx = 0$ for all r in $R(\mathfrak{g}, K)$. Since X is approximately unital, $x = 0$. Thus j is one-one.

To check circumstances under which j is onto, let us realize $X^\vee$ as an $R(L)$ module in the form given in Lemma 2.28. If x is in X, then the γth factor of $j(x)$ on the right side of (2.30) is $j(x)|_{U(\mathfrak{g})\otimes R(K)_\gamma}$, and this corresponds under (2.31) to the element $j(x)|_{1\otimes R(K)_\gamma}$ of $\mathrm{Hom}_{R(K)}(R(K)_\gamma, X)$, which in turn corresponds under (2.32) to the member $j(x)(1 \otimes \chi_\gamma)$ of X_γ. By definition of j, this element is just $(1 \otimes \chi_\gamma)(x)$. Thus, under our identification of $X^\vee$ as in Lemma 2.28,

$$ j(x) \longleftrightarrow \prod_{\gamma \in \widehat{K}} \chi_\gamma(x). $$

Hence j maps onto the subset $\bigoplus_{\gamma \in \widehat{K}} X_\gamma$ of $X^\vee$. From the relationship of $\mathcal{F}^\vee(X)$ to $X^\vee$, we see that $\mathcal{F}^\vee(X) = \left(\prod_{\gamma \in \widehat{K}} X_\gamma \right)_L$ as a vector space. Thus j is onto if and only if $\bigoplus_{\gamma \in \widehat{K}} X_\gamma = \left(\prod_{\gamma \in \widehat{K}} X_\gamma \right)_L$ as a vector space. Each side may be decomposed as a direct sum over $\tau \in \widehat{L}$ of the τ isotypic subspace, the subspaces for a particular τ being $\bigoplus_{\gamma \in \widehat{K}} (X_\gamma)_\tau$ and $\prod_{\gamma \in \widehat{K}} (X_\gamma)_\tau$, respectively. These spaces are equal if and only if there are at most finitely many terms, as we wished to show. The sufficiency of condition (i) is a consequence of classical Frobenius reciprocity, and the sufficiency of condition (ii) is immediate.

Proposition 2.34 (Frobenius reciprocity). The functor P is left adjoint to $\mathcal{F}^\vee$. That is, there is a natural isomorphism

$$ \mathrm{Hom}_{R(\mathfrak{g},K)}(P(V), X) \cong \mathrm{Hom}_{R(\mathfrak{h},L)}(V, \mathcal{F}^\vee(X)) $$

for V in $\mathcal{C}(\mathfrak{h}, L)$ and X in $\mathcal{C}(\mathfrak{g}, K)$.

PROOF.

$\mathrm{Hom}_{R(\mathfrak{g},K)}(P(V), X)$
$$ = \mathrm{Hom}_{R(\mathfrak{g},K)}(R(\mathfrak{g}, K) \otimes_{R(\mathfrak{h},L)} V, X) $$
$$ \cong \mathrm{Hom}_{R(\mathfrak{h},L)}(V, \mathrm{Hom}_{R(\mathfrak{g},K)}(R(\mathfrak{g}, K), X)) \quad \text{by (C.20)} $$
$$ = \mathrm{Hom}_{R(\mathfrak{h},L)}(V, \mathrm{Hom}_{R(\mathfrak{g},K)}(R(\mathfrak{g}, K), X)_L) $$
$$ = \mathrm{Hom}_{R(\mathfrak{h},L)}(V, \mathcal{F}^\vee(X)). $$

In the same way as with Corollaries 2.23, 2.25, and 2.26, we obtain the following three corollaries.

Corollary 2.35. P carries projectives to projectives.

Corollary 2.36. If V is in $\mathcal{C}(\mathfrak{k}, K)$, then $P_{\mathfrak{k}, K}^{\mathfrak{g}, K}(V)$ is projective in $\mathcal{C}(\mathfrak{g}, K)$.

REMARK. A projective in $\mathcal{C}(\mathfrak{g}, K)$ of the form $P_{\mathfrak{k}, K}^{\mathfrak{g}, K}(V)$ is called a **standard projective**.

Corollary 2.37.

(a) If X is in $\mathcal{C}(\mathfrak{g}, K)$, then X is a quotient of a standard projective. Consequently $\mathcal{C}(\mathfrak{g}, K)$ has enough projectives.

(b) If X is projective in $\mathcal{C}(\mathfrak{g}, K)$, then X is a direct summand of a standard projective.

3. Constructions within $\mathcal{C}(\mathfrak{g}, K)$

We begin with closure properties for $(\mathfrak{g}, K)$ modules under $\otimes_{\mathbb{C}}$ and $\mathrm{Hom}_{\mathbb{C}}$. If V and W are in $\mathcal{C}(\mathfrak{g}, K)$, we obtain representations of $\mathfrak{g}$ and K on $V \otimes_{\mathbb{C}} W$ by the usual definitions

$$(2.38) \qquad \begin{aligned} X(x \otimes y) &= Xx \otimes y + x \otimes Xy && \text{for } X \in \mathfrak{g} \\ k(x \otimes y) &= kx \otimes ky && \text{for } k \in K. \end{aligned}$$

Properties (i) and (ii) of a $(\mathfrak{g}, K)$ module in (1.86) are clear for $V \otimes_{\mathbb{C}} W$. For (iii), if X is in $\mathfrak{g}$, then

$$\begin{aligned} (\mathrm{Ad}(k)X)&(x \otimes y) \\ &= (\mathrm{Ad}(k)X)x \otimes y + x \otimes (\mathrm{Ad}(k)X)y \\ &= k(X(k^{-1}x)) \otimes y + x \otimes k(X(k^{-1}y)) && \text{by (iii) for } V \text{ and } W \\ &= k(X(k^{-1}x) \otimes k^{-1}y + k^{-1}x \otimes X(k^{-1}y)) \\ &= k(X(k^{-1}(x \otimes y))). \end{aligned}$$

Thus (iii) holds for $V \otimes_{\mathbb{C}} W$, and $V \otimes_{\mathbb{C}} W$ is in $\mathcal{C}(\mathfrak{g}, K)$.

The situation for $\mathrm{Hom}_{\mathbb{C}}$ is more complicated. If V and W are in $\mathcal{C}(\mathfrak{g}, K)$, we obtain representations of $\mathfrak{g}$ and K on $\mathrm{Hom}_{\mathbb{C}}(V, W)$ by the definitions

$$(2.39) \qquad \begin{aligned} (X\varphi)(x) &= X(\varphi(x)) - \varphi(Xx) && \text{for } X \in \mathfrak{g} \\ (k\varphi)(x) &= k(\varphi(k^{-1}x)) && \text{for } k \in K. \end{aligned}$$

But the members of $\mathrm{Hom}_{\mathbb{C}}(V, W)$ need not be K finite, i.e., property (i) in (1.86) need not hold. For example, if $\mathfrak{g} = \mathfrak{k}$ and $V = \bigoplus_{\gamma \in \widehat{K}} V_{\gamma}$ and

$W = \mathbb{C}$ (the trivial representation), then $\operatorname{Hom}_\mathbb{C}(V, \mathbb{C}) = \prod_{\gamma \in \widehat{K}} V_\gamma$, and this is K finite if and only if $V_\gamma \neq 0$ for only finitely many $\gamma \in \widehat{K}$.

The way around this difficulty is to define it away. We have made $\operatorname{Hom}_\mathbb{C}(V, W)$ into a representation space for K (as well as for $\mathfrak{g}$), and we shall work with the K finite subspace $\operatorname{Hom}_\mathbb{C}(V, W)_K$, which was defined in (1.17). *Notice that this use of* $(\cdot)_K$ *is different from the one in* §§1–2. In §§1–2, $(\cdot)_K$ was applied to a left $R(K)$ module; now it is applied to a representation of K. As was mentioned after Proposition 1.55, use of the same notation for two notions should not cause any difficulty in this instance.

We require an analog of (2.2) for the group-representation version of $(\cdot)_K$. To work within the context of Appendix C, we use the abstract **group algebra** $\mathbb{C}[K]$ of K over $\mathbb{C}$, which made an appearance in the proof of Proposition 1.18. This algebra has a basis $\{v_k\}$ parametrized by the elements of K, and its multiplication law is $v_k v_{k'} = v_{kk'}$. It is immediate that the unital left $\mathbb{C}[K]$ modules are the arbitrary representations of K. The category $C(\mathbb{C}[K])$ of all unital left $\mathbb{C}[K]$ modules is a good category, and the results of Appendix C apply. If M is in $C(\mathbb{C}[K])$, then M_K is in $C(\mathfrak{k}, K)$, and $M \mapsto M_K$ is a functor from $C(\mathbb{C}[K])$ to $C(\mathfrak{k}, K)$ called the K **finite part**, the effect on maps being given by restriction.

In defining a functor in this way, there is one thing to check: If $\varphi : M \to N$ is a map in $C(\mathbb{C}[K])$, then φ carries M_K to N_K. To check this statement, we use Proposition 1.18a to write $M_K = \bigoplus_\alpha M_\alpha$ with each M_α an irreducible finite-dimensional smooth representation of K. Schur's Lemma says that $\varphi|_{M_\alpha}$ is zero or invertible. In the latter case, the image $N_\alpha = \varphi(M_\alpha)$ is equivalent with a smooth representation and hence is smooth. (That is, the linear map $\varphi|_{M_\alpha}$ is smooth, being between finite-dimensional spaces, and therefore the action of K on N_α is smooth, being given by $kx = \varphi(k\varphi^{-1}(x))$ for $k \in K$ and $x \in N_\alpha$.) Hence $N_\alpha \subseteq N_K$, and we conclude that $\varphi(M_K) \subseteq N_K$.

In place of (2.2) this version of K finite part satisfies

(2.40) $(\cdot)_K$ is covariant and left exact.

To verify (2.40), let

$$0 \longrightarrow A \overset{\psi}{\longrightarrow} B \overset{\varphi}{\longrightarrow} C \longrightarrow 0$$

be exact in $C(\mathbb{C}[K])$, and let

$$0 \longrightarrow A_K \overset{\psi|_{A_K}}{\longrightarrow} B_K \overset{\varphi|_{B_K}}{\longrightarrow} C_K$$

be the image complex under the functor of K finite part. The map $\psi|_{A_K}$ is one-one since it is a restriction of the one-one map ψ, and we know that image$(\psi|_{A_K}) \subseteq \ker(\varphi|_{B_K})$. If b is in $\ker(\varphi|_{B_K})$, then $b = \psi(a)$ for some $a \in A$ since b is in $\ker\varphi$. Since ψ is one-one,

$$\dim\langle Ka\rangle = \dim\psi(\langle Ka\rangle) = \dim\langle K\psi(a)\rangle = \dim\langle Kb\rangle,$$

and the right side is finite because b is K finite. The linear map ψ exhibits the representation of K on $\langle Ka\rangle$ as equivalent with the smooth representation of K on $\langle Kb\rangle$, and therefore the representation of K on $\langle Ka\rangle$ is smooth. Thus a is K finite, and (2.40) is proved.

We cannot conclude exactness in (2.40). In fact, with $K = S^1$, let $e : \mathbb{C}[K] \to \mathbb{C}$ be the sum-of-coefficients mapping. This is a $\mathbb{C}[K]$ module map if K acts by the left regular representation on $\mathbb{C}[K]$ and by the trivial representation on $\mathbb{C}$. If $\mathbb{C}_0[K]$ denotes the kernel of e, we have an exact sequence

$$0 \longrightarrow \mathbb{C}_0[K] \longrightarrow \mathbb{C}[K] \overset{e}{\longrightarrow} \mathbb{C} \longrightarrow 0.$$

The claim is that the complex of K finite parts is

$$0 \longrightarrow 0 \longrightarrow 0 \longrightarrow \mathbb{C} \longrightarrow 0,$$

which is not exact. In fact, a K finite subspace of $\mathbb{C}[K]$ must be fully reducible into one-dimensional invariant subspaces, since K is compact abelian, and it is easy to see that any one-dimensional K invariant subspace of functions on K must consist of multiples of a character. But the characters are not in $\mathbb{C}[K]$, and hence $\mathbb{C}[K]_K = 0$.

Because we do not have a conclusion of exactness in (2.40), our approach to proving properties of $\mathrm{Hom}_{\mathbb{C}}(V, W)_K$ will be a little indirect. First let us check that $\mathrm{Hom}_{\mathbb{C}}(V, W)_K$ is indeed a $(\mathfrak{g}, K)$ module.

Proposition 2.41. If V and W are in $\mathcal{C}(\mathfrak{g}, K)$, then $\mathrm{Hom}_{\mathbb{C}}(V, W)_K$ is a $\mathfrak{g}$ invariant subspace of $\mathrm{Hom}_{\mathbb{C}}(V, W)$ and is in $\mathcal{C}(\mathfrak{g}, K)$.

PROOF. For $k \in K$, $X \in \mathfrak{g}$, and $\varphi \in \mathrm{Hom}_{\mathbb{C}}(V, W)$, we have

(2.42) $$kX\varphi = (\mathrm{Ad}(k)X)(k\varphi) \in \mathfrak{g}(K\varphi)$$

because

$$((\mathrm{Ad}(k)X)(k\varphi))(x)$$
$$= (\mathrm{Ad}(k)X)(k\varphi(x)) - (k\varphi)((\mathrm{Ad}(k)X)x)$$
$$= (\mathrm{Ad}(k)X)(k(\varphi(k^{-1}x))) - k(\varphi(k^{-1}(\mathrm{Ad}(k)X)x))$$
$$= k(X(\varphi(k^{-1}x))) - k(\varphi(X(k^{-1}x))) \quad \text{by (ii) for } V \text{ and } W$$
$$= k(X\varphi(k^{-1}x)) = kX\varphi(x).$$

The set on the right of (2.42) lies in a finite-dimensional space if φ is K finite. In other words, if φ is K finite, so is $X\varphi$. Thus $\mathrm{Hom}_{\mathbb{C}}(V, W)_K$ is $\mathfrak{g}$ invariant.

For $\mathrm{Hom}_{\mathbb{C}}(V, W)_K$, property (i) of a $(\mathfrak{g}, K)$ module in (1.86) is now true by definition, (ii) is trivial, and (iii) is the equality in (2.42). Thus $\mathrm{Hom}_{\mathbb{C}}(V, W)_K$ is a $(\mathfrak{g}, K)$ module.

REMARK. In the special case that $W = \mathbb{C}$ with trivial $(\mathfrak{g}, K)$ action, we denote the $(\mathfrak{g}, K)$ module $\mathrm{Hom}_{\mathbb{C}}(V, \mathbb{C})_K$ by V^c and call V^c the K finite **contragredient** of V.

Now let us fix one of V and W and make the passages to $V \otimes_{\mathbb{C}} W$ and $\mathrm{Hom}_{\mathbb{C}}(V, W)_K$ into functors carrying $\mathcal{C}(\mathfrak{g}, K)$ into itself. To do so, we have to exhibit the effect on maps.

1) With $F(V) = V \otimes_{\mathbb{C}} W$, define $F(\varphi) = \varphi \otimes 1$. Then it is a standard fact about tensor products of vector spaces that F is a covariant functor. Similar remarks apply to $F(W) = V \otimes_{\mathbb{C}} W$.

2) With $F(V) = \mathrm{Hom}_{\mathbb{C}}(V, W)_K$, define $F(\varphi) = \mathrm{Hom}(\varphi, 1)$. In seeing that F is a contravariant functor, we need to check that if φ is a $(\mathfrak{g}, K)$ map from V into V', then $\mathrm{Hom}(\varphi, 1)$ is a $(\mathfrak{g}, K)$ map from $\mathrm{Hom}_{\mathbb{C}}(V', W)_K$ into $\mathrm{Hom}_{\mathbb{C}}(V, W)_K$. Certainly $\mathrm{Hom}(\varphi, 1)$ is a $(\mathfrak{g}, K)$ map from $\mathrm{Hom}_{\mathbb{C}}(V', W)$ into $\mathrm{Hom}_{\mathbb{C}}(V, W)$ and therefore from $\mathrm{Hom}_{\mathbb{C}}(V', W)_K$ into $\mathrm{Hom}_{\mathbb{C}}(V, W)$. Since $\mathrm{Hom}(\varphi, 1)$ is a K map, it carries K finite elements to K finite elements. Therefore the image must be in $\mathrm{Hom}_{\mathbb{C}}(V, W)_K$. The other properties of F are standard, and thus F is a contravariant functor. (One instance is worth special note. Expanding on the notion of K finite contragredient of a $(\mathfrak{g}, K)$ module, we call the functor

$$V \mapsto V^c = \mathrm{Hom}_{\mathbb{C}}(V, \mathbb{C})_K$$

the **contragredient** functor on $\mathcal{C}(\mathfrak{g}, K)$.)

3) With $F(W) = \mathrm{Hom}_{\mathbb{C}}(V, W)_K$, define $F(\varphi) = \mathrm{Hom}(1, \varphi)$. The same kind of argument as in (2) shows that F is a covariant functor.

We shall work toward a proof that all these functors are exact. Because (2.40) does not give us exactness of $(\cdot)_K$, we digress briefly in order to develop some associativity formulas. The first associativity formula requires two lemmas. Recall for a $(\mathfrak{g}, K)$ module V that the vector space of $(\mathfrak{g}, K)$ invariants was defined in (2.15). It is a subspace of V.

Lemma 2.43. If V and W are $(\mathfrak{g}, K)$ modules, then the vector space of invariants of $\mathrm{Hom}_{\mathbb{C}}(V, W)_K$ is the subspace $\mathrm{Hom}_{R(\mathfrak{g}, K)}(V, W)$.

PROOF. An element $\varphi \in \mathrm{Hom}_{\mathbb{C}}(V, W)_K$ is invariant if and only if

$$0 = X\varphi(x) = X(\varphi(x)) - \varphi(Xx) \qquad \text{for } X \in \mathfrak{g}$$

and $\qquad \varphi(x) = k\varphi(x) = k(\varphi(k^{-1}x)) \qquad \text{for } k \in K,$

i.e., if and only if φ is $\mathfrak{g}$ linear and K linear. Moreover, a K invariant element is automatically K finite. Hence the space of invariants is $\mathrm{Hom}_{\mathfrak{g},k}(V, W)$, and this is the same as $\mathrm{Hom}_{R(\mathfrak{g},K)}(V, W)$ by Theorem 1.117.

Lemma 2.44. Let representations of K on vector spaces V and W be given, and define a representation of K on $\mathrm{Hom}_{\mathbb{C}}(V, W)$ by $(k\varphi)(x) = k(\varphi(k^{-1}x))$ for $k \in K$. If φ is K finite in $\mathrm{Hom}_{\mathbb{C}}(V, W)$ and x is K finite in V, then $\varphi(x)$ is K finite in W.

PROOF. We have $k(\varphi(x)) = (k\varphi)(kx) \subseteq (K\varphi)(Kx)$. Now $\langle K\varphi \rangle$ and $\langle Kx \rangle$ are finite dimensional, say with bases $\{\varphi_i\}$ and $\{x_j\}$. Then $K(\varphi(x))$ is contained in the finite-dimensional space $\sum_{i,j} \mathbb{C}\varphi_i(x_j)$. Since $k(\varphi(x)) = (k\varphi)(kx)$, the map $k \mapsto k(\varphi(x))$ is smooth. Therefore $\varphi(x)$ is K finite.

Proposition 2.45. If A, B, and C are in $C(\mathfrak{g}, K)$, then

$$\mathrm{Hom}_{\mathbb{C}}(A \otimes_{\mathbb{C}} B, C)_K \cong \mathrm{Hom}_{\mathbb{C}}(A, \mathrm{Hom}_{\mathbb{C}}(B, C)_K)_K$$

as modules in $C(\mathfrak{g}, K)$, and this isomorphism is natural in A, B, and C.

PROOF. We start from the natural isomorphism of complex vector spaces

$$(2.46) \qquad \mathrm{Hom}_{\mathbb{C}}(A \otimes_{\mathbb{C}} B, C) \cong \mathrm{Hom}_{\mathbb{C}}(A, \mathrm{Hom}_{\mathbb{C}}(B, C))$$

of (C.19), which is given by $\Phi \mapsto \varphi$ with $\varphi(a)(b) = \Phi(a \otimes b)$. This isomorphism respects the actions of $X \in \mathfrak{g}$ and $k \in K$ because

$$\begin{aligned}
X\varphi(a)(b) &= (X(\varphi(a)))(b) - \varphi(Xa)(b) \\
&= X(\varphi(a)(b)) - \varphi(a)(Xb) - \varphi(Xa)(b) \\
&= X(\Phi(a \otimes b)) - \Phi(a \otimes Xb) - \Phi(Xa \otimes b) \\
&= X(\Phi(a \otimes b)) - \Phi(X(a \otimes b)) \\
&= X\Phi(a \otimes b)
\end{aligned}$$

shows $X\Phi$ corresponds to $X\varphi$ and because

$$
\begin{aligned}
k\varphi(a)(b) &= (k(\varphi(k^{-1}a)))(b) \\
&= k(\varphi(k^{-1}a)(k^{-1}b)) \\
&= k(\Phi(k^{-1}a \otimes k^{-1}b)) \\
&= k(\Phi(k^{-1}(a \otimes b))) \\
&= (k\Phi)(a \otimes b)
\end{aligned}
$$

shows $k\Phi$ corresponds to $k\varphi$.

Since the K actions correspond, we can take the K finite part of both sides of (2.46), obtaining

$$
\operatorname{Hom}_{\mathbb{C}}(A \otimes_{\mathbb{C}} B,\ C)_K \cong \operatorname{Hom}_{\mathbb{C}}(A,\ \operatorname{Hom}_{\mathbb{C}}(B, C))_K.
$$

Applying Lemma 2.44 to the outer Hom on the right side, we can put the missing $(\cdot)_K$ in place and obtain the proposition.

Corollary 2.47. If A, B, and C are in $C(\mathfrak{g}, K)$, then

$$
\operatorname{Hom}_{R(\mathfrak{g},K)}(A \otimes_{\mathbb{C}} B,\ C) \cong \operatorname{Hom}_{R(\mathfrak{g},K)}(A,\ \operatorname{Hom}_{\mathbb{C}}(B, C)_K)
$$

as complex vector spaces, and this isomorphism is natural in A, B, and C.

REMARK.   The corollary says that $(\cdot) \otimes_{\mathbb{C}} B$ is left adjoint to $\operatorname{Hom}_{\mathbb{C}}(B, \cdot)$. But it says a little more, since it gives also naturality in B.

PROOF. In Proposition 2.45, the subsets of $(\mathfrak{g}, K)$ invariants of the two sides must correspond. Lemma 2.43 identifies these, and the corollary follows.

There are corresponding formulas for triple tensor products. Again we begin with a lemma. Recall for a $(\mathfrak{g}, K)$ module V that the vector space of $(\mathfrak{g}, K)$ coinvariants was defined in (2.15). It is a quotient space of V.

Lemma 2.48. If V and W are $(\mathfrak{g}, K)$ modules, then the vector space of coinvariants of $V \otimes_{\mathbb{C}} W$ is the quotient space $V \otimes_{R(\mathfrak{g},K)} W$, where V is converted from a left $R(\mathfrak{g}, K)$ module to a right $R(\mathfrak{g}, K)$ module by the definition

(2.49) $xr = r^t x$ for $r \in R(\mathfrak{g}, K)$, $x \in V$.

PROOF. According to (2.15), the space of coinvariants of $V \otimes_{\mathbb{C}} W$ is the quotient of $V \otimes_{\mathbb{C}} W$ by the linear span

(2.50a) $\langle X(x \otimes y), (1 - k)(x \otimes y) \mid X \in \mathfrak{g}, k \in K, x \in V, y \in W \rangle.$

If we convert V into a right $R(\mathfrak{g}, K)$ module, then the right actions of $\mathfrak{g}$ and K on V are given by

$$ xX = X^t x = -Xx \quad \text{and} \quad xk = x\delta_k = \delta_k^t x = \delta_{k^{-1}} x = k^{-1} x $$

for $X \in \mathfrak{g}$ and $k \in K$. Therefore the action on $V \otimes_{\mathbb{C}} W$ is given by

$$ X(x \otimes y) = Xx \otimes y + x \otimes Xy = -xX \otimes y + x \otimes Xy $$

and

$$ (1 - k)(x \otimes y) = x \otimes y - kx \otimes ky = (xk^{-1})k \otimes y - (xk^{-1}) \otimes ky. $$

From these formulas it is apparent that (2.50a) equals

(2.50b)
$\langle xX \otimes y - x \otimes Xy, xk \otimes y - x \otimes ky \mid X \in \mathfrak{g}, k \in K, x \in V, y \in W \rangle.$

Applying Theorem 1.119, we obtain the lemma.

Proposition 2.51. If A, B, and C are in $\mathcal{C}(\mathfrak{g}, K)$, then

$$ (A \otimes_{\mathbb{C}} B) \otimes_{\mathbb{C}} C \cong A \otimes_{\mathbb{C}} (B \otimes_{\mathbb{C}} C) $$

as modules in $\mathcal{C}(\mathfrak{g}, K)$, and this isomorphism is natural in A, B, and C.

PROOF. We start from the natural isomorphism of vector spaces

$$ (A \otimes_{\mathbb{C}} B) \otimes_{\mathbb{C}} C \cong A \otimes_{\mathbb{C}} (B \otimes_{\mathbb{C}} C) $$

of (C.21), which is given by $(a \otimes b) \otimes c \mapsto a \otimes (b \otimes c)$. This isomorphism respects the actions of $X \in \mathfrak{g}$ and $k \in K$ because we find

$$ X((a \otimes b) \otimes c) = Xa \otimes b \otimes c + a \otimes Xb \otimes c + a \otimes b \otimes Xc $$
$$ = X(a \otimes (b \otimes c)) $$

and $k((a \otimes b) \otimes c) = ka \otimes kb \otimes kc = k(a \otimes (b \otimes c)).$

The proposition follows.

Corollary 2.52. If A, B, and C are in $\mathcal{C}(\mathfrak{g}, K)$, then

$$(A \otimes_{\mathbb{C}} B) \otimes_{R(\mathfrak{g}, K)} C \cong A \otimes_{R(\mathfrak{g}, K)} (B \otimes_{\mathbb{C}} C)$$

as complex vector spaces, and this isomorphism is natural in A, B, and C. Here $A \otimes_{\mathbb{C}} B$ on the left side and A on the right side have been converted into right $R(\mathfrak{g}, K)$ modules by (2.49).

PROOF. In Proposition 2.51, the quotient spaces of $(\mathfrak{g}, K)$ coinvariants of the two sides must correspond. Lemma 2.48 identifies these, and the corollary follows.

Our digression is complete, and we now prove the main result on constructions within $\mathcal{C}(\mathfrak{g}, K)$.

Proposition 2.53. Let V be in $\mathcal{C}(\mathfrak{g}, K)$. Then
  (a) the covariant functor $W \mapsto V \otimes_{\mathbb{C}} W$ on $\mathcal{C}(\mathfrak{g}, K)$ is exact, and it sends projectives to projectives.
  (b) the contravariant functor $W \mapsto \mathrm{Hom}_{\mathbb{C}}(W, V)_K$ on $\mathcal{C}(\mathfrak{g}, K)$ is exact, and it sends projectives to injectives.
  (c) the covariant functor $W \mapsto \mathrm{Hom}_{\mathbb{C}}(V, W)_K$ on $\mathcal{C}(\mathfrak{g}, K)$ is exact, and it sends injectives to injectives.

PROOF. The exactness in (a) is from (C.15), but the exactness in (b) and (c) needs proof because (2.40) has not given us the exactness we would like for $(\cdot)_K$. Let us prove exactness for (c), omitting the analogous argument for (b). If

$$(2.54) \qquad\qquad 0 \longrightarrow A \overset{\psi}{\longrightarrow} B \overset{\varphi}{\longrightarrow} C \longrightarrow 0$$

is exact in $\mathcal{C}(\mathfrak{g}, K)$, then the sequence

$$0 \to \mathrm{Hom}_{\mathbb{C}}(V, A)_K \xrightarrow{\mathrm{Hom}(1, \psi)} \mathrm{Hom}_{\mathbb{C}}(V, B)_K \xrightarrow{\mathrm{Hom}(1, \varphi)} \mathrm{Hom}_{\mathbb{C}}(V, C)_K$$

is exact in $\mathcal{C}(\mathfrak{g}, K)$ as a consequence of (C.13) and (2.40). We are to show that $\mathrm{Hom}(1, \varphi)$ is onto. By Proposition 1.18b, (2.54) splits in $\mathcal{C}(\mathfrak{k}, K)$. If $\varphi' : C \to B$ is a $(\mathfrak{k}, K)$ map with $\varphi\varphi' = 1_C$, then

$$\mathrm{Hom}(1, \varphi)\mathrm{Hom}(1, \varphi') = 1_{\mathrm{Hom}_{\mathbb{C}}(V, C)_K}$$

in $\mathcal{C}(\mathfrak{k}, K)$, and it follows that $\mathrm{Hom}(1, \varphi)$ is onto. This proves the exactness in (c).

Each part of the proposition asserts something about projectives or injectives. In (a), let W be projective. We use Corollary 2.47. The functor

$$X \mapsto \operatorname{Hom}_{R(\mathfrak{g},K)}(W \otimes_{\mathbb{C}} V, X) \cong \operatorname{Hom}_{R(\mathfrak{g},K)}(W, \operatorname{Hom}_{\mathbb{C}}(V, X)_K)$$

is naturally isomorphic to the composition of $X \mapsto \operatorname{Hom}_{\mathbb{C}}(V, X)_K$, which we have just seen is exact, and $U \mapsto \operatorname{Hom}_{R(\mathfrak{g},K)}(W, U)$, which is exact by (C.14). Hence it is exact, and a second application of (C.14) says $W \otimes_{\mathbb{C}} V$ is projective.

In (b), let W be projective. The functor

$$X \mapsto \operatorname{Hom}_{R(\mathfrak{g},K)}(X, \operatorname{Hom}_{\mathbb{C}}(W, V)_K) \cong \operatorname{Hom}_{R(\mathfrak{g},K)}(X \otimes_{\mathbb{C}} W, V)$$
$$\cong \operatorname{Hom}_{R(\mathfrak{g},K)}(W, \operatorname{Hom}_{\mathbb{C}}(X, V)_K)$$

is naturally isomorphic to the composition of $X \mapsto \operatorname{Hom}_{\mathbb{C}}(X, V)_K$ and $U \mapsto \operatorname{Hom}_{R(\mathfrak{g},K)}(W, U)$, both of which are exact. Hence $\operatorname{Hom}_{\mathbb{C}}(W, V)_K$ is injective by (C.14).

In (c), let W be injective. The functor

$$X \mapsto \operatorname{Hom}_{R(\mathfrak{g},K)}(X, \operatorname{Hom}_{\mathbb{C}}(V, W)_K) \cong \operatorname{Hom}_{R(\mathfrak{g},K)}(X \otimes_{\mathbb{C}} V, W)$$

is naturally isomorphic to the composition of $X \mapsto X \otimes_{\mathbb{C}} V$ and $U \mapsto \operatorname{Hom}_{R(\mathfrak{g},K)}(U, W)$, both of which are exact. Hence $\operatorname{Hom}_{\mathbb{C}}(V, W)_K$ is injective by (C.14). This completes the proof.

There is one relatively minor additional result on constructions within $\mathcal{C}(\mathfrak{g}, K)$ that we shall need to use.

Proposition 2.55. If V and W are $(\mathfrak{g}, K)$ modules, there is an injection of $(\mathfrak{g}, K)$ modules

$$V^c \otimes_{\mathbb{C}} W \hookrightarrow \operatorname{Hom}_{\mathbb{C}}(V, W)_K,$$

and this is natural in V and W. If V is finite dimensional, then $\operatorname{Hom}_{\mathbb{C}}(V, W)$ is automatically locally K finite, and the natural inclusion is a natural isomorphism:

$$V^c \otimes_{\mathbb{C}} W = V^* \otimes_{\mathbb{C}} W \cong \operatorname{Hom}_{\mathbb{C}}(V, W) = \operatorname{Hom}_{\mathbb{C}}(V, W)_K.$$

PROOF. We start from the natural vector-space injection

$$V^* \otimes_{\mathbb{C}} W \hookrightarrow \operatorname{Hom}_{\mathbb{C}}(V, W),$$

which is given by $\varphi \otimes y \mapsto \Phi$ with $\Phi(x) = \varphi(x)y$. This injection respects the actions of $X \in \mathfrak{g}$ and $k \in K$. In fact, $X(\varphi \otimes y) \mapsto \Phi'$ with

$$\Phi'(x) = X\varphi(x)y + \varphi(x)Xy = -\varphi(Xx)y + \varphi(x)Xy$$
$$= -\Phi(Xx) + X(\Phi(x)) = (X\Phi)(x).$$

Thus $X(\varphi \otimes y) \mapsto X\Phi$. Similarly $k(\varphi \otimes y) = k\varphi \otimes ky$ corresponds to

$$(k\varphi)(x)ky = \varphi(k^{-1}x)ky = k(\Phi(k^{-1}x)) = (k\Phi)(x).$$

Restricting the injection to the subspace $V^c \otimes_{\mathbb{C}} W$ of $V^* \otimes_{\mathbb{C}} W$, we obtain an injection $V^c \otimes_{\mathbb{C}} W \hookrightarrow \operatorname{Hom}_{\mathbb{C}}(V, W)$.

Now $V^c \otimes_{\mathbb{C}} W$ is locally K finite, and its K equivariant image must be locally K finite. Hence the image is contained in the K finite part of $\operatorname{Hom}_{\mathbb{C}}(V, W)$.

If V is finite dimensional, then $V^c = V^*$, and vector-space theory shows that $V^* \otimes_{\mathbb{C}} W$ maps onto $\operatorname{Hom}_{\mathbb{C}}(V, W)$. Since $V^* \otimes_{\mathbb{C}} W$ is locally K finite, the image $\operatorname{Hom}_{\mathbb{C}}(V, W)$ must be locally K finite, i.e., must equal its K finite part. This completes the proof.

4. Special Properties of P and I in Examples

Let us return to the examples discussed in §1. We shall prove the assertions made there and also obtain some additional properties.

1a. Functors pro and ind

Example 1a concerned an inclusion of pairs $(\mathfrak{h}, L) \hookrightarrow (\mathfrak{g}, L)$. In this context ind and pro are defined by (2.12). If V is in $\mathcal{C}(\mathfrak{h}, L)$, it is routine to check that $\operatorname{ind}_{\mathfrak{h},L}^{\mathfrak{g},L}(V)$ and $\operatorname{pro}_{\mathfrak{h},L}^{\mathfrak{g},L}(V)$ are in $\mathcal{C}(\mathfrak{g}, L)$. If $\varphi : V \to W$ is a map in $\operatorname{Hom}_{R(\mathfrak{h},L)}(V, W)$, then we can define $\operatorname{ind}(\varphi) = 1 \otimes \varphi$ and $\operatorname{pro}(\varphi) = \operatorname{Hom}(1, \varphi)$. With these definitions, ind and pro are covariant functors. The main result will be that P and I for this situation are naturally isomorphic with ind and pro, respectively, and are exact. But first we prove a lemma. If $\mathfrak{q}$ is a subspace of $\mathfrak{g}$, we let $S(\mathfrak{q})$ be the symmetric algebra of $\mathfrak{q}$ and we let $\sigma : S(\mathfrak{q}) \to U(\mathfrak{g})$ be the symmetrization map.

Lemma 2.56. If $(\mathfrak{h}, L) \hookrightarrow (\mathfrak{g}, L)$ is an inclusion of pairs and $\mathfrak{q}$ is an $\mathrm{Ad}(L)$ invariant susbspace of $\mathfrak{g}$ such that $\mathfrak{g} = \mathfrak{h} \oplus \mathfrak{q}$, then

$$R(\mathfrak{g}, L) \cong S(\mathfrak{q}) \otimes_{\mathbb{C}} R(\mathfrak{h}, L)$$

under the map from right to left given by

$$q \otimes (u \otimes T) \mapsto \sigma(q)u \otimes T \quad \text{for } q \in S(\mathfrak{q}),\ u \in U(\mathfrak{h}),\ T \in R(L).$$

The isomorphism respects the right action of $R(\mathfrak{h}, L)$ (with $R(\mathfrak{h}, L)$ acting trivially on $S(\mathfrak{q})$). In addition, the left-regular representation of L on $R(\mathfrak{g}, L)$ corresponds to $\mathrm{Ad} \otimes l$ on $S(\mathfrak{q}) \otimes_{\mathbb{C}} R(\mathfrak{h}, L)$.

PROOF. We can write

$$R(\mathfrak{g}, L) = U(\mathfrak{g}) \otimes_{U(\mathfrak{l})} R(L) \cong (S(\mathfrak{q}) \otimes_{\mathbb{C}} U(\mathfrak{h})) \otimes_{U(\mathfrak{l})} R(L)$$
$$\cong S(\mathfrak{q}) \otimes_{\mathbb{C}} (U(\mathfrak{h}) \otimes_{U(\mathfrak{l})} R(L)) = S(\mathfrak{q}) \otimes_{\mathbb{C}} R(\mathfrak{h}, L),$$

with the first isomorphism given within $U(\mathfrak{g})$ from right to left by the composition of $\sigma \otimes 1$ and multiplication and with the second isomorphism given by (C.21). The composite map is then the one in the statement of the lemma. The chain of isomorphisms certainly respects the right action of L, and the corresponding left actions by L are

$$l \to \mathrm{Ad} \otimes l \to (\mathrm{Ad} \otimes \mathrm{Ad}) \otimes l \to \mathrm{Ad} \otimes (\mathrm{Ad} \otimes l) \to \mathrm{Ad} \otimes l$$

since symmetrization commutes with $\mathrm{Ad}(L)$. We need to show that the right $\mathfrak{h}$ module structure corresponds throughout. Let X be in $\mathfrak{h}$. A general element of $R(\mathfrak{g}, L) = U(\mathfrak{g}) \otimes_{U(\mathfrak{l})} R(L)$ is a sum of terms of the form $su \otimes T$ with $s \in \sigma(S(\mathfrak{q})), u \in U(\mathfrak{h})$, and $T \in R(L)$. This element corresponds to a sum of terms $\sigma^{-1}(s) \otimes (u \otimes T)$ in $S(\mathfrak{q}) \otimes_{\mathbb{C}} R(\mathfrak{h}, L)$. In vector notation in $R(\mathfrak{g}, L)$,

$$(su \otimes T)X = su(\mathrm{Ad}(\cdot)X) \otimes T.$$

This corresponds to a sum of terms $\sigma^{-1}(s) \otimes (u(\mathrm{Ad}(\cdot)X) \otimes T)$ in $S(\mathfrak{q}) \otimes_{\mathbb{C}} R(\mathfrak{h}, L)$, hence to a sum of terms $\sigma^{-1}(s) \otimes (u \otimes T)X$ since $\mathrm{Ad}(\cdot)X$ takes values in $U(\mathfrak{h})$.

Proposition 2.57. If $(\mathfrak{h}, L) \hookrightarrow (\mathfrak{g}, L)$ is an inclusion of pairs, then

(a) $P_{\mathfrak{h},L}^{\mathfrak{g},L}$ is naturally isomorphic to $\mathrm{ind}_{\mathfrak{h},L}^{\mathfrak{g},L}$,

(b) $I_{\mathfrak{h},L}^{\mathfrak{g},L}$ is naturally isomorphic to $\mathrm{pro}_{\mathfrak{h},L}^{\mathfrak{g},L}$,

(c) $\mathrm{ind}_{\mathfrak{h},L}^{\mathfrak{g},L}$ and $\mathrm{pro}_{\mathfrak{h},L}^{\mathfrak{g},L}$ are exact,

(d) $\mathcal{F}_{\mathfrak{g},L}^{\mathfrak{h},L}$ sends projectives to projectives and injectives to injectives,

(e) $(\mathcal{F}^{\vee})_{\mathfrak{g},L}^{\mathfrak{h},L} = \mathcal{F}_{\mathfrak{g},L}^{\mathfrak{h},L}$.

REMARK. As the proof will show, parts (a), (b), and (e) remain valid whenever i_{gp} is one-one onto; it is not necessary to assume that $(\mathfrak{h}, L) \to (\mathfrak{g}, L)$ be an inclusion. This extension is of interest for special semidirect product maps of pairs.

PROOF. Part (e) is a special case of (i) in Proposition 2.33. We prove the other parts in order.

(a) We shall prove that $\operatorname{ind}_{\mathfrak{h}, L}^{\mathfrak{g}, L}$ is left adjoint to $\mathcal{F}_{\mathfrak{g}, L}^{\mathfrak{h}, L}$, which is naturally isomorphic to $(\mathcal{F}^{\vee})_{\mathfrak{g}, L}^{\mathfrak{h}, L}$ by (e). Since $P_{\mathfrak{h}, L}^{\mathfrak{g}, L}$ is left adjoint to $(\mathcal{F}^{\vee})_{\mathfrak{g}, L}^{\mathfrak{h}, L}$ by Proposition 2.34, conclusion (a) then follows from (C.22). If V is in $\mathcal{C}(\mathfrak{h}, L)$ and X is in $\mathcal{C}(\mathfrak{g}, L)$, Theorem 1.117 gives

$$\operatorname{Hom}_{R(\mathfrak{g}, L)}(\operatorname{ind}_{\mathfrak{h}, L}^{\mathfrak{g}, L}(V), X)$$

$$= \operatorname{Hom}_{R(\mathfrak{g}, L)}(U(\mathfrak{g}) \otimes_{U(\mathfrak{h})} V, X)$$

$$= \operatorname{Hom}_{\mathfrak{g}, L}(U(\mathfrak{g}) \otimes_{U(\mathfrak{h})} V, X)$$

$$(2.58) \qquad = \operatorname{Hom}_{\mathfrak{g}}(U(\mathfrak{g}) \otimes_{U(\mathfrak{h})} V, X) \cap \operatorname{Hom}_{L}(U(\mathfrak{g}) \otimes_{U(\mathfrak{h})} V, X).$$

By (C.19)

$$(2.59) \quad \operatorname{Hom}_{U(\mathfrak{g})}(U(\mathfrak{g}) \otimes_{U(\mathfrak{h})} V, X) = \operatorname{Hom}_{U(\mathfrak{h})}(V, \operatorname{Hom}_{U(\mathfrak{g})}(U(\mathfrak{g}), X)).$$

Meanwhile, $\operatorname{Hom}_{L}(U(\mathfrak{g}) \otimes_{U(\mathfrak{h})} V, X)$ is the set of L invariants of $\operatorname{Hom}_{\mathbb{C}}(U(\mathfrak{g}) \otimes_{U(\mathfrak{h})} V, X)$, by Lemma 2.43. Let us check that the isomorphism of (C.20) given by

$$(2.60) \qquad \operatorname{Hom}_{\mathbb{C}}(U(\mathfrak{g}) \otimes_{U(\mathfrak{h})} V, X) \cong \operatorname{Hom}_{U(\mathfrak{h})}(V, \operatorname{Hom}_{\mathbb{C}}(U(\mathfrak{g}), X))$$

respects the action of L if L acts on the right side as it does on $\operatorname{Hom}_{\mathbb{C}}(V, \operatorname{Hom}_{\mathbb{C}}(U(\mathfrak{g}), X))$. If Φ on the left corresponds to φ on the right, then $\varphi(v)(u) = \Phi(u \otimes v)$ for $u \in U(\mathfrak{g})$ and $v \in V$. For $k \in L$, (2.13a) gives

$$(k\Phi)(u \otimes v) = k(\Phi(k^{-1}(u \otimes v))) = k(\Phi(\operatorname{Ad}(k)^{-1}u \otimes k^{-1}v))$$
$$= k(\varphi(k^{-1}v)(\operatorname{Ad}(k)^{-1}u)) = k(\varphi(k^{-1}v))(u) = (k\varphi)(v)(u).$$

Thus the asserted actions of L correspond on the two sides of (2.60).
If φ on the right side of (2.60) is L invariant, we thus have

$$\varphi(v)(1) = (k\varphi)(v)(1) = k(\varphi(k^{-1}v)(1)),$$

which we summarize as

$$(2.61) \qquad\qquad k(\varphi(v)(1)) = \varphi(kv)(1).$$

In (2.59) the right side is isomorphic to $\mathrm{Hom}_{U(\mathfrak{h})}(V, X)$ by $\varphi \mapsto \varphi'$ with $\varphi'(v) = \varphi(v)(1)$. What (2.61) shows is that L invariance for such an element φ translates into the condition that φ' is L linear. Hence the right side of (2.58) is

$$\cong \mathrm{Hom}_{\mathfrak{h},L}(V, X) \cong \mathrm{Hom}_{R(\mathfrak{h},L)}(V, X),$$

and we check by inspection that the various isomorphisms are natural in X. This proves the desired adjoint relation, and (a) follows.

(b) We shall prove that $\mathrm{pro}_{\mathfrak{h},L}^{\mathfrak{g},L}$ is right adjoint to $\mathcal{F}_{\mathfrak{g},L}^{\mathfrak{h},L}$, and then (b) will follow from Proposition 2.21 and (C.22). If V is in $\mathcal{C}(\mathfrak{h}, L)$ and X is in $\mathcal{C}(\mathfrak{g}, L)$, Theorem 1.117 gives

$$
\begin{aligned}
\mathrm{Hom}_{R(\mathfrak{g},L)}(X, \mathrm{pro}_{\mathfrak{h},L}^{\mathfrak{g},L}(V)) &= \mathrm{Hom}_{R(\mathfrak{g},L)}(X, \mathrm{Hom}_{U(\mathfrak{h})}(U(\mathfrak{g}), V)_L) \\
&= \mathrm{Hom}_{\mathfrak{g},L}(X, \mathrm{Hom}_{U(\mathfrak{h})}(U(\mathfrak{g}), V)_L) \\
&= \mathrm{Hom}_{\mathfrak{g}}(X, \mathrm{Hom}_{U(\mathfrak{h})}(U(\mathfrak{g}), V)) \\
&\quad \cap \mathrm{Hom}_L(X, \mathrm{Hom}_{U(\mathfrak{h})}(U(\mathfrak{g}), V)_L).
\end{aligned}
$$

(2.62)

By (C.19)

$$(2.63) \quad \mathrm{Hom}_{U(\mathfrak{g})}(X, \mathrm{Hom}_{U(\mathfrak{h})}(U(\mathfrak{g}), V)) = \mathrm{Hom}_{U(\mathfrak{h})}(U(\mathfrak{g}) \otimes_{U(\mathfrak{g})} X, V).$$

Meanwhile,

$$\mathrm{Hom}_L(X, \mathrm{Hom}_{U(\mathfrak{h})}(U(\mathfrak{g}), V)_L) \cong \mathrm{Hom}_L(X, \mathrm{Hom}_{U(\mathfrak{h})}(U(\mathfrak{g}), V))$$

is the set of L invariants of $\mathrm{Hom}_{\mathbb{C}}(X, \mathrm{Hom}_{U(\mathfrak{h})}(U(\mathfrak{g}), V))$, by Lemma 2.43. We readily check using (2.13b) that the isomorphism of (C.20) given by

$$(2.64) \qquad \mathrm{Hom}_{\mathbb{C}}(X, \mathrm{Hom}_{U(\mathfrak{h})}(U(\mathfrak{g}), V)) \cong \mathrm{Hom}_{U(\mathfrak{h})}(U(\mathfrak{g}) \otimes_{\mathbb{C}} X, V)$$

respects the action of L if L acts on the right side as it does on $\mathrm{Hom}_{\mathbb{C}}(U(\mathfrak{g}) \otimes_{\mathbb{C}} X, V)$.

If φ on the right side of (2.64) is L invariant, we thus have

$$\varphi(1 \otimes x) = (k\varphi)(1 \otimes x) = k(\varphi(1 \otimes k^{-1}x)),$$

which we summarize as

$$(2.65) \qquad\qquad k(\varphi(1 \otimes x)) = \varphi(1 \otimes kx).$$

In (2.63) the right side is isomorphic to $\mathrm{Hom}_{U(\mathfrak{h})}(X, V)$ by $\varphi \mapsto \varphi'$ with $\varphi'(x) = \varphi(1 \otimes x)$. What (2.65) shows is that L invariance for such an element φ translates into the condition that φ' is L linear. Hence the right side of (2.62) is

$$\cong \mathrm{Hom}_{\mathfrak{h}, L}(X, V) \cong \mathrm{Hom}_{R(\mathfrak{h}, L)}(X, V),$$

and we check by inspection that the various isomorphisms are natural in X.

(c) Let q be an $\mathrm{Ad}(L)$ invariant vector-space complement to $\mathfrak{h}$ in $\mathfrak{g}$. Then we have vector-space isomorphisms

$$
\begin{aligned}
P_{\mathfrak{h},L}^{\mathfrak{g},L}(V) &= R(\mathfrak{g}, L) \otimes_{R(\mathfrak{h},L)} V \\
&\cong (S(\mathfrak{q}) \otimes_{\mathbb{C}} R(\mathfrak{h}, L)) \otimes_{R(\mathfrak{h},L)} V &&\text{by Lemma 2.56} \\
&\cong S(\mathfrak{q}) \otimes_{\mathbb{C}} (R(\mathfrak{h}, L) \otimes_{R(\mathfrak{h},L)} V) &&\text{by (C.21)} \\
&\cong S(\mathfrak{q}) \otimes_{\mathbb{C}} V &&\text{by (2.18a).}
\end{aligned}
$$

By (C.15), $P_{\mathfrak{h},L}^{\mathfrak{g},L}$ is exact.

For $I_{\mathfrak{h},L}^{\mathfrak{g},L}$, we have $R(L)$ isomorphisms

$$
\begin{aligned}
I_{\mathfrak{h},L}^{\mathfrak{g},L}(V) &= \mathrm{Hom}_{R(\mathfrak{h},L)}(R(\mathfrak{g}, L), V)_L \\
&\cong \mathrm{Hom}_{R(\mathfrak{h},L)}(R(\mathfrak{h}, L) \otimes_{\mathbb{C}} S(\mathfrak{q}), V)_L &&\text{by Lemma 2.56} \\
&\cong \mathrm{Hom}_{R(\mathfrak{h},L)}(R(\mathfrak{h}, L), \mathrm{Hom}_{\mathbb{C}}(S(\mathfrak{q}), V)_L)_L &&\text{by Corollary 2.47} \\
&\cong \mathrm{Hom}_{\mathbb{C}}(S(\mathfrak{q}), V)_L &&\text{by (2.18b).}
\end{aligned}
$$

By Proposition 2.53c, $I_{\mathfrak{h},L}^{\mathfrak{g},L}$ is exact.

(d) If X is injective in $\mathcal{C}(\mathfrak{g}, L)$ and V is arbitrary in $\mathcal{C}(\mathfrak{h}, L)$, then (e) and Proposition 2.34 give natural isomorphisms

$$
\begin{aligned}
\mathrm{Hom}_{R(\mathfrak{h},L)}(V, \mathcal{F}_{\mathfrak{g},L}^{\mathfrak{h},L}(X)) &\cong \mathrm{Hom}_{R(\mathfrak{h},L)}(V, (\mathcal{F}^{\vee})_{\mathfrak{g},L}^{\mathfrak{h},L}(X)) \\
&\cong \mathrm{Hom}_{R(\mathfrak{g},L)}(P_{\mathfrak{h},L}^{\mathfrak{g},L}(V), X).
\end{aligned}
$$

As a functor in V, the right side is isomorphic to the composition of two exact functors, namely $V \mapsto P_{\mathfrak{h},L}^{\mathfrak{g},L}(V)$ and $U \mapsto \mathrm{Hom}_{R(\mathfrak{g},L)}(U, X)$, by (c) and (C.14). Hence it is exact, and (C.14) shows that $\mathcal{F}_{\mathfrak{g},L}^{\mathfrak{h},L}(X)$ is injective. An analogous argument shows that $\mathcal{F}_{\mathfrak{g},L}^{\mathfrak{h},L}(X)$ carries projectives to projectives. This completes the proof.

1b. Functors Π and Γ

Example 1b concerned an inclusion of pairs $(\mathfrak{g}, L) \hookrightarrow (\mathfrak{g}, K)$. In this context we may write Π and Γ for P and I, respectively. The main result is that Π and Γ for $(\mathfrak{k}, L) \hookrightarrow (\mathfrak{k}, K)$ are obtained by applying a forgetful functor to the effect of Π and Γ for $(\mathfrak{g}, L) \hookrightarrow (\mathfrak{g}, K)$. We begin with another lemma.

Lemma 2.66. Let $(\mathfrak{g}, L) \hookrightarrow (\mathfrak{g}, K)$ be an inclusion of pairs. Then

$$R(K) \otimes_{R(\mathfrak{k}, L)} R(\mathfrak{g}, L) \cong R(\mathfrak{g}, K)$$

under convolution on L:

$$T_1 \otimes (T_2 \otimes u) \mapsto (T_1 *_L T_2) \otimes u \quad \text{for } T_1 \in R(K), \; T_2 \in R(L), \; u \in U(\mathfrak{g}).$$

The isomorphism carries the K action $l \otimes 1$ on the left side to the left-regular representation of K on $R(\mathfrak{g}, K)$ on the right side, and it carries the right $R(\mathfrak{g}, L)$ module structure on the left side to the right $R(\mathfrak{g}, L)$ module structure of $R(\mathfrak{g}, K)$ on the right side. Here $T_1 *_L T_2$ is the push-forward of $T_1 \times T_2$ by the group action map $K \times L \to K$ and is given by

$$\langle T_1 *_L T_2, \; f \rangle = \int_{k_1 \in K} \int_{k_2 \in L} f(k_1 k_2) \, dT_1(k_1) \, dT_2(k_2).$$

PROOF. We apply Lemma 2.56 twice, once with $(\mathfrak{k}, L) \hookrightarrow (\mathfrak{g}, L)$ and $\mathfrak{g} = \mathfrak{k} \oplus \mathfrak{p}$ to obtain

$$(2.67) \qquad R(\mathfrak{g}, L) \cong R(\mathfrak{k}, L) \otimes_{\mathbb{C}} S(\mathfrak{p}) \quad \text{as left } R(\mathfrak{k}, L) \text{ modules,}$$

and once with $(\mathfrak{k}, K) \hookrightarrow (\mathfrak{g}, K)$ and the same $\mathfrak{g} = \mathfrak{k} \oplus \mathfrak{p}$ to obtain

$$(2.68) \qquad R(\mathfrak{g}, K) \cong R(\mathfrak{k}, K) \otimes_{\mathbb{C}} S(\mathfrak{p}) \quad \text{as left } R(\mathfrak{k}, K) \text{ modules.}$$

Then we have

$$
\begin{aligned}
R(K) \otimes_{R(\mathfrak{k}, L)} R(\mathfrak{g}, L) &\cong R(K) \otimes_{R(\mathfrak{k}, L)} (R(\mathfrak{k}, L) \otimes_{\mathbb{C}} S(\mathfrak{p})) && \text{by (2.67)} \\
&\cong (R(K) \otimes_{R(\mathfrak{k}, L)} R(\mathfrak{k}, L)) \otimes_{\mathbb{C}} S(\mathfrak{p}) && \text{by (C.21)} \\
&\cong R(K) \otimes_{\mathbb{C}} S(\mathfrak{p}) && \text{by (2.18a)} \\
&\cong R(\mathfrak{k}, K) \otimes_{\mathbb{C}} S(\mathfrak{p}) && \text{by (2.3)} \\
&\cong R(\mathfrak{g}, K) && \text{by (2.68).}
\end{aligned}
$$

Let us identify the composite isomorphism. Let $T_1 \in R(K), T_2 \in R(L)$, $u_K \in U(\mathfrak{k})$, and $p \in S(\mathfrak{p})$. Under the above isomorphisms we have

$$\begin{aligned}
T_1 \otimes (T_2 \otimes u_K \sigma(p)) &\mapsto T_1 \otimes ((T_2 \otimes u_K) \otimes p) \\
&\mapsto (T_1 \otimes (T_2 \otimes u_K)) \otimes p \\
&\mapsto ((T_1 *_L T_2) \otimes u_K) \otimes p \\
&\mapsto ((T_1 *_L T_2) * \partial(u_K)) \otimes p \\
&\mapsto (T_1 *_L T_2) \otimes u_K \sigma(p).
\end{aligned}$$

Hence the composite map is

$$T_1 \otimes (T_2 \otimes u) \mapsto (T_1 *_L T_2) \otimes u,$$

as asserted. Since u appears on the right end of each side, the right $\mathfrak{g}$ action corresponds. For $k_2 \in L$, the effect on the left side is

$$r(k_2)(T_1 \otimes (T_2 \otimes u)) = T_1 \otimes r(k_2)(T_2 \otimes u) = T_1 \otimes r(k_2)T_2 \otimes \mathrm{Ad}(k_2)u.$$

The effect on the right side is

$$\begin{aligned}
r(k_2)((T_1 *_L T_2) \otimes u) &= r(k_2)(T_1 *_L T_2) \otimes \mathrm{Ad}(k_2)u \\
&= ((T_1 *_L T_2) *_K \delta_{k_2^{-1}}) \otimes \mathrm{Ad}(k_2)u \\
&= (T_1 *_L (T_2 *_L \delta_{k_2^{-1}})) \otimes \mathrm{Ad}(k_2)u \\
&= (T_1 *_L r(k_2)T_2) \otimes \mathrm{Ad}(k_2)u,
\end{aligned}$$

and the effects on the two sides correspond.

Proposition 2.69. If $(\mathfrak{g}, L) \hookrightarrow (\mathfrak{g}, K)$ is an inclusion of pairs, then
  (a) for V in $\mathcal{C}(\mathfrak{g}, L)$ there are natural isomorphisms

$$\Pi(V) \cong R(K) \otimes_{R(\mathfrak{k}, L)} V$$
$$\Gamma(V) \cong \mathrm{Hom}_{R(\mathfrak{k}, L)}(R(K), V)_K$$

  as locally K finite representations of K, and
  (b) there are natural isomorphisms

$$\Pi_{\mathfrak{k}, L}^{\mathfrak{k}, K} \circ \mathcal{F}_{\mathfrak{g}, L}^{\mathfrak{k}, L} \cong \mathcal{F}_{\mathfrak{g}, K}^{\mathfrak{k}, K} \circ \Pi_{\mathfrak{g}, L}^{\mathfrak{g}, K}$$
$$\Gamma_{\mathfrak{k}, L}^{\mathfrak{k}, K} \circ \mathcal{F}_{\mathfrak{g}, L}^{\mathfrak{k}, L} \cong \mathcal{F}_{\mathfrak{g}, K}^{\mathfrak{k}, K} \circ \Gamma_{\mathfrak{g}, L}^{\mathfrak{g}, K}$$

of functors from $\mathcal{C}(\mathfrak{g}, L)$ to $\mathcal{C}(\mathfrak{k}, K)$.

PROOF. For (a), let V be in $C(\mathfrak{g}, L)$. Then we have natural isomorphisms in $C(\mathfrak{k}, K)$ given by

$$
\begin{aligned}
P_{\mathfrak{g},L}^{\mathfrak{g},K}(V) &= R(\mathfrak{g}, K) \otimes_{R(\mathfrak{g},L)} V \\
&\cong (R(K) \otimes_{R(\mathfrak{k},L)} R(\mathfrak{g}, L)) \otimes_{R(\mathfrak{g},L)} V && \text{by Lemma 2.66} \\
&\cong R(K) \otimes_{R(\mathfrak{k},L)} (R(\mathfrak{g}, L) \otimes_{R(\mathfrak{g},L)} V) && \text{by (C.21)} \\
&\cong R(K) \otimes_{R(\mathfrak{k},L)} V && \text{by (2.18a)}
\end{aligned}
$$

and by

$$
\begin{aligned}
I_{\mathfrak{g},L}^{\mathfrak{g},K}(V) &= \operatorname{Hom}_{R(\mathfrak{g},L)}(R(\mathfrak{g}, K), V)_K \\
&\cong \operatorname{Hom}_{R(\mathfrak{g},L)}(R(\mathfrak{g}, L) \otimes_{R(\mathfrak{k},L)} R(K), V)_K && \text{by Lemma 2.66} \\
&\cong \operatorname{Hom}_{R(\mathfrak{k},L)}(R(K), \operatorname{Hom}_{R(\mathfrak{g},L)}(R(\mathfrak{g}, L), V)_L)_K && \text{by (C.20)} \\
&\cong \operatorname{Hom}_{R(\mathfrak{k},L)}(R(K), V)_K && \text{by (2.18b)}.
\end{aligned}
$$

This proves (a), and (b) is just a restatement of (a).

Examples 1a and 1b say something about the most general inclusion of pairs $(\mathfrak{h}, L) \hookrightarrow (\mathfrak{g}, K)$. Such an inclusion is a composition

$$(\mathfrak{h}, L) \hookrightarrow (\mathfrak{g}, L) \qquad \text{followed by} \qquad (\mathfrak{g}, L) \hookrightarrow (\mathfrak{g}, K).$$

Combining Proposition 2.19 with the propositions of this section, we obtain

$$P_{\mathfrak{h},L}^{\mathfrak{g},K} \cong P_{\mathfrak{g},L}^{\mathfrak{g},K} \circ P_{\mathfrak{h},L}^{\mathfrak{g},L} \cong \Pi_{\mathfrak{g},L}^{\mathfrak{g},K} \circ \operatorname{ind}_{\mathfrak{h},L}^{\mathfrak{g},L}$$

and
$$I_{\mathfrak{h},L}^{\mathfrak{g},K} \cong I_{\mathfrak{g},L}^{\mathfrak{g},K} \circ I_{\mathfrak{h},L}^{\mathfrak{g},L} \cong \Gamma_{\mathfrak{g},L}^{\mathfrak{g},K} \circ \operatorname{pro}_{\mathfrak{h},L}^{\mathfrak{g},L}.$$

Consequently the study of P and I for a general inclusion of pairs reduces to a large extent to properties of P and I in the special cases in Examples 1a and 1b.

1c. Zuckerman functor with K connected

Example 1c concerned an inclusion of pairs $(\mathfrak{g}, L) \hookrightarrow (\mathfrak{g}, K)$ with K connected. If V is in $C(\mathfrak{g}, L)$, let $\Gamma'(V)$ be the subspace of V spanned by all vectors x such that $U(\mathfrak{k})x$ is finite dimensional and the action of $\mathfrak{k}$ on $U(\mathfrak{k})x$ is the differential of some action of K. We can make Γ' into a functor by defining it on maps as follows: If $\varphi : V \to W$ is a map, then $\Gamma'(\varphi) : \Gamma'(V) \to \Gamma'(W)$ will be the restriction of φ to $\Gamma'(V)$.

To make sense of this definition, we have to show that $\varphi(\Gamma'(V)) \subseteq \Gamma'(W)$. Let us prove first that $\ker(\varphi|_{\Gamma'(V)})$ is an invariant subspace for the locally K finite representation of K on $\Gamma'(V)$. If x is in $\ker(\varphi|_{\Gamma'(V)})$, let V_0 be the finite-dimensional K invariant subspace generated by x. Since K is connected, $V_0 = U(\mathfrak{k})x$. The $\mathfrak{k}$ equivariance of φ forces $U(\mathfrak{k})x$ to be in $\ker(\varphi|_{\Gamma'(V)})$. Hence $V_0 \subseteq \ker(\varphi|_{\Gamma'(V)})$, and $\ker(\varphi|_{\Gamma'(V)})$ is K invariant.

Now we can show that $\varphi(\Gamma'(V)) \subseteq \Gamma'(W)$. If $w \in \varphi(\Gamma'(V))$ is in W and k is in K, let $v \in \varphi^{-1}(w) \cap \Gamma'(V)$ and define $kw = \varphi(kv)$. This definition makes sense because two such v's differ by a member of $\ker(\varphi|_{\Gamma'(V)})$ and we can apply the result of the previous paragraph. If V_0 is a finite-dimensional K invariant subspace of $\Gamma'(V)$ containing v, it is easy to check that this definition gives a representation of K on $\varphi(V_0)$ whose differential agrees with the given representation of $\mathfrak{k}$ on W. Hence w is in $\Gamma'(W)$.

Thus Γ' is a covariant functor from $C(\mathfrak{g}, L)$ to $C(\mathfrak{g}, K)$, and one readily checks that Γ' is left exact. The main result below is that this functor of "largest K finite subspace" is naturally isomorphic with Γ. As a consequence, we can see easily that Γ is not always exact. This conclusion should be contrasted with Proposition 2.57c.

EXAMPLE. Let the pair $(\mathfrak{k}, K)$ be obtained from the circle group $\{e^{i\theta}\}$. Form the two-dimensional representation of $\mathfrak{k} \cong \mathbb{C}$ on $\mathbb{C}^2$ given by $1 \in \mathbb{C} \mapsto \begin{pmatrix} 0 & 1 \\ 0 & 0 \end{pmatrix}$, let ψ be the inclusion of the $\mathfrak{k}$ invariant subspace $\mathbb{C}\begin{pmatrix} 1 \\ 0 \end{pmatrix}$ into $\mathbb{C}^2$ and let φ be the quotient map. Then the sequence

$$0 \longrightarrow \mathbb{C} \overset{\psi}{\longrightarrow} \mathbb{C}^2 \overset{\varphi}{\longrightarrow} \mathbb{C} \longrightarrow 0$$

in $C(\mathfrak{k}, \{1\})$ is exact. Applying Γ' as above, we obtain the complex

$$0 \longrightarrow \mathbb{C} \overset{\Gamma'(\psi)}{\longrightarrow} \mathbb{C} \overset{\Gamma'(\varphi)}{\longrightarrow} \mathbb{C} \longrightarrow 0$$

in $C(\mathfrak{k}, K)$, and this is not exact. (If the middle module were $\mathbb{C}^2$, $\mathbb{C}^2$ would be fully reducible under K and hence also $\mathfrak{k}$, but it is not.) Therefore $\Gamma' : C(\mathfrak{k}, \{1\}) \to C(\mathfrak{k}, K)$ is not exact, and it will follow from the next proposition that Γ is not exact.

Proposition 2.70. If $(\mathfrak{g}, L) \hookrightarrow (\mathfrak{g}, K)$ is an inclusion of pairs and K is connected, then Γ' is naturally isomorphic with Γ.

PROOF. By Proposition 2.21 and (C.22), it is enough to prove that Γ' is right adjoint to $\mathcal{F}_{\mathfrak{g},K}^{\mathfrak{g},L}$, i.e., that

$$(2.71) \qquad \mathrm{Hom}_{\mathfrak{g},K}(X, \Gamma'(V)) \cong \mathrm{Hom}_{\mathfrak{g},L}(\mathcal{F}_{\mathfrak{g},K}^{\mathfrak{g},L}(X), V)$$

naturally for $X \in \mathcal{C}(\mathfrak{g}, K)$ and $V \in \mathcal{C}(\mathfrak{g}, L)$. In fact, the isomorphism is really an equality. If φ is a $(\mathfrak{g}, K)$ map of X into $\Gamma'(V)$, then φ is a map of X into V since $\Gamma'(V) \subseteq V$, and φ respects the $\mathfrak{g}$ action and the L action. Hence the left side of (2.71) is contained in the right side. In the reverse direction if φ is a $(\mathfrak{g}, L)$ map of $\mathcal{F}_{\mathfrak{g},K}^{\mathfrak{g},L}(X)$ into V, then φ carries X into V and respects the $\mathfrak{g}$ action. Also $\varphi(\Gamma'(\mathcal{F}_{\mathfrak{g},K}^{\mathfrak{g},L}(X))) \subseteq \Gamma'(V)$ since Γ' is a functor. But $\Gamma'(\mathcal{F}_{\mathfrak{g},K}^{\mathfrak{g},L}(X)) = X$. Thus φ carries X into $\Gamma'(V)$. Since $\mathfrak{g} \supseteq \mathfrak{k}$, φ respects the $\mathfrak{k}$ action. If X_0 is a finite-dimensional K invariant subspace of X, then $\mathfrak{k}$ acts by the (complexified) differential of the action by K in X_0 and $\varphi(X_0)$, and φ respects this $\mathfrak{k}$ action. Since K is connected, φ is K equivariant from X_0 to $\varphi(X_0)$. Hence φ is a $(\mathfrak{g}, K)$ map. Thus the right side of (2.71) is contained in the left side, and the proof is complete.

1d. Finite groups

Example 1d concerned an inclusion of pairs $(0, L) \hookrightarrow (0, K)$ with L and K finite groups. If (ρ, V) is a finite-dimensional representation of L, the classical induced representation of K is $\pi = \text{induced}_L^K(\rho)$, acting on the space

(2.72a) $\{f : K \to V \mid f(lk) = \rho(l)(f(k)) \text{ for } k \in K, l \in L\}$

by

(2.72b) $(\pi(k_0)f)(k) = f(kk_0)$.

Then induced_L^K is a covariant functor, carrying an L map $\varphi : (\rho, V) \to (\rho', V')$ to the K map Φ with $\Phi(f)(k) = \varphi(f(k))$.

Proposition 2.73. If $(0, L) \hookrightarrow (0, K)$ is an inclusion of pairs, then induced_L^K is the same as $I_{0,L}^{0,K}$.

PROOF. $R(K)$ and $R(L)$ are just the complex group algebras of K and L. If V is in $\mathcal{C}(0, L)$, the definitions make

$$I_{0,L}^{0,K}(V) = \text{Hom}_{R(L)}(R(K), V)_K = \text{Hom}_{R(L)}(R(K), V).$$

The right side is exactly the space (2.72a), and the K action is exactly what is given in (2.72b).

1e. Subpairs of finite index

Example 1e concerned an inclusion of pairs $(\mathfrak{g}, K_1) \hookrightarrow (\mathfrak{g}, K)$, where K_1 has finite index in K. We shall show that P and I are naturally isomorphic to a third functor and that all three functors are exact.

The third functor, called induced, is a variant of **classical induction**. If V is a $(\mathfrak{g}, K_1)$ module, then

$$(2.74a) \quad \mathrm{induced}_{\mathfrak{g}, K_1}^{\mathfrak{g}, K} V$$
$$= \{ f : K \to V \mid f(kk_1) = k_1^{-1}(f(k)) \text{ for } k_1 \in K_1, \, k \in K \}.$$

We let K act by

$$(2.74b) \qquad (kf)(k') = f(k^{-1}k') \qquad \text{for } k \text{ and } k' \text{ in } K,$$

and we let $\mathfrak{g}$ act by

$$(2.74c) \qquad (Xf)(k) = (\mathrm{Ad}(k)^{-1}X)(f(k)) \qquad \text{for } k \in K, \, X \in \mathfrak{g}.$$

It is easy to check that $\mathrm{induced}_{\mathfrak{g}, K_1}^{\mathfrak{g}, K} V$ is a $(\mathfrak{g}, K)$ module.

Proposition 2.75. (Frobenius reciprocity). Let $(\mathfrak{g}, K_1) \hookrightarrow (\mathfrak{g}, K)$ be an inclusion of pairs, and suppose K_1 has finite index in K. Let V be a $(\mathfrak{g}, K_1)$ module, and let W be a $(\mathfrak{g}, K)$ module.

(a) Define

$$e_V : \mathrm{induced}_{\mathfrak{g}, K_1}^{\mathfrak{g}, K} V \longrightarrow V$$

to be evaluation at 1. Then

$$(2.76a) \qquad \mathrm{Hom}_{\mathfrak{g}, K}(W, \mathrm{induced}_{\mathfrak{g}, K_1}^{\mathfrak{g}, K} V) \cong \mathrm{Hom}_{\mathfrak{g}, K_1}(\mathcal{F}_{\mathfrak{g}, K}^{\mathfrak{g}, K_1} W, V)$$

under the map $\Phi \mapsto e_V \circ \Phi$, and this isomorphism is natural in W and in V.

(b) Define

$$j_V : V \longrightarrow \mathrm{induced}_{\mathfrak{g}, K_1}^{\mathfrak{g}, K} V$$

to be the map sending $v \in V$ to the function on K given by

$$j_V(v)(k) = \begin{cases} k^{-1}v & \text{for } k \in K_1 \\ 0 & \text{for } k \notin K_1. \end{cases}$$

Then

$$(2.76b) \qquad \mathrm{Hom}_{\mathfrak{g}, K}(\mathrm{induced}_{\mathfrak{g}, K_1}^{\mathfrak{g}, K} V, W) \cong \mathrm{Hom}_{\mathfrak{g}, K_1}(V, \mathcal{F}_{\mathfrak{g}, K}^{\mathfrak{g}, K_1} W)$$

under the map $\Phi \mapsto \Phi \circ j_V$, and this isomorphism is natural in W and in V.

PROOF. For (a), let $w \in W$, $k_1 \in K_1$, and $X \in \mathfrak{g}$. Then

$$k_1(e_V \Phi w) = k_1[(\Phi w)(1)] = (\Phi w)(k_1^{-1})$$
$$= (k_1(\Phi w))(1) = (\Phi(k_1 w))(1) = e_V \Phi(k_1 w)$$

and

$$X(e_V \Phi w) = X[(\Phi w)(1)] = (X(\Phi w))(1) = (\Phi(Xw))(1) = e_V \Phi(Xw).$$

Thus $e_V \Phi$ is in the right side of (2.76a), and $e_V \circ (\cdot)$ carries the left side of (2.76a) to the right side.

Let us see that $e_V \circ (\cdot)$ is an isomorphism. To see that it is one-one, suppose $e_V \Phi w = 0$ for all $w \in W$. Then $(\Phi w)(1) = 0$ for all w. Applying this conclusion to $w = k^{-1}w'$, we obtain

$$0 = (\Phi w)(1) = \Phi(k^{-1}w')(1) = k^{-1}(\Phi w')(1) = (\Phi w')(k).$$

Since $k \in K$ is arbitrary, $\Phi w' = 0$. Since w' is arbitrary, $\Phi = 0$. Thus e_V is one-one.

To see that $e_V \circ (\cdot)$ is onto, let φ be in the right side of (2.76a). Define

$$(\Phi w)(k) = \varphi(k^{-1}w) \qquad \text{for } w \in W, \ k \in K.$$

Then

$$(\Phi w)(kk_1) = \varphi(k_1^{-1}k^{-1}w) = k_1^{-1}(\varphi(k^{-1}w)) = k_1^{-1}((\Phi w)(k))$$

shows that Φw is in induced$_{\mathfrak{g},K_1}^{\mathfrak{g},K}$ V. The map Φ is a $(\mathfrak{g}, K)$ map because

$$k(\Phi w)(k') = \Phi w(k^{-1}k') = \varphi(k'^{-1}kw) = \Phi(kw)(k')$$

and

$$X(\Phi w)(k') = (\mathrm{Ad}(k')^{-1}X)(\Phi w(k')) = (\mathrm{Ad}(k')^{-1}X)(\varphi(k'^{-1}w))$$
$$= \varphi((\mathrm{Ad}(k')^{-1}X)(k'^{-1}w)) = \varphi(k'^{-1}Xw) = \Phi(Xw)(k').$$

Finally Φ maps onto φ because

$$(e\Phi)w = (\Phi w)(1) = \varphi(1w) = \varphi(w).$$

Thus e_V yields the isomorphism (2.76a).

The naturality in the W variable amounts to the trivial statement that $e_V \circ (\Phi \circ F) = (e_V \circ \Phi) \circ F$ if $F : W' \to W$ is a $(\mathfrak{g}, K)$ map. Naturality in the V variable comes down to the statement that

$$H(e_V(f)) = e_{V'}(\text{induced}(H)(f))$$

for any $(\mathfrak{g}, K_1)$ map $H : V \to V'$ and any f in $\text{induced}_{\mathfrak{g}, K_1}^{\mathfrak{g}, K} V$; here each side is just $H(f(1))$.

The proof of (b) proceeds in somewhat the same spirit. We readily check that j_V carries V into $\text{induced}_{\mathfrak{g}, K_1}^{\mathfrak{g}, K} V$ and is a K_1 map. If Φ is in the left side of (2.76b), then it follows that $\Phi \circ j_V$ is in the right side of (2.76b).

To see that $(\cdot) \circ j_V$ is an isomorphism, let $\{k_j\}$ be a set of coset representatives of K/K_1. Then any f in the induced space decomposes as $f = \sum k_j j_V(f(k_j))$. If $\Phi \circ j_V = 0$, then $0 = k_j \Phi(j_V(f(k_j))) = \Phi(k_j j_V(f(k_j)))$, and so $\Phi(f) = 0$. Since f is arbitrary, $\Phi \mapsto \Phi \circ j_V$ is one-one. If φ is given in the right side of (2.76b) and if we define Φ by $\Phi(f) = \sum k_j \varphi(f(k_j))$, then we find that Φ is in the left side of (2.76b) and $\Phi \circ j_V = \varphi$. Hence $\Phi \mapsto \Phi \circ j_V$ is onto.

The naturality is established as for (a), and the proof is complete.

Proposition 2.77. Let $(\mathfrak{g}, K_1) \hookrightarrow (\mathfrak{g}, K)$ be an inclusion of pairs, and suppose K_1 has finite index in K. Then the functors $P_{\mathfrak{g}, K_1}^{\mathfrak{g}, K}$, $I_{\mathfrak{g}, K_1}^{\mathfrak{g}, K}$, and $\text{induced}_{\mathfrak{g}, K_1}^{\mathfrak{g}, K}$ are naturally isomorphic exact functors from $C(\mathfrak{g}, K_1)$ to $C(\mathfrak{g}, K)$.

PROOF. The natural isomorphism of $I_{\mathfrak{g}, K_1}^{\mathfrak{g}, K}$ with $\text{induced}_{\mathfrak{g}, K_1}^{\mathfrak{g}, K}$ follows from (C.32) and Frobenius reciprocity (Propositions 2.21 and 2.75a, respectively). The natural isomorphism of $P_{\mathfrak{g}, K_1}^{\mathfrak{g}, K}$ with $\text{induced}_{\mathfrak{g}, K_1}^{\mathfrak{g}, K}$ follows similarly from Propositions 2.34, 2.33, and 2.75b. Hence $P_{\mathfrak{g}, K_1}^{\mathfrak{g}, K} \cong I_{\mathfrak{g}, K_1}^{\mathfrak{g}, K}$. Since $P_{\mathfrak{g}, K_1}^{\mathfrak{g}, K}$ is left exact and $I_{\mathfrak{g}, K_1}^{\mathfrak{g}, K}$ is right exact, (C.11) shows that these functors are exact.

Corollary 2.78. Let $(\mathfrak{g}, K_2) \hookrightarrow (\mathfrak{g}, K_1) \hookrightarrow (\mathfrak{g}, K)$ be inclusions of pairs such that K_2 has finite index in K. For V in $C(\mathfrak{g}, K_2)$, there is a natural isomorphism

$$\text{induced}_{\mathfrak{g}, K_1}^{\mathfrak{g}, K} (\text{induced}_{\mathfrak{g}, K_2}^{\mathfrak{g}, K_1}(V)) \cong \text{induced}_{\mathfrak{g}, K_2}^{\mathfrak{g}, K}(V).$$

PROOF. This follows immediately from Propositions 2.77 and 2.19.

If $(\mathfrak{g}, K)$ is a pair and V is a $(\mathfrak{g}, K)$ module, we say that a subspace W is **invariant** (or $(\mathfrak{g}, K)$ **invariant**) if it is both a $U(\mathfrak{g})$ submodule and an invariant subspace for the representation of K. We say that V is **irreducible** if the only invariant subspaces are 0 and V. Our objective in the remainder of this section is to study the effect of the functor (2.74) on irreducibility. We proceed in two steps. The first step is to simplify

$$(2.79) \qquad \mathrm{Hom}_{\mathfrak{g}, K}(\mathrm{induced}_{\mathfrak{g}, K_1}^{\mathfrak{g}, K}(V), \ \mathrm{induced}_{\mathfrak{g}, K_1}^{\mathfrak{g}, K}(V)).$$

The technique is known as **Mackey theory**, and the results are given in the next proposition and corollary.

Proposition 2.80. Let $(\mathfrak{g}, K)$ be a pair, let K_1 and K_2 be two subgroups of finite index in K, and form the inclusions of pairs $(\mathfrak{g}, K_1) \hookrightarrow (\mathfrak{g}, K)$ and $(\mathfrak{g}, K_2) \hookrightarrow (\mathfrak{g}, K)$. Let V be a $(\mathfrak{g}, K_1)$ module. For k in K, let kV be the $(\mathfrak{g}, K_2 \cap kK_1k^{-1})$ module whose underlying space is V, whose action by $X \in \mathfrak{g}$ is the usual action of $\mathrm{Ad}(k^{-1})X$ on V, and whose action by $k' \in K_2 \cap kK_1k^{-1}$ is the usual action of $k^{-1}k'k$ on V. Then

$$(2.81) \qquad \mathcal{F}_{\mathfrak{g}, K}^{\mathfrak{g}, K_2}(\mathrm{induced}_{\mathfrak{g}, K_1}^{\mathfrak{g}, K}(V)) \cong \bigoplus_{\substack{\text{double cosets} \\ K_2 k K_1}} \mathrm{induced}_{\mathfrak{g}, K_2 \cap kK_1k^{-1}}^{\mathfrak{g}, K_2}(kV).$$

PROOF. When the action of K on $\mathrm{induced}_{\mathfrak{g}, K_1}^{\mathfrak{g}, K}(V)$ given in (2.74a) is restricted to K_2, the conditions on f on the distinct double cosets $K_2 k K_1$ in $K_2 \backslash K / K_1$ are independent of one another. Hence the left side of (2.81) is isomorphic as a $(\mathfrak{g}, K_2)$ module to

$$(2.82) \qquad \bigoplus_{K_2 k K_1} \{f \in \mathrm{induced}_{\mathfrak{g}, K_1}^{\mathfrak{g}, K}(V) \mid f \text{ has support in } K_2 k K_1\}.$$

Write π for the action of K_1 on V. For f in the kth term of (2.82), define

$$f_0(k_2) = f(k_2 k) \qquad \text{when } k_2 \text{ is in } K_2.$$

If k' is in the subgroup $K_2 \cap kK_1k^{-1}$ of K_2, then $k^{-1}k'k$ is in K_1 and we have

$$f_0(k_2 k') = f(k_2 k' k) = f(k_2 k \cdot k^{-1}k'k)$$
$$= \pi(k^{-1}k'^{-1}k)(f(k_2 k)) = \pi(k^{-1}k'^{-1}k)(f_0(k_2)).$$

Hence f_0 is in the space of $\mathrm{induced}_{\mathfrak{g}, K_2 \cap kK_1k^{-1}}^{\mathfrak{g}, K_2}(kV)$. On the other hand, we can reconstruct an f from any member f_0 of this induced space by the formula

$$f(k_2 k k_1) = \pi(k_1)^{-1}(f(k_2 k)) = \pi(k_1)^{-1}(f_0(k_2)).$$

Hence $f \mapsto f_0$ is a vector-space isomorphism from the kth term of (2.82) onto the space of $\mathrm{induced}_{\mathfrak{g}, K_2 \cap kK_1k^{-1}}^{\mathfrak{g}, K_2}(kV)$. It is clear that this isomorphism respects the $(\mathfrak{g}, K_2)$ action, and the proposition follows.

Corollary 2.83. Let $(\mathfrak{g}, K)$ be a pair, let K_1 and K_2 be two subgroups of finite index in K, and form the inclusions of pairs $(\mathfrak{g}, K_1) \hookrightarrow (\mathfrak{g}, K)$ and $(\mathfrak{g}, K_2) \hookrightarrow (\mathfrak{g}, K)$. If V_1 is a $(\mathfrak{g}, K_1)$ module and V_2 is a $(\mathfrak{g}, K_2)$ module, then

$$\mathrm{Hom}_{\mathfrak{g}, K}(\mathrm{induced}_{\mathfrak{g}, K_1}^{\mathfrak{g}, K}(V_1), \mathrm{induced}_{\mathfrak{g}, K_2}^{\mathfrak{g}, K}(V_2))$$
$$\cong \bigoplus_{\substack{\text{double cosets} \\ K_2 k K_1}} \mathrm{Hom}_{\mathfrak{g}, K_2 \cap k K_1 k^{-1}}(k V_1, V_2).$$

PROOF. The left side of the identity is

$$\cong \mathrm{Hom}_{\mathfrak{g}, K_2}(\mathcal{F}_{\mathfrak{g}, K}^{\mathfrak{g}, K_2}(\mathrm{induced}_{\mathfrak{g}, K_1}^{\mathfrak{g}, K}(V_1)), \ V_2) \qquad \text{by Proposition 2.75a}$$

$$\cong \bigoplus_{K_2 k K_1} \mathrm{Hom}_{\mathfrak{g}, K_2}(\mathrm{induced}_{\mathfrak{g}, K_2 \cap k K_1 k^{-1}}^{\mathfrak{g}, K_2}(k V_1)), \ V_2) \qquad \text{by Proposition 2.80}$$

$$\cong \bigoplus_{K_2 k K_1} \mathrm{Hom}_{\mathfrak{g}, K_2 \cap k K_1 k^{-1}}(k V_1, V_2) \qquad \text{by Proposition 2.75b.}$$

This proves the corollary.

This completes our discussion of Mackey theory. The second step in the analysis of reducibility is to show the extent to which (2.79) measures reducibility. Since $(\mathfrak{g}, K)$ modules are the same as unital $R(\mathfrak{g}, K)$ modules, the definitions and results of §A.2 and §A.3 are applicable. We can therefore speak of "completely reducible" modules of "finite length" in the categories $\mathcal{C}(\mathfrak{g}, K)$, $\mathcal{C}(\mathfrak{g}, K_1)$, and $\mathcal{C}(\mathfrak{g})$.

When $\mathrm{induced}_{\mathfrak{g}, K_1}^{\mathfrak{g}, K}(V)$ is completely reducible and has finite length, (2.79) does measure reducibility. (Namely the invariant subspaces are the images of the projection operators in (2.79).) What we need is conditions under which this complete reducibility is valid. For now, we give some preliminary results. In Chapter VII, under the hypothesis that $(\mathfrak{g}, K)$ is a "reductive pair," we shall obtain additional results that will complete the present discussion.

Proposition 2.84. Let $(\mathfrak{g}, K)$ be a pair, and suppose V is an irreducible $(\mathfrak{g}, K)$ module that has an irreducible $U(\mathfrak{g})$ submodule. Then V is completely reducible as a $U(\mathfrak{g})$ module and has finite length.

PROOF. If K_0 denotes the identity component of K, then K/K_0 is compact and discrete, hence finite. Let W be an irreducible $U(\mathfrak{g})$ submodule, and let $\{k_j\}_{j=1}^n$ be coset representatives for K/K_0. Each space $k_j W$ is $U(\mathfrak{g})$ invariant and irreducible, since

$$u(k_j W) = k_j (\mathrm{Ad}(k_j^{-1})u) W \subseteq k_j W \qquad \text{for } u \in U(\mathfrak{g}).$$

The irreducibility of V as a $(\mathfrak{g}, K)$ module therefore implies that

$$\sum_{j=1}^{n} k_j W = V.$$

This equation exhibits V as the finite sum of irreducible $U(\mathfrak{g})$ modules, and Corollary A.10b shows that V is completely reducible and has finite length.

Proposition 2.85. Let $(\mathfrak{g}, K_1) \hookrightarrow (\mathfrak{g}, K)$ be an inclusion of pairs, and suppose K_1 has finite index in K. Let V be a $(\mathfrak{g}, K_1)$ module. For each $k \in K$, let kV be the $(\mathfrak{g}, K_0)$ module with underlying space V, with $X \in \mathfrak{g}$ acting as $\mathrm{Ad}(k)^{-1}X$, and with $k_0 \in K_0$ acting as $k^{-1}k_0k$. If $\{k_j\}$ is a set of coset representatives for K/K_1, then

$$\mathcal{F}_{\mathfrak{g},K}^{\mathfrak{g},K_0}(\mathrm{induced}_{\mathfrak{g},K_1}^{\mathfrak{g},K} V) \cong \bigoplus_j k_j V.$$

Consequently if V is completely reducible and has finite length as a $U(\mathfrak{g})$ module, then $\mathcal{F}_{\mathfrak{g},K}^{\mathfrak{g},K_0}(\mathrm{induced}_{\mathfrak{g},K_1}^{\mathfrak{g},K} V)$ is completely reducible and has finite length as a $U(\mathfrak{g})$ module.

PROOF. This is a special case of Proposition 2.80.

Corollary 2.86. Let $(\mathfrak{g}, K_1) \hookrightarrow (\mathfrak{g}, K)$ be an inclusion of pairs, and suppose K_1 has finite index in K. Let V be a $(\mathfrak{g}, K_1)$ module. Then V is $U(\mathfrak{g})$ isomorphic with a direct summand of

$$\mathcal{F}_{\mathfrak{g},K}^{\mathfrak{g},K_0}(\mathrm{induced}_{\mathfrak{g},K_1}^{\mathfrak{g},K} V).$$

PROOF. We are allowed to take one of the coset representatives k_j in Proposition 2.85 to be 1.

Proposition 2.87. Let $(\mathfrak{g}, K_1) \hookrightarrow (\mathfrak{g}, K)$ be an inclusion of pairs, and suppose K_1 has finite index in K.

(a) If V is a $(\mathfrak{g}, K)$ module, then V is $(\mathfrak{g}, K)$ isomorphic with a $(\mathfrak{g}, K)$ submodule of $\mathrm{induced}_{\mathfrak{g},K_0}^{\mathfrak{g},K}(\mathcal{F}_{\mathfrak{g},K}^{\mathfrak{g},K_0}(V))$.

(b) If V is a $(\mathfrak{g}, K_1)$ module, then $\mathrm{induced}_{\mathfrak{g},K_1}^{\mathfrak{g},K}(V)$ is $(\mathfrak{g}, K)$ isomorphic with a $(\mathfrak{g}, K)$ submodule of $\mathrm{induced}_{\mathfrak{g},K_0}^{\mathfrak{g},K}(\mathcal{F}_{\mathfrak{g},K_1}^{\mathfrak{g},K_0}(V))$.

PROOF.

(a) By Frobenius reciprocity (Proposition 2.75a),

$$\mathrm{Hom}_{\mathfrak{g},K}(V,\ \mathrm{induced}_{\mathfrak{g},K_0}^{\mathfrak{g},K}(\mathcal{F}_{\mathfrak{g},K}^{\mathfrak{g},K_0}(V))) \cong \mathrm{Hom}_{\mathfrak{g},K_0}(\mathcal{F}_{\mathfrak{g},K}^{\mathfrak{g},K_0}(V),\ \mathcal{F}_{\mathfrak{g},K}^{\mathfrak{g},K_0}(V)).$$

The identity map on the right side corresponds to a nonzero $(\mathfrak{g}, K)$ map φ on the left side. If S denotes its kernel, then the naturality in Proposition 2.75a says that the inclusion $S \subseteq V$ leads to a commutative diagram

$$\begin{array}{ccc}
\mathrm{Hom}_{\mathfrak{g},K}(V,\ \mathrm{induced}(\mathcal{F}(V))) & \longrightarrow & \mathrm{Hom}_{\mathfrak{g},K_0}(\mathcal{F}(V),\mathcal{F}(V)) \\
\downarrow & & \downarrow \\
\mathrm{Hom}_{\mathfrak{g},K}(S,\ \mathrm{induced}(\mathcal{F}(V))) & \longrightarrow & \mathrm{Hom}_{\mathfrak{g},K_0}(\mathcal{F}(S),\mathcal{F}(V))
\end{array}$$

in which the vertical maps are restrictions. The member φ of the top left maps to $\varphi|_S = 0$ on the lower left, while the identity on the upper right maps to the inclusion of $\mathcal{F}(S)$ in $\mathcal{F}(V)$ on the lower right. Thus, on the bottom row, 0 corresponds to the inclusion of $\mathcal{F}(S)$ in $\mathcal{F}(V)$, and we must have $\mathcal{F}(S) = 0$. So $S = 0$, and φ is one-one.

(b) By (a) and the exactness of $\mathrm{induced}_{\mathfrak{g},K_1}^{\mathfrak{g},K}$ given in Proposition 2.77, $\mathrm{induced}_{\mathfrak{g},K_1}^{\mathfrak{g},K}(V)$ is $(\mathfrak{g}, K)$ isomorphic with a $(\mathfrak{g}, K)$ submodule of

$$\mathrm{induced}_{\mathfrak{g},K_1}^{\mathfrak{g},K}(\mathrm{induced}_{\mathfrak{g},K_0}^{\mathfrak{g},K_1}(\mathcal{F}_{\mathfrak{g},K_1}^{\mathfrak{g},K_0}(V))).$$

The latter $(\mathfrak{g}, K)$ module is isomorphic with $\mathrm{induced}_{\mathfrak{g},K_0}^{\mathfrak{g},K}(\mathcal{F}_{\mathfrak{g},K_1}^{\mathfrak{g},K_0}(V))$ by Corollary 2.78, and the result follows.

In Chapter VII we shall specialize these considerations to "reductive pairs" $(\mathfrak{g}, K)$. In that case, irreducible $(\mathfrak{g}, K)$ modules automatically have irreducible $U(\mathfrak{g})$ submodules and hence, by Proposition 2.84, are completely reducible and have finite length as $U(\mathfrak{g})$ modules. Conversely a $(\mathfrak{g}, K)$ module that is completely reducible and has finite length as a $U(\mathfrak{g})$ module is completely reducible as a $(\mathfrak{g}, K)$ module. Armed with these facts, we shall prove for each irreducible $(\mathfrak{g}, K_1)$ module V that $\mathrm{induced}_{\mathfrak{g},K_1}^{\mathfrak{g},K}(V)$ is completely reducible and has finite length as a $(\mathfrak{g}, K)$ module. Hence Mackey theory (Corollary 2.83) identifies the reducibility.

2. Trivial map of pairs

Example 2 concerned a map of pairs $(\mathfrak{g}, K) \to (0, \{1\})$. The main result is that P and I are isomorphic with the functors of coinvariants and invariants, which are defined in (2.15).

Proposition 2.88. For a map of pairs $(\mathfrak{g}, K) \to (0, \{1\})$, if V is in $C(\mathfrak{g}, K)$, then there are complex vector-space isomorphisms

$$P_{\mathfrak{g},K}^{0,\{1\}}(V) \cong V_{\mathfrak{g},K} \qquad (= \text{coinvariants})$$

and $\qquad\qquad I_{\mathfrak{g},K}^{0,\{1\}}(V) \cong V^{\mathfrak{g},K} \qquad (= \text{invariants}),$

and these isomorphisms are natural in V.

PROOF. The algebra $R(0, \{1\})$ is just $\mathbb{C}$. Application of Lemmas 2.43 and 2.48 gives natural isomorphisms

$$P_{\mathfrak{g},K}^{0,\{1\}}(V) = \mathbb{C} \otimes_{R(\mathfrak{g},K)} V = (\mathbb{C} \otimes_{\mathbb{C}} V)_{\mathfrak{g},K} \cong V_{\mathfrak{g},K}$$

and $\qquad I_{\mathfrak{g},K}^{0,\{1\}}(V) = \mathrm{Hom}_{R(\mathfrak{g},K)}(\mathbb{C}, V) = (\mathrm{Hom}_{\mathbb{C}}(\mathbb{C}, V))^{\mathfrak{g},K} \cong V^{\mathfrak{g},K},$

and the result follows.

3. Semidirect product map of pairs

Example 3 concerned a semidirect product map of pairs $(\mathfrak{q}, L) \to (\mathfrak{l}, L)$ with $\mathfrak{q} = \mathfrak{l} \oplus \mathfrak{u}$ and with associated inclusion $(\mathfrak{u}, \{1\}) \hookrightarrow (\mathfrak{q}, L)$. The main result for this case is essentially that P and I for the semidirect product turn out to be the same as for the map $(\mathfrak{u}, \{1\}) \to (0, \{1\})$ except that an L action remains in place.

Proposition 2.89. If $(\mathfrak{q}, L) \to (\mathfrak{l}, L)$ is a special semidirect product map of pairs with $\mathfrak{q} = \mathfrak{l} \oplus \mathfrak{u}$ and if $(\mathfrak{u}, \{1\}) \hookrightarrow (\mathfrak{q}, L)$ is the associated inclusion, then

(a) there are natural isomorphisms

$$\mathcal{F}_{\mathfrak{l},L}^{0,\{1\}} \circ P_{\mathfrak{q},L}^{\mathfrak{l},L} \cong (\,\cdot\,)_{\mathfrak{u}} \circ \mathcal{F}_{\mathfrak{q},L}^{\mathfrak{u},\{1\}}$$

$$\mathcal{F}_{\mathfrak{l},L}^{0,\{1\}} \circ I_{\mathfrak{q},L}^{\mathfrak{l},L} \cong (\,\cdot\,)^{\mathfrak{u}} \circ \mathcal{F}_{\mathfrak{q},L}^{\mathfrak{u},\{1\}}$$

of functors from $C(\mathfrak{q}, L)$ to $C(0, \{1\})$; if π denotes the action of L on V, then the actions of L on $V_{\mathfrak{u}} = \mathbb{C} \otimes_{U(\mathfrak{u})} V$ and $V^{\mathfrak{u}} = \mathrm{Hom}_{U(\mathfrak{u})}(\mathbb{C}, V)$ are $1 \otimes \pi$ and $\mathrm{Hom}(1, \pi)$, respectively, and

(b) $P_{\mathfrak{u},\{1\}}^{\mathfrak{q},L}$ and $I_{\mathfrak{u},\{1\}}^{\mathfrak{q},L}$ are exact.

REMARK. Actually the significant result is not exactly Proposition 2.89a, but the extension of it to derived functors that appears below as Proposition 2.130. Proposition 2.130 does not use Proposition 2.89a in its proof, and Proposition 2.89a should be regarded only as motivation.

PROOF.
(a) For $V \in \mathcal{C}(\mathfrak{q}, L)$, we map

$$P_{\mathfrak{q},L}^{\mathfrak{l},L}(V) = R(L) \otimes_{R(\mathfrak{q},L)} V \qquad \text{to} \qquad V_{\mathfrak{u}} = \mathbb{C} \otimes_{U(\mathfrak{u})} V$$

by

(2.90a) $S \otimes v \mapsto 1 \otimes Sv$ for $S \in R(L)$, $v \in V$.

The inverse is

(2.90b) $1 \otimes v \mapsto \chi_A \otimes v$, where $\chi_A \in R(L)$ and $\chi_A v = v$.

We shall check that (2.90a) and (2.90b) are well defined. Then it is clear that they are inverses of each other and that the left action by L on $R(L)$ transforms to the action by $1 \otimes \pi$ on $V_{\mathfrak{u}}$. So (a) will follow.

Initially (2.90a) is defined from $R(L) \otimes_{\mathbb{C}} V$ to $\mathbb{C} \otimes_{U(\mathfrak{u})} V$, and we are to check that $X \in \mathfrak{u}$ and $l \in L$ imply that the images of

$$SX \otimes v - S \otimes Xv \qquad \text{and} \qquad Sl \otimes v - S \otimes lv,$$

namely $1 \otimes (-SXv)$ and $1 \otimes (S\delta_l)v - 1 \otimes S(lv)$,

are 0 in $\mathbb{C} \otimes_{U(\mathfrak{u})} V$. These images are indeed 0: By (1.21a) we have

$$1 \otimes (-SXv) = -\int_L (1 \otimes kXv) \, dS(k) = -\int_L (1 \otimes kXk^{-1}kv) \, dS(k)$$
$$= -\int_L (1 \cdot (kXk^{-1}) \otimes kv) \, dS(k) = 0.$$

Also if $\chi_A v = v$, then

$$1 \otimes (S\delta_l)v = 1 \otimes (S\delta_l)(\chi_A v) = 1 \otimes ((S\delta_l)\chi_A)v$$
$$= 1 \otimes (S(\delta_l \chi_A))v = 1 \otimes S((\delta_l \chi_A)v)$$
$$= 1 \otimes S((l(l)\chi_A)v) = 1 \otimes S(l(\chi_A v)) = 1 \otimes S(lv).$$

In the reverse direction, (2.90b) is defined initially from $\mathbb{C} \otimes_{\mathbb{C}} V$ to $R(L) \otimes_{R(\mathfrak{q},L)} V$, and we are to check that $X \in \mathfrak{u}$ implies that the image of

$$1 \cdot X \otimes v - 1 \otimes Xv = -1 \otimes Xv,$$

namely $-\chi_A \otimes Xv$ for $\chi_A \in R(L)$ with $\chi_A Xv = Xv$, is 0 in $R(L) \otimes_{R(\mathfrak{q},L)} V$. Indeed, it equals $-\chi_A X \otimes v$ in $R(L) \otimes_{R(\mathfrak{q},L)} V$, and this is 0. This completes the proof of (a) for the P functor.

For V in $\mathcal{C}(\mathfrak{q}, L)$, we map

$$I_{\mathfrak{q}, L}^{\mathfrak{l}, L}(V) = \mathrm{Hom}_{R(\mathfrak{q}, L)}(R(L), V)_L \qquad \text{to} \qquad V^{\mathfrak{u}} = \mathrm{Hom}_{U(\mathfrak{u})}(\mathbb{C}, V)$$

by $\Phi \mapsto \varphi$, where $\varphi(1) = \Phi(\chi_A)$ with χ_A chosen so that $\chi_A \Phi = \Phi$. The inverse is $\varphi \mapsto \Phi$ with $\Phi(S) = S(\varphi(1))$. Computations like those for (a) show that $\Phi \mapsto \varphi$ and $\varphi \mapsto \Phi$ are well defined. To see that they are inverses, let Φ be given with $\chi_A \Phi = \Phi$, and let the combined maps be $\Phi \mapsto \varphi \mapsto \Phi'$. The definition of the action of $R(L)$ on Φ, together with the $R(L)$ linearity of Φ, makes

$$\Phi(S) = \chi_A \Phi(S) = \Phi(S\chi_A) = S(\Phi(\chi_A)) = S(\varphi(1)) = \Phi'(S).$$

So $\Phi' = \Phi$. If φ is given and the combined maps are $\varphi \mapsto \Phi \mapsto \varphi'$, choose χ_A with $\chi_A(\varphi(1)) = \varphi(1)$. Then

$$\chi_A \Phi(S) = \Phi(S\chi_A) = S\chi_A(\varphi(1)) = S(\varphi(1)) = \Phi(S),$$

and hence

$$\varphi'(1) = \Phi(\chi_A) = \chi_A(\varphi(1)) = \varphi(1).$$

So $\varphi' = \varphi$, and $\varphi \mapsto \Phi$ is inverse to $\Phi \mapsto \varphi$.

To identify the action by L, let Φ be given and let $l \in L$. If Φ maps to φ, then $l\Phi$ maps to φ' with

$$\begin{aligned}
\varphi'(1) = (l\Phi)(\chi_A) &= \Phi(\chi_A \delta_l) \\
&= \Phi(\delta_l \chi_A) && \text{by (1.52)} \\
&= l(\Phi(\chi_A)) \\
&= l(\varphi(1)),
\end{aligned}$$

as asserted. This completes the proof of (a) for the I functor.

(b) By Lemma 2.56, the inclusion of pairs $(\mathfrak{l}, L) \hookrightarrow (\mathfrak{q}, L)$ gives us vector-space isomorphisms

$$\text{(2.91a)} \qquad R(\mathfrak{q}, L) \cong R(L) \otimes_{\mathbb{C}} S(\mathfrak{u}) \cong R(L) \otimes_{\mathbb{C}} U(\mathfrak{u}),$$

the maps being

$$\text{(2.91b)} \qquad S \otimes u_{\mathfrak{l}} u_{\mathfrak{u}} \mapsto S\partial(u_{\mathfrak{l}}) \otimes \sigma^{-1}(u_{\mathfrak{u}}) \mapsto S\partial(u_{\mathfrak{l}}) \otimes u_{\mathfrak{u}}.$$

The composite isomorphism respects the left action by L, and (2.91b) shows that it respects the right $U(\mathfrak{u})$ module structure. If V is in $\mathcal{C}(\mathfrak{u}, \{1\})$, we therefore have vector-space isomorphisms

$$\begin{aligned}
P_{\mathfrak{u}, \{1\}}^{\mathfrak{q}, L}(V) = R(\mathfrak{q}, L) \otimes_{U(\mathfrak{u})} V & \\
\cong (R(L) \otimes_{\mathbb{C}} U(\mathfrak{u})) \otimes_{U(\mathfrak{u})} V && \text{by (2.91)} \\
\cong R(L) \otimes_{\mathbb{C}} (U(\mathfrak{u}) \otimes_{U(\mathfrak{u})} V) && \text{by (C.21)} \\
\cong R(L) \otimes_{\mathbb{C}} V.
\end{aligned}$$

Then $P_{\mathfrak{u},\{1\}}^{\mathfrak{q},L}$ is exact by (C.15). Similarly we have isomorphisms of left $R(L)$ modules

$$
\begin{aligned}
I_{\mathfrak{u},\{1\}}^{\mathfrak{q},L}(V) &= \mathrm{Hom}_{U(\mathfrak{u})}(R(\mathfrak{q}, L), \, V)_L \\
&\cong \mathrm{Hom}_{U(\mathfrak{u})}(U(\mathfrak{u}) \otimes_{\mathbb{C}} R(L), \, V)_L \qquad \text{by (2.91)} \\
&\cong \mathrm{Hom}_{U(\mathfrak{u})}(U(\mathfrak{u}), \, \mathrm{Hom}_{\mathbb{C}}(R(L), V))_L \quad \text{by (C.17)} \\
&\cong \mathrm{Hom}_{\mathbb{C}}(R(L), V)_L.
\end{aligned}
$$

By Proposition 2.53c, $I_{\mathfrak{u},\{1\}}^{\mathfrak{q},L}$ is exact.

4. Covering map of pairs

Example 4 concerned a map of pairs $i : (\mathfrak{g}, \tilde{K}) \to (\mathfrak{g}, K)$, where i_{alg} is the identity and i_{gp} is onto with finite kernel. This is the kind of situation that arises with covering groups. It is apparent that the forgetful functor $\mathcal{F}_{\mathfrak{h},T}^{\mathfrak{h},\tilde{T}}$ just lifts $(\mathfrak{h}, T)$ modules to $(\mathfrak{h}, \tilde{T})$ modules. Proposition 2.33 shows that the pseudoforgetful functor $(\mathcal{F}^{\vee})_{\mathfrak{h},T}^{\mathfrak{h},\tilde{T}}$ is the same as $\mathcal{F}_{\mathfrak{h},T}^{\mathfrak{h},\tilde{T}}$, and the proposition below will show that P and I are naturally isomorphic to a third functor and that all three are exact.

The third functor is the **averaging functor** $A_{\mathfrak{g},\tilde{K}}^{\mathfrak{g},K}$:

$$
A_{\mathfrak{g},\tilde{K}}^{\mathfrak{g},K}(V) = \{v \in V \mid \ker(i_{\mathrm{gp}})v = v\} = \left\{ \int_{\ker(i_{\mathrm{gp}})} \tilde{k}v \, d\tilde{k} \,\Big|\, v \in V \right\}.
$$

Here it is to be understood that the Haar measure $d\tilde{k}$ on the finite group $\ker(i_{\mathrm{gp}})$ has total mass 1.

Proposition 2.92. Let $(\mathfrak{g}, \tilde{K}) \to (\mathfrak{g}, K)$ be a map of pairs in which i_{alg} is the identity and i_{gp} is onto with finite kernel. Then the corresponding P and I functors are naturally isomorphic with the averaging functor $A_{\mathfrak{g},\tilde{K}}^{\mathfrak{g},K}$ and are exact.

REMARK. Consequently $\mathcal{F}_{\mathfrak{g},K}^{\mathfrak{g},\tilde{K}}$ followed by either $P_{\mathfrak{g},\tilde{K}}^{\mathfrak{g},K}$ or $I_{\mathfrak{g},\tilde{K}}^{\mathfrak{g},K}$ is naturally isomorphic with the identity functor.

PROOF. Let V be a $(\mathfrak{g}, \tilde{K})$ module, and let W be a $(\mathfrak{g}, K)$ module. Then

$$
\mathrm{Hom}_{\mathfrak{g},K}(A_{\mathfrak{g},\tilde{K}}^{\mathfrak{g},K}(V), W) \cong \mathrm{Hom}_{\mathfrak{g},\tilde{K}}(V, \mathcal{F}_{\mathfrak{g},K}^{\mathfrak{g},\tilde{K}}(W)),
$$

the isomorphism from left to right being given by lifting. Moreover the isomorphism is natural in V and W. Now Frobenius reciprocity

(Propositions 2.34 and 2.33) shows that $P^{\mathfrak{g},K}_{\mathfrak{g},\tilde{K}}$ satisfies a similar identity. By (A.22) it follows that $P^{\mathfrak{g},K}_{\mathfrak{g},\tilde{K}} \cong A^{\mathfrak{g},K}_{\mathfrak{g},\tilde{K}}$. A similar argument using Proposition 2.21 shows that $I^{\mathfrak{g},K}_{\mathfrak{g},\tilde{K}} \cong A^{\mathfrak{g},K}_{\mathfrak{g},\tilde{K}}$. These isomorphisms shows that $A^{\mathfrak{g},K}_{\mathfrak{g},\tilde{K}}$ is left exact and right exact, hence exact. Therefore $P^{\mathfrak{g},K}_{\mathfrak{g},\tilde{K}}$ and $I^{\mathfrak{g},K}_{\mathfrak{g},\tilde{K}}$ are exact.

5. Mackey Isomorphisms

In this section we shall prove suitable formulations of the facts that the P functor commutes with tensor product and the I functor commutes with Hom. When P and I are written out with $R(\mathfrak{g}, K)$ in place, these results appear at first glance to be associativity formulas similar to those in (C.19), (C.20), (C.21), Propositions 2.45 and 2.51, and Corollaries 2.47 and 2.52. In those earlier results, the mapping implementing the isomorphism is fairly simple; the three variables are mapped to the three variables with no twisting or mixing. But the isomorphisms in this section are more subtle because the three variables get twisted.

For purposes of motivation, we shall formulate the isomorphisms first for the case of finite groups, which is the special case discussed in Example 1d in §§1 and 4. These isomorphisms were introduced for finite groups by G. Mackey, who later generalized them to locally compact groups. Accordingly we call the generalizations to the P and I functors **Mackey isomorphisms**.

Let L and K be two finite groups with $L \subseteq K$. We have seen that this situation corresponds to an inclusion of pairs $(0, L) \hookrightarrow (0, K)$. We suppose that we are given representations of L on V and of K on X, but we shall suppress the names of the representations. The representation of K on $\mathrm{induced}^K_L(V)$ is just $I(V)$, according to Proposition 2.73, and is described in (2.72). With the names of the representations suppressed, we can identify

$$I(V) = \{f : K \to V \mid f(lk) = l(f(k)) \text{ for } k \in L, \, l \in L\}$$

$$(k_0 f)(k) = f(kk_0).$$

We shall describe an isomorphism

$$\Phi : \mathrm{Hom}_{\mathbb{C}}(X, I(V)) \xrightarrow{\sim} I(\mathrm{Hom}_{\mathbb{C}}(X, V)).$$

If φ is in $\mathrm{Hom}_{\mathbb{C}}(X, I(V))$, then $\Phi(\varphi)$ is the member of $I(\mathrm{Hom}_{\mathbb{C}}(X, V))$ given by

(2.93a) $$(\Phi(\varphi))(k)(x) = \varphi(k^{-1}x)(k).$$

The factor of k^{-1} on the right side is what makes the map twisted; it is also what makes it K equivariant since

$$\Phi(k_0\varphi)(k)(x) = k_0\varphi(k^{-1}x)(k) = k_0(\varphi(k_0^{-1}k^{-1}x))(k)$$
$$= \varphi(k_0^{-1}k^{-1}x)(kk_0) = \Phi(\varphi)(kk_0)(x) = k_0(\Phi(\varphi))(k)(x).$$

To see that Φ is invertible, we simply exhibit Φ^{-1}. It is given by

(2.93b) $$\Phi^{-1}(f)(x)(k) = f(k)(kx).$$

For the case of finite groups, the P functor is easily described, too. The Hecke algebra $R(K)$ is just the group ring $\mathbb{C}[K]$, and we can write $R(K) \otimes_{\mathbb{C}} V$ as generated by tensors $k \otimes v$ if we write k in place of the basic element (k) in $\mathbb{C}[K]$. The isomorphism is

$$\Phi : P(X \otimes_{\mathbb{C}} V) \xrightarrow{\sim} X \otimes_{\mathbb{C}} P(V)$$

and is given by

(2.94a) $$\Phi(k \otimes (x \otimes v)) = kx \otimes (k \otimes v).$$

The factor of k on the x on the right side is what makes the map twisted. For K equivariance, we have

$$\Phi(k_0(k \otimes (x \otimes v))) = \Phi(k_0k \otimes x \otimes v) = k_0kx \otimes (k_0k \otimes v)$$
$$= k_0(kx \otimes (k \otimes v)) = k_0(\Phi(k \otimes (x \otimes v))).$$

The inverse is given by

(2.94b) $$\Phi^{-1}(x \otimes (k \otimes v)) = k \otimes (k^{-1}x \otimes v).$$

Let us consider the general case of a map of pairs $i : (\mathfrak{h}, L) \to (\mathfrak{g}, K)$. Then $R(\mathfrak{g}, K)$ is no longer a group algebra, and we need to reinterpret (2.93) and (2.94). The easiest case to understand is when $L = K = \{1\}$. Then $R(\mathfrak{g}, K) = U(\mathfrak{g})$. In (2.93a), $\Phi(\varphi)$ becomes a function on $U(\mathfrak{g})$ instead of K, and it is to be expected that the Hom action of $U(\mathfrak{g})$ will replace the simple expression on the right side of (2.93a). Similar remarks apply to (2.94a).

Theorem 2.95 (Mackey isomorphism). Let $i : (\mathfrak{h}, L) \to (\mathfrak{g}, K)$ be a map of pairs, let X be in $\mathcal{C}(\mathfrak{g}, K)$, and let V be in $\mathcal{C}(\mathfrak{h}, L)$. Then there is a unique $\mathcal{C}(\mathfrak{g}, K)$ isomorphism

(2.96a) $$\Phi : \mathrm{Hom}_{\mathbb{C}}(X, I(V))_K \xrightarrow{\sim} I(\mathrm{Hom}_{\mathbb{C}}(\mathcal{F}(X), V)_L)$$

such that

(2.96b) $$\Phi(\varphi)(T)(x) = (T\varphi)(x)(\chi_A)$$

whenever T is in $R(K)$, x is in X, and the finite set $A \subseteq \widehat{K}$ is chosen so that χ_A fixes $T\varphi \otimes x$. Moreover, this isomorphism is natural in X and V. In terms of the application of distributions to vector-valued functions as in (B.17), the inverse satisfies

(2.96c) $$\Phi^{-1}(\varphi)(x)(T) = \int_K \varphi(k\chi_A)(kx) \, dT(k).$$

PROOF. For uniqueness, if u is in $U(\mathfrak{g})$, then we have

$$\Phi(\varphi)(T \otimes u)(x) = \Phi(\varphi)((T \otimes 1)u)(x)$$
$$= (u(\Phi(\varphi)))(T \otimes 1)(x) = \Phi(u\varphi)(T)(x),$$

and the right side is determined by (2.96b). For existence, let U be in $\mathcal{C}(\mathfrak{g}, K)$. Then we have natural isomorphisms (in X, V, and U)

$$\mathrm{Hom}_{R(\mathfrak{g},K)}(U, \mathrm{Hom}_{\mathbb{C}}(X, I(V))_K)$$

$$\begin{aligned}
&\cong \mathrm{Hom}_{R(\mathfrak{g},K)}(U \otimes_{\mathbb{C}} X, I(V)) && \text{by Corollary 2.47}\\
&\cong \mathrm{Hom}_{R(\mathfrak{h},L)}(\mathcal{F}(U \otimes_{\mathbb{C}} X), V) && \text{by Proposition 2.21}\\
&\cong \mathrm{Hom}_{R(\mathfrak{h},L)}(\mathcal{F}(U) \otimes_{\mathbb{C}} \mathcal{F}(X), V)\\
&\cong \mathrm{Hom}_{R(\mathfrak{h},L)}(\mathcal{F}(U), \mathrm{Hom}_{\mathbb{C}}(\mathcal{F}(X), V)_L) && \text{by Corollary 2.47}\\
&\cong \mathrm{Hom}_{R(\mathfrak{g},K)}(U, I(\mathrm{Hom}_{\mathbb{C}}(\mathcal{F}(X), V)_L)) && \text{by Proposition 2.21.}
\end{aligned}$$

By (C.22), we obtain a natural isomorphism of the form (2.96a).

Let us track down the formula for Φ and see that Φ satisfies (2.96b). By (C.22), Φ is the member of the last Hom above, when $U = \mathrm{Hom}_{\mathbb{C}}(X, I(V))_K$, that corresponds to the identity in the first Hom. Before specializing U, let $\Phi_1, \dots, \Phi_6$ be corresponding maps in the six Hom's above. If $u \in U$, $r \in R(\mathfrak{g}, K)$, and $x \in X$, then

$$\Phi_6(u)(r)(x) = \Phi_5(ru)(x)$$
$$\Phi_5(u)(x) = \Phi_4(u \otimes x) = \Phi_3(u \otimes x)$$
$$\Phi_3(u \otimes x) = \Phi_2(u \otimes x)(\chi_A) \text{ if } \chi_A(u \otimes x) = u \otimes x$$
$$\Phi_2(u \otimes x)(r) = \Phi_1(u)(x)(r).$$

Thus

$$\Phi_6(u)(r)(x) = \Phi_5(ru)(x) = \Phi_3(ru \otimes x)$$
$$= \Phi_2(ru \otimes x)(\chi_A) \quad \text{if } \chi_A(ru \otimes x) = ru \otimes x$$
$$= \Phi_1(ru)(x)(\chi_A).$$

Now let us specialize U. Fix $u = \varphi$ in $U = \mathrm{Hom}_{\mathbb{C}}(X, I(V))_K$. For T in $R(K)$, our computations give

$$\Phi(\varphi)(T)(x) = \Phi_6(\varphi)(T)(x)$$
$$= \Phi_1(T\varphi)(x)(\chi_A) \qquad \text{if } \chi_A(T\varphi \otimes x) = T\varphi \otimes x$$
$$= (T\varphi)(x)\chi_A.$$

Thus the constructed isomorphism Φ satisfies (2.96b).

For the inverse, we enlarge one of the above formulas by using $T \in R(K)$ in place of χ_A:

$$\Phi_3(T(u \otimes x)) = \Phi_2(u \otimes x)(T).$$

Applying (1.21a) to the left-regular representation of K on $R(K)$, we can write $T = \int_K k\chi_A \, dT(k)$ if $T\chi_A = T$, and then we obtain

$$\Phi_2(u \otimes x)(T) = \int_K \Phi_3(k\chi_A(u \otimes x)) \, dT(k) = \int_K \Phi_3(ku \otimes kx) \, dT(k)$$

if $\chi_A(u \otimes x) = u \otimes x$. Consequently

$$\Phi_1(u)(x)(T) = \int_K \Phi_6(u)(k\chi_A)(kx) \, dT(k)$$

if $\chi_A u = u$. Specializing U to be $I(\mathrm{Hom}_{\mathbb{C}}(X, I(V))_L)$ and taking $u = \varphi$ and $\Phi_6 = 1$, we obtain (2.96c).

Corollary 2.97. Let $i : (\mathfrak{h}, L) \to (\mathfrak{g}, K)$ be a map of pairs, let V be in $\mathcal{C}(\mathfrak{h}, L)$, and let F be a finite-dimensional member of $\mathcal{C}(\mathfrak{g}, K)$. Then there is a $\mathcal{C}(\mathfrak{g}, K)$ isomorphism

$$\Phi : F \otimes I(V) \xrightarrow{\sim} I(\mathcal{F}(F) \otimes V)$$

natural in F and V.

PROOF. We put $X = F^*$ in Theorem 2.95 and apply Proposition 2.55. Then we have

$$F \otimes I(V) \cong F^{**} \otimes I(V) \cong \mathrm{Hom}_{\mathbb{C}}(F^*, I(V)) \cong I(\mathrm{Hom}_{\mathbb{C}}(\mathcal{F}(F^*), V))$$
$$\cong I(\mathcal{F}(F^*)^* \otimes_{\mathbb{C}} V) \cong I(\mathcal{F}(F) \otimes_{\mathbb{C}} V),$$

and the result follows.

The corresponding Mackey isomorphism for the P functor is more difficult to prove because P is adjoint to $\mathcal{F}^\vee$ rather than to a forgetful functor. An argument similar to that in Theorem 2.95 is blocked by the question whether $\mathrm{Hom}_\mathbb{C}(\mathcal{F}(X), \mathcal{F}^\vee(U))_L \cong \mathcal{F}^\vee(\mathrm{Hom}_\mathbb{C}(X, U)_K)$. With the aid of Lemma 2.28, it is not hard to settle this question affirmatively on the level of $(\mathfrak{l}, L)$ modules, but the $\mathfrak{h}$ action is not transparent.

Instead we shall take a different approach. A map of pairs $i : (\mathfrak{h}, L) \to (\mathfrak{g}, K)$ factors as a composition

(2.98)
$$i : (\mathfrak{h}, L) \to (i(\mathfrak{h}), i(L)), \quad (i(\mathfrak{h}), i(L)) \hookrightarrow (\mathfrak{g}, i(L)), \quad (\mathfrak{g}, i(L)) \hookrightarrow (\mathfrak{g}, K),$$

and Mackey isomorphisms are consistent with composition. For the first two maps (which we can compose into one), $\mathcal{F}^\vee = \mathcal{F}$ by (i) in Proposition 2.33, and there is no problem. In the third case the formula for the Mackey isomorphism is global, and we establish its properties directly. It is this third case that we isolate in the following lemma.

Lemma 2.99. Let $(\mathfrak{g}, M) \hookrightarrow (\mathfrak{g}, K)$ be an inclusion of pairs, let X be in $C(\mathfrak{g}, K)$, and let V be in $C(\mathfrak{g}, M)$. Then the formula

(2.100a)
$$\Phi(T \otimes (x \otimes v)) = \int_K (kx \otimes (k\chi_A \otimes v)) \, dT(k)$$

(for $T \in R(K), x \in X, v \in V$, and a sufficiently large finite subset $A \subseteq \widehat{K}$) defines a $C(\mathfrak{g}, K)$ isomorphism

(2.100b)
$$\Phi : \Pi(\mathcal{F}(X) \otimes_\mathbb{C} V) \xrightarrow{\sim} X \otimes_\mathbb{C} \Pi(V)$$

natural in X and in V. Moreover, Φ^{-1} is given by

(2.101)
$$\Phi^{-1}(x \otimes (S \otimes v)) = \int_K (k\chi_B \otimes (k^{-1}x \otimes v)) \, dS(k)$$

if $B \subseteq \widehat{K}$ is sufficiently large.

PROOF. To see that (2.100a) stabilizes for large A, let us expand (2.100a) in scalar notation. Suppose that X_0 is a K invariant subspace of X containing x and that π is the representation of K on X_0. Let $\{x_i\}$ be a basis of X_0, and let $\{x_i^*\}$ be the dual basis of the dual space. Then the right side is

$$= \sum_i \int_K (\langle \pi(k)x, x_i^* \rangle x_i \otimes k\chi_A \otimes v) \, dT(k)$$

$$= \sum_i \int_K (x_i \otimes k\chi_A \otimes v) \, dT_i(k) \quad \text{with } T_i = \langle \pi(\cdot)x, x_i^* \rangle T$$

$$= \sum_i x_i \otimes T_i \chi_A \otimes v,$$

and this stabilizes as

$$(2.102a) \qquad \Phi(T \otimes (x \otimes v)) = \sum_i x_i \otimes (T_i \otimes v), \qquad T_i = \langle \pi(\cdot)x, x_i^* \rangle T$$

if A is large enough so that $T_i \chi_A = T_i$ for all i. Similarly (2.101) stabilizes as

$$(2.102b) \quad \Phi^{-1}(x \otimes (S \otimes v)) = \sum_j S_j \otimes (x_j \otimes v), \qquad S_j = \langle \pi(\cdot)^{-1}x, x_j^* \rangle S$$

when B is large enough so that $S_j \chi_B = S_j$ for all j.

Thus Φ is a well-defined map

$$\Phi : R(K) \otimes_\mathbb{C} (\mathcal{F}(X) \otimes_\mathbb{C} V) \to X \otimes_\mathbb{C} \Pi(V).$$

By Proposition 2.69a and Theorem 1.119, we obtain a well-defined linear map (2.100b) if we prove that Φ carries

$$T\partial(Y) \otimes (x \otimes v) - T \otimes Y(x \otimes v) \quad \text{and} \quad Tm \otimes (x \otimes v) - T \otimes m(x \otimes v)$$

into 0 if $Y \in \mathfrak{k}$ and $m \in M$. In the case of $Y \in \mathfrak{k}$, it is enough to treat $Y \in \mathfrak{k}_0$. Let $\tilde{Y}$ be the corresponding left-invariant vector field: $\tilde{Y}f(x) = \frac{d}{dt}f(x \exp tY)_{t=0}$. Then we have

$$\langle T\partial(Y), f \rangle = \int_K \int_K f(kl) \, dT(k) \, d\partial(Y)(l) = \langle T, \tilde{Y}f \rangle$$

and

$$\Phi(T\partial(Y) \otimes (x \otimes v))$$

$$= \int_K (kx \otimes k\chi_A \otimes v) \, dT\partial(Y)(k)$$

$$= \int_K \tfrac{d}{dt}(k(\exp tY)x \otimes k(\exp tY)\chi_A \otimes v)_{t=0} \, dT(k)$$

$$= \int_K (kYx \otimes k\chi_A \otimes v) \, dT(k) + \int_K (kx \otimes kY\chi_A \otimes v) \, dT(k)$$

$$= \Phi(T \otimes Yx \otimes v) + \int_K (kx \otimes k\chi_A Y \otimes v) \, dT(k) \qquad \text{by (1.52)}$$

$$= \Phi(T \otimes Yx \otimes v) + \int_K (kx \otimes k\chi_A \otimes Yv) \, dT(k)$$

$$= \Phi(T \otimes Yx \otimes v) + \Phi(T \otimes x \otimes Yv)$$

$$= \Phi(T \otimes Y(x \otimes v)).$$

Also

$$\Phi(Tm \otimes (x \otimes v)) = \int_K (kmx \otimes km\chi_A \otimes v)\, dT(k)$$

$$= \int_K (kmx \otimes k\chi_A m \otimes v)\, dT(k)$$

$$= \int_K (kmx \otimes k\chi_A \otimes mv)\, dT(k)$$

$$= \Phi(T \otimes mx \otimes mv)$$

$$= \Phi(T \otimes m(x \otimes v)).$$

Thus Φ is a map between the spaces in (2.100b).

Similarly (2.101) descends to a map in the reverse direction. To check that Φ and Φ^{-1} are really inverses, we use (2.102) to write

$$\Phi^{-1}\Phi(T \otimes x \otimes v) = \sum_i \Phi^{-1}(x_i \otimes \langle \pi(\cdot)x, x_i^* \rangle T \otimes v)$$

$$= \sum_{i,j} \langle \pi(\cdot)^{-1}x_i, x_j^* \rangle \langle \pi(\cdot)x, x_i^* \rangle T \otimes x_j \otimes v$$

$$= \sum_j \langle \pi(\cdot)^{-1}\pi(\cdot)x, x_j^* \rangle T \otimes x_j \otimes v$$

$$= T \otimes x \otimes v$$

and

$$\Phi\Phi^{-1}(x \otimes S \otimes v) = \sum_j \Phi(\langle \pi(\cdot)^{-1}x, x_j^* \rangle S \otimes x_j \otimes v)$$

$$= \sum_{i,j} x_i \otimes \langle \pi(\cdot)x_j, x_i^* \rangle \langle \pi(\cdot)^{-1}x, x_j^* \rangle S \otimes v$$

$$= \sum_i x_i \otimes \langle \pi(\cdot)\pi(\cdot)^{-1}x, x_i^* \rangle S \otimes v$$

$$= x \otimes S \otimes v.$$

The main step is to check that Φ is a $(\mathfrak{g}, K)$ map. If $k_0 \in K$ is given, then

$$\Phi(k_0(T \otimes x \otimes v)) = \Phi(k_0 T \otimes x \otimes v)$$

$$= \int_K (kx \otimes k\chi_A \otimes v)\, d(k_0 T)(k)$$

$$= \int_K (k_0 kx \otimes k_0 k\chi_A \otimes v)\, dT(k)$$

$$= k_0(\Phi(T \otimes x \otimes v)).$$

If $Y \in \mathfrak{g}$ is given, then

$$
\begin{aligned}
\Phi(Y(T \otimes x \otimes v)) &= \Phi((Y \otimes T) \otimes x \otimes v) \\
&= \Phi((T \otimes \mathrm{Ad}(\,\cdot\,)^{-1}Y) \otimes x \otimes v) \\
&= \Phi(T \otimes (\mathrm{Ad}(\,\cdot\,)^{-1}Y)(x \otimes v)) \\
&= \Phi(T \otimes (\mathrm{Ad}(\,\cdot\,)^{-1}Y)x \otimes v) \\
&\quad + \Phi(T \otimes x \otimes (\mathrm{Ad}(\,\cdot\,)^{-1}Y)v) \\
&= \int_K (k(\mathrm{Ad}(k)^{-1}Y)x \otimes k\chi_A \otimes v)\, dT(k) \\
&\quad + \int_K (kx \otimes k\chi_A \otimes (\mathrm{Ad}(k)^{-1}Y)v)\, dT(k) \\
&= \int_K (Ykx \otimes k\chi_A \otimes v)\, dT(k) \\
&\quad + \int_K (kx \otimes (k\chi_A \otimes \mathrm{Ad}(k)^{-1}Y) \otimes v)\, dT(k) \\
&= \mathrm{I} + \mathrm{II}, \text{ say.}
\end{aligned}
$$

Here

$$
\begin{aligned}
\mathrm{II} &= \int_K (kx \otimes (\chi_A k \otimes \mathrm{Ad}(k)^{-1}Y) \otimes v)\, dT(k) \\
&= \int_K (kx \otimes (\chi_A \otimes Y)k \otimes v)\, dT(k).
\end{aligned}
$$

Let us expand $kx = \sum_j \langle kx, x_j^* \rangle x_j$ and put $T_j = \langle \,\cdot\, x, x_j^* \rangle T$. Choose $B \subseteq \widehat{K}$ large enough so that $\chi_B T_j = T_j$ for all j. Then

$$
\begin{aligned}
\mathrm{II} &= \sum_j \int_K (x_j \otimes (\chi_A \otimes Y)k \otimes v)\, dT_j(k) \\
&= \sum_j x_j \otimes (\chi_A \otimes Y)T_j \otimes v \\
&= \sum_j x_j \otimes (\chi_A \otimes Y)\chi_B T_j \otimes v \\
&= \sum_j x_j \otimes \chi_A(Y \otimes \chi_B)T_j \otimes v.
\end{aligned}
$$

The set A is still at our disposal, and we choose it so that $\chi_A(Y \otimes \chi_B) =$

$Y \otimes \chi_B$. Then

$$\mathrm{II} = \sum_j x_j \otimes (Y \otimes \chi_B) T_j \otimes v$$

$$= \sum_j \int_K (x_j \otimes (Y \otimes \chi_B) k \otimes v) \, dT_j(k)$$

$$= \int_K (kx \otimes (Y \otimes \chi_B) k \otimes v) \, dT(k),$$

and

$$\mathrm{I} + \mathrm{II} = Y \int_K (kx \otimes \chi_B k \otimes v) \, dT(k)$$

$$= Y \int_K (kx \otimes k\chi_B \otimes v) \, dT(k) \qquad \text{by (1.52)}$$

$$= Y(\Phi(T \otimes x \otimes v)).$$

This completes the proof of the lemma.

Theorem 2.103 (Mackey isomorphism). Let $i : (\mathfrak{h}, L) \to (\mathfrak{g}, K)$ be a map of pairs, let X be in $C(\mathfrak{g}, K)$, and let V be in $C(\mathfrak{h}, L)$. Then there is a unique $C(\mathfrak{g}, K)$ isomorphism

(2.104a) $$\Phi : P(\mathcal{F}(X) \otimes_{\mathbb{C}} V) \xrightarrow{\sim} X \otimes_{\mathbb{C}} P(V)$$

such that

(2.104b) $$\Phi(T \otimes (x \otimes v)) = \int_K (kx \otimes (k\chi_A \otimes v)) \, dT(k)$$

for $T \in R(K)$, $x \in K$, $v \in V$, and all finite $A \subseteq \widehat{K}$ that are sufficiently large. Moreover, this isomorphism is natural in X and in V, and

(2.105) $$\Phi^{-1}(x \otimes (S \otimes v)) = \int_K (k\chi_B \otimes (k^{-1}x \otimes v)) \, dS(k)$$

for all B sufficiently large.

PROOF. For uniqueness, if u is in $U(\mathfrak{g})$, then we have

$$\Phi((u \otimes T) \otimes (x \otimes v)) = \Phi(u(T \otimes x \otimes v)) = u(\Phi(T \otimes x \otimes v)),$$

and the right side is determined by (2.104b).

For existence we decompose the map of pairs i as in (2.98). If we let $M = i(L)$, then we have

(2.106)
$$\mathcal{F}^{\mathfrak{h},L}_{\mathfrak{g},K} = \mathcal{F}^{\mathfrak{h},L}_{\mathfrak{g},M} \circ \mathcal{F}^{\mathfrak{g},M}_{\mathfrak{g},K}$$

$$P^{\mathfrak{g},K}_{\mathfrak{h},L} \cong P^{\mathfrak{g},K}_{\mathfrak{g},M} \circ P^{\mathfrak{g},M}_{\mathfrak{h},L} \qquad \text{by Proposition 2.19.}$$

Lemma 2.99 proves the proposition for $P^{\mathfrak{g},K}_{\mathfrak{g},M}$. We shall prove it now for $P^{\mathfrak{g},M}_{\mathfrak{h},L}$ and then for the composition.

By (i) in Proposition 2.33, $(\mathcal{F}^{\vee})^{\mathfrak{h},L}_{\mathfrak{g},M} \cong \mathcal{F}^{\mathfrak{h},L}_{\mathfrak{g},M}$. For X' and U' in $C(\mathfrak{g}, M)$ and V in $C(\mathfrak{h}, L)$, we have natural isomorphisms

$$\mathrm{Hom}_{R(\mathfrak{g},M)}(P^{\mathfrak{g},M}_{\mathfrak{h},L}(\mathcal{F}^{\mathfrak{h},L}_{\mathfrak{g},M}(X') \otimes_{\mathbb{C}} V), \, U')$$

$$\cong \mathrm{Hom}_{R(\mathfrak{h},L)}(\mathcal{F}^{\mathfrak{h},L}_{\mathfrak{g},M}(X') \otimes_{\mathbb{C}} V, \, \mathcal{F}^{\mathfrak{h},L}_{\mathfrak{g},M}(U')) \qquad \text{by Proposition 2.34}$$

$$\cong \mathrm{Hom}_{R(\mathfrak{h},L)}(V, \, \mathrm{Hom}_{\mathbb{C}}(\mathcal{F}^{\mathfrak{h},L}_{\mathfrak{g},M}(X'), \, \mathcal{F}^{\mathfrak{h},L}_{\mathfrak{g},M}(U'))_L) \qquad \text{by Corollary 2.47}$$

$$\cong \mathrm{Hom}_{R(\mathfrak{h},L)}(V, \, \mathcal{F}^{\mathfrak{h},L}_{\mathfrak{g},M}(\mathrm{Hom}_{\mathbb{C}}(X', U')_M))$$

$$\cong \mathrm{Hom}_{R(\mathfrak{g},M)}(P^{\mathfrak{g},M}_{\mathfrak{h},L}(V), \, \mathrm{Hom}_{\mathbb{C}}(X', U')_M) \qquad \text{by Proposition 2.34}$$

$$\cong \mathrm{Hom}_{R(\mathfrak{g},M)}(X' \otimes_{\mathbb{C}} P^{\mathfrak{g},M}_{\mathfrak{h},L}(V), \, U') \qquad \text{by Corollary 2.47.}$$

By (C.22) we obtain a natural isomorphism

(2.107) $$\Psi : P^{\mathfrak{g},M}_{\mathfrak{h},L}(\mathcal{F}^{\mathfrak{h},L}_{\mathfrak{g},M}(X') \otimes_{\mathbb{C}} V) \xrightarrow{\sim} X' \otimes_{\mathbb{C}} P^{\mathfrak{g},M}_{\mathfrak{h},L}(V)$$

by taking $U' = X' \otimes_{\mathbb{C}} P^{\mathfrak{g},M}_{\mathfrak{h},L}(V)$ and defining Ψ to be the member of the first Hom above that corresponds to 1 in the last Hom. For general U', let $\Psi_1, \ldots, \Psi_6$ be corresponding members of the six Hom's. For $x \in X'$, $v \in V$, and $r \in R(\mathfrak{g}, M)$, we have

$$\Psi_1(r \otimes (x \otimes v)) = r(\Psi_2(x \otimes v))$$
$$\Psi_2(x \otimes v) = \Psi_3(v)(x) = \Psi_4(v)(x)$$
$$\Psi_4(v)(x) = \Psi_5((\chi_A \otimes v)(x) \quad \text{if } \chi_A(\Psi_4(v)(x)) = \Psi_4(v)(x)$$
$$\Psi_5(r \otimes v)(x) = \Psi_6(x \otimes (r \otimes v)).$$

For the special U', $\Psi_2(x \otimes v) = \Psi_6(x \otimes (\chi_A \otimes v)) = x \otimes (\chi_A \otimes v)$. If T is in $R(K)$, we obtain

(2.108)
$$\Psi(T \otimes (x \otimes v)) = T(\Psi_2(x \otimes v)) = T(x \otimes (\chi_A \otimes v))$$
$$= \int_K (kx \otimes k\chi_A \otimes v) \, dT(k)$$

as required. A formula for Ψ^{-1} as in (2.105) follows similarly by writing Ψ_6 in terms of Ψ_1 and taking Ψ_1 to be the identity.

Finally we can consider the composition. From (2.106) we have

$$
P_{\mathfrak{h},L}^{\mathfrak{g},K}(\mathcal{F}_{\mathfrak{g},K}^{\mathfrak{h},L}(X) \otimes_{\mathbb{C}} V)
$$

$$
\cong P_{\mathfrak{g},M}^{\mathfrak{g},K}(P_{\mathfrak{h},L}^{\mathfrak{g},M}(\mathcal{F}_{\mathfrak{g},M}^{\mathfrak{h},L}(\mathcal{F}_{\mathfrak{g},K}^{\mathfrak{g},M}(X))) \otimes_{\mathbb{C}} V))
$$

$$
\cong P_{\mathfrak{g},M}^{\mathfrak{g},K}(\mathcal{F}_{\mathfrak{g},K}^{\mathfrak{g},M}(X)_{\mathbb{C}} P_{\mathfrak{h},L}^{\mathfrak{g},M}(V)) \qquad\qquad \text{by (2.107)}
$$

$$
\cong X \otimes_{\mathbb{C}} P_{\mathfrak{g},M}^{\mathfrak{g},K}(P_{\mathfrak{h},L}^{\mathfrak{g},M}(V)) \qquad\qquad \text{by Lemma 2.99}
$$

$$
\cong X \otimes_{\mathbb{C}} P_{\mathfrak{h},L}^{\mathfrak{g},K}(V) \qquad\qquad\qquad \text{by (2.106),}
$$

and this is (2.104a). To get an explicit formula, let $S \in R(K)$, $T \in R(M)$, $x \in X$, and $v \in V$. Then

$$
ST \otimes (x \otimes v)
$$

$$
\leftrightarrow S \otimes (T \otimes (x \otimes v))
$$

$$
\leftrightarrow \int_M (S \otimes mx \otimes m\chi_A \otimes v)\, dT(m) \qquad\qquad \text{by (2.108)}
$$

$$
\leftrightarrow \int_M \int_K (kmx \otimes k\chi_{A'} \otimes (m\chi_A \otimes v))\, dS(k)\, dT(m) \qquad \text{by Lemma 2.99}
$$

$$
\leftrightarrow \int_M \int_K (kmx \otimes k\chi_{A'} m\chi_A \otimes v)\, dS(k)\, dT(m)
$$

$$
= \int_M \int_K (kmx \otimes km\chi_{A'} \otimes v)\, dS(k)\, dT(m) \qquad\qquad \begin{array}{l}\text{by (1.52)} \\ \text{for } A \text{ large}\end{array}
$$

$$
= \int_K (mx \otimes m\chi_{A'} \otimes v)\, dST(m).
$$

This proves (2.104b), and (2.105) is derived similarly from (2.101) and (2.108).

Corollary 2.109. Let $i : (\mathfrak{h}, L) \to (\mathfrak{g}, K)$ be a map of pairs, and let X and U be in $C(\mathfrak{g}, K)$. Then there is a $C(\mathfrak{h}, L)$ isomorphism

$$
\Psi : \mathrm{Hom}_{\mathbb{C}}(\mathcal{F}(X), \mathcal{F}^{\vee}(U))_L \xrightarrow{\sim} \mathcal{F}^{\vee}(\mathrm{Hom}_{\mathbb{C}}(X, U)_K)
$$

natural in X and in U.

PROOF. For V in $C(\mathfrak{h}, L)$, we have natural isomorphisms

$$\text{Hom}_{R(\mathfrak{h}, L)}(V, \text{Hom}_{\mathbb{C}}(\mathcal{F}(X), \mathcal{F}^{\vee}(U))_L)$$

$$\cong \text{Hom}_{R(\mathfrak{h}, L)}(\mathcal{F}(X) \otimes_{\mathbb{C}} V, \mathcal{F}^{\vee}(U)) \qquad \text{by Corollary 2.47}$$

$$\cong \text{Hom}_{R(\mathfrak{g}, K)}(P(\mathcal{F}(X) \otimes_{\mathbb{C}} V), U) \qquad \text{by Proposition 2.34}$$

$$\cong \text{Hom}_{R(\mathfrak{g}, K)}(X \otimes_{\mathbb{C}} P(V), U) \qquad \text{by Theorem 2.103}$$

$$\cong \text{Hom}_{R(\mathfrak{g}, K)}(P(V), \text{Hom}_{\mathbb{C}}(X, U)_K) \qquad \text{by Corollary 2.47}$$

$$\cong \text{Hom}_{R(\mathfrak{h}, L)}(V, \mathcal{F}^{\vee}(\text{Hom}_{\mathbb{C}}(X, U)_K)) \qquad \text{by Proposition 2.34.}$$

Then the result follows from (C.22).

6. Derived Functors of P and I

In this section we shall introduce the derived functors of P and I and some of their special cases that have been considered in this chapter. The relevant background in homological algebra is summarized in §§3–4 of Appendix C. Fix a map of pairs $(\mathfrak{h}, L) \to (\mathfrak{g}, K)$. The functors P and I go from $C(\mathfrak{h}, L)$ to $C(\mathfrak{g}, K)$, and $C(\mathfrak{h}, L)$ has enough projectives and enough injectives, by Corollaries 2.37 and 2.26. Consequently derived functors are meaningful.

The functor P is covariant and right exact, by Proposition 2.9, and its derived functors

$$(2.110a) \qquad\qquad P_j : C(\mathfrak{h}, L) \to C(\mathfrak{g}, K)$$

are obtained on modules by applying P to projective resolutions in $C(\mathfrak{h}, L)$ and taking homology, by (C.24). The functor I is covariant and left exact, by Proposition 2.11, and its derived functors

$$(2.110b) \qquad\qquad I^j : C(\mathfrak{h}, L) \to C(\mathfrak{g}, K)$$

are obtained on modules by applying I to injective resolutions in $C(\mathfrak{h}, L)$ and taking cohomology, by (C.24). The effects of P_j and I^j on maps are given in (C.25). By (C.26), we have natural isomorphisms

$$(2.111) \qquad\qquad \begin{aligned} P_0 &\cong P \\ I^0 &\cong I \end{aligned}$$

Special cases of P and I were considered in detail in §4. For an inclusion of pairs $(\mathfrak{h}, L) \hookrightarrow (\mathfrak{g}, L)$, Proposition 2.57 showed that P and I are naturally isomorphic to ind and pro, respectively, as defined in (2.12). The functors ind and pro are exact, according to Proposition 2.57c, and hence they have

$$(2.112) \qquad P_j = 0 \text{ and } I^j = 0 \text{ for } j > 0.$$

In the situation of a subpair of finite index as in Example 1e of §4, Proposition 2.77 shows that P and I are exact, and again (2.112) holds. The same thing happens in Example 4 of §4 when i_{alg} is the identity and i_{gp} is onto with finite kernel, according to Proposition 2.92.

For an inclusion of pairs $(\mathfrak{g}, L) \hookrightarrow (\mathfrak{g}, K)$, P and I are the Bernstein functor Π and the Zuckerman functor Γ, respectively. These need not be exact. Their derived functors are denoted Π_j and Γ^j.

For a map of pairs $(\mathfrak{g}, K) \to (0, \{1\})$, P and I are naturally isomorphic to the coinvariants functor $(\,\cdot\,)_{\mathfrak{g}, K}$ and the invariants functor $(\,\cdot\,)^{\mathfrak{g}, K}$, respectively, according to Proposition 2.88. The derived functors are called **relative Lie algebra homology** $H_j(\mathfrak{g}, K; \cdot)$ and **relative Lie algebra cohomology** $H^j(\mathfrak{g}, K; \cdot)$. When $K = \{1\}$, we may drop K from the notation and refer to **Lie algebra homology** $H_j(\mathfrak{g}, \cdot)$ and **Lie algebra cohomology** $H^j(\mathfrak{g}, \cdot)$.

We summarize these special cases as follows:

$$(\Pi_{\mathfrak{g},L}^{\mathfrak{g},K})_j = \text{Bernstein derived functors}$$

$$(\Gamma_{\mathfrak{g},L}^{\mathfrak{g},K})^j = \text{Zuckerman derived functors}$$

(2.113)
$$H_j(\mathfrak{g}, K; \cdot) = \text{relative Lie algebra homology}$$

$$H^j(\mathfrak{g}, K; \cdot) = \text{relative Lie algebra cohomology}$$

$$H_j(\mathfrak{g}, \cdot) = \text{Lie algebra homology}$$

$$H^j(\mathfrak{g}, \cdot) = \text{Lie algebra cohomology}.$$

The 0th derived functors are summarized by

$$\Pi_0 \cong \Pi$$

$$\Gamma^0 \cong \Gamma$$

(2.114)
$$H_0(\mathfrak{g}, K; \cdot) = (\,\cdot\,)_{\mathfrak{g}, K}$$

$$H^0(\mathfrak{g}, K; \cdot) = (\,\cdot\,)^{\mathfrak{g}, K}$$

$$H_0(\mathfrak{g}, \cdot) = (\,\cdot\,)_{\mathfrak{g}}$$

$$H^0(\mathfrak{g}, \cdot) = (\,\cdot\,)^{\mathfrak{g}}.$$

In Propositions 2.69 and 2.89, we obtained identities relating P and I with forgetful functors. The next proposition shows how Proposition 2.69 extends to P_j and I^j. The extension of Proposition 2.89 will be given in Proposition 2.130.

Proposition 2.115. For an inclusion of pairs $(\mathfrak{g}, L) \hookrightarrow (\mathfrak{g}, K)$, let $\Pi = \Pi_{\mathfrak{g},L}^{\mathfrak{g},K}$ and $\Gamma = \Gamma_{\mathfrak{g},L}^{\mathfrak{g},K}$, and let $\Pi^K = \Pi_{\mathfrak{k},L}^{\mathfrak{k},K}$ and $\Gamma_K = \Gamma_{\mathfrak{k},L}^{\mathfrak{k},K}$. Then there are natural isomorphisms

$$\Pi_j^K \circ \mathcal{F}_{\mathfrak{g},L}^{\mathfrak{k},K} \cong \mathcal{F}_{\mathfrak{g},K}^{\mathfrak{k},K} \circ \Pi_j$$

and
$$\Gamma_K^j \circ \mathcal{F}_{\mathfrak{g},L}^{\mathfrak{k},K} \cong \mathcal{F}_{\mathfrak{g},K}^{\mathfrak{k},K} \circ \Gamma^j$$

for each $j \geq 0$.

PROOF. For the first identity, we start from the isomorphism

$$\Pi^K \circ \mathcal{F}_{\mathfrak{g},L}^{\mathfrak{k},L} \cong \mathcal{F}_{\mathfrak{g},K}^{\mathfrak{k},K} \circ \Pi$$

in Proposition 2.69b and form $(\cdot)_j$ of both sides. Proposition 2.57d shows that the exact functor $\mathcal{F}_{\mathfrak{g},L}^{\mathfrak{k},L}$ sends projectives to projectives, and (C.28a1) allows us to conclude that $(\Pi^K \circ \mathcal{F}_{\mathfrak{g},L}^{\mathfrak{k},L})_j \cong \Pi_j^K \circ \mathcal{F}_{\mathfrak{g},L}^{\mathfrak{k},L}$. On the other side, (C.27a) allows us to conclude that $(\mathcal{F}_{\mathfrak{g},K}^{\mathfrak{k},K} \circ \Pi)_j \cong \mathcal{F}_{\mathfrak{g},K}^{\mathfrak{k},K} \circ \Pi_j$. Thus the isomorphism involving Π follows. The isomorphism involving Γ is proved similarly, using (C.28a2) and (C.27a).

The remainder of this section will consist of a discussion of a functor Ext. Lemma 2.43 shows that the invariants functor from $\mathcal{C}(\mathfrak{g}, K)$ to $\mathcal{C}(0, \{1\})$ is naturally isomorphic to $V \mapsto \mathrm{Hom}_{\mathfrak{g},K}(\mathbb{C}, V)$, and we are writing the derived functors as $H^j(\mathfrak{g}, K; V)$. We define

(2.116) $\mathrm{Ext}_{\mathfrak{g},K}^j(U, \cdot) : \mathcal{C}(\mathfrak{g}, K) \to \mathcal{C}(0, \{1\})$

to be the jth derived functor of $\mathrm{Hom}_{\mathfrak{g},K}(U, \cdot)$ if U is in $\mathcal{C}(\mathfrak{g}, K)$.

Proposition 2.117. $\mathrm{Ext}_{\mathfrak{g},K}^j(\cdot, V)$ is the jth derived functor of $\mathrm{Hom}_{\mathfrak{g},K}(\cdot, V)$ if V is in $\mathcal{C}(\mathfrak{g}, K)$, and also

$$\mathrm{Ext}_{\mathfrak{g},K}^j(U, V) \cong H^j(\mathfrak{g}, K; \mathrm{Hom}_{\mathbb{C}}(U, V)_K)$$

naturally for U and V in $\mathcal{C}(\mathfrak{g}, K)$.

REMARK. The first statement of the proposition extends to any good category with enough projectives and enough injectives, but we shall give a proof special to $C(\mathfrak{g}, K)$.

PROOF. $\mathrm{Hom}_{\mathfrak{g},K}(U, \cdot)$ is the composition of $\mathrm{Hom}_{\mathbb{C}}(U, \cdot)_K$ and the invariants functor. The first of these, by Proposition 2.53c, is exact and sends injectives to injectives. By (C.28a2), the derived functors satisfy

$$\mathrm{Ext}_{\mathfrak{g},K}^j(U, \cdot) \cong H^j(\mathfrak{g}, K; \cdot) \circ \mathrm{Hom}_{\mathbb{C}}(U, \cdot)_K$$

naturally in U. Thus

(2.118a) $$\mathrm{Ext}_{\mathfrak{g},K}^j(U, V) \cong H^j(\mathfrak{g}, K; \mathrm{Hom}_{\mathbb{C}}(U, V)_K)$$

naturally in U and V.

Similarly $\mathrm{Hom}_{\mathfrak{g},K}(\cdot, V)$ is the composition of $\mathrm{Hom}_{\mathbb{C}}(\cdot, V)_K$ and the invariants functor. The first of these, by Proposition 2.53b, is exact and sends projectives to injectives. By (C.27a), the derived functors of $\mathrm{Hom}_{\mathfrak{g},K}(\cdot, V)$, call them $\mathrm{Hom}_{\mathfrak{g},K}^j(\cdot, V)$, satisfy

$$\mathrm{Hom}_{\mathfrak{g},K}^j(\cdot, V) \cong H^j(\mathfrak{g}, K; \cdot) \circ \mathrm{Hom}_{\mathbb{C}}(\cdot, V)_K$$

naturally in V. Thus

(2.118b) $$\mathrm{Hom}_{\mathfrak{g},K}^j(U, V) \cong H^j(\mathfrak{g}, K; \mathrm{Hom}_{\mathbb{C}}(U, V)_K)$$

naturally in U and V. Combining (2.118a) and (2.118b), we obtain the proposition.

Theorem 2.119 (Shapiro's Lemma). If $(\mathfrak{h}, L) \hookrightarrow (\mathfrak{g}, L)$ is an inclusion of pairs, then

$$\mathrm{Ext}_{\mathfrak{g},L}^j(P_{\mathfrak{h},L}^{\mathfrak{g},L}(V), X) \cong \mathrm{Ext}_{\mathfrak{h},L}^j(V, \mathcal{F}_{\mathfrak{g},L}^{\mathfrak{h},L}(X))$$

and $$\mathrm{Ext}_{\mathfrak{g},L}^j(X, I_{\mathfrak{h},L}^{\mathfrak{g},L}(V)) \cong \mathrm{Ext}_{\mathfrak{h},L}^j(\mathcal{F}_{\mathfrak{g},L}^{\mathfrak{h},L}(X), V)$$

naturally for V in $C(\mathfrak{h}, L)$ and for X in $C(\mathfrak{g}, L)$.

REMARK. By Proposition 2.57, we can write ind in place of P above, and we can write pro in place of I.

PROOF. By Frobenius reciprocity and part (i) of Proposition 2.33,

$$\operatorname{Hom}_{\mathfrak{g},L}(\,\cdot\,, X) \circ P^{\mathfrak{g},L}_{\mathfrak{h},L} \cong \operatorname{Hom}_{\mathfrak{h},L}(\,\cdot\,, \mathcal{F}^{\mathfrak{h},L}_{\mathfrak{g},L}(X))$$

naturally in X. On the left side, $P^{\mathfrak{g},L}_{\mathfrak{h},L}$ is exact by Proposition 2.57c, and it sends projectives to projectives by Corollary 2.35. Passing to derived functors and using (C.28a1) and the identification of Ext in Proposition 2.117, we obtain the first conclusion.

For the second conclusion, Frobenius reciprocity gives

$$\operatorname{Hom}_{\mathfrak{g},L}(X, \,\cdot\,) \circ I^{\mathfrak{g},L}_{\mathfrak{h},L} \cong \operatorname{Hom}_{\mathfrak{h},L}(\mathcal{F}^{\mathfrak{h},L}_{\mathfrak{g},L}(X), \,\cdot\,).$$

On the left side, $I^{\mathfrak{g},L}_{\mathfrak{h},L}$ is exact by Proposition 2.57c, and it sends injectives to injectives by Corollary 2.23. Thus if we pass to derived functors, (C.28a2) allows us to complete the proof.

Theorem 2.119 is limited by the fact that it deals with only inclusions of pairs for which i_{gp} is the identity. In the proof we used this hypothesis on the pairs in the form of the exactness of $P^{\mathfrak{g},L}_{\mathfrak{h},L}$ and $I^{\mathfrak{g},L}_{\mathfrak{h},L}$. For a general map of pairs, P and I need not be exact, and the appropriate tool for analyzing the composition of functors is a spectral sequence. We shall not use spectral sequences until Chapter V, and we postpone the generalization of Theorem 2.119 until then.

7. Standard Resolutions

Fix a pair $(\mathfrak{g}, K)$. In this section we shall introduce concrete projective and injective resolutions of each module in $\mathcal{C}(\mathfrak{g}, K)$.

The group $\operatorname{Ad}(K)$ operates on $\mathfrak{g}/\mathfrak{k}$ and hence on $\bigwedge^n(\mathfrak{g}/\mathfrak{k})$. Thus $\bigwedge^n(\mathfrak{g}/\mathfrak{k})$ becomes a $(\mathfrak{k}, K)$ module. By Corollary 2.36

(2.120a) $$X_n = P^{\mathfrak{g},K}_{\mathfrak{k},K}(\textstyle\bigwedge^n(\mathfrak{g}/\mathfrak{k})) = R(\mathfrak{g}, K) \otimes_{R(K)} \textstyle\bigwedge^n(\mathfrak{g}/\mathfrak{k})$$

is a projective in $\mathcal{C}(\mathfrak{g}, K)$. If $\mathfrak{p}$ denotes an $\operatorname{Ad}(K)$ invariant complementary subspace to $\mathfrak{k}$ in $\mathfrak{g}$, then we can write X_n also as

(2.120b) $$X_n = R(\mathfrak{g}, K) \otimes_{R(K)} \textstyle\bigwedge^n\mathfrak{p}.$$

Proposition 2.57a allows us to write X_n with the more manageable algebra $U(\mathfrak{g})$ in place of $R(\mathfrak{g}, K)$:

(2.120c) $$X_n = U(\mathfrak{g}) \otimes_{U(\mathfrak{k})} \textstyle\bigwedge^n(\mathfrak{g}/\mathfrak{k})$$

or

$$(2.120d) \qquad\qquad X_n = U(\mathfrak{g}) \otimes_{U(\mathfrak{k})} \textstyle\bigwedge^n \mathfrak{p}.$$

In (2.120c) and (2.120d), $\mathfrak{g}$ acts by left multiplication in the $U(\mathfrak{g})$ factor alone, and K acts by $\mathrm{Ad} \otimes \pi$, where π is the representation of K on the alternating tensors.

Let us realize X_n as in (2.120c). The **Koszul resolution** in $\mathcal{C}(\mathfrak{g}, K)$ is the sequence

$$(2.121a) \quad 0 \longleftarrow \mathbb{C} \overset{\epsilon}{\longleftarrow} X_0 \overset{\partial_0}{\longleftarrow} X_1 \overset{\partial_1}{\longleftarrow} X_2 \overset{\partial_2}{\longleftarrow} \cdots \longleftarrow X_{\dim(\mathfrak{g}/\mathfrak{k})} \longleftarrow 0$$

with $\epsilon : U(\mathfrak{g}) \otimes_{U(\mathfrak{k})} \mathbb{C} \to \mathbb{C}$ given as

$$(2.121b) \qquad\qquad \epsilon = \text{projection to constant term of } U(\mathfrak{g})$$

and with $\partial_{n-1} : X_n \to X_{n-1}$ given by

$$(2.121c)$$
$$\partial_{n-1}(u \otimes Y_1 \wedge \cdots \wedge Y_n)$$

$$= \sum_{l=1}^{n} (-1)^{l+1} (u\tilde{Y}_l \otimes Y_1 \wedge \cdots \wedge \widehat{Y}_l \wedge \cdots \wedge Y_n)$$

$$+ \sum_{r<s} (-1)^{r+s} (u \otimes ([\tilde{Y}_r, \tilde{Y}_s] + \mathfrak{k}) \wedge Y_1 \wedge \cdots \wedge \widehat{Y}_r \wedge \cdots \wedge \widehat{Y}_s \wedge \cdots \wedge Y_n),$$

where $Y_1, \ldots, Y_n$ are in $\mathfrak{g}/\mathfrak{k}$ and $\tilde{Y}_1, \ldots, \tilde{Y}_n$ are representatives in $\mathfrak{g}$.

It will have to be proved that (2.121c) is well defined and is indeed a resolution. Once (2.121c) is well defined, we may want to make a particular choice of representatives. Let $\mathfrak{g} = \mathfrak{k} \oplus \mathfrak{p}$ as above, and let $\mathcal{P}$ be the projection of $\mathfrak{g}$ onto $\mathfrak{p}$ along $\mathfrak{k}$. If we realize X_n as in (2.120d), then the formula for ∂_{n-1} becomes

$$(2.121d)$$
$$\partial_{n-1}(u \otimes Y_1 \wedge \cdots \wedge Y_n)$$

$$= \sum_{l=1}^{n} (-1)^{l+1} (uY_l \otimes Y_1 \wedge \cdots \wedge \widehat{Y}_l \wedge \cdots \wedge Y_n)$$

$$+ \sum_{r<s} (-1)^{r+s} (u \otimes \mathcal{P}[Y_r, Y_s] \wedge Y_1 \wedge \cdots \wedge \widehat{Y}_r \wedge \cdots \wedge \widehat{Y}_s \wedge \cdots \wedge Y_n),$$

where $Y_1, \ldots, Y_n$ are in $\mathfrak{p}$.

If we realize X_n as in (2.120b), we can write ∂_{n-1} with the aid of an element $\chi_A \in R(K)$ such that χ_A is the identity on $\bigwedge^k \mathfrak{g}$ for all k. The formula is

(2.121e)
$$\partial_{n-1}((u \otimes \chi_A) \otimes Y_1 \wedge \cdots \wedge Y_n)$$

$$= \sum_{l=1}^{n} (-1)^{l+1}((uY_l \otimes \chi_A) \otimes Y_1 \wedge \cdots \wedge \widehat{Y_l} \wedge \cdots \wedge Y_n)$$

$$+ \sum_{r<s}(-1)^{r+s}(u \otimes \chi_A) \otimes \mathcal{P}[Y_r, Y_s] \wedge Y_1 \wedge \cdots \wedge \widehat{Y_r} \wedge \cdots \wedge \widehat{Y_s} \wedge \cdots \wedge Y_n.$$

Theorem 2.122. In (2.121) the maps ϵ and ∂_{n-1} are well defined independently of the choice of the representatives $\tilde{Y}_1, \ldots, \tilde{Y}_n$, and the passage to the tensor product over $U(\mathfrak{k})$ is well defined. Moreover, the maps ϵ and ∂_{n-1} are $(\mathfrak{g}, K)$ maps, and (2.121a) is a projective resolution of $\mathbb{C}$.

The proofs that (2.121) is well defined and (2.121a) is a complex are computational and will be given in §8. We know already that the X_n's are projective, and it is clear by inspection that the maps ϵ and ∂_{n-1} are $(\mathfrak{g}, K)$ maps. The main content of Theorem 2.122 is the exactness of (2.121a), which will be proved in §§9–10.

Let V be in $\mathcal{C}(\mathfrak{g}, K)$, and let $N = \dim(\mathfrak{g}/\mathfrak{k}) = \dim \mathfrak{p}$. Applying (a) and (b) of Proposition 2.53 to the Koszul resolution, we obtain a projective resolution

(2.123)
$$0 \longleftarrow V \longleftarrow X_0 \otimes_{\mathbb{C}} V \longleftarrow X_1 \otimes_{\mathbb{C}} V \longleftarrow \cdots \longleftarrow X_N \otimes_{\mathbb{C}} V \longleftarrow 0$$

and an injective resolution

(2.124)
$$0 \longrightarrow V \longrightarrow \mathrm{Hom}_{\mathbb{C}}(X_0, V)_K \longrightarrow \mathrm{Hom}_{\mathbb{C}}(X_1, V)_K \longrightarrow$$

$$\cdots \longrightarrow \mathrm{Hom}_{\mathbb{C}}(X_N, V)_K \longrightarrow 0.$$

These are called the **standard resolutions** of V.

Corollary 2.125.

(a) If $i : (\mathfrak{h}, L) \to (\mathfrak{g}, K)$ is a map of pairs, then $(P_{\mathfrak{h},L}^{\mathfrak{g},K})_j = 0$ and $(I_{\mathfrak{h},L}^{\mathfrak{g},K})^j = 0$ for $j > \dim(\mathfrak{h}/\mathfrak{l})$.

(b) If $(\mathfrak{g}, L) \hookrightarrow (\mathfrak{g}, K)$ is an inclusion of pairs, then $(\Pi_{\mathfrak{g},L}^{\mathfrak{g},K})_j = 0$ and $(\Gamma_{\mathfrak{g},L}^{\mathfrak{g},K})^j = 0$ for $j > \dim(\mathfrak{k}/\mathfrak{l})$.

PROOF.   (a) $P_j(V)$ and $I^j(V)$ are computed from projective and injective resolutions of V in $\mathcal{C}(\mathfrak{h}, L)$. If we use the standard resolutions, we get 0 for $j > \dim(\mathfrak{h}/\mathfrak{l})$.

(b) Proposition 2.115 says we may compute $\mathcal{F}_{\mathfrak{g},K}^{\mathfrak{k},K}$ of $\Pi_j(V)$ and $\Gamma^j(V)$ by using Π_j^K and Γ_K^j. By (a), $\Pi_j^K = 0$ and $\Gamma_K^j = 0$ for $j > \dim(\mathfrak{k}/\mathfrak{l})$.

Let us now make relative Lie algebra homology more explicit. By Proposition 2.88 this is the instance of the P functor for the map of pairs $(\mathfrak{g}, K) \to (0, \{1\})$, and P reduces to the coinvariants functor. We consider the standard projective resolution (2.123) of V. A term is $X_n \otimes_{\mathbb{C}} V$, and we write $X_n = R(\mathfrak{g}, K) \otimes_{R(K)} \bigwedge^n \mathfrak{p}$ as in (2.123b). Applying the coinvariants functor to the resolution, we have

$$
\begin{aligned}
(X_n \otimes_{\mathbb{C}} V)_{\mathfrak{g},K} &\cong X_n \otimes_{R(\mathfrak{g},K)} V && \text{by Lemma 2.48} \\
&= \left(\textstyle\bigwedge^n \mathfrak{p} \otimes_{R(K)} R(\mathfrak{g}, K)\right) \otimes_{R(\mathfrak{g},K)} V \\
&\cong \textstyle\bigwedge^n \mathfrak{p} \otimes_{R(K)} (R(\mathfrak{g}, K) \otimes_{R(\mathfrak{g},K)} V) && \text{by (C.21)} \\
&\cong \textstyle\bigwedge^n \mathfrak{p} \otimes_{R(K)} V && \text{by Lemma 2.17} \\
&= \textstyle\bigwedge^n \mathfrak{p} \otimes_K V.
\end{aligned}
$$

Thus the complex from which we obtain relative Lie algebra homology is

(2.126a)
$$
0 \longleftarrow \textstyle\bigwedge^0 \mathfrak{p} \otimes_K V \xleftarrow{\partial} \textstyle\bigwedge^1 \mathfrak{p} \otimes_K V \xleftarrow{\partial} \cdots \xleftarrow{\partial} \textstyle\bigwedge^N \mathfrak{p} \otimes_K V \longleftarrow 0,
$$

and we readily check that the operator ∂ is given on $\bigwedge^n \mathfrak{p} \otimes_K V$ by

(2.126b)
$$
\begin{aligned}
\partial(Y_1 \wedge \cdots \wedge Y_n \otimes v) \\
= \sum_{l=1}^{n} (-1)^l (Y_1 \wedge \cdots \wedge \widehat{Y_l} \wedge \cdots \wedge Y_n \otimes Y_l v) \\
+ \sum_{r<s} (-1)^{r+s} (P[Y_r, Y_s] \wedge Y_1 \wedge \cdots \wedge \widehat{Y_r} \wedge \cdots \wedge \widehat{Y_s} \wedge \cdots \wedge Y_n \otimes v).
\end{aligned}
$$

In fact, in verifying (2.126b), it is convenient to write the isomorphism $(X_n \otimes_{\mathbb{C}} V)_{\mathfrak{g}, K} \cong \bigwedge^n \mathfrak{p} \otimes_K V$ as

$$V \otimes_{R(\mathfrak{g}, K)} R(\mathfrak{g}, K) \otimes_K \bigwedge^n \mathfrak{p} \xrightarrow{\sim} V \otimes_K \bigwedge^n \mathfrak{p}.$$

The operator ∂ on $X_n = R(\mathfrak{g}, K) \otimes_K \bigwedge^n \mathfrak{p}$ leads to $1 \otimes \partial$ on the left side, and then we collapse $V \otimes_{R(\mathfrak{g}, K)} R(\mathfrak{g}, K)$ by the induced right $R(\mathfrak{g}, K)$ module structure on V to get the corresponding action of ∂ on the right side. Specifically we make the following correspondences for the terms of (2.121e):

$$v \otimes (u \otimes \chi_A) \otimes (Y_1 \wedge \cdots \wedge Y_n) \mapsto v u \chi_A \otimes (Y_1 \wedge \cdots \wedge Y_n)$$

$$\sum (-1)^{l+1} (v \otimes (u \tilde{Y}_l \otimes \chi_A) \otimes Y\text{'s}) \mapsto \sum (-1)^{l+1} (v u \tilde{Y}_l \chi_A \otimes Y\text{'s})$$

$$\sum (-1)^{r+s} (v \otimes (u \otimes \chi_A) \otimes Y\text{'s}) \mapsto \sum (-1)^{r+s} (v u \chi_A \otimes Y\text{'s}).$$

We take u to be 1, drop the χ_A's (because A is large), and regard V again as a left $R(\mathfrak{g}, K)$ module by changing $v\tilde{Y}_l$ to $-\tilde{Y}_l v$ in the middle line. Then (2.126b) follows.

Next let us make relative Lie algebra cohomology more explicit. By Proposition 2.88 this is the instance of the I functor for the map of pairs $(\mathfrak{g}, K) \to (0, \{1\})$, and I reduces to the invariants functor. We consider the standard injective resolution (2.124) of V. A term is $\mathrm{Hom}_{\mathbb{C}}(X_n, V)_K$, and we write again X_n as in (2.120b). Applying the invariants functor to the resolution, we have

$(\mathrm{Hom}_{\mathbb{C}}(X_n, V)_K)^{\mathfrak{g}, K}$

$\quad = \mathrm{Hom}_{R(\mathfrak{g}, K)}(X_n, V)$ by Lemma 2.43

$\quad = \mathrm{Hom}_{R(\mathfrak{g}, K)}(R(\mathfrak{g}, K) \otimes_{R(K)} \bigwedge^n \mathfrak{p}, V)$

$\quad \cong \mathrm{Hom}_{R(K)}(\bigwedge^n \mathfrak{p}, \mathrm{Hom}_{R(\mathfrak{g}, K)}(R(\mathfrak{g}, K), V))$ by (C.20)

$\quad = \mathrm{Hom}_{R(K)}(\bigwedge^n \mathfrak{p}, \mathrm{Hom}_{R(\mathfrak{g}, K)}(R(\mathfrak{g}, K), V)_K)$

$\quad \cong \mathrm{Hom}_{R(K)}(\bigwedge^n \mathfrak{p}, V)$ by Lemma 2.17

$\quad = \mathrm{Hom}_K(\bigwedge^n \mathfrak{p}, V).$

Thus the complex from which we obtain relative Lie algebra cohomology is

(2.127a)

$$0 \longrightarrow \mathrm{Hom}_K(\textstyle\bigwedge^0 \mathfrak{p}, V) \xrightarrow{\;d\;} \mathrm{Hom}_K(\textstyle\bigwedge^1 \mathfrak{p}, V) \xrightarrow{\;d\;}$$

$$\cdots \xrightarrow{\;d\;} \mathrm{Hom}_K(\textstyle\bigwedge^N \mathfrak{p}, V) \longrightarrow 0,$$

and we readily check that the operator d is given on $\mathrm{Hom}_K(\bigwedge^{n-1}\mathfrak{p}, V)$ by

(2.127b)
$$d\lambda(Y_1 \wedge \cdots \wedge Y_n)$$

$$= \sum_{l=1}^n (-1)^{l+1} Y_l(\lambda(Y_1 \wedge \cdots \wedge \widehat{Y_l} \wedge \cdots \wedge Y_n))$$

$$+ \sum_{r<s} (-1)^{r+s} \lambda(\mathcal{P}[Y_r, Y_s] \wedge Y_1 \wedge \cdots \wedge \widehat{Y_r} \wedge \cdots \wedge \widehat{Y_s} \wedge \cdots \wedge Y_n).$$

The argument is similar to that for (2.126b).

More generally we can make $(P_{\mathfrak{h},L}^{\mathfrak{g},K})_n$ and $(I_{\mathfrak{h},L}^{\mathfrak{g},K})^n$ more explicit for any map of pairs $i : (\mathfrak{h}, L) \to (\mathfrak{g}, K)$. Let $\mathfrak{q}$ be an $\mathrm{Ad}(L)$ invariant complement to $\mathfrak{l}$ in $\mathfrak{h}$. If we apply $P_{\mathfrak{h},L}^{\mathfrak{g},K}$ to the standard projective resolution (2.123) of V in $\mathcal{C}(\mathfrak{h}, L)$, we can simplify the terms as follows:

$$P_{\mathfrak{h},L}^{\mathfrak{g},K}(X_n \otimes_{\mathbb{C}} V)$$

$$\begin{aligned}
&= R(\mathfrak{g}, K) \otimes_{R(\mathfrak{h},L)} ((R(\mathfrak{h}, L) \otimes_L \textstyle\bigwedge^n \mathfrak{q}) \otimes_{\mathbb{C}} V) && \text{by (2.120b)} \\
&\cong (R(\mathfrak{g}, K) \otimes_{\mathbb{C}} V) \otimes_{R(\mathfrak{h},L)} (R(\mathfrak{h}, L) \otimes_L \textstyle\bigwedge^n \mathfrak{q}) && \text{by Corollary 2.52} \\
&\cong ((R(\mathfrak{g}, K) \otimes_{\mathbb{C}} V) \otimes_{R(\mathfrak{h},L)} R(\mathfrak{h}, L)) \otimes_L \textstyle\bigwedge^n \mathfrak{q} && \text{by (C.21)} \\
&\cong (R(\mathfrak{g}, K) \otimes_{\mathbb{C}} V) \otimes_L \textstyle\bigwedge^n \mathfrak{q} && \text{by Lemma 2.17} \\
&\cong R(\mathfrak{g}, K) \otimes_L (\textstyle\bigwedge^n \mathfrak{q} \otimes_{\mathbb{C}} V) && \text{by Corollary 2.52.}
\end{aligned}$$

In these steps, $R(\mathfrak{g}, K)$ is a right $(\mathfrak{h}, L)$ module at each step, while V is converted from a left $(\mathfrak{h}, L)$ module to a right module at the second step and then back to a left module at the last step. Thus the complex whose n^{th} homology is $(P_{\mathfrak{h},L}^{\mathfrak{g},K})_n(V)$ is

(2.128a)

$$0 \longleftarrow R(\mathfrak{g}, K) \otimes_L (\textstyle\bigwedge^0 \mathfrak{q} \otimes_{\mathbb{C}} V) \xleftarrow{\;\partial\;} R(\mathfrak{g}, K) \otimes_L (\textstyle\bigwedge^1 \mathfrak{q} \otimes_{\mathbb{C}} V) \xleftarrow{\;\partial\;}$$

$$\cdots \xleftarrow{\;\partial\;} R(\mathfrak{g}, K) \otimes_L (\textstyle\bigwedge^N \mathfrak{q} \otimes_{\mathbb{C}} V) \longleftarrow 0,$$

where $N = \dim(\mathfrak{h}/\mathfrak{l})$ and where ∂ is given on $R(\mathfrak{g}, K) \otimes_L (\bigwedge^n \mathfrak{q} \otimes_{\mathbb{C}} V)$ by

(2.128b)
$$\partial(r \otimes (Y_1 \wedge \cdots \wedge Y_n \otimes v))$$

$$= \sum_{l=1}^n (-1)^{l+1} (rY_l \otimes (Y_1 \wedge \cdots \wedge \widehat{Y_l} \wedge \cdots \wedge Y_n \otimes v))$$

$$+ \sum_{l=1}^n (-1)^l (r \otimes (Y_1 \wedge \cdots \wedge \widehat{Y_l} \wedge \cdots \wedge Y_n \otimes Y_l v))$$

$$+ \sum_{s<t} (-1)^{s+t} (r \otimes \mathcal{P}[Y_s, Y_t] \wedge Y_1 \wedge \cdots \wedge \widehat{Y_s} \wedge \cdots \wedge \widehat{Y_t} \wedge \cdots \wedge Y_n \otimes v).$$

Similarly if we apply $I_{\mathfrak{h},L}^{\mathfrak{g},K}$ to the standard injective resolution (2.124) of V in $\mathcal{C}(\mathfrak{h}, L)$, we can simplify the terms as follows:

$I_{\mathfrak{h},L}^{\mathfrak{g},K} (\mathrm{Hom}_{\mathbb{C}}(X_n, V)_L)$

$= \mathrm{Hom}_{\mathfrak{h},L}(R(\mathfrak{g}, K), \mathrm{Hom}_{\mathbb{C}}(R(\mathfrak{h}, L) \otimes_L \bigwedge^n \mathfrak{q}, V)_L)_K$ by (2.120b)

$\cong \mathrm{Hom}_{\mathfrak{h},L}(R(\mathfrak{g}, K) \otimes_{\mathbb{C}} (R(\mathfrak{h}, L) \otimes_L \bigwedge^n \mathfrak{q}), V)_K$ by Corollary 2.47

$\cong \mathrm{Hom}_{\mathfrak{h},L}(R(\mathfrak{h}, L) \otimes_L \bigwedge^n \mathfrak{q}, \mathrm{Hom}_{\mathbb{C}}(R(\mathfrak{g}, K), V)_L)_K$ by Corollary 2.47

$\cong \mathrm{Hom}_L(\bigwedge^n \mathfrak{q}, \mathrm{Hom}_{\mathfrak{h},L}(R(\mathfrak{h}, L), \mathrm{Hom}_{\mathbb{C}}(R(\mathfrak{g}, K), V)_L))_K$ by (C.20)

$\cong \mathrm{Hom}_L(\bigwedge^n \mathfrak{q}, \mathrm{Hom}_{\mathfrak{h},L}(R(\mathfrak{h}, L), \mathrm{Hom}_{\mathbb{C}}(R(\mathfrak{g}, K), V)_L)_L)_K$

$\cong \mathrm{Hom}_L(\bigwedge^n \mathfrak{q}, \mathrm{Hom}_{\mathbb{C}}(R(\mathfrak{g}, K), V)_L)_K$ by Lemma 2.17

$\cong \mathrm{Hom}_L(R(\mathfrak{g}, K) \otimes_{\mathbb{C}} \bigwedge^n \mathfrak{q}, V)_K$ by Corollary 2.47

$\cong \mathrm{Hom}_L(R(\mathfrak{g}, K), \mathrm{Hom}_{\mathbb{C}}(\bigwedge^n \mathfrak{q}, V))_K$ by Proposition 2.55.

Throughout these steps, $R(\mathfrak{g}, K)$ and V are left $(\mathfrak{h}, L)$ modules, and $R(\mathfrak{h}, L)$ is a left $(\mathfrak{h}, L)$ module and right L module. Thus the complex whose n^{th} cohomology is $(I_{\mathfrak{h},L}^{\mathfrak{g},K})^n(V)$ is

(2.129a)
$$0 \longrightarrow \mathrm{Hom}_L(R(\mathfrak{g}, K), \mathrm{Hom}_{\mathbb{C}}(\bigwedge^0 \mathfrak{q}, V))_K \overset{d}{\longrightarrow}$$

$$\mathrm{Hom}_L(R(\mathfrak{g}, K), \mathrm{Hom}_{\mathbb{C}}(\bigwedge^1 \mathfrak{q}, V))_K \overset{d}{\longrightarrow}$$

$$\cdots \overset{d}{\longrightarrow} \mathrm{Hom}_L(R(\mathfrak{g}, K), \mathrm{Hom}_{\mathbb{C}}(\bigwedge^N \mathfrak{q}, V))_K \longrightarrow 0,$$

where $N = \dim(\mathfrak{h}/\mathfrak{l})$ and where d is given on

$$\mathrm{Hom}_L(R(\mathfrak{g}, K), \mathrm{Hom}_{\mathbb{C}}(\bigwedge^{n-1} \mathfrak{q}, V))_K$$

by

(2.129b)
$$d\lambda(r)(Y_1 \wedge \cdots \wedge Y_n)$$

$$= \sum_{l=1}^{n} (-1)^l \lambda(Y_l r)(Y_1 \wedge \cdots \wedge \widehat{Y_l} \wedge \cdots \wedge Y_n)$$

$$+ \sum_{l=1}^{n} (-1)^{l+1} Y_l(\lambda(r)(Y_1 \wedge \cdots \wedge \widehat{Y_l} \wedge \cdots \wedge Y_n))$$

$$+ \sum_{s<t} (-1)^{s+t} \lambda(r)(\mathcal{P}[Y_s, Y_t] \wedge Y_1 \wedge \cdots \wedge \widehat{Y_s} \wedge \cdots \wedge \widehat{Y_t} \wedge \cdots \wedge Y_n).$$

Here r is in $R(\mathfrak{g}, K)$ and $Y_1, \ldots, Y_n$ are in $\mathfrak{q}$.

Proposition 2.130. If $(\mathfrak{q}, L) \to (\mathfrak{l}, L)$ is a special semidirect product map of pairs with $\mathfrak{q} = \mathfrak{l} \oplus \mathfrak{u}$ and if $(\mathfrak{u}, \{1\}) \hookrightarrow (\mathfrak{q}, L)$ is the associated inclusion, then

$$\mathcal{F}_{\mathfrak{l},L}^{0,\{1\}} \circ (P_{\mathfrak{q},L}^{\mathfrak{l},L})_j(V) \cong H_j(\mathfrak{u}, \mathcal{F}_{\mathfrak{q},L}^{\mathfrak{u},\{1\}}(V))$$

and
$$\mathcal{F}_{\mathfrak{l},L}^{0,\{1\}} \circ (I_{\mathfrak{q},L}^{\mathfrak{l},L})^j(V) \cong H^j(\mathfrak{u}, \mathcal{F}_{\mathfrak{q},L}^{\mathfrak{u},\{1\}}(V))$$

naturally for $V \in \mathcal{C}(\mathfrak{q}, L)$. That is, $(P_{\mathfrak{q},L}^{\mathfrak{l},L})_j(V)$ and $(I_{\mathfrak{q},L}^{\mathfrak{l},L})^j(V)$ may be computed, apart from the L action, as $H_j(\mathfrak{u}, V)$ and $H^j(\mathfrak{u}, V)$, respectively. The action of L is given on the level of complexes $\bigwedge^j \mathfrak{u} \otimes_{\mathbb{C}} V$ in (2.126) and $\mathrm{Hom}_{\mathbb{C}}(\bigwedge^j \mathfrak{u}, V)$ in (2.127) by

$$l(\xi \otimes v) = \mathrm{Ad}(l)\xi \otimes lv \qquad \text{for } \xi \in \bigwedge^j \mathfrak{u}, \ v \in V$$

and $\quad (l\lambda)(\xi) = l(\lambda(\mathrm{Ad}(l)^{-1}\xi)) \quad$ for $\xi \in \bigwedge^j \mathfrak{u}, \ \lambda \in \mathrm{Hom}_{\mathbb{C}}(\bigwedge^j \mathfrak{u}, V)$.

PROOF. The complex (2.128) for $(P_{\mathfrak{q},L}^{\mathfrak{l},L})_j(V)$ has j^{th} term

$$R(L) \otimes_L (\bigwedge^j \mathfrak{u} \otimes_{\mathbb{C}} V),$$

and this is isomorphic as a vector space to the term $\bigwedge^j \mathfrak{u} \otimes_{\mathbb{C}} V$ in the complex (2.126) for $(P_{\mathfrak{u},\{1\}}^{0,\{1\}})_j(V)$, by Lemma 2.17. The isomorphism carries the element $T \otimes (\xi \otimes v)$ to $T(\xi \otimes v)$.

Let $\xi = Y_1 \wedge \cdots \wedge Y_j \in \bigwedge^j \mathfrak{u}$ and $v \in V$ be given, and choose a finite subset $A \subseteq \widehat{K}$ so that $\chi_A(\xi \otimes v) = \xi \otimes v$. Using (2.128b), we can compute $\partial(\chi_A \otimes (\xi \otimes v))$. If A is sufficiently large, then χ_A acts as the identity on the $\bigwedge^j \mathfrak{u} \otimes_{\mathbb{C}} V$ part of each term of (2.128b), and the resulting sum matches the right side of (2.126b). Therefore our isomorphism of complexes is a chain map. This proves the isomorphism for $(P_{\mathfrak{q},L}^{l,L})_j(V)$. Tracking down what happens to the action by L under this isomorphism, we obtain $l(\xi \otimes v) = \mathrm{Ad}(l)\xi \otimes lv$, as asserted.

The argument for $(I_{\mathfrak{q},L}^{l,L})^j(V)$ is similar, using the complexes (2.129) and (2.127) and the isomorphism

$$\mathrm{Hom}_L(R(L), \mathrm{Hom}_{\mathbb{C}}(\textstyle\bigwedge^j \mathfrak{u}, V))_L \cong \mathrm{Hom}_{\mathbb{C}}(\textstyle\bigwedge^j \mathfrak{u}, V)$$

of Lemma 2.17.

8. Koszul Resolution as a Complex

The remainder of the chapter is devoted to the proof of Theorem 2.122. In this section we shall prove that the maps of (2.121) are well defined and that (2.121a) is a complex. We work in the good category $\mathcal{C}(\mathfrak{g}, K)$ with modules $X_n = U(\mathfrak{g}) \otimes_{\mathfrak{k}} (\bigwedge^n (\mathfrak{g}/\mathfrak{k}))$ and maps ϵ and ∂_n. The group K will not play any role; only the Lie algebras $\mathfrak{g}$ and $\mathfrak{k}$ are relevant.

In the category $\mathcal{C}(\mathfrak{g}, \{1\})$, the corresponding modules are $U(\mathfrak{g}) \otimes_{\mathfrak{k}} \bigwedge^n \mathfrak{g}$, and we let $\tilde{\epsilon}$ and $\tilde{\partial}_n$ be the corresponding maps. The idea is to relate ϵ and ∂_n to $\tilde{\epsilon}$ and $\tilde{\partial}_n$. Then it will follow automatically that ϵ and ∂_n are well defined. We shall write out the relationship only for ∂_n and $\tilde{\partial}_n$, omitting the easy details for ϵ and $\tilde{\epsilon}$. A relatively clean computation will give the results $\tilde{\epsilon}\tilde{\partial}_0 = 0$ and $\tilde{\partial}_{n-2}\tilde{\partial}_{n-1} = 0$, and it will follow immediately that $\epsilon\partial_0 = 0$ and $\partial_{n-2}\partial_{n-1} = 0$.

The formula for ∂_{n-1} is

(2.131)
$$\tilde{\partial}_{n-1}(u \otimes Y_1 \wedge \cdots \wedge Y_n)$$

$$= \sum_{l=1}^{n} (-1)^{l+1} (uY_l \otimes Y_1 \wedge \cdots \wedge \widehat{Y_l} \wedge \cdots \wedge Y_n)$$

$$+ \sum_{r<s} (-1)^{r+s} (u \otimes [Y_r, Y_s] \wedge Y_1 \wedge \cdots \wedge \widehat{Y_r} \wedge \cdots \wedge \widehat{Y_s} \wedge \cdots \wedge Y_n)$$

with $u \in U(\mathfrak{g})$ and $Y_1, \ldots, Y_n \in \mathfrak{g}$. It is a familiar computation (which we omit) that the right side is alternating in $Y_1, \ldots, Y_n$, so that $\tilde{\partial}_{n-1}$ is well defined with alternating tensors in place. Thus $\tilde{\partial}_{n-1}$ is well defined as a map

(2.132) $$\tilde{\partial}_{n-1} : U(\mathfrak{g}) \otimes_{\mathbb{C}} \textstyle\bigwedge^n \mathfrak{g} \to U(\mathfrak{g}) \otimes_{\mathbb{C}} \textstyle\bigwedge^{n-1} \mathfrak{g}.$$

Let us see that $\tilde{\partial}_{n-1}$ descends to a map

(2.133) $$\tilde{\partial}'_{n-1} : U(\mathfrak{g}) \otimes_{\mathfrak{k}} \textstyle\bigwedge^n \mathfrak{g} \to U(\mathfrak{g}) \otimes_{\mathfrak{k}} \textstyle\bigwedge^{n-1} \mathfrak{g},$$

where $\mathfrak{k}$ acts on $U(\mathfrak{g})$ by right translation and on $\bigwedge \mathfrak{g}$ by adjoint. For $T \in \mathfrak{k}$, we are to show that

(2.134a) $$uT \otimes Y_1 \wedge \cdots \wedge Y_n$$

and

(2.134b) $$\sum_{j=1}^{n} u \otimes Y_1 \wedge \cdots \wedge [T, Y_j] \wedge \cdots \wedge Y_n$$

have the same image in $U(\mathfrak{g}) \otimes_{\mathfrak{k}} \bigwedge^{n-1} \mathfrak{g}$ under $\tilde{\partial}_{n-1}$ (when followed by passage to the quotient).

Under $\tilde{\partial}_{n-1}$, (2.134a) maps to

(2.135)
$$\sum_{l=1}^{n} (-1)^{l+1} (uTY_l \otimes Y_1 \wedge \cdots \wedge \widehat{Y_l} \wedge \cdots \wedge Y_n)$$

$$+ \sum_{r<s} (-1)^{r+s} (uT \otimes [Y_r, Y_s] \wedge Y_1 \wedge \cdots \wedge \widehat{Y_r} \wedge \cdots \wedge \widehat{Y_s} \wedge \cdots \wedge Y_n)$$

$$= \sum_{l=1}^{n} (-1)^{l+1} u[T, Y_l] \otimes Y_1 \wedge \cdots \wedge \widehat{Y_l} \wedge \cdots \wedge Y_n \tag{a}$$

$$+ \sum_{l=1}^{n} (-1)^{l+1} uY_l \otimes (\operatorname{ad} T)(Y_1 \wedge \cdots \wedge \widehat{Y_l} \wedge \cdots \wedge Y_n) \tag{b}$$

$$+ \sum_{r<s} (-1)^{r+s} (u \otimes [T, [Y_r, Y_s]]$$

$$\wedge Y_1 \wedge \cdots \wedge \widehat{Y_r} \wedge \cdots \wedge \widehat{Y_s} \wedge \cdots \wedge Y_n) \tag{c}$$

$$+ \sum_{r<s} (-1)^{r+s} (u \otimes [Y_r, Y_s]$$

$$\wedge (\operatorname{ad} T)(Y_1 \wedge \cdots \wedge \widehat{Y_r} \wedge \cdots \wedge \widehat{Y_s} \wedge \cdots \wedge Y_n)), \tag{d}$$

and (2.134b) maps to

(2.136)

$$\sum_{\substack{l=1 \\ (j=l)}}^{n} (-1)^{l+1} u[T, Y_l] \otimes Y_1 \wedge \cdots \wedge \widehat{Y_l} \wedge \cdots \wedge Y_n \qquad \text{(a)}$$

$$+ \sum_{l=1}^{n} \sum_{j \neq l} (-1)^{l+1} u Y_l \otimes Y_1 \wedge \cdots \wedge \widehat{Y_l} \wedge \cdots \wedge [T, Y_j] \wedge \cdots \wedge Y_n \quad \text{(b)}$$

$$+ \sum_{\substack{r<s \\ (j=r)}} (-1)^{r+s} (u \otimes [[T, Y_r], Y_s]$$

$$\wedge Y_1 \wedge \cdots \wedge \widehat{Y_r} \wedge \cdots \wedge \widehat{Y_s} \wedge \cdots \wedge Y_n) \qquad \text{(c)}$$

$$+ \sum_{\substack{r<s \\ (j=s)}} (-1)^{r+s} (u \otimes [Y_r, [T, Y_s]]$$

$$\wedge Y_1 \wedge \cdots \wedge \widehat{Y_r} \wedge \cdots \wedge \widehat{Y_s} \wedge \cdots \wedge Y_n) \qquad \text{(d)}$$

$$+ \sum_{r<s} \sum_{j \notin \{r,s\}} (-1)^{r+s} (u \otimes [Y_r, Y_s]$$

$$\wedge Y_1 \wedge \cdots \wedge \widehat{Y_r} \wedge \cdots \wedge \widehat{Y_s} \wedge \cdots \wedge [T, Y_j] \wedge \cdots \wedge Y_n). \text{ (e)}$$

Now (a) in (2.135) equals (a) in (2.136), (b) equals (b), and (d) equals (e). Also (c) in (2.135) equals the sum of (c) and (d) in (2.136), by the Jacobi identity. Hence (2.134a) and (2.134b) have the same image in $U(\mathfrak{g}) \otimes_{\mathfrak{k}} \bigwedge^{n-1}\mathfrak{g}$, and ∂_{n-1} descends to the map $\tilde{\partial}'_{n-1}$ in (2.133).

Now let $\bigwedge^n_1\mathfrak{g}$ be the alternating tensors of rank n spanned by monomials having at least one factor in $\mathfrak{k}$. This is an ad $\mathfrak{k}$ invariant subspace of $\bigwedge^n\mathfrak{g}$. Let us see that $\tilde{\partial}'_{n-1}$ restricts to a map

(2.137) $$\tilde{\partial}'_{n-1} : U(\mathfrak{g}) \otimes_{\mathfrak{k}} \bigwedge^n_1\mathfrak{g} \to U(\mathfrak{g}) \otimes_{\mathfrak{k}} \bigwedge^{n-1}_1\mathfrak{g}.$$

In fact, in (2.131) let us suppose, without loss of generality, that Y_1 is in $\mathfrak{k}$. The only terms on the right side of (2.131) that are not obviously in $U(\mathfrak{g}) \otimes_{\mathfrak{k}} \bigwedge^{n-1}_1\mathfrak{g}$ are

$$u Y_1 \otimes Y_2 \wedge \cdots \wedge Y_n \qquad \qquad \text{(from } l = 1\text{)}$$

$$+ \sum_{1<s} (-1)^{1+s} (u \otimes [Y_1, Y_s] \wedge Y_2 \wedge \cdots \wedge \widehat{Y_s} \wedge \cdots \wedge Y_n) \qquad \text{(from } r = 1\text{)},$$

and this is 0 since Y_1 is in $\mathfrak{k}$. Hence $\tilde{\partial}'_{n-1}$ is well defined as a map of the form (2.137).

Because of (2.133) and (2.137), $\tilde{\partial}'_{n-1}$ induces a map ∂'_{n-1} between quotients. We use the isomorphism

$$(2.138) \qquad (\textstyle\bigwedge^n \mathfrak{g})/(\bigwedge^n_1 \mathfrak{g}) \cong \bigwedge^n(\mathfrak{g}/\mathfrak{k}).$$

The functor $V \mapsto U(\mathfrak{g}) \otimes_{\mathfrak{k}} V$ is exact by (C.13) since it is naturally isomorphic to $V \mapsto S(\mathfrak{p}) \otimes_{\mathbb{C}} V$. By (C.11), ∂'_{n-1} can be regarded as a map

$$\partial'_{n-1} : U(\mathfrak{g}) \otimes_{\mathfrak{k}} \textstyle\bigwedge^n(\mathfrak{g}/\mathfrak{k}) \to U(\mathfrak{g}) \otimes_{\mathfrak{k}} \bigwedge^{n-1}(\mathfrak{g}/\mathfrak{k}).$$

Comparing the formulas (2.131) and (2.121c), we see that ∂'_{n-1} coincides with ∂_{n-1}. Therefore ∂_{n-1} is well defined.

If $\varphi_n : \bigwedge^n \mathfrak{g} \to \bigwedge^n(\mathfrak{g}/\mathfrak{k})$ is the map leading to (2.138), what we have just seen is that $\partial_{n-1}\varphi_n = \varphi_{n-1}\tilde{\partial}'_{n-1}$. Hence we can conclude $\partial_{n-2}\partial_{n-1} = 0$ if we can prove $\tilde{\partial}'_{n-2}\tilde{\partial}'_{n-1} = 0$. Similarly we can conclude $\tilde{\partial}'_{n-2}\tilde{\partial}'_{n-1} = 0$ if we can prove $\tilde{\partial}_{n-2}\tilde{\partial}_{n-1} = 0$. Analogous remarks apply to $\epsilon\partial_0$ and $\tilde{\epsilon}\tilde{\partial}_0$.

Let us consider $\tilde{\epsilon}\tilde{\partial}_0$. In (2.131) for $n = 1$, the term $\sum_{r<s}$ on the right side is 0, and $\tilde{\epsilon}$ annihilates each member of the sum $\sum_{l=1}^n$. Hence $\tilde{\epsilon}\tilde{\partial}_0 = 0$.

For $\tilde{\partial}_{n-2}\tilde{\partial}_{n-1}$, it is enough to consider

$$\tilde{\partial}_{n-2}\tilde{\partial}_{n-1}(1 \otimes Y_1 \wedge \cdots \wedge Y_n)$$

$$= \sum_{i=1}^n (-1)^{i+1}\tilde{\partial}_{n-2}(Y_i \otimes Y_1 \wedge \cdots \wedge \widehat{Y_i} \wedge \cdots \wedge Y_n)$$

$$+ \sum_{k<l} (-1)^{k+l}\tilde{\partial}_{n-2}(1 \otimes [Y_k, Y_l] \wedge Y_1 \wedge \cdots \wedge \widehat{Y_k} \wedge \cdots \wedge \widehat{Y_l} \wedge \cdots \wedge Y_n)$$

$$= \mathrm{I} + \mathrm{II}, \quad \text{say.}$$

Here

$$\mathrm{I} = \sum_{i=1}^n \sum_{j<i} (-1)^{i+1+j+1} Y_i Y_j \otimes Y_1 \wedge (\text{omit } j, i) \wedge Y_n$$

$$+ \sum_{i=1}^n \sum_{j>i} (-1)^{i+1+j} Y_i Y_j \otimes Y_1 \wedge (\text{omit } i, j) \wedge Y_n$$

$$+ \sum_i \sum_{\substack{r<s \\ \neq i}} (-1)^{r+s+\delta+i+1} Y_i \otimes [Y_r, Y_s] \wedge Y_1 \wedge (\text{omit } r, s, i) \wedge Y_n$$

$$\text{where } \delta = 1 \text{ if } r < i < s \text{ and } \delta = 0 \text{ otherwise}$$

$$= \sum_{j<i} (-1)^{i+j} [Y_i, Y_j] \otimes Y_1 \wedge (\text{omit } j, i) \wedge Y_n$$

$$+ \sum_i \sum_{\substack{r<s \\ \neq i}} (-1)^{r+s+\delta+i+1} Y_i \otimes [Y_r, Y_s] \wedge Y_1 \wedge (\text{omit } r, s, i) \wedge Y_n$$

$$= \text{I A} + \text{I B},$$

and

$$\text{II} = \sum_{k<l} (-1)^{k+l} [Y_k, Y_l] \otimes Y_1 \wedge (\text{omit } k, l) \wedge Y_n$$

$$+ \sum_{k<l} \sum_{i \neq k,l} (-1)^{k+l+i+\delta} Y_i \otimes [Y_k, Y_l] \wedge Y_1 \wedge (\text{omit } k, l, i) \wedge Y_n$$

$$\text{where } \delta = 1 \text{ if } r < i < s$$

$$+ \sum_{k<l} (-1)^{k+l} \sum_{\substack{r<s \\ \neq k,l}} (-1)^{r+s+\sigma} 1 \otimes [Y_r, Y_s] \wedge [Y_k, Y_l]$$

$$\wedge Y_1 \wedge (\text{omit } k, l, r, s) \wedge Y_n$$

$$\text{where } \sigma = 1 \text{ if exactly one of } k$$
$$\text{and } l \text{ is between } r \text{ and } s$$

$$+ \sum_{k<l} (-1)^{k+l} \sum_{j \neq k,l} (-1)^{j+\tau} 1 \otimes [[Y_k, Y_l], Y_j]$$

$$\wedge Y_1 \wedge (\text{omit } j, k, l) \wedge Y_n$$

$$\text{where } \tau = 1 \text{ if } k < j < l$$

$$= \text{II A} + \text{II B} + \text{II C} + \text{II D}.$$

Now I A cancels with II A, and I B cancels with II B. Thus we are left with II C and II D. For II C, the (k, l) and (r, s) pairs contribute together to two terms, and σ is the same for both; thus the terms of II C cancel in pairs.

In II D, fix three indices $a < b < c$. The corresponding terms come from

$k = a,$	$l = b,$	$j = c$	with term	$(-1)^{a+b+c} [[Y_a, Y_b], Y_c]$
$k = a,$	$l = c,$	$j = b$	with term	$(-1)^{a+b+c+1} [[Y_a, Y_c], Y_b]$
$k = b,$	$l = c,$	$j = a$	with term	$(-1)^{a+b+c} [[Y_b, Y_c], Y_a],$

and these terms add to 0 by the Jacobi identity. Thus II D is 0, and the proof that $\tilde{\partial}_{n-2} \tilde{\partial}_{n-1} = 0$ is complete. This concludes the proof that the maps in (2.121) are well defined and that (2.121a) is a complex.

9. Reduction of Exactness for the Koszul Resolution

To complete the proof of Theorem 2.122, we must prove the exactness of the Koszul resolution. In this section we shall reduce the proof of exactness to the special case of the theorem in which $\mathfrak{g}$ is abelian and $K = \{1\}$. We assume the theorem in that case for the present, and we work with (2.121) in a general $C(\mathfrak{g}, K)$.

We take X_n to be given by (2.120d) and ϵ and ∂_{n-1} to be given by (2.121d). For $j \geq 0$, let $U_j(\mathfrak{g})$ be the finite-dimensional subspace of $U(\mathfrak{g})$ of elements of order $\leq j$. For $0 \leq n \leq k$, let $X_n^{(k)}$ be the image of $U_{k-n}(\mathfrak{g}) \otimes_{\mathbb{C}} \bigwedge^n \mathfrak{p}$ in $U(\mathfrak{g}) \otimes_{\mathfrak{k}} \bigwedge^n \mathfrak{p}$.

The formulas (2.121b) and (2.121d) for ϵ and ∂_{n-1} make sense on $U_{k-n}(\mathfrak{g}) \otimes_{\mathbb{C}} \bigwedge^n \mathfrak{p}$ (although we do not necessarily have a complex on this level). Thus it is clear that that ∂ maps $X_n^{(k)}$ into $X_{n-1}^{(k)}$. Hence we have a complex of vector spaces

$$(2.139) \qquad 0 \longleftarrow \mathbb{C} \xleftarrow{\ \epsilon\ } X_0^{(k)} \xleftarrow{\ \partial_0\ } X_1^{(k)} \xleftarrow{\ \partial_1\ } \cdots \longleftarrow X_k^{(k)} \longleftarrow 0.$$

We prove exactness of (2.139) for $k \geq 1$ by induction on k. To see that exactness of (2.121a) follows, let x be in $\ker \partial_n \subseteq X_{n+1}$. Then x is in $X_{n+1}^{(k)}$ for large enough k, and $x = \partial_{n+1}\tilde{x}$ for some $\tilde{x}$ in $X_{n+2}^{(k)} \subseteq X_{n+2}$. So exactness of (2.121a) will follow from exactness of (2.139) and the fact proved in §8 that (2.121a) is at least a complex.

Let $N = \dim \mathfrak{p}$, let $x_1, \ldots, x_N$ be a basis of $\mathfrak{p}$, and let $\{\omega_r\}$ be a basis of $\bigwedge^n \mathfrak{p}$. We shall show that the images in the quotient of the elements

$$x^j \otimes \omega_r = x_1^{j_1} \cdots x_N^{j_N} \otimes \omega_r, \qquad \sum j_i \leq k - n,$$

are a basis of $X_n^{(k)}$ as a vector space. It is clear from the Poincaré-Birkhoff-Witt Theorem that they span. Let us see that they are linearly independent. If a linear combination $\sum_{j,r} c_{jr} x^j \otimes \omega_r$ of such elements descends to 0, it is a linear combination of elements

$$x^l u \otimes \omega' - x^l \otimes u\omega'$$

with $u \in U(\mathfrak{k})$ of constant term 0, $\omega' \in \bigwedge^n \mathfrak{p}$, $l = (l_1, \ldots, l_N)$, and $\sum l_i \leq k - n$. Thus we have an equality of the form

$$(2.140) \qquad \sum_{j,r} c_{jr} x^j \otimes \omega_r = \sum_m c_m (x^{l_m} u_m \otimes \omega'_m - x^{l_m} \otimes u_m \omega'_m)$$

with $\{x^{l_m} \otimes \omega'_m\}$ linearly independent in $U(\mathfrak{g}) \otimes_{\mathbb{C}} \bigwedge^n \mathfrak{p}$ as m varies. Since each u_m has constant term 0, the Poincaré-Birkhoff-Witt Theorem implies that

$$\left(\sum_m \mathbb{C} x^{l_m} u_m\right) \cap \left(\sum_l \mathbb{C} x^l\right) = 0.$$

Therefore (2.140) implies that

$$\sum_m c_m (x^{l_m} u_m \otimes \omega'_m) = 0,$$

and we deduce that all c_m are 0. Hence the images of the elements $x^j \otimes \omega_r$ with $\sum j_i \le k - n$ are a basis of $X_n^{(k)}$.

Returning to (2.139), fix $k \ge 1$. We have just seen that the elements $x^j \otimes 1$ with $\sum j_i \le k$ give a basis of $X_0^{(k)}$. Since ϵ annihilates all of these but $1 \otimes 1$, the elements $x^j \otimes 1$ with $1 \le \sum j_i \le k$ give a basis of $\ker \epsilon \subseteq X_0^{(k)}$. Meanwhile the elements $x^{j'} \otimes x_m$ with $\sum j'_i \le k - 1$ give a basis of $X_1^{(k)}$. If $\sum j'_i \ge 1$ and if M is the last index i with $j'_i > 0$, then

$$\partial_0(x_1^{j'_1} \cdots x_M^{j'_M - 1} \otimes x_M) = x_1^{j'_1} \cdots x_M^{j'_M} = x^{j'};$$

hence image $\partial_0|_{X_1^{(k)}} = \ker \epsilon|_{X_0^{(k)}}$. Thus the exactness of (2.139) is equivalent with the exactness of

$$(2.141) \qquad 0 \longleftarrow \sum_{1 \le \sum j_i \le k} \mathbb{C} x^j \otimes 1 \xleftarrow{\partial_0} X_1^{(k)} \xleftarrow{\partial_1} \cdots \longleftarrow X_k^{(k)} \longleftarrow 0.$$

For $k = 1$ this complex is

$$0 \longleftarrow \mathfrak{p} \otimes_{\mathbb{C}} \mathbb{C} \xleftarrow{\partial_0} \mathbb{C} \otimes_{\mathbb{C}} \mathfrak{p} \longleftarrow 0$$

with $\partial_0(1 \otimes x) = x \otimes 1$, and the complex is exact.

Proceeding inductively, suppose (2.141) is exact for some $k - 1 > 0$, and consider (2.141) for k. We embed (2.141) in the diagram (2.142). In (2.142) the rows are exact by construction, the left column of vertical maps is exact by inductive hypothesis, the middle column of maps is a complex by the part of Theorem 2.122 proved in §8, the left column of squares commutes by construction, and the maps $\bar{\partial}_{n-1}$ are obtained by passage to the quotient (so that the right column of squares commutes). Certainly the right column of maps is a complex, being the image of a complex by passage to the quotient. We shall prove that this column is exact. Then it will follow that the middle column is exact. (In fact, (C.31) gives for this situation a long exact homology sequence whose modules occur in triples 0, H, 0, where H is the homology of the middle column; by exactness the homology of the middle column must be 0.)

(2.142)

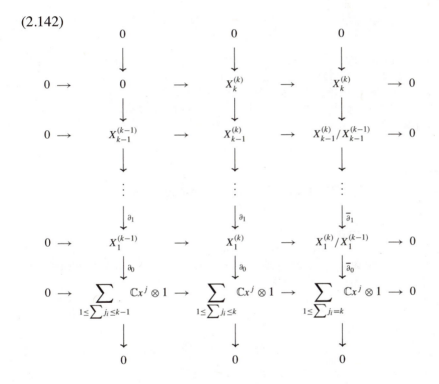

From our bases of $X_n^{(k)}$ and $X_n^{(k-1)}$, we see that a basis of $X_n^{(k)}/X_n^{(k-1)}$ is the image of all $x^j \otimes \omega_r$ with $\sum j_i = k - n$. Hence

$$X_n^{(k)}/X_n^{(k-1)} \cong S^{k-n}(\mathfrak{p}) \otimes_{\mathbb{C}} \bigwedge^n \mathfrak{p}$$

and

$$X_k^{(k)} \cong \mathbb{C} \otimes_{\mathbb{C}} \bigwedge^k \mathfrak{p}.$$

With these isomorphisms in place, we want to identify $\bar{\partial}_{n-1}$. So let

$\sum j_i = k - n$. Then

$$\bar{\partial}_{n-1}(x^j \otimes x_{i_1} \wedge \cdots \wedge x_{i_n})$$

$$= \sum_{l=1}^{n} (-1)^{l+1} (x^j x_{i_l} \otimes x_{i_1} \wedge \cdots \wedge \widehat{x}_{i_l} \wedge \cdots x_{i_n})$$

$$+ \sum_{r<s} (-1)^{r+s} (x^j \otimes [x_{i_r}, x_{i_s}] \wedge x_{i_1} \wedge \cdots \wedge \widehat{x}_{i_r} \wedge \cdots \wedge \widehat{x}_{i_s} \wedge \cdots \wedge x_{i_n})$$

$$\equiv \sum_{l=1}^{n} (-1)^{l+1} (x^j x_{i_l} \otimes x_{i_1} \wedge \cdots \wedge \widehat{x}_{i_l} \wedge \cdots x_{i_n}) \mod X_{n-1}^{(k-1)}.$$

Thus the right column of (2.142) is isomorphic to that part of the standard complex for $\mathcal{C}(\mathfrak{p}, \{1\})$ (with $\mathfrak{p}$ made into an abelian Lie algebra) in which the $U(\mathfrak{p}) = S(\mathfrak{p})$ coefficients are homogeneous of degree $k - n$. This sequence is exact by Theorem 2.122 for $\mathcal{C}(\mathfrak{p}, \{1\})$, which we are assuming for this section, and the induction is complete.

10. Exactness in the Abelian Case

The final step in the proof of Theorem 2.122 is to show that the Koszul resolution is exact when $\mathfrak{g}$ is abelian and $K = \{1\}$. Several proofs are possible, and we give only one of them. For a discussion of other proofs, see the Notes for Chapter II.

Let us take $\mathfrak{g} = \mathbb{C}^N$. Then $U(\mathbb{C}^n) = S(\mathbb{C}^N)$ is generated by a basis $x_1, \ldots, x_N$ of $\mathbb{C}^N$, and we may thus regard $U(\mathbb{C}^N)$ as the space of polynomials

$$U(\mathbb{C}^N) = \mathbb{C}[x_1, \ldots, x_N].$$

Let us write U for the left $U(\mathbb{C}^N)$ module $\mathbb{C}[x_1, \ldots, x_N]$. Let J_k be the submodule

$$J_k = \sum_{i=1}^{k} U(\mathbb{C}^N) x_i.$$

This is the space of polynomials vanishing for $x_1 = \cdots = x_k = 0$, and thus

$$U/J_k \cong \mathbb{C}[x_{k+1}, \ldots, x_N].$$

In particular, we have

$$U/J_N \cong \mathbb{C}.$$

The operation of ∂_{n-1} on $U \otimes_{\mathbb{C}} \bigwedge^n \mathbb{C}^N$ is given from (2.121d) as

$$\partial_{n-1}(u \otimes x_{i_1} \wedge \cdots \wedge x_{i_n}) = \sum_{j=1}^{n}(-1)^{j+1}ux_{i_j} \otimes x_{i_1} \wedge \cdots \wedge \widehat{x_{i_j}} \wedge \cdots \wedge x_{i_n}.$$

Within $U \otimes_{\mathbb{C}} \bigwedge^n \mathbb{C}^N$, the elements whose $\bigwedge^n \mathbb{C}^N$ part involves only $x_1, \ldots, x_k$ form a $U(\mathbb{C}^N)$ submodule $U \otimes_{\mathbb{C}} \bigwedge^n(x_1, \ldots, x_k)$, and ∂_{n-1} carries this into the corresponding submodule of $U \otimes_{\mathbb{C}} \bigwedge^{n-1} \mathbb{C}^N$. Thus we obtain a sequence

(2.143)
$$0 \longleftarrow U/J_k \overset{\epsilon_k}{\longleftarrow} U \otimes_{\mathbb{C}} \mathbb{C} \overset{\partial_0}{\longleftarrow} U \otimes_{\mathbb{C}} \bigwedge^1(x_1, \ldots, x_k)$$
$$\overset{\partial_1}{\longleftarrow} \cdots \longleftarrow U \otimes_{\mathbb{C}} \bigwedge^k(x_1, \ldots, x_k) \longleftarrow 0.$$

that is a complex as a result of the argument in §8. Here the map ϵ_k carries $u \otimes 1$ to $u + J_k$. To prove the theorem for $\mathfrak{g}$ abelian, we shall prove by induction on k that the complex (2.143) is exact, the result for $k = N$ being the result we seek.

If $k = 0$, the complex in question is

$$0 \longleftarrow U/J_0 \overset{\epsilon_0}{\longleftarrow} U \otimes_{\mathbb{C}} \mathbb{C} \overset{\partial_0}{\longleftarrow} 0$$

with $J_0 = 0$. Since ϵ_0 is an isomorphism, this is exact.

Suppose inductively that (2.143) is exact with $k - 1$ in place of k. We are to prove it is exact as it stands. Let

$$X_n^{(k)} = U \otimes_{\mathbb{C}} \bigwedge^n(x_1, \ldots, x_k).$$

In (2.143), ϵ_k is defined so as to be onto. For exactness at $U \otimes_{\mathbb{C}} \mathbb{C}$, we note that

$$\ker \epsilon_k = J_k = \left\{ \sum_{i=1}^{k} u_i x_i \right\}$$

and
$$\text{image } \partial_0 = \left\{ \partial_0 \left(\sum_{i=1}^{k} u_i \otimes x_i \right) \right\} = \left\{ \sum_{i=1}^{k} u_i x_i \right\}.$$

Thus the exactness of (2.143) is equivalent with the exactness of

(2.144)
$$0 \longleftarrow J_k (\subseteq X_0^{(k)}) \overset{\partial_0}{\longleftarrow} X_1^{(k)} \overset{\partial_1}{\longleftarrow} \cdots \overset{\partial_{k-1}}{\longleftarrow} X_k^{(k)} \longleftarrow 0.$$

We embed (2.144) in the diagram (2.145).

(2.145)

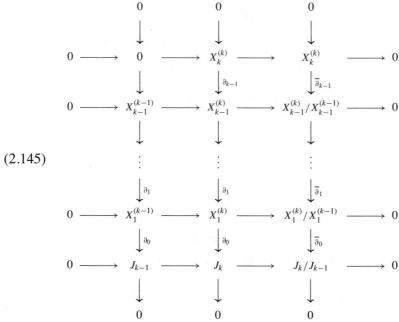

The first column of vertical maps is (2.144) for $k-1$ and is exact. The second column of vertical maps is (2.144) for k and is a complex. In each row the horizontal maps are the inclusion and quotient mappings. Thus the rows are exact and the left column of squares commutes. The third column of vertical maps exists as induced by passage to the quotient, and hence the second column of squares commutes. Moreover the third column is a complex, being obtained from a complex by passage to the quotient. We shall prove that the third column is exact, and then the exactness of (2.143) will follow from the long exact homology sequence (C.31), just as in the analogous situation in §9. Thus consider the diagram (2.146) below.

In (2.146) each horizontal mapping R_k is a version of right multiplication by x_k:

$$R_k(u + J_{k-1}) = ux_k + J_{k-1} \qquad\qquad \text{on } U/J_{k-1}$$

$$R_k(u \otimes x_{i_1} \wedge \cdots \wedge x_{i_n}) = u \otimes x_{i_1} \wedge \cdots \wedge x_{i_n} \wedge x_k + X_{n+1}^{(k-1)} \text{ on } X_n^{(k-1)}.$$

The left column of (2.146) is (2.143) for $k-1$ and is exact. The right column is the third vertical of (2.145), whose exactness is to be proved. At the least we know that this column is a complex.

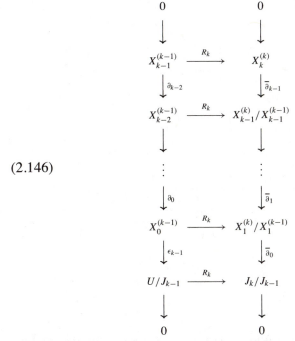

(2.146)

We shall prove that the horizontal maps of (2.146) are isomorphisms onto and that the squares commute. Then it will follow that the second column of (2.146) is exact, and our induction will go through.

First we consider the horizontal maps. The elements $u x_k + \sum_{i=1}^{k-1} u_i x_i$ exhaust J_k and thus $R_k(u) = u x_k$ carries U / J_{k-1} onto J_k / J_{k-1}. This map is one-one because if $u x_k$ is in J_{k-1}, then $u x_k$ vanishes for $x_1 = \cdots = x_{k-1} = 0$; dividing by x_k, we see that u vanishes for $x_1 = \cdots = x_{k-1} = 0$, so that u is in J_{k-1}. The other horizontal maps involve

$$x_{i_1} \wedge \cdots \wedge x_{i_n} \mapsto x_{i_1} \wedge \cdots \wedge x_{i_n} \wedge x_k \qquad \text{for } 1 \leq i_1 \leq \cdots \leq i_n \leq k - 1.$$

The images here, together with all alternating tensors from $\bigwedge^{n+1}(x_1, \ldots, x_{k-1})$, give all of $X_{n+1}^{(k)}$, so that R_k carries $X_n^{(k-1)}$ onto $X_{n+1}^{(k)} / X_{n+1}^{(k-1)}$. Clearly it is one-one. Thus R_k is an isomorphism in all cases.

Now we examine the squares in (2.146). For the bottom square, we have

$$R_k \circ \epsilon_{k-1}(u \otimes 1) = R_k(u + J_{k-1}) = u x_k + J_{k-1}$$

and

$$\bar{\partial}_0 \circ R_k(u \otimes 1) = \bar{\partial}_0(u \otimes x_k + X_1^{(k-1)}) = u x_k + J_{k-1}.$$

Thus the bottom square commutes. For the other squares, we have

$$R_k \circ \partial_{n-1}(u \otimes x_{i_1} \wedge \cdots \wedge x_{i_n})$$

$$= R_k \left(\sum_{j=1}^{n} (-1)^{j+1} u x_{i_j} \otimes x_{i_1} \wedge \cdots \wedge \widehat{x}_{i_j} \wedge \cdots \wedge x_{i_n} \right)$$

$$= \sum_{j=1}^{n} (-1)^{j+1} u x_{i_j} \otimes x_{i_1} \wedge \cdots \wedge \widehat{x}_{i_j} \wedge \cdots \wedge x_{i_n} \wedge x_k + X_n^{(k-1)}$$

on elements of $X_n^{(k-1)}$. Also

$$\overline{\partial}_{n-1} \circ R_k(u \otimes x_{i_1} \wedge \cdots \wedge x_{i_n})$$

$$= \overline{\partial}_{n-1}(u \otimes x_{i_1} \wedge \cdots \wedge x_{i_n} \wedge x_k + X_n^{(k-1)})$$

$$= \sum_{j=1}^{n} (-1)^{j+1} u x_{i_j} \otimes x_{i_1} \wedge \cdots \wedge \widehat{x}_{i_j} \wedge \cdots \wedge x_{i_n} \wedge x_k$$

$$+ (-1)^n u x_k \otimes x_{i_1} \wedge \cdots \wedge x_{i_n} + X_n^{(k-1)}$$

$$= \sum_{j=1}^{n} (-1)^{j+1} u x_{i_j} \otimes x_{i_1} \wedge \cdots \wedge \widehat{x}_{i_j} \wedge \cdots \wedge x_{i_n} \wedge x_k + X_n^{(k-1)}.$$

Thus the other squares commute, our induction goes through, and the proof is complete when $\mathfrak{g}$ is abelian.

CHAPTER III

DUALITY THEOREM

The derived functors of P and I exhibit two kinds of duality. The first kind, known as Easy Duality, is valid for any map of pairs and is the assertion that $P_j(V)^c \cong I^j(V^c)$ for all j. The second kind, known as Hard Duality, is valid if the algebra part of the map of pairs is onto. Hard Duality relates the derived functors of P with those of I for essentially the same module, but in a complementary degree. Special cases include Poincaré duality for relative Lie algebra homology and cohomology, Zuckerman duality for the Bernstein and Zuckerman functors, and the isomorphism of P with I for subpairs of finite index and for covering maps of pairs.

1. Easy Duality

We come to the first of the five main theorems of the book. The Duality Theorem relates the derived functors of P with those of I in two ways. We take up the simpler of those ways in this section.

Theorem 3.1 (Easy Duality). Let $i : (\mathfrak{h}, L) \rightarrow (\mathfrak{g}, K)$ be a map of pairs. For V in $\mathcal{C}(\mathfrak{h}, L)$ and $j \geq 0$, there are $\mathcal{C}(\mathfrak{g}, K)$ isomorphisms

$$(I_{\mathfrak{h},L}^{\mathfrak{g},K})^j(V^c) \cong (P_{\mathfrak{h},L}^{\mathfrak{g},K})_j(V)^c$$

natural in V. Here $(\cdot)^c$ refers to the L finite or K finite contragredient.

PROOF. We have

$$P(V)^c = \mathrm{Hom}_{\mathbb{C}}(P(V), \mathbb{C})_K = \mathrm{Hom}_{\mathbb{C}}(R(\mathfrak{g}, K) \otimes_{R(\mathfrak{h},L)} V, \mathbb{C})_K.$$

On the right side we convert V into a right $R(\mathfrak{h}, L)$ module $\tilde{V}$ by means of the transpose map on $R(\mathfrak{h}, L)$, and we convert $R(\mathfrak{g}, K)$ into a left $R(\mathfrak{h}, L)$ module and a right $R(\mathfrak{g}, K)$ module. Then this Hom becomes

$$= \mathrm{Hom}_{\mathbb{C}}(\tilde{V} \otimes_{R(\mathfrak{h},L)} R(\mathfrak{g}, K), \mathbb{C})_K.$$

If we regard V also as a left $\mathbb{C}$ module, then we can rewrite this Hom, via (C.20), as

$$\cong \mathrm{Hom}_{R(\mathfrak{h},L)}(R(\mathfrak{g}, K), \mathrm{Hom}_{\mathbb{C}}(\tilde{V}, \mathbb{C}))_K.$$

Now we convert $\tilde{V}$ back into V, and this Hom becomes

$$
\begin{aligned}
&= \mathrm{Hom}_{R(\mathfrak{h},L)}(R(\mathfrak{g},K),\ \mathrm{Hom}_{\mathbb{C}}(V,\mathbb{C})_L)_K \\
&= \mathrm{Hom}_{R(\mathfrak{h},L)}(R(\mathfrak{g},K),\ V^c)_K = I(V^c).
\end{aligned}
$$

All the steps are natural in V. Thus

$$
I \circ (\cdot)^c \cong (\cdot)^c \circ P.
$$

On the left side, $(\cdot)^c$ is exact and sends projectives to injectives, by Proposition 2.53b. On the right side, $(\cdot)^c$ is exact. Taking the jth derived functor of both sides and applying (C.28b2) and (C.27), we obtain the conclusion of the theorem.

Corollary 3.2.
(a) If $(\mathfrak{h}, L) \hookrightarrow (\mathfrak{g}, L)$ is an inclusion of pairs, then

$$
\mathrm{pro}_{\mathfrak{h},L}^{\mathfrak{g},L}(V^c) \cong (\mathrm{ind}_{\mathfrak{h},L}^{\mathfrak{g},L}(V))^c
$$

naturally for V in $\mathcal{C}(\mathfrak{h}, L)$.
(b) If $(\mathfrak{g}, L) \hookrightarrow (\mathfrak{g}, K)$ is an inclusion of pairs and if $j \geq 0$, then

$$
(\Gamma_{\mathfrak{g},L}^{\mathfrak{g},K})^j(V^c) \cong (\Pi_{\mathfrak{g},L}^{\mathfrak{g},K})_j(V)^c
$$

naturally for V in $\mathcal{C}(\mathfrak{g}, L)$.
(c) If $(\mathfrak{g}, K)$ is a pair, then

$$
H^j(\mathfrak{g}, K; V^c) \cong H_j(\mathfrak{g}, K; V)^*
$$

naturally for V in $\mathcal{C}(\mathfrak{g}, K)$.

PROOF. These are specializations of Theorem 3.1 to Examples 1a, 1b, and 2 in §II.4. See Propositions 2.57 and 2.88, as well as (2.113). In (a) the higher derived functors are 0, and there is no point in listing them. In (c) the map of pairs is $(\mathfrak{g}, K) \to (0, \{1\})$.

2. Statement of Hard Duality

In this section we shall formulate the second kind of duality, which relates the derived functors of P to those of I for essentially the same module but in a complementary degree.

For such a result some restriction is needed on the map of pairs $i : (\mathfrak{h}, L) \to (\mathfrak{g}, K)$. In fact, consider the simple case in which the map of pairs is $i : (0, \{1\}) \to (\mathfrak{g}, \{1\})$ with $\mathfrak{g} \neq 0$. Then $P(\mathbb{C}) \cong U(\mathfrak{g})$ has countable vector-space dimension and $P_j(\mathbb{C}) = 0$ for $j > 0$. Also $I(\mathbb{C}) \cong U(\mathfrak{g})^*$ has uncountable vector-space dimension and $I^j(\mathbb{C}) = 0$ for $j > 0$. So $I^0(\mathbb{C})$ cannot match anything close to $P_j(\mathbb{C})$ for any $j \geq 0$.

The condition to impose on $i : (\mathfrak{h}, L) \to (\mathfrak{g}, K)$ is that i_{alg} carry $\mathfrak{h}$ onto $\mathfrak{g}$. The full relevance of this condition will be more apparent in the next section. We define

$$(3.3a) \qquad \mathfrak{h}_K = i_{\text{alg}}^{-1}(\mathfrak{k}).$$

This is a subalgebra of $\mathfrak{h}$ containing $\mathfrak{l}$ and stabilized by $\text{Ad}(L)$, by (ii) of (2.6c). Hence $(\mathfrak{h}_K, L)$ is a pair. Let $\mathfrak{c}$ be an $\text{Ad}(L)$ invariant complement of $\mathfrak{l}$ in $\mathfrak{h}_K$, so that

$$(3.3b) \qquad \mathfrak{h}_K = \mathfrak{l} \oplus \mathfrak{c}, \qquad \text{all stable under } \text{Ad}(L),$$

and let

$$(3.3c) \qquad m = \dim \mathfrak{c} = \dim(\mathfrak{h}_K / \mathfrak{l}).$$

We shall make the top exterior power $\bigwedge^m \mathfrak{c}$ into an $(\mathfrak{h}, L)$ module. The L action is given by Ad, and we have to define the $\mathfrak{h}$ action.

Since $\bigwedge^m \mathfrak{c}$ is one dimensional, an $\mathfrak{h}$ action on $\bigwedge^m \mathfrak{c}$ amounts to a linear functional on $\mathfrak{h}$ that vanishes on $[\mathfrak{h}, \mathfrak{h}]$. Let $\mathfrak{s} = \ker(i_{\text{alg}})$. Then $\mathfrak{h}$ acts by ad on the one-dimensional space $\bigwedge^{\dim \mathfrak{s}} \mathfrak{s}$ since $\mathfrak{s}$ is an ideal in $\mathfrak{h}$. The linear functional for the action on $\bigwedge^{\dim \mathfrak{s}} \mathfrak{s}$ is the one we use to define an $\mathfrak{h}$ action on $\bigwedge^m \mathfrak{c}$. Its value on $H \in \mathfrak{h}$ is $\text{Tr}_{\mathfrak{s}}(\text{ad} H)$, where $\text{Tr}_{\mathfrak{s}}$ denotes the trace of the restriction to $\mathfrak{s}$.

Thus we have an $\mathfrak{h}$ action and an L action. These are compatible, according to the next lemma. We postpone the proof of the lemma to the end of this section.

Lemma 3.4. Under the assumption that the map of pairs $i : (\mathfrak{h}, L) \to (\mathfrak{g}, K)$ has $i_{\text{alg}}(\mathfrak{h}) = \mathfrak{g}$, the L action on $\bigwedge^m \mathfrak{c}$ given by Ad and the $\mathfrak{h}$ action constructed to correspond to $\bigwedge^{\dim \mathfrak{s}} \mathfrak{s}$ make $\bigwedge^m \mathfrak{c}$ into an $(\mathfrak{h}, L)$ module.

Theorem 3.5 (Hard Duality). Let $i : (\mathfrak{h}, L) \to (\mathfrak{g}, K)$ be a map of pairs such that i_{alg} carries $\mathfrak{h}$ onto $\mathfrak{g}$. Define $\mathfrak{h}_K$, $\mathfrak{c}$, and m as in (3.3), and let $P = P_{\mathfrak{h},L}^{\mathfrak{g},K}$ and $I = I_{\mathfrak{h},L}^{\mathfrak{g},K}$.

(a) For $j > m$, $P_j = 0$ and $I^j = 0$.

(b) For $0 \leq j \leq m$ and V in $\mathcal{C}(\mathfrak{h}, L)$, there are $(\mathfrak{g}, K)$ isomorphisms

$$P_j(V \otimes_{\mathbb{C}} (\textstyle\bigwedge^m \mathfrak{c})^*) \cong I^{m-j}(V)$$

natural in V.

(c) For $0 \leq j \leq m$ and V in $\mathcal{C}(\mathfrak{h}, L)$, there are $(\mathfrak{g}, K)$ isomorphisms

$$P_j(V \otimes_{\mathbb{C}} \textstyle\bigwedge^m \mathfrak{c})^c \cong P_{m-j}(V^c)$$

and
$$I^j(V \otimes_{\mathbb{C}} \textstyle\bigwedge^m \mathfrak{c})^c \cong I^{m-j}(V^c)$$

natural in V.

REMARKS.

1) We shall prove (a) in §3. (See the remark with Corollary 3.26 in that section.) Notice that the standard resolutions of §II.7 give the vanishing of P_j and I^j only for $j > \dim(\mathfrak{h}/\mathfrak{l})$, which in general is greater than m.

2) Result (c) is an immediate consequence of (b) and Theorem 3.1. But (b) is not a formal consequence of (c) and Theorem 3.1, since the double contragredient functor does not generally agree with the identity functor. It is (b), and not merely (c), that is needed for our second main theorem, the Signature Theorem.

3) The main result is (b), whose proof will take the rest of this chapter. We begin by proving the isomorphism on the level of $(\mathfrak{k}, K)$ modules. This step is carried out by an explicit computation in §§3–4. To pass to an isomorphism on the level of $(\mathfrak{g}, K)$ modules, we first consider two opposite extreme cases:

(i) $i_{\mathrm{gp}} : L \to K$ is onto;
(ii) i is an inclusion of pairs $(\mathfrak{g}, L) \hookrightarrow (\mathfrak{g}, K)$.

The proofs in Cases (i) and (ii) occupy §§5–7. In §8 we show how to put the two special cases together to handle the general case; one step in this argument involves a spectral sequence, and we consequently postpone this step to §V.11.

4) If i_{alg} is one-one (as well as onto) and if $i_{\mathrm{alg}}(\mathfrak{l}) = \mathfrak{k}$, then $m = 0$. Two situations where these conditions hold are a covering map of pairs and an inclusion for a subpair of finite index. Both were considered in §II.4, and Theorem 3.5 in these situations follows from Propositions 2.92 and 2.77, respectively. In fact, a covering map of pairs is what Case (i) in Remark 3 becomes when $m = 0$, while an inclusion of a subpair of finite index is what Case (ii) becomes when $m = 0$. The general case for $m = 0$ follows by combining Proposition 2.92 and 2.77.

Corollary 3.6 (Poincaré duality). Let $(\mathfrak{g}, K)$ be a pair, and let $m = \dim(\mathfrak{g}/\mathfrak{k})$. For $0 \leq j \leq m$ and V in $\mathcal{C}(\mathfrak{g}, K)$, there are vector-space isomorphisms

$$H_j(\mathfrak{g}, K; V \otimes_{\mathbb{C}} (\textstyle\bigwedge^m (\mathfrak{g}/\mathfrak{k}))^*) \cong H^{m-j}(\mathfrak{g}, K; V)$$

$$H_j(\mathfrak{g}, K; V \otimes_{\mathbb{C}} \textstyle\bigwedge^m (\mathfrak{g}/\mathfrak{k}))^* \cong H_{m-j}(\mathfrak{g}, K; V^c)$$

$$H^j(\mathfrak{g}, K; V \otimes_{\mathbb{C}} \textstyle\bigwedge^m (\mathfrak{g}/\mathfrak{k}))^* \cong H^{m-j}(\mathfrak{g}, K; V^c)$$

natural in V.

PROOF. This is the special case of Theorem 3.5 in which the map of pairs is $(\mathfrak{g}, K) \to (0, \{1\})$. (Only the argument in §§3–4 is needed for the proof.) See Example 2 in §II.4, especially Proposition 2.88, and also (2.113).

Corollary 3.7 (Zuckerman duality). Let $(\mathfrak{g}, L) \hookrightarrow (\mathfrak{g}, K)$ be an inclusion of pairs, and let $m = \dim(\mathfrak{k}/\mathfrak{l})$. For $0 \leq j \leq m$ and V in $\mathcal{C}(\mathfrak{g}, L)$, there are $\mathcal{C}(\mathfrak{g}, K)$ isomorphisms

$$\Pi_j(V \otimes_{\mathbb{C}} (\textstyle\bigwedge^m (\mathfrak{k}/\mathfrak{l}))^*) \cong \Gamma^{m-j}(V)$$

$$(\Pi_j(V \otimes_{\mathbb{C}} \textstyle\bigwedge^m (\mathfrak{k}/\mathfrak{l})))^c \cong \Pi_{m-j}(V^c)$$

$$(\Gamma^j(V \otimes_{\mathbb{C}} \textstyle\bigwedge^m (\mathfrak{k}/\mathfrak{l})))^c \cong \Gamma^{m-j}(V^c)$$

natural in V.

PROOF. This is the special case of Theorem 3.5 in which the map of pairs is the inclusion $(\mathfrak{g}, L) \hookrightarrow (\mathfrak{g}, K)$. (This is Case (ii) in Remark 3 after the statement of the theorem.) See Example 1b in §II.4.

Corollary 3.8. Let $(\mathfrak{q}, L) \to (\mathfrak{l}, L)$ be a special semidirect product map of pairs with $\mathfrak{q} = \mathfrak{l} \oplus \mathfrak{u}$, let $(\mathfrak{u}, \{1\}) \hookrightarrow (\mathfrak{q}, L)$ be the associated inclusion, and let $m = \dim \mathfrak{u}$. For $0 \leq j \leq m$ and V in $\mathcal{C}(\mathfrak{q}, L)$, there are $\mathcal{C}(L)$ isomorphisms

$$H_j(\mathfrak{u}, V \otimes_{\mathbb{C}} (\textstyle\bigwedge^m \mathfrak{u})^*) \cong H^{m-j}(\mathfrak{u}, V)$$

$$H_j(\mathfrak{u}, V \otimes_{\mathbb{C}} \textstyle\bigwedge^m \mathfrak{u})^* \cong H_{m-j}(\mathfrak{u}, V^c)$$

$$H^j(\mathfrak{u}, V \otimes_{\mathbb{C}} \textstyle\bigwedge^m \mathfrak{u})^* \cong H^{m-j}(\mathfrak{u}, V^c)$$

natural in V.

PROOF. This is the special case of Theorem 3.5 in which the map of pairs is $(\mathfrak{q}, L) \to (\mathfrak{l}, L)$. (Only the argument in §§3–4 is needed for the proof.) See Example 3 in §II.4, especially Proposition 2.89, and also Proposition 2.130.

The treatment of semidirect product maps can be extended significantly. A (general) **semidirect product map** is a map of pairs $(q, B) \to (\mathfrak{l}, B)$ in which $q = \mathfrak{l} \oplus u$ with u an ideal, $\mathfrak{l}$ and u are stable under $\mathrm{Ad}(B)$, $i_{\mathrm{alg}} : q \to \mathfrak{l}$ is the projection, and $i_{\mathrm{gp}} : B \to B$ is the identity. We call $(u, \{1\}) \hookrightarrow (q, B)$ the **associated inclusion**. We shall see that a general semidirect product map of pairs leads to u homology and cohomology with an $(\mathfrak{l}, B)$ structure in place. This fact generalizes Propositions 2.89 and 2.130. Although the extension of Proposition 2.89 given below in Proposition 3.9 is fairly easy, the extension of Proposition 2.130 given in Proposition 3.12 requires new ideas and will not be proved until §5.

Proposition 3.9. If $(q, B) \to (\mathfrak{l}, B)$ is a semidirect product map of pairs with $q = \mathfrak{l} \oplus u$ and if $(u, \{1\}) \hookrightarrow (q, B)$ is the associated inclusion, then

(a) there are natural isomorphisms

$$\mathcal{F}_{\mathfrak{l},B}^{0,\{1\}} \circ P_{q,B}^{\mathfrak{l},B} \cong (\,\cdot\,)_u \circ \mathcal{F}_{q,B}^{u,\{1\}}$$

$$\mathcal{F}_{\mathfrak{l},B}^{0,\{1\}} \circ I_{q,B}^{\mathfrak{l},B} \cong (\,\cdot\,)^u \circ \mathcal{F}_{q,B}^{u,\{1\}}$$

of functors from $\mathcal{C}(q, B)$ to $\mathcal{C}(0, \{1\})$; if π denotes the action of $(\mathfrak{l}, B)$ on V, then the actions of $(\mathfrak{l}, B)$ on $V_u = \mathbb{C} \otimes_{U(u)} V$ and $V^u = \mathrm{Hom}_{U(u)}(\mathbb{C}, V)$ are $1 \otimes \pi$ and $\mathrm{Hom}(1, \pi)$, respectively, and

(b) $P_{u,\{1\}}^{q,B}$ and $I_{u,\{1\}}^{q,B}$ are exact.

PROOF.

(a) We imitate the proof of Proposition 2.89a. We shall be content with a sketch since Proposition 3.9a is a special case of Proposition 3.12 below. For $V \in \mathcal{C}(q, B)$, we map

$$P_{q,B}^{\mathfrak{l},B}(V) = R(\mathfrak{l}, B) \otimes_{R(q,B)} V \qquad \text{to} \qquad V_u = \mathbb{C} \otimes_{U(u)} V$$

by

$$(u \otimes S) \otimes v \mapsto 1 \otimes u(Sv) \qquad \text{for } u \otimes S \in U(\mathfrak{l}) \otimes_{U(b)} R(B), \ v \in V.$$

The map is defined initially on $R(\mathfrak{l}, B) \otimes_{\mathbb{C}} V$, and we are to check that $X \in u$ and $l \in B$ imply that the images of

$$(u \otimes S)X \otimes v - (u \otimes S) \otimes Xv \qquad \text{and} \qquad (u \otimes S)l \otimes v - (u \otimes S) \otimes lv,$$

namely

$$-1 \otimes u(SXv) \qquad \text{and} \qquad 1 \otimes u(S\delta_l(v)) - 1 \otimes u(S(lv)),$$

are 0 in $\mathbb{C} \otimes_{U(\mathfrak{u})} V$. The second of these is handled as in Proposition 2.89. For the first, we have

(3.10)
$$-1 \otimes uSXv = -\int_B (1 \otimes ukXv) \, dS(k) = -\int_B (1 \otimes u(kXk^{-1})kv) \, dS(k).$$

Since $\mathfrak{u}$ is an ideal in $\mathfrak{q}$, a little computation with the Poincaré-Birkhoff-Witt Theorem shows that $u(kXk^{-1})$ is in $U(\mathfrak{u}) \otimes_{\mathbb{C}} U(\mathfrak{l})$ with 0 constant term for the $U(\mathfrak{u})$ component of each summand. Each $U(\mathfrak{u})$ component moves across the tensor product sign in (3.10) to act on 1 and give 0. Thus (3.10) is 0. The rest of the proof of (a) proceeds similarly.

(b) By Lemma 2.56, the inclusion of pairs $(\mathfrak{l}, B) \hookrightarrow (\mathfrak{q}, B)$ gives us vector-space isomorphisms

(3.11a) $$R(\mathfrak{q}, B) \cong R(\mathfrak{l}, B) \otimes_{\mathbb{C}} S(\mathfrak{u}) \cong R(\mathfrak{l}, B) \otimes_{\mathbb{C}} U(\mathfrak{u}),$$

the maps being

(3.11b) $$S \otimes u_{\mathfrak{l}} u_{\mathfrak{u}} \mapsto (S \otimes u_{\mathfrak{l}}) \otimes \sigma(u_{\mathfrak{u}}) \mapsto (S \otimes u_{\mathfrak{l}}) \otimes u_{\mathfrak{u}}.$$

The composite isomorphism respects the left action by $\mathfrak{l}$ and B, and (3.11b) shows that it respects the right $U(\mathfrak{u})$ module structure. If V is in $\mathcal{C}(\mathfrak{u}, \{1\})$, we therefore have vector-space isomorphisms

$$
\begin{aligned}
P_{\mathfrak{u},\{1\}}^{\mathfrak{q},B}(V) &= R(\mathfrak{q}, B) \otimes_{U(\mathfrak{u})} V \\
&\cong (R(\mathfrak{l}, B) \otimes_{\mathbb{C}} U(\mathfrak{u})) \otimes_{U(\mathfrak{u})} V && \text{by (3.11)} \\
&\cong R(\mathfrak{l}, B) \otimes_{\mathbb{C}} (U(\mathfrak{u}) \otimes_{U(\mathfrak{u})} V) && \text{by (C.21)} \\
&\cong R(\mathfrak{l}, B) \otimes_{\mathbb{C}} V.
\end{aligned}
$$

Then $P_{\mathfrak{u},\{1\}}^{\mathfrak{q},B}$ is exact by (C.15). Similarly we have isomorphisms of left $R(B)$ modules

$$
\begin{aligned}
I_{\mathfrak{u},\{1\}}^{\mathfrak{q},B}(V) &= \operatorname{Hom}_{U(\mathfrak{u})}(R(\mathfrak{q}, L), V)_B \\
&\cong \operatorname{Hom}_{U(\mathfrak{u}}(U(\mathfrak{u}) \otimes_{\mathbb{C}} R(\mathfrak{l}, B), V)_B && \text{by (3.11)} \\
&\cong \operatorname{Hom}_{U(\mathfrak{u})}(U(\mathfrak{u}), \operatorname{Hom}_{\mathbb{C}}(R(\mathfrak{l}, B), V))_B && \text{by (C.17)} \\
&\cong \operatorname{Hom}_{\mathbb{C}}(R(\mathfrak{l}, B), V)_B.
\end{aligned}
$$

By Proposition 2.53c, $I_{\mathfrak{u},\{1\}}^{\mathfrak{q},B}$ is exact.

Proposition 3.12. If $(q, B) \to (\mathfrak{l}, B)$ is a semidirect product map of pairs with $q = \mathfrak{l} \oplus \mathfrak{u}$ and if $(\mathfrak{u}, \{1\}) \hookrightarrow (q, B)$ is the associated inclusion, then

$$\mathcal{F}^{0,\{1\}}_{\mathfrak{l},B} \circ (P^{\mathfrak{l},B}_{q,B})_j(V) \cong H_j(\mathfrak{u}, \, \mathcal{F}^{\mathfrak{u},\{1\}}_{q,B}(V))$$

and

$$\mathcal{F}^{0,\{1\}}_{\mathfrak{l},B} \circ (I^{\mathfrak{l},B}_{q,B})^j(V) \cong H^j(\mathfrak{u}, \, \mathcal{F}^{\mathfrak{u},\{1\}}_{q,B}(V))$$

naturally for $V \in \mathcal{C}(q, B)$. That is, $(P^{\mathfrak{l},B}_{q,B})_j(V)$ and $(I^{\mathfrak{l},B}_{q,B})^j(V)$ may be computed, apart from the $(\mathfrak{l}, B)$ action, as $H_j(\mathfrak{u}, V)$ and $H^j(\mathfrak{u}, V)$, respectively. The actions of $\mathfrak{l}$ and B are given on the level of complexes $\bigwedge^j \mathfrak{u} \otimes_{\mathbb{C}} V$ in (2.126) and $\mathrm{Hom}_{\mathbb{C}}(\bigwedge^j \mathfrak{u}, V)$ in (2.127) by

$$X(\xi \otimes v) = (\mathrm{ad}\, X)\xi \otimes v + \xi \otimes Xv$$
$$l(\xi \otimes v) = \mathrm{Ad}(l)\xi \otimes lv$$

and

$$X\lambda(\xi) = X(\lambda(\xi)) - \lambda((\mathrm{ad}\, X)\xi)$$
$$l\lambda(\xi) = l(\lambda(\mathrm{Ad}(l)^{-1}\xi))$$

for $X \in \mathfrak{l}, l \in B, \xi \in \bigwedge^j \mathfrak{u}, v \in V$, and $\lambda \in \mathrm{Hom}_{\mathbb{C}}(\bigwedge^j \mathfrak{u}, V)$.

REMARKS. The difficulty with imitating the proof of Proposition 2.130 is that the terms of the complex (2.128) for $(P^{\mathfrak{l},B}_{q,L})_j(V)$ are not isomorphic to the terms of the complex (2.126) for $(P^{0,\{1\}}_{\mathfrak{u},\{1\}})_j(V)$, and similarly for the I functor. Instead of proving Proposition 3.12 at this time, we shall prove a generalization as Proposition 3.41 in §5.

In any event, a general semidirect product map of pairs is another situation in which Hard Duality is applicable.

Corollary 3.13. Let $(q, B) \to (\mathfrak{l}, B)$ be a semidirect product map of pairs with $q = \mathfrak{l} \oplus \mathfrak{u}$, let $(\mathfrak{u}, \{1\}) \hookrightarrow (q, B)$ be the associated inclusion, and let $m = \dim \mathfrak{u}$. For $0 \le j \le m$ and V in $\mathcal{C}(q, B)$, there are $\mathcal{C}(\mathfrak{l}, B)$ isomorphisms

$$H_j(\mathfrak{u}, \, V \otimes_{\mathbb{C}} (\textstyle\bigwedge^m \mathfrak{u})^*) \cong H^{m-j}(\mathfrak{u}, V)$$

$$H_j(\mathfrak{u}, \, V \otimes_{\mathbb{C}} \textstyle\bigwedge^m \mathfrak{u})^* \cong H_{m-j}(\mathfrak{u}, V^c)$$

$$H^j(\mathfrak{u}, \, V \otimes_{\mathbb{C}} \textstyle\bigwedge^m \mathfrak{u})^* \cong H^{m-j}(\mathfrak{u}, V^c)$$

natural in V.

PROOF. This is the special case of Theorem 3.5 in which the map of pairs is $(q, B) \to (\mathfrak{l}, B)$. (This is an instance of Case (i) in Remark 3 after the statment of the theorem.)

PROOF OF LEMMA 3.4. For $\eta \in \bigwedge^m \mathfrak{c}$, we are to prove that

(3.14a) $\qquad (\mathrm{Ad}(l)H)l\eta = lH\eta \qquad\qquad$ for $l \in L$, $H \in \mathfrak{h}$

(3.14b) $\qquad\qquad\qquad X\eta = (\mathrm{ad}\,X)\eta \qquad$ for $X \in \mathfrak{l}$.

Equation (3.14a) follows from the computation

$$(\mathrm{Ad}(l)H)l\eta = \mathrm{Tr}_\mathfrak{s}(\mathrm{ad}(\mathrm{Ad}(l)H))l\eta = \mathrm{Tr}_\mathfrak{s}(\mathrm{Ad}(l)(\mathrm{ad}\,H)\mathrm{Ad}(l)^{-1})l\eta$$
$$= \mathrm{Tr}_\mathfrak{s}(\mathrm{ad}\,H)l\eta = l\,\mathrm{Tr}_\mathfrak{s}(\mathrm{ad}\,H)\eta = lH\eta.$$

For (3.14b) we shall prove first that

(3.15) $\qquad\qquad \mathrm{Tr}_{\mathfrak{h}_K}(\mathrm{ad}\,H) = \mathrm{Tr}_\mathfrak{s}(\mathrm{ad}\,H) \qquad$ for $H \in \mathfrak{h}_K$.

In fact, choose a vector subspace $\mathfrak{a}$ of $\mathfrak{h}_K$ such that

(3.16) $\qquad\qquad\qquad\qquad \mathfrak{h}_K = \mathfrak{s} \oplus \mathfrak{a}.$

Then i carries $\mathfrak{a}$ one-one onto $\mathfrak{k}$ since $i(\mathfrak{h}) = \mathfrak{g}$. If $\{X_j\}$ is a basis of $\mathfrak{a}$ and we write $(\mathrm{ad}\,H)X_j = \sum_k c_{kj}X_k \mod \mathfrak{s}$, then we have

$$[i(H), i(X_j)] = \sum_k c_{kj}\,i(X_k)$$

and hence

(3.17) $\qquad\qquad \mathrm{Tr}_\mathfrak{k}(\mathrm{ad}\,i(H)) = \sum_j c_{jj} = \mathrm{Tr}_{\mathfrak{h}_K}(\mathrm{ad}\,H) - \mathrm{Tr}_\mathfrak{s}(\mathrm{ad}\,H).$

But $\mathrm{Tr}_\mathfrak{k}(\mathrm{ad}\,X) = 0$ for all $X \in \mathfrak{k}$ since K is compact. Hence (3.17) implies (3.15).

The compactness of L implies that

(3.18) $\qquad\qquad\qquad \mathrm{Tr}_\mathfrak{l}(\mathrm{ad}\,X) = 0 \qquad$ for $X \in \mathfrak{l}$.

Combining this equality with (3.15), we see that

$$\mathrm{Tr}_\mathfrak{s}(\mathrm{ad}\,X) = \mathrm{Tr}_{\mathfrak{h}_K/\mathfrak{l}}(\mathrm{ad}\,X) \qquad \text{for } X \in \mathfrak{l},$$

and this is (3.14b).

3. Complexes for Computing P_j and I^j

For a general map of pairs $i : (\mathfrak{h}, L) \to (\mathfrak{g}, K)$, (2.128a) and (2.129a) are complexes from which one can compute $P_j(V)$ and $I^j(V)$ for V in $\mathcal{C}(\mathfrak{h}, L)$. Under the assumption that i_{alg} carries $\mathfrak{h}$ onto $\mathfrak{g}$, shorter complexes will serve this purpose. They have the disadvantage that they compute only the K action, the $\mathfrak{g}$ action being lost. In this section we exhibit these complexes.

Proposition 3.19. Let $i : (\mathfrak{h}, L) \to (\mathfrak{g}, K)$ be a map of pairs such that $i_{\mathrm{alg}}(\mathfrak{h}) = \mathfrak{g}$. Put $\mathfrak{h}_K = i_{\mathrm{alg}}^{-1}(\mathfrak{k})$, so that $(\mathfrak{h}_K, L)$ is a pair, $(\mathfrak{h}_K, L) \hookrightarrow (\mathfrak{h}, L)$ is an inclusion of pairs, and $i : (\mathfrak{h}_K, L) \to (\mathfrak{k}, K)$ is a map of pairs. Then

$$R(K) \otimes_{R(\mathfrak{h}_K, L)} R(\mathfrak{h}, L) \cong R(\mathfrak{g}, K)$$

under convolution on L:

$$T_1 \otimes (T_2 \otimes u) \mapsto (T_1 *_L T_2) \otimes i_{\mathrm{alg}}(u)$$

for $T_1 \in R(K)$, $T_2 \in R(L)$, $u \in U(\mathfrak{h})$. This isomorphism carries the K action $l \otimes 1$ on the left side to the left-regular representation of K on $R(\mathfrak{g}, K)$ on the right side, and it carries the right $R(\mathfrak{h}, L)$ module structure on the left side to the right $R(\mathfrak{h}, L)$ module structure of $R(\mathfrak{g}, K)$ on the right side.

REMARK. This result generalizes Lemma 2.66, where the special case of an inclusion with $\mathfrak{h} = \mathfrak{g}$ is treated.

PROOF. Let $\mathfrak{q}$ be an $\mathrm{Ad}(L)$ invariant complement to $\mathfrak{h}_K$ in $\mathfrak{h}$, and let $\mathfrak{p}$ be an $\mathrm{Ad}(K)$ invariant complement to $\mathfrak{k}$ in $\mathfrak{g}$, so that

(3.20) $\mathfrak{h} = \mathfrak{h}_K \oplus \mathfrak{q}$

(3.21a) $\mathfrak{g} = \mathfrak{k} \oplus \mathfrak{p}.$

Since $\mathfrak{h}_K = i_{\mathrm{alg}}^{-1}(\mathfrak{k}) \supseteq i_{\mathrm{alg}}^{-1}(0) = \ker(i_{\mathrm{alg}})$,

(3.21b) i_{alg} maps $\mathfrak{q}$ one-one into $\mathfrak{g}$.

Application of i_{alg} to (3.20) gives $\mathfrak{g} = \mathfrak{k} + i_{\mathrm{alg}}(\mathfrak{q})$ since $i_{\mathrm{alg}}(\mathfrak{h}) = \mathfrak{g}$. To see that this sum is a direct sum, we argue as follows: Let $X \in \mathfrak{k} \cap i_{\mathrm{alg}}(\mathfrak{q})$

with $X = i_{\mathrm{alg}}(Z)$ for some $Z \in \mathfrak{q}$. Then $Z \in i_{\mathrm{alg}}^{-1}(X) \subseteq i_{\mathrm{alg}}^{-1}(\mathfrak{k}) = \mathfrak{h}_K$ and $Z \in \mathfrak{q}$; hence $Z = 0$. Therefore

$$(3.22) \qquad \mathfrak{g} = \mathfrak{k} \oplus i_{\mathrm{alg}}(\mathfrak{q}),$$

and (ii) of (2.6c) shows that each term is $\mathrm{Ad}(L)$ invariant. Comparing (3.21) and (3.22), we see that there is an L equivariant isomorphism

$$(3.23) \qquad j : \mathfrak{q} \xrightarrow{\sim} \mathfrak{p}.$$

We apply Lemma 2.56 twice, once with $(\mathfrak{h}_K, L) \hookrightarrow (\mathfrak{h}, L)$ and $\mathfrak{h} = \mathfrak{h}_K \oplus \mathfrak{q}$ to obtain

$$(3.24) \qquad R(\mathfrak{h}, L) \cong R(\mathfrak{h}_K, L) \otimes_{\mathbb{C}} S(\mathfrak{q}) \qquad \text{as left } R(\mathfrak{h}_K, L) \text{ modules,}$$

and once with $(\mathfrak{k}, K) \hookrightarrow (\mathfrak{g}, K)$ and $\mathfrak{g} = \mathfrak{k} \oplus \mathfrak{p}$ to obtain

$$(3.25) \qquad R(\mathfrak{g}, K) \cong R(K) \otimes_{\mathbb{C}} S(\mathfrak{p}) \qquad \text{as left } R(K) \text{ modules.}$$

Then we have

$$R(K) \otimes_{R(\mathfrak{h}_K, L)} R(\mathfrak{h}, L)$$

$$\begin{aligned}
&\cong R(K) \otimes_{R(\mathfrak{h}_K, L)} (R(\mathfrak{h}_K, L) \otimes_{\mathbb{C}} S(\mathfrak{q})) && \text{by (3.24)} \\
&\cong (R(K) \otimes_{R(\mathfrak{h}_K, L)} R(\mathfrak{h}_K, L)) \otimes_{\mathbb{C}} S(\mathfrak{q}) && \text{by (C.21)} \\
&\cong R(K) \otimes_{\mathbb{C}} S(\mathfrak{q}) && \text{by (2.18a)} \\
&\cong R(K) \otimes_{\mathbb{C}} S(\mathfrak{p}) && \text{by (3.23)} \\
&\cong R(\mathfrak{g}, K) && \text{by (3.25).}
\end{aligned}$$

These isomorphisms respect the left actions by K. Tracking down the explicit composite isomorphism as in the proof of Lemma 2.66, we see that the right action of $R(\mathfrak{h}, L)$ on $R(K) \otimes_{R(\mathfrak{h}_K, L)} R(\mathfrak{h}, L)$ corresponds to the right action of $R(\mathfrak{h}, L)$ on $R(\mathfrak{g}, K)$.

Corollary 3.26. If $i : (\mathfrak{h}, L) \to (\mathfrak{g}, K)$ is a map of pairs with i_{alg} carrying $\mathfrak{h}$ onto $\mathfrak{g}$, then

(a) for V in $\mathcal{C}(\mathfrak{h}, L)$ there are natural isomorphisms

$$P(V) \cong R(K) \otimes_{R(\mathfrak{h}_K, L)} V$$

$$I(V) \cong \mathrm{Hom}_{R(\mathfrak{h}_K, L)}(R(K), V)_K$$

as locally K finite representations of K,

(b) there are natural isomorphisms

$$P^{\mathfrak{k},K}_{\mathfrak{h}_K,L} \circ \mathcal{F}^{\mathfrak{h}_K,L}_{\mathfrak{h},L} \cong \mathcal{F}^{\mathfrak{k},K}_{\mathfrak{g},K} \circ P^{\mathfrak{g},K}_{\mathfrak{h},L}$$

$$I^{\mathfrak{k},K}_{\mathfrak{h}_K,L} \circ \mathcal{F}^{\mathfrak{h}_K,L}_{\mathfrak{h},L} \cong \mathcal{F}^{\mathfrak{k},K}_{\mathfrak{g},K} \circ I^{\mathfrak{g},K}_{\mathfrak{h},L}$$

of functors from $\mathcal{C}(\mathfrak{h}, L)$ to $\mathcal{C}(\mathfrak{k}, K)$, and
(c) for each $j \geq 0$, there are natural isomorphisms

$$(P^{\mathfrak{k},K}_{\mathfrak{h}_K,L})_j \circ \mathcal{F}^{\mathfrak{h}_K,L}_{\mathfrak{h},L} \cong \mathcal{F}^{\mathfrak{k},K}_{\mathfrak{g},K} \circ (P^{\mathfrak{g},K}_{\mathfrak{h},L})_j$$

$$(I^{\mathfrak{k},K}_{\mathfrak{h}_K,L})^j \circ \mathcal{F}^{\mathfrak{h}_K,L}_{\mathfrak{h},L} \cong \mathcal{F}^{\mathfrak{k},K}_{\mathfrak{g},K} \circ (I^{\mathfrak{g},K}_{\mathfrak{h},L})^j$$

of functors from $\mathcal{C}(\mathfrak{h}, L)$ to $\mathcal{C}(\mathfrak{k}, K)$.

REMARK. Theorem 3.5a is an immediate consequence of conclusion (c) and Corollary 2.125a for the map of pairs $i : (\mathfrak{h}_K, L) \to (\mathfrak{k}, K)$. (This argument generalizes the proof of Corollary 2.125b.)

PROOF. Part (a) follows from Proposition 3.19 in the same way that Proposition 2.69 follows from Lemma 2.66, and (b) is just a restatement of (a).

For the first identity in (c), we take the jth derived functor of both sides of the first identity in (b). Since each forgetful functor is exact and $\mathcal{F}^{\mathfrak{h}_K,L}_{\mathfrak{h},L}$ sends projectives to projectives (Proposition 2.57d), the result follows from (C.27a) and (C.28a1).

For the second identity in (c), we argue similarly from the jth derived functor of both sides of the second identity in (b), using Proposition 2.57d and (C.27a) and (C.28a2).

Let us combine the complexes (2.128) and (2.129) for the map of pairs $i : (\mathfrak{h}_K, L) \to (\mathfrak{k}, K)$ with the conclusions in (b) and (c) of Corollary 3.26. Writing $\mathcal{F} = \mathcal{F}^{\mathfrak{h}_K,L}_{\mathfrak{h},L}$ and $\tilde{\mathcal{F}} = \mathcal{F}^{\mathfrak{k},K}_{\mathfrak{g},K}$, we see that the complex whose n^{th} homology is $\tilde{\mathcal{F}}((P^{\mathfrak{g},K}_{\mathfrak{h},L})_n(V))$ is

(3.27)

$$0 \longleftarrow R(K) \otimes_L (\textstyle\bigwedge^0 \mathfrak{c} \otimes_{\mathbb{C}} \mathcal{F}(V)) \overset{\partial}{\longleftarrow} R(K) \otimes_L (\textstyle\bigwedge^1 \mathfrak{c} \otimes_{\mathbb{C}} \mathcal{F}(V)) \overset{\partial}{\longleftarrow}$$

$$\cdots \overset{\partial}{\longleftarrow} R(K) \otimes_L (\textstyle\bigwedge^m \mathfrak{c} \otimes_{\mathbb{C}} \mathcal{F}(V)) \longleftarrow 0,$$

with ∂ given by (2.128b) except for obvious notational changes. Similarly the complex whose n^{th} cohomology is $\tilde{\mathcal{F}}((I_{\mathfrak{h},L}^{\mathfrak{g},K})^n(V))$ is

$$0 \longrightarrow \operatorname{Hom}_L(R(K), \operatorname{Hom}_{\mathbb{C}}(\textstyle\bigwedge^0\mathfrak{c}, \mathcal{F}(V)))_K \xrightarrow{d}$$

(3.28) $$\operatorname{Hom}_L(R(K), \operatorname{Hom}_{\mathbb{C}}(\textstyle\bigwedge^1\mathfrak{c}, \mathcal{F}(V)))_K \xrightarrow{d}$$

$$\cdots \xrightarrow{d} \operatorname{Hom}_L(R(K), \operatorname{Hom}_{\mathbb{C}}(\textstyle\bigwedge^m\mathfrak{c}, \mathcal{F}(V)))_K \longrightarrow 0,$$

with d given by (2.129b) except for obvious notational changes.

4. Hard Duality as a K Isomorphism

In this section we shall establish the duality in Theorem 3.5b as an isomorphism in $\mathcal{C}(\mathfrak{k}, K)$, postponing consideration of the $\mathfrak{g}$ equivariance to §§5–8. To prove Theorem 3.5b, we shall exhibit an explicit $(\mathfrak{k}, K)$ isomorphism on the level of complexes that descends to an isomorphism on the level of homology and cohomology. The isomorphism will involve, for the first time, a choice of normalization of Haar measure on K. Although one can argue that total mass 1 might be a natural choice, the counting measure in the case of finite groups is equally natural. We use the notation in the statement of Theorem 3.5. The relevant complexes are (3.27) and (3.28), which are complexes in the category $\mathcal{C}(\mathfrak{k}, K)$.

To begin with, we lift ∂ from

(3.29a) $$R(K) \otimes_L (\textstyle\bigwedge^n\mathfrak{c} \otimes_{\mathbb{C}} \mathcal{F}(V) \otimes_{\mathbb{C}} (\textstyle\bigwedge^m\mathfrak{c})^*)$$

to

(3.29b) $$R(K) \otimes_{\mathbb{C}} \textstyle\bigwedge^n\mathfrak{c} \otimes_{\mathbb{C}} \mathcal{F}(V) \otimes_{\mathbb{C}} (\textstyle\bigwedge^m\mathfrak{c})^*,$$

where the formula for ∂ is still given by (2.128b) but we no longer expect $\partial^2 = 0$. If π denotes the representation of L on V, then L acts on (3.29b) by $r \otimes \operatorname{Ad} \otimes \pi \otimes \operatorname{Ad}^*$, and we can recover (3.29a), apart from a canonical isomorphism, as the subspace of L invariants of (3.29b). (In fact, (3.29a) is really the quotient space of L coinvariants of (3.29b). This is the quotient of (3.29b) by the sum of all L types but the trivial one. By Proposition 1.18b it is canonically isomorphic with the trivial L type, which is the space of invariants.)

Similarly we extend d from

(3.30a) $\mathrm{Hom}_L(R(K), \mathrm{Hom}_{\mathbb{C}}(\bigwedge^k \mathfrak{c}, \mathcal{F}(V)))_K$

to

(3.30b) $\mathrm{Hom}_{\mathbb{C}}(R(K), \mathrm{Hom}_{\mathbb{C}}(\bigwedge^k \mathfrak{c}, \mathcal{F}(V)))_K,$

using (2.129b) as a definition and no longer expecting that $d^2 = 0$. The group L acts on (3.30b) by $\mathrm{Hom}(l, \mathrm{Hom}(\mathrm{Ad}, \pi))$, and we recover (3.30a) as the space of L invariants of (3.30b).

The spaces (3.29b) and (3.30b) have K actions that are given, respectively, by $l \otimes 1 \otimes 1 \otimes 1$ and $\mathrm{Hom}(r, 1)$. These commute with the L actions and reduce to the usual K actions on (3.29a) and (3.30a).

We give the duality isomorphism initially on the level of a map

(3.31a) $\begin{aligned} C(K)\, dk \otimes_{\mathbb{C}} \bigwedge^n \mathfrak{c} \otimes_{\mathbb{C}} \mathcal{F}(V) \otimes_{\mathbb{C}} (\bigwedge^m \mathfrak{c})^* \\ \longrightarrow \mathrm{Hom}_{\mathbb{C}}(R(K), \mathrm{Hom}_{\mathbb{C}}(\bigwedge^{m-n} \mathfrak{c}, \mathcal{F}(V)))_K. \end{aligned}$

Here we have replaced $R(K)$ in the domain by $C(K)\, dk$, using the isomorphism (1.34) determined by our particular normalization of Haar measure dk on K. Namely for $f\, dk \in C(K)\, dk$, $\xi \in \bigwedge^n \mathfrak{c}$, $v \in V$, $\varepsilon \in (\bigwedge^m \mathfrak{c})^*$, $T \in R(K)$, and $\gamma \in \bigwedge^{m-n} \mathfrak{c}$, the map (3.31a) is given by

(3.31b) $\begin{aligned} f\, dk \otimes \xi \otimes v \otimes \varepsilon &\mapsto \lambda_{f \otimes \xi \otimes v \otimes \varepsilon} \\ \lambda_{f \otimes \xi \otimes v \otimes \varepsilon}(T)(\gamma) &= \varepsilon(\xi \wedge \gamma)\langle T, f^t \rangle v. \end{aligned}$

To see that λ is a vector-space isomorphism, we use the following lemma.

Lemma 3.32. Let U be a trivial K module, and let K act on $C(K)$ and $R(K)$, respectively, by l and r. Then the linear map

$$C(K)\, dk \otimes_{\mathbb{C}} U \longrightarrow \mathrm{Hom}_{\mathbb{C}}(R(K), U)_K$$

$$f\, dk \otimes u \mapsto \lambda'_{f \otimes u} \quad \text{with} \quad \lambda'_{f \otimes u}(T) = \langle T, f^t \rangle u$$

is a K equivariant isomorphism.

PROOF. The computation

$$\begin{aligned} (k\lambda'_{f \otimes u})(T) &= k(\lambda'_{f \otimes u}(r(k)^{-1}T)) = \lambda'_{f \otimes u}(r(k)^{-1}T) = \langle r(k)^{-1}T, f^t \rangle \\ &= \langle T, r(k)f^t \rangle = \langle T, (l(k)f)^t \rangle = \lambda'_{l(k)f \otimes u}(T) \end{aligned}$$

shows that λ' is K equivariant. Therefore it is enough to prove that λ' is an isomorphism on each K type.

If φ is in $\operatorname{Hom}_{\mathbb{C}}(R(K), U)_K$, then $(k\varphi)(T) = \varphi(r(k)^{-1}T)$ for $k \in K$, since K acts trivially on U, and we can write the right side as $\varphi(T\delta_k)$ by Proposition 1.20. If γ is in $\widehat{K}$, then Proposition 1.29 gives

$$\chi_\gamma \varphi(T) = \varphi(T\chi_\gamma) = \varphi(r(\chi_{\gamma^*})T).$$

For φ in $\operatorname{Hom}_{\mathbb{C}}(R(K), U)_\gamma$, the left side is $\varphi(T)$. Therefore φ vanishes on all K types of $R(K)$ other than $R(K)_{\gamma^*}$, and it follows that restriction of members of $\operatorname{Hom}_{\mathbb{C}}(R(K), U)_\gamma$ to $R(K)_{\gamma^*}$ is an isomorphism:

$$\operatorname{Hom}_{\mathbb{C}}(R(K), U)_\gamma \cong \operatorname{Hom}_{\mathbb{C}}(R(K)_{\gamma^*}, U).$$

By Proposition 2.55 the Hom on the right is

$$\cong (R(K)_{\gamma^*})^* \otimes_{\mathbb{C}} U \cong (R(K)^*)_\gamma \otimes_{\mathbb{C}} U.$$

Thus it is enough to prove that $(R(K)^*)_\gamma \cong (C(K)\,dk)_\gamma$, hence to prove that $R(K)^* \cong C(K)\,dk$. In other words, there is no loss of generality in taking $U = \mathbb{C}$.

Let $\lambda_f'' = \lambda_{f'\otimes 1}'$, so that $\lambda_f''(T) = \langle T, f \rangle$. Then λ'' is K equivariant from $(r, C(K))$ to $(\operatorname{Hom}(r, 1), \operatorname{Hom}_{\mathbb{C}}(R(K), \mathbb{C})_K)$. For $\gamma \in \widehat{K}$, the γ subspaces, by (1.28) and (1.36), are respectively $C(K)_\gamma$ and $\operatorname{Hom}_{\mathbb{C}}(R(K)_\gamma, \mathbb{C})$, and Proposition 1.39c shows that λ'' exhibits these as isomorphic. Since $C(K)$ and $\operatorname{Hom}_{\mathbb{C}}(R(K), \mathbb{C})_K$ are locally K finite, λ'' is an isomorphism. Hence λ' is an isomorphism.

Returning to (3.31), we can write λ as an obvious composition

$$C(K)\,dk \otimes_{\mathbb{C}} (\textstyle\bigwedge^n \mathfrak{c} \otimes_{\mathbb{C}} \mathcal{F}(V) \otimes_{\mathbb{C}} (\bigwedge^m \mathfrak{c})^*)$$

(3.33)
$$\longrightarrow C(K)\,dk \otimes_{\mathbb{C}} \operatorname{Hom}_{\mathbb{C}}(\textstyle\bigwedge^{m-n} \mathfrak{c}, \mathcal{F}(V))$$
$$\longrightarrow \operatorname{Hom}_{\mathbb{C}}(R(K), \operatorname{Hom}_{\mathbb{C}}(\textstyle\bigwedge^{m-n} \mathfrak{c}, \mathcal{F}(V)))_K.$$

The first map is an isomorphism by Proposition 2.55, and the second map is an isomorphism by Lemma 3.32. Therefore λ is a vector-space isomorphism.

Each map in (3.33) is K equivariant (the second one by Lemma 3.32), and therefore the isomorphism λ is K equivariant.

The map λ respects the L actions because, for $k \in L$,

$$
\begin{aligned}
k\lambda_{f\otimes\xi\otimes v\otimes\varepsilon}(T)(\gamma) &= k(\lambda_{f\otimes\xi\otimes v\otimes\varepsilon}(l(k^{-1})T))(\gamma) \\
&= \pi(\lambda_{f\otimes\xi\otimes v\otimes\varepsilon}(l(k^{-1})T)(\mathrm{Ad}(k^{-1})\gamma)) \\
&= \pi(\varepsilon(\xi \wedge \mathrm{Ad}(k^{-1})\gamma)\langle l(k^{-1})T, f^t\rangle v) \\
&= k\varepsilon(\mathrm{Ad}(k)\xi \wedge \gamma)\langle T, (r(k)f)^t\rangle\pi(v) \\
&= \lambda_{r(k)f\otimes\mathrm{Ad}(k)\xi\otimes\pi(k)\otimes\varepsilon}(T)(\gamma).
\end{aligned}
$$

Passing to the L invariants, we therefore obtain a K equivariant isomorphism

$$
(3.34) \qquad
\begin{aligned}
\lambda_n : C(K)\, dk \otimes_L (\textstyle\bigwedge^n\mathfrak{c} \otimes_{\mathbb{C}} \mathcal{F}(V) \otimes_{\mathbb{C}} (\textstyle\bigwedge^m\mathfrak{c})^*) \\
\longrightarrow \mathrm{Hom}_L(R(K), \mathrm{Hom}_{\mathbb{C}}(\textstyle\bigwedge^{m-n}\mathfrak{c}, \mathcal{F}(V)))_K.
\end{aligned}
$$

To complete the proof of Theorem 3.5b within the category $\mathcal{C}(\mathfrak{k}, K)$, we shall prove that λ_n descends to a map of homology to cohomology. This descent will be possible if we show that

$$
(3.35) \qquad\qquad \lambda_{n-1} \circ \partial = (-1)^n d \circ \lambda_n.
$$

We use two lemmas.

Lemma 3.36. If $Z_1, \ldots, Z_{m+1}$ are in $\mathfrak{c}$, then

$$
\sum_{l=1}^{m+1} (-1)^l \varepsilon(Z_1 \wedge \cdots \wedge \widehat{Z_l} \wedge \cdots \wedge Z_{m+1})Z_l = 0 \quad \text{in } \mathfrak{c}
$$

and

$$
\sum_{s<t} (-1)^{s+t}\varepsilon(\mathcal{P}[Z_s, Z_t] \wedge Z_1 \wedge \cdots \wedge \widehat{Z_s} \wedge \cdots \wedge \widehat{Z_t} \wedge \cdots \wedge Z_{m+1}) = 0
$$

in $\mathbb{C}$.

PROOF. In each case the left side is alternating $(m + 1)$-linear on an m-dimensional space and hence is 0.

Lemma 3.37. Let $Z_1, \ldots, Z_{m+1}$ be in $\mathfrak{c}$ and let $0 \le k \le m + 1$. For $s < t$, define $\zeta_{st} = Z_1 \wedge \cdots \wedge \widehat{Z}_s \wedge \cdots \wedge \widehat{Z}_t \wedge \cdots \wedge Z_{m+1}$. Then

$$\sum_{l=1}^{k} (-1)^l Z_l \varepsilon (Z_1 \wedge \cdots \wedge \widehat{Z}_l \wedge \cdots \wedge Z_{m+1})$$
$$= -2 \sum_{s<t\le k} (-1)^{s+t} \varepsilon(\mathcal{P}[Z_s, Z_t] \wedge \zeta_{st}) - \sum_{s\le k<t} (-1)^{s+t} \varepsilon(\mathcal{P}[Z_s, Z_t] \wedge \zeta_{st}).$$

Here $Z_l \varepsilon$ refers to the action of $\mathfrak{h}_K$ on $(\bigwedge^m \mathfrak{c})^*$ contragredient to that in Lemma 3.4.

PROOF. Let η be in $\bigwedge^m \mathfrak{c}$. If H is in $\mathfrak{h}_K$, then

(3.38a) $$H\eta = \mathrm{Tr}_{\mathfrak{s}}(\mathrm{ad}\, H)\eta = \mathrm{Tr}_{\mathfrak{h}_K}(\mathrm{ad}\, H)\eta$$

by definition and (3.15). Let $\bigwedge^{\mathrm{top}}$ refer to the top-degree alternating tensors on a vector space. If τ is in $\bigwedge^{\mathrm{top}}\mathfrak{l}$, then $\eta \wedge \tau$ is in $\bigwedge^{\mathrm{top}}\mathfrak{h}_K$, and we have

(3.38b) $$\mathrm{Tr}_{\mathfrak{h}_K}(\mathrm{ad}\, H)(\eta \wedge \tau) = (\mathrm{ad}\, H)(\eta \wedge \tau) = (\mathrm{ad}\, H)\eta \wedge \tau + \eta \wedge (\mathrm{ad}\, H)\tau.$$

Let us show that

(3.39) $$\eta \wedge (\mathrm{ad}\, H)\tau = 0.$$

In fact, there are two cases. If H is in $\mathfrak{l}$, then

$$(\mathrm{ad}\, H) = \mathrm{Tr}_{\mathfrak{l}}(\mathrm{ad}\, H)\tau = 0$$

because L is compact. If H is in $\mathfrak{c}$, then $[H, X]$ is in $\mathfrak{c}$ for each factor $X \in \mathfrak{l}$ of τ. Hence each term of $\eta \wedge (\mathrm{ad}\, H)\tau$ has $m + 1$ factors in $\mathfrak{c}$ and must be 0. This proves (3.39). Combining (3.39) with (3.38), we see that

$$H\eta \wedge \tau = (\mathrm{ad}\, H) \wedge \tau.$$

Passing to the dual, we therefore have

$$H\varepsilon(H_1 \wedge \cdots \wedge H_m) = -\sum_{j=1}^{m} \varepsilon(H_1 \wedge \cdots \wedge \mathcal{P}[H, H_j] \wedge \cdots \wedge H_m)$$

for H in $\mathfrak{h}_K$ and $H_1, \ldots, H_m$ in $\mathfrak{c}$. Then the left side in the statement of the lemma is

$$= \sum_{l=1}^{k} \left\{ \sum_{j<l} (-1)^{l+1} \varepsilon(Z_1 \wedge \cdots \wedge \mathcal{P}[Z_l, Z_j] \wedge \cdots \wedge \widehat{Z}_l \wedge \cdots \wedge Z_{m+1}) \right.$$
$$\left. + \sum_{j>l} (-1)^{l+1} \varepsilon(Z_1 \wedge \cdots \wedge \widehat{Z}_l \wedge \cdots \wedge \mathcal{P}[Z_l, Z_j] \wedge \cdots \wedge Z_{m+1}) \right\}$$
$$= \sum_{j<l \leq k} (-1)^{l+j} \varepsilon(\mathcal{P}[Z_l, Z_j] \wedge \zeta_{jl}) + \sum_{l<j \leq k} (-1)^{l+j+1} \varepsilon(\mathcal{P}[Z_l, Z_j] \wedge \zeta_{lj})$$
$$+ \sum_{l \leq k < j} (-1)^{l+j+1} \varepsilon(\mathcal{P}[Z_l, Z_j] \wedge \zeta_{lj}),$$

and this equals the right side in the statement of the lemma.

Returning to (3.35), we work with (3.31a) rather than the map after passage to L invariants. Write

$$\xi = X_1 \wedge \cdots \wedge X_n$$
$$\xi_l = X_1 \wedge \cdots \wedge \widehat{X}_l \wedge \cdots \wedge X_n$$
$$\xi_{st} = X_1 \wedge \cdots \wedge \widehat{X}_s \wedge \cdots \wedge \widehat{X}_t \wedge \cdots \wedge X_n \quad \text{for } s < t$$
$$\gamma = Y_1 \wedge \cdots \wedge Y_n$$
$$\gamma_l = Y_1 \wedge \cdots \wedge \widehat{Y}_l \wedge \cdots \wedge Y_n$$
$$\gamma_{st} = Y_1 \wedge \cdots \wedge \widehat{Y}_s \wedge \cdots \wedge \widehat{Y}_t \wedge \cdots \wedge Y_n \quad \text{for } s < t.$$

Then

$$\lambda_{\partial(f \otimes \xi \otimes v \otimes \varepsilon)}(T)(\gamma) - (-1)^n d\lambda_{f \otimes \xi \otimes v \otimes \varepsilon}(T)(\gamma)$$
$$= \left\{ \sum_{l=1}^{n} (-1)^{l+1} \lambda_{fi(X_l) \otimes \xi_l \otimes v \otimes \varepsilon}(T)(\gamma) + \sum_{l=1}^{n} (-1)^l \lambda_{f \otimes \xi_l \otimes X_l(v \otimes \varepsilon)}(T)(\gamma) \right.$$
$$\left. + \sum_{s<t} (-1)^{s+t} \lambda_{f \otimes (\mathcal{P}[X_s, X_t] \wedge \xi_{st}) \otimes v \otimes \varepsilon}(T)(\gamma) \right\}$$
$$- (-1)^n \left\{ \sum_{l=0}^{m-n} (-1)^{l+1} \lambda_{f \otimes \xi \otimes v \otimes \varepsilon}(i(Y_l)T)(\gamma_l) \right.$$
$$+ \sum_{l=0}^{m-n} (-1)^l Y_l(\lambda_{f \otimes \xi \otimes v \otimes \varepsilon}(T)(\gamma_l))$$
$$\left. + \sum_{s<t} (-1)^{s+t} \lambda_{f \otimes \xi \otimes v \otimes \varepsilon}(T)(\mathcal{P}[Y_s, Y_t] \wedge \gamma_{st}) \right\}$$

(3.40)

$$= \left\{ \sum_{l=1}^{n} (-1)^{l+1} \varepsilon(\xi_l \wedge \gamma) \langle T, (f \cdot i(X_l))^t \rangle v \right. \tag{a}$$

$$+ \sum_{l=1}^{n} (-1)^l \varepsilon(\xi_l \wedge \gamma) \langle T, f^t \rangle X_l v \tag{b}$$

$$+ \sum_{l=1}^{n} (-1)^l X_l \varepsilon(\xi_l \wedge \gamma) \langle T, f^t \rangle v \tag{c}$$

$$+ \sum_{s<t} (-1)^{s+t} \varepsilon(\mathcal{P}[X_s, X_t] \wedge \xi_{st} \wedge \gamma) \langle T, f^t \rangle v \tag{d}$$

$$- (-1)^n \left\{ \sum_{l=0}^{m-n} (-1)^{l+1} \varepsilon(\xi \wedge \gamma_l) \langle i(Y_l) T, f^t \rangle v \right. \tag{e}$$

$$+ \sum_{l=0}^{m-n} (-1)^l \varepsilon(\xi \wedge \gamma_l) \langle T, f^t \rangle Y_l v \tag{f}$$

$$+ \sum_{s<t} (-1)^{s+t} \varepsilon(\xi \wedge \mathcal{P}[Y_s, Y_t] \wedge \gamma_{st}) \langle T, f^t \rangle v \right\}. \tag{g}$$

Terms (a) and (e) in (3.40) add to 0 by the first formula in Lemma 3.36, as do (b) and (f). For (c) we substitute from Lemma 3.37, and the sum of (d) and (g) with the substituted terms is the product of $-\langle T, f^t \rangle v$ by the left side of the second formula in Lemma 3.36. Lemma 3.36 says this expression is 0, and the proof of (3.35) is complete.

5. Proof of 𝔤 Equivariance in Case (i)

The remainder of the chapter gives the proof of 𝔤 equivariance in Theorem 3.5b, starting with Cases (i) and (ii) described in Remark 3 of that theorem. Case (i) is conceptually easier, and we begin with that.

The map of pairs is assumed to have i_{alg} and i_{gp} onto. Let 𝔰 be the kernel of i_{alg} and M be the kernel of i_{gp}, so that $(\mathfrak{s}, M) \hookrightarrow (\mathfrak{h}, L)$ is an inclusion of pairs. Referring to (i) of (2.6c), we see that $\mathfrak{s} \cap \mathfrak{l} = \mathfrak{m}$. Also (ii) of (2.6c) shows that $\mathrm{Ad}(L)$ carries 𝔰 to itself.

For each X in 𝔤, let $\tilde{X}$ be a member of 𝔥 with $i_{\mathrm{alg}}(\tilde{X}) = X$ and $\mathrm{Ad}(m)\tilde{X} = \tilde{X}$ for all $m \in M$. To see that $\tilde{X}$ exists, we first pick $X_1 \in \mathfrak{h}$ with $i_{\mathrm{alg}}(X_1) = X$; we can do so since i_{alg} is onto 𝔤. Then we put $\tilde{X} = \int_M \mathrm{Ad}(m) X_1 \, dm$, where dm is the normalized Haar measure on M.

The element $\tilde{X}$ is fixed by $\mathrm{Ad}(M)$, and $i_{\mathrm{alg}}(\tilde{X}) = X$ since (ii) of (2.6) shows that

$$i_{\mathrm{alg}}(\mathrm{Ad}(m)X_1) = \mathrm{Ad}(i_{\mathrm{gp}}(m))i_{\mathrm{alg}}(X_1) = X$$

for each $m \in M$.

For each k in K, choose $\tilde{k}$ in L with $i_{\mathrm{gp}}(\tilde{k}) = k$.

Proposition 3.41. In the above notation

(3.42)
$$\mathcal{F}_{\mathfrak{g},K}^{0,\{1\}} \circ (P_{\mathfrak{h},L}^{\mathfrak{g},K})_j(V) \cong H_j(\mathfrak{s}, M; \mathcal{F}_{\mathfrak{h},L}^{\mathfrak{s},M}(V))$$
$$\mathcal{F}_{\mathfrak{g},K}^{0,\{1\}} \circ (I_{\mathfrak{h},L}^{\mathfrak{g},K})^j(V) \cong H^j(\mathfrak{s}, M; \mathcal{F}_{\mathfrak{h},L}^{\mathfrak{s},M}(V))$$

naturally for $V \in \mathcal{C}(\mathfrak{h}, L)$. That is, $(P_{\mathfrak{h},L}^{\mathfrak{g},K})_j(V)$ and $(I_{\mathfrak{h},L}^{\mathfrak{g},K})^j(V)$ may be computed, apart from the $(\mathfrak{g}, K)$ action, as $H_j(\mathfrak{s}, M; V)$ and $H^j(\mathfrak{s}, M; V)$, respectively. The effects of the $\mathfrak{g}$ and K actions may be given on the level of complexes $\bigwedge^j(\mathfrak{s}/\mathfrak{m}) \otimes_M V$ in (2.126) and $\mathrm{Hom}_M(\bigwedge^j(\mathfrak{s}/\mathfrak{m}), V)$ in (2.127) by the linear extension of

(3.43a)
$$X(\xi \otimes v) = (\mathrm{ad}\,\tilde{X})\xi \otimes v + \xi \otimes \tilde{X}v$$
$$k(\xi \otimes v) = \mathrm{Ad}(\tilde{k})\xi \otimes \tilde{k}v$$

and by

(3.43b)
$$X\lambda(\xi) = \tilde{X}(\lambda(\xi)) - \lambda((\mathrm{ad}\,\tilde{X})\xi)$$
$$k\lambda(\xi) = \tilde{k}(\lambda(\mathrm{Ad}(\tilde{k})^{-1}\xi))$$

for $X \in \mathfrak{g}, k \in K, \xi \in \bigwedge^j(\mathfrak{s}/\mathfrak{m}), v \in V$, and $\lambda \in \mathrm{Hom}_M(\bigwedge^j(\mathfrak{s}/\mathfrak{m}), V)$.

REMARKS.

1) In (3.43a) and (3.43b), $(\mathrm{ad}\,\tilde{X})\xi$ and $\mathrm{Ad}(\tilde{k})\xi$ are meaningful for ξ in $\bigwedge^j(\mathfrak{s}/\mathfrak{m})$. First of all, $\mathrm{ad}\,\tilde{X}$ acts on $\mathfrak{s}$ since $\mathfrak{s}$ is an ideal in $\mathfrak{h}$. Also if Z is in $\mathfrak{m}$, then the identity $\mathrm{Ad}(m)\tilde{X} = \tilde{X}$ leads by differentiation to $[\tilde{X}, Z] = 0$. In other words, $\mathrm{ad}\,\tilde{X}$ acts by 0 on $\mathfrak{m}$. Hence the action descends to $\mathfrak{s}/\mathfrak{m}$, and then $\mathrm{ad}\,\tilde{X}$ acts on $\bigwedge^j(\mathfrak{s}/\mathfrak{m})$. Next, we have observed that $\mathrm{Ad}(L)$ carries $\mathfrak{s}$ to itself. Also M is normal in L, and consequently $\mathrm{Ad}(L)\mathfrak{m} \subseteq \mathfrak{m}$. Then it follows that $\mathrm{Ad}(\tilde{k})$ acts on $\bigwedge^j(\mathfrak{s}/\mathfrak{m})$.

2) This proposition reduces to Proposition 3.12 in the special case of a semidirect product mapping of pairs $(\mathfrak{q}, B) \to (\mathfrak{l}, B)$ with $\mathfrak{q} = \mathfrak{l} \oplus \mathfrak{u}$. In fact, we can put $\mathfrak{g} = \mathfrak{l}$, $\mathfrak{h} = \mathfrak{q}$, and $L = K = B$. The kernel $\mathfrak{s}$ of i_{alg} is $\mathfrak{u}$, and i_{gp} is the identity. Since i_{alg} is just the projection of $\mathfrak{q}$ on $\mathfrak{l}$ along $\mathfrak{u}$, we can take $X \mapsto \tilde{X}$ to be the identity map of $\mathfrak{l}$ to itself (realized as a

subalgebra of $\mathfrak{q}$). Also $k \mapsto \tilde{k}$ is the identity map on B. Then (3.42) and (3.43) are exactly the conclusions of Proposition 3.12.

3) The proof of Proposition 3.41 will occupy most of the remainder of this section. First we prove (3.42), then we prove (3.43) for the K actions, then we show how Proposition 3.41 implies Case (i) of Theorem 3.5b, and finally we prove (3.43a) and (3.43b) for the $\mathfrak{g}$ actions.

NOTATION IN THE PROOF. The $\mathrm{Ad}(L)$ invariant complement $\mathfrak{c}$ of $\mathfrak{l}$ in $\mathfrak{h}_K$, given in (3.3b), may be taken to be of a special form. In fact, since $\mathfrak{l}$ maps onto $\mathfrak{k}$, it follows that $\mathfrak{h}_K = \mathfrak{l} + \mathfrak{s}$. Here both $\mathfrak{l}$ and $\mathfrak{s}$ are invariant subspaces for $\mathrm{Ad}(L)$, and thus we can choose the $\mathrm{Ad}(L)$ invariant subspace $\mathfrak{c}$ so that $\mathfrak{c} \subseteq \mathfrak{s}$, as well as so that $\mathfrak{h}_K = \mathfrak{l} \oplus \mathfrak{c}$.

Since $\mathfrak{m} \subseteq \mathfrak{l}$, $\mathfrak{c} \cap \mathfrak{m} = 0$. On the other hand, $\mathfrak{s} \cap \mathfrak{l} = \mathfrak{m}$ implies that

$$\dim \mathfrak{c} + \dim \mathfrak{m} = \dim(\mathfrak{l} + \mathfrak{s}) - \dim \mathfrak{l} + \dim \mathfrak{m}$$
$$= \dim \mathfrak{s} - \dim(\mathfrak{s} \cap \mathfrak{l}) + \dim \mathfrak{m}$$
$$= \dim \mathfrak{s}.$$

Thus $\mathfrak{s} = \mathfrak{c} \oplus \mathfrak{m}$, and we see that $\mathfrak{c} \cong \mathfrak{s}/\mathfrak{m}$ as M modules.

PROOF OF (3.42). First let us show that

$$(3.44) \qquad \mathcal{F}_{\mathfrak{l},K}^{0,\{1\}} \circ (P_{\mathfrak{h}_K,L}^{\mathfrak{l},K})_j(V) \cong (P_{\mathfrak{s},M}^{0,\{1\}})_j \circ \mathcal{F}_{\mathfrak{h}_K,L}^{\mathfrak{s},M}(V).$$

According to (2.128), the left side is computed from a complex with terms

$$(3.45\text{a}) \qquad R(K) \otimes_L (\textstyle\bigwedge^j \mathfrak{c} \otimes_\mathbb{C} V),$$

while the right side is computed from a complex with terms

$$(3.45\text{b}) \qquad \textstyle\bigwedge^j \mathfrak{c} \otimes_M V.$$

We shall exhibit (3.45a) and (3.45b) as isomorphic by a map that commutes with ∂ and therefore passes to homology, and then (3.44) follows.

In fact, let us show for any L module W that

$$(3.45\text{c}) \qquad R(K) \otimes_L W \cong \mathbb{C} \otimes_M W.$$

We begin by making $\mathbb{C} \otimes_M W$ into a K module. The space $\mathbb{C} \otimes_M W$ may be identified with the subspace of M invariants of the L module $\mathbb{C} \otimes_\mathbb{C} W \cong W$, by Lemma 2.48. Since M is normal in L, $\mathbb{C} \otimes_M W$ is an L

invariant subspace for which the action passes to the quotient $L/M \cong K$.
We prove (3.45c) as an isomorphism of K modules.

If X is any finite-dimensional K module, we have

$$\mathrm{Hom}_K(R(K) \otimes_L W, X)$$
$$= \mathrm{Hom}_K(P_{\mathfrak{l},L}^{\mathfrak{k},K}(W), X)$$
$$\cong \mathrm{Hom}_L(W, \mathcal{F}_{\mathfrak{k},K}^{\mathfrak{l},L}(X)) \qquad \text{by Propositions 2.33 and 2.34}$$
$$\cong \mathrm{Hom}_L(\mathbb{C} \otimes_{\mathbb{C}} W, \mathcal{F}_{\mathfrak{k},K}^{\mathfrak{l},L}(X))$$
$$\cong \mathrm{Hom}_L(\mathbb{C} \otimes_M W, \mathcal{F}_{\mathfrak{k},K}^{\mathfrak{l},L}(X)) \qquad \begin{array}{l}\text{by composition with projection}\\ \text{on the trivial } M \text{ type}\end{array}$$
$$\cong \mathrm{Hom}_K(\mathbb{C} \otimes_M W, X).$$

This isomorphism extends to arbitrary X in $\mathcal{C}(\mathfrak{k}, K)$ and is natural in
X. Therefore (3.45c) follows from (C.22). From the proof of (C.22),
one sees that the map from right to left in (3.45c) is the member of
$\mathrm{Hom}_K(\mathbb{C} \otimes_M W, R(K) \otimes_L W)$ that corresponds to the identity map
on $R(K) \otimes_L W$. Unwinding our isomorphisms, we find that the map
is given by taking $1 \otimes w$ into $(i_{\mathrm{gp}})_*(\chi_A) \otimes w$, where χ_A is a member
of the approximate identity of $R(L)$ with $\chi_A w = w$. In the case that
$W = \bigwedge^j \mathfrak{c} \otimes_{\mathbb{C}} V$, we readily check that this map commutes with ∂, and
(3.44) follows.

Next we recall from Corollary 3.26c that

$$(3.46) \qquad \mathcal{F}_{\mathfrak{g},K}^{\mathfrak{k},K} \circ (P_{\mathfrak{h},L}^{\mathfrak{g},K})_j \cong (P_{\mathfrak{h}_K,L}^{\mathfrak{k},K})_j \circ \mathcal{F}_{\mathfrak{h},L}^{\mathfrak{h}_K,L}.$$

Therefore

$$\mathcal{F}_{\mathfrak{g},K}^{0,\{1\}} \circ (P_{\mathfrak{h},L}^{\mathfrak{g},K})_j \cong \mathcal{F}_{\mathfrak{k},K}^{0,\{1\}} \circ \mathcal{F}_{\mathfrak{g},K}^{\mathfrak{k},K} \circ (P_{\mathfrak{h},L}^{\mathfrak{g},K})_j$$
$$\cong \mathcal{F}_{\mathfrak{k},K}^{0,\{1\}} \circ (P_{\mathfrak{h}_K,L}^{\mathfrak{k},K})_j \circ \mathcal{F}_{\mathfrak{h},L}^{\mathfrak{h}_K,L} \qquad \text{by (3.46)}$$
$$\cong (P_{\mathfrak{s},M}^{0,\{1\}})_j \circ \mathcal{F}_{\mathfrak{h}_K,L}^{\mathfrak{s},M} \circ \mathcal{F}_{\mathfrak{h},L}^{\mathfrak{h}_K,L} \qquad \text{by (3.44)}$$
$$\cong (P_{\mathfrak{s},M}^{0,\{1\}})_j \circ \mathcal{F}_{\mathfrak{h},L}^{\mathfrak{s},M},$$

which is the desired isomorphism for P.

For the I functor, the identity that replaces (3.45c) is

$$\mathrm{Hom}_L(R(K), W)_K \cong \mathrm{Hom}_M(\mathbb{C}, W),$$

and we deduce from it a version of (3.44). Also Corollary 3.26c contains
a version of (3.46). Thus the above argument works also for the I functor.

PROOF OF (3.43) FOR K ACTIONS. According to Corollary 3.26c and (3.27), the K action on $(P_{\mathfrak{h},L}^{\mathfrak{g},K})_j(V)$ can be computed from the left K action on the complex whose j^{th} term is $R(K) \otimes_L (\bigwedge^j \mathfrak{c} \otimes_{\mathbb{C}} \mathcal{F}(V))$, with $\mathcal{F} = \mathcal{F}_{\mathfrak{h},L}^{\mathfrak{h}_K,L}$, and we saw in the previous part of the proof how to make K act compatibly on the isomorphic complex whose j^{th} term is $\bigwedge^j \mathfrak{c} \otimes_M \mathcal{F}(V)$. Namely the action is given by restricting the L action on $\bigwedge^j \mathfrak{c} \otimes_{\mathbb{C}} \mathcal{F}(V)$ to the subspace of M invariants, which we can identify with $\bigwedge^j \mathfrak{c} \otimes_M \mathcal{F}(V)$, and the action then descends to $K \cong L/M$. The element $k \in K$ therefore acts on $\bigwedge^j \mathfrak{c} \otimes_M \mathcal{F}(V)$ by the linear extension of $\xi \otimes v \mapsto \mathrm{Ad}(\tilde{k})\xi \otimes \tilde{k}v$, as required. This establishes the K action for $(P_{\mathfrak{h},L}^{\mathfrak{g},K})_j(V)$, and $(I_{\mathfrak{h},L}^{\mathfrak{g},K})^j(V)$ is handled similarly.

PROOF THAT PROPOSITION 3.41 IMPLIES CASE (i) OF THEOREM 3.5b. Let us apply (3.43a) with V replaced by $V \otimes_{\mathbb{C}} (\bigwedge^m \mathfrak{c})^*$. If ε is in $(\bigwedge^m \mathfrak{c})^*$, we obtain

$$X(\xi \otimes v \otimes \varepsilon) = (\mathrm{ad}\,\tilde{X})\xi \otimes v \otimes \varepsilon + \xi \otimes \tilde{X}v \otimes \varepsilon + \xi \otimes v \otimes \tilde{X}\varepsilon,$$

where $\tilde{X}\varepsilon$ refers to the action of $\mathfrak{h}$ on $(\bigwedge^m \mathfrak{c})^*$ given in Lemma 3.4. Since $\mathrm{ad}\,\tilde{X}$ is 0 on $\mathfrak{m}$, the action in that lemma matches the contragredient of the action of $\tilde{X}$ on $\bigwedge^m(\mathfrak{s}/\mathfrak{m})$ given in Remark 1. Thus we can write

$$X(\xi \otimes v \otimes \varepsilon) = \tilde{X}(\xi \otimes v \otimes \varepsilon),$$

and the meaning of the right side is unambiguous.

Referring to (3.31), we are to show that

(3.47) $$X\lambda_{f \otimes \xi \otimes v \otimes \varepsilon} = \lambda_{X(f \otimes \xi \otimes v \otimes \varepsilon)}.$$

The formulas (3.43) define actions of X on

$$\bigwedge^n \mathfrak{c} \otimes_{\mathbb{C}} V \otimes (\bigwedge^m \mathfrak{c})^* \qquad \text{and} \qquad \mathrm{Hom}_{\mathbb{C}}(\bigwedge^n \mathfrak{c}, V).$$

Before attempting to prove (3.47), we have to reinterpret these actions on the isomorphic spaces

(3.48a) $$R(K) \otimes_K (\bigwedge^n \mathfrak{c} \otimes_{\mathbb{C}} V \otimes_{\mathbb{C}} (\bigwedge^m \mathfrak{c})^*)$$

and

(3.48b) $$\mathrm{Hom}_K(R(K), \mathrm{Hom}_{\mathbb{C}}(\bigwedge^n \mathfrak{c}, V))_K.$$

For the action of X in (3.48a), we have

$$S \otimes \xi \otimes v \otimes \varepsilon \cong S(\xi \otimes v \otimes \varepsilon).$$

Thus we are to take

$$
\begin{aligned}
X(S \otimes \xi \otimes v \otimes \varepsilon) &\cong X(S(\xi \otimes v \otimes \varepsilon)) \\
&= X \int_K k(\xi \otimes v \otimes \varepsilon)\, dS(k) \\
&= \int_K \tilde{X} k(\xi \otimes v \otimes \varepsilon)\, dS(k) \\
&= \int_K k(\mathrm{Ad}(k)^{-1} \tilde{X})(\xi \otimes v \otimes \varepsilon)\, dS(k).
\end{aligned}
$$

In the usual way, let us write $\mathrm{Ad}(k)^{-1} \tilde{X} = \sum_i \langle \mathrm{Ad}(k)^{-1} \tilde{X}, \tilde{X}_i^* \rangle \tilde{X}_i$ and $S_i = \langle \mathrm{Ad}(\cdot)^{-1} \tilde{X}, \tilde{X}_i^* \rangle S$. Then the above is

$$
\begin{aligned}
&= \sum_i \int_K k \tilde{X}_i (\xi \otimes v \otimes \varepsilon)\, dS_i(k) \\
&= \sum_i S_i(\tilde{X}_i(\xi \otimes v \otimes \varepsilon)) \\
&\cong \sum_i S_i \otimes \tilde{X}_i(\xi \otimes v \otimes \varepsilon).
\end{aligned}
$$

So

(3.49) $$X(f\, dk \otimes \xi \otimes v \otimes \varepsilon) = \sum_i f_i\, dk \otimes \tilde{X}_i(\xi \otimes v \otimes \varepsilon),$$

where $f_i = \langle \mathrm{Ad}(\cdot)^{-1} \tilde{X}, \tilde{X}_i^* \rangle f$.

For the action of X in (3.48b), let $\lambda' \in \mathrm{Hom}_{\mathbb{C}}(\bigwedge^n \mathfrak{c}, V)$ be given. The corresponding member of (3.48b) is λ with

$$\lambda(T)(\gamma) \cong (T\lambda')(\gamma) = \int_K (k\lambda')(\gamma)\, dT(k) = \int_K k(\lambda'(k^{-1}\gamma))\, dT(k).$$

Then the member $X\lambda$ corresponding to $X\lambda'$ has

$$
\begin{aligned}
X\lambda(T)(\gamma) &= \int_K k(X\lambda'(k^{-1}\gamma))\, dT(k) \\
&= \int_K k\tilde{X}(\lambda'(k^{-1}\gamma))\, dT(k) - \int_K k(\lambda'(\tilde{X}k^{-1}\gamma))\, dT(k) \\
&= \int_K (\mathrm{Ad}(k)\tilde{X})(k(\lambda'(k^{-1}\gamma)))\, dT(k) - \int_K k(\lambda'(k^{-1}\mathrm{Ad}(k)\tilde{X}\gamma))\, dT(k) \\
&= \int_K (\mathrm{Ad}(k)\tilde{X})((k\lambda')(\gamma))\, dT(k) - \int_K (k\lambda')((\mathrm{Ad}(k)\tilde{X})\gamma)\, dT(k).
\end{aligned}
$$

Let us write $T_i = \langle \text{Ad}(\cdot)\tilde{X}, \tilde{X}_i^* \rangle T$ (with no inverse on $\text{Ad}(\cdot)$). Then the above is

$$= \sum_i \int_K \tilde{X}_i((k\lambda')(\gamma))\, dT_i(k) - \sum_i \int_K (k\lambda')(\tilde{X}_i\gamma)\, dT_i(k)$$

$$= \sum_i \tilde{X}_i((T_i\lambda')(\gamma)) - \sum_i (T_i\lambda')(\tilde{X}_i\gamma)$$

(3.50) $$= \sum_i \tilde{X}_i((\lambda(T_i))(\gamma)) - \sum_i \lambda(T_i)(\tilde{X}_i\gamma).$$

From (3.31b) we have

$$\lambda_{f\otimes\xi\otimes v\otimes\varepsilon}(T)(\gamma) = \varepsilon(\xi \wedge \gamma)\langle T, f^t \rangle v.$$

Therefore (3.49) gives

$$\lambda_{X(f\otimes\xi\otimes v\otimes\varepsilon)}(T)(\gamma) = \sum_i \lambda_{f_i\otimes\tilde{X}_i(\xi\otimes v\otimes\varepsilon)}(T)(\gamma)$$

$$= \sum_i \varepsilon(\tilde{X}_i\xi \wedge \gamma)\langle T, f_i^t \rangle v$$

$$+ \sum_i \varepsilon(\xi \wedge \gamma)\langle T, f_i^t \rangle \tilde{X}_i v$$

$$+ \sum_i (\tilde{X}_i\varepsilon)(\xi \wedge \gamma)\langle T, f_i^t \rangle v.$$

Here $f_i^t = \langle \text{Ad}(\cdot)\tilde{X}, \tilde{X}_i^* \rangle f^t$, and $\langle T, f_i^t \rangle = \langle T_i, f^t \rangle$. Also $(\tilde{X}_i\varepsilon)(\xi \wedge \gamma) = -\varepsilon(\tilde{X}_i\xi \wedge \gamma) - \varepsilon(\xi \wedge \tilde{X}_i\gamma)$. Thus the above expression is

(3.51) $$= \sum_i \langle T_i, f^t \rangle \varepsilon(\xi \wedge \gamma)\tilde{X}_i v - \sum_i \langle T_i, f^t \rangle \varepsilon(\xi \wedge \tilde{X}_i\gamma)v.$$

On the other hand, (3.50) gives

$$X\lambda_{f\otimes\xi\otimes v\otimes\varepsilon}(T)(\gamma)$$

$$= \sum_i \tilde{X}_i((\lambda_{f\otimes\xi\otimes v\otimes\varepsilon}(T_i))(\gamma)) - \sum_i \lambda_{f\otimes\xi\otimes v\otimes\varepsilon}(T_i)(\tilde{X}_i\gamma)$$

(3.52) $$= \sum_i \varepsilon(\xi \wedge \gamma)\langle T_i, f^t \rangle \tilde{X}_i v - \sum_i \varepsilon(\xi \wedge \tilde{X}_i\gamma)\langle T_i, f^t \rangle v.$$

The equality of (3.51) and (3.52) for all T and γ means that (3.47) holds. This completes the proof that Proposition 3.41 implies Case (i) of Theorem 3.5b.

PROOF OF (3.43a) FOR $\mathfrak{g}$ ACTION. First we suppose that $K = \{1\}$ and $L = M$. Let $X_j^{\mathfrak{h},M} = U(\mathfrak{h}) \otimes_{U(\mathfrak{m})} \bigwedge^j(\mathfrak{h}/\mathfrak{m})$ be the jth term (2.120) of the Koszul resolution (2.121) in $\mathcal{C}(\mathfrak{h}, M)$. Each $X_j^{\mathfrak{h},M}$ is an $(\mathfrak{h}, M)$ module with $\mathfrak{h}$ acting only in the $U(\mathfrak{h})$ factor and with M acting by $\text{Ad} \otimes \text{Ad}$, and V is of course an $(\mathfrak{h}, M)$ module. We make $U(\mathfrak{g})$ into an $(\mathfrak{h}, M)$ module by the definitions

$$m(u) = u \qquad \text{and} \qquad H(u) = -u(i_{\text{alg}}(H))$$

for $m \in M$, $H \in \mathfrak{h}$, and $u \in U(\mathfrak{g})$. To see that $U(\mathfrak{g})$ is indeed an $(\mathfrak{h}, M)$ module, we observe that the differential of the M action is 0, which coincides with the $\mathfrak{m}$ action inherited from $\mathfrak{h}$ since $i_{\text{alg}}(\mathfrak{m}) = 0$; also

$$(\text{Ad}(m)H)(mu) = (\text{Ad}(m)H)u = -u(i_{\text{alg}}(\text{Ad}(m)H))$$
$$= -u(i_{\text{alg}}(H)) = H(u) = m(H(u)).$$

Consequently $U(\mathfrak{g}) \otimes X_j^{\mathfrak{h},M} \otimes V$ is a well-defined $(\mathfrak{h}, M)$ module.

We form the diagram

(3.53)
$$0 \leftarrow U(\mathfrak{g}) \otimes V \leftarrow U(\mathfrak{g}) \otimes X_0^{\mathfrak{h},M} \otimes V \leftarrow U(\mathfrak{g}) \otimes X_1^{\mathfrak{h},M} \otimes V \leftarrow \cdots$$

$$\quad 1 \Big\downarrow 0 \qquad\qquad \tilde{\psi}_0 \Big\downarrow \tilde{\varphi}_0 \qquad\qquad \tilde{\psi}_1 \Big\downarrow \tilde{\varphi}_1$$

$$0 \leftarrow U(\mathfrak{g}) \otimes V \leftarrow U(\mathfrak{g}) \otimes X_0^{\mathfrak{h},M} \otimes V \leftarrow U(\mathfrak{g}) \otimes X_1^{\mathfrak{h},M} \otimes V \leftarrow \cdots$$

in $\mathcal{C}(\mathfrak{h}, M)$. The rows are the standard projective resolutions (2.123) of $U(\mathfrak{g}) \otimes_{\mathbb{C}} V$ in $\mathcal{C}(\mathfrak{h}, M)$. We regard $X \in \mathfrak{g}$ as fixed, and we take $\tilde{\psi}_j$ and $\tilde{\varphi}_j$ to be the $(\mathfrak{h}, M)$ maps defined by the linear extensions of

$$\tilde{\psi}_j(u_{\mathfrak{g}} \otimes u_{\mathfrak{h}} \otimes \xi \otimes v) = Xu_{\mathfrak{g}} \otimes u_{\mathfrak{h}} \otimes \xi \otimes v$$
(3.54)
$$\tilde{\varphi}_j(u_{\mathfrak{g}} \otimes u_{\mathfrak{h}} \otimes \xi \otimes v) = -u_{\mathfrak{g}} \otimes u_{\mathfrak{h}} \tilde{X} \otimes \xi \otimes v + u_{\mathfrak{g}} \otimes u_{\mathfrak{h}} \otimes (\text{ad}\,\tilde{X})\xi \otimes v$$

for $u_{\mathfrak{g}} \in U(\mathfrak{g})$, $u_{\mathfrak{h}} \in U(\mathfrak{h})$, $\xi \in \bigwedge^j(\mathfrak{h}/\mathfrak{m})$, and $v \in V$. It is easy to see that the system $\tilde{\psi}_j$ is a chain map covering the $(\mathfrak{h}, M)$ map $u_{\mathfrak{g}} \otimes v \mapsto Xu_{\mathfrak{g}} \otimes v$ of $U(\mathfrak{g}) \otimes_{\mathbb{C}} V$ to itself. A little computation shows that the system $\tilde{\varphi}_j$ is a chain map, and it is clear that it covers the 0 map on $U(\mathfrak{g}) \otimes_{\mathbb{C}} V$. Consequently the systems $\tilde{\psi}_j$ and $\tilde{\psi}_j + \tilde{\varphi}_j$ are chain maps covering the $(\mathfrak{h}, M)$ map $u_{\mathfrak{g}} \otimes v \mapsto Xu_{\mathfrak{g}} \otimes v$ of $U(\mathfrak{g}) \otimes_{\mathbb{C}} V$ to itself, and (C.6) shows that $\tilde{\psi}_j$ and $\tilde{\psi}_j + \tilde{\varphi}_j$ are homotopic in $\mathcal{C}(\mathfrak{h}, M)$.

To the diagram (3.53), we apply the coinvariants functor $(\cdot)_{\mathfrak{h},M}$ and drop the effect on $U(\mathfrak{g}) \otimes V$, obtaining a diagram in $\mathcal{C}(0)$. Then $U(\mathfrak{g}) \otimes X_j^{\mathfrak{h},M} \otimes V$ gets replaced by

(3.55)
$$
\begin{aligned}
(U(\mathfrak{g}) \otimes X_j^{\mathfrak{h},M} &\otimes V)_{\mathfrak{h},M} \\
&\cong ((U(\mathfrak{g}) \otimes X_j^{\mathfrak{h},M} \otimes V)_{\mathfrak{h}})_M && \text{by (2.15b)} \\
&\cong ((U(\mathfrak{g}) \otimes V) \otimes_{U(\mathfrak{h})} X_j^{\mathfrak{h},M})_M && \text{by Lemma 2.48} \\
&= ((U(\mathfrak{g}) \otimes V) \otimes_{U(\mathfrak{h})} U(\mathfrak{h}) \otimes_{U(\mathfrak{m})} {\textstyle\bigwedge}^j(\mathfrak{h}/\mathfrak{m}))_M && \text{by (A.7)} \\
&\cong ((U(\mathfrak{g}) \otimes V) \otimes_{U(\mathfrak{m})} {\textstyle\bigwedge}^j(\mathfrak{h}/\mathfrak{m}))_M \\
&\cong (U(\mathfrak{g}) \otimes V) \otimes_M {\textstyle\bigwedge}^j(\mathfrak{h}/\mathfrak{m}) \\
&\cong U(\mathfrak{g}) \otimes ({\textstyle\bigwedge}^j(\mathfrak{h}/\mathfrak{m}) \otimes_M V) && \text{since M acts trivially} \\
& && \text{on $U(\mathfrak{g})$.}
\end{aligned}
$$

If we write ψ_j and φ_j for the maps that replace $\tilde{\psi}_j$ and $\tilde{\varphi}_j$, then the diagram in $\mathcal{C}(0)$ is

(3.56)
$$
0 \leftarrow U(\mathfrak{g}) \otimes ({\textstyle\bigwedge}^0(\mathfrak{h}/\mathfrak{m}) \otimes_M V) \leftarrow U(\mathfrak{g}) \otimes ({\textstyle\bigwedge}^1(\mathfrak{h}/\mathfrak{m}) \otimes_M V) \leftarrow \cdots
$$
$$
\qquad\qquad \psi_0 \downarrow \varphi_0 \qquad\qquad\qquad\qquad \psi_1 \downarrow \varphi_1
$$
$$
0 \leftarrow U(\mathfrak{g}) \otimes ({\textstyle\bigwedge}^0(\mathfrak{h}/\mathfrak{m}) \otimes_M V) \leftarrow U(\mathfrak{g}) \otimes ({\textstyle\bigwedge}^1(\mathfrak{h}/\mathfrak{m}) \otimes_M V) \leftarrow \cdots
$$

Under the isomorphisms (3.55), linear combinations of elements $u_{\mathfrak{g}} \otimes (1 \otimes \xi) \otimes v$ that are in $(U(\mathfrak{g}) \otimes X_j^{\mathfrak{h},M} \otimes V)_{\mathfrak{h},M}$ map by

$$
\begin{aligned}
u_{\mathfrak{g}} \otimes (1 \otimes \xi) \otimes v &\cong u_{\mathfrak{g}} \otimes (1 \otimes \xi) \otimes v \\
&\cong u_{\mathfrak{g}} \otimes v \otimes (1 \otimes \xi) \\
&\cong u_{\mathfrak{g}} \otimes v \otimes 1 \otimes \xi \\
&\cong u_{\mathfrak{g}} \otimes v \otimes \xi \\
&\cong u_{\mathfrak{g}} \otimes v \otimes \xi \\
&\cong u_{\mathfrak{g}} \otimes \xi \otimes v
\end{aligned}
$$

onto $U(\mathfrak{g}) \otimes ({\textstyle\bigwedge}^j(\mathfrak{h}/\mathfrak{m}) \otimes_M V)$. We apply $\tilde{\varphi}_j$ to the lift of this result and track down what happens as we retrace (3.55) from left to right, skipping trivial steps:

$$u_\mathfrak{g} \otimes (1 \otimes \xi) \otimes v \mapsto -u_\mathfrak{g} \otimes (\tilde{X} \otimes \xi) \otimes v + u_\mathfrak{g} \otimes (1 \otimes (\operatorname{ad} \tilde{X})\xi) \otimes v$$
$$\cong -u_\mathfrak{g} \otimes v \otimes \tilde{X} \otimes \xi + u_\mathfrak{g} \otimes v \otimes 1 \otimes (\operatorname{ad} \tilde{X})\xi$$
$$\cong \tilde{X}(u_\mathfrak{g} \otimes v) \otimes \xi + u_\mathfrak{g} \otimes v \otimes (\operatorname{ad} \tilde{X})\xi$$
$$\cong -u_\mathfrak{g} X \otimes v \otimes \xi + u_\mathfrak{g} \otimes \tilde{X}v \otimes \xi + u_\mathfrak{g} \otimes v \otimes (\operatorname{ad} \tilde{X})\xi$$
$$\cong -u_\mathfrak{g} X \otimes \xi \otimes v + u_\mathfrak{g} \otimes \xi \otimes \tilde{X}v + u_\mathfrak{g} \otimes (\operatorname{ad} \tilde{X})\xi \otimes v.$$

Thus the map φ_j in (3.56) is given by

$$\varphi_j(u_\mathfrak{g} \otimes \xi \otimes v) = -u_\mathfrak{g} X \otimes \xi \otimes v + u_\mathfrak{g} \otimes \xi \otimes \tilde{X}v + u_\mathfrak{g} \otimes (\operatorname{ad} \tilde{X})\xi \otimes v.$$

Similarly we track down the effect of ψ_j and find that it just gives left multiplication by X in the $U(\mathfrak{g})$ factor:

$$\psi_j(u_\mathfrak{g} \otimes \xi \otimes v) = X u_\mathfrak{g} \otimes \xi \otimes v.$$

Therefore

(3.57) $(\psi_j + \varphi_j)(u_\mathfrak{g} \otimes \xi \otimes v)$
$$= (\operatorname{ad} X)u_\mathfrak{g} \otimes \xi \otimes v + u_\mathfrak{g} \otimes (\operatorname{ad} \tilde{X})\xi \otimes v + u_\mathfrak{g} \otimes \xi \otimes \tilde{X}v.$$

Since the systems $\tilde{\psi}_j$ and $\tilde{\psi}_j + \tilde{\varphi}_j$ are homotopic in the $C(\mathfrak{h}, M)$ diagram (3.53), the systems ψ_j and $\psi_j + \varphi_j$ are homotopic in the $C(0)$ diagram (3.56).

Meanwhile, $(P_{\mathfrak{h},M}^{\mathfrak{g},\{1\}})_j(V)$ is computed from the complex

$$P_{\mathfrak{h},M}^{\mathfrak{g},\{1\}}(X_j^{\mathfrak{h},M} \otimes V) = U(\mathfrak{g}) \otimes_{R(\mathfrak{h},M)} (X_j^{\mathfrak{h},M} \otimes V)$$
(3.58)
$$\cong (U(\mathfrak{g}) \otimes X_j^{\mathfrak{h},M} \otimes V)_{\mathfrak{h},M}$$
$$\cong U(\mathfrak{g}) \otimes (\textstyle\bigwedge^j(\mathfrak{h}/\mathfrak{m}) \otimes_M V) \qquad \text{by (3.55)},$$

and the action of $X \in \mathfrak{g}$ on the left side carries over to the map ψ_j on the right side. Because of the homotopy, (C.4) shows that the action by $X \in \mathfrak{g}$ on $(P_{\mathfrak{h},\{1\}}^{\mathfrak{g},\{1\}})_j(V)$ is induced in (3.55) at the level of complexes also by $\psi_j + \varphi_j$.

Remembering that $\mathfrak{s}$ is the kernel of i_{alg}, let

$$X_j^{\mathfrak{s},M} = U(\mathfrak{s}) \otimes_{U(\mathfrak{m})} \textstyle\bigwedge^j(\mathfrak{s}/\mathfrak{m})$$

be the jth term of the Koszul resolution in $C(\mathfrak{s}, M)$. Also let $\tilde{I}_j : X_j^{\mathfrak{s},M} \to X_j^{\mathfrak{h},M}$ be given by inclusion in each factor. Then we obtain the diagram

(3.59)
$$
\begin{array}{ccccccc}
0 & \leftarrow & V & \leftarrow & X_0^{\mathfrak{s},M} \otimes V & \leftarrow & X_1^{\mathfrak{s},M} \otimes V & \leftarrow & \cdots \\
& & \downarrow^1 & & \downarrow^{\tilde{I}_0 \otimes 1} & & \downarrow^{\tilde{I}_1 \otimes 1} & & \\
0 & \leftarrow & V & \leftarrow & X_0^{\mathfrak{h},M} \otimes V & \leftarrow & X_1^{\mathfrak{h},M} \otimes V & \leftarrow & \cdots
\end{array}
$$

in $C(\mathfrak{s}, M)$; here the top row is the standard projective resolution of V in $C(\mathfrak{s}, M)$, and the bottom row is the standard projective resolution of V in $C(\mathfrak{h}, M)$. The latter is also a projective resolution of V in $C(\mathfrak{s}, M)$. In fact, it remains exact in $C(\mathfrak{s}, M)$. To see that it contains $C(\mathfrak{s}, M)$ projectives, let us write $\mathfrak{h} = \mathfrak{s} \oplus \mathfrak{b}$, where $\mathfrak{b}$ is an $\text{Ad}(L)$ invariant subspace of $\mathfrak{h}$. Then we have isomorphisms

$$
\begin{aligned}
X_j^{\mathfrak{h},M} &\cong R(\mathfrak{h}, M) \otimes_M \textstyle\bigwedge^j (\mathfrak{h}/\mathfrak{m}) & \text{by (2.120)} \\
&\cong (R(\mathfrak{s}, M) \otimes_{\mathbb{C}} S(\mathfrak{b})) \otimes_M \textstyle\bigwedge^j (\mathfrak{h}/\mathfrak{m}) & \text{by Lemma 2.56} \\
&\cong R(\mathfrak{s}, M) \otimes_M (S(\mathfrak{b}) \otimes_{\mathbb{C}} \textstyle\bigwedge^j (\mathfrak{h}/\mathfrak{m})) & \text{by Corollary 2.52}
\end{aligned}
$$

as $R(\mathfrak{s}, M)$ modules. Hence Corollary 2.36 shows that $X_j^{\mathfrak{h},M}$ is projective in $\mathbb{C}(\mathfrak{s}, M)$, and Proposition 2.53a shows that the bottom row of (3.59) is a projective resolution in $C(\mathfrak{s}, M)$. Inspection of the definitions shows that (3.59) commutes, i.e., that the system $\tilde{I}_j \otimes 1$ is a chain map over the identity on V. By (C.6), there exist vertical $C(\mathfrak{s}, M)$ maps

(3.60)
$$
\begin{array}{c}
X_j^{\mathfrak{s}} \otimes V \\
\tilde{\alpha}_j \uparrow \\
X_j^{\mathfrak{h}} \otimes V
\end{array}
$$

that are inverses up to homotopy in $C(\mathfrak{s}, M)$.

We apply $P_{\mathfrak{s},M}^{0,\{1\}}$ to (3.59), dropping the effect on V at the left. The modules $X_j^{\mathfrak{s},M} \otimes_{\mathbb{C}} V$ become $\bigwedge^j (\mathfrak{s}/\mathfrak{m}) \otimes_M V$. On the bottom row, we have

$$
\begin{aligned}
P_{\mathfrak{s},M}^{0,\{1\}} \mathcal{F}_{\mathfrak{h},M}^{\mathfrak{s},M} & (X_j^{\mathfrak{h},M} \otimes_{\mathbb{C}} V) \\
&\cong \mathcal{F}_{\mathfrak{g},\{1\}}^{0,\{1\}} P_{\mathfrak{h},M}^{\mathfrak{g},\{1\}} (X_j^{\mathfrak{h},M} \otimes_{\mathbb{C}} V) & \text{by Corollary 3.26b} \\
&\cong \mathcal{F}_{\mathfrak{g},\{1\}}^{0,\{1\}} (U(\mathfrak{g}) \otimes_{\mathbb{C}} (\textstyle\bigwedge^j (\mathfrak{h}/\mathfrak{m}) \otimes_M V)) & \text{by (3.58).}
\end{aligned}
$$

Dropping forgetful functors, we obtain the commutative diagram

(3.61)

$$0 \leftarrow \qquad {\textstyle\bigwedge}^0(\mathfrak{s}/\mathfrak{m}) \otimes_M V \qquad \leftarrow \qquad {\textstyle\bigwedge}^1(\mathfrak{s}/\mathfrak{m}) \otimes_M V \qquad \leftarrow \cdots$$

$$\downarrow I_0 \qquad\qquad\qquad\qquad \downarrow I_1$$

$$0 \leftarrow U(\mathfrak{g}) \otimes ({\textstyle\bigwedge}^0(\mathfrak{h}/\mathfrak{m}) \otimes_M V) \leftarrow U(\mathfrak{g}) \otimes ({\textstyle\bigwedge}^1(\mathfrak{h}/\mathfrak{m}) \otimes_M V) \leftarrow \cdots$$

in $\mathcal{C}(0)$. A little check shows that the vertical maps I_j satisfy

$$I_j(\xi \otimes v) = 1 \otimes \xi \otimes v.$$

Also the maps $\tilde{\alpha}_j$ of (3.60) descend to give to a system of upward-pointing vertical maps that invert the system of maps I_j up to homotopy. Hence the maps I_j induce isomorphisms on homology.

Define η_j from $\bigwedge^j(\mathfrak{s}/\mathfrak{m}) \otimes_M V$ to itself by the linear extension of

$$(3.62) \qquad\qquad \eta_j(\xi \otimes v) = (\mathrm{ad}\,\tilde{X})\xi \otimes v + \xi \otimes \tilde{X}v.$$

We have seen that the action by $X \in \mathfrak{g}$ on $(P_{\mathfrak{h},M}^{\mathfrak{g},\{1\}})_j(V)$ is given in terms of the complex in the bottom row of (3.61) by $\psi_j + \varphi_j$. Since

$$(3.63) \qquad\qquad (\psi_j + \varphi_j) \circ I_j = I_j \circ \eta_j,$$

it follows that the action by $X \in \mathfrak{g}$ on $(P_{\mathfrak{h},M}^{\mathfrak{g},\{1\}})_j(V)$ is given in terms of the complex in the top row of (3.61) by η_j. This proves the formula for the $\mathfrak{g}$ action by X in (3.43a), provided $K = \{1\}$ and $L = M$.

Now suppose that K and L are general in (3.43a), still subject to the condition $i_{\mathrm{gp}}(L) = K$. The difficulty with imitating the above proof is that there is an obstruction to defining maps φ_j satisfying the analog of (3.63). Instead we shall obtain (3.43a) for general K as a consequence of the result for $K = \{1\}$.

Let

$$X_j^{\mathfrak{h},L} = R(\mathfrak{h}, L) \otimes_L {\textstyle\bigwedge}^j(\mathfrak{h}/\mathfrak{l}) \cong U(\mathfrak{h}) \otimes_{U(\mathfrak{l})} {\textstyle\bigwedge}^j(\mathfrak{h}/\mathfrak{l})$$

be the jth term of the Koszul resolution in $\mathcal{C}(\mathfrak{h}, L)$. If we write

$$\mathfrak{h} = \mathfrak{h}_K \oplus \mathfrak{b} = \mathfrak{l} \oplus \mathfrak{c} \oplus \mathfrak{b}$$

with each summand $\mathrm{Ad}(L)$ invariant, then we see that

$$\mathfrak{h}/\mathfrak{l} \cong \mathfrak{c} \oplus \mathfrak{b}$$

as L modules. Hence we can replace $\bigwedge^j(\mathfrak{h}/\mathfrak{l})$ by $\bigwedge^j(\mathfrak{c} \oplus \mathfrak{b})$ in $X_j^{\mathfrak{h},L}$ whenever we want. We form the diagram

(3.64)

$$
\begin{array}{ccc}
X_j^{\mathfrak{s},M} & & \\
\tilde{\iota}_j \downarrow & \searrow & \\
X_j^{\mathfrak{h},M} & \xrightarrow{\tilde{Q}_j} & X_j^{\mathfrak{h},L}
\end{array}
$$

in $C(\mathfrak{s}, M)$, in which the horizontal map is passage to the quotient and the diagonal map is "inclusion" followed by passage to the quotient. This diagram commutes, and each system of maps as a function of j is a chain map over the identity on $\mathbb{C}$.

We saw earlier that $X_j^{\mathfrak{h},M}$ is projective in $C(\mathfrak{s}, M)$. Let us prove that $X_j^{\mathfrak{h},L}$ is projective in $C(\mathfrak{s}, M)$. To do so, we begin by proving that

$$U(\mathfrak{h}_K) \cong U(\mathfrak{s}) \otimes_{U(\mathrm{m})} U(\mathfrak{l})$$

as left $U(\mathfrak{s})$ modules and right $U(\mathfrak{l})$ modules. In fact, if we recall that $\mathfrak{h}_K = \mathfrak{s} + \mathfrak{l}$ with $\mathfrak{s} \cap \mathfrak{l} = \mathrm{m}$, then we see that the multiplication map from $U(\mathfrak{s}) \times U(\mathfrak{l})$ to $U(\mathfrak{h}_K)$ extends to a linear map of $U(\mathfrak{s}) \otimes_{\mathbb{C}} U(\mathfrak{l})$ into $U(\mathfrak{h}_K)$ that descends to a linear map of $U(\mathfrak{s}) \otimes_{U(\mathrm{m})} U(\mathfrak{l})$ into $U(\mathfrak{h}_K)$. This linear map is onto by the existence half of the Poincaré-Birkhoff-Witt Theorem since $\mathfrak{h}_K = \mathfrak{l} + \mathfrak{s}$. The map respects the left $U(\mathfrak{s})$ modules structures and the right $U(\mathfrak{l})$ modules structures. To see that it is one-one, we write $\mathfrak{s} = \mathfrak{c} \oplus \mathrm{m}$ and $\mathfrak{l} = \mathrm{m} \oplus \mathfrak{l}'$ for some subspace $\mathfrak{l}'$. Let $u_{\mathfrak{c}}$, u_{m} or u'_{m}, and $u_{\mathfrak{l}'}$ be monomials built from ordered bases of $\mathfrak{c}$, m, and $\mathfrak{l}'$, respectively. If a linear combination of terms $u_{\mathfrak{c}} u_{\mathrm{m}} \otimes u'_{\mathrm{m}} u_{\mathfrak{l}'}$ maps to 0, then the same linear combination of $u_{\mathfrak{c}} u_{\mathrm{m}} u'_{\mathrm{m}} \otimes u_{\mathfrak{l}'}$ maps to 0, and the uniqueness half of the Poincaré-Birkhoff-Witt Theorem for $U(\mathfrak{h}_K)$ shows that the linear combination itself is 0. This proves the isomorphism for the universal enveloping algebras. Bringing in the compact groups, we then have

$$R(\mathfrak{h}_K, L) \cong U(\mathfrak{h}_K) \otimes_{U(\mathfrak{l})} R(L) \qquad \text{as left } (\mathfrak{h}_K, L) \text{ modules}$$
$$\text{and right } L \text{ modules}$$

$$\cong U(\mathfrak{s}) \otimes_{U(\mathrm{m})} U(\mathfrak{l}) \otimes_{U(\mathfrak{l})} R(L) \qquad \text{from above}$$

$$\cong U(\mathfrak{s}) \otimes_{U(\mathrm{m})} R(L) \qquad \text{as left } (\mathfrak{s}, M) \text{ modules}$$
$$\text{and right } L \text{ modules}$$

$$\cong U(\mathfrak{s}) \otimes_{U(\mathrm{m})} R(M) \otimes_M R(L)$$

$$\cong R(\mathfrak{s}, M) \otimes_M R(L).$$

Finally we have $(\mathfrak{s}, M)$ isomorphisms

$$
\begin{aligned}
X_j^{\mathfrak{h},L} &= R(\mathfrak{h}, L) \otimes_L \textstyle\bigwedge^j(\mathfrak{c} \oplus \mathfrak{b}) \\
&\cong (R(\mathfrak{h}_K, L) \otimes_{\mathbb{C}} S(\mathfrak{b})) \otimes_L \textstyle\bigwedge^j(\mathfrak{c} \oplus \mathfrak{b}) && \text{by Lemma 2.56} \\
&\cong R(\mathfrak{h}_K, L) \otimes_L (S(\mathfrak{b}) \otimes_{\mathbb{C}} \textstyle\bigwedge^j(\mathfrak{c} \oplus \mathfrak{b})) && \text{by Corollary 2.52} \\
&\cong R(\mathfrak{s}, M) \otimes_M R(L) \otimes_L (S(\mathfrak{b}) \otimes_{\mathbb{C}} \textstyle\bigwedge^j(\mathfrak{c} \oplus \mathfrak{b})) && \text{from above} \\
&\cong R(\mathfrak{s}, M) \otimes_M (S(\mathfrak{b}) \otimes_{\mathbb{C}} \textstyle\bigwedge^j(\mathfrak{c} \oplus \mathfrak{b})) \\
&= P_{\mathfrak{m},M}^{\mathfrak{s},M}(S(\mathfrak{b}) \otimes_{\mathbb{C}} \textstyle\bigwedge^j(\mathfrak{c} \oplus \mathfrak{b})),
\end{aligned}
$$

and Corollary 2.36 shows that $X_j^{\mathfrak{h},L}$ is projective in $\mathcal{C}(\mathfrak{s}, M)$.

Thus $X_j^{\mathfrak{s},M}$, $X_j^{\mathfrak{h},M}$, and $X_j^{\mathfrak{h},L}$ are all projective in $\mathcal{C}(\mathfrak{s}, M)$. Tensoring (3.64) with V over $\mathbb{C}$, we obtain a commutative diagram

(3.65)

$$
\begin{array}{ccc}
X_j^{\mathfrak{s},M} \otimes_{\mathbb{C}} V & & \\
{\scriptstyle \tilde{I}_j \otimes 1}\Big\downarrow & \searrow & \\
X_j^{\mathfrak{h},M} \otimes_{\mathbb{C}} V & \xrightarrow[{\scriptstyle \tilde{Q}_j \otimes 1}]{} & X_j^{\mathfrak{h},L} \otimes_{\mathbb{C}} V
\end{array}
$$

in $\mathcal{C}(\mathfrak{s}, M)$ whose modules, as j varies, are projective resolutions of V and whose maps, as j varies, are chain maps over the identity on V. By (C.6), there exist chain maps in the inverse directions that are inverse maps up to homotopy in $\mathcal{C}(\mathfrak{s}, M)$.

Shortly we shall apply $P_{\mathfrak{s},M}^{0,\{1\}}$ to the system of diagrams (3.65). But first we need to know the isomorphism

$$
(3.66) \qquad P_{\mathfrak{h},L}^{\mathfrak{g},K}(X_j^{\mathfrak{h},L} \otimes_{\mathbb{C}} V) \cong R(\mathfrak{g}, K) \otimes_L (\textstyle\bigwedge^j(\mathfrak{h}/\mathfrak{l}) \otimes_{\mathbb{C}} V),
$$

which is an instance of the computation preceding (2.128a). The action by X on $(P_{\mathfrak{h},K}^{\mathfrak{g},K})_j(V)$ is induced on the level of complexes by left-by-X on the left side of this isomorphism and corresponds to the $\mathcal{C}(0)$ map

$$
(3.67) \qquad \psi_j'((u_{\mathfrak{g}} \otimes T) \otimes \xi \otimes v) = (X u_{\mathfrak{g}} \otimes T) \otimes \xi \otimes v
$$

on the right side. Also we have

$$
\begin{aligned}
P_{\mathfrak{s},M}^{0,\{1\}} \mathcal{F}_{\mathfrak{h},L}^{\mathfrak{s},M}&(X_j^{\mathfrak{h},L} \otimes_{\mathbb{C}} V) \\
&\cong \mathcal{F}_{\mathfrak{g},K}^{0,\{1\}} P_{\mathfrak{h},L}^{\mathfrak{g},K}(X_j^{\mathfrak{h},L} \otimes_{\mathbb{C}} V) && \text{by (3.42)} \\
&\cong \mathcal{F}_{\mathfrak{g},K}^{0,\{1\}}(R(\mathfrak{g}, K) \otimes_L (\textstyle\bigwedge^j(\mathfrak{h}/\mathfrak{l}) \otimes_{\mathbb{C}} V)) && \text{by (3.66).}
\end{aligned}
$$

Thus, if we drop forgetful functors and apply $P^{0,\{1\}}_{\mathfrak{s},M}$ to the system of diagrams (3.65), we obtain the system of commutative diagrams

(3.68)

$$
\begin{array}{c}
\bigwedge^j(\mathfrak{s}/\mathfrak{m}) \otimes_M V \\
\scriptstyle{I_j}\big\downarrow \qquad\qquad\qquad\searrow \\
U(\mathfrak{g}) \otimes_{\mathbb{C}} (\bigwedge^j(\mathfrak{h}/\mathfrak{m}) \otimes_M V) \xrightarrow[\;\;Q_j\;\;]{} R(\mathfrak{g}, K) \otimes_L (\bigwedge^j(\mathfrak{h}/\mathfrak{l}) \otimes_{\mathbb{C}} V)
\end{array}
$$

in which each system of maps is a chain map in $\mathcal{C}(0)$ with an inverse up to homotopy.

In homology each of the maps in (3.68) becomes a $\mathcal{C}(0)$ isomorphism. We have seen that the system of maps η_j in (3.62) of $\bigwedge^j(\mathfrak{s}/\mathfrak{m}) \otimes_M V$ to itself corresponds to the system $\psi_j + \varphi_j$ of maps of $U(\mathfrak{g}) \otimes_{\mathbb{C}} (\bigwedge^j(\mathfrak{h}/\mathfrak{m}) \otimes_M V)$ to itself. This in turn is homotopic to the system ψ_j of maps of $U(\mathfrak{g}) \otimes_{\mathbb{C}} (\bigwedge^j(\mathfrak{h}/\mathfrak{m}) \otimes_M V)$ to itself, which corresponds to the system ψ'_j of maps of $R(\mathfrak{g}, K) \otimes_L (\bigwedge^j(\mathfrak{h}/\mathfrak{l}) \otimes_{\mathbb{C}} V)$ to itself given in (3.67). By (C.4), under the isomorphism on homology from $\bigwedge^j(\mathfrak{s}/\mathfrak{m}) \otimes_M V$ to homology from $R(\mathfrak{g}, K) \otimes_L (\bigwedge^j(\mathfrak{h}/\mathfrak{l}) \otimes_{\mathbb{C}} V)$, the effect of η_j corresponds to the effect of ψ'_j. The latter is the action by X on $(P^{\mathfrak{g},K}_{\mathfrak{h},K})_j(V)$, and the formula in (3.43a) for the action by X follows.

PROOF OF (3.43b) FOR $\mathfrak{g}$ ACTION. Again we start with the special case that $K = \{1\}$ and $L = M$. The diagram (3.53) is replaced by

(3.69)

$$
\begin{array}{ccccccc}
0 & \to & \mathrm{Hom}_{\mathbb{C}}(U(\mathfrak{g}), V) & \to & \mathrm{Hom}_{\mathbb{C}}(U(\mathfrak{g}) \otimes X^{\mathfrak{h},M}_0, V) & \to & \mathrm{Hom}_{\mathbb{C}}(U(\mathfrak{g}) \otimes X^{\mathfrak{h},M}_1, V) \\
 & & \scriptstyle{1}\big\downarrow\scriptstyle{0} & & \scriptstyle{\tilde{\psi}_0}\big\downarrow\scriptstyle{\tilde{\varphi}_0} & & \scriptstyle{\tilde{\psi}_1}\big\downarrow\scriptstyle{\tilde{\varphi}_1} \\
0 & \to & \mathrm{Hom}_{\mathbb{C}}(U(\mathfrak{g}), V) & \to & \mathrm{Hom}_{\mathbb{C}}(U(\mathfrak{g}) \otimes X^{\mathfrak{h},M}_0, V) & \to & \mathrm{Hom}_{\mathbb{C}}(U(\mathfrak{g}) \otimes X^{\mathfrak{h},M}_1, V)
\end{array}
$$

in $\mathcal{C}(\mathfrak{h}, M)$, with $\mathfrak{h}$ acting on the left of $U(\mathfrak{g})$, on the left of the $U(\mathfrak{h})$ factor of $U(\mathfrak{h}) \otimes_{U(\mathfrak{m})} \bigwedge^j(\mathfrak{h}/\mathfrak{m})$, and on V. The group M acts trivially on $U(\mathfrak{g})$. Formulas (3.54) become

(3.70)

$$
\tilde{\psi}_j \lambda(u_\mathfrak{g} \otimes u_\mathfrak{h} \otimes \xi) = \lambda(u_\mathfrak{g} X \otimes u_\mathfrak{h} \otimes \xi)
$$

$$
\tilde{\varphi}_j(u_\mathfrak{g} \otimes u_\mathfrak{h} \otimes \xi) = \lambda(u_\mathfrak{g} \otimes u_\mathfrak{h}\tilde{X} \otimes \xi) - \lambda(u_\mathfrak{g} \otimes u_\mathfrak{h} \otimes (\mathrm{ad}\,\tilde{X})\xi),
$$

and $\tilde{\psi}_j$ is homotopic to $\tilde{\psi}_j + \tilde{\varphi}_j$.

We apply the invariants functor $(\,\cdot\,)^{\mathfrak{h},M}$ to the diagram (3.69), replacing the computation (3.55) by

(3.71)
$$
\begin{aligned}
(\mathrm{Hom}_{\mathbb{C}}(U(\mathfrak{g}) \otimes X_j^{\mathfrak{h},M},\, V))^{\mathfrak{h},M} \\
\cong (\mathrm{Hom}_{\mathfrak{h}}(U(\mathfrak{g}) \otimes X_j^{\mathfrak{h},M},\, V))^{M} \\
\cong (\mathrm{Hom}_{\mathfrak{h}}(U(\mathfrak{h}) \otimes_{U(\mathfrak{m})} \textstyle\bigwedge^j(\mathfrak{h}/\mathfrak{m}),\, \mathrm{Hom}_{\mathbb{C}}(U(\mathfrak{g}),\, V)))^{M} \\
\cong (\mathrm{Hom}_{\mathfrak{m}}(\textstyle\bigwedge^j(\mathfrak{h}/\mathfrak{m}),\, \mathrm{Hom}_{\mathfrak{h}}(U(\mathfrak{h}),\, \mathrm{Hom}_{\mathbb{C}}(U(\mathfrak{g}),\, V))))^{M} \\
\cong \mathrm{Hom}_{M}(\textstyle\bigwedge^j(\mathfrak{h}/\mathfrak{m}),\, \mathrm{Hom}_{\mathbb{C}}(U(\mathfrak{g}),\, V)) \\
\cong \mathrm{Hom}_{\mathbb{C}}(U(\mathfrak{g}),\, \mathrm{Hom}_{M}(\textstyle\bigwedge^j(\mathfrak{h}/\mathfrak{m}),\, V)) \quad \text{since } M \text{ acts trivially} \\
\text{on } U(\mathfrak{g}),
\end{aligned}
$$

and we arrive at the diagram

(3.72)
$$
\begin{array}{ccccc}
0 & \to & \mathrm{Hom}_{\mathbb{C}}(U(\mathfrak{g}), \mathrm{Hom}_{M}(\textstyle\bigwedge^0(\mathfrak{h}/\mathfrak{m}), V)) & \to & \mathrm{Hom}_{\mathbb{C}}(U(\mathfrak{g}), \mathrm{Hom}_{M}(\textstyle\bigwedge^1(\mathfrak{h}/\mathfrak{m}), V)) \\
& & \psi_0 \downarrow \varphi_0 & & \psi_1 \downarrow \varphi_1 \\
0 & \to & \mathrm{Hom}_{\mathbb{C}}(U(\mathfrak{g}), \mathrm{Hom}_{M}(\textstyle\bigwedge^0(\mathfrak{h}/\mathfrak{m}), V)) & \to & \mathrm{Hom}_{\mathbb{C}}(U(\mathfrak{g}), \mathrm{Hom}_{M}(\textstyle\bigwedge^1(\mathfrak{h}/\mathfrak{m}), V))
\end{array}
$$

in place of (3.56). Then we can calculate from (3.70) and (3.71) that

$$
\begin{aligned}
\varphi_j \lambda(u_{\mathfrak{g}}(\xi)) &= \tilde{X}(\lambda(u_{\mathfrak{g}}(\xi))) - \lambda(Xu_{\mathfrak{g}}(\xi)) - \lambda(u_{\mathfrak{g}}((\mathrm{ad}\,\tilde{X})\xi)) \\
\psi_j \lambda(u_{\mathfrak{g}}(\xi)) &= \lambda(u_{\mathfrak{g}} X(\xi))
\end{aligned}
$$

and that

(3.73)
$$
(\psi_j + \varphi_j)\lambda(u_{\mathfrak{g}}(\xi)) = \tilde{X}(\lambda(u_{\mathfrak{g}}(\xi))) - \lambda(((\mathrm{ad}\,X)u_{\mathfrak{g}})(\xi)) - \lambda(u_{\mathfrak{g}}((\mathrm{ad}\,\tilde{X})\xi)).
$$

Since the systems $\tilde{\psi}_j$ and $\tilde{\psi}_j + \tilde{\varphi}_j$ are homotopic in the $C(\mathfrak{h}, M)$ diagram (3.69), the systems ψ_j and $\psi_j + \varphi_j$ are homotopic in the $C(0)$ diagram (3.72).

Meanwhile, $(I_{\mathfrak{h},M}^{\mathfrak{g},\{1\}})^j(V)$ is computed from the complex

$$
\begin{aligned}
I_{\mathfrak{h},M}^{\mathfrak{g},\{1\}}\mathrm{Hom}_{\mathbb{C}}(X_j^{\mathfrak{h},M}, V)) &= \mathrm{Hom}_{\mathfrak{h},M}(U(\mathfrak{g}), \mathrm{Hom}_{\mathbb{C}}(X_j^{\mathfrak{h},M}, V)) \\
&\cong (\mathrm{Hom}_{\mathbb{C}}(U(\mathfrak{g}) \otimes X_j^{\mathfrak{h},M}, V))^{\mathfrak{h},M} \\
&\cong \mathrm{Hom}_{\mathbb{C}}(U(\mathfrak{g}), \mathrm{Hom}_{M}(\textstyle\bigwedge^j(\mathfrak{h}/\mathfrak{m}), V)) \quad \text{by (3.71)},
\end{aligned}
$$

and the action of $X \in \mathfrak{g}$ on the left side carries over to the map ψ_j of the right side to itself. Thus our homotopy shows that the action by $X \in \mathfrak{g}$ on $(I_{\mathfrak{h},M}^{\mathfrak{g},\{1\}})^j(V)$ is induced in (3.72) at the level of complexes by $\psi_j + \varphi_j$.

Next we set up as an analog of (3.59) the diagram

$$
\begin{array}{ccccccc}
0 & \to & V & \to & \mathrm{Hom}_{\mathbb{C}}(X_0^{\mathfrak{h},M}, V) & \to & \mathrm{Hom}_{\mathbb{C}}(X_1^{\mathfrak{h},M}, V) & \to & \cdots \\
& & \downarrow{\scriptstyle 1} & & \downarrow{\scriptstyle \tilde{R}_0 \otimes 1} & & \downarrow{\scriptstyle \tilde{R}_1 \otimes 1} & & \\
0 & \to & V & \to & \mathrm{Hom}_{\mathbb{C}}(X_0^{\mathfrak{s},M}, V) & \to & \mathrm{Hom}_{\mathbb{C}}(X_1^{\mathfrak{s},M}, V) & \to & \cdots
\end{array}
$$

in $\mathcal{C}(\mathfrak{s}, M)$, in which the $\tilde{R}_j$'s are restrictions. Since, as we saw in an earlier part of the proof, $X_j^{\mathfrak{h},M}$ is projective in $\mathcal{C}(\mathfrak{s}, M)$, both rows are injective resolutions in $\mathcal{C}(\mathfrak{s}, M)$. We apply $I_{\mathfrak{s},M}^{0,\{1\}}$ and obtain from (3.71) the diagram

$$
\begin{array}{ccccc}
0 & \to & \mathrm{Hom}_{\mathbb{C}}(U(\mathfrak{g}), \mathrm{Hom}_M(\textstyle\bigwedge^0(\mathfrak{h}/\mathfrak{m}), V)) & \to & \mathrm{Hom}_{\mathbb{C}}(U(\mathfrak{g}), \mathrm{Hom}_M(\textstyle\bigwedge^1(\mathfrak{h}/\mathfrak{m}), V)) \\
& & \downarrow{\scriptstyle R_0} & & \downarrow{\scriptstyle R_1} \\
0 & \to & \mathrm{Hom}_M(\textstyle\bigwedge^0(\mathfrak{s}/\mathfrak{m}), V) & \to & \mathrm{Hom}_M(\textstyle\bigwedge^1(\mathfrak{s}/\mathfrak{m}), V)
\end{array}
$$

Here the maps R_j are given by restriction and induce isomorphisms on cohomology. Restricting $\psi_j + \varphi_j$ from (3.73) by taking $u_{\mathfrak{g}} = 1$, we obtain the desired formula for the action by X in (3.43b), provided $K = \{1\}$ and $L = M$. The passage to general K is accomplished by the same kind of reduction as with (3.43a), and then the argument for (3.43b) is complete.

6. Motivation for $\mathfrak{g}$ Equivariance in Case (ii)

For this section and the next we specialize to the case of an inclusion of pairs $(\mathfrak{g}, L) \hookrightarrow (\mathfrak{g}, K)$. The Hard Duality map takes the form

$$(3.74) \qquad \lambda : (P_{\mathfrak{g},L}^{\mathfrak{g},K})_j(V \otimes_{\mathbb{C}} (\textstyle\bigwedge^m \mathfrak{c})^*) \longrightarrow (I_{\mathfrak{g},L}^{\mathfrak{g},K})^j(V),$$

and we proved in §4 that it was a K equivariant isomorphism. To complete the proof of Theorem 3.5b in Case (ii), we need to prove that (3.74) is $\mathfrak{g}$ equivariant. The difficulty in doing so is that the complexes used to define λ are complexes of K modules, not complexes of $(\mathfrak{g}, K)$ modules.

What we shall do for each of the K modules in question is impose artificial definitions of actions of $\mathfrak{g}$ that descend to the correct action in homology or cohomology. When we considered Case (i) in §5, we were able to work with a single member X of $\mathfrak{g}$ at a time, but that approach does not seem to work for an inclusion of pairs. The key question is how to recognize the correct action at the homology or cohomology level when we have it.

We begin with a general remark: A Lie algebra action can be formulated in two different ways so as to become a map in the category. In fact, let $\mathfrak{g}$ be a Lie algebra, and let M be a $\mathfrak{g}$ module. Then multiplication as a map of $\mathfrak{g} \times M$ to M is bilinear and extends to a linear map

$$(3.75) \qquad\qquad \mu : \mathfrak{g} \otimes_{\mathbb{C}} M \to M.$$

Alternatively we can define

$$(3.76) \qquad\qquad \mu : M \to \operatorname{Hom}_{\mathbb{C}}(\mathfrak{g}, M)$$

by $\mu(v)(X) = Xv$. The striking fact is that μ is $\mathfrak{g}$ equivariant in both cases. This equivariance follows from the two computations

$$\begin{aligned}
\mu(X(Y \otimes v)) &= \mu([X, Y] \otimes v) + \mu(Y \otimes Xv) \\
&= [X, Y]v + YXv = XYv = X\mu(Y \otimes v) \qquad \text{for (3.75)} \\
X(\mu(v))(Y) &= X(\mu(v)(Y)) - \mu(v)[X, Y] \\
&= X(Yv) - [X, Y]v = YXv = \mu(Xv)(Y) \qquad \text{for (3.76).}
\end{aligned}$$

Moreover, if M is a $(\mathfrak{g}, K)$ module, then μ is K equivariant as well, because

$$\begin{aligned}
\mu(k(X \otimes v)) &= \mu(\operatorname{Ad}(k)X \otimes kv) \\
&= (\operatorname{Ad}(k)X)kv = kXv = k\mu(X \otimes v) \qquad \text{for (3.75)} \\
k(\mu(v))(X) &= k(\mu(v)(\operatorname{Ad}(k)^{-1}X)) \\
&= k((\operatorname{Ad}(k)^{-1}X)v) = Xkv = \mu(kv)(X) \qquad \text{for (3.76).}
\end{aligned}$$

Let us come back to the situation in Theorem 3.5b, writing Π and Γ as usual for $P_{\mathfrak{g},L}^{\mathfrak{g},K}$ and $I_{\mathfrak{g},L}^{\mathfrak{g},K}$. For V in $\mathcal{C}(\mathfrak{g}, L)$, we can use (3.75) and (3.76) to give a description of multiplication in $\Pi(V)$, in $\Gamma(V)$, and in the derived functor modules. We write μ with a subscript to emphasize where multiplication is taking place.

Proposition 3.77 (Enright-Wallach). Let $(\mathfrak{g}, L) \hookrightarrow (\mathfrak{g}, K)$ be an inclusion of pairs, and let V be in $\mathcal{C}(\mathfrak{g}, L)$. Then the $\mathcal{C}(\mathfrak{g}, K)$ diagrams

(3.78a)

$$
\begin{array}{ccc}
\mathfrak{g} \otimes_{\mathbb{C}} \Pi(V) & \xrightarrow{\;\;\Phi^{-1}\;\;} & \Pi(\mathfrak{g} \otimes_{\mathbb{C}} V) \\
{\scriptstyle \mu_{\Pi(V)}} \downarrow & \diagup & {\scriptstyle \Pi(\mu_V)} \\
\Pi(V) & &
\end{array}
$$

and

(3.78b)

$$
\begin{array}{ccc}
\mathrm{Hom}_{\mathbb{C}}(\mathfrak{g}, \Gamma(V)) & \xleftarrow{\;\;\Psi^{-1}\;\;} & \Gamma(\mathrm{Hom}_{\mathbb{C}}(\mathfrak{g}, V) \\
{\scriptstyle \mu_{\Gamma(V)}} \uparrow & \diagup & {\scriptstyle \Gamma(\mu_V)} \\
\Gamma(V) & &
\end{array}
$$

are commutative, and their maps are natural in V. Here Φ and Ψ are Mackey isomorphisms relative to $(\mathfrak{g}, L)$ and $(\mathfrak{g}, K)$. The diagrams remain commutative when Π is replaced by Π_j and Γ is replaced by Γ^j; the maps Φ^{-1} and Ψ^{-1} get replaced by the induced maps Φ_j^{-1} and Ψ_j^{-1} on the quotients.

PROOF. For (3.78a), let $T \in R(K) \subseteq R(\mathfrak{g}, K)$, $X \in \mathfrak{g}$, and $v \in V$. We trace the effects of the maps in question on the element $X \otimes (T \otimes v)$ of $\mathfrak{g} \otimes_{\mathbb{C}} \Pi(V)$. On the one hand,

$$
\begin{aligned}
\mu_{\Pi(V)}(X \otimes (T \otimes v)) &= (X \otimes T) \otimes v \\
&= (T \otimes \mathrm{Ad}(\cdot)^{-1} X) \otimes v \\
&= T \otimes (\mathrm{Ad}(\cdot)^{-1} X) v,
\end{aligned}
$$

the last step since the tensor product in $\Pi(V)$ is over $\mathfrak{g}$ (and L). On the other hand

$$
\begin{aligned}
\Pi(\mu_V) &\circ \Phi^{-1}(X \otimes (T \otimes v)) \\
&= \Pi(\mu_V)\left(\int_K (k\chi_A \otimes (\mathrm{Ad}(k)^{-1} X \otimes v))\, dT(k) \right) \\
&= \int_K \Pi(\mu_V)(k\chi_A \otimes (\mathrm{Ad}(k)^{-1} X \otimes v))\, dT(k) \\
&= \int_K (k\chi_A \otimes (\mathrm{Ad}(k)^{-1} X) v)\, dT(k).
\end{aligned}
$$

If $\{X_j\}$ is a basis of $\mathfrak{g}$, we write $\mathrm{Ad}(k)^{-1}X = \sum_j \langle \mathrm{Ad}(k)^{-1}X, X_j^* \rangle X_j$ and put $T_j = \langle \mathrm{Ad}(\cdot)^{-1}X, X_j^* \rangle T$. Then the above expression is

$$= \sum_j \int_K (k\chi_A \otimes X_j v)\, dT_j(k)$$

$$= \sum_j (T_j \chi_A \otimes X_j v).$$

If A is chosen large enough so that $T_j \chi_A = T_j$ for all j, this is

$$= \sum_j (T_j \otimes X_j v)$$

$$= T \otimes (\mathrm{Ad}(\cdot)^{-1}X)v.$$

Hence $\mu_{\Pi(V)}$ equals $\Pi(\mu_V) \circ \Phi^{-1}$ on elements $X \otimes (T \otimes V)$. The linear span of such elements is all of $\mathfrak{g} \otimes_{\mathbb{C}} \Pi(V)$, by Proposition 2.69a, and the commutativity of (3.78a) follows. The naturality of Φ^{-1} in V was observed in Theorem 2.103, and the naturality of the other two maps is clear.

For (3.78b), let $\varphi \in \Gamma(V)$, $T \in R(K) \subseteq R(\mathfrak{g}, K)$, and $X \in \mathfrak{g}$. We trace the effects of the maps in question on the element φ of $\Gamma(V)$. On the one hand,

$$\mu_{\Gamma(V)}(\varphi)(X)(T) = (X\varphi)(T)$$

$$= \varphi(T \otimes X)$$

$$= \varphi(\mathrm{Ad}(\cdot)X \otimes T)$$

$$= \mathrm{Ad}(\cdot)X(\varphi(T)),$$

the last step since φ commutes with $\mathfrak{g}$ (and L). On the other hand,

$$\Psi^{-1} \circ \Gamma(\mu_V)\varphi(X)(T) = \int_K \Gamma(\mu_V)\varphi(k\chi_A)(\mathrm{Ad}(k)X)\, dT(k)$$

$$= \int_K (\mathrm{Ad}(k)X)(\varphi(k\chi_A))\, dT(k).$$

If $\{X_j\}$ is a basis of $\mathfrak{g}$, we write $\mathrm{Ad}(k)X = \sum_j \langle \mathrm{Ad}(k)X, X_j^* \rangle X_j$ and put $T_j = \langle \mathrm{Ad}(\cdot)X, X_j^* \rangle T$. Then the above expression is

$$= \sum_j \int_K X_j(\varphi(k\chi_A))\, dT_j(k)$$

$$= \sum_j X_j(\varphi(T_j \chi_A)).$$

If A is chosen large enough so that $T_j \chi_A = T_j$ for all j, this is

$$= \sum_j X_j(\varphi(T_j))$$

$$= \text{Ad}(\,\cdot\,)X(\varphi(T)).$$

Hence $\mu_{\Gamma(V)}\varphi(X)$ equals $\Phi^{-1}\circ\Gamma(\mu_V)\varphi(X)$ on $R(K)$, hence everywhere by Proposition 2.69a. The commutativity of (3.78b) follows. The naturality of Ψ^{-1} in V was observed in Theorem 2.95, and the naturality of the other two maps is clear.

Let us now consider the passage in ( ) from Π to Π_j. We replace V by the members V_n of a projective resolution in $\mathcal{C}(\mathfrak{g}, L)$, obtaining a diagram (3.78a) for each V_n. The naturality of the maps of (3.78a) in V implies that $\{\mu_{\Pi(V_n)}\}$, $\{\Phi^{-1}\}$, and $\{\Pi(\mu_{V_n})\}$ are chain maps. Hence we get a commutative diagram in homology. The jth homology of $\Pi(V_n)$ is $\Pi_j(V)$. Since $\mathfrak{g} \otimes_{\mathbb{C}} V_n$ is a projective resolution of $\mathfrak{g} \otimes_{\mathbb{C}} V$ by Proposition 2.53a, the jth homology of $\mathfrak{g} \otimes_{\mathbb{C}} V_n$ is $\Pi_j(\mathfrak{g} \otimes_{\mathbb{C}} V)$. Finally $\mathfrak{g} \otimes_{\mathbb{C}} (\,\cdot\,)$ is an exact functor by Proposition 2.53a, and it sends homology to homology by (C.11). Thus $\mathfrak{g} \otimes_{\mathbb{C}} \Pi(V_n)$ has homology $\mathfrak{g} \otimes_{\mathbb{C}} \Pi_j(V)$. Consequently (3.78a) is valid with Π replaced by Π_j. A similar argument shows that (3.78b) remains valid if Γ is replaced by Γ^j. This completes the proof.

Put

$$\Pi^K = P_{\mathfrak{k},L}^{\mathfrak{k},K}, \qquad \Gamma_K = I_{\mathfrak{k},L}^{\mathfrak{k},K}, \qquad \mathcal{F} = \mathcal{F}_{\mathfrak{g},L}^{\mathfrak{k},L}, \qquad \tilde{\mathcal{F}} = \mathcal{F}_{\mathfrak{g},K}^{\mathfrak{k},K}.$$

We can apply $\tilde{\mathcal{F}}$ to the diagrams (3.78) for Π_j and Γ^j, and Corollary 3.26 shows that the resulting diagrams are

(3.79a)

$$
\begin{array}{ccc}
\mathfrak{g} \otimes_{\mathbb{C}} \Pi_j^K(\mathcal{F}V) & \xrightarrow{\;\;\Phi_j^{-1}\;\;} & \Pi_j^K(\mathfrak{g} \otimes_{\mathbb{C}} \mathcal{F}V) \\
{\scriptstyle \mu_{\Pi_j(V)}}\downarrow & \swarrow & \;\;{\scriptstyle \Pi_j^K(\mathcal{F}\mu_V)} \\
\Pi_j^K(\mathcal{F}V) & &
\end{array}
$$

and

(3.79b)

$$
\begin{array}{ccc}
\text{Hom}_{\mathbb{C}}(\mathfrak{g}, \Gamma_K^j(\mathcal{F}V)) & \xleftarrow{\;\;\Psi_j^{-1}\;\;} & \Gamma_K^j(\text{Hom}_{\mathbb{C}}(\mathfrak{g}, \mathcal{F}V)) \\
{\scriptstyle \mu_{\Gamma^j(V)}}\uparrow & \nearrow & \;\;{\scriptstyle \Gamma_K^j(\mathcal{F}\mu_V)} \\
\Gamma_K^j(\mathcal{F}V) & &
\end{array}
$$

These are diagrams in the category $C(\mathfrak{k}, K)$, and we should bear in mind the steps in their construction: We started from the diagrams with $j = 0$, given in (3.78), we replaced V by a projective or injective resolution in $C(\mathfrak{g}, L)$, and we passed to homology or cohomology. The projective or injective resolution in $C(\mathfrak{g}, L)$ is also a projective or injective resolution in $C(\mathfrak{k}, L)$, and we can ask whether an arbitrary projective or injective resolution in $C(\mathfrak{k}, L)$ will work in its place. The answer is that we do not have an interpretation of $\mu_{\Pi(V_n)}$ or $\mu_{\Gamma(V_n)}$ when V_n is merely a $(\mathfrak{k}, L)$ module, and we can make no sense of (3.79). But there is something else we can do. Using a member V_n of a resolution in (3.79) for $j = 0$ in place of V, let us *define* the vertical maps at the left in (3.79) so that the diagrams commute. Then the induced map on homology or cohomology ought to be multiplication in $\Pi_j^K(\mathcal{F}V)$ or $\Gamma_K^j(\mathcal{F}V)$. The first step is to compute the vertical maps; we do so when the resolution in question is the standard projective or injective resolution of V in $C(\mathfrak{k}, L)$.

Proposition 3.80 (Duflo-Vergne). Let $(\mathfrak{g}, L) \hookrightarrow (\mathfrak{g}, K)$ be an inclusion of pairs, and let V be in $C(\mathfrak{g}, L)$. Regard $\bigwedge^n \mathfrak{c}$ as a trivial $(\mathfrak{g}, L)$ module, and make $\mathcal{F}V$ into a $(\mathfrak{g}, L)$ module so that its action is the same as in V. Let Φ_0 and Ψ_0 be Mackey isomorphisms relative to $(\mathfrak{l}, L)$ and $(\mathfrak{k}, K)$. Then the $C(\mathfrak{k}, K)$ maps α_n and β_n that make the diagrams

(3.81a)

$$\mathfrak{g} \otimes_{\mathbb{C}} (R(K) \otimes_L (\textstyle\bigwedge^n \mathfrak{c} \otimes_{\mathbb{C}} \mathcal{F}V)) \xrightarrow{\Phi_0^{-1}} R(K) \otimes_L (\mathfrak{g} \otimes_{\mathbb{C}} \textstyle\bigwedge^n \mathfrak{c} \otimes_{\mathbb{C}} \mathcal{F}V)$$

$$\alpha_n \downarrow \qquad\qquad 1 \otimes \mu_{(\wedge^n \mathfrak{c} \otimes \mathcal{F}V)}$$

$$R(K) \otimes_L (\textstyle\bigwedge^n \mathfrak{c} \otimes_{\mathbb{C}} \mathcal{F}V)$$

and

(3.81b)

$$\begin{array}{ccc}
\mathrm{Hom}_{\mathbb{C}}(\mathfrak{g}, & \xleftarrow{\Psi_0^{-1}} & \mathrm{Hom}_L(R(K), \\
\quad \mathrm{Hom}_L(R(K), \mathrm{Hom}_{\mathbb{C}}(\textstyle\bigwedge^n \mathfrak{c}, \mathcal{F}V))_K & & \quad \mathrm{Hom}_{\mathbb{C}}(\mathfrak{g}, \mathrm{Hom}_{\mathbb{C}}(\textstyle\bigwedge^n \mathfrak{c}, \mathcal{F}V)))_K \\
\beta_n \uparrow & & \mathrm{Hom}(1, \mu_{\mathrm{Hom}(\wedge^n \mathfrak{c}, \mathcal{F}V)}) \\
\mathrm{Hom}_L(R(K), \mathrm{Hom}_{\mathbb{C}}(\textstyle\bigwedge^n \mathfrak{c}, \mathcal{F}V))_K & &
\end{array}$$

commute are given by

$$\alpha_n(X \otimes (T \otimes w)) = T \otimes (\mathrm{Ad}(\cdot)^{-1}X)w \qquad \begin{array}{l}\text{for } X \in \mathfrak{g}, T \in R(K), \\ \text{and } w \in \textstyle\bigwedge^n \mathfrak{c} \otimes \mathcal{F}V\end{array}$$

$$\beta_n(\varphi)(X)(T) = \mathrm{Ad}(\cdot)X(\varphi(T)) \qquad \begin{array}{l}\text{for } \varphi \in \mathrm{domain}(\beta_n), X \in \mathfrak{g}, \\ \text{and } T \in R(K).\end{array}$$

REMARK. In the definition of α_n, we follow the convention adopted in (3.29) and use L invariants rather than L coinvariants to define $\otimes_L$. It has to be understood then that a monomial $X \otimes (T \otimes w)$ is not necessarily in the space, and the formula should really be written with finite sums of such monomials.

PROOF. In the case of α_n, we have

$$\alpha_n(X \otimes (T \otimes w)) = (1 \otimes \mu) \circ \Phi_0^{-1}(X \otimes (T \otimes w))$$
$$= (1 \otimes \mu) \int_K (k\chi_A \otimes (\mathrm{Ad}(k)^{-1}X \otimes w)) \, dT(k)$$
$$= \int_K (1 \otimes \mu)(k\chi_A \otimes (\mathrm{Ad}(k)^{-1}X \otimes w)) \, dT(k)$$
$$= \int_K (k\chi_A \otimes (\mathrm{Ad}(k)^{-1}X)w) \, dT(k).$$

Now we argue as in the proof of (3.78a). If $\{X_j\}$ is a basis of $\mathfrak{g}$, we write $T_j = \langle \mathrm{Ad}(\cdot)^{-1}X, X_j^* \rangle T$ and $\mathrm{Ad}(k)^{-1}X = \sum_j \langle \mathrm{Ad}(k)^{-1}X, X_j^* \rangle X_j$. If $A \subseteq \widehat{K}$ is chosen large enough so that $T_j\chi_A = T_j$ for all j, then the above expression is

$$= T \otimes (\mathrm{Ad}(\cdot)^{-1}X)w$$

as required.

In the case of β_n, we argue similarly. We have

$$\beta_n(\varphi)(X)(T) = \Psi_0^{-1} \circ \mathrm{Hom}(-)(\varphi)(X)(T)$$
$$= \int_K \mathrm{Hom}(-)(\varphi)(k\chi_A)(\mathrm{Ad}(k)X) \, dT(k)$$
$$= \int_K (\mathrm{Ad}(k)X)\varphi(k\chi_A)) \, dT(k),$$

and we can argue as in the proof of (3.78b). If $\{X_j\}$ is a basis of $\mathfrak{g}$, we write $\mathrm{Ad}(k)X = \sum_j \langle \mathrm{Ad}(k)X, X_j^* \rangle X_j$ and $T_j = \langle \mathrm{Ad}(\cdot)X, X_j^* \rangle T$. If A is chosen large enough so that $T_j\chi_A = T_j$ for all j, then the above expression is

$$= \mathrm{Ad}(\cdot)X(\varphi(T))$$

as required. This completes the proof.

The path to completing the proof of Theorem 3.5b is now clear. We use α_n and β_n to make artificial definitions of actions of $\mathfrak{g}$ at the level of complexes, prove that these actions respect duality, and prove that they descend to homology and cohomology and match the actions given by (3.79). As in Case (i) in §5, we do not get a match of actions on the level of complexes, only a match up to homotopy.

7. Proof of $\mathfrak{g}$ Equivariance in Case (ii)

As in §6, let $(\mathfrak{g}, L) \hookrightarrow (\mathfrak{g}, K)$ be an inclusion of pairs, and let $\mathfrak{c}$ be a complement to $\mathfrak{l}$ in $\mathfrak{k}$. In this section we shall prove that the Hard Duality map (3.74) is $\mathfrak{g}$ equivariant. In the notation used before Lemma 3.4, $\mathfrak{s}$ is 0; thus $\bigwedge^m \mathfrak{c}$ is a trivial $\mathfrak{g}$ module (though not necessarily a trivial K module).

Motivated by Proposition 3.80, we define Lie algebra actions of $\mathfrak{g}$ on (3.29b) and (3.30b) by

$$(3.82a) \qquad X^\#(f\, dk \otimes \xi \otimes v \otimes \varepsilon) = f(\cdot)\, dk \otimes \xi \otimes (\mathrm{Ad}(\cdot)^{-1}X)v \otimes \varepsilon$$

for $X \in \mathfrak{g}$, $f \in C(K)$, $\xi \in \bigwedge^n \mathfrak{c}$, $v \in V$, and $\varepsilon \in (\bigwedge^m \mathfrak{c})^*$, and by

$$(3.82b) \qquad (X^\# \varphi)(T)(\eta) = (\mathrm{Ad}(\cdot)X)(\varphi(T)(\eta))$$

for $X \in \mathfrak{g}$, $\varphi \in$ (3.30b), $T \in R(K)$, and $\eta \in \bigwedge^{m-n} \mathfrak{c}$.

To see that these definitions descend to (3.29a) and (3.30a), we check that each $X^\#$ commutes with the action of L: For (3.82a), L acts by $r \otimes \mathrm{Ad} \otimes \pi \otimes \mathrm{Ad}^*$. If $k_0 \in L$ and $X \in \mathfrak{g}$, we have

$$
\begin{aligned}
k_0 X^\#(f\, dk \otimes \xi \otimes v \otimes \varepsilon) &= k_0(f(\cdot)\, dk \otimes \xi \otimes (\mathrm{Ad}(\cdot)^{-1}X)v \otimes \varepsilon) \\
&= f(\cdot k_0)dk \otimes \mathrm{Ad}(k_0)\xi \otimes k_0(\mathrm{Ad}(\cdot k_0)^{-1}X)v \otimes k_0\varepsilon \\
&= f(\cdot k_0)\, dk \otimes \mathrm{Ad}(k_0)\xi \otimes (\mathrm{Ad}(\cdot)^{-1}X)k_0 v \otimes k_0\varepsilon \\
&= X^\#(f(\cdot k_0)\, dk \otimes \mathrm{Ad}(k_0)\xi \otimes k_0 v \otimes k_0\varepsilon) \\
&= X^\# k_0(f\, dk \otimes \xi \otimes v \otimes \varepsilon).
\end{aligned}
$$

For (3.82b), L acts by $\mathrm{Hom}(l, \mathrm{Hom}(\mathrm{Ad}, \pi))$. If $k_0 \in L$ and $X \in \mathfrak{g}$, we have

$$
\begin{aligned}
(X^\# k_0 \varphi)(T)(\eta) &= (\mathrm{Ad}(\cdot)X)(k_0\varphi(T)(\eta)) \\
&= (\mathrm{Ad}(k_0 \cdot)X)k_0(\varphi(k_0^{-1}T)(k_0^{-1}\eta)) \\
&= k_0((\mathrm{Ad}(\cdot)X)(\varphi(k_0^{-1}T)(k_0^{-1}\eta))) \\
&= k_0((X^\# \varphi)(k_0^{-1}T)(k_0^{-1}\eta)) \\
&= (k_0 X^\# \varphi)(T)(\eta).
\end{aligned}
$$

Thus the Lie algebra actions of $\mathfrak{g}$ defined by (3.82) descend to (3.29a) and (3.30a). It makes sense to define

$$\alpha_n(X \otimes y) = X^{\#}(y) \qquad \text{for } y \text{ in (3.29a)}$$

and

$$\beta_{m-n}(\varphi)(X) = X^{\#}(\varphi) \qquad \text{for } \varphi \text{ in (3.30a)}.$$

Proposition 3.80 identifies

$$\alpha_n = (1 \otimes \mu_{(\wedge^n\mathfrak{c}\otimes\mathcal{F}V\otimes(\wedge^m\mathfrak{c})^*)}) \circ \Phi^{-1}$$

and

$$\beta_{m-n} = \Psi^{-1} \circ \mathrm{Hom}(1, \mu_{\mathrm{Hom}(\wedge^{m-n}\mathfrak{c},\mathcal{F}V)}).$$

Each of the factors of α_n and of β_n is a chain map as n varies. Consequently $\{\alpha_n\}$ and $\{\beta_n\}$ are chain maps.

Proposition 3.83. Put

$$V_n^{\Pi} = \textstyle\bigwedge^n \mathfrak{c} \otimes_{\mathbb{C}} \mathcal{F}V \otimes_{\mathbb{C}} (\textstyle\bigwedge^m \mathfrak{c})^*$$
$$V_n^{\Gamma} = \mathrm{Hom}_{\mathbb{C}}(\textstyle\bigwedge^n \mathfrak{c}, \mathcal{F}V)_K.$$

Then the $\mathcal{C}(\mathfrak{k}, K)$ diagrams

(3.84a)

$$\left\{ \begin{array}{c} \ker(1 \otimes \partial) \text{ in} \\ \mathfrak{g} \otimes_{\mathbb{C}} (R(K) \otimes_L V_n^{\Pi}) \end{array} \right\} \xrightarrow{\alpha_n} \left\{ \begin{array}{c} \ker \partial \text{ in} \\ R(K) \otimes_L V_n^{\Pi} \end{array} \right\}$$

$$\downarrow \qquad\qquad\qquad\qquad \downarrow$$

$$\mathfrak{g} \otimes_{\mathbb{C}} \Pi_n^K(\mathcal{F}(V \otimes_{\mathbb{C}} (\textstyle\bigwedge^m \mathfrak{c})^*)) \xrightarrow{\mu} \Pi_n^K(\mathcal{F}(V \otimes_{\mathbb{C}} (\textstyle\bigwedge^m \mathfrak{c})^*))$$

and

(3.84b)

$$\left\{ \begin{array}{c} \ker(1 \otimes d) \text{ in} \\ \mathrm{Hom}_{\mathbb{C}}(\mathfrak{g}, \mathrm{Hom}_L(R(K), V_n^{\Gamma})_K) \end{array} \right\} \xleftarrow{\beta_n} \left\{ \begin{array}{c} \ker d \text{ in} \\ \mathrm{Hom}_L(R(K), V_n^{\Gamma})_K \end{array} \right\}$$

$$\downarrow \qquad\qquad\qquad\qquad \downarrow$$

$$\mathrm{Hom}_{\mathbb{C}}(\mathfrak{g}, \Gamma_K^n(\mathcal{F}V)) \xleftarrow{\mu} \Gamma_K^n(\mathcal{F}V)$$

commute.

PROOF. We give the argument for (3.84a), the argument for (3.84b) being similar. Let W_n and W_n^K be the standard projective resolutions of $V \otimes_{\mathbb{C}} (\bigwedge^m \mathfrak{c})^*$ in $\mathcal{C}(\mathfrak{g}, L)$ and $\mathcal{C}(\mathfrak{k}, L)$, namely

(3.85a) $W_n = (R(\mathfrak{g}, L) \otimes_L \bigwedge^n (\mathfrak{g}/\mathfrak{l})) \otimes_{\mathbb{C}} (V \otimes_{\mathbb{C}} (\bigwedge^m \mathfrak{c})^*)$

(3.85b) $W_n^K = (R(\mathfrak{k}, L) \otimes_L \bigwedge^n (\mathfrak{k}/\mathfrak{l})) \otimes_{\mathbb{C}} (V \otimes_{\mathbb{C}} (\bigwedge^m \mathfrak{c})^*).$

These are both projective resolutions of $V \otimes_{\mathbb{C}} (\bigwedge^m \mathfrak{c})^*$ in $\mathcal{C}(\mathfrak{k}, L)$. The inclusions $R(\mathfrak{k}, L) \subseteq R(\mathfrak{g}, L)$ and $\bigwedge^n (\mathfrak{k}/\mathfrak{l}) \subseteq \bigwedge^n (\mathfrak{g}/\mathfrak{l})$ induce $\mathcal{C}(\mathfrak{k}, L)$ inclusions

(3.86) $\mathrm{inc}_n : W_n^K \to W_n$

that form a chain map over the identity on $V \otimes_{\mathbb{C}} (\bigwedge^m \mathfrak{c})^*$.

In (3.84a) the definition of μ is that it is the map on homology induced by the multiplication mapping

(3.87) $\mu_{\Pi(W_n)} : \mathfrak{g} \otimes_{\mathbb{C}} \Pi(W_n) \to \Pi(W_n).$

Consider the diagram

(3.88)

$$
\begin{array}{ccc}
\mathfrak{g} \otimes_{\mathbb{C}} (R(K) \otimes_L V_n^{\Pi}) & \xrightarrow{\ \alpha_n\ } & R(K) \otimes_L V_n^{\Pi} \\
{\scriptstyle 1 \otimes I_n} \downarrow & & \downarrow {\scriptstyle I_n} \\
\mathfrak{g} \otimes_{\mathbb{C}} \Pi^K(W_n^K) & & \Pi^K(W_n^K) \\
{\scriptstyle 1 \otimes J_n} \downarrow & & \downarrow {\scriptstyle J_n} \\
\mathfrak{g} \otimes_{\mathbb{C}} \Pi(W_n) & \xrightarrow{\ \mu_{\Pi(W_n)}\ } & \Pi(W_n)
\end{array}
$$

in which I_n is the isomorphism in $\mathcal{C}(\mathfrak{k}, K)$ given before (2.128a) and J_n is induced by inc_n. On each side the composite vertical, restricted to the kernel of $1 \otimes \partial$ or ∂ and followed by passage to homology, is the corresponding vertical of (3.84a). The commutativity of (3.84a) will follow from (C.4) if we prove that (3.88) commutes up to homotopy.

To prove that (3.88) commutes up to homotopy, we use Propositions 3.77 and 3.80 to factor the horizontal maps so that (3.88) is the union of a left half

$$\mathfrak{g} \otimes_{\mathbb{C}} (R(K) \otimes_L V_n^\Pi) \xrightarrow{\Phi_0^{-1}} R(K) \otimes_L (\mathfrak{g} \otimes_{\mathbb{C}} V_n^\Pi)$$

$$\downarrow \qquad\qquad\qquad\qquad \downarrow$$

(3.89) $\mathfrak{g} \otimes_{\mathbb{C}} \Pi^K (W_n^K) \qquad\qquad \Pi^K (\mathfrak{g} \otimes_{\mathbb{C}} W_n^K)$

$$\downarrow \qquad\qquad\qquad\qquad \downarrow$$

$$\mathfrak{g} \otimes_{\mathbb{C}} \Pi(W_n) \xrightarrow{\Phi^{-1}} \Pi(\mathfrak{g} \otimes_{\mathbb{C}} W_n)$$

and a right half

$$R(K) \otimes_L (\mathfrak{g} \otimes_{\mathbb{C}} V_n^\Pi) \longrightarrow R(K) \otimes_L V_n^\Pi$$

$$\downarrow \qquad\qquad\qquad\qquad \downarrow$$

(3.90) $\Pi^K (\mathfrak{g} \otimes_{\mathbb{C}} W_n^K) \qquad\qquad \Pi^K (W_n^K)$

$$\downarrow \qquad\qquad\qquad\qquad \downarrow$$

$$\Pi(\mathfrak{g} \otimes_{\mathbb{C}} W_n) \longrightarrow \Pi(W_n).$$

It is enough to prove that (3.89) and (3.90) both commute up to homotopy.

Let us have a formula for $J_n I_n$ available. Tracking down the definition, we see that

$$I_n(T \otimes \xi \otimes v \otimes \varepsilon) = T \otimes (\chi_B \otimes \xi \otimes v \otimes \varepsilon)$$

for $T \in R(K)$, $\chi_B \in R(L) \subseteq R(\mathfrak{g}, L)$ with $T \chi_B = T$, $\xi \in \bigwedge^n (\mathfrak{k}/\mathfrak{l})$, $v \in V$, and $\varepsilon \in (\bigwedge^m \mathfrak{c})^*$. Then $J_n I_n$ has the same formula as I_n.

Let us prove that (3.89) actually commutes. Right and then down in (3.89) gives

$$X \otimes (T \otimes \xi \otimes v \otimes \varepsilon) \mapsto \int_K (k\chi_A \otimes \mathrm{Ad}(k)^{-1} X \otimes \xi \otimes v \otimes \varepsilon) \, dT(k)$$

$$\mapsto \int_K k\chi_A \otimes (\mathrm{Ad}(k)^{-1} X \otimes \chi_B \otimes \xi \otimes v \otimes \varepsilon) \, dT(k),$$

where B is large enough so that $(k\chi_A)\chi_B = k\chi_A$. Down and then right gives

$$X \otimes (T \otimes \xi \otimes v \otimes \varepsilon) \mapsto X \otimes (T \otimes (\chi_B \otimes \xi \otimes v \otimes \varepsilon))$$

$$\mapsto \int_K (k\chi_A \otimes \mathrm{Ad}(k)^{-1} X \otimes \chi_B \otimes \xi \otimes v \otimes \varepsilon) \, dT(k),$$

if $(T)\chi_B = T$. The two expressions are equal for B large enough, and therefore (3.89) commutes.

Let us prove that (3.90) commutes up to homotopy. We can write $W_n = X_n \otimes_{\mathbb{C}} (V \otimes_{\mathbb{C}} (\bigwedge^m \mathfrak{c})^*)$, where $\{X_n\}$ is the Koszul resolution in $\mathcal{C}(\mathfrak{g}, L)$. Consider the two maps

$$\mathfrak{g} \otimes_{\mathbb{C}} X_n \otimes_{\mathbb{C}} (V \otimes_{\mathbb{C}} (\textstyle\bigwedge^m \mathfrak{c})^*) \longrightarrow X_n \otimes_{\mathbb{C}} (V \otimes_{\mathbb{C}} (\textstyle\bigwedge^m \mathfrak{c})^*)$$

given by

$$\mu_1 = 1_{X_n} \otimes (\mathfrak{g}\text{ multiplication in } V \otimes_{\mathbb{C}} (\textstyle\bigwedge^m \mathfrak{c})^*)$$

$$\mu_2 = (\mathfrak{g}\text{ multiplication in } X_n) \otimes 1.$$

Their sum $\mu = \mu_1 + \mu_2$ is $\mathfrak{g}$ multiplication in W_n. Relative to the standard differentials, μ_1 is a chain map because ∂ does not act in the last coordinate, and μ_2 is a chain map because ∂ commutes with $\mathfrak{g}$. We readily check that μ_1 is a chain map over the $\mathfrak{g}$ multiplication map in $V \otimes_{\mathbb{C}} (\bigwedge^m \mathfrak{c})^*$ and that μ_2 is a chain map over 0. Hence $\mu = \mu_1 + \mu_2$ and μ_1 are chain maps over the same thing. Since they act between projective resolutions, (C.6) says that μ and μ_1 are homotopic. Therefore $\Pi(\mu)$ and $\Pi(\mu_1)$ are homotopic.

Let (3.90′) be the diagram (3.90) with the bottom map changed from $\Pi(\mu)$ to $\Pi(\mu_1)$. We shall prove that (3.90′) actually commutes. In fact, in (3.90′), right and then down gives

$$T \otimes (X \otimes \xi \otimes v \otimes \varepsilon) \mapsto T \otimes \xi \otimes X(v \otimes \varepsilon)$$
$$\mapsto T \otimes (\chi_B \otimes \xi \otimes X(v \otimes \varepsilon))$$

if $(T)\chi_B = T$. Down and then right gives

$$T \otimes (X \otimes \xi \otimes v \otimes \varepsilon) \mapsto T \otimes (\chi_B \otimes X \otimes \xi \otimes v \otimes \varepsilon)$$
$$\mapsto T \otimes (\chi_B \otimes \xi \otimes X(v \otimes \varepsilon))$$

if $(T)\chi_B = T$. The two expressions are equal, and therefore (3.90′) commutes. Thus (3.90) commutes up to homotopy, (3.88) commutes up to homotopy, and the proof of Proposition 3.83 is complete.

Now we can complete the proof of Theorem 3.5b in Case (ii). In (3.31) we have

$$\beta_n(X, \lambda_{f \otimes \xi \otimes v \otimes \varepsilon})(T)(\gamma) = X(\lambda_{f \otimes \xi \otimes v \otimes \varepsilon}(T)(\gamma))$$
$$= \varepsilon(\xi \wedge \gamma)\langle T, f' \rangle Xv$$
$$= \lambda_{f \otimes \xi \otimes Xv \otimes \varepsilon}(T)(\gamma).$$

Since

$$\alpha_n(X, \, f \, dk \otimes \xi \otimes v \otimes \varepsilon) = f \, dk \otimes \xi \otimes Xv \otimes \varepsilon,$$

we see that

$$\beta_n(X, \lambda_y) = \lambda_{\alpha_n(X,y)}$$

for $y \in C(K) \, dk \otimes_L (\bigwedge^n \mathfrak{c} \otimes_{\mathbb{C}} \mathcal{F}V \otimes_{\mathbb{C}} (\bigwedge^m \mathfrak{c})^*)$. Suppose y is in the kernel of ∂. Let us apply the commutativity of (3.84), writing $[\,\cdot\,]$ for a homology or cohomology element. Then we have

$$\lambda_{X[y]} = [\lambda_{\alpha_n(X,y)}] = [\beta_n(X, \lambda_y)] = X[\lambda_y].$$

Thus (3.31) is $\mathfrak{g}$ equivariant.

8. Proof of Hard Duality in the General Case

In this section we show how to combine Cases (i) and (ii) to complete the proof of Theorem 3.5b in the general case. One step in the argument involves a spectral sequence. Since spectral sequences will not otherwise be used in this book until the end of Chapter V, we postpone this step to §V.11.

Let $i_{\mathrm{gp}}(L) = K' \subseteq K$. Then our map of pairs $(\mathfrak{h}, L) \to (\mathfrak{g}, K)$ factors as

$$(\mathfrak{h}, L) \to (\mathfrak{g}, K') \hookrightarrow (\mathfrak{g}, K),$$

with the first factor satisfying the hypotheses of Case (i) and the second factor satisfying the hypotheses of Case (ii).

The indices m for the factors are

$$m_1 = \dim(i_{\mathrm{alg}}^{-1}(\mathfrak{k}')/\mathfrak{l}) \qquad \text{and} \qquad m_2 = \dim(\mathfrak{k}/\mathfrak{k}'),$$

and they satisfy the key identity

$$(3.91) \qquad\qquad m = m_1 + m_2.$$

In fact, the hypothesis that i_{alg} is onto makes $i_{\mathrm{alg}}^{-1}(\mathfrak{k})/i_{\mathrm{alg}}^{-1}(\mathfrak{k}') \cong \mathfrak{k}/\mathfrak{k}'$ via the map i_{alg}, and thus we have

$$m_1 + m_2 = \dim(i_{\mathrm{alg}}^{-1}(\mathfrak{k}')) - \dim \mathfrak{l} + \dim(i_{\mathrm{alg}}^{-1}(\mathfrak{k})) - \dim(i_{\mathrm{alg}}^{-1}(\mathfrak{k}')) = m.$$

The vector spaces $\mathfrak{c}_1$ and $\mathfrak{c}_2$ for the factors are

$$\mathfrak{c}_1 = i_{\mathrm{alg}}^{-1}(\mathfrak{k}')/\mathfrak{l} \qquad \text{and} \qquad \mathfrak{c}_2 = \mathfrak{k}/\mathfrak{k}',$$

and (3.91) shows that their dimensions add to the dimension of

$$\mathfrak{c} = i_{\mathrm{alg}}^{-1}(\mathfrak{k})/\mathfrak{l}.$$

The top exterior powers $\bigwedge^{m_1}\mathfrak{c}_1$ and $\bigwedge^{m}\mathfrak{c}$ are $(\mathfrak{h}, L)$ modules by Lemma 3.4, and $\bigwedge^{m_2}\mathfrak{c}_2$ is a $(\mathfrak{g}, K')$ module. We make $\bigwedge^{m_2}\mathfrak{c}_2$ into an $(\mathfrak{h}, L)$ module by pulling back the $(\mathfrak{g}, K')$ action via $(i_{\mathrm{alg}}, i_{\mathrm{gp}})$. Then the claim is that

(3.92) $\qquad \bigwedge^{m}\mathfrak{c} \cong (\bigwedge^{m_1}\mathfrak{c}_1) \otimes_{\mathbb{C}} (\bigwedge^{m_2}\mathfrak{c}_2)$ as $(\mathfrak{h}, L)$ modules.

In fact, let us first consider the actions by L. The identity

$$i_{\mathrm{alg}}(\mathrm{Ad}(l)(X + i_{\mathrm{alg}}^{-1}(\mathfrak{k}'))) = \mathrm{Ad}(i_{\mathrm{gp}}(l))(i_{\mathrm{alg}}(X) + \mathfrak{k}')$$

for $X \in i_{\mathrm{alg}}^{-1}(\mathfrak{k})$ and $l \in L$ shows that the action of L on $\mathfrak{c}_2$ via the lift from K_1 matches the adjoint action of L on $i_{\mathrm{alg}}^{-1}(\mathfrak{k})/i_{\mathrm{alg}}^{-1}(\mathfrak{k}')$. Then

$$\mathfrak{c} \cong \mathfrak{c}_1 \oplus \mathfrak{c}_2$$

as L modules, and (3.92) is therefore an isomorphism of L modules. Now let us consider the $\mathfrak{h}$ actions. The $\mathfrak{h}$ action on $\bigwedge^{m_2}\mathfrak{c}_2$ is the lift of the trivial $\mathfrak{g}$ action and is trivial, while the $\mathfrak{h}$ action on $\bigwedge^{m}\mathfrak{c}$ is defined by the same prescription in terms of $\mathfrak{s} = \ker(i_{\mathrm{alg}})$ as the $\mathfrak{h}$ action on $\bigwedge^{m_1}\mathfrak{c}_1$. Thus (3.92) is an isomorphism of $U(\mathfrak{h})$ modules.

By Theorem 3.5a, $(I_{\mathfrak{h},L}^{\mathfrak{g},K'})^j = 0$ for $j > m_1$ and $(I_{\mathfrak{g},K'}^{\mathfrak{g},K})^j = 0$ for $j > m_2$. Combining this fact with the identity (3.91), we shall prove in §V.11 that

(3.93) $\qquad (I_{\mathfrak{g},K'}^{\mathfrak{g},K})^{m_2}(I_{\mathfrak{h},L}^{\mathfrak{g},K'})^{m_1}(V) \cong (I_{\mathfrak{h},L}^{\mathfrak{g},K})^{m}(V)$

naturally for V in $\mathcal{C}(\mathfrak{h}, L)$. Then we have

(3.94)
$(I_{\mathfrak{h},L}^{\mathfrak{g},K})^{m}(V)$

$\begin{aligned}
&\cong (I_{\mathfrak{g},K'}^{\mathfrak{g},K})^{m_2}(I_{\mathfrak{h},L}^{\mathfrak{g},K'})^{m_1}(V) && \text{by (3.93)} \\
&\cong (I_{\mathfrak{g},K'}^{\mathfrak{g},K})^{m_2}(P_{\mathfrak{h},L}^{\mathfrak{g},K'}(V \otimes (\textstyle\bigwedge^{m_1}\mathfrak{c}_1)^*)) && \text{by Case (i)} \\
&\cong P_{\mathfrak{g},K'}^{\mathfrak{g},K}(P_{\mathfrak{h},L}^{\mathfrak{g},K'}(V \otimes (\textstyle\bigwedge^{m_1}\mathfrak{c}_1)^*) \otimes (\textstyle\bigwedge^{m_2}\mathfrak{c}_2)^*) && \text{by Case (ii)} \\
&\cong P_{\mathfrak{g},K'}^{\mathfrak{g},K}(P_{\mathfrak{h},L}^{\mathfrak{g},K'}(V \otimes (\textstyle\bigwedge^{m_1}\mathfrak{c}_1)^* \otimes (\textstyle\bigwedge^{m_2}\mathfrak{c}_2)^*)) && \text{by Corollary 2.97} \\
&\cong P_{\mathfrak{g},K'}^{\mathfrak{g},K}(P_{\mathfrak{h},L}^{\mathfrak{g},K'}(V \otimes (\textstyle\bigwedge^{m}\mathfrak{c})^*)) && \text{by (3.92)} \\
&\cong P_{\mathfrak{h},L}^{\mathfrak{g},K}(V \otimes (\textstyle\bigwedge^{m}\mathfrak{c})^*) && \text{by Proposition 2.19.}
\end{aligned}$

The isomorphism (3.94) proves Theorem 3.5b for $j = 0$. To handle general j, we make use of a cohomological technique known as **dimension shifting**. Namely we shall make use of a long exact cohomology sequence in which every third term is 0, so that we obtain isomorphisms in cohomology that cross from one dimension to the next.

By Corollary 2.37a, V is an $(\mathfrak{h}, L)$ quotient of some $(\mathfrak{h}, L)$ projective module P. Writing M for the kernel, we have an exact sequence

(3.95a) $$0 \longrightarrow M \longrightarrow P \longrightarrow V \longrightarrow 0.$$

Let us tensor with $(\bigwedge^m \mathfrak{c})^*$, denoting the resulting modules M', P', and V'. Then

(3.95b) $$0 \longrightarrow M' \longrightarrow P' \longrightarrow V' \longrightarrow 0.$$

is exact, and P' is projective. We apply $I_{\mathfrak{h},L}^{\mathfrak{g},K}$ and $P_{\mathfrak{h},L}^{\mathfrak{g},K}$ to (3.95a) and (3.95b) respectively, obtaining from (C.36c) and (C.36a) the respective long exact sequences

(3.96a)
$$\longrightarrow (I_{\mathfrak{h},L}^{\mathfrak{g},K})^{m-j}(P) \longrightarrow (I_{\mathfrak{h},L}^{\mathfrak{g},K})^{m-j}(V)$$
$$\longrightarrow (I_{\mathfrak{h},L}^{\mathfrak{g},K})^{m-j+1}(M) \longrightarrow (I_{\mathfrak{h},L}^{\mathfrak{g},K})^{m-j+1}(P) \longrightarrow$$

and

(3.96b)
$$\longrightarrow (P_{\mathfrak{h},L}^{\mathfrak{g},K})_j(P') \longrightarrow (P_{\mathfrak{h},L}^{\mathfrak{g},K})_j(V')$$
$$\longrightarrow (P_{\mathfrak{h},L}^{\mathfrak{g},K})_{j-1}(M') \longrightarrow (P_{\mathfrak{h},L}^{\mathfrak{g},K})_{j-1}(P') \longrightarrow$$

Since P' is projective,

(3.97a) $$(P_{\mathfrak{h},L}^{\mathfrak{g},K})_i(P') = 0 \qquad \text{for } 0 < i \leq m.$$

We shall prove below that

(3.97b) $$(I_{\mathfrak{h},L}^{\mathfrak{g},K})^i(P) = 0 \qquad \text{for } 0 \leq i < m.$$

Accepting this fact for the moment, we prove the formula of Theorem 3.5b (simultaneously for all $(\mathfrak{h}, L)$ modules) by induction on j, the case $j = 0$ being (3.94).

For $j = 1$, we assemble (3.96a) and (3.96b) into a diagram

$$0 \longrightarrow (I_{\mathfrak{h},L}^{\mathfrak{g},K})^{m-1}(V) \longrightarrow (I_{\mathfrak{h},L}^{\mathfrak{g},K})^m(M) \longrightarrow (I_{\mathfrak{h},L}^{\mathfrak{g},K})^m(P)$$

(3.98) $\downarrow$ $\downarrow$

$$0 \longrightarrow (P_{\mathfrak{h},L}^{\mathfrak{g},K})_1(V') \longrightarrow P_{\mathfrak{h},L}^{\mathfrak{g},K}(M') \longrightarrow P_{\mathfrak{h},L}^{\mathfrak{g},K}(P')$$

in which the top row takes account of (3.97b), the bottom row takes account of (3.97a), and the vertical maps are the isomorphisms (3.94). The square at the right commutes. In fact, (3.93) is natural in V, and it follows step-by-step that (3.94) is natural in V. This naturality gives the asserted commutativity. A little diagram chase then shows that (3.98) induces an isomorphism $(I_{\mathfrak{h},L}^{\mathfrak{g},K})^{m-1}(V) \to (P_{\mathfrak{h},L}^{\mathfrak{g},K})_1(V')$, and this completes the proof of the formula of Theorem 3.5b for $j = 1$.

Let $j > 1$, and assume the formula for $j - 1$. For the index j, the end terms in each of (3.96a) and (3.96b) are 0 by (3.97). Thus we obtain isomorphisms

$$\begin{aligned}
(I_{\mathfrak{h},L}^{\mathfrak{g},K})^{m-j}(V) &\cong (I_{\mathfrak{h},L}^{\mathfrak{g},K})^{m-j+1}(M) &&\text{by (3.96a)}\\
&\cong (P_{\mathfrak{h},L}^{\mathfrak{g},K})_{j-1}(M') &&\text{by inductive hypothesis}\\
&\cong (P_{\mathfrak{h},L}^{\mathfrak{g},K})_j(V') &&\text{by (3.96b),}
\end{aligned}$$

and the induction is complete.

Finally we have to prove (3.97b). We have

$$\begin{aligned}
\mathcal{F}_{\mathfrak{g},K}^{\mathfrak{k},K}(I_{\mathfrak{h},L}^{\mathfrak{g},K})^{m-j}(P) &\cong (I_{\mathfrak{h}_K,L}^{\mathfrak{k},K})^{m-j}\mathcal{F}_{\mathfrak{h},L}^{\mathfrak{h}_K,L}(P) &&\text{by Corollary 3.26c}\\
&\cong (P_{\mathfrak{h}_K,L}^{\mathfrak{k},K})_j\mathcal{F}_{\mathfrak{h},L}^{\mathfrak{h}_K,L}(P') &&\text{by the result of §4}\\
&\cong \mathcal{F}_{\mathfrak{g},K}^{\mathfrak{k},K}(P_{\mathfrak{h},L}^{\mathfrak{g},K})_j(P') &&\text{by Corollary 3.26c,}
\end{aligned}$$

and this is 0 by (3.97a).

CHAPTER IV

REDUCTIVE PAIRS

The theory that classifies the irreducible representations of compact connected Lie groups by means of maximal tori and highest weights has a suitable generalization to compact groups that are disconnected. Maximal tori are replaced by "large Cartan subgroups," and highest weights are replaced by the representation types of a Cartan subgroup in spaces of highest-weight vectors.

Compact disconnected groups play a role in the structure of general reductive groups whose identity components have finite center. Reductive groups yield pairs with special properties, and it is possible to characterize the pairs that arise in this way. For a noncompact reductive group, there may be more than one isomorphism class of Cartan subalgebras.

Properties of irreducible finite-dimensional representations of reductive groups follow from the classification theorem for compact disconnected groups. They are given in terms of Cartan subgroups that are as noncompact as possible, and again the representation types of this kind of Cartan subgroup in spaces of highest-weight vectors are the relevant parameters.

Parabolic subalgebras provide a framework for a generalization of Cartan subgroups and highest weights to Levi subgroups and their action in spaces of invariants in finite-dimensional representations.

The Harish-Chandra isomorphism is an algebra isomorphism from the center of the universal enveloping algebra of a reductive Lie algebra to the algebra of Weyl group invariants in the symmetric algebra of a Cartan subalgebra. It is the first tool for working algebraically with infinite-dimensional representations of a reductive Lie algebra. In particular it yields an invariant called the "infinitesimal character" in the case of an irreducible representation.

Use of the Harish-Chandra isomorphism leads to Kostant's Theorem, which computes the Lie algebra homology or cohomology of a representation of a compact group with respect to certain nilpotent subalgebras. In turn Kostant's Theorem yields various explicit cohomological realizations of irreducible representations of compact groups.

1. Review of Cartan-Weyl Theory

Cartan-Weyl theory parametrizes the irreducible (continuous finite-dimensional) representations of a compact connected Lie group in terms of their highest weights. There are two aspects to the theory — a center-independent one that classifies irreducible representations of a complex semisimple Lie algebra and a center-dependent one that incorporates the appropriate integrality conditions for the passage to groups. We shall summarize the theory in this section, emphasizing the center-dependent part of the theory while merging the two aspects and omitting most proofs. For references to proofs and examples, see the Notes for Chapter IV.

Let G be a compact connected Lie group, and let $\mathfrak{g}_0$ be its Lie algebra as usual. Since G is compact, there exists an $\mathrm{Ad}(G)$ invariant inner product on $\mathfrak{g}_0$. We fix such an inner product and write $\langle \cdot, \cdot \rangle$ for its negative, so that $\langle \cdot, \cdot \rangle$ is negative definite.

The Lie algebra $\mathfrak{g}_0$ of G is **reductive** in the sense that $\mathfrak{g}_0$ is fully reducible under ad $\mathfrak{g}_0$. It follows that

$$(4.1) \qquad\qquad \mathfrak{g}_0 = Z_{\mathfrak{g}_0} \oplus [\mathfrak{g}_0, \mathfrak{g}_0],$$

where $Z_{\mathfrak{g}_0}$ is the center of $\mathfrak{g}_0$ and where the commutator subalgebra $[\mathfrak{g}_0, \mathfrak{g}_0]$ is semisimple. This sum is an orthogonal sum with respect to $\langle \cdot, \cdot \rangle$. Let Z_G be the center of G. Corresponding to (4.1) is a decomposition into a product of analytic subgroups

$$(4.2) \qquad\qquad G = (Z_G)_0 G_{ss},$$

where $(Z_G)_0$ is the identity component of Z_G. The subgroups $(Z_G)_0$ and G_{ss} commute, but their product may fail to be direct. The Peter-Weyl Theorem shows that G has a faithful matrix representation, and one can deduce as a consequence that G_{ss} is a closed (hence compact) subgroup of G.

Theorem 4.3 (Weyl's Theorem). The universal covering group of a compact semsimple Lie group is compact.

In (4.2) the universal cover of G_{ss} is compact, by Theorem 4.3, and the universal cover of $(Z_G)_0$ is a Euclidean group. The product of these universal covers maps homomorphically onto G, and thus we obtain the following result.

Corollary 4.4. A compact connected Lie group has a finite cover that is the product of a torus and a compact simply connected semisimple group.

Corollary 4.4 implies that much of the representation theory of G follows from knowing the theory for tori and the theory for compact simply connected semisimple groups. The irreducible representations of a torus T are one dimensional, since T is abelian, and they correspond, under passage to logarithms, to a lattice in $i\mathfrak{t}_0^*$. The irreducible representations of a compact simply connected semisimple group correspond to the irreducible finite-dimensional representations of the Lie algebra and can be classified algebraically.

Let G be a compact connected Lie group. If $\mathfrak{t}_0$ is a maximal abelian subalgebra of the Lie algebra $\mathfrak{g}_0$ of G, then the analytic subgroup T corresponding to $\mathfrak{t}_0$ is a maximal torus of G. A continuous finite-dimensional representation Φ of G is automatically smooth. The restriction $\Phi|_T$ is the direct sum of irreducible representations of T, by Proposition 1.18a. These must be one dimensional, since T is abelian, and each must be a character of the form ξ_λ, where

$$\lambda \in i\mathfrak{t}_0^*$$

and $\qquad\qquad \xi_\lambda(\exp H) = e^{\lambda(H)} \qquad$ for $H \in \mathfrak{t}_0$.

The linear functionals λ determined from Φ in this way are called the **weights** of Φ, and the corresponding simultaneous eigenspaces are called the **weight spaces** of Φ. Members of these eigenspaces are called **weight vectors**. The set of weights does not depend seriously on the choice of $\mathfrak{t}_0$, as a result of the next theorem.

Theorem 4.5. In a compact connected Lie group G, any two maximal abelian subspaces of $\mathfrak{g}_0$ can be mapped onto one another via $\mathrm{Ad}(G)$. Consequently any two maximal tori in G are conjugate.

Continuing with T as above, we say that a linear functional λ on $\mathfrak{t}$ is **analytically integral** if there exists a character ξ_λ of T such that $\xi_\lambda(\exp H) = e^{\lambda(H)}$ for all $H \in \mathfrak{t}_0$. It is equivalent to assume that $\lambda(H) \in 2\pi i\mathbb{Z}$ whenever $\exp H = 1$. Such a linear functional is necessarily imaginary-valued on $\mathfrak{t}_0$ and may be regarded as a member of $i\mathfrak{t}_0^*$. The weights of any finite-dimensional representation are examples of analytically integral elements. The set of analytically integral members of $i\mathfrak{t}_0^*$ encodes information about the center of G by virtue of the following theorem.

Theorem 4.6. In a compact connected Lie group G,
  (a) the centralizer of a torus is always connected
  (b) a maximal torus is always a maximal abelian subgroup
  (c) the center of G is contained in every maximal torus
  (d) the exponential map is onto G.

The $\mathbb{R}$ bilinear form $\langle\,\cdot\,,\cdot\,\rangle$ on $\mathfrak{g}_0$ extends to a $\mathbb{C}$ bilinear form on $\mathfrak{g}$. From there, it restricts to bilinear forms on $\mathfrak{t}$ and $\mathfrak{t}_0$ and induces bilinear forms on $\mathfrak{t}^*$, $\mathfrak{t}_0^*$, and $i\mathfrak{t}_0^*$. On $i\mathfrak{t}_0^*$ this form is positive definite.

Now let Δ be the set of roots of $(\mathfrak{g}, \mathfrak{t})$. Each root is in $i\mathfrak{t}_0^*$ and is analytically integral. A member λ of $\mathfrak{t}^*$ is **algebraically integral** if $2\langle\lambda, \alpha\rangle/\langle\alpha, \alpha\rangle$ is in $\mathbb{Z}$ for each $\alpha \in \Delta$. Analytically integral implies algebraically integral.

The choice of a basis of $i\mathfrak{t}_0^*$ determines a lexicographic ordering of $i\mathfrak{t}_0^*$. Fixing such an ordering, let Δ^+ be the set of positive roots. A member λ of $\mathfrak{t}^*$ is said to be **dominant** if $2\langle\lambda, \alpha\rangle/\langle\alpha, \alpha\rangle \geq 0$ for all $\alpha \in \Delta^+$. The **highest weight** of a finite-dimensional representation Φ is the largest weight in the ordering.

Theorem 4.7 (Theorem of the Highest Weight). Let G be a compact connected Lie group. Apart from equivalence, the irreducible representations Φ of G stand in one-one correspondence with the dominant, analytically integral linear functionals λ on $\mathfrak{t}$, the correspondence being that λ is the highest weight of Φ_λ. The highest weight λ of Φ_λ has these properties:

(a) λ depends only on Δ^+ and not on the particular lexicographic ordering that defines Δ^+

(b) the weight space V_λ corresponding to λ is one dimensional

(c) each root vector for a positive root annihilates the members of V_λ, and the members of V_λ are the only vectors with this property

(d) every weight of Φ_λ is of the form $\lambda - \sum_{\alpha\in\Delta^+} n_\alpha\alpha$ with the n_α integers ≥ 0

(e) every dominant weight μ of Φ_λ other than λ has $|\mu| < |\lambda|$.

Furthermore if G is semisimple and simply connected, then every algebraically integral linear functional on $\mathfrak{t}$ is analytically integral, so that the correspondence in this case is between irreducible representations of G and dominant, algebraically integral, linear functionals on $\mathfrak{t}$.

The hard step in the proof of Theorem 4.7 is the construction of an irreducible representation of G with given highest weight. This construction is carried out first by an algebraic argument in the semisimple simply connected case. Passage to general G is aided by Corollary 4.4 and Theorem 4.6c.

A connected component of the set $i\mathfrak{t}_0 - \cup_{\alpha\in\Delta} \ker \alpha$ is called a **Weyl chamber** in $i\mathfrak{t}_0$. The subset of $H \in i\mathfrak{t}_0$ such that $\alpha(H) > 0$ for all $\alpha \in \Delta^+$ is one such component and is called the **positive** Weyl chamber relative to Δ^+.

If α is a root, then the linear transformation s_α of $\mathfrak{t}^*$ given by

$$s_\alpha(\lambda) = \lambda - \frac{2\langle \lambda, \alpha \rangle}{\langle \alpha, \alpha \rangle} \alpha$$

is called the **root reflection** in α. The transformation group of $\mathfrak{t}^*$ generated by the root reflections is called the **Weyl group** $W(\mathfrak{g}, \mathfrak{t})$. It acts on $\mathfrak{t}$ by duality: $(w\lambda)(H) = \lambda(w^{-1}H)$.

Theorem 4.8. Each root reflection carries Δ into itself and therefore maps Weyl chambers to Weyl chambers. The Weyl group $W(\mathfrak{g}, \mathfrak{t})$ is a finite group and is simply transitive on the set of Weyl chambers. Consequently if Δ^+ and $(\Delta^+)'$ are the positive systems obtained from two lexicographic orderings, there exists one and only one member w of $W(\mathfrak{g}, \mathfrak{t})$ such that $w\Delta^+ = (\Delta^+)'$.

Corollary 4.9. If $\lambda \in \mathfrak{t}^*$ is real on $i\mathfrak{t}_0^*$, then there exists $w \in W(\mathfrak{g}, \mathfrak{t})$ with $w\lambda$ dominant.

The **Weyl group** $W(G, T)$ is defined as the normalizer divided by the centralizer

$$W(G, T) = N_G(T)/Z_G(T),$$

and it acts as a transformation group of T, hence of $\mathfrak{t}$ and $\mathfrak{t}^*$. In acting on $\mathfrak{t}^*$, it permutes the roots and maps analytically integral elements to analytically integral elements. Each root reflection in $W(\mathfrak{g}, \mathfrak{t})$ can be implemented by Ad of an element of $N_G(T)$ and hence corresponds to a member of $W(G, T)$. Thus $W(\mathfrak{g}, \mathfrak{t}) \subseteq W(G, T)$.

Theorem 4.10. For G compact connected, the Weyl groups $W(\mathfrak{g}, \mathfrak{t})$ and $W(G, T)$ coincide. Consequently in an irreducible finite-dimensional representation of G with highest weight λ,

   (a) the set of weights, together with their multiplicities, is invariant under the Weyl group
   (b) any weight μ has $|\mu| \leq |\lambda|$, with equality only if $\mu = w\lambda$ for some w in the Weyl group.

The weights of longest possible length in an irreducible finite-dimensional representation are said to be **extreme**. In view of Theorem 4.10b, these weights are exactly the Weyl-group transforms of the highest weight.

It follows from Theorems 4.8 and 4.10 that the choice of ordering in Theorem 4.7 affects matters only up to conjugacy in G.

One of the techniques in the proof of Theorem 4.7 occurs repeatedly in applications, and we shall isolate it in three lemmas. For $\alpha \in \Delta$, let $\mathfrak{g}_\alpha$ be the root space corresponding to α. Put

$$(4.11) \qquad \mathfrak{n} = \sum_{\alpha \in \Delta^+} \mathfrak{g}_\alpha \quad \text{and} \quad \bar{\mathfrak{n}} = \sum_{\alpha \in \Delta^+} \mathfrak{g}_{-\alpha}.$$

Lemma 4.12. If V is a finite-dimensional $U(\mathfrak{g})$ module, then

$$V = V^{\mathfrak{n}} \oplus \bar{\mathfrak{n}} V.$$

PROOF. Since $\mathfrak{n}$ and $\bar{\mathfrak{n}}$ are contained in $[\mathfrak{g}, \mathfrak{g}]$ by (4.1), we may replace $\mathfrak{g}$ by $[\mathfrak{g}, \mathfrak{g}]$. Changing notation, we may assume that $\mathfrak{g}$ is semisimple. Then V is a direct sum of irreducible representations, and there is no loss of generality in assuming that V is irreducible, say of highest weight λ.

With V irreducible, choose nonzero root vectors e_α for every root α, and let $h_1, \ldots, h_l$ be a basis of $\mathfrak{t}$. By the Poincaré-Birkhoff-Witt Theorem, $U(\mathfrak{g})$ is spanned by all elements

$$e_{-\beta_1} \cdots e_{-\beta_p} h_{i_1} \cdots h_{i_q} e_{\alpha_1} \cdots e_{\alpha_r},$$

where the α_i and β_j are positive roots, not necessarily distinct. Since V is irreducible, V is spanned by all elements

$$e_{-\beta_1} \cdots e_{-\beta_p} h_{i_1} \cdots h_{i_q} e_{\alpha_1} \cdots e_{\alpha_r} v$$

with v in V_λ. Since V_λ is annihilated by $\mathfrak{n}$, such an element is 0 unless $r = 0$. The space V_λ is mapped into itself by $\mathfrak{t}$, and we conclude that V is spanned by all elements

$$e_{-\beta_1} \cdots e_{-\beta_p} v$$

with v in V_λ. If $p > 0$, such an element is in $\bar{\mathfrak{n}} V$ and has weight less than λ, while if $p = 0$, it is in V_λ. Consequently

$$V = V_\lambda \oplus \bar{\mathfrak{n}} V.$$

Theorem 4.7c shows that $V^{\mathfrak{n}}$ is just the λ weight space of V, and the lemma follows.

Lemma 4.13. If V is any finite-dimensional $U(\mathfrak{g})$ module, then the natural map

$$V^{\mathfrak{n}} \longrightarrow V/(\bar{\mathfrak{n}} V)$$

is an isomorphism of $U(\mathfrak{t})$ modules. In particular, if V is irreducible of highest weight λ, then $V/(\bar{\mathfrak{n}} V)$ is one dimensional of weight λ.

PROOF. This is immediate from Lemma 4.12.

Lemma 4.14. Suppose V is a finite-dimensional $U(\mathfrak{g})$ module generated by a subspace V_0 such that

(a) $\mathfrak{n}$ annihilates V_0, and

(b) $\mathfrak{t}$ maps V_0 into itself.

Then $V_0 = V^{\mathfrak{n}}$.

PROOF. Certainly $V_0 \subseteq V^{\mathfrak{n}}$. By Lemma 4.12, it is enough to prove that $V = V_0 + \bar{\mathfrak{n}}V$. Here we are assuming that V_0 generates V. Making use of the notation of Lemma 4.12, but without passing to the irreducible case, we see as a consequence that V is spanned by all elements

$$e_{-\beta_1} \cdots e_{-\beta_p} h_{i_1} \cdots h_{i_q} e_{\alpha_1} \cdots e_{\alpha_r} v$$

with v in V_0. By (a), such an element is 0 unless $r = 0$. Consequently (b) shows that V is spanned by all elements

$$e_{-\beta_1} \cdots e_{-\beta_p} v$$

with v in V_0. If $p > 0$, such an element is in $\bar{\mathfrak{n}}V$, and if $p = 0$, it is in V_0. Consequently

$$V = V_0 + \bar{\mathfrak{n}}V,$$

and the result follows.

It is instructive to describe the representation Φ_λ of Theorem 4.7 in terms of the constructions in Chapters I and II. With $\mathfrak{n}$ and $\bar{\mathfrak{n}}$ as in (4.11), put

$$\mathfrak{b} = \mathfrak{t} \oplus \mathfrak{n} \qquad \text{and} \qquad \bar{\mathfrak{b}} = \mathfrak{t} \oplus \bar{\mathfrak{n}}.$$

Since λ is analytically integral, we can make T act on $\mathbb{C}$ by ξ_λ; we use $\mathbb{C}_\lambda$ as notation for the resulting $(\mathfrak{t}, T)$ module. We can extend $\mathbb{C}_\lambda$ to a $(\bar{\mathfrak{b}}, T)$ module by having $\bar{\mathfrak{n}}$ act as 0; this extension is nothing more than $\mathcal{F}_{\mathfrak{t},T}^{\bar{\mathfrak{b}},T}(\mathbb{C}_\lambda)$, in the notation of Example 3 of §II.1. A realization of Φ_λ is as the $(\mathfrak{g}, G)$ module

$$(4.15) \qquad I_{\bar{\mathfrak{b}},T}^{\mathfrak{g},G}(\mathcal{F}_{\mathfrak{t},T}^{\bar{\mathfrak{b}},T}(\mathbb{C}_\lambda)).$$

If we adjust notation and regard $\mathbb{C}_\lambda$ as a $(\bar{\mathfrak{b}}, T)$ module, the module (4.15) can be rewritten as

$$\Gamma_{\mathfrak{g},T}^{\mathfrak{g},G}(\mathrm{pro}_{\bar{\mathfrak{b}},T}^{\mathfrak{g},T}(\mathbb{C}_\lambda)),$$

which looks like an algebraic analog of the realization of Φ_λ in the Borel-Weil Theorem. (See the Introduction.)

For a proof that (4.15) indeed realizes Φ_λ, let V be an irreducible $(\mathfrak{g}, G)$ module of highest weight μ. Then

(4.16)
$$\mathrm{Hom}_{\mathfrak{g},G}(V, I_{\bar{\mathfrak{b}},T}^{\mathfrak{g},G}(\mathcal{F}_{\mathfrak{t},T}^{\bar{\mathfrak{b}},T}(\mathbb{C}_\lambda)))$$

$$\cong \mathrm{Hom}_{\bar{\mathfrak{b}},T}(V, \mathcal{F}_{\mathfrak{t},T}^{\bar{\mathfrak{b}},T}(\mathbb{C}_\lambda)) \qquad \text{by Proposition 2.21}$$

$$\cong \mathrm{Hom}_{\bar{\mathfrak{b}},T}(V, (\mathcal{F}^\vee)_{\mathfrak{t},T}^{\bar{\mathfrak{b}},T}(\mathbb{C}_\lambda)) \qquad \text{by Proposition 2.33}$$

$$\cong \mathrm{Hom}_{\mathfrak{t},T}(P_{\bar{\mathfrak{b}},T}^{\mathfrak{t},T}(V), \mathbb{C}_\lambda) \qquad \text{by Proposition 2.34}$$

$$\cong \mathrm{Hom}_{\mathfrak{t},T}(V_{\bar{\mathfrak{n}}}, \mathbb{C}_\lambda) \qquad \text{by Proposition 2.89a}$$

$$\cong \mathrm{Hom}_{\mathfrak{t},T}(V^{\mathfrak{n}}, \mathbb{C}_\lambda) \qquad \text{by Lemma 4.13.}$$

Since $V^{\mathfrak{n}} \cong \mathbb{C}_\mu$ by Theorem 4.7c, the right side of (4.16) is one dimensional if $\mu = \lambda$ and is 0 if $\mu \neq \lambda$. Since (4.15) is fully reducible, it is determined by the multiplicities of its irreducible constituents, and it follows that (4.15) is irreducible with highest weight λ.

There is an alternate realization of Φ_λ that does not relate to the Borel-Weil Theorem but that will be useful in §2. Specifically Φ_λ is given by the $(\mathfrak{g}, G)$ module

(4.17)
$$P_{\mathfrak{b},T}^{\mathfrak{g},G}(\mathcal{F}_{\mathfrak{t},T}^{\mathfrak{b},T}(\mathbb{C}_\lambda)).$$

For a proof, again let V be an irreducible $(\mathfrak{g}, G)$ module of highest weight μ. Then

(4.18)
$$\mathrm{Hom}_{\mathfrak{g},G}(P_{\mathfrak{b},T}^{\mathfrak{g},G}(\mathcal{F}_{\mathfrak{t},T}^{\mathfrak{b},T}(\mathbb{C}_\lambda)), V)$$

$$\cong \mathrm{Hom}_{\mathfrak{b},T}(\mathcal{F}_{\mathfrak{t},T}^{\mathfrak{b},T}(\mathbb{C}_\lambda), (\mathcal{F}^\vee)_{\mathfrak{g},G}^{\mathfrak{b},T}(V)) \qquad \text{by Proposition 2.34}$$

$$\cong \mathrm{Hom}_{\mathfrak{b},T}(\mathcal{F}_{\mathfrak{t},T}^{\mathfrak{b},T}(\mathbb{C}_\lambda), V) \qquad \text{by Proposition 2.33}$$

$$\cong \mathrm{Hom}_{\mathfrak{t},T}(\mathbb{C}_\lambda, I_{\mathfrak{b},T}^{\mathfrak{t},T}(V)) \qquad \text{by Proposition 2.21}$$

$$\cong \mathrm{Hom}_{\mathfrak{t},T}(\mathbb{C}_\lambda, V^{\mathfrak{n}}) \qquad \text{by Proposition 2.89b.}$$

Since $V^{\mathfrak{n}} \cong \mathbb{C}_\mu$ by Theorem 4.7c, the right side of (4.18) is one dimensional if $\mu = \lambda$ and is 0 if $\mu \neq \lambda$. Since (4.17) is fully reducible, it follows that (4.17) is irreducible with highest weight λ.

2. Cartan-Weyl Theory for Disconnected Groups

Let G be a compact Lie group, possibly disconnected, and let G_0 be its identity component. In this section we shall give a parametrization of the irreducible representations of G.

EXAMPLE. Let $G = O(2)$. There are two one-dimensional representations of $O(2)$, namely the trivial representation and the determinant, and the other irreducible representations are two dimensional. They are parametrized by an integer $n \geq 1$ and are given, up to equivalence, by

$$\begin{pmatrix} \cos\theta & \sin\theta \\ -\sin\theta & \cos\theta \end{pmatrix} \rightarrow \begin{pmatrix} \cos n\theta & \sin n\theta \\ -\sin n\theta & \cos n\theta \end{pmatrix} \quad \text{and} \quad \begin{pmatrix} 1 & 0 \\ 0 & -1 \end{pmatrix} \rightarrow \begin{pmatrix} 1 & 0 \\ 0 & -1 \end{pmatrix}.$$

Returning to a general G, fix a maximal torus T_0 of G_0, let Δ be the system of roots for $(\mathfrak{g}, \mathfrak{t})$, and let $\mathfrak{g}_\alpha$ be the root space for the root α. Introduce a lexicographic ordering for $i\mathfrak{t}_0^*$, let Δ^+ be the corresponding system of positive roots, and define

$$(4.19) \qquad \mathfrak{n} = \sum_{\alpha \in \Delta^+} \mathfrak{g}_\alpha \qquad \text{and} \qquad \mathfrak{b} = \mathfrak{t} \oplus \mathfrak{n}.$$

The centralizer $Z_G(\mathfrak{t}_0)$ is sometimes called a "small Cartan subgroup" of G, and the normalizer $N_G(\mathfrak{b})$ is sometimes called a "large Cartan subgroup." We shall use only the latter group, and we call

$$(4.20) \qquad T = N_G(\mathfrak{b})$$

the **Cartan subgroup** of G corresponding to $\mathfrak{t}$ and Δ^+. The group T normalizes $\mathfrak{t}$ and $\mathfrak{n}$ separately. Since it normalizes T, it acts on $\mathfrak{t}$ and $\mathfrak{t}^*$, sending roots to roots and analytically integral members of $\mathfrak{t}^*$ to analytically integral members.

Note that

$$(4.21) \qquad T_0 \subseteq Z_G(\mathfrak{t}_0) \subseteq N_G(\mathfrak{b}).$$

If G is connected, equality holds in (4.21): The first inclusion becomes an equality by Theorem 4.6a, and the second inclusion becomes an equality by Theorems 4.8 and 4.10.

EXAMPLES.

1) Let $G = O(2)$. For this group, T_0 is $SO(2)$, Δ is empty, $Z_G(t_0)$ is $SO(2)$, and T is $O(2)$.

2) Let $G = O(2n)$ with $n > 1$. For this group T_0 is the usual product of n circles, each imbedded as an $SO(2)$, and $Z_G(t_0)$ equals T_0. The group T has two components. Its exact form depends on the choice of Δ^+, but the nonidentity component arises from an outer automorphism of the Dynkin diagram of $SO(2n)$, which is of type D_n.

3) Let $G = O(2n + 1)$. For this group T_0 is a product of n circles, and $Z_G(t_0)$ has two components, consisting of T_0 and the product of T_0 with the matrix $-I$. The group T coincides with $Z_G(t_0)$, no matter how Δ^+ is chosen.

Proposition 4.22. If G is compact, then T_0 has finite index in T. Every element of G is the product of a member of T with a member of G_0, and $T \cap G_0 = T_0$.

PROOF. A curve from 1 in $T = N_G(\mathfrak{b})$ must lie in G_0, hence in $N_{G_0}(\mathfrak{b})$. We have just seen that this latter group equals T_0. Thus $N_G(\mathfrak{b})$ has Lie algebra t_0. The group $N_G(\mathfrak{b})/T_0$ is therefore compact and 0-dimensional, hence finite.

Let x be in G. Since $\mathrm{Ad}(x)^{-1}t_0$ is maximal abelian in $\mathfrak{g}_0$, Theorem 4.5 gives us an element $y \in G_0$ with $\mathrm{Ad}(y)\mathrm{Ad}(x)^{-1}t_0 = t_0$. Then yx^{-1} normalizes t_0, and it must act on the roots, sending Δ^+ to another positive system $(\Delta^+)'$. By Theorems 4.8 and 4.10, we can choose $z \in N_{G_0}(T_0)$ such that $z(\Delta^+)' = \Delta^+$. Then $zyx^{-1} = n$ is in $T = N_G(\mathfrak{b})$, and $x = n^{-1}(zy)$ exhibits x as in TG_0.

Finally $T \cap G_0 = N_{G_0}(\mathfrak{b})$, and we have seen that this group coincides with T_0. Hence $T \cap G_0 = T_0$.

In an irreducible representation of G_0 on a space V, the highest-weight space coincides with the subspace V^n of n invariants, by Theorem 4.7c; the space V^n is a representation space for T_0. Since any finite-dimensional representation of G_0 is fully reducible, it follows for a general V that a decomposition of V^n into irreducible representations under T_0 yields a decomposition of V into irreducible representations under G_0. Specifically if

$$(4.23a) \qquad\qquad V^n = \bigoplus_{i=1}^{m} \mathbb{C}v_i$$

with $\mathbb{C}v_i$ irreducible under T_0, then

$$(4.23b) \qquad\qquad V = \bigoplus_{i=1}^{m} U(\mathfrak{g})v_i$$

with $U(\mathfrak{g})v_i$ irreducible under G_0.

Now let V be the space for an irreducible representation of G, and decompose V^n and V under T_0 and G_0 as in (4.23). If λ_i is the weight of v_i under T_0 in (4.23a), then λ_i is Δ^+ dominant by Theorem 4.7.

Lemma 4.24. If V is irreducible under G, then V^n is irreducible under T.

PROOF. Assuming the contrary, let W be a proper nonzero T invariant subspace of V^n. Since $T \supseteq T_0$, we can write $W = \bigoplus \mathbb{C}w_i$ with $\mathbb{C}w_i$ irreducible under T_0. By Lemma 4.14,

$$V' = \bigoplus_{i=1}^{l} U(\mathfrak{g})w_i$$

is a proper nonzero G_0 invariant subspace of V. If t is in T, then

$$t\left(\sum_i u_i w_i\right) = \sum_i (\mathrm{Ad}(t)u_i) t w_i = \sum_i (\mathrm{Ad}(t)u_i) \sum_j c_{ji} w_j$$

is in V', and hence V' is T invariant. By Proposition 4.22, V' is G invariant, in contradiction to the irreducibility of V.

The representation of T on V^n has the special property that all its weights λ_i under T_0 are Δ^+ dominant. We call such a representation of T a **dominant representation**. The action of T on analytically integral members of $\mathfrak{t}^*$ gives an action of T directly on the weights of V^n, and the irreducibility forces these weights to lie in a single orbit. For t in $\mathfrak{t}$, α in Δ^+, and λ dominant in $\mathfrak{t}^*$, we have

$$\langle t\lambda, \alpha \rangle = \langle \lambda, t^{-1}\alpha \rangle \geq 0.$$

Thus an irreducible representation of T is dominant as soon as one of its weights is dominant.

It is easy to see that the multiplicities of the weights of an irreducible representation of T are all the same. However, the common multiplicity need not be 1, even in cases of interest. (For example, take $G = T$ to be a nonabelian finite group and V to be an irreducible representation of dimension greater than 1.)

Theorem 4.25. Let G be a compact Lie group with Cartan subgroup T relative to a maximal abelian subalgebra $\mathfrak{t}_0$ of $\mathfrak{g}_0$ and a positive system of roots Δ^+. Apart from equivalence, the irreducible representations (Φ, V) of G stand in one-one correspondence with the irreducible dominant representations π of T, the correspondence being that π is the restriction of $\Phi|_T$ to V^n.

PROOF. Everything has been proved above except that the correspondence $\Phi \to \pi$ is one-one and onto. Let us now show that the correspondence is one-one. Let (Φ_1, V_1) and (Φ_2, V_2) be irreducible representations with the property that there is a T isomorphism $e : V_1^n \to V_2^n$. If

$$V_1^n = \bigoplus_{i=1}^{m} \mathbb{C}v_i, \qquad v_i \text{ of weight } \lambda_i,$$

is a decomposition of V_1^n under T_0, then V_2^n decomposes under T_0 as

$$V_2^n = \bigoplus_{i=1}^{m} \mathbb{C}e(v_i), \qquad e(v_i) \text{ of weight } \lambda_i.$$

The spaces $U(\mathfrak{g})v_i$ and $U(\mathfrak{g})e(v_i)$ are irreducible under G_0 with common highest weight λ_i, and we let $E : U(\mathfrak{g})v_i \to U(\mathfrak{g})e(v_i)$ be a G_0 isomorphism with $E(v_i) = e(v_i)$. Since the spaces $U(\mathfrak{g})v_i$ are linearly independent as i varies, we obtain a G_0 isomorphism E from the sum

$$V_1 = \bigoplus_{i=1}^{m} U(\mathfrak{g})v_i$$

to the sum $\qquad \qquad V_2 = \bigoplus_{i=1}^{m} U(\mathfrak{g})e(v_i),$

and E is an extension of the T isomorphism $e : V_1^n \to V_2^n$. If $t \in T$ is given, we have

$$
\begin{aligned}
Et\left(\sum u_i v_i\right) &= E\left(\sum (\mathrm{Ad}(t)u_i)(tv_i)\right) = \sum (\mathrm{Ad}(t)u_i)E(tv_i) \\
&= \sum (\mathrm{Ad}(t)u_i)e(tv_i) = \sum (\mathrm{Ad}(t)u_i)te(v_i) \\
&= t\sum u_i e(v_i) = tE\left(\sum u_i v_i\right).
\end{aligned}
$$

Thus E commutes with the action of T. By Proposition 4.22, E commutes with the action of G. Thus the correspondence is one-one.

To prove that the correspondence is onto, let (π, W) be an irreducible dominant representation of T. Let λ be a weight of W, and let (τ, V_0) be an irreducible representation of G_0 with highest weight λ and with v_0 as a nonzero highest-weight vector. Form the classically induced representation

$$V' = \mathrm{induced}_{G_0}^{G}(V_0),$$

as defined in (2.74). The members of the space for V' are functions $f : G \to V_0$ with

(4.26) $\qquad f(gg_0) = \tau(g_0)^{-1} f(g) \qquad$ for all $g_0 \in G_0$, $g \in G$,

and G acts by the left-regular representation.

According to Proposition 4.22, every element of G is of the form tg_0 with $t \in T$ and $g_0 \in G_0$. For $X \in \mathfrak{g}$ and $f \in V'$, (2.74c) and (2.74b) give

$$Xf(tg_0) = \tau(g_0)^{-1} \tau(\mathrm{Ad}(t)^{-1}X)(f(t)).$$

Applying this formula to a root vector for a positive root, we see that f is in $(V')^{\mathfrak{n}}$ if and only if $f(T) \subseteq \mathbb{C}v_0$.

Let us establish an isomorphism of T representations

(4.27) $\qquad\qquad (V')^{\mathfrak{n}} \cong \mathrm{induced}_{T_0}^T(\mathbb{C}_\lambda).$

Let $t_1, \dots, t_n$ be coset representatives for T/T_0. According to Proposition 4.22, the t_i are also coset representatives for G/G_0. If values $f(t_i)$ are given for all i, (4.26) uniquely determines a member of $(V')^{\mathfrak{n}}$ with these values, and the version of (4.26) for the right side of (4.27) uniquely determines a member of that space. Hence we get a one-one onto map as in (4.27). The T actions correspond, and thus we have the claimed isomorphism.

By classical Frobenius reciprocity, (4.27) yields

$$(V')^{\mathfrak{n}} \cong \bigoplus_{W' \in \widehat{T}} [W'|_{T_0} : \mathbb{C}_\lambda] \, W'.$$

Since $W|_{T_0}$ contains $\mathbb{C}_\lambda$, W is isomorphic with a direct summand of $(V')^{\mathfrak{n}}$. Let V be the smallest G invariant subspace of V' containing this copy of W. The space V is just the subspace $U(\mathfrak{g})W$. By Lemma 4.14 its space of $\mathfrak{n}$ invariants is exactly W, and therefore it is irreducible. This completes the proof that the correspondence $\Phi \to \pi$ is onto.

Corollary 4.28. Let G be a compact Lie group with Cartan subgroup T relative to a maximal abelian subalgebra $\mathfrak{t}_0$ of $\mathfrak{g}_0$ and a positive system of roots Δ^+. If (π, W) is an irreducible dominant representation of T, then the corresponding irreducible representation of G is given by either of the formulas

$$V = P_{\mathfrak{b},T}^{\mathfrak{g},G}(\mathcal{F}_{\mathfrak{t},T}^{\mathfrak{b},T}(W))$$

and

$$V = I_{\bar{\mathfrak{b}},T}^{\mathfrak{g},G}(\mathcal{F}_{\mathfrak{t},T}^{\bar{\mathfrak{b}},T}(W)).$$

PROOF. By Theorem 4.25, let V' be the irreducible representation of G corresponding to an irreducible dominant representation (π', W') of T. By the same calculation as in (4.18),

$$\mathrm{Hom}_{\mathfrak{g},G}(P_{\mathfrak{b},T}^{\mathfrak{g},G}(\mathcal{F}_{\mathfrak{t},T}^{\mathfrak{b},T}(W)),\ V') \cong \mathrm{Hom}_{\mathfrak{t},T}(W,\ (V')^{\mathfrak{n}}) \cong \mathrm{Hom}_{\mathfrak{t},T}(W,\ W').$$

Hence the only irreducible representation of G, up to equivalence, that occurs in $P_{\mathfrak{b},T}^{\mathfrak{g},G}(\mathcal{F}_{\mathfrak{t},T}^{\mathfrak{b},T}(W))$ is V, and it occurs with multiplicity 1. This proves the first identity, and the second follows from the isomorphisms

$$\mathrm{Hom}_{\mathfrak{g},G}(V',\ I_{\bar{\mathfrak{b}},T}^{\mathfrak{g},G}(\mathcal{F}_{\mathfrak{t},T}^{\bar{\mathfrak{b}},T}(W))) \cong \mathrm{Hom}_{\mathfrak{t},T}((V')_{\bar{\mathfrak{n}}},\ W) \cong \mathrm{Hom}_{\mathfrak{t},T}((V')^{\mathfrak{n}},\ W)$$

that are proved in the same way as in (4.16).

3. Reductive Groups and Reductive Pairs

In the representation theory of semisimple Lie groups, representations that are induced from closed subgroups (especially parabolic subgroups) play an important role. As a result, one is quickly forced to enlarge the class of groups under study to allow for some disconnectedness and for a positive-dimensional center. Groups in the enlarged class are always called "reductive," but their characterizing properties vary from author to author. We shall use the following definition, which incorporates a certain amount of structure theory that could be proved from weaker hypotheses.

DEFINITION 4.29. A **reductive group** for our purposes is actually a 4-tuple $(G, K, \theta, \langle \cdot, \cdot \rangle)$ consisting of a real Lie group G, a compact subgroup K of G, a Lie algebra involution θ of $\mathfrak{g}_0$, and a nondegenerate, $\mathrm{Ad}(G)$ invariant, θ invariant, bilinear form $\langle \cdot, \cdot \rangle$ on $\mathfrak{g}_0$ such that

  (i) $\mathfrak{g}_0$ is a **reductive Lie algebra** (i.e., $\mathfrak{g}_0$ is fully reducible under $\mathrm{ad}\,\mathfrak{g}_0$)
 (ii) the decomposition of $\mathfrak{g}_0$ into $+1$ and -1 eigenspaces under θ is $\mathfrak{g}_0 = \mathfrak{k}_0 \oplus \mathfrak{p}_0$, where $\mathfrak{k}_0$ is the Lie algebra of K
(iii) multiplication, as a map from $K \times \exp\mathfrak{p}_0$ into G, is a diffeomorphism onto
 (iv) $\mathfrak{k}_0$ and $\mathfrak{p}_0$ are orthogonal under $\langle \cdot, \cdot \rangle$, and $\langle \cdot, \cdot \rangle$ is positive definite on $\mathfrak{p}_0$ and negative definite on $\mathfrak{k}_0$.

If also

(v) every automorphism $\mathrm{Ad}(g)$ of $\mathfrak{g}$ is **inner** for $g \in G$, i.e., is given
by $\mathrm{Ad}(x)$ for some $x \in G_{\mathbb{C}}$, where $G_{\mathbb{C}}$ is any complex connected
Lie group with Lie algebra $\mathfrak{g}$,

then we say that G is in the **Harish-Chandra class**. Since (iii) already
implies that G has finitely many components, this usage of "Harish-
Chandra class" is consistent with customary usage.

EXAMPLES.

1) $G = K =$ any compact Lie group, $\theta = 1, \langle \cdot , \cdot \rangle$ equal to the negative
of a K invariant inner product on $\mathfrak{k}_0$.

2) $G =$ any connected semisimple Lie group with finite center, K
maximal compact, $\theta =$ Cartan involution, $\langle \cdot , \cdot \rangle$ equal to the Killing
form.

3) $G = GL(n, \mathbb{R})$, $K = O(n)$, $\theta =$ negative transpose, $\langle \cdot , \cdot \rangle$ equal to
the trace form.

4) $G =$ real points of a connected reductive algebraic group defined
over $\mathbb{R}$.

5) $G =$ normalizer in $G_{\mathbb{C}}$ of $\mathfrak{g}_0$, where $G_{\mathbb{C}}$ is a complex connected
semisimple Lie group and $\mathfrak{g}_0$ is a real form of its Lie algebra $\mathfrak{g}$.

Examples 2–5 are in the Harish-Chandra class, but Example 1 need
not be. For example, $O(n)$ is not in the Harish-Chandra class if n is
even. In view of Example 3, this fact about $O(n)$ illustrates a problem
with confining our attention to the Harish-Chandra class: Even if G is
in the class, K may not be. People who work with algebraic groups,
studying groups defined over locally compact fields other than $\mathbb{R}$, will
not be bothered by this difficulty; in the theory of algebraic groups,
there is no reason to expect K to have the same kinds of nice properties
that G does. But people who work with representation theory of Lie
groups have reason for concern; if the goal is an understanding of the
representation theory of all Lie groups of type I (perhaps with finitely
many components), then reductive groups form one of the building
blocks for the theory, and G and K should be considered on the same
footing.

Anyway we shall work with all reductive groups when it is convenient
to do so, and we shall restrict to those in the Harish-Chandra class when
necessary.

Let G be a reductive group. Since $\mathfrak{g}_0$ is reductive, $\mathfrak{g}_0$ is the direct sum
of ideals

$$\mathfrak{g}_0 = Z_{\mathfrak{g}_0} \oplus [\mathfrak{g}_0, \mathfrak{g}_0]$$

with $[\mathfrak{g}_0, \mathfrak{g}_0]$ semisimple. Enlarging the scope of the standard terminology used in Example 2, we extend θ complex linearly to $\mathfrak{g}$ and call it the **Cartan involution** of $\mathfrak{g}_0$ and of $\mathfrak{g}$. Also we call $\langle \cdot, \cdot \rangle$ the **invariant bilinear form** for G. This form will be restricted, complexified, and transferred to dual vector spaces without change of notation.

The involution $\theta : \mathfrak{g}_0 \to \mathfrak{g}_0$ lifts to an involution $\theta : G \to G$ such that $\theta = 1$ on K. The involution $\theta : G \to G$ is called the **Cartan involution** of G and is given by the formula

$$\theta(kx) = kx^{-1} \qquad \text{for } k \in K \text{ and } x \in \exp \mathfrak{p}_0.$$

By (iii), θ is well defined on G. To see that θ is multiplicative, we take as known that θ is multiplicative on G_0. We note from $\mathrm{Ad}(k)\mathfrak{p}_0 \subseteq \mathfrak{p}_0$ that

$$\theta(kxk^{-1}) = k\theta(x)k^{-1} \qquad \text{for } k \in K \text{ and } x \in G.$$

Property (iii) shows that $G = KG_0$ and that

$$\theta(kg_0) = k\theta(g_0) \qquad \text{for } k \in K \text{ and } g_0 \in G_0.$$

If kg_0 and $k'g_0'$ are given, then

$$\begin{aligned}
\theta(kg_0k'g_0') &= \theta(kk'k'^{-1}g_0k'g_0') = kk'\theta(k'^{-1}g_0k'g_0') \\
&= kk'\theta(k'^{-1}g_0k')\theta(g_0') = kk'(k'^{-1}\theta(g_0)k')\theta(g_0') \\
&= k\theta(g_0)k'\theta(g_0') = \theta(kg_0)\theta(k'g_0').
\end{aligned}$$

Hence θ is an involution of G.

The constructions in this book are associated with pairs, not with groups, and it is appropriate to formulate the notion of "reductive group" in terms of pairs. If G is a reductive group with complexified Lie algebra $\mathfrak{g}$, then $(\mathfrak{g}, K)$ is a reductive pair in the following sense.

DEFINITION 4.30. A **reductive pair** $(\mathfrak{g}, K)$ is actually a tuple $((\mathfrak{g}, K), \mathfrak{g}_0, \theta, \langle \cdot, \cdot \rangle)$ consisting of a pair $(\mathfrak{g}, K)$, a real form $\mathfrak{g}_0$ of $\mathfrak{g}$, a Lie algebra involution θ of $\mathfrak{g}_0$, and a nondegenerate bilinear form $\langle \cdot, \cdot \rangle$ on $\mathfrak{g}_0$ that is $\mathrm{Ad}(K)$ invariant and is skew symmetric under $\mathrm{ad}\,\mathfrak{g}_0$. Moreover, it is assumed that

(i) $\mathfrak{g}_0$ is a reductive Lie algebra
(ii) the decomposition of $\mathfrak{g}_0$ into $+1$ and -1 eigenspaces under θ is $\mathfrak{g}_0 = \mathfrak{k}_0 \oplus \mathfrak{p}_0$, where $\mathfrak{k}_0$ is the Lie algebra of K
(iii) $\mathfrak{k}_0$ and $\mathfrak{p}_0$ are orthogonal under $\langle \cdot, \cdot \rangle$, and $\langle \cdot, \cdot \rangle$ is positive definite on $\mathfrak{p}_0$ and negative definite on $\mathfrak{k}_0$.

Proposition 4.31. Each reductive pair $(\mathfrak{g}, K)$ arises from a reductive group G, and the reductive group G is unique up to isomorphism.

PROOF OF UNIQUENESS.   Let G_1 and G_2 be reductive groups that lead to the same pair $(\mathfrak{g}, K)$. By (iii) in Definition 4.29, $G_1 = K(G_1)_0$. Multiplication in G_1 is determined by multiplication within K and within $(G_1)_0$, and by the formula for conjugation of elements of K on $(G_1)_0$. The latter is determined by the action of $\mathrm{Ad}(K)$ on $\mathfrak{g}_0$, which is a property of the pair $(\mathfrak{g}, K)$. Consequently $G_1 \cong G_2$ will follow if we prove that $(G_1)_0 \cong (G_2)_0$.

Let $\tilde{G}_0$ be a simply connected group with Lie algebra $\mathfrak{g}_0$, and let $\tilde{K}_0$ be the analytic subgroup corresponding to $\mathfrak{k}_0$. Since $\tilde{G}_0$ and $(G_1)_0$ both have $\mathfrak{g}_0$ as Lie algebra, there exists a covering homomorphism $\varphi : \tilde{G}_0 \to (G_1)_0$, and $\varphi(\tilde{K}_0) = K_0$. The kernel Z of φ on $\tilde{G}_0$ is contained in $\tilde{K}_0$ and is thus determined by the reductive pair. Since $\tilde{G}_0$ is determined by the pair, so is $\tilde{G}_0/Z \cong (G_1)_0$. Arguing similarly with G_2, we obtain $(G_1)_0 \cong (G_2)_0$.

PROOF OF EXISTENCE.   First we construct a connected reductive group G_0 that leads to the pair $(\mathfrak{g}, K_0)$. Let $\tilde{G}_0$ be a simply connected group with Lie algebra $\mathfrak{g}_0$, and let $\tilde{K}_0$ be the analytic subgroup corresponding to $\mathfrak{k}_0$. Since $\tilde{K}_0$ and K_0 have the same Lie algebra, there exists a covering homomorphism $\varphi : \tilde{K}_0 \to K_0$. Let Z be the kernel of φ. By definition of a pair, K_0 has a representation Ad on $\mathfrak{g}$ with differential $\mathrm{ad}\,\mathfrak{k}_0 \subseteq \mathrm{ad}\,\mathfrak{g}$, and by construction, the representation Ad of $\tilde{K}_0$ on $\mathfrak{g}$ has differential $\mathrm{ad}\,\mathfrak{k}_0$. Therefore the representation of $\tilde{K}_0$ on $\mathfrak{g}$ descends to K_0, and Z must act trivially. Consequently $Z \subseteq Z_{\tilde{G}_0}$, and it makes sense to define $G_0 = \tilde{G}_0/Z$. Property (iii) for G_0 in Definition 4.29 follows from the general theory of connected reductive groups, and the $\mathrm{Ad}(G_0)$ invariance of $\langle \cdot , \cdot \rangle$ follows from the skew symmetry of $\langle \cdot , \cdot \rangle$ under $\mathrm{ad}\,\mathfrak{g}_0$. Thus G_0 is a reductive group that leads to the pair $(\mathfrak{g}, K_0)$.

In the general case, we first construct G_0 as above. Then we attempt to lift $\mathrm{Ad}(k) : \mathfrak{g}_0 \to \mathfrak{g}_0$, for $k \in K$, to a homomorphism (hence isomorphism) $\mathfrak{g}_0 \to \mathfrak{g}_0^k$ of G_0. In terms of the simply connected cover $\tilde{G}_0$, we at least get an isomorphism of $\tilde{G}_0$ with differential $\mathrm{Ad}(k)$. If we follow this isomorphism with the covering homomorphism $\varphi : \tilde{G}_0 \to G_0$, then the kernel of the composition is contained in $\tilde{K}_0$ since $\mathrm{Ad}(k)(\mathfrak{k}_0) \subseteq \mathfrak{k}_0$. So the question of descent to G_0 is one of descent to K_0. But conjugation of K_0 by k has differential $\mathrm{Ad}(k)$. Thus the descent is possible from $\tilde{K}_0$ to K_0. It follows that $\mathfrak{g}_0 \to \mathfrak{g}_0^k$ is well defined.

Now let

$$G = K \times_{K_0} G_0 = \{(k, g_0) \mid k \in K, \ g_0 \in G\}/ \sim,$$

where

$$(kk_0, g_0) \sim (k, k_0 g_0) \qquad \text{for } k \in K, \; g_0 \in G_0, \; k_0 \in K_0.$$

The product formula is

$$(k, g_0)(k', g_0') = (kk', g_0^{k'^{-1}} g_0').$$

As a topological space, $K \times_{K_0} G_0$ gets the quotient topology and is just a union of finitely many components of type G_0. Easy calculation shows that the product formula is well defined and associative, and it follows that G is the required reductive group.

4. Cartan Subpairs

Let $(\mathfrak{g}, K)$ be a reductive pair in the sense of Definition 4.30, and let G be the corresponding reductive group given by Proposition 4.31. In this section we shall introduce a notion of Cartan subgroup for G that generalizes the definition (4.20) in the compact case and that translates into a notion of Cartan subpair for $(\mathfrak{g}, K)$. When G is in the Harish-Chandra class, the definitions depend only on the Lie algebra of the Cartan subgroup. But for general G, the definitions depend also on a choice of positive system of roots, as in (4.20). We therefore begin with notation for roots.

A θ **stable Cartan subalgebra** $\mathfrak{h}_0$ of $\mathfrak{g}_0$ is a maximal abelian subalgebra of $\mathfrak{g}_0$ such that $\theta(\mathfrak{h}_0) = \mathfrak{h}_0$. The complexification $\mathfrak{h}$ of $\mathfrak{h}_0$ is a Cartan subalgebra of $\mathfrak{g}$, and $\mathfrak{h}_0$ has a direct-sum decomposition

$$\mathfrak{h}_0 = \mathfrak{t}_0 \oplus \mathfrak{a}_0 \qquad \text{with } \mathfrak{t}_0 = \mathfrak{h}_0 \cap \mathfrak{k}_0 \text{ and } \mathfrak{a}_0 = \mathfrak{h}_0 \cap \mathfrak{p}_0.$$

The set of roots of $\mathfrak{h}$ in $\mathfrak{g}$ is written $\Delta(\mathfrak{g}, \mathfrak{h}) \subseteq \mathfrak{h}^*$. If α is a root, then $\mathfrak{g}_\alpha$ denotes the corresponding root space. A member of $\mathfrak{g}_\alpha$ is called a **root vector** for the root α; we shall often use e_α to denote such a vector. The **root-space decomposition** of $\mathfrak{g}$ relative to $\mathfrak{h}$ is

$$\mathfrak{g} = \mathfrak{h} \oplus \bigoplus_{\alpha \in \Delta(\mathfrak{g}, \mathfrak{h})} \mathfrak{g}_\alpha.$$

A root $\alpha \in \Delta(\mathfrak{g}, \mathfrak{h})$ is called **real** if $\alpha|_\mathfrak{t} = 0$, **imaginary** if $\alpha|_\mathfrak{a} = 0$, and **complex** if it is neither real nor imaginary. The Cartan involution acts on $\mathfrak{h}^*$ by duality by $(\theta\lambda)(h) = \lambda(\theta h)$ since $\mathfrak{h}$ is θ stable. Complex conjugation of $\mathfrak{g}$ relative to $\mathfrak{g}_0$ (denoted by a bar) acts on $\mathfrak{h}^*$ by $\bar{\lambda}(h) = \overline{\lambda(\bar{h})}$ since $\mathfrak{h}$ is a complexification.

Proposition 4.32. Let $(\mathfrak{g}, K)$ be a reductive pair, and let $\mathfrak{h}_0$ be a θ stable Cartan subalgebra of $\mathfrak{g}_0$. Then

(a) the roots take real values on $\mathfrak{a}_0$ and purely imaginary values on $\mathfrak{t}_0$ (so that a root is real, imaginary, or complex as its values on $\mathfrak{h}_0$ are)

(b) $\Delta(\mathfrak{g}, \mathfrak{h})$ is θ stable and

$$\alpha \text{ is real} \Longleftrightarrow \theta\alpha = -\alpha$$

$$\alpha \text{ is imaginary} \Longleftrightarrow \theta\alpha = \alpha$$

$$\alpha \text{ is complex} \Longleftrightarrow \theta\alpha \neq \pm\alpha$$

(c) $\Delta(\mathfrak{g}, \mathfrak{h})$ is invariant under conjugation, and $\bar{\alpha} = -\theta\alpha$.

PROOF.

(a) The bilinear form $\langle \cdot, \cdot \rangle$ is negative definite on the real form $\mathfrak{t}_0 \oplus i\mathfrak{p}_0$ of $\mathfrak{g}$. If h is in $\mathfrak{t}_0$, then the skew symmetric transformation $\mathrm{ad}\, h$ maps $\mathfrak{t}_0 \oplus i\mathfrak{p}_0$ into itself and thus has imaginary eigenvalues. Similarly if h is in $\mathfrak{a}_0$, then $\mathrm{ad}\, ih$ maps $\mathfrak{t}_0 \oplus i\mathfrak{p}_0$ into itself and has imaginary eigenvalues; hence $\mathrm{ad}\, h$ has real eigenvalues.

(b) If e_α is a root vector for the root α and if h is in $\mathfrak{h}$, then

$$[h, \theta e_\alpha] = \theta[\theta h, e_\alpha] = \alpha(\theta h)\theta e_\alpha = (\theta\alpha(h))\theta e_\alpha$$

shows that $\theta\alpha$ is a root. Hence $\Delta(\mathfrak{g}, \mathfrak{h})$ is θ stable. The root $\theta\alpha$ equals α on $\mathfrak{t}_0$ and equals $-\alpha$ on $\mathfrak{a}_0$; thus the equivalences follow.

(c) If e_α is a root vector for the root α and if h is in $\mathfrak{h}$, then

$$[h, \bar{e}_\alpha] = \overline{[\bar{h}, e_\alpha]} = \overline{\alpha(\bar{h})e_\alpha} = \bar{\alpha}(h)\bar{e}_\alpha$$

shows that $\bar{\alpha}$ is a root. The root $\bar{\alpha}$ equals α on $\mathfrak{a}_0$ and equals $-\alpha$ on $\mathfrak{t}_0$, by (a), and hence $\bar{\alpha} = -\theta\alpha$.

When α is imaginary, (b) of the proposition says that $\theta\alpha = \alpha$. Then the one-dimensionality of $\mathfrak{g}_\alpha$ forces $\mathfrak{g}_\alpha \subseteq \mathfrak{k}$ or $\mathfrak{g}_\alpha \subseteq \mathfrak{p}$, and we call α **compact imaginary** or **noncompact imaginary** in the respective cases.

By (a) of the proposition, the roots are real on $\mathfrak{a}_0 \oplus i\mathfrak{t}_0$. Thus a lexicographic ordering of $\mathfrak{a}_0 \oplus i\mathfrak{t}_0$ or its dual determines a positive system $\Delta^+(\mathfrak{g}, \mathfrak{h})$ within the set of roots $\Delta(\mathfrak{g}, \mathfrak{h})$. Relative to a fixed choice of $\Delta^+(\mathfrak{g}, \mathfrak{h})$, we define

$$(4.33) \qquad \mathfrak{n} = \bigoplus_{\alpha \in \Delta^+(\mathfrak{g}, \mathfrak{h})} \mathfrak{g}_\alpha \qquad \text{and} \qquad \mathfrak{b} = \mathfrak{h} \oplus \mathfrak{n}.$$

These are subalgebras of $\mathfrak{g}$ with $[\mathfrak{h}, \mathfrak{n}] \subseteq \mathfrak{n}$. The subalgebra $\mathfrak{b}$ is called the **Borel subalgebra** of $\mathfrak{g}$ determined by $\mathfrak{h}$ and $\Delta^+(\mathfrak{g}, \mathfrak{h})$. A connected component of the set

$$(\mathfrak{a}_0 \oplus i\mathfrak{t}_0) - \bigcup_{\alpha \in \Delta(\mathfrak{g}, \mathfrak{h})} \ker \alpha$$

is called a **Weyl chamber** in $\mathfrak{a}_0 \oplus i\mathfrak{t}_0$. The subset of $h \in \mathfrak{a}_0 \oplus i\mathfrak{t}_0$ such that $\alpha(h) > 0$ for all $\alpha \in \Delta^+(\mathfrak{g}, \mathfrak{h})$ is one such component and is called the **positive** Weyl chamber relative to $\Delta^+(\mathfrak{g}, \mathfrak{h})$.

If α is a root, then the linear transformation s_α of $\mathfrak{h}^*$ given by

$$(4.34) \qquad\qquad s_\alpha(\lambda) = \lambda - \frac{2\langle \lambda, \alpha \rangle}{\langle \alpha, \alpha \rangle} \alpha$$

is called the **root reflection** in α. The transformation group of $\mathfrak{h}^*$ generated by the root reflections is called the **Weyl group** $W(\mathfrak{g}, \mathfrak{h})$. It acts on $\mathfrak{h}$ by duality: $(w\lambda)(h) = \lambda(w^{-1}h)$.

Theorem 4.35. Each root reflection carries $\Delta(\mathfrak{g}, \mathfrak{h})$ into itself and therefore maps Weyl chambers to Weyl chambers. The Weyl group $W(\mathfrak{g}, \mathfrak{h})$ is a finite group and is simply transitive on the set of Weyl chambers. Consequently if Δ^+ and $(\Delta^+)'$ are the positive systems obtained from two lexicographic orderings, there exists one and only one member w of $W(\mathfrak{g}, \mathfrak{h})$ such that $w\Delta^+ = (\Delta^+)'$.

With the θ stable Cartan subalgebra $\mathfrak{h}_0 = \mathfrak{t}_0 \oplus \mathfrak{a}_0$ fixed, we shall introduce the notion of a corresponding θ **stable Cartan subgroup** H. The group H will depend on $\mathfrak{h}_0$ and also on the choice of a positive system $\Delta^+(\mathfrak{g}, \mathfrak{h})$. If $\mathfrak{b}$ denotes the Borel subalgebra associated to $\Delta^+(\mathfrak{g}, \mathfrak{h})$, then H is defined as the intersection of normalizers

$$(4.36a) \qquad\qquad H = N_G(\mathfrak{b}) \cap N_G(\theta\mathfrak{b}).$$

This definition of θ stable Cartan subgroup is consistent with the definition (4.20) in the compact case. The symmetry of the right side of (4.36a) in θ forces H to be θ stable. As a consequence of (4.36a), we have

$$(4.36b) \qquad\qquad H = N_G(\bar{\mathfrak{b}}) \cap N_G(\theta\bar{\mathfrak{b}}),$$

where the bar refers to the conjugation of $\mathfrak{g}$ with respect to $\mathfrak{g}_0$. Define

$$(4.36c) \qquad T = N_K(\mathfrak{b}) = N_K(\theta\mathfrak{b}) \qquad \text{and} \qquad A = \exp \mathfrak{a}_0.$$

Lemma 4.37. If $\mathfrak{h}_0$ is a θ stable Cartan subalgebra and A is as in (4.36c), then

$$N_G(\mathfrak{h}) = N_K(\mathfrak{h})A$$

as a semidirect product with A normal. Also

$$Z_G(\mathfrak{h}) = Z_K(\mathfrak{h})A$$

as a direct product.

PROOF. The subgroup A is closed by (iii) in Definition 4.29, and also $N_K(\mathfrak{h}) \cap A = 1$ by the same property. It is clear that $N_K(\mathfrak{h})A \subseteq N_G(\mathfrak{h})$, and we are left with showing the reverse inclusion. Let $k \exp X$ be the decomposition according to (iii) of a member of $N_G(\mathfrak{h})$. Since $\mathfrak{h}$ is θ stable, $\theta(k \exp X) = k(\exp X)^{-1}$ is in $N_G(\mathfrak{h})$ and hence so is

$$(k(\exp X)^{-1})^{-1}k \exp X = \exp 2X.$$

In other words, the positive definite transformation $\mathrm{Ad}(\exp 2X)$ on $\mathfrak{g}_0$ leaves $\mathfrak{h}_0$ stable. Its logarithm, namely $\mathrm{ad}\, X$, must leave $\mathfrak{h}_0$ stable, and hence X is in $N_{\mathfrak{g}_0}(\mathfrak{h}_0) = \mathfrak{h}_0$. Since X is assumed to be in $\mathfrak{p}_0$, X is in $\mathfrak{a}_0$. Then $\exp X$ is in A, and k must be in $N_K(\mathfrak{h})$. This proves the first equality.

For the second we have

$$Z_K(\mathfrak{h})A \subseteq Z_G(\mathfrak{h}) \subseteq N_G(\mathfrak{h}) = N_K(\mathfrak{h})A.$$

If ka is the decomposition of an element of $Z_G(\mathfrak{h})$ according to $N_K(\mathfrak{h})A$, then a centralizes $\mathfrak{h}$ and hence so must k. We conclude that $Z_K(\mathfrak{h})A = Z_G(\mathfrak{h})$.

Proposition 4.38. If $\mathfrak{h}_0$ is a θ stable Cartan subalgebra and H, T, and A are as in (4.36), then

(a) $H = TA$ as a semidirect product with A normal
(b) $(H, T, \theta, \langle\,\cdot\,,\,\cdot\,\rangle)$ is a reductive group, and $((\mathfrak{h}, T), \mathfrak{h}_0, \theta, \langle\,\cdot\,,\,\cdot\,\rangle)$ is the corresponding reductive pair
(c) $(\mathfrak{h}, T) \hookrightarrow (\mathfrak{g}, K)$ is an inclusion of pairs
(d) $H = Z_G(\mathfrak{h})$ in the special case that G is in the Harish-Chandra class; hence in this case H is independent of the choice of a positive system in $\Delta(\mathfrak{g}, \mathfrak{h})$.

PROOF. From (4.36a) and (4.36b), we have

$$H = N_G(\mathfrak{b}) \cap N_G(\theta\mathfrak{b})$$
$$= N_G(\mathfrak{b}) \cap N_G(\theta\mathfrak{b}) \cap N_G(\bar{\mathfrak{b}}) \cap N_G(\theta\bar{\mathfrak{b}})$$
$$\subseteq N_G(\mathfrak{b} \cap \theta\mathfrak{b} \cap \bar{\mathfrak{b}} \cap \theta\bar{\mathfrak{b}})$$

(4.39a) $= N_G(\mathfrak{h}).$

Therefore

$$N_K(\mathfrak{b})A = N_K(\mathfrak{b})A \cap N_K(\theta\mathfrak{b})A \qquad \text{by (4.36c)}$$
$$\subseteq N_G(\mathfrak{b}) \cap N_G(\theta\mathfrak{b})$$
$$\subseteq N_G(\mathfrak{h}) \qquad\qquad\qquad\text{by (4.39a)}$$

(4.39b) $= N_K(\mathfrak{h})A \qquad\qquad\qquad\text{by Lemma 4.37.}$

If ka is the decomposition of an element of $N_G(\mathfrak{b}) \cap N_G(\theta\mathfrak{b})$ according to $N_K(\mathfrak{h})A$, then a normalizes $\mathfrak{b}$ and hence so must k. In other words, $N_K(\mathfrak{b})A = N_G(\mathfrak{b}) \cap N_G(\theta\mathfrak{b})$. This proves (a). Conclusion (b) is immediate from (a), and (c) is trivial.

For (d) we note that an element of $Z_K(\mathfrak{h})$ normalizes every root space. Hence (a) gives

$$Z_G(\mathfrak{h}) = Z_K(\mathfrak{h})A \subseteq N_K(\mathfrak{b})A = H.$$

For the reverse inclusion, we certainly have $A \subseteq Z_G(\mathfrak{h})$. Also any element of $N_K(\mathfrak{b})$ normalizes $\mathfrak{h}$, by (4.39a), and carries $\Delta^+(\mathfrak{g}, \mathfrak{h})$ to itself. Since any element g in a group in the Harish-Chandra class must act with $\mathrm{Ad}(g)$ inner on $\mathfrak{g}$, the element of $N_K(\mathfrak{b})$ acts on $\mathfrak{h}$ by an element w of $W(\mathfrak{g}, \mathfrak{h})$ preserving $\Delta^+(\mathfrak{g}, \mathfrak{h})$. By Theorem 4.8, $w = 1$; that is, the element centralizes $\mathfrak{h}$. This proves (d).

EXAMPLE. Let $G = GL(2, \mathbb{R})$ and $K = O(2)$ as in Example 3 of §3. The subalgebras

$$\mathfrak{h}_1 = \left\{ \begin{pmatrix} a & 0 \\ 0 & b \end{pmatrix} \right\} \qquad \text{and} \qquad \mathfrak{h}_2 = \left\{ \begin{pmatrix} c & d \\ -d & c \end{pmatrix} \right\}$$

are θ stable Cartan subalgebras. The corresponding Cartan subgroups are independent of the choices of positive systems and are given by

$$H_1 = T_1 A_1 \qquad \text{and} \qquad H_2 = T_2 A_2,$$

where

$$T_1 = \left\{ \begin{pmatrix} \pm 1 & 0 \\ 0 & \pm 1 \end{pmatrix} \right\}, \qquad A_1 = \left\{ \begin{pmatrix} e^s & 0 \\ 0 & e^t \end{pmatrix} \right\},$$
$$T_2 = \left\{ \begin{pmatrix} \cos\theta & \sin\theta \\ -\sin\theta & \cos\theta \end{pmatrix} \right\}, \qquad A_2 = \left\{ \begin{pmatrix} e^t & 0 \\ 0 & e^t \end{pmatrix} \right\}.$$

The group H_1 meets every component of G, but H_2 does not. Notice also that $\mathfrak{t}_2$ is a Cartan subalgebra of $\mathfrak{k}_0$ but T_2 is not a Cartan subgroup of K.

A reductive pair arising from a θ stable Cartan subalgebra $\mathfrak{h}_0$ by Proposition 4.38b will be called a **Cartan subpair**. It is determined from $((\mathfrak{g}, K), \mathfrak{g}_0, \theta, \langle \cdot , \cdot \rangle)$ by specifying $\mathfrak{h}_0$ and $T = N_K(\mathfrak{b})$. When it is given, the θ stable Cartan subgroup H can be reconstructed from $H = TA$, with A as in (4.36c).

Suppose $H \subseteq G$ is a θ stable Cartan subgroup as in (4.36). The **Weyl group** of H in G, denoted $W(G, H)$, is defined as the normalizer divided by centralizer

(4.40a) $$W(G, H) = N_G(\mathfrak{h})/Z_G(\mathfrak{h}).$$

It is the same to define $W(G, H)$ by

(4.40b) $$W(G, H) = N_K(\mathfrak{h})/Z_K(\mathfrak{h}).$$

In fact, Lemma 4.37 gives

$$N_G(\mathfrak{h}) = N_K(\mathfrak{h})A \qquad \text{and} \qquad Z_G(\mathfrak{h}) = Z_K(\mathfrak{h})A.$$

Thus (4.40b) follows from (4.40a). The group $W(G, H)$ is a group of automorphisms of $\mathfrak{h}$, and the examples below shows that neither $W(G, H) \subseteq W(\mathfrak{g}, \mathfrak{h})$ nor $W(G, H) \supseteq W(\mathfrak{g}, \mathfrak{h})$ is universally true. However, $W(G, H) \subseteq W(\mathfrak{g}, \mathfrak{h})$ if G is of Harish-Chandra class.

EXAMPLES.

1) With $G = SL(2, \mathbb{R})$ and $K = SO(2)$, $H = SO(2)$ is a θ stable Cartan subgroup. The group $W(\mathfrak{g}, \mathfrak{h})$ has the two elements ± 1, but $W(G, H)$ has only the one element $+1$.

2) With $G = K = O(2n)$ and $n > 1$, a θ stable Cartan subgroup has two components, according to Example 2 of §2. The Cartan subgroup is not abelian, and hence the members of its nonidentity component descend to a nontrivial member of $W(G, H)$. This element corresponds to a nontrivial outer automorphism of G_0 and is not in $W(\mathfrak{g}, \mathfrak{h})$.

There are two extreme cases for θ stable Cartan subalgebras. At one extreme a θ stable Cartan subalgebra $\mathfrak{h}_0$ is called **maximally split** if $\mathfrak{a}_0$ is maximal abelian in $\mathfrak{p}_0$. In this case we say also that any corresponding θ stable Cartan subgroup and the corresponding Cartan subpair are **maximally split**. Maximally split θ stable Cartan subalgebras always exist; we simply start with a maximal abelian subspace $\mathfrak{a}_0$ of $\mathfrak{p}_0$ and adjoin a maximal abelian subspace $\mathfrak{t}_0$ of $Z_K(\mathfrak{a}_0)$. In the maximally split case, θ stable Cartan subgroups have the special properties given in Proposition 4.42 below. We begin by recalling their special properties when G is connected.

Lemma 4.41. Suppose G is connected and $\mathfrak{h}_0 = \mathfrak{t}_0 \oplus \mathfrak{a}_0$ is a maximally split θ stable Cartan subalgebra. Let $H = Z_G(\mathfrak{h})$ be the corresponding θ stable Cartan subgroup. Then

(a) there are no noncompact imaginary roots
(b) the set of nonzero restrictions to $\mathfrak{a}_0$ of members of $\Delta(\mathfrak{g}, \mathfrak{h})$ is a (possibly nonreduced) root system $\Delta(\mathfrak{g}_0, \mathfrak{a}_0)$
(c) the Weyl group of $\Delta(\mathfrak{g}_0, \mathfrak{a}_0)$ is given by

$$N_G(A)/Z_G(A) \qquad \text{and} \qquad N_K(A)/Z_K(A)$$

(d)

$$W(G, H) = \{w \in W(\mathfrak{g}, \mathfrak{h}) \mid \theta w = w\theta\} = \{w \in W(\mathfrak{g}, \mathfrak{h}) \mid w(\mathfrak{a}_0) = \mathfrak{a}_0\}.$$

In addition, any two maximally split θ stable Cartan subalgebras are conjugate via K.

Proposition 4.42. For general reductive G, let $\mathfrak{h}_0$ be a maximally split θ stable Cartan subalgebra, and fix a positive system $\Delta^+(\mathfrak{g}, \mathfrak{h})$ so that the complex conjugate of any nonimaginary positive root is positive. Then the corresponding θ stable Cartan subgroup $H = TA$ has the following properties:

(a) T meets every component of G
(b) $W(G, H)$ is generated by $W(G_0, H \cap G_0)$ and $T/Z_K(\mathfrak{h})$.

REMARKS. The group $W(G_0, H \cap G_0)$ in conclusion (b) is given as an explicit subgroup of the Weyl group $W(\mathfrak{g}, \mathfrak{h})$ by Lemma 4.41d. The conclusions in (b) can be strengthened to say that $W(G, H)$ is the semdirect product of the two mentioned subgroups, with the first one normal. In particular, T acts on $W(G_0, H \cap G_0)$.

PROOF.
(a) Let $g \in G$ be given. By (iii) of Definition 4.29, we may assume $g \in K$. Then $\mathrm{Ad}(g)\mathfrak{h}_0$ and $\mathfrak{h}_0$ are two maximally split θ stable Cartan subalgebras, and Lemma 4.41 gives us $k_0 \in K_0$ with $\mathrm{Ad}(k_0)\mathrm{Ad}(g)\mathfrak{h}_0 = \mathfrak{h}_0$. Hence $k_0 g$ is in $N_K(\mathfrak{h})$.

Changing notation, we may assume that our given element in G is an element $k \in N_K(\mathfrak{h})$. The set of nonzero restrictions to $\mathfrak{a}_0$ of the members of $\Delta(\mathfrak{g}, \mathfrak{h})$ is a root system $\Delta(\mathfrak{g}_0, \mathfrak{a}_0)$, by Lemma 4.41b, and our special choice of $\Delta^+(\mathfrak{g}, \mathfrak{h})$ makes $\Delta^+(\mathfrak{g}_0, \mathfrak{a}_0)$ well defined by restriction. Then $\Delta^+(\mathfrak{g}_0, \mathfrak{a}_0)$ and $\mathrm{Ad}(k)\Delta^+(\mathfrak{g}_0, \mathfrak{a}_0)$ are two positive systems for $\Delta(\mathfrak{g}_0, \mathfrak{a}_0)$. By Lemma 4.41c and Theorem 4.8, we can choose $k_0 \in K_0$ with $\mathrm{Ad}(k_0)\mathrm{Ad}(k)\Delta^+(\mathfrak{g}_0, \mathfrak{a}_0) = \Delta^+(\mathfrak{g}_0, \mathfrak{a}_0)$.

Let $M = Z_K(\mathfrak{a}_0)$. This is a closed subgroup of K with Lie algebra

$$\mathfrak{m}_0 = Z_{\mathfrak{k}_0}(\mathfrak{a}_0).$$

Since $k_0 k$ normalizes K and $\mathfrak{a}_0$, it normalizes $\mathfrak{m}_0$. The subspace $\mathfrak{t}_0$ of $\mathfrak{m}_0$ is maximal abelian in $\mathfrak{m}_0$ since $\mathfrak{h}_0$ is maximal abelian in $\mathfrak{g}_0$. By Theorem 4.5, there exists $m_0 \in M_0$ with

$$\mathrm{Ad}(m_0)\mathrm{Ad}(k_0 k)\mathfrak{t}_0 = \mathfrak{t}_0,$$

and Theorems 4.8 and 4.10 allow us to assume that $\mathrm{Ad}(m_0)\Delta^+(\mathfrak{m}_0, \mathfrak{t}_0) = \Delta^+(\mathfrak{m}_0, \mathfrak{t}_0)$. Then the element $m_0 k_0 k$ normalizes $\mathfrak{h}$ and carries $\Delta^+(\mathfrak{g}, \mathfrak{h})$ to itself. Hence $m_0 k_0 k$ is in T.

(b) Let $k \in K$ represent a member of $W(G, H)$, i.e., let $k \in N_K(\mathfrak{h})$. Then the proof of (a) produces $m_0 k_0 \in K_0$ so that $m_0 k_0 k$ normalizes $\mathfrak{h}$ and is in T. The element $m_0 k_0$ is in $N_{K_0}(\mathfrak{h})$ and hence represents a member of $W(G_0, H \cap G_0)$. The result follows.

At the other extreme, a θ stable Cartan subalgebra $\mathfrak{h}_0$ is called **maximally compact** if $\mathfrak{t}_0$ is maximal abelian in $\mathfrak{k}_0$. In this case we say that the corresponding θ stable Cartan subgroup and the corresponding Cartan subpair are **maximally compact**. Maximally compact θ stable Cartan subalgebras always exist; we simply start with a maximal abelian subspace $\mathfrak{t}_0$ of $\mathfrak{k}_0$ and adjoin a subspace $\mathfrak{a}_0$ of $\mathfrak{p}_0$ that is maximal with respect to the properties of being abelian and commuting with $\mathfrak{t}_0$. The analogs of Lemma 4.41 and Proposition 4.42 are as follows.

Lemma 4.43. Suppose G is connected and $\mathfrak{h}_0 = \mathfrak{t}_0 \oplus \mathfrak{a}_0$ is a maximally compact θ stable Cartan subalgebra. Let $H = Z_G(\mathfrak{h})$ be the corresponding θ stable Cartan subgroup. Then

(a) there are no real roots
(b) $N_G(\mathfrak{t}) = N_G(\mathfrak{h})$
(c) the members of $\Delta(\mathfrak{k}, \mathfrak{t})$ are precisely the restrictions to $\mathfrak{t}$ of all the complex and compact imaginary roots in $\Delta(\mathfrak{g}, \mathfrak{h})$
(d) T and H are connected
(e) $W(G, H)$ is given by

$$\{w \in W(\mathfrak{g}, \mathfrak{h}) \mid \theta w = w\theta \text{ and } w|_{\mathfrak{t}} \text{ belongs to } W(\mathfrak{k}, \mathfrak{t}) = W(K, T)\},$$

and restriction to T defines an isomorphism $W(G, H) \cong W(K, T)$.

In addition, any two maximally compact θ stable Cartan subalgebras are conjugate via K. Conversely if there are no real roots, then the θ stable Cartan subalgebra $\mathfrak{h}_0$ is maximally compact.

Proposition 4.44. For general reductive G, let $\mathfrak{h}_0$ be a maximally compact θ stable Cartan subalgebra, and fix a positive system $\Delta^+(\mathfrak{g}, \mathfrak{h})$ that is stable under θ. Then the corresponding θ stable Cartan subgroup $H = TA$ has the following properties:

(a) $H = N_G(\mathfrak{b})$

(b) the subalgebra $\mathfrak{b} \cap \mathfrak{k}$ of $\mathfrak{k}$ is a Borel subalgebra relative to the Cartan subalgebra $\mathfrak{t}_0$ of $\mathfrak{k}_0$, and the corresponding Cartan subgroup is $T' = N_K(\mathfrak{b} \cap \mathfrak{k})$

(c) $W(G, H)$ is generated by $W(G_0, H \cap G_0)$ and $T'/Z_K(\mathfrak{h})$, where T' is as in (b).

REMARKS. It is always true that $T \subseteq T'$. The example with $G = GL(2, \mathbb{R})$ and $T = T_2$ earlier in this section shows that the inclusion can be strict, even if G is in the Harish-Chandra class. As with Proposition 4.42b, the conclusion in (c) can be strengthened to say that $W(G, H)$ is the semidirect product of the two mentioned subgroups, with the first one normal.

PROOF.

(a) Our special choice of $\Delta^+(\mathfrak{g}, \mathfrak{h})$ makes $\theta\mathfrak{b} = \mathfrak{b}$. Thus

$$H = N_G(\mathfrak{b}) \cap N_G(\theta\mathfrak{b}) = N_G(\mathfrak{b}).$$

(b) The root system $\Delta(\mathfrak{k}, \mathfrak{t})$ is given by Lemma 4.43c. Let us show that our special choice of $\Delta^+(\mathfrak{g}, \mathfrak{h})$ makes $\Delta^+(\mathfrak{k}, \mathfrak{t})$ well defined by restriction. In fact, in obvious notation, let $\alpha = \alpha_{\mathfrak{t}} + \alpha_{\mathfrak{a}}$ and $\beta = \beta_{\mathfrak{t}} + \beta_{\mathfrak{a}}$ be in $\Delta(\mathfrak{g}, \mathfrak{h})$. If $\alpha_{\mathfrak{t}} = \beta_{\mathfrak{t}}$, then

$$\langle \alpha, \beta + \theta\beta \rangle = 2|\alpha_{\mathfrak{t}}|^2 > 0,$$

and we must have $\langle \alpha, \beta \rangle > 0$ or $\langle \alpha, \theta\beta \rangle > 0$. Thus $\alpha - \beta$ or $\alpha - \theta\beta$ is in $\Delta(\mathfrak{g}, \mathfrak{h}) \cup \{0\}$. By Lemma 4.43a, $\alpha = \beta$ or $\alpha = \theta\beta$. So the only other root having the same restriction to $\mathfrak{t}$ as α is $\theta\alpha$. Since α and $\theta\alpha$ are both positive or both negative, $\Delta^+(\mathfrak{k}, \mathfrak{t})$ is well defined.

Now that $\Delta^+(\mathfrak{k}, \mathfrak{t})$ is well defined, the corresponding Borel subalgebra of $\mathfrak{b}$ is then $\mathfrak{b} \cap \mathfrak{k}$, and our definitions make $T' = N_K(\mathfrak{b} \cap \mathfrak{k})$ the corresponding Cartan subgroup.

(c) Let $k \in K$ represent a member of $W(G, H)$, i.e., let $k \in N_K(\mathfrak{h})$. Applying Proposition 4.22 to K, we can write $k = k_0 t$ with $k_0 \in K_0$ and $t \in T'$. The element t is in $N_K(\mathfrak{b} \cap \mathfrak{k})$, which is contained in $N_K(\mathfrak{t})$, and hence k_0 is in $N_{K_0}(\mathfrak{t})$. By Lemma 4.43b, k_0 is in $N_{G_0}(\mathfrak{h})$ and thus represents a member of $W(G_0, H \cap G_0)$. Consequently $k = k_0 t$ is the required decomposition.

To conclude this section, let us amplify Lemma 4.43c by describing in more detail the relationship between roots of $\mathfrak{g}$ and roots of $\mathfrak{k}$. Thus let $\mathfrak{h}_0 = \mathfrak{t}_0 \oplus \mathfrak{a}_0$ be a maximally compact θ stable Cartan subalgebra of $\mathfrak{g}_0$. For $\alpha \in \Delta(\mathfrak{g}, \mathfrak{h})$, let e_α be a nonzero root vector for α. Then the root-space decomposition of $\mathfrak{g}$ can be regrouped as

$$\mathfrak{g} = \left(\mathfrak{t} \oplus \underbrace{\bigoplus}_{\substack{\alpha \text{ compact} \\ \text{imaginary}}} \mathbb{C}e_\alpha \oplus \underbrace{\bigoplus}_{\substack{\alpha, \theta\alpha \\ \text{complex}}} \mathbb{C}(e_\alpha + \theta e_\alpha) \right)$$

(4.45a)

$$\oplus \left(\mathfrak{a} \oplus \underbrace{\bigoplus}_{\substack{\alpha \text{ noncompact} \\ \text{imaginary}}} \mathbb{C}e_\alpha \oplus \underbrace{\bigoplus}_{\substack{\alpha, \theta\alpha \\ \text{complex}}} \mathbb{C}(e_\alpha - \theta e_\alpha) \right),$$

and the sums within the large parentheses are $\mathfrak{k}$ and $\mathfrak{p}$, respectively. Even more, the sum within the first pair of large parentheses is the root-space decomposition of $\mathfrak{k}$ with respect to $\mathfrak{t}$, and the nonzero weights must therefore have multiplicity 1.

The decomposition in the second set of large parentheses is the decomposition of $\mathfrak{p}$ according to weights of $\mathfrak{t}$. From the fact that the nonzero weights of $\mathfrak{k}$ have multiplicity 1, we can deduce the same conclusion about $\mathfrak{p}$. In fact, two pairs of complex roots cannot contribute to the same weight of $\mathfrak{p}$ because of multiplicity 1 in $\mathfrak{k}$. If a complex root were to have the same restriction to $\mathfrak{t}$ as a noncompact imaginary root, then the difference would have to be a root supported on $\mathfrak{a}$, in contradiction to Lemma 4.43a. Thus the nonzero $\mathfrak{t}$ weights are as follows:

for $\mathfrak{k}$: $\begin{cases} \alpha|_\mathfrak{t} & \text{for each compact imaginary root} \\ \alpha|_\mathfrak{t} & \text{for each pair } \{\alpha, \theta\alpha\} \text{ of complex roots} \end{cases}$

for $\mathfrak{k}$: $\begin{cases} \alpha|_\mathfrak{t} & \text{for each noncompact imaginary root} \\ \alpha|_\mathfrak{t} & \text{for each pair } \{\alpha, \theta\alpha\} \text{ of complex roots,} \end{cases}$

and all the multiplicities are 1.

To analyze further the bracket relations among the basis vectors in (4.45a), we introduce a slightly unwieldy notational convention. A $(\mathfrak{t}, \theta)$ **form** is a pair $\gamma = (\beta, \varepsilon)$ consisting of a linear functional $\beta \in \mathfrak{t}^*$ and a number $\varepsilon = \pm 1$. A $(\mathfrak{t}, \theta)$ **weight vector** for γ is an element $e \in \mathfrak{g}$ with the property that

$$[h, e] = \beta(h)e \quad \text{for all } h \in \mathfrak{t}^*, \qquad \theta e = \varepsilon e.$$

That is, e is a $\mathfrak{t}$ weight vector of weight β, and e belongs to $\mathfrak{k}$ or to $\mathfrak{p}$ according as ε is $+1$ or -1. If there exists a nonzero $(\mathfrak{t}, \theta)$ weight vector for γ, we call γ a $(\mathfrak{t}, \theta)$ **weight**. We write $\mathfrak{g}_\gamma$ for the space of all $(\mathfrak{t}, \theta)$ weight vectors of weight γ. Explicitly the compilation of $\mathfrak{t}$ weights in the previous paragraph shows that

$$
\mathfrak{g}_{(\beta,\varepsilon)} =
\begin{cases}
\mathbb{C}e_\alpha & \text{if } \varepsilon = +1 \text{ and } \beta = \alpha|_\mathfrak{t} \text{ for some} \\
& \text{compact imaginary root } \alpha \\
\mathbb{C}e_\alpha & \text{if } \varepsilon = -1 \text{ and } \beta = \alpha|_\mathfrak{t} \text{ for some} \\
& \text{noncompact imaginary root } \alpha \\
\mathbb{C}(e_\alpha + \varepsilon\theta e_\alpha) & \text{if } \beta = \alpha|_\mathfrak{t} \text{ for some complex root } \alpha \\
\mathfrak{t} & \text{if } \beta = 0 \text{ and } \varepsilon = +1 \\
\mathfrak{a} & \text{if } \beta = 0 \text{ and } \varepsilon = -1.
\end{cases}
$$

In particular, $\mathfrak{g}_{(\beta,\varepsilon)}$ has dimension at most one whenever $\beta \neq 0$. We add $(\mathfrak{t}, \theta)$ weights by adding the first coordinate and multiplying the second:

$$(\beta, \varepsilon) + (\beta', \varepsilon') = (\beta + \beta', \varepsilon\varepsilon').$$

With this convention, $[\mathfrak{g}_\gamma, \mathfrak{g}_{\gamma'}] \subseteq \mathfrak{g}_{\gamma+\gamma'}$. The zero $(\mathfrak{t}, \theta)$ weight is $(0, 1)$ with weight space $\mathfrak{t}$, and the negative of $\gamma = (\beta, \varepsilon)$ is $-\gamma = (-\beta, \varepsilon)$. Consequently if $\gamma \neq \gamma'$, we see that

(4.45b) $[\mathfrak{g}_\gamma, \mathfrak{g}_{\gamma'}]$ is in $\mathfrak{a} \oplus \displaystyle\sum_{\gamma''|_\mathfrak{t} \neq 0} \mathfrak{g}_{\gamma''}.$

If $\mathfrak{g}_{\gamma''}$ is in the last sum, then $\mathfrak{g}_{\gamma''} = [\mathfrak{t}, \mathfrak{g}_{\gamma''}]$.

It is often convenient to abuse notation slightly by writing simply γ in place of $\gamma|_\mathfrak{t}$ Thus if $\gamma = (\beta, \varepsilon)$, we shall write $\gamma(h) = \beta(h)$ for $h \in \mathfrak{t}$. Also we define $\langle \gamma, \gamma' \rangle = \langle \beta, \beta' \rangle$ if $\gamma = (\beta, \varepsilon)$ and $\gamma' = (\beta', \varepsilon')$.

For each nonzero $(\mathfrak{t}, \theta)$ weight $\gamma = (\gamma|_\mathfrak{t}, \varepsilon)$ with $\gamma|_\mathfrak{t} \neq 0$, let e_γ be a nonzero $(\mathfrak{t}, \theta)$ weight vector. The claim is that we can make these choices so that

(4.45c) $[e_\gamma, e_{-\gamma}] = h'_\gamma$ in $\mathfrak{t}$ corresponds to $2|\gamma|^{-2}\gamma|_\mathfrak{t}$ in $\mathfrak{t}^*$.

Furthermore, without compromising this normalization, we can arrange that

(4.45d) $\bar{e}_\gamma = -\varepsilon e_{-\gamma}.$

In fact, if $\langle \cdot, \cdot \rangle_\mathfrak{g}$ is the invariant bilinear form on $\mathfrak{g}$ given in Definition 4.29, then $\langle \mathfrak{g}_\gamma, \mathfrak{g}_{-\gamma'} \rangle_\mathfrak{g} = 0$ unless $\gamma = \gamma'$. Because the form is nondegenerate, it follows that $\langle e_\gamma, e_{-\gamma} \rangle_\mathfrak{g} \neq 0$. To achieve (4.45c), we choose the basis vectors so that

$$\langle e_\gamma, e_{-\gamma} \rangle = \frac{2}{|\gamma|^2}.$$

If $h \in \mathfrak{t}$ is given, then

$$\langle h, [e_\gamma, e_{-\gamma}] \rangle_{\mathfrak{g}} = -\langle [e_\gamma, h], e_{-\gamma} \rangle_{\mathfrak{g}}$$
$$= \langle [h, e_\gamma], e_{-\gamma} \rangle_{\mathfrak{g}} = \langle \gamma(h) e_\gamma, e_{-\gamma} \rangle_{\mathfrak{g}} = \frac{2\gamma(h)}{|\gamma|^2},$$

and nondegeneracy implies (4.45c).

Now let us adjust the normalization so that (4.45d) holds as well. Since the weights of $\mathfrak{t}$ take purely imaginary values on $\mathfrak{t}_0$, $\bar{\mathfrak{g}}_\gamma = \mathfrak{g}_{-\gamma}$. Write $\bar{e}_\gamma = c e_{-\gamma}$, and let $e_\gamma = x_\gamma + i y_\gamma$ with x_γ and y_γ in $\mathfrak{g}_0$. Then

$$2c/|\gamma|^2 = \langle e_\gamma, c e_{-\gamma} \rangle = \langle e_\gamma, \bar{e}_\gamma \rangle = \langle x_\gamma, x_\gamma \rangle + \langle y_\gamma, y_\gamma \rangle.$$

If $\varepsilon = +1$, then x_γ and y_γ belong to $\mathfrak{k}_0$, the right side is negative, and $c < 0$. In this case if we define

$$e'_\gamma = (-c)^{-1/2} e_\gamma \qquad \text{and} \qquad e'_{-\gamma} = (-c)^{1/2} e_{-\gamma},$$

then $[e'_\gamma, e'_{-\gamma}] = [e_\gamma, e_{-\gamma}]$ (so that (4.45c) is preserved), and $\{e'_\gamma, e'_{-\gamma}\}$ satisfies (4.45d). Similarly if $\varepsilon = -1$, we find that c is positive. In this case if we define

$$e'_\gamma = c^{-1/2} e_\gamma \qquad \text{and} \qquad e'_{-\gamma} = c^{1/2} e_{-\gamma},$$

then $[e'_\gamma, e'_{-\gamma}] = [e_\gamma, e_{-\gamma}]$ (so that (4.45c) is preserved), and $\{e'_\gamma, e'_{-\gamma}\}$ satisfies (4.45d).

5. Finite-Dimensional Representations

If $(\mathfrak{g}, K)$ is a pair and V is a $(\mathfrak{g}, K)$ module, we recall from §II.4 that V is **irreducible** as a $(\mathfrak{g}, K)$ module if the only $(\mathfrak{g}, K)$ invariant subspaces are 0 and V. In this section we shall study the irreducible finite-dimensional $(\mathfrak{g}, K)$ modules for a reductive pair. By Proposition 2.84, any irreducible finite-dimensional $(\mathfrak{g}, K)$ module is the direct sum of finitely many irreducible $U(\mathfrak{g})$ modules.

Let G be a Lie group with identity component G_0 and with Lie algebra $\mathfrak{g}_0$ and complexification $\mathfrak{g}$, and suppose that K is a compact subgroup of G meeting every component of G, i.e., $G = K G_0$. Certainly $(\mathfrak{g}, K)$ is a pair. If (π, V) is a (continuous) finite-dimensional representation of G, then π is smooth and V becomes a $(\mathfrak{g}, K)$ module from the corresponding action of $\mathfrak{g}$ and the restriction $\pi|_K$. For a converse, we need to impose further conditions on K, and we shall be content to work with reductive groups.

Proposition 4.46. Let $(\mathfrak{g}, K)$ be a reductive pair, and let G be the corresponding reductive group. If V is a finite-dimensional $(\mathfrak{g}, K)$ module, then V comes from exactly one representation π of G on V.

PROOF. The uniqueness of π is clear since $G = KG_0$ and since the action of $\mathfrak{g}_0$ determines the action of G_0. Let $\tilde{G}$ be a simply connected covering group of G_0, and let $\varphi : \tilde{G} \to G_0$ be the covering homomorphism. Since $\tilde{G}$ is simply connected, the representation of $\mathfrak{g}_0$ on V is the differential of a representation $\tilde{\pi}$ of $\tilde{G}$ on V. Then

$$(4.47a) \qquad \frac{d}{dt}\tilde{\pi}(\exp_{\tilde{G}} tX)v|_{t=0} = Xv \qquad \text{for } X \in \mathfrak{g}_0,\ v \in V.$$

Let $\tilde{K}$ be the subgroup $\varphi^{-1}(K_0)$ of $\tilde{G}$. The special feature of the reductive case is that $\tilde{K}$ is connected (and actually simply connected). Since V is a $(\mathfrak{g}, K)$ module,

$$(4.47b) \qquad \frac{d}{dt}(\exp_G tX)v|_{t=0} = Xv \qquad \text{for } X \in \mathfrak{k}_0,\ v \in V.$$

Let $c_1(t)$ and $c_2(t)$ be the curves in V given for fixed $X \in \mathfrak{k}_0$ and $v \in V$ by

$$c_1(t) = \tilde{\pi}(\exp_{\tilde{G}} tX)v \qquad \text{and} \qquad c_2(t) = (\exp_G tX)v.$$

From (4.47) we obtain

$$c_1'(t) = Xc_1(t) \qquad \text{and} \qquad c_2'(t) = Xc_2(t).$$

Since $c_1(0) = c_2(0) = v$, it follows from the uniqueness theorem for systems of ordinary differential equations that $c_1(t) = c_2(t)$ for all $t \in \mathbb{R}$. Since K_0 is compact and $\tilde{K}$ is connected, Theorem 4.6d and Corollary 4.4 show that exp is onto $\tilde{K}$. Therefore

$$\tilde{\pi}(k)v = \varphi(k)v \qquad \text{for } k \in \tilde{K}.$$

Taking $k \in \ker \varphi$, we see that $\tilde{\pi}$ is trivial on $\ker \varphi$. Hence $\tilde{\pi}$ descends to a representation π of G_0 on V with differential the given action of $\mathfrak{g}_0$.

One of the properties of a $(\mathfrak{g}, K)$ module is that K_0 acts with differential equal to the restriction to $\mathfrak{k}_0$ of the action of $\mathfrak{g}_0$. Consequently $\pi(k_0)v = k_0 v$ for $k_0 \in K_0$. Therefore we get a consistent definition of π on G if we let

$$(4.48) \qquad \pi(kg_0)v = k\pi(g_0)v \qquad \text{for } k \in K,\ g_0 \in G_0,\ v \in V.$$

Moreover, the restriction of π to K coincides with the given K action.

The continuity of π on G comes from the continuity of π on G_0. To see that π is a representation, we use the identity

$$(\mathrm{Ad}(k)X)kv = kXv \qquad \text{for } k \in K,\ X \in \mathfrak{g}_0,\ v \in V$$

valid in any $(\mathfrak{g}, K)$ module. It implies

$$\pi(k(\exp tX)k^{-1})\pi(k)v = \pi(\exp t\,\mathrm{Ad}(k)X)kv$$

$$= \sum_{n=0}^{\infty} \frac{t^n}{n!}\,(\mathrm{Ad}(k)X)^n kv$$

$$= \sum_{n=0}^{\infty} \frac{t^n}{n!}\, kX^n v$$

$$= \pi(k)\pi(\exp tX)v.$$

Therefore

$$(4.49) \qquad \pi(kg_0k^{-1})\pi(k) = \pi(k)\pi(g_0) \qquad \text{for } k \in K,\ g_0 \in G_0.$$

If k and k' are in K and g_0 and g_0' are in G_0, we then have

$$
\begin{aligned}
\pi(kg_0)\pi(k'g_0') &= \pi(k)\pi(k')\pi(k')^{-1}\pi(g_0)\pi(k')\pi(g_0') && \text{by (4.48)} \\
&= \pi(kk')\pi(k'^{-1}g_0k')\pi(g_0') && \text{by (4.49)} \\
&= \pi(kk')\pi(k'^{-1}g_0k'g_0') && \\
&= \pi(kk'k'^{-1}g_0k'g_0') && \text{by (4.48)} \\
&= \pi(kg_0k'g_0'),
\end{aligned}
$$

and π is the required representation.

Let $(\mathfrak{g}, K)$ be a reductive pair, and suppose that V is an irreducible finite-dimensional $(\mathfrak{g}, K)$ module. Proposition 2.84 shows that $V = \bigoplus_{j=1}^{n} V_j$ with V_j an irreducible $U(\mathfrak{g})$ module. An understanding of V_j can be reduced to the considerations in §1 by use of the "unitary trick," which we now describe. Since $\mathfrak{g}$ is reductive, $\mathfrak{g}$ decomposes as

$$\mathfrak{g} = Z_{\mathfrak{g}} \oplus [\mathfrak{g}, \mathfrak{g}],$$

where $Z_{\mathfrak{g}}$ is the center and $[\mathfrak{g}, \mathfrak{g}]$ is semisimple. Each member of $Z_{\mathfrak{g}}$ acts by scalars in V_j, by Schur's Lemma, and hence V_j is irreducible under

$[\mathfrak{g}, \mathfrak{g}]$. In other words, we may assume without loss of generality that $\mathfrak{g}$ is semisimple.

With $\mathfrak{g}$ semisimple, let $\mathfrak{u}_0 = \mathfrak{k}_0 \oplus i\mathfrak{p}_0$. This is a real semisimple Lie algebra called the **compact form** of $\mathfrak{g}_0$; its adjoint group is compact. By Theorem 4.3 a simply connected group U with Lie algebra $\mathfrak{u}_0$ is compact.

Under the assumption that $\mathfrak{g}$ is semisimple, any complex-linear finite-dimensional representation of $\mathfrak{g}$ restricts to $\mathfrak{u}_0$ and then lifts to a representation of the compact group U. A finite-dimensional representation of U is fully reducible, and the theory of §1 is applicable. This process of passing to U and using facts about the representation theory of compact groups is called the **unitary trick**.

Now let us return to the case of a reductive pair $(\mathfrak{g}, K)$. Let $(\mathfrak{h}, T)$ be a Cartan subpair, with $\mathfrak{h}_0 = \mathfrak{t}_0 \oplus \mathfrak{a}_0$ and with T built from a positive system $\Delta^+(\mathfrak{g}, \mathfrak{h})$. Here

$$\mathfrak{h}_0 = Z_{\mathfrak{g}_0} \oplus ([\mathfrak{g}_0, \mathfrak{g}_0] \cap \mathfrak{h}_0),$$

and $[\mathfrak{g}_0, \mathfrak{g}_0] \cap (\mathfrak{t}_0 \oplus i\mathfrak{a}_0)$ is a maximal abelian subspace of the compact form of $[\mathfrak{g}_0, \mathfrak{g}_0]$. Using the unitary trick, we see that Cartan-Weyl theory, as outlined in §1, is applicable. Thus an irreducible complex-linear finite-dimensional representation V_j of $\mathfrak{g}$ is a direct sum of its weight spaces relative to $\mathfrak{h}$. Its highest weight is dominant and algebraically integral, and the highest weight determines V_j up to equivalence. Any dominant, algebraically integral form is the highest weight of some irreducible complex-linear finite-dimensional representation of $\mathfrak{g}$.

Returning to the irreducible finite-dimensional $(\mathfrak{g}, K)$ module $V = \bigoplus_{j=1}^{n} V_j$, we can attempt to imitate the theory of §2. There are two difficulties. One is that $\mathfrak{t}_0$ is not necessarily maximal abelian in $\mathfrak{k}_0$; hence T does not fully capture integrality conditions. The other is that T does not necessarily meet every component of K; hence T does not cope adequately with the disconnectedness of K. We shall therefore be content with some partial analogs of Theorem 4.25.

Let v_j be a nonzero highest-weight vector of V_j relative to $\Delta^+(\mathfrak{g}, \mathfrak{h})$. The above theory for $\mathfrak{g}$ gives

$$V = \bigoplus_{j=1}^{n} U(\mathfrak{g}) v_j$$

with $U(\mathfrak{g}) v_j$ irreducible under G_0. If $\mathfrak{b} = \mathfrak{h} \oplus \mathfrak{n}$ is the Borel subalgebra corresponding to $\Delta^+(\mathfrak{g}, \mathfrak{h})$, then the Theorem of the Highest Weight (Theorem 4.7) gives

$$V^{\mathfrak{n}} = \bigoplus_{j=1}^{n} \mathbb{C} v_j.$$

The subspace V^n of V is invariant under T. In fact, let $t \in T$, $X \in \mathfrak{n}$, and $v \in V^n$. Then $\mathrm{Ad}(t)^{-1}X$ is in $\mathfrak{n}$, and

$$X t v = t(\mathrm{Ad}(t)^{-1}X)v = 0.$$

Hence tv is in V^n. Thus V^n is an $(\mathfrak{h}, T)$ module.

In general, V^n need not be irreducible because T need not meet every component of K.

Lemma 4.50. Let $(\mathfrak{g}, K)$ be a reductive pair, and let $(\mathfrak{h}, T)$ be a Cartan subpair such that T meets every component of K, i.e., $K = T K_0$. If V is an irreducible finite-dimensional $(\mathfrak{g}, K)$ module, then V^n is an irreducible $(\mathfrak{h}, T)$ module.

REMARK. By Proposition 4.42, the condition on T is satisfied if $(\mathfrak{h}, T)$ is maximally split and the complex conjugate of each nonimaginary positive root is positive. The setting of §2 is of this kind.

PROOF. Assuming the contrary, let W be a proper nonzero $(\mathfrak{h}, T)$ invariant subspace of V^n. By Lemma 4.14, $U(\mathfrak{g})W$ is not all of V. On the other hand, $U(\mathfrak{g})W$ is T invariant, since

$$t\Big(\sum_i u_i w_i\Big) = \sum_i (\mathrm{Ad}(t)u_i) t w_i = \sum_i (\mathrm{Ad}(t)u_i) \sum_j c_{ji} w_j$$

is in $U(\mathfrak{g})W$. Also it is K_0 invariant, since it is a $U(\mathfrak{k})$ submodule. By assumption $K = T K_0$. Thus $U(\mathfrak{g})W$ is K invariant and is a proper nonzero $(\mathfrak{g}, K)$ invariant subspace, in contradiction to the irreducibility of V.

As in the compact case, the irreducible $(\mathfrak{h}, T)$ module V^n has the special property that all its weights are $\Delta^+(\mathfrak{g}, \mathfrak{h})$ dominant. We call such an $(\mathfrak{h}, T)$ module **dominant**.

Theorem 4.51. Let $(\mathfrak{g}, K)$ be a reductive pair, and let $(\mathfrak{h}, T)$ be a Cartan subpair such that T meets every component of K, i.e., $K = T K_0$. If V is an irreducible finite-dimensional $(\mathfrak{g}, K)$ module, then the $(\mathfrak{h}, T)$ module V^n is irreducible and dominant. If W is another irreducible finite-dimensional $(\mathfrak{g}, K)$ module and $W^n \cong V^n$ as $(\mathfrak{h}, T)$ modules, then $W \cong V$ as $(\mathfrak{g}, K)$ modules.

REMARK. By Proposition 4.42, the condition on T is satisfied if $(\mathfrak{h}, T)$ is maximally split and the complex conjugate of each nonimaginary positive root is positive. The setting of §2 is of this kind.

PROOF. Lemma 4.50 shows that V^n is irreducible, and we have noted that V^n is dominant. The argument that $V \rightarrow V^n$ is one-one up to isomorphism is substantially the same as the proof of the corresponding assertion for the compact case in Theorem 4.25.

Theorem 4.52. Let $(\mathfrak{g}, K)$ be a reductive pair, and let $(\mathfrak{h}, T)$ be a Cartan subpair. If M is a finite-dimensional irreducible dominant $(\mathfrak{h}, T)$ module whose weights are algebraically integral, then there exists an irreducible finite-dimensional $(\mathfrak{g}, T)$ module V such that $V^n \cong M$ as an $(\mathfrak{h}, T)$ module. The module V may be taken to be a $(\mathfrak{g}, K)$ module if the Cartan subpair $(\mathfrak{h}, T)$ has the properties that $(\mathfrak{h}, T)$ is maximally compact, $\Delta^+(\mathfrak{g}, \mathfrak{h})$ is θ stable, and T meets every component of K.

REMARKS.

1) Let $(\mathfrak{g}, K) = (\mathfrak{sl}(2, \mathbb{R}), SO(2))$, let $\mathfrak{h}$ be diagonal, let $T = \{\pm 1\}$, and let α be the positive root. If $\mathfrak{h}$ acts on $\mathbb{C}$ by α and T acts by the nontrivial character, we get an $(\mathfrak{h}, T)$ module, and the theorem yields a 3-dimensional irreducible $(\mathfrak{g}, T)$ module. But the action by T is not the same as in the unique 3-dimensional $(\mathfrak{g}, K)$ module, and hence the constructed $(\mathfrak{g}, T)$ module cannot be made into a $(\mathfrak{g}, K)$ module.

2) For Theorem 4.52, unlike with Theorem 4.51, the optimal result comes when $(\mathfrak{h}, T)$ is maximally compact, rather than maximally split. It is not automatic that T meet every component of K when $(\mathfrak{h}, T)$ is maximally compact, and we have to assume this extra condition. In the presence of a θ stable $\Delta^+(\mathfrak{g}, \mathfrak{h})$, it is equivalent to assume that T is a Cartan subgroup of K, i.e., $T = T'$ in the notation of Proposition 4.44.

PROOF. Let λ be a weight of M, and let V_0 be an irreducible finite-dimensional $U(\mathfrak{g})$ module with highest weight λ and a nonzero highest-weight vector v_0. Write $\mathbb{C}_\lambda = \mathbb{C}v_0$. Since λ is integral for T_0 and since any weight of V_0 differs from λ by a sum of roots, V_0 is a $(\mathfrak{g}, T_0)$ module. Motivated by the proof of Theorem 4.25, define a $(\mathfrak{g}, T)$ module V' by

$$V' = \text{induced}_{\mathfrak{g}, T_0}^{\mathfrak{g}, T}(V_0).$$

By (2.74c) and Theorem 4.7c,

$$(V')^n = \{f \in \text{induced}_{\mathfrak{g}, T_0}^{\mathfrak{g}, T}(V_0) \mid f(T) \subseteq (V_0)^n\}$$
$$= \text{induced}_{\mathfrak{t}, T_0}^{\mathfrak{t}, T}((V_0)^n)$$
$$= \text{induced}_{\mathfrak{t}, T_0}^{\mathfrak{t}, T}(\mathbb{C}_\lambda).$$

By (2.74), V' as a T module is given by $\text{induced}_{\mathfrak{t}, T_0}^{\mathfrak{t}, T}(V_0|_{T_0})$. Since $\mathbb{C}_\lambda$ is a direct summand of $V_0|_{T_0}$, $\text{induced}_{\mathfrak{t}, T_0}^{\mathfrak{t}, T}(\mathbb{C}_\lambda)$ is a direct summand of induced$_{t,T_0}^{t,T}(V_0|_{T_0})$. Frobenius reciprocity (Proposition 2.75a) shows that the decomposition of induced$_{t,T_0}^{t,T}(\mathbb{C}_\lambda)$ according to Proposition 1.18 is

$$\text{induced}_{t,T_0}^{t,T}(\mathbb{C}_\lambda) \cong \bigoplus_{W' \in \widehat{T}} [W'|_{T_0} : \mathbb{C}_\lambda]\, W'.$$

Since $M|_{T_0}$ contains $\mathbb{C}_\lambda$, this formula shows that we can regard M as a T submodule of induced$_{t,T_0}^{t,T}(\mathbb{C}_\lambda)$. With this identification in place, we then have

(4.53) $M \subseteq (V')^n.$

Let V be the smallest $U(\mathfrak{g})$ submodule of V' containing M. Since M is T invariant, V is a $(\mathfrak{g}, T)$ module. Applying Lemma 4.14 to the generating space M and using (4.53) to verify hypothesis (a), we see that $M = V^n$. It follows that V is irreducible as a $(\mathfrak{g}, T)$ module. This proves the first conclusion of the theorem.

Now suppose that the Cartan subpair $(\mathfrak{h}, T)$ has the properties that $(\mathfrak{h}, T)$ is maximally compact, $\Delta^+(\mathfrak{g}, \mathfrak{h})$ is θ stable, and T meets every component of K. Let λ, ξ_λ, and V be as above. We shall make V into a $(\mathfrak{g}, K_0)$ module. Let G be the reductive group corresponding to $(\mathfrak{g}, K)$ under Proposition 4.31, let $\tilde{G}$ be a simply connected cover of G_0, let $\varphi : \tilde{G} \to G_0$ be the covering homomorphism, and let $\tilde{T} = \varphi^{-1}(T_0)$. The group $\tilde{T}$ is connected. The $U(\mathfrak{g})$ module structure on V lifts to an irreducible representation $\tilde{G}$ on V. The kernel of φ consists of certain members of the center $Z_{\tilde{G}}$ and lies in $\tilde{T}$. If x is in $\ker \varphi$, write $x = \exp_{\tilde{T}} h$ with $h \in \mathfrak{t}_0$. Then x acts in V as a scalar, and x acts on v_0 by $e^{\lambda(h)} = \xi_\lambda(\exp_{T_0} h)$. From

$$1 = \varphi(x) = \varphi(\exp_{\tilde{T}} h) = \exp_{T_0} h,$$

we see that x acts by 1 in V. Therefore the representation of $\tilde{G}$ on V descends to a representation of G_0 on V, and this representation of G_0 gives us the desired $(\mathfrak{g}, K_0)$ module structure.

We now have a $(\mathfrak{g}, K_0)$ module structure on V and, from the first part of the proof, a $(\mathfrak{g}, T)$ module structure. On T_0, both the K_0 representation and the T representation differentiate to the $\mathfrak{t}_0$ part of the $\mathfrak{g}$ action. Hence the two module structures are compatible. Since $K = T K_0$ by assumption and since $T \cap K_0 = T_0$ by Lemma 4.43d, we obtain a well-defined K representation that extends the K_0 and T representations if we put

$$(t k_0) v = t(k_0(v)) \qquad \text{for } t \in T,\ k_0 \in K_0,\ v \in V.$$

If u is in $U(\mathfrak{g})$, we have

$$(\mathrm{Ad}(tk_0)u)v = (\mathrm{Ad}(t)\mathrm{Ad}(k_0)u)v = t(\mathrm{Ad}(k_0)u)t^{-1}v$$
$$= tk_0uk_0^{-1}t^{-1}v = (tk_0)u(tk_0)^{-1}v.$$

Hence the $U(\mathfrak{g})$ module structure on V and the K representation combine to make V into a $(\mathfrak{g}, K)$ module.

6. Parabolic Subpairs

In this section we take up the subjects of parabolic subalgebras, Levi subgroups, and parabolic subpairs. Let $\mathfrak{g}$ be a complex reductive Lie algebra. In standard usage a "parabolic subalgebra" of $\mathfrak{g}$ is any Lie subalgebra of $\mathfrak{g}$ containing a Borel subalgebra $\mathfrak{b} = \mathfrak{h} \oplus \mathfrak{n}$. Here it is understood that $\mathfrak{h}$ is a Cartan subalgebra, that a system $\Delta^+(\mathfrak{g}, \mathfrak{h})$ of positive roots has been specified within $\Delta(\mathfrak{g}, \mathfrak{h})$, and that $\mathfrak{n}$ is the nilpotent subalgebra

$$\mathfrak{n} = \bigoplus_{\alpha \in \Delta^+(\mathfrak{g},\mathfrak{h})} \mathfrak{g}_\alpha.$$

For our purposes we shall restrict this notion a little. We start from a reductive pair $(\mathfrak{g}, K)$, and we define a **parabolic subalgebra** of $\mathfrak{g}$ to be any Lie subalgebra of $\mathfrak{g}$ containing a Borel subalgebra $\mathfrak{b}$ built from a Cartan subpair $(\mathfrak{h}, T)$. For the time being, we shall consider $(\mathfrak{h}, T)$ and $\mathfrak{b}$ to be fixed. Any parabolic subalgebra $\mathfrak{q}$ containing $\mathfrak{b}$ is of the form

$$(4.54) \qquad\qquad \mathfrak{q} = \mathfrak{h} \oplus \bigoplus_{\alpha \in \Gamma} \mathfrak{g}_\alpha,$$

where Γ is a subset of $\Delta(\mathfrak{g}, \mathfrak{h})$ containing $\Delta^+(\mathfrak{g}, \mathfrak{h})$. The extreme cases are $\mathfrak{q} = \mathfrak{b}$ (with $\Gamma = \Delta^+(\mathfrak{g}, \mathfrak{h})$) and $\mathfrak{q} = \mathfrak{g}$ (with $\Gamma = \Delta(\mathfrak{g}, \mathfrak{h})$).

To provide examples of parabolic subalgebras, we recall the notion of a "simple root." Relative to the positive system $\Delta^+(\mathfrak{g}, \mathfrak{h})$, a root $\alpha \in \Delta^+$ is **simple** if it is not the sum of two positive roots. We take the following result as known.

Proposition 4.55.

(a) The simple roots form a vector space basis for the subspace of $\mathfrak{h}^*$ that vanishes on $\mathfrak{h} \cap Z_{\mathfrak{g}}$. In the expansion of any positive root in terms of the simple roots, the coefficients are integers ≥ 0.

(b) If α and β are distinct simple roots, then $\beta - \alpha$ is not a root. Hence $\langle \alpha, \beta \rangle \leq 0$.

(c) If β is any positive root that is not simple, then there exists a simple root α such that $\beta - \alpha$ is a (positive) root. Consequently a root vector of β is an iterated bracket of root vectors for simple roots.

(d) The root reflections in the simple roots generate the Weyl group.

To obtain an example of a parabolic subalgebra, we fix a subset Π of simple roots and let

(4.56) $\Gamma = \Delta^+(\mathfrak{g}, \mathfrak{h}) \cup \{\alpha \in \Delta(\mathfrak{g}, \mathfrak{h}) \mid \alpha \in \mathrm{span}(\Pi)\}.$

Then (4.54) is a parabolic subalgebra containing the fixed Borel subalgebra $\mathfrak{b}$. (Closure under brackets follows from the fact that if α and β are in Γ and if $\alpha + \beta$ is a root, then $\alpha + \beta$ is in Γ.) All examples are of this form, according to Proposition 4.57 below. With Γ as in (4.54), define $-\Gamma$ to be the set of negatives of the members of Γ.

Proposition 4.57. The parabolic subalgebras $\mathfrak{q}$ containing $\mathfrak{b}$ are parametrized by the set of subsets of simple roots; the one corresponding to a subset Π is of the form (4.54) with Γ as in (4.56).

PROOF. If $\mathfrak{q}$ is given, we define $\Gamma(\mathfrak{q})$ to be the Γ in (4.54), and we define $\Pi(\mathfrak{q})$ to be the set of simple roots in the linear span of $\Gamma(\mathfrak{q}) \cap -\Gamma(\mathfrak{q})$. Then $\mathfrak{q} \mapsto \Pi(\mathfrak{q})$ is a map from parabolic subalgebras $\mathfrak{q}$ containing $\mathfrak{b}$ to subsets of simple roots. In the reverse direction, if Π is given, we define $\Gamma(\Pi)$ to be the Γ in (4.56), and then $\mathfrak{q}(\Pi)$ is defined by means of (4.54). We have seen that $\mathfrak{q}(\Pi)$ is a subalgebra, and thus $\Pi \mapsto \mathfrak{q}(\Pi)$ is a map from subsets of simple roots to parabolic subalgebras containing $\mathfrak{b}$.

To complete the proof we have to show that these two maps are inverse to one another. To see that $\Pi(\mathfrak{q}(\Pi)) = \Pi$, we observe that

$$\{\alpha \in \Delta(\mathfrak{g}, \mathfrak{h}) \mid \alpha \in \mathrm{span}(\Pi)\}$$

is closed under negatives. Therefore (4.56) gives

$\Gamma(\Pi) \cap -\Gamma(\Pi)$
$$= (\Delta^+(\mathfrak{g}, \mathfrak{h}) \cup \{\alpha \in \Delta(\mathfrak{g}, \mathfrak{h}) \mid \alpha \in \mathrm{span}(\Pi)\})$$
$$\cap (-\Delta^+(\mathfrak{g}, \mathfrak{h}) \cup \{\alpha \in \Delta(\mathfrak{g}, \mathfrak{h}) \mid \alpha \in \mathrm{span}(\Pi)\})$$
$$= (\Delta^+(\mathfrak{g}, \mathfrak{h}) \cap -\Delta^+(\mathfrak{g}, \mathfrak{h})) \cup \{\alpha \in \Delta(\mathfrak{g}, \mathfrak{h}) \mid \alpha \in \mathrm{span}(\Pi)\}$$
$$= \{\alpha \in \Delta(\mathfrak{g}, \mathfrak{h}) \mid \alpha \in \mathrm{span}(\Pi)\}.$$

The simple roots in the span of the right side are the members of Π, by the independence in Proposition 4.55a, and it follows that $\Pi(q(\Pi)) = \Pi$.

To see that $q(\Pi(q)) = q$, we are to show that $\Gamma(\Pi(q)) = \Gamma(q)$. Since $\Delta^+(\mathfrak{g}, \mathfrak{h}) \subseteq \Gamma(q)$, the inclusion $\Gamma(\Pi(q)) \subseteq \Gamma(q)$ will follow if we show that

$$(4.58) \qquad \{\alpha \in \Delta(\mathfrak{g}, \mathfrak{h}) \mid \alpha \in \mathrm{span}(\Pi(q))\} \subseteq \Gamma(q).$$

Since $\Gamma(q) = \Delta^+(\mathfrak{g}, \mathfrak{h}) \cup (\Gamma(q) \cap -\Gamma(q))$, the inclusion $\Gamma(\Pi(q)) \supseteq \Gamma(q)$ will follow if we show that

$$(4.59) \qquad \Gamma(q) \cap -\Gamma(q) \subseteq \Gamma(\Pi(q)).$$

Let us first prove (4.58). The positive members of the left side of (4.58) are elements of the right side, since $\mathfrak{b} \subseteq q$. Any negative root in the left side is a negative-integer combination of the members of $\Pi(q)$, by Proposition 4.55a. Let $-\alpha$ be such a root. By Proposition 4.55c, a nonzero root vector $e_{-\alpha}$ for $-\alpha$ is an iterated bracket of root vectors for the negatives of the simple roots contributing to α. The individual factors are in q; since q is a subalgebra, the iterated bracket $e_{-\alpha}$ is in q. Thus $-\alpha$ is in $\Gamma(q)$, and (4.58) is proved.

Finally let us prove (4.59). Let $-\alpha = \sum (-n_i)\alpha_i$ be the expansion of a negative root in $\Gamma(q)$ in terms of simple roots. The assertion is that each α_i for which $n_i > 0$ is in $\Pi(q)$, i.e., has $-\alpha_i \in \Gamma(q)$. We prove this assertion by induction on $\sum n_i$, the case of sum 1 being trivial. If the sum $\sum n_i$ for α is > 1, then $\alpha = \beta + \gamma$ with β and γ in $\Delta^+(\mathfrak{g}, \mathfrak{h})$. The root vectors $e_{-\alpha}$ and e_β are in q, and hence so is their bracket, which is a nonzero multiple of $e_{-\gamma}$. Similarly $e_{-\alpha}$ and e_γ are in q, and hence so is $e_{-\beta}$. Thus $-\gamma$ and $-\beta$ are in $\Gamma(q)$. By induction the constituent simple roots of β and γ are in $\Pi(q)$, and thus the same thing is true of α. This completes the proof.

Now define

$$(4.60a) \qquad \mathfrak{l} = \mathfrak{h} \oplus \bigoplus_{\alpha \in \Gamma \cap -\Gamma} \mathfrak{g}_\alpha \quad \text{and} \quad \mathfrak{u} = \bigoplus_{\substack{\alpha \in \Gamma, \\ \alpha \notin -\Gamma}} \mathfrak{g}_\alpha,$$

so that

$$(4.60b) \qquad \mathfrak{q} = \mathfrak{l} \oplus \mathfrak{u}.$$

Corollary 4.61. Relative to a parabolic subalgebra q containing $\mathfrak{b}$,

(a) $\mathfrak{l}$ and $\mathfrak{u}$ are subalgebras of q, and $\mathfrak{u}$ is an ideal in q
(b) $\mathfrak{l}$ is reductive, and $\mathfrak{u}$ is nilpotent
(c) $\Delta(\mathfrak{l}, \mathfrak{h}) = \Gamma \cap -\Gamma \subseteq \Delta(\mathfrak{g}, \mathfrak{h})$.

PROOF. By Proposition 4.57, let $\mathfrak{q}$ be built from Π by means of (4.56) and (4.54). Then (a) is clear. In (b), we have $\mathfrak{u} \subseteq \mathfrak{n}$, and hence $\mathfrak{u}$ is nilpotent. Since $\Gamma \cap -\Gamma$ is a root system, it follows that $\mathfrak{l}$ is reductive. This proves the other part of (b), as well as (c).

In the decomposition (4.60) of $\mathfrak{q}$, $\mathfrak{l}$ is called the **Levi factor** and $\mathfrak{u}$ is called the **nilpotent radical**. The nilpotent radical can be characterized in terms of $\mathfrak{q}$ as the radical of the symmetric bilinear form $\langle \cdot, \cdot \rangle|_{\mathfrak{q} \times \mathfrak{q}}$. Because of our assumptions that make $\mathfrak{h}$ the complexification of a θ stable subspace $\mathfrak{h}_0$, the Levi factor has a characterization. Namely we recall that $-\alpha = \theta \bar{\alpha}$ for every root α. If we define

$$(4.62\text{a}) \qquad \mathfrak{q}^- = \theta \bar{\mathfrak{q}} \qquad \text{and} \qquad \mathfrak{u}^- = \theta \bar{\mathfrak{u}},$$

then

$$(4.62\text{b}) \qquad \mathfrak{q}^- = \mathfrak{l} \oplus \mathfrak{u}^-,$$

where

$$(4.62\text{c}) \qquad \mathfrak{u}^- = \bigoplus_{\substack{\alpha \in \Gamma, \\ \alpha \notin -\Gamma}} \mathfrak{g}_{-\alpha}.$$

(The subalgebra $\mathfrak{q}^-$ is a parabolic subalgebra containing the Borel subalgebra $\mathfrak{b}^- = \mathfrak{h} \oplus \mathfrak{n}^-$.) The characterization of the Levi factor of $\mathfrak{q}$ is

$$(4.63) \qquad \mathfrak{l} = \mathfrak{q} \cap \mathfrak{q}^-.$$

Let us notice also the important identity

$$(4.64) \qquad \mathfrak{g} = \mathfrak{u}^- \oplus \mathfrak{l} \oplus \mathfrak{u}.$$

Parabolic subalgebras have been defined relative to a particular θ stable Cartan subalgebra $\mathfrak{h}_0$ of $\mathfrak{g}_0$. Until now, we have fixed $\mathfrak{h}_0$ in our study. However, if a subalgebra $\mathfrak{q}$ of $\mathfrak{g}$ contains two θ stable Cartan subalgebras $\mathfrak{h}_0$ and $\mathfrak{h}_0'$, then $\mathfrak{q}$ is parabolic relative to both $\mathfrak{h}_0$ and $\mathfrak{h}_0'$ or to neither. To see this, let $\mathfrak{q}$ be parabolic relative to $\mathfrak{h}_0$, and let $\mathfrak{h}_0' \subseteq \mathfrak{q}$. Since $\mathfrak{h}_0'$ is in $\mathfrak{g}_0$ and is θ stable, $\theta \bar{\mathfrak{h}}' = \mathfrak{h}'$. Thus $\mathfrak{h}' \subseteq \mathfrak{q}^-$, and (4.63) shows that $\mathfrak{h}' \subseteq \mathfrak{l}$. Let $G_{\mathbb{C}}$ be a complex group with Lie algebra $\mathfrak{g}$, and let $L_{\mathbb{C}}$ be the analytic subgroup with Lie algebra $\mathfrak{l}$. Any two Cartan subalgebras of $\mathfrak{l}$ are conjugate by an inner automorphism, and thus there exists $l \in L_{\mathbb{C}}$ with $\mathrm{Ad}(l)\mathfrak{h} = \mathfrak{h}'$. The map $\mathrm{Ad}(l)$ carries $\mathfrak{l}$ to $\mathfrak{l}$ and $\mathfrak{u}$ to $\mathfrak{u}$, by Corollary 4.61a. Also $\mathfrak{l}$ carries $\Delta(\mathfrak{g}, \mathfrak{h})$ in obvious fashion to $\Delta(\mathfrak{g}, \mathfrak{h}')$. If $\mathfrak{b} = \mathfrak{h} \oplus \mathfrak{n}$ is a Borel subalgebra relative to $\mathfrak{h}$ such that $\mathfrak{q} \supseteq \mathfrak{b}$ and if $\mathfrak{b}$ corresponds to $\Delta^+(\mathfrak{g}, \mathfrak{h})$, then $\mathrm{Ad}(l)\mathfrak{b} = \mathfrak{h}' \oplus \mathrm{Ad}(l)\mathfrak{n}$ is the Borel subalgebra corresponding to $\Delta^+(\mathfrak{g}, \mathfrak{h}') = l\Delta^+(\mathfrak{g}, \mathfrak{h})$. Since $\mathfrak{q} \supseteq \mathrm{Ad}(l)\mathfrak{b}$, $\mathfrak{q}$ is parabolic relative to $\mathfrak{h}'_0$.

Since parabolic subalgebras do not depend so rigidly on the initial Cartan subalgebra, it is reasonable to look for constructions and characterizations that avoid using a full Cartan subalgebra. We shall provide constructions and characterizations in terms of centralizers and eigenvalues.

We begin with some notation. In the background will be an unspecified Cartan subalgebra, such as $\mathfrak{h}$. We suppose that V is a finite-dimensional fully reducible representation of $\mathfrak{h}$, and we denote by $\Delta(V)$ the set of weights of $\mathfrak{h}$ in V. Some examples are

$$\Delta(\mathfrak{g}) = \Delta(\mathfrak{g}, \mathfrak{h}) \cup \{0\}$$
$$\Delta(\mathfrak{n}) = \Delta^+(\mathfrak{g}, \mathfrak{h})$$
$$\Delta(\mathfrak{q}) = \Gamma \cup \{0\}$$
$$\Delta(\mathfrak{l}) = (\Gamma \cap -\Gamma) \cup \{0\}$$
$$\Delta(\mathfrak{u}) = \{\alpha \in \Gamma \mid -\alpha \notin \Gamma\}.$$

For each weight $\omega \in \Delta(V)$, let m_ω be the multiplicity of ω. We define

$$(4.65) \qquad \qquad \delta(V) = \tfrac{1}{2} \sum_{\omega \in \Delta(V)} m_\omega \omega,$$

half the sum of the weights with multiplicities counted. Often we abbreviate

$$(4.66) \qquad \qquad \delta = \delta(\mathfrak{n}) = \tfrac{1}{2} \sum_{\alpha \in \Delta^+(\mathfrak{g}, \mathfrak{h})} \alpha.$$

In the next proposition we give two elementary properties of δ. Proposition 4.68 generalizes this kind of result considerably.

Proposition 4.67. If α is a simple root, then s_α permutes the positive roots other than α. Consequently with δ as in (4.66),

(a) $s_\alpha \delta = \delta - \alpha$ for each simple root α
(b) $2\langle \delta, \alpha \rangle / |\alpha|^2 = 1$ for each simple root α.

PROOF. Let $\alpha_1, \ldots, \alpha_l$ be the simple roots, and say $\alpha_1 = \alpha$. If $\beta = \sum n_i \alpha_i$ is a positive root other than α, then

$$s_\alpha \beta = \sum_{i \geq 1} n_i s_\alpha \alpha_i = \sum_{i \geq 1} \left(\alpha_i - \frac{2\langle \alpha_i, \alpha_1 \rangle}{|\alpha|^2} \alpha_1 \right) = c\alpha_1 + \sum_{i \geq 2} n_i \alpha_i.$$

For $i \geq 2$, at least one n_i is > 0. By Proposition 4.55a, $s_\alpha \beta$ is positive. Hence s_α permutes the positive roots other than α. Taking the sum of those roots, we have $s_\alpha(2\delta - \alpha) = 2\delta - \alpha$. Since $s_\alpha \alpha = -\alpha$, (a) follows. Since $s_\alpha \delta = \delta - \frac{2\langle \delta, \alpha \rangle}{|\alpha|^2} \alpha$, (b) follows from (a).

Proposition 4.68. Let V be a finite-dimensional fully reducible representation of $\mathfrak{g}$, and let Λ be a subset of $\Delta(V)$. Suppose that α is a root such that $\lambda \in \Lambda$ and $\alpha + \lambda \in \Delta(V)$ together imply $\alpha + \lambda \in \Lambda$. Then $\left\langle \sum_{\lambda \in \Lambda} m_\lambda \lambda, \alpha \right\rangle \geq 0$. Strict inequality holds when $V = \mathfrak{g}$ and α is in Λ and $-\alpha$ is not in Λ.

PROOF. Since we cannot have simultaneously $\lambda \in \Lambda$, $s_\alpha \lambda \notin \Lambda$, and $\langle \lambda, \alpha \rangle < 0$, we see that

$$\sum_{\lambda \in \Lambda} m_\lambda \lambda = \sum_{\substack{\lambda \in \Lambda, \\ \langle \lambda, \alpha \rangle < 0}} m_\lambda(\lambda + s_\alpha \lambda) + \sum_{\substack{\lambda \in \Lambda, \\ \langle \lambda, \alpha \rangle = 0}} m_\lambda \lambda + \sum_{\substack{\lambda \in \Lambda, s_\alpha \lambda \notin \Lambda, \\ \langle \lambda, \alpha \rangle > 0}} m_\lambda \lambda.$$

The inner product of α with the first two sums on the right is 0, and the inner product of α with the third sum is term-by-term positive. This proves the first assertion. When $V = \mathfrak{g}$, if $\alpha \in \Lambda$ and $-\alpha \notin \Lambda$, then α occurs in the third sum and gives a positive inner product. This proves the second assertion.

Corollary 4.69. Let $\mathfrak{q}$ be a parabolic subalgebra. If α is in $\Delta^+(\mathfrak{g}, \mathfrak{h})$, then

$$\langle \delta(\mathfrak{u}), \alpha \rangle \quad \text{is} \quad \begin{cases} = 0 & \text{if } \alpha \in \Delta(\mathfrak{l}, \mathfrak{h}) \\ > 0 & \text{if } \alpha \in \Delta(\mathfrak{u}). \end{cases}$$

PROOF. In Proposition 4.68, let $V = \mathfrak{g}$ and $\Lambda = \Delta(\mathfrak{u})$. If α is in $\Delta(\mathfrak{l}, \mathfrak{h})$, the proposition applies to α and $-\alpha$ and gives $\langle \delta(\mathfrak{u}), \alpha \rangle = 0$. If α is in $\Delta(\mathfrak{u})$, then $-\alpha$ is not in Λ and the proposition gives $\langle \delta(\mathfrak{u}), \alpha \rangle > 0$.

Proposition 4.70. Fix a θ stable Cartan subalgebra $\mathfrak{h}_0$ of $\mathfrak{g}_0$.

(a) Suppose $h_1, \ldots, h_m$ is a linearly independent set in $\mathfrak{h}$ on which all roots are real. Call a simultaneous eigenvalue of $\operatorname{ad} h_1, \ldots, \operatorname{ad} h_m$ positive or 0 or negative if it positive or 0 or negative lexicographically. Define

$\mathfrak{u} = $ sum of simultaneous eigenspaces of $\{\operatorname{ad} h_j\}$ for positive simultaneous eigenvalues

$\mathfrak{l} = Z_\mathfrak{g}(\mathbb{C}h_1 + \cdots + \mathbb{C}h_m) = $ simultaneous eigenspace of $\{\operatorname{ad} h_j\}$ for simultaneous eigenvalue 0

$\mathfrak{u}^- = $ sum of simultaneous eigenspaces of $\{\operatorname{ad} h_j\}$ for negative simultaneous eigenvalues.

Then $\mathfrak{q} = \mathfrak{l} \oplus \mathfrak{u}$ is a parabolic subalgebra with Levi factor $\mathfrak{l}$ and nilpotent radical $\mathfrak{u}$, and $\mathfrak{q}^- = \mathfrak{l} \oplus \mathfrak{u}^-$.

(b) Conversely if $\mathfrak{q} = \mathfrak{l} \oplus \mathfrak{u}$ is a parabolic subalgebra of $\mathfrak{g}$, then there exists $h \in \mathfrak{h}$ such that all roots are real on h and

   $\mathfrak{u}$ = sum of eigenspaces of ad h for positive eigenvalues

   $\mathfrak{l} = Z_{\mathfrak{g}}(h)$ = eigenspace of ad h for eigenvalue 0

   $\mathfrak{u}^-$ = sum of eigenspaces of ad h for negative eigenvalues.

REMARK. This proposition enables us to identify parabolic subalgebras using just a portion of $\mathfrak{h}$. Part (a) gives a construction of parabolic subalgebras, and part (b) characterizes them as always coming from this construction.

PROOF.

(a) We can extend $h_1, \ldots, h_m$ to a basis $h_1, \ldots, h_l$ of $\mathfrak{h}$ on which all roots are real. The simultaneous eigenspaces become the root spaces and 0. If the corresponding notion of positivity is used in this situation, the sum of the root spaces for positive roots is $\mathfrak{n}$, the 0 eigenspace is $\mathfrak{h}$, and the sum of the root spaces for negative roots is $\mathfrak{n}^-$. Thus we obtain a Borel subalgebra $\mathfrak{b} = \mathfrak{h} \oplus \mathfrak{n}$ such that $\mathfrak{l} \oplus \mathfrak{u} \supseteq \mathfrak{b}$.

It remains to prove that $\mathfrak{q} = \mathfrak{l} \oplus \mathfrak{u}$ is a subalgebra of $\mathfrak{g}$ and that $\mathfrak{l}$ and $\mathfrak{u}$ have the stated properties. Let ad h act on X and Y with eigenvalues λ_1 and λ_2. By the Jacobi identity

$$(\operatorname{ad} h)[X, Y] = [(\operatorname{ad} h)X, Y] + [X, (\operatorname{ad} h)Y] = \lambda_1[X, Y] + \lambda_2[X, Y].$$

So ad h acts on $[X, Y]$ with eigenvalue $\lambda_1 + \lambda_2$. From this fact it readily follows that $\mathfrak{q}$ is a subalgebra. The other properties follow from our construction of the Borel subalgebra and from the identification of the roots.

(b) Let $h_{\delta(\mathfrak{u})}$ be the member of $\mathfrak{h}$ that corresponds to $\delta(\mathfrak{u})$ under $\langle \cdot, \cdot \rangle$. Corollary 4.69 gives

(4.71) $\alpha(h_{\delta(\mathfrak{u})})$ is $\begin{cases} = 0 & \text{if } \alpha \in \Delta(\mathfrak{l}) \\ > 0 & \text{if } \alpha \in \Delta(\mathfrak{u}). \end{cases}$

It follows that $\alpha(h_{\delta(\mathfrak{u})})$ is < 0 if $\alpha \in \Delta(\mathfrak{u}^-)$. Then (b) follows.

Proposition 4.72. Fix a θ stable Cartan subalgebra $\mathfrak{h}_0$ of $\mathfrak{g}_0$, a positive system $\Delta^+(\mathfrak{g}, \mathfrak{h})$, and a corresponding parabolic subalgebra $\mathfrak{q} = \mathfrak{l} \oplus \mathfrak{u}$ of $\mathfrak{g}$. Let V be a finite-dimensional irreducible $U(\mathfrak{g})$ module with highest weight λ and nonzero highest-weight vector v_0. Then

   (a) V is fully reducible as a $U(\mathfrak{l})$ module
   (b) the $U(\mathfrak{l})$ module $U(\mathfrak{l})v_0$ is irreducible and coincides with the subspace $V^{\mathfrak{u}}$ of $\mathfrak{u}$ invariants.

PROOF.

(a) The reductive Lie algebra $\mathfrak{l}$ decomposes as $Z_{\mathfrak{l}} \oplus [\mathfrak{l}, \mathfrak{l}]$ with $[\mathfrak{l}, \mathfrak{l}]$ semisimple. Since $Z_{\mathfrak{l}} \subseteq \mathfrak{h}$, $Z_{\mathfrak{l}}$ acts fully reducibly. Each simultaneous eigenspace under $Z_{\mathfrak{l}}$ is a representation space for a complex-linear finite-dimensional representation of $[\mathfrak{l}, \mathfrak{l}]$ and is fully reducible under $[\mathfrak{l}, \mathfrak{l}]$ by the unitary trick.

(b) To see that $V^{\mathfrak{u}}$ is a $U(\mathfrak{l})$ module, let $v \in V^{\mathfrak{u}}$, $X \in \mathfrak{u}$, and $Y \in \mathfrak{l}$. Since $[X, Y]$ is in $\mathfrak{u}$, we have

$$X(Yv) = [X, Y]v + YXv = 0 + 0 = 0.$$

Hence Yv is in $V^{\mathfrak{u}}$, and $V^{\mathfrak{u}}$ is a $U(\mathfrak{l})$ module.

By (a) we can write $V^{\mathfrak{u}} = V_1 \oplus \cdots \oplus V_k$ as a direct sum of irreducible $U(\mathfrak{l})$ modules. Let v_1 be a nonzero weight vector in V_1, say with weight μ. We use (4.64) to form an ordered basis of $\mathfrak{g}$ consisting of a basis of $\mathfrak{u}^-$, followed by a basis of $\mathfrak{l}$, followed by a basis of $\mathfrak{u}$. By the Poincaré-Birkhoff-Witt Theorem, the ordered monomials in this basis form a basis of $U(\mathfrak{g})$. Since V is irreducible, $U(\mathfrak{g})v_1 = V$. Members of $\mathfrak{u}$ act on v_1 by 0, and members of $\mathfrak{l}$ keep v_1 in V_1. Thus

$$(4.73) \qquad\qquad V = U(\mathfrak{u}^-)U(\mathfrak{l})v_1.$$

Consider the element $h = h_{\delta(\mathfrak{u})}$ given by (4.71) in Proposition 4.70. The proposition implies that h commutes with $\mathfrak{l}$ and hence that h acts by the scalar $\mu(h)$ on $U(\mathfrak{l})v_1$. The proposition implies also that if any nontrivial monomial in $U(\mathfrak{u}^-)$ is applied to a member of $U(\mathfrak{l})v_1$, then h acts by an eigenvalue $< \mu(h)$. By (4.73), $\mu(h)$ is the maximum eigenvalue of h on V, and that maximum is attained on $V_1 = U(\mathfrak{l})v_1$ and only there. We could have chosen any V_i for this argument in place of V_1, and we conclude that V_1 is the only such V_i. That is, $V^{\mathfrak{u}} = V_1$. This proves that $V^{\mathfrak{u}}$ is irreducible under $U(\mathfrak{l})$. Since v_0 is in $V^{\mathfrak{u}}$, $U(\mathfrak{l})v_0 = V^{\mathfrak{u}}$.

Not all parabolic subalgebras in $\mathfrak{g}$ will be of interest to us. The interesting ones are those satisfying the equivalent conditions of Proposition 4.74. We call such parabolic subalgebras **germane**.

Proposition 4.74. Let $\mathfrak{q} = \mathfrak{l} \oplus \mathfrak{u}$ be a parabolic subalgebra, and let Γ be the correponding set of roots given by (4.54). Then the following conditions on $\mathfrak{q}$ are equivalent:

(a) $\mathfrak{l}$ is the complexification of $\mathfrak{l}_0 = \mathfrak{l} \cap \mathfrak{g}_0$

(b) $\mathfrak{l}$ is closed under conjugation (of $\mathfrak{g}$ with respect to $\mathfrak{g}_0$)

(c) $\Gamma \cap -\Gamma$ is closed under bar

(d) $\Gamma \cap -\Gamma$ is closed under θ

(e) $\mathfrak{l}$ is θ stable.

PROOF.

(a) $\Longleftrightarrow$ (b) This is a general fact about real vector spaces and their complexifications.

(b) $\Longleftrightarrow$ (c) This follows from the formulas $\bar{\mathfrak{h}} = \mathfrak{h}$ and $\overline{\mathfrak{g}_\alpha} = \mathfrak{g}_{\bar{\alpha}}$.

(c) $\Longleftrightarrow$ (d) This follows since $-\alpha = \theta\bar{\alpha}$ and since $\Gamma \cap -\Gamma$ is closed under negatives.

(d) $\Longleftrightarrow$ (e) This follows from the formulas $\theta\mathfrak{h} = \mathfrak{h}$ and $\theta\mathfrak{g}_\alpha = \mathfrak{g}_{\theta\alpha}$.

For a germane parabolic subalgebra, $\mathfrak{l}_0$ is θ stable since θ and bar commute. Thus we can write

$$(4.75) \qquad\qquad \mathfrak{l}_0 = (\mathfrak{l}_0 \cap \mathfrak{k}_0) \oplus (\mathfrak{l}_0 \cap \mathfrak{p}_0).$$

There are two extreme cases for the notion of "germane." One is that the parabolic subalgebra $\mathfrak{q}$ is **real**, i.e., $\mathfrak{q}$ is the complexification of $\mathfrak{q}_0 = \mathfrak{q} \cap \mathfrak{g}_0$ (or equivalently $\mathfrak{q}$ is closed under conjugation). In this case $\mathfrak{q}^-$ is real, and hence so is $\mathfrak{l} = \mathfrak{q} \cap \mathfrak{q}^-$. The other extreme case is that the parabolic subalgebra $\mathfrak{q}$ is θ stable, i.e., $\theta\mathfrak{q} = \mathfrak{q}$. In this case $\theta\mathfrak{q}^- = \mathfrak{q}^-$, and hence $\theta\mathfrak{l} = \mathfrak{l}$. Therefore real parabolic subalgebras and θ stable parabolic subalgebras are always germane. The counterpart of Proposition 4.70 for germane parabolic subalgebras is the following.

Proposition 4.76. Fix a θ stable Cartan subalgebra $\mathfrak{h}_0 = \mathfrak{t}_0 \oplus \mathfrak{a}_0$ of $\mathfrak{g}_0$.

(a) Suppose $h_1, \ldots, h_m$ is a linearly independent set in $i\mathfrak{t}_0 \cup \mathfrak{a}_0$. Call a simultaneous eigenvalue of $\operatorname{ad} h_1, \ldots, \operatorname{ad} h_m$ positive or 0 or negative if it positive or 0 or negative lexicographically. Define

> $\mathfrak{u}$ = sum of simultaneous eigenspaces of $\{\operatorname{ad} h_j\}$ for positive simultaneous eigenvalues
>
> $\mathfrak{l} = Z_\mathfrak{g}(\mathbb{C}h_1 + \cdots + \mathbb{C}h_m)$ = simultaneous eigenspace of $\{\operatorname{ad} h_j\}$ for simultaneous eigenvalue 0
>
> $\mathfrak{u}^-$ = sum of simultaneous eigenspaces of $\{\operatorname{ad} h_j\}$ for negative simultaneous eigenvalues.

Then $\mathfrak{q} = \mathfrak{l} \oplus \mathfrak{u}$ is a germane parabolic subalgebra with Levi factor $\mathfrak{l}$ and nilpotent radical $\mathfrak{u}$, and $\mathfrak{q}^- = \mathfrak{l} \oplus \mathfrak{u}^-$. The subalgebra $\mathfrak{q}$ is real if all h_j are in $\mathfrak{a}_0$, and it is θ stable if all h_j are in $i\mathfrak{t}_0$.

(b) Conversely if $\mathfrak{q} = \mathfrak{l} \oplus \mathfrak{u}$ is a real or θ stable parabolic subalgebra of $\mathfrak{g}$, then there exists h in $i\mathfrak{t}_0 \cup \mathfrak{a}_0$ such that

> $\mathfrak{u}$ = sum of eigenspaces of $\operatorname{ad} h$ for positive eigenvalues
>
> $\mathfrak{l} = Z_\mathfrak{g}(h)$ = eigenspace of $\operatorname{ad} h$ for eigenvalue 0
>
> $\mathfrak{u}^-$ = sum of eigenspaces of $\operatorname{ad} h$ for negative eigenvalues.

The element h may be taken in $\mathfrak{a}_0$ if $\mathfrak{q}$ is real, and it may be taken in $i\mathfrak{t}_0$ if $\mathfrak{q}$ is θ stable.

PROOF.

(a) By Proposition 4.70a, $\mathfrak{q} = \mathfrak{l} \oplus \mathfrak{u}$ is a parabolic subalgebra. From the form of $\mathfrak{l}$ in that proposition, we see that $\mathfrak{l}$ is closed under conjugation. Hence $\mathfrak{q}$ is germane. If all h_j are in $\mathfrak{a}_0$, the only conditions on the roots defining $\mathfrak{q}$ are on their values on $\mathfrak{a}_0$, where they are real. The conjugates of these roots satisfy the same conditions, and hence $\mathfrak{q}$ is real. A similar argument shows that if all h_j are in $i\mathfrak{t}_0$, then $\mathfrak{q}$ is θ stable.

(b) We use the element $h = h_{\delta(\mathfrak{u})}$ of (4.71). This is the half sum of members of $i\mathfrak{t}_0 \oplus \mathfrak{a}_0$ and hence is in $i\mathfrak{t}_0 \oplus \mathfrak{a}_0$. All we have to see is that h is in $\mathfrak{a}_0$ if $\mathfrak{q}$ is real and that h is in $i\mathfrak{t}_0$ if $\mathfrak{q}$ is θ stable. If $\mathfrak{q}$ is real, then $\mathfrak{q} = \bar{\mathfrak{q}}$ and $\mathfrak{u} = \bar{\mathfrak{u}}$. Hence $\delta(\mathfrak{u}) = \delta(\bar{\mathfrak{u}}) = \overline{\delta(\mathfrak{u})}$, and $h_{\delta(\mathfrak{u})} = \overline{h_{\delta(\mathfrak{u})}}$. Since $h_{\delta(\mathfrak{u})}$ is in $i\mathfrak{t}_0 \oplus \mathfrak{a}_0$, it must be in $\mathfrak{a}_0$. Similarly if $\mathfrak{q}$ is θ stable, then $\theta\mathfrak{u} = \mathfrak{u}$. Hence $\delta(\mathfrak{u}) = \theta(\delta(\mathfrak{u}))$, and $h_{\delta(\mathfrak{u})} = \theta h_{\delta(\mathfrak{u})} = -\overline{h_{\delta(\mathfrak{u})}}$. Since $h_{\delta(\mathfrak{u})}$ is in $i\mathfrak{t}_0 \oplus \mathfrak{a}_0$, it must be in $i\mathfrak{t}_0$. This completes the proof.

For germane parabolic subalgebras $\mathfrak{q} = \mathfrak{l} \oplus \mathfrak{u}$, we shall introduce the notion of a corresponding "Levi subgroup" L. The context is a reductive pair $(\mathfrak{g}, K)$, and we let G be the corresponding reductive group given by Proposition 4.31. The **Levi subgroup** L is the closed subgroup of G given by

$$(4.77a) \qquad L = N_G(\mathfrak{q}) \cap N_G(\theta\mathfrak{q}).$$

In the case that $\mathfrak{q}$ is a Borel subalgebra, this definition of Levi subgroup reduces to the definition of Cartan subgroup in (4.31a). The symmetry of the right side of (4.77a) in θ forces L to be θ stable. As a consequence of (4.77a), we have

$$(4.77b) \qquad L = N_G(\bar{\mathfrak{q}}) \cap N_G(\theta\bar{\mathfrak{q}}).$$

Observe from (4.77a) that

$$(4.77c) \qquad L \cap K = N_K(\mathfrak{q}) = N_K(\theta\mathfrak{q}).$$

Proposition 4.78. If $\mathfrak{q}$ is a germane parabolic subalgebra and L is the corresponding Levi subgroup, then

(a) L has Lie algebra $\mathfrak{l}_0$, and $\mathfrak{l}_0 = N_{\mathfrak{g}_0}(\mathfrak{l}_0)$

(b) $L \cap K$ normalizes the analytic subgroup L_0 of G with Lie algebra $\mathfrak{l}_0$

(c) $L = (L \cap K)L_0$

(d) $(L, L \cap K, \theta, \langle \cdot, \cdot \rangle)$ and $((\mathfrak{l}, L \cap K), \mathfrak{l}_0, \theta, \langle \cdot, \cdot \rangle)$ are a reductive group and its corresponding reductive pair

(e) $(\mathfrak{l}, L \cap K) \hookrightarrow (\mathfrak{q}, L \cap K) \hookrightarrow (\mathfrak{g}, K)$ are inclusions of pairs

(f) L normalizes $\mathfrak{u}, \bar{\mathfrak{u}}, \theta\mathfrak{u},$ and $\theta\bar{\mathfrak{u}}$.

PROOF. For (a), let $\tilde{\mathfrak{l}}_0$ be the Lie algebra of L. Then (4.77a) and (4.77b) give

$$\tilde{\mathfrak{l}}_0 = N_{\mathfrak{g}_0}(\mathfrak{q}) \cap N_{\mathfrak{g}_0}(\theta\mathfrak{q}) \cap N_{\mathfrak{g}_0}(\bar{\mathfrak{q}}) \cap N_{\mathfrak{g}_0}(\theta\bar{\mathfrak{q}})$$
$$\subseteq N_{\mathfrak{g}_0}(\mathfrak{q} \cap \theta\mathfrak{q} \cap \bar{\mathfrak{q}} \cap \theta\bar{\mathfrak{q}}) = N_{\mathfrak{g}_0}(\mathfrak{l}) = N_{\mathfrak{g}_0}(\mathfrak{l}_0).$$

Let $\mathfrak{h}_0$ be a θ stable Cartan subalgebra contained in $\mathfrak{q}$. We have $\mathfrak{h}_0 \subseteq \tilde{\mathfrak{l}}_0$ clearly. Hence

(4.79) $\mathfrak{h} \subseteq \tilde{\mathfrak{l}} \subseteq N_{\mathfrak{g}}(\mathfrak{l}).$

We shall show that

(4.80) $N_{\mathfrak{g}}(\mathfrak{l}) \subseteq \mathfrak{l},$

so that we can conclude $\mathfrak{l}_0 = N_{\mathfrak{g}_0}(\mathfrak{l}_0)$. Since $\mathfrak{h} \subseteq N_{\mathfrak{g}}(\mathfrak{l})$, $N_{\mathfrak{g}}(\mathfrak{l})$ is of the form

$$N_{\mathfrak{g}}(\mathfrak{l}) = \mathfrak{h} \oplus \bigoplus_{\alpha \in \Sigma} \mathfrak{g}_\alpha$$

for some subset $\Sigma \subseteq \Delta(\mathfrak{g}, \mathfrak{h})$. Let $X = h + \sum_{\alpha \in \Sigma} e_\alpha$ be the corresponding decomposition of a member of $N_{\mathfrak{g}}(\mathfrak{l})$, and let h_1 be in $\mathfrak{h} \subseteq \mathfrak{l}$. Then

$$[h_1, X] = \sum_{\alpha \in \Sigma} \alpha(h_1)e_\alpha \in \mathfrak{l}.$$

Since the e_α's are arbitrary, $\alpha(h_1)$ can be nonzero only for $\alpha \in \Gamma \cap -\Gamma$. Since h_1 is arbitrary, $\Sigma \subseteq \Gamma \cap -\Gamma$. This proves (4.80), and thus (4.79) gives $\tilde{\mathfrak{l}} \subseteq \mathfrak{l}$ and $\tilde{\mathfrak{l}}_0 \subseteq \mathfrak{l}_0$. But every element of $\mathfrak{l}$ normalizes $\mathfrak{q}$ and $\theta\mathfrak{q}$. Thus $\mathfrak{l}_0 \subseteq \tilde{\mathfrak{l}}_0$ and we conclude that $\tilde{\mathfrak{l}}_0 = \mathfrak{l}_0$.

For (b), conclusion (a) identifies the analytic subgroup in question as the identity component of L. Since the identity component is normal, (b) follows.

For (d), let us observe that $((\mathfrak{l}, L \cap K), \mathfrak{l}_0, \theta, \langle \cdot, \cdot \rangle)$ is a reductive pair in the sense of Definition 4.30. We saw in (4.75) that $\mathfrak{l}_0$ is θ stable. Since $\langle \cdot, \cdot \rangle$ is definite on $\mathfrak{k}_0$ and $\mathfrak{p}_0$, it is definite on $\mathfrak{l}_0 \cap \mathfrak{k}_0$ and $\mathfrak{l}_0 \cap \mathfrak{p}_0$, and (4.75) shows it is nondegenerate on $\mathfrak{l}_0$. The other properties are clear, and hence we have a reductive pair. To prove that $(L, L \cap K, \theta, \langle \cdot, \cdot \rangle)$ is the corresponding reductive group, we have to prove that the Cartan decomposition in G of an element of L takes place in L. Thus let $l = k \exp X$ with $l \in L$, $k \in K$, and $X \in \mathfrak{p}_0$. Then $(\theta l)^{-1}l = \exp 2X$ belongs to L. Now $\mathrm{Ad}(\exp 2X)$ is a positive definite transformation on $\mathfrak{g}_0$ whose complex-linear extension to $\mathfrak{g}$ preserves $\mathfrak{q}$ and $\theta\mathfrak{q}$. Hence its logarithm ad $2X$ preserves $\mathfrak{q}$ and $\theta\mathfrak{q}$. Then X is in $N_{\mathfrak{g}_0}(\mathfrak{q}) \cap N_{\mathfrak{g}_0}(\theta\mathfrak{q})$, and (a) shows that X is in $\mathfrak{l}_0$. Consequently $\exp X$ and k are in L. This proves (d).

Then (c) follows from the fact that L is a reductive group, and (e) is clear. Finally (f) for $\mathfrak{u}$ follows from the fact that $\mathfrak{u}$ is the radical of $\langle \cdot, \cdot \rangle_{\mathfrak{q} \times \mathfrak{q}}$, and (f) is proved similarly for $\bar{\mathfrak{u}}$, $\theta\mathfrak{u}$, and $\theta\bar{\mathfrak{u}}$.

We call $(\mathfrak{q}, L \cap K)$ a **parabolic subpair** of $(\mathfrak{g}, K)$, and we call the reductive pair $(\mathfrak{l}, L \cap K)$ the corresponding **Levi subpair**. Relationships among modules for the pairs $(\mathfrak{g}, K)$, $(\mathfrak{q}, L \cap K)$, and $(\mathfrak{l}, L \cap K)$ will be a principal topic of interest in the remainder of the book.

Let $(\mathfrak{h}, T)$ be a Cartan subpair of $(\mathfrak{g}, K)$, and let $\mathfrak{b}$ be a corresponding Borel subalgebra of $\mathfrak{g}$. If $\mathfrak{q} \supseteq \mathfrak{b}$ is a germane parabolic subalgebra, it does not follow that $L \supseteq T$. The example below shows that this inclusion may fail even in the most opportune-looking cases. A positive result is Proposition 4.81 below, which relates Cartan subpairs of $(\mathfrak{g}, K)$ with Cartan subpairs of $(\mathfrak{l}, L \cap K)$.

EXAMPLE. Let G be the two-component compact group given as a semidirect product of $\{1, \gamma\}$ and $G_0 = SU(3)$. Define $\gamma g_0 \gamma^{-1} = \overline{g_0}$ for $g_0 \in G_0$. We may take $\mathfrak{h}_0$ to be the two-dimensional diagonal subalgebra of $\mathfrak{g}_0 = \mathfrak{su}(3)$, and then T has two components. The root system in question is of type A_2, and members of the nonidentity component of T interchange the two simple roots. If the simple roots are α_1 and α_2, let $\mathfrak{q}$ be the parabolic subalgebra with roots $\pm\alpha_1$ and α_2. Members of T that lie in L must permute the roots of $\mathfrak{l}$. Since α_2 is not a root of $\mathfrak{l}$, the nonidentity component of T is not contained in L.

The subgroup L is connected and thus does not meet both components of G. Yet $\mathfrak{h}_0$ is both maximally split and maximally θ stable, and also $\mathfrak{q}$ is θ stable. An analog of this example can be constructed with $G_0 = SL(3, \mathbb{R})$, and again L does not meet both components of G. This time $\mathfrak{h}_0$ can be chosen maximally split or maximally θ stable, and also $\mathfrak{q}$ is real.

Proposition 4.81. Let $(\mathfrak{h}, T)$ be a Cartan subpair of $(\mathfrak{g}, K)$, and let $\mathfrak{b}$ be a corresponding Borel subalgebra. Let $\mathfrak{q} \supseteq \mathfrak{b}$ be a germane parabolic subalgebra, and let $(\mathfrak{h}, T_L)$ be the Cartan subpair of $(\mathfrak{l}, L \cap K)$ corresponding to the Borel subalgebra $\mathfrak{l} \cap \mathfrak{b}$ of $\mathfrak{l}$. Then $T_L = L \cap T$. If T_L meets every component of K, then $T_L = T$.

PROOF. The definitions are

$$T_L = N_{L \cap K}(\mathfrak{l} \cap \mathfrak{b}) \qquad \text{and} \qquad T = N_K(\mathfrak{b}),$$

and L normalizes $\mathfrak{l}$ and $\mathfrak{u}$. Also

$$\mathfrak{b} = (\mathfrak{l} \cap \mathfrak{b}) \oplus \mathfrak{u}.$$

Then

$$T_L \subseteq N_K(\mathfrak{l} \cap \mathfrak{b}) \cap N_K(\mathfrak{u}) \subseteq N_K((\mathfrak{l} \cap \mathfrak{b}) \oplus \mathfrak{u}) = N_K(\mathfrak{b}) = T,$$

and $\quad L \cap T = L \cap N_K(\mathfrak{b}) = N_{L \cap K}(\mathfrak{b}) \subseteq N_{L \cap K}(\mathfrak{l} \cap \mathfrak{b}) = T_L.$

Thus $T_L = L \cap T$.

If T_L meets every component of K, we have $K = T_L K_0$. If t is in T, we can therefore write $t = t_L k_0$ with $t_L \in T_L$ and $k_0 \in K_0$. Since $T \supseteq T_L$ from above, k_0 must be in $T \cap K_0$. By Proposition 4.38d, k_0 is in $Z_{K_0}(\mathfrak{h}_0)$. But $Z_{K_0}(\mathfrak{h}_0) \subseteq N_K(\mathfrak{q}) = L \cap K$. Therefore k_0 is in $L \cap T$, and so is t. Since $T_L = L \cap T$, $T_L = T$.

Now we examine the role of parabolic subalgebras in finite-dimensional representations. For the compact case Theorem 4.83 will give a tidy generalization of the Theorem of the Highest Weight. In the noncompact case we encounter the same difficulties as in Theorems 4.51 and 4.52. Our analogs will be Theorems 4.85 and 4.86.

Lemma 4.82. Let G be a compact Lie group, and let $(\mathfrak{q}, L)$ be a parabolic subpair of $(\mathfrak{g}, G)$. If V is any finite-dimensional $U(\mathfrak{g})$ module, then

(a) $V = V^{\mathfrak{u}} \oplus \bar{\mathfrak{u}} V$

(b) the natural map $V^{\mathfrak{u}} \to V/(\bar{\mathfrak{u}} V)$ is an isomorphism of $U(\mathfrak{l})$ modules.

If V is a $(\mathfrak{g}, G)$ module, then the map in (b) is an $(\mathfrak{l}, L)$ isomorphism.

PROOF. Since all roots are fixed by θ, we have $\mathfrak{u}^- = \bar{\mathfrak{u}}$. For (a) and (b), we may replace $\mathfrak{g}$ by $[\mathfrak{g}, \mathfrak{g}]$ since $\mathfrak{u}$ and $\bar{\mathfrak{u}}$ are in $[\mathfrak{g}, \mathfrak{g}]$. Changing notation, we may assume that $\mathfrak{g}$ is semisimple. Thus V is a direct sum of irreducible representations, and there is no loss of generality in assuming that V is irreducible, say of highest weight λ.

With V irreducible, we argue as in the proof of Lemma 4.12, using a Poincaré-Birkhoff-Witt basis of $U(\mathfrak{g})$ built from root vectors in $\bar{\mathfrak{u}}$, root vectors in $\mathfrak{l}$ together with members of $\mathfrak{t}$, and root vectors in $\mathfrak{u}$. Each such vector is an eigenvector under $\operatorname{ad} h_{\delta(\mathfrak{u})}$, and the eigenvalues are negative, zero, and positive in the three cases, by Proposition 4.70b. Using this eigenvalue as a substitute for weight in the proof of Lemma 4.12, we see that

$$V = U(\mathfrak{l}) V_\lambda \oplus \bar{\mathfrak{u}} V.$$

Since $U(\mathfrak{l})V_\lambda \subseteq V^{\mathfrak{u}}$, (a) will follow if we show that $V^{\mathfrak{u}} \cap \bar{\mathfrak{u}}V = 0$. The above argument leads to the conclusion that $U(\mathfrak{l})V_\lambda$ equals the $\lambda(h_{\delta(\mathfrak{u})})$ eigenspace under the action of $h_{\delta(\mathfrak{u})}$ and that $h_{\delta(\mathfrak{u})}$ acts diagonally on $\bar{\mathfrak{u}}V$ with eigenvalues $< \lambda(h_{\delta(\mathfrak{u})})$. If $V^{\mathfrak{u}} \cap \bar{\mathfrak{u}}V \neq 0$, we can find a nonzero weight vector v in $V^{\mathfrak{u}} \cap \bar{\mathfrak{u}}V$ with $h_{\delta(\mathfrak{u})}$ acting by an eigenvalue $< \lambda(h_{\delta(\mathfrak{u})})$. Repeating the above argument with $\mathbb{C}v$ in place of V_λ, we obtain $V = U(\mathfrak{l})v \oplus \bar{\mathfrak{u}}V$. But this decomposition does not allow for $\lambda(h_{\delta(\mathfrak{u})})$ as an eigenvalue of $h_{\delta(\mathfrak{u})}$. We conclude that $U(\mathfrak{l})V_\lambda = V^{\mathfrak{u}}$, and (a) follows.

Conclusion (b) is immediate from (a). When G acts on V, L acts on both $V^{\mathfrak{u}}$ and $\bar{\mathfrak{u}}V$. To see this action in the case of $V^{\mathfrak{u}}$, let $v \in V^{\mathfrak{u}}, l \in L$, and $X \in \mathfrak{u}$. Then $\mathrm{Ad}(l^{-1})X$ is in $\mathfrak{u}$, and

$$X(lv) = l(\mathrm{Ad}(l^{-1})X)v = 0.$$

Hence lv is in $V^{\mathfrak{u}}$, and $V^{\mathfrak{u}}$ is an $(\mathfrak{l}, L)$ module. Similarly $\bar{\mathfrak{u}}V$ is an $(\mathfrak{l}, L)$ module. Thus $V = V^{\mathfrak{u}} \oplus \bar{\mathfrak{u}}V$ respects the action by $(\mathfrak{l}, L)$, and therefore $V^{\mathfrak{u}} \to V/\bar{\mathfrak{u}}V$ is an $(\mathfrak{l}, L)$ map.

Theorem 4.83. Let G be a compact Lie group, and let $(\mathfrak{q}, L)$ be a parabolic subpair of $(\mathfrak{g}, G)$ such that L meets every component of G. Fix a maximal abelian subspace $\mathfrak{t}_0$ of $\mathfrak{l}_0$, let $\mathfrak{b}$ be a Borel subalgebra of $\mathfrak{g}$ such that $\mathfrak{t} \subseteq \mathfrak{b} \subseteq \mathfrak{q}$, and let $\Delta^+(\mathfrak{g}, \mathfrak{t})$ be the corresponding positive system. Let $(\mathfrak{t}, T)$ and $(\mathfrak{t}, T_L)$ be the corresponding Cartan subpairs of $(\mathfrak{g}, G)$ and $(\mathfrak{l}, L)$.

(a) $T = T_L$.

(b) If V is an irreducible finite-dimensional $(\mathfrak{g}, G)$ module, then $V^{\mathfrak{u}}$ is an irreducible $(\mathfrak{l}, L)$ module, and all of the highest weights of the representation of L_0 on $V^{\mathfrak{u}}$ are dominant for $\Delta^+(\mathfrak{g}, \mathfrak{t})$.

(c) If V and W are irreducible finite-dimensional $(\mathfrak{g}, G)$ modules such that the associated irreducible $(\mathfrak{l}, L)$ modules $V^{\mathfrak{u}}$ and $W^{\mathfrak{u}}$ are equivalent, then V and W are equivalent.

(d) If M is an irreducible finite-dimensional $(\mathfrak{l}, L)$ module all of whose highest weights under L_0 are dominant for $\Delta^+(\mathfrak{g}, \mathfrak{t})$, then there exists an irreducible finite-dimensional $(\mathfrak{g}, G)$ module V with $V^{\mathfrak{u}} \cong M$ as $(\mathfrak{l}, L)$ modules.

(e) The irreducible finite-dimensional $(\mathfrak{g}, G)$ module in (d) is given by either of the formulas

$$V = P_{\mathfrak{q},L}^{\mathfrak{g},G}(\mathcal{F}_{\mathfrak{l},L}^{\mathfrak{q},L}(M))$$

and

$$V = I_{\bar{\mathfrak{q}},L}^{\mathfrak{g},G}(\mathcal{F}_{\mathfrak{l},L}^{\bar{\mathfrak{q}},L}(M)).$$

REMARKS.

1) This theorem generalizes Theorem 4.25 and Corollary 4.28, which effectively treat the case that q is a Borel subalgebra. In those results T played the role that L plays here, and Proposition 4.22 established that T meets every component of G.

2) If M is an irreducible finite-dimensional $(\mathfrak{l}, L)$ module, either all of its highest weights under L_0 are dominant for $\Delta^+(\mathfrak{g}, \mathfrak{t})$ or none are, by the remarks following Lemma 4.24. In the first case, (d) says that $M \cong V^{\mathfrak{u}}$ for some irreducible $(\mathfrak{g}, G)$ module V, while in the second case, (b) says M is not of this form. When $M \cong V^{\mathfrak{u}}$, M and V have the same highest weights.

PROOF.

(a) Proposition 4.22 gives $L = T_L L_0$. Since by assumption L meets every component of G, $G = LG_0$. Thus

$$G = LG_0 = T_L L_0 G_0 = T_L G_0.$$

Then $T = T_L$ by Proposition 4.81.

(b) An argument at the end of the proof of Lemma 4.82 shows that $V^{\mathfrak{u}}$ is an $(\mathfrak{l}, L)$ module.

Write $\mathfrak{b} = \mathfrak{h} \oplus \mathfrak{n}$. Since V is an irreducible $(\mathfrak{g}, G)$ module, Theorem 4.25 applied to G says that $V^{\mathfrak{n}}$ is an irreducible $(\mathfrak{t}, T)$ module. Now

$$(4.84) \qquad (V^{\mathfrak{u}})^{\mathfrak{l} \cap \mathfrak{n}} = V^{(\mathfrak{l} \cap \mathfrak{n}) \oplus \mathfrak{u}} = V^{\mathfrak{n}}.$$

Since $T_L = T$, the space of $\mathfrak{l} \cap \mathfrak{n}$ invariants of $V^{\mathfrak{u}}$ is an irreducible $(\mathfrak{t}, T_L)$ module. Hence Theorem 4.25 applied to L allows us to conclude that $V^{\mathfrak{u}}$ is an irreducible $(\mathfrak{l}, L)$ module.

The highest weights of the $(\mathfrak{l}, L)$ module $V^{\mathfrak{u}}$ are the weights of $V^{\mathfrak{n}}$, and these are highest weights of the $(\mathfrak{g}, G)$ module V. Thus they are $\Delta^+(\mathfrak{g}, \mathfrak{t})$ dominant.

(c) In the notation of (b), if $V^{\mathfrak{u}}$ and $W^{\mathfrak{u}}$ are equivalent $(\mathfrak{l}, L)$ modules, then $V^{\mathfrak{n}}$ and $W^{\mathfrak{n}}$ are equivalent $(\mathfrak{t}, T_L)$ modules (by (4.84)) and hence equivalent $(\mathfrak{t}, T)$ modules (since $T_L = T$). By Theorem 4.25, V and W are equivalent $(\mathfrak{g}, G)$ modules.

(d) By Lemma 4.24, $M^{\mathfrak{l} \cap \mathfrak{n}}$ is an irreducible $(\mathfrak{t}, T_L)$ module, hence an irreducible $(\mathfrak{t}, T)$ module since $T_L = T$. By assumption the weights of $M^{\mathfrak{l} \cap \mathfrak{n}}$ are dominant for $\Delta^+(\mathfrak{g}, \mathfrak{t})$. By Theorem 4.25, there exists an irreducible finite-dimensional $(\mathfrak{g}, G)$ module V with $V^{\mathfrak{n}} \cong M^{\mathfrak{l} \cap \mathfrak{n}}$ as $(\mathfrak{t}, T)$ modules. By (b), $V^{\mathfrak{u}}$ is an irreducible $(\mathfrak{l}, L)$ module, and it must contain $V^{\mathfrak{n}} \cong M^{\mathfrak{l} \cap \mathfrak{n}}$. Since $M^{\mathfrak{l} \cap \mathfrak{n}}$ generates the irreducible $(\mathfrak{l}, L)$ module M, we conclude that $V^{\mathfrak{u}} \cong M$ as $(\mathfrak{l}, L)$ modules.

(e) This is proved in the same way as Corollary 4.28. Let us recall the argument for the P functor. If W is an irreducible finite-dimensional $(\mathfrak{g}, G)$ module, then

$$\operatorname{Hom}_{\mathfrak{g},G}(P_{\mathfrak{q},L}^{\mathfrak{g},G}(\mathcal{F}_{\mathfrak{l},L}^{\mathfrak{q},L}(M)),\ W) \cong \operatorname{Hom}_{\mathfrak{q},L}(\mathcal{F}_{\mathfrak{l},L}^{\mathfrak{q},L}(M),\ (\mathcal{F}^{\vee})_{\mathfrak{g},G}^{\mathfrak{q},L}(W))$$

$$\cong \operatorname{Hom}_{\mathfrak{q},L}(\mathcal{F}_{\mathfrak{l},L}^{\mathfrak{q},L}(M),\ W)$$

$$\cong \operatorname{Hom}_{\mathfrak{l},L}(M,\ I_{\mathfrak{q},L}^{\mathfrak{l},L}(W))$$

$$\cong \operatorname{Hom}_{\mathfrak{l},L}(M,\ W^{\mathfrak{u}}),$$

the last of these isomorphisms following from Proposition 2.89a. Since $P_{\mathfrak{q},L}^{\mathfrak{g},G}(\mathcal{F}_{\mathfrak{l},L}^{\mathfrak{q},L}(M))$ is fully reducible, the above chain of isomorphisms allows us to determine the multiplicity of each irreducible W in it. For $W = V$ the multiplicity is 1, by the irreducibility in (b). For any inequivalent W, the multiplicity is 0, by (c).

For the I functor, we argue similarly, using the chain of isomorphisms

$$\operatorname{Hom}_{\mathfrak{g},G}(W,\ I_{\bar{\mathfrak{q}},L}^{\mathfrak{g},G}(\mathcal{F}_{\mathfrak{l},L}^{\bar{\mathfrak{q}},L}(M))) \cong \operatorname{Hom}_{\bar{\mathfrak{q}},L}(W,\ \mathcal{F}_{\mathfrak{l},L}^{\bar{\mathfrak{q}},L}(M))$$

$$\cong \operatorname{Hom}_{\bar{\mathfrak{q}},L}(W,\ (\mathcal{F}^{\vee})_{\mathfrak{l},L}^{\bar{\mathfrak{q}},L}(M))$$

$$\cong \operatorname{Hom}_{\mathfrak{l},L}(P_{\bar{\mathfrak{q}},L}^{\mathfrak{l},L}(W),\ M)$$

$$\cong \operatorname{Hom}_{\mathfrak{l},L}(W_{\bar{\mathfrak{u}}},\ M)$$

and employing Proposition 2.89a to justify the last step. The right side is isomorphic to $\operatorname{Hom}_{\mathfrak{l},L}(W^{\mathfrak{u}}, M)$ by Lemma 4.82, and the proof is complete.

Theorem 4.85. Let $(\mathfrak{g}, K)$ be a reductive pair, and let $(\mathfrak{q}, L \cap K)$ be a parabolic subpair such that $L \cap K$ meets every component of K. If V is an irreducible finite-dimensional $(\mathfrak{g}, K)$ module, then the $(\mathfrak{l}, L \cap K)$ module $V^{\mathfrak{u}}$ is irreducible. If W is another irreducible finite-dimensional $(\mathfrak{g}, K)$ module and $W^{\mathfrak{u}} \cong V^{\mathfrak{u}}$ as $(\mathfrak{l}, L \cap K)$ modules, then $W \cong V$ as $(\mathfrak{g}, K)$ modules.

PROOF. Let $\mathfrak{h}_0 \subseteq \mathfrak{l}_0$ be a maximally split Cartan subalgebra, and form $\Delta(\mathfrak{g}, \mathfrak{h})$. By declaring the members of $\Delta(\mathfrak{u})$ to be positive and by choosing a positive system $\Delta^+(\mathfrak{l}, \mathfrak{h})$ so that the conjugate of each nonimaginary positive root of $\mathfrak{l}$ is positive, we determine a positive system $\Delta^+(\mathfrak{g}, \mathfrak{h})$ and a corresponding Borel subalgebra $\mathfrak{b}$ of $\mathfrak{g}$ with $\mathfrak{h} \subseteq \mathfrak{b} \subseteq \mathfrak{q}$. Let $(\mathfrak{h}, T)$ and $(\mathfrak{h}, T_L)$ be the corresponding Cartan subpairs of $(\mathfrak{g}, K)$ and $(\mathfrak{l}, L \cap K)$. We are assuming that $K = (L \cap K)K_0$, and Proposition 4.42a gives $L = T_L L_0$ and hence $L \cap K = T_L(L \cap K)_0$. Therefore $K = T_L K_0$. By Proposition 4.81, $T = T_L$.

Write $\mathfrak{b} = \mathfrak{h} \oplus \mathfrak{n}$. Since V is an irreducible $(\mathfrak{g}, K)$ module, Theorem 4.51 applied to $(\mathfrak{g}, K)$ says that $V^{\mathfrak{n}}$ is an irreducible $(\mathfrak{h}, T)$ module. Now $(V^{\mathfrak{u}})^{\mathfrak{l} \cap \mathfrak{n}} = V^{\mathfrak{n}}$, and hence the space of $\mathfrak{l} \cap \mathfrak{n}$ invariants of $V^{\mathfrak{u}}$ is an irreducible $(\mathfrak{h}, T)$ module. Theorem 4.51 applied to $(\mathfrak{l}, L \cap K)$ allows us to conclude that $V^{\mathfrak{u}}$ is an irreducible $(\mathfrak{l}, L \cap K)$ module.

If $W^{\mathfrak{u}} \cong V^{\mathfrak{u}}$ as $(\mathfrak{l}, L \cap K)$ modules, then $(W^{\mathfrak{u}})^{\mathfrak{l} \cap \mathfrak{n}} \cong (V^{\mathfrak{u}})^{\mathfrak{l} \cap \mathfrak{n}}$ as $(\mathfrak{h}, T)$ modules and hence $W^{\mathfrak{n}} \cong V^{\mathfrak{n}}$ as $(\mathfrak{h}, T)$ modules. By Theorem 4.51, $W \cong V$ as $(\mathfrak{g}, K)$ modules.

Theorem 4.86. Let $(\mathfrak{g}, K)$ be a reductive pair, and let $(\mathfrak{q}, L \cap K)$ be a parabolic subpair such that $L \cap K$ meets every component of K. Let $\mathfrak{b} = \mathfrak{h} \oplus \mathfrak{n}$ be a Borel subalgebra with $\mathfrak{h} \subseteq \mathfrak{b} \subseteq \mathfrak{q}$, let $\Delta^{+}(\mathfrak{g}, \mathfrak{h})$ be the corresponding positive system, and let $(\mathfrak{h}, T)$ and $(\mathfrak{h}, T_L)$ be the corresponding Cartan subpairs of $(\mathfrak{g}, K)$ and $(\mathfrak{l}, L \cap K)$, respectively. Assume that $T \subseteq L$, that T meets every component of $L \cap K$, that $(\mathfrak{h}, T_L)$ is maximally compact in $(\mathfrak{l}, L \cap K)$, and that $\Delta^{+}(\mathfrak{l}, \mathfrak{h})$ is θ stable. If M is an irreducible finite-dimensional $(\mathfrak{l}, L \cap K)$ module whose highest weights are algebraically integral and $\Delta^{+}(\mathfrak{g}, \mathfrak{h})$ dominant, then there exists an irreducible finite-dimensional $(\mathfrak{g}, L \cap K)$ module V such that $V^{\mathfrak{u}} \cong M$ as $(\mathfrak{l}, L \cap K)$ modules. The module V may be taken to be a $(\mathfrak{g}, K)$ module if the Cartan subpair $(\mathfrak{h}, T)$ has the additional properties that $(\mathfrak{h}, T)$ is maximally compact in $(\mathfrak{g}, K)$ and that $\Delta^{+}(\mathfrak{g}, \mathfrak{h})$ is θ stable.

PROOF. Let $(\mathfrak{h}, T_L)$ be the Cartan subpair of $(\mathfrak{l}, L \cap K)$ corresponding to $\Delta^{+}(\mathfrak{l}, \mathfrak{h})$. Since $T \subseteq L$, Proposition 4.81 shows that $T = T_L$. By Theorem 4.51, $M^{\mathfrak{l} \cap \mathfrak{n}}$ is an irreducible dominant $(\mathfrak{h}, T)$ module whose weights are algebraically integral. By Theorem 4.52, there exists an irreducible finite-dimensional $(\mathfrak{g}, T)$ module V such that $V^{\mathfrak{n}} \cong M^{\mathfrak{l} \cap \mathfrak{n}}$ as $(\mathfrak{h}, T)$ modules.

The $(\mathfrak{g}, T)$ module V is fully reducible as an $(\mathfrak{l}, T)$ module, say as $V = \bigoplus V_i$. We may assume that V_1 is the $U(\mathfrak{l})$ span of $V^{\mathfrak{n}}$. Within each V_i, we have a representation of T on $V_i^{\mathfrak{l} \cap \mathfrak{n}}$, and this must be irreducible since a nontrivial invariant subspace would generate a nontrivial subspace of V_i invariant under $(\mathfrak{l}, T)$. Theorem 4.52 then shows that the $(\mathfrak{l}, T)$ module structure on V_i extends to an $(\mathfrak{l}, L \cap K)$ module structure. Hence V acquires an $(\mathfrak{l}, L \cap K)$ module structure. This structure is consistent with the $(\mathfrak{g}, T)$ module structure on V, and hence V is a $(\mathfrak{g}, L \cap K)$ module.

The $(\mathfrak{l}, L \cap K)$ module V_1 and the given M both have spaces of $\mathfrak{l} \cap \mathfrak{n}$ invariants isomorphic to $M^{\mathfrak{l} \cap \mathfrak{n}}$. By Theorem 4.51, $V_1 \cong M$ as an $(\mathfrak{l}, L \cap K)$ module. Since V_1 is generated by $V^{\mathfrak{n}}$ over $U(\mathfrak{l})$ and since $V^{\mathfrak{u}}$ is $U(\mathfrak{l})$ invariant, the inclusion $V^{\mathfrak{n}} \subseteq V^{\mathfrak{u}}$ implies that $V_1 \subseteq V^{\mathfrak{u}}$. The same argument as in Theorem 4.85 shows that $V^{\mathfrak{u}}$ is irreducible as an $(\mathfrak{l}, L \cap K)$ module, and therefore $V_1 = V^{\mathfrak{u}}$, i.e., $V^{\mathfrak{u}} \cong M$.

Now suppose also that $(\mathfrak{h}, T)$ is maximally compact in $(\mathfrak{g}, K)$ and that $\Delta^+(\mathfrak{g}, \mathfrak{h})$ is θ stable. Theorem 4.52 then gives V the structure of a $(\mathfrak{g}, K_0)$ module. But V is already a $(\mathfrak{g}, L \cap K)$ module, and the actions of K_0 and $(L \cap K)_0$ both differentiate to the restriction of the $\mathfrak{g}_0$ action. Hence the $(\mathfrak{g}, K_0)$ and $(\mathfrak{g}, L \cap K)$ module structures on V are compatible. By assumption, $K = (L \cap K)K_0$; thus V is a $(\mathfrak{g}, K)$ module.

7. Harish-Chandra Isomorphism

Let $\mathfrak{g}$ be a complex Lie algebra, and let $U(\mathfrak{g})$ be its universal enveloping algebra as usual. As a consequence of the generalization of Schur's Lemma given in Proposition 4.87 below, the center $Z(\mathfrak{g})$ of $U(\mathfrak{g})$ acts by scalars in any irreducible $U(\mathfrak{g})$ module, even an infinite-dimensional one. The resulting homomorphism $\chi : Z(\mathfrak{g}) \to \mathbb{C}$ is the first serious algebraic invariant of an irreducible representation and is called the **infinitesimal character**. This invariant is most useful in situations where $Z(\mathfrak{g})$ can be shown to be large.

In the case that $\mathfrak{g}$ is reductive, let $\mathfrak{h}$ be a Cartan subalgebra, and let $S(\mathfrak{h})^W$ be the subalgebra of Weyl-group invariants (under $W = W(\mathfrak{g}, \mathfrak{h})$) of the symmetric algebra of $\mathfrak{h}$. In this section we shall introduce and establish the "Harish-Chandra isomorphism" $\gamma : Z(\mathfrak{g}) \to S(\mathfrak{h})^W$. In the next section we shall see an indication how this isomorphism allows us to work with infinitesimal characters when $\mathfrak{g}$ is reductive.

Proposition 4.87 (Dixmier). Let $\mathfrak{g}$ be a complex Lie algebra, and let V be an irreducible $U(\mathfrak{g})$ module. Then the only $U(\mathfrak{g})$ linear maps $E : V \to V$ are the scalars.

PROOF. If $v \neq 0$ is in V, then the irreducibility implies that $V = U(\mathfrak{g})v$. Hence the dimension of V is countable. Then the result follows from Proposition A.12c.

Let us analyze $Z(\mathfrak{g})$ in the reductive case. Fix a complex reductive Lie algebra $\mathfrak{g}$ and a Cartan subalgebra $\mathfrak{h}$. It will be helpful in part of the development to think of $\mathfrak{g}$ as the complexification of the Lie algebra of a compact group. More precisely we can find a compact connected Lie group G whose real Lie algebra has complexification isomorphic to $\mathfrak{g}$. Furthermore we can choose the isomorphism in such a way that $\mathfrak{h}$ is the complexification of a real subalgebra $\mathfrak{h}_0$ of $\mathfrak{g}_0$. Then the corresponding analytic subgroup H of G is a maximal torus. In the special case that $\mathfrak{h}$

arises as the complexification of a θ stable Cartan subalgebra of a real reductive Lie algebra, the existence of G is a consequence of the "unitary trick" described in §5. For the general case, see Helgason [1962], 155.

Let $W = W(\mathfrak{g}, \mathfrak{h})$ be the Weyl group of the root system $\Delta(\mathfrak{g}, \mathfrak{h})$. For each root α, let e_α be a nonzero root vector. Let $\langle \cdot, \cdot \rangle$ be a nonsingular $\mathbb{C}$ bilinear form on $\mathfrak{g}$ such that each ad X acts by a skew transformation and such that $\langle \cdot, \cdot \rangle$ is negative definite on $\mathfrak{g}_0$. We can take $\langle \cdot, \cdot \rangle$ to be the Killing form of $\mathfrak{g}$ if $\mathfrak{g}$ is semisimple. In any case this form sets up an isomorphism of $\mathfrak{h}$ with $\mathfrak{h}^*$, and we let h_α be the member of $\mathfrak{h}$ that corresponds to α. The form is positive definite on the real span of the h_α's.

Introduce a positive system $\Delta^+ = \Delta^+(\mathfrak{g}, \mathfrak{h})$, let $\mathfrak{b} = \mathfrak{h} \oplus \mathfrak{n}$ be the associated Borel subalgebra, and define

$$\mathfrak{n}^- = \sum_{\alpha \in \Delta^+} \mathbb{C}e_{-\alpha}.$$

Enumerate the positive roots as $\beta_1, \dots, \beta_k$, and let $h_1, \dots, h_l$ be a basis of $\mathfrak{h}$ over $\mathbb{C}$. Then the monomials

$$(4.88) \qquad e_{-\beta_1}^{q_1} \cdots e_{-\beta_k}^{q_k} h_1^{m_1} \cdots h_l^{m_l} e_{\beta_1}^{p_1} \cdots e_{\beta_k}^{p_k}$$

are a basis of $U(\mathfrak{g})$ over $\mathbb{C}$, by the Poincaré-Birkhoff-Witt Theorem.

The Harish-Chandra mapping is motivated by considering how an element $z \in Z(\mathfrak{g})$ acts in an irreducible complex-linear finite-dimensional representation with highest weight λ. The action is by scalars, by Proposition 4.87, and we compute those scalars by testing the action on a nonzero highest-weight vector v_0. If we expand z in terms of the above basis of $U(\mathfrak{g})$ and consider the effect of the term (4.88), there are two possibilities. One is that some p_j is > 0, and then the term acts as 0. The other is that all p_j are 0. In this case, as we shall see in Proposition 4.89b below, all q_j are 0. The $U(\mathfrak{h})$ part acts on v_0 by the scalar

$$\lambda(h_1)^{m_1} \cdots \lambda(h_l)^{m_l},$$

and that is the total effect of the term. Hence we can compute the effect of z if we can extract those terms in the expansion relative to the basis (4.88) such that only the $U(\mathfrak{h})$ part is present.

Thus define

$$\mathcal{H} = U(\mathfrak{h}), \qquad \mathcal{P} = \sum_{\alpha \in \Delta^+} U(\mathfrak{g})e_\alpha, \qquad \mathcal{N} = \sum_{\alpha \in \Delta^+} e_{-\alpha}U(\mathfrak{g}).$$

Since $\mathfrak{h}$ is abelian, $\mathcal{H}$ may be regarded as the symmetric algebra $S(\mathfrak{h})$ of $\mathfrak{h}$. The action of the Weyl group sending $\mathfrak{h}$ to $\mathfrak{h} \subseteq \mathcal{H}$ extends, by the universal mapping property of a symmetric algebra, to an action of W on $\mathcal{H}$ by automorphisms.

Proposition 4.89.

(a) $U(\mathfrak{g}) = \mathcal{H} \oplus (\mathcal{P} + \mathcal{N})$

(b) Any member of $Z(\mathfrak{g})$ has its $\mathcal{P} + \mathcal{N}$ component in $\mathcal{P}$.

PROOF.

(a) The fact that $U(\mathfrak{g}) = \mathcal{H} + (\mathcal{P} + \mathcal{N})$ follows by the Poincaré-Birkhoff-Witt Theorem from the fact that the elements (4.88) span $U(\mathfrak{g})$. Fix the basis (4.88). For any nonzero element of $U(\mathfrak{g})e_\alpha$ with $\alpha \in \Delta^+$, write out the $U(\mathfrak{g})$ factor in terms of the basis (4.88), and consider a single term of the product, say

$$(4.90) \qquad ce_{-\beta_1}^{q_1} \cdots e_{-\beta_k}^{q_k} h_1^{m_1} \cdots h_l^{m_l} e_{\beta_1}^{p_1} \cdots e_{\beta_k}^{p_k} e_\alpha.$$

The factor $e_{\beta_1}^{p_1} \cdots e_{\beta_k}^{p_k} e_\alpha$ is in $U(\mathfrak{n})$ and has no constant term. By the Poincaré-Birkhoff-Witt Theorem, we can rewrite it as a linear combination of terms $e_{\beta_1}^{r_1} \cdots e_{\beta_k}^{r_k}$ with $r_1 + \cdots + r_k > 0$. Putting

$$ce_{-\beta_1}^{q_1} \cdots e_{-\beta_k}^{q_k} h_1^{m_1} \cdots h_l^{m_l}$$

in place on the left of each term, we see that (4.90) is a linear combination of terms (4.88) with $p_1 + \cdots + p_k > 0$. Similarly any member of $\mathcal{N}$ is a linear combination of terms (4.88) with $q_1 + \cdots + q_k > 0$. Thus any member of $\mathcal{P} + \mathcal{N}$ is a linear combination of terms (4.88) with $p_1 + \cdots + p_k > 0$ or $q_1 + \cdots + q_k > 0$. Any member of $\mathcal{H}$ has $p_1 + \cdots + p_k = 0$ and $q_1 + \cdots + q_k = 0$ in every term of its expansion, and thus (a) follows.

(b) The monomials (4.88) are a basis of $U(\mathfrak{g})$ of weight vectors for ad $\mathfrak{h}$, the weight of (4.88) being

$$(4.91) \qquad -q_1\beta_1 - \cdots - q_k\beta_k + p_1\beta_1 + \cdots + p_k\beta_k.$$

Any member z of $Z(\mathfrak{g})$ satisfies $[h, z] = hz - zh = 0$ for $h \in \mathfrak{h}$ and thus is of weight 0. Hence its expansion in terms of the basis (4.88) involves only terms of weight 0. In the proof of (a) we saw that any member of $\mathcal{P} + \mathcal{N}$ has each term with $p_1 + \cdots + p_k > 0$ or $q_1 + \cdots + q_k > 0$. Since the p's and q's are constrained by the condition that (4.91) equal 0, each term must have both $p_1 + \cdots + p_k > 0$ and $q_1 + \cdots + q_k > 0$. Hence each term is in $\mathcal{P}$.

Let $\gamma_\mathfrak{n}'$ be the projection of $Z(\mathfrak{g})$ into the $\mathcal{H}$ term in Proposition 4.89a. We have seen that this projection is related to the computation of the effect of members of $Z(\mathfrak{g})$ on highest-weight vectors. Harish-Chandra found that a slight adjustment of γ'_n leads to a more symmetric formula. From (4.66), we recall the definition

$$\delta = \tfrac{1}{2} \sum_{\alpha \in \Delta^+} \alpha.$$

Define a linear map $\tau_n : \mathfrak{h} \to \mathcal{H}$ by

(4.92) $\tau_n(h) = h - \delta(h)1,$

and extend τ_n to an algebra automorphism of $\mathcal{H}$ by the universal mapping property for symmetric algebras. The **Harish-Chandra map** γ is defined by

(4.93) $\gamma = \tau_n \circ \gamma'_n$

as a mapping of $Z(\mathfrak{g})$ into $\mathcal{H}$.

Any element $\lambda \in \mathfrak{h}^*$ defines an algebra homomorphism $\lambda : \mathcal{H} \to \mathbb{C}$ with $\lambda(1) = 1$, because the universal mapping property of symmetric algebras allows one to extend $\lambda : \mathfrak{h} \to \mathbb{C}$ to $\mathcal{H}$. In terms of this extension, the maps γ and γ'_n are related by

(4.94) $\lambda(\gamma(z)) = (\lambda - \delta)(\gamma'_n(z))$ for $z \in Z(\mathfrak{g})$, $\lambda \in \mathfrak{h}^*$.

Theorem 4.95 (Harish-Chandra). The mapping γ in (4.94) is an algebra isomorphism of $Z(\mathfrak{g})$ onto the algebra of Weyl-group invariants in $\mathcal{H}$, namely

$$\mathcal{H}^W = \{h \in \mathcal{H} \mid wh = h \text{ for all } w \in W\},$$

and it does not depend on the choice of the positive system Δ^+.

EXAMPLE. $\mathfrak{g} = \mathfrak{sl}(2, \mathbb{C})$. Let $\Omega = \tfrac{1}{2}h^2 + ef + fe$.
Here we use the notation

(4.96) $h = \begin{pmatrix} 1 & 0 \\ 0 & -1 \end{pmatrix}, \qquad e = \begin{pmatrix} 0 & 1 \\ 0 & 0 \end{pmatrix}, \qquad f = \begin{pmatrix} 0 & 0 \\ 1 & 0 \end{pmatrix}.$

We readily check that Ω is in $Z(\mathfrak{g})$ by seeing that Ω commutes with h, e, and f. (Or we can appeal to Proposition 4.119 below.) Let us agree that e corresponds to the positive root α. Then $ef = fe + [e, f] = fe + h$ implies

$$\Omega = \tfrac{1}{2}h^2 + ef + fe = (\tfrac{1}{2}h^2 + h) + 2fe \in \mathcal{H} \oplus \mathcal{P}.$$

Hence

$$\gamma'_{\mathfrak{n}}(\Omega) = \tfrac{1}{2}h^2 + h.$$

Now $\delta(h) = \tfrac{1}{2}\alpha \begin{pmatrix} 1 & 0 \\ 0 & -1 \end{pmatrix} = 1$, and so

$$\tau_{\mathfrak{n}}(h) = h - 1.$$

Thus

$$\gamma(\Omega) = \tfrac{1}{2}(h-1)^2 + (h-1) = \tfrac{1}{2}h^2 - \tfrac{1}{2}.$$

The nontrivial element of the two-element Weyl group acts on $\mathcal{H}$ by sending h to $-h$, and we see that $\gamma(\Omega)$ is invariant under this transformation.

The proof of Theorem 4.95 will occupy the remainder of this section and will take five steps. Two of the steps will have lemmas embedded in them.

PROOF THAT image$(\gamma) \subseteq \mathcal{H}^W$.

We shall make use of the P functor from §II.1. Fix $\lambda \in \mathfrak{h}^*$ and define a one-dimensional complex-linear representation of $\mathfrak{b}$ on a space $\mathbb{C}_{\lambda-\delta} \cong \mathbb{C}$ by

$$Xz = \begin{cases} (\lambda - \delta)(X)z & \text{for } X \in \mathfrak{h} \\ 0 & \text{for } X \in \mathfrak{n}. \end{cases}$$

(Here z is in $\mathbb{C}$.) Then $\mathbb{C}_{\lambda-\delta}$ is a $U(\mathfrak{b})$ module, and we let

$$V(\lambda) = P^{\mathfrak{g},\{1\}}_{\mathfrak{b},\{1\}}(\mathbb{C}_{\lambda-\delta}) = U(\mathfrak{g}) \otimes_{U(\mathfrak{b})} \mathbb{C}_{\lambda-\delta}.$$

The $U(\mathfrak{g})$ module $V(\lambda)$ is called a **Verma module**. We shall derive some of its properties.

First we have $\mathfrak{g} = \mathfrak{n}^- \oplus \mathfrak{b}$ as vector spaces and hence

$$U(\mathfrak{g}) \cong U(\mathfrak{n}^-) \otimes_{\mathbb{C}} U(\mathfrak{b})$$

as a left $U(\mathfrak{n}^-)$ module and right $U(\mathfrak{b})$ module. Consequently we have a $U(\mathfrak{n}^-)$ isomorphism

$$\begin{aligned} (4.97) \qquad V(\lambda) &\cong (U(\mathfrak{n}^-) \otimes_{\mathbb{C}} U(\mathfrak{b})) \otimes_{U(\mathfrak{b})} \mathbb{C}_{\lambda-\delta} \\ &\cong U(\mathfrak{n}^-) \otimes_{\mathbb{C}} (U(\mathfrak{b}) \otimes_{U(\mathfrak{b})} \mathbb{C}_{\lambda-\delta}) \cong U(\mathfrak{n}^-) \otimes_{\mathbb{C}} \mathbb{C}_{\lambda-\delta}. \end{aligned}$$

Next let us compute the effect of $z \in Z(\mathfrak{g})$ on the element $1 \otimes 1$. For $X \in \mathfrak{n}$, we have

$$(4.98\text{a}) \qquad X(1 \otimes 1) = X \otimes 1 = 1 \otimes X(1) = 0$$

so that $U(\mathfrak{g})\mathfrak{n}$ acts as 0. For $X \in \mathfrak{h}$, we have

(4.98b) $$X(1 \otimes 1) = 1 \otimes X(1) = (\lambda - \delta)(X)(1 \otimes 1),$$

so that $\gamma'_\mathfrak{n}(z)$ acts as $(\lambda - \delta)(\gamma'_\mathfrak{n}(z))$. By Proposition 4.89b, z therefore acts as $(\lambda - \delta)(\gamma'_\mathfrak{n}(z))$ on $1 \otimes 1$, and (4.94) says that z acts as $\lambda(\gamma(z))$ on $1 \otimes 1$. The most general element of $V(\lambda)$ is $u \otimes 1$ with $u \in U(\mathfrak{g})$, and z acts on this by

$$z(u \otimes 1) = zu(1 \otimes 1) = uz(1 \otimes 1) = \lambda(\gamma(z))u(1 \otimes 1) = \lambda(\gamma(z))(u \otimes 1).$$

That is,

(4.99) $\qquad\qquad z$ acts in $V(\lambda)$ by the scalar $\lambda(\gamma(z))$.

Lemma 4.100. Let λ be in $\mathfrak{h}^*$, let α be a simple root, let e_α be a nonzero root vector, and suppose that $m = 2\langle \lambda, \alpha \rangle/|\alpha|^2$ is an integer ≥ 0. Then the $U(\mathfrak{g})$ submodule $M = U(\mathfrak{g})(e_{-\alpha})^m(1 \otimes 1)$ of $V(\lambda)$ is isomorphic to $V(s_\alpha \lambda)$.

PROOF. We may assume that $m > 0$. The vector $v = (e_{-\alpha})^m(1 \otimes 1)$ is not 0, as a consequence of (4.97). Let h be in $\mathfrak{h}$, and apply h to v. Commuting h past each factor of $(e_{-\alpha})^m$ one step at a time, we see that h acts on v by the scalar

$$-m\alpha(h) + (\lambda - \delta)(h) = \left(\lambda - \frac{2\langle \lambda, \alpha \rangle}{|\alpha|^2} \alpha \right)(h) - \delta(h) = (s_\alpha \lambda - \delta)(h).$$

Therefore the linear extension of $1 \to v$ maps $\mathbb{C}_{s_\alpha \lambda - \delta}$ into the submodule M while respecting the $\mathfrak{h}$ action.

We shall check that $1 \to v$ is actually a $U(\mathfrak{b})$ map of $\mathbb{C}_{s_\alpha \lambda - \delta}$ into $V(\lambda)$. We are to see that each e_β with $\beta \in \Delta^+$ maps v to 0. Since a general e_β with $\beta \in \Delta^+$ is an iterated bracket of e_β's with β simple (Proposition 4.55c), it is enough to prove that $e_\beta v = 0$ for β simple. For $\beta \neq \alpha$, $[e_\beta, e_{-\alpha}] = 0$ since $\beta - \alpha$ is not a root (Proposition 4.55b) and is not 0. Thus e_β commutes with $(e_{-\alpha})^m$. Since $e_\beta(1 \otimes 1) = 0$ by (4.98a), $e_\beta v = 0$.

Let us now show that $e_\alpha v = 0$. By (4.98a), we have

$$\begin{aligned} e_\alpha v &= e_\alpha(e^m_{-\alpha})(1 \otimes 1) \\ &= [e_\alpha, e^m_{-\alpha}](1 \otimes 1) + e^m_{-\alpha} e_\alpha(1 \otimes 1) \\ &= [e_\alpha, e^m_{-\alpha}](1 \otimes 1). \end{aligned}$$

Apart from normalizations, we can now make the computation in $\mathfrak{sl}(2, \mathbb{C})$. With h, e, and f as in (4.96), we write

$$[e, f^m] = (Rf)^m e - (Lf)^m e,$$

where Rf is right multiplication by f and Lf is left multiplication by f. Now $Rf = Lf - \text{ad } f$, and the terms on the right commute. By the binomial theorem,

$$(Rf)^m e = \sum_{j=0}^{m} \binom{m}{j} (Lf)^{m-j}(-\text{ad } f)^j e$$

$$= (Lf)^m e + m(Lf)^{m-1}(-\text{ad } f)e + \tfrac{m(m-1)}{2}(Lf)^{m-2}(-\text{ad } f)^2 e,$$

with the higher-order terms vanishing on the right since $(\text{ad } f)^3 e = 0$. The above expression is

$$= (Lf)^m e + m(Lf)^{m-1}h + \tfrac{m(m-1)}{2} f^{m-2}(-2f)$$

$$= (Lf)^m e + mf^{m-1}(h - (m-1)1).$$

Thus

(4.101) $$[e, f^m] = (Rf)^m e - (Lf)^m e = mf^{m-1}(h - (m-1)1).$$

Back in $U(\mathfrak{g})$, we may assume that the root vectors corresponding to e and f are normalized so that $[e_\alpha, e_{-\alpha}] = 2|\alpha|^{-2}h_\alpha$. Then (4.101) and (4.98b) show that

$$[e_\alpha, (e_{-\alpha})^m](1 \otimes 1) = m(e_{-\alpha})^{m-1}(2|\alpha|^{-2}h_\alpha - (m-1))(1 \otimes 1)$$

$$= m\left(\frac{2\langle \lambda - \delta, \alpha\rangle}{|\alpha|^2} - (m-1)\right)(e_{-\alpha})^{m-1}(1 \otimes 1)$$

$$= 0,$$

the last step holding since $2\langle \delta, \alpha\rangle/|\alpha|^2 = 1$ by Proposition 4.67b. We conclude that $e_\alpha v = 0$ and thus that $1 \to v$ gives a $U(\mathfrak{b})$ map of $\mathbb{C}_{s_\alpha\lambda - \delta}$ into $V(\lambda)$.

By Frobenius reciprocity (Propositions 2.33(i) and 2.34), we have an explicit isomorphism

$$\text{Hom}_{\mathfrak{b}}(\mathbb{C}_{s_\alpha\lambda - \delta}, M) \cong \text{Hom}_{\mathfrak{g}}(P_{\mathfrak{b},\{1\}}^{\mathfrak{g},\{1\}}(\mathbb{C}_{s_\alpha\lambda - \delta}), M),$$

and thus we obtain a $U(\mathfrak{g})$ map of $V(s_\alpha\lambda)$ into M carrying $1 \otimes 1$ into v. Then

$$V(s_\alpha\lambda) = U(\mathfrak{g})(1 \otimes 1) \qquad \text{maps to} \qquad U(\mathfrak{g})v = M,$$

and our map is onto. A general element of $V(s_\alpha\lambda)$ is of the form $u \otimes 1$ with $u \in U(\mathfrak{n}^-)$, by (4.97), and this maps to $uv = u(e_{-\alpha})^m(1 \otimes 1)$. If $u \neq 0$, then $u(e_{-\alpha})^m$ is a nonzero member of $U(\mathfrak{n}^-)$, and its effect on $1 \otimes 1$ is nonzero, again by (4.97). Thus our map is one-one, and the lemma is proved.

Now we can give the proof that image$(\gamma) \subseteq U(\mathfrak{h})^W$. Since members of $\mathcal{H}$ are determined by the effect of all $\lambda \in \mathfrak{h}^*$ on them, we need to prove that

$$\lambda(w(\gamma(z))) = \lambda(\gamma(z))$$

for all $\lambda \in \mathfrak{h}^*$ and $w \in W$. In other words, we need to see that every $w \in W$ has

(4.102) $(w^{-1}\lambda)(\gamma(z)) = \lambda(\gamma(z))$

and it is enough to handle w equal to a reflection in a simple root, by Proposition 4.55d. Moreover each side for fixed z is a polynomial in λ, and thus it is enough to prove (4.102) for λ dominant integral.

Form the Verma module $V(\lambda)$. We know from (4.99) that z acts in $V(\lambda)$ by the scalar $\lambda(\gamma(z))$. Also z acts in $V(s_\alpha\lambda)$ by the scalar $(s_\alpha\lambda)(\gamma(z))$. Since $2\langle\lambda, \alpha\rangle/|\alpha|^2$ is an integer ≥ 0, Lemma 4.100 says that $V(s_\alpha\lambda)$ is isomorphic to a (clearly nonzero) $U(\mathfrak{g})$ submodule of $V(\lambda)$. Thus the two scalars must match, and (4.102) is proved.

PROOF THAT γ DOES NOT DEPEND ON THE CHOICE OF Δ^+.

This step will make use of the connected compact group G introduced after Proposition 4.87. Let $\mathfrak{h} \oplus \mathfrak{n}^\#$ be the Borel subalgebra corresponding to a second positive system $\Delta^{+\#}$, and let $\delta^\#$ and $\gamma^\#$ be defined relative to $\Delta^{+\#}$ by (4.66) and (4.93). Similarly define $\mathcal{P}^\# = \sum_{\alpha\in\Delta^{+\#}} U(\mathfrak{g})e_\alpha$. By Theorem 4.8 there exists a unique element $w \in W(\mathfrak{g}, \mathfrak{h})$ so that $w\Delta^+ = \Delta^{+\#}$. By Theorem 4.10 we can choose $x \in G$ normalizing $\mathfrak{h}$ such that $w = \mathrm{Ad}(x)|_\mathfrak{h}$.

Any member z of $Z(\mathfrak{g})$ satisfies $Xz = zX$ for $X \in \mathfrak{g}_0$. Consequently $(\mathrm{ad}\, X)z = 0$ for $X \in \mathfrak{g}_0$, and hence $\mathrm{Ad}(g)z = z$ for $g \in G$. In particular $\mathrm{Ad}(x)z = z$. Application of $\mathrm{Ad}(x)$ to

$$z - \gamma'_\mathfrak{n}(z) \in \mathcal{P}$$

then gives $z - \mathrm{Ad}(x)(\gamma'_\mathfrak{n}(z)) \in \mathcal{P}^\#$.

Since $\mathrm{Ad}(x)|_\mathcal{H} = w$, we can conclude that

$$\gamma'_{\mathfrak{n}^\#}(z) = w(\gamma'_\mathfrak{n}(z)).$$

Since $w\delta = \delta^\#$, each $\lambda \in \mathfrak{h}^*$ has

$$\begin{aligned}
\lambda(\gamma^\#(z)) &= (\lambda - \delta^\#)(\gamma'_{\mathfrak{n}^\#}(z)) = (\lambda - w\delta)(w(\gamma'_\mathfrak{n}(z))) \\
&= w^{-1}(\lambda - w\delta)(\gamma'_\mathfrak{n}(z)) = (w^{-1}\lambda - \delta)(\gamma'_\mathfrak{n}(z)) \\
&= w^{-1}\lambda(\gamma(z)) = \lambda(w(\gamma(z))).
\end{aligned}$$

Thus $\gamma^\#(z) = w(\gamma(z))$. Since γ has image in $\mathcal{H}^W$, $w(\gamma(z)) = \gamma(z)$, and we see that γ does not depend on the choice of Δ^+.

PROOF THAT γ IS MULTIPLICATIVE.

Since τ_n is an algebra isomorphism, we need to show that

(4.103) $$\gamma_n'(z_1 z_2) = \gamma_n'(z_1)\gamma_n'(z_2).$$

We have

$$z_1 z_2 - \gamma_n'(z_1)\gamma_n'(z_2) = z_1(z_2 - \gamma_n'(z_2)) + \gamma_n'(z_2)(z_1 - \gamma_n'(z_1)),$$

which is in $\mathcal{P}$, and therefore (4.103) follows.

PROOF THAT γ IS ONE-ONE.

Again we make use of the compact connected group G introduced after Proposition 4.87. As a consequence of the Peter-Weyl Theorem, G has a faithful matrix representation. We fix such a representation, calling it the **standard representation**.

If $\gamma(z) = 0$, then $\gamma_n'(z) = 0$ also, and hence $\lambda(\gamma_n'(z)) = 0$ for every highest weight λ. This is the scalar by which z acts in the irreducible finite-dimensional representation of highest weight λ. Since every finite-dimensional representation of G is the direct sum of irreducibles, we have $\pi(z) = 0$ for every finite-dimensional representation of G.

Regard π as unitary, and let $(\pi(g)u, v)$ be a matrix coefficient of π. If X is in $\mathfrak{g}_0$ and we make X act by left-invariant differentiation, we obtain

$$X(\pi(g)v_1, v_2) = \tfrac{d}{dt}(\pi(g \exp tX)v_1, v_2)|_{t=0}$$
$$= \tfrac{d}{dt}(\pi(\exp tX)v_1, \pi(g)^* v_2)|_{t=0}$$
$$= (\pi(g)\pi(X)v_1, v_2).$$

Iterating this equality, we obtain

(4.104) $$u(\pi(g)v_1, v_2) = (\pi(g)\pi(u)v_1, v_2)$$

for every $u \in U(\mathfrak{g})$. Taking $u = z$, we see that z, as a left-invariant differential operator, annihilates every matrix coefficient of every finite-dimensional representation of G. This holds for the k-fold tensor product of the standard representation of G with itself, for all $k \geq 1$, and in this case the matrix coefficients at x are the kth degree monomials in the entries of the matrix x. It holds also for the complex conjugate of any representation and for the trivial representation. Thus z, as a differential operator, annihilates all monomials in the real and imaginary parts of the coordinates, and $z = 0$ as a differential operator. It is known that the map of $U(\mathfrak{g})$ into left-invariant differential operators is one-one, and hence $z = 0$ in $U(\mathfrak{g})$. Consequently γ is one-one.

PROOF THAT γ IS ONTO.

To prove that γ is onto $\mathcal{H}^W$, we need a supply of members of $Z(\mathfrak{g})$. As we shall see, the following lemma provides sufficiently many members of $Z(\mathfrak{g})$ for our purposes.

Lemma 4.105. Let π be a finite-dimensional representation of a compact connected Lie group G, and let $\langle \cdot, \cdot \rangle$ be the negative of an $\mathrm{Ad}(G)$ invariant inner product on $\mathfrak{g}_0$, so that the complexification of $\langle \cdot, \cdot \rangle$ is nondegenerate on $\mathfrak{g}$. If X_i is a basis of $\mathfrak{g}$ over $\mathbb{C}$, let $\tilde{X}_i$ be the dual basis relative to $\langle \cdot, \cdot \rangle$, i.e., the basis with

$$\langle \tilde{X}_i, X_j \rangle = \delta_{ij},$$

where δ_{ij} is 1 if $i = j$ and is 0 if $i \neq 0$. Fix an integer $n \geq 1$ and define

$$z = \sum_{i_1,\dots,i_n} \mathrm{Tr}\,\pi(X_{i_1} \cdots X_{i_n}) \tilde{X}_{i_1} \cdots \tilde{X}_{i_n}$$

as a member of $U(\mathfrak{g})$. Then

(a) z is independent of the choice of basis X_i
(b) $\mathrm{Ad}(g)z = z$ for all $g \in G$
(c) z is in $Z(\mathfrak{g})$.

REMARK. For an application in the next section, it is worth noting that the only relevant properties of the form $\mathrm{Tr}\,\pi(X_{i_1} \cdots X_{i_n})$ above are that it is n-multilinear and is $\mathrm{Ad}(G)$ invariant.

PROOF.

(a) Let a second basis X_i' be given by means of a nonsingular complex matrix (a_{ij}) as

$$X_j' = \sum_m a_{mj} X_m.$$

Let (b_{ij}) be the inverse of the matrix (a_{ij}), and define

$$\tilde{X}_i' = \sum_l b_{il} \tilde{X}_l.$$

Then

$$\langle \tilde{X}_i', X_j' \rangle = \sum_{l,m} b_{il} a_{mj} \langle \tilde{X}_l, X_m \rangle = \sum_l b_{il} a_{lj} = \delta_{ij}.$$

Thus $\tilde{X}'_i$ is the dual basis of X'_j. The element to consider is

$$
\begin{aligned}
z' &= \sum_{i_1,\ldots,i_n} \operatorname{Tr} \pi(X'_{i_1} \cdots X'_{i_n}) \tilde{X}'_{i_1} \cdots \tilde{X}'_{i_n} \\
&= \sum_{m_1,\ldots,m_n} \sum_{l_1,\ldots,l_n} \sum_{i_1,\ldots,i_n} a_{m_1 i_1} \cdots a_{m_n i_n} \operatorname{Tr} \pi(X_{m_1} \cdots X_{m_n}) \\
&\qquad\qquad\qquad\qquad\qquad \times \, b_{i_1 l_1} \cdots b_{i_n l_n} \tilde{X}_{l_1} \cdots \tilde{X}_{l_n} \\
&= \sum_{m_1,\ldots,m_n} \sum_{l_1,\ldots,l_n} \delta_{m_1 l_1} \cdots \delta_{m_n l_n} \operatorname{Tr} \pi(X_{m_1} \cdots X_{m_n}) \tilde{X}_{l_1} \cdots \tilde{X}_{l_n} \\
&= \sum_{l_1,\ldots,l_n} \operatorname{Tr} \pi(X_{l_1} \cdots X_{l_n}) \tilde{X}_{l_1} \cdots \tilde{X}_{l_n} \\
&= z.
\end{aligned}
$$

This proves (a).

(b) Fix $g \in G$. In (a) let the new basis be $X'_i = \operatorname{Ad}(g)X_i$. By assumption

$$
\langle \operatorname{Ad}(g)X, \operatorname{Ad}(g)Y \rangle = \langle X, Y \rangle.
$$

Thus

$$
\langle \operatorname{Ad}(g)\tilde{X}_i, X'_j \rangle = \langle \tilde{X}_i, \operatorname{Ad}(g)^{-1} X'_j \rangle = \langle \tilde{X}_i, X_j \rangle = \delta_{ij},
$$

and we conclude that $\tilde{X}'_i = \operatorname{Ad}(g)\tilde{X}_i$. Consequently

$$
\begin{aligned}
\operatorname{Ad}(g)z &= \sum_{i_1,\ldots,i_n} \operatorname{Tr} \pi(X_{i_1} \cdots X_{i_n}) \operatorname{Ad}(g)(\tilde{X}_{i_1} \cdots \tilde{X}_{i_n}) \\
&= \sum_{i_1,\ldots,i_n} \operatorname{Tr}(\pi(g)\pi(X_{i_1} \cdots X_{i_n})\pi(g)^{-1}) \tilde{X}'_{i_1} \cdots \tilde{X}'_{i_n} \\
&= \sum_{i_1,\ldots,i_n} \operatorname{Tr} \pi(X'_{i_1} \cdots X'_{i_n}) \tilde{X}'_{i_1} \cdots \tilde{X}'_{i_n},
\end{aligned}
$$

and this equals z, by (a). This proves (b).

(c) Differentiating the equality $\operatorname{Ad}(\exp tX)z = z$ given in (b), we obtain $(\operatorname{ad} X)z = 0$ and $Xz = zX$. Then (c) follows.

Returning to the proof that γ is onto $\mathcal{H}^W$, we shall once again make use of the compact connected group G introduced after Proposition 4.87. Write $U_n(\mathfrak{g})$ for the subspace of $U(\mathfrak{g})$ of elements of degree $\leq n$, and let $\mathcal{H}_n$ and $\mathcal{H}_n^W$ be the subspaces of $\mathcal{H}$ and $\mathcal{H}^W$ of elements homogeneous of degree n. It is clear from the Poincaré-Birkhoff-Witt Theorem that

$$
(4.106) \qquad\qquad \gamma(Z(\mathfrak{g}) \cap U_n(\mathfrak{g})) \subseteq \bigoplus_{d=0}^{n} \mathcal{H}_d^W.
$$

Let λ be any dominant integral member of $\mathfrak{h}^*$, and let π_λ be the irreducible finite-dimensional representation of G with highest weight λ. Let $\Lambda(\lambda)$ be the weights of π_λ, repeated as often as their multiplicities. In Lemma 4.105, let X_i be the ordered basis dual to one consisting of a basis $h_1, \ldots, h_l$ of $\mathfrak{h}$ followed by the root vectors e_α. The lemma says that the following element z is in $Z(\mathfrak{g})$:

$$
\begin{aligned}
z &= \sum_{i_1,\ldots,i_n} \operatorname{Tr}\pi(\tilde{X}_{i_1} \cdots \tilde{X}_{i_n}) X_{i_1} \cdots X_{i_n} \\
&= \sum_{\substack{i_1,\ldots,i_n, \\ \text{all} \leq l}} \operatorname{Tr}\pi(\tilde{h}_{i_1} \cdots \tilde{h}_{i_n}) h_{i_1} \cdots h_{i_n} \\
&\quad + \sum_{\substack{j_1,\ldots,j_n, \\ \text{at least one} > l}} \operatorname{Tr}\pi(\tilde{X}_{j_1} \cdots \tilde{X}_{j_n}) X_{j_1} \cdots X_{j_n}.
\end{aligned}
$$

In the second sum on the right side of the equality, some factor of $X_{j_1} \cdots X_{j_n}$ is a root vector. Commuting the factors into their positions to match terms with the basis vectors (4.88) of $U(\mathfrak{g})$, we see that

$$
X_{j_1} \cdots X_{j_n} \equiv u \mod U_{n-1}(\mathfrak{g}) \qquad \text{with } u \in \mathcal{P} + \mathcal{N},
$$

i.e.,
$$
X_{j_1} \cdots X_{j_n} \equiv 0 \mod \left(\bigoplus_{d=0}^{n-1} \mathcal{H}_d \oplus (\mathcal{P} + \mathcal{N}) \right).
$$

Application of $\gamma_{\mathfrak{n}}'$ to z therefore gives

$$
\gamma_{\mathfrak{n}}'(z) \equiv \sum_{\substack{i_1,\ldots,i_n, \\ \text{all} \leq l}} \operatorname{Tr}\pi(\tilde{h}_{i_1} \cdots \tilde{h}_{i_n}) h_{i_1} \cdots h_{i_n} \mod \left(\bigoplus_{d=0}^{n-1} \mathcal{H}_d \right).
$$

The automorphism $\tau_{\mathfrak{n}}$ of $\mathcal{H}$ affects elements only modulo lower-order terms, and thus

$$
\gamma(z) \equiv \sum_{\substack{i_1,\ldots,i_n, \\ \text{all} \leq l}} \operatorname{Tr}\pi(\tilde{h}_{i_1} \cdots \tilde{h}_{i_n}) h_{i_1} \cdots h_{i_n} \mod \left(\bigoplus_{d=0}^{n-1} \mathcal{H}_d \right)
$$

$$
= \sum_{\mu \in \Lambda(\lambda)} \sum_{\substack{i_1,\ldots,i_n, \\ \text{all} \leq l}} \mu(\tilde{h}_{i_1}) \cdots \mu(\tilde{h}_{i_n}) h_{i_1} \cdots h_{i_n} \mod \left(\bigoplus_{d=0}^{n-1} \mathcal{H}_d \right).
$$

Now

(4.107)
$$\sum_i \mu(\tilde{h}_i) h_i = h_\mu$$

since

$$\left\langle \sum_i \mu(\tilde{h}_i) h_i, \ \tilde{h}_j \right\rangle = \mu(\tilde{h}_j) = \langle h_\mu, \tilde{h}_j \rangle \qquad \text{for all } j.$$

Thus

$$\gamma(z) \equiv \sum_{\mu \in \Lambda(\lambda)} (h_\mu)^n \quad \mathrm{mod} \ \left(\bigoplus_{d=0}^{n-1} \mathcal{H}_d \right).$$

The set of weights of π_λ, together with their multiplicities, is invariant under W, by Theorem 4.10. Hence $\sum_{\mu \in \Lambda(\lambda)} (h_\mu)^n$ is in $\mathcal{H}^W$, and we can write

(4.108)
$$\gamma(z) \equiv \sum_{\mu \in \Lambda(\lambda)} (h_\mu)^n \quad \mathrm{mod} \ \left(\bigoplus_{d=0}^{n-1} \mathcal{H}_d^W \right).$$

To prove that γ is onto $\mathcal{H}^W$, we show that the image of γ contains $\bigoplus_{d=0}^m \mathcal{H}_d^W$ for every m. For $m = 0$, we have $\gamma(1) = 1$, and there is nothing further to prove. Assuming the result for $n = m - 1$, we see from (4.108) that we can choose $z_1 \in Z(\mathfrak{g})$ with

(4.109)
$$\gamma(z - z_1) = \sum_{\mu \in \Lambda(\lambda)} (h_\mu)^n.$$

To complete the induction, we shall show that

(4.110)
$$\text{the elements} \ \sum_{\mu \in \Lambda(\lambda)} (h_\mu)^n \ \text{span} \ \mathcal{H}_n^W.$$

Let $\Lambda_D(\lambda)$ be the set of dominant weights of $\pi(\lambda)$, repeated according to their multiplicities. Since again the set of weights, together with their multiplicities, is invariant under W, we can rewrite the right side of (4.110) as

(4.111)
$$= \sum_{\mu \in \Lambda_D(\lambda)} c_\mu \sum_{w \in W} (h_{w\mu})^n,$$

where c_μ^{-1} is the order of the stablizer of μ in W. We know that π_λ contains the weight λ with multiplicity 1. Equation (4.109) shows that the elements (4.111) are in the image of γ in $\mathcal{H}_n^W$. To complete the induction, it is thus enough to show that

(4.112)   the elements (4.111) span $\mathcal{H}_n^W$.

We do so by showing that

(4.113a)   the span of all elements (4.111) includes all
elements $\sum_{w \in W} (h_{wv})^n$ for v dominant integral

(4.113b) $\mathcal{H}_n^W$ is spanned by all elements $\sum_{w \in W} (h_{wv})^n$
for v dominant integral.

To prove (4.113a), note that the set of dominant integral v in a compact set is finite because the integral points correspond to the characters of the maximal torus $\exp \mathfrak{h}_0$ and form a lattice in $i\mathfrak{h}_0$. Hence it is permissible to induct on $|v|$. The trivial case for the induction is $|v| = 0$. Suppose inductively that (4.113a) has been proved for all dominant integral v with $|v| < |\lambda|$. If μ is any dominant weight of π_λ other than λ, then $|\mu| < |\lambda|$ by Theorem 4.7e. Thus the expression (4.111) involving λ is the sum of $c_\lambda \sum_{w \in W} (h_{w\lambda})^n$ and a linear combination of terms for which (4.113a) is assumed by induction already to be proved. Since $c_\lambda \neq 0$, (4.113a) holds for $\sum_{w \in W} (h_{w\lambda})^n$. This completes the induction and the proof of (4.113a).

To prove (4.113b), it is enough (by summing over $w \in W$) to prove that

(4.113c) $\mathcal{H}_n$ is spanned by all elements $(h_v)^n$
for v dominant integral,

and we do so by induction on n. The trivial case of the induction is $n = 0$.

We can choose dominant integral forms λ_i for $1 \leq i \leq \dim \mathfrak{h}$ such that $\{\lambda_i\}$ is a $\mathbb{C}$ basis for $\mathfrak{h}^*$. Since the λ_i's span $\mathfrak{h}^*$, the h_{λ_i} span $\mathfrak{h}$. Consequently the nth degree monomials in the h_{λ_i} span $\mathcal{H}_n$.

Assuming (4.113c) inductively for $n - 1$, we now prove it for n. Let $v_1, \ldots, v_n$ be dominant integral. It is enough to show that the monomial $h_{v_1} \cdots h_{v_n}$ is a linear combination of elements $(h_v)^n$ with v dominant integral. By the induction hypothesis,

$$(h_{v_1} \cdots h_{v_{n-1}})h_{v_n} = \sum_v c_v h_v^{n-1} h_{v_n},$$

and it is enough to show that $h_v^{n-1}h_{v'}$ is a linear combination of terms $(h_{v+rv'})^n$ with $r \geq 0$ in $\mathbb{Z}$. By the invertibility of a Vandermonde matrix, choose constants $c_1, \ldots, c_n$ with

$$
\begin{pmatrix}
1 & 1 & 1 & \cdots & 1 \\
1 & 2 & 2^2 & \cdots & 2^{n-1} \\
1 & 3 & 3^2 & \cdots & 3^{n-1} \\
 & & \vdots & & \\
1 & n & n^2 & \cdots & n^{n-1}
\end{pmatrix}
\begin{pmatrix}
c_1 \\ c_2 \\ c_3 \\ \vdots \\ c_n
\end{pmatrix}
=
\begin{pmatrix}
0 \\ 1 \\ 0 \\ \vdots \\ 0
\end{pmatrix}.
$$

Then

$$
\sum_{j=1}^{n} c_j (h_{v+jv'})^n = \sum_{j=1}^{n} c_j (h_v + j h_{v'})^n
$$

$$
= \sum_{k=0}^{n} \binom{n}{k} h_v^{n-k} h_{v'}^{k} \sum_{j=1}^{n} c_j j^k
$$

$$
= n h_v^{n-1} h_{v'}.
$$

Thus $h_v^{n-1}h_{v'}$ has the required expansion, and the induction is complete. This proves (4.113c), and consequently γ is onto $\mathcal{H}^W$.

8. Infinitesimal Character

Fix a complex reductive Lie algebra $\mathfrak{g}$. We say that a $U(\mathfrak{g})$ module V "has an infinitesimal character" if $Z(\mathfrak{g})$ acts by scalars in V. In this case the **infinitesimal character** of V is the homomorphism $\chi : Z(\mathfrak{g}) \to \mathbb{C}$ with $\chi(z)$ equal to the scalar by which z acts. Proposition 4.87 says that every irreducible $U(\mathfrak{g})$ module has an infinitesimal character.

The Harish-Chandra isomorphism of §7 allows us to determine explicitly all possible infinitesimal characters. Let $\mathfrak{h}$ be a Cartan subalgebra of $\mathfrak{g}$. If λ is in $\mathfrak{h}^*$, then λ is defined on the element $\gamma(z)$ of $\mathcal{H}$, and we can obtain a homomorphism $\chi_\lambda : Z(\mathfrak{g}) \to \mathbb{C}$ by setting

(4.114) $\chi_\lambda(z) = \lambda(\gamma(z))$.

Theorem 4.115. If $\mathfrak{g}$ is a reductive Lie algebra and $\mathfrak{h}$ is a Cartan subalgebra, then every homomorphism of $Z(\mathfrak{g})$ into $\mathbb{C}$ sending 1 into 1 is of the form χ_λ for some $\lambda \in \mathfrak{h}^*$. If λ' and λ are in $\mathfrak{h}^*$, then $\chi_{\lambda'} = \chi_\lambda$ if and only if λ' and λ are in the same orbit under the Weyl group $W = W(\mathfrak{g}, \mathfrak{h})$.

PROOF. Let $\chi : Z(\mathfrak{g}) \to \mathbb{C}$ be a homomorphism with $\chi(1) = 1$. By Theorem 4.95, γ carries $Z(\mathfrak{g})$ onto $\mathcal{H}^W$, and therefore $\gamma(\ker \chi)$ is an ideal in $\mathcal{H}^W$. Let us check that the corresponding ideal $I = \mathcal{H}\gamma(\ker \chi)$ in $\mathcal{H}$ is proper. Assuming the contrary, suppose $u_1, \ldots, u_n$ in $\mathcal{H}$ and $h_1, \ldots, h_n$ in $\gamma(\ker \chi)$ are such that $\sum_i u_i h_i = 1$. Application of $w \in W$ gives $\sum_i (w u_i) h_i = 1$. Summing on w, we obtain

$$\sum_i \left(\sum_{w \in W} w u_i \right) h_i = |W|.$$

Since $\sum_{w \in W} w u_i$ is in $\mathcal{H}^W$, we can apply $\chi \circ \gamma^{-1}$ to both sides. Since $\chi(1) = 1$, the result is

$$\sum_i \chi \left(\gamma^{-1} \left(\sum_{w \in W} w u_i \right) \right) \chi(\gamma^{-1}(h_i)) = |W|.$$

But the left side is 0 since $\chi(\gamma^{-1}(h_i)) = 0$ for all i, and we have a contradiction. We conclude that the ideal I is proper.

By Zorn's Lemma, extend I to a maximal ideal $\tilde{I}$ of $\mathcal{H}$. The Hilbert Nullstellensatz tells us that there is some $\lambda \in \mathfrak{h}^*$ with

$$\tilde{I} = \{ h \in \mathcal{H} \mid \lambda(h) = 0 \}.$$

Since $\gamma(\ker \chi) \subseteq I \subseteq \tilde{I}$, we have $\chi_\lambda(z) = \lambda(\gamma(z)) = 0$ for all $z \in \ker \chi$. In other words, $\chi(z) = \chi_\lambda(z)$ for $z \in \ker \chi$ and for $z = 1$. These z's span $\mathcal{H}^W$, and hence $\chi = \chi_\lambda$.

If λ' and λ are in the same orbit under W, say $\lambda' = w\lambda$, then the identity $w(\gamma(z)) = \gamma(z)$ for $w \in W$ forces

$$\chi_{\lambda'}(z) = \lambda'(\gamma(z)) = \lambda'(w(\gamma(z))) = w^{-1}\lambda'(\gamma(z)) = \lambda(\gamma(z)) = \chi_\lambda(z).$$

Finally suppose λ' and λ are not in the same orbit under W. Choose a polynomial p on $\mathfrak{h}^*$ that is 1 on $W\lambda$ and 0 on $W\lambda'$. The polynomial p on $\mathfrak{h}^*$ is nothing more than an element h of $\mathcal{H}$ with

$$(4.116) \qquad w\lambda(h) = 1 \quad \text{and} \quad w\lambda'(h) = 0 \quad \text{for all } w \in W.$$

The element $\tilde{h}$ of $\mathcal{H}$ with $\tilde{h} = |W|^{-1} \sum_{w \in W} wh$ is in $\mathcal{H}^W$ and satisfies the same properties (4.116) as h. By Theorem 4.95 we can choose $z \in Z(\mathfrak{g})$ with $\gamma(z) = \tilde{h}$. Then $\chi_\lambda(z) = \lambda(\gamma(z)) = \lambda(\tilde{h}) = 1$ while $\chi_{\lambda'}(z) = 0$. Hence $\chi_{\lambda'} \neq \chi_\lambda$.

Now suppose that V is a $U(\mathfrak{g})$ module with infinitesimal character χ. By Theorem 4.115, $\chi = \chi_\lambda$ for some $\lambda \in \mathfrak{h}^*$. We often abuse notation and say that V has **infinitesimal character** λ. The element λ is determined up to the operation of the Weyl group, again by Theorem 4.115.

EXAMPLES.

1) Let V be a finite-dimensional irreducible $U(\mathfrak{g})$ module with highest weight λ. We saw in §7 that $z \in Z(\mathfrak{g})$ acts by the scalar $\lambda(\gamma_n'(z))$ in V. By (4.94) this scalar equals $(\lambda + \delta)(\gamma(z))$. Therefore V has infinitesimal character $\lambda + \delta$.

2) If λ is in $\mathfrak{h}^*$, then the Verma module $V(\lambda)$ defined in §7 has infinitesimal character λ, by (4.99).

Proposition 4.117. Let $(\mathfrak{g}, K)$ be a reductive pair, and suppose K_1 has finite index in K. Let V be a $(\mathfrak{g}, K_1)$ module with infinitesimal character χ. Then each $z \in Z(\mathfrak{g})$ acts diagonably on $\mathrm{induced}_{\mathfrak{g}, K_1}^{\mathfrak{g}, K}(V)$, the eigenvalues being $\chi(\mathrm{Ad}(k_j)z)$ for $\{k_j^{-1}\}$ a set of coset representatives of K/K_1.

REMARKS. The $(\mathfrak{g}, K)$ module $\mathrm{induced}_{\mathfrak{g}, K_1}^{\mathfrak{g}, K}(V)$ was defined in (2.74). It is isomorphic with $P_{\mathfrak{g}, K_1}^{\mathfrak{g}, K}(V)$ and $I_{\mathfrak{g}, K_1}^{\mathfrak{g}, K}(V)$, according to Proposition 2.77. When $(\mathfrak{g}, K)$ is in the Harish-Chandra class, $\mathrm{Ad}(k_j)z = z$ for all k_j. In this case the present proposition concludes that $\mathrm{induced}_{\mathfrak{g}, K_1}^{\mathfrak{g}, K}(V)$ has infinitesimal character χ from the fact that V has infinitesimal character χ.

PROOF. Proposition 2.85 says that $\mathrm{induced}_{\mathfrak{g}, K_1}^{\mathfrak{g}, K}(V)$, as a $U(\mathfrak{g})$ module, is isomorphic to $\bigoplus_j k_j^{-1}V$. The action of $z \in Z(\mathfrak{g})$ on $k_j^{-1}V$, according to that proposition, is by the usual action of $\mathrm{Ad}(k_j)z$ on V. Since $\mathrm{Ad}(k_j)z$ is in $Z(\mathfrak{g})$, it acts by $\chi(\mathrm{Ad}(k_j)z)$ on $k_j^{-1}V$. The result follows.

Lemma 4.105 constructed some explicit members of $Z(\mathfrak{g})$. One of these, the "Casimir element," is of particular interest. To work within the hypotheses of Lemma 4.105, we should assume that G is a compact connected Lie group and that $\langle \cdot, \cdot \rangle$ is the negative of an $\mathrm{Ad}(G)$ invariant inner product on $\mathfrak{g}_0$. We extend $\langle \cdot, \cdot \rangle$ to $\mathfrak{g}$ so as to be $\mathbb{C}$ bilinear. Let X_i be any basis of $\mathfrak{g}$ over $\mathbb{C}$, and let $\tilde{X}_i$ be the dual basis relative to $\langle \cdot, \cdot \rangle$. The **Casimir element** Ω is defined by

$$(4.118) \qquad \Omega = \sum_{i,j} \langle X_i, X_j \rangle \tilde{X}_i \tilde{X}_j.$$

Lemma 4.105 and its remark give us the following result.

Proposition 4.119. In a complex reductive Lie algebra $\mathfrak{g}$, the Casimir element Ω is defined independently of the basis X_i and is a member of $Z(\mathfrak{g})$.

REMARK. Actually the compact group G appeared in the definition of Casimir element only to simplify the proof of Lemma 4.105. We can do without G. It is enough to have a complex reductive Lie algebra $\mathfrak{g}$ and a nondegenerate symmetric $\mathbb{C}$ bilinear form $\langle \cdot, \cdot \rangle$ on $\mathfrak{g}$ such that ad $\mathfrak{g}$ acts by skew transformations. Then (4.118) is meaningful, and the proposition is still valid.

EXAMPLE. For $\mathfrak{g} = \mathfrak{sl}(2, \mathbb{C})$, let $\langle X, Y \rangle = \operatorname{Tr}(XY)$. Let the basis X_i consist of h, e, f as in (4.96). Then we find that $\Omega = \frac{1}{2}h^2 + ef + fe$, the same element of $Z(\mathfrak{sl}(2, \mathbb{C}))$ considered in the example after the statement of Theorem 4.95. That example showed that $\gamma(\Omega) = \frac{1}{2}h^2 - \frac{1}{2}$. Now $\tilde{h} = \frac{1}{2}h$, and (4.107) gives

$$\tfrac{1}{2}\lambda(h)^2 = \lambda(\lambda(h)\tilde{h}) = \lambda(h_\lambda) = \langle \lambda, \lambda \rangle.$$

Hence

$$\chi_\lambda(\Omega) = \lambda(\gamma(\Omega)) = \tfrac{1}{2}\lambda(h)^2 - \tfrac{1}{2} = \langle \lambda, \lambda \rangle - \tfrac{1}{2}.$$

The subtracted term $\frac{1}{2}$ on the right can be regarded as $\langle \delta, \delta \rangle$.

Proposition 4.120. Let $\mathfrak{g}$ be a complex reductive Lie algebra. Relative to any Cartan subalgebra $\mathfrak{h}$ of $\mathfrak{g}$,

$$\chi_\lambda(\Omega) = \langle \lambda, \lambda \rangle - \langle \delta, \delta \rangle,$$

where $\delta = \delta(\Delta^+)$ is the half sum of the positive roots for an arbitrary choice of Δ^+.

REMARK. If Δ^+ is changed, then δ is changed by a Weyl group element and $\langle \delta, \delta \rangle$ is not changed. Similarly χ_λ has to depend only on the Weyl group orbit of λ, according to Theorem 4.115; indeed, $\langle \lambda, \lambda \rangle$ does not change if λ is replaced by another element in the same orbit.

PROOF. Form $\Delta(\mathfrak{g}, \mathfrak{h})$, and normalize root vectors so that $[e_\alpha, e_{-\alpha}] = h_\alpha$ for all roots α. Fix a positive system $\Delta^+ = \Delta^+(\mathfrak{g}, \mathfrak{h})$, and let $h_1, \ldots, h_l$ be a basis of $\mathfrak{h}$. Using a basis consisting of the e_α's and the h_i's, we easily check that

$$\Omega = \sum_i h_i \tilde{h}_i + \sum_{\alpha \in \Delta^+} (e_\alpha e_{-\alpha} + e_{-\alpha} e_\alpha)$$

$$= \sum_i h_i \tilde{h}_i + 2h_\delta + \sum_{\alpha \in \Delta^+} 2e_{-\alpha} e_\alpha.$$

Hence

$$\gamma'_{\mathfrak{n}}(\Omega) = \sum_i h_i \tilde{h}_i + 2h_\delta$$

and

$$\gamma(\Omega) = \sum_i (h_i - \delta(h_i))(\tilde{h}_i - \delta(\tilde{h}_i)) + 2(h_\delta - \delta(h_\delta))$$

$$= \sum_i h_i \tilde{h}_i - 2h_\delta + \langle \delta, \delta \rangle + 2(h_\delta - \langle \delta, \delta \rangle)$$

by (4.107). Thus

(4.121) $$\gamma(\Omega) = \sum_i h_i \tilde{h}_i - \langle \delta, \delta \rangle.$$

To compute $\chi_\lambda(\Omega)$, we apply λ to both sides of this equation, again using (4.107), and the proposition follows.

The Harish-Chandra map γ and the auxiliary map $\gamma'_{\mathfrak{n}}$ have generalizations in which the Borel subalgebra is replaced by a parabolic subalgebra. In keeping with our restricted definition of parabolic subalgebra in §6, let $(\mathfrak{g}, K)$ be a reductive pair, let $\mathfrak{h}_0$ be a θ stable Cartan subalgebra of $\mathfrak{g}_0$, let $\mathfrak{b} = \mathfrak{h} \oplus \mathfrak{n}$ be a Borel subalgebra built from $\mathfrak{h}$, and let $\mathfrak{q} = \mathfrak{l} \oplus \mathfrak{u}$ be a parabolic subalgebra of $\mathfrak{g}$ containing $\mathfrak{b}$.

In Proposition 4.89a we saw that

(4.122) $$U(\mathfrak{g}) = U(\mathfrak{h}) \oplus (\mathfrak{n}^- U(\mathfrak{g}) + U(\mathfrak{g})\mathfrak{n}).$$

If we use the element $h_{\delta(\mathfrak{u})} \in \mathfrak{h}$ of Proposition 4.70b to measure effects of root vectors, we immediately obtain parts (a) and (b) of the following generalization.

Lemma 4.123.
(a) $U(\mathfrak{g}) = U(\mathfrak{l}) \oplus (\mathfrak{u}^- U(\mathfrak{g}) + U(\mathfrak{g})\mathfrak{u})$
(b) Any member of $Z(\mathfrak{g})$ has its $\mathfrak{u}^- U(\mathfrak{g}) + U(\mathfrak{g})\mathfrak{u}$ component in $U(\mathfrak{g})\mathfrak{u}$.
(c) Any member of $Z(\mathfrak{g})$ has its $U(\mathfrak{l})$ component in $Z(\mathfrak{l})$.

PROOF OF (c). Let z be in $Z(\mathfrak{g})$, and write, in obvious notation,

$$z = u_{\mathfrak{l}} + \sum \bar{u}_i x_i + \sum y_j u_j.$$

If X is in $\mathfrak{l}$, then

$$0 = (\text{ad } X)z = (\text{ad } X)u_{\mathfrak{l}} + \sum ((\text{ad } X)\bar{u}_i)x_i + \sum \bar{u}_i ((\text{ad } X)x_i)$$
$$+ \sum ((\text{ad } X)y_j)u_j + \sum y_j ((\text{ad } X)u_j),$$

and we see from the directness of the sum in (a) that $(\text{ad } X)u_{\mathfrak{l}} = 0$. Since $X \in \mathfrak{l}$ is arbitrary, we conclude that $u_{\mathfrak{l}}$ is in $Z(\mathfrak{l})$.

Let $\mu'_{\mathfrak{u}} : U(\mathfrak{g}) \to U(\mathfrak{l})$ be the projection of $U(\mathfrak{g})$ into the $U(\mathfrak{l})$ component in Lemma 4.123a. By Lemma 4.123c, $\mu'_{\mathfrak{u}}$ carries $Z(\mathfrak{g})$ into $Z(\mathfrak{l})$. The mapping $\mu'_{\mathfrak{u}}$ is a generalization of $\gamma'_{\mathfrak{n}}$. For a comparable generalization of $\tau_{\mathfrak{n}} : U(\mathfrak{h}) \to U(\mathfrak{h})$, we shall construct a mapping $\tau_{\mathfrak{u}} : U(\mathfrak{l}) \to U(\mathfrak{l})$. The member $\delta(\mathfrak{u})$ of $\mathfrak{h}^*$ is orthogonal to $\Delta(\mathfrak{l}, \mathfrak{h})$ and hence is dominant and algebraically integral for $\mathfrak{l}$. Therefore $\delta(\mathfrak{u})$ is the (highest) weight of a one-dimensional representation $\pi_{\delta(\mathfrak{u})}$ of $\mathfrak{l}$ on a space that we may identify with the multiples of 1 in $U(\mathfrak{l})$. For X in $\mathfrak{l}$, we define

(4.124) $$\tau_{\mathfrak{u}}(X) = X - \pi_{\delta(\mathfrak{u})}(X)1$$

as a member of $U(\mathfrak{l})$. Then we extend $\tau_{\mathfrak{u}}$ to an algebra map of the tensor algebra over $\mathfrak{l}$. For X and Y in $\mathfrak{l}$, we have

$$\tau_{\mathfrak{u}}(X)\tau_{\mathfrak{u}}(Y) - \tau_{\mathfrak{u}}(Y)\tau_{\mathfrak{u}}(X)$$
$$= (X - \pi_{\delta(\mathfrak{u})}(X)1)(Y - \pi_{\delta(\mathfrak{u})}(Y)1)$$
$$\quad - (Y - \pi_{\delta(\mathfrak{u})}(Y)1)(X - \pi_{\delta(\mathfrak{u})}(X)1)$$
$$= [X, Y]$$
$$\quad - (\pi_{\delta(\mathfrak{u})}(X)1)Y - X(\pi_{\delta(\mathfrak{u})}(Y)1) + (\pi_{\delta(\mathfrak{u})}(Y)1)X + Y(\pi_{\delta(\mathfrak{u})}(X)1)$$
$$\quad + (\pi_{\delta(\mathfrak{u})}(X)1)(\pi_{\delta(\mathfrak{u})}(Y)1) - (\pi_{\delta(\mathfrak{u})}(Y)1)(\pi_{\delta(\mathfrak{u})}(X)1)$$
$$= [X, Y] + (\pi_{\delta(\mathfrak{u})}(X)\pi_{\delta(\mathfrak{u})}(Y) - \pi_{\delta(\mathfrak{u})}(Y)\pi_{\delta(\mathfrak{u})}(X))1$$
$$= \tau_{\mathfrak{u}}[X, Y]1,$$

the last step holding since $\pi_{\delta(\mathfrak{u})}$ is a representation of $\mathfrak{l}$. Hence $\tau_{\mathfrak{u}}$ descends to a representation of $U(\mathfrak{l})$ carrying 1 into 1. Repeating the construction for $X \mapsto X + \pi_{\delta(\mathfrak{u})}(X)1$, we obtain a two-sided inverse of $\tau_{\mathfrak{u}}$ on $U(\mathfrak{l})$. Hence $\tau_{\mathfrak{u}}$ is an automorphism of $U(\mathfrak{l})$.

If we specialize $\tau_{\mathfrak{u}}$ to $U(\mathfrak{h})$, then it is clear from (4.124) that $\tau_{\mathfrak{u}}$ carries $U(\mathfrak{h})$ to itself. Combining the formula (4.124) with the identity $\delta = \delta(\mathfrak{l} \cap \mathfrak{n}) + \delta(\mathfrak{u})$, we obtain

(4.125) $$\tau_{\mathfrak{n}} = \tau_{\mathfrak{l} \cap \mathfrak{n}} \circ \tau_{\mathfrak{u}} = \tau_{\mathfrak{u}} \circ \tau_{\mathfrak{l} \cap \mathfrak{n}} \qquad \text{on } U(\mathfrak{h}).$$

Since $\tau_{\mathfrak{u}}$ is an automorphism of $U(\mathfrak{l})$, $\tau_{\mathfrak{u}}$ maps $Z(\mathfrak{l})$ to itself. To understand this mapping, we write $\mathfrak{l} = Z_{\mathfrak{l}} \oplus [\mathfrak{l}, \mathfrak{l}]$ as the sum of ideals, with $Z_{\mathfrak{l}}$ equal to the center of $\mathfrak{l}$; here $Z_{\mathfrak{l}} \subseteq \mathfrak{h}$ since $\mathfrak{h} \subseteq \mathfrak{l}$ and $\mathfrak{h}$ is maximal abelian in $\mathfrak{l}$. Then

$$U(\mathfrak{l}) = U(Z_{\mathfrak{l}}) \otimes_{\mathbb{C}} U([\mathfrak{l}, \mathfrak{l}])$$

with $U(Z_{\mathfrak{l}})$ central in $U(\mathfrak{l})$. Hence

(4.126) $$Z(\mathfrak{l}) = U(Z_{\mathfrak{l}}) \otimes_{\mathbb{C}} Z([\mathfrak{l}, \mathfrak{l}]).$$

On $[\mathfrak{l}, \mathfrak{l}]$, $\tau_{\delta(u)}$ is 0. Hence τ_u is the identity on $U([\mathfrak{l}, \mathfrak{l}])$. Since $U(Z_\mathfrak{l}) \subseteq U(\mathfrak{h})$, τ_u is given on $U(Z_\mathfrak{l})$ by the extension of the linear map

$$(4.127) \qquad \tau_{u,1}(h) = h - \delta(u)(h)1$$

of $Z_\mathfrak{l} \to U(Z_\mathfrak{l})$ to an algebra automorphism of $U(Z_\mathfrak{l})$. Therefore τ_u is given on (4.126) by $\tau_u = \tau_{u,1} \otimes 1$.

The mapping $\mu_{\mathfrak{g}/\mathfrak{l}} : Z(\mathfrak{g}) \to Z(\mathfrak{l})$ is given by

$$(4.128) \qquad \mu_{\mathfrak{g}/\mathfrak{l}} = \tau_u \circ \mu_u'.$$

In the proposition below we shall identify $\mu_{\mathfrak{g}/\mathfrak{l}}$ as $\gamma_\mathfrak{l}^{-1} \circ \gamma$, where $\gamma_\mathfrak{l}$ is the Harish-Chandra isomorphism for $\mathfrak{l}$; therefore $\mu_{\mathfrak{g}/\mathfrak{l}}$ is independent of u.

Proposition 4.129. The mappings μ_u' and $\mu_{\mathfrak{g}/\mathfrak{l}}$ satisfy

(a) $\gamma_\mathfrak{n}' = \gamma_{\mathfrak{l}\cap\mathfrak{n}}' \circ \mu_u'$ on $Z(\mathfrak{g})$
(b) $\gamma_{\mathfrak{l}\cap\mathfrak{n}}' \circ \tau_u = \tau_u \circ \gamma_{\mathfrak{l}\cap\mathfrak{n}}'$ on $Z(\mathfrak{l})$
(c) $\mu_{\mathfrak{g}/\mathfrak{l}} = \gamma_\mathfrak{l}^{-1} \circ \gamma$ on $Z(\mathfrak{g})$.

REMARKS. In (a) and (b), $\gamma_{\mathfrak{l}\cap\mathfrak{n}}'$ is γ for $\mathfrak{l}$ with the system of positive roots $\Delta^+(\mathfrak{g}, \mathfrak{h}) \cap \Delta(\mathfrak{l})$. In (c), $\gamma_\mathfrak{l}^{-1} \circ \gamma$ is well defined since image$(\gamma) \subseteq$ image$(\gamma_\mathfrak{l})$: The Weyl-group invariants for $\mathfrak{g}$ in $\mathfrak{h}$ are contained in the Weyl-group invariants for $\mathfrak{l}$ in $\mathfrak{h}$.

PROOF.
(a) If we follow the decomposition of Lemma 4.123a with the decomposition according to

$$U(\mathfrak{l}) = U(\mathfrak{h}) \oplus ((\mathfrak{l}\cap\mathfrak{n}^-)U(\mathfrak{l}) + U(\mathfrak{l})(\mathfrak{l}\cap\mathfrak{n})),$$

we obtain a decomposition consistent with (4.122), and the result is (a).

(b) By (4.126) it is enough to treat $U(Z_\mathfrak{l})$ and $Z([\mathfrak{l}, \mathfrak{l}])$ separately. If z is in $Z([\mathfrak{l}, \mathfrak{l}])$, then $\gamma_{\mathfrak{l}\cap\mathfrak{n}}'(z)$ is in $U(\mathfrak{h} \cap [\mathfrak{l}, \mathfrak{l}])$ and τ_u is the identity on this. Also $\tau_u(z) = z$, and the result is that (b) holds on $Z([\mathfrak{l}, \mathfrak{l}])$. On $U(Z_\mathfrak{l})$, $\gamma_{\mathfrak{l}\cap\mathfrak{n}}'$ is the identity, and then (b) holds on $U(Z_\mathfrak{l})$ as well.

(c) We have

$$\begin{aligned}
\gamma &= \tau_\mathfrak{n} \circ \gamma_\mathfrak{n}' && \text{by (4.93)} \\
&= \tau_{\mathfrak{l}\cap\mathfrak{n}} \circ \tau_u \circ \gamma_{\mathfrak{l}\cap\mathfrak{n}}' \circ \mu_u' && \text{by (4.125) and (a)} \\
&= \tau_{\mathfrak{l}\cap\mathfrak{n}} \circ \gamma_{\mathfrak{l}\cap\mathfrak{n}}' \circ \tau_u \circ \mu_u' && \text{by (b)} \\
&= \gamma_\mathfrak{l} \circ \mu_{\mathfrak{g}/\mathfrak{l}} && \text{by (4.93) and (4.128).}
\end{aligned}$$

9. Kostant's Theorem

Let G be a compact connected Lie group, let T be a maximal torus, let $\mathfrak{b} = \mathfrak{t} \oplus \mathfrak{n}$ be a Borel subalgebra, and let $\Delta^+(\mathfrak{g}, \mathfrak{t})$ be the corresponding positive system of roots. The special case of Kostant's Theorem given in Theorem 4.135 below computes the cohomology $H^k(\mathfrak{n}, V)$ for each irreducible finite-dimensional representation V of G. Here the action on V is understood to be restricted so that V is regarded as an $\mathfrak{n}$ module. Use of the standard complex (2.127a) shows that this cohomology vanishes for $k > \dim \mathfrak{n}$.

When $k = 0$, the cohomology is simply the subspace $V^{\mathfrak{n}}$, according to (2.114), and the Theorem of the Highest Weight (Theorem 4.7c) identifies $V^{\mathfrak{n}}$ as the highest-weight space. The highest-weight space is not simply a vector space but is actually a representation space for T. In the same way Kostant's Theorem provides more information about $H^k(\mathfrak{n}, V)$ than just its identification as a vector space: It gives an action of T on $H^k(\mathfrak{n}, V)$ in the process. We have already seen how such an action by T can arise. Specifically in Chapter II we carried along the example of a special semidirect product map of pairs and proved in Proposition 2.130 that

$$(4.130) \qquad \mathcal{F}_{\mathfrak{t},T}^{0,\{1\}} \circ (I_{\mathfrak{b},T}^{\mathfrak{t},T})^k(V) \cong H^k(\mathfrak{n}, \mathcal{F}_{\mathfrak{b},T}^{\mathfrak{n},\{1\}}(V))$$

naturally for $V \in \mathcal{C}(\mathfrak{b}, T)$. In other words, if we start with any $(\mathfrak{b}, T)$ module V and apply the kth derived functor of $I_{\mathfrak{b},T}^{\mathfrak{t},T}$, then we obtain the cohomology $H^k(\mathfrak{n}, V)$ with an action by T in place. The proof of Proposition 2.130 explains the isomorphism (4.130) concretely. The module $(I_{\mathfrak{b},T}^{\mathfrak{t},T})^k(V)$ is computed from a complex (2.129a) with terms

$$(4.131a) \qquad \mathrm{Hom}_T(R(T), \mathrm{Hom}_{\mathbb{C}}(\textstyle\bigwedge^k \mathfrak{n}, V)),$$

while the cohomology $H^k(\mathfrak{n}, V)$ is computed from a complex (2.127a) with terms

$$(4.131b) \qquad \mathrm{Hom}_{\mathbb{C}}(\textstyle\bigwedge^k \mathfrak{n}, V).$$

The isomorphism (2.18b) of (4.131a) with (4.131b) commutes with the operator d, by inspection; hence the isomorphism passes to cohomology and yields (4.130). Concretely the action of T on (4.131b) is therefore given by

$$(4.132) \qquad (t\omega)(\xi) = t(\omega(\mathrm{Ad}(t)^{-1}(\xi))) \qquad \text{for } \xi \in \textstyle\bigwedge^k \mathfrak{n}.$$

The statement of Kostant's Theorem involves the notion of "length" of members of the Weyl group. Using Theorem 4.10, we write

$$W = W(G, T) = W(\mathfrak{g}, \mathfrak{t})$$

for the Weyl group. For $w \in W$, let

(4.133)
$$\Delta^+(w) = \{\alpha \in \Delta^+(\mathfrak{g}, \mathfrak{t}) \mid w^{-1}\alpha < 0\}$$
$$l(w) = |\Delta^+(w)|.$$

The integer $l(w)$ is the **length** of w. We check easily that $l(w) = l(w^{-1})$.

Lemma 4.134. If w is in W, then

$$\sum_{\beta \in \Delta^+(w)} \beta = \delta - w\delta.$$

PROOF. We write

$$\delta = \tfrac{1}{2}\sum\{\beta \mid \beta > 0, \ w^{-1}\beta > 0\} + \tfrac{1}{2}\sum\{\beta \mid \beta > 0, \ w^{-1}\beta < 0\}$$

and

$$\begin{aligned}
w\delta &= \tfrac{1}{2}w\sum\{\alpha \mid \alpha > 0, \ w\alpha > 0\} + \tfrac{1}{2}w\sum\{\alpha \mid \alpha > 0, \ w\alpha < 0\} \\
&= \tfrac{1}{2}\sum\{w\alpha \mid \alpha > 0, \ w\alpha > 0\} + \tfrac{1}{2}\sum\{w\alpha \mid \alpha > 0, \ w\alpha < 0\} \\
&= \tfrac{1}{2}\sum\{\beta \mid w^{-1}\beta > 0, \ \beta > 0\} + \tfrac{1}{2}\sum\{\eta \mid w^{-1}\eta > 0, \ \eta < 0\} \\
&\qquad\qquad\qquad\qquad \text{under } \beta = w\alpha \text{ and } \eta = w\alpha \\
&= \tfrac{1}{2}\sum\{\beta \mid w^{-1}\beta > 0, \ \beta > 0\} - \tfrac{1}{2}\sum\{\beta \mid w^{-1}\beta < 0, \ \beta > 0\} \\
&\qquad\qquad\qquad\qquad \text{under } \beta = -\eta.
\end{aligned}$$

Subtracting, we obtain

$$\delta - w\delta = \sum\{\beta \mid \beta > 0, \ w^{-1}\beta < 0\}$$

as required.

From this lemma we see that $w\delta - \delta$ is always analytically integral. This fact makes the conclusion meaningful in the following statement.

Theorem 4.135 (Kostant's Theorem, special case). Let G be a compact connected Lie group, let T be a maximal torus, and let $\mathfrak{b} = \mathfrak{t} \oplus \mathfrak{n}$ be the Borel subalgebra corresponding to a positive system $\Delta^+(\mathfrak{g}, \mathfrak{t})$. Let V be an irreducible finite-dimensional representation of G with highest weight λ, and let $0 \leq k \leq \dim \mathfrak{n}$. As a representation space for T, $H^k(\mathfrak{n}, V)$ is the direct sum of all the one-dimensional representations of T with weights

$$w(\lambda + \delta) - \delta \quad \text{such that } w \in W \text{ and } l(w) = k,$$

all occurring with multiplicity 1.

We shall come to the proof later in this section. For $k = 0$, the theorem is simple. The only member of the Weyl group with length 0 is $w = 1$. The weight is therefore $1(\lambda + \delta) - \delta = \lambda$, and the theorem says that $H^0(\mathfrak{n}, V)$ is a one-dimensional representation space of T with weight λ. This conclusion is in agreement with what we noted above, namely that $H^0(\mathfrak{n}, V) \cong V^{\mathfrak{n}}$ is the highest-weight space of V.

The other simple case of the theorem is $k = \dim \mathfrak{n}$. There is a unique element w_0 of W with $w_0 \Delta^+(\mathfrak{g}, \mathfrak{t}) = -\Delta^+(\mathfrak{g}, \mathfrak{t})$, and w_0 is the unique element of length $\dim \mathfrak{n}$. The lowest weight of V is $w_0 \lambda$, and we can see directly that

$$
\begin{aligned}
H^{\dim \mathfrak{n}}(\mathfrak{n}, V) &\cong (I_{\mathfrak{b}, T}^{\mathfrak{t}, T})^{\dim \mathfrak{n}}(V) \\
&\cong (P_{\mathfrak{b}, T}^{\mathfrak{t}, T})_0(V \otimes_{\mathbb{C}} (\textstyle\bigwedge^{\dim \mathfrak{n}} \mathfrak{n})^*) && \text{by Theorem 3.5b} \\
&\cong (V \otimes_{\mathbb{C}} (\textstyle\bigwedge^{\dim \mathfrak{n}} \mathfrak{n})^*)_{\mathfrak{n}} && \text{by Proposition 3.9a} \\
&\cong V_{\mathfrak{n}} \otimes \mathbb{C}_{-2\delta} \\
&\cong V^{\bar{\mathfrak{n}}} \otimes \mathbb{C}_{-2\delta} && \text{by Lemma 4.13} \\
&\cong \mathbb{C}_{w_0 \lambda} \otimes \mathbb{C}_{-2\delta} && \text{from the lowest weight} \\
&= \mathbb{C}_{w_0 \lambda - 2\delta} \\
&= \mathbb{C}_{w_0(\lambda + \delta) - \delta},
\end{aligned}
$$

in agreement with what is predicted by the theorem.

The proof of Theorem 4.135 in other degrees of cohomology lies deeper and divides into two steps. The first step, which is just a combinatorial calculation with weights, is to see that the weight $w(\lambda + \delta) - \delta$ occurs in $\mathrm{Hom}_{\mathbb{C}}(\bigwedge^{l(w)} \mathfrak{n}, V)$ with multiplicity 1 and in all other $\mathrm{Hom}_{\mathbb{C}}(\bigwedge^k \mathfrak{n}, V)$ with multiplicity 0. Then d of its occurrence in $\mathrm{Hom}_{\mathbb{C}}(\bigwedge^{l(w)} \mathfrak{n}, V)$ must be 0, and its occurrence cannot arise from d applied to $\mathrm{Hom}_{\mathbb{C}}(\bigwedge^{l(w)-1} \mathfrak{n}, V)$. Hence the weight passes to cohomology with multiplicity 1 in degree $l(w)$ and with multiplicity 0 in other degrees. The second step in the proof is to show that no other weights occur. This step will use the center of the universal enveloping algebra and will be carried out is §10.

It is possible to identify explicitly the one-dimensional subspace of $\mathrm{Hom}_{\mathbb{C}}(\bigwedge^{l(w)}\mathfrak{n}, V)$ of weight $w(\lambda + \delta) - \delta$. Namely for each $\alpha \in \Delta^+(\mathfrak{g}, \mathfrak{t})$, let e_α be a nonzero root vector for α. Enumerate $\Delta^+(w)$ as $\beta_1, \ldots, \beta_{l(w)}$, and let $v_{w\lambda}$ be a nonzero vector in V of weight $w\lambda$. (This vector is unique up to a scalar, since a representative of w^{-1} in the normalizer $N_G(\mathfrak{t})$ carries it to a highest-weight vector.) Define ω in $\mathrm{Hom}_{\mathbb{C}}(\bigwedge^{l(w)}\mathfrak{n}, V)$ by

(4.136)

$$\omega(e_{\alpha_1} \wedge \cdots \wedge e_{\alpha_{l(w)}}) = \begin{cases} v_{w\lambda} & \text{if } \alpha_1 = \beta_1, \ldots, \alpha_{l(w)} = \beta_{l(w)} \\ 0 & \text{if } \{\alpha_1, \ldots, \alpha_{l(w)}\} \neq \{\beta_1, \ldots, \beta_{l(w)}\}. \end{cases}$$

Then $\mathbb{C}\omega$ is a one-dimensional T invariant subspace, and Lemma 4.134 allows us to compute its weight as

$$w\lambda - \sum_{\beta \in \Delta^+(w)} \beta = w\lambda - (\delta - w\delta) = w(\lambda + \delta) - \delta.$$

With G still compact and connected, the general case of Kostant's Theorem replaces T by a Levi subgroup L. Let $\mathfrak{q} = \mathfrak{l} \oplus \mathfrak{u}$ be a parabolic subalgebra with $\mathfrak{t} \subseteq \mathfrak{b} \subseteq \mathfrak{q}$, and let L be the corresponding Levi subgroup (4.77). Inclusions (4.21) show that the maximal torus T is a Cartan subgroup of G, Theorem 4.83a says that $T = T_L$, and Proposition 4.42a says that T_L meets every component of L; hence L is connected. Define

(4.137) $$W^1 = \{w \in W \mid \Delta^+(w) \subseteq \Delta(\mathfrak{u})\}.$$

In this situation, $(\mathfrak{q}, L) \to (\mathfrak{l}, L)$ is a special semidirect product map of pairs. By Proposition 2.130, L acts on $H^k(\mathfrak{u}, V)$ whenever V is a $(\mathfrak{q}, L)$ module. The action is given on the level of the complex whose j^{th} term is

(4.138a) $$\mathrm{Hom}_{\mathbb{C}}(\bigwedge^k\mathfrak{u}, V)$$

by the formula

(4.138b) $$(l\omega)(\xi) = l(\omega(\mathrm{Ad}(l)^{-1}\xi)).$$

Theorem 4.139 (Kostant's Theorem, general case). Let G be a compact connected Lie group, let T be a maximal torus, and let $\mathfrak{b} = \mathfrak{t} \oplus \mathfrak{n}$

be the Borel subalgebra corresponding to a positive system $\Delta^+(\mathfrak{g}, \mathfrak{t})$. Let $\mathfrak{q} = \mathfrak{l} \oplus \mathfrak{u}$ be a parabolic subalgebra with $\mathfrak{t} \subseteq \mathfrak{b} \subseteq \mathfrak{q}$, let L be the corresponding Levi subgroup, and let W^1 be as in (4.137). Let V be an irreducible finite-dimensional representation of G with highest weight λ, and let $0 \le k \le \dim \mathfrak{u}$. As a representation space for L, $H^k(\mathfrak{u}, V)$ is the direct sum of all the irreducible finite-dimensional representations of L with highest weights

$$w(\lambda + \delta) - \delta \quad \text{such that } w \in W^1 \text{ and } l(w) = k,$$

all occurring with multiplicity 1.

We shall prove this theorem and the special case (Theorem 4.135) together shortly, carrying out the same two steps that we outlined for the special case. The first step will be carried out in this section, and the second step will be carried out in §10. The only real difference between the proofs of the general case and the special case is a small complication in the first step: We have to show that the member ω of $\operatorname{Hom}_{\mathbb{C}}(\bigwedge^{l(w)} \mathfrak{u}, V)$ given in (4.136) is a highest-weight vector under the representation of L, provided w is in W^1.

Lemma 4.140. Let λ be dominant, and let S be a subset of $\Delta^+(\mathfrak{g}, \mathfrak{t})$. If w is in W, then

$$(4.141) \qquad \left| w(\lambda + \delta) - \delta + \sum_{\alpha \in S} \alpha \right|^2 \ge |\lambda|^2,$$

with equality only if $S = \Delta^+(w)$.

PROOF. Since $\delta - w\delta = \sum_{\beta \in \Delta^+(w)} \beta$ by Lemma 4.134, we have

$$(4.142)$$

$$\left| w(\lambda + \delta) - \delta + \sum_{\alpha \in S} \alpha \right|^2 - |w\lambda|^2$$

$$= \left| w\lambda + \sum_{\alpha \in S} \alpha - \sum_{\beta \in \Delta^+(w)} \beta \right|^2 - |w\lambda|^2$$

$$= 2 \left\langle w\lambda, \sum_{\alpha \in S} \alpha - \sum_{\beta \in \Delta^+(w)} \beta \right\rangle + \left| \sum_{\alpha \in S} \alpha - \sum_{\beta \in \Delta^+(w)} \beta \right|^2.$$

In considering $\sum_{\alpha \in S} \alpha - \sum_{\beta \in \Delta^+(w)} \beta$, let us think of canceling the roots that match in the two sums. Let η be a root that survives this cancellation.

If η comes from the first term, then η is positive but η is not in $\Delta^+(w)$. Thus $w^{-1}\eta > 0$. Since λ is dominant by assumption and δ is dominant by Proposition 4.67b, we have

$$(4.143) \qquad \langle w\lambda, \eta \rangle = \langle \lambda, w^{-1}\eta \rangle \geq 0 \quad \text{and} \quad \langle w\delta, \eta \rangle = \langle \delta, w^{-1}\eta \rangle > 0.$$

If η is some $-\beta$ with β from the second term, then β is in $\Delta^+(w)$ and satisfies $w^{-1}\beta < 0$. Thus $\eta = -\beta$ has $w^{-1}\eta > 0$, and the dominance of λ and δ again gives (4.143). In other words, every η of interest satisfies (4.143).

Adding the λ inequalities of (4.143), we see that the first term on the right side of (4.142) is ≥ 0, and thus (4.141) follows. Adding the δ inequalities of (4.143), we find that $|\sum_{\alpha \in S} \alpha - \sum_{\beta \in \Delta^+(w)} \beta|^2$ is nonzero unless the roots in S coincide with those in $\Delta^+(w)$. Referring to the second term on the right side of (4.142), we see that the inequality in (4.141) is strict unless $S = \Delta^+(w)$. This proves the lemma.

PROOF OF FIRST PART OF THEOREM 4.139. We compute the multiplicity of $w(\lambda + \delta) - \delta$ in $\mathrm{Hom}_{\mathbb{C}}(\bigwedge^k \mathfrak{u}, V)$, showing that it is 1 if $k = l(w)$ and is 0 otherwise. To show that these weights are all highest for $\mathfrak{l}$, we show that $w(\lambda + \delta) - \delta + \alpha_0$ is not a weight if w is in W^1 and α_0 is in $\Delta^+(\mathfrak{l}, \mathfrak{t})$.

Thus let μ be a weight of $\mathrm{Hom}_{\mathbb{C}}(\bigwedge^k \mathfrak{u}, V)$. The weights of $\bigwedge^k \mathfrak{u}$ are of the form $\sum_{\alpha \in S} \alpha$, where S is a subset of k distinct elements of $\Delta(\mathfrak{u}) \subseteq \Delta^+(\mathfrak{g}, \mathfrak{t})$, and each such weight has multiplicity equal to the number of such S's yielding the same sum. Thus μ is of the form

$$(4.144) \qquad \mu = \eta - \sum_{\alpha \in S} \alpha,$$

where η is a weight of V and S is a subset of k elements of $\Delta(\mathfrak{u})$. Moreover, the multiplicity of μ is the sum, over all η and S satisfying (4.144), of the multiplicity of η.

We are interested in $\mu = w(\lambda + \delta) - \delta$. Thus we consider the relation

$$(4.145) \qquad w(\lambda + \delta) - \delta = \eta - \sum_{\alpha \in S} \alpha.$$

Then Lemma 4.140 says that $|\eta|^2 \geq |\lambda|^2$, while Theorems 4.7e and 4.10 say that $|\eta|^2 \leq |\lambda|^2$. Hence $|\eta|^2 = |\lambda|^2$, and Lemma 4.140 thus says $S = \Delta^+(w)$. Then $\sum_{\alpha \in S} \alpha = \delta - w\delta$, so that (4.145) gives $\eta = w\lambda$.

In other words, the only possible η is $w\lambda$ (which has multiplicity 1 in V) and the only possible S is $\Delta^+(w)$. Hence $w(\lambda + \delta) - \delta$ has multiplicity 1 if $k = l(w)$, and it has multiplicity 0 otherwise.

Finally let us see that $w(\lambda + \delta) - \delta$ is the weight of a highest-weight vector in the representation of L on $\mathrm{Hom}_{\mathbb{C}}(\bigwedge^k \mathfrak{u}, V)$. Assuming the contrary, suppose that α_0 is in $\Delta^+(\mathfrak{l}, \mathfrak{t})$ and that $w(\lambda + \delta) - \delta + \alpha_0$ is a weight. Then

$$w(\lambda + \delta) - \delta + \alpha_0 = \eta - \sum_{\alpha \in S} \alpha,$$

which is an instance of (4.145) with S replaced by $S \cup \{\alpha_0\}$. (The root α_0 does not occur in S since α_0 is in $\Delta^+(\mathfrak{l}, \mathfrak{t})$, not in $\Delta(\mathfrak{u})$.) By the above argument this situation forces $S \cup \{\alpha_0\} = \Delta^+(w)$, in contradiction to the assumption that w is in W^1. This completes the proof.

The proof of the second part of Theorem 4.139 will be given in §10. Before coming to that proof, we make two digressions. The first is to give some remarks about what happens when G is disconnected. If V is an irreducible finite-dimensional representation of G, then we can decompose V into a direct sum of irreducibles under G_0. Since $H^k(\mathfrak{u}, V)$ splits as a corresponding direct sum, Theorem 4.139 then describes $H^k(\mathfrak{u}, V)$ as a representation of L_0. Let us indicate without proof how these representations of L_0 fit together as representations of L. The representation of L on $H^k(\mathfrak{u}, V)$ is still described via Theorem 4.83 by the representation of T_L on the $\mathfrak{l} \cap \mathfrak{n}$ invariants of $H^k(\mathfrak{u}, V)$, and this space is still given by Theorem 4.139. So it is necessary only to describe the decomposition of this representation of T_L into irreducibles.

Recall from the remarks with Proposition 4.42 that T acts on $W = W(G_0, T_0)$. The subgroup T_L preserves the subset W^1. Fix w in W^1, and write $T_{L,w}$ for the isotropy subgroup at w of the action of T_L on W^1. The highest-weight space $V^{\mathfrak{n}}$ of V is an irreducible representation λ of T that can be restricted to $T_{L,w}$. Choose a representative x for w in the normalizer $N_{G_0}(T_0)$. Then x normalizes $T_{L,w}$, and so acts on representations thereof; this action depends only on w, not on the choice of representative. Thus $w\lambda$ is a well-defined representation of $T_{L,w}$.

Meanwhile, we can interpret $w\delta - \delta$ as the one-dimensional representation of $T_{L,w}$ on the top exterior power of $\mathfrak{n} \cap \mathrm{Ad}(x)^{-1}\mathfrak{n}$. It makes sense to speak of the tensor product of the representation $w\lambda$ and this one-dimensional representation, and we denote it $w(\lambda + \delta) - \delta$. The representation of T_L on the $\mathfrak{l} \cap \mathfrak{n}$ invariants of $H^k(\mathfrak{u}, V)$ is

$$\bigoplus_{\substack{\text{orbits of } T_L \\ \text{on length } k \\ \text{elements of } W^1}} \mathrm{induced}_{T_{L,w}}^{T_L} (w(\lambda + \delta) - \delta).$$

Nothing simplifies very much even if L meets every component of G. A typical example is $G = O(4)$ with $L = T$; here T meets both components of G. The Weyl group has order 4, and T permutes the two elements of length 1. One can check that the details of the above decomposition are complicated to state, even in this ostensibly simple situation.

The second digression from the second part of the proof of Theorem 4.139 is to give a different application of the length function of (4.133). This application, known as **Chevalley's Lemma**, will be used repeatedly starting in Chapter VII.

Proposition 4.146 (Chevalley). Fix v in $i\mathfrak{t}_0^*$ and let

$$W_0 = \{w \in W(\mathfrak{g}, \mathfrak{t}) \mid wv = v\}.$$

Then W_0 is generated by the root reflections s_α such that $\langle v, \alpha \rangle = 0$.

PROOF. Assuming as we may that $v \neq 0$, choose an ordering using an orthogonal basis with v first. Then $\langle v, \beta \rangle > 0$ implies $\beta > 0$. Arguing by contradiction, choose $w \in W_0$ with $l(w)$ as small as possible so that w is not a product of root reflections s_α with $\langle v, \alpha \rangle = 0$. Since $w \neq 1$, Theorem 4.8 and Proposition 4.55a yield a simple root γ such that $w\gamma < 0$. Then $\langle v, \gamma \rangle \geq 0$ since γ is positive. If $\langle v, \gamma \rangle > 0$, then

$$\langle v, w\gamma \rangle = \langle wv, w\gamma \rangle = \langle v, \gamma \rangle > 0,$$

which is a contradiction. Hence $\langle v, \gamma \rangle = 0$. That is, s_γ is in W_0. But then ws_γ is in W_0.

One of the roots in $\Delta^+(w^{-1})$ is γ. Any other root β in $\Delta^+(w^{-1})$ is $s_\gamma(s_\gamma\beta)$ with $s_\gamma\beta > 0$ by Proposition 4.67 and with

$$(s_\gamma w^{-1})^{-1}(s_\gamma\beta) = w\beta < 0;$$

thus such a root β is in $s_\gamma\Delta^+(s_\gamma w^{-1})$. Similarly every root in $s_\gamma\Delta^+(s_\gamma w^{-1})$ is in $\Delta^+(w^{-1})$. Therefore

$$\Delta^+(w^{-1}) = s_\gamma\Delta^+(s_\gamma w^{-1}) \cup \{\gamma\}$$

disjointly, and it follows that

$$l(w) = l(w^{-1}) > l(s_\gamma w^{-1}) = l(ws_\gamma).$$

From the minimality condition on w, ws_γ is a product of root reflections s_α with $\langle v, \alpha \rangle = 0$, and therefore the same thing is true of w. This statement provides the desired contradiction and completes the proof.

10. Casselman-Osborne Theorem

In this section we give the proof of the second part of Kostant's Theorem, Theorem 4.139. Temporarily we continue with the notation $G, \mathfrak{b} = \mathfrak{t} \oplus \mathfrak{n}, T, \mathfrak{n}^- = \bar{\mathfrak{n}}, \mathfrak{q} = \mathfrak{l} \oplus \mathfrak{u}, \mathfrak{u}^- = \bar{\mathfrak{u}}, L, \Delta^+(\mathfrak{g}, \mathfrak{t}), W$, etc., as in §9.

We shall bring to bear the Harish-Chandra isomorphism. Let V be a $(\mathfrak{g}, L)$ module, not necessarily finite dimensional. We saw already in §9 how $H^k(\mathfrak{u}, V)$ acquires an $(\mathfrak{l}, L)$ module structure. Although $\mathfrak{g}$ does not act in any evident way on $H^k(\mathfrak{u}, V)$, we can introduce an action of $Z(\mathfrak{g})$ as follows. If z is in $Z(\mathfrak{g})$, let z act on $\mathrm{Hom}_{\mathbb{C}}(\bigwedge^k \mathfrak{u}, V)$ by acting on the V values:

$$(4.147) \qquad (z\omega)(\xi) = z(\omega(\xi)) \qquad \text{for } \xi \in \textstyle\bigwedge^k \mathfrak{u}.$$

Referring to the formula for d in (2.127b), we see that d commutes with the action by z. Therefore the action by z descends to the cohomology $H^k(\mathfrak{u}, V)$. In this way, $H^k(\mathfrak{u}, V)$ has two actions of interest on it, one by $\mathfrak{l}$ from the $(\mathfrak{l}, L)$ module structure and one by $Z(\mathfrak{g})$ from (4.147).

These two actions are related by the Casselman-Osborne Theorem given below. We shall state the result as Theorem 4.149 and then immediately show how it finishes the proof of Kostant's Theorem.

Actually the Casselman-Osborne Theorem is not limited to the setting with a compact connected Lie group G. Let $(\mathfrak{g}, K)$ be a reductive pair, and let $(\mathfrak{q}, L \cap K)$ be a parabolic subpair. If we write $\mathfrak{q} = \mathfrak{l} \oplus \mathfrak{u}$ with $\mathfrak{l}$ as Levi factor and $\mathfrak{u}$ as nilpotent radical, then $(\mathfrak{q}, L \cap K) \to (\mathfrak{l}, L \cap K)$ is a semidirect product map of pairs in the sense of §III.2. According to Proposition 3.12, if V is any $(\mathfrak{g}, L \cap K)$ module, then $H^k(\mathfrak{u}, V)$ acquires the structure of an $(\mathfrak{l}, L \cap K)$ module, with $\mathfrak{l}$ acting on $\mathrm{Hom}_{\mathbb{C}}(\bigwedge^k \mathfrak{u}, V)$ by

$$(4.148) \qquad X\lambda(\xi) = X(\lambda(\xi)) - \lambda((\mathrm{ad}\, X)\xi).$$

Also $H^k(\mathfrak{u}, V)$ becomes a $Z(\mathfrak{g})$ module by the action in (4.147). The Casselman-Osborne Theorem relates these two actions.

Theorem 4.149 (Casselman-Osborne). Let $(\mathfrak{g}, K)$ be a reductive pair, and let $(\mathfrak{q}, L \cap K)$ be a parabolic subpair with $\mathfrak{q} = \mathfrak{l} \oplus \mathfrak{u}$. If V is a $(\mathfrak{g}, L \cap K)$ module, then the $Z(\mathfrak{g})$ and $(\mathfrak{l}, L \cap K)$ actions on $H^k(\mathfrak{u}, V)$ are related by the formula

$$z\omega = \mu'_{\mathfrak{u}}(z)\omega$$

for all $z \in Z(\mathfrak{g})$ and $\omega \in H^k(\mathfrak{u}, V)$.

REMARK. The compact groups may be suppressed in the statement of the theorem, and the result is still valid. In other words, $\mathfrak{g}$ is to be a complex reductive Lie algebra, V a $U(\mathfrak{g})$ module, and $\mathfrak{q}$ a parabolic subalgebra of $\mathfrak{g}$. The action of $\mathfrak{l}$ on $H^k(\mathfrak{u}, V)$ is still given by (4.148), and the conclusion of the theorem remains valid.

PROOF OF SECOND PART OF THEOREM 4.139. Let us return to the notation of §9. We are to show that the only $\Delta^+(\mathfrak{l}, \mathfrak{t})$ highest weights in $H^k(\mathfrak{u}, V)$ are the ones of the form $w(\lambda + \delta) - \delta$ with $w \in W^1$. For V an irreducible finite-dimensional representation of G with highest weight λ, we saw in §7 that each $z \in Z(\mathfrak{g})$ acts on V by the scalar $\lambda(\gamma_\mathfrak{n}'(z))$. In $H^k(\mathfrak{u}, V)$, the action of z comes from the action of z on the V values of members of $\operatorname{Hom}_\mathbb{C}(\bigwedge^k \mathfrak{u}, V)$, and hence z acts on $H^k(\mathfrak{u}, V)$ by the scalar $\lambda(\gamma_\mathfrak{n}'(z))$.

If ν is the highest weight of a representation of L occurring in $H^k(\mathfrak{u}, V)$, then any member z_1 of $Z(\mathfrak{l})$ acts on the corresponding representation space by the scalar $\nu(\gamma_{\mathfrak{l} \cap \mathfrak{n}}'(z_1))$. Applying this fact to $z_1 = \mu_\mathfrak{u}'(z)$ and using Theorem 4.149 and Proposition 4.129a, we see that

$$\lambda(\gamma_\mathfrak{n}'(z)) = \nu(\gamma_{\mathfrak{l} \cap \mathfrak{n}}'(\mu_\mathfrak{u}'(z))) = \nu(\gamma_\mathfrak{n}'(z))$$

for all $z \in Z(\mathfrak{g})$. By (4.94) and (4.114),

$$\chi_{\lambda + \delta}(z) = \chi_{\nu + \delta}(z)$$

for all $z \in Z(\mathfrak{g})$. Then Theorem 4.115 gives

$$\nu + \delta = w(\lambda + \delta)$$

for some $w \in W$. Since $\nu + \delta$ is dominant for $\Delta^+(\mathfrak{l}, \mathfrak{t})$ by Proposition 4.67b, so is $w(\lambda + \delta)$. Thus, for each α in $\Delta^+(\mathfrak{l}, \mathfrak{t})$,

$$\langle \lambda + \delta, w^{-1}\alpha \rangle = \langle w(\lambda + \delta), \alpha \rangle > 0,$$

and $\Delta^+(w) \cap \Delta^+(\mathfrak{l}, \mathfrak{t})$ is empty. By (4.137), w is in W^1. Therefore $\nu = w(\lambda + \delta) - \delta$ with $w \in W^1$, as required.

Our objective in the remainder of this section will be to prove Theorem 4.149, using a cohomological technique known as **dimension shifting**. We shall make use of a long exact cohomology sequence in which every third term is 0, so that we obtain isomorphisms of cohomology that cross from one dimension to the next.

The long exact sequence in question is the following. First suppose that we have an exact sequence

(4.150) $$0 \longrightarrow V \xrightarrow{\psi} U \xrightarrow{\varphi} V' \longrightarrow 0$$

in the category $\mathcal{C}(\mathfrak{u})$. Applying (C.36c) to the invariants functor $(\cdot)^{\mathfrak{u}}$, we obtain the long exact sequence

(4.151)

$$0 \longrightarrow H^0(\mathfrak{u}, V) \longrightarrow H^0(\mathfrak{u}, U) \longrightarrow H^0(\mathfrak{u}, V')$$
$$\longrightarrow H^1(\mathfrak{u}, V) \longrightarrow H^1(\mathfrak{u}, U) \longrightarrow H^1(\mathfrak{u}, V')$$
$$\longrightarrow H^2(\mathfrak{u}, V) \longrightarrow H^2(\mathfrak{u}, U) \longrightarrow H^2(\mathfrak{u}, V')$$
$$\longrightarrow \cdots \longrightarrow H^{\dim \mathfrak{u}}(\mathfrak{u}, U) \longrightarrow H^{\dim \mathfrak{u}}(\mathfrak{u}, V') \longrightarrow 0,$$

because the Lie algebra cohomology functors are the derived functors of the invariants functor. If (4.150) is actually an exact sequence in $\mathcal{C}(\mathfrak{g})$, then each cohomology space in (4.151) possesses actions by $U(\mathfrak{l})$ and $Z(\mathfrak{g})$ from (4.148) and (4.147). In order to make use of (4.151), we need to know that the maps in (4.151) commute with these actions.

In the case of $U(\mathfrak{l})$, this commutativity is automatic: The sequence (4.150) is exact in $\mathcal{C}(\mathfrak{q})$. Application of the functor $I_{\mathfrak{q},\{1\}}^{\mathfrak{l},\{1\}}$ leads to an exact sequence in $\mathcal{C}(\mathfrak{l})$ such that (4.151) results by forgetting the $U(\mathfrak{l})$ module structure, by Proposition 3.12. Therefore (4.151) respects the $U(\mathfrak{l})$ action.

In the case of $Z(\mathfrak{g})$, a little argument is needed. We regard (4.150) as exact in $\mathcal{C}(\mathfrak{g})$. We recall how the long exact sequence in (4.151) can be obtained. The standard injective resolutions of V, U, and V' in $\mathcal{C}(\mathfrak{u})$ may be assembled as a diagram of the form (C.31). Application of $(\cdot)^{\mathfrak{u}}$ to this diagram leads to the diagram (4.152) below with exact columns. There is then a prescription (see §V.5 of Knapp [1988]) for computing the maps of the long exact sequence from this diagram. The maps that preserve degree in (4.151) are quotients of restrictions of $\mathrm{Hom}(1, \psi)$ and $\mathrm{Hom}(1, \varphi)$, and these commute with $Z(\mathfrak{g})$ because ψ and φ commute with the action by $\mathfrak{g}$.

The maps that increase degree in (4.151) come from the connecting homomorphism for (4.152). To define the connecting homomorphism ρ, we start with ω in $\mathrm{Hom}_{\mathbb{C}}(\bigwedge^{k-1}\mathfrak{u}, V')$ such that $d\omega = 0$. We choose

ξ in $\mathrm{Hom}_\mathbb{C}(\bigwedge^{k-1}\mathfrak{u}, U)$ with $\mathrm{Hom}(1, \varphi)\xi = \omega$. Then $d\xi$ is of the form $\mathrm{Hom}(1, \psi)\eta$ for some η in $\mathrm{Hom}(\bigwedge^k\mathfrak{u}, V)$, and we define $\rho(\omega)$ to be the class of η in $H^k(\mathfrak{u}, V)$.

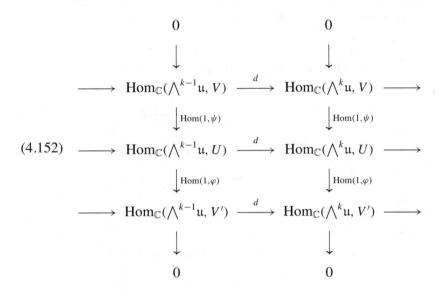

(4.152)

Now let us consider $z\omega$ with $z \in Z(\mathfrak{g})$. We have $d(z\omega) = z(d\omega) = 0$. The element $z\omega$ has $\mathrm{Hom}(1, \varphi)(z\xi) = z(\mathrm{Hom}(1, \varphi)\xi) = z\omega$. Moreover,

$$\mathrm{Hom}(1, \psi)(z\eta) = z(\mathrm{Hom}(1, \psi)\eta) = z(d\xi) = d(z\xi).$$

Thus $\rho(z\omega)$ is the class of $z\eta$ in $H^k(\mathfrak{u}, V)$. This statement says that ρ commutes with the action of $Z(\mathfrak{g})$, and we conclude that (4.151) respects the $Z(\mathfrak{g})$ action.

PROOF OF THEOREM 4.149. We show by induction on k that the identity

$$z\omega = \mu'_{\mathfrak{u}}(z)\omega, \qquad \text{for } z \in Z(\mathfrak{g}),$$

holds on $H^k(\mathfrak{u}, V)$ for all $(\mathfrak{g}, L \cap K)$ modules V.

For $k = 0$, $H^0(\mathfrak{u}, V) \cong V^{\mathfrak{u}}$ as a $U(\mathfrak{l})$ module. Both z and $\mu'_{\mathfrak{u}}(z)$ act by the restriction of their actions on V. Since Lemma 4.123b says that $z - \mu'_{\mathfrak{u}}(z)$ is in $U(\mathfrak{g})\mathfrak{u}$, we see that $(z - \mu'_{\mathfrak{u}}(z))\omega = 0$ for ω in $H^0(\mathfrak{u}, V)$. This is the desired conclusion for $k = 0$.

Now assume the result for degree $k - 1$. The beginning of the standard injective resolution of V (in the category $\mathcal{C}(\mathfrak{g})$) gives us an exact sequence

$$(4.153) \qquad 0 \longrightarrow V \xrightarrow{\ \psi\ } \mathrm{Hom}_{\mathbb{C}}(U(\mathfrak{g}), V) \xrightarrow{\ \varphi\ } Q \longrightarrow 0$$

where Q is a suitable $U(\mathfrak{g})$ module. Regarding (4.153) as an exact sequence in $\mathcal{C}(\mathfrak{u})$, we obtain from (4.151) an exact sequence

$$(4.154) \qquad H^{k-1}(\mathfrak{u}, Q) \xrightarrow{\ \rho\ } H^k(\mathfrak{u}, V) \longrightarrow H^k(\mathfrak{u}, \mathrm{Hom}_{\mathbb{C}}(U(\mathfrak{g}), V)),$$

and we have seen that this exact sequence respects the $U(\mathfrak{l})$ and $Z(\mathfrak{g})$ actions.

As vector spaces $\mathfrak{g} = \mathfrak{q}^- \oplus \mathfrak{u}$. Thus as $\mathfrak{u}$ modules,

$$\mathrm{Hom}_{\mathbb{C}}(U(\mathfrak{g}), V) \cong \mathrm{Hom}_{\mathbb{C}}(S(\mathfrak{q}^-) \otimes_{\mathbb{C}} U(\mathfrak{u}), V)$$
$$\cong \mathrm{Hom}_{\mathbb{C}}(U(\mathfrak{u}), \mathrm{Hom}_{\mathbb{C}}(S(\mathfrak{q}^-), V)),$$

where $\mathfrak{u}$ acts trivially on the symmetric algebra $S(\mathfrak{q}^-)$. The right side here is injective in $\mathcal{C}(\mathfrak{u})$, by Proposition 2.53b, and therefore $I = \mathrm{Hom}_{\mathbb{C}}(U(\mathfrak{g}), V)$ is injective in $\mathcal{C}(\mathfrak{u})$.

Since I is injective, $0 \to I \to I \to 0$ is an injective resolution of I, and it follows that $H^l(\mathfrak{u}, I) = 0$ for $l > 0$. In particular, $H^k(\mathfrak{u}, I) = 0$, and the exactness of (4.154) therefore says that the connecting homomorphism ρ is onto $H^k(\mathfrak{u}, V)$. Given ω in $H^k(\mathfrak{u}, V)$, choose $\tilde{\omega}$ in $H^{k-1}(\mathfrak{u}, Q)$ with $\rho(\tilde{\omega}) = \omega$. By inductive hypothesis, $z\tilde{\omega} = \mu'_{\mathfrak{u}}(z)\tilde{\omega}$. Applying ρ to both sides and using the compatibility of the actions of $U(\mathfrak{l})$ and $Z(\mathfrak{g})$ with long exact sequences, we obtain

$$z\omega = z\rho(\tilde{\omega}) = \rho(z\tilde{\omega}) = \rho(\mu'_{\mathfrak{u}}(z)\tilde{\omega}) = \mu'_{\mathfrak{u}}(z)\rho(\tilde{\omega}) = \mu'_{\mathfrak{u}}(z)\omega.$$

This completes the induction and the proof.

11. Algebraic Analog of Bott-Borel-Weil Theorem

To conclude this chapter, we shall apply our results to give realizations of the irreducible representations of compact Lie groups in homology and cohomology spaces. We continue with the notation G, $\mathfrak{b} = \mathfrak{t} \oplus \mathfrak{n}$, T, $\mathfrak{n}^- = \bar{\mathfrak{n}}$, $\mathfrak{q} = \mathfrak{l} \oplus \mathfrak{u}$, $\mathfrak{u}^- = \bar{\mathfrak{u}}$, L, $\Delta^+(\mathfrak{g}, \mathfrak{t})$, W, etc., as in §§9–10.

For motivation let us relate matters to the analytic setting discussed in the Introduction. Suppose G is connected. If we start from a one-dimensional representation of T (say with weight λ on the space $\mathbb{C}_\lambda$), then we can extend λ to $\bar{\mathfrak{b}}$ by making it 0 on the subalgebra $\bar{\mathfrak{n}}$, and this extension is just $\mathcal{F}_{\mathfrak{t},T}^{\bar{\mathfrak{b}},T}(\mathbb{C}_\lambda)$. As noted in the Introduction, $I_{\bar{\mathfrak{b}},T}^{\mathfrak{g},T}(\mathcal{F}_{\mathfrak{t},T}^{\bar{\mathfrak{b}},T}(\mathbb{C}_\lambda))$ will be related to the Taylor coefficients of locally defined holomorphic functions having domain in a complexification $G_\mathbb{C}$ of G and satisfying a transformation law given by λ. Let $\bar{B}$ be the analytic subgroup of $G_\mathbb{C}$ with Lie algebra $\bar{\mathfrak{b}}$. These locally defined holomorphic functions may be viewed also as locally defined holomorphic sections of the holomorphic line bundle over $G/\bar{B}$ associated to the representation $\mathbb{C}_\lambda$. The functor $I_{\mathfrak{g},T}^{\mathfrak{g},G}$ is analogous with the extraction of global holomorphic sections from these Taylor coefficients, and it is traditional to write Γ in place of $I_{\mathfrak{g},T}^{\mathfrak{g},G}$ to emphasize the connection with sections. The functor Γ has been discussed in Chapters II and III. It is not exact and has nonzero derived functors (obtained from an injective resolution), and these are analogous to extracting global $(0, k)$ forms and passing to cohomology. The Bott-Borel-Weil Theorem locates the irreducible representations of G in such spaces of forms. We shall be more precise about this point in the remarks attached to Corollary 4.160.

In the algebraic setting, the technique for identifying a derived functor representation of this kind is to reduce spaces of G maps into this representation to spaces of T maps into a representation that can be identified by means of Kostant's Theorem (Theorem 4.139). This reduction technique is important, since its points of success and failure are where the theory begins in Chapter V in the case of noncompact groups.

For the reduction technique we shall work with a general compact Lie group G, the parabolic subalgebra $\mathfrak{q}$, and the Levi subgroup L. Let

$$S = \dim \mathfrak{u}.$$

The spaces

$$\bigwedge\nolimits^{\mathrm{top}} \mathfrak{u} = \bigwedge\nolimits^{S} \mathfrak{u} \qquad \text{and} \qquad \bigwedge\nolimits^{\mathrm{top}} \bar{\mathfrak{u}} = \bigwedge\nolimits^{S} \bar{\mathfrak{u}}$$

are one-dimensional representation spaces for L with unique weights $2\delta(\mathfrak{u})$ and $2\delta(\bar{\mathfrak{u}})$, respectively. Write

$$\Pi = P^{\mathfrak{g},G}_{\mathfrak{g},L} \qquad \text{and} \qquad \Gamma = I^{\mathfrak{g},G}_{\mathfrak{g},L},$$

and let Π_j and Γ^j be the derived functors of Π and Γ.

Theorem 4.155. Let G be a compact Lie group, let $\mathfrak{t}_0$ be a maximal abelian subspace of $\mathfrak{g}_0$, let $\mathfrak{b} = \mathfrak{t} \oplus \mathfrak{n}$ be the Borel subalgebra corresponding to a positive system $\Delta^+(\mathfrak{g}, \mathfrak{t})$, let $\mathfrak{q} = \mathfrak{l} \oplus \mathfrak{u}$ be a parabolic subalgebra with $\mathfrak{t} \subseteq \mathfrak{b} \subseteq \mathfrak{q}$, and let L be the corresponding Levi subgroup. For $j \geq 0$,

$$\operatorname{Hom}_G(\Pi_j(P^{\mathfrak{g},L}_{\mathfrak{q},L}(\mathcal{F}^{\mathfrak{q},L}_{\mathfrak{l},L}(Z))), V) \cong \operatorname{Hom}_L(Z, H^j(\mathfrak{u}, V))$$

and

$$\operatorname{Hom}_G(V, \Gamma^j(I^{\mathfrak{g},L}_{\bar{\mathfrak{q}},L}(\mathcal{F}^{\bar{\mathfrak{q}},L}_{\mathfrak{l},L}(Z)))) \cong \operatorname{Hom}_L(H_j(\bar{\mathfrak{u}}, V), Z)$$
$$\cong \operatorname{Hom}_L(H^{S-j}(\bar{\mathfrak{u}}, V) \otimes_{\mathbb{C}} {\textstyle\bigwedge}^{\text{top}}\bar{\mathfrak{u}}, Z)$$

naturally for all finite-dimensional Z in $\mathcal{C}(\mathfrak{l}, L)$ and all finite-dimensional V in $\mathcal{C}(\mathfrak{g}, G)$.

REMARK. For $j > S$, $H^{S-j}(\bar{\mathfrak{u}}, V)$ should be interpreted as 0.

PROOF. For the first identity we write down all the steps together, and we justify them afterward. We have

$$\operatorname{Hom}_{\mathfrak{g},G}(\Pi_j(P^{\mathfrak{g},L}_{\mathfrak{q},L}(\mathcal{F}^{\mathfrak{q},L}_{\mathfrak{l},L}(Z))), V)$$

(4.156a) $$\cong \operatorname{Ext}^j_{\mathfrak{g},L}(P^{\mathfrak{g},L}_{\mathfrak{q},L}(\mathcal{F}^{\mathfrak{q},L}_{\mathfrak{l},L}(Z)), (\mathcal{F}^\vee)^{\mathfrak{g},L}_{\mathfrak{g},G}(V))$$

(4.156b) $$\cong \operatorname{Ext}^j_{\mathfrak{g},L}(P^{\mathfrak{g},L}_{\mathfrak{q},L}(\mathcal{F}^{\mathfrak{q},L}_{\mathfrak{l},L}(Z)), V)$$

(4.156c) $$\cong \operatorname{Ext}^j_{\mathfrak{q},L}(\mathcal{F}^{\mathfrak{q},L}_{\mathfrak{l},L}(Z), V)$$

(4.156d) $$\cong \operatorname{Hom}_{\mathfrak{l},L}(Z, H^j(\mathfrak{u}, V)).$$

Step (b) is by Proposition 2.33, V being finite dimensional, and step (c) is by Shapiro's Lemma (Theorem 2.119).

Step (a) is proved as follows: The two functors

$$\operatorname{Hom}_{\mathfrak{g},G}(\cdot, V) \circ P^{\mathfrak{g},G}_{\mathfrak{g},L} \qquad \text{and} \qquad \operatorname{Hom}_{\mathfrak{g},L}(\cdot, (\mathcal{F}^\vee)^{\mathfrak{g},L}_{\mathfrak{g},G}(V))$$

are naturally isomorphic by Frobenius reciprocity (Proposition 2.34), and hence their derived functors are naturally isomorphic. Since

$\mathrm{Hom}_{\mathfrak{g},G}(\,\cdot\,,V)$ is exact by Lemma 2.4 and (C.14), (C.27a) and Proposition 2.117 give

$$\mathrm{Hom}_{\mathfrak{g},G}(\,\cdot\,,V)\circ(P_{\mathfrak{g},L}^{\mathfrak{g},G})_j \cong (\mathrm{Hom}_{\mathfrak{g},G}(\,\cdot\,,V)\circ P_{\mathfrak{g},L}^{\mathfrak{g},G})_j$$

(4.157)
$$\cong \mathrm{Hom}_{\mathfrak{g},L}(\,\cdot\,,(\mathcal{F}^{\vee})_{\mathfrak{g},G}^{\mathfrak{g},L}(V))_j$$

$$\cong \mathrm{Ext}_{\mathfrak{g},L}^{j}(\,\cdot\,,(\mathcal{F}^{\vee})_{\mathfrak{g},G}^{\mathfrak{g},L}(V)).$$

Then (a) follows.

Step (d) is handled similarly. The two functors

$$\mathrm{Hom}_{\mathfrak{l},L}(Z,\,\cdot\,)\circ I_{\mathfrak{q},L}^{\mathfrak{l},L} \qquad \text{and} \qquad \mathrm{Hom}_{\mathfrak{q},L}(\mathcal{F}_{\mathfrak{l},L}^{\mathfrak{q},L}(Z),\,\cdot\,)$$

are naturally isomorphic by Frobenius reciprocity (Proposition 2.21), and hence their derived functors are naturally isomorphic. Since $\mathrm{Hom}_{\mathfrak{l},L}(Z,\,\cdot\,)$ is exact, (C.27a) and (2.116) give

$$\mathrm{Hom}_{\mathfrak{l},L}(Z,\,\cdot\,)\circ(I_{\mathfrak{q},L}^{\mathfrak{l},L})^{j} \cong (\mathrm{Hom}_{\mathfrak{l},L}(Z,\,\cdot\,)\circ(I_{\mathfrak{q},L}^{\mathfrak{l},L}))^{j}$$

(4.158)
$$\cong \mathrm{Hom}_{\mathfrak{q},L}(\mathcal{F}_{\mathfrak{l},L}^{\mathfrak{q},L}(Z),\,\cdot\,)^{j}$$

$$= \mathrm{Ext}_{\mathfrak{q},L}^{j}(\mathcal{F}_{\mathfrak{l},L}^{\mathfrak{q},L}(Z),\,\cdot\,).$$

To complete the argument, we have only to recall from Proposition 3.12 that

$$(I_{\mathfrak{q},L}^{\mathfrak{l},L})^{j}(V) \cong H^{j}(\mathfrak{u},V)$$

if we suppress forgetful functors in the notation. Then (d) follows.

For the second identity in the theorem, the steps are

$$\mathrm{Hom}_{\mathfrak{g},G}(V,\,\Gamma^{j}(I_{\bar{\mathfrak{q}},L}^{\mathfrak{g},L}(\mathcal{F}_{\mathfrak{l},L}^{\bar{\mathfrak{q}},L}(Z))))$$

(4.159a)
$$\cong \mathrm{Ext}_{\mathfrak{g},L}^{j}(V,\,I_{\bar{\mathfrak{q}},L}^{\mathfrak{g},L}(\mathcal{F}_{\mathfrak{l},L}^{\bar{\mathfrak{q}},L}(Z)))$$

(4.159b)
$$\cong \mathrm{Ext}_{\bar{\mathfrak{q}},L}^{j}(V,\,\mathcal{F}_{\mathfrak{l},L}^{\bar{\mathfrak{q}},L}(Z))$$

(4.159c)
$$\cong \mathrm{Hom}_{\mathfrak{l},L}(H_{j}(\bar{\mathfrak{u}},V),\,Z)$$

(4.159d)
$$\cong \mathrm{Hom}_{\mathfrak{l},L}(H^{S-j}(\bar{\mathfrak{u}},\,V\otimes_{\mathbb{C}}\textstyle\bigwedge^{\mathrm{top}}\bar{\mathfrak{u}}),\,Z)$$

(4.159e)
$$\cong \mathrm{Hom}_{\mathfrak{l},L}(H^{S-j}(\bar{\mathfrak{u}},V)\otimes_{\mathbb{C}}\textstyle\bigwedge^{\mathrm{top}}\bar{\mathfrak{u}},\,Z).$$

Step (a) is handled as in (4.158) but with changed notation, step (b) is by Shapiro's Lemma, step (c) is handled as in (4.157) but with changed notation and use of Proposition 2.33 and the formula

$$(P_{\bar{\mathfrak{q}},L}^{\mathfrak{l},L})_{j}(V) \cong H_{j}(\bar{\mathfrak{u}},V)$$

from Proposition 3.12, and step (d) is by Hard Duality (Theorem 3.5b). For step (e) we use that the $\bar{\mathfrak{u}}$ action on $\bigwedge^{\mathrm{top}}\bar{\mathfrak{u}}$ is trivial. In fact, §III.2 defines this action by a certain linear functional on $\bar{\mathfrak{q}}$ that vanishes on $[\bar{\mathfrak{q}},\bar{\mathfrak{q}}]$. Since $[\mathfrak{t},\bar{\mathfrak{u}}]=\bar{\mathfrak{u}}$, $\bar{\mathfrak{u}}$ is contained in $[\bar{\mathfrak{q}},\bar{\mathfrak{q}}]$ and must act trivially on $\bigwedge^{\mathrm{top}}\bar{\mathfrak{u}}$. This completes the proof.

In Theorem 4.155 the functors Π_j and Γ^j are 0 when j exceeds $\dim_{\mathbb{C}}(\mathfrak{g}/\mathfrak{l}) = \dim_{\mathbb{C}}\mathfrak{u} + \dim_{\mathbb{C}}\bar{\mathfrak{u}} = 2S$, by Corollary 2.125. Isomorphisms (4.156a-c) and (4.159a-b) imply that $\Pi_j(P_{\mathfrak{q},L}^{\mathfrak{g},L}(\mathcal{F}_{\mathfrak{l},L}^{\mathfrak{q},L}(Z)))$ and $\Gamma^j(I_{\bar{\mathfrak{q}},L}^{\mathfrak{g},L}(\mathcal{F}_{\mathfrak{l},L}^{\bar{\mathfrak{q}},L}(Z)))$ are 0 when j exceeds S, since $\text{Ext}_{\mathfrak{q},L}^j$ and $\text{Ext}_{\bar{\mathfrak{q}},L}^j$ are both the 0 functor for $j > S$.

Corollary 4.160. Let G be a compact connected Lie group, let $\mathfrak{t}_0$ be a maximal abelian subspace of $\mathfrak{g}_0$, let $\mathfrak{b} = \mathfrak{t} \oplus \mathfrak{n}$ be the Borel subalgebra corresponding to a positive system $\Delta^+(\mathfrak{g}, \mathfrak{t})$, let $\mathfrak{q} = \mathfrak{l} \oplus \mathfrak{u}$ be a parabolic subalgebra with $\mathfrak{t} \subseteq \mathfrak{b} \subseteq \mathfrak{q}$, and let L be the corresponding Levi subgroup. Let Z be an irreducible $(\mathfrak{l}, L)$ module with $\Delta^+(\mathfrak{l}, \mathfrak{t})$ highest weight λ.

(a) If $\lambda + \delta$ is orthogonal to some root, then $\Pi_j(P_{\mathfrak{q},L}^{\mathfrak{g},L}(\mathcal{F}_{\mathfrak{l},L}^{\mathfrak{q},L}(Z)))$ and $\Gamma^j(I_{\bar{\mathfrak{q}},L}^{\mathfrak{g},L}(\mathcal{F}_{\mathfrak{l},L}^{\bar{\mathfrak{q}},L}(Z)))$ are both 0 for all $j \geq 0$.

(b) If $\lambda + \delta$ is orthogonal to no root, put

$$(4.161) \qquad q = \#\{\alpha \in \Delta^+(\mathfrak{g}, \mathfrak{t}) \mid \langle \lambda + \delta, \alpha \rangle < 0\}.$$

There exists a unique element $w \in W$ such that $w^{-1}(\lambda + \delta)$ is $\Delta^+(\mathfrak{g}, \mathfrak{t})$ dominant; this w has length q and is in the subset W^1 defined in (4.137). The representations $\Pi_j(P_{\mathfrak{q},L}^{\mathfrak{g},L}(\mathcal{F}_{\mathfrak{l},L}^{\mathfrak{q},L}(Z)))$ and $\Gamma^j(I_{\bar{\mathfrak{q}},L}^{\mathfrak{g},L}(\mathcal{F}_{\mathfrak{l},L}^{\bar{\mathfrak{q}},L}(Z)))$ are 0 if $j \neq q$. For $j = q$,

$$\Pi_q(P_{\mathfrak{q},L}^{\mathfrak{g},L}(\mathcal{F}_{\mathfrak{l},L}^{\mathfrak{q},L}(Z))) \quad \text{and} \quad \Gamma^q(I_{\bar{\mathfrak{q}},L}^{\mathfrak{g},L}(\mathcal{F}_{\mathfrak{l},L}^{\bar{\mathfrak{q}},L}(Z)))$$

are irreducible representations of G, and both have highest weight $\mu = w^{-1}(\lambda + \delta) - \delta$.

REMARKS.

1) If λ is dominant relative to $\Delta^+(\mathfrak{g}, \mathfrak{t})$, then $q = 0$ in (4.161) and the corollary reduces to Theorem 4.83e.

2) If $L = T$ and λ is dominant relative to $\Delta^+(\mathfrak{g}, \mathfrak{t})$, then the corollary identifies $\Gamma(I_{\bar{\mathfrak{q}},L}^{\mathfrak{g},L}(\mathcal{F}_{\mathfrak{l},L}^{\bar{\mathfrak{q}},L}(\mathbb{C}_\lambda)))$ as an irreducible representation of G with highest weight λ. This is an algebraic analog of the analytic statement that the representation of G on the global holomorphic sections of the holomorphic line bundle

$$(4.162) \qquad G_{\mathbb{C}} \times_{\bar{B}} \mathbb{C}_\lambda \to G_{\mathbb{C}}/\bar{B}$$

is irreducible with highest weight λ. The analytic statement is true and is known as the **Borel-Weil Theorem**.

3) If $L = T$ and λ is not dominant, the corollary gives

$$\Gamma(I^{\mathfrak{g},L}_{\bar{\mathfrak{q}},L}(\mathcal{F}^{\bar{\mathfrak{q}},L}_{\mathfrak{l},L}(\mathbb{C}_\lambda))) = 0.$$

In this case, assuming $\lambda + \delta$ is not orthogonal to any root, we instead get a nontrivial representation in a space $\Gamma^q(I^{\mathfrak{g},L}_{\bar{\mathfrak{q}},L}(\mathcal{F}^{\bar{\mathfrak{q}},L}_{\mathfrak{l},L}(\mathbb{C}_\lambda)))$ that is an algebraic analog of the space of $(0, q)$-form sections of the line bundle (4.162). Corollary 4.160 is an algebraic analog of the statement that the representation of G in the $(0, q)^{\mathrm{th}}$ space of Dolbeault cohomology sections of (4.162) is irreducible with highest weight $\mu = w(\lambda + \delta) - \delta$. The analytic statement is true and is known as the **Bott-Borel-Weil Theorem**.

4) Both the Borel-Weil Theorem and the Bott-Borel-Weil Theorem are valid for general L (if G is connected), and Corollary 4.160 provides an algebraic analog.

5) When G is disconnected, it does not simplify much (except in degrees 0 and S) to assume that L meets every component of G. The $(\mathfrak{l}, L)$ module Z may have several $\Delta^+(\mathfrak{l}, \mathfrak{t})$ highest weights λ, but (4.161) will yield the same number q for all of them. Moreover a common element $w \in W^1$ will make all $w(\lambda+\delta)$ dominant for $\Delta^+(\mathfrak{g}, \mathfrak{t})$. Something interesting happens when one (or equivalently all) of these $w(\lambda + \delta)$ is orthogonal to no roots. In this case we regard $Z^{\mathfrak{l} \cap \mathfrak{n}}$ as an irreducible representation Λ with the λ's as weights. In terms of the notation at the end of §9, one can show that Λ is induced from an irreducible representation Λ_w of $T_{L,w}$ and can use Λ_w to construct an explicit generalization of μ in the corollary. The generalization of μ may be reducible even if $T_L = T$. We omit the details.

PROOF. The result will be proved by calculating the multiplicity of each irreducible representation of G in the representation in question. Let V be an irreducible representation of G with highest weight ν.

Let us consider $\Pi_j(P^{\mathfrak{g},L}_{\mathfrak{q},L}(\mathcal{F}^{\mathfrak{q},L}_{\mathfrak{l},L}(Z)))$. The first identity in Theorem 4.155 shows that the multiplicity of V in this representation equals the dimension of

$$\mathrm{Hom}_L(Z, H^j(\mathfrak{u}, V)),$$

i.e., the multiplicity of the irreducible representation of L with highest weight λ in $H^j(\mathfrak{u}, V)$. By Theorem 4.139, this multiplicity is 0 or 1. It is 1 if and only if there is an element $w \in W^1$ of length j such that

$$w(\nu + \delta) - \delta = \lambda,$$

i.e.,

(4.163) $$w^{-1}(\lambda + \delta) = \nu + \delta.$$

Since ν is dominant for $\Delta^+(\mathfrak{g}, \mathfrak{t})$, Proposition 4.67b shows that $\nu + \delta$ is orthogonal to no root. Hence if w exists as in (4.163), $\lambda + \delta$ is orthogonal to no root. This proves (a) for $\Pi_j(P_{\mathfrak{q},L}^{\mathfrak{g},L}(\mathcal{F}_{\mathfrak{l},L}^{\mathfrak{q},L}(Z)))$.

Under the assumption that $\lambda + \delta$ is orthogonal to no root, there exists a unique $w \in W$ such that $w^{-1}(\lambda + \delta)$ is dominant. Let us check that w is in W^1. If α is in $\Delta^+(w)$, then

(4.164) $$\langle \lambda + \delta, \alpha \rangle = \langle w(\nu + \delta), \alpha \rangle = \langle \nu + \delta, w^{-1}\alpha \rangle < 0.$$

Since $\lambda + \delta$ is $\Delta^+(\mathfrak{l}, \mathfrak{t})$ dominant, we conclude that α in in $\Delta(\mathfrak{u})$. Thus w is in W^1.

Hence V occurs in $\Pi_j(P_{\mathfrak{q},L}^{\mathfrak{g},L}(\mathcal{F}_{\mathfrak{l},L}^{\mathfrak{q},L}(Z)))$ if and only if ν satisfies (4.163) for this w and j is the length of w. Thus $\Pi_j(P_{\mathfrak{q},L}^{\mathfrak{g},L}(\mathcal{F}_{\mathfrak{l},L}^{\mathfrak{q},L}(Z)))$ is irreducible of highest weight $w^{-1}(\lambda + \delta) - \delta$ if $j = l(w)$ and it is 0 otherwise. To complete the proof for $\Pi_j(P_{\mathfrak{q},L}^{\mathfrak{g},L}(\mathcal{F}_{\mathfrak{l},L}^{\mathfrak{q},L}(Z)))$, we need to see that $l(w) = q$. If α is in $\Delta^+(\mathfrak{g}, \mathfrak{t})$, then

$$\langle \lambda + \delta, \alpha \rangle = \langle w(\nu + \delta, \alpha \rangle = \langle \nu + \delta, w^{-1}\alpha \rangle.$$

Since $\nu + \delta$ is $\Delta^+(\mathfrak{g}, \mathfrak{t})$ dominant, this is < 0 if and only if α is in $\Delta^+(w)$. Referring to (4.161), we see that $q = |\Delta^+(w)| = l(w)$.

Let us consider $\Gamma^j(I_{\bar{\mathfrak{q}},L}^{\mathfrak{g},L}(\mathcal{F}_{\mathfrak{l},L}^{\bar{\mathfrak{q}},L}(Z)))$. The second identity in Theorem 4.155 shows that the multiplicity of V in this representation equals the multiplicity of the irreducible representation of L with highest weight λ in $H^{S-j}(\bar{\mathfrak{u}}, V) \otimes_{\mathbb{C}} \bigwedge^{\text{top}} \bar{\mathfrak{u}}$, i.e., the multiplicity of the irreducible representation of L with highest weight $\lambda + 2\delta(\mathfrak{u})$ in $H^{S-j}(\bar{\mathfrak{u}}, V)$.

Again we shall apply Theorem 4.139, but this time we use $\bar{\mathfrak{b}}$, $\bar{\mathfrak{q}}$, and $-\Delta^+(\mathfrak{g}, \mathfrak{t})$ in place of $\mathfrak{b}$, $\mathfrak{q}$, and $\Delta^+(\mathfrak{g}, \mathfrak{t})$. The length function and W^1 are unchanged. The change of positive system means that we seek the multiplicity of the irreducible representation of L with lowest weight $\lambda + 2\delta(\mathfrak{u})$ (relative to $-\Delta^+(\mathfrak{g}, \mathfrak{t})$) in $H^{S-j}(\bar{\mathfrak{u}}, V)$. If w_L denotes the element of $W(\mathfrak{l}, \mathfrak{t})$ that sends $\Delta^+(\mathfrak{l}, \mathfrak{t})$ to $-\Delta^+(\mathfrak{l}, \mathfrak{t})$, we then seek the multiplicity of the irreducible representation of L with highest weight $w_L(\lambda + 2\delta(\mathfrak{u})) = w_L\lambda + 2\delta(\mathfrak{u})$ (relative to $-\Delta^+(\mathfrak{l}, \mathfrak{t})$) in $H^{S-j}(\bar{\mathfrak{u}}, V)$. Also V now has lowest weight ν and therefore has highest weight $w_G\nu$, where w_G is the element of $W(\mathfrak{g}, \mathfrak{t})$ that sends $\Delta^+(\mathfrak{g}, \mathfrak{t})$ to $-\Delta^+(\mathfrak{g}, \mathfrak{t})$.

The multiplicity in question is 0 or 1. It is 1 if and only if there is an element $s \in W^1$ of length $S - j$ such that

(4.165) $$s(w_G\nu - \delta) + \delta = w_L\lambda + 2\delta(\mathfrak{u}).$$

Applying w_L to both sides and using the identity $w_L^2 = 1$, we obtain

$$w_L s w_G(\nu + \delta) + \delta - 2\delta(\mathfrak{l} \cap \mathfrak{n}) = \lambda + 2\delta(\mathfrak{u})$$

and hence

$$w_L s w_G (\nu + \delta) = \lambda + \delta$$

or

(4.166) $$(w_L s w_G)^{-1}(\lambda + \delta) = \nu + \delta.$$

As with (4.163), this equation shows that if s exists, then $\lambda + \delta$ is orthogonal to no root. This proves (a) for $\Gamma^j(I_{\bar{\mathfrak{q}},L}^{\mathfrak{g},L}(\mathcal{F}_{\mathfrak{l},L}^{\bar{\mathfrak{q}},L}(Z)))$.

Put $w = w_L s w_G$. Let us check that $w \in W^1$ if and only if $s \in W^1$. Suppose s is in W^1. If α is in $\Delta^+(w)$, then $\alpha > 0$ and $w^{-1}\alpha < 0$. Hence $w_G s^{-1} w_L \alpha < 0$ and $s^{-1} w_L \alpha > 0$. If also α is in $\Delta^+(\mathfrak{l}, \mathfrak{t})$, then $w_L \alpha < 0$ and $-w_L \alpha$ is in $\Delta^+(\mathfrak{l}, \mathfrak{t})$. But the inequalities $-w_L \alpha > 0$ and $s^{-1}(-w_L \alpha) < 0$ together give a contradiction since s is in W^1. Thus $s \in W^1$ implies $w \in W^1$. Writing $s = w_L w w_G$ and repeating the argument, we see that $w \in W^1$ implies $s \in W^1$.

Under the assumption that $\lambda + \delta$ is orthogonal to no root, there exists a unique $w \in W$ such that $w^{-1}(\lambda + \delta)$ is dominant, and the argument with (4.164) shows that w is in W^1. Defining s by $w = w_L s w_G$, we see from the previous paragraph that s is in W^1.

Hence V occurs in $\Gamma^j(I_{\bar{\mathfrak{q}},L}^{\mathfrak{g},L}(\mathcal{F}_{\mathfrak{l},L}^{\bar{\mathfrak{q}},L}(Z)))$ if and only if ν satisfies (4.166) for this w with $l(s) = S - j$, if and only if ν satisfies (4.165) for this s with $l(s) = S - j$. Thus $\Gamma^j(I_{\bar{\mathfrak{q}},L}^{\mathfrak{g},L}(\mathcal{F}_{\mathfrak{l},L}^{\bar{\mathfrak{q}},L}(Z)))$ is irreducible of highest weight $w^{-1}(\lambda + \delta) - \delta$ if $S - j = l(s)$ and it is 0 otherwise. To complete the proof for $\Gamma^j(I_{\bar{\mathfrak{q}},L}^{\mathfrak{g},L}(\mathcal{F}_{\mathfrak{l},L}^{\bar{\mathfrak{q}},L}(Z)))$, we need to see that this j is $j = q$. From the argument for $\Pi_j(P_{\mathfrak{q},L}^{\mathfrak{g},L}(\mathcal{F}_{\mathfrak{l},L}^{\mathfrak{q},L}(Z)))$ we know that $q = l(w)$. Hence it is enough to prove that $l(w) + l(s) = S$, i.e., that

(4.167) $$l(w_L s w_G) + l(s) = |\Delta(\mathfrak{u})|.$$

For any element $x \in W(G, T), l(x) = |\Delta(\mathfrak{n})| - l(x w_G)$. Hence it is enough to prove that

$$l(w_L s) = |\Delta(\mathfrak{l} \cap \mathfrak{n})| + l(s).$$

But $\Delta^+(w_L s) = w_L \Delta^+(s) \cup \Delta(\mathfrak{l} \cap \mathfrak{n})$ disjointly, and hence this equality and (4.167) follow. This completes the proof of the corollary.

In the next chapter we shall make computations of this sort in the context of the representation theory of noncompact reductive groups. The results that we shall generalize will be *variants* of (4.156), (4.159), and Corollary 4.160; we shall not generalize directly the results that we have stated. The difficulty is that for noncompact groups the representations in intermediate degrees of homology and cohomology tend to be either messy or uninteresting. To obtain interesting representations, we must arrange that they should appear in the highest-possible degree of homology or cohomology.

We give the variants for the compact case now. For purposes of motivation, an important new ingredient is a changed order of events. It will help to think of G as connected, although we shall drop this assumption shortly. We are given $\mathfrak{l}_0$, possibly constructed from Proposition 4.76 as a centralizer. Since G is connected, L is going to be the analytic subgroup of G with Lie algebra $\mathfrak{l}_0$. Inside $\mathfrak{l}_0$ we are given a maximal abelian subspace $\mathfrak{t}_0$, and we are given a positive system $\Delta^+(\mathfrak{l}, \mathfrak{t})$ and an irreducible representation Z of L with highest weight λ. We then choose $\Delta^+(\mathfrak{g}, \mathfrak{t})$ containing $\Delta^+(\mathfrak{l}, \mathfrak{t})$ to make λ dominant. If we take $\mathfrak{u}$ to be generated by root vectors for positive roots outside $\Delta^+(\mathfrak{l}, \mathfrak{t})$, then $\mathfrak{q} = \mathfrak{l} \oplus \mathfrak{u}$ is a θ stable parabolic subalgebra of $\mathfrak{g}$, and we can proceed as earlier.

But to arrange for the nonzero representations to occur in degree S, we make one further change, reversing the roles of $\mathfrak{q}$ and $\bar{\mathfrak{q}}$. Specifically the representation built from Π will now involve $\bar{\mathfrak{q}}$, and the representation built from Γ will now involve $\mathfrak{q}$.

In practice, we just take all these data as given. Thus we have $\Delta^+(\mathfrak{g}, \mathfrak{t})$ and $\mathfrak{q}$ as usual, and the highest weight λ of an irreducible representation of L is assumed dominant for $\Delta^+(\mathfrak{g}, \mathfrak{t})$. With this setting, we can allow G to be disconnected. Once $\Delta^+(\mathfrak{g}, \mathfrak{t})$ and $\mathfrak{q}$ are in place, L is well defined. (Recall that L can depend on the choice of $\Delta^+(\mathfrak{g}, \mathfrak{t})$ if G is disconnected.) Once L is well defined, we can speak of an irreducible representation of L whose highest weights are dominant.

With G possibly disconnected, let Z be in $\mathcal{C}(\mathfrak{l}, L)$. Define

$$(4.168) \qquad\qquad Z^{\#} = Z \otimes_{\mathbb{C}} \textstyle\bigwedge^{\mathrm{top}} \mathfrak{u}.$$

Recall from Proposition 4.78f that $\bigwedge^{\mathrm{top}} \mathfrak{u}$ is a one-dimensional representation of L with unique weight $2\delta(\mathfrak{u})$. Then $Z^{\#}$ is in $\mathcal{C}(\mathfrak{l}, L)$. Let $\mathcal{L}_j(Z)$ and $\mathcal{R}^j(Z)$ be the members of $\mathcal{C}(\mathfrak{g}, G)$ given by

$$(4.169a) \qquad \mathcal{L}_j(Z) = \Pi_j(P^{\mathfrak{g},L}_{\bar{\mathfrak{q}},L}(\mathcal{F}^{\bar{\mathfrak{q}},L}_{\mathfrak{l},L}(Z^{\#}))) = (P^{\mathfrak{g},G}_{\bar{\mathfrak{q}},L})_j(\mathcal{F}^{\bar{\mathfrak{q}},L}_{\mathfrak{l},L}(Z^{\#}))$$

$$(4.169b) \qquad \mathcal{R}^j(Z) = \Gamma^j(I^{\mathfrak{g},L}_{\mathfrak{q},L}(\mathcal{F}^{\mathfrak{q},L}_{\mathfrak{l},L}(Z^{\#}))) = (I^{\mathfrak{g},G}_{\mathfrak{q},L})^j(\mathcal{F}^{\mathfrak{q},L}_{\mathfrak{l},L}(Z^{\#})).$$

(The second inequality in each case follows from Propositions 2.19 and 2.57, along with (C.28a).)

With V as a finite-dimensional representation of G, the variants of

(4.156) and (4.159) are

$$\mathrm{Hom}_{\mathfrak{g},G}(\mathcal{L}_j(Z), V)$$

$$= \mathrm{Hom}_{\mathfrak{g},G}(\Pi_j(P_{\bar{\mathfrak{q}},L}^{\mathfrak{g},L}(\mathcal{F}_{\mathfrak{l},L}^{\bar{\mathfrak{q}},L}(Z^{\#}))), V)$$

(4.170a) $$\cong \mathrm{Ext}_{\mathfrak{g},L}^j(P_{\bar{\mathfrak{q}},L}^{\mathfrak{g},L}(\mathcal{F}_{\mathfrak{l},L}^{\bar{\mathfrak{q}},L}(Z^{\#})), (\mathcal{F}^{\vee})_{\mathfrak{g},G}^{\mathfrak{g},L}(V))$$

(4.170b) $$\cong \mathrm{Ext}_{\mathfrak{g},L}^j(P_{\bar{\mathfrak{q}},L}^{\mathfrak{g},L}(\mathcal{F}_{\mathfrak{l},L}^{\bar{\mathfrak{q}},L}(Z^{\#})), V)$$

(4.170c) $$\cong \mathrm{Ext}_{\bar{\mathfrak{q}},L}^j(\mathcal{F}_{\mathfrak{l},L}^{\bar{\mathfrak{q}},L}(Z^{\#}), V)$$

(4.170d) $$\cong \mathrm{Hom}_{\mathfrak{l},L}(Z^{\#}, H^j(\bar{\mathfrak{u}}, V))$$

(4.170e) $$\cong \mathrm{Hom}_{\mathfrak{l},L}(Z^{\#}, H_{S-j}(\bar{\mathfrak{u}}, V \otimes_{\mathbb{C}} \textstyle\bigwedge^{\mathrm{top}}\mathfrak{u}))$$

(4.170f) $$\cong \mathrm{Hom}_{\mathfrak{l},L}(Z \otimes_{\mathbb{C}} \textstyle\bigwedge^{\mathrm{top}}\mathfrak{u}, H_{S-j}(\bar{\mathfrak{u}}, V) \otimes_{\mathbb{C}} \textstyle\bigwedge^{\mathrm{top}}\mathfrak{u})$$

(4.170g) $$\cong \mathrm{Hom}_{\mathfrak{l},L}(Z, H_{S-j}(\bar{\mathfrak{u}}, V))$$

and

$$\mathrm{Hom}_{\mathfrak{g},G}(V, \mathcal{R}^j(Z))$$

$$= \mathrm{Hom}_{\mathfrak{g},G}(V, \Gamma^j(I_{\mathfrak{q},L}^{\mathfrak{g},L}(\mathcal{F}_{\mathfrak{l},L}^{\mathfrak{q},L}(Z^{\#}))))$$

(4.171a) $$\cong \mathrm{Ext}_{\mathfrak{g},L}^j(V, I_{\mathfrak{q},L}^{\mathfrak{g},L}(\mathcal{F}_{\mathfrak{l},L}^{\mathfrak{q},L}(Z^{\#})))$$

(4.171b) $$\cong \mathrm{Ext}_{\mathfrak{q},L}^j(V, \mathcal{F}_{\mathfrak{l},L}^{\mathfrak{q},L}(Z^{\#}))$$

(4.171c) $$\cong \mathrm{Hom}_{\mathfrak{l},L}(H_j(\mathfrak{u}, V), Z^{\#})$$

(4.171d) $$\cong \mathrm{Hom}_{\mathfrak{l},L}(H^{S-j}(\mathfrak{u}, V \otimes_{\mathbb{C}} \textstyle\bigwedge^{\mathrm{top}}\mathfrak{u}), Z^{\#})$$

(4.171e) $$\cong \mathrm{Hom}_{\mathfrak{l},L}(H^{S-j}(\mathfrak{u}, V) \otimes_{\mathbb{C}} \textstyle\bigwedge^{\mathrm{top}}\mathfrak{u}, Z \otimes_{\mathbb{C}} \textstyle\bigwedge^{\mathrm{top}}\mathfrak{u}).$$

(4.171f) $$\cong \mathrm{Hom}_{\mathfrak{l},L}(H^{S-j}(\mathfrak{u}, V), Z).$$

Steps (a) through (d) of (4.170) are proved as with (4.156), and step (e) is by Hard Duality (Theorem 3.5b). Steps (a) through (e) of (4.171) are proved as with (4.159). So that they are not lost in the middle, let us isolate the beautiful and powerful formulas (4.170d) and (4.171c):

(4.172a) $$\quad \mathrm{Hom}_{\mathfrak{g},G}(\mathcal{L}_j(Z), V) \cong \mathrm{Hom}_{\mathfrak{l},L}(Z^{\#}, H^j(\bar{\mathfrak{u}}, V))$$

(4.172b) $$\quad \mathrm{Hom}_{\mathfrak{g},G}(V, \mathcal{R}^j(Z)) \cong \mathrm{Hom}_{\mathfrak{l},L}(H_j(\mathfrak{u}, V), Z^{\#}).$$

We can apply the computations (4.170) and (4.171) as follows.

Proposition 4.173. Let G be a compact Lie group, let $\mathfrak{t}_0$ be a maximal abelian subspace of $\mathfrak{g}_0$, let $\mathfrak{b} = \mathfrak{t} \oplus \mathfrak{n}$ be the Borel subalgebra corresponding to a positive system $\Delta^+(\mathfrak{g}, \mathfrak{t})$, let $\mathfrak{q} = \mathfrak{l} \oplus \mathfrak{u}$ be a parabolic subalgebra with $\mathfrak{t} \subseteq \mathfrak{b} \subseteq \mathfrak{q}$, and let L be the corresponding Levi subgroup. Suppose that L meets every component of G. Let Z be an irreducible $(\mathfrak{l}, L)$ module whose $\Delta^+(\mathfrak{l}, \mathfrak{t})$ highest weights are $\Delta^+(\mathfrak{g}, \mathfrak{t})$ dominant. Then the representations $\mathcal{L}_j(Z)$ and $\mathcal{R}^j(Z)$ are 0 if $j \neq S$. For $j = S$,

$$\mathcal{L}_S(Z) \qquad \text{and} \qquad \mathcal{R}^S(Z)$$

are irreducible representations of G, and they are characterized by having Z as the representation of L in their spaces of $\mathfrak{u}$ invariants.

PROOF. Working with Kostant's Theorem (Theorem 4.139) for the group G_0, we readily see from (4.170d) and (4.171f) that $\mathcal{L}_j(Z)$ and $\mathcal{R}^j(Z)$ are 0 for $j \neq S$. For $j = S$, (4.170g) and (4.171f) give

(4.174a) $\mathrm{Hom}_G(\mathcal{L}_S(Z), V) \cong \mathrm{Hom}_L(Z, V_{\bar{\mathfrak{u}}})$

(4.174b) $\mathrm{Hom}_G(V, \mathcal{R}^S(Z)) \cong \mathrm{Hom}_L(V^{\mathfrak{u}}, Z).$

The conclusion about $\mathcal{R}^S(Z)$ follows immediately from (4.174b) and Theorem 4.83. For $\mathcal{L}_S(Z)$, we combine (4.174a) and Theorem 4.83 with the isomorphism $V_{\bar{\mathfrak{u}}} \cong V^{\mathfrak{u}}$ given in Lemma 4.82.

CHAPTER V

COHOMOLOGICAL INDUCTION

If $(\mathfrak{g}, K)$ is a reductive pair and $(\mathfrak{l}, L \cap K)$ is the Levi pair built from a θ stable parabolic subalgebra of $\mathfrak{g}$, "cohomological induction" refers to any of several functors for passing from $(\mathfrak{l}, L \cap K)$ modules to $(\mathfrak{g}, K)$ modules. These are of two kinds: $\mathcal{L}_j$ is based on P, and $\mathcal{R}^j$ is based on I. In this abstract it is assumed that L meets every component of K.

Under cohomological induction if the $(\mathfrak{l}, L \cap K)$ module has an infinitesimal character, then the $(\mathfrak{g}, K)$ module has an infinitesimal character, and its parameter is a simple translate of the parameter for the $(\mathfrak{l}, L \cap K)$ module. The first attack on cohomologically induced modules is to investigate what K types can occur. By exploiting a filtration on the induced module at an intermediate step of the construction, one obtains an upper bound on multiplicities of K types. A first consequence is a vanishing theorem for cohomologically induced modules above the key degree S of interest. A second consequence is that one can use an Euler-Poincaré principle in order to get an exact value, for each K type, of the alternating sum of the multiplicities in the various degrees. More precise information is available for K types in the "bottom layer." A bottom-layer map controls their multiplicities precisely.

Under a dominance condition on the infinitesimal character of the $(\mathfrak{l}, L \cap K)$ module, one obtains a vanishing theorem below the key degree S of interest. The proof uses the Duality Theorem. A stronger vanishing theorem is available when the $(\mathfrak{l}, L \cap K)$ module is one dimensional. In the situation of each theorem, the functors $\mathcal{L}_S$ and $\mathcal{R}^S$ yield isomorphic modules.

Some information about the $(\mathfrak{g}, K)$ structure of cohomologically induced modules comes from spectral sequences. Frobenius reciprocity for the P and I functors leads to general spectral sequences involving the derived functors of P and I, and these in turn lead to two pairs of fundamental spectral sequences for cohomological induction, the members of a pair having a common abutment.

The Hochschild-Serre spectral sequences relate $\mathfrak{u}$ homology and cohomology to $\mathfrak{u} \cap \mathfrak{k}$ homology and cohomology. From them and Kostant's Theorem, it follows that the $\mathfrak{u}$ homology and cohomology of an admissible $(\mathfrak{g}, K)$ module are admissible as $(\mathfrak{l}, L \cap K)$ modules.

1. Setting

Cohomological induction is a generalization to reductive pairs of the construction for compact groups introduced in (4.169) and used to give algebraic analogs of the Bott-Borel-Weil Theorem. The Introduction explains its analytic origins.

327

The setting will be the following starting in §4 of this chapter. We let $(\mathfrak{g}, K)$ be a reductive pair, and we work with a θ stable parabolic subalgebra $\mathfrak{q} = \mathfrak{l} \oplus \mathfrak{u}$. It is implicit in our definition of "parabolic subalgebra" that $\mathfrak{q}$ contains a θ stable Cartan subalgebra $\mathfrak{h}_0$ of $\mathfrak{g}_0$. Since $\mathfrak{q}$ is θ stable, the set of roots in $\Delta(\mathfrak{g}, \mathfrak{h})$ contributing to $\mathfrak{q}$ is stable under θ and so is its set of negatives. By Proposition 4.74c, $\mathfrak{q}$ is germane. Consequently we automatically obtain from $\mathfrak{q}$ a parabolic subpair $(\mathfrak{q}, L \cap K)$. The definition of L in (4.77) shows that L does not depend on the Cartan subalgebra $\mathfrak{h}_0$. Since $\mathfrak{q}$ and $\mathfrak{u}$ are θ stable, we have

$$(5.1a) \qquad\qquad \mathfrak{q}^- = \bar{\mathfrak{q}} \qquad \text{and} \qquad \mathfrak{u}^- = \bar{\mathfrak{u}}.$$

In particular,

$$(5.1b) \qquad\qquad \mathfrak{g} = \bar{\mathfrak{u}} \oplus \mathfrak{l} \oplus \mathfrak{u}.$$

Let Z be in $\mathcal{C}(\mathfrak{l}, L \cap K)$, and define

$$(5.2) \qquad\qquad Z^\# = Z \otimes_{\mathbb{C}} \textstyle\bigwedge^{\text{top}} \mathfrak{u}.$$

Since $\mathfrak{u}$ is an $(\mathfrak{l}, L \cap K)$ module, $\bigwedge^{\text{top}} \mathfrak{u}$ is a one-dimensional $(\mathfrak{l}, L \cap K)$ module; its unique weight relative to $\mathfrak{h}$ is $2\delta(\mathfrak{u})$. Then $Z^\#$ is in $\mathcal{C}(\mathfrak{l}, L \cap K)$. Let $\mathcal{L}_j(Z)$ and $\mathcal{R}^j(Z)$ be the members of $\mathcal{C}(\mathfrak{g}, K)$ given by

$$(5.3a) \quad \mathcal{L}_j(Z) = \Pi_j(P_{\bar{\mathfrak{q}}, L \cap K}^{\mathfrak{g}, L \cap K}(\mathcal{F}_{\mathfrak{l}, L \cap K}^{\bar{\mathfrak{q}}, L \cap K}(Z^\#))) = (P_{\bar{\mathfrak{q}}, L \cap K}^{\mathfrak{g}, K})_j(\mathcal{F}_{\mathfrak{l}, L \cap K}^{\bar{\mathfrak{q}}, L \cap K}(Z^\#))$$

$$(5.3b) \quad \mathcal{R}^j(Z) = \Gamma^j(I_{\mathfrak{q}, L \cap K}^{\mathfrak{g}, L \cap K}(\mathcal{F}_{\mathfrak{l}, L \cap K}^{\mathfrak{q}, L \cap K}(Z^\#))) = (I_{\mathfrak{q}, L \cap K}^{\mathfrak{g}, K})^j(\mathcal{F}_{\mathfrak{l}, L \cap K}^{\mathfrak{q}, L \cap K}(Z^\#)).$$

The functors $\mathcal{L}_j$ and $\mathcal{R}^j$ are the functors of **cohomological induction**. We say that $\mathcal{L}_j(Z)$ and $\mathcal{R}^j(Z)$ are **cohomologically induced** from Z. To simplify the notation, we shall often abbreviate

$$(5.4) \qquad Z_{\bar{\mathfrak{q}}}^\# = \mathcal{F}_{\mathfrak{l}, L \cap K}^{\bar{\mathfrak{q}}, L \cap K}(Z^\#) \qquad \text{and} \qquad Z_{\mathfrak{q}}^\# = \mathcal{F}_{\mathfrak{l}, L \cap K}^{\mathfrak{q}, L \cap K}(Z^\#).$$

Let

$$(5.5) \qquad S = \dim(\mathfrak{u} \cap \mathfrak{k}) \qquad \text{and} \qquad R = \dim(\mathfrak{u} \cap \mathfrak{p}),$$

so that

$$\textstyle\bigwedge^{\text{top}} \mathfrak{u} = \bigwedge^{R+S} \mathfrak{u}.$$

By Corollary 2.125b, Π_j and Γ^j are 0 for $j > 2S$; hence $\mathcal{L}_j$ and $\mathcal{R}^j$ are 0 for $j > 2S$.

In Theorem 5.35 we shall prove the deeper result that $\mathcal{L}_j$ and $\mathcal{R}^j$ are 0 for $j > S$. We shall be especially interested in identifying $\mathcal{L}_S(Z)$ and $\mathcal{R}^S(Z)$. Since $\mathcal{L}_j$ and $\mathcal{R}^j$ are 0 for easy reasons when $j > 2S$, $\mathcal{L}_S(Z)$ and $\mathcal{R}^S(Z)$ are the cohomologically induced modules in what is effectively the middle degree of interest.

EXAMPLES.

1) Let G be a compact Lie group, and let Z be an irreducible $(\mathfrak{l}, L)$ module whose $\Delta^+(\mathfrak{l}, \mathfrak{t})$ highest weights are $\Delta^+(\mathfrak{g}, \mathfrak{t})$ dominant.

(a) Assume that L meets every component of G. Then Proposition 4.173 shows that $\mathcal{L}_j(Z)$ and $\mathcal{R}^j(Z)$ are 0 for $j \neq S$. For $j = S$ they are irreducible, and they are characterized by having Z as the representation of L in their respective spaces of $\mathfrak{u}$ invariants. In particular, $\mathcal{L}_S(Z) \cong \mathcal{R}^S(Z)$.

(b) Suppose G is the semidirect product of a two-element group with $SU(3)$ such that $SU(3)$ is normal and the nontrivial element γ of the two-element group acts by complex conjugation. With standard notation, let $\mathfrak{t}_0$ be the diagonal subalgebra of $\mathfrak{g}_0$, and let

$$\Delta^+(\mathfrak{g}, \mathfrak{t}) = \{e_i - e_j \mid 1 \leq i < j \leq 3\}.$$

Define $\mathfrak{q}$ to have $\Delta(\mathfrak{l}, \mathfrak{t}) = \{\pm(e_1 - e_2)\}$ and $\Delta(\mathfrak{u}) = \{e_2 - e_3, e_1 - e_3\}$, so that $S = 2$. One can show that $L \subseteq G_0$. Let V_1 be the adjoint representation of G_0 on $\mathfrak{g}$, and define $Z = (V_1)^{\mathfrak{u}}$. Then Z is an irreducible $(\mathfrak{l}, L)$ module, and V_1 is recovered as

$$V_1 \cong (P^{\mathfrak{g},G_0}_{\bar{\mathfrak{q}},L})_2(Z^{\#}_{\bar{\mathfrak{q}}}) \cong (I^{\mathfrak{g},G_0}_{\mathfrak{q},L})_2(Z^{\#}_{\mathfrak{q}}).$$

As we shall observe in (5.8) below,

$$\mathcal{L}_2(Z) = (P^{\mathfrak{g},G}_{\bar{\mathfrak{q}},L})_2(Z^{\#}_{\bar{\mathfrak{q}}}) \cong P^{\mathfrak{g},G}_{\mathfrak{g},G_0}(P^{\mathfrak{g},G_0}_{\bar{\mathfrak{q}},L})_2(Z^{\#}_{\bar{\mathfrak{q}}}) = P^{\mathfrak{g},G}_{\mathfrak{g},G_0}(V_1).$$

Now $\gamma V_1 \cong V_1$ because the nontrivial outer automorphism of $SU(3)$ fixes the highest weight of V_1. Thus Propositions 2.77 and 2.85 show that $\mathcal{F}^{\mathfrak{g},G_0}_{\mathfrak{g},G}(\text{induced}^{\mathfrak{g},G}_{\mathfrak{g},G_0}(V_1)) \cong V_1 \oplus V_1$. Referring to Corollary 2.83, we see that $\mathcal{L}_2(Z)$ is reducible. As a consequence of (5.8) and Proposition 2.77, $\mathcal{R}^2(Z) \cong \mathcal{L}_2(Z)$. Hence $\mathcal{R}^2(z)$ is reducible, too.

2) Let $(\mathfrak{g}, K)$ be a reductive pair for which G is connected but is allowed to be noncompact. Lemma 5.10 below will show that L is connected. If we choose $\mathfrak{h}_0$ to be a maximally compact θ stable Cartan subalgebra of $\mathfrak{l}_0$, then Lemma 5.9 will show that $\mathfrak{h}_0$ is maximally compact as a θ stable Cartan subalgebra of $\mathfrak{g}_0$. Proposition 4.38d and Lemma 4.43d show that the group T that makes $(\mathfrak{h}, T)$ into a Cartan subpair is connected, no matter what the positive system. The group T satisfies the hypotheses on it in Theorem 4.52 relative to the reductive pair $(\mathfrak{l}, L \cap K)$. Also we can choose a positive system $\Delta^+(\mathfrak{l}, \mathfrak{h})$ that is θ stable by taking $i\mathfrak{t}_0$ before

$\mathfrak{a}_0$ in the ordering; to see that $\Delta^+(\mathfrak{l}, \mathfrak{h})$ is θ stable, we have only to recall from Lemma 4.43 that $\Delta(\mathfrak{l}, \mathfrak{h})$ has no real roots. Thus Theorem 4.52 is applicable to $(\mathfrak{l}, L \cap K)$. Let λ be an analytically integral member of $\mathfrak{h}^*$ that is orthogonal to all members of $\Delta(\mathfrak{l})$, and let $\mathbb{C}_\lambda$ be the corresponding $(\mathfrak{h}, T)$ module. By Theorem 4.52, there is an irreducible $(\mathfrak{l}, L \cap K)$ module V with highest weight λ. Since λ is orthogonal to all members of $\Delta(\mathfrak{l})$, V is one dimensional, i.e., $V = \mathbb{C}_\lambda$. Thus $\mathbb{C}_\lambda$ becomes a one-dimensional representation of L. We define $(\mathfrak{g}, K)$ modules by

$$(5.6) \qquad A_{\mathfrak{q}}(\lambda) = \mathcal{L}_S(\mathbb{C}_\lambda) \qquad \text{and} \qquad A^{\mathfrak{q}}(\lambda) = \mathcal{R}^S(\mathbb{C}_\lambda).$$

We have noted that Theorem 5.35 will prove $\mathcal{L}_j(\mathbb{C}_\lambda)$ and $\mathcal{R}^j(\mathbb{C}_\lambda)$ to be 0 for $j > S$. For $j < S$, we shall see in Theorem 5.99 that $\mathcal{L}_j(\mathbb{C}_\lambda)$ and $\mathcal{R}^j(\mathbb{C}_\lambda)$ are 0 if λ satisfies a suitable dominance condition. Therefore the only degree of interest is $j = S$. Under the same dominance condition on λ, Theorem 5.99 will establish an isomorphism in degree S: $A_{\mathfrak{q}}(Z) \cong A^{\mathfrak{q}}(Z)$. Moreover, in Chapter VIII we shall prove that the isomorphic modules $A_{\mathfrak{q}}(\lambda)$ and $A^{\mathfrak{q}}(\lambda)$ are irreducible, and in Chapter IX, under the additional assumption that $\lambda|_{\mathfrak{a}_0}$ is purely imaginary, we shall prove that they are the underlying $(\mathfrak{g}, K)$ module of a unitary representation. Subsequently we shall locate them in a classification of irreducible $(\mathfrak{g}, K)$ modules.

3) Let $G = O(2m, 2n)$ with $m > 1$ and $n > 1$. The maximal compact subgroup is $K = O(2m) \times O(2n)$, and it follows that G has four components. A maximal abelian subspace $\mathfrak{t}_0$ of $\mathfrak{k}_0$ is a maximally compact θ stable Cartan subalgebra of $\mathfrak{g}_0$. In standard notation, the roots are given by

$$\Delta(\mathfrak{g}, \mathfrak{t}) = \{\pm e_i \pm e_j \mid 1 \le i < j \le m + n\}$$

and

$$\Delta(\mathfrak{k}, \mathfrak{t}) = \{\pm e_i \pm e_j \mid 1 \le i < j \le m \text{ or } m + 1 \le i < j \le n\}.$$

Let us fix the Borel subalgebra $\mathfrak{b}$ determined by

$$\Delta^+(\mathfrak{g}, \mathfrak{t}) = \{e_i \pm e_j \mid 1 \le i < j \le m + n\}$$

and consider parabolic subalgebras $\mathfrak{q}$ containing $\mathfrak{b}$. Each such is θ stable, since θ is 1 on $\mathfrak{t}$ and therefore is 1 on every root. According to Proposition 4.57, there is one $\mathfrak{q}$ for each subset of simple roots. Let us consider how many components the corresponding L has. We shall see in Lemma 5.10 that $L \cap G_0$ is connected, so that the problem is to decide how many components of G are met by L. Since any two maximal abelian subspaces of $\mathfrak{k}_0$ are conjugate via K_0, any member of $N_K(\mathfrak{q})$ is the product of a member of $N_{K_0}(\mathfrak{q})$ by a member of $N_K(\mathfrak{q})$ that normalizes $\mathfrak{t}$. Only the latter element is of interest, and the number of components of L is therefore

$$\frac{\#\{w \in W(K, T_0) \mid w\Delta(\mathfrak{u}) \subseteq \Delta(\mathfrak{u})\}}{\#\{w \in W(K_0, T_0) \mid w\Delta(\mathfrak{u}) \subseteq \Delta(\mathfrak{u})\}}.$$

This can be 1, 2, or 4:

m and n arbitrary	and	$\Delta^+(\mathfrak{l}, \mathfrak{t}) = \emptyset$	gives 4
$m = n = 2$	and	$\Delta^+(\mathfrak{l}, \mathfrak{t}) = \{e_3 \pm e_4\}$	gives 2
$m = n = 2$	and	$\Delta^+(\mathfrak{l}, \mathfrak{t}) = \{e_3 - e_4\}$	gives 1.

We can calculate $N_K(\mathfrak{q} \cap \mathfrak{k})$ similarly as

$$\frac{\#\{w \in W(K, T_0) \mid w\Delta(\mathfrak{u} \cap \mathfrak{k}) \subseteq \Delta(\mathfrak{u} \cap \mathfrak{k})\}}{\#\{w \in W(K_0, T_0) \mid w\Delta(\mathfrak{u} \cap \mathfrak{k}) \subseteq \Delta(\mathfrak{u} \cap \mathfrak{k})\}}.$$

This also can be 1, 2, or 4:

m and n arbitrary	and	$\Delta^+(\mathfrak{l}, \mathfrak{t}) = \emptyset$	gives 4
$m = n = 2$	and	$\Delta^+(\mathfrak{l}, \mathfrak{t}) = \{e_3 - e_4\}$	gives 2
$m = n = 2$	and	$\Delta^+(\mathfrak{l}, \mathfrak{t}) = \{e_1 - e_2, e_3 - e_4\}$	gives 1.

From the definition we always have $N_K(\mathfrak{q} \cap \mathfrak{k}) \supseteq N_K(\mathfrak{q})$. The example in the above tables in which $\Delta^+(\mathfrak{l}, \mathfrak{t}) = \{e_3 - e_4\}$ shows that the inequality can be strict. In this case it follows that although $\mathfrak{q} \cap \mathfrak{k}$ is a θ stable parabolic subalgebra of $\mathfrak{k}$ (by Proposition 4.44b), $L \cap K$ is too small for $(\mathfrak{q} \cap \mathfrak{k}, L \cap K)$ to be the corresponding parabolic subpair of $(\mathfrak{k}, K)$.

Some of the results about cohomological induction will depend critically on introducing the assumption

(5.7) L meets every component of G, i.e., $K = (L \cap K)K_0$.

By way of illustration, Example 1 shows that we get a conclusion of irreducibility in the compact case under the assumption (5.7) and that irreducibility may fail without this assumption. On the other hand Example 3 indicates that insistence on (5.7) would exclude some interesting, naturally arising examples. This state of affairs complicates matters, and we shall try to avoid assuming (5.7) whenever possible.

When (5.7) is needed for the best possible result, sometimes a useful conclusion is still possible in the general case. Our standard technique for salvaging something in the general case is indicated in Example 1b and is as follows: For any reductive G and Levi subgroup L, the assumption (5.7) is satisfied for the subgroup $G_1 = LG_0$. In terms of reductive pairs, (5.7) holds for $(\mathfrak{g}, (L \cap K)K_0)$, which is a subpair of finite index in $(\mathfrak{g}, K)$. Now the functors $P^{\mathfrak{g},K}_{\mathfrak{g},(L \cap K)K_0}$ and $I^{\mathfrak{g},K}_{\mathfrak{g},(L \cap K)K_0}$ are naturally isomorphic and exact, by Proposition 2.77. By (C.27),

$$
\begin{aligned}
\mathcal{L}_j(Z) &= (P^{\mathfrak{g},K}_{\bar{\mathfrak{q}},L \cap K})_j(Z^{\#}_{\bar{\mathfrak{q}}}) \cong P^{\mathfrak{g},K}_{\mathfrak{g},(L \cap K)K_0}(P^{\mathfrak{g},(L \cap K)K_0}_{\bar{\mathfrak{q}},L \cap K})_j(Z^{\#}_{\bar{\mathfrak{q}}}) \\
\mathcal{R}^j(Z) &= (I^{\mathfrak{g},K}_{\mathfrak{q},L \cap K})^j(Z^{\#}_{\mathfrak{q}}) \cong I^{\mathfrak{g},K}_{\mathfrak{g},(L \cap K)K_0}(I^{\mathfrak{g},(L \cap K)K_0}_{\mathfrak{q},L \cap K})^j(Z^{\#}_{\mathfrak{q}}).
\end{aligned}
$$
(5.8)

The right sides of these identities are $P^{\mathfrak{g},K}_{\mathfrak{g},(L \cap K)K_0}$ and $I^{\mathfrak{g},K}_{\mathfrak{g},(L \cap K)K_0}$ applied to $\mathcal{L}_j(Z)$ and $\mathcal{R}^j(Z)$ for $(\mathfrak{g}, K_0(L \cap K))$. Hence results about $\mathcal{L}_j(Z)$ and $\mathcal{R}^j(Z)$ for G often follow by combining results about LG_0 with results about $P^{\mathfrak{g},K}_{\mathfrak{g},(L \cap K)K_0}$ and $I^{\mathfrak{g},K}_{\mathfrak{g},(L \cap K)K_0}$. Many properties of the latter functors can be understood through Mackey theory, as presented in §II.4.

Example 3 points to a second difficulty. It shows that when $(\mathfrak{q}, L \cap K)$ is a parabolic subpair of $(\mathfrak{g}, K)$, it is not necessarily true that $(\mathfrak{q} \cap \mathfrak{k}, L \cap K)$ is a parabolic subpair of $(\mathfrak{k}, K)$. This is an awkward circumstance, because one of the techniques for analyzing cohomological induction within $(\mathfrak{g}, K)$ will be to analyze an analog within $(\mathfrak{k}, K)$. Therefore some authors find it convenient to enlarge the definition of cohomological induction to allow replacing the Levi factor L by certain kinds of subgroups of finite index in L. This issue will arise in connection with the bottom-layer map in §6. We shall avoid the problem, however, by assuming in that section that L meets every component of G.

The remainder of this section is devoted to various structure-theoretic lemmas, all in the setting of a reductive pair $(\mathfrak{g}, K)$ and a θ stable parabolic subalgebra $\mathfrak{q} = \mathfrak{l} \oplus \mathfrak{u}$. The first two of these were used in discussing Example 2 above.

Lemma 5.9. If $\mathfrak{h}_0 = \mathfrak{t}_0 \oplus \mathfrak{a}_0$ is a maximally compact θ stable Cartan subalgebra of $\mathfrak{l}_0$, then $\mathfrak{h}_0$ is maximally compact as a θ stable Cartan subalgebra of $\mathfrak{g}_0$.

PROOF. By Lemma 4.43, $\Delta(\mathfrak{l}, \mathfrak{h})$ has no real roots. Hence if we take $i\mathfrak{t}_0$ before $\mathfrak{a}_0$ in defining an ordering, we can arrange that $\Delta^+(\mathfrak{l}, \mathfrak{h})$ is θ stable. We can then define

$$
\Delta^+(\mathfrak{g}, \mathfrak{h}) = \Delta(\mathfrak{u}) \cup \Delta^+(\mathfrak{l}, \mathfrak{h}),
$$

and the result will be a positive system for $\mathfrak{g}$. This system is θ stable and, being a positive system, contains no pair $\pm\alpha$. Therefore $\Delta^+(\mathfrak{g}, \mathfrak{h})$ contains no real roots. By Lemma 4.43, $\mathfrak{h}_0$ is maximally compact in $\mathfrak{g}_0$.

Lemma 5.10. If G is connected, then L is connected.

PROOF. Since L is reductive, we are to prove that $L \cap K$ is connected. Let $\mathfrak{h}_0 = \mathfrak{t}_0 \oplus \mathfrak{a}_0$ be a maximally compact θ stable Cartan subalgebra of $\mathfrak{l}_0$, and form a θ stable Borel subalgebra $\mathfrak{b}$ with $\mathfrak{h} \subseteq \mathfrak{b} \subseteq \mathfrak{q}$. By Proposition 4.44b, $\mathfrak{b} \cap (\mathfrak{l} \cap \mathfrak{k})$ is a Borel subalgebra of $\mathfrak{l} \cap \mathfrak{k}$ relative to the Cartan subalgebra $\mathfrak{t}_0$ of $\mathfrak{l}_0 \cap \mathfrak{k}_0$, and the corresponding Cartan subgroup is $T' = N_{L \cap K}(\mathfrak{b} \cap \mathfrak{l} \cap \mathfrak{k})$. By Proposition 4.22,

$$L \cap K = T'(L \cap K)_0.$$

Therefore it is enough to prove that T' is connected. The members of T' normalize $\mathfrak{u}$, being in L, and they normalize $\mathfrak{k}$, being in K. Hence

$$T' = N_{L \cap K}(\mathfrak{b} \cap \mathfrak{l} \cap \mathfrak{k}) \cap N_{L \cap K}(\mathfrak{u} \cap \mathfrak{k})$$
$$\subseteq N_{L \cap K}((\mathfrak{b} \cap \mathfrak{l}) + \mathfrak{u}) \cap \mathfrak{k}) = N_{L \cap K}(\mathfrak{b} \cap \mathfrak{k}) \subseteq N_K(\mathfrak{b} \cap \mathfrak{k}).$$

Lemma 5.9 shows that $\mathfrak{h}_0$ is a maximally compact θ stable Cartan subalgebra of $\mathfrak{g}_0$, and Proposition 4.44b allows us to deduce that $\mathfrak{b} \cap \mathfrak{k}$ is a Borel subalgebra of $\mathfrak{k}$ relative to the Cartan subalgebra $\mathfrak{t}_0$ of $\mathfrak{k}_0$. Since K is connected by assumption, the remark following (4.21) shows that $N_K(\mathfrak{b} \cap \mathfrak{k}) = T_0$. Therefore $T' \subseteq T_0$. Since T' has $\mathfrak{t}_0$ as Lie algebra, we conclude that $T' = T_0$. Hence T' is connected.

Lemma 5.11. Let $\mathfrak{h}_0 = \mathfrak{t}_0 \oplus \mathfrak{a}_0$ be a maximally compact θ stable Cartan subalgebra of $\mathfrak{l}$. Then

(a) the elements $h_{\delta(\mathfrak{u} \cap \mathfrak{k})}$ and $h_{\delta(\mathfrak{u} \cap \mathfrak{p})}$ are fixed by $\mathrm{Ad}(L \cap K)$, and
(b) the element $h_{\delta(\mathfrak{u})}$ is fixed by $\mathrm{Ad}(L \cap K)$, its centralizer in $\mathfrak{g}$ is $\mathfrak{l}$, and $\mathfrak{u}$ is the sum of the eigenspaces in $\mathfrak{g}$ of $\mathrm{ad}\, h_{\delta(\mathfrak{u})}$ for positive eigenvalues.

PROOF. By Proposition 4.44, $\mathfrak{t}$ is a Cartan subalgebra of $\mathfrak{l} \cap \mathfrak{k}$. We form the one-dimensional representations $\bigwedge^R(\mathfrak{u} \cap \mathfrak{p})$ and $\bigwedge^S(\mathfrak{u} \cap \mathfrak{k})$ of $L \cap K$. Let us review the computation of the weight of each. In the case of $\bigwedge^R(\mathfrak{u} \cap \mathfrak{p})$, if $X_1, \ldots, X_{\dim(\mathfrak{u} \cap \mathfrak{p})}$ are vectors forming a basis of $\mathfrak{u} \cap \mathfrak{p}$, then $X_1 \wedge \cdots \wedge X_{\dim(\mathfrak{u} \cap \mathfrak{p})}$ is a one-element basis of the one-dimensional space $\bigwedge^R(\mathfrak{u} \cap \mathfrak{p})$. It is clear for $h \in \mathfrak{t}$ that

$$(\mathrm{ad}\, h)(X_1 \wedge \cdots \wedge X_{\dim(\mathfrak{u} \cap \mathfrak{p})}) = 2\delta(\mathfrak{u} \cap \mathfrak{p})(h)(X_1 \wedge \cdots \wedge X_{\dim(\mathfrak{u} \cap \mathfrak{p})})$$

if the X_j are weight vectors under $\mathfrak{t}$, and therefore $\bigwedge^R(\mathfrak{u} \cap \mathfrak{p})$ has the unique weight $2\delta(\mathfrak{u} \cap \mathfrak{p})$. Similarly $\bigwedge^S(\mathfrak{u} \cap \mathfrak{k})$ has the unique weight $2\delta(\mathfrak{u} \cap \mathfrak{k})$. Since the representations $\bigwedge^R(\mathfrak{u} \cap \mathfrak{p})$ and $\bigwedge^S(\mathfrak{u} \cap \mathfrak{k})$ are one dimensional, we have

$$(5.12) \qquad \langle 2\delta(\mathfrak{u} \cap \mathfrak{p}), \alpha \rangle = \langle 2\delta(\mathfrak{u} \cap \mathfrak{k}), \alpha \rangle = 0 \qquad \text{for } \alpha \in \Delta^+(\mathfrak{l} \cap \mathfrak{k}, \mathfrak{t}).$$

(Alternatively we can deduce (5.12) from Proposition 4.68.)

Since $\mathfrak{t} \subseteq \mathfrak{l} \cap \mathfrak{k}$, the members $h_{\delta(\mathfrak{u} \cap \mathfrak{p})}$ and $h_{\delta(\mathfrak{u} \cap \mathfrak{k})}$ of $i\mathfrak{t}_0$ are therefore in the center of $\mathfrak{l} \cap \mathfrak{k}$. Fix l in $L \cap K$. We shall prove that $\mathrm{Ad}(l)$ fixes $h_{\delta(\mathfrak{u} \cap \mathfrak{p})}$; a similar argument proves that $\mathrm{Ad}(l)$ fixes $h_{\delta(\mathfrak{u} \cap \mathfrak{k})}$. Since $\mathrm{Ad}(l)$ normalizes $\mathfrak{l}$ and $\mathfrak{k}$, $\mathrm{Ad}(l)h_{\delta(\mathfrak{u} \cap \mathfrak{p})}$ is in $\mathfrak{l} \cap \mathfrak{k}$. Also if X is in $\mathfrak{l} \cap \mathfrak{k}$, then

$$[\mathrm{Ad}(l)h_{\delta(\mathfrak{u} \cap \mathfrak{p})}, X] = \mathrm{Ad}(l)[h_{\delta(\mathfrak{u} \cap \mathfrak{p})}, \mathrm{Ad}(l)^{-1}X] = 0$$

since $\mathrm{Ad}(l)^{-1}X$ is in $\mathfrak{l} \cap \mathfrak{k}$. It follows that $\mathrm{Ad}(l)h_{\delta(\mathfrak{u} \cap \mathfrak{p})}$ is in the center of $\mathfrak{l} \cap \mathfrak{k}$, hence is in $\mathfrak{t}$. Thus $\mathrm{Ad}(l)h_{\delta(\mathfrak{u} \cap \mathfrak{p})}$ is in $i\mathfrak{t}_0$.

Applying our weight computation, we see that $\mathrm{ad}(\mathrm{Ad}(l)h_{\delta(\mathfrak{u} \cap \mathfrak{p})})$ acts on $\bigwedge^R(\mathfrak{u} \cap \mathfrak{p})$ by the scalar $\langle h_{2\delta(\mathfrak{u} \cap \mathfrak{p})}, \mathrm{Ad}(l)h_{\delta(\mathfrak{u} \cap \mathfrak{p})} \rangle$. On the other hand, if we recompute the action of

$$\mathrm{ad}(\mathrm{Ad}(l)h_{\delta(\mathfrak{u} \cap \mathfrak{p})}) = \mathrm{Ad}(l)\mathrm{ad}\, h_{\delta(\mathfrak{u} \cap \mathfrak{p})}\mathrm{Ad}(l)^{-1}$$

using the basis $\mathrm{Ad}(l)X_1, \ldots, \mathrm{Ad}(l)X_{\dim(\mathfrak{u} \cap \mathfrak{p})}$ of $\mathfrak{u} \cap \mathfrak{p}$, we find that $\mathrm{ad}(\mathrm{Ad}(l)h_{\delta(\mathfrak{u} \cap \mathfrak{p})})$ acts on $\bigwedge^R(\mathfrak{u} \cap \mathfrak{p})$ by the scalar $\langle h_{2\delta(\mathfrak{u} \cap \mathfrak{p})}, h_{\delta(\mathfrak{u} \cap \mathfrak{p})} \rangle$. Therefore

$$\langle h_{2\delta(\mathfrak{u} \cap \mathfrak{p})}, \mathrm{Ad}(l)h_{\delta(\mathfrak{u} \cap \mathfrak{p})} \rangle = \langle h_{2\delta(\mathfrak{u} \cap \mathfrak{p})}, h_{\delta(\mathfrak{u} \cap \mathfrak{p})} \rangle.$$

Since $\langle \mathrm{Ad}(l)h_{\delta(\mathfrak{u} \cap \mathfrak{p})}, \mathrm{Ad}(l)h_{\delta(\mathfrak{u} \cap \mathfrak{p})} \rangle = \langle h_{\delta(\mathfrak{u} \cap \mathfrak{p})}, h_{\delta(\mathfrak{u} \cap \mathfrak{p})} \rangle$ and since $\langle \cdot, \cdot \rangle$ is positive definite on $i\mathfrak{t}_0$, the equality part of the Schwarz inequality allows us to conclude that

$$\mathrm{Ad}(l)h_{\delta(\mathfrak{u} \cap \mathfrak{p})} = h_{\delta(\mathfrak{u} \cap \mathfrak{p})}.$$

Similarly $\mathrm{Ad}(L \cap K)$ fixes $h_{\delta(\mathfrak{u} \cap \mathfrak{k})}$, and (a) follows. For (b), $h_{\delta(\mathfrak{u})}$ is the sum of $h_{\delta(\mathfrak{u} \cap \mathfrak{k})}$ and $h_{\delta(\mathfrak{u} \cap \mathfrak{p})}$, with 0 component in $\mathfrak{a}$, and hence is fixed by $\mathrm{Ad}(L \cap K)$ as a result of (a). The rest follows from the proof of Proposition 4.76b.

2. Effect on Infinitesimal Character

As we have already mentioned, the assumption (5.7) will play a crucial role in parts of the theory. When the assumption is in force, the results of this section show that the P and I functors for passing from $(\mathfrak{l}, L \cap K)$ to $(\mathfrak{g}, K)$ are intertwined by forgetful functors with the P and I functors for passing from $(\mathfrak{l}, (L \cap K)_0)$ to $(\mathfrak{g}, K_0)$. A first consequence is that cohomological induction has a simple effect on infinitesimal characters (see §IV.8) when (5.7) is in force.

Several factors play a role in this analysis, and we shall take several steps to understand the role of each factor. The first factor is the component structure of the subgroup L. In analyzing this structure, we need to assume only that L is the underlying group of a reductive subpair of the reductive pair $(\mathfrak{g}, K)$.

Throughout this section we assume that $(\mathfrak{g}, K)$ is a reductive pair with underlying group G.

Proposition 5.13. Let $(\mathfrak{l}, L \cap K)$ be a reductive subpair of $(\mathfrak{g}, K)$ with underlying group $L \subseteq G$. If L meets every component of G, then the homomorphism

$$(L \cap K)/(L \cap K_0) \to K/K_0$$

induced by inclusion is one-one onto. In the special case that L is the Levi subgroup of a θ stable parabolic subalgebra of $\mathfrak{g}$, $(L \cap K)_0 = L \cap K_0$; hence under the assumption (5.7) the homomorphism

$$(L \cap K)/(L \cap K)_0 \to K/K_0$$

induced by inclusion is one-one onto.

PROOF. Condition (iii) of Definition 4.29 shows that $L \cap K$ meets every component of K, and hence $(L \cap K)/(L \cap K_0)$ maps onto K/K_0. The kernel is $(L \cap K) \cap K_0 = L \cap K_0$, and hence the map is an isomorphism. If $\mathfrak{q}$ is a θ stable parabolic subalgebra of $\mathfrak{g}$ and if L is its Levi subgroup, certainly $(L \cap K)_0 \subseteq L \cap K_0$. On the other hand, if l is in $L \cap K_0$, then l is in $N_{K_0}(\mathfrak{q})$, which is the group $L \cap K$ for G_0. By Lemma 5.10 this is connected. Hence l is in $(L \cap K)_0$, and we obtain the equality $(L \cap K)_0 = L \cap K_0$. The result follows.

Proposition 5.14. Let $(\mathfrak{l}, L \cap K)$ be a reductive subpair of $(\mathfrak{g}, K)$ with underlying group $L \subseteq G$. If L meets every component of G, then

(5.15a) $$R(K) \cong R(K_0) \otimes_{R(L \cap K_0)} R(L \cap K)$$

under the mapping of right side to left side given by

$$(5.15b) \quad T_1 \otimes T_2 \mapsto T \text{ with } \langle T, f \rangle = \int_{K_0 \times (L \cap K)} f(k_0 l) \, dT_1(k_0) \, dT_2(l).$$

This isomorphism respects the left action by K_0 and the right action by $L \cap K$. Consequently

(a) $R(\mathfrak{g}, K) \cong R(\mathfrak{g}, K_0) \otimes_{R(\mathfrak{g}, L \cap K_0)} R(\mathfrak{g}, L \cap K)$ under an isomorphism that respects the left and right actions by $\mathfrak{g}$, the left action by K_0, and the right action by $L \cap K$,

(b) $\mathcal{F}_{\mathfrak{g}, K}^{\mathfrak{g}, K_0} \circ P_{\mathfrak{g}, L \cap K}^{\mathfrak{g}, K} \cong P_{\mathfrak{g}, L \cap K_0}^{\mathfrak{g}, K_0} \circ \mathcal{F}_{\mathfrak{g}, L \cap K}^{\mathfrak{g}, L \cap K_0}$,

(c) $\mathcal{F}_{\mathfrak{g}, K}^{\mathfrak{g}, K_0} \circ I_{\mathfrak{g}, L \cap K}^{\mathfrak{g}, K} \cong I_{\mathfrak{g}, L \cap K_0}^{\mathfrak{g}, K_0} \circ \mathcal{F}_{\mathfrak{g}, L \cap K}^{\mathfrak{g}, L \cap K_0}$.

In these conclusions, $L \cap K_0$ may be replaced by $(L \cap K)_0$ if L is the Levi subgroup of a θ stable parabolic subalgebra of $\mathfrak{g}$.

PROOF. Let $m : K_0 \times (L \cap K) \to K$ be the multiplication map. In (5.15b), T is given in the notation of (B.26) by

$$T = m_*(T_1 \times T_2),$$

and thus (5.15b) gives us a map

$$(5.16a) \qquad \mathcal{E}'(K_0) \otimes_{\mathbb{C}} \mathcal{E}'(L \cap K) \to \mathcal{E}'(K).$$

The members T_0 of $\mathcal{E}'(L \cap K_0)$ act on the right of $\mathcal{E}'(K_0)$ and the left of $\mathcal{E}'(L \cap K)$ by the formulas

$$\langle T_1 T_0, h \rangle = \int_{K_0 \times L \cap K_0} h(k_0 l_0) \, dT_1(k_0) \, dT_0(l_0)$$

and $\qquad \langle T_0 T_2, h \rangle = \int_{(L \cap K_0) \times (L \cap K)} h(l_0 l) \, dT_0(l_0) \, dT_2(l),$

and we find that $T_1 T_0 \otimes T_2$ and $T_1 \otimes T_0 T_2$ map to the same element of $\mathcal{E}'(K)$, namely

(5.16b)

$$\langle m_*(T_1 T_0 \times T_2), f \rangle = \int_{K_0 \times (L \cap K_0) \times (L \cap K)} f(k_0 l_0 l) \, dT_1(k_0) \, dT_0(l_0) \, dT_2(l)$$

$$= \langle m_*(T_1 \times T_0 T_2), f \rangle.$$

Hence (5.16a) descends to a map

$$(5.16c) \qquad \mathcal{E}'(K_0) \otimes_{\mathcal{E}'(L \cap K_0)} \mathcal{E}'(L \cap K) \to \mathcal{E}'(K).$$

Under (5.16a), we readily check that $R(K_0) \otimes_{\mathbb{C}} R(L \cap K)$ maps to $R(K)$. Consequently the argument that passes from the map (5.16a) to the map (5.16c) also gives us a compatible map

(5.16d) $$R(K_0) \otimes_{R(L \cap K_0)} R(L \cap K) \to R(K).$$

Let $p_0^{L \cap K}$ and p_0^K be the respective characteristic functions of $L \cap K_0$ and K_0. Then $p_0^{L \cap K} R(L \cap K)$ and $p_0^K R(K)$ are subalgebras of $R(L \cap K)$ and $R(K)$ isomorphic to $R(L \cap K_0)$ and $R(K_0)$. Let us use a subscript 0 on a member of $R(L \cap K)$ or $R(K)$ to indicate the product with $p_0^{L \cap K}$ or p_0^K. If we make these identifications, then we see that we can rewrite (5.16) as the map $T_1 \otimes T_2 \mapsto (T_1)T_2$, where $T_1 \in R(K_0)$ has been identified with a member of $R(K)$ and acted upon by the member T_2 of $R(L \cap K)$.

Let us construct an inversion formula for (5.16d). Let $\Lambda = \{l\}$ be a set of coset representatives of $L \cap K_0$ in $L \cap K$. Since L meets every component of G, Proposition 5.13 is applicable and shows that Λ is also a set of coset representatives of K_0 in K. If $T \in \mathcal{E}'(K)$ is given, we can write

(5.17a) $$T = \sum_{l \in \Lambda} (T \delta_{l^{-1}})_0 \delta_l.$$

Assuming that T is in $R(K)$, choose a finite set $A \subseteq (L \cap K_0)\hat{\ }$ with $(T \delta_{l^{-1}})_0 \chi_A = (T \delta_{l^{-1}})_0$ for all $l \in \Lambda$. Then (5.17a) becomes

(5.17b) $$T = \sum_{l \in \Lambda} (T \delta_{l^{-1}})_0 (\chi_A \delta_l),$$

and our inversion formula is

(5.17c) $$T \mapsto \sum_{l \in \Lambda} (T \delta_{l^{-1}})_0 \otimes \chi_A \delta_l.$$

Application of the map (5.16a) to the right side of (5.17c) gives the right side of (5.17a). In other words, (5.17c) followed by (5.16a) is the identity.

In the reverse direction, let $T_1 \in R(K)$ and $T_2 \in R(L \cap K)$ be given. Then

(5.18a) $$T_2 = \sum_{l \in \Lambda} (T_2 \delta_{l^{-1}})_0 \delta_l = \sum_{l \in \Lambda} (T_2 \delta_{l^{-1}})_0 \chi_A \delta_l$$

for sufficiently large A, and

(5.18b)
$$\langle T_1 (T_2 \delta_{l^{-1}})_0, f \rangle = \int_{K_0 \times (L \cap K)} f(k_0 l') p_0^{L \cap K}(l') \, dT_1(k_0) \, d(T_2 \delta_{L^{-1}})(l')$$
$$= \int_{K_0 \times (L \cap K)} f(k_0 l') p_0^K(k_0 l') \, dT_1(k_0) \, d(T_2 \delta_{L^{-1}})(l')$$
$$= \langle (((T_1)T_2)\delta_{l^{-1}})_0, f \rangle,$$

the second equality following from Proposition 5.13. Hence

$$
\begin{aligned}
m_*(T_1 \times T_2) &= \sum_{l \in \Lambda} m_*(T_1 \times (T_2\delta_{l-1})_0 \chi_A \delta_l) &&\text{by (5.18a)} \\
&= \sum_{l \in \Lambda} m_*(T_1(T_2\delta_{l-1})_0 \times \chi_A \delta_l) &&\text{by (5.16b)} \\
&= \sum_{l \in \Lambda} m_*((((T_1)T_2)\delta_{l-1})_0 \times \chi_A \delta_l) &&\text{by (5.18b)} \\
&= \sum_{l \in \Lambda} (((T_1)T_2)\delta_{l-1})_0 \times \chi_A \delta_l.
\end{aligned}
$$

The right side here is the decomposition (5.17b) for $m_*(T_1 \times T_2)$. Hence if we apply (5.16) to $T_1 \otimes T_2$ and then apply (5.17), we obtain

$$
\begin{aligned}
\sum_{l \in \Lambda} (((T_1)T_2)\delta_{l-1})_0 \otimes \chi_A \delta_l &= \sum_{l \in \Lambda} T_1(T_2\delta_{l-1})_0 \otimes \chi_A \delta_l &&\text{by (5.18b)} \\
&= \sum_{l \in \Lambda} T_1 \otimes (T_2\delta_{l-1})_0 \chi_A \delta_l \\
&= T_1 \otimes T_2 &&\text{by (5.18a).}
\end{aligned}
$$

In other words, (5.16) followed by (5.17) is the identity. This proves that (5.15) is an isomorphism. Clearly (5.15) respects the left action by K_0 and the right action by $L \cap K$.

Applying $U(\mathfrak{g}) \otimes_{U(\mathfrak{k})} (\,\cdot\,)$ to our isomorphism (5.15a), we obtain an isomorphism

$$
R(\mathfrak{g}, K) \cong R(\mathfrak{g}, K_0) \otimes_{R(L \cap K_0)} R(L \cap K).
$$

Let us use Corollary 3.26 in the context of the inclusion of pairs $(\mathfrak{g}, L \cap K_0) \hookrightarrow (\mathfrak{g}, L \cap K)$, taking the $(\mathfrak{g}, L \cap K_0)$ module V to be $R(\mathfrak{g}, K_0)$. Then the result is

$$
R(\mathfrak{g}, K_0) \otimes_{R(L \cap K_0)} R(L \cap K) \cong R(\mathfrak{g}, K_0) \otimes_{R(\mathfrak{g}, L \cap K_0)} R(\mathfrak{g}, L \cap K).
$$

Hence

$$
(5.19) \qquad R(\mathfrak{g}, K) \cong R(\mathfrak{g}, K_0) \otimes_{R(\mathfrak{g}, L \cap K_0)} R(\mathfrak{g}, L \cap K).
$$

By construction the isomorphism (5.19) respects the left action by $\mathfrak{g}$ and the right action by $L \cap K$. It follows by passage from Lie algebras to Lie groups that (5.19) respects also the left action by K_0. (Or one can appeal to the first formula of (1.118b).)

To complete the proof of (a), we have to show that (5.19) respects the right action by $\mathfrak{g}$. For $X \in \mathfrak{g}$, $T_1 \in R(K_0)$, and $T_2 \in R(L \cap K_0)$, let us write $\mathrm{Ad}(\cdot)X = \sum \langle \mathrm{Ad}(\cdot)X, X_i^* \rangle X_i$ in the usual way. Then our identifications give

$$
\begin{aligned}
(T_1 \otimes T_2)X &= T_1 \otimes (T_2 \otimes X) \\
&= T_1 \otimes (\mathrm{Ad}(\cdot)X \otimes T_2) \\
&= (T_1 \otimes \mathrm{Ad}(\cdot)X) \otimes T_2 \\
&= \sum_i (T_1 \otimes X_i) \otimes \langle \mathrm{Ad}(\cdot)X, X_i^* \rangle T_2 \\
&= \sum_{i,j} (X_j \otimes \langle \mathrm{Ad}(\cdot)X_i, X_j^* \rangle T_1) \otimes \langle \mathrm{Ad}(\cdot)X, X_i^* \rangle T_2.
\end{aligned}
$$

Also, within $R(\mathfrak{g}, K)$, we have

$$
T_1 T_2 \otimes X = \mathrm{Ad}(\cdot)X \otimes T_1 T_2 = \sum_j X_j \otimes \langle \mathrm{Ad}(\cdot)X, X_j^* \rangle T_1 T_2.
$$

To check that (5.19) respects the right $\mathfrak{g}$ actions, it is therefore enough to check that

(5.20) $$\sum_i \langle \mathrm{Ad}(\cdot)X_i, X_j^* \rangle T_1 \otimes \langle \mathrm{Ad}(\cdot)X, X_i^* \rangle T_2$$

maps under (5.15) to $\langle \mathrm{Ad}(\cdot)X, X_j^* \rangle T_1 T_2$. To do so, we simply compute the effect of the image of (5.20) on f:

$$
\begin{aligned}
&\sum_i \int_{K_0 \times (L \cap K)} f(k_0 l)\, d(\langle \mathrm{Ad}(k_0)X_i, X_j^* \rangle T_1)(k_0)\, d(\langle \mathrm{Ad}(l)X, X_i^* \rangle T_2)(l) \\
&= \int_{K_0 \times (L \cap K)} f(k_0 l) \left(\sum_i \langle \mathrm{Ad}(k_0)X_i, X_j^* \rangle \langle \mathrm{Ad}(l)X, X_i^* \rangle \right) dT_1(k_0)\, dT_2(l) \\
&= \int_{K_0 \times (L \cap K)} f(k_0 l) \langle \mathrm{Ad}(k_0 l)X, X_j^* \rangle\, dT_1(k_0)\, dT_2(l) \\
&= \langle \mathrm{Ad}(\cdot)X, X_j^* \rangle T_1 T_2, f \rangle.
\end{aligned}
$$

This is the required answer and completes the proof of (a) of the proposition.

For (b) we apply $(\cdot) \otimes_{R(\mathfrak{g}, L \cap K)} V$ to the two sides of (a) and obtain

$$
\begin{aligned}
P_{\mathfrak{g}, L \cap K}^{\mathfrak{g}, K}(V) &\cong R(\mathfrak{g}, K_0) \otimes_{R(\mathfrak{g}, L \cap K_0)} R(\mathfrak{g}, L \cap K) \otimes_{R(\mathfrak{g}, L \cap K)} V \\
&\cong R(\mathfrak{g}, K_0) \otimes_{R(\mathfrak{g}, L \cap K_0)} V,
\end{aligned}
$$

and the result is (b). The isomorphism (b) is natural in V.

For (c) we write (a) in the form

$$R(\mathfrak{g}, K) \cong R(\mathfrak{g}, L \cap K) \otimes_{R(\mathfrak{g}, L \cap K_0)} R(\mathfrak{g}, K_0)$$

and apply $\mathrm{Hom}_{R(\mathfrak{g}, L \cap K)}(\,\cdot\,, V)_K = \mathrm{Hom}_{R(\mathfrak{g}, L \cap K)}(\,\cdot\,, V)_{K_0}$ to the two sides. We obtain

$I^{\mathfrak{g}, K}_{\mathfrak{g}, L \cap K}(V)$

$\cong \mathrm{Hom}_{R(\mathfrak{g}, L \cap K)}(R(\mathfrak{g}, L \cap K) \otimes_{R(\mathfrak{g}, L \cap K_0)} R(\mathfrak{g}, K_0), V)_{K_0}$

$\cong \mathrm{Hom}_{R(\mathfrak{g}, L \cap K_0)}(R(\mathfrak{g}, K_0), \mathrm{Hom}_{R(\mathfrak{g}, L \cap K)} R(\mathfrak{g}, L \cap K), V))_{K_0}$

$\cong \mathrm{Hom}_{R(\mathfrak{g}, L \cap K_0)}(R(\mathfrak{g}, K_0), \mathrm{Hom}_{R(\mathfrak{g}, L \cap K)} R(\mathfrak{g}, L \cap K), V)_{L \cap K_0})_{K_0}$

$\cong \mathrm{Hom}_{R(\mathfrak{g}, L \cap K_0)}(R(\mathfrak{g}, K_0), \mathrm{Hom}_{R(\mathfrak{g}, L \cap K)} R(\mathfrak{g}, L \cap K), V)_{L \cap K})_{K_0}$

$\cong \mathrm{Hom}_{R(\mathfrak{g}, L \cap K_0)}(R(\mathfrak{g}, K_0), V)_{K_0}$,

and the result is (c). The isomorphism (c) is natural in V.

If L is the Levi subgroup of a θ stable parabolic subalgebra of $\mathfrak{g}$, then Proposition 5.13 shows that $L \cap K_0 = (L \cap K)_0$. This completes the proof.

Now let us turn to infinitesimal characters. For the moment we continue to work with the underlying group L of a reductive subpair of $(\mathfrak{g}, K)$. Theorem 4.115 shows that infinitesimal characters of $U(\mathfrak{g})$ modules may be given as members of $\mathfrak{h}^*$ in terms of the Harish-Chandra isomorphism γ.

Theorem 5.21. Let $(\mathfrak{l}, L \cap K)$ be a reductive subpair of $(\mathfrak{g}, K)$ with underlying group L, and let W be a $(\mathfrak{g}, L \cap K)$ module with infinitesimal character λ.

(a) If L meets every component of G, then the $(\mathfrak{g}, K)$ modules $(\Pi^{\mathfrak{g}, K}_{\mathfrak{g}, L \cap K})_j(W)$ and $(\Gamma^{\mathfrak{g}, K}_{\mathfrak{g}, L \cap K})^j(W)$ have infinitesimal character λ.

(b) If $(\mathfrak{g}, K)$ is in the Harish-Chandra class, then the $(\mathfrak{g}, K)$ modules $(\Pi^{\mathfrak{g}, K}_{\mathfrak{g}, L \cap K})_j(W)$ and $(\Gamma^{\mathfrak{g}, K}_{\mathfrak{g}, L \cap K})^j(W)$ have infinitesimal character λ.

(c) Without either of the special assumptions in (a) and (b), the $(\mathfrak{g}, K)$ modules $(\Pi^{\mathfrak{g}, K}_{\mathfrak{g}, L \cap K})_j(W)$ and $(\Gamma^{\mathfrak{g}, K}_{\mathfrak{g}, L \cap K})^j(W)$ are acted upon diagonably by $Z(\mathfrak{g})$, the simultaneous eigenvalues on $z \in Z(\mathfrak{g})$ being $\chi_\lambda(\mathrm{Ad}(k_j)z)$ for $\{k_j^{-1}\}$ a set of coset representatives of $K/((L \cap K)K_0)$.

PROOF. We give the argument for $\Pi_j = (\Pi^{\mathfrak{g}, K}_{\mathfrak{g}, L \cap K})_j(W)$, the argument for $(\Gamma^{\mathfrak{g}, K}_{\mathfrak{g}, L \cap K})^j(W)$ being similar. We prove the assertions in the order (a), (c), (b).

(a) Write $\mathcal{F}$ for $\mathcal{F}_{\mathfrak{g},L\cap K}^{\mathfrak{g},L\cap K_0}$, suppose that z is in $Z(\mathfrak{g})$, and let P be any $(\mathfrak{g}, L\cap K)$ module. Since z is central, the action of z on $\mathcal{F}P$ is a $(\mathfrak{g}, L\cap K_0)$ map of $\mathcal{F}P$ into itself. Write Π_0 for $P_{\mathfrak{g},L\cap K_0}^{\mathfrak{g},K_0}$. We claim that the action of z on $\Pi_0(\mathcal{F}P)$ is just $\Pi_0(z)$. In fact,

$$\Pi_0(\mathcal{F}P) = R(\mathfrak{g}, K_0) \otimes_{R(\mathfrak{g},L\cap K_0)} \mathcal{F}P$$

with $\Pi_0(z)$ given as the map $1 \otimes z$. Meanwhile z acts on $\Pi_0(\mathcal{F}P)$ by $z \otimes 1$, and it is easy to check that these actions are equal by using (1.116) and the fact that $\mathrm{Ad}(K_0)$ fixes z.

In our given $(\mathfrak{g}, L\cap K)$ module W, let z act by a scalar $c(z)$, and let

$$0 \longleftarrow W \longleftarrow P_0 \longleftarrow P_1 \longleftarrow \cdots$$

be a projective resolution of W in $\mathcal{C}(\mathfrak{g}, L\cap K)$. The diagram

(5.22a)

$$
\begin{array}{ccccccc}
0 & \longleftarrow & \mathcal{F}W & \longleftarrow & \mathcal{F}P_0 & \longleftarrow & \mathcal{F}P_1 & \longleftarrow & \cdots \\
& & \downarrow z & & \downarrow z & & \downarrow z & & \\
0 & \longleftarrow & \mathcal{F}W & \longleftarrow & \mathcal{F}P_0 & \longleftarrow & \mathcal{F}P_1 & \longleftarrow & \cdots
\end{array}
$$

is commutative since the horizontal maps are $U(\mathfrak{g})$ maps, and it is a diagram of $(\mathfrak{g}, L\cap K_0)$ maps since z is central. Applying Π_0 and using the result of the previous paragraph, we obtain a commutative diagram

(5.22b)

$$
\begin{array}{ccccccc}
0 & \longleftarrow & \Pi_0(\mathcal{F}P_0) & \longleftarrow & \Pi_0(\mathcal{F}P_1) & \longleftarrow & \cdots \\
& & \downarrow z & & \downarrow z & & \\
0 & \longleftarrow & \Pi_0(\mathcal{F}P_0) & \longleftarrow & \Pi_0(\mathcal{F}P_1) & \longleftarrow & \cdots
\end{array}
$$

in $\mathcal{C}(\mathfrak{g}, K_0)$.

The diagram (5.22a) exhibits the system $\{z\}$ as a chain map over z, and it is clear that a system of scalar multiplications by $c(z)$ is another chain map over z, since z acts by $c(z)$ on W. By (C.6), (5.22a) is homotopic to

$$
\begin{array}{ccccccc}
0 & \longleftarrow & \mathcal{F}W & \longleftarrow & \mathcal{F}P_0 & \longleftarrow & \mathcal{F}P_1 & \longleftarrow & \cdots \\
& & \downarrow z & & \downarrow c(z) & & \downarrow c(z) & & \\
0 & \longleftarrow & \mathcal{F}W & \longleftarrow & \mathcal{F}P_0 & \longleftarrow & \mathcal{F}P_1 & \longleftarrow & \cdots
\end{array}
$$

Application of Π shows that (5.22b) is homotopic to the diagram

$$0 \longleftarrow \Pi_0(\mathcal{F}P_0) \longleftarrow \Pi_0(\mathcal{F}P_1) \longleftarrow \cdots$$

(5.22c) $\qquad\qquad c(z)\Big\downarrow \qquad\qquad c(z)\Big\downarrow$

$$0 \longleftarrow \Pi_0(\mathcal{F}P_0) \longleftarrow \Pi_0(\mathcal{F}P_1) \longleftarrow \cdots$$

in $C(\mathfrak{g}, K_0)$. By Proposition 5.14b we can rewrite (5.22b) and (5.22c) as homotopic diagrams

$$0 \longleftarrow \Pi(P_0) \longleftarrow \Pi(P_1) \longleftarrow \cdots$$

$$z\Big\downarrow \qquad\qquad z\Big\downarrow$$

$$0 \longleftarrow \Pi(P_0) \longleftarrow \Pi(P_1) \longleftarrow \cdots$$

and

$$0 \longleftarrow \Pi(P_0) \longleftarrow \Pi(P_1) \longleftarrow \cdots$$

$$c(z)\Big\downarrow \qquad\qquad c(z)\Big\downarrow$$

$$0 \longleftarrow \Pi(P_0) \longleftarrow \Pi(P_1) \longleftarrow \cdots$$

in $C(\mathfrak{g}, K_0)$, provided we suppress occurrences of $\mathcal{F}_{\mathfrak{g},K}^{\mathfrak{g},K_0}$. Passing to homology and using (C.4), we see that z acts by $c(z)$ on $\Pi_j(W)$. Combining this fact with the fact that W has infinitesimal character λ, we obtain the conclusion that $\Pi_j(W)$ has infinitesimal character λ.

(c) Let $\Pi_j'(W)$ be $\Pi_j(W)$ for the group $G_1 = LG_0$. Using Proposition 2.77 to argue as for (5.8), we have

(5.23) $\qquad\qquad \Pi_j(W) \cong \text{induced}_{\mathfrak{g},(L\cap K)K_0}^{\mathfrak{g},K}(\Pi_j'(W))$.

By (a), $\Pi_j'(W)$ has infinitesimal character λ. Proposition 4.117 and (5.23) therefore show that $Z(\mathfrak{g})$ acts diagonably on $\Pi_j(W)$, the simultaneous eigenvalues of $z \in Z(\mathfrak{g})$ being $\chi_\lambda(\text{Ad}(k_j)z)$. This proves (c) for $\Pi_j(W)$.

(b) If $(\mathfrak{g}, K)$ is in the Harish-Chandra class, then $\text{Ad}(k_j)z = z$ for all $k_j \in K$. Thus (b) follows immediately from (c).

The other factor in infinitesimal characters is the effect of ind and pro. If $\mathfrak{l}$ is the Levi factor of a parabolic subalgebra of $\mathfrak{g}$ and if $\mathfrak{h}$ is a Cartan subalgebra of $\mathfrak{l}$, then we can realize infinitesimal characters for both $U(\mathfrak{g})$ and $U(\mathfrak{l})$ as members of $\mathfrak{h}^*$, using the Harish-Chandra isomorphism γ in the case of $U(\mathfrak{g})$ modules and $\gamma_\mathfrak{l}$ in the case of $U(\mathfrak{l})$ modules.

Theorem 5.24. Let $\mathfrak{q} = \mathfrak{l} \oplus \mathfrak{u}$ be a germane parabolic subalgebra of $\mathfrak{g}$, and let L be any closed subgroup of G with Lie algebra $\mathfrak{l}_0$ such that $L \subseteq N_G(\mathfrak{q}) \cap N_G(\theta\mathfrak{q})$. If Z' is an $(\mathfrak{l}, L \cap K)$ module with infinitesimal character λ, then

    (a) $\operatorname{ind}_{\mathfrak{q}, L \cap K}^{\mathfrak{g}, L \cap K}(\mathcal{F}_{\mathfrak{l}, L \cap K}^{\mathfrak{q}, L \cap K}(Z'))$ is a $(\mathfrak{g}, L \cap K)$ module with infinitesimal character $\lambda + \delta(\mathfrak{u})$

    (b) $\operatorname{pro}_{\mathfrak{q}, L \cap K}^{\mathfrak{g}, L \cap K}(\mathcal{F}_{\mathfrak{l}, L \cap K}^{\mathfrak{q}, L \cap K}(Z'))$ is a $(\mathfrak{g}, L \cap K)$ module with infinitesimal character $\lambda - \delta(\mathfrak{u})$.

PROOF. For purposes of understanding the $U(\mathfrak{g})$ action, Proposition 2.57a says that we can make the identification

$$P_{\mathfrak{q}, L \cap K}^{\mathfrak{g}, L \cap K}(\mathcal{F}_{\mathfrak{l}, L \cap K}^{\mathfrak{q}, L \cap K}(Z')) \cong U(\mathfrak{g}) \otimes_{U(\mathfrak{q})} \mathcal{F}_{\mathfrak{l}, L \cap K}^{\mathfrak{q}, L \cap K}(Z').$$

If z is in $Z(\mathfrak{g})$, write

$$z = \mu_{\mathfrak{u}}'(z) + \sum x_i u_i \qquad \text{with } x_i \in U(\mathfrak{g}) \text{ and } u_i \in \mathfrak{u},$$

according to the decomposition in Lemma 4.123a. (Here the $\bar{\mathfrak{u}}U(\mathfrak{g})$ terms are not needed, according to Lemma 4.123b.) Then

$$
\begin{aligned}
z(x \otimes v) \\
&= zx \otimes v \\
&= xz \otimes v \\
&= x\mu_{\mathfrak{u}}'(z) \otimes v + \sum x x_i u_i \otimes v \\
&= x \otimes \mu_{\mathfrak{u}}'(z)v + \sum x x_i \otimes u_i v \\
&= x \otimes \mu_{\mathfrak{u}}'(z)v && \text{since } \mathfrak{u} \text{ annihilates} \\
& && \mathcal{F}_{\mathfrak{l}, L \cap K}^{\mathfrak{q}, L \cap K}(Z^{\#}) \\
&= (\lambda(\gamma_{\mathfrak{l}}(\mu_{\mathfrak{u}}'(z))))(x \otimes v) \\
&= (\lambda(\tau_{\mathfrak{l} \cap \mathfrak{n}} \circ \gamma_{\mathfrak{l} \cap \mathfrak{n}}' \circ \mu_{\mathfrak{u}}'(z)))(x \otimes v) && \text{by (4.93) for } \mathfrak{l} \\
&= (\lambda(\tau_{\mathfrak{u}}^{-1} \circ \tau_{\mathfrak{n}} \circ \gamma_{\mathfrak{n}}'(z)))(x \otimes v) && \text{by Proposition 4.129a} \\
&= (\lambda(\tau_{\mathfrak{u}}^{-1} \circ \gamma(z)))(x \otimes v) && \text{by (4.93) for } \mathfrak{g} \\
&= ((\lambda + \delta(\mathfrak{u}))(\gamma(z)))(x \otimes v).
\end{aligned}
$$

So z acts on the $(\mathfrak{g}, L \cap K)$ module $P_{\mathfrak{q}, L \cap K}^{\mathfrak{g}, L \cap K}(\mathcal{F}_{\mathfrak{l}, L \cap K}^{\mathfrak{q}, L \cap K}(Z'))$ with infinitesimal character $\lambda + \delta(\mathfrak{u})$. This proves (a), and the proof of (b) is similar.

Corollary 5.25. Let $\mathfrak{q} = \mathfrak{l} \oplus \mathfrak{u}$ be a θ stable parabolic subalgebra of $\mathfrak{g}$, let L be the corresponding Levi subgroup, and let z be an $(\mathfrak{l}, L \cap K)$ module with infinitesimal character λ.

(a) If L meets every component of G, then the $(\mathfrak{g}, K)$ modules $\mathcal{L}_j(Z)$ and $\mathcal{R}^j(Z)$ have infinitesimal character $\lambda + \delta(\mathfrak{u})$.

(b) If $(\mathfrak{g}, K)$ is in the Harish-Chandra class, then the $(\mathfrak{g}, K)$ modules $\mathcal{L}_j(Z)$ and $\mathcal{R}^j(Z)$ have infinitesimal character $\lambda + \delta(\mathfrak{u})$.

(c) Without either of the assumptions in (a) and (b), the $(\mathfrak{g}, K)$ modules $\mathcal{L}_j(Z)$ and $\mathcal{R}^j(Z)$ are acted upon diagonably by $Z(\mathfrak{g})$, the simultaneous eigenvalues on $z \in Z(\mathfrak{g})$ being $\chi_\lambda(\mathrm{Ad}(k_j)z)$ for $\{k_j^{-1}\}$ a set of coset representatives of $K/((L \cap K)K_0)$.

PROOF. First we check readily that the $(\mathfrak{l}, L \cap K)$ module $Z^\#$ has infinitesimal character $\lambda + 2\delta(\mathfrak{u})$. By Theorem 5.24 (either with (a) applied to $\bar{\mathfrak{q}}$ or with (b) applied to $\mathfrak{q}$), the $(\mathfrak{g}, L \cap K)$ modules $\mathrm{ind}_{\bar{\mathfrak{q}}, L \cap K}^{\mathfrak{g}, L \cap K}(\mathcal{F}_{\mathfrak{l}, L \cap K}^{\bar{\mathfrak{q}}, L \cap K}(Z^\#))$ and $\mathrm{pro}_{\mathfrak{q}, L \cap K}^{\mathfrak{g}, L \cap K}(\mathcal{F}_{\mathfrak{l}, L \cap K}^{\mathfrak{q}, L \cap K}(Z^\#))$ have infinitesimal character $\lambda + \delta(\mathfrak{u})$. Then the theorem follows by applying Theorem 5.21.

EXAMPLE. Let the reductive pair $(\mathfrak{g}, K)$ have G connected. The $(\mathfrak{g}, K)$ modules $A_\mathfrak{q}(\lambda)$ and $A^\mathfrak{q}(\lambda)$ defined in (5.6) have infinitesimal character $\lambda + \delta$. In fact, they are cohomologically induced from $Z = \mathbb{C}_\lambda$, which, according to the first example after Theorem 4.115, has infinitesimal character $\lambda + \delta(\mathfrak{l} \cap \mathfrak{n})$ relative to $\mathfrak{l}$. By Corollary 5.25 the cohomologically induced modules have infinitesimal character $(\lambda + \delta(\mathfrak{l} \cap \mathfrak{n})) + \delta(\mathfrak{u}) = \lambda + \delta$.

3. Preliminary Lemmas

In this section we collect two technical lemmas concerning the P and I functors. Let $(\mathfrak{g}, B)$ be a pair. Suppose that $\mathfrak{m}$ and $\mathfrak{m}'$ are $\mathrm{Ad}(B)$ invariant complex Lie subalgebras of $\mathfrak{g}$. Then we have inclusions of pairs

$$(\mathfrak{m} \cap \mathfrak{m}', B) \hookrightarrow (\mathfrak{m}, B) \hookrightarrow (\mathfrak{g}, B)$$

and
$$(\mathfrak{m} \cap \mathfrak{m}', B) \hookrightarrow (\mathfrak{m}', B) \hookrightarrow (\mathfrak{g}, B),$$

and we refer to $(\mathfrak{m}, B)$, $(\mathfrak{m}', B)$, and $(\mathfrak{m} \cap \mathfrak{m}', B)$ as **compatible pairs** within $\mathfrak{g}$.

The first lemma concerns the P functor and will be used later in two contexts: one when $\mathfrak{m} = \mathfrak{l}$ and $\mathfrak{m}' = \bar{\mathfrak{q}}$, and one when $\mathfrak{m} = \mathfrak{q}$ and $\mathfrak{m}' = \bar{\mathfrak{q}}$. Recall that $\sigma : S(\mathfrak{g}) \to U(\mathfrak{g})$ is the symmetrization map.

Lemma 5.26. Let $(\mathfrak{g}, B)$ be a pair, let $\mathfrak{m}$ and $\mathfrak{m}'$ be Lie subalgebras of $\mathfrak{g}$, and suppose that $(\mathfrak{m}, B)$, $(\mathfrak{m}', B)$, and $(\mathfrak{m} \cap \mathfrak{m}', B)$ are compatible pairs.

(a) The map

$$(5.27) \qquad \Phi : P^{\mathfrak{m}, B}_{\mathfrak{m} \cap \mathfrak{m}', B}(Z) \to P^{\mathfrak{g}, B}_{\mathfrak{m}', B}(Z)$$

given by $\Phi(r \otimes z) = r \otimes z$ is a one-one $(\mathfrak{m}, B)$ map natural for Z in $\mathcal{C}(\mathfrak{m}', B)$.

(b) The map Φ in (5.27) is an $(\mathfrak{m}, B)$ isomorphism if $\mathfrak{m} + \mathfrak{m}' = \mathfrak{g}$.

(c) Composition of Φ with Frobenius reciprocity yields an isomorphism

$$(5.28) \qquad \mathrm{Hom}_{\mathfrak{g}, B}(P^{\mathfrak{g}, B}_{\mathfrak{m}', B}(Z), I^{\mathfrak{g}, B}_{\mathfrak{m}, B}(Z')) \cong \mathrm{Hom}_{\mathfrak{m} \cap \mathfrak{m}', B}(Z, Z')$$

that is natural for Z in $\mathcal{C}(\mathfrak{m}', B)$ and Z' in $\mathcal{C}(\mathfrak{m}, B)$, provided $\mathfrak{m} + \mathfrak{m}' = \mathfrak{g}$.

PROOF.
(a) By definition

$$(5.29a) \qquad P^{\mathfrak{g}, B}_{\mathfrak{m}', B}(Z) = R(\mathfrak{g}, B) \otimes_{R(\mathfrak{m}', B)} Z$$

$$(5.29b) \qquad P^{\mathfrak{m}, B}_{\mathfrak{m} \cap \mathfrak{m}', B}(Z) = R(\mathfrak{m}, B) \otimes_{R(\mathfrak{m} \cap \mathfrak{m}', B)} Z.$$

Since $R(\mathfrak{m} \cap \mathfrak{m}', B) \subseteq R(\mathfrak{m}, B) \subseteq R(\mathfrak{g}, B)$, we have a natural $R(\mathfrak{m}, B)$ map Φ from (5.29b) to (5.29a) given by $r \otimes z \to r \otimes z$. We shall prove Φ is one-one.

Let $\mathfrak{c}$ be an $\mathrm{Ad}(B)$ invariant vector subspace with $\mathfrak{m} = \mathfrak{c} \oplus (\mathfrak{m} \cap \mathfrak{m}')$, and let $\mathfrak{s} \supseteq \mathfrak{c}$ be an $\mathrm{Ad}(B)$ invariant vector subspace with $\mathfrak{g} = \mathfrak{s} \oplus \mathfrak{m}'$. To prove that Φ is one-one, we shall exhibit vector-space isomorphisms

$$\Phi_1 : S(\mathfrak{c}) \otimes_{\mathbb{C}} Z \to (5.29b)$$

and

$$\Phi_2 : S(\mathfrak{s}) \otimes_{\mathbb{C}} Z \to (5.29a)$$

such that $\Phi_2^{-1} \circ \Phi \circ \Phi_1$ is $\iota \otimes 1$, where $\iota : S(\mathfrak{c}) \to S(\mathfrak{s})$ denotes inclusion. In fact, if we regard $S(\mathfrak{c})$ as a trivial $R(\mathfrak{m} \cap \mathfrak{m}', B)$ module, then Φ_1 is the composition

$$\begin{aligned}
S(\mathfrak{c}) \otimes_{\mathbb{C}} Z &\cong S(\mathfrak{c}) \otimes_{\mathbb{C}} (R(\mathfrak{m} \cap \mathfrak{m}', B) \otimes_{R(\mathfrak{m} \cap \mathfrak{m}', B)} Z) &&\text{by (2.18a)} \\
&\cong (S(\mathfrak{c}) \otimes_{\mathbb{C}} R(\mathfrak{m} \cap \mathfrak{m}', B)) \otimes_{R(\mathfrak{m} \cap \mathfrak{m}', B)} Z &&\text{by (C.21)} \\
&\cong R(\mathfrak{m}, B) \otimes_{R(\mathfrak{m} \cap \mathfrak{m}', B)} Z &&\text{by Lemma 2.56}
\end{aligned}$$

and is given explicitly by $\Phi_1(r \otimes z) = (\sigma(r) \otimes \chi_A) \otimes z$ for $r \in S(\mathfrak{c})$ and $z \in Z$; here A is a finite subset of $\widehat{B}$ large enough so that $\chi_A z = z$. The map Φ_2 is given similarly. Since $\sigma(\mathfrak{c}) \subseteq \sigma(\mathfrak{s})$, $\Phi_2^{-1} \circ \Phi \circ \Phi_1(r \otimes z) = r \otimes z$, and thus $\Phi_2^{-1} \circ \Phi \circ \Phi_1 = \iota \otimes 1$ as asserted.

(b) If $\mathfrak{m} + \mathfrak{m}' = \mathfrak{g}$, then we may take $\mathfrak{c} = \mathfrak{s}$ in the proof of (a), by a count of dimensions. Then ι is an isomorphism, and so is $\Phi_2^{-1} \circ \Phi \circ \Phi_1 = \iota \otimes 1$. Hence Φ is an isomorphism.

(c) Frobenius reciprocity for I (Proposition 2.21) gives

$$\mathrm{Hom}_{\mathfrak{g},B}(P_{\mathfrak{m}',B}^{\mathfrak{g},B}(Z), I_{\mathfrak{m},B}^{\mathfrak{g},B}(Z')) \cong \mathrm{Hom}_{\mathfrak{m},B}(P_{\mathfrak{m}',B}^{\mathfrak{g},B}(Z), Z')$$

$$\cong \mathrm{Hom}_{\mathfrak{m},B}(P_{\mathfrak{m} \cap \mathfrak{m}',B}^{\mathfrak{m},B}(Z), Z') \quad \text{by (b)},$$

and this is

$$\cong \mathrm{Hom}_{\mathfrak{m} \cap \mathfrak{m}',B}(Z, Z')$$

by Frobenius reciprocity for P (Proposition 2.34) and Proposition 2.33(i).

The second lemma is an analog for the I functor. Only the counterpart of Lemma 5.26a will be needed, and it will be used later only in the context that $\mathfrak{m} = \mathfrak{k}$ and $\mathfrak{m}' = \mathfrak{q}$.

Lemma 5.30. Let $\mathfrak{m}$ and $\mathfrak{m}'$ be Lie subalgebras of $\mathfrak{g}$, and suppose that $(\mathfrak{m}, B)$, $(\mathfrak{m}', B)$, and $(\mathfrak{m} \cap \mathfrak{m}', B)$ are compatible pairs. Then

(a) the map

$$\Phi : I_{\mathfrak{m}',B}^{\mathfrak{g},B}(Z) \to I_{\mathfrak{m} \cap \mathfrak{m}',B}^{\mathfrak{m},B}(Z)$$

given by restriction $\Phi(\varphi)(r) = \varphi(r)$ from $R(\mathfrak{g}, B)$ to $R(\mathfrak{m}, B)$ is a surjective $(\mathfrak{m}, B)$ map natural for Z in $\mathcal{C}(\mathfrak{m}', B)$

(b) the map Φ in (a) is an $(\mathfrak{m}, B)$ isomorphism if $\mathfrak{m} + \mathfrak{m}' = \mathfrak{g}$.

PROOF.

(a) What needs proof is that Φ is onto. Let $\mathfrak{c}$ be an $\mathrm{Ad}(B)$ invariant vector space with $\mathfrak{m} = \mathfrak{c} \oplus (\mathfrak{m} \cap \mathfrak{m}')$, and let $\mathfrak{s} \supseteq \mathfrak{c}$ be an $\mathrm{Ad}(B)$ invariant vector space with $\mathfrak{g} = \mathfrak{s} \oplus \mathfrak{m}'$. To prove that Φ is onto, we shall exhibit vector-space isomorphisms

$$\Phi_1 : \mathrm{Hom}_{\mathfrak{m}',B}(R(\mathfrak{g}, B), Z)_B \to \mathrm{Hom}_{\mathbb{C}}(S(\mathfrak{s}), Z)_B$$

and

$$\Phi_2 : \mathrm{Hom}_{\mathfrak{m} \cap \mathfrak{m}',B}(R(\mathfrak{m}, B), Z)_B \to \mathrm{Hom}_{\mathbb{C}}(S(\mathfrak{c}), Z)_B$$

such that $\Phi_2 \circ \Phi \circ \Phi_1^{-1}$ is restriction from $S(\mathfrak{s})$ to $S(\mathfrak{c})$, which is onto. In fact, Φ_1 is the composition

$$\text{Hom}_{\mathfrak{m}',B}(R(\mathfrak{g},B),Z)_B$$
$$\cong \text{Hom}_{R(\mathfrak{m}',B)}(S(\mathfrak{s}) \otimes_{\mathbb{C}} R(\mathfrak{m}',B),Z)_B \qquad \text{by Lemma 2.56}$$
$$\cong \text{Hom}_{\mathbb{C}}(S(\mathfrak{s}), \text{Hom}_{R(\mathfrak{m}',B)}(R(\mathfrak{m}',B),Z)_B)_B \quad \text{by (C.20)}$$
$$\cong \text{Hom}_{\mathbb{C}}(S(\mathfrak{s}),Z)_B \qquad\qquad\qquad\qquad \text{by (2.18b)}$$

and is given explicitly by $\Phi_1(\varphi)(s) = \varphi(\sigma(s) \otimes \chi_A)$, where A is large enough so that $\chi_A \varphi = \varphi$. The map Φ_2 is given similarly, and $(\Phi_2 \circ \Phi \circ \Phi_1^{-1})(\varphi)(r) = \varphi(r)$ as asserted.

(b) If $\mathfrak{m} + \mathfrak{m}' = \mathfrak{g}$, then we may take $\mathfrak{c} = \mathfrak{s}$ in the proof of (a), by a count of dimensions. Then $\Phi_2 \circ \Phi \circ \Phi_1^{-1}$ is the identity, and Φ is an isomorphism.

4. Upper Bound on Multiplicities of K Types

For the remainder of the chapter, we return to the setting of a reductive pair and a θ stable parabolic subalgebra as described in §1. In this section we shall analyze $\mathcal{L}_j(Z)$ and $\mathcal{R}^j(Z)$ and give upper bounds for the multiplicities of their K types. The idea will be to try to imitate the theory of §IV.11 as much as possible, trying to push through (4.170) for $\mathcal{L}_j(Z)$ and (4.171) for $\mathcal{R}^j(Z)$. Unfortunately these arguments do not apply in our new wider setting; they merely provide a frame of reference.

It turns out that we can get some information by using K analogs of the arguments with Hom's in §IV.11, and we can get further information by using K analogs of the modules $\mathcal{L}_j(Z)$ and $\mathcal{R}^j(Z)$ themselves. These K analogs accomplish two things: (1) they provide some tools for working with noncompact $(\mathfrak{g}, K)$ and (2) they provide precise information about some basic K types that occur in $\mathcal{L}_j(Z)$ and $\mathcal{R}^j(Z)$ (those in the "bottom layer"). We shall postpone consideration of the bottom layer to §6, concentrating on (1) for the moment. In concentrating on (1), we shall not need to make the assumption (5.7) that L meets every component of G.

To simplify the notation in the definitions of $\mathcal{L}_j(Z)$ and $\mathcal{R}^j(Z)$ a little, we again abbreviate

$$(5.31) \qquad Z_{\bar{\mathfrak{q}}}^{\#} = \mathcal{F}_{\mathfrak{l},L\cap K}^{\bar{\mathfrak{q}},L\cap K}(Z^{\#}) \qquad \text{and} \qquad Z_{\mathfrak{q}}^{\#} = \mathcal{F}_{\mathfrak{l},L\cap K}^{\mathfrak{q},L\cap K}(Z^{\#}).$$

We begin with the K analogs of the arguments with Hom's. Let V be a finite-dimensional $(\mathfrak{k}, K)$ module. In view of §IV.11 the computations we would like to make are

$$\operatorname{Hom}_K(\mathcal{L}_j(Z), V) = \operatorname{Hom}_{\mathfrak{k},K}(\Pi_j(P^{\mathfrak{g},L\cap K}_{\bar{\mathfrak{q}},L\cap K}(Z^{\#}_{\bar{\mathfrak{q}}})), V)$$

$$(5.32a) \qquad \overset{?}{\cong} \operatorname{Ext}^j_{\mathfrak{k},L\cap K}(P^{\mathfrak{g},L\cap K}_{\bar{\mathfrak{q}},L\cap K}(Z^{\#}_{\bar{\mathfrak{q}}}), (\mathcal{F}^{\vee})^{\mathfrak{k},L\cap K}_{\mathfrak{k},K}(V))$$

$$(5.32b) \qquad \overset{?}{\cong} \operatorname{Ext}^j_{\mathfrak{k},L\cap K}(P^{\mathfrak{g},L\cap K}_{\bar{\mathfrak{q}},L\cap K}(Z^{\#}_{\bar{\mathfrak{q}}}), V)$$

$$(5.32c) \qquad \overset{?}{\cong} \operatorname{Ext}^j_{\bar{\mathfrak{q}}\cap\mathfrak{k},L\cap K}(Z^{\#}_{\bar{\mathfrak{q}}}, V)$$

$$(5.32d) \qquad \overset{?}{\cong} \operatorname{Hom}_{L\cap K}(Z^{\#}, H^j(\bar{\mathfrak{u}}\cap\mathfrak{k}, V))$$

and

$$\operatorname{Hom}_K(V, \mathcal{R}^j(Z)) = \operatorname{Hom}_{\mathfrak{k},K}(V, \Gamma^j(I^{\mathfrak{g},L\cap K}_{\mathfrak{q},L\cap K}(Z^{\#}_{\mathfrak{q}})))$$

$$(5.33a) \qquad \overset{?}{\cong} \operatorname{Ext}^j_{\mathfrak{k},L\cap K}(V, I^{\mathfrak{g},L\cap K}_{\mathfrak{q},L\cap K}(Z^{\#}_{\mathfrak{q}}))$$

$$(5.33b) \qquad \overset{?}{\cong} \operatorname{Ext}^j_{\mathfrak{q}\cap\mathfrak{k},L\cap K}(V, Z^{\#}_{\mathfrak{q}})$$

$$(5.33c) \qquad \overset{?}{\cong} \operatorname{Hom}_{L\cap K}(H_j(\mathfrak{u}\cap\mathfrak{k}, V), Z^{\#}).$$

Steps (5.32a) and (5.33a) are actually valid. With (5.32a) the argument is that

$$\mathcal{F}^{\mathfrak{k},K}_{\mathfrak{g},K}\mathcal{L}_j(Z) = \mathcal{F}^{\mathfrak{k},K}_{\mathfrak{g},K}\Pi_j(P^{\mathfrak{g},L\cap K}_{\bar{\mathfrak{q}},L\cap K}(Z^{\#}_{\bar{\mathfrak{q}}})) \cong (P^{\mathfrak{k},K}_{\mathfrak{k},L\cap K})_j\mathcal{F}^{\mathfrak{k},L\cap K}_{\mathfrak{g},L\cap K}(P^{\mathfrak{g},L\cap K}_{\bar{\mathfrak{q}},L\cap K}(Z^{\#}_{\bar{\mathfrak{q}}}))$$

by Proposition 2.115. The functors

$$\operatorname{Hom}_K(\,\cdot\,, V) \circ P^{\mathfrak{k},K}_{\mathfrak{k},L\cap K} \qquad \text{and} \qquad \operatorname{Hom}_{\mathfrak{k},L\cap K}(\,\cdot\,, (\mathcal{F}^{\vee})^{\mathfrak{k},L\cap K}_{\mathfrak{k},K}(V))$$

are naturally isomorphic by Frobenius reciprocity (Proposition 2.34), and hence their derived functors are naturally isomorphic. Since $\operatorname{Hom}_K(\,\cdot\,, V)$ is exact by Lemma 2.4 and (C.14), (C.27a) and Proposition 2.117 give

$$\operatorname{Hom}_K(\,\cdot\,, V) \circ (P^{\mathfrak{k},K}_{\mathfrak{k},L\cap K})_j \cong (\operatorname{Hom}_K(\,\cdot\,, V) \circ P^{\mathfrak{k},K}_{\mathfrak{k},L\cap K})_j$$

$$\cong \operatorname{Hom}_{\mathfrak{k},L\cap K}(\,\cdot\,, (\mathcal{F}^{\vee})^{\mathfrak{k},L\cap K}_{\mathfrak{k},K}(V))_j$$

$$\cong \operatorname{Ext}^j_{\mathfrak{k},L\cap K}(\,\cdot\,, (\mathcal{F}^{\vee})^{\mathfrak{k},L\cap K}_{\mathfrak{k},K}(V)).$$

Therefore

$$\operatorname{Hom}_K(\mathcal{F}_{\mathfrak{g},K}^{\mathfrak{k},K}\mathcal{L}_j(Z), V)$$
$$\cong \operatorname{Hom}_K((P_{\mathfrak{k},L\cap K}^{\mathfrak{k},K})_j(\mathcal{F}_{\mathfrak{g},L\cap K}^{\mathfrak{k},L\cap K}(P_{\bar{\mathfrak{q}},L\cap K}^{\mathfrak{g},L\cap K}(Z_{\bar{\mathfrak{q}}}^{\#}))), V)$$
$$\cong \operatorname{Ext}_{\mathfrak{k},L\cap K}^{j}(\mathcal{F}_{\mathfrak{g},L\cap K}^{\mathfrak{k},L\cap K}(P_{\bar{\mathfrak{q}},L\cap K}^{\mathfrak{g},L\cap K}(Z_{\bar{\mathfrak{q}}}^{\#})), (\mathcal{F}^{\vee})_{\mathfrak{k},K}^{\mathfrak{k},L\cap K}(V)).$$

Suppressing forgetful functors, we obtain (5.32a). The proof of (5.33a) is similar.

Isomorphism (5.32b) is immediate from Proposition 2.33. It is (5.32c) and (5.33b) that are not generally valid. However, (5.32d) and (5.33c) are valid in all cases; these are results about the compact case and follow from (4.156d) and (4.159c). The critical facts that are used in their proofs are that $\bar{u}$ acts trivially in $Z_{\bar{\mathfrak{q}}}^{\#}$ and $\mathfrak{u}$ acts trivially in $Z_{\mathfrak{q}}^{\#}$.

In short, (5.32) is valid except at step (c), and (5.33) is valid except at step (b). In Theorem 5.35 below, it is these steps that will occupy most of our attention.

To obtain substitutes for (5.32c) and (5.33b), we shall exploit the long exact sequences for $\operatorname{Ext}_{\mathfrak{k},L\cap K}^{j}$. The technique involves filtering $P_{\bar{\mathfrak{q}},L\cap K}^{\mathfrak{g},L\cap K}(Z_{\bar{\mathfrak{q}}}^{\#})$ and $I_{\mathfrak{q},L\cap K}^{\mathfrak{g},L\cap K}(Z_{\mathfrak{q}}^{\#})$ by subspaces with actions by $\mathfrak{k}$ and $L\cap K$. For this purpose, let $n \geq 0$ and recall that $U_n(\mathfrak{g})$ is the subspace of all elements of $U(\mathfrak{g})$ of degree $\leq n$. Define

$$U_{n,\mathfrak{k}}(\mathfrak{g}) = U(\mathfrak{k})U_n(\mathfrak{g})$$
$$= \text{span in } U(\mathfrak{g}) \text{ of all } uX_1 \cdots X_r \text{ with } r \leq n,\ X_i \in \mathfrak{p},\ u \in U(\mathfrak{k}).$$

The spaces $U_{n,\mathfrak{k}}(\mathfrak{g})$ are $\mathfrak{k}$ modules under left multiplication and are $L\cap K$ modules under adjoint. Define $U_{n,\mathfrak{k}}(\mathfrak{u})$ and $U_{n,\mathfrak{k}}(\bar{\mathfrak{u}})$ similarly, e.g.,

$$U_{n,\mathfrak{k}}(\mathfrak{u}) = U(\mathfrak{u}\cap\mathfrak{k})U_n(\mathfrak{u})$$
$$= \text{span of all } uX_1 \cdots X_r \text{ with } r \leq n,\ X_i \in \mathfrak{u}\cap\mathfrak{p},\ u \in U(\mathfrak{u}\cap\mathfrak{k});$$

$U_{n,\mathfrak{k}}(\mathfrak{u})$ and $U_{n,\mathfrak{k}}(\bar{\mathfrak{u}})$ are $(\mathfrak{l}\cap\mathfrak{k}, L\cap K)$ modules under adjoint. We let $U_{-1,\mathfrak{k}}(\mathfrak{u}) = U_{-1,\mathfrak{k}}(\bar{\mathfrak{u}}) = U_{-1,\mathfrak{k}}(\mathfrak{g}) = 0$.

Lemma 5.34. For $n \geq 0$,

$$U_{n,\mathfrak{k}}(\mathfrak{g}) \subseteq U_{n,\mathfrak{k}}(\mathfrak{u}) \otimes_{\mathbb{C}} U(\bar{\mathfrak{q}})$$

and

$$U_{n,\mathfrak{k}}(\mathfrak{g}) \subseteq U_{n,\mathfrak{k}}(\bar{\mathfrak{u}}) \otimes_{\mathbb{C}} U(\mathfrak{q})$$

if the right sides are identified with subsets of $U(\mathfrak{g})$ via the multiplication maps.

PROOF. For the first identity we have

$$
\begin{aligned}
U_{n,\mathfrak{k}} &= U(\mathfrak{k})U_n(\mathfrak{g}) \\
&= U(\mathfrak{u}\cap\mathfrak{k})U(\bar{\mathfrak{q}}\cap\mathfrak{k})U_n(\mathfrak{g}) \\
&= U(\mathfrak{u}\cap\mathfrak{k})U_n(\mathfrak{g})U(\bar{\mathfrak{q}}\cap\mathfrak{k}) \qquad \text{since } U_n(\mathfrak{g}) \text{ is stable under ad} \\
&\subseteq U(\mathfrak{u}\cap\mathfrak{k})U_n(\mathfrak{u})U_n(\bar{\mathfrak{q}})U(\bar{\mathfrak{q}}\cap\mathfrak{k}) \\
&\subseteq U(\mathfrak{u}\cap\mathfrak{k})U_n(\mathfrak{u})U(\bar{\mathfrak{q}}) \\
&= U_{n,\mathfrak{k}}(\mathfrak{u})U(\bar{\mathfrak{q}}).
\end{aligned}
$$

The second identity follows by applying bar to the first.

Theorem 5.35.

(a) For $j > S$, $\mathcal{L}_j(Z) = 0$. For $0 \le j \le S$ and for any finite-dimensional $(\mathfrak{k}, K)$ module V,

$$(5.36a) \quad \dim\operatorname{Hom}_K(\mathcal{L}_j(Z), V)$$

$$\le \sum_{n=0}^{\infty} \dim\operatorname{Hom}_{L\cap K}(S^n(\mathfrak{u}\cap\mathfrak{p}) \otimes_{\mathbb{C}} Z^{\#}, \, H^j(\bar{\mathfrak{u}}\cap\mathfrak{k}, V)).$$

Here S^n refers to the symmetric tensors homogeneous of degree n.

(b) For $j > S$, $\mathcal{R}^j(Z) = 0$. For $0 \le j \le S$ and for any finite-dimensional $(\mathfrak{k}, K)$ module V,

$$(5.36b) \quad \dim\operatorname{Hom}_K(V, \mathcal{R}^j(Z))$$

$$\le \sum_{n=0}^{\infty} \dim\operatorname{Hom}_{L\cap K}(H_j(\mathfrak{u}\cap\mathfrak{k}, V), \, S^n(\mathfrak{u}\cap\mathfrak{p}) \otimes_{\mathbb{C}} Z^{\#}).$$

REMARKS.

1) The motivation is clearer for (b) than for (a) and comes from the analytic analog of the Γ functor mentioned in the Introduction. The filtration M_n of M in the proof of (b) given below is an analog of the "order-of-vanishing filtration" in the analytic setting. In the analogy with sheaves on G/L, M_n plays the role of germs of sections vanishing to order n along $K/(L\cap K)$. The proof of (a) has no analytic counterpart and should be regarded as a dualized version of the proof of (b); we include the argument since it is not completely obvious how to dualize the proof of (b).

2) The inequalities in (5.36a) and (5.36b) can be strict. An example will be given in §7.

3) A certain part of the proof of each of (a) and (b) can be reformulated as a spectral sequence argument. We return to this point in §9.

PROOF. We begin with some preliminary facts. As we saw in connection with (5.32) and (5.33),

$$(5.37a) \qquad \operatorname{Hom}_K(\mathcal{L}_j(Z), V) \cong \operatorname{Ext}^j_{\mathfrak{k}, L \cap K}(P^{\mathfrak{g}, L \cap K}_{\bar{\mathfrak{q}}, L \cap K}(Z^{\#}_{\bar{\mathfrak{q}}}), V)$$

and

$$(5.37b) \qquad \operatorname{Hom}_K(V, \mathcal{R}^j(Z)) \cong \operatorname{Ext}^j_{\mathfrak{k}, L \cap K}(V, I^{\mathfrak{g}, L \cap K}_{\mathfrak{q}, L \cap K}(Z^{\#}_{\mathfrak{q}})).$$

By Lemma 5.11 the element $h_{\delta(\mathfrak{u})}$ of $i \mathfrak{t}_0$ is fixed by $\operatorname{Ad}(L \cap K)$. Therefore $h_{\delta(\mathfrak{u})}$ acts as a scalar in every $L \cap K$ type, and $Z^{\#}$ is fully reducible under $h_{\delta(\mathfrak{u})}$. Since $h_{\delta(\mathfrak{u})}$ commutes with $\mathfrak{l}$ (also by Lemma 5.11), $Z^{\#}$ splits into the finite or countable direct sum of its eigenspaces under $h_{\delta(\mathfrak{u})}$, each of which is an $(\mathfrak{l}, L \cap K)$ module.

In the case of $\mathcal{L}_j(Z)$, since V is finite dimensional, the Ext on the right of (5.37a) splits as a corresponding sum, and the Hom's on the right of (5.36a) split as a corresponding sum. Thus we may assume from the outset that $h_{\delta(\mathfrak{u})}$ acts as a scalar in $Z^{\#}$. In the case of $\mathcal{R}^j(Z)$, similar remarks apply to (5.37b) and (5.36b), and again we may assume from the outset that $h_{\delta(\mathfrak{u})}$ acts as a scalar in $Z^{\#}$.

Finally we shall use one more conclusion from Lemma 5.11: $\operatorname{ad} h_{\delta(\mathfrak{u})}$ acts fully reducibly on $\mathfrak{u}$, and all the eigenvalues are $\geq \varepsilon > 0$.

Now we are ready to prove (b). Define a $(\mathfrak{g}, L \cap K)$ module M by

$$M = I^{\mathfrak{g}, L \cap K}_{\mathfrak{q}, L \cap K}(Z^{\#}_{\mathfrak{q}}) \cong \operatorname{pro}^{\mathfrak{g}, L \cap K}_{\mathfrak{q}, L \cap K}(Z^{\#}_{\mathfrak{q}}) = \operatorname{Hom}_{\mathfrak{q}}(U(\mathfrak{g}), Z^{\#}_{\mathfrak{q}})_{L \cap K},$$

the isomorphism following from Proposition 2.57b. Here exceptionally we regard the $\mathfrak{q}$ invariance in the right member as meaning that $\varphi(uX) = -X(\varphi(u))$ for $u \in U(\mathfrak{g})$ and $X \in \mathfrak{q}$. Then $U(\mathfrak{g})$ acts exceptionally on the left side of the $U(\mathfrak{g})$ variable, the action being $u_0 \varphi(u) = \varphi(u^t_0 u)$. (This realization is isomorphic to the usual realization under the map that takes the transpose of the $U(\mathfrak{g})$ variable.) As usual with the pro functor, $L \cap K$ acts in both variables of $\operatorname{Hom}_{\mathfrak{q}}$, as in (2.13b). From the decomposition $\mathfrak{g} = \bar{\mathfrak{u}} \oplus \mathfrak{q}$, we have

$$M \cong \operatorname{Hom}_{\mathfrak{q}}(U(\mathfrak{g}), Z^{\#}_{\mathfrak{q}})_{L \cap K}$$
$$\cong \operatorname{Hom}_{\mathfrak{q}}(U(\bar{\mathfrak{u}}) \otimes_{\mathbb{C}} U(\mathfrak{q}), Z^{\#}_{\mathfrak{q}})_{L \cap K}$$
$$\cong \operatorname{Hom}_{\mathbb{C}}(U(\bar{\mathfrak{u}}), \operatorname{Hom}_{\mathfrak{q}}(U(\mathfrak{q}), Z^{\#}_{\mathfrak{q}}))_{L \cap K}$$
$$\cong \operatorname{Hom}_{\mathbb{C}}(U(\bar{\mathfrak{u}}), Z^{\#}_{\mathfrak{q}})_{L \cap K}.$$

For $n \geq -1$, let

$$(5.38a) \qquad M_n = \{\varphi \in M \mid \varphi|_{U_{n, \mathfrak{k}}(\mathfrak{g})} = 0\}$$

as a $(\mathfrak{k}, L \cap K)$ submodule of M. Since we defined $U_{-1,\mathfrak{k}}(\mathfrak{g})$ to be 0, $M_{-1} = M$. To see that M_n is a $(\mathfrak{k}, L \cap K)$ submodule of M, recall that $U_{n,\mathfrak{k}}(\mathfrak{g})$ is stable under left multiplication by $\mathfrak{k}$ in $U(\mathfrak{g})$. For this reason the action of $\mathfrak{k}$ on M preserves M_n. Similarly the fact that $U_{n,\mathfrak{k}}(\mathfrak{g})$ is stable under the action of $L \cap K$ makes M_n an $L \cap K$ submodule of M.

Let us see that

$$(5.38b) \qquad M_n \cong \mathrm{Hom}_{\mathbb{C}}(U(\bar{\mathfrak{u}})/U_{n,\mathfrak{k}}(\bar{\mathfrak{u}}), \, Z_{\mathfrak{q}}^{\#})_{L \cap K}$$

as $(\mathfrak{l} \cap \mathfrak{k}, L \cap K)$ modules, with the action by $L \cap K$ given in both variables as in (2.13b) The isomorphism is defined by restricting $\varphi \in M_n$ to $U(\bar{\mathfrak{u}})$ and writing φ as a map of $U(\bar{\mathfrak{u}})/U_{n,\mathfrak{k}}(\bar{\mathfrak{u}})$ since $\varphi|_{U_{n,\mathfrak{k}}(\bar{\mathfrak{u}})} = 0$. In fact, Lemma 5.34 gives

$$\{\varphi \in M \mid \varphi|_{U_{n,\mathfrak{k}}(\bar{\mathfrak{u}})} = 0\} \supseteq \{\varphi \in M \mid \varphi|_{U_{n,\mathfrak{k}}(\mathfrak{g})} = 0\} = M_n$$
$$\supseteq \{\varphi \in M \mid \varphi|_{U_{n,\mathfrak{k}}(\bar{\mathfrak{u}}) \otimes_{\mathbb{C}} U(\mathfrak{q})} = 0\}$$
$$\supseteq \{\varphi \in M \mid \varphi|_{U_{n,\mathfrak{k}}(\bar{\mathfrak{u}})} = 0\},$$

the last step following since φ is a $U(\mathfrak{q})$ map. The end members here are equal, and hence

$$M_n \cong \{\varphi \in \mathrm{Hom}_{\mathbb{C}}(U(\bar{\mathfrak{u}}), Z_{\mathfrak{q}}^{\#})_{L \cap K} \mid \varphi|_{U_{n,\mathfrak{k}}(\bar{\mathfrak{u}})} = 0\}.$$

Then (5.38b) follows.

Differentiating the $L \cap K$ action in (5.38b), we see that $\mathfrak{l} \cap \mathfrak{k}$ acts by Hom of the adjoint action into the action on $Z_{\mathfrak{q}}^{\#}$. Here each member of $U(\bar{\mathfrak{u}})/U_{n,\mathfrak{k}}(\bar{\mathfrak{u}})$ is represented by a sum of monomials involving at least $n + 1$ factors from $\bar{\mathfrak{u}}$, and thus it has only eigenvalues $\leq -(n+1)\varepsilon$ under $\mathrm{ad}\, h_{\delta(\mathfrak{u})}$. Since $h_{\delta(\mathfrak{u})}$ acts as a scalar a in $Z_{\mathfrak{q}}^{\#}$, the eigenvalues of $h_{\delta(\mathfrak{u})}$ on M_n are $\geq (n+1)\varepsilon + a$ and therefore tend to $+\infty$ with n. Now

$$\mathrm{Ext}_{\mathfrak{k},L \cap K}^{j}(V, M_n) \cong H^j(\mathfrak{k}, L \cap K; \mathrm{Hom}_{\mathbb{C}}(V, M_n)_{L \cap K})$$

may be computed from the complex

$$\mathrm{Hom}_{L \cap K}(\textstyle\bigwedge^{j}(\mathfrak{k}/(\mathfrak{l} \cap \mathfrak{k})), \mathrm{Hom}_{\mathbb{C}}(V, M_n)_{L \cap K}),$$

and the cochains must respect eigenvalues under the action by $h_{\delta(\mathfrak{u})}$. Since M_n has only large eigenvalues for large n, we see that

$$(5.39) \qquad\qquad \mathrm{Ext}_{\mathfrak{k},L \cap K}^{j}(V, M_N) = 0$$

for some N and all j. Plugging (5.39) into the long exact sequence of (C.36c) for $\operatorname{Ext}^j_{\mathfrak{k}, L\cap K}(V, \cdot)$, namely

$$\cdots \longrightarrow \operatorname{Ext}^{j-1}_{\mathfrak{k}, L\cap K}(V, M/M_N) \longrightarrow \operatorname{Ext}^j_{\mathfrak{k}, L\cap K}(V, M_N)$$

$$\longrightarrow \operatorname{Ext}^j_{\mathfrak{k}, L\cap K}(V, M) \longrightarrow \operatorname{Ext}^j_{\mathfrak{k}, L\cap K}(V, M/M_N) \longrightarrow \cdots,$$

we conclude that

$$(5.40) \qquad \operatorname{Ext}^j_{\mathfrak{k}, L\cap K}(V, M) \cong \operatorname{Ext}^j_{\mathfrak{k}, L\cap K}(V, M/M_N)$$

for this N and all j.

From the long exact sequence for $\operatorname{Ext}(V, \cdot)$, it follows trivially that

$$(5.41) \qquad \dim \operatorname{Ext}^j(V, A) \le \dim \operatorname{Ext}^j(V, A/B) + \dim \operatorname{Ext}^j(V, B).$$

If C is a submodule of B, we can apply this inequality similarly to $\operatorname{Ext}^j(V, B)$ and proceed inductively. The result is a subadditivity formula for filtrations. If A has a finite filtration, then $\dim \operatorname{Ext}^j(V, A)$ is $\le$ the sum of $\dim \operatorname{Ext}^j$ for the various subquotients. Applying this fact to the filtration

$$M_N \subseteq M_{N-1} \subseteq \cdots \subseteq M_0 \subseteq M_{-1} = M$$

and using (5.39), we obtain

$$\dim \operatorname{Ext}^j_{\mathfrak{k}, L\cap K}(V, M) \le \sum_{n=0}^{N} \dim \operatorname{Ext}^j_{\mathfrak{k}, L\cap K}(V, M_{n-1}/M_n)$$

$$(5.42) \qquad\qquad = \sum_{n=0}^{\infty} \dim \operatorname{Ext}^j_{\mathfrak{k}, L\cap K}(V, M_{n-1}/M_n).$$

Next let us establish the $(\mathfrak{k}, L \cap K)$ isomorphism

$$(5.43) \qquad M_{n-1}/M_n \cong \operatorname{Hom}_{\mathfrak{q}\cap\mathfrak{k}}(U(\mathfrak{k}), \operatorname{Hom}_{\mathbb{C}}(S^n(\mathfrak{g}/(\mathfrak{q}+\mathfrak{k})), Z^{\#}_{\mathfrak{q}}))_{L\cap K}.$$

Here $\mathfrak{q} \cap \mathfrak{k}$ acts exceptionally on the right of $U(\mathfrak{k})$, and it acts on the inner Hom by means of the adjoint action on the symmetric algebra and the natural action on $Z^{\#}_{\mathfrak{q}}$. To establish (5.43) we begin with the map

$$(5.44a) \qquad M_{n-1} \to \operatorname{Hom}_{\mathbb{C}}(U(\mathfrak{k}), \operatorname{Hom}_{\mathbb{C}}(T^n(\mathfrak{g}), Z^{\#}_{\mathfrak{q}}))$$

given by $\varphi \mapsto \Phi$ with

$$(5.44b) \qquad \Phi(u)(X_1 \cdots X_n) = \varphi(uX_1 \cdots X_n) \qquad \text{for } u \in U(\mathfrak{k}), X_i \in \mathfrak{g}.$$

(Here $T^n(\mathfrak{g})$ is the nth tensor power of $\mathfrak{g}$.) We make the right side of (5.44) into a $\mathfrak{k}$ module by the left action on $U(\mathfrak{k})$, and we make $L \cap K$ act by the adjoint actions on $U(\mathfrak{k})$ and $T(\mathfrak{g})$ and by the given action on $Z_{\mathfrak{q}}^{\#}$. Then (5.44) is easily seen to respect the actions of $\mathfrak{k}$ and $L \cap K$. The image of M_{n-1} must therefore be contained in the $L \cap K$ finite part of the right side.

To pass from $T^n(\mathfrak{g})$ in (5.44) to $S^n(\mathfrak{g})$, we check that $\Phi(u)$ is independent of the order of the X_i. In fact,

$$\varphi(uX_1 \cdots X_n) - \varphi(uX_1 \cdots X_{i-1}X_{i+1}X_iX_{i+2} \cdots X_n)$$
$$= \varphi(uX_1 \cdots X_{i-1}[X_i, X_{i+1}]X_{i+2} \cdots X_n).$$

The argument of φ on the right side belongs to $U_{n-1,\mathfrak{k}}(\mathfrak{g})$. Since φ is in M_{n-1}, the right side is 0. It follows that (5.44) provides a well-defined map

$$(5.45) \qquad M_{n-1} \to \operatorname{Hom}_{\mathbb{C}}(U(\mathfrak{k}), \operatorname{Hom}_{\mathbb{C}}(S^n(\mathfrak{g}), Z_{\mathfrak{q}}^{\#}))_{L \cap K}.$$

Next we shall show that $\Phi(u)(X_1 \cdots X_n) = 0$ if one of the X_i belongs to $\mathfrak{k}$ or to $\mathfrak{q}$. For the case of $\mathfrak{k}$, we may assume from the symmetry of $\Phi(u)$ in its n arguments that X_1 is in $\mathfrak{k}$. Then

$$\Phi(u)(X_1 \cdots X_n) = \varphi((uX_1)X_2 \cdots X_n).$$

The argument of φ on the right side belongs to $U_{n-1,\mathfrak{k}}(\mathfrak{g})$. Since φ is in M_{n-1}, the right side is 0. For the case of $\mathfrak{q}$, we may assume that X_n is in $\mathfrak{q}$. Then

$$\Phi(u)(X_1 \cdots X_n) = \varphi((uX_1 \cdots X_{n-1})X_n) = -X_n\varphi(uX_1 \cdots X_{n-1})$$

since φ is a $U(\mathfrak{q})$ map. The argument of φ on the right side belongs to $U_{n-1,\mathfrak{k}}(\mathfrak{g})$. Since φ is in M_{n-1}, the right side is 0. It follows that (5.45) may be reinterpreted as a map

$$(5.46) \qquad M_{n-1} \to \operatorname{Hom}_{\mathbb{C}}(U(\mathfrak{k}), \operatorname{Hom}_{\mathbb{C}}(S^n(\mathfrak{g}/(\mathfrak{q} + \mathfrak{k})), Z_{\mathfrak{q}}^{\#}))_{L \cap K}.$$

Finally we show that Φ respects the $\mathfrak{q} \cap \mathfrak{k}$ actions. To do so, suppose we are given $u \in U(\mathfrak{k})$, $X \in \mathfrak{q} \cap \mathfrak{k}$, and $X_i \in \mathfrak{g}$. Then

$$\begin{aligned}
\Phi(uX)(X_1 \cdots X_n) &= \varphi(uXX_1 \cdots X_n) \\
&= \varphi(u(\operatorname{ad} X)(X_1 \cdots X_n)) + \varphi(uX_1 \cdots X_nX) \\
&= \varphi(u(\operatorname{ad} X)(X_1 \cdots X_n)) - X\varphi(uX_1 \cdots X_n) \\
&= \Phi(u)(\operatorname{ad} X)(X_1 \cdots X_n)) - X(\Phi(u)(X_1 \cdots X_n)) \\
&= (X\Phi(u))(X_1 \cdots X_n),
\end{aligned}$$

as we wished to show.

So (5.46) is actually a map

(5.47) $M_{n-1} \to \text{Hom}_{\mathfrak{q} \cap \mathfrak{k}}(U(\mathfrak{k}), \text{Hom}_{\mathbb{C}}(S^n(\mathfrak{g}/(\mathfrak{q} + \mathfrak{k})), Z_{\mathfrak{q}}^{\#}))_{L \cap K}.$

From the original definition (5.44), it is clear that the kernel of (5.47) is precisely M_n. Thus (5.47) descends to a one-one map of $(\mathfrak{k}, L \cap K)$ modules from left to right in (5.43).

To see that the map from left to right in (5.43) is onto, notice that

(5.48) $\mathfrak{g}/(\mathfrak{q} + \mathfrak{k}) \cong \overline{\mathfrak{u}} \cap \mathfrak{p}$ as an $L \cap K$ module.

Hence (5.38b) provides an isomorphism of $L \cap K$ modules

M_{n-1}/M_n

$\cong \text{Hom}_{\mathbb{C}}(U_{n,\mathfrak{k}}(\overline{\mathfrak{u}})/U_{n-1,\mathfrak{k}}(\overline{\mathfrak{u}}), Z_{\mathfrak{q}}^{\#})_{L \cap K}$

$\cong \text{Hom}_{\mathbb{C}}(U(\overline{\mathfrak{u}} \cap \mathfrak{k}) \otimes_{\mathbb{C}} S^n(\overline{\mathfrak{u}} \cap \mathfrak{p}), Z_{\mathfrak{q}}^{\#})_{L \cap K}$

$\cong \text{Hom}_{\mathbb{C}}(U(\overline{\mathfrak{u}} \cap \mathfrak{k}), \text{Hom}_{\mathbb{C}}(S^n(\overline{\mathfrak{u}} \cap \mathfrak{p}), Z_{\mathfrak{q}}^{\#}))_{L \cap K}$

$\cong \text{Hom}_{\mathbb{C}}(U(\overline{\mathfrak{u}} \cap \mathfrak{k}), \text{Hom}_{\mathbb{C}}(S^n(\mathfrak{g}/(\mathfrak{q} + \mathfrak{k})), Z_{\mathfrak{q}}^{\#}))_{L \cap K}$

$\cong \text{Hom}_{\mathfrak{q} \cap \mathfrak{k}}(U(\overline{\mathfrak{u}} \cap \mathfrak{k}) \otimes_{\mathbb{C}} U(\mathfrak{q} \cap \mathfrak{k}), \text{Hom}_{\mathbb{C}}(S^n(\mathfrak{g}/(\mathfrak{q} + \mathfrak{k})), Z_{\mathfrak{q}}^{\#}))_{L \cap K}$

$\cong \text{Hom}_{\mathfrak{q} \cap \mathfrak{k}}(U(\mathfrak{k}), \text{Hom}_{\mathbb{C}}(S^n(\mathfrak{g}/(\mathfrak{q} + \mathfrak{k})), Z_{\mathfrak{q}}^{\#}))_{L \cap K}.$

These $L \cap K$ isomorphisms are compatible with the map (5.43). Thus the $(\mathfrak{k}, L \cap K)$ map (5.43) is an isomorphism as well.

Combining (5.37b), (5.42), and (5.43) and using Proposition 2.57b for $\mathfrak{k}$ instead of $\mathfrak{g}$, we obtain

$\dim \text{Hom}_K(V, \mathcal{R}^j(Z))$

$\leq \sum_{n=0}^{\infty} \dim \text{Ext}^j_{\mathfrak{k}, L \cap K}(V, I^{\mathfrak{k}, L \cap K}_{\mathfrak{q} \cap \mathfrak{k}, L \cap K}(\text{Hom}_{\mathbb{C}}(S^n(\mathfrak{g}/(\mathfrak{q} + \mathfrak{k})), Z_{\mathfrak{q}}^{\#})))$

for the nontrivial action of $\mathfrak{u} \cap \mathfrak{k}$ on $S^n(\mathfrak{g}/(\mathfrak{q} + \mathfrak{k}))$. By Shapiro's Lemma (Theorem 2.119), the right side here is

(5.49) $= \sum_{n=0}^{\infty} \dim \text{Ext}^j_{\mathfrak{q} \cap \mathfrak{k}, L \cap K}(V, \text{Hom}_{\mathbb{C}}(S^n(\mathfrak{g}/(\mathfrak{q} + \mathfrak{k})), Z_{\mathfrak{q}}^{\#})).$

The $(\mathfrak{q} \cap \mathfrak{k}, L \cap K)$ module $S^n(\mathfrak{g}/(\mathfrak{q} + \mathfrak{k}))$ is filtered according to the eigenvalues of $\text{ad}\, h_{\delta(\mathfrak{u})}$, since $\mathfrak{u} \cap \mathfrak{k}$ increases eigenvalues. We apply our subadditivity formula for Ext (the one extending (5.41)) to (5.49): The respective quotients have 0 action by $\mathfrak{u} \cap \mathfrak{k}$, and we can use (5.48) to reassemble them as a direct sum and see that (5.49) is

$$(5.50) \qquad \leq \sum_{n=0}^{\infty} \dim \mathrm{Ext}^j_{\mathfrak{q} \cap \mathfrak{k}, L \cap K}(V, \mathrm{Hom}_{\mathbb{C}}(S^n(\bar{\mathfrak{u}} \cap \mathfrak{p}), Z^{\#}_{\bar{\mathfrak{q}}}))$$

with trivial $\mathfrak{u} \cap \mathfrak{k}$ action on $S^n(\bar{\mathfrak{u}} \cap \mathfrak{p})$.

We have now concluded the step in the argument that substitutes for (5.33a), and we have incorporated the analog of (5.33b). To complete the proof, we use the established isomorphism (5.33c) to obtain

$$\mathrm{Ext}^j_{\mathfrak{q} \cap \mathfrak{k}, L \cap K}(V, W) \cong \mathrm{Hom}_{\mathfrak{l} \cap \mathfrak{k}, L \cap K}(H_j(\mathfrak{u} \cap \mathfrak{k}, V), W)$$

for any $(\mathfrak{q} \cap \mathfrak{k}, L \cap K)$ module W for which $\mathfrak{u} \cap \mathfrak{k}$ acts trivially. Applying this isomorphism with $W = \mathrm{Hom}_{\mathbb{C}}(S^n(\bar{\mathfrak{u}} \cap \mathfrak{p}), Z^{\#}_{\bar{\mathfrak{q}}})$, we obtain

$$\dim \mathrm{Hom}_K(V, \mathcal{R}^j(Z))$$
$$\leq \sum_{n=0}^{\infty} \dim \mathrm{Hom}_{L \cap K}(H_j(\mathfrak{u} \cap \mathfrak{k}, V), \mathrm{Hom}_{\mathbb{C}}(S^n(\bar{\mathfrak{u}} \cap \mathfrak{p}), Z^{\#})).$$

Since $\bar{\mathfrak{u}}^* \cong \mathfrak{u}$ as $(\mathfrak{l}, L \cap K)$ modules, we have

$$\mathrm{Hom}_{\mathbb{C}}(S^n(\bar{\mathfrak{u}} \cap \mathfrak{p}), Z^{\#}) \cong S^n(\bar{\mathfrak{u}} \cap \mathfrak{p})^* \otimes_{\mathbb{C}} Z^{\#} \cong S^n(\mathfrak{u} \cap \mathfrak{p}) \otimes_{\mathbb{C}} Z^{\#}.$$

Then (5.36b) follows. If $j > S$, then $H_j(\mathfrak{u} \cap \mathfrak{k}, V) = 0$, and the right side of (5.36b) is 0. Since $\mathcal{R}^j(Z)$ is fully reducible as a $(\mathfrak{k}, K)$ module and since V can be an arbitrary irreducible $(\mathfrak{k}, K)$ module, it follows that $\mathcal{R}^j(Z) = 0$ for $j > S$. This completes the proof of (b) in the theorem.

The proof of (a) is somewhat dual to the above argument, but the proof of the analog of (5.43) deserves some explanation. We outline the steps for (a), giving details only for the analog of (5.43). Put

$$M = P^{\mathfrak{g}, L \cap K}_{\bar{\mathfrak{q}}, L \cap K}(Z^{\#}_{\bar{\mathfrak{q}}}) \cong \mathrm{ind}^{\mathfrak{g}, L \cap K}_{\bar{\mathfrak{q}}, L \cap K}(Z^{\#}_{\bar{\mathfrak{q}}}) = U(\mathfrak{g}) \otimes_{U(\bar{\mathfrak{q}})} Z^{\#}_{\bar{\mathfrak{q}}},$$

the isomorphism following from Proposition 2.57a. For $n \geq -1$, let

$$(5.51\mathrm{a}) \qquad M_n = \text{image in } M \text{ of } U_{n,\mathfrak{k}}(\mathfrak{g}) \otimes_{\mathbb{C}} Z^{\#}_{\bar{\mathfrak{q}}}$$

as a $(\mathfrak{k}, L \cap K)$ module. We check using Lemma 5.34 that

$$(5.51\mathrm{b}) \qquad M_n \cong U_{n,\mathfrak{k}}(\mathfrak{u}) \otimes_{\mathbb{C}} Z^{\#}_{\bar{\mathfrak{q}}}$$

as $(\mathfrak{l} \cap \mathfrak{k}, L \cap K)$ modules; the isomorphism embeds the right side into $U(\mathfrak{g}) \otimes_{\mathbb{C}} Z_{\bar{\mathfrak{q}}}^{\#}$ and then passes to the quotient. From (5.51b) we obtain

$$M/M_n \cong (U(\mathfrak{u})/U_{n,\mathfrak{k}}(\mathfrak{u})) \otimes_{\mathbb{C}} Z_{\bar{\mathfrak{q}}}^{\#},$$

and $\mathfrak{l} \cap \mathfrak{k}$ acts by the tensor-product action, with ad in the first factor. This formula enables us to see that the eigenvalues of $\operatorname{ad} h_{\delta(\mathfrak{u})}$ on M/M_n tend to $+\infty$ with n. Consequently

$$(5.52) \qquad \operatorname{Ext}^j_{\mathfrak{k},L\cap K}(M/M_N, V) = 0$$

for some N and all j. Plugging (5.52) into the long exact sequence of (C.36a) for $\operatorname{Ext}^j_{\mathfrak{k},L\cap K}(\,\cdot\,, V)$, we conclude that

$$(5.53) \qquad \operatorname{Ext}^j_{\mathfrak{k},L\cap K}(M, V) \cong \operatorname{Ext}^j_{\mathfrak{k},L\cap K}(M_N, V)$$

for this N and all j. The same kind of argument that combines (5.40) and (5.41) to prove (5.42) can be used to combine (5.53) and

$$(5.54) \qquad \dim \operatorname{Ext}^j(A, V) \le \dim \operatorname{Ext}^j(A/B, V) + \dim \operatorname{Ext}^j(B, V).$$

to prove

$$(5.55) \qquad \dim \operatorname{Ext}^j_{\mathfrak{k},L\cap K}(M, V) \le \sum_{n=0}^{\infty} \dim \operatorname{Ext}^j_{\mathfrak{k},L\cap K}(M_n/M_{n-1}, V).$$

Now let us establish the $(\mathfrak{k}, L \cap K)$ isomorphism

$$(5.56) \qquad M_n/M_{n-1} \cong U(\mathfrak{k}) \otimes_{U(\bar{\mathfrak{q}} \cap \mathfrak{k})} (S^n(\mathfrak{g}/(\bar{\mathfrak{q}} + \mathfrak{k})) \otimes_{\mathbb{C}} Z_{\bar{\mathfrak{q}}}^{\#}).$$

To establish (5.56) we begin with the map

$$(5.57a) \qquad U(\mathfrak{k}) \otimes_{\mathbb{C}} (T^n(\mathfrak{g}) \otimes_{\mathbb{C}} Z_{\bar{\mathfrak{q}}}^{\#}) \to M_n/M_{n-1}$$

given by

$$(5.57b) \qquad u \otimes (X_1 \cdots X_n \otimes z) \mapsto u X_1 \cdots X_n \otimes z + M_{n-1}$$

for $u \in U(\mathfrak{k})$, $X_i \in \mathfrak{g}$, $z \in Z_{\bar{\mathfrak{q}}}^{\#}$. We make the right side of (5.57) into a $\mathfrak{k}$ module by the left action on $U(\mathfrak{k})$, and we make $L \cap K$ act by the adjoint actions on $U(\mathfrak{k})$ and $T(\mathfrak{g})$ and by the given action on $Z_{\bar{\mathfrak{q}}}^{\#}$. Then (5.57) is easily seen to respect the actions of $\mathfrak{k}$ and $L \cap K$.

We check readily that the image is independent of the order of the X_i, and then (5.57) provides a well-defined map

$$(5.58) \qquad U(\mathfrak{k}) \otimes_{\mathbb{C}} (S^n(\mathfrak{g}) \otimes_{\mathbb{C}} Z_{\bar{\mathfrak{q}}}^{\#}) \to M_n/M_{n-1}.$$

Next we verify that $uX_1 \cdots X_n \otimes z$ is in M_{n-1} if one of the X_i belongs to $\mathfrak{k}$ or to $\bar{\mathfrak{q}}$. It follows that (5.58) may be reinterpreted as a map

$$(5.59) \qquad U(\mathfrak{k}) \otimes_{\mathbb{C}} (S^n(\mathfrak{g}/(\bar{\mathfrak{q}} + \mathfrak{k})) \otimes_{\mathbb{C}} Z_{\bar{\mathfrak{q}}}^{\#}) \to M_n/M_{n-1}.$$

To see that (5.59) descends to the tensor product over $U(\bar{\mathfrak{q}} \cap \mathfrak{k})$, suppose we are given $u \in U(\mathfrak{k})$, $X \in \bar{\mathfrak{q}} \cap \mathfrak{k}$, and $X_i \in \mathfrak{g}$. Then

$$\begin{aligned}
uX \otimes (X_1 \cdots X_n \otimes z) &- u \otimes (X(X_1 \cdots X_n \otimes z) \\
&= uX \otimes X_1 \cdots X_n \otimes z - u \otimes X_1 \cdots x_n \otimes Xz \\
&\mapsto uXX_1 \cdots X_n \otimes z - uX_1 \cdots X_n \otimes Xz + M_{n-1} \\
&= uXX_1 \cdots X_n \otimes z - uX_1 \cdots X_n X \otimes z + M_{n-1} \\
&= M_{n-1}.
\end{aligned}$$

Thus (5.59) descends to a map

$$(5.60) \qquad U(\mathfrak{k}) \otimes_{U(\bar{\mathfrak{q}} \cap \mathfrak{k})} (S^n(\mathfrak{g}/(\bar{\mathfrak{q}} + \mathfrak{k})) \otimes_{\mathbb{C}} Z_{\bar{\mathfrak{q}}}^{\#}) \to M_n/M_{n-1}.$$

From the original definition (5.57), it is clear that this map is onto. Thus (5.60) gives a map from the right side of (5.56) onto the left side.

To see that the map from right to left in (5.56) is one-one, notice that

$$(5.61) \qquad \mathfrak{g}/(\bar{\mathfrak{q}} + \mathfrak{k}) \cong \mathfrak{u} \cap \mathfrak{p} \qquad \text{as an } L \cap K \text{ module.}$$

Hence (5.51b) provides an isomorphism of $L \cap K$ modules

$$\begin{aligned}
M_n/M_{n-1} &\cong (U_{n,\mathfrak{k}}(\mathfrak{u})/U_{n-1,\mathfrak{k}}(\mathfrak{u})) \otimes_{\mathbb{C}} Z_{\bar{\mathfrak{q}}}^{\#} \\
&\cong U(\mathfrak{u} \cap \mathfrak{k}) \otimes_{\mathbb{C}} S^n(\mathfrak{u} \cap \mathfrak{p}) \otimes_{\mathbb{C}} Z_{\bar{\mathfrak{q}}}^{\#} \\
&\cong U(\mathfrak{u} \cap \mathfrak{k}) \otimes_{\mathbb{C}} S^n(\mathfrak{g}/(\bar{\mathfrak{q}} + \mathfrak{k})) \otimes_{\mathbb{C}} Z_{\bar{\mathfrak{q}}}^{\#} \\
&\cong U(\mathfrak{u} \cap \mathfrak{k}) \otimes_{\mathbb{C}} U(\bar{\mathfrak{q}} \cap \mathfrak{k}) \otimes_{U(\bar{\mathfrak{q}} \cap \mathfrak{k})} (S^n(\mathfrak{g}/(\bar{\mathfrak{q}} + \mathfrak{k})) \otimes_{\mathbb{C}} Z_{\bar{\mathfrak{q}}}^{\#}) \\
&\cong U(\mathfrak{k}) \otimes_{U(\bar{\mathfrak{q}} \cap \mathfrak{k})} (S^n(\mathfrak{g}/(\bar{\mathfrak{q}} + \mathfrak{k})) \otimes_{\mathbb{C}} Z_{\bar{\mathfrak{q}}}^{\#}).
\end{aligned}$$

These $L \cap K$ isomorphisms are compatible with the map (5.56). Thus the $(\mathfrak{k}, L \cap K)$ map (5.56) is an isomorphism as well.

Combining (5.37a), (5.55), and (5.56) and using Proposition 2.57a for $\mathfrak{k}$ instead of $\mathfrak{g}$, we obtain

$$\dim \mathrm{Hom}_K(\mathcal{L}_j(Z), V)$$

$$\leq \sum_{n=0}^{\infty} \dim \mathrm{Ext}^j_{\mathfrak{k}, L \cap K}(P^{\mathfrak{k}, L \cap K}_{\bar{\mathfrak{q}} \cap \mathfrak{k}, L \cap K}(S^n(\mathfrak{g}/(\bar{\mathfrak{q}} + \mathfrak{k})) \otimes_{\mathbb{C}} Z^{\#}_{\bar{\mathfrak{q}}}), V)$$

for the nontrivial action of $\bar{\mathfrak{u}} \cap \mathfrak{k}$ on $S^n(\mathfrak{g}/(\bar{\mathfrak{q}} + \mathfrak{k}))$. By Shapiro's Lemma (Theorem 2.119), the right side here is

$$(5.62) \qquad = \sum_{n=0}^{\infty} \dim \mathrm{Ext}^j_{\bar{\mathfrak{q}} \cap \mathfrak{k}, L \cap K}(S^n(\mathfrak{g}/(\bar{\mathfrak{q}} + \mathfrak{k})) \otimes_{\mathbb{C}} Z^{\#}_{\bar{\mathfrak{q}}}, V).$$

The $(\bar{\mathfrak{q}} \cap \mathfrak{k}, L \cap K)$ module $S^n(\mathfrak{g}/(\bar{\mathfrak{q}} + \mathfrak{k}))$ is filtered according to the eigenvalues of $\mathrm{ad}\, h_{\delta(\mathfrak{u})}$, since $\bar{\mathfrak{u}} \cap \mathfrak{k}$ decreases eigenvalues. We apply our subadditivity formula for Ext (the one extending (5.54)) to (5.62): The respective quotients have 0 action by $\bar{\mathfrak{u}} \cap \mathfrak{k}$, and we can use (5.61) to reassemble them as a direct sum and see that (5.62) is

$$(5.63) \qquad \leq \sum_{n=0}^{\infty} \dim \mathrm{Ext}^j_{\bar{\mathfrak{q}} \cap \mathfrak{k}, L \cap K}(S^n(\mathfrak{u} \cap \mathfrak{p}) \otimes_{\mathbb{C}} Z^{\#}_{\bar{\mathfrak{q}}}, V)$$

with trivial $\bar{\mathfrak{u}} \cap \mathfrak{k}$ action on $S^n(\mathfrak{u} \cap \mathfrak{p})$.

We have now concluded the step in the argument that substitutes for (5.32b), and we have incorporated the analog of (5.32c). To complete the proof, we use the established isomorphism (5.32d) to obtain

$$\mathrm{Ext}^j_{\bar{\mathfrak{q}} \cap \mathfrak{k}, L \cap K}(W, V) \cong \mathrm{Hom}_{\mathfrak{l} \cap \mathfrak{k}, L \cap K}(W, H^j(\bar{\mathfrak{u}} \cap \mathfrak{k}, V))$$

for any $(\bar{\mathfrak{q}} \cap \mathfrak{k}, L \cap K)$ module W for which $\bar{\mathfrak{u}} \cap \mathfrak{k}$ acts trivially. Applying this isomorphism with $W = S^n(\mathfrak{u} \cap \mathfrak{p}) \otimes_{\mathbb{C}} Z^{\#}_{\bar{\mathfrak{q}}}$, we obtain (5.38a). Since $H^j(\bar{\mathfrak{u}} \cap \mathfrak{k}, V) = 0$ for $j > S$, we see from (5.38a) that $\mathcal{L}_j(Z) = 0$ for $j > S$, by the same argument as at the end of the proof of (b). This completes the proof of the theorem.

5. An Euler-Poincaré Principle for K Types

In the situation of Theorem 5.35 we shall apply the Euler-Poincaré Principle of §C.5 to deduce that equality holds when the alternating sum on j is taken of the two sides of (5.36a) or (5.36b). The proof involves the serious technical point of showing that we are not subtracting infinities, and for this purpose we shall assume that Z is admissible and that the element $h_{\delta(\mathfrak{u})}$ in the center of $\mathfrak{l}$ acts as a scalar in Z. The latter condition is of course satisfied if Z has an infinitesimal character.

Theorem 5.64. Suppose that Z is an admissible $(\mathfrak{l}, L \cap K)$ module and that the element $h_{\delta(\mathfrak{u})}$ acts as a scalar in Z. Then $\mathcal{L}_j(Z)$ and $\mathcal{R}^j(Z)$ are admissible $(\mathfrak{g}, K)$ modules for all $j \geq 0$, and, for any finite-dimensional $(\mathfrak{k}, K)$ module V,

$$\sum_{j=0}^{S} (-1)^j \dim \operatorname{Hom}_K(\mathcal{L}_j(Z), V)$$

$$= \sum_{j=0}^{S} (-1)^j \sum_{n=0}^{\infty} \dim \operatorname{Hom}_{L \cap K}(S^n(\mathfrak{u} \cap \mathfrak{p}) \otimes_{\mathbb{C}} Z^{\#}, H^j(\bar{\mathfrak{u}} \cap \mathfrak{k}, V))$$

and

$$\sum_{j=0}^{S} (-1)^j \dim \operatorname{Hom}_K(V, \mathcal{R}^j(Z))$$

$$= \sum_{j=0}^{S} (-1)^j \sum_{n=0}^{\infty} \dim \operatorname{Hom}_{L \cap K}(H_j(\mathfrak{u} \cap \mathfrak{k}, V), S^n(\mathfrak{u} \cap \mathfrak{p}) \otimes_{\mathbb{C}} Z^{\#}).$$

In each case the terms on the right side are 0 for n sufficiently large.

PROOF. Referring to the proof of Theorem 5.35 and remembering that $h_{\delta(\mathfrak{u})}$ is assumed to act as a scalar in Z, we see that the terms on the right side of (5.36a) and (5.36b) are 0 for n sufficiently large. Hence there are only finitely many nonzero terms on each right side. Within each term, $H^j(\bar{\mathfrak{u}} \cap \mathfrak{k})$ and $H_j(\mathfrak{u} \cap \mathfrak{k})$ are finite dimensional since they can be obtained from standard complexes involving finite-dimensional modules. The admissibility of Z therefore forces each term on the right side of (5.36a) or (5.36b) to be finite. It follows from Theorem 5.35 that $\dim \operatorname{Hom}_K(\mathcal{L}_j(Z), V)$ and $\dim \operatorname{Hom}_K(V, \mathcal{R}^j(Z))$ are finite for any finite-dimensional V, and therefore $\mathcal{L}_j(Z)$ and $\mathcal{R}^j(Z)$ are admissible as $(\mathfrak{g}, K)$ modules.

Let us prove the identity involving $\dim \operatorname{Hom}_K(\mathcal{L}_j(Z), V)$, the proof of the other identity being similar. Put

$$M = P_{\bar{\mathfrak{q}}, L \cap K}^{\mathfrak{g}, L \cap K}(Z_{\bar{\mathfrak{q}}}^{\#}),$$

and define M_n by (5.51). By (5.37a) and (5.53), we have

$$(5.65) \qquad \operatorname{Hom}_K(\mathcal{L}_j(Z), V) \cong \operatorname{Ext}_{\mathfrak{k}, L \cap K}^j(M, V) \cong \operatorname{Ext}_{\mathfrak{k}, L \cap K}^j(M_N, V)$$

for sufficiently large N (depending on Z). We form the filtration

(5.66) $M_N \supseteq M_{N-1} \supseteq \cdots \supseteq M_0 \supseteq M_{-1} = 0$

in $C(\mathfrak{k}, L \cap K)$. From the proof of (5.36a), we see that

(5.67)

$\dim \operatorname{Ext}^j_{\mathfrak{k}, L \cap K}(M_n/M_{n-1}, V)$

$= \dim \operatorname{Ext}^j_{\mathfrak{k}, L \cap K}(P^{\mathfrak{k}, L \cap K}_{\bar{\mathfrak{q}} \cap \mathfrak{k}, L \cap K}(S^n(\mathfrak{g}/(\bar{\mathfrak{q}} + \mathfrak{k})) \otimes_{\mathbb{C}} Z^{\#}_{\bar{\mathfrak{q}}}), V)$ by (5.56)

$= \dim \operatorname{Ext}^j_{\bar{\mathfrak{q}} \cap \mathfrak{k}, L \cap K}(S^n(\mathfrak{g}/(\bar{\mathfrak{q}} + \mathfrak{k})) \otimes_{\mathbb{C}} Z^{\#}_{\bar{\mathfrak{q}}}, V)$ by Theorem 2.119

$\leq \dim \operatorname{Ext}^j_{\bar{\mathfrak{q}} \cap \mathfrak{k}, L \cap K}(S^n(\mathfrak{u} \cap \mathfrak{p}) \otimes_{\mathbb{C}} Z^{\#}_{\bar{\mathfrak{q}}}, V)$ with a trivial action of $\bar{\mathfrak{u}} \cap \mathfrak{k}$ on $S^n(\mathfrak{u} \cap \mathfrak{p})$

$= \dim \operatorname{Hom}_{L \cap K}(S^n(\mathfrak{u} \cap \mathfrak{p}) \otimes_{\mathbb{C}} Z^{\#}_{\bar{\mathfrak{q}}}, H^j(\bar{\mathfrak{u}} \cap \mathfrak{k}, V))$ by (5.32d)

$< \infty,$

the last step holding since $S^n(\mathfrak{u} \cap \mathfrak{p})$ and $H^j(\bar{\mathfrak{u}} \cap \mathfrak{k}, V)$ are finite dimensional and $Z^{\#}_{\bar{\mathfrak{q}}}$ is admissible.

We shall apply Proposition C.44. We cannot take C to be $C(\mathfrak{k}, L \cap K)$, because the result is not a small category. Instead we take the modules of C to be all $(\mathfrak{k}, L \cap K)$ modules with a suitably bounded cardinality. The functors in the proposition will be $F_j = \operatorname{Ext}^j_{\mathfrak{k}, L \cap K}(\cdot, V)$, carrying C to a small good category C' of complex vector spaces. The collection S consists of all modules S in C for which $F_j(S)$ is finite dimensional for all j. Each M_n/M_{n-1} is in S by (5.67). If we take C'' in Proposition C.44 to be the good subcategory of all finite-dimensional members of C', then the hypotheses of that proposition are satisfied. Taking dim of both sides of conclusion (c) of that proposition and using (5.65), we obtain

$$\sum_{j=0}^{S} (-1)^j \dim \operatorname{Hom}_K(\mathcal{L}_j(Z), V)$$

$$= \sum_{j=0}^{S} \sum_{n=0}^{\infty} (-1)^j \dim \operatorname{Ext}^j_{\mathfrak{k}, L \cap K}(P^{\mathfrak{k}, L \cap K}_{\bar{\mathfrak{q}} \cap \mathfrak{k}, L \cap K}(S^n(\mathfrak{g}/(\bar{\mathfrak{q}} + \mathfrak{k})) \otimes_{\mathbb{C}} Z^{\#}_{\bar{\mathfrak{q}}}), V).$$

Then we can apply Theorem 2.119 to rewrite the right side as

$$(5.68) \quad = \sum_{j=0}^{S} \sum_{n=0}^{\infty} (-1)^j \dim \operatorname{Ext}^j_{\bar{\mathfrak{q}} \cap \mathfrak{k}, L \cap K}(S^n(\mathfrak{g}/(\bar{\mathfrak{q}} + \mathfrak{k})) \otimes_{\mathbb{C}} Z^{\#}_{\bar{\mathfrak{q}}}, V),$$

for the nontrivial action of $\bar{u} \cap \mathfrak{k}$ on $S^n(\mathfrak{g}/(\bar{\mathfrak{q}} + \mathfrak{k}))$.

Fix $n \geq 0$. The $(\bar{\mathfrak{q}} \cap \mathfrak{k}, L \cap K)$ module $S^n(\mathfrak{g}/(\bar{\mathfrak{q}}+\mathfrak{k}))$ is filtered according to the eigenvalues of $\operatorname{ad} h_{\delta(\mathfrak{u})}$, since $\bar{u} \cap \mathfrak{k}$ decreases eigenvalues. Let us write this filtration as

$$S^n(\mathfrak{g}/(\bar{\mathfrak{q}} + \mathfrak{k})) = S_r \supseteq S_{r-1} \supseteq \cdots \supseteq S_0 \supseteq S_{-1} = 0,$$

with $\bar{u} \cap \mathfrak{k}$ acting trivially on each S_i/S_{i-1}. Arguing with Proposition C.44 as above, we obtain

$$(5.69) \quad \sum_{j=0}^{S} (-1)^j \dim \operatorname{Ext}^j_{\bar{\mathfrak{q}} \cap \mathfrak{k}, L \cap K} (S^n(\mathfrak{g}/(\bar{\mathfrak{q}} + \mathfrak{k})) \otimes_{\mathbb{C}} Z^\#_{\bar{\mathfrak{q}}}, V)$$

$$= \sum_{j=0}^{S} \sum_{i=0}^{r} (-1)^j \dim \operatorname{Ext}^j_{\bar{\mathfrak{q}} \cap \mathfrak{k}, L \cap K} (S_i/S_{i-1} \otimes_{\mathbb{C}} Z^\#_{\bar{\mathfrak{q}}}, V).$$

In (5.69) we read from left to right, using the nontrivial action of $\bar{u} \cap \mathfrak{k}$ on $S^n(\mathfrak{g}/(\bar{\mathfrak{q}}+\mathfrak{k}))$. But, by using the $L \cap K$ isomorphism $\mathfrak{u} \cap \mathfrak{p} \cong \mathfrak{g}/(\bar{\mathfrak{q}} + \mathfrak{k})$ of (5.61), we may regard the S_j as a filtration of $S^n(\mathfrak{u} \cap \mathfrak{p})$. Then we see that the right side of (5.69) is

$$= \sum_{j=0}^{S} (-1)^j \dim \operatorname{Ext}^j_{\bar{\mathfrak{q}} \cap \mathfrak{k}, L \cap K} (S^n(\mathfrak{u} \cap \mathfrak{p}) \otimes_{\mathbb{C}} Z^\#_{\bar{\mathfrak{q}}}, V).$$

Therefore (5.68) is

$$= \sum_{j=0}^{S} \sum_{n=0}^{\infty} (-1)^j \dim \operatorname{Ext}^j_{\bar{\mathfrak{q}} \cap \mathfrak{k}, L \cap K} (S^n(\mathfrak{u} \cap \mathfrak{p}) \otimes_{\mathbb{C}} Z^\#_{\bar{\mathfrak{q}}}, V)$$

for the trivial action of $\bar{u} \cap \mathfrak{k}$ on $S^n(\mathfrak{u} \cap \mathfrak{p})$, and (5.32d) shows that this is

$$= \sum_{j=0}^{S} \sum_{n=0}^{\infty} (-1)^j \dim \operatorname{Hom}_{L \cap K} (S^n(\mathfrak{u} \cap \mathfrak{p}) \otimes_{\mathbb{C}} Z^\#_{\bar{\mathfrak{q}}}, H^j(\bar{u} \cap \mathfrak{k}, V)).$$

The theorem follows.

6. Bottom-Layer Map

As mentioned in §4, some information about $\mathcal{L}_j(Z)$ and $\mathcal{R}^j(Z)$ comes by introducing K analogs $\mathcal{L}_j^K(Z)$ and $\mathcal{R}_K^j(Z)$ and relating them to $\mathcal{L}_j(Z)$ and $\mathcal{R}^j(Z)$. In order to obtain the most clearly defined results, we shall assume throughout this section that (5.7) holds, i.e., that L meets every component of G.

According to Lemma 5.9 and Proposition 4.44, we can intersect our various Lie algebras and groups with $\mathfrak{k}$ and K and obtain valid reductive and parabolic pairs for study. Specifically we know that $(\mathfrak{k}, K)$ is a reductive pair, and the proposition tells us that $\mathfrak{q} \cap \mathfrak{k}$ is a θ stable parabolic subalgebra of $\mathfrak{k}$. Because $L \cap K$ meets every component of G and we always have $N_K(\mathfrak{q} \cap \mathfrak{k}) \supseteq N_K(\mathfrak{q})$, we must have $N_K(\mathfrak{q} \cap \mathfrak{k}) = N_K(\mathfrak{q})$. Therefore the parabolic and Levi subpairs are respectively $(\mathfrak{q} \cap \mathfrak{k}, L \cap K)$ and $(\mathfrak{l} \cap \mathfrak{k}, L \cap K)$, and the hypothesis (5.7) remains valid for the Levi subgroup of K. In this setting we continue to write $Z_{\bar{\mathfrak{q}}}^{\#}$ and $Z_{\mathfrak{q}}^{\#}$ for the K analogs

$$\mathcal{F}_{\bar{\mathfrak{q}}, L\cap K}^{\bar{\mathfrak{q}} \cap \mathfrak{k}, L\cap K}(Z_{\bar{\mathfrak{q}}}^{\#}) = \mathcal{F}_{\mathfrak{l}, L\cap K}^{\bar{\mathfrak{q}} \cap \mathfrak{k}, L\cap K}(Z^{\#})$$

and

$$\mathcal{F}_{\mathfrak{q}, L\cap K}^{\mathfrak{q} \cap \mathfrak{k}, L\cap K}(Z_{\mathfrak{q}}^{\#}) = \mathcal{F}_{\mathfrak{l}, L\cap K}^{\mathfrak{q} \cap \mathfrak{k}, L\cap K}(Z^{\#}).$$

In particular, the superscript $(\,\cdot\,)^{\#}$ continues to refer to the tensor product with $\bigwedge^{\text{top}} \mathfrak{u}$ (and does not refer to tensoring with $\bigwedge^{\text{top}}(\mathfrak{u} \cap \mathfrak{k})$). As in Proposition 2.115, define

$$\Pi_j^K = (P_{\mathfrak{k}, L\cap K}^{\mathfrak{k}, K})_j \qquad \text{and} \qquad \Gamma_K^j = (I_{\mathfrak{k}, L\cap K}^{\mathfrak{k}, K})^j.$$

The K analogs of $\mathcal{L}_j(Z)$ and $\mathcal{R}^j(Z)$ are then

(5.70)
$$\mathcal{L}_j^K(Z) = \Pi_j^K (P_{\bar{\mathfrak{q}} \cap \mathfrak{k}, L\cap K}^{\mathfrak{k}, L\cap K}(Z_{\bar{\mathfrak{q}}}^{\#}))$$

$$\mathcal{R}_K^j(Z) = \Gamma_K^j (P_{\mathfrak{q} \cap \mathfrak{k}, L\cap K}^{\mathfrak{k}, L\cap K}(Z_{\mathfrak{q}}^{\#})).$$

These are functors from $\mathcal{C}(\mathfrak{l} \cap \mathfrak{k}, L \cap K)$ to $\mathcal{C}(\mathfrak{k}, K)$.

Proposition 5.71. For $j > S$, $\mathcal{L}_j^K(Z) = 0$ and $\mathcal{R}_K^j(Z) = 0$. For $0 \le j \le S$ and for any finite-dimensional $(\mathfrak{k}, K)$ module V,

$$\text{Hom}_K(\mathcal{L}_j^K(Z), V) \cong \text{Hom}_{L\cap K}(Z^{\#}, H^j(\bar{\mathfrak{u}} \cap \mathfrak{k}, V))$$

and

$$\text{Hom}_K(V, \mathcal{R}_K^j(Z)) \cong \text{Hom}_{L\cap K}(H_j(\mathfrak{u} \cap \mathfrak{k}, V), Z^{\#}).$$

PROOF. The first identity is an instance of (4.172a); the meaning of $Z^{\#}$ is different in (4.172a), but that fact does not affect the argument. The second identity is an instance of (4.172b).

If $j > S$, both right sides are 0 for all V. Since $\mathcal{L}_j^K(Z)$ and $\mathcal{R}_K^j(Z)$ are fully reducible as $(\mathfrak{k}, K)$ modules, it follows that $\mathcal{L}_j^K(Z)$ and $\mathcal{R}_K^j(Z)$ are 0 for $j > S$.

Corollary 5.72. For any irreducible $(\mathfrak{k}, K)$ module V, let $V^{\mathfrak{u} \cap \mathfrak{k}}$ be the representation of $L \cap K$ in the subspace of $\mathfrak{u} \cap \mathfrak{k}$ invariants. Then

$$(5.73a) \qquad \mathrm{Hom}_K(\mathcal{L}_S^K(Z), V) \cong \mathrm{Hom}_{L \cap K}(Z \otimes_{\mathbb{C}} {\textstyle\bigwedge}^R(\mathfrak{u} \cap \mathfrak{p}), \ V^{\mathfrak{u} \cap \mathfrak{k}})$$

and

$$(5.73b) \qquad \mathrm{Hom}_K(V, \mathcal{R}_K^S(Z)) \cong \mathrm{Hom}_{L \cap K}(V^{\mathfrak{u} \cap \mathfrak{k}}, \ Z \otimes_{\mathbb{C}} {\textstyle\bigwedge}^R(\mathfrak{u} \cap \mathfrak{p})).$$

REMARKS. If Z is irreducible as an $(\mathfrak{l} \cap \mathfrak{k}, L \cap K)$ module, then so is $Z \otimes_{\mathbb{C}} {\textstyle\bigwedge}^R(\mathfrak{u} \cap \mathfrak{p})$, since ${\textstyle\bigwedge}^R(\mathfrak{u} \cap \mathfrak{p})$ is one dimensional. Under the assumption (5.7), Corollary 5.72 and Remark 2 for Theorem 4.83 identify $\mathcal{L}_S^K(Z)$. Namely $\mathcal{L}_S^K(Z)$ is 0 unless the highest weights of $Z \otimes_{\mathbb{C}} {\textstyle\bigwedge}^R(\mathfrak{u} \cap \mathfrak{p})$ are dominant for K. When they are dominant for K, $\mathcal{L}_S^K(Z)$ is the irreducible representation V of K given in Theorem 4.83 such that

$$V^{\mathfrak{u} \cap \mathfrak{k}} \cong Z \otimes_{\mathbb{C}} {\textstyle\bigwedge}^R(\mathfrak{u} \cap \mathfrak{p}).$$

In this case if the highest weights of Z are $\{\mu_L\}$, then the highest weights of $\mathcal{L}_S^K(Z)$ are $\{\mu_G\}$ with

$$\mu_G = \mu_L + 2\delta(\mathfrak{u} \cap \mathfrak{p}),$$

and their respective multiplicities match.

PROOF. For the first conclusion we take $j = S$ in the first identity of Proposition 5.71 and continue with

$$\mathrm{Hom}_K(\mathcal{L}_S^K(Z), V) \cong \mathrm{Hom}_{L \cap K}(Z^{\#}, H^S(\bar{\mathfrak{u}} \cap \mathfrak{k}, V))$$

$$\cong \mathrm{Hom}_{L \cap K}(Z^{\#}, H_0(\bar{\mathfrak{u}} \cap \mathfrak{k}, V \otimes_{\mathbb{C}} {\textstyle\bigwedge}^S(\bar{\mathfrak{u}} \cap \mathfrak{k})^*))$$

$$\text{by Corollary 3.13}$$

$$\cong \mathrm{Hom}_{L \cap K}(Z^{\#}, H_0(\bar{\mathfrak{u}} \cap \mathfrak{k}, V) \otimes_{\mathbb{C}} {\textstyle\bigwedge}^S(\bar{\mathfrak{u}} \cap \mathfrak{k})^*)$$

since ${\textstyle\bigwedge}^S(\bar{\mathfrak{u}} \cap \mathfrak{k})^*$ is a trivial $\bar{\mathfrak{u}} \cap \mathfrak{k}$ module. Now

$$H_0(\bar{\mathfrak{u}} \cap \mathfrak{k}, V) = V \otimes_{\bar{\mathfrak{u}} \cap \mathfrak{k}} \mathbb{C} \cong V^{\mathfrak{u} \cap \mathfrak{k}}$$

by Lemma 4.82. Since

$$Z^{\#} = Z \otimes_{\mathbb{C}} \bigwedge{}^{R+S} \mathfrak{u} \cong \bigwedge{}^{R}(\mathfrak{u} \cap \mathfrak{p}) \otimes_{\mathbb{C}} \bigwedge{}^{S}(\mathfrak{u} \cap \mathfrak{k}) \cong \bigwedge{}^{R}(\mathfrak{u} \cap \mathfrak{p}) \otimes_{\mathbb{C}} \bigwedge{}^{S}(\bar{\mathfrak{u}} \cap \mathfrak{k})^{*},$$

the first conclusion follows.

For the second conclusion, we take $j = S$ in the second identity of Proposition 5.71 and continue with

$$\begin{aligned} \operatorname{Hom}_{K}(V, \mathcal{R}_{K}^{S}(Z)) &\cong \operatorname{Hom}_{L \cap K}(H_{S}(\mathfrak{u} \cap \mathfrak{k}, V), Z^{\#}) \\ &\cong \operatorname{Hom}_{L \cap K}(H^{0}(\mathfrak{u} \cap \mathfrak{k}, V) \otimes_{\mathbb{C}} \bigwedge{}^{S}(\mathfrak{u} \cap \mathfrak{k}), Z^{\#}) \end{aligned}$$

by Corollary 3.13. Since $Z^{\#} \cong \bigwedge{}^{R}(\mathfrak{u} \cap \mathfrak{p}) \otimes_{\mathbb{C}} \bigwedge{}^{S}(\mathfrak{u} \cap \mathfrak{k})$ and $H^{0}(\mathfrak{u} \cap \mathfrak{k}, V) \cong V^{\mathfrak{u} \cap \mathfrak{k}}$, the second conclusion follows.

We refer to the K types of $\mathcal{L}_{S}(Z)$ that appear in $\mathcal{L}_{S}^{K}(Z)$ as the **bottom layer** of K types of $\mathcal{L}_{S}(Z)$. To justify this terminology, we shall introduce a **bottom-layer map** of $(\mathfrak{k}, K)$ modules

$$\mathcal{B}_{Z} : \mathcal{L}_{S}^{K}(Z) \to \mathcal{L}_{S}(Z)$$

with the properties (under the assumption (5.7)) that

   (i) $\mathcal{B}_{Z}$ is one-one on $\mathcal{L}_{S}^{K}(Z)$
   (ii) if τ is a K type in $\mathcal{L}_{S}^{K}(Z)$ and if $h_{\delta(\mathfrak{u})}$ acts by a scalar in Z, then $\mathcal{B}_{Z}$ maps the τ subspace of $\mathcal{L}_{S}^{K}(Z)$ one-one onto the τ subspace of $\mathcal{L}_{S}(Z)$.

Actually we shall construct $(\mathfrak{k}, K)$ maps $\mathcal{B}_{Z} : \mathcal{L}_{j}^{K}(Z) \to \mathcal{L}_{j}(Z)$ for all j, but we shall obtain properties (i) and (ii) only for $j = S$. (In the absence of (5.7), we would obtain only conclusion (i).)

To construct the maps, we use Lemma 5.26, taking $\mathfrak{m} = \mathfrak{k}, \mathfrak{m}' = \bar{\mathfrak{q}}$, and $Z = Z_{\bar{\mathfrak{q}}}^{\#}$ and obtain a one-one $(\mathfrak{k}, L \cap K)$ map

$$(5.74) \qquad \beta_{Z} : P_{\bar{\mathfrak{q}} \cap \mathfrak{k}, L \cap K}^{\mathfrak{k}, L \cap K}(Z_{\bar{\mathfrak{q}}}^{\#}) \to P_{\bar{\mathfrak{q}}, L \cap K}^{\mathfrak{g}, L \cap K}(Z_{\bar{\mathfrak{q}}}^{\#}).$$

(To be quite precise, we ought to include a forgetful functor $\mathcal{F}_{\mathfrak{g}, L \cap K}^{\mathfrak{k}, L \cap K}$ with the image.) Recall from Proposition 2.115 that

$$(5.75) \qquad \Pi_{j}^{K} \circ \mathcal{F}_{\mathfrak{g}, L \cap K}^{\mathfrak{k}, L \cap K} \cong \mathcal{F}_{\mathfrak{g}, K}^{\mathfrak{k}, K} \circ \Pi_{j}.$$

Because of this isomorphism, it is meaningful to form $\mathcal{B}_{Z} = \Pi_{j}^{K}(\beta_{Z})$ from (5.74); the result is our desired map

$$(5.76) \qquad \mathcal{B}_{Z} : \mathcal{L}_{j}^{K}(Z) \to \mathcal{L}_{j}(Z).$$

Similarly we introduce a **bottom-layer map**

$$\tilde{\mathcal{B}}_Z : \mathcal{R}^S(Z) \to \mathcal{R}_K^S(Z)$$

with the properties (under the assumption (5.7)) that

  (i) $\tilde{\mathcal{B}}_Z$ is onto $\mathcal{R}_K^S(Z)$
  (ii) if τ is a K type in $\mathcal{R}_K^S(Z)$ and if $h_\delta(\mathfrak{u})$ acts by a scalar in Z, then $\tilde{\mathcal{B}}_Z$ maps the τ subspace of $\mathcal{R}^S(Z)$ one-one onto the τ subspace of $\mathcal{R}_K^S(Z)$.

In Lemma 5.30, we take $\mathfrak{m} = \mathfrak{k}$, $\mathfrak{m}' = \mathfrak{q}$, and $Z = Z_\mathfrak{q}^\#$ and obtain a surjective $(\mathfrak{k}, L \cap K)$ map

$$(5.77) \qquad \tilde{\beta}_Z : P_{\mathfrak{q}, L \cap K}^{\mathfrak{g}, L \cap K}(Z_\mathfrak{q}^\#) \to P_{\mathfrak{q} \cap \mathfrak{k}, L \cap K}^{\mathfrak{k}, L \cap K}(Z_\mathfrak{q}^\#).$$

(This time, to be precise, we ought to include a forgetful functor $\mathcal{F}_{\mathfrak{g}, L \cap K}^{\mathfrak{k}, L \cap K}$ with the domain.) Using the isomorphism

$$(5.78) \qquad \Gamma_K^j \circ \mathcal{F}_{\mathfrak{g}, L \cap K}^{\mathfrak{k}, L \cap K} \cong \mathcal{F}_{\mathfrak{g}, K}^{\mathfrak{k}, K} \circ \Gamma^j$$

from Proposition 2.115, we can form $\tilde{\mathcal{B}}_Z = \Gamma_K^j(\tilde{\beta}_Z)$ from (5.77); the result is our desired map

$$(5.79) \qquad \tilde{\mathcal{B}}_Z : \mathcal{R}^j(Z) \to \mathcal{R}_K^j(Z).$$

Theorem 5.80. Suppose $h_{\delta(\mathfrak{u})}$ acts by a scalar in the $(\mathfrak{l}, L \cap K)$ module Z, and suppose L meets every component of G.

  (a) If τ is a K type in $\mathcal{L}_S^K(Z)$, then the bottom-layer map $\mathcal{B}_Z$ maps the τ subspace of $\mathcal{L}_S^K(Z)$ one-one onto the τ subspace of $\mathcal{L}_S(Z)$.

  (b) If τ is a K type in $\mathcal{R}_K^S(Z)$, then the bottom-layer map $\tilde{\mathcal{B}}_Z$ maps the τ subspace of $\mathcal{R}^S(Z)$ one-one onto the τ subspace of $\mathcal{R}_K^S(Z)$.

PROOF. The proofs of the two statements are rather similar, and we shall prove only (a). The argument for (a) consists of a refinement of the proof of Theorem 5.35. In the notation of that proof, we have $U_{0, \mathfrak{k}}(\mathfrak{g}) = U(\mathfrak{k})$ and therefore

$$M = P_{\bar{\mathfrak{q}}, L \cap K}^{\mathfrak{g}, L \cap K}(Z_{\bar{\mathfrak{q}}}^\#)$$

and

$$M_0 = P_{\bar{\mathfrak{q}} \cap \mathfrak{k}, L \cap K}^{\mathfrak{k}, L \cap K}(Z_{\bar{\mathfrak{q}}}^\#).$$

The map β_Z of (5.74) gives us an exact sequence

$$0 \longrightarrow M_0 \overset{\beta_Z}{\longrightarrow} M \longrightarrow M/M_0 \longrightarrow 0$$

in $\mathcal{C}(\mathfrak{g}, L \cap K)$, and we consequently obtain an exact sequence

$$\longrightarrow \Pi_{S+1}^K(M/M_0) \longrightarrow \Pi_S^K(M_0) \xrightarrow{\mathcal{B}_Z} \Pi_S^K(M) \longrightarrow \Pi_S^K(M/M_0) \longrightarrow$$

from the long exact sequence (C.36a) for Π_j^K. Applying the exact functor $\mathrm{Hom}_K(\,\cdot\,, V)$, we obtain an exact sequence

(5.81)
$$\longleftarrow \mathrm{Hom}_K(\Pi_{S+1}^K(M/M_0), V) \longleftarrow \mathrm{Hom}_K(\Pi_S^K(M_0), V)$$
$$\longleftarrow \mathrm{Hom}_K(\Pi_S^K(M), V) \longleftarrow \mathrm{Hom}_K(\Pi_S^K(M/M_0), V) \longleftarrow .$$

Let V be irreducible of type τ, with τ in the bottom layer. Then (5.81) shows that the corollary will follow from knowing

(5.82a) $\mathrm{Hom}_K(\Pi_{S+1}^K(M/M_0), V) = 0$

(5.82b) $\mathrm{Hom}_K(\Pi_S^K(M/M_0), V) = 0.$

The same lengthy argument that gives (5.63) gives

$$\dim \mathrm{Hom}_K(\Pi_i^K(M/M_0), V)$$
$$\leq \sum_{n=1}^{\infty} \dim \mathrm{Ext}_{\bar{\mathfrak{q}} \cap \mathfrak{k}, L \cap K}^i(S^n(\mathfrak{u} \cap \mathfrak{p}) \otimes_{\mathbb{C}} Z_{\bar{\mathfrak{q}}}^{\#}, V).$$

Since the right side is 0 for $i > S$, (5.82a) follows. For $i = S$, we can apply (5.32d) to the right side to obtain

(5.83) $\dim \mathrm{Hom}_K(\Pi_S^K(M/M_0), V)$
$$\leq \sum_{n=1}^{\infty} \dim \mathrm{Hom}_{L \cap K}(S^n(\mathfrak{u} \cap \mathfrak{p}) \otimes_{\mathbb{C}} Z \otimes_{\mathbb{C}} \bigwedge^R(\mathfrak{u} \cap \mathfrak{p}), V^{\mathfrak{u} \cap \mathfrak{k}}).$$

Because $L \cap K$ meets every component of K, Theorem 4.83b says that $L \cap K$ acts irreducibly in $V^{\mathfrak{u} \cap \mathfrak{k}}$. Let μ_G be a highest weight. If the right side of (5.83) is nonzero, there must be a highest weight μ_L of Z and some weight ω of some $S^n(\mathfrak{u} \cap \mathfrak{p})$ with $n \geq 1$ such that

(5.84) $\omega + \mu_L + 2\delta(\mathfrak{u} \cap \mathfrak{p}) = \mu_G.$

(Here we are using that every highest weight of a tensor product $V_1 \otimes_{\mathbb{C}} V_2$ for a compact connected Lie group is the sum of a weight of V_1 and a highest weight of V_2; this fact is readily seen from the Weyl Character

Formula, for example.) Meanwhile τ is in the bottom layer, and (5.73a) must be nonzero. Thus there must be a highest weight μ'_L of Z such that

$$\mu'_L + 2\delta(\mathfrak{u} \cap \mathfrak{p}) = \mu_G.$$

Hence $\omega = \mu'_L - \mu_L$.

But $h_{\delta(\mathfrak{u})}$ acts in Z by a scalar, say c. Then we have

$$\omega(h_{\delta(\mathfrak{u})}) = \mu'_L(h_{\delta(\mathfrak{u})}) - \mu_L(h_{\delta(\mathfrak{u})}) = c - c = 0,$$

in contradiction to the fact that ω is a nonempty sum of weights of $\mathfrak{u}$ and ad $h_{\delta(\mathfrak{u})}$ acts with positive eigenvalues in $\mathfrak{u}$. We conclude that the right side of (5.83) is 0, and so is (5.82b). This completes the proof.

Corollary 5.85. Suppose $h_{\delta(\mathfrak{u})}$ acts by a scalar in the $(\mathfrak{l}, L \cap K)$ module Z, and suppose L meets every component of G. For any irreducible $(\mathfrak{k}, K)$ module V in the bottom layer, let $V^{\mathfrak{u} \cap \mathfrak{k}}$ be the representation of $L \cap K$ in the subspace of $\mathfrak{u} \cap \mathfrak{k}$ invariants. Then

$$\mathrm{Hom}_K(\mathcal{L}_S(Z), V) \cong \mathrm{Hom}_{L \cap K}(Z \otimes_{\mathbb{C}} \textstyle\bigwedge^R(\mathfrak{u} \cap \mathfrak{p}), V^{\mathfrak{u} \cap \mathfrak{k}})$$

and $\mathrm{Hom}_K(V, \mathcal{R}^S(Z)) \cong \mathrm{Hom}_{L \cap K}(V^{\mathfrak{u} \cap \mathfrak{k}}, Z \otimes_{\mathbb{C}} \bigwedge^R(\mathfrak{u} \cap \mathfrak{p})).$

PROOF. Combine Theorem 5.80 and Corollary 5.72.

In view of the remarks after the statement of Corollary 5.72, Corollary 5.85 sets up a correspondence between all K types in the bottom layer of $\mathcal{L}_S(Z)$ (or $\mathcal{R}^S(Z)$) and certain $L \cap K$ types of Z, and the multiplicities match. The formula

(5.86a) $\mu_G = \mu_L + 2\delta(\mathfrak{u} \cap \mathfrak{p})$

matches the highest weights, and

(5.86b) $\pi_G = \pi_L \otimes \textstyle\bigwedge^R(\mathfrak{u} \cap \mathfrak{p})$

matches the respective representations of a common Cartan subgroup of $L \cap K$ and K in the sums of weight spaces for the highest weights. The correspondence is onto those $L \cap K$ types of Z for which the right side of (5.86a) is K dominant, as a result of Theorem 4.83d. The correspondence is implemented by the bottom-layer map.

The name "bottom layer" comes about as follows. Assume that $h_{\delta(\mathfrak{u})}$ acts in Z by a scalar, say c. If μ_L is the highest weight of an $L \cap K$ type in Z, then $\mu_L(h_{\delta(\mathfrak{u})}) = c$. The above proofs show that if μ_G is the highest weight of a K type in $\mathcal{L}_S(Z)$, then

$$\mu_G(h_{\delta(\mathfrak{u})}) \geq c + 2\delta(\mathfrak{u} \cap \mathfrak{p})(h_{\delta(\mathfrak{u})}),$$

and equality holds if and only if the K type is in the bottom layer. [In fact, (5.86a) shows that equality holds if the K type is in the bottom layer, and (5.84) shows that strict inequality holds if the K type is not in the bottom layer.]

Condition (5.7), that L meet every component of K, is essential for some of the results of this section, as the example below shows.

EXAMPLE. With $G = SL^{\pm}(2, \mathcal{R})$, we have $K = O(2)$, and we can arrange that $L = SO(2)$. Condition (5.7) is not satisfied. Except for the two representations that are trivial on L, the irreducible representations of K are of dimension 2, with $V^{\mathfrak{u} \cap \mathfrak{k}} = V$. The group $L \cap K$ acts reducibly in $V^{\mathfrak{u} \cap \mathfrak{k}}$ in the case of all the two-dimensional representations.

Let Z be one dimensional on L, with weight $t\alpha$, where $t \in \frac{1}{2}\mathbb{Z}$ and where α is the positive root (defining $\mathfrak{q}$). Then $P_{\bar{\mathfrak{q}}, L \cap K}^{\mathfrak{g}, L \cap K}(Z_{\bar{\mathfrak{q}}}^{\#}) = \text{ind}_{\bar{\mathfrak{q}}}^{\mathfrak{g}}(Z_{\bar{\mathfrak{q}}}^{\#})$ is an upside-down Verma module with weights $(t + 1)\alpha$, $(t + 2)\alpha$, $(t + 3)\alpha$, ... In this example, we have $S = 0$, and (5.37a) is applicable and gives

(5.87) $\qquad \text{Hom}_K(\mathcal{L}_0(Z), V) \cong \text{Hom}_{\mathfrak{k}, L \cap K}(P_{\bar{\mathfrak{q}}, L \cap K}^{\mathfrak{g}, L \cap K}(Z_{\bar{\mathfrak{q}}}^{\#}), V)$.

If $t \leq -\frac{3}{2}$, the K type with weights $(t+1)\alpha$ and $-(t+1)\alpha$ is in the bottom layer and has multiplicity 1 in $\mathcal{L}_0^K(Z)$. On the other hand, from (5.87) we see that it has multiplicity 2 in $\mathcal{L}_0(Z)$.

7. Vanishing Theorem

We saw in §4 that $\mathcal{L}_j(Z)$ and $\mathcal{R}^j(Z)$ are 0 if $j > S$. Our objective in this section is to show that $\mathcal{L}_j(Z)$ and $\mathcal{R}^j(S)$ are 0 also if $j < S$, provided Z is admissible and has an infinitesimal character with a certain dominance property. In Chapter VIII we shall use deeper properties of infinitesimal characters to prove a stronger theorem.

The idea of the proof, at least initially, is to use Hard Duality on the $(\mathfrak{k}, K)$ level to reduce the new vanishing theorem to what was proved in

§4. To do so, we use Lemma 5.26 to relate $P_{\bar{\mathfrak{q}},L\cap K}^{\mathfrak{g},L\cap K}(Z_{\bar{\mathfrak{q}}}^{\#})$ to $I_{\mathfrak{q},L\cap K}^{\mathfrak{g},L\cap K}(Z_{\mathfrak{q}}^{\#})$. Namely recall from Lemma 5.26c that we have a natural isomorphism

$$(5.88\text{a}) \qquad \operatorname{Hom}_{\mathfrak{g},L\cap K}(P_{\bar{\mathfrak{q}},L\cap K}^{\mathfrak{g},L\cap K}(Z_{\bar{\mathfrak{q}}}^{\#}),\ I_{\mathfrak{q},L\cap K}^{\mathfrak{g},L\cap K}(Z_{\mathfrak{q}}^{\#})) \cong \operatorname{Hom}_{\mathfrak{l},L\cap K}(Z^{\#},Z^{\#}),$$

which we can write also as

$$(5.88\text{b}) \quad \operatorname{Hom}_{\mathfrak{g},L\cap K}(\operatorname{ind}_{\bar{\mathfrak{q}},L\cap K}^{\mathfrak{g},L\cap K}(Z_{\bar{\mathfrak{q}}}^{\#}),\ \operatorname{pro}_{\mathfrak{q},L\cap K}^{\mathfrak{g},L\cap K}(Z_{\mathfrak{q}}^{\#})) \cong \operatorname{Hom}_{\mathfrak{l},L\cap K}(Z^{\#},Z^{\#}).$$

Let

$$(5.89) \qquad \varphi_Z^G : \operatorname{ind}_{\bar{\mathfrak{q}},L\cap K}^{\mathfrak{g},L\cap K}(Z_{\bar{\mathfrak{q}}}^{\#}) \to \operatorname{pro}_{\mathfrak{q},L\cap K}^{\mathfrak{g},L\cap K}(Z_{\mathfrak{q}}^{\#})$$

be the member of the left side of (5.88b) that corresponds to the identity on the right side. For later reference we record an explicit formula for φ_Z^G that uses the generalized Harish-Chandra mapping $\mu_{\bar{\mathfrak{u}}}'$ discussed in §IV.8. Proposition 5.93 will then tell how φ_Z^G can be used to prove a vanishing theorem.

Proposition 5.90. The map φ_Z^G in (5.89) is given by

$$\varphi_Z^G(x \otimes z)(y) = \mu_{\bar{\mathfrak{u}}}'(yx)z$$

for $x \in U(\mathfrak{g})$, $y \in U(\mathfrak{g})$, and $z \in Z^{\#}$.

PROOF. The argument that verifies (5.88b) may be summarized as

$$\operatorname{Hom}_{\mathfrak{g},L\cap K}(\operatorname{ind}_{\bar{\mathfrak{q}},L\cap K}^{\mathfrak{g},L\cap K}(Z_{\bar{\mathfrak{q}}}^{\#}),\ \operatorname{pro}_{\mathfrak{q},L\cap K}^{\mathfrak{g},L\cap K}(Z_{\mathfrak{q}}^{\#}))$$
$$\cong \operatorname{Hom}_{\mathfrak{q},L\cap K}(\operatorname{ind}_{\bar{\mathfrak{q}},L\cap K}^{\mathfrak{g},L\cap K}(Z_{\bar{\mathfrak{q}}}^{\#}),\ Z_{\mathfrak{q}}^{\#})$$
$$\cong \operatorname{Hom}_{\mathfrak{q},L\cap K}(P_{\mathfrak{l},L\cap K}^{\mathfrak{q},L\cap K}(Z_{\bar{\mathfrak{q}}}^{\#}),\ Z_{\mathfrak{q}}^{\#})$$
$$\cong \operatorname{Hom}_{\mathfrak{l},L\cap K}(Z^{\#},Z^{\#}).$$

If $\varphi, \varphi_1, \varphi_2$, and φ_3 are corresponding members of these four Hom's, then we have

$$\varphi(1 \otimes z)(1) = \varphi_1(1 \otimes z) = \varphi_2(1 \otimes z) = \varphi_3(z).$$

Taking φ_3 to be the identity, we obtain

$$(5.91) \qquad \varphi_Z^G(1 \otimes z)(1) = z.$$

Thus

$$\varphi_Z^G(x \otimes z)(y) = (y(\varphi_Z^G(x \otimes z)))(1)$$
$$= \varphi_Z^G(y(x \otimes z))(1) \qquad \text{since } \varphi_Z^G \text{ is } U(\mathfrak{g}) \text{ linear}$$
$$(5.92) \qquad = \varphi_Z^G(yx \otimes z)(1).$$

If y is in $U(\mathfrak{g})$ and X is in $\mathfrak{u}$, then (5.91) gives

$$\varphi_Z^G(Xy \otimes z)(1) = \varphi_Z^G(y \otimes z)(X) = X(\varphi_Z^G(x \otimes z)(1)) = 0,$$

while if $\bar{X}$ is in $\bar{\mathfrak{u}}$, then

$$\varphi_Z^G(y\bar{X} \otimes z)(1) = \varphi_Z^G(y \otimes \bar{X}z)(1) = \varphi_Z^G(0)(1) = 0.$$

Thus (5.91) gives

$$\varphi_Z^G(x \otimes z)(1) = \varphi_Z^G(\mu_{\bar{\mathfrak{u}}}'(x) \otimes z)(1) = \varphi_Z^G(1 \otimes \mu_{\bar{\mathfrak{u}}}'(x)z)(1) = \mu_{\bar{\mathfrak{u}}}'(x)z,$$

and the proposition follows by combining this relation with (5.92).

Proposition 5.93. If the map φ_Z^G in (5.89) is one-one onto, then $\mathcal{L}_j(Z)$ and $\mathcal{R}^j(Z)$ are 0 for $j < S$ and are isomorphic for $j = S$.

PROOF. Under our hypotheses we have

$$(5.94) \qquad \mathrm{ind}_{\bar{\mathfrak{q}}, L \cap K}^{\mathfrak{g}, L \cap K}(Z_{\bar{\mathfrak{q}}}^\#) \cong \mathrm{pro}_{\mathfrak{q}, L \cap K}^{\mathfrak{g}, L \cap K}(Z_{\mathfrak{q}}^\#).$$

We apply Hard Duality in the context of the inclusion of pairs $(\mathfrak{g}, L \cap K) \hookrightarrow (\mathfrak{g}, K)$. (Although the full strength of Hard Duality will be used when $j = S$, only the duality on the $(\mathfrak{k}, K)$ level will be needed for the vanishing statement.) In the notation of (3.3), we have

$$\mathfrak{c} = \mathfrak{k}/(\mathfrak{l} \cap \mathfrak{k}) \cong (\mathfrak{u} \cap \mathfrak{k}) \oplus (\bar{\mathfrak{u}} \oplus \mathfrak{k})$$

and $m = 2S$. If $0 \le j < S$, then $S < 2S - j \le 2S$ and

$$\Pi_j(\mathrm{ind}_{\bar{\mathfrak{q}}, L \cap K}^{\mathfrak{g}, L \cap K}(Z_{\bar{\mathfrak{q}}}^\#))$$

$$\begin{aligned}
&\cong \Pi_j(\mathrm{pro}_{\mathfrak{q}, L \cap K}^{\mathfrak{g}, L \cap K}(Z_{\mathfrak{q}}^\#)) && \text{by (5.94)} \\
&\cong \Gamma^{2S-j}(\mathrm{pro}_{\mathfrak{q}, L \cap K}^{\mathfrak{g}, L \cap K}(Z_{\mathfrak{q}}^\#) \otimes_{\mathbb{C}} \textstyle\bigwedge^{2S}\mathfrak{c}) && \text{by Theorem 3.5b} \\
&\cong \Gamma^{2S-j}(\mathrm{pro}_{\mathfrak{q}, L \cap K}^{\mathfrak{g}, L \cap K}(Z \otimes \textstyle\bigwedge^{2S}\mathfrak{c})_{\mathfrak{q}}^\#) && \text{by Corollary 2.97} \\
&= 0 && \text{by Theorem 5.35.}
\end{aligned}$$

Also

$$\Gamma^j(\mathrm{pro}_{\mathfrak{q}, L \cap K}^{\mathfrak{g}, L \cap K}(Z_{\mathfrak{q}}^\#))$$

$$\begin{aligned}
&\cong \Gamma^j(\mathrm{ind}_{\bar{\mathfrak{q}}, L \cap K}^{\mathfrak{g}, L \cap K}(Z_{\bar{\mathfrak{q}}}^\#)) && \text{by (5.94)} \\
&\cong \Pi_{2S-j}(\mathrm{ind}_{\bar{\mathfrak{q}}, L \cap K}^{\mathfrak{g}, L \cap K}(Z_{\bar{\mathfrak{q}}}^\#) \otimes_{\mathbb{C}} (\textstyle\bigwedge^{2S}\mathfrak{c})^*) && \text{by Theorem 3.5b} \\
&\cong \Pi_{2S-j}(\mathrm{ind}_{\bar{\mathfrak{q}}, L \cap K}^{\mathfrak{g}, L \cap K}(Z \otimes_{\mathbb{C}} (\textstyle\bigwedge^{2S}\mathfrak{c})^*)_{\bar{\mathfrak{q}}}^\#) && \text{by Theorem 2.103} \\
&= 0 && \text{by Theorem 5.35,}
\end{aligned}$$

and the conclusion follows for $j < S$.

If $j = S$, each of the above computations is valid up to the last line and gives the conclusion of the proposition if it is shown that $\bigwedge^{2S} \mathfrak{c}$ is a trivial $(\mathfrak{g}, L \cap K)$ module.

Referring to the definition of the action before Lemma 3.4, we see that $\mathfrak{g}$ acts trivially on $\bigwedge^{2S} \mathfrak{c}$ (since $\mathfrak{s} = 0$). The action by $L \cap K$ is by Ad, and

$$\bigwedge\nolimits^{2S} \mathfrak{c} \cong \bigwedge\nolimits^{\mathrm{top}} (\mathfrak{u} \cap \mathfrak{k}) \otimes_{\mathbb{C}} \bigwedge\nolimits^{\mathrm{top}} (\bar{\mathfrak{u}} \cap \mathfrak{k}).$$

Under our invariant bilinear form $\langle \cdot, \cdot \rangle$, the root space for any root α is nonsingularly paired with the root space for $-\alpha$, and thus $\bar{\mathfrak{u}}^* \cong \mathfrak{u}$ canonically as vector spaces. The invariance of $\langle \cdot, \cdot \rangle$ under $\mathrm{Ad}(L \cap K)$ implies that this isomorphism respects the action of $L \cap K$. Thus

(5.95) $\bar{\mathfrak{u}}^c \cong \mathfrak{u}$ as $(\mathfrak{l} \cap \mathfrak{k}, L \cap K)$ modules.

Similarly $(\bar{\mathfrak{u}} \cap \mathfrak{k})^c \cong \mathfrak{u} \cap \mathfrak{k}$, and hence $\bigwedge^{\mathrm{top}} (\bar{\mathfrak{u}} \cap \mathfrak{k})^c \cong \bigwedge^{\mathrm{top}} (\mathfrak{u} \cap \mathfrak{k})$. It follows that $\bigwedge^{2S} \mathfrak{c}$ is a trivial module, and the proof of the proposition is complete.

In applying Proposition 5.93, we shall see that the kernel of φ_Z^G is somewhat manageable. If it is 0, then $\mathrm{ind}_{\bar{\mathfrak{q}}, L \cap K}^{\mathfrak{g}, L \cap K}(Z_{\bar{\mathfrak{q}}}^{\#})$ embeds in $\mathrm{pro}_{\mathfrak{q}, L \cap K}^{\mathfrak{g}, L \cap K}(Z_{\mathfrak{q}}^{\#})$. Determining the image, however, is not so easy without some additional hypothesis on Z.

Proposition 5.96.

(a) If Z is admissible as an $L \cap K$ module, then

$$\mathrm{ind}_{\bar{\mathfrak{q}}, L \cap K}^{\mathfrak{g}, L \cap K}(Z_{\bar{\mathfrak{q}}}^{\#}) \qquad \text{and} \qquad \mathrm{pro}_{\mathfrak{q}, L \cap K}^{\mathfrak{g}, L \cap K}(Z_{\mathfrak{q}}^{\#})$$

are isomorphic as $L \cap K$ modules.

(b) If Z is admissible as an $L \cap K$ module and has an infinitesimal character, then $\mathrm{ind}_{\bar{\mathfrak{q}}, L \cap K}^{\mathfrak{g}, L \cap K}(Z_{\bar{\mathfrak{q}}}^{\#})$ is admissible as an $L \cap K$ module.

(c) If Z is admissible as an $L \cap K$ module and has an infinitesimal character and if φ_Z^G is one-one, then φ_Z^G is onto.

PROOF.

(a) Since Z is admissible, $Z^{\#} = ((Z^{\#})^c)^c$. Therefore we have $L \cap K$ isomorphisms

$$\mathrm{pro}_{\mathfrak{q}, L \cap K}^{\mathfrak{g}, L \cap K}(Z_{\mathfrak{q}}^{\#}) \cong \mathrm{pro}_{\mathfrak{q}, L \cap K}^{\mathfrak{g}, L \cap K}(((Z_{\mathfrak{q}}^{\#})^c)^c)$$

$$\cong \mathrm{ind}_{\mathfrak{q}, L \cap K}^{\mathfrak{g}, L \cap K}((Z_{\bar{\mathfrak{q}}}^{\#})^c)^c \qquad \text{by Easy Duality}$$

$$\cong (U(\bar{\mathfrak{u}}) \otimes_{\mathbb{C}} (Z^{\#})^c)^c \qquad \text{by the Poincaré-Birkhoff-Witt Theorem}$$

$$\cong U(\bar{\mathfrak{u}})^c \otimes_{\mathbb{C}} ((Z^{\#})^c)^c$$

(5.97) $\cong U(\bar{\mathfrak{u}})^c \otimes_{\mathbb{C}} Z^{\#}.$

By (5.95), $\bar{u}^c \cong u$ as $(\mathfrak{l} \cap \mathfrak{k}, L \cap K)$ modules. The element $\operatorname{ad} h_{\delta(u)}$ acts fully reducibly on $U(u)$, and its eigenspaces are finite dimensional by (4.71). Thus

$$(5.98) \qquad U(\bar{u})^c \cong U(u) \qquad \text{as } (\mathfrak{l} \cap \mathfrak{k}, L \cap K) \text{ modules},$$

and (5.97) is

$$\cong U(u) \otimes_{\mathbb{C}} Z^{\#} \cong \operatorname{ind}_{\bar{\mathfrak{q}}, L \cap K}^{\mathfrak{g}, L \cap K}(Z_{\bar{\mathfrak{q}}}^{\#})$$

as an $L \cap K$ module.

(b) From Lemma 5.10 it follows that $h_{\delta(u)}$, which is in the center of $\mathfrak{l} \cap \mathfrak{k}$, acts as a scalar in every $L \cap K$ type. Let $h_{\delta(u)}$ act by the scalar c_0 in $Z^{\#}$. Since $U(u)$ is fully reducible under $\operatorname{ad} h_{\delta(u)}$, so is

$$\operatorname{ind}_{\bar{\mathfrak{q}}, L \cap K}^{\mathfrak{g}, L \cap K}(Z_{\bar{\mathfrak{q}}}^{\#}) \cong U(u) \otimes_{\mathbb{C}} Z^{\#},$$

the eigenspace for eigenvalue c being the tensor product with $Z^{\#}$ of the subspace of $U(u)$ where $\operatorname{ad} h_{\delta(u)}$ acts by $c - c_0$. The latter subspace is finite dimensional. Therefore any $L \cap K$ type of $U(u) \otimes_{\mathbb{C}} Z^{\#}$ is contained in the tensor product of a finite-dimensional subspace of $U(u)$ by all of $Z^{\#}$. Since $Z^{\#}$ is admissible, it follows that the $L \cap K$ type occurs only finitely often in $U(u) \otimes_{\mathbb{C}} Z^{\#}$.

(c) For any $L \cap K$ type, the multiplicity is finite in $\operatorname{ind}_{\bar{\mathfrak{q}}, L \cap K}^{\mathfrak{g}, L \cap K}(Z_{\bar{\mathfrak{q}}}^{\#})$, by (b), and is equal to the multiplicity for $\operatorname{pro}_{\mathfrak{q}, L \cap K}^{\mathfrak{g}, L \cap K}(Z_{\mathfrak{q}}^{\#})$. Since φ_Z^G respects the $L \cap K$ actions, φ_Z^G is one-one on an $L \cap K$ type if and only if it is onto. This completes the proof.

A member of $\mathfrak{h}^*$ is called **real** if it is real-valued on $i\mathfrak{t}_0 \oplus \mathfrak{a}_0$. In terms of this notion, we can speak of the **real** and **imaginary parts** of members of $\mathfrak{h}^*$. (WARNING: "Real root," as defined in §IV.4, is a different notion. All roots are real in the present sense.) The Weyl group of $\Delta(\mathfrak{g}, \mathfrak{h})$ maps real elements to real elements.

Theorem 5.99. Suppose that the $(\mathfrak{l}, L \cap K)$ module Z is admissible and has an infinitesimal character $\lambda \in \mathfrak{h}^*$ such that

$$\langle \operatorname{Re} \lambda + \delta(u), \alpha \rangle \geq 0 \qquad \text{for all } \alpha \in \Delta(u).$$

Then

(a) any nonzero $(\mathfrak{g}, L \cap K)$ submodule of $\operatorname{ind}_{\bar{\mathfrak{q}}, L \cap K}^{\mathfrak{g}, L \cap K}(Z_{\bar{\mathfrak{q}}}^{\#})$ has nonzero intersection with $1 \otimes_{\mathbb{C}} Z^{\#}$, and

(b) $\mathcal{L}_j(Z)$ and $\mathcal{R}^j(Z)$ are 0 for $j < S$

(c) $\mathcal{L}_S(Z) \cong \mathcal{R}^S(Z)$.

In proving Theorem 5.99, we shall use the following lemma, whose proof is deferred to the end of §VII.8. No circularity is involved since the results of this section will not be applied until Chapter VIII. Observe that the lemma is trivial for Z finite dimensional.

Lemma 5.100. Suppose that the nonzero member Z of $\mathcal{C}(\mathfrak{l})$ has an infinitesimal character and that the nonzero member F of $\mathcal{C}(\mathfrak{l})$ is finite dimensional. Then $F \otimes_{\mathbb{C}} Z$ has a nonzero $U(\mathfrak{l})$ submodule with an infinitesimal character.

PROOF OF THEOREM 5.99.

(a) Introduce $w \in W(\mathfrak{l}, \mathfrak{h})$ so that Re $w\lambda$ is dominant relative to $\Delta^+(\mathfrak{l}, \mathfrak{h})$. For $\alpha \in \Delta(\mathfrak{u})$, we have

$$\mathrm{Re}\langle w\lambda + \delta(\mathfrak{u}), \alpha \rangle = \mathrm{Re}\langle w(\lambda + \delta(\mathfrak{u})), \alpha \rangle = \mathrm{Re}\langle \lambda + \delta(\mathfrak{u}), w^{-1}\alpha \rangle \geq 0$$

by hypothesis. For $\beta \in \Delta^+(\mathfrak{l}, \mathfrak{h})$, we have

$$\mathrm{Re}\langle w\lambda + \delta(\mathfrak{u}), \beta \rangle = \mathrm{Re}\langle w\lambda, \beta \rangle + \langle \delta(\mathfrak{u}), \beta \rangle = \mathrm{Re}\langle w\lambda, \beta \rangle \geq 0.$$

Changing notation, we see that there is no loss of generality in assuming that Re $\lambda + \delta(\mathfrak{u})$ is dominant relative to $\Delta^+(\mathfrak{g}, \mathfrak{h})$.

Let $M = \mathrm{ind}_{\bar{\mathfrak{q}}, L \cap K}^{\mathfrak{g}, L \cap K}(Z_{\bar{\mathfrak{q}}}^{\#})$, so that

$$M \cong U(\mathfrak{u}) \otimes_{\mathbb{C}} Z^{\#}$$

as an $(\mathfrak{l}, L \cap K)$ module. Let N be a nonzero $(\mathfrak{g}, L \cap K)$ submodule of M. Since Z has an infinitesimal character, $h_{\delta(\mathfrak{u})} = h_{\delta(\mathfrak{u})}$ acts by a scalar in $Z^{\#}$, say by $c_0 + ib$ with c_0 and b real. From (4.71) we see that $h_{\delta(\mathfrak{u})}$ acts fully reducibly on M, the $c_0 + ib$ eigenspace of $h_{\delta(\mathfrak{u})}$ in M is $1 \otimes_{\mathbb{C}} Z^{\#}$, and the other eigenvalues lie in a discrete set of complex numbers with real parts $> c_0$ and with imaginary parts b. Since $h_{\delta(\mathfrak{u})}$ carries N to itself, N is the sum of its intersections with the eigenspaces in M under $h_{\delta(\mathfrak{u})}$. Let c be the smallest value such that the eigenspace for $c + ib$ meets N, and let N_0 be this eigenspace. We have $c \geq c_0$.

Arguing by contradiction, suppose N does not meet $1 \otimes_{\mathbb{C}} Z^{\#}$. Then we have

$$(5.101) \qquad\qquad\qquad c > c_0.$$

The eigenspace N_0 is an $(\mathfrak{l}, L \cap K)$ module as a consequence of Lemma 5.10, and it coincides with $F \otimes_{\mathbb{C}} Z^{\#}$, where F is the subspace of $U(\mathfrak{u})$ where $\mathrm{ad}\, h_{\delta(\mathfrak{u})}$ acts by the scalar $c - c_0$. By Lemma 5.100, $F \otimes_{\mathbb{C}} Z^{\#}$ has a nonzero $\mathfrak{l}$ submodule with an infinitesimal character, say ν. Then there exists some nonzero $x \in F \otimes_{\mathbb{C}} Z^{\#}$ with

$$(5.102a) \qquad z_{\mathfrak{l}} x = \nu(\gamma_{\mathfrak{l}}(z_{\mathfrak{l}})) x \qquad \text{for all } z_{\mathfrak{l}} \in Z(\mathfrak{l}).$$

Taking $z_{\mathfrak{l}} = h_{\delta(\mathfrak{u})}$, we have in particular

$$(5.102b) \qquad c + ib = \langle \nu, \delta(\mathfrak{u}) \rangle.$$

Meanwhile Theorem 5.24a shows that M has infinitesimal character $\lambda + \delta(\mathfrak{u})$. Thus

$$(5.103) \qquad zx = (\lambda + \delta(\mathfrak{u}))(\gamma(z)) \qquad \text{for all } z \in Z(\mathfrak{g}).$$

Since x is in N_0 and $\bar{\mathfrak{u}}$ decreases eigenvalues under $h_{\delta(\mathfrak{u})}$, we have $\bar{\mathfrak{u}}(N_0) = 0$. Therefore $z \in Z(\mathfrak{g})$ implies

$$
\begin{aligned}
zx &= \mu'_{\bar{\mathfrak{u}}}(z)x \\
&= \nu(\gamma_{\mathfrak{l}}(\mu'_{\bar{\mathfrak{u}}}(z)))x & \text{by (5.102a)} \\
&= \nu(\tau_{\mathfrak{l} \cap \bar{\mathfrak{n}}} \circ \gamma'_{\mathfrak{l} \cap \bar{\mathfrak{n}}} \circ \mu'_{\bar{\mathfrak{u}}}(z))x & \text{by (4.93)} \\
&= \nu(\tau_{\mathfrak{l} \cap \bar{\mathfrak{n}}} \circ \gamma'_{\bar{\mathfrak{n}}}(z))x & \text{by Proposition 4.129a} \\
&= \nu(\tau_{\bar{\mathfrak{u}}}^{-1}(\gamma(z)))x & \text{by (4.93)} \\
&= (\nu - \delta(\mathfrak{u}))(\gamma(z))x.
\end{aligned}
$$

Comparing this result with (5.103), we see that $\chi_{\nu - \delta(\mathfrak{u})} = \chi_{\lambda + \delta(\mathfrak{u})}$. By Theorem 4.115,

$$(5.104) \qquad \nu - \delta(\mathfrak{u}) = s(\lambda + \delta(\mathfrak{u}))$$

for some $s \in W(\mathfrak{g}, \mathfrak{h})$. Taking the real part of both sides and forming the inner product with $\delta(\mathfrak{u})$ gives

$$
\begin{aligned}
c - |\delta(\mathfrak{u})|^2 &= \operatorname{Re}\langle \nu - \delta(\mathfrak{u}), \delta(\mathfrak{u}) \rangle & \text{by (5.102b)} \\
&= \operatorname{Re}\langle s(\lambda + \delta(\mathfrak{u})), \delta(\mathfrak{u}) \rangle \\
&\leq \operatorname{Re}\langle \lambda + \delta(\mathfrak{u}), \delta(\mathfrak{u}) \rangle & \text{since } \operatorname{Re} \lambda + \delta(\mathfrak{u}) \text{ is} \\
& & \text{assumed dominant} \\
&= \operatorname{Re}\langle \lambda + 2\delta(\mathfrak{u}), \delta(\mathfrak{u}) \rangle - |\delta(\mathfrak{u})|^2 \\
&= c_0 - |\delta(\mathfrak{u})|^2.
\end{aligned}
$$

Thus $c \leq c_0$, in contradiction to (5.101).

(b,c) Let us prove that φ_Z^G is one-one. Arguing by contradiction, suppose that the kernel N of φ_Z^G is nonzero. Since N is a $(\mathfrak{g}, L \cap K)$ submodule of M, (a) shows that it meets $1 \otimes_{\mathbb{C}} Z^{\#}$. If $1 \otimes z$ is a nonzero element in the intersection, then $\varphi_Z^G(1 \otimes z)$ is 0 since $1 \otimes z$ is in N. On the other hand, Proposition 5.90 gives

$$\varphi_Z^G(1 \otimes z)(1) = z,$$

contradiction. We conclude that φ_Z^G is one-one. By Proposition 5.96c, φ_Z^G is onto. Then $\mathcal{L}_j(Z)$ and $\mathcal{R}^j(Z)$ are 0 for $j < S$ by Proposition 5.93, and also $\mathcal{L}_S(Z) \cong \mathcal{R}^S(Z)$.

Corollary 5.105. If an irreducible admissible $(\mathfrak{l}, L \cap K)$ module Z has an infinitesimal character $\lambda \in \mathfrak{h}^*$ such that $\langle \operatorname{Re} \lambda + \delta(\mathfrak{u}), \alpha \rangle \geq 0$ for all $\alpha \in \Delta(\mathfrak{u})$, then $\operatorname{ind}_{\bar{\mathfrak{q}}, L \cap K}^{\mathfrak{g}, L \cap K}(Z_{\bar{\mathfrak{q}}}^{\#})$ is irreducible as a $(\mathfrak{g}, L \cap K)$ module.

PROOF. Theorem 5.99a shows that any nonzero $(\mathfrak{g}, L \cap K)$ submodule N of $M = \operatorname{ind}_{\bar{\mathfrak{q}}, L \cap K}^{\mathfrak{g}, L \cap K}(Z_{\bar{\mathfrak{q}}}^{\#})$ meets $1 \otimes_{\mathbb{C}} Z^{\#}$. If $1 \otimes z$ is a nonzero element of the intersection, then $R(\mathfrak{l}, L \cap K)(1 \otimes z) = 1 \otimes_{\mathbb{C}} Z^{\#}$ since $Z^{\#}$ is irreducible. Hence $N \supseteq 1 \otimes_{\mathbb{C}} Z^{\#}$. Applying $U(\mathfrak{g})$, we see that $N \supseteq M$. Hence $N = M$.

Let us now apply the above considerations to the modules $A_{\mathfrak{q}}(\lambda)$ and $A^{\mathfrak{q}}(\lambda)$ defined in (5.6). We assume that the reductive pair $(\mathfrak{g}, K)$ has G connected. Since $\mathbb{C}_\lambda$ has infinitesimal character $\lambda + \delta(\mathfrak{l} \cap \mathfrak{n})$, Theorem 5.99b shows that

$$(5.106a) \qquad \mathcal{L}_j(\mathbb{C}_\lambda) = 0 \qquad \text{and} \qquad \mathcal{R}^j(\mathbb{C}_\lambda) = 0 \qquad \text{for } j < S$$

and

$$(5.106b) \qquad \mathcal{L}_S(\mathbb{C}_\lambda) \cong \mathcal{R}^S(\mathbb{C}_\lambda),$$

provided

$$(5.107) \qquad \operatorname{Re}\langle \lambda + \delta, \alpha \rangle \geq 0 \qquad \text{for all } \alpha \in \Delta(\mathfrak{u}).$$

Under these circumstances we can apply Theorem 5.64 and obtain

(5.108a)
$$\dim \operatorname{Hom}_K(\mathcal{L}_S(\mathbb{C}_\lambda), V)$$
$$= \sum_{j=0}^{S} (-1)^{S-j} \sum_{n=0}^{\infty} \dim \operatorname{Hom}_{L \cap K}(S^n(\mathfrak{u} \cap \mathfrak{p}) \otimes_{\mathbb{C}} \mathbb{C}_{\lambda+2\delta(\mathfrak{u})}, H^j(\bar{\mathfrak{u}} \cap \mathfrak{k}, V))$$

(5.108b)
$$\dim \operatorname{Hom}_K(V, \mathcal{R}^S(\mathbb{C}_\lambda))$$
$$= \sum_{j=0}^{S} (-1)^{S-j} \sum_{n=0}^{\infty} \dim \operatorname{Hom}_{L \cap K}(H_j(\mathfrak{u} \cap \mathfrak{k}, V), S^n(\mathfrak{u} \cap \mathfrak{p}) \otimes_{\mathbb{C}} \mathbb{C}_{\lambda+2\delta(\mathfrak{u})})$$

for any finite-dimensional $(\mathfrak{k}, K)$ module V. These formulas are called the **Blattner multiplicity formulas**. The right sides can be evaluated explicitly by means of Kostant's Theorem.

EXAMPLE. For $G = SU(2, 1)$, we shall show that the inequalities in Theorem 5.35 can be strict. Let $\mathfrak{h}_0$ be a maximally compact θ stable Cartan subalgebra, let

$$\Delta^+(\mathfrak{g}, \mathfrak{h}) = \{e_1 - e_2, e_2 - e_3, e_1 - e_3\},$$

and assume that $e_1 - e_2$ is compact and that the other two positive roots are noncompact. Take $\mathfrak{q} = \mathfrak{b}$, so that $S = 1$ and $L = T$. Here $\delta = e_1 - e_3$. If we define

$$\lambda = -e_1 + e_3,$$

then λ is integral and $\lambda + \delta = 0$ is dominant. Hence $\mathbb{C}_\lambda$ is defined, and (5.106) and (5.107) show that $\mathcal{L}_0(\mathbb{C}_\lambda) = 0$. In Theorem 5.35a, let V be the one-dimensional K module with unique weight $e_1 + e_2 - 2e_3$. Then

$$(e_2 - e_3) + (e_1 - e_3) = e_1 + e_2 - 2e_3$$

exhibits

$$\mathrm{Hom}_T(S^1(\mathfrak{u} \cap \mathfrak{p}) \otimes_{\mathbb{C}} \mathbb{C}_\lambda^\#, \ H^0(\bar{\mathfrak{u}} \cap \mathfrak{k}, V))$$

as nonzero. Consequently the inequality (5.36a) for $j = 0$ has 0 on the left side and at least 1 on the right side. Thus equality fails in (5.36a).

In the case of $A_\mathfrak{q}(\lambda)$ and $A^\mathfrak{q}(\lambda)$, the final steps of the proof of Theorem 5.99 can be adjusted to give a different condition in place of (5.107). This variant of Theorem 5.99 will be generalized in Chapter XII.

Theorem 5.109. In the case of $A_\mathfrak{q}(\lambda)$ and $A^\mathfrak{q}(\lambda)$, the conclusions (5.106) and (5.108) remain valid if

$$(5.110) \qquad \mathrm{Re}\langle \lambda + \delta(\mathfrak{u}), \alpha \rangle \geq 0 \qquad \text{for all } \alpha \in \Delta(\mathfrak{u}).$$

REMARKS. This theorem is an improvement over Theorem 5.99 for $A_\mathfrak{q}(\lambda)$ and $A^\mathfrak{q}(\lambda)$ in the sense that (5.107) implies (5.110) when λ is orthogonal to $\Delta(\mathfrak{l}, \mathfrak{h})$. In fact, it is enough to see this implication for α simple. If the simple root α is in $\Delta(\mathfrak{l}, \mathfrak{h})$, then the left side of (5.110) is 0, and (5.110) is valid. If the simple root α is in $\Delta(\mathfrak{u})$, then it is distinct from all the simple roots of $\Delta(\mathfrak{l}, \mathfrak{h})$, and hence $\langle \delta(\mathfrak{l} \cap \mathfrak{n}), \alpha \rangle \leq 0$ by Proposition 4.55b. Subtracting this inequality from (5.107), we obtain (5.110).

PROOF. We start from (5.104), but with λ replaced by the infinitesimal character of $\mathbb{C}_\lambda$, namely $\lambda + \delta(\mathfrak{l} \cap \mathfrak{n})$. Thus

$$\nu - \delta(\mathfrak{u}) = s(\lambda + \delta(\mathfrak{l} \cap \mathfrak{n}) + \delta(\mathfrak{u})).$$

To simplify notation, we shall regard λ as real; then ν is real also. (For complex λ, we replace λ and ν by their real parts in all norms and estimates below.) We have

(5.111) $$|\nu - \delta(\mathfrak{u})|^2 = |\lambda + \delta(\mathfrak{l} \cap \mathfrak{n}) + \delta(\mathfrak{u})|^2.$$

Let us write $\mathfrak{h}^* = \mathfrak{h}_1^* \oplus \mathfrak{h}_2^*$, where $\mathfrak{h}_1^*$ is the dual of $\mathfrak{h} \cap Z_{\mathfrak{l}}$ and $\mathfrak{h}_2^*$ is the dual of $\mathfrak{h} \cap [\mathfrak{l}, \mathfrak{l}]$. A member β of $\mathfrak{h}^*$ decomposes correspondingly as $\beta = \beta_1 + \beta_2$. If we define

$$|\beta|_1^2 = |\beta_1|^2 \qquad \text{and} \qquad |\beta|_2^2 = |\beta_2|^2,$$

then we have

$$|\beta|^2 = |\beta|_1^2 + |\beta|_2^2.$$

In the proof of Theorem 5.99, $Z^{\#}$ is now one dimensional. Therefore the nonzero $\mathfrak{l}$ submodule of $F \otimes_{\mathbb{C}} Z^{\#}$ with infinitesimal character ν is finite dimensional, and it follows that

$$|\nu|_2^2 \geq |\delta(\mathfrak{l} \cap \mathfrak{n})|^2.$$

Since $\mathbb{C}_\lambda$ is one dimensional, λ is orthogonal to $\Delta(\mathfrak{l}, \mathfrak{h})$. Thus

$$|\nu - \delta(\mathfrak{u})|_2^2 = |\nu|_2^2 \geq |\delta(\mathfrak{l} \cap \mathfrak{n})|^2 = |\lambda + \delta(\mathfrak{l} \cap \mathfrak{n}) + \delta(\mathfrak{u})|_2^2.$$

Taking (5.111) into account, we see that

(5.112) $$|\nu - \delta(\mathfrak{u})|_1^2 \leq |\lambda + \delta(\mathfrak{l} \cap \mathfrak{n}) + \delta(\mathfrak{u})|_1^2 = |\lambda + \delta(\mathfrak{u})|^2.$$

Let x be a nonzero highest-weight vector of the $\mathfrak{l}$ submodule of $F \otimes_{\mathbb{C}} Z^{\#}$ with infinitesimal character ν. Then x has weight $\nu - \delta(\mathfrak{l} \cap \mathfrak{n})$, and this weight must be related to the weight of $1 \otimes \mathbb{C}_\lambda^{\#}$ by

$$\nu - \delta(\mathfrak{l} \cap \mathfrak{n}) = \lambda + 2\delta(\mathfrak{u}) + \sum_{\alpha \in \Delta(\mathfrak{u})} n_\alpha \alpha \qquad \text{with } n_\alpha \geq 0.$$

Subtracting $\delta(\mathfrak{u})$ from both sides and applying $|\cdot|_1^2$, we obtain

$$
\begin{aligned}
|v - \delta(\mathfrak{u})|_1^2 &= \left| (\lambda + \delta(\mathfrak{u})) + \left(\sum_{\alpha \in \Delta(\mathfrak{u})} n_\alpha \alpha \right)_1 \right|^2 \\
&= |\lambda + \delta(\mathfrak{u})|^2 + 2\left\langle \lambda + \delta(\mathfrak{u}), \sum_{\alpha \in \Delta(\mathfrak{u})} n_\alpha \alpha \right\rangle + \left| \sum_{\alpha \in \Delta(\mathfrak{u})} n_\alpha \alpha \right|_1^2 \\
&\geq |\lambda + \delta(\mathfrak{u})|^2 + \left| \sum_{\alpha \in \Delta(\mathfrak{u})} n_\alpha \alpha \right|_1^2,
\end{aligned}
$$

the inequality following from (5.110). Now $\left(\sum_{\alpha \in \Delta(\mathfrak{u})} n_\alpha \alpha \right)_1$ is not 0 unless all n_α are 0, since (4.71) says that $\langle \alpha, \delta(\mathfrak{u}) \rangle > 0$ for all $\alpha \in \Delta(\mathfrak{u})$. Hence

$$
|v - \delta(\mathfrak{u})|_1^2 > |\lambda + \delta(\mathfrak{u})|^2
$$

unless all n_α are 0. From (5.112) we conclude that all n_α are 0, and it follows that the vector x is in $1 \otimes \mathbb{C}_\lambda^\#$. In other words, the submodule N in the proof of Theorem 5.99 meets $1 \otimes \mathbb{C}_\lambda^\#$, and we can repeat the rest of the argument for that theorem to complete the proof of Theorem 5.109.

EXAMPLE. Let G be the identity component of $SO(4, 1)$. Fix a Cartan subalgebra $\mathfrak{h}_0$ of $\mathfrak{g}_0$ that lies in $\mathfrak{k}_0$, and realize $\Delta(\mathfrak{g}, \mathfrak{h})$ as a root system of type B_2. We define $\mathfrak{l}$ and $\mathfrak{u}$ so that $\Delta^+(\mathfrak{l}, \mathfrak{h}) = \{e_1 - e_2\}$ and $\Delta(\mathfrak{u}) = \{e_1, e_2, e_1 + e_2\}$. Then $\delta(\mathfrak{u}) = e_1 + e_2$, while $\delta = \frac{3}{2}e_1 + \frac{1}{2}e_2$. If we take $\lambda = -e_1 - e_2$, then λ is orthogonal to $\Delta(\mathfrak{l}, \mathfrak{h})$ and does not satisfy (5.107), because $\langle \lambda + \delta, e_2 \rangle < 0$. However, $\lambda + \delta(\mathfrak{u}) = 0$, and thus λ does satisfy (5.110).

8. Fundamental Spectral Sequences

In §§4–7 we obtained information about $\mathcal{L}_j(Z)$ and $\mathcal{R}^j(Z)$ by using various K analogs of constructs in §IV.11. To get information about the constructs themselves, we shall use spectral sequences. The reader is referred to Appendix D for background on this topic; special attention should be given to the language in which results are stated, which is explained after (D.27).

The guiding computations are those in (5.32) and (5.33), except that the Hom's and Ext's are to be taken relative to pairs $(\mathfrak{g}, K)$, $(\bar{\mathfrak{q}}, L \cap K)$, $(\mathfrak{q}, L \cap K)$, and $(\mathfrak{l}, L \cap K)$, rather than their intersections with $(\mathfrak{k}, K)$.

The starting point for our investigation is what happens to Frobenius reciprocity for P and I when we pass to derived functors of P and I. We work in the context of Chapter II, namely with a general map of pairs $(\mathfrak{h}, L) \to (\mathfrak{g}, K)$. The good categories $\mathcal{C}(\mathfrak{h}, L)$ and $\mathcal{C}(\mathfrak{g}, K)$ are closed under countable direct sums, and the results of Appendix D are therefore applicable.

Proposition 5.113 (Frobenius reciprocity spectral sequences). Let $(\mathfrak{h}, L) \to (\mathfrak{g}, K)$ be a map of pairs, let V be in $\mathcal{C}(\mathfrak{h}, L)$, and let X be in $\mathcal{C}(\mathfrak{g}, K)$.

(a) There exists a first-quadrant spectral sequence

$$E_r^{p,q} \implies \mathrm{Ext}_{\mathfrak{h},L}^{p+q}(V, \, (\mathcal{F}^{\vee})_{\mathfrak{g},K}^{\mathfrak{h},L}(X))$$

with E_2 term $\qquad E_2^{p,q} = \mathrm{Ext}_{\mathfrak{g},K}^{p}((P_{\mathfrak{h},L}^{\mathfrak{g},K})_q(V), \, X).$

The differential on $E_r^{p,q}$ has bidegree $(r, 1 - r)$.

(b) There exists a first-quadrant spectral sequence

$$E_r^{p,q} \implies \mathrm{Ext}_{\mathfrak{h},L}^{p+q}(\mathcal{F}_{\mathfrak{g},K}^{\mathfrak{h},L}(X), \, V)$$

with E_2 term $\qquad E_2^{p,q} = \mathrm{Ext}_{\mathfrak{g},K}^{p}(X, \, (I_{\mathfrak{h},L}^{\mathfrak{g},K})^q(V)).$

The differential on $E_r^{p,q}$ has bidegree $(r, 1 - r)$.

PROOF. (a) Frobenius reciprocity (Proposition 2.34) gives

$$\mathrm{Hom}_{\mathfrak{g},K}(\,\cdot\,, X) \circ P_{\mathfrak{h},L}^{\mathfrak{g},K} \cong \mathrm{Hom}_{\mathfrak{h},L}(\,\cdot\,, \, (\mathcal{F}^{\vee})_{\mathfrak{g},K}^{\mathfrak{h},L}(X)),$$

and $P_{\mathfrak{h},L}^{\mathfrak{g},K}$ carries projectives to projectives, by Corollary 2.35. Applying Variation (2) of the Grothendieck Spectral Sequence (Theorem D.51) for derived functors of a composition, we obtain (a).

(b) Frobenius reciprocity (Proposition 2.21) gives

$$\mathrm{Hom}_{\mathfrak{g},K}(X, \,\cdot\,) \circ I_{\mathfrak{h},L}^{\mathfrak{g},K} \cong \mathrm{Hom}_{\mathfrak{h},L}(\mathcal{F}_{\mathfrak{g},K}^{\mathfrak{h},L}(X), \,\cdot\,),$$

and $I_{\mathfrak{h},L}^{\mathfrak{g},K}$ carries injectives to injectives, by Corollary 2.23. Applying Variation (5) of the Grothendieck Spectral Sequence, we obtain (b).

Corollary 5.114. Let $(\mathfrak{h}, L) \to (\mathfrak{g}, K)$ be a map of pairs, and let V be in $\mathcal{C}(\mathfrak{h}, L)$.

(a) If q_0 is the least q for which $(P_{\mathfrak{h},L}^{\mathfrak{g},K})_q(V) \neq 0$, then q_0 is also the least q for which there is a $(\mathfrak{g}, K)$ module X_0 such that

$$\mathrm{Ext}_{\mathfrak{h},L}^q(V, (\mathcal{F}^{\vee})_{\mathfrak{g},K}^{\mathfrak{h},L}(X_0)) \neq 0.$$

If X is any $(\mathfrak{g}, K)$ module, then

$$\mathrm{Hom}_{\mathfrak{g},K}((P_{\mathfrak{h},L}^{\mathfrak{g},K})_{q_0}(V), X) \cong \mathrm{Ext}_{\mathfrak{h},L}^{q_0}(V, (\mathcal{F}^{\vee})_{\mathfrak{g},K}^{\mathfrak{h},L}(X)).$$

(b) If q_0 is the least q for which $(I_{\mathfrak{h},L}^{\mathfrak{g},K})^q(V) \neq 0$, then q_0 is also the least q for which there is a $(\mathfrak{g}, K)$ module X_0 such that

$$\mathrm{Ext}_{\mathfrak{h},L}^q(\mathcal{F}_{\mathfrak{g},K}^{\mathfrak{h},L}(X_0), V) \neq 0.$$

If X is any $(\mathfrak{g}, K)$ module, then

$$\mathrm{Hom}_{\mathfrak{g},K}(X, (I_{\mathfrak{h},L}^{\mathfrak{g},K})^{q_0}(V)) \cong \mathrm{Ext}_{\mathfrak{h},L}^{q_0}(\mathcal{F}_{\mathfrak{g},K}^{\mathfrak{h},L}(X), V).$$

REMARKS. The proof will show also that, in (a), if $(P_{\mathfrak{h},L}^{\mathfrak{g},K})_q(V) = 0$ for all q, then
$$\mathrm{Ext}_{\mathfrak{h},L}^q(V, (\mathcal{F}^{\vee})_{\mathfrak{g},K}^{\mathfrak{h},L}(X)) = 0$$
for all $X \in \mathcal{C}(\mathfrak{g}, K)$. Similarly, in (b), if $(I_{\mathfrak{h},L}^{\mathfrak{g},K})^q(V) = 0$ for all q, then

$$\mathrm{Ext}_{\mathfrak{h},L}^q(\mathcal{F}_{\mathfrak{g},K}^{\mathfrak{h},L}(X), V) = 0$$

for all $X \in \mathcal{C}(\mathfrak{g}, K)$.

PROOF. We prove (a), the argument for (b) being similar. Fix X in $\mathcal{C}(\mathfrak{g}, K)$. For $q < q_0$ we have $(P_{\mathfrak{h},L}^{\mathfrak{g},K})_q(V) = 0$ by hypothesis. Therefore, in the notation of Proposition 5.113a, we obtain

$$(5.115) \quad E_r^{p,q} \cong \begin{cases} \mathrm{Hom}_{\mathfrak{g},K}((P_{\mathfrak{h},L}^{\mathfrak{g},K})_{q_0}(V), X) & \text{if } q = q_0 \text{ and } p = 0 \\ 0 & \text{if } q < q_0 \end{cases}$$

for $r = 2$. We shall prove by induction on r that (5.115) remains valid for all $r \geq 2$, the case $r = 2$ being the base case for the induction. Assume (5.115) for some r with $r \geq 2$. The module $E_{r+1}^{p,q}$ is isomorphic to a quotient of a submodule of $E_r^{p,q}$ and is 0 by (5.115) if $q < q_0$. The module E_{r+1}^{0,q_0} is isomorphic with the cohomology of the complex

$$(5.116) \qquad\qquad E_r^{-r,q_0-1+r} \longrightarrow E_r^{0,q_0} \longrightarrow E_r^{r,q_0-r},$$

since Proposition 5.113a says that the differential has bidegree $(r, 1-r)$. Since $q_0 - r < q_0$, (5.115) shows that $E_r^{r,q_0-r} = 0$. Since $-r < 0$ and since the spectral sequence in Proposition 5.113a is first quadrant, $E_r^{-r,q_0-1+r} = 0$. Therefore the cohomology of (5.116) is E_r^{0,q_0}, and we have $E_{r+1}^{0,q_0} \cong E_r^{0,q_0}$. This completes the induction, proving (5.115) for all $r \geq 2$. Hence

$$(5.117) \qquad\qquad E_2^{p,q} \cong \cdots \cong E_r^{p,q} \cong \cdots$$

for $(p,q) = (0, q_0)$ and for all (p,q) with $q < q_0$. By Proposition D.26b, the spectral sequence is convergent. Because of (5.117) and (5.115), we thus have

$$(5.118)$$
$$E_\infty^{p,q} \cong E_2^{p,q} \cong \begin{cases} \mathrm{Hom}_{\mathfrak{g},K}((P_{\mathfrak{h},L}^{\mathfrak{g},K})_{q_0}(V), \, X) & \text{if } q = q_0 \text{ and } p = 0 \\ 0 & \text{if } q < q_0 \end{cases}$$

The second line of (5.118), together with (D.28) and Proposition 5.113a, imply

$$(5.119\text{a}) \qquad 0 = \bigoplus_{p+q'=q} E_\infty^{p,q'} \cong \mathrm{Gr}\,\mathrm{Ext}_{\mathfrak{h},L}^q(V, \, (\mathcal{F}^\vee)_{\mathfrak{g},K}^{\mathfrak{h},L}(X))$$

for $q' < q_0$. Also they imply

$$(5.119\text{b}) \qquad\begin{aligned} E_\infty^{0,q_0} &= \bigoplus_{p+q=q_0} E_\infty^{p,q} \cong \mathrm{Gr}\,\mathrm{Ext}_{\mathfrak{h},L}^{q_0}(V, \, (\mathcal{F}^\vee)_{\mathfrak{g},K}^{\mathfrak{h},L}(X)) \\ &\cong \mathrm{Ext}_{\mathfrak{h},L}^{q_0}(V, \, (\mathcal{F}^\vee)_{\mathfrak{g},K}^{\mathfrak{h},L}(X)). \end{aligned}$$

Combining (5.119) and the first line of (5.118), we obtain

$$\mathrm{Ext}_{\mathfrak{h},L}^q(V, \, (\mathcal{F}^\vee)_{\mathfrak{g},K}^{\mathfrak{h},L}(X)) \cong \begin{cases} \mathrm{Hom}_{\mathfrak{g},K}((P_{\mathfrak{h},L}^{\mathfrak{g},K})_{q_0}(V), \, X) & \text{if } q = q_0 \\ 0 & \text{if } q < q_0. \end{cases}$$

Choosing $X_0 = (P_{\mathfrak{h},L}^{\mathfrak{g},K})_{q_0}(V)$, we obtain all the conclusions of (a). This proves the corollary.

The style of proof in the corollary is called a **corner argument**. It works with any Grothendieck spectral sequence in the presence of the kind of vanishing condition assumed in the corollary. Evidently it is applicable in a wider class of situations as well, and we shall see an example in Corollary 5.123.

Let us now return to the hypotheses in §1, with $(\mathfrak{g}, K)$ a reductive pair, etc. Repeated applications of Proposition 5.113 yield the following **fundamental spectral sequences** for cohomological induction.

Theorem 5.120. Let X be in $\mathcal{C}(\mathfrak{g}, K)$.

(a) There exist two first-quadrant spectral sequences with differential of bidegree $(r, 1 - r)$, with respective E_2 terms

$$E_2^{p,q} = \operatorname{Ext}_{\mathfrak{g}, K}^p(\mathcal{L}_q(Z), X)$$

and $$E_2^{p,q} = \operatorname{Ext}_{\mathfrak{l}, L \cap K}^p(Z^{\#}, H^q(\bar{\mathfrak{u}}, (\mathcal{F}^{\vee})_{\mathfrak{g}, K}^{\bar{\mathfrak{q}}, L \cap K}(X))),$$

and with a common abutment

$$E_r^{p,q} \Longrightarrow \operatorname{Ext}_{\bar{\mathfrak{q}}, L \cap K}^{p+q}(\mathcal{F}_{\mathfrak{l}, L \cap K}^{\bar{\mathfrak{q}}, L \cap K}(Z^{\#}), (\mathcal{F}^{\vee})_{\mathfrak{g}, K}^{\bar{\mathfrak{q}}, L \cap K}(X)).$$

(b) There exist two first-quadrant spectral sequences with differential of bidegree $(r, 1 - r)$, with respective E_2 terms

$$E_2^{p,q} = \operatorname{Ext}_{\mathfrak{g}, K}^p(X, \mathcal{R}^q(Z))$$

and $$E_2^{p,q} = \operatorname{Ext}_{\mathfrak{l}, L \cap K}^p(H_q(\mathfrak{u}, \mathcal{F}_{\mathfrak{g}, K}^{\mathfrak{q}, L \cap K}(X)), Z^{\#}),$$

and with a common abutment

$$E_r^{p,q} \Longrightarrow \operatorname{Ext}_{\mathfrak{q}, L \cap K}^{p+q}(\mathcal{F}_{\mathfrak{g}, K}^{\mathfrak{q}, L \cap K}(X), (\mathcal{F}^{\vee})_{\mathfrak{l}, L \cap K}^{\mathfrak{q}, L \cap K}(Z^{\#})).$$

REMARKS. In (b) the abutment is written with $\mathcal{F}^{\vee}$ in order to facilitate comparing (b) with (a). The "pseudo-forgetful" functor $\mathcal{F}^{\vee}$ may be replaced in (b) by the forgetful functor $\mathcal{F}$, by Proposition 2.33.

The presence of pseudo-forgetful functors in (a) is more troublesome. An infinite-dimensional $(\mathfrak{g}, K)$ module is rarely admissible as an $L \cap K$ module, and for this reason $\mathcal{F}_{\mathfrak{g}, K}^{\bar{\mathfrak{q}}, L \cap K}(X)$ is usually not isomorphic to $(\mathcal{F}^{\vee})_{\mathfrak{g}, K}^{\bar{\mathfrak{q}}, L \cap K}(X)$ under the inclusion j of Proposition 2.33. In Corollary 5.141 in §10 we shall show that j does induce an isomorphism on $\bar{\mathfrak{u}}$ cohomology. Consequently the pseudo-forgetful functors $\mathcal{F}^{\vee}$ in (a) also may be replaced by forgetful functors. This remark applies equally to Corollaries 5.121, 5.122, and 5.123 below.

PROOF.

(a) First we apply Proposition 5.113a to the inclusion of pairs $(\bar{\mathfrak{q}}, L \cap K) \hookrightarrow (\mathfrak{g}, K)$, taking $V = \mathcal{F}_{\mathfrak{l},L \cap K}^{\bar{\mathfrak{q}},L \cap K}(Z^{\#})$, and we obtain the first spectral sequence. Then we apply Proposition 5.113b to the map of pairs $(\bar{\mathfrak{q}}, L \cap K) \to (\mathfrak{l}, L \cap K)$, taking $X = Z^{\#}$ and $V = (\mathcal{F}^{\vee})_{\mathfrak{g},K}^{\bar{\mathfrak{q}},L \cap K}(X)$. The E_2 term in this case is

$$E_2^{p,q} = \mathrm{Ext}_{\mathfrak{l},L \cap K}^{p}(Z^{\#}, (I_{\bar{\mathfrak{q}},L \cap K}^{\mathfrak{l},L \cap K})^{q}(\mathcal{F}^{\vee})_{\mathfrak{g},K}^{\bar{\mathfrak{q}},L \cap K}(X))$$
$$= \mathrm{Ext}_{\mathfrak{l},L \cap K}^{p}(Z^{\#}, H^{q}(\bar{\mathfrak{u}}, (\mathcal{F}^{\vee})_{\mathfrak{g},K}^{\bar{\mathfrak{q}},L \cap K}(X))),$$

by Proposition 3.12, and thus we obtain the second spectral sequence.

(b) First we apply Proposition 5.113b to the inclusion of pairs $(\mathfrak{q}, L \cap K) \hookrightarrow (\mathfrak{g}, K)$, taking $V = (\mathcal{F}^{\vee})_{\mathfrak{l},L \cap K}^{\mathfrak{q},L \cap K}(Z^{\#})$, and we obtain the first spectral sequence. Then we apply Proposition 5.113a to the map of pairs $(\mathfrak{q}, L \cap K) \to (\mathfrak{l}, L \cap K)$, taking $X = Z^{\#}$ and $V = \mathcal{F}_{\mathfrak{g},K}^{\mathfrak{q},L \cap K}(X)$. The E_2 term in this case is

$$E_2^{p,q} = \mathrm{Ext}_{\mathfrak{l},L \cap K}^{p}((P_{\mathfrak{q},L \cap K}^{\mathfrak{l},L \cap K})_q(\mathcal{F}_{\mathfrak{g},K}^{\mathfrak{q},L \cap K}((X)), Z^{\#})$$
$$= \mathrm{Ext}_{\mathfrak{l},L \cap K}^{p}(H_q(\mathfrak{u}, (\mathcal{F}_{\mathfrak{g},K}^{\mathfrak{q},L \cap K}(X)), Z^{\#}),$$

by Proposition 3.12, and thus we obtain the second spectral sequence.

Corollary 5.121. Let X be in $\mathcal{C}(\mathfrak{g}, K)$.

(a) If $\mathcal{L}_q(Z) = 0$ for $q \neq q_0$, then there is a first-quadrant spectral sequence

$$E_r^{p,q} \Longrightarrow \mathrm{Ext}_{\mathfrak{g},K}^{p+q-q_0}(\mathcal{L}_{q_0}(Z), X)$$

with differential of bidegree $(r, 1-r)$ and with E_2 term

$$E_2^{p,q} = \mathrm{Ext}_{\mathfrak{l},L \cap K}^{p}(Z^{\#}, H^q(\bar{\mathfrak{u}}, (\mathcal{F}^{\vee})_{\mathfrak{g},K}^{\bar{\mathfrak{q}},L \cap K}(X)).$$

(b) If $\mathcal{R}^q(Z) = 0$ for $q \neq q_0$, then there is a first-quadrant spectral sequence

$$E_r^{p,q} \Longrightarrow \mathrm{Ext}_{\mathfrak{g},K}^{p+q-q_0}(X, \mathcal{R}^{q_0}(Z))$$

with differential of bidegree $(r, 1-r)$ and with E_2 term

$$E_2^{p,q} = \mathrm{Ext}_{\mathfrak{l},L \cap K}^{p}(H_q(\mathfrak{u}, \mathcal{F}_{\mathfrak{g},K}^{\mathfrak{q},L \cap K}(X)), Z^{\#}).$$

PROOF. We prove (a), the argument for (b) being similar. Under our hypothesis the first spectral sequence of Theorem 5.120a has $E_2^{p,q} = 0$

for $q \neq q_0$. Since the bidegree is $(r, 1 - r)$, it follows that the spectral sequence collapses and degenerates, and we read off from (D.28) and Theorem 5.120a that

$$\text{Ext}^p_{\mathfrak{g}, K}(\mathcal{L}_{q_0}(Z), X) = E_2^{p, q_0}$$
$$\cong E_\infty^{p, q_0}$$
$$\cong \text{Ext}^{p+q_0}_{\bar{\mathfrak{q}}, L \cap K}(\mathcal{F}^{\bar{\mathfrak{q}}, L \cap K}_{\mathfrak{l}, L \cap K}(Z^\#), (\mathcal{F}^\vee)^{\bar{\mathfrak{q}}, L \cap K}_{\mathfrak{g}, K}(X)).$$

In other words, the abutment of the second spectral sequence of Theorem 5.120a has

$$\text{Ext}^{p+q}_{\bar{\mathfrak{q}}, L \cap K}(\mathcal{F}^{\bar{\mathfrak{q}}, L \cap K}_{\mathfrak{l}, L \cap K}(Z^\#), (\mathcal{F}^\vee)^{\bar{\mathfrak{q}}, L \cap K}_{\mathfrak{g}, K}(X)) \cong \text{Ext}^{p+q-q_0}_{\mathfrak{g}, K}(\mathcal{L}_{q_0}(Z), V).$$

Substituting this value of the abutment into the statement of existence of the second spectral sequence, we obtain the statement of (a) in the corollary.

Corollary 5.122. Suppose L is compact and X is in $\mathcal{C}(\mathfrak{g}, K)$.

(a) There is a first-quadrant spectral sequence

$$E_r^{p,q} \implies \text{Hom}_L(Z^\#, H^{p+q}(\bar{\mathfrak{u}}, (\mathcal{F}^\vee)^{\bar{\mathfrak{q}}, L \cap K}_{\mathfrak{g}, K}(X))$$

with differential of bidegree $(r, 1 - r)$ and with E_2 term

$$E_2^{p,q} = \text{Ext}^p_{\mathfrak{g}, K}(\mathcal{L}_q(Z), X).$$

(b) There is a first-quadrant spectral sequence

$$E_r^{p,q} \implies \text{Hom}_L(H_{p+q}(\mathfrak{u}, \mathcal{F}^{\mathfrak{q}, L \cap K}_{\mathfrak{g}, K}(X)), Z^\#)$$

with differential of bidegree $(r, 1 - r)$ and with E_2 term

$$E_2^{p,q} = \text{Ext}^p_{\mathfrak{g}, K}(X, \mathcal{R}^q(Z)).$$

PROOF. We prove (a), the argument for (b) being similar. Since L is compact, Lemma 2.4 and (C.14) show that $\text{Hom}_L(\cdot, \cdot)$ is exact in each variable. Therefore

$$\text{Ext}^p_{\mathfrak{l}, L}(Z^\#, H^q(\bar{\mathfrak{u}}, (\mathcal{F}^\vee)^{\bar{\mathfrak{q}}, L \cap K}_{\mathfrak{g}, K}(X)) = 0 \qquad \text{for } p > 0.$$

In Theorem 5.120a, the second spectral sequence thus has

$$E_2^{p,q} = 0 \qquad \text{for } p > 0.$$

It follows immediately that the second spectral sequence collapses and degenerates, and the discussion after (D.28) shows that $E_2^{0,q}$ equals the abutment. That is,

$$\mathrm{Hom}_L(Z^\#,\ H^q(\bar{\mathfrak{u}},\ (\mathcal{F}^\vee)_{\mathfrak{g},K}^{\bar{\mathfrak{q}},L\cap K}(X))$$

$$\cong \mathrm{Ext}_{\bar{\mathfrak{q}},L\cap K}^q(\mathcal{F}_{\mathfrak{l},L\cap K}^{\bar{\mathfrak{q}},L\cap K}(Z^\#),\ (\mathcal{F}^\vee)_{\mathfrak{g},K}^{\bar{\mathfrak{q}},L\cap K}(X)).$$

Hence the first spectral sequence of Theorem 5.120a has

$$E_r^{p,q} \implies \mathrm{Hom}_L(Z^\#,\ H^{p+q}(\bar{\mathfrak{u}},\ (\mathcal{F}^\vee)_{\mathfrak{g},K}^{\bar{\mathfrak{q}},L\cap K}(X))),$$

and the result follows.

Corollary 5.123. Suppose L is compact and Z is an irreducible representation of L.

(a) Let q_0 be the least q such that there is a $(\mathfrak{g}, K)$ module X with $H^q(\bar{\mathfrak{u}},\ (\mathcal{F}^\vee)_{\mathfrak{g},K}^{\bar{\mathfrak{q}},L}(X))$ containing $Z^\#$. Then q_0 is also the least q such that $\mathcal{L}_q(Z) \neq 0$, and

$$\mathrm{Hom}_{\mathfrak{g},K}(\mathcal{L}_{q_0}(Z), X) \cong \mathrm{Hom}_L(Z^\#,\ H^{q_0}(\bar{\mathfrak{u}},\ (\mathcal{F}^\vee)_{\mathfrak{g},L}^{\bar{\mathfrak{q}},L}(X))).$$

(If q_0 does not exist, then $\mathcal{L}_q(Z) = 0$ for all q.)

(b) Let q_0 be the least q such that there is a $(\mathfrak{g}, K)$ module X with $H_q(\mathfrak{u},\ \mathcal{F}_{\mathfrak{g},K}^{q,L}(X))$ containing $Z^\#$. Then q_0 is also the least q such that $\mathcal{R}^q(Z) \neq 0$, and

$$\mathrm{Hom}_{\mathfrak{g},K}(X, \mathcal{R}^{q_0}(Z)) \cong \mathrm{Hom}_L(H_{q_0}(\mathfrak{u},\ \mathcal{F}_{\mathfrak{g},L}^{q,L}(X)),\ Z^\#).$$

(If q_0 does not exist, then $\mathcal{R}^q(Z) = 0$ for all q.)

PROOF. Each part follows from Corollary 5.121 by a corner argument (in the style of the proof of Corollary 5.114).

9. Spectral Sequences for Analysis of K Types

Derived functors of a filtered module yield spectral sequences that are useful in relating ind and pro from $\mathfrak{q}$ or $\bar{\mathfrak{q}}$ to ind and pro from $\mathfrak{q} \cap \mathfrak{k}$ or $\bar{\mathfrak{q}} \cap \mathfrak{k}$. For the first of these, recall from the proof of Theorem 5.35a that we formed a filtration of

$$M = \mathrm{ind}_{\bar{\mathfrak{q}},L\cap K}^{\mathfrak{g},L\cap K}(Z_{\bar{\mathfrak{q}}}^\#)$$

by $(\mathfrak{k}, L \cap K)$ modules

$$M_n = \text{image in } M \text{ of } U_{n,\mathfrak{k}}(\mathfrak{g}) \otimes_{\mathbb{C}} Z_{\bar{\mathfrak{q}}}^{\#}.$$

Here $U_{n,\mathfrak{k}}(\mathfrak{g})$ was defined just before Lemma 5.34. Under the assumption that $h_{\delta(\mathfrak{u})}$ acts in Z by a scalar and that V is a finite-dimensional K module, the filtration

(5.124) $$0 = M_{-1} \subseteq M_0 \subseteq M_1 \subseteq \ldots M$$

had the property that

$$\text{Ext}^j_{\mathfrak{k}, L \cap K}(M/M_N, V) = \text{Ext}^j_{\mathfrak{k}, L \cap K}(M_{N+1}/M_N, V) = 0$$

for all large N. By (5.32a),

$$\text{Hom}_K(\Pi^K_j(M/M_N), V) = \text{Hom}_K(\Pi^K_j(M_{N+1}/M_N), V) = 0$$

for such N. Note, however, that "large N" depends on V; the filtration (5.124) usually cannot be replaced by a finite filtration independent of V. This circumstance makes the use of spectral sequences more cumbersome.

Anyway, (5.56) showed that

$$M_p/M_{p-1} \cong \text{ind}^{\mathfrak{k}}_{\bar{\mathfrak{q}} \cap \mathfrak{k}}(S^p(\mathfrak{g}/(\bar{\mathfrak{q}} + \mathfrak{k})) \otimes_{\mathbb{C}} Z_{\bar{\mathfrak{q}}}^{\#}),$$

a special case being the formula

$$M_0 \cong \text{ind}^{\mathfrak{k}}_{\bar{\mathfrak{q}} \cap \mathfrak{k}}(Z_{\bar{\mathfrak{q}}}^{\#}).$$

Applying Proposition D.57b, we obtain the following result for the Π functor.

Proposition 5.125. Let Z be an $(\mathfrak{l}, L \cap K)$ module. For every integer $N \geq 0$, there exists a convergent spectral sequence

$$E_r^{p,q} \Longrightarrow \Pi^K_{p+q}(M_N)$$

in $\mathcal{C}(\mathfrak{k}, K)$ with a differential of bidegree $(-r, r-1)$ and with E_1 term

$$E_1^{p,q} = \Pi^K_{p+q}(\text{ind}^{\mathfrak{k}}_{\bar{\mathfrak{q}} \cap \mathfrak{k}}(S^p(\mathfrak{g}/(\bar{\mathfrak{q}} + \mathfrak{k})) \otimes_{\mathbb{C}} Z_{\bar{\mathfrak{q}}}^{\#}))$$

for $0 \leq p \leq N$ and $q \geq -p$. Suppose that $h_{\delta(\mathfrak{u})}$ acts in Z by a scalar and that V is any finite-dimensional $(\mathfrak{k}, K)$ module. Then for all N sufficiently large,

(5.126) $$\text{Hom}_K(\Pi^K_j(M_N), V) \cong \text{Hom}_K(\Pi^K_j(\text{ind}^{\mathfrak{g}, L \cap K}_{\bar{\mathfrak{q}}, L \cap K}(Z_{\bar{\mathfrak{q}}}^{\#})), V),$$

the isomorphism being induced by the inclusion $M_N \to \text{ind}^{\mathfrak{g}, L \cap K}_{\bar{\mathfrak{q}}, L \cap K}(Z_{\bar{\mathfrak{q}}}^{\#})$.

REMARKS. By Proposition 2.115, $\Pi^K_{p+q} \circ \mathcal{F}^{\mathfrak{k}, L \cap K}_{\mathfrak{g}, L \cap K} \cong \mathcal{F}^{\mathfrak{k}, K}_{\mathfrak{g}, K} \circ \Pi_{p+q}$. Hence the first argument of the last Hom is $\mathcal{L}_{p+q}(Z)$ but is to be viewed in the category $\mathcal{C}(\mathfrak{k}, K)$. Thus the proposition effectively relates $\text{Hom}_K(\mathcal{L}^K_q(\cdot), V)$ to $\text{Hom}_K(\mathcal{L}_q(Z), V)$ via a spectral sequence that depends on V.

Proposition 5.125 does not help very much in simplifying the proof of Theorem 5.35a. However, it does help to predict some other results. For example, on a formal level, Theorem 5.64 follows by applying (D.30) to Proposition 5.125. (The spectral sequence does not give a quick complete proof of Theorem 5.64, however, because the finite-dimensionality of various Ext's still has to be addressed.) Also Proposition 5.125 points to construction of the bottom-layer map in Theorem 5.80a.

To construct the bottom-layer map $\mathcal{B}_Z : \mathcal{L}_q^K(Z) \to \mathcal{L}_q(Z)$ on a K type V in this setting, we choose N large enough for (5.126) to hold. Then we observe the vanishing result

$$(5.127) \qquad E_1^{p,q} = 0 \qquad \text{for } p + q > \dim(\mathfrak{u} \cap \mathfrak{k}) = S.$$

(This follows from Proposition 5.71 and is an instance of the computation (4.170) in the compact case.) Since the differential has bidegree $(-r, 1 - r)$, $E_r^{0,q}$ maps to $E_r^{-r,1-r+q}$, which is 0. Hence $E_2^{0,q}$ is a quotient of $E_1^{0,q}$, $E_3^{0,q}$ is a quotient of $E_2^{0,q}$, and so on. Because of the convergence, we conclude that $E_\infty^{0,q}$ is a quotient of $E_1^{0,q}$. From the location of $E_\infty^{0,q}$ in the abutment, one can check that $E_\infty^{0,q}$ is a submodule of $\Pi_q^K(M_N)$. Thus the quotient map of $E_1^{0,q}$ to $E_\infty^{0,q}$ gives us a map of $\mathcal{L}_q^K(Z)$ into $\Pi_q^K(M_N)$. By unwinding the definitions, we can see that this map is $\mathcal{B}_Z$ on the K type V.

In this setting it is easy to see that $\mathcal{B}_Z$ is one-one in degree S. (This part of the conclusion was easy also in the proof of Theorem 5.80a.) Namely the kernel of $E_1^{0,S} \to E_\infty^{0,S}$ comes from the images of the maps $E_r^{r,S+1-r} \to E_r^{0,S}$; these images are 0 because (5.127) shows that $E_r^{r,S+1-r}$ is 0.

A similar analysis is available for the Γ functor. The tools are the proof of Theorem 5.35b, the isomorphism (5.43), and Proposition D.57a. The filtration M_n is as in (5.38), and the conclusion is as follows.

Proposition 5.128. Let Z be an $(\mathfrak{l}, L \cap K)$ module. For every integer $N \geq 0$, there exists a convergent spectral sequence

$$E_r^{p,q} \Longrightarrow \Gamma_K^{p+q}(M/M_N)$$

in $\mathcal{C}(\mathfrak{k}, K)$ with a differential of bidegree $(r, 1 - r)$ and with E_1 term

$$E_1^{p,q} = \Gamma_K^{p+q}(\text{pro}_{\mathfrak{q}\cap\mathfrak{k}}^{\mathfrak{k}}(\text{Hom}_{\mathbb{C}}(S^p(\mathfrak{g}/(\mathfrak{q} + \mathfrak{k})), Z_\mathfrak{q}^\#)))$$

for $0 \leq p \leq N$ and $q \geq -p$. Suppose that $h_{\delta(\mathfrak{u})}$ acts in Z by a scalar and that V is any finite-dimensional $(\mathfrak{k}, K)$ module. Then for all N sufficiently large,

$$\text{Hom}_K(V, \Gamma_K^j(M/M_N)) \cong \text{Hom}_K(V, \Gamma_K^j(\text{pro}_{\mathfrak{q}, L\cap K}^{\mathfrak{g}, L\cap K}(Z_\mathfrak{q}^\#))),$$

the isomorphism being induced by the quotient

$$\text{pro}_{q,L\cap K}^{\mathfrak{g},L\cap K}(Z_q^{\#}) \to M/M_N.$$

10. Hochschild-Serre Spectral Sequences

In this section we give results in the spirit of Propositions 5.125 and 5.128 that relate $\mathfrak{u}$ homology and cohomology to $\mathfrak{u} \cap \mathfrak{k}$ homology and cohomology. The proofs will not use general derived-functor arguments but will instead rely on computations with standard resolutions.

We begin by selecting in $\bigwedge(\mathfrak{u}\cap\mathfrak{p})$ a sequence $\{V_p\}_{p=0}^N$ of $L\cap K$ invariant subspaces such that

(5.129)

    (a) $V_0 = \bigwedge^0(\mathfrak{u} \cap \mathfrak{p}) = \mathbb{C}$

    (b) $V_p \subseteq \bigwedge^{r(p)}(\mathfrak{u} \cap \mathfrak{p})$, i.e., the members of V_p are homogeneous of some degree $r(p) \leq p$; also $p' \leq p$ implies $r(p') \leq r(p)$

    (c) $(\mathfrak{u}\cap\mathfrak{k})V_p \subseteq \langle V_0, V_1, \ldots, V_{p-1}\rangle.$

EXAMPLE. In $G = SO(4, 1)$, write

$$\Delta(\mathfrak{g}, \mathfrak{h}) = \{\pm e_1 \pm e_2, \ \pm e_1, \ \pm e_2\}$$

as usual, with the long roots compact and the short roots noncompact. Take $L = T$ and

$$\Delta^+(\mathfrak{g}, \mathfrak{h}) = \{e_1 \pm e_2, \ e_1, \ e_2\},$$

so that $\Delta(\mathfrak{u} \cap \mathfrak{p}) = \{e_1, e_2\}$. If we let e_α be a nonzero root vector for the root α, then we can take

$$V_0 = \langle 1\rangle, \quad V_1 = \langle e_{e_1}\rangle, \quad V_2 = \langle e_{e_2}\rangle, \quad V_3 = \langle e_{e_1} \wedge e_{e_2}\rangle.$$

Each V_p is T invariant, and

$$r(0) = 0, \quad r(1) = r(2) = 1, \quad r(3) = 2.$$

Hence (5.129a) and (5.129b) are satisfied. Property (5.129c) comes down to the fact that

$$e_{e_1-e_2}(V_2) \subseteq \langle V_0, V_1\rangle,$$

which follows from the relation $[e_{e_1-e_2}, e_{e_2}] = ce_{e_1}.$

To prove that (5.129) can always be achieved, we decompose $\bigwedge^n(\mathfrak{u}\cap\mathfrak{p})$ into a direct sum of irreducible $L \cap K$ invariant subspaces $U_1, \ldots, U_m$. Let $h_{\delta(\mathfrak{u})}$ act in U_j with the eigenvalue $d_j > 0$, and let $h_{\delta(\mathfrak{u})}$ act in $\mathfrak{u} \cap \mathfrak{k}$ fully reducibly with the eigenvalues $c_i > 0$. Then $h_{\delta(\mathfrak{u})}$ acts in $(\mathfrak{u} \cap \mathfrak{k})U_j$ fully reducibly with eigenvalues of the form $c_i + d_j$. Thus if we arrange that $d_1 \geq \cdots \geq d_m$, we obtain

$$(\mathfrak{u} \cap \mathfrak{k})U_j \subseteq \langle U_1, \ldots, U_{j-1}\rangle.$$

When the U_j's are ordered in this way, the sequence of V_p's can be formed by concatenating the various sequences of U_j's, with their degrees increasing.

Theorem 5.130 (Hochschild-Serre spectral sequences). Let X be a $(\mathfrak{q}, L \cap K)$ module, and let V_p be as in (5.129).

(a) There exists a convergent spectral sequence

$$E_r^{p,q} \implies H_{p+q}(\mathfrak{u}, X)$$

in $\mathcal{C}(\mathfrak{l} \cap \mathfrak{k}, L \cap K)$ with a differential of bidegree $(-r, r-1)$ and with E_1 term

$$E_1^{p,q} = H_{p+q-r(p)}(\mathfrak{u} \cap \mathfrak{k}, X) \otimes_{\mathbb{C}} V_p.$$

(b) There exists a convergent spectral sequence

$$E_r^{p,q} \implies H^{p+q}(\mathfrak{u}, X)$$

in $\mathcal{C}(\mathfrak{l} \cap \mathfrak{k}, L \cap K)$ with a differential of bidegree $(r, 1-r)$ and with E_1 term

$$E_1^{p,q} = H^{p+q-r(p)}(\mathfrak{u} \cap \mathfrak{k}, X) \otimes_{\mathbb{C}} V_p^*.$$

PROOF.

(a) Recall from Proposition 3.12 that the derived functors of $P_{\mathfrak{q},L\cap K}^{\mathfrak{l},L\cap K}$ on X compute the $\mathfrak{u}$ homology of X with an $(\mathfrak{l}, L \cap K)$ action in place. Moreover, these derived functors may be computed from the complex in $\mathcal{C}(\mathfrak{q}, L \cap K)$ whose n^{th} term is $\bigwedge^n \mathfrak{u} \otimes_{\mathbb{C}} X$, the differential ∂ being given by (2.126). Put $A = \bigwedge \mathfrak{u} \otimes_{\mathbb{C}} X$, and regard (A, ∂) as a module with differential in the category $\mathcal{C}(\mathfrak{q}\cap\mathfrak{k}, L\cap K)$. The module A is graded, with

$$^nA = \bigwedge^n \mathfrak{u} \otimes_{\mathbb{C}} X.$$

Let

$$(5.131) \qquad W^p = \bigoplus_{p' \leq p} (\bigwedge(\mathfrak{u} \cap \mathfrak{k}) \otimes_{\mathbb{C}} V_{p'}) \qquad \text{and} \qquad A^p = W^p \otimes_{\mathbb{C}} X$$

so that

$$0 = A^{-1} \subseteq A^0 \subseteq A^1 \subseteq \cdots \subseteq A^N = A$$

is an increasing filtration of A by $(\mathfrak{q} \cap \mathfrak{k}, L \cap K)$ modules, by (5.129c). We shall prove that

(5.132) $$\partial A^p \subseteq A^p.$$

The spectral sequence of the filtration converges and abuts on $H(A)$. In fact, only finitely many indices (p, q) are involved, and the proof of Proposition D.26′ works after adjustments in notation. Thus we have

(5.133a) $$E_r^{p,q} \implies H_{p+q}(\mathfrak{u}, X)$$

by (D.16), the bidegree is $(-r, r - 1)$, and the E_1 term is

(5.133b) $$E_1^{p,q} = H(^{p+q}A^p / {}^{p+q}A^{p-1})$$

by (D.9b). We shall prove also that

(5.134) $$(^nA^p / {}^nA^{p-1}, \partial) \cong ((\textstyle\bigwedge^{n-r(p)}(\mathfrak{u} \cap \mathfrak{k}) \otimes_{\mathbb{C}} X) \otimes_{\mathbb{C}} V_p, \partial_{\mathfrak{k}} \otimes 1),$$

where $\partial_{\mathfrak{k}}$ is the differential for computing $\mathfrak{u} \cap \mathfrak{k}$ homology from $\bigwedge(\mathfrak{u} \cap \mathfrak{k}) \otimes_{\mathbb{C}} X$. Combining (5.133) and (5.134), we obtain the conclusion (a) of the theorem. To prove (5.132), we form the sequence of isomorphisms

$$
\begin{aligned}
(\textstyle\bigwedge^{n-r(p)}(\mathfrak{u} \cap \mathfrak{k}) & \otimes_{\mathbb{C}} X) \otimes_{\mathbb{C}} V_p \\
&\cong (\textstyle\bigwedge^{n-r(p)}(\mathfrak{u} \cap \mathfrak{k}) \otimes_{\mathbb{C}} V_p) \otimes_{\mathbb{C}} X \\
&\cong (\textstyle\bigwedge^n \mathfrak{u} \cap (W^p / W^{p-1})) \otimes_{\mathbb{C}} X \\
&\cong ((\textstyle\bigwedge^n \mathfrak{u} \cap W^p) \otimes_{\mathbb{C}} X) / ((\textstyle\bigwedge^n \mathfrak{u} \cap W^{p-1}) \otimes_{\mathbb{C}} X) \\
&\cong {}^nA^p / {}^nA^{p-1},
\end{aligned}
$$

writing $^n\tau^p$ both for the first isomorphism of the sequence and for the composition of all four isomorphisms. For $\eta \in \bigwedge^{n-r(p)}(\mathfrak{u} \cap \mathfrak{k}), \xi \in V_p$, and $x \in X$, we shall show that

(5.135) $$\partial(^n\tau^p((\eta \otimes x) \otimes \xi)) - {}^{n-1}\tau^p(\partial_{\mathfrak{k}}(\eta \otimes x) \otimes \xi) \quad \text{is in } {}^{n-1}A^{p-1}.$$

Since $\partial_{\mathfrak{k}}(\eta \otimes x)$ is in $\bigwedge^{n-1-r(p)}(\mathfrak{u} \cap \mathfrak{k}) \otimes_{\mathbb{C}} X$, it follows that $^{n-1}\tau^p(\partial_{\mathfrak{k}}(\eta \otimes x) \otimes \xi)$ is in $^nA^p$. Then (5.135) shows that $\partial(^n\tau^p((\eta \otimes x) \otimes \xi))$ is in $^nA^p$. Consequently (5.132) holds. Since $^n\tau^p$ is an $L \cap K$ isomorphism onto $^n A^p / {}^n A^{p-1}$, (5.135) also proves (5.134). In other words, the theorem will follow if we prove (5.135).

For simplicity of exposition, let us write

$$\eta = Y_1 \wedge \cdots \wedge Y_{n-r(p)} \qquad \text{and} \qquad \xi = X_1 \wedge \cdots \wedge X_{r(p)}$$

even though η and ξ should really be taken to be linear combinations of such terms. Then

(5.136)
$$\partial(^n \tau^p((\eta \otimes x) \otimes \xi)) = \partial((\eta \otimes \xi) \otimes x)$$

(a)
$$= \sum_l (-1)^l (Y_1 \wedge \cdots \wedge \widehat{Y_l} \wedge \cdots \wedge Y_{n-r(p)} \wedge \xi \otimes Y_l x)$$

(b)
$$+ \sum_l (-1)^{l+n-r(p)} (\eta \wedge X_1 \wedge \cdots \wedge \widehat{X_l} \wedge \cdots \wedge X_{r(p)} \otimes X_l x)$$

(c)
$$+ \sum_{r<s} (-1)^{r+s} ([Y_r, Y_s] \wedge Y_1 \wedge \cdots \wedge \widehat{Y_r} \wedge \cdots$$
$$\wedge \widehat{Y_s} \wedge \cdots \wedge Y_{n-r(p)} \wedge \xi \otimes x)$$

(d)
$$+ \sum_{r<s} (-1)^{r+s} ([X_r, X_s] \wedge \eta \wedge X_1 \wedge \cdots \wedge \widehat{X_r} \wedge \cdots$$
$$\wedge \widehat{X_s} \wedge \cdots \wedge X_{r(p)} \otimes x)$$

(e)
$$+ \sum_{r<s} (-1)^{r+s+n-r(p)} ([Y_r, X_s] \wedge Y_1 \wedge \cdots \wedge \widehat{Y_r} \wedge \cdots \wedge Y_{n-r(p)}$$
$$\wedge X_1 \wedge \cdots \wedge \widehat{X_s} \wedge \cdots \wedge X_{r(p)} \otimes x).$$

Meanwhile $^{n-1}\tau^p(\partial_{\mathfrak{k}}(\eta \otimes x) \otimes \xi)$ is the sum of (a) and (c). Therefore (5.135) is the sum of (b), (d), and (e). For each of these, let us regroup the alternating tensors according to the decomposition

$$\bigwedge^{n-1}\mathfrak{u} = \bigoplus \bigwedge^{n-1-r(p')}(\mathfrak{u} \cap \mathfrak{k}) \otimes_{\mathbb{C}} V_{p'}.$$

In (b), the p''s that occur have $r(p') = r(p) - 1$ and therefore $p' < p$, by (5.129b). Hence these terms are in $^{n-1}A^{p-1}$. In (d), the p''s that occur have $r(p') = r(p) - 1$ or $r(p') = r(p) - 2$. Thus these p''s have $p' < p$, and the terms are in $^{n-1}A^{p-1}$.

For fixed r, the contribution to (5.136e) is

$$= \sum_s (-1)^{s+1} (Y_1 \wedge \cdots \wedge \widehat{Y_r} \wedge \cdots \wedge Y_{n-r(p)}$$
$$\wedge [Y_r, X_s] \wedge X_1 \wedge \cdots \wedge \widehat{X_s} \wedge \cdots \wedge X_{r(p)} \otimes x)$$

$$= Y_1 \wedge \cdots \wedge \widehat{Y_r} \wedge \cdots \wedge Y_{n-r(p)} \wedge (\operatorname{ad} Y_r)\eta \otimes x,$$

and this corresponds only to p''s with $p' < p$, by (5.129c). Hence (5.136b), (5.136d), and (5.136e) are all in $^{n-1}A^{p-1}$, and (5.135) holds. This completes the proof of (a) of the theorem.

(b) The argument proceeds along similar lines, and we just sketch it. The $\mathfrak{u}$ cohomology of X, with an $(\mathfrak{l}, L \cap K)$ action in place, may be computed from the complex in $\mathcal{C}(\mathfrak{q}, L \cap K)$ whose n^{th} term is $\text{Hom}_{\mathbb{C}}(\bigwedge^n \mathfrak{u}, X)$, the differential d being given by (2.127). Put $A = \text{Hom}_{\mathbb{C}}(\bigwedge \mathfrak{u}, X)$, and regard (A, d) as a module with differential in the category $\mathcal{C}(\mathfrak{q} \cap \mathfrak{k}, L \cap K)$. The module A is graded, with

$$^n A = \text{Hom}_{\mathbb{C}}(\textstyle\bigwedge^n \mathfrak{u}, X).$$

With W^p as in (5.131), let

$$A^p = \{f \in A \mid f(W^{p-1}) = 0\}.$$

Then

$$A = A^0 \supseteq A^1 \supseteq \cdots \supseteq A^N \supseteq A^{N+1} = 0$$

is a decreasing filtration of A by $(\mathfrak{q} \cap \mathfrak{k}, L \cap K)$ modules. The main step is to prove that

(5.137) $$d A^p \subseteq A^p.$$

The spectral sequence of the filtration converges and has

(5.138a) $$E_r^{p,q} \Longrightarrow H^{p+q}(\mathfrak{u}, X),$$

the bidegree is $(r, 1 - r)$, and the E_1 term is

(5.138b) $$E_1^{p,q} = H(^{p+q}A^p / ^{p+q}A^{p+1}).$$

The other thing that needs proof is that

(5.139) $$(^n A^p / ^n A^{p+1}, d) \cong (\text{Hom}_{\mathbb{C}}(\textstyle\bigwedge^{n-r(p)}(\mathfrak{u} \cap \mathfrak{k}), X) \otimes_{\mathbb{C}} V_p^*, d_{\mathfrak{k}} \otimes 1),$$

where $d_{\mathfrak{k}}$ is the differential for computing $\mathfrak{u} \cap \mathfrak{k}$ cohomology from $\text{Hom}_{\mathbb{C}}(\bigwedge(\mathfrak{u} \cap \mathfrak{k}), X)$. Combining (5.138) and (5.139), we obtain the conclusion (b) of the theorem.

To prove (5.137), one begins by establishing (5.139) as an $L \cap K$ isomorphism of modules. Then one proves an analog of (5.135) and shows that (5.137) and the full (5.139) follow from it. The heart of the argument is a computation similar to (5.136) and what follows it. This completes the proof of (b) of the theorem.

We give two applications of Theorem 5.130. The first is to the issue of admissibility of $\mathfrak{u}$ homology and cohomology.

Corollary 5.140. If X is admissible as a $(\mathfrak{g}, K)$ module, then $H_n(\mathfrak{u}, X)$ and $H^n(\mathfrak{u}, X)$ are admissible as $(\mathfrak{l}, L \cap K)$ modules.

PROOF. In the case of $H_n(\mathfrak{u}, X)$, it is enough to prove that $H_n(\mathfrak{u}, X)$ is admissible as an $(\mathfrak{l}, (L \cap K)_0)$ module. If φ is an irreducible representation of K_0 and Φ is an irreducible representation of K, then φ occurs in $\Phi|_{K_0}$ if and only if Φ occurs in $\mathrm{induced}_{K_0}^K \varphi$, by Frobenius reciprocity. Since $\mathrm{induced}_{K_0}^K \varphi$ is finite dimensional, φ occurs in the restriction to K_0 of only finitely many irreducible representations of K. Consequently the admissibility of X as a $(\mathfrak{g}, K)$ module implies the admissibility of X as a $(\mathfrak{g}, K_0)$ module.

Under this condition, we shall prove that $E_1^{p,q}$ in Theorem 5.130a is admissible as an $(L \cap K)_0$ module. Since $E_\infty^{p,q}$ is a quotient of a submodule of $E_1^{p,q}$, $E_\infty^{p,q}$ is then admissible. Summing this conclusion over $p + q = n$ and applying the theorem, we see that $H_n(\mathfrak{u}, X)$ is admissible.

Let (μ, V_μ) be an $(L \cap K)_0$ type, and write

$$V_\mu \otimes_{\mathbb{C}} V_p^* \cong \bigoplus_{(\tau, V_\tau) \in (L \cap K)_0^{\wedge}} n_\tau \tau.$$

Use of characters shows that the multiplicities of $(L \cap K)_0$ types satisfy

$$[H_n(\mathfrak{u} \cap \mathfrak{k}, X) \otimes_{\mathbb{C}} V_p : \mu] = \sum n_\tau [H_n(\mathfrak{u} \cap \mathfrak{k}, X) : \tau].$$

Hence $E_1^{p,q}$ will be admissible as an $(L \cap K)_0$ module if $H_n(\mathfrak{u} \cap \mathfrak{k}, X)$ is. Decompose X under K_0 as $X \cong \bigoplus_{(\varphi, V_\varphi) \in \widehat{K_0}} m_\varphi \varphi$, and write

$$H_n(\mathfrak{u} \cap \mathfrak{k}, X) = \sum m_\varphi H_n(\mathfrak{u} \cap \mathfrak{k}, V_\varphi).$$

By Corollary 3.13

$$H_n(\mathfrak{u} \cap \mathfrak{k}, V_\varphi) \cong H^{S-n}(\mathfrak{u} \cap \mathfrak{k}, V_\varphi \otimes_{\mathbb{C}} \textstyle\bigwedge^S(\mathfrak{u} \cap \mathfrak{k})).$$

Kostant's Theorem (Theorem 4.139) identifies the right side as a finite sum of certain $(L \cap K)_0$ types with highest weights of the form $w(\lambda + \delta(\mathfrak{n} \cap \mathfrak{k})) - \delta(\mathfrak{n} \cap \mathfrak{k})$, all with multiplicity 1. Here λ is the highest weight of $V_\varphi \otimes_{\mathbb{C}} \bigwedge^S(\mathfrak{u} \cap \mathfrak{k})$, and w is in the Weyl group. Since an $(L \cap K)_0$ type can occur for only one φ, it follows that $H_n(\mathfrak{u} \cap \mathfrak{k}, X)$ is admissible as an $(L \cap K)_0$ module.

This completes the argument for $H_n(\mathfrak{u}, X)$. The argument for $H^n(\mathfrak{u}, X)$ is similar except that Theorem 5.130b is used in place of Theorem 5.130a and it is not necessary to invoke Corollary 3.13.

The second application of Theorem 5.130 is to the isomorphism promised in the remarks after Theorem 5.120. For notational simplicity we phrase the result in terms of $\mathfrak{u}$ instead of $\bar{\mathfrak{u}}$.

Corollary 5.141. If X is in $\mathcal{C}(\mathfrak{g}, K)$, then the one-one map

$$j : \mathcal{F}_{\mathfrak{g},K}^{\mathfrak{q},L \cap K}(X) \to (\mathcal{F}^\vee)_{\mathfrak{g},K}^{\mathfrak{q},L \cap K}(X)$$

of Proposition 2.33 induces an isomorphism of $(\mathfrak{l}, L \cap K)$ modules

$$j^n : H^n(\mathfrak{u}, \mathcal{F}_{\mathfrak{g},K}^{\mathfrak{q},L \cap K}(X)) \to H^n(\mathfrak{u}, (\mathcal{F}^\vee)_{\mathfrak{g},K}^{\mathfrak{q},L \cap K}(X))$$

in every degree n.

REMARKS. Because cohomology is a functor from $\mathcal{C}(\mathfrak{q}, L \cap K)$ to $\mathcal{C}(\mathfrak{l}, L \cap K)$, j^n is a map of $(\mathfrak{l}, L \cap K)$ modules. To prove that j^n is an isomorphism, we need only show that it is an isomorphism when regarded as a map in $\mathcal{C}(\mathfrak{l} \cap \mathfrak{k}, L \cap K)$. For this purpose we shall compute the domain and the range using the Hochschild-Serre spectral sequence of Theorem 5.130b. We begin with two lemmas.

Lemma 5.142. Let A and B be modules for a ring R, and suppose that A and B are endowed with filtrations

$$0 = A^{-1} \subseteq A^0 \subseteq \cdots \subseteq A^N = A$$

$$0 = B^{-1} \subseteq B^0 \subseteq \cdots \subseteq B^N = B$$

by R submodules. Let $j : A \to B$ be an R module map such that $j(A^p) \subseteq B^p$ for all p, and let j^p be the corresponding map of consecutive quotients given by

$$j^p : A^p/A^{p-1} \to B^p/B^{p-1} \qquad \text{for } p = 0, 1, \dots, N.$$

If every map j^p is an isomorphism, then j is an isomorphism.

PROOF. We use induction on N, the base case $N = 0$ being trivial. Suppose $N > 0$. We then have a commutative diagram

$$
\begin{array}{ccccccccc}
0 & \longrightarrow & A^{N-1} & \longrightarrow & A & \longrightarrow & A/A^{N-1} & \longrightarrow & 0 \\
& & \downarrow{\scriptstyle j|_{A^{N-1}}} & & \downarrow{\scriptstyle j} & & \downarrow{\scriptstyle j^N} & & \\
0 & \longrightarrow & B^{N-1} & \longrightarrow & B & \longrightarrow & B/B^{N-1} & \longrightarrow & 0
\end{array}
$$

with exact rows. The inductive hypothesis shows that $j|_{A^{N-1}}$ maps A^{N-1} isomorphically onto B^{N-1}, and j^N is an isomorphism by hypothesis. By (C.31), j is an isomorphism. This completes the induction.

If $(\mathfrak{g}, H)$ is a pair, recall from (2.1) that $\tilde{C}(\mathfrak{g}, H)$ denotes the category of all $R(\mathfrak{g}, H)$ modules, approximately unital or not.

Lemma 5.143. Suppose $(\mathfrak{g}, H)$ is a pair, Y is in $C(\mathfrak{g}, H)$, and Z is in $\tilde{C}(\mathfrak{g}, H)$. Then the natural inclusion $Z_H \subseteq Z$ induces an isomorphism of $(\mathfrak{g}, H)$ modules

$$Y \otimes_{\mathbb{C}} (Z_H) \to (Y \otimes_{\mathbb{C}} Z)_H.$$

PROOF. The inclusion $Z_H \subseteq Z$ induces a one-one map of $R(\mathfrak{g}, H)$ modules $Y \otimes_{\mathbb{C}} (Z_H) \to Y \otimes_{\mathbb{C}} Z$, and the image is contained in $(Y \otimes_{\mathbb{C}} Z)_H$. What must be checked is that the image is all of $(Y \otimes_{\mathbb{C}} Z)_H$. Since the action of $\mathfrak{g}$ does not affect the definitions of any of these modules and maps as sets and functions, we may assume that $\mathfrak{g} = \mathfrak{h}$. Suppose $u = \sum y_i \otimes z_i$ is in $(Y \otimes Z)_H$. Since the sum is finite, we can find a finite-dimensional H invariant subspace E of Y containing all the y_i. To show that u is in the image of $Y \otimes_{\mathbb{C}} (Z_H)$, it suffices to replace Y by the submodule E. Thus we may assume that Y is finite dimensional. Now if F is any finite-dimensional representation of H, we have

$\mathrm{Hom}_H(F, Y \otimes_{\mathbb{C}} Z) \cong \mathrm{Hom}_H(F \otimes_{\mathbb{C}} Y^*, Z)$

$\cong \mathrm{Hom}_H(F \otimes_{\mathbb{C}} Y^*, Z_H)$ since $F \otimes Y^*$ is H finite

$\cong \mathrm{Hom}_H(F, Y \otimes_{\mathbb{C}} (Z_H)).$

The images of members of $\mathrm{Hom}_H(F, Y \otimes_{\mathbb{C}} Z)$ span $(Y \otimes_{\mathbb{C}} Z)_H$, but they are always contained in $Y \otimes_{\mathbb{C}} (Z_H)$ because of the isomorphism. The lemma follows.

PROOF OF COROLLARY 5.141. The map j induces $L \cap K$ maps

$$(5.144) \quad j_1^{p,q} : H^{p+q-r(p)}(\mathfrak{u} \cap \mathfrak{k}, \mathcal{F}_{\mathfrak{g},K}^{\mathfrak{q},L \cap K}(X)) \otimes V_p^*$$

$$\to H^{p+q-r(p)}(\mathfrak{u} \cap \mathfrak{k}, (\mathcal{F}^\vee)_{\mathfrak{g},K}^{\mathfrak{q},L \cap K}(X)) \otimes V_p^*.$$

We propose to show that each $j_1^{p,q}$ is an isomorphism.

Granting this conclusion for the moment, let us finish the proof of the theorem. Appendix D does not address the naturality of the construction of a spectral sequence, but we can check by inspection that the maps $j_1^{p,q}$ respect the differentials and hence induce recursively

$$j_r^{p,q} : E_r^{p,q}(\mathcal{F}_{\mathfrak{g},K}^{\mathfrak{q},L \cap K}(X)) \to E_r^{p,q}((\mathcal{F}^\vee)_{\mathfrak{g},K}^{\mathfrak{q},L \cap K}(X)).$$

We deduce first that all the maps $j_r^{p,q}$ are isomorphisms for $r \geq 1$. Since the Hochschild-Serre spectral sequences in Theorem 5.130 converge, the maps $j_r^{p,q}$ induce an isomorphism $j_\infty^{p,q}$ between the spaces $E_\infty^{p,q}$. Now we apply Lemma 5.142 to the filtrations of $H^n(\mathfrak{u}, \mathcal{F}_{\mathfrak{g},K}^{q,L\cap K}(X))$ and $H^n(\mathfrak{u}, (\mathcal{F}^\vee)_{\mathfrak{g},K}^{q,L\cap K}(X))$ used in the spectral sequences. The map j^n respects the filtrations (by the naturality of their construction), and the induced maps on subquotients are just the various $j_\infty^{p,q}$. Since we have seen that the $j_\infty^{p,q}$ are all isomorphisms, it follows from the lemma that j^n is an isomorphism as well, as we wished to show.

It remains to show that $j_1^{p,q}$ in (5.144) is an isomorphism. From (5.144) it is apparent that $j_1^{p,q}$ is unaffected by the Lie algebra action outside $\mathfrak{k}$. Thus we may assume from the outset that j in Proposition 2.33 is a one-one map

$$j : \mathcal{F}_{\mathfrak{k},K}^{q\cap\mathfrak{k},L\cap K}(X) \to (\mathcal{F}^\vee)_{\mathfrak{k},K}^{q\cap\mathfrak{k},L\cap K}(X),$$

and we are to show that the induced map of $(\mathfrak{l}\cap\mathfrak{k}, L\cap K)$ modules

$$(5.145) \qquad j^n : H^n(\mathfrak{u}\cap\mathfrak{k}, \mathcal{F}_{\mathfrak{k},K}^{q\cap\mathfrak{k},L\cap K}(X)) \to H^n(\mathfrak{u}\cap\mathfrak{k}, (\mathcal{F}^\vee)_{\mathfrak{k},K}^{q\cap\mathfrak{k},L\cap K}(X))$$

is an isomorphism.

For each $\gamma \in \widehat{K}$, write X_γ for the corresponding isotypic subspace of X. Then

$$X \cong \bigoplus_{\gamma\in\widehat{K}} X_\gamma.$$

It is clear from the standard complex (2.127) that taking cohomology commutes with direct sums. Hence the domain space in (5.145) has

$$(5.146) \qquad H^n(\mathfrak{u}\cap\mathfrak{k}, \mathcal{F}_{\mathfrak{k},K}^{q\cap\mathfrak{k},L\cap K}(X)) \cong \bigoplus_{\gamma\in\widehat{K}} H^n(\mathfrak{u}\cap\mathfrak{k}, \mathcal{F}_{\mathfrak{k},K}^{q\cap\mathfrak{k},L\cap K}(X_\gamma)).$$

For the range space in (5.145), recall from (2.27) and Lemma 2.28 that

$$(\mathcal{F}^\vee)_{\mathfrak{k},K}^{q\cap\mathfrak{k},L\cap K}(X) = \left(\prod_{\gamma\in\widehat{K}} X_\gamma\right)_{L\cap K}.$$

The cohomology of this module is computed from the standard complex

$$(\textstyle\bigwedge^n(\mathfrak{u}\cap\mathfrak{k}))^* \otimes_{\mathbb{C}} \left(\prod_{\gamma\in\widehat{K}} X_\gamma\right)_{L\cap K},$$

which Lemma 5.143 allows us to rewrite as

$$\cong \left((\textstyle\bigwedge^n(\mathfrak{u}\cap\mathfrak{k}))^* \otimes_{\mathbb{C}} \prod_{\gamma\in\widehat{K}} X_\gamma\right)_{L\cap K}.$$

Since tensoring with a finite-dimensional vector space commutes with direct products, this module is

$$\cong \Big(\prod_{\gamma \in \widehat{K}} (\textstyle\bigwedge^n(\mathfrak{u} \cap \mathfrak{k}))^* \otimes_{\mathbb{C}} X_\gamma \Big)_{L \cap K}.$$

Now the $L \cap K$ finite-part functor on $\tilde{\mathcal{C}}(\mathfrak{l} \cap \mathfrak{k}, L \cap K)$ is exact by (2.2). Hence the cohomology of the $L \cap K$ finite part of a complex is the $L \cap K$ finite part of the cohomology. Consequently the cohomology of $\mathcal{F}^\vee(X)$ is the $L \cap K$ finite part of the cohomology of the complex

$$\prod_{\gamma \in \widehat{K}} (\textstyle\bigwedge^n(\mathfrak{u} \cap \mathfrak{k}))^* \otimes_{\mathbb{C}} X_\gamma.$$

Since the cohomology of a direct product of complexes is the direct product of the cohomologies, we conclude that

$$H^n(\mathfrak{u} \cap \mathfrak{k}, (\mathcal{F}^\vee)_{\mathfrak{k},K}^{\mathfrak{q} \cap \mathfrak{k}, L \cap K}(X)) \cong \Big(\prod_{\gamma \in \widehat{K}} H^n(\mathfrak{u} \cap \mathfrak{k}, X_\gamma) \Big)_{L \cap K}.$$

The factors here are essentially computed by Kostant's Theorem, Theorem 4.139. (Remarks on the case of disconnected K may be found after its proof.) What matters for us is that for each irreducible representation τ of $L \cap K$, there are only finitely many irreducible representations $\gamma_1, \ldots, \gamma_m$ of K with the property that τ occurs in $H^n(\mathfrak{u} \cap \mathfrak{k}, X_\gamma)$. (If K is connected, there is only one such γ.) If we use a subscript τ to denote extraction of the τ isotypic subspace, we therefore have

$$H^n(\mathfrak{u} \cap \mathfrak{k}, (\mathcal{F}^\vee)_{\mathfrak{k},K}^{\mathfrak{q} \cap \mathfrak{k}, L \cap K}(X))_\tau \cong \Big(\prod_{i=1}^m H^n(\mathfrak{u} \cap \mathfrak{k}, X_{\gamma_i}) \Big)_\tau.$$

A finite direct product is isomorphic to the direct sum. Thus we can rewrite the right side as

$$\cong \Big(\bigoplus_{i=1}^m H^n(\mathfrak{u} \cap \mathfrak{k}, X_{\gamma_i}) \Big)_\tau \cong \Big(\bigoplus_{\gamma \in \widehat{K}} H^n(\mathfrak{u} \cap \mathfrak{k}, X_\gamma) \Big)_\tau.$$

By (5.146) the right side is isomorphic to $H^n(\mathfrak{u} \cap \mathfrak{k}, \mathcal{F}_{\mathfrak{k},K}^{\mathfrak{q} \cap \mathfrak{k}, L \cap K}(X))_\tau$. Hence the natural inclusion induces an isomorphism

$$H^n(\mathfrak{u} \cap \mathfrak{k}, \mathcal{F}_{\mathfrak{k},K}^{\mathfrak{q} \cap \mathfrak{k}, L \cap K}(X))_\tau \cong H^n(\mathfrak{u} \cap \mathfrak{k}, (\mathcal{F}^\vee)_{\mathfrak{k},K}^{\mathfrak{q} \cap \mathfrak{k}, L \cap K}(X))_\tau.$$

Since τ is arbitrary, (5.145) is an isomorphism. Hence (5.144) is an isomorphism, and the proof is complete.

11. Composite P Functors and I Functors

Our final use of spectral sequences in this chapter is in connection with the composition of maps of pairs

$$(5.147) \qquad (\mathfrak{j}, M) \to (\mathfrak{h}, L) \to (\mathfrak{g}, K).$$

We shall work with the corresponding I functors, but similar conclusions are valid also for the P functors. Proposition 2.19 gives a natural isomorphism

$$(5.148) \qquad I_{\mathfrak{j},M}^{\mathfrak{g},K} \cong I_{\mathfrak{h},L}^{\mathfrak{g},K} \circ I_{\mathfrak{j},M}^{\mathfrak{h},L},$$

but it is not generally true that

$$(5.149) \qquad (I_{\mathfrak{j},M}^{\mathfrak{g},K})^{p+q} \overset{?}{\cong} (I_{\mathfrak{h},L}^{\mathfrak{g},K})^q \circ (I_{\mathfrak{j},M}^{\mathfrak{h},L})^p.$$

Instead the two sides are related by the Grothendieck Spectral Sequence (Theorem D.51). Use of this spectral sequence assumes that $I_{\mathfrak{j},M}^{\mathfrak{h},L}$ carries injectives to injectives; this hypothesis is verified by Corollary 2.23.

Besides (5.148), there is one interesting case where (5.149) is valid, and it is of importance in the proof of Hard Duality (Theorem 3.5).

Proposition 5.150. For a composition of maps of pairs (5.147), suppose that $(I_{\mathfrak{j},M}^{\mathfrak{h},L})^p = 0$ for $p > m_1$ and $(I_{\mathfrak{h},L}^{\mathfrak{g},K})^q = 0$ for $q > m_2$. Then

$$(I_{\mathfrak{h},L}^{\mathfrak{g},K})^{m_2} \circ (I_{\mathfrak{j},M}^{\mathfrak{h},L})^{m_1}(V) \cong (I_{\mathfrak{j},M}^{\mathfrak{g},K})^{m_1+m_2}(V)$$

naturally for V in $\mathcal{C}(\mathfrak{j}, M)$.

PROOF. In view of (5.148) and Corollary 2.23, the Grothendieck Spectral Sequence (Theorem D.51) gives a first-quadrant spectral sequence $E_r^{p,q} \Longrightarrow (I_{\mathfrak{j},M}^{\mathfrak{g},K})^{p+q}(V)$ with E_2 term $E_2^{p,q} \cong (I_{\mathfrak{h},L}^{\mathfrak{g},K})^q \circ (I_{\mathfrak{j},M}^{\mathfrak{h},L})^p(V)$ such that the differential has bidegree $(r, 1-r)$. Our assumptions imply that $E_2^{p,q} = 0$ if $p > m_1$ or $q > m_2$. Hence $E_r^{p,q} = 0$ for $r \geq 2$ when $p > m_1$ or $q > m_2$. Because the differential has bidegree $(r, 1-r)$, an easy induction on r shows that $E_r^{p,q} \cong E_2^{p,q}$ when $p = m_1$ or $q = m_2$. On the other hand, the spectral sequence is first quadrant and therefore converges. Thus $E_2^{p,q} \cong E_\infty^{p,q}$ if $p+q = m_1+m_2$. When $p+q = m_1+m_2$ and $(p, q) \neq (m_1, m_2)$, we have $E_\infty^{p,q} = 0$. Thus the abutment in degree $m_1 + m_2$ is just $E_2^{m_1,m_2}$, and we have an isomorphism

$$(I_{\mathfrak{h},L}^{\mathfrak{g},K})^{m_2} \circ (I_{\mathfrak{j},M}^{\mathfrak{h},L})^{m_1}(V) \cong E_2^{m_1,m_2} \cong (I_{\mathfrak{j},M}^{\mathfrak{g},K})^{m_1+m_2}(V).$$

This proves the required isomorphism.

There is still the matter of naturality in V. Suppose $\varphi : V \to V'$ is a map in $\mathcal{C}(\mathfrak{j}, M)$. It is necessary to go over the arguments in §D.3 to see that versions of the map φ can be implemented throughout the argument, at least with diagrams commuting up to homotopy. Carrying through a number of details that we omit, we find that the isomorphism in the proposition is natural in V.

The hypotheses of the proposition apply in the setting of §III.8, where we have a composition of maps of pairs

$$(\mathfrak{h}, L) \to (\mathfrak{g}, K') \hookrightarrow (\mathfrak{g}, K)$$

with $i_{\mathrm{alg}}(\mathfrak{h}) = \mathfrak{g}$ and $i_{\mathrm{gp}}(L) = K'$. Since $m = m_1 + m_2$ by (3.91), Proposition 5.150 establishes the natural isomorphism (3.93). This completes the proof of Hard Duality.

CHAPTER VI

SIGNATURE THEOREM

The Signature Theorem relates positivity of an invariant Hermitian form for a module Z with positivity of an invariant Hermitian form for the cohomologically induced module $\mathcal{L}_S(Z)$, but just on the K types of the bottom layer. A precise statement of the theorem requires a definition of a group-theoretic notion of signature, with respect to a compact Lie group, of an invariant Hermitian form on an isotypic subspace of a representation of the compact group.

Hermitian forms, and sesquilinear forms more generally, may be converted into the framework of good categories and modules as maps from a module to its Hermitian dual. Combining use of this framework with both Easy and Hard Duality, one associates to any invariant sesquilinear form for an $(\mathfrak{l}, L \cap K)$ module Z an invariant sesquilinear form for $\mathcal{L}_S(Z)$ called the associated "Shapovalov form." The equality of the signature of the given form on each $L \cap K$ type with the signature of the Shapovalov form on the corresponding K type of the bottom layer is the subject of the Signature Theorem.

The proof consists in reducing matters to a K analog, so that Z can be replaced by an irreducible $(\mathfrak{l} \cap \mathfrak{k}, L \cap K)$ module, and in making an explicit computation for the irreducible case.

1. Setting

This chapter contains the second of the five main theorems of the book. The Signature Theorem relates positivity of an invariant Hermitian form for a module Z with positivity of an invariant Hermitian form for the cohomologically induced module $\mathcal{L}_S(Z)$, but just on the K types of the bottom layer. Part of the power of the theorem comes from the absence of any hypothesis of dominance for the infinitesimal character of Z.

Except in §2 the setting will be the following throughout the chapter. We let $(\mathfrak{g}, K)$ be a reductive pair, with G as the corresponding reductive group. Let $\mathfrak{q} = \mathfrak{l} \oplus \mathfrak{u}$ be a θ stable parabolic subalgebra (containing, by definition, some θ stable Cartan subalgebra $\mathfrak{h}_0$ of $\mathfrak{g}_0$), and let L be the corresponding Levi subgroup of G. Then $(\mathfrak{q}, L \cap K)$ is the corresponding parabolic subpair. The statement and proof of the Signature Theorem are built around the bottom-layer map, and consequently we assume in §§1–6 that

$$(6.1) \qquad L \text{ meets every component of } G, \text{ i.e., } K = (L \cap K)K_0.$$

Since $\mathfrak{q}$ and $\mathfrak{u}$ are θ stable, we have $\mathfrak{q}^- = \bar{\mathfrak{q}}$ and $\mathfrak{u}^- = \bar{\mathfrak{u}}$.

We denote by Z a module in $\mathcal{C}(\mathfrak{l}, L \cap K)$, and we define

$$(6.2) \qquad\qquad Z^{\#} = Z \otimes_{\mathbb{C}} \bigwedge^{\text{top}} \mathfrak{u}$$

as another module in $\mathcal{C}(\mathfrak{l}, L \cap K)$. As in Chapter V, we let

$$(6.3) \qquad\qquad S = \dim(\mathfrak{u} \cap \mathfrak{k}) \qquad \text{and} \qquad R = \dim(\mathfrak{u} \cap \mathfrak{p}).$$

The cohomologically induced modules $\mathcal{L}_j(Z)$ and $\mathcal{R}^j(Z)$ are defined in (5.3). In the Signature Theorem we shall assume that $\mathcal{L}_S(Z)$ is admissible as an $(\mathfrak{g}, K)$ module and that $h_{\delta(\mathfrak{u})}$ acts by a scalar in Z. By Theorem 5.64 it is enough to assume that Z is an admissible $(\mathfrak{l}, L \cap K)$ module and that $h_{\delta(\mathfrak{u})}$ acts as a scalar in Z.

To simplify the notation we shall systematically use the abbreviations

$$(6.4a) \qquad Z_{\bar{\mathfrak{q}}}^{\#} = \mathcal{F}_{\mathfrak{l}, L \cap K}^{\bar{\mathfrak{q}}, L \cap K}(Z^{\#}) \qquad \text{and} \qquad Z_{\mathfrak{q}}^{\#} = \mathcal{F}_{\mathfrak{l}, L \cap K}^{\mathfrak{q}, L \cap K}(Z^{\#}).$$

As we observed in §V.6, we can intersect $\mathfrak{g}$, $\mathfrak{q}$, $\bar{\mathfrak{q}}$, $\mathfrak{u}$, $\bar{\mathfrak{u}}$, and $\mathfrak{l}$ with $\mathfrak{k}$ and obtain valid reductive and parabolic pairs for study. Specifically $(\mathfrak{k}, K)$ is a reductive pair, and $\mathfrak{q} \cap \mathfrak{k}$ is a θ stable parabolic subalgebra of $\mathfrak{k}$. The parabolic and Levi subpairs are respectively $(\mathfrak{q} \cap \mathfrak{k}, L \cap K)$ and $(\mathfrak{l} \cap \mathfrak{k}, L \cap K)$, and (6.1) remains valid. However, we continue to write $Z_{\bar{\mathfrak{q}}}^{\#}$ and $Z_{\mathfrak{q}}^{\#}$ for the K analogs

$$(6.4b) \qquad Z_{\bar{\mathfrak{q}}}^{\#} = \mathcal{F}_{\mathfrak{l}, L \cap K}^{\bar{\mathfrak{q}} \cap \mathfrak{k}, L \cap K}(Z^{\#}) \qquad \text{and} \qquad Z_{\mathfrak{q}}^{\#} = \mathcal{F}_{\mathfrak{l}, L \cap K}^{\mathfrak{q} \cap \mathfrak{k}, L \cap K}(Z^{\#}).$$

The K analogs $\mathcal{L}_j^K(Z)$ and $\mathcal{R}_K^j(Z)$ of $\mathcal{L}_j(Z)$ and $\mathcal{R}^j(Z)$ are defined in (5.70). The K types of the **bottom layer** are defined to be those occurring in $\mathcal{L}_S^K(Z)$. These are identified in the remarks with Corollary 5.72. If ρ_L is any irreducible representation of $L \cap K$ occurring in Z such that the highest weights of $\rho_L \otimes_{\mathbb{C}} \bigwedge^R(\mathfrak{u} \cap \mathfrak{p})$ are K dominant, then the representation ρ_G of K given by Theorem 4.83 and having its space of $\mathfrak{u} \cap \mathfrak{k}$ invariants of type $\rho_L \otimes_{\mathbb{C}} \bigwedge^R(\mathfrak{u} \cap \mathfrak{p})$ is in the bottom layer; moreover, no other K types are in the bottom layer.

The bottom-layer map

$$(6.5) \qquad\qquad \mathcal{B}_Z : \mathcal{L}_j^K(Z) \to \mathcal{L}_j(Z)$$

was introduced in (5.74) and (5.76) and relates these modules. For $j = S$ in the case that $h_{\delta(\mathfrak{u})}$ acts by a scalar, Theorem 5.80a says that $\mathcal{B}_Z$ is one-one and that it is onto the sum of the full isotypic subspaces for the K types of the bottom layer.

According to the end of §V.6, when $h_{\delta(\mathfrak{u})}$ acts in Z by a scalar, we can recognize the K types of the bottom layer as follows: Let μ_L be a highest weight of any $L \cap K$ type of Z, and let μ_G be a highest weight of a K type of $\mathcal{L}_S(Z)$. Then

$$(6.6) \qquad \mu_G(h_{\delta(\mathfrak{u})}) \geq \mu_L(h_{\delta(\mathfrak{u})}) + 2\delta(\mathfrak{u} \cap \mathfrak{p})(h_{\delta(\mathfrak{u})}),$$

and equality holds if and only if the K type in question is in the bottom layer.

2. Hermitian Dual and Signature

For this section only, we relax the assumptions in §1. Our objectives are to define a Hermitian dual functor in a suitable setting, to discuss invariant sesquilinear forms, and to define a group-theoretic notion of "signature" for a Hermitian form invariant under a compact Lie group.

For now, we let $(\mathfrak{g}, B)$ be any pair such that $\mathfrak{g}$ is the complexification of a real Lie algebra $\mathfrak{g}_0$ with $\mathfrak{b}_0 \subseteq \mathfrak{g}_0$ and with $\mathrm{Ad}(B)$ normalizing $\mathfrak{g}_0$.

We begin with some discussion of Hermitian duals. We define conjugate linear automorphisms (denoted by bars) of $U(\mathfrak{g})$ and $R(B)$ as follows: For X in $\mathfrak{g}$, $X \mapsto \bar{X}$ is conjugation of $\mathfrak{g}$ with respect to $\mathfrak{g}_0$, and this map is extended to $U(\mathfrak{g})$ as a conjugate linear automorphism. For T in $R(B)$, $\bar{T}$ denotes the complex conjugate distribution to T with $\langle \bar{T}, f \rangle = \overline{\langle T, \bar{f} \rangle}$. (The B finiteness of $\bar{T}$ is clear since if $A \subseteq \hat{B}$ is a finite subset such that the convolution $\chi_A T$ equals T, then also $\bar{\chi}_A \bar{T} = \bar{T}$.)

As in (1.3) we denote the right and left regular representations of $\mathfrak{b}$ on $R(B)$ by r and l. Then we have

$$\overline{r(X)T} = r(\bar{X})\bar{T} \qquad \text{and} \qquad \overline{l(X)T} = l(\bar{X})\bar{T}$$

for $X \in \mathfrak{b}$ and $T \in R(B)$, and it follows by the same argument as for Proposition 1.122 that conjugation makes sense on $R(\mathfrak{g}, B)$ and is a conjugate linear automorphism. Conjugation commutes with the canonical transpose mapping on $R(\mathfrak{g}, B)$, which we continue to denote by $(\cdot)^t$.

If V is a complex vector space, we let $\bar{V}$ be the same underlying real vector space with multiplication by i in $\bar{V}$ the negative of multiplication by i in V. Denoting multiplication by i in V and $\bar{V}$ by J_V and $J_{\bar{V}}$, respectively, we have $J_{\bar{V}} = -J_V$. If $L : V \to V$ is a complex-linear map, then L remains complex linear as a map $L : \bar{V} \to \bar{V}$. We denote by ι the identity map from V to $\bar{V}$ or from $\bar{V}$ to V; ι is conjugate linear.

If V possesses some distinguished **real form**, i.e., a real subspace V_0 such that $V = V_0 \oplus iV_0$, then V acquires a corresponding **conjugation mapping** (denoted by bar) that is 1 on V_0 and is -1 on V_0. This conjugation mapping is conjugate linear on V. Composing it with ι, we see that a real form of V yields a canonical complex-linear isomorphism of V onto $\bar{V}$. In this case, if U is a complex subspace of V, then $\bar{U}$ has two interpretations, one as the image of U under conjugation and one as the set U with an altered definition of multiplication by i. The two need not be the same but are canonically isomorphic (by conjugation). Unless we specify otherwise, when we write $\bar{U}$ in the setting of a subspace of a space V with a distinguished real form, we shall mean the image of U under conjugation.

Let q be a complex Lie subalgebra of $\mathfrak{g}$, and suppose that $\mathfrak{b}_0 \subseteq \mathfrak{q}$ and that $\mathrm{Ad}(B)$ normalizes q, thus making $(\mathfrak{q}, B)$ a pair. On the level of vector spaces, the situation is as in the preceding paragraph. Then $\bar{\mathfrak{q}}$ is a complex Lie subalgebra of $\mathfrak{g}$, and $(\bar{\mathfrak{q}}, B)$ is a pair. Moreover, $\overline{R(\mathfrak{q}, B)} = R(\bar{\mathfrak{q}}, B)$ if both are regarded as subsets of $R(\mathfrak{g}, B)$. If V is in $\mathcal{C}(\mathfrak{q}, B)$, we can make $\bar{V}$ into a member of $\mathcal{C}(\bar{\mathfrak{q}}, B)$ by defining $sv = \iota(\bar{s}(\iota v))$ for $s \in R(\bar{\mathfrak{q}}, B)$. Let us check that this definition is complex linear in s:

$$(is)v = \iota(-i\bar{s})\iota v = \iota(-J_V\bar{s})\iota v = J_{\bar{V}}(\iota\bar{s}\iota v) = J_{\bar{V}}sv.$$

It is clear that the resulting **conjugation functor** $V \to \bar{V}$ is a covariant exact functor from $\mathcal{C}(\mathfrak{q}, B)$ to $\mathcal{C}(\bar{\mathfrak{q}}, B)$ and that it send projectives to projectives and injectives to injectives.

As in the remark after Proposition 2.41, we defined the **contragredient functor** $V \to V^c$ from $\mathcal{C}(\mathfrak{q}, B)$ to itself by $V^c = \mathrm{Hom}_{\mathbb{C}}(V, \mathbb{C})_B$. The results of §II.3 showed that this functor is contravariant and exact, and it sends projectives to injectives. Contragredient commutes with conjugation in the sense that there is a natural $\mathcal{C}(\bar{\mathfrak{q}}, B)$ isomorphism $\overline{V^c} \cong (\bar{V})^c$ given by

$$(6.7) \qquad \Phi(\varphi)(v) = (\iota\varphi)(\iota v) \qquad \text{for } \varphi \in \overline{V^c} \text{ and } v \in \bar{V}.$$

We define the **Hermitian dual functor** $V \to V^h$ from $\mathcal{C}(\mathfrak{q}, B)$ to $\mathcal{C}(\bar{\mathfrak{q}}, B)$ to be the composition of conjugation and contragredient, i.e., $V^h = (\bar{V})^c$. This functor is contravariant and exact, and it sends projectives to injectives. If $\beta : V \to W$ is a map, then $\beta^h : W^h \to V^h$ is given by

$$(6.8) \qquad\qquad\qquad \beta^h = \iota\beta^c\iota.$$

There is a natural inclusion $V \hookrightarrow V^{hh}$. This map is an isomorphism if V is finite dimensional, or more generally if V has each B type of finite multiplicity.

Next suppose that $q = \bar{q}$, and let V be in $\mathcal{C}(q, B)$. Then it is easy to check that the following three classes of objects correspond:

(6.9)
  (a) $R(q, B)$ invariant sesquilinear forms on $V \times V$ (linear in first, conjugate linear in second)
  (b) $R(q, B)$ invariant complex-bilinear forms on $V \times \bar{V}$
  (c) maps in $\mathrm{Hom}_{q, B}(V, V^h)$.

Here the invariance condition in (a) is that

$$\langle rv, w \rangle = \langle v, \bar{r}^t w \rangle \qquad \text{for } r \in R(q, K),\ v \in V,\ w \in V,$$

while the invariance condition in (b) is that

$$\langle rv, w \rangle = \langle v, r^t w \rangle \qquad \text{for } r \in R(q, K),\ v \in V,\ w \in \bar{V}.$$

The correspondence between (a) and (b) is obtained by inserting ι into the second coordinate. The correspondence between (b) and (c) is given by $\langle v, w \rangle = \psi(v)(w)$ for $v \in V$ and $w \in \bar{V}$; thus the correspondence between (a) and (c) is given by

$$(6.10) \qquad \langle v, w \rangle = \psi(v)(\iota w) \qquad \text{for } v \in V \text{ and } w \in V.$$

If ψ is a map as in (c), then ψ^h maps V^{hh} to V^h. We say that ψ is **formally Hermitian** if the restriction of ψ^h to V is equal to ψ. In the correspondence between (a) and (c) in (6.9), invariant Hermitian forms correspond to formally Hermitian maps; this is immediate from the definitions (6.8) and (6.10).

Proposition 6.11. If V is an irreducible finite-dimensional $(\mathfrak{b}, B)$ module, then V has a unique invariant sesquilinear form up to a scalar, and any such form is a complex multiple of a Hermitian inner product.

REMARK. This result for Hermitian forms is a special case of Proposition 6.12 below. We isolate Proposition 6.11 because it provides part of the motivation for the explicit computation to be done at the end of §6.

PROOF. Since V is irreducible, so is V^h. By Schur's Lemma, the dimension of $\mathrm{Hom}_B(V, V^h)$ is ≤ 1. On the other hand, we know that V has an invariant Hermitian inner product. Hence the result follows from the equivalence of (a) and (c) in (6.9).

Proposition 6.12. Let V be a finite-dimensional $(\mathfrak{b}, B)$ module that is isotypic of a single B type τ, and suppose that $\langle \cdot, \cdot \rangle$ is an invariant Hermitian form on $V \times V$. Then there exists an orthogonal decomposition

$$(6.13) \qquad\qquad V = V_+ \oplus V_0 \oplus V_-$$

into invariant subspaces such that $\langle \cdot, \cdot \rangle$ is positive definite on V_+, is 0 on V_0, and is negative definite on V_-. Moreover, if

$$V = V'_+ \oplus V'_0 \oplus V'_-$$

is another such decomposition, then $V_0 = V'_0$ and also the multiplicities of τ in V'_+ and V'_- match respectively the multiplicities of τ in V_+ and V_-.

REMARKS.

1) We assemble the multiplicities of τ in V_+, V_0, and V_- into a triple (n_+, n_0, n_-). This triple is called the **signature** of the form $\langle \cdot, \cdot \rangle$ on V with respect to τ.

2) Before coming to the proof, we review some linear algebra. Let R be a Hermitian matrix operating on $\mathbb{C}^n$. By the finite-dimensional spectral theorem, $\mathbb{C}^n$ splits as an orthogonal sum

$$\mathbb{C}^n = V_+ \oplus V_0 \oplus V_-$$

relative to the usual Hermitian inner product on $\mathbb{C}^n$, where

$$(6.14) \qquad \begin{aligned} V_+ &= \text{sum of eigenspaces of } R \text{ for positive eigenvalues} \\ V_0 &= \text{zero eigenspace of } R \\ V_- &= \text{sum of eigenspaces of } R \text{ for negative eigenvalues.} \end{aligned}$$

The triple $(\dim V_+, \dim V_0, \dim V_-)$ is called the **signature** of the Hermitian matrix R. In similar fashion we can define the **signature** of a Hermitian operator on a finite-dimensional Hermitian inner-product space.

3) If R is a Hermitian matrix and A is a nonsingular matrix, let us show that $A^* R A$ has the same signature as R. In fact, let

$$\mathbb{C}^n = W_+ \oplus W_0 \oplus W_-$$

be the orthogonal decomposition of $\mathbb{C}^n$ given in Remark 2 for $A^* R A$. Certainly $\dim V_0 = \dim W_0$. If $w \neq 0$ is in $(W_0 \oplus W_-) \cap A^{-1} V_+$, then

$\langle A^* R A w, w \rangle \leq 0$ since $A^* R A$ is negative semidefinite on $W_0 \oplus W_-$. On the other hand, we can write $w = A^{-1} v$ with $v \in V_+$, and then

$$\langle A^* R A w, w \rangle = \langle A^* R v, A^{-1} v \rangle = \langle R v, v \rangle > 0,$$

contradiction. Thus

$$(W_0 \oplus W_-) \cap A^{-1} V_+ = 0,$$

and we must have

$$\dim W_0 + \dim W_- + \dim V_+ \leq n.$$

Adding $\dim W_+ - n$ to both sides, we obtain

$$\dim V_+ \leq \dim W_+.$$

Since the roles of V_+ and W_+ are interchanged if we replace A by A^{-1}, equality must hold. Similarly $\dim V_- = \dim W_-$. Thus $A^* R A$ and R have the same signature.

4) Now we can define the "vector-space signature" of a Hermitian form on a finite-dimensional complex vector space V. We fix a vector-space isomorphism $\Phi : V \to \mathbb{C}^n$. Then Hermitian forms on V correspond bijectively to Hermitian matrices R by the formula

$$\langle v, w \rangle_{R, \Phi} = \langle R \Phi v, \Phi w \rangle_{\mathbb{C}^n}.$$

Here the form on the right is the standard inner product on $\mathbb{C}^n$. We define the **vector-space signature** of $\langle \cdot, \cdot \rangle_{R, \Phi}$ to be the signature of the matrix R. To see that this is well defined, let $\Phi' : V \to \mathbb{C}^n$ be another isomorphism, and let S be the matrix of $\langle \cdot, \cdot \rangle_{R, \Phi}$ with respect to the isomorphism Φ'. Then

$$\langle R \Phi v, \Phi w \rangle_{\mathbb{C}^n} = \langle S \Phi' v, \Phi' w \rangle_{\mathbb{C}^n}.$$

Since Φ and Φ' are both isomorphisms, there is an invertible matrix A so that $A \Phi' = \Phi$. Then

$$\langle R A \Phi' v, A \Phi' w \rangle_{\mathbb{C}^n} = \langle S \Phi' v, \Phi' w \rangle_{\mathbb{C}},$$

and thus

$$\langle A^* R A \Phi' v, \Phi' w \rangle_{\mathbb{C}^n} = \langle S \Phi' v, \Phi' w \rangle_{\mathbb{C}^n}.$$

As v and w run through V, $\Phi'v$ and $\Phi'w$ run through $\mathbb{C}^n$. Hence $S = A^*RA$. By Remark 3, S has the same signature as R, and vector-space signature is therefore well defined.

5) It is often convenient to rephrase the definition of vector-space signature as follows. Fix a positive definite Hermitian inner product $\langle\,\cdot\,,\,\cdot\,\rangle_{PD}$ on V. Then Hermitian operators $T \in \text{End}_{\mathbb{C}} V$ correspond bijectively to Hermitian forms $\langle\,\cdot\,,\,\cdot\,\rangle_T$ on V by the formula

$$\langle v, w\rangle_T = \langle Tv, w\rangle_{PD}.$$

We claim that the vector-space signature of $\langle\,\cdot\,,\,\cdot\,\rangle_T$ is just the signature of the operator T in the sense of Remark 2, i.e., the numbers of positive, zero, and negative eigenvalues of T. To see this, fix a vector space isomorphism $\Phi : V \to \mathbb{C}^n$ carrying $\langle\,\cdot\,,\,\cdot\,\rangle_{PD}$ to the standard inner product on $\mathbb{C}^n$:

$$\langle v, w\rangle_{PD} = \langle \Phi v, \Phi w\rangle_{\mathbb{C}^n}.$$

Then

$$\langle v, w\rangle_T = \langle Tv, w\rangle_{PD} = \langle \Phi Tv, \Phi w\rangle_{\mathbb{C}^n} = \langle \Phi T\Phi^{-1}\Phi v, \Phi w\rangle_{\mathbb{C}^n}.$$

The matrix of $\langle\,\cdot\,,\,\cdot\,\rangle_T$ with respect to the isomorphism Φ is therefore $\Phi T\Phi^{-1}$, which has the same eigenvalues as T.

PROOF. For existence, let $\langle\,\cdot\,,\,\cdot\,\rangle_{PD}$ be an *invariant* Hermitian inner product on V. As in Remark 5, we define

$$\langle v, w\rangle = \langle Rv, w\rangle_{PD} \qquad \text{for all } v \text{ and } w \text{ in } V,$$

and then R is a Hermitian operator. Also the invariance of both forms implies

$$\langle bRv, w\rangle_{PD} = \langle Rv, b^{-1}w\rangle_{PD} = \langle v, b^{-1}w\rangle = \langle bv, w\rangle = \langle Rbv, w\rangle_{PD}$$

for $b \in B$, and it follows that R commutes with each $b \in B$. If we define V_+, V_0, and V_- as in (6.14), then we readily obtain (6.13).

For uniqueness of multiplicities, we appeal to Remark 5. In an invariant decomposition the multiplicities are obtained by dividing the vector-space dimensions by $\dim \tau$. Moreover, the space V_0 itself is unique, being the radical of the given form. This completes the proof.

3. Hermitian Duality Relative to P and I

We now return to the setting of §1. In this section we shall give some formulas connecting the functors P, I, conjugation, and Hermitian dual.

First, for our $(\mathfrak{l}, L \cap K)$ module Z, the $(\mathfrak{q}, L \cap K)$ module $Z_\mathfrak{q}$ and the $(\bar{\mathfrak{q}}, L \cap K)$ module $Z_{\bar{\mathfrak{q}}}$ are related by

$$(6.15) \qquad (Z^h)_\mathfrak{q} \cong (Z_{\bar{\mathfrak{q}}})^h.$$

If we have an invariant sesquilinear form $\langle \cdot, \cdot \rangle_L$ on Z, then the equivalence of (a) and (c) in (6.9) gives us an $(\mathfrak{l}, L \cap K)$ map $\zeta_Z : Z \to Z^h$, by the formula (6.10). We can regard ζ_Z as a $(\mathfrak{q}, L \cap K)$ map $\zeta_Z : Z_\mathfrak{q} \to (Z^h)_\mathfrak{q}$, and (6.15) allows us to interpret ζ_Z as a map

$$(6.16) \qquad \zeta_Z : Z_\mathfrak{q} \to (Z_{\bar{\mathfrak{q}}})^h.$$

Next the module

$$(6.17a) \qquad P_{\mathfrak{q}, L\cap K}^{\mathfrak{g}, L\cap K}(\overline{Z_{\bar{\mathfrak{q}}}}) = R(\mathfrak{g}, L \cap K) \otimes_{R(\mathfrak{q}, L\cap K)} \overline{Z_{\bar{\mathfrak{q}}}}$$

has the internal law $sq \otimes \iota z = s \otimes \iota \bar{q} z$ and the external action $r(s \otimes \iota z) = rs \otimes \iota z$, while the module

$$(6.17b) \qquad \overline{P_{\bar{\mathfrak{q}}, L\cap K}^{\mathfrak{g}, L\cap K}(Z_{\bar{\mathfrak{q}}})} = \overline{R(\mathfrak{g}, L \cap K) \otimes_{R(\bar{\mathfrak{q}}, L\cap K)} Z_{\bar{\mathfrak{q}}}}$$

has the internal law $\iota(s\bar{q} \otimes z) = \iota(s \otimes \bar{q} z)$ and the external action

$$r(\iota(s \otimes z)) = \iota \bar{r}(s \otimes z) = \iota(\bar{r} s \otimes z);$$

(6.17a) and (6.17b) are isomorphic under the map $s \otimes \iota z \mapsto \iota(\bar{s} \otimes z)$, and thus

$$(6.18) \qquad P_{\mathfrak{q}, L\cap K}^{\mathfrak{g}, L\cap K}(\overline{Z_{\bar{\mathfrak{q}}}}) \cong \overline{P_{\bar{\mathfrak{q}}, L\cap K}^{\mathfrak{g}, L\cap K}(Z_{\bar{\mathfrak{q}}})}$$

naturally in Z. Similarly

$$(6.19) \qquad I_{\bar{\mathfrak{q}}, L\cap K}^{\mathfrak{g}, L\cap K}(\overline{Z_\mathfrak{q}}) \cong \overline{I_{\mathfrak{q}, L\cap K}^{\mathfrak{g}, L\cap K}(Z_\mathfrak{q})}$$

naturally in Z. Meanwhile, Easy Duality (Theorem 3.1) gives

$$(6.20) \qquad I_{\mathfrak{q}, L\cap K}^{\mathfrak{g}, L\cap K}(V^c) \cong P_{\mathfrak{q}, L\cap K}^{\mathfrak{g}, L\cap K}(V)^c$$

naturally for V in $\mathcal{C}(\mathfrak{q}, L \cap K)$. Combining these relations, we obtain

(6.21a)
$$
\begin{aligned}
[P_{\bar{\mathfrak{q}},L\cap K}^{\mathfrak{g},L\cap K}(Z_{\bar{\mathfrak{q}}})]^h &= \overline{[P_{\bar{\mathfrak{q}},L\cap K}^{\mathfrak{g},L\cap K}(Z_{\bar{\mathfrak{q}}})]^c} \\
&\cong [P_{\mathfrak{q},L\cap K}^{\mathfrak{g},L\cap K}(\overline{Z_{\bar{\mathfrak{q}}}})]^c \\
&\cong I_{\mathfrak{q},L\cap K}^{\mathfrak{g},L\cap K}((\overline{Z_{\bar{\mathfrak{q}}}})^c) \\
&= I_{\mathfrak{q},L\cap K}^{\mathfrak{g},L\cap K}((Z_{\bar{\mathfrak{q}}})^h).
\end{aligned}
$$

If φ is a member of the right side, then the corresponding member $e(\varphi)$ of the left side has

(6.21b)
$$ e(\varphi)(\iota(s \otimes z)) = \varphi(\bar{s}^t)(\iota z). $$

Note the transpose in this formula; it comes from the occurrence of transpose in the proof of Theorem 3.1.

The above formulas have analogs for the passage from $(\mathfrak{g}, L \cap K)$ modules to $(\mathfrak{g}, K)$ modules. Arguing as for (6.18), we obtain

(6.22)
$$ P_{\mathfrak{g},L\cap K}^{\mathfrak{g},K}(\bar{V}) \cong \overline{P_{\mathfrak{g},L\cap K}^{\mathfrak{g},K}(V)} $$

naturally for V in $\mathcal{C}(\mathfrak{g}, L \cap K)$, the isomorphism being $s \otimes \iota v \mapsto \iota(\bar{s} \otimes v)$. A similar identity holds for $I_{\mathfrak{g},L\cap K}^{\mathfrak{g},K}$. As usual (see (2.14)), we write

$$ \Pi = P_{\mathfrak{g},L\cap K}^{\mathfrak{g},K} \qquad \text{and} \qquad \Gamma = I_{\mathfrak{g},L\cap K}^{\mathfrak{g},K} $$

and denote the corresponding derived functors by Π_j and Γ^j. Since conjugation is exact and sends projectives to projectives and injectives to injectives, we obtain

(6.23a)
$$ \Pi_j(\bar{V}) \cong \overline{\Pi_j(V)} \qquad \text{and} \qquad \Gamma^j(\bar{V}) \cong \overline{\Gamma^j(V)} $$

naturally for V in $\mathcal{C}(\mathfrak{g}, L \cap K)$. Recall from Proposition 2.115 that the derived functors

$$ \Pi_j^K = (P_{\mathfrak{k},L\cap K}^{\mathfrak{k},K})_j \qquad \text{and} \qquad \Gamma_K^j = (I_{\mathfrak{k},L\cap K}^{\mathfrak{k},K})^j $$

are related to Π_j and Γ^j via forgetful functors:

(6.23b)
$$
\begin{aligned}
\Pi_j^K \circ \mathcal{F}_{\mathfrak{g},L\cap K}^{\mathfrak{k},L\cap K} &\cong \mathcal{F}_{\mathfrak{g},K}^{\mathfrak{k},K} \circ \Pi_j \\
\Gamma_K^j \circ \mathcal{F}_{\mathfrak{g},L\cap K}^{\mathfrak{k},L\cap K} &\cong \mathcal{F}_{\mathfrak{g},K}^{\mathfrak{k},K} \circ \Gamma^j.
\end{aligned}
$$

Thus $\Pi_j(V)$ and $\Gamma^j(V)$ can be computed as $(\mathfrak{k}, K)$ modules from standard complexes in $\mathcal{C}(\mathfrak{k}, K)$.

Now Easy Duality (Corollary 3.2a) gives us

$$(6.24) \qquad\qquad \Pi_j(V)^c \cong \Gamma^j(V^c)$$

naturally in V. Combining this relation with (6.23), we obtain

$$(6.25) \qquad\qquad \Pi_j(V)^h \cong \Gamma^j(V^h)$$

naturally for V in $\mathcal{C}(\mathfrak{g}, L \cap K)$.

If $\mathcal{E}$ denotes the isomorphism from right to left in (6.25), we can give an explicit formula for $\mathcal{E}$ on the level of complexes. By (6.23b) and (2.128), $\Pi_j(V)$ can be computed as a $(\mathfrak{k}, K)$ module from the complex

$$(6.26a) \qquad\qquad R(K) \otimes_{L \cap K} (\textstyle\bigwedge^j \mathfrak{c} \otimes_{\mathbb{C}} V),$$

and $\Gamma^j(V)$ can be computed as a $(\mathfrak{k}, K)$ module from the complex

$$(6.26b) \qquad\qquad \operatorname{Hom}_{L \cap K}(R(K), \operatorname{Hom}_{\mathbb{C}}(\textstyle\bigwedge^j \mathfrak{c}, V))_K.$$

Here $\mathfrak{c}$ is an $\mathfrak{l} \cap \mathfrak{k}$ invariant complement of $\mathfrak{l} \cap \mathfrak{k}$ in $\mathfrak{k}$, and we take it to be $\mathfrak{c} = (\mathfrak{u} \cap \mathfrak{k}) \oplus (\bar{\mathfrak{u}} \cap \mathfrak{k})$. (Actually there is no other choice.) In (6.26a), K acts by l on $R(K)$, but in (6.26b), K acts by r on $R(K)$.

Lemma 6.27. The isomorphism $\mathcal{E} : \Gamma^j(V^h) \to \Pi_j(V)^h$ of (6.25), as a map on the level of complexes

$$(6.28a) \quad \mathcal{E} : \operatorname{Hom}_{L \cap K}(R(K), \operatorname{Hom}_{\mathbb{C}}(\textstyle\bigwedge^j \mathfrak{c}, V^h))_K$$
$$\to \overline{(R(K) \otimes_{L \cap K} (\textstyle\bigwedge^j \mathfrak{c} \otimes_{\mathbb{C}} V))^c}$$

is given by

$$(6.28b) \qquad\qquad \mathcal{E}(\varphi)(\iota(T \otimes \xi \otimes v)) = \varphi(\bar{T}^t)(\bar{\xi})(\iota v).$$

REMARK. Note the transpose in (6.28b).

PROOF. First we track down (6.24) on the level of complexes. For $j = 0$, if we write $\mathcal{E}_0$ for the isomorphism $\Gamma(V^c) \xrightarrow{\sim} \Pi(V)^c$, then inspection of the proof of Theorem 3.1 shows that $\mathcal{E}_0$ is given by

$$\mathcal{E}_0(\varphi)(T \otimes v) = \varphi(T^t)(v).$$

The transpose comes from the conversion of $R(\mathfrak{g}, K)$ in that proof (here $= R(K)$) from a K module on one side to a K module on the other side by the transpose map.

The passage to the jth derived functors is now dictated by naturality of $\mathcal{E}_0$, according to (C.26). We let $\{X_j\}$ be the Koszul resolution in $\mathcal{C}(\mathfrak{k}, L \cap K)$ and replace V in the above computation by $X_j \otimes_{\mathbb{C}} V$. If we write a typical monomial as $x_j \otimes v$, then the definition

$$\mathcal{E}_0(\varphi)(T \otimes x_j \otimes v) = \varphi(T^t)(x_j \otimes v)$$

gives a chain map

$$\mathcal{E}_0 : \Gamma((X_j \otimes_{\mathbb{C}} V)^c) \overset{\sim}{\to} \Pi(X_j \otimes_{\mathbb{C}} V)^c$$

that descends to cohomology as the map implementing (6.24) from right to left.

We still have to interpret $\mathcal{E}_0$ in terms of our standard complexes. We have

$\Gamma((X_j \otimes_{\mathbb{C}} V)^c)$

$\cong \operatorname{Hom}_{\mathfrak{k}, L \cap K}(R(K), \operatorname{Hom}_{\mathbb{C}}((R(\mathfrak{k}, L \cap K) \otimes_{L \cap K} (\bigwedge^j \mathfrak{c})) \otimes_{\mathbb{C}} V, \mathbb{C})_{L \cap K})_K$

$\cong \operatorname{Hom}_{\mathfrak{k}, L \cap K}(R(K), \operatorname{Hom}_{\mathbb{C}}(R(\mathfrak{k}, L \cap K) \otimes_{L \cap K} (\bigwedge^j \mathfrak{c}), V^c)_{L \cap K})_K,$

and the computation before (2.129) shows that this is

$$\cong \operatorname{Hom}_{L \cap K}(R(K), \operatorname{Hom}_{\mathbb{C}}(\bigwedge^j \mathfrak{c}, V^c))_K.$$

Also the contragredient of the computation before (2.128) shows that

$$\Pi(X_j \otimes_{\mathbb{C}} V)^c \cong (R(K) \otimes_{L \cap K} (\bigwedge^j \mathfrak{c} \otimes_{\mathbb{C}} V))^c.$$

Tracking down the effect on $\mathcal{E}_0$ of these isomorphisms, we find that $\mathcal{E}_0$ on the level of complexes is given by the chain map

$$\mathcal{E}_0 : \operatorname{Hom}_{L \cap K}(R(K), \operatorname{Hom}_{\mathbb{C}}(\bigwedge^j \mathfrak{c}, V^c))_K \to (R(K) \otimes_{L \cap K} (\bigwedge^j \mathfrak{c} \otimes_{\mathbb{C}} V))^c$$

with

$$\mathcal{E}_0(\varphi)(T \otimes \xi \otimes v) = \varphi(T^t)(\xi)(v).$$

Now $\mathcal{E}$ is a composition

$$\Gamma^j(V^h) = \Gamma^j((\bar{V})^c) \cong \Pi_j(\bar{V})^c \cong (\overline{\Pi_j(V)})^c = \Pi_j(V)^h.$$

The first isomorphism is implemented by $\mathcal{E}_0$ but with v replaced by ιv, and the second isomorphism is implemented by conjugating T^t and ξ. Hence $\mathcal{E}$ is given as in (6.28b).

4. Statement of Signature Theorem

Before stating the Signature Theorem, we shall define the "Shapovalov form" on $\mathcal{L}_S(Z)$ corresponding to an invariant sesquilinear form on Z.

We start from an invariant sesquilinear form on Z and construct the corresponding $\zeta_Z : Z \to Z^h$ given by (6.10). Since $\mathfrak{u}^c \cong \bar{\mathfrak{u}}$, we obtain $\mathfrak{u}^h \cong \mathfrak{u}$ and $(\bigwedge^{\text{top}}\mathfrak{u})^h \cong \bigwedge^{\text{top}}\mathfrak{u}$. Then $\zeta_{Z^\#} = \zeta_Z \otimes_{\mathbb{C}} 1$ maps $Z^\#$ into

$$ Z^h \otimes_{\mathbb{C}} \textstyle\bigwedge^{\text{top}}\mathfrak{u} \cong Z^h \otimes_{\mathbb{C}} (\textstyle\bigwedge^{\text{top}}\mathfrak{u})^h \cong (Z \otimes_{\mathbb{C}} \textstyle\bigwedge^{\text{top}}\mathfrak{u})^h = (Z^\#)^h. $$

As in (6.16), we can interpret $\zeta_{Z^\#}$ as a map

$$ \zeta_{Z^\#} : Z_{\mathfrak{q}}^\# \to (Z_{\bar{\mathfrak{q}}}^\#)^h. $$

Consistently with (5.89), let

$$ (6.29\text{a}) \qquad \varphi_Z^G : P_{\bar{\mathfrak{q}}, L\cap K}^{\mathfrak{g}, L\cap K}(Z_{\bar{\mathfrak{q}}}^\#) \to I_{\mathfrak{q}, L\cap K}^{\mathfrak{g}, L\cap K}(Z_{\mathfrak{q}}^\#) $$

be the member of the left side of the isomorphism

$$ (6.29\text{b}) \qquad \text{Hom}_{\mathfrak{g}, L\cap K}(P_{\bar{\mathfrak{q}}, L\cap K}^{\mathfrak{g}, L\cap K}(Z_{\bar{\mathfrak{q}}}^\#), I_{\mathfrak{q}, L\cap K}^{\mathfrak{g}, L\cap K}(Z_{\mathfrak{q}}^\#)) \cong \text{Hom}_{\mathfrak{l}, L\cap K}(Z^\#, Z^\#) $$

that corresponds to the identity on the right side. Although φ_Z^G need not be an isomorphism (cf. Proposition 5.93), it is not 0 since it does not correspond to 0 under (6.29b); Proposition 5.90 gives an explicit formula for it. Finally let

$$ (6.30) \qquad e_Z^G : I_{\mathfrak{q}, L\cap K}^{\mathfrak{g}, L\cap K}((Z_{\bar{\mathfrak{q}}}^\#)^h) \to [P_{\bar{\mathfrak{q}}, L\cap K}^{\mathfrak{g}, L\cap K}(Z_{\bar{\mathfrak{q}}}^\#)]^h $$

be the isomorphism from right to left in (6.21a) when Z is replaced by $Z^\#$. Put

$$ (6.31\text{a}) \qquad \psi_Z^G = e_Z^G \circ I_{\mathfrak{q}, L\cap K}^{\mathfrak{g}, L\cap K}(\zeta_{Z^\#}) \circ \varphi_Z^G. $$

Then

$$ (6.31\text{b}) \qquad \psi_Z^G : P_{\bar{\mathfrak{q}}, L\cap K}^{\mathfrak{g}, L\cap K}(Z_{\bar{\mathfrak{q}}}^\#) \to [P_{\bar{\mathfrak{q}}, L\cap K}^{\mathfrak{g}, L\cap K}(Z_{\bar{\mathfrak{q}}}^\#)]^h $$

defines an invariant sesquilinear form on $P_{\bar{\mathfrak{q}}, L\cap K}^{\mathfrak{g}, L\cap K}(Z_{\bar{\mathfrak{q}}}^\#)$, by (6.10).

If the invariant sesquilinear form on Z is nondegenerate, then $\zeta_{Z^\#}$ is an isomorphism, and it follows that ψ_Z^G cannot be the 0 map. In fact, φ_Z^G in (6.31a) is not 0, and e_Z^G and $I_{\mathfrak{q}, L\cap K}^{\mathfrak{g}, L\cap K}(\zeta_{Z^\#})$ are isomorphisms.

To carry our form to $(\mathfrak{g}, K)$ modules, we shall bring in the Hard Duality Theorem (Theorem 3.5) in the form of the isomorphism

$$\Pi_j(V \otimes_{\mathbb{C}} (\textstyle\bigwedge^{2S}\mathfrak{c})^*) \cong \Gamma^{2S-j}(V)$$

naturally for V in $\mathcal{C}(\mathfrak{g}, L \cap K)$. In the general setting of Chapter III, $\bigwedge^{\text{top}}\mathfrak{c}$ is a trivial $\mathfrak{g}$ module but may not be a trivial module for the compact group. However, in our present setting,

$$\textstyle\bigwedge^{\text{top}}\mathfrak{c} \cong \bigwedge^{\text{top}}(\mathfrak{u} \cap \mathfrak{k}) \otimes_{\mathbb{C}} \bigwedge^{\text{top}}(\bar{\mathfrak{u}} \cap \mathfrak{k})$$

is a trivial $L \cap K$ module. Thus $(\bigwedge^{2S}\mathfrak{c})^*$ pulls out, and we have a natural isomorphism

$$(6.32) \qquad \Pi_j(V) \otimes_{\mathbb{C}} (\textstyle\bigwedge^{2S}\mathfrak{c})^* \cong \Gamma^{2S-j}(V) \qquad \text{for } V \in \mathcal{C}(\mathfrak{g}, L \cap K).$$

When $V = P^{\mathfrak{g},L \cap K}_{\bar{\mathfrak{q}},L \cap K}(Z^{\#}_{\bar{\mathfrak{q}}})$, let us write $\mathcal{D}^G_Z$ for the isomorphism from left to right in (6.32). Also let $\mathcal{E}^G_Z$ be the isomorphism from right to left in (6.25) for this V. Define

$$(6.33) \qquad\qquad \Psi^G_Z = \mathcal{E}^G_Z \circ \Gamma^{2S-j}(\psi^G_Z) \circ \mathcal{D}^G_Z$$

as a $(\mathfrak{g}, K)$ map

$$\Psi^G_Z : \mathcal{L}_j(Z) \otimes_{\mathbb{C}} (\textstyle\bigwedge^{2S}\mathfrak{c})^* \to [\mathcal{L}_{2S-j}(Z)]^h.$$

We shall be mostly interested in the case $j = S$ and therefore suppress the index j in the notation Ψ^G_Z. In any event, in the case that $j = S$, if we fix a choice of a member of $(\bigwedge^{2S}\mathfrak{c})^*$, then Ψ^G_Z provides us (via (6.10)) with an invariant sesquilinear form on $\mathcal{L}_S(Z)$. Below we shall fix such a choice $\varepsilon_0 \in (\bigwedge^{2S}\mathfrak{c})^*$ depending only on G and $\mathfrak{q}$, and the resulting form will be what we call the **Shapovalov form** constructed from the form on Z given by ζ_Z.

To define ε_0, we let ξ be any nonzero element of $\bigwedge^S(\bar{\mathfrak{u}} \cap \mathfrak{k})$, so that $\xi \wedge \bar{\xi}$ is a nonzero element of $\bigwedge^{\text{top}}\mathfrak{c}$. Then we define ε_0 to be any nonzero element of $(\bigwedge^{\text{top}}\mathfrak{c})^*$ such that $\varepsilon_0(\xi \wedge \bar{\xi}) > 0$. This definition is independent of ξ, and ε_0 is uniquely defined up to a positive constant.

At this stage it is not apparent that the Shapovalov form is nonzero, but (as a consequence of the Signature Theorem) it will be nonzero if $\mathcal{L}_S(Z)$ has at least one K type in the bottom layer. If the form on Z is nondegenerate and if φ^G_Z is an isomorphism (as it is under the hypotheses of Theorem 5.99), then Ψ^G_Z is an isomorphism and the Shapovalov form is nondegenerate. This latter situation will be of interest for the Unitarizability Theorem in Chapter IX, but these special hypotheses are not needed for the Signature Theorem.

We shall give an explicit formula for the Shapovalov form on the level of complexes in (6.45), and in §IX.2 we shall compute some explicit examples using this formula.

Let V be an admissible $(\mathfrak{g}, K)$ module, let τ be a K type, and let V_τ be the sum of all K invariant subspaces of V of type τ. If $\langle\,\cdot\,,\cdot\,\rangle_G$ is an invariant Hermitian form on V, then Proposition 6.12 showed that V_τ splits as a direct sum $V_+ \oplus V_0 \oplus V_-$ of K invariant subspaces such that $\langle\,\cdot\,,\cdot\,\rangle_G$ is positive definite on V_+, zero on V_0, and negative definite on V_-. We defined the triple of multiplicities (n_+, n_0, n_-) of τ in these subspaces to be the signature of $\langle\,\cdot\,,\cdot\,\rangle_G|_{V_\tau \times V_\tau}$ with respect to τ. We shall refer to this triple also as the **signature** of $\langle\,\cdot\,,\cdot\,\rangle_G$ on the K type τ.

Theorem 6.34 (Signature Theorem). Suppose Z is an $(\mathfrak{l}, L \cap K)$ module in which $h_{\delta(\mathfrak{u})}$ acts by a scalar, and suppose Z possesses an invariant Hermitian form $\langle\,\cdot\,,\cdot\,\rangle_L$. If $\mathcal{L}_S(Z)$ is admissible, then the corresponding Shapovalov form $\langle\,\cdot\,,\cdot\,\rangle_G$ on $\mathcal{L}_S(Z)$ is Hermitian. On each K type in the bottom layer of $\mathcal{L}_S(Z)$, the signature of $\langle\,\cdot\,,\cdot\,\rangle_G$ matches the signature of $\langle\,\cdot\,,\cdot\,\rangle_L$ on the $L \cap K$ type of Z that corresponds under the remarks following Corollary 5.85.

In broad detail the proof of Theorem 6.34 will proceed in two steps. First we use the bottom-layer map to reduce matters to an analog of Theorem 6.34 for K. In the analog of Theorem 6.34, a form for an $L \cap K$ module Z is carried from Z to $\mathcal{L}_S^K(Z)$ via a map Ψ_Z^K similar to (6.33). This first step enables us to replace Z by an $(\mathfrak{l} \cap \mathfrak{k}, L \cap K)$ module and then to reduce to the case of an irreducible $(\mathfrak{l} \cap \mathfrak{k}, L \cap K)$ module. Second we combine an explicit computation of the Shapovalov form for K with our multiplicity results in §V.6 to see that the signature is preserved in passing from $L \cap K$ to K. We carry out these two steps in the next two sections.

5. Comparison of Shapovalov Forms on K and G

In this section we use the bottom layer map to reduce Theorem 6.34 to an analogous statement for K. We can repeat all of our constructions in §4 with $\mathfrak{g}, \mathfrak{q}, \bar{\mathfrak{q}}, \mathfrak{u}, \bar{\mathfrak{u}}$, and $\mathfrak{l}$ all replaced by their intersections with $\mathfrak{k}$.

Doing so, we obtain as analogs of $\varphi_Z^G, e_Z^G, \psi_Z^G, \mathcal{D}_Z^G, \mathcal{E}_Z^G$, and Ψ_Z^G the maps $\varphi_Z^K, e_Z^K, \psi_Z^K, \mathcal{D}_Z^K, \mathcal{E}_Z^K$, and Ψ_Z^K, respectively. The map Ψ_Z^K is a $(\mathfrak{k}, K)$ map

$$\Psi_Z^K : \mathcal{L}_j^K(Z) \otimes_{\mathbb{C}} (\textstyle\bigwedge^{2S}\mathfrak{c})^* \to [\mathcal{L}_{2S-j}^K(Z)]^h.$$

For $j = S$ and with ε_0 as in §4, Ψ_Z^K defines by (6.10) a Shapovalov form $\langle \cdot, \cdot \rangle_K$ on $\Pi_S^K(P_{\bar{\mathfrak{q}}, L \cap K}^{\mathfrak{k}, L \cap K}(Z_{\bar{\mathfrak{q}}}^{\#}))$. The main result of this section is as follows.

Proposition 6.35. For $0 \leq j \leq 2S$ and with ε fixed at ε_0 as in §4, the diagram

(6.36)

$$
\begin{array}{ccc}
\mathcal{L}_j^K(Z) & \xrightarrow{\;\;\mathcal{B}_Z\;\;} & \mathcal{L}_j(Z) \\[2pt]
{\scriptstyle \Psi_Z^K}\big\downarrow & & \big\downarrow{\scriptstyle \Psi_Z^G} \\[4pt]
[\mathcal{L}_{2S-j}^K(Z)]^h & \xleftarrow{\;(\mathcal{B}_Z)^h\;} & [\mathcal{L}_{2S-j}(Z)]^h
\end{array}
$$

commutes. Consequently the Shapovalov forms on $\mathcal{L}_S^K(Z)$ and $\mathcal{L}_S(Z)$ are related by

(6.37)
$$\langle x, y \rangle_K = \langle \mathcal{B}_Z x, \mathcal{B}_Z y \rangle_G.$$

Schematically the proof of Proposition 6.35 consists in replacing (6.36) by the larger diagram (6.38) and showing that each of the three small rectangles commutes. Then the outer rectangle of (6.38) commutes, and this is the same as the rectangle in (6.36).

(6.38)

$$
\begin{array}{ccc}
\Pi_j^K(\text{ind}_{\bar{\mathfrak{q}}\cap\mathfrak{k}, L\cap K}^{\mathfrak{k}, L\cap K}(Z_{\bar{\mathfrak{q}}}^{\#})) \otimes (\textstyle\bigwedge^{2S}\mathfrak{c})^* & \xrightarrow{\;\;\mathcal{B}_Z\;\;} & \Pi_j(\text{ind}_{\bar{\mathfrak{q}}, L\cap K}^{\mathfrak{g}, L\cap K}(Z_{\bar{\mathfrak{q}}}^{\#})) \otimes (\textstyle\bigwedge^{2S}\mathfrak{c})^* \\[6pt]
{\scriptstyle \mathcal{D}_Z^K}\big\downarrow & & \big\downarrow{\scriptstyle \mathcal{D}_Z^G} \\[6pt]
\Gamma_K^{2S-j}(\text{ind}_{\bar{\mathfrak{q}}\cap\mathfrak{k}, L\cap K}^{\mathfrak{k}, L\cap K}(Z_{\bar{\mathfrak{q}}}^{\#})) & \longrightarrow & \Gamma^{2S-j}(\text{ind}_{\bar{\mathfrak{q}}, L\cap K}^{\mathfrak{g}, L\cap K}(Z_{\bar{\mathfrak{q}}}^{\#})) \\[6pt]
{\scriptstyle \Gamma^{2S-j}(\psi_Z^K)}\big\downarrow & & \big\downarrow{\scriptstyle \Gamma^{2S-j}(\psi_Z^G)} \\[6pt]
\Gamma_K^{2S-j}((\text{ind}_{\bar{\mathfrak{q}}\cap\mathfrak{k}, L\cap K}^{\mathfrak{k}, L\cap K}(Z_{\bar{\mathfrak{q}}}^{\#}))^h) & \longleftarrow & \Gamma^{2S-j}((\text{ind}_{\bar{\mathfrak{q}}, L\cap K}^{\mathfrak{g}, L\cap K}(Z_{\bar{\mathfrak{q}}}^{\#}))^h) \\[6pt]
{\scriptstyle \mathcal{E}_Z^K}\big\downarrow & & \big\downarrow{\scriptstyle \mathcal{E}_Z^G} \\[6pt]
[\Pi_{2S-j}^K(\text{ind}_{\bar{\mathfrak{q}}\cap\mathfrak{k}, L\cap K}^{\mathfrak{k}, L\cap K}(Z_{\bar{\mathfrak{q}}}^{\#}))]^h & \xleftarrow{\;(\mathcal{B}_Z)^h\;} & [\mathcal{L}_{2S-j}(\text{ind}_{\bar{\mathfrak{q}}, L\cap K}^{\mathfrak{g}, L\cap K}(Z_{\bar{\mathfrak{q}}}^{\#}))]^h
\end{array}
$$

For each of the small rectangles, the commutativity represents a general fact that we shall verify on the level of complexes. We shall use the resulting formulas again in §6. The commutativity of the first of the three small rectangles says that the Hard Duality maps for K and G are related by forgetful functors; the verification with complexes will take place in (6.43) and (6.44) below. The commutativity of the third of the three small rectangles says that the Easy Duality maps for K and G are related by forgetful functors (and also that Easy Duality is natural as an isomorphism); its verification with complexes has already taken place in Lemma 6.27.

The heart of the matter is the commutativity of the middle rectangle, seen on the level of complexes. This rectangle is Γ^{2S-j} of the diagram

(6.39)

$$
\begin{array}{ccc}
\mathrm{ind}_{\bar{\mathfrak{q}}\cap\mathfrak{k},L\cap K}^{\mathfrak{k},L\cap K}(Z_{\bar{\mathfrak{q}}}^{\#}) & \longrightarrow & \mathrm{ind}_{\bar{\mathfrak{q}},L\cap K}^{\mathfrak{g},L\cap K}(Z_{\bar{\mathfrak{q}}}^{\#}) \\
\psi_Z^K \downarrow & & \downarrow \psi_Z^G \\
(\mathrm{ind}_{\bar{\mathfrak{q}}\cap\mathfrak{k},L\cap K}^{\mathfrak{k},L\cap K}(Z_{\bar{\mathfrak{q}}}^{\#}))^h & \longrightarrow & (\mathrm{ind}_{\bar{\mathfrak{q}},L\cap K}^{\mathfrak{g},L\cap K}(Z_{\bar{\mathfrak{q}}}^{\#}))^h
\end{array}
$$

and the two lemmas that follow will address this problem.

Recall the definition and properties of $\mu_{\bar{\mathfrak{u}}}'$ from §IV.8.

Lemma 6.40. The map ψ_Z^G of (6.31) satisfies

(6.41) $$\psi_Z^G(x \otimes z)(\iota(x' \otimes z')) = \langle \mu_{\bar{\mathfrak{u}}}'(\bar{x}'^t x)z, z' \rangle_L^{\#}$$

for x and x' in $U(\mathfrak{g})$ and z and z' in $Z_{\bar{\mathfrak{q}}}^{\#}$. Here $\langle \cdot, \cdot \rangle_L^{\#}$ is the form for $Z^{\#}$ corresponding to $\langle \cdot, \cdot \rangle_L$ for Z.

PROOF. The left side of (6.41) is just

$$
\begin{aligned}
&= (e_Z^G \circ I_{\mathfrak{q},L\cap K}^{\mathfrak{g},L\cap K}(\zeta_{Z^{\#}}) \circ \varphi_Z^G)(x \otimes z)(\iota(x' \otimes z')) \\
&= (I_{\mathfrak{q},L\cap K}^{\mathfrak{g},L\cap K}(\zeta_{Z^{\#}}) \circ \varphi_Z^G)(x \otimes z)(\bar{x}'^t)(\iota z') && \text{by (6.21b)} \\
&= \zeta_{Z^{\#}}(\varphi_Z^G(x \otimes z)(\bar{x}'^t))(\iota z') \\
&= \zeta_{Z^{\#}}(\mu_{\bar{\mathfrak{u}}}'(\bar{x}'^t x)z)(\iota z') && \text{by Proposition 5.90} \\
&= \langle \mu_{\bar{\mathfrak{u}}}'(\bar{x}'^t x)z, z' \rangle_L^{\#} && \text{by (6.10),}
\end{aligned}
$$

and this is the right side of (6.41).

Observe that $\mu_{\bar{\mathfrak{u}}\cap\mathfrak{k}}'$ is the restriction of $\mu_{\bar{\mathfrak{u}}}'$ to $U(\mathfrak{k})$. Applying Lemma 6.40 once to G and once to K, we obtain

$$\psi_Z^K(x \otimes z)(\iota(x' \otimes z')) = \psi_Z^G(x \otimes z)(\iota(x' \otimes z'))$$

for x and x' in $U(\mathfrak{k})$ and z and z' in $Z_{\bar{\mathfrak{q}}}^{\#}$. Consequently the middle rectangle of (6.38) commutes. We shall make this commutativity visible on the level of complexes.

Lemma 6.42. Let $\beta : V \to W$ be a $(\mathfrak{g}, L \cap K)$ map between $(\mathfrak{g}, L \cap K)$ modules. Then $\Gamma_K^{2S-j}(\beta)$ can be computed as $\mathrm{Hom}(1, \mathrm{Hom}(1, \beta))$ on the level of the complexes

$$\mathrm{Hom}_{L \cap K}(R(K), \mathrm{Hom}_{\mathbb{C}}(\textstyle\bigwedge^{2S-j} \mathfrak{c}, V))_K$$

and $\qquad \mathrm{Hom}_{L \cap K}(R(K), \mathrm{Hom}_{\mathbb{C}}(\textstyle\bigwedge^{2S-j} \mathfrak{c}, W))_K.$

PROOF. Let X_j be the Koszul resolution in $\mathcal{C}(\mathfrak{k}, L \cap K)$. Then

$$\mathrm{Hom}(1, \beta) : \mathrm{Hom}_{\mathbb{C}}(X_j, V)_{L \cap K} \to \mathrm{Hom}_{\mathbb{C}}(X_j, W)_{L \cap K}$$

is a cochain map over $\beta : V \to W$, the associated complexes being injective resolutions. Applying Γ_K, we see that $\Gamma_K^j(\beta)$ is the map induced on the jth cohomology by the cochain map

$$\mathrm{Hom}(1, \beta) : \Gamma_K(\mathrm{Hom}_{\mathbb{C}}(X_j, V)_{L \cap K}) \to \Gamma_K(\mathrm{Hom}_{\mathbb{C}}(X_j, W)_{L \cap K}).$$

The sequence of isomorphisms preceding (2.129a) shows how to interpret this cochain map as a map between the standard complexes

$$\mathrm{Hom}_{L \cap K}(R(K), \mathrm{Hom}_{\mathbb{C}}(\textstyle\bigwedge^{2S-j} \mathfrak{c}, V))_K$$

and $\qquad \mathrm{Hom}_{L \cap K}(R(K), \mathrm{Hom}_{\mathbb{C}}(\textstyle\bigwedge^{2S-j} \mathfrak{c}, W))_K.$

Tracking down that sequence of isomorphisms, we find that the resulting map is $\mathrm{Hom}(1, \mathrm{Hom}(1, \beta))$, as asserted.

To see the commutativity of the top rectangle in (6.38), we need to compare $\mathcal{D}_Z^G$ with $\mathcal{D}_Z^K$. According to §III.4, $\mathcal{D}_Z^G$ is first given as a K map

(6.43a) $\quad \mathcal{D}_Z^G : C(K) \otimes_{\mathbb{C}} \textstyle\bigwedge^j \mathfrak{c} \otimes_{\mathbb{C}} V) \otimes_{\mathbb{C}} (\textstyle\bigwedge^{2S} \mathfrak{c})^*$

$$\to \mathrm{Hom}_{\mathbb{C}}(R(K), \mathrm{Hom}_{\mathbb{C}}(\textstyle\bigwedge^{2S-j} \mathfrak{c}, V))_K.$$

Here K acts by l on $C(K)$ in the domain and by r on $R(K)$ in the range. The formula is

(6.43b)
$$\mathcal{D}(f \otimes \xi \otimes v \otimes \varepsilon) = \lambda_{f \otimes \xi \otimes v \otimes \varepsilon}$$

$$\lambda_{f \otimes \xi \otimes v \otimes \varepsilon}(T')(\gamma) = \varepsilon(\xi \wedge \gamma)\langle T', f'\rangle v.$$

It will be more convenient for us to work with $R(K)$ in place of $C(K)$, using the isomorphism $C(K) \xrightarrow{\sim} R(K)$ given by $f \mapsto f \, dk$. This isomorphism depends on a choice of normalization of Haar measure dk, and we shall take dk to have total mass 1. Then $\mathcal{D}_Z^G$ is given as a map

(6.43c) $\quad \mathcal{D}_Z^G : R(K) \otimes_{\mathbb{C}} \textstyle\bigwedge^j \mathfrak{c} \otimes_{\mathbb{C}} V) \otimes_{\mathbb{C}} (\textstyle\bigwedge^{2S} \mathfrak{c})^*$

$$\to \mathrm{Hom}_{\mathbb{C}}(R(K), \mathrm{Hom}_{\mathbb{C}}(\textstyle\bigwedge^{2S-j} \mathfrak{c}, V))_K,$$

and the formula is

(6.43d)
$$\mathcal{D}_Z^G(T \otimes \xi \otimes v \otimes \varepsilon) = \lambda_{T \otimes \xi \otimes v \otimes \varepsilon}$$
$$\lambda_{T \otimes \xi \otimes v \otimes \varepsilon}(T')(\gamma) = \varepsilon(\xi \wedge \gamma)\langle T', T'\rangle v.$$

Here it is understood that $\langle T', T'\rangle = \langle T', f'\rangle$ if $T = f\,dk$, i.e.,

$$\langle T', T'\rangle = \int_K f'(k) f(k^{-1})\,dk \qquad \text{if } T' = f'\,dk \text{ and } T = f\,dk.$$

Then $\mathcal{D}_Z^G$ restricts to the $L \cap K$ invariants as a map between complexes

(6.44a)
$$(R(K) \otimes_{L \cap K} (\textstyle\bigwedge^j \mathfrak{c} \otimes_{\mathbb{C}} V)) \otimes_{\mathbb{C}} (\textstyle\bigwedge^{2S} \mathfrak{c})^*$$

and

(6.44b)
$$\operatorname{Hom}_{L \cap K}(R(K), \operatorname{Hom}_{\mathbb{C}}(\textstyle\bigwedge^{2S-j} \mathfrak{c}, V))_K,$$

the boundary/coboundary operators being given as in (2.128) and (2.129). We know from (2.128) and (2.129) that $\Pi_j^K(V)$ is given by the homology of (6.44a) and $\Gamma_K^j(V)$ is given by the cohomology of (6.44b). Also (3.35) shows that $\mathcal{D}_Z^G$ is a chain map, so that $\mathcal{D}_Z^G$ implements a K map of $\Pi_j^K(V)$ to $\Gamma_K^j(V)$.

Now let us compute the vertical composition

$$\Psi_Z^G = \mathcal{E}_Z^G \circ \Gamma^{2S-j}(\psi_Z^G) \circ \mathcal{D}_Z^G$$

in (6.38) by putting together (6.43) and Lemmas 6.27, 6.42, and 6.40. The relevant V in those results is

$$V = P_{\bar{\mathfrak{q}}, L \cap K}^{\mathfrak{g}, L \cap K}(Z_{\bar{\mathfrak{q}}}^\#) = \operatorname{ind}_{\bar{\mathfrak{q}}, L \cap K}^{\mathfrak{g}, L \cap K}(Z_{\bar{\mathfrak{q}}}^\#).$$

Write $v = u \otimes z$ with $u \in U(\mathfrak{g})$ and $z \in Z_{\bar{\mathfrak{q}}}^\#$ for a typical monomial in V. We first make the computation as if the domain of our maps is somewhat larger than in the complexes of the lemmas, and then we interpret matters. With this understanding, we start with elements $T \otimes \xi \otimes (u \otimes z) \otimes \varepsilon_0$ and

$T' \otimes \xi' \otimes (u' \otimes z')$ as in the notation of (6.43d). Then

$$\Psi_Z^G(T \otimes \xi \otimes (u \otimes z) \otimes \varepsilon_0)(\iota(T' \otimes \xi' \otimes (u' \otimes z')))$$

$$= \mathcal{E}_Z^G \circ \Gamma_K^{2S-j}(\psi_Z^G) \circ \mathcal{D}_Z^G(T \otimes \xi \otimes (u \otimes z) \otimes \varepsilon_0)(\iota(T' \otimes \xi' \otimes (u' \otimes z')))$$

by (6.33)

$$= (\Gamma_K^{2S-j}(\psi_Z^G) \circ \mathcal{D}_Z^G(T \otimes \xi \otimes (u \otimes z) \otimes \varepsilon_0))(\overline{T'^t})(\bar{\xi}')(\iota(u' \otimes z'))$$

by (6.28b)

$$= \psi_Z^G(\mathcal{D}_Z^G(T \otimes \xi \otimes (u \otimes z) \otimes \varepsilon_0)(\overline{T'^t})(\bar{\xi}'))(\iota(u' \otimes z'))$$

by Lemma 6.42

$$= \psi_Z^G(\varepsilon_0(\xi \wedge \bar{\xi}')\langle \overline{T'^t}, T^t \rangle (u \otimes z))(\iota(u' \otimes z'))$$

by (6.43d)

$$= \varepsilon_0(\xi \wedge \bar{\xi}')\langle \overline{T'}, T \rangle \psi_Z^G(u \otimes z))(\iota(u' \otimes z')).$$

Hence Lemma 6.40 gives

(6.45a) $\Psi_Z^G(T \otimes \xi \otimes (u \otimes z) \otimes \varepsilon_0)(\iota(T' \otimes \xi' \otimes (u' \otimes z')))$

$$= \varepsilon_0(\xi \wedge \bar{\xi}')\langle \overline{T'}, T \rangle \langle \mu_{\bar{u}}'(\bar{u}'^t u)z, z' \rangle_L^\#.$$

The subtle point here is that the formula (6.43) for $\mathcal{D}_Z^G$, with ε fixed as ε_0, descends from $R(K) \otimes_{\mathbb{C}} (\bigwedge^j \mathfrak{c} \otimes_{\mathbb{C}} V)$ to $R(K) \otimes_{L \cap K} (\bigwedge^j \mathfrak{c} \otimes_{\mathbb{C}} V)$ not by passage to the quotient but by restriction to the $L \cap K$ invariants. Thus in order to interpret (6.45) as a formula on $R(K) \otimes_{\mathbb{C}} (\bigwedge^j \mathfrak{c} \otimes_{\mathbb{C}} V)$ that allows passage to the quotient, we must first average $T \otimes \xi \otimes (u \otimes z)$ and $T' \otimes \xi' \otimes (u' \otimes z')$ separately with respect to $L \cap K$.

With this proviso, Ψ_Z^G gives an explicit formula for the Shapovalov form on G. Namely the form lifts to

(6.45b) $\langle T \otimes \xi \otimes (u \otimes z), T' \otimes \xi' \otimes (u' \otimes z') \rangle_G$

$$= \varepsilon_0(\xi \wedge \bar{\xi}')\langle \overline{T'}, T \rangle \langle \mu_{\bar{u}}'(\bar{u}'^t u)z, z' \rangle_L^\#$$

on the level of complexes.

PROOF OF PROPOSITION 6.35. Let u be in $U(\mathfrak{k})$. Lemma 6.42 allows us to compute $\mathcal{B}_Z$ on the level of complexes as

$$\mathcal{B}_Z(T \otimes \xi \otimes (u \otimes z)) = (1 \otimes 1 \otimes \beta_Z)(T \otimes \xi \otimes (u \otimes z))$$

$$= T \otimes \xi \otimes \beta_Z(u \otimes z)$$

$$= T \otimes \xi \otimes (u \otimes z).$$

Here u on the right is interpreted as a member of $U(\mathfrak{g})$. That is, $\mathcal{B}_Z$ is an inclusion on the level of complexes. Bringing in (6.8), we obtain

$$
\begin{aligned}
(\mathcal{B}_Z)^h \circ \Psi_Z^G \circ \mathcal{B}_Z(T \otimes \xi \otimes (u \otimes z) \otimes \varepsilon_0)(\iota(T' \otimes \xi' \otimes (u' \otimes z'))) \\
= \Psi_Z^G \circ \mathcal{B}_Z(T \otimes \xi \otimes (u \otimes z) \otimes \varepsilon_0)(\iota\mathcal{B}_Z(T' \otimes \xi' \otimes (u' \otimes z'))) \\
= \Psi_Z^G(T \otimes \xi \otimes (u \otimes z) \otimes \varepsilon_0)(\iota(T' \otimes \xi' \otimes (u' \otimes z'))).
\end{aligned}
$$

This expression is given by the right side of (6.45), and, in turn, the right side of (6.45) is

$$
= \Psi_Z^K(T \otimes \xi \otimes (u \otimes z) \otimes \varepsilon_0)(\iota(T' \otimes \xi' \otimes (u' \otimes z')))
$$

since $\mu'_{\bar{\mathfrak{u}}}|_{U(\mathfrak{k})} = \mu'_{\bar{\mathfrak{u}} \cap \mathfrak{k}}$. This proves (6.36), and (6.37) is an immediate consequence of (6.36) and (6.10).

6. Preservation of Positivity from $L \cap K$ to K

In this section we shall complete the proof of Theorem 6.34. We continue with the setting of §1.

Lemma 6.46. Under the assumption that $\langle \cdot, \cdot \rangle_L$ is Hermitian for Z, $\langle \cdot, \cdot \rangle_G$ is Hermitian for $\mathcal{L}_S(Z)$.

PROOF. Certainly the form $\langle \cdot, \cdot \rangle_L^\#$ on $Z^\#$ will be Hermitian. From (6.45b) we may work with $\langle \cdot, \cdot \rangle_G$ on the level of complexes. The most general element of

$$
R(K) \otimes_{L \cap K} (\textstyle\bigwedge^S \mathfrak{c} \otimes_{\mathbb{C}} \operatorname{ind}_{\bar{\mathfrak{q}}, L \cap K}^{\mathfrak{g}, L \cap K}(Z_{\bar{\mathfrak{q}}}^\#))
$$

is any linear combination of elements

$$
\int_{L \cap K} \{r(m)T \otimes \operatorname{Ad}(m)\xi \otimes \operatorname{Ad}(m)u \otimes mz\} \, dm
$$

with $T \in R(K)$, $\xi \in \bigwedge^S \mathfrak{c}$, $u \in U(\mathfrak{u})$, and $z \in Z_{\bar{\mathfrak{q}}}^\#$; and $\langle \cdot, \cdot \rangle_G$ for two such elements is computed by linearity from (6.45b). Hence it is enough to see that the right side of (6.45b) itself is Hermitian. Thus it is enough to see that

(6.47) $$ \overline{\varepsilon_0(\xi \wedge \bar{\xi}')} = \varepsilon_0(\xi' \wedge \bar{\xi}) $$

and that

(6.48)
$$\overline{\langle \mu'_{\bar{u}}(\bar{u}''u)z, z'\rangle^{\#}_L} = \langle \mu'_{\bar{u}}(\bar{u}'u')z', z\rangle^{\#}_L.$$

Formula (6.47) claims that the sesquilinear form $\langle \xi, \xi'\rangle = \varepsilon_0(\xi \wedge \bar{\xi}')$ is Hermitian. To verify this assertion, it is enough to treat the case in which $\xi = \xi_1 \wedge \xi_2, \xi' = \xi'_1 \wedge \xi'_2, \xi_1 \in \bigwedge^a(\bar{u} \cap \mathfrak{k}), \xi_2 \in \bigwedge^b(u \cap \mathfrak{k}), \xi'_1 \in \bigwedge^c(\bar{u} \cap \mathfrak{k}),$ $\xi'_2 \in \bigwedge^d(u \cap \mathfrak{k}), a + b = S, c + d = S$. Both sides of (6.47) are 0 unless $a = c$, and thus we may assume $b = S - a, c = a, d = S - a$. Moreover, our definition of ε_0 makes (6.47) valid if $a = S$. For general a, we have

$$\begin{aligned}
\overline{\langle \xi, \xi'\rangle} &= \overline{\varepsilon_0(\xi_1 \wedge \xi_2 \wedge \bar{\xi}'_1 \wedge \bar{\xi}'_2)} \\
&= (-1)^{S(S-a)}\overline{\varepsilon_0(\xi_1 \wedge \bar{\xi}'_2 \wedge \xi_2 \wedge \bar{\xi}'_1)} \\
&= (-1)^{S(S-a)}\overline{\langle \xi_1 \wedge \bar{\xi}'_2, \bar{\xi}_2 \wedge \xi'_1\rangle} \\
&= (-1)^{S(S-a)}\langle \bar{\xi}_2 \wedge \xi'_1, \xi_1 \wedge \bar{\xi}'_2\rangle \qquad \text{by the special case } a = S \\
&= (-1)^{S(S-a)}\varepsilon_0(\bar{\xi}_2 \wedge \xi'_1 \wedge \bar{\xi}_1 \wedge \xi'_2) \\
&= (-1)^{(S+a)(S-a)+S(S-a)}\varepsilon_0(\xi'_1 \wedge \bar{\xi}_1 \wedge \xi'_2 \wedge \bar{\xi}_2) \\
&= (-1)^{a(S-a)+a(S-a)}\varepsilon_0(\xi'_1 \wedge \xi'_2 \wedge \bar{\xi}_1 \wedge \bar{\xi}_2) \\
&= \langle \xi', \xi\rangle,
\end{aligned}$$

and (6.47) is proved.

Consider (6.48). Since $\langle \cdot, \cdot\rangle^{\#}_L$ is Hermitian, it is enough to see that

(6.49)
$$\mu'_{\bar{u}}(\bar{x}^t) = \overline{\mu'_{\bar{u}}(x)}^t$$

for x in $U(\mathfrak{g})$. Let x be a monomial in a Poincaré-Birkhoff-Witt basis relative to

$$U(\mathfrak{g}) \cong U(u) \otimes_{\mathbb{C}} U(\mathfrak{l}) \otimes_{\mathbb{C}} U(\bar{u}).$$

If there are nontrivial u or $\bar{u}$ factors present, both sides of (6.49) are 0. Otherwise x is in $U(\mathfrak{l})$, and both sides of (6.49) reduce to $\bar{x}^t$. Thus (6.48) is proved, and $\langle \cdot, \cdot\rangle_G$ is Hermitian.

Proposition 6.50. Let W be an $(\mathfrak{l} \cap \mathfrak{k}, L \cap K)$ module and define

$W^{\#} = W \otimes_{\mathbb{C}} \bigwedge^{\text{top}} u,$	an $(\mathfrak{l} \cap \mathfrak{k}, L \cap K)$ module
$W^{\#}_{\bar{q}} = W^{\#}$ with $\bar{u} \cap \mathfrak{k}$ acting by 0,	a $(\bar{q} \cap \mathfrak{k}, L \cap K)$ module.

Let $\langle \cdot, \cdot \rangle_{L \cap K}$ be an invariant Hermitian form on W, and let $\langle \cdot, \cdot \rangle_K$ be the corresponding (Hermitian) Shapovalov form on

$$\mathcal{L}_S^K(W) = \Pi_S^K(P_{\bar{\mathfrak{q}} \cap \mathfrak{k}, L \cap K}^{\mathfrak{k}, L \cap K}(W_{\bar{\mathfrak{q}}}^{\#})).$$

Suppose that an $L \cap K$ type τ_L of W has the property that all the highest weights of $\tau_L \otimes_{\mathbb{C}} \bigwedge^R(\mathfrak{u} \cap \mathfrak{p})$ are K dominant. Let τ_G be the K type of $\mathcal{L}_S^K(W)$ that corresponds to $\tau_L \otimes_{\mathbb{C}} \bigwedge^R(\mathfrak{u} \cap \mathfrak{p})$ by the construction of Theorem 4.83e. Then the signature of $\langle \cdot, \cdot \rangle_{L \cap K}$ on the τ_L subspace of W equals the signature of $\langle \cdot, \cdot \rangle_K$ on the τ_G subspace of $\mathcal{L}_S^K(W)$.

REMARKS.

1) The Shapovalov form is Hermitian by Lemma 6.46.

2) If W is irreducible of type τ_L, then Corollary 5.72 shows that $\mathcal{L}_S^K(W)$ is irreducible of type τ_G. In this situation the map φ_W^K of (6.29) is an isomorphism, by inspection of the proof of (b,c) of Theorem 5.99. If the form $\langle \cdot, \cdot \rangle_{L \cap K}$ is nondegenerate, then it follows that ψ_W^K in (6.31) is an isomorphism. Hence Ψ_W^K in (6.33) is an isomorphism, and $\langle \cdot, \cdot \rangle_K$ is nondegenerate.

PROOF. Our results for G apply also to K (since K satisfies our axioms) and $\langle \cdot, \cdot \rangle_K$ is Hermitian by Lemma 6.46. Let $\langle \cdot, \cdot \rangle_{L \cap K}^{\#}$ be the tensor-product form for $W^{\#}$, so that $\langle \cdot, \cdot \rangle_K$ can be read off from (6.45) with G replaced by K. Write W and $\mathcal{L}_S^K(W)$ as orthogonal invariant sums

$$W = W_{\tau_L} \oplus W'$$
$$\mathcal{L}_S^K(W) = \mathcal{L}_S^K(W)_{\tau_G} \oplus \mathcal{L}_S^K(W)',$$

where W' is the sum of the remaining $L \cap K$ isotypic subspaces and $\mathcal{L}_S^K(W)'$ is the sum of the remaining K isotypic subspaces. Then Corollary 5.72 shows that

$$\mathcal{L}_S^K(W_{\tau_L}) = \mathcal{L}_S^K(W)_{\tau_G}.$$

Changing notation, we may assume that W has the single $L \cap K$ type τ_L. Next, using Proposition 6.12, write W as an orthogonal $L \cap K$ invariant sum
$$W = W_+ \oplus W_0 \oplus W_-$$

according to the signature of $\langle \cdot, \cdot \rangle_{L \cap K}$. Since (6.45b) for K shows that orthogonal $L \cap K$ invariant subspaces are carried by $\mathcal{L}_S^K$ to subspaces that are orthogonal under $\langle \cdot, \cdot \rangle_K$, we obtain an orthogonal K invariant decomposition

$$\mathcal{L}_S^K(W) = \mathcal{L}_S^K(W_+) \oplus \mathcal{L}_S^K(W_0) \oplus \mathcal{L}_S^K(W_-).$$

Since $\mathcal{L}_S^K(W_0)$ is in the radical of $\langle \cdot, \cdot \rangle_K$ by (6.45b), it is enough to handle W_+.

Changing notation a second time, we may assume that $\langle \cdot, \cdot \rangle_{L \cap K}$ is positive definite on W. Then we can write W as an orthogonal sum of irreducible $L \cap K$ invariant subspaces of type τ_L, with $\langle \cdot, \cdot \rangle_{L \cap K}$ positive definite on each. Again (6.45b) for K shows that $\mathcal{L}_S^K$ carries these spaces to orthogonal spaces.

Changing notation a third time, we may assume that W is irreducible of type τ_L and $\langle \cdot, \cdot \rangle_{L \cap K}$ is positive definite. Then $\langle \cdot, \cdot \rangle_{L \cap K}^{\#}$ is positive definite. Corollary 5.72 shows that $\mathcal{L}_S^K(W)$ is irreducible. Hence the signature is detected by the inner product of any nonzero element in homology with itself. We shall exhibit such an element where the inner product is positive, and then the proof will be complete.

The element that we shall use is the homology class of the average over $L \cap K$ of

$$T \otimes \xi \otimes (1 \otimes (w \otimes \eta)),$$

where $T = \bar{f} \, dk$ for the function $f = d_{\tau_G} \chi_{\tau_G}$, ξ is nonzero in $\bigwedge^S(\bar{u} \cap \mathfrak{k})$, 1 is in $U(\mathfrak{k})$, w is arbitrary nonzero in W, and η is nonzero in $\bigwedge^{\text{top}} u$. Here d_{τ_G} and χ_{τ_G} are the degree and the character of τ_G. Thus we use

$$v = \int_{L \cap K} \{\bar{f}(\cdot \, m) \otimes \text{Ad}(m)\xi \otimes 1 \otimes mw \otimes \text{Ad}(m)\eta\} \, dm.$$

Let us write $\xi = Y_1 \wedge \cdots \wedge Y_S$ with all Y_i in $\bar{u} \cap \mathfrak{k}$. According to (2.128), the boundary operator is given as the linear extension of

(6.51)
$$\partial(\bar{f} \, dk \otimes \xi \otimes (1 \otimes w \otimes \eta))$$

$$= \sum_{i=1}^{S} (-1)^i (\bar{f} \, dk \otimes Y_1 \wedge \cdots \wedge \widehat{Y_i} \wedge \cdots \wedge Y_S \otimes Y_i(1 \otimes w \otimes \eta))$$

$$+ \sum_{i=1}^{S} (-1)^{i+1} (r(Y_i) \bar{f} \, dk \otimes Y_1 \wedge \cdots \wedge \widehat{Y_i} \wedge \cdots \wedge Y_S \otimes (1 \otimes w \otimes \eta))$$

$$+ \sum_{i<j} (-1)^{i+j} (\bar{f} \, dk \otimes \mathcal{P}[Y_i, Y_j] \wedge Y_1 \wedge \cdots \wedge \widehat{Y_i} \wedge \cdots \wedge \widehat{Y_j} \wedge \cdots \wedge Y_S$$

$$\otimes (1 \otimes w \otimes \eta)).$$

We shall show that $\partial v = 0$. In the first sum on the right of (6.51), we have

$$Y_i(1 \otimes w \otimes \eta) = Y_i \otimes w \otimes \eta = 1 \otimes Y_i(w \otimes \eta),$$

and this is 0 since $\bar{u} \cap \mathfrak{k}$ acts as 0 in $W^{\#}$. In the third sum on the right of (6.51), the alternating tensor is 0 since $\bar{u} \cap \mathfrak{k}$ is nilpotent and $S = \dim(\bar{u} \cap \mathfrak{k})$. Thus ∂v is the integral over $L \cap K$ of the m transform of the second term on the right of (6.51), for $m \in L \cap K$. This is a linear combination of

(6.52)
$$\int_{L \cap K} (r(\mathrm{Ad}(m)Y_i)r(m)f \, dk) \otimes \mathrm{Ad}(m)(Y_1 \wedge \cdots \wedge \widehat{Y_i} \wedge \cdots \wedge Y_S)$$
$$\otimes (1 \otimes m(w \otimes \eta)) \, dm$$
$$= \int_{L \cap K} (r(m)r(Y_i)f \, dk) \otimes \mathrm{Ad}(m)(Y_1 \wedge \cdots \wedge \widehat{Y_i} \wedge \cdots \wedge Y_S)$$
$$\otimes (1 \otimes m(w \otimes \eta)) \otimes dm.$$

Since W is irreducible under $L \cap K$, $h_{\delta(u)}$ acts as a scalar (Lemma 5.11). Hence all highest weights μ_L of τ_L have $\mu_L(h_{\delta(u)})$ the same, say c. Any highest weight μ_G of τ_G is of the form

$$\mu_G = \mu_L + 2\delta(u \cap p)$$

for some highest weight μ_L of τ_L, and any weight v of τ_G is of the form

$$v = \mu_G - \sum_{\beta \in \Delta(n \cap \mathfrak{k})} n_\beta \beta, \qquad n_\beta \geq 0,$$

for some μ_G. Thus

$$v(h_{\delta(u)}) = \mu_G(h_{\delta(u)}) - \sum_{\beta \in \Delta(u \cap \mathfrak{k})} n_\beta \beta(h_{\delta(u)})$$
$$= c + 2\delta(u \cap p)(h_{\delta(u)}) - \sum_{\beta \in \Delta(u \cap \mathfrak{k})} n_\beta \beta(h_{\delta(u)}).$$

The function f is a sum of functions of various weights v of this form (under r), and $\bar{f}$ is a sum of functions whose weights are the negatives of these. The same is true of $r(Y_i)\bar{f}$. The smallest value of one of these weights on $h_{\delta(u)}$ occurs when all n_β are 0 and is

$$-c - 2\delta(u \cap p)(h_{\delta(u)}).$$

The expression (6.52) will be 0 by Schur orthogonality unless the tensor product in question contains the trivial representation of $L \cap K$. Then some weight contributing to $r(Y_i)\bar{f}$ must be the negative of a highest weight of

$$\textstyle\bigwedge^{S-1}(\bar{\mathfrak{u}} \cap \mathfrak{k}) \otimes_{\mathbb{C}} W \otimes_{\mathbb{C}} \bigwedge^{\mathrm{top}}\mathfrak{u}.$$

Hence some weight of $\bigwedge^{S-1}(\bar{\mathfrak{u}} \cap \mathfrak{k})$ plus the highest weight of $W \otimes \bigwedge^{\mathrm{top}}\mathfrak{u}$, namely some $\mu_L + 2\delta(\mathfrak{u})$, must match the negative of a weight contributing to $r(Y_i)\bar{f}$. On $h_{\delta(\mathfrak{u})}$ this relation says

$$c + 2\delta(\mathfrak{u})(h_{\delta(\mathfrak{u})}) + \sum_{\substack{\text{set of } S-1 \\ \gamma\text{'s in } \Delta(\bar{\mathfrak{u}}\cap\mathfrak{k})}} \gamma(h_{\delta(\mathfrak{u})}) \le c + 2\delta(\mathfrak{u} \cap \mathfrak{p})(h_{\delta(\mathfrak{u})}).$$

In other words,

$$2\delta(\mathfrak{u} \cap \mathfrak{k})(h_{\delta(\mathfrak{u})}) \le \sum_{\substack{\text{set of } S-1 \\ \beta\text{'s in } \Delta(\mathfrak{u}\cap\mathfrak{k})}} \beta(h_{\delta(\mathfrak{u})}).$$

This is a contradiction since $\beta(h_{\delta(\mathfrak{u})}) > 0$ for every $\beta \in \Delta(\mathfrak{u})$, by (4.71). Therefore (6.52) is 0, and $\partial v = 0$.

To complete the proof, it is enough to show that $\langle v, v \rangle_K$ is a positive multiple of $\langle w, w \rangle_{L \cap K}$. (Then it follows that $v \ne 0$ in homology.) If we let $\chi_{2\delta(\mathfrak{u}\cap\mathfrak{p})}$ be the character of the one-dimensional representation of $L \cap K$ on $\bigwedge^R(\mathfrak{u} \cap \mathfrak{p})$, then (6.45b) gives

$$\langle v, v \rangle_G = \int_{L \cap K} \int_{L \cap K} \int_K \varepsilon_0(\mathrm{Ad}(m)\xi \wedge \mathrm{Ad}(m')\bar{\xi})\overline{f(km)}f(km')$$
$$\times \langle mw, m'w \rangle_{L\cap K} \langle m\eta, m'\eta \rangle\, dk\, dm\, dm'$$

$$= \int_{L \cap K} \int_K \varepsilon_0(\mathrm{Ad}(m)\xi \wedge \bar{\xi})\overline{f(k)}f(km^{-1})\langle mw, w \rangle_{L\cap K}$$
$$\times \langle m\eta, \eta \rangle\, dk\, dm \quad \text{after } k \mapsto km^{-1}$$
$$\text{and } m'^{-1}m \mapsto m$$

$$= \varepsilon_0(\xi \wedge \bar{\xi})\langle \eta, \eta \rangle \int_{L\cap K} \int_K \chi_{2\delta(\mathfrak{u}\cap\mathfrak{p})}(m)\langle mw, w \rangle_{L\cap K}\rangle_{L\cap K}$$
$$\times f(k^{-1})f(km^{-1})\, dk\, dm \quad \text{since } \bar{f} = f^t$$

$$= \varepsilon_0(\xi \wedge \bar{\xi})\langle \eta, \eta \rangle \int_{L\cap K} \chi_{2\delta(\mathfrak{u}\cap\mathfrak{p})}(m)\langle mw, w \rangle_{L\cap K} f(m^{-1})\, dm$$
$$\text{since } f = d_{\tau_G}\chi_{\tau_G}$$

$$(6.53) \qquad = \varepsilon_0(\xi \wedge \bar{\xi})\langle \eta, \eta \rangle \int_{L\cap K} \chi_{2\delta(\mathfrak{u}\cap\mathfrak{p})}(m)\langle mw, w \rangle_{L\cap K}\, \overline{f(m)}\, dm.$$

Apart from $\overline{f(m)}$, the integrand is in the $L \cap K$ type $\tau_L \otimes_{\mathbb{C}} \bigwedge^R(\mathfrak{u} \cap \mathfrak{p})$. Meanwhile $f(m)$ is a linear combination of characters of $L \cap K$, and the coefficient of $\chi_{\tau_L \otimes \wedge^R(\mathfrak{u} \cap \mathfrak{p})}$ is d_{τ_G}. If d_{τ_L} denotes the degree of τ_L, (6.53) thus reduces to

$$= \varepsilon_0 (\xi \wedge \bar{\xi}) \langle \eta, \eta \rangle (d_{\tau_G}/d_{\tau_L}) \langle w, w \rangle_{L \cap K},$$

by Schur orthogonality. Because of our choice of ε_0, the coefficient of $\langle w, w \rangle_{L \cap K}$ is positive, and the proof of Proposition 6.50 is complete.

PROOF OF THEOREM 6.34. Lemma 6.46 shows that the Shapovalov form is Hermitian. Let τ_G be a K type in the bottom layer of $\mathcal{L}_S(Z)$, and let τ_L be the $L \cap K$ type of Z such that $\tau_L \otimes_{\mathbb{C}} \bigwedge^R(\mathfrak{u} \cap \mathfrak{p})$ is the representation in the $\mathfrak{u} \cap \mathfrak{k}$ invariants of τ_G (see Theorem 4.83e). Regard Z as an $(\mathfrak{l} \cap \mathfrak{k}, L \cap K)$ module, and apply Proposition 6.50 to see that $\langle \cdot, \cdot \rangle_L$ has the same signature on the τ_L subspace of Z that $\langle \cdot, \cdot \rangle_K$ has on the τ_G subspace of $\mathcal{L}_S^K(Z)$. Theorem 5.80 says that the bottom-layer map $\mathcal{B}_Z$ carries the τ_G subspace of $\mathcal{L}_S^K(Z)$ one-one onto the τ_G subspace of $\mathcal{L}_S(Z)$, and Proposition 6.35 says that $\mathcal{B}_Z$ preserves inner products. Thus $\langle \cdot, \cdot \rangle_K$ has the same signature on the τ_G subspace of $\mathcal{L}_S^K(Z)$ that $\langle \cdot, \cdot \rangle_G$ has on the τ_G subspace of $\mathcal{L}_S(Z)$.

7. Signature Theorem for K Badly Disconnected

In this section we investigate Hermitian forms on cohomologically induced modules without the assumption (6.1). The setting is otherwise that of §1. We define $Z^{\#}$ as in (5.2) or (6.2), $Z_{\bar{\mathfrak{q}}}^{\#}$ and $Z_{\mathfrak{q}}^{\#}$ as in (5.4) or (6.4), and the cohomologically induced modules

$$\mathcal{L}_j(Z), \quad \mathcal{R}^j(Z), \quad \mathcal{L}_j^K(Z), \quad \mathcal{R}_K^j(Z)$$

as in (5.3) and (5.70).

Just as in (5.76) and (6.5), there are bottom-layer maps

(6.54) $$\mathcal{B}_Z : \mathcal{L}_j^K(Z) \to \mathcal{L}_j(Z).$$

It is still true that $\mathcal{B}_Z$ is injective when $j = S$, and we call the image the **bottom layer** of $\mathcal{L}_S(Z)$. The consequence for $\mathcal{B}_Z$ of dropping the assumption (6.1) is that $\mathcal{B}_Z$ for $j = S$ need no longer map onto the full isotypic subspaces for the K types of the bottom layer. These facts were already noted in §V.6.

We want to define the Shapovalov form on $\mathcal{L}_S(Z)$ corresponding to an invariant sesquilinear form $\langle \cdot, \cdot \rangle_L$ on Z. The definition in §§3–4 carries over without change to the present setting, including the formulas on the level of complexes of all the maps involved. Thus we get an invariant sesquilinear form $\langle \cdot, \cdot \rangle_G$ on $\mathcal{L}_S(Z)$. Similarly any $(\mathfrak{l} \cap \mathfrak{k}, L \cap K)$ invariant sesquilinear form $\langle \cdot, \cdot \rangle_{L \cap K}$ on Z induces a K invariant sesquilinear form $\langle \cdot, \cdot \rangle_K$ on $\mathcal{L}_S^K(Z)$. Here is a version of the Signature Theorem for these forms.

Theorem 6.55. Suppose Z is an $(\mathfrak{l}, L \cap K)$ module with an invariant Hermitian form $\langle \cdot, \cdot \rangle_L$.

(a) The Shapovalov form $\langle \cdot, \cdot \rangle_G$ on $\mathcal{L}_S(Z)$ is Hermitian.

(b) If $\langle \cdot, \cdot \rangle_{L \cap K}$ is defined as the form $\langle \cdot, \cdot \rangle_L$, then the Shapovalov forms $\langle \cdot, \cdot \rangle_K$ and $\langle \cdot, \cdot \rangle_G$ are related by

$$\langle x, y \rangle_K = \langle \mathcal{B}_Z x, \mathcal{B}_Z y \rangle_G.$$

(c) If $\langle \cdot, \cdot \rangle_{L \cap K}$ is any positive definite $L \cap K$ invariant Hermitian form on Z, then the Shapovalov form $\langle \cdot, \cdot \rangle_K$ is positive definite on $\mathcal{L}_S^K(Z)$.

REMARKS. For (a) the proof of Lemma 6.46 carries over without change. For (b) the proof of Proposition 6.35 applies. Part (c) requires some preparation. This part is a version of Proposition 6.50, whose statement and proof made use of the assumption (6.1). We shall reduce matters to that case.

Before proving (c), we need to understand explicitly the nature of the Shapovalov forms away from $K_1 = (L \cap K)K_0$. We begin with an analysis of Hermitian forms in the context of §II.4.

Temporarily let us suppose that $(\mathfrak{g}, K_1) \hookrightarrow (\mathfrak{g}, K)$ is an arbitrary inclusion of pairs such that K_1 has finite index in K. Let $\mathfrak{g}_0$ be a real form of $\mathfrak{g}$ containing $\mathfrak{k}_0$ and preserved by $\mathrm{Ad}(K)$, so that we may speak of invariant Hermitian forms on $(\mathfrak{g}, K)$ or $(\mathfrak{g}, K_1)$ modules. Let V be a $(\mathfrak{g}, K_1)$ module endowed with an invariant sesquilinear form $\langle \cdot, \cdot \rangle_1$, and write

$$\zeta_1 : V \to V^h$$

for the corresponding map. To imitate matters in §§3–4, we start with (6.31a), replacing φ_Z^G and e_Z^G by the identity and taking ζ_1 to play the roles of $I_{\mathfrak{q}, L \cap K}^{\mathfrak{g}, L \cap K}(\zeta_{Z^\#})$ and ψ_Z^G. Let $\mathcal{D}_V^G$ be the Hard Duality mapping

$$\mathcal{D}_V^G : P_{\mathfrak{g}, K_1}^{\mathfrak{g}, K}(V) \to I_{\mathfrak{g}, K_1}^{\mathfrak{g}, K}(V)$$

of Theorem 3.5, and let $\mathcal{E}_Z^G$ be the Easy Duality mapping

$$\mathcal{E}_V^G : I_{\mathfrak{g},K_1}^{\mathfrak{g},K}(V^h) \to P_{\mathfrak{g},K_1}^{\mathfrak{g},K}(V)^h$$

of Theorem 3.1. Putting $S = j = 0$ in (6.33), we obtain

(6.56a)
$$\Psi_V^G = \mathcal{E}_V^G \circ \Gamma_{\mathfrak{g},K_1}^{\mathfrak{g},K}(\zeta_1) \circ \mathcal{D}_V^G$$

as a $(\mathfrak{g}, K)$ map

(6.56b)
$$\Psi_V^G : P_{\mathfrak{g},K_1}^{\mathfrak{g},K}(V) \to [P_{\mathfrak{g},K_1}^{\mathfrak{g},K}(V)]^h.$$

Because of the similarity with the construction in §§3–4, it is reasonable to call the resulting form $\langle \cdot, \cdot \rangle_P$ on $P_{\mathfrak{g},K_1}^{\mathfrak{g},K}(V)$ the **Shapovalov form** induced by $\langle \cdot, \cdot \rangle_1$.

Recall from Proposition 2.77 that the functor $P_{\mathfrak{g},K_1}^{\mathfrak{g},K}$ is naturally isomorphic with induced$_{\mathfrak{g},K_1}^{\mathfrak{g},K}$. Because of (C.22) this isomorphism is given by the maps

(6.57a)
$$\Phi_V : P_{\mathfrak{g},K_1}^{\mathfrak{g},K}(V) \to \text{induced}_{\mathfrak{g},K_1}^{\mathfrak{g},K}(V)$$

belonging to the left side of

(6.57b)
$$\text{Hom}_{\mathfrak{g},K}(P_{\mathfrak{g},K_1}^{\mathfrak{g},K}(V), \text{induced}_{\mathfrak{g},K_1}^{\mathfrak{g},K}(V)) \cong \text{Hom}_{\mathfrak{g},K_1}(V, \text{induced}_{\mathfrak{g},K_1}^{\mathfrak{g},K}(V))$$

that correspond to j_V in the right side. (For the definition of

(6.57c)
$$j_V : V \to \text{induced}_{\mathfrak{g},K_1}^{\mathfrak{g},K}(V),$$

see the statement of Proposition 2.75.)

We can reinterpret the Shapovalov form $\langle \cdot, \cdot \rangle_P$ induced by $\langle \cdot, \cdot \rangle_1$ as a form $\langle \cdot, \cdot \rangle$ on induced$_{\mathfrak{g},K_1}^{\mathfrak{g},K}(V)$ by means of the map

$$(\Phi_V^h)^{-1} \circ \Psi_V^G \circ \Phi_V^{-1} : \text{induced}_{\mathfrak{g},K_1}^{\mathfrak{g},K}(V) \to [\text{induced}_{\mathfrak{g},K_1}^{\mathfrak{g},K}(V)]^h.$$

Lemma 6.58. If A is a finite subset of $\widehat{K}_0$, extend the member χ_A of the approximate identity of $R(K_0)$ to a member χ_A^e of $R(K)$ by defining its value on $f \in C(K)$ to be $\langle \chi_A, f|_{K_0}\rangle = \int_{K_0} f(k_0)\, d\chi_A(k_0)$. If A is K stable (under the action of K on $\widehat{K}_0$), then χ_A^e is a member of the approximate identity of $R(K)$. In this case if $f \in C(K)$ is given and if A is large enough, then $\chi_A^e * f = f$.

PROOF. The extension χ_A^e is obtained by putting χ_A equal to 0 off the subgroup K_0 of K. Thus the formula $\chi_A * \chi_A$ in $R(K_0)$ leads to $\chi_A^e * \chi_A^e = \chi_A^e$ in $R(K)$.

The condition that A be K stable implies that $l(k)r(k)\chi_A^e = \chi_A^e$ for all $k \in K$. Now any $T \in R(K)$ with $T * T = T$ and with $l(k)r(k)T = T$ for all $k \in K$ is a member of the approximate identity. In fact, the invariance condition $l(k)r(k)T = T$ implies that T is the product of a linear combination of irreducible characters by normalized Haar measure on K. Substituting such a linear combination into the identity $T * T = T$, we see that the coefficient of each contributing irreducible character has to be the degree of that character. Thus T has the required form.

In particular, χ_A^e is a member of the approximate identity of $R(K)$. Let k_j be coset representatives of K/K_0. If $f \in C(K)$ is given, then each $l(k_j^{-1})f|_{K_0}$ is in $C(K_0)$. Choose a K invariant finite subset $A \subseteq \widehat{K_0}$ so that $\chi_A * (l(k_j^{-1})f|_{K_0}) = l(k_j^{-1})f|_{K_0}$ for all j. Then

$$\chi_A^e * f(k_j k_0) = l(k_j^{-1})(\chi_A^e * f)(k_0) = (\chi_A^e * l(k_j^{-1})f)(k_0)$$
$$= (\chi_A * (l(k_j^{-1})f|_{K_0}))(k_0) = l(k_j^{-1})f|_{K_0}(k_0) = f(k_j k_0)$$

for $k_0 \in K_0$, and we see that $\chi_A^e * f = f$.

Lemma 6.59. When the Shapovalov form $\langle \cdot, \cdot \rangle_P$ induced by $\langle \cdot, \cdot \rangle_1$ is reinterpreted as a form $\langle \cdot, \cdot \rangle$ on $\text{induced}_{\mathfrak{g},K_1}^{\mathfrak{g},K}(V)$, the form is given by

$$\langle F_1, F_2 \rangle = [K : K_1] \sum_{k \in K/K_1} \langle F_1(k), F_2(k) \rangle_1.$$

Consequently if the form $\langle \cdot, \cdot \rangle_1$ on V is Hermitian, then the Shapovalov form $\langle \cdot, \cdot \rangle_P$ on $P(V)$ is Hermitian; if $\langle \cdot, \cdot \rangle_1$ is positive definite, then the Shapovalov form $\langle \cdot, \cdot \rangle_P$ is positive definite.

PROOF. Write dk and dk_1 for the normalized Haar measures on K and K_1. As in (6.45b), the form on $P_{\mathfrak{g},K_1}^{\mathfrak{g},K}(V)$ is obtained from the form

$$\langle T \otimes v, T' \otimes v' \rangle_P = \langle \overline{T'}, T \rangle \langle v, v' \rangle_1 \qquad \text{for } T, T' \in R(K) \text{ and } v, v' \in V$$

on $R(K) \otimes_{\mathbb{C}} V$ by passage to the K_1 invariants. The $(\mathfrak{g}, K)$ module $P_{\mathfrak{g},K_1}^{\mathfrak{g},K}(V)$ is generated by elements

$$(6.60) \qquad \int_{K_1} (r(k_1)T \otimes k_1 v) \, dk_1,$$

and we thus have

(6.61)
$$\left\langle \int_{K_1} (r(k_1)T \otimes k_1 v)\, dk_1, \int_{K_1} (r(k_1)T' \otimes k_1 v')\, dk_1 \right\rangle_P$$

$$= \int_{K_1} \int_{K_1} \langle \overline{r(k_1')T'}, r(k_1)T \rangle \langle k_1 v, k_1' v' \rangle_1 \, dk_1 \, dk_1'$$

$$= \int_{K_1} \langle \overline{T'}, r(k_1)T \rangle \langle k_1 v, v' \rangle_1 \, dk_1 \qquad \text{after } k_1'^{-1} k_1 \to k_1.$$

Meanwhile, we readily check that Φ_V in (6.57) is given by

$$\Phi_V(T \otimes v)(k) = (T(j_V(v)))(k) \qquad \text{for } T \in R(K),\ v \in V,\ k \in K.$$

Let k_i be coset representatives of K/K_1. According to the proof of Proposition 2.75, any $F \in \text{induced}_{\mathfrak{g},K_1}^{\mathfrak{g},K}(V)$ satisfies the identity $F = \sum k_i\, j_V(F(k_i))$. Let χ_A^e be a member of the approximate identity of $R(K)$ of the form in Lemma 6.58, with $A \subseteq \widehat{K}_0$ large enough so that χ_A^e fixes $j_V(F(k_i))$ for all i. Then we have

$$F = \sum_i (l(k_i)\chi_A^e)(j_V(F(k_i))) = \Phi_V\left(\sum_i l(k_i)\chi_A^e \otimes F(k_i) \right),$$

and thus

(6.62)
$$\Phi_V^{-1} F = \sum_i l(k_i)\chi_A^e \otimes F(k_i).$$

If F_1 and F_2 are in $\text{induced}_{\mathfrak{g},K_1}^{\mathfrak{g},K}(V)$, we can choose the finite set $A \subseteq \widehat{K}_0$ large enough so that χ_A^e fixes $j_V(F_1(k_i))$ and $j_V(F_2(k_i))$ for all i. In order to apply (6.61), we need to apply the averaging process of (6.60) to each term of (6.62) for F_1 and F_2. Then (6.61) gives

$$\langle \Phi_V^{-1} F_1, \Phi_V^{-1} F_2 \rangle_P$$

$$= \sum_{i,j} \int_{K_1} \langle l(k_j)\overline{\chi_A^e}, r(k_1)l(k_i)\chi_A^e \rangle \langle k_1 F_1(k_i), F_2(k_j) \rangle_1 \, dk_1.$$

If the distribution χ_A^e is given as a function by $\chi_A^e(k)\, dk$, then the first factor of the integrand is

$$\int_K l(k_i k_1^{-1})\chi_A^e(k)\overline{l(k_j)\chi_A^e(k)}\, dk = \int_K \chi_A^e(k_1 k_i^{-1} k)\overline{\chi_A^e(k_j^{-1} k)}\, dk$$

$$= \int_K \chi_A^e(k_1 k_i^{-1} k)\chi_A^e(k^{-1} k_j)\, dk$$

$$= \chi_A^e(k_1 k_i^{-1} k_j),$$

and this vanishes for $i \neq j$. Therefore

$$(6.63) \qquad \langle \Phi_V^{-1} F_1, \Phi_V^{-1} F_2 \rangle_P = \sum_i \int_{K_1} \chi_A^e(k_1) \langle k_1 F_1(k_i), F_2(k_i) \rangle_1 \, dk_1.$$

The condition that χ_A^e fix $j_V(F_1(k_i))$ implies that

$$F_1(k_i) = j_V(F_1(k_i))(1) = \chi_A^e * (j_V(F_1(k_i)))(1)$$

$$= \int_{K_0} j_V(F_1(k_i))(k_0^{-1}) \, d\chi_A(k_0)$$

$$= \int_{K_0} k_0 F_1(k_i) \, d\chi_A(k_0).$$

The expression $k F_1(k_i)$ is defined for $k \in K_1$, and we extend it to $k \in K$ by defining it to be 0 for $k \notin K_1$. Then the above expression is

$$= \int_K k F_1(k_i) \, d\chi_A^e(k)$$

$$= \int_K \chi_A^e(k)(k F_1(k_i)) \, dk$$

$$= [K : K_1]^{-1} \int_{K_1} \chi_A^e(k_1)(k_1 F_1(k_i)) \, dk_1.$$

Substituting into (6.63), we see that

$$\langle \Phi_V^{-1} F_1, \Phi_V^{-1} F_2 \rangle_P = [K : K_1] \sum_i \langle F_1(k_i), F_2(k_i) \rangle_1,$$

and the lemma follows.

We return now to the setting of §1, but without the assumption (6.1). Let $K_1 = (L \cap K)K_0$, and write $\mathcal{L}_{j,1}$ for the cohomological induction functors for $(\mathfrak{g}, K_1)$ attached to $(\mathfrak{q}, L \cap K)$. Recall from (5.8) that

$$(6.64) \qquad \begin{aligned} \mathcal{L}_j(Z) &= P_{\mathfrak{g}, K_1}^{\mathfrak{g}, K} \mathcal{L}_{j,1}(Z), \\ \mathcal{L}_j^K(Z) &= P_{\mathfrak{k}, K_1}^{\mathfrak{k}, K} \mathcal{L}_{j,1}^K(Z). \end{aligned}$$

Lemma 6.65. In the setting of Theorem 6.55, write $\langle \cdot, \cdot \rangle_{G_1}$ for the Shapovalov form on $\mathcal{L}_{S,1}(Z)$ induced by $\langle \cdot, \cdot \rangle_L$, and similarly write $\langle \cdot, \cdot \rangle_{K_1}$ for the Shapovalov form on $\mathcal{L}_{S,1}^K(Z)$ induced by $\langle \cdot, \cdot \rangle_{L \cap K}$. Under the identifications (6.64), the Shapovalov form $\langle \cdot, \cdot \rangle_G$ on $\mathcal{L}_S(Z)$ is induced from $\langle \cdot, \cdot \rangle_{G_1}$ by the construction of (6.56), and similarly the Shapovalov form $\langle \cdot, \cdot \rangle_K$ on $\mathcal{L}_S^K(Z)$ is induced from $\langle \cdot, \cdot \rangle_{K_1}$ by the same construction. Consequently the form $\langle \cdot, \cdot \rangle_G$ is positive definite if and only if $\langle \cdot, \cdot \rangle_{G_1}$ is positive definite. Similarly the form $\langle \cdot, \cdot \rangle_K$ is positive definite if and only if $\langle \cdot, \cdot \rangle_{K_1}$ is positive definite.

PROOF. We give the argument for the forms relative to G, the other case being essentially identical. The composite form is given on

$$R(K) \otimes_{\mathbb{C}} R((L \cap K)K_0) \otimes_{\mathbb{C}} \textstyle\bigwedge^j \mathfrak{c} \otimes_{\mathbb{C}} \operatorname{ind}_{\bar{\mathfrak{q}}, L \cap K}^{\mathfrak{g}, L \cap K}(Z_{\bar{\mathfrak{q}}}^{\#})$$

by

$$\begin{aligned}
&\langle T \otimes t \otimes \xi \otimes (u \otimes z), T' \otimes t' \otimes \xi' \otimes (u' \otimes z')\rangle_{\text{composite}}\\
&= \langle \overline{T'}, T\rangle\langle t \otimes \xi \otimes (u \otimes z),\ t' \otimes \xi' \otimes (u' \otimes z')\rangle_{G_1}\\
&= \langle \overline{T'}, T\rangle\langle \overline{t'}, t\rangle \varepsilon_0(\xi \wedge \bar{\xi}')\langle \mu'_{\bar{u}}(\bar{u}''u)z, z'\rangle_L^{\#}\\
&= \langle \overline{T'} \otimes \overline{t'}, T \otimes t\rangle \varepsilon_0(\xi \wedge \bar{\xi}')\langle \mu'_{\bar{u}}(\bar{u}''u)z, z'\rangle_L^{\#}.
\end{aligned}$$

We restrict from $R(K) \otimes_{\mathbb{C}} R((L \cap K)K_0)$ to the $(L \cap K)K_0$ invariants, identifying these invariants with $R(K)$, and the result is the composite form on

$$R(K) \otimes_{\mathbb{C}} \textstyle\bigwedge^j \mathfrak{c} \otimes_{\mathbb{C}} \operatorname{ind}_{\bar{\mathfrak{q}}, L \cap K}^{\mathfrak{g}, L \cap K}(Z_{\bar{\mathfrak{q}}}^{\#}).$$

Let dk and dk_1 be normalized Haar measures on K and $(L \cap K)K_0$. To make $T \otimes t$ into an $(L \cap K)K_0$ invariant, we form the average

$$(6.66a) \qquad \int_{(L \cap K)K_0} r(k_1)T \otimes l(k_1)t \, dk_1,$$

which acts on a function $H \in C(K \times (L \cap K)K_0)$ as

$$(6.66b) \qquad \int_{K \times (L \cap K)K_0 \times (L \cap K)K_0} H(kk_1'^{-1}, k_1'k_1) \, dT(k) \, dt(k_1) \, dk_1'.$$

To identify the distribution (6.66b) with a member of $R(K)$, we transfer the action into the first coordinate of H. Expression (6.66b) equals

$$\int_{K \times (L \cap K)K_0} \left(\int_{(L \cap K)K_0} H(kk_1k_1'^{-1}, k_1') \, dk_1' \right) dT(k) \, dt(k_1),$$

and thus (6.66) gets identified with $T * t$.

To see that the composite form matches the form for G, we have thus to show that the pairing of two members (6.66) of $R(K) \otimes R((L \cap K)K_0)$ matches the pairing in $R(K)$ of the two corresponding distributions $T * t$. Let $\overline{T'} \otimes \overline{t'}$ be given, and write $\overline{T'} = \overline{F(k)} \, dk$ and $\overline{t'} = \overline{f(k_1)} \, dk_1$. Then the pairing of the two members (6.66) is

$$
\begin{aligned}
&= \int_{(L\cap K)K_0} \langle r(k_1'')\overline{T'} \otimes l(k_1'')\overline{t'},\, r(k_1')T \otimes l(k_1')t \rangle \, dk_1' \, dk_1'' \\
&= \int_{(L\cap K)K_0} \langle \overline{T'} \otimes \overline{t'},\, r(k_1')T \otimes l(k_1')t \rangle \, dk_1' \\
&= \int_{K\times(L\cap K)K_0\times(L\cap K)K_0} \overline{F(kk_1'^{-1})f(k_1'k_1)} \, dT(k)\, dt(k_1)\, dk_1' \\
&= \int_{K\times(L\cap K)K_0} \overline{F * f(kk_1)} \, dT(k)\, dt(k_1) \\
&= \langle T * t,\, \overline{F * f} \rangle,
\end{aligned}
$$

and this equals $\langle \overline{T'} * \overline{t'},\, T * t \rangle$. Hence the composite form matches the form for G.

The second stage of the composition is identified by Lemma 6.59, whose formula makes it clear that positivity and nonpositivity are preserved in the second stage. The lemma follows.

PROOF OF THEOREM 6.55c. Lemma 6.65 allows us to reduce to the case $K = (L \cap K)K_0$, and this is exactly the case treated by Proposition 6.50.

CHAPTER VII

TRANSLATION FUNCTORS

For $\mathfrak{g}$ reductive, a $Z(\mathfrak{g})$ finite $U(\mathfrak{g})$ module is one that is annihilated by an ideal of finite codimension in $Z(\mathfrak{g})$. Such a module is a direct sum of submodules each having a "generalized infinitesimal character," i.e., having $(z - \chi(z)1)^n$ act as 0 for some n and some homomorphism χ of $Z(\mathfrak{g})$ into $\mathbb{C}$.

The translation functors are obtained by composing the operation of tensor product with a finite-dimensional representation and the operation of projection according to a given generalized infinitesimal character. It has to be proved that tensoring with a finite-dimensional representation carries $Z(\mathfrak{g})$ finite modules to $Z(\mathfrak{g})$ finite modules, and this result takes considerable preparation. One needs Chevalley's theory of invariants under finite groups generated by reflections, Kostant's decomposition of the symmetric algebra of $\mathfrak{g}^*$ as the tensor product of invariants and harmonics, and a theorem of Duflo-Dixmier characterizing the annihilators of Verma modules. A result along the way is that the functors of $\mathfrak{u}$ homology and cohomology carry $Z(\mathfrak{g})$ finite modules to $Z(\mathfrak{l})$ finite modules.

Under a condition called "integral dominance," together with no decrease in non-singularity, the translation functors denoted by ψ carry irreducible $U(\mathfrak{g})$ modules to irreducible $U(\mathfrak{g})$ modules or 0. Under integral dominance and no increase in nonsingularity, the translation functors ψ carry nonzero modules to nonzero modules. In the equisingular case, the ψ functor is an isomorphism between categories of modules with an infinitesimal character. With no decrease in singularity, ψ cannot map two irreducible modules to the same nonzero module unless the irreducible modules are isomorphic.

The theory gives further results when specialized to $(\mathfrak{g}, K)$ modules. When K is connected, the functors ψ carry $(\mathfrak{g}, K)$ modules to $(\mathfrak{g}, K)$ modules. When K is disconnected, one obtains a theory by deriving results from the connected case. It is assumed that the finite-dimensional representation used in defining ψ has a unique highest weight and that weight has multiplicity 1. Moreover, it is assumed that the parameters for ψ are fixed by K. One obtains the same kinds of results about irreducibility and nonvanishing as for $U(\mathfrak{g})$ modules. In the dominant equisingular case, the composition of ψ with one set of parameters and ψ with the reversed set of parameters is naturally equivalent to the identity functor on modules of finite length.

The theory may be applied to cohomological induction and to $\mathfrak{u}$ homology and cohomology. Under conditions of integral dominance and relative nonsingularity, the translation functors ψ commute with all these functors up to shifts in parameters.

1. Motivation and Examples

A characteristic feature of the representation theory of real reductive groups is that many representations appear in nice families indexed by some real and some integer parameters, with the qualitative features of the representations exhibiting a kind of periodicity in some or all of the parameters. This kind of periodicity can be viewed as arising from an interaction with finite-dimensional representations and from the parametrization of irreducible finite-dimensional representations by the Theorem of the Highest Weight.

Our goal in this chapter is to introduce the "translation functors," which have evolved from work of Bernstein-Gelfand-Gelfand, Jantzen, Zuckerman, and others. Use of the translation functors helps to explain the existence of such periodicity in families of representations and turns it into a valuable technical tool: Properties proved first only for "generic" representations will sometimes automatically be inherited by more singular ones.

The first application of the functors will be in Chapter VIII to the question of the preservation of irreducibility under cohomological induction. A direct attack on irreducibility handles representations with a dominant infinitesimal character that is far from singular, but the argument runs into obstructions when the infinitesimal character is singular or nearly singular. Use of the translation functors will enable us to get around these obstructions, deducing irreducibility in singular and nearly singular cases from irreducibility in the generic case.

The periodicity in families of representations arises from functors that allow one to pass between representations in a family whose parameters differ by a period. We shall work with functors denoted $\psi_\lambda^{\lambda'}$. Let us describe them approximately at this stage. The indices λ and λ' refer to infinitesimal characters, and, on a first reading, both λ and λ' may be taken to be in the same Weyl chamber, say the dominant one. It is assumed that the difference $\lambda' - \lambda$ is algebraically integral, so that it is an extreme weight of an irreducible finite-dimensional representation, say F, which is unique up to isomorphism. The functor $\psi_\lambda^{\lambda'}$ is applied to representations of infinitesimal character λ in two steps—an operation of tensoring by F, followed by an operation of projecting according to λ'. The necessary background to get $\psi_\lambda^{\lambda'}$ well defined is rather lengthy and will take through §7. The functors will be defined in §8, and the remainder of the chapter will be devoted to handling their effect on irreducibility. We shall begin the rigorous development of these functors in §2, considering only examples for the remainder of this section. Although the rigorous development will treat $U(\mathfrak{g})$ modules and $(\mathfrak{g}, K)$ modules, the examples for now will all be in the context of group representations.

EXAMPLE 1. The family of irreducible representations of a compact connected Lie group G.

This family is parametrized, of course, by the highest weights, which by Theorem 4.7 are all of the dominant analytically integral forms on a Cartan subalgebra $\mathfrak{t}_0$ of $\mathfrak{g}_0$. Let us take irreducible representations V^μ and V^λ with respective highest weights μ and λ, regarding V^λ as a member of the family with infinitesimal character $\lambda + \delta$ and regarding V^μ as a finite-dimensional representation that we are using in tensor products. The decomposition of the tensor product $V^\lambda \otimes_{\mathbb{C}} V^\mu$ is determined by the decomposition of the product $\chi_\lambda \chi_\mu$ of the characters χ_λ and χ_μ, which is known to be

$$(7.1) \qquad \chi_\lambda \chi_\mu = \sum_{\substack{\mu' = \text{weight} \\ \text{of } V^\mu}} m_\mu(\mu') \operatorname{sgn}(\lambda + \delta + \mu') \chi_{(\lambda + \delta + \mu')^\sim - \delta},$$

where

$$m_\mu(\mu') = \text{multiplicity of weight } \mu' \text{ in } V^\mu$$

$$(\lambda + \delta + \mu')^\sim = \lambda + \delta + \mu' \text{ made dominant by some } s \in W(\mathfrak{g}, \mathfrak{t})$$

$$\operatorname{sgn}(\lambda + \delta + \mu') = \begin{cases} 0 & \text{if } \lambda + \delta + \mu' \text{ is singular} \\ \det s & \text{otherwise, with } s \in W(\mathfrak{g}, \mathfrak{t}) \text{ as above.} \end{cases}$$

(See Knapp [1986], 111–112.)

The extreme weights of V^μ are $w\mu$ with w in $W(\mathfrak{g}, \mathfrak{t})$, and the claim is that $V^{\lambda + w\mu}$ occurs exactly once in $V^\lambda \otimes_{\mathbb{C}} V^\mu$ if $\lambda + w\mu$ is dominant. In fact, $\mu' = w\mu$ contributes $\chi_{\lambda + w\mu}$ to the right side of (7.1). To see that there is no other contribution, suppose μ' contributes. Then

$$(\lambda + \delta + \mu')^\sim - \delta = \lambda + w\mu.$$

We can rewrite this as $(\lambda + \delta + \mu')^\sim = \lambda + \delta + w\mu$ and then as

$$\lambda + \delta + \mu' = s(\lambda + \delta + w\mu)$$

for some $s \in W(\mathfrak{g}, \mathfrak{t})$. Subtracting $\lambda + \delta$ from both sides and computing the length gives

$$(7.2) \qquad \begin{aligned} |\mu'|^2 &= |s(\lambda + \delta + w\mu) - (\lambda + \delta)|^2 \\ &= |\lambda + \delta + w\mu|^2 - 2\langle s(\lambda + \delta + w\mu), \lambda + \delta \rangle + |\lambda + \delta|^2. \end{aligned}$$

In the middle term on the right, we have

$$s(\lambda + \delta + w\mu) = \lambda + \delta + w\mu - \sum_{\alpha > 0} n_\alpha \alpha, \qquad n_\alpha \geq 0,$$

since $\lambda + \delta + w\mu$ is dominant nonsingular, and all the n_α's are 0 only if $s = 1$. Since $\lambda + \delta$ is dominant nonsingular, (7.2) is

$$\geq |\lambda + \delta + w\mu|^2 - 2\langle\lambda + \delta + w\mu, \lambda + \delta\rangle + |\lambda + \delta|^2$$
$$= |(\lambda + \delta + w\mu) - (\lambda + \delta)|^2$$
$$= |w\mu|^2,$$

with equality only if $s = 1$. But Theorem 4.7e shows that $|\mu'|^2 \leq |w\mu|^2$, and equality is forced. Thus $s = 1$, and we see that $\mu' = w\mu$.

Therefore $V^{\lambda+w\mu}$ indeed occurs exactly once in $V^\lambda \otimes_{\mathbb{C}} V^\mu$ if $\lambda + w\mu$ is dominant. In terms of translation functors, this fact will be written

(7.3) $$\psi_{\lambda+\delta}^{\lambda+\delta+w\mu}(V^\lambda) \cong V^{\lambda+w\mu}.$$

The lower index indicates that we started from infinitesimal character $\lambda + \delta$, which corresponds to highest weight λ. We tensored with the irreducible finite-dimensional representation with the difference of the indices $w\mu$ as extreme weight, thus with V^μ. Finally we projected according to the top index $\lambda + \delta + w\mu$ as infinitesimal character. The result was $V^{\lambda+w\mu}$ with multiplicity 1.

EXAMPLE 2. The principal series of $G = SL(2, \mathbb{R})$.
Let MAN be the upper triangular subgroup, with

$$M = \pm\begin{pmatrix} 1 & 0 \\ 0 & 1 \end{pmatrix}, \quad A = \left\{\begin{pmatrix} e^t & 0 \\ 0 & e^{-t} \end{pmatrix}\right\}, \quad N = \left\{\begin{pmatrix} 1 & x \\ 0 & 1 \end{pmatrix}\right\}.$$

Let σ be either of the one-dimensional representations of M, trivial or nontrivial, and define a linear functional ρ on $\mathfrak{a}$ by $\rho\begin{pmatrix} t & 0 \\ 0 & -t \end{pmatrix} = t$. Let ν be any linear functional on $\mathfrak{a}$. The principal series consists of the representations $U(\sigma, \nu, \cdot)$ defined as follows: The space for $U(\sigma, \nu, \cdot)$ is

(7.4a)
$$V^{\sigma,\nu} = \{f : G \to \mathbb{C} \text{ of class } C^\infty \mid f(xman) = e^{-(\nu+\rho)\log a}\sigma(m)^{-1}f(x)\},$$

and the action is

(7.4b) $$(U(\sigma, \nu, g)f)(x) = f(g^{-1}x).$$

The space $V^{\sigma,\nu}$ becomes a $U(\mathfrak{g})$ module through differentiation of the G action. We write

(7.4c) $$U(\sigma, \nu, \cdot) = \text{induced}_{MAN}^G \pi_{\sigma,\nu},$$

where $\pi_{\sigma,\nu}$ is the one-dimensional representation of MAN given by

(7.4d)
$$\pi_{\sigma,\nu}(man) = e^{\nu \log a}\sigma(m).$$

The definitions (7.4) should be compared with the corresponding construction for finite groups given in (2.72).

We shall consider the tensor product of a finite-dimensional representation with $U(\sigma, \nu, \cdot)$, using one particular finite-dimensional representation for an illustration. Let (π, F) be the standard representation of $SL(2, \mathbb{R})$ on $\mathbb{C}^2$. One easily establishes that

(7.5a) $\Phi : (\text{induced}_{MAN}^G \pi_{\sigma,\nu}) \otimes_{\mathbb{C}} \pi \rightarrow \text{induced}_{MAN}^G (\pi_{\sigma,\nu} \otimes \pi|_{MAN})$

given by

(7.5b)
$$\Phi(f \otimes v)(g) = f(g) \otimes \pi(g)^{-1}v$$

is a G isomorphism onto. (Compare with the Mackey isomorphism (2.93).) The subspace $\mathbb{C}\begin{pmatrix} 1 \\ 0 \end{pmatrix}$ for $\pi|_{MAN}$ is an invariant subspace, and $\pi|_{MAN}$ acts on it by $e^\rho \otimes \sigma_0$, where σ_0 is the nontrivial character of M. Consequently

$$\pi_{\sigma,\nu} \otimes \mathbb{C}\begin{pmatrix} 1 \\ 0 \end{pmatrix} \cong \pi_{\sigma\sigma_0, \nu+\rho}.$$

It is not hard to see that the construction called "induced" in (7.4) is an exact functor, and it follows from (7.5) that

(7.6)
$$(\text{induced}_{MAN}^G \pi_{\sigma,\nu}) \otimes_{\mathbb{C}} \pi$$

has an invariant subspace isomorphic to the space for $U(\sigma\sigma_0, \nu + \rho, \cdot)$. Analyzing $\pi|_{MAN}$ further, we see similarly that (7.6) has as quotient the space for $U(\sigma\sigma_0, \nu - \rho, \cdot)$. Thus up to an isomorphism (7.5), we obtain a finite filtration (in the sense of §A.3)

(7.7a)
$$V^{\sigma,\nu} \otimes_{\mathbb{C}} F \supseteq V^{\sigma\sigma_0, \nu+\rho} \supseteq 0$$

with successive quotients $V^{\sigma\sigma_0, \nu-\rho}$ and $V^{\sigma\sigma_0, \nu+\rho}$ that are principal series.

If we choose $K = SO(2)$ in G, then the subspace $\mathfrak{a}_0$ of $\mathfrak{g}_0$ is a θ stable Cartan subalgebra, and one can show that $U(\sigma, \nu, \cdot)$ has infinitesimal character ν. In (7.7a), $V^{\sigma\sigma_0, \nu+\rho}$ is thus contained in the subspace where $Z(\mathfrak{g})$ acts by $\chi_{\nu+\rho}$. If $\chi_{\nu-\rho} \neq \chi_{\nu+\rho}$ (i.e., if $\nu \neq 0$), then the subspace of $V^{\sigma,\nu} \otimes_{\mathbb{C}} F$ where $Z(\mathfrak{g})$ acts by $\chi_{\nu-\rho}$ is a $\mathfrak{g}$ invariant complement for $V^{\sigma\sigma_0, \nu+\rho}$. Consequently the inclusion in (7.7a) splits:

(7.7b)
$$V^{\sigma,\nu} \otimes_{\mathbb{C}} F \cong V^{\sigma\sigma_0, \nu-\rho} \oplus V^{\sigma\sigma_0, \nu+\rho}.$$

In this case $Z(\mathfrak{g})$ has two eigenspaces in $V^{\sigma,\nu} \otimes_{\mathbb{C}} F$, with respective eigenvalues $\chi_{\nu+\rho}$ and $\chi_{\nu-\rho}$. The eigenvalue χ_ν was the original eigenvalue for $V^{\sigma,\nu}$; the significance of ρ and $-\rho$ is that they are the weights of (π, F). The formulas analogous to (7.3) are

(7.8)
$$\psi_\nu^{\nu+\rho}(V^{\sigma,\nu}) \cong V^{\sigma\sigma_0,\nu+\rho} \qquad \text{if } \nu \neq 0$$
$$\psi_\nu^{\nu-\rho}(V^{\sigma,\nu}) \cong V^{\sigma\sigma_0,\nu-\rho} \qquad \text{if } \nu \neq 0.$$

If $\nu = 0$, the conclusion is more complicated: $\psi_0^\rho(V^{\sigma,0})$ and $\psi_0^{-\rho}(V^{\sigma,0})$ are isomorphic, and each has a two-step filtration with $V^{\sigma\sigma_0,-\rho}$ as quotient and $V^{\sigma\sigma_0,\rho}$ as subrepresentation.

EXAMPLE 3. A family including part of the discrete series of $SL(2, \mathbb{R})$. To simplify the notation, we work with the conjugate group

$$G = SU(1, 1) = \left\{ \begin{pmatrix} \alpha & \beta \\ \bar{\beta} & \bar{\alpha} \end{pmatrix}, \ |\alpha|^2 - |\beta|^2 = 1 \right\}.$$

The subgroup $K = \left\{ \begin{pmatrix} e^{i\theta} & 0 \\ 0 & e^{-i\theta} \end{pmatrix} \right\}$ is maximal compact in G. For $n \geq 1$, let $\mathcal{D}_n$ be the representation whose space $\mathcal{H}_n$ is all analytic functions in the unit disc $\{|z| < 1\}$ and whose action by $g = \begin{pmatrix} \alpha & \beta \\ \bar{\beta} & \bar{\alpha} \end{pmatrix}$ is

(7.9)
$$(\mathcal{D}_n(g)f)(z) = (-\bar{\beta}z + \alpha)^{-n} f\left(\frac{\bar{\alpha}z - \beta}{-\bar{\beta}z + \alpha} \right).$$

Then $\mathcal{H}_n$ becomes a $U(\mathfrak{g})$ module by differentiation of the G action. (It is customary to consider only those functions in $\mathcal{H}_n$ for which a certain norm is finite, but use of the full space $\mathcal{H}_n$ has the advantage of making every vector smooth.)

The subspace of polynomials in $\mathcal{H}_n$ is the subspace of $\mathcal{H}_n$ of K finite vectors in the sense of (1.17). Hence it is a $(\mathfrak{g}, K)$ module. The functions $\{z^k\}_{k=0}^\infty$ form a basis of this subspace, and

(7.10)
$$\mathcal{D}_n \begin{pmatrix} e^{i\theta} & 0 \\ 0 & e^{-i\theta} \end{pmatrix} (z^k) = e^{-(n+2k)i\theta} z^k.$$

Also the diagonal subalgebra $\mathfrak{h}$ of $\mathfrak{g}$ is a Cartan subalgebra, and (7.10) shows that this $(\mathfrak{g}, K)$ module is isomorphic to the Verma module $V(-(n-1)\delta)$. The infinitesimal character is $\chi_{-(n-1)\delta} = \chi_{(n-1)\delta}$, by (4.99).

Let us tensor $\mathcal{D}_n$ with the standard representation (π, F) of $SU(1, 1)$ on $\mathbb{C}^2$. On the level of $(\mathfrak{g}, K)$ modules, it is apparent that the weights of

(7.11)
$$\mathcal{D}_n \otimes_{\mathbb{C}} \pi$$

are

$$-(n-1)\delta \qquad \text{with multiplicity 1}$$
$$-(n-1+2k)\delta, \text{ for } k \geq 1, \qquad \text{with multiplicity 2.}$$

Working with Verma modules, we are led to suspect that $\mathcal{D}_{n-1}$ is a subrepresentation of (7.11) and that $\mathcal{D}_{n+1}$ is what is left over. In fact, if we isolate the subspaces where $Z(\mathfrak{g})$ acts with eigenvalues $\chi_{(n-2)\delta}$ and $\chi_{n\delta}$, we find that

$$\mathcal{D}_n \otimes_{\mathbb{C}} \pi \cong \mathcal{D}_{n-1} \oplus \mathcal{D}_{n+1},$$

at least on the level of $(\mathfrak{g}, K)$ modules. (Actually such a decomposition is valid even on the level of topological group representations, but this point will not concern us.) The eigenvalue $\chi_{(n-1)\delta}$ was the original eigenvalue for $\mathcal{D}_n$, and the new eigenvalues are $\chi_{(n-2)\delta}$ and $\chi_{n\delta}$; the significance of the increments $-\delta$ and δ is that they are the weights of (π, F). The formulas analogous to (7.3) are

(7.12)
$$\psi^{n\delta}_{(n-1)\delta}(\mathcal{D}_n) \cong \mathcal{D}_{n+1} \qquad \text{as } (\mathfrak{g}, K) \text{ modules if } n \geq 2$$

$$\psi^{(n-2)\delta}_{(n-1)\delta}(\mathcal{D}_n) \cong \mathcal{D}_{n-1} \qquad \text{as } (\mathfrak{g}, K) \text{ modules if } n \geq 2.$$

2. Generalized Infinitesimal Character

In this section we shall enlarge the notion that a representation has an infinitesimal character. Let $\mathfrak{g}$ be a complex reductive Lie algebra. For much of the chapter we shall work only with $U(\mathfrak{g})$ modules. Only at the end of the chapter (and in a few other special circumstances) will we interpret our results for $(\mathfrak{g}, K)$ modules, where $(\mathfrak{g}, K)$ is a reductive pair. Let $\mathfrak{h}$ be a Cartan subalgebra of $\mathfrak{g}$.

More knowledge will be assumed about $\mathfrak{g}$ and $\mathfrak{h}$ in this chapter than was spelled out in Chapter IV. We take as known that Cartan subalgebras exist and are abelian, that the theory of roots and the Weyl group extends to this situation from the situation in Chapter IV, and that $\mathfrak{h}$ can be used to construct a compact form $\mathfrak{g}_0$ of $\mathfrak{g}$. (Once the compact form has been constructed, we have established that the present situation is no more general than that in Chapter IV.)

The $U(\mathfrak{g})$ module V is said to be $Z(\mathfrak{g})$ **finite** if the ideal

(7.13) $$I = \text{Ann}_{Z(\mathfrak{g})}(V) = \{z \in Z(\mathfrak{g}) \mid zv = 0 \text{ for all } v \in V\}$$

has finite codimension in $Z(\mathfrak{g})$.

EXAMPLES.

1) If V has infinitesimal character $\lambda \in \mathfrak{h}^*$, then $\mathrm{Ann}_{Z(\mathfrak{g})}(V)$ in (7.13) includes all $z - \chi_\lambda(z)1$ for $z \in Z(\mathfrak{g})$. Hence $\mathrm{Ann}_{Z(\mathfrak{g})}(V)$ has codimension 1 in $Z(\mathfrak{g})$, and V is $Z(\mathfrak{g})$ finite.

2) Let $(\mathfrak{g}, K)$ be a reductive pair, and let V be a $(\mathfrak{g}, K)$ module. Suppose that V is admissible as a K module and is finitely generated as a $U(\mathfrak{g})$ module. Then V is $Z(\mathfrak{g})$ finite. In fact, let a finite set of generators $v_1, \ldots, v_n$ be specified, and let $V_0 = \bigoplus V_\tau$ be a finite sum of K isotypic subspaces that contains all the generators. If the pair $(\mathfrak{g}, K)$ is in the Harish-Chandra class, then each V_τ is invariant under $Z(\mathfrak{g})$. To see this, let z be in $Z(\mathfrak{g})$, and let v be in V_τ. With notation as in (1.23), we have $\chi_\tau v = v$, where

$$\chi_\tau v = d_\tau \int_K \overline{\Theta(k)} kv \, dk.$$

Then

$$\chi_\tau z v = d_\tau \int_K \overline{\Theta(k)} (\mathrm{Ad}(k)z) kv \, dk$$

$$= d_\tau \int_K \overline{\Theta(k)} z kv \, dk \qquad \text{since } \mathrm{Ad}(k)z = z \text{ when } (\mathfrak{g}, K) \text{ is}$$
$$\qquad\qquad\qquad\qquad\qquad\qquad \text{in the Harish-Chandra class}$$

$$= zv,$$

and zv is in V_τ. Thus V_τ is invariant under $Z(\mathfrak{g})$, and so is V_0. (For $(\mathfrak{g}, K)$ not in the Harish-Chandra class, we should enlarge V_0 by adjoining the finitely many $V_{\tau'}$ such that $\tau'|_{K_0} \cong \tau|_{K_0}$ for some τ contributing to V_0. This enlarged V_0 is invariant under $Z(\mathfrak{g})$.)

If $J = \mathrm{Ann}_{Z(\mathfrak{g})}(V_0)$, then $Z(\mathfrak{g})/J$ maps one-one into $\mathrm{End}_{\mathbb{C}} V_0$, and it follows that J has finite codimension in $Z(\mathfrak{g})$. The most general element of V is of the form

$$v = u_1 v_1 + \cdots + u_n v_n \qquad \text{with all } u_j \in U(\mathfrak{g}).$$

If z is in J, then

$$zv = zu_1 v_1 + \cdots + zu_n v_n = u_1 zv_1 + \cdots + u_n zv_n = 0.$$

Hence z is in (7.13). We conclude that J equals the ideal $\mathrm{Ann}_{Z(\mathfrak{g})}(V)$ given in (7.13). Consequently the ideal $\mathrm{Ann}_{Z(\mathfrak{g})}(V)$ has finite codimension in $Z(\mathfrak{g})$, and V is $Z(\mathfrak{g})$ finite.

We now prove that any $Z(\mathfrak{g})$ finite $U(\mathfrak{g})$ module is the direct sum of finitely many $U(\mathfrak{g})$ submodules each of which nearly has an infinitesimal character.

Lemma 7.14. Let R be a commutative associative algebra with identity over $\mathbb{C}$. Suppose $\{\chi_1, \ldots, \chi_k\}$ is a finite set of distinct algebra homomorphisms of R into $\mathbb{C}$ (understood to carry the identity to 1), and suppose $\chi_0 : R \to \mathbb{C}$ is an algebra homomorphism not belonging to the set. Then there exists $r \in R$ such that $\chi_0(r) = 1$ and $\chi_i(r) = 0$ for $1 \le i \le k$. Consequently if constants c_i are given for $0 \le i \le k$, then there exists $r \in R$ with $\chi_i(r) = c_i$ for $0 \le i \le k$.

PROOF. Since $\chi_0 \ne \chi_i$, there is an element r_i with $\chi_0(r_i) \ne \chi_i(r_i)$. Then

$$r = \prod_{i=1}^{k} (\chi_0(r_i) - \chi_i(r_i))^{-1}(r_i - \chi_i(r_i))$$

has the required properties for the first conclusion. The second conclusion follows by linearity.

Lemma 7.15. Suppose R is a commutative associative algebra with identity over $\mathbb{C}$, V is a finite-dimensional complex vector space, and $\{v_1, \ldots, v_N\}$ is a basis of V. Suppose that l is a representation of R on V that is upper triangular in this basis:

$$l(r)v_j = \sum_{i=1}^{j} l(r)_{ij} v_i.$$

Let $\chi_i : R \to \mathbb{C}$ be the ith diagonal entry: $\chi_i(r) = l(r)_{ii}$. Then the set of distinct homomorphisms among the χ_i coincides with the set of simultaneous eigenvalues of $l(r)$ for r in R.

REMARK. It follows that the set of distinct χ_i is an invariant of the representation l not depending on the basis that exhibits l as upper triangular.

PROOF. Suppose $\chi : R \to \mathbb{C}$ is an algebra homomorphism different from the χ_i. Let $v_0 = \sum c_i v_i$ be a simultaneous eigenvector for χ. By Lemma 7.14, choose $r_0 \in R$ with $\chi(r_0) = 1$ and $\chi_i(r_0) = 0$ for all i. Then

$$\sum_{i=1}^{N} c_i v_i = v_0 = \chi(r_0)v_0$$

$$= l(r_0)v_0 = \sum_{i=1}^{N} c_i l(r_0)v_i$$

$$= \sum_{i=1}^{N} \sum_{k=1}^{i} c_i l(r_0)_{ki} v_k$$

$$= \sum_{i=1}^{N} c_i \chi_i(r_0) v_i + \sum_{i=1}^{N} \sum_{k=1}^{i-1} c_i l(r_0)_{ki} v_k$$

$$= \sum_{i=1}^{N} \sum_{k=1}^{i-1} c_i l(r_0)_{ki} v_k.$$

We want to show that v_0 is 0. Assuming the contrary, let m be the largest index with $c_m \neq 0$. Then the coefficient of v_m on the left side of this equation is c_m, while the coefficient of v_m on the right side is $\sum_{i=m+1}^{N} c_i l(r_0)_{mi}$. The latter expression is 0 since $c_i = 0$ for $i \geq m+1$. Thus $c_m = 0$, and we have a contradiction. We conclude that $v_0 = 0$ and that χ is not a simultaneous eigenvalue.

Conversely we are to show that each χ_i is a simultaneous eigenvalue. Possibly by using $\langle v_1, \ldots, v_j \rangle$ in place of V, we may assume that χ_N is different from $\chi_1, \ldots, \chi_{N-1}$ and that we are to exhibit χ_N as a simultaneous eigenvalue. By Lemma 7.14, choose $r_0 \in R$ with $\chi_N(r_0) = 1$ and $\chi_i(r_0) = 0$ for $1 \leq i \leq N-1$. The generalized eigenspace for $l(r_0)$ and the eigenvalue 1 is then one dimensional, say $\mathbb{C}v_0$ with $v_0 \neq 0$. Since R is commutative, $l(R)v_0 \subseteq \mathbb{C}v_0$. Thus v_0 is a simultaneous eigenvector. Let χ be its simultaneous eigenvalue.

Write $v_0 = \sum_{i=1}^{N} c_i v_i$. If $c_N = 0$, then v_0 lies in a subspace on which $l(r_0)$ acts nilpotently, whereas $l(r_0)v_0 = \chi_N(r_0)v_0 = v_0$. So $c_N \neq 0$. For r in R, we have

$$\chi(r)(c_1 v_1 + \cdots + c_N v_N) = l(r)(c_1 v_1 + \cdots + c_N v_N)$$
$$= c_1 l(r) v_1 + \cdots + c_N l(r) v_N$$
$$= c_1 \chi_1(r) v_1 + \cdots + c_N \chi_N(r) v_N.$$

Comparing the coefficients of v_N, we see that $\chi(r) = \chi_N(r)$. Therefore χ_N occurs as a simultaneous eigenvalue.

Lemma 7.16. Let R be a finite-dimensional commutative associative algebra with identity over $\mathbb{C}$. For each algebra homomorphism $\chi : R \to \mathbb{C}$ (understood to carry the identity to 1), let R_χ be defined to be

(7.17a)
$$= \{ r' \in R \mid (r - \chi(r)1)^n r' = 0 \text{ for all } r \in R \text{ and for some } n = n(r, r') \}.$$

Then R_χ is an ideal, and

(7.17b)
$$R = \bigoplus_{\chi} R_\chi.$$

PROOF. Let $\dim_{\mathbb{C}} R = N$. If we regard R as an abelian Lie algebra and we regard left multiplication with $l(r)r' = rr'$ as an N-dimensional representation of the Lie algebra R on the vector space R, then Lie's Theorem gives us a basis in which l becomes simultaneously upper triangular. The various diagonal entries of $l(r)$, as r varies, define homomorphisms $\chi : R \to \mathbb{C}$, and we let $\chi_1, \ldots, \chi_k$ be the distinct ones. By Lemma 7.14, choose $r_0 \in R$ such that $\chi_j(r_0) = j$ for $1 \le j \le k$, and let R_j be the generalized eigenspace for $l(r_0)$ and the eigenvalue j. The existence of Jordan form says that the vector space R splits as

$$(7.18) \qquad\qquad R = \bigoplus_{j=1}^{k} R_j.$$

Since the algebra R is commutative, $l(R)$ carries each R_j into itself. We can choose a basis of each R_j such that $l(R)$ is upper triangular on R_j, and the union of these bases is a basis of the vector space R such that $l(R)$ is upper triangular. By Lemma 7.15 the simultaneous diagonal entries are still the χ_j, perhaps occurring with different multiplicities and in a different order. On the space R_j, however, each diagonal entry for $l(r_0)$ is j, and this fact forces the simultaneous diagonal entry to be χ_j. Therefore

$$(r - \chi_j(r)1)^{\dim R_j} r' = 0$$

for all $r' \in R_j$ and $r \in R$. Consequently $R_j \subseteq R_{\chi_j}$. But the reverse inclusion $R_{\chi_j} \subseteq R_j$ is immediate from the definitions, and hence $R_j = R_{\chi_j}$. Substituting into (7.18), we see that

$$(7.19) \qquad\qquad R = \bigoplus_{j=1}^{k} R_{\chi_j}.$$

Since R is commutative, R_χ is an ideal. Suppose that $R_\chi \ne 0$ for some χ not in the list $\chi_1, \ldots, \chi_k$. By Lemma 7.14, choose $r \in R$ with $\chi_j(r) = 0$ for $1 \le j \le k$ and with $\chi(r) = 1$. From (7.17a) we see that $l(r)$ is nilpotent on R_{χ_j}. By (7.19), $l(r)$ is nilpotent on all of R. On the other hand, (7.17a) shows that $l(r)$ has only 1 as generalized eigenvalue on R_χ. Hence $R_\chi = 0$. Thus (7.19) yields (7.17b).

If V is a $Z(\mathfrak{g})$ finite $U(\mathfrak{g})$ module and if $\chi : Z(\mathfrak{g}) \to \mathbb{C}$ is a homomorphism, we define

$$P_\chi(V) = \{v \in V \mid (z - \chi(z)1)^n v = 0 \text{ for all } z \in Z(\mathfrak{g}) \text{ and some } n\},$$

calling $P_\chi(V)$ the χ **primary component** of V. This is a $U(\mathfrak{g})$ module. If $\chi = \chi_\lambda$ for some $\lambda \in \mathfrak{h}^*$, we may write $P_\lambda(V)$ in place of $P_\chi(V)$.

Proposition 7.20. If V is a $Z(\mathfrak{g})$ finite $U(\mathfrak{g})$ module, then V is the direct sum of its primary components and only finitely many of them are nonzero. If $P_\chi(V)$ is nonzero only for $\chi = \chi_1, \ldots, \chi_k$, then there exists N such that

(7.21)
$$(z_1 - \chi_1(z_1)1)^N \cdots (z_k - \chi_k(z_k)1)^N v = 0 \quad \text{for all } z_1, \ldots, z_k \in Z(\mathfrak{g})$$
$$\text{and } v \in V.$$

If V has an infinitesimal character $\lambda \in \mathfrak{h}^*$, then $P_\lambda(V)$ is the only nonzero primary component.

REMARKS. The direct-sum decomposition in the proposition is called the **primary decomposition** of V relative to $Z(\mathfrak{g})$, and we can define a corresponding projection operator P_χ (or P_λ) on $P_\chi(V)$. If V is a $Z(\mathfrak{g})$ finite $(\mathfrak{g}, K)$ module with K disconnected, the decomposition is still valid, but the individual primary components need not be $(\mathfrak{g}, K)$ modules. However, K acts on $Z(\mathfrak{g})$ with finite orbits and hence on the homomorphisms of $Z(\mathfrak{g})$ into $\mathbb{C}$, again with finite orbits; the sum of all the primary components of V corresponding to a single orbit under K is a $(\mathfrak{g}, K)$ module.

PROOF. Let $I = \text{Ann}_{Z(\mathfrak{g})}(V)$ be the ideal (7.13), and define $R = Z(\mathfrak{g})/I$. Since $Iv = 0$ for all v, the action of $Z(\mathfrak{g})$ on V factors through R. Applying Lemma 7.16 to R, we decompose 1 according to (7.17b) as

(7.22)
$$1 = \sum_\chi e_\chi.$$

Since each R_χ is an ideal, $e_\chi^2 = e_\chi$ and $e_\chi e_{\chi'} = 0$ for $\chi \neq \chi'$. Each homomorphism $\chi : R \to \mathbb{C}$ lifts to a homomorphism $\chi : Z(\mathfrak{g}) \to \mathbb{C}$, and we use the same symbol χ for the lift.

From the properties of the e_χ, $V = \bigoplus_\chi e_\chi V$. We shall prove that $e_\chi = P_\chi(V)$. We know from (7.17a) that

(7.23)
$$(r - \chi(r)1)^N e_\chi = 0$$

for all $r \in R$ if $N = \dim R$, and it follows that

(7.24a)
$$e_\chi(V) \subseteq P_\chi(V).$$

Conversely suppose $P_\chi(V) \neq 0$. We show that χ descends to R and that

(7.24b)
$$P_\chi(V) \subseteq e_\chi(V).$$

Fix z in I. From the definition of $P_\chi(V)$, there is some $v \neq 0$ in V with $(z - \chi(z)1)^n v = 0$ for some n. Replacing v by some $(z - \chi(z)1)^m v$ and changing notation, we may assume that $v \neq 0$ and $(z - \chi(z)1)v = 0$, i.e., $zv = \chi(z)v$. Since z is in I, $zv = 0$. Thus $\chi(z) = 0$ and z is in $\ker \chi$. Hence $I \subseteq \ker \chi$ and χ descends to R.

From

$$P_\chi(V) = \sum_{\chi'} e_{\chi'}(P_\chi(V)) \subseteq e_\chi(V) + \sum_{\chi' \neq \chi} e_{\chi'}(P_\chi(V)),$$

we see that (7.24b) will follow if we show that $e_{\chi'}(P_\chi(V)) = 0$ whenever $\chi' \neq \chi$. Now $e_{\chi'}(P_\chi(V)) \subseteq P_\chi(V)$ since $P_\chi(V)$ is a $Z(\mathfrak{g})$ submodule, and $e_{\chi'}(P_\chi(V)) \subseteq P_{\chi'}(V)$ by (7.24a). Thus (7.24b) will follow if we show that

(7.25) $$P_\chi(V) \cap P_{\chi'}(V) = 0$$

for $\chi' \neq \chi$. Arguing by contradiction, let v be nonzero in the left side of (7.25). Choose $z_0 \in Z(\mathfrak{g})$ with $\chi(z_0) \neq \chi'(z_0)$, and choose m and n as small as possible with

$$(z_0 - \chi(z_0)1)^m v = (z_0 - \chi'(z_0)1)^n v = 0.$$

Put $v' = (z_0 - \chi(z_0)1)^{m-1} v$. This is not 0, but

$$z_0 v' = \chi(z_0)v' \qquad \text{and} \qquad (z_0 - \chi'(z_0)1)^n v' = 0.$$

Since $z_0 - \chi'(z_0)1$ then acts as $(\chi(z_0) - \chi'(z_0))1$ on v', we obtain

$$(\chi(z_0) - \chi'(z_0)1)^n v' = 0$$

and $\chi(z_0) = \chi'(z_0)$, contradiction. This proves the first conclusion.

For (7.21), we observe from (7.23) and (7.22) that

$$(r_1 - \chi_1(r_1))^N \cdots (r_k - \chi_k(r_k))^N = 0$$

in R for all $r_1, \ldots, r_k \in R$. Reinterpreting this identity in terms of $Z(\mathfrak{g})/\mathrm{Ann}_{Z(\mathfrak{g})}(V)$, we obtain (7.21).

If V has an infinitesimal character, say λ, then the homomorphism $\chi_\lambda : Z(\mathfrak{g}) \to \mathbb{C}$ has kernel $\mathrm{Ann}_{Z(\mathfrak{g})}(V)$, and it follows that the algebra R above is isomorphic to $\mathbb{C}$. Consequently χ_λ is the unique algebra homomorphism $\chi : R \to \mathbb{C}$, and the proof above shows that $P_{\chi_\lambda}(V)$ is the unique primary component.

We say that a $Z(\mathfrak{g})$ finite $U(\mathfrak{g})$ module V has **generalized infinitesimal character** $\lambda \in \mathfrak{h}^*$ if it has just one primary component and the corresponding χ is χ_λ. In any $Z(\mathfrak{g})$ finite $U(\mathfrak{g})$ module, the primary component $P_\lambda(V)$ has generalized infinitesimal character λ. According to Proposition 7.20, if V has infinitesimal character λ, then V has generalized infinitesimal character λ. In symbols a $Z(\mathfrak{g})$ finite $U(\mathfrak{g})$ module V has generalized infinitesimal character λ if there is some n such that

(7.26a) $(z - \chi_\lambda(z)1)^n v = 0$ for all $z \in Z(\mathfrak{g})$ and $v \in V$.

In Corollary 7.32 we shall see that (7.26a) automatically implies that V is $Z(\mathfrak{g})$ finite; therefore (7.26a) can be used by itself to define when the $U(\mathfrak{g})$ module V has a generalized infinitesimal character.

If N is the dimension of the algebra R in the proof of Proposition 7.20, then

(7.26b)
$$\prod_{i=1}^{N} (z_i - \chi_\lambda(z_i))v = 0 \quad \text{for all } N \text{ tuples } \{z_i\} \text{ in } Z(\mathfrak{g}) \text{ and for all } v \in V.$$

Corollary 7.27. If a $Z(\mathfrak{g})$ finite $U(\mathfrak{g})$ module V has generalized infinitesimal character $\lambda \in \mathfrak{h}^*$, then the subspace

$$U = \{v \in V \mid zv = \chi_\lambda(z)v \text{ for all } z \in Z(\mathfrak{g})\}$$

is not 0.

PROOF. By (7.26b), there exists n such that

$$\prod_{i=1}^{n} (z_i - \chi_\lambda(z_i)1)v = 0$$

for all n tuples $\{z_i\}$ in $Z(\mathfrak{g})$ and all $v \in V$ but some

$$v' = \prod_{i=1}^{n-1} (z_i - \chi_\lambda(z_i)1)v$$

is not 0. Then v' is a nonzero element in U.

Proposition 7.28 (Wigner's Lemma). Let U and V be $U(\mathfrak{g})$ modules having respective generalized infinitesimal characters λ and λ', where $\chi_\lambda \neq \chi_{\lambda'}$. Then $\text{Ext}_{\mathfrak{g}}^n(U, V) = 0$ for all n.

PROOF. Let $X_n \otimes_{\mathbb{C}} U$ be the standard projective resolution (2.123) of U in $\mathcal{C}(\mathfrak{g})$. Fix z in $Z(\mathfrak{g})$. We construct from z two chain maps of $X_n \otimes_{\mathbb{C}} U$ to itself. One chain map is given by having z act on $X_n \otimes_{\mathbb{C}} U$ by $1 \otimes z$. This commutes with $U(\mathfrak{g}) \otimes 1$ trivially and with $1 \otimes U(\mathfrak{g})$ since z is central, and it commutes with the differential since the differential acts only in X_n. The other chain map is given by having z act on $X_n \otimes_{\mathbb{C}} U$ via the $U(\mathfrak{g})$ action on the tensor product. This commutes with $U(\mathfrak{g})$ since z is central, and it commutes with the differential since the differential is a $U(\mathfrak{g})$ map.

Both chain maps are maps over the action of z on U. By (C.6), the two chain maps are homotopic. Applying $\mathrm{Hom}_{\mathfrak{g}}(\,\cdot\,, V)$ to all the modules and maps in question, we obtain homotopic chain maps of $\mathrm{Hom}_{\mathbb{C}}(\bigwedge^n \mathfrak{g} \otimes_{\mathbb{C}} U, V)$ to itself. In the first case the map still acts in U, and in the second case the map acts in V. By (C.4) the two chain maps pass to equal maps on homology $\mathrm{Ext}_{\mathfrak{g}}^n(U, V)$.

By Lemma 7.14, choose $z \in Z(\mathfrak{g})$ with $\chi_\lambda(z) = 0$ and $\chi_{\lambda'}(z) = 1$. Since U has generalized infinitesimal character λ, z^l acts by 0 in U for some l. Hence our first map corresponding to z^l is 0 on $\mathrm{Ext}_{\mathfrak{g}}^n(U, V)$. Our second map is induced by z^l on V, which is invertible since $(z^l - 1)^N V = 0$ implies

$$z^l(z^{l(N-1)} - \binom{N}{1} z^{l(N-2)} + \cdots + (-1)^{N-1} \binom{N}{N-1})) = (-1)^{N-1}.$$

Hence our second map is invertible on $\mathrm{Ext}_{\mathfrak{g}}^n(U, V)$. Since the two maps are equal, $\mathrm{Ext}_{\mathfrak{g}}^n(U, V) = 0$.

Corollary 7.29. If F is an irreducible nontrivial finite-dimensional $U(\mathfrak{g})$ module, then $H_n(\mathfrak{g}, F) = 0$ and $H^n(\mathfrak{g}, F) = 0$ for all n.

PROOF. These are special cases of Proposition 7.28 in which U or V is $\mathbb{C}$. The irreducible finite-dimensional F has an infinitesimal character that is different from that of $\mathbb{C}$.

3. Chevalley's Structure Theorem for $Z(\mathfrak{g})$

In Chapter V we addressed admissibility for cohomologically induced modules and for $\mathfrak{u}$ homology and cohomology, and we addressed the existence of infinitesimal characters for cohomologically induced modules. When $(\mathfrak{g}, K)$ is a reductive pair, it is not true, however, that the $\mathfrak{u}$ homology and cohomology of a $(\mathfrak{g}, K)$ module with an infinitesimal character under $Z(\mathfrak{g})$ has an infinitesimal character under $Z(\mathfrak{l})$. (The case that G is compact connected already provides counterexamples.) What is true is that the $\mathfrak{u}$ homology and cohomology are $Z(\mathfrak{l})$ finite. We shall prove this result in the next section.

But first we need some further structure theory for $Z(\mathfrak{g})$. Fix a complex reductive Lie algebra $\mathfrak{g}$, and let $\mathfrak{h}$ be a Cartan subalgebra. Recall that the Harish-Chandra isomorphism carries $Z(\mathfrak{g})$ onto the algebra $U(\mathfrak{h})^W$ of Weyl-group invariants in $U(\mathfrak{h})$. In the following theorem, when there is an underlying reductive pair $(\mathfrak{g}, K)$ and $\mathfrak{h}$ is the complexification of a θ stable Cartan subalgebra $\mathfrak{h}_0 = \mathfrak{t}_0 \oplus \mathfrak{a}_0$ of $\mathfrak{g}_0$, we should think of the real inner-product space V as $i\mathfrak{t}_0 \oplus \mathfrak{a}_0$ and the complexification $V_{\mathbb{C}}$ as $\mathfrak{h}$. A **reflection** on V is an orthogonal transformation of order 2 whose -1 eigenspace has dimension 1. A **finite reflection group** W is a finite subgroup of $O(V)$ generated by reflections. Let $|W|$ be its order. We define

$\mathcal{P} =$ complex-valued polynomials on V (the algebra of functions generated by members of V^*, i.e., the symmetric algebra of V^*)

$\mathcal{I} =$ members of $\mathcal{P}$ fixed by W.

The algebra $\mathcal{P}$ is graded by degree, and we let $\mathcal{P}_d$ be the subspace of elements homogeneous of degree d. The subalgebra $\mathcal{I}$ inherits the grading, and we let $\mathcal{I}_d = \mathcal{P}_d \cap \mathcal{I}$. Let

$$\mathcal{P}^+ = \bigoplus_{d>0} \mathcal{P}_d, \text{ an ideal in } \mathcal{P}$$

$$\mathcal{I}^+ = \bigoplus_{d>0} \mathcal{I}_d, \text{ an ideal in } \mathcal{I}$$

$$\mathcal{J} = \mathcal{P}\mathcal{I}^+, \text{ an ideal in } \mathcal{P} \text{ contained in } \mathcal{P}^+.$$

Recall that members $p_1, \ldots, p_m$ of $\mathcal{P}$ are said to be **algebraically independent** if the algebra homomorphism of $\mathbb{C}[X_1, \ldots, X_m]$ into $\mathcal{P}$ given by $X_j \mapsto p_j$ is one-one.

Theorem 7.30 (Chevalley). Let W be a finite reflection group in $O(V)$, where V is a real inner-product space of finite dimension l.

(a) There are l algebraically independent homogeneous elements $p_1, \ldots, p_l$ such that $\mathcal{I} = \mathbb{C}[p_1, \ldots, p_l]$. Their degrees $d_1, \ldots, d_l$ satisfy

(7.31a) $$\sum_{j=1}^{l} (d_j - 1) = \text{number of reflections in } W$$

(7.31b) $$\prod_{j=1}^{l} d_j = |W|.$$

(b) $\mathcal{P}$ is a free $\mathcal{I}$ module of rank $|W|$. More precisely, let $\mathcal{H}$ be a graded subspace of $\mathcal{P}$ such that $\mathcal{P} = \mathcal{J} \oplus \mathcal{H}$; then dim $\mathcal{H} = |W|$, and the map $p \otimes h \mapsto ph$ for $p \in \mathcal{I}$ and $h \in \mathcal{H}$ extends to a linear isomorphism of $\mathcal{I} \otimes_{\mathbb{C}} \mathcal{H}$ onto $\mathcal{P}$.

We shall prove this theorem after obtaining some lemmas. But before giving the lemmas, we derive as a corollary of (a) the fact noted in §2 that "generalized infinitesimal character" does not need to assume $Z(\mathfrak{g})$ finiteness.

Corollary 7.32. Let $U(\mathfrak{g})$ be a complex reductive Lie algebra, let $\mathfrak{h}$ be a Cartan subalgebra, let λ be in $\mathfrak{h}^*$, and fix $n > 0$. Then the ideal in $Z(\mathfrak{g})$ generated by

$$\{(z - \chi_\lambda(z)1)^n \mid z \in Z(\mathfrak{g})\}$$

has finite codimension in $Z(\mathfrak{g})$.

PROOF. The Harish-Chandra isomorphism and Theorem 7.30a allow us to regard $Z(\mathfrak{g})$ as a polynomial algebra $\mathbb{C}[X_1, \ldots, X_l]$. On this algebra, χ_λ is a homomorphism into $\mathbb{C}$ and is therefore a point evaluation, by the universal mapping property of symmetric algebras. Without loss of generality we may take the point in question to be the origin. Then the given ideal is generated by the nth powers of all members of $\mathbb{C}[X_1, \ldots, X_l]$ without constant term. By (4.113c) the ideal is generated by all elements homogeneous of degree n. Therefore the ideal contains all elements homgeneous of degree $\geq n$, and it has finite codimension.

We turn to the proof of Theorem 7.30, beginning with some lemmas.

Lemma 7.33. The ideal $\mathcal{J}$ in $\mathcal{P}$ is finitely generated, and the generators may be taken to be homogeneous elements in $\mathcal{I}^+$. If $\{p_1, \ldots, p_m\}$ is a set of homogeneous generators of $\mathcal{J}$ taken from $\mathcal{I}^+$, then the algebra with identity generated by $p_1, \ldots, p_m$ is $\mathcal{I}$.

PROOF. The ring $\mathcal{P}$ is Noetherian by the Hilbert Basis Theorem, and hence every ideal is finitely generated. Among all ideals in $\mathcal{P}$ having a finite set of homogeneous generators from within $\mathcal{I}^+$, choose one $\mathcal{J}_0$ that is maximal under inclusion. We can write $\mathcal{J}_0 = \mathcal{P}S$ with S a finite set of homogeneous elements in $\mathcal{I}^+$. Here $\mathcal{J} \supseteq \mathcal{J}_0$. If $p \in \mathcal{I}^+$ were in $\mathcal{J}$ but not $\mathcal{J}_0$, then one of its homogeneous components q would fail to be in $\mathcal{J}_0$. So we could adjoin q to S and contradict maximality. We conclude that S generates $\mathcal{J}$.

Fix such a generating set $S = \{p_1, \ldots, p_m\}$. We prove by induction on $d \geq 0$ that the algebra with identity generated by S contains $\mathcal{I}_d$. For $d = 0$ the element 1 handles $\mathcal{I}_0$. Let p be in $\mathcal{I}_d$ with $d > 0$, and write

(7.34a) $p = q_1 p_1 + \cdots + q_m p_m$ with $q_j \in \mathcal{P}_{d - \deg p_j}$.

If we define $\dot{q}_j = |W|^{-1} \sum_{w \in W} w q_j$, then the average of the equation (7.34a) with $w \in W$ applied is

(7.34b) $p = \dot{q}_1 p_1 + \ldots \dot{q}_m p_m$ with $\dot{q}_j \in \mathcal{I}_{d - \deg p_j}$.

Since $d - \deg p_j < d$, the inductive hypothesis implies that $\dot{q}_j$ is in the algebra with identity generated by S, and hence so is p.

Lemma 7.35. Let $p_1, \ldots, p_m$ be members of $\mathcal{I}$ such that p_1 is not in $\sum_{j=2}^{m} \mathcal{I} p_j$. If $q_1, \ldots, q_m$ are homogeneous elements in $\mathcal{P}$ with

(7.36a) $q_1 p_1 + \cdots + q_m p_m = 0,$

then q_1 is in $\mathcal{J}$.

PROOF. We induct on $\deg q_1$. As in (7.34), we have

(7.36b) $\dot{q}_1 p_1 + \cdots + \dot{q}_m p_m = 0.$

The base case of the induction is $\deg q_1 = 0$. In this case $\dot{q}_1 = q_1$; if $q_1 \neq 0$, (7.36b) gives

$$p_1 = -\dot{q}_1^{-1} (\dot{q}_2 p_2 + \cdots + \dot{q}_m p_m),$$

in contradiction to our assumption. The inductive step is $\deg q_1 = d \geq 1$. Let $s \in W$ be a reflection, and let λ be a nonzero linear functional whose kernel is the $+1$ eigenspace of s. From (7.36a) we have $\sum_{j=1}^{m} (q_j - s q_j) p_j = 0$. But $q_j - s q_j = \lambda r_j$ with $\deg r_j < \deg q_j$. Thus $\sum_{j=1}^{m} \lambda r_j p_j = 0$ and hence $\sum_{j=1}^{m} r_j p_j = 0$. Since $\deg r_1 < d$, the inductive hypothesis says that r_1 is in $\mathcal{J}$. Hence $q_1 \equiv s q_1 \mod \mathcal{J}$. Since s is an arbitrary reflection and the reflections generate W, $q_1 \equiv w q_1 \mod \mathcal{J}$ for all $w \in W$. Averaging these congruences, we obtain $q_1 \equiv \dot{q}_1 \mod \mathcal{J}$. But $\dot{q}_1$ is a member of $\mathcal{I}$ homogeneous of degree > 0 and is therefore in $\mathcal{J}$. Hence $q_1 \equiv 0 \mod \mathcal{J}$, and the induction is complete.

Lemma 7.37. Suppose that $p_1, \dots, p_m$ are homogeneous elements in $\mathcal{I}^+$ that generate $\mathcal{I}$, and suppose that $p_i \notin \sum_{j \neq i} \mathcal{I} p_j$ for $1 \leq i \leq m$. Then the p_i are algebraically independent.

PROOF. Let $d_j = \deg p_j$. Suppose on the contrary that the p_j are not algebraically independent. Then there is some nonzero P in $\mathbb{C}[X_1, \dots, X_m]$ with $P(p_1, \dots, p_m) = 0$ in $\mathcal{I}$. We may assume that the degree of P is as small as possible and, since the p_j are homogeneous, that there is some $d > 0$ such that every monomial $X_1^{k_1} \cdots X_m^{k_m}$ contributing to P has

$$(7.38) \qquad\qquad \sum k_j d_j = d.$$

Let $P_j = \dfrac{\partial P}{\partial X_j}$ and $q_j = P_j(p_1, \dots, p_m)$ for $1 \leq j \leq m$. Since q_j is a polynomial in $p_1, \dots, p_m$, it is in $\mathcal{I}$. Also the minimality of $\deg P$ implies that $q_j = 0$ if and only if $P_j = 0$. Since P is of degree > 0, some P_j is $\neq 0$ and therefore some q_j is $\neq 0$. By (7.38), q_j is homogeneous of degree $d - d_j$.

Let us renumber the indices so that $q_1, \dots, q_t$ are $\neq 0$ and $q_{t+1}, \dots, q_m$ are 0. Within $\{1, \dots, t\}$ let S be a minimal subset such that $\sum_{j \in S} \mathcal{I} q_j = \sum_{j=1}^{t} \mathcal{I} q_j$. Renumber the q_i's corresponding to S first, calling them $q_1, \dots, q_s$. The minimality of S forces

$$(7.39) \qquad\qquad q_i \notin \sum_{\substack{1 \leq j \leq s, \\ j \neq i}} \mathcal{I} q_j \qquad \text{for } i \leq s,$$

while the identity $\sum_{j=1}^{s} \mathcal{I} q_j = \sum_{j=1}^{t} \mathcal{I} q_j$ forces relations of the form

$$(7.40) \qquad\qquad q_{s+j} = \sum_{i=1}^{s} r_{ji} q_i \qquad \text{for } 1 \leq j \leq t - s$$

with r_{ji} in $\mathcal{I}$. Since q_i is homogeneous of degree $d - d_i$, there is no loss of generality in assuming that r_{ji} is homogeneous with

$$(d - d_{s+j}) = \deg r_{ji} + (d - d_i),$$

i.e.,

$$(7.41) \qquad\qquad \deg r_{ji} + d_{s+j} = d_i.$$

Let $x_1, \dots, x_l$ be Euclidean coordinates for V. The relation

$$P(p_1, \dots, p_m) = 0$$

is an identity on V, and we can differentiate it with respect to x_k to obtain

$$\sum_{j=1}^{t} q_j \frac{\partial p_j}{\partial x_k} = 0 \qquad \text{for } 1 \le k \le l.$$

Substitution from (7.40) gives

$$\sum_{i=1}^{s} q_i \left(\frac{\partial p_i}{\partial x_k} + \sum_{j=1}^{t-s} r_{ji} \frac{\partial p_{s+j}}{\partial x_k} \right) = 0 \qquad \text{for } 1 \le k \le l.$$

Here the q_i's are in $\mathcal{I}$, they satisfy (7.39), and the terms in parentheses are homogeneous by (7.41). Thus Lemma 7.35 is applicable. The lemma allows us to conclude that

$$\frac{\partial p_i}{\partial x_k} + \sum_{j=1}^{t-s} r_{ji} \frac{\partial p_{s+j}}{\partial x_k} \qquad \text{is in } \mathcal{P}\mathcal{I}^+ \text{ for } i \le s \text{ and all } k.$$

Multiplying by x_k and adding, we see from Euler's identity

$$(7.42) \qquad \sum x_k \frac{\partial h}{\partial x_k} = d'h \qquad \text{for } h \text{ homogeneous of degree } d'$$

that

$$d_i p_i + \sum_{j=1}^{t-s} r_{ji} d_{s+j} p_{s+j} \qquad \text{is in } \mathcal{P}^+\mathcal{I}^+ \text{ for } i \le s.$$

Since the $p_{i'}$ generate $\mathcal{P}\mathcal{I}^+$, Lemma 7.33 shows that $\mathcal{I}^+ = \sum_{i'=1}^{m} \mathcal{I} p_{i'}$. Thus there exist $u_{ii'} \in \mathcal{P}^+$ for $i \le s$ and $i' \le m$ such that

$$(7.43) \qquad d_i p_i + \sum_{j=1}^{t-s} r_{ji} d_{s+j} p_{s+j} = \sum_{i'=1}^{m} u_{ii'} p_{i'}.$$

By (7.41) the left side of (7.43) is homogeneous of degree d_i. Thus we can discard terms on the right side of (7.43) that do not have this homogeneity. Since $u_{ii'}$ has 0 constant term, we may assume, in other words, that $u_{ii'}$ is homogeneous of degree $d_i - d_{i'}$ if $d_i - d_{i'} > 0$ and that $u_{ii'} = 0$ if $d_i - d_{i'} \le 0$. Taking $i = 1$, we obtain from (7.43) a relation

$$p_1 = \sum_{i'=2}^{m} v_{i'} p_{i'}$$

since $u_{11} = 0$. Applying $w \in W$ to this relation and averaging, we obtain

$$p_1 = \sum_{i'=2}^{m} \bar{v}_{i'} p_{i'},$$

which contradicts the hypothesis in the lemma. This completes the proof.

PROOF OF THEOREM 7.30a EXCEPT FOR (7.31). By Lemma 7.33 we can find homogeneous elements $p_1, \ldots, p_m$ in $\mathcal{I}^+$ such that the algebra with identity generated by $p_1, \ldots, p_m$ is $\mathcal{I}$. Let S be a minimal subset of $\{1, \ldots, m\}$ such that $\sum_{j \in S} \mathcal{I} p_j = \sum_{j=1}^m \mathcal{I} p_j \ (= \mathcal{I}^+)$, and reorder indices so that $S = \{1, \ldots, s\}$. Then

$$p_i \notin \sum_{\substack{1 \le j \le s, \\ j \ne i}} \mathcal{I} p_j \qquad \text{for } 1 \le i \le s$$

and the algebra with identity generated by $p_1, \ldots, p_s$ is $\mathcal{I}$. By Lemma 7.37, the p_i for $i \le s$ are algebrically independent. Therefore $\mathcal{I} = \mathbb{C}[p_1, \ldots, p_s]$.

To complete the proof, we show that $s = l$. The transcendence degree of the field of quotients $\mathbb{C}(p_1, \ldots, p_s)$ is s. If we write $\mathcal{P} = \mathbb{C}[\lambda_1, \ldots, \lambda_l]$ for a basis $\lambda_1, \ldots, \lambda_l$ of V^*, then the field of quotients $\mathbb{C}(\lambda_1, \ldots, \lambda_l)$ is an extension field of $\mathbb{C}(p_1, \ldots, p_s)$ of transcendence degree l over $\mathbb{C}$. We show that this extension is algebraic over $\mathbb{C}(p_1, \ldots, p_s)$, and then it follows that $l = s$.

It is enough to prove that each element q of $\mathcal{P}$ is algebraic over $\mathbb{C}(p_1, \ldots, p_s)$. All we have to do is observe that the polynomial

$$\prod_{w \in W} (X - wq) = X^{|W|} + r_{|W|-1} X^{|W|-1} + \cdots + r_0,$$

which has coefficients in $\mathcal{I}$ and has q as one of its roots, exhibits q as algebraic.

PROOF OF THEOREM 7.30b EXCEPT FOR $\dim \mathcal{H} = |W|$. Since $\mathcal{J}$ is graded, we can choose a graded subspace $\mathcal{H} = \bigoplus_d \mathcal{H}_d$ of $\mathcal{P}$ such that $\mathcal{P} = \mathcal{J} \oplus \mathcal{H}$. The map $\tau(p, h) = ph$ of $\mathcal{I} \times \mathcal{H}$ into $\mathcal{P}$ is bilinear and therefore extends to a linear map $\tau : \mathcal{I} \otimes_{\mathbb{C}} \mathcal{H} \to \mathcal{P}$. We prove that τ is one-one onto.

To see that τ is onto, let $\mathcal{P}'$ be the image. Since $\mathcal{I}$ is an algebra, $\mathcal{I}\mathcal{P}' \subseteq \mathcal{P}'$. We prove inductively that $\mathcal{P}_d \subseteq \mathcal{P}'$ for all $d \ge 0$. For $d = 0$, $\mathcal{H}_0 = \mathbb{C}$ and thus $\mathcal{P}_0 \subseteq \mathcal{P}'$. For the inductive step, let $d \ge 1$ be given. It is enough to prove that $\mathcal{J}_d \subseteq \mathcal{P}'$. Let $p_1, \ldots p_l$ be as in (a) of the theorem. Since $\mathcal{I}^+ = \sum \mathcal{I} p_i$, we have $\mathcal{J} = \sum \mathcal{P} \mathcal{I} p_i = \sum \mathcal{P} p_i$. If $r \in \mathcal{J}_d$ is given, we can write $r = \sum q_i p_i$ with $q_i \in \mathcal{P}$. Without loss of generality, we may assume that q_i is homogeneous and that it has degree $d - d_i$, where $d_i = \deg p_i$. By inductive hypothesis each q_i is in $\mathcal{P}'$; since p_i is in $\mathcal{I}$ and $\mathcal{I}\mathcal{P}' \subseteq \mathcal{P}'$, it follows that r is in $\mathcal{P}'$. Thus τ is onto.

To prove that τ is one-one, suppose that

$$(7.44) \qquad \tau\left(\sum_{i=1}^{m} q_i \otimes h_i\right) = 0, \quad \text{i.e.,} \quad \sum_{i=1}^{m} q_i h_i = 0.$$

We may assume that the q_i are homogeneous members of $\mathcal{I}$ and that the h_i are homogeneous and linearly independent in $\mathcal{H}$. Also we may assume that

$$(7.45) \qquad \deg q_1 = \cdots = \deg q_s < \deg q_{s+1} \le \cdots \le \deg q_m.$$

We are to prove that all q_i are 0. Assuming the contrary, we may assume that m is as small as possible among all counterexamples. We show that q_1 is not in $\sum_{i=2}^{m} \mathcal{I} q_i$. Assume the contrary. Taking into account the degrees in (7.45) and the homogeneity, we see that $q_1 = \sum_{i=2}^{s} c_i q_i$ with c_i constant. Hence

$$\sum_{i=2}^{s} q_i (c_i h_1 + h_i) + \sum_{i=s+1}^{m} q_i h_i = 0.$$

This is another nontrivial relation of the form (7.44) and contradicts the minimality of m.

We conclude that q_1 is not in $\sum_{i=2}^{m} \mathcal{I} q_i$. By Lemma 7.35, h_1 is in $\mathcal{J}$. Thus h_1 is in $\mathcal{J} \cap \mathcal{H} = 0$, in contradiction to the linear independence of $h_1, \ldots, h_m$. This completes the proof of (b) except for the relation $\dim \mathcal{H} = |W|$.

PROOF THAT $\dim \mathcal{H} = |W|$ IN THEOREM 7.30b. We shall use generating functions. If $\mathcal{E} = \bigoplus_{d \ge 0} \mathcal{E}_d$ is a graded subspace of $\mathcal{P}$, the **Poincaré series** of $\mathcal{E}$, denoted $P_{\mathcal{E}}(t)$, is the formal power series

$$P_{\mathcal{E}}(t) = \sum_{d=0}^{\infty} (\dim \mathcal{E}_d) t^d.$$

The Poincaré series of a direct sum is the sum of the Poincaré series. If $\mathcal{E}$ and $\mathcal{E}'$ are given and if

$$(7.46a) \qquad (\mathcal{E}\mathcal{E}')_d \cong \bigoplus_{k=0}^{d} (\mathcal{E}_k \otimes_{\mathbb{C}} \mathcal{E}'_{d-k}),$$

then

$$(7.46b) \qquad P_{\mathcal{E}\mathcal{E}'}(t) = P_{\mathcal{E}}(t) P_{\mathcal{E}'}(t).$$

From (7.46) it follows inductively on the number of variables that

(7.47)
$$P_{\mathcal{P}}(t) = \frac{1}{(1-t)^l}.$$

Consequently $P_{\mathcal{E}}(t)$ is convergent for $|t| < 1$. If $p_1, \ldots, p_l$ are homogeneous of degrees $d_j > 0$ and are algebraically independent, let $\mathcal{E} = \mathbb{C}[p_1, \ldots, p_l]$. Then it follows from (7.46) that

(7.48)
$$P_{\mathcal{E}}(t) = \prod_{j=1}^{l} \frac{1}{(1-t^{d_j})}.$$

We apply these observations to the decomposition $\mathcal{P} \cong \mathcal{I} \otimes_{\mathbb{C}} \mathcal{H}$. We know that $\mathcal{I} \cong \mathbb{C}[p_1, \ldots, p_l]$ with the p_j homogeneous of degree $d_j > 0$ and algebraically independent. Formulas (7.46) through (7.48) give

$$\frac{1}{(1-t)^l} = P_{\mathcal{P}}(t) = P_{\mathcal{I}}(t) P_{\mathcal{H}}(t) = P_{\mathcal{H}}(t) \prod_{j=1}^{l} \frac{1}{(1-t^{d_j})}.$$

Hence

(7.49)
$$P_{\mathcal{H}}(t) = \prod_{j=1}^{l} (1 + t + \cdots + t^{d_j - 1}).$$

Since $P_{\mathcal{H}}$ is then a polynomial, it follows that $\dim \mathcal{H} < \infty$. We can evaluate $\dim \mathcal{H}$ as $P_{\mathcal{H}}(1)$, and therefore it is enough to prove that

(7.50)
$$\lim_{t \uparrow 1} P_{\mathcal{H}}(t)^{-1} = |W|^{-1}.$$

For any member L of the orthogonal group $O(V)$, let L_d be the corresponding linear mapping on $\mathcal{P}_d$. Let us observe that

(7.51)
$$(\det(1-tL))^{-1} = \sum_{d=0}^{\infty} (\operatorname{Tr} L_d) t^d.$$

In fact, in a basis of $V_{\mathbb{C}}$ in which L is diagonal with entries $\{\alpha_j\}_{j=1}^{l}$, the left side of (7.51) is

$$\prod_{j=1}^{l} \frac{1}{1 - \alpha_j t} = \prod_{j=1}^{l} (1 + \alpha_j t + \alpha_j^2 t^2 + \ldots),$$

and the coefficient of t^d on the right side is just $\operatorname{Tr} L_d$. This proves (7.51). On $\mathcal{P}_d$ the linear transformation $|W|^{-1} \sum_{w \in W} w$ is the projection on $\mathcal{I}_d$ along the subspace that transforms by the other irreducible representations of W. Its trace is thus the dimension of $\mathcal{I}_d$. Hence

(7.52)
$$P_{\mathcal{I}}(t) = \sum_{d=0}^{\infty} (\dim \mathcal{I}_d) t^d = |W|^{-1} \sum_{d=0}^{\infty} \sum_{w \in W} (\operatorname{Tr} w_d) t^d$$
$$= |W|^{-1} \sum_{w \in W} (\det(1 - tw))^{-1}.$$

For $w \in W$, let $\alpha_{w,i}$ be the eigenvalues of w, repeated according to their multiplicities. Then (7.52) gives

(7.53) $$P_{\mathcal{H}}(t)^{-1} = P_{\mathcal{I}}(t)(1 - t)^l = |W|^{-1} \sum_{w \in W} \frac{\prod_{i=1}^{l} (1 - t)}{\prod_{i=1}^{l} (1 - t\alpha_{w,i})}.$$

Each factor $\dfrac{1 - t}{1 - t\alpha_{w,i}}$ is bounded as $t \uparrow 1$, and the product tends to 0 if some $\alpha_{w,i}$ is different from 1. Hence the only term of $\sum_{w \in W}$ that makes a contribution as $t \uparrow 1$ is $w = 1$. The summand in this case is 1, and (7.50) follows. This completes the proof of (b).

PROOF OF (7.31) IN THEOREM 7.30a. Putting $t = 1$ in (7.49) and comparing the result with (7.50), we obtain (7.31b). Thus we have only to prove (7.31a). We differentiate (7.53) and put $t = 1$. In the expression

(7.54)
$$\frac{\prod_{i=1}^{l} (1 - t)}{\prod_{i=1}^{l} (1 - t\alpha_{w,i})},$$

the numerator has a 0 of order l at $t = 1$, and the denominator has a 0 of order equal to the multiplicity of 1 as an eigenvalue of w. Thus the derivative of (7.54) at $t = 1$ is 0 if this multiplicity is not $l - 1$. When the multiplicity is $l - 1$, w is a reflection, the other $\alpha_{w,i}$ is -1, and (7.54) reduces to $(1 + t)^{-1}$, which has derivative $-\frac{1}{2}$ at $t = 1$. Hence

$$\tfrac{d}{dt} P_{\mathcal{H}}(t)^{-1}|_{t=1} = |W|^{-1} \sum_{w=\text{reflection}} (-\tfrac{1}{2}).$$

The left side here, by (7.50) is

$$= -\frac{P'_{\mathcal{H}}(1)}{P_{\mathcal{H}}(1)^2} = -P_{\mathcal{H}}(1)^{-1}\tfrac{d}{dt}\log P_{\mathcal{H}}(t)|_{t=1}$$

$$= -|W|^{-1}\sum_{j=1}^{l}\frac{1 + 2t + \cdots + (d_j - 1)t^{d_j-2}}{1 + t + \cdots + t^{d_j-1}}\bigg|_{t=1}$$

$$= -|W|^{-1}\sum_{j=1}^{l}\frac{(d_j - 1)d_j/2}{d_j}$$

$$= |W|^{-1}\sum_{j=1}^{l}(-\tfrac{1}{2})(d_j - 1).$$

Thus (7.31a) follows.

4. $Z(\mathfrak{l})$ Finiteness of $\mathfrak{u}$ Homology and Cohomology

Let $(\mathfrak{g}, K)$ be a reductive pair, and let $(\mathfrak{q}, L \cap K)$ be a parabolic subpair with $\mathfrak{q} = \mathfrak{l} \oplus \mathfrak{u}$. Then the map $(\mathfrak{q}, L \cap K) \to (\mathfrak{l}, L \cap K)$ is a semidirect product map of pairs in the sense of §III.2. According to Proposition 3.12, if V is any $(\mathfrak{g}, L \cap K)$ module, then $H_j(\mathfrak{u}, V)$ and $H^j(\mathfrak{u}, V)$ acquire the structure of $(\mathfrak{l}, L \cap K)$ modules. As we noted at the start of §3, even if V is a $(\mathfrak{g}, K)$ module with an infinitesimal character under $Z(\mathfrak{g})$, it does not necessarily follow that the $\mathfrak{u}$ homology and cohomology of V have infinitesimal characters under $Z(\mathfrak{l})$. However, the $\mathfrak{u}$ homology and cohomology will be $Z(\mathfrak{l})$ finite, as the main result of this section will show.

Before coming to that result, let us make some observations about the action of $Z(\mathfrak{g})$ on a contragredient. If V is a given $U(\mathfrak{g})$ module, then the action of $U(\mathfrak{g})$ on V^* is given by

$$\langle uv^*, v\rangle = \langle v^*, u^t v\rangle \qquad \text{for } u \in U(\mathfrak{g}),\ v^* \in V^*,\ v \in V.$$

Consequently $\text{Ann}_{Z(\mathfrak{g})}(V^*) = \text{Ann}_{Z(\mathfrak{g})}(V)^t$, and V^* is $Z(\mathfrak{g})$ finite if V is. Let us prove that

$$(7.55) \qquad\qquad \chi_\lambda(z^t) = \chi_{-\lambda}(z) \qquad \text{for } z \in Z(\mathfrak{g}).$$

From this identity it follows that if V is $Z(\mathfrak{g})$ finite, then so is V^*; moreover, if V has generalized infinitesimal character λ, then V^* has generalized infinitesimal character $-\lambda$.

To prove (7.55), we use standard notation and observe from the definition and Proposition 4.89b that

$$\gamma'_n(z^t) = \gamma'_{n^-}(z)^t.$$

Then two applications of (4.94) give

$$\chi_\lambda(z^t) = \lambda(\gamma(z^t)) = (\lambda - \delta)(\gamma'_n(z^t)) = (\lambda - \delta)(\gamma'_{n^-}(z)^t)$$
$$= (-\lambda + \delta)(\gamma'_{n^-}(z)) = -\lambda(\gamma(z)) = \chi_{-\lambda}(z),$$

and (7.55) follows.

Theorem 7.56. Let $(\mathfrak{g}, K)$ be a reductive pair, and let $(\mathfrak{q}, L \cap K)$ be a parabolic subpair with $\mathfrak{q} = \mathfrak{l} \oplus \mathfrak{u}$. If V is a $Z(\mathfrak{g})$ finite $(\mathfrak{g}, L \cap K)$ module, then $H_j(\mathfrak{u}, V)$ and $H^j(\mathfrak{u}, V)$ are $Z(\mathfrak{l})$ finite $(\mathfrak{l}, L \cap K)$ modules. In addition, let $\mathfrak{h}$ be a θ stable Cartan subalgebra of $\mathfrak{l}$, and choose a positive system $\Delta^+(\mathfrak{g}, \mathfrak{h})$ containing $\Delta(\mathfrak{u})$. Suppose that the χ primary components of V relative to $Z(\mathfrak{g})$ are nonzero only for $\chi = \chi_\lambda$ with

$$\lambda \in \{\lambda_1, \ldots, \lambda_r\} \subseteq \mathfrak{h}^*.$$

Then the χ primary components of $H^j(\mathfrak{u}, V)$ relative to $Z(\mathfrak{l})$ can be nonzero only for $\chi = \chi_\nu$ with ν of the form

$$\nu = w\lambda_i - \delta(\mathfrak{u}) \qquad \text{for some } w \in W(\mathfrak{g}, \mathfrak{h}) \text{ and some } i,$$

while the χ primary components of $H_j(\mathfrak{u}, V)$ relative to $Z(\mathfrak{l})$ can be nonzero only for $\chi = \chi_\nu$ with ν of the form

$$\nu = w\lambda_i + \delta(\mathfrak{u}) \qquad \text{for some } w \in W(\mathfrak{g}, \mathfrak{h}) \text{ and some } i.$$

REMARK. The compact groups may be suppressed in the statement of the theorem, and the result is still valid. In other words, $\mathfrak{g}$ is to be a complex reductive Lie algebra, V is to be a $Z(\mathfrak{g})$ finite $U(\mathfrak{g})$ module, and $\mathfrak{q}$ is to be a parabolic subalgebra of $\mathfrak{g}$. Then $H_j(\mathfrak{u}, V)$ and $H^j(\mathfrak{u}, V)$ are $Z(\mathfrak{l})$ finite $U(\mathfrak{l})$ modules, and the assertions about the primary components are still valid.

PROOF. We use notation as in §IV.8. First we consider $H^j(\mathfrak{u}, V)$. Let V be annihilated by an ideal I in $Z(\mathfrak{g})$ of finite codimension. The action of $z \in Z(\mathfrak{g})$ at the level of cocyles in (4.147) descends to $H^j(\mathfrak{u}, V)$ and is given there, according to Theorem 4.149, by

$$z\omega = \mu'_\mathfrak{u}(z)\omega \qquad \text{for } \omega \in H^j(\mathfrak{u}, V).$$

Hence $\mu'_{\mathfrak{u}}(I)$ annihilates $H^j(\mathfrak{u}, V)$, and so does $Z(\mathfrak{l})\mu'_{\mathfrak{u}}(I)$. To prove the first conclusion of the theorem, we shall prove that this ideal has finite codimension in $Z(\mathfrak{l})$.

Let $W = W(\mathfrak{g}, \mathfrak{h})$ and $W_L = W(\mathfrak{l}, \mathfrak{h})$ be the Weyl groups of $\mathfrak{g}$ and $\mathfrak{l}$. Since $\mu_{\mathfrak{g}/\mathfrak{l}} = \tau_{\mathfrak{u}} \circ \mu'_{\mathfrak{u}}$ by (4.128) and since $\tau_{\mathfrak{u}}$ is an algebra automorphism of $Z(\mathfrak{l})$, it is enough to prove that $Z(\mathfrak{l})\mu_{\mathfrak{g}/\mathfrak{l}}(I)$ has finite codimension in $Z(\mathfrak{l})$. Since $\gamma_{\mathfrak{l}} : Z(\mathfrak{l}) \to U(\mathfrak{h})^{W_L}$ is an isomorphism by Theorem 4.95, it is enough to prove that $\gamma_{\mathfrak{l}}(Z(\mathfrak{l})\mu_{\mathfrak{g}/\mathfrak{l}}(I))$ has finite codimension in $U(\mathfrak{h})^{W_L}$. But

$$\gamma_{\mathfrak{l}}(Z(\mathfrak{l})\mu_{\mathfrak{g}/\mathfrak{l}}(I)) = \gamma_{\mathfrak{l}}(Z(\mathfrak{l}))\gamma_{\mathfrak{l}}(\mu_{\mathfrak{g}/\mathfrak{l}}(I)) = U(\mathfrak{h})^{W_L}\gamma(I)$$

by Proposition 4.129c. Hence it is enough to prove that $U(\mathfrak{h})^{W_L}\gamma(I)$ has finite codimension in $U(\mathfrak{h})^{W_L}$.

Since $\gamma(I)$ has finite codimension in $U(\mathfrak{h})^W$, we can choose $f_1, \dots, f_m$ in $U(\mathfrak{h})^W$ such that

$$(7.57) \qquad U(\mathfrak{h})^W = \gamma(I) + \sum_{j=1}^{m} \mathbb{C}f_j.$$

From Theorem 7.30b we know that $U(\mathfrak{h})$ is a finitely generated $U(\mathfrak{h})^W$ module. Since $U(\mathfrak{h})^W$ is Noetherian (by Lemma 7.33 or Theorem 7.30a), the $U(\mathfrak{h})^W$ submodule $U(\mathfrak{h})^{W_L}$ is finitely generated. Let us write

$$(7.58) \qquad U(\mathfrak{h})^{W_L} = \sum_{k=1}^{n} U(\mathfrak{h})^W u_k \qquad \text{with } u_k \in U(\mathfrak{h})^{W_L}.$$

Combining (7.57) and (7.58), we obtain

$$U(\mathfrak{h})^{W_L} = \sum_{k=1}^{n} u_k\gamma(I) + \sum_{k=1}^{n}\sum_{j=1}^{m} \mathbb{C}f_j u_k.$$

Here $u_k\gamma(I)$ is contained in $U(\mathfrak{h})^{W_L}\gamma(I)$, and thus $U(\mathfrak{h})^{W_L}\gamma(I)$ has finite codimension in $U(\mathfrak{h})^{W_L}$.

To limit the primary components relative to $Z(\mathfrak{l})$, suppose $\nu \in \mathfrak{h}^*$ is such that the χ_ν primary component of $H^j(\mathfrak{u}, V)$ relative to $Z(\mathfrak{l})$ is nonzero. We prove that ν is of the form $w\lambda_i - \delta(\mathfrak{u})$ for some $w \in W$ and some i. Assuming the contrary, we can find a member p of $U(\mathfrak{h})$, which we regard as a polynomial on $\mathfrak{h}^*$, such that

$$(7.59a) \qquad\qquad p(\mu + \delta(\mathfrak{u})) \neq 0$$
$$(7.59b) \qquad p(w\lambda_i) = 0 \qquad \text{for all } w \text{ and } i.$$

Averaging p over W, we may assume that p is in $U(\mathfrak{h})^W$. By Theorem 4.95, choose $z_0 \in Z(\mathfrak{g})$ with $\gamma(z_0) = p$. The element z_0 satisfies $\chi_{\lambda_i}(z_0) = 0$ for all i. Since V is $Z(\mathfrak{g})$ finite, there is some $n > 0$ such that

$$\prod_{i=1}^{r} (z - \chi_{\lambda_i}(z)1)^n$$

acts as 0 on V, for all $z \in Z(\mathfrak{g})$. Then z_0^{nr} acts as 0 on V. Replacing p by p^{nr} does not affect (7.59), and thus we may assume that z_0 annihilates V. By Theorem 4.149, $\mu_{\mathfrak{u}}'(z_0)$ annihilates $H^j(\mathfrak{u}, V)$. Since the χ_ν primary component is assumed nonzero, we must have $\nu(\gamma_\mathfrak{l}(\mu_{\mathfrak{u}}'(z_0))) = 0$. But

$$
\begin{aligned}
\nu(\gamma_\mathfrak{l}(\mu_{\mathfrak{u}}'(z_0))) &= \nu(\gamma_\mathfrak{l} \tau_{\mathfrak{u}}^{-1} \mu_{\mathfrak{g}/\mathfrak{l}}(z_0)) && \text{by (4.128)} \\
&= \nu(\tau_{\mathfrak{u}}^{-1} \gamma_\mathfrak{l} \mu_{\mathfrak{g}/\mathfrak{l}}(z_0)) && \text{by Proposition 4.129b} \\
&= (\nu + \delta(\mathfrak{u}))(\gamma_\mathfrak{l} \mu_{\mathfrak{g}/\mathfrak{l}}(z_0)) && \\
&= (\nu + \delta(\mathfrak{u}))(\gamma(z_0)) && \text{by Proposition 4.129c} \\
&= p(\nu + \delta(\mathfrak{u})).
\end{aligned}
$$

Hence $p(\nu + \delta(\mathfrak{u})) = 0$, in contradiction to (7.59a).

This completes the proof of the assertions about cohomology $H^j(\mathfrak{u}, V)$. To handle homology $H_j(\mathfrak{u}, V)$, one approach is to go over the above argument, including the material in §IV.8, and make the appropriate adjustments for homology. A simpler approach is to use Easy Duality (Theorem 3.1 and Proposition 3.12). The $Z(\mathfrak{g})$ finiteness of V implies that of V^c, by the discussion before the statement of the present theorem. Thus we have

$$H_j(\mathfrak{u}, V)^c \cong H^j(\mathfrak{u}, V^c).$$

Taking contragredients, we obtain

$$H_j(\mathfrak{u}, V) \hookrightarrow (H_j(\mathfrak{u}, V)^c)^c \cong H^j(\mathfrak{u}, V^c)^c.$$

Since $H^j(\mathfrak{u}, V^c)$ is $Z(\mathfrak{l})$ finite, so are $H^j(\mathfrak{u}, V^c)^c$ and $H_j(\mathfrak{u}, V)$.

To complete the proof for homology, we need only track down the specific parameters. If the parameters for V relative to $Z(\mathfrak{g})$ are $\{\lambda_i\}$, (7.55) shows that those for V^c are $\{-\lambda_i\}$. The result for cohomology says that the parameters for $H^j(\mathfrak{u}, V^c)$ relative to $Z(\mathfrak{l})$ are $-w\lambda_i - \delta(\mathfrak{u})$, and (7.55) shows that the parameters for $H_j(\mathfrak{u}, V^c)^c$ (and hence also for $H_j(\mathfrak{u}, V)$) are $w\lambda_i + \delta(\mathfrak{u})$.

5. Invariants in the Symmetric Algebra

Let $\mathfrak{g}$ be a complex semisimple Lie algebra. The goal of §§5–7 is to obtain a theorem of Dixmier and Duflo identifying the two-sided ideal in $U(\mathfrak{g})$ that acts as 0 in a Verma module. This result will allow us to show that the tensor product of a finite-dimensional $U(\mathfrak{g})$ module with a $Z(\mathfrak{g})$ finite $U(\mathfrak{g})$ module is $Z(\mathfrak{g})$ finite. As a consequence we shall be able to define the translation functors $\psi_\lambda^{\lambda'}$.

The result about Verma modules requires some elaborate preparation that will not be used otherwise until §X.1. The reader may prefer to take the result on faith, skip now to §7, read only until the end of the statement of Theorem 7.118 in that section, and resume reading with §8.

The main result of §6 states that $U(\mathfrak{g})$ is a free $Z(\mathfrak{g})$ module, and it shows how to obtain a free basis. Only at the last moment does $U(\mathfrak{g})$ actually enter the argument. Most of the time we work with the symmetric algebra $S(\mathfrak{g})$, which is canonically isomorphic with the associated graded algebra obtained from $U(\mathfrak{g})$ and its usual filtration. We write $S(\mathfrak{g}) = \bigoplus_{d=0}^{\infty} S^d(\mathfrak{g})$ for the grading.

In the present section we study invariants in the symmetric algebra. Let us recall some facts about $S(\mathfrak{g})$. Let G be a (complex) Lie group with Lie algebra $\mathfrak{g}$. The group G acts holomorphically on $\mathfrak{g}$ by Ad, and the action extends uniquely from $\mathfrak{g}$ to an action by algebra automorphisms on $S(\mathfrak{g})$; we denote this action by Ad also. Differentiating in G, we obtain a complex-linear action ad of $\mathfrak{g}$ on $S(\mathfrak{g})$. Let

$$(7.60a) \qquad \qquad \mathcal{I} = S(\mathfrak{g})^G$$

be the algebra of invariants under $\mathrm{Ad}(G)$ or $\mathrm{ad}\,\mathfrak{g}$. This is a graded subalgebra, and we let $\mathcal{I}_d$ be the dth graded subspace. Let

$$(7.60b) \qquad \qquad \mathcal{I}_+ = \bigoplus_{d>0} \mathcal{I}_d.$$

The Killing form B provides a canonical isomorphism $X \mapsto \lambda_X$ of $\mathfrak{g}$ onto $\mathfrak{g}^*$ via $\lambda_X(Y) = B(X, Y)$. Because B is invariant, this isomorphism carries the action Ad of G on $\mathfrak{g}$ to Ad^* on $\mathfrak{g}^*$ (the contragredient of the adjoint action):

$$(\mathrm{Ad}^*(g)\lambda_X)(Y) = \lambda_X(\mathrm{Ad}(g)^{-1}Y) = B(X, \mathrm{Ad}(g)^{-1}Y)$$
$$= B(\mathrm{Ad}(g)X, Y) = \lambda_{\mathrm{Ad}(g)X}(Y).$$

We shall refer to the action of G on $\mathfrak{g}^*$ as the **coadjoint action** of G. In view of its equivalence with the adjoint action, we may denote it by Ad or Ad^*, depending on our emphasis.

The isomorphism $\mathfrak{g} \cong \mathfrak{g}^*$ extends to an algebra isomorphism $S(\mathfrak{g}) \cong S(\mathfrak{g}^*)$, and the adjoint and coadjoint actions by G and $\mathfrak{g}$ extend compatibly. Let

(7.61a) $$\mathcal{I}' = S(\mathfrak{g}^*)^G$$

be the invariants in $S(\mathfrak{g}^*)$ under $Ad(G)$ or $ad\,\mathfrak{g}$, and let

(7.61b) $$\mathcal{I}'_+ = \bigoplus_{d>0} \mathcal{I}'_d.$$

Members of $\mathfrak{g}^* \cong S^1(\mathfrak{g}^*)$ are linear functions from $\mathfrak{g}$ into $\mathbb{C}$, and general members of $S(\mathfrak{g}^*)$ may be regarded as (holomorphic) polynomial functions from $\mathfrak{g}$ into $\mathbb{C}$. Those in $S^d(\mathfrak{g}^*)$ are homogeneous of degree d. If $\mathfrak{g}_0$ is a compact real form of $\mathfrak{g}$, we may regard $S(\mathfrak{g}^*)$ also as all complex-valued polynomials on $\mathfrak{g}_0$.

Kostant's theory of harmonics, to be introduced in §6, gives a tensor-product decomposition of $S(\mathfrak{g}^*)$. The notion of harmonic enters through the following classical example.

EXAMPLE. Let $\mathfrak{g}_0 = \mathfrak{so}(3)$, $\mathfrak{g} = \mathfrak{so}(3,\mathbb{C}) \cong \mathfrak{sl}(2,\mathbb{C})$, $K = SO(3)$, and $G = SO(3,\mathbb{C})$. We can identify $\mathfrak{so}(3)$ with $\mathbb{R}^3$ by $\begin{pmatrix} 0 & a & c \\ -a & 0 & b \\ -c & -b & 0 \end{pmatrix} \leftrightarrow (a,b,c)$. The algebra $S(\mathfrak{g}^*)$ may be regarded as the algebra of complex-valued polynomials on $\mathbb{R}^3$, and $\mathcal{I}'$ consists of all polynomials in $|x|^2$. We denote by $S(\mathfrak{g}_0^*)$ the real algebra of all real-valued polynomials on $\mathbb{R}^3$; this is a real form of $S(\mathfrak{g}^*)$. To any $p \in S(\mathfrak{g}^*)$ we make correspond a differential operator $\partial(p)$ with constant coefficients by requiring that $\partial(\cdot)$ is linear and that

$$\partial(x_1^{k_1} x_2^{k_2} x_3^{k_3}) = \frac{\partial^{k_1+k_2+k_3}}{\partial x_1^{k_1} \partial x_2^{k_2} \partial x_3^{k_3}}.$$

If p and q are in $S^d(\mathfrak{g}^*)$, then $\partial(q)p$ is a constant polynomial, and we define $\langle p, q \rangle$ to be that constant. On $S^d(\mathfrak{g}^*) \cap S(\mathfrak{g}_0^*)$, $\langle \cdot, \cdot \rangle$ is a K invariant inner product with the monomials as an orthogonal basis; hence $\langle \cdot, \cdot \rangle$ is an invariant nondegenerate symmetric bilinear form on $S^d(\mathfrak{g}^*)$. A member p of $S(\mathfrak{g}^*)$ is **harmonic** if $\partial(|x^2|)p = 0$, i.e., if

$$\frac{\partial^2}{\partial x_1^2} + \frac{\partial^2}{\partial x_2^2} + \frac{\partial^2}{\partial x_3^2}$$

acts as 0 on p. In this case we write $p \in \mathcal{H}'$. We let $\mathcal{H}'_d$ be the space of harmonic polynomials homogeneous of degree d; $\mathcal{H}'_d$ is a K invariant subspace of $S(\mathfrak{g}^*)$, and $\mathcal{H}'_d \cap S(\mathfrak{g}_0^*)$ is a real form of it. Let us establish an orthogonal direct-sum decomposition

(7.62a) $$S^d(\mathfrak{g}_0^*) = (\mathcal{H}'_d \cap S(\mathfrak{g}_0^*)) \oplus |x|^2 S^{d-2}(\mathfrak{g}_0^*),$$

which, because of the identification $\mathcal{I}' = \mathbb{C}[|x|^2]$, then yields

(7.62b) $$S(\mathfrak{g}^*) = \mathcal{H}' \oplus S(\mathfrak{g}^*)\mathcal{I}'_+.$$

(In fact, if p is in $S^d(\mathfrak{g}_0^*)$ and q is in $S^{d-2}(\mathfrak{g}_0^*)$, then

$$\langle \partial(|x|^2)p, q \rangle = \partial(q)\partial(|x|^2)p = \partial(|x|^2 q)p = \langle p, |x|^2 q \rangle.$$

If p is harmonic, this shows that p is orthogonal to $|x|^2 S^{d-2}(\mathfrak{g}_0^*)$. Conversely if p is orthogonal to $|x|^2 S^{d-2}(\mathfrak{g}_0^*)$, then $\partial(|x|^2)p = 0$, as we see by taking $q = \partial(|x|^2)p$.) Iterating (7.62a), we see that each member p of $S^d(\mathfrak{g}_0^*)$ decomposes uniquely as

(7.63a) $$p = h_d + |x|^2 h_{d-2} + |x|^4 h_{d-4} + \cdots$$

with $h_d, h_{d-2}, h_{d-4}, \ldots$ homogeneous harmonic of the indicated degrees. Again taking into account the identification $\mathcal{I}' = \mathbb{C}[|x|^2]$, we can rewrite (7.63a) as

(7.63b) $$S(\mathfrak{g})^* \cong \mathcal{H}' \otimes_{\mathbb{C}} \mathcal{I}',$$

the isomorphism being given by multiplication.

For the generalization in §6 to other $\mathfrak{g}$, the passage from (7.62) to (7.63) will be much more subtle because of the presence of multiple generators of $\mathcal{I}'$. In working with (7.62), we shall need a characterization of $S(\mathfrak{g}^*)\mathcal{I}'_+$; it will not be enough merely to identify $\mathcal{H}'$ as the kernel of $\partial(\mathcal{I}'_+)$. If $X = \begin{pmatrix} 0 & a & c \\ -a & 0 & b \\ -c & -b & 0 \end{pmatrix}$ is a member of $\mathfrak{g}$, we readily check that its characteristic polynomial is

$$\lambda^3 + (a^2 + b^2 + c^2)\lambda.$$

Hence X is a nilpotent matrix if and only if $a^2 + b^2 + c^2 = 0$. If we regard $a^2 + b^2 + c^2$ as the holomorphic extension to $\mathfrak{g}$ of the polynomial on $\mathfrak{g}_0$ that we have been calling $|x|^2$, we see that $|x|^2$ vanishes on X in $\mathfrak{g}$ if and only if X is nilpotent. Thus any polynomial $|x|^2 q$ in $S(\mathfrak{g}^*)$ vanishes on all nilpotents. Conversely the ideal of all polynomials vanishing on all nilpotents is the principal ideal $(|x|^2)$ since $|x|^2$ is prime; this conclusion uses the Hilbert Nullstellensatz. In other words, $S(\mathfrak{g}^*)\mathcal{I}'_+$ is characterized as the ideal of all polynomials in $S(\mathfrak{g}^*)$ vanishing on all nilpotents.

The second approach to the above example, via nilpotents, is how we shall proceed. We shall take the following definitions and facts as known. The Lie algebra $\mathfrak{g}$ continues to be assumed complex semisimple.

(7.64a)   For X in $\mathfrak{g}$, let $\mathfrak{g}_{0,X}$ be the generalized eigenspace in $\mathfrak{g}$ for the linear transformation ad X and the eigenvalue 0. The minimum dimension of $\mathfrak{g}_{0,X}$, for X in $\mathfrak{g}$, is the **rank** l of $\mathfrak{g}$. An element X for which dim $\mathfrak{g}_{0,X} = l$ is called **generic**. The set of generic elements is nonempty open dense.

(7.64b)   If X in $\mathfrak{g}$ is generic, then $\mathfrak{g}_{0,X}$ is a Cartan subalgebra of $\mathfrak{g}$.

(7.64c)   For X in $\mathfrak{g}$, let $\mathfrak{g}^X$ be the centralizer of X in $\mathfrak{g}$. The minimum dimension of $\mathfrak{g}^X$, for X in $\mathfrak{g}$, is the **index** of $\mathfrak{g}$. An element X for which dim $\mathfrak{g}^X$ is the index is called **regular**. The set of regular elements is nonempty open dense.

(7.64d)   For any X, $\mathfrak{g}^X \subseteq \mathfrak{g}_{0,X}$. Equality holds if X is generic. Hence generic elements are regular, and the rank equals the index.

(7.64e)   Each Cartan subalgebra of $\mathfrak{g}$ is of the form $\mathfrak{g}_{0,X}$ for some X in $\mathfrak{g}$. Any two Cartan subalgebras are conjugate via $\mathrm{Ad}(G)$.

(7.64f)   Let $\mathfrak{g} = \mathfrak{n}^- \oplus \mathfrak{h} \oplus \mathfrak{n}$ be the triangular decomposition of $\mathfrak{g}$ obtained by fixing a positive system of roots $\Delta^+(\mathfrak{g}, \mathfrak{h})$ and taking $\mathfrak{n}$ to be the sum of the root spaces for the positive roots. Let N^-, H, and N be the corresponding analytic subgroups of G. Then $N^- H N$ is an open dense subset of G. If K is a maximal compact subgroup of G (and therefore also a compact form of G), then $G = KHN$.

Now let us return to $S(\mathfrak{g}^*)$. Again we may regard members of $S(\mathfrak{g}^*)$ as holomorphic polynomials from $\mathfrak{g}$ into $\mathbb{C}$. As such, they may be restricted to any subspace of $\mathfrak{g}$ to yield holomorphic polynomials on that subspace.

Theorem 7.65 (Chevalley). If $\mathfrak{h}$ is a Cartan subalgebra of $\mathfrak{g}$ and if $W = W(\mathfrak{g}, \mathfrak{h})$ is the Weyl group, then the restriction map

$$i : S(\mathfrak{g}^*) \to S(\mathfrak{h}^*)$$

yields an algebra isomorphism of $S(\mathfrak{g}^*)^G$ onto the algebra $S(\mathfrak{h}^*)^W$ of W invariants in $S(\mathfrak{h}^*)$. Moreover, for any $d > 0$, $S^d(\mathfrak{g}^*)^G$ is the linear span of all functions $X \mapsto \mathrm{Tr}(\pi(X)^d)$, where π varies over all finite-dimensional (complex-linear) representations of $\mathfrak{g}$.

REMARKS. If $\mathfrak{h}_0$ denotes the real subspace of $\mathfrak{h}$ on which all roots are real, then Theorem 7.30 is applicable to the action of W on $\mathfrak{h}_0$. That theorem tells the structure of $S(\mathfrak{h}^*)^W$ and, when combined with Theorem 7.65, tells the structure of $S(\mathfrak{g}^*)^G$. Since the number of reflections in W equals half the number of roots, relations (7.31) become

(7.66)
$$\sum_{j=1}^{l} d_j = \tfrac{1}{2}(l + \dim \mathfrak{g})$$

$$\prod_{j=1}^{l} d_j = |W|.$$

PROOF. Without loss of generality, we may assume that G is simply connected. If g is in G, then $\pi(g)$ is defined and

$$\mathrm{Tr}(\pi(\mathrm{Ad}(g)X)^d) = \mathrm{Tr}((\pi(g)\pi(X)\pi(G)^{-1})^d)$$
$$= \mathrm{Tr}(\pi(g)\pi(X)^d\pi(g)^{-1}) = \mathrm{Tr}(\pi(X)^d).$$

Hence

(7.67) $X \mapsto \mathrm{Tr}(\pi(X)^d)$ is in $S^d(\mathfrak{g}^*)^G$.

Also (4.110) shows that

(7.68)   the linear span of all polynomials $X \mapsto \mathrm{Tr}(\pi(X)^d)$ on $\mathfrak{h}$ equals $S(\mathfrak{h}^*)^W$.

By (7.68), i is an algebra homomorphism of $S(\mathfrak{g}^*)^G$ into $S(\mathfrak{h}^*)^W$. To see that i is one-one, let $p \in S(\mathfrak{g}^*)^G$ have $i(p) = 0$. Then p vanishes on $\mathrm{Ad}(G)\mathfrak{h}$. This set is dense in $\mathfrak{g}$ by (7.64abe), and hence $p = 0$.

To see that i is onto, let L be the linear span of all polynomials $X \mapsto \mathrm{Tr}(\pi(X)^d)$ on $\mathfrak{g}$. Then $L \subseteq S^d(\mathfrak{g}^*)^G$ by (7.67), and (7.68) gives

$$S(\mathfrak{h}^*)^W = i(L) \subseteq i(S^d(\mathfrak{g}^*)^G) \subseteq S(\mathfrak{h}^*)^W.$$

Hence $i(S^*(\mathfrak{g}^*)^G) = S(\mathfrak{h}^*)^W$. Also $i(L) = i(S^d(\mathfrak{g}^*)^G)$, and we conclude that $L = S^d(\mathfrak{g}^*)^G$, since i is one-one. This completes the proof.

The remainder of this section translates Theorem 7.65 into a result about $S(\mathfrak{g})$. We shall not need Theorem 7.71 in this book, but we include it since it is usually proved in the course of establishing the Harish-Chandra isomorphism of §IV.7.

Lemma 7.69. Fix a positive system of roots $\Delta^+(\mathfrak{g}, \mathfrak{h})$, and introduce $\mathfrak{n}$ and $\mathfrak{n}^-$ relative to it. Let α and β be the isomorphisms

$$\alpha : S(\mathfrak{g}) \to S(\mathfrak{g}^*) \qquad \text{and} \qquad \beta : S(\mathfrak{h}) \to S(\mathfrak{h}^*)$$

induced by the Killing form of $\mathfrak{g}$. Let $j : S(\mathfrak{g}) \to S(\mathfrak{h})$ be the projection according to the decomposition

$$(7.70) \qquad\qquad S(\mathfrak{g}) = S(\mathfrak{h}) \oplus (S(\mathfrak{g})\mathfrak{n} + S(\mathfrak{g})\mathfrak{n}^-),$$

and let $i : S(\mathfrak{g}^*) \to S(\mathfrak{h}^*)$ be the restriction homomorphism. Then $i \circ \alpha = \beta \circ j$ on $S(\mathfrak{g})$.

REMARKS. The map j is an analog for symmetric algebras of the $\gamma'_{\mathfrak{n}}$ factor in the Harish-Chandra homomorphism for universal enveloping algebras. The above direct-sum decomposition of $S(\mathfrak{g})$ is a version of Proposition 4.89a, but the proof of (7.70) is immediate because of the commutativity of $S(\mathfrak{g})$. The projection j is obviously an algebra homomorphism.

PROOF. Since α, β, i, and j are algebra homomorphisms, it suffices to prove that $i(\alpha(X)) = \beta(j(X))$ for $X \in \mathfrak{g}$. For arbitrary $h \in \mathfrak{h}$, we have

$$
\begin{aligned}
\langle i(\alpha(X)), h \rangle &= \langle \alpha(X), h \rangle \\
&= B(X, h) \\
&= B(j(X), h) \qquad \text{since } B(\mathfrak{n} + \mathfrak{n}^-, \mathfrak{h}) = 0 \\
&= \langle \beta(j(X)), h \rangle.
\end{aligned}
$$

Thus $i(\alpha(X)) = \beta(j(X))$ as required.

Theorem 7.71 (Chevalley). Fix a positive system of roots $\Delta^+(\mathfrak{g}, \mathfrak{h})$, and introduce $\mathfrak{n}$ and $\mathfrak{n}^-$ relative to it. Let $j : S(\mathfrak{g}) \to S(\mathfrak{h})$ be the projection according to the decomposition (7.70). Then j restricts to an algebra isomorphism of $S(\mathfrak{g})^G$ onto $S(\mathfrak{h})^W$.

PROOF. The restriction $i : S(\mathfrak{g}^*)^G \to S(\mathfrak{h}^*)^W$ is one-one by Theorem 7.65. Since Lemma 7.69 gives $j = \beta^{-1} \circ i \circ \alpha$, it follows that j is one-one. The map α is G equivariant and carries $S(\mathfrak{g})^G$ onto $S(\mathfrak{g}^*)^G$, and Theorem 7.65 says that i carries $S(\mathfrak{g}^*)^G$ onto $S(\mathfrak{h}^*)^W$. The map β is W equivariant, and β^{-1} carries $S(\mathfrak{h}^*)^W$ onto $S(\mathfrak{h})^W$. Thus $j = \beta^{-1} \circ i \circ \alpha$ carries $S(\mathfrak{g})^G$ onto $S(\mathfrak{h})^W$.

6. Kostant's Theory of Harmonics

We continue with notation as in §5. Our goal in this section is to obtain a tensor-product decomposition $S(\mathfrak{g})^* \cong \mathcal{H}' \otimes_{\mathbb{C}} \mathcal{I}'$ generalizing (7.63b) and to use it to exhibit $U(\mathfrak{g})$ as a free $Z(\mathfrak{g})$ module. Our restriction of the setting to semisimple $\mathfrak{g}$ avoids inessential complications in the proofs. After the statement of Theorem 7.118 in §7, we shall remark on what happens when $\mathfrak{g}$ is reductive.

The decomposition generalizing (7.62b) is a straightforward extension of the example in §5, and we start with that. To any $p \in S(\mathfrak{g}^*)$ we make correspond a differential operator $\partial(p)$ on $S(\mathfrak{g}^*)$ with constant coefficients by requiring that

(7.72)

  (a) $\partial(\lambda)\lambda' = -\lambda'(X_\lambda) = -\lambda(X_{\lambda'}) = -B(X_\lambda, X_{\lambda'})$,
       where X_λ and $X_{\lambda'}$ are the members of $\mathfrak{g}$ that
       correspond to λ and λ' in $\mathfrak{g}^*$

  (b) $q \mapsto \partial(\lambda)q$ is a derivation if λ is in $\mathfrak{g}^* = S^1(\mathfrak{g}^*)$

  (c) $p \mapsto \partial(p)$ is an algebra homomorphism.

If p and q are in $S^d(\mathfrak{g}^*)$, then $\partial(q)p$ is a constant polynomial, and we define $\langle p, q \rangle$ to be that constant.

Lemma 7.73. Let $\mathfrak{g}_0$ be a compact form of $\mathfrak{g}$, and let K be the corresponding (compact) analytic subgroup of G. Let $\mathfrak{g}_0^*$ be the real subspace of elements of $\mathfrak{g}^*$ that are real-valued on $\mathfrak{g}_0$, so that $S^d(\mathfrak{g}_0^*)$ is a real form of $S^d(\mathfrak{g}^*)$. Then

  (a) $\partial(gq)(gp) = g(\partial(q)p)$ for $g \in G$ and $p, q \in S(\mathfrak{g}^*)$
  (b) $\partial(q)p$ is in $S(\mathfrak{g}_0^*)$ if p and q are in $S(\mathfrak{g}_0^*)$
  (c) $\langle \cdot, \cdot \rangle$ is an inner product on $S^d(\mathfrak{g}_0^*)$
  (d) $\mathcal{I}' \cap S(\mathfrak{g}_0^*)$ is a real form of $\mathcal{I}'$
  (e) $\langle p, rq \rangle = \langle \partial(r)p, q \rangle$ if p, r, and q are homogeneous in $S(\mathfrak{g}^*)$ of respective degrees d, k, and $d - k$.

PROOF.
  (a) The conclusion is valid if p and q are in $\mathfrak{g}^*$, by (7.72a). It extends to be valid for general $p \in S(\mathfrak{g}^*)$ by (7.72b) and then for general $q \in S(\mathfrak{g}^*)$ by (7.72c).

  (b, c) The Killing form B is negative definite on $\mathfrak{g}_0$. Choose an orthonormal basis $X_1, \ldots, X_n$ of $\mathfrak{g}_0$ relative to $-B$, and let $\lambda_1, \ldots, \lambda_n$ be the corresponding elements of $\mathfrak{g}_0^*$. The monomials

$$\lambda_1^{j_1} \cdots \lambda_n^{j_n} \qquad \text{with } j_1 + \cdots + j_n = d$$

form a basis of $S^d(\mathfrak{g}_0^*)$, and the properties (7.72) force

(7.74)
$$\langle \lambda_1^{i_1} \cdots \lambda_n^{i_n}, \lambda_1^{j_1} \cdots \lambda_n^{j_n} \rangle = \begin{cases} j_1! \cdots j_n! & \text{if } i_1 = j_1, \ldots, i_n = j_n \\ 0 & \text{otherwise.} \end{cases}$$

Then (b) and (c) follow.

(d) Let p be in $\mathcal{I}' \cap S^d(\mathfrak{g}^*)$, and write $p = p_1 + ip_2$ with p_1 and p_2 in $S^d(\mathfrak{g}_0^*)$. Since K maps $S^d(\mathfrak{g}_0^*)$ into itself, p_1 and p_2 are K invariant. They are thus K invariant as members of $S^d(\mathfrak{g}^*)$. Passing to $\mathfrak{g}_0$, then to $\mathfrak{g}$, and finally to G, we see that p_1 and p_2 are G invariant as members of $S^d(\mathfrak{g}^*)$.

(e) We have $\langle p, rq \rangle = \partial(rq)p = \partial(q)\partial(r)p = \langle \partial(r)p, q \rangle$. This completes the proof.

A member p of $S(\mathfrak{g}^*)$ is **harmonic** if $\partial(\mathcal{I}_+')p = 0$. The subspace of harmonics is denoted $\mathcal{H}'$. This is a graded subspace, and it is invariant under G by Lemma 7.73. Let $\mathcal{H}_d' = \mathcal{H}' \cap S^d(\mathfrak{g}^*)$. By (a) and (c) of the lemma, $\mathcal{H}_d' \cap S^d(\mathfrak{g}_0^*)$ is a real form of $\mathcal{H}_d'$.

Proposition 7.75. $S(\mathfrak{g}^*) = \mathcal{H}' \oplus \mathcal{I}_+'S(\mathfrak{g}^*)$.

PROOF. The subspaces in question are graded, and we consider their intersections with $S^d(\mathfrak{g}^*)$. Lemma 7.73 shows that it is enough to prove that

(7.76) $S^d(\mathfrak{g}_0^*) = (\mathcal{H}' \cap S^d(\mathfrak{g}_0^*)) \oplus (\mathcal{I}_+'S(\mathfrak{g}^*) \cap S^d(\mathfrak{g}_0^*))$.

We shall show that the two spaces on the right are orthogonal complements in $S^d(\mathfrak{g}_0^*)$ relative to $\langle \cdot, \cdot \rangle$, which is an inner product by (b) of the lemma. Let p, r, and q be homogeneous in $S(\mathfrak{g}_0^*)$ of respective degrees $d, k > 0$, and $d - k$, and suppose that r, as a member of $S(\mathfrak{g}^*)$, is G invariant. By (d) of the lemma, we have $\langle p, rq \rangle = \langle \partial(r)p, q \rangle$. If p is harmonic, we obtain $\langle p, rq \rangle = 0$; hence the two spaces on the right of (7.76) are orthogonal. Conversely if p is orthogonal to the second space, we obtain $\langle \partial(r)p, q \rangle = 0$; taking $q = \partial(r)p$, we see that $\partial(r)p = 0$ and p is harmonic.

Corollary 7.77. $S(\mathfrak{g}^*) = \mathcal{H}'\mathcal{I}'$ in the sense that every member of $S(\mathfrak{g}^*)$ is a sum of products of members of $\mathcal{H}'$ by members of $\mathcal{I}'$.

PROOF. The argument is based on Proposition 7.75 and proceeds in the same way as the proof of the corresponding part of Theorem 7.30b.

We do not automatically get the appropriate independence result that would allow us to conclude that $S(\mathfrak{g}^*) \cong \mathcal{H}' \otimes_{\mathbb{C}} \mathcal{I}'$. To get this stronger result, we shall examine the ideal $\mathcal{I}'_+ S(\mathfrak{g}^*)$ in some detail. The desired tensor-product relation will follow soon after we show in Proposition 7.106 that the ideal $\mathcal{I}'_+ S(\mathfrak{g}^*)$ is prime.

In the semisimple Lie algebra $\mathfrak{g}$, an element X is called **nilpotent** if it satisfies the equivalent conditions of Proposition 7.78 below. Because of (a) in the proposition, the set $\mathfrak{N}$ of nilpotent elements is an affine algebraic subset of $\mathfrak{g}$, being the set where the characteristic polynomial of ad X reduces to $\lambda^{\dim \mathfrak{g}}$.

Proposition 7.78. If X is in the semisimple Lie algebra $\mathfrak{g}$, then the following conditions on X are equivalent:

(a) ad X is a nilpotent endomorphism of $\mathfrak{g}$
(b) $\pi(X)$ is a nilpotent endomorphism for some one-one finite-dimensional representation π of $\mathfrak{g}$
(c) $\pi'(X)$ is a nilpotent endomorphism for every finite-dimensional representation π' of $\mathfrak{g}$.

PROOF. In (b), let π act on V. We may drop the π, regarding $\mathfrak{g}$ as a subalgebra of $\mathfrak{gl}(V)$. Let us prove that (a) and (b) are equivalent. We begin with the identity

$$(7.79) \qquad (\operatorname{ad} X)^n Y = \sum_{k=0}^{n} (-1)^k \binom{n}{k} X^{n-k} Y X^k.$$

Suppose (b) holds, so that $X^m = 0$ for some m. As soon as $n \geq 2m$, every term on the right side of (7.79) is 0; hence ad X is nilpotent, and (a) holds.

Conversely suppose (a) holds, so that ad X is nilpotent on $\mathfrak{g}$. We are to show that X is nilpotent on V. Since $\mathfrak{g}$ acts fully reducibly on V, write $V = \bigoplus_i V_i$ with each V_i invariant and irreducible under $\mathfrak{g}$. By the Jordan decomposition in $\mathfrak{gl}(V)$, write

$$(7.80) \qquad X = s + n$$

with $s \in \mathfrak{gl}(V)$ diagonable, $n \in \mathfrak{gl}(V)$ nilpotent, and $sn = ns$. We shall prove that $s = 0$. Here X carries each V_i to itself, and so do s and n. Since X is in $\mathfrak{g}$, it is a sum of brackets and therefore has trace 0 in its action on each V_i. Also n is nilpotent and has trace 0. Therefore

$$(7.81) \qquad \operatorname{Tr}_{V_i} s = 0 \qquad \text{for all } i.$$

Meanwhile, from (7.80) we have

(7.82) $\mathrm{ad}_{\mathfrak{gl}(V)}\, X = \mathrm{ad}_{\mathfrak{gl}(V)}\, s + \mathrm{ad}_{\mathfrak{gl}(V)}\, n,$

and $\mathrm{ad}_{\mathfrak{gl}(V)}\, n$ is nilpotent by (7.79). Also $\mathrm{ad}_{\mathfrak{gl}(V)}\, s$ is semisimple since if s has eigenvalues s_i, then $\mathrm{ad}_{\mathfrak{gl}(V)}\, s$ acts diagonably with eigenvalues $s_i - s_j$. Moreover,

$$[\mathrm{ad}_{\mathfrak{gl}(V)}\, s,\; \mathrm{ad}_{\mathfrak{gl}(V)}\, n] = \mathrm{ad}_{\mathfrak{gl}(V)}\, [s, n] = 0,$$

so that (7.82) is the Jordan decomposition of $\mathrm{ad}_{\mathfrak{gl}(V)}\, X$.

Let $N'(\mathfrak{g})$ be the subset of the normalizer of $\mathfrak{g}$ in $\mathfrak{gl}(V)$ given by

$$N'(\mathfrak{g}) = \{Z \in \mathfrak{gl}(V) \mid [Z, \mathfrak{g}] \subseteq \mathfrak{g} \text{ and } Z(V_i) \subseteq V_i \text{ for all } i\}.$$

This is a Lie subalgebra of $\mathfrak{gl}(V)$, and $\mathfrak{g}$ is an ideal in $N'(\mathfrak{g})$. Since every finite-dimensional representation of $\mathfrak{g}$ is fully reducible, $N'(\mathfrak{g}) = \mathfrak{g} \oplus \mathfrak{g}'$ with $[\mathfrak{g}, \mathfrak{g}'] \subseteq \mathfrak{g}'$. On the other hand, $[\mathfrak{g}, \mathfrak{g}'] \subseteq \mathfrak{g}$ since $\mathfrak{g}$ is an ideal in $N'(\mathfrak{g})$. Thus $[\mathfrak{g}, \mathfrak{g}'] = 0$. From the Jordan decomposition (7.82), it follows that $\mathrm{ad}_{\mathfrak{gl}(V)}\, s$ is a polynomial in $\mathrm{ad}_{\mathfrak{gl}(V)}\, X$ and hence carries $\mathfrak{g}$ into itself. Thus s is in $N'(\mathfrak{g})$, and so is n. Consequently

(7.83) $\mathrm{ad}_{N'(\mathfrak{g})}\, X = \mathrm{ad}_{N'(\mathfrak{g})}\, s + \mathrm{ad}_{N'(\mathfrak{g})}\, n.$

The left side carries $\mathfrak{g}$ into itself and acts as 0 on $\mathfrak{g}'$ since $[\mathfrak{g}, \mathfrak{g}'] = 0$. Since $\mathrm{ad}_{\mathfrak{g}}\, X$ is nilpotent, $\mathrm{ad}_{N'(\mathfrak{g})}\, X$ is then nilpotent. By the uniqueness of the Jordan decomposition, $\mathrm{ad}_{N'(\mathfrak{g})}\, s = 0$. Therefore s is in the center of $N'(\mathfrak{g})$. Since $\mathfrak{g}$ acts irreducibly on each V_i, so does $N'(\mathfrak{g})$. By Schur's Lemma, s is scalar on each V_i. Then (7.81) shows that $s = 0$, and (b) holds.

In other words, (a) and (b) are equivalent. Since (c) clearly implies (b), it remains to prove that (b) implies (c). Let $\pi'(X)$ act on V', and put $\mathfrak{g}' = \mathfrak{g}/\ker \pi'$. We can regard $\mathfrak{g}'$ as a subalgebra of $\mathfrak{gl}(V')$ via π'. Since X is nilpotent on V, the implication (b) $\Rightarrow$ (a) shows that $\mathrm{ad}_{\mathfrak{g}}\, X$ is nilpotent on $\mathfrak{g}$. Hence $\mathrm{ad}_{\mathfrak{g}'}\, \pi'(X)$ is nilpotent on $\pi'(\mathfrak{g}) = \mathfrak{g}'$, and the implication (a) $\Rightarrow$ (b) shows that $\pi'(X)$ is nilpotent on V'. Thus (b) implies (c).

Let $\mathfrak{h}'$ be a Cartan subalgebra of $\mathfrak{g}$. If $\Delta^+(\mathfrak{g}, \mathfrak{h}')$ is a system of positive roots and $\mathfrak{n}'$ is the sum of the root spaces for the positive roots, then $\mathfrak{h}' \oplus \mathfrak{n}'$ is the most general Borel subalgebra of $\mathfrak{g}$. If $\mathfrak{h} \oplus \mathfrak{n}$ is a fixed Borel subalgebra, then $\mathfrak{h}'$ is conjugate to $\mathfrak{h}$ via $\mathrm{Ad}(G)$, and the conjugation carries $\mathfrak{n}'$ to the version $\mathfrak{n}''$ of $\mathfrak{n}$ appropriate to some system of positive roots for $\Delta(\mathfrak{g}, \mathfrak{h})$ that is possibly different from $\Delta^+(\mathfrak{g}, \mathfrak{h})$. By Theorems 4.8 and 4.10, $\mathfrak{n}''$ is conjugate to $\mathfrak{n}$ by a member of $\mathrm{Ad}(G)$ normalizing $\mathfrak{h}$. We conclude that

(7.84)   Any two Borel subalgebras of $\mathfrak{g}$ are conjugate via $\mathrm{Ad}(G)$.

Consequently we shall fix a single Borel subalgebra $\mathfrak{h} \oplus \mathfrak{n}$ and refer other Borel subalgebras to it by conjugacy.

Proposition 7.85. Any element of $\mathfrak{g}$ is contained in a Borel subalgebra.

PROOF. Let X be given. We shall prove that X is conjugate to a member of $\mathfrak{h} \oplus \mathfrak{n}$. By (7.64a), we can choose a sequence X_n of generic elements with $X_n \to X$. By (7.64b) and (7.64e), there exist elements g_n in G with $\mathrm{Ad}(g_n)X_n$ in $\mathfrak{h}$. By (7.64f) we can write $g_n = b_n k_n$ with $b_n \in HN$ and $k_n \in K$. As a result, $\mathrm{Ad}(k_n)X_n$ is in $\mathfrak{b}$. Since K is compact, we can select a subsequence $\{k_{n_j}\}$ of $\{k_n\}$ convergent to some $k \in K$. Then $\mathrm{Ad}(k_{n_j})X_{n_j}$ converges to $\mathrm{Ad}(k)X$, and we conclude that $\mathrm{Ad}(k)X$ is in $\mathfrak{b}$.

Corollary 7.86. Fix a system $\Delta^+(\mathfrak{g}, \mathfrak{h})$ of positive roots, and introduce $\mathfrak{n}$ relative to it. Then $\mathrm{Ad}(G)\mathfrak{n} = \mathfrak{N}$.

PROOF. Since $\mathrm{ad}(\mathrm{Ad}(g)X) = \mathrm{Ad}(g)(\mathrm{ad}\,X)\mathrm{Ad}(g)^{-1}$, it follows that $\mathrm{Ad}(G)$ carries $\mathfrak{n}$ into $\mathfrak{N}$. In the reverse direction, let $X \in \mathfrak{N}$ be given. By Proposition 7.85 we can find $g \in G$ with $\mathrm{Ad}(g)^{-1}X \in \mathfrak{h} \oplus \mathfrak{n}$. It is enough to prove that $\mathrm{Ad}(g)^{-1}X$ is actually in $\mathfrak{n}$. Assuming the contrary, write $\mathrm{Ad}(g)^{-1}X = h + n$ with $h \neq 0$. Choose α simple with $\alpha(h) \neq 0$. If Y is a nonzero root vector for α, then

$$(\mathrm{ad}(\mathrm{Ad}(g)^{-1}X))^k Y \equiv \alpha(h)^k Y \quad \mathrm{mod} \bigoplus_{\substack{\beta > 0 \\ \beta \neq \alpha}} \mathfrak{g}_\beta,$$

in contradiction to the nilpotence of $\mathrm{ad}(\mathrm{Ad}(g)^{-1}X)$. Hence $h = 0$, and $\mathrm{Ad}(g)^{-1}X$ is in $\mathfrak{n}$.

Lemma 7.87. Let l be the rank of $\mathfrak{g}$, let $\Delta^+(\mathfrak{g}, \mathfrak{h})$ be a positive system of roots, and let Π be the set of simple roots in $\Delta^+(\mathfrak{g}, \mathfrak{h})$. Define h to be the unique element of $\mathfrak{h}$ such that $\alpha(h) = 2$ for all $\alpha \in \Pi$, let H_α be the element of $\mathfrak{h}$ corresponding to $\alpha \in \mathfrak{h}^*$, and define constants a_α by $h = \sum_{\alpha \in \Pi} a_\alpha H_\alpha$. For each $\alpha \in \Pi$, let b_α and c_α be constants such that $b_\alpha c_\alpha = a_\alpha$, and choose root vectors $X_\alpha \in \mathfrak{g}_\alpha$ and $X_{-\alpha} \in \mathfrak{g}_{-\alpha}$ such that $[X_\alpha, X_{-\alpha}] = H_\alpha$. Define $e = \sum_{\alpha \in \Pi} b_\alpha X_\alpha$, $f = \sum_{\alpha \in \Pi} c_\alpha X_{-\alpha}$, and $\mathfrak{s} = \mathbb{C}e + \mathbb{C}h + \mathbb{C}f$.

(a) The elements e and f are regular. Also $[h, e] = 2e$, $[h, f] = -2f$, and $[e, f] = h$. Hence $\mathfrak{s} \cong \mathfrak{sl}(2, \mathbb{C})$.

(b) Let $\mathfrak{g} = \mathfrak{a}_1 \oplus \cdots \oplus \mathfrak{a}_n$ be the decomposition of $\mathfrak{g}$ under $\mathrm{ad}\,\mathfrak{s}$ into simple modules, and, without loss of generality, let $\dim \mathfrak{a}_i = r_i + 1$ with $r_1 \leq \cdots \leq r_n$. Then $n = l$.

(c) Let $p_1, \ldots, p_l$ be homogeneous generators of $S(\mathfrak{g}^*)^G$ that are algebraically independent and have respective degrees $d_1, \ldots, d_l$ with $d_1 \leq \cdots \leq d_l$. Then $d_j = 1 + \frac{1}{2}r_j$ for $1 \leq j \leq l$.

(d) The differentials of $p_1, \ldots, p_l$ are linearly independent at every point of $e + \mathfrak{g}^f$.

PROOF.

(a) Direct calculation gives

$$[h, e] = \sum_{\alpha \in \Pi} b_\alpha \alpha(h) X_\alpha = 2e,$$

$$[h, f] = \sum_{\alpha \in \Pi} c_\alpha(-\alpha)(h) X_\alpha = -2f,$$

$$[e, f] = \sum_{\alpha, \beta \in \Pi} b_\alpha c_\beta [X_\alpha, X_{-\beta}] = \sum_{\alpha \in \Pi} b_\alpha c_\alpha [X_\alpha, X_{-\alpha}] = \sum_{\alpha \in \Pi} a_\alpha H_\alpha = h.$$

Hence $\mathfrak{s} \cong \mathfrak{sl}(2, \mathbb{C})$. On any root vector X_β, $[h, X_\beta] = \beta(h) X_\beta$; here $\beta(h)$ is an even integer, since every root is an integral combination of simple roots. Hence every eigenvalue of $\operatorname{ad} h$ is an even integer. In each irreducible $\mathfrak{s}$ module of this kind, the 0 eigenspaces of h, e, and f are all one dimensional. Therefore

(7.88) $n = \dim \mathfrak{g}^h = \dim \mathfrak{g}^e = \dim \mathfrak{g}^f.$

Since no root vanishes on h, $\mathfrak{g}^h = \mathfrak{h}$. Thus h is generic, hence regular. By (7.88), e and f are regular.

(b) In (7.88), $n = \dim \mathfrak{g}^h = \dim \mathfrak{h} = l$.

(c) From the representation theory of $\mathfrak{sl}(2, \mathbb{C})$, we know that $\mathfrak{g} = [e, \mathfrak{g}] \oplus \mathfrak{g}^f$; moreover $\mathfrak{g}^f$ has a basis $y_1, \ldots, y_l$ such that $[h, y_i] = -r_i y_i$ for $1 \leq i \leq l$. Define a map $\psi : G \times \mathbb{C}^l \to \mathfrak{g}$ by

(7.89) $\psi(g, \xi_1, \ldots, \xi_l) = \operatorname{Ad}(g)(e + \xi_1 y_1 + \cdots + \xi_l y_l).$

Let $d\psi_{\xi_1, \ldots, \xi_l}$ be the differential of ψ at $(1, \xi_1, \ldots, \xi_l)$. Differentiating (7.89), we obtain

(7.90)
$d\psi_{\xi_1, \ldots, \xi_l}(z, \eta_1, \ldots, \eta_l) = (\operatorname{ad} z)(e + \xi_1 y_1 + \cdots + \xi_l y_l) + \eta_1 y_1 + \cdots + \eta_l y_l$

for $z \in \mathfrak{g}$ and $(\eta_1, \ldots, \eta_l) \in \mathbb{C}^l$. Therefore the image of $d\psi_{0, \ldots, 0}$ is $[\mathfrak{g}, e] + \mathbb{C} y_1 + \cdots + \mathbb{C} y_l = \mathfrak{g}$. Hence $\psi(G \times \mathbb{C}^l) = \operatorname{Ad}(G)(e + \mathfrak{g}^f)$ is an open neighborhood of e in $\mathfrak{g}$. Consequently the mapping $p \mapsto p|_{e+\mathfrak{g}^f}$ of $S(\mathfrak{g}^*)^G$ into the algebra of polynomial functions on $e + \mathfrak{g}^f$ is one-one. Later we shall use the following consequence of this fact:

(7.91)   For the homogeneous generators $p_1, \ldots, p_l$, the polynomials $p_i|_{e+\mathfrak{g}^f}$ on $e + \mathfrak{g}^f$ are algebraically independent.

Let p be an element of $S(\mathfrak{g}^*)^G$ homogeneous of degree d. Let us take $u = e + \xi_1 y_1 + \cdots + \xi_l y_l$ and rewrite u by (7.90) as

$$
\begin{aligned}
u &= (e - \tfrac{1}{2}\xi_1 r_1 y_1 - \cdots - \tfrac{1}{2}\xi_l r_l y_l) + (1 + \tfrac{1}{2}r_1)\xi_1 y_1 + \cdots + (1 + \tfrac{1}{2}r_l)\xi_l y_l \\
&= \tfrac{1}{2}(\operatorname{ad} h)(e + \xi_1 y_1 + \cdots + \xi_l y_l) + (1 + \tfrac{1}{2}r_1)\xi_1 y_1 + \cdots + (1 + \tfrac{1}{2}r_l)\xi_l y_l \\
&= d\psi_{\xi_1,\ldots,\xi_l}(\tfrac{1}{2}h, \ (1 + \tfrac{1}{2}r_1)\xi_1, \ldots, \ (1 + \tfrac{1}{2}r_l)\xi_l).
\end{aligned}
$$

Then

(7.92)
$$
\begin{aligned}
\frac{d}{dt}\, & p(u + tu)|_{t=0} \\
&= \partial(u)\,p(e + \xi_1 y_1 + \cdots + \xi_l y_l) \\
&= [d\psi_{\xi_1,\ldots,\xi_l}(\tfrac{1}{2}h, \ (1 + \tfrac{1}{2}r_1)\xi_1, \ldots, \ (1 + \tfrac{1}{2}r_l)\xi_l)\,p](e + \xi_1 y_1 + \cdots + \xi_l y_l) \\
&= [(\tfrac{1}{2}h, \ (1 + \tfrac{1}{2}r_1)\xi_1, \ldots, \ (1 + \tfrac{1}{2}r_l)\xi_l)(p \circ \psi)](1, \xi_1, \ldots, \xi_l).
\end{aligned}
$$

Define a polynomial q in l complex variables $\zeta_1, \ldots, \zeta_l$ by

$$
q(\zeta_1, \ldots, \zeta_l) = p(e + \zeta_1 y_1 + \cdots + \zeta_l y_l).
$$

The function $p \circ \psi$ is defined on $G \times \mathbb{C}^l$, is independent of the first variable because p is G invariant, and satisfies

$$
(p \circ \psi)(1, (\zeta_1, \ldots, \zeta_l)) = q(\zeta_1, \ldots, \zeta_l).
$$

From (7.92), we obtain

$$
\begin{aligned}
\frac{d}{dt}\, & p(u + tu)|_{t=0} \\
&= (1 + \tfrac{1}{2}r_1)\xi_1 \frac{\partial q}{\partial \xi_1}(\xi_1, \ldots, \xi_l) + \cdots + (1 + \tfrac{1}{2}r_l)\xi_l \frac{\partial q}{\partial \xi_l}(\xi_1, \ldots, \xi_l).
\end{aligned}
$$

Now let us apply Euler's identity (7.42), which we write in the form

(7.93)
$$
\frac{d}{dt}\, p(u + tu)|_{t=0} = d\, p(u).
$$

Combining (7.92) and (7.93), we obtain

(7.94)
$$
d\, q(\zeta_1, \ldots, \zeta_l) = \sum_{i=1}^{l} (1 + \tfrac{1}{2}r_i)\zeta_i \frac{\partial q}{\partial \zeta_i}(\zeta_1, \ldots, \zeta_l).
$$

In particular, let $q_1, \ldots, q_l$ be the polynomial functions obtained in this way from $p_1, \ldots, p_l$. What (7.94) says is that the polynomial q_j is a linear combination

$$q_j = \sum_k c_{jk} \zeta_1^{m_{1k}} \cdots \zeta_l^{m_{lk}}$$

of monomials in $\zeta_1, \ldots, \zeta_l$ with exponents such that

(7.95) $$\sum_{i=1}^{l} (1 + \tfrac{1}{2} r_i) m_{ik} = d_j.$$

Let us prove that

(7.96) $$1 + \tfrac{1}{2} r_j \leq d_j \qquad \text{for } 1 \leq j \leq r.$$

Arguing by contradiction, suppose that $d_{j_0} < 1 + \tfrac{1}{2} r_{j_0}$ for a certain j_0. Since $d_1 \leq \cdots \leq d_l$, we obtain $d_j < 1 + \tfrac{1}{2} r_{j_0}$ for $j \leq j_0$. Since $r_1 \leq \cdots \leq r_l$, (7.95) shows that $j \leq j_0 \leq i$ implies

$$(1 + \tfrac{1}{2} r_{j_0}) m_{ik} \leq (1 + \tfrac{1}{2} r_i) m_{ik} \leq d_j < 1 + \tfrac{1}{2} r_{j_0}.$$

Hence $m_{ik} = 0$ for $i \geq j_0$, and q_j depends only on $\zeta_1, \ldots, \zeta_{j_0-1}$. Therefore $q_1, \ldots, q_{j_0}$ are algebraically dependent, in contradiction to (7.91). Hence (7.96) follows. Consequently

$$
\begin{aligned}
d_1 + \cdots + d_l &\geq l + \tfrac{1}{2}(r_1 + \cdots + r_l) && \text{by (7.96)} \\
&= \tfrac{1}{2} l + \tfrac{1}{2}((r_1 + 1) + \cdots + (r_l + 1)) \\
&= \tfrac{1}{2}(l + \dim \mathfrak{g}) && \text{from (b)} \\
&= d_1 + \cdots + d_l && \text{by (7.66).}
\end{aligned}
$$

Thus the inequalities (7.96) are all equalities, and (c) is proved.

(d) Using that equality holds in (7.96), we can rewrite (7.95) as

(7.97) $$d_1 m_{1k} + \cdots + d_l m_{lk} = d_j.$$

If $j' > j$ is such that $d_{j'} > d_j$, then (7.97) forces $m_{j'k} = 0$. Let us rephrase this condition in terms of the distinct elements $\mu_1 < \cdots < \mu_p$ of the set $\{d_1, \ldots, d_l\}$. For $1 \leq s \leq p$, let C_s be the set of indices j such that $d_j = \mu_s$. Condition (7.97) proves, for $j \in C_s$, that

$$q_j(\zeta_1, \ldots, \zeta_l) = \sum_{j' \in C_s} \alpha_{jj'} \zeta_{j'} + q_j^*,$$

where the $\alpha_{jj'}$ are scalars and where q_j^* depends only on those ζ_i such that i is in $\cup_{t \leq s-1} C_t$. Let A_s be the matrix $(\alpha_{jj'})_{j, j' \in C_s}$. Then the Jacobian determinant of $q_1, \ldots, q_l$ with respect to $\zeta_1, \ldots, \zeta_l$ is $\prod_{s=1}^{p} \det(A_s)$. In particular it is a constant. Since (7.91) noted that $q_1, \ldots, q_l$ are algebraically independent, this constant is not zero. This proves (d).

Lemma 7.98. Let e be as in Lemma 7.87. For each root β, let X_β be a nonzero root vector for the root β. Then $\mathrm{Ad}(HN)e$ consists of all elements

$$e' = \sum_{\beta > 0} x_\beta X_\beta \qquad \text{with } x_\alpha \neq 0 \text{ for } \alpha \text{ simple.}$$

This is a dense subset of $\mathfrak{n}$, and consequently $\mathrm{Ad}(G)e$ is dense in $\mathfrak{N}$.

REMARK. The regular elements in $\mathfrak{N}$ are called **principal nilpotents**. One can show that $\mathrm{Ad}(G)e$ exhausts the principal nilpotents.

PROOF. If β is a positive root, we can expand β in terms of simple roots as $\beta = \sum_{\alpha \in \Pi} n_\alpha \alpha$, and we let $l(\beta) = \sum_{\alpha \in \Pi} n_\alpha$. Define $\mathfrak{n}_k$ to be the sum of the root spaces $\mathfrak{g}_\beta$ for all $\beta > 0$ with $l(\beta) = k$. Then e is in $\mathfrak{n}_1$ and $[\mathfrak{n}_i, \mathfrak{n}_j] \subseteq \mathfrak{n}_{i+j}$. We shall show that the map

$$(7.99) \qquad \mathrm{ad}\, e : \mathfrak{n}_k \to \mathfrak{n}_{k+1}$$

is onto. Thus consider

$$(7.100) \qquad \mathrm{ad}\, e : \bigoplus_{k \geq 1} \mathfrak{n}_k \to \bigoplus_{k \geq 2} \mathfrak{n}_k.$$

On all of $\mathfrak{g}$, $\mathrm{ad}\, e$ has an l-dimensional kernel, by Lemma 7.87a. Therefore the kernel of (7.100) has dimension $\leq l$, and we obtain

$$\dim(\text{domain}) = l + \dim\left(\bigoplus_{k \geq 2} \mathfrak{n}_k\right)$$

$$\geq l + \dim(\text{image})$$

$$\geq \dim(\text{kernel}) + \dim(\text{image}).$$

Since the left side and the right side are actually equal, we conclude that equality holds in both places where $\geq$ occurs. From the first of these conclusions, we see that (7.100) is onto. It follows that (7.99) is onto for every k.

Using (7.99), we shall show that

$$(7.101) \qquad \mathrm{Ad}(N)e = e + \bigoplus_{k \geq 2} \mathfrak{n}_k.$$

Let m be the maximum value of $l(\beta)$, and let $e + X_2 + X_3 + \cdots + X_m$ be given with $X_k \in \mathfrak{n}_k$. Since (7.99) is onto for $k = 1$, choose $Y_1 \in \mathfrak{n}_1$ with $[Y_1, e] = X_2$. Then

$$\mathrm{Ad}(\exp Y_1)e = e + (\mathrm{ad}\, Y_1)e + \tfrac{1}{2}(\mathrm{ad}\, Y_1)^2 e + \dots$$

$$= e + X_2 + X_3' + X_4' + \dots$$

with $X'_k \in \mathfrak{n}_k$. Since (7.99) is onto for $k = 2$, choose $Y_2 \in \mathfrak{n}_2$ with $[Y_2, e] = X_3 - X'_3$. Then

$$
\begin{aligned}
\mathrm{Ad}(\exp Y_2)(e + X_2 + X'_3 + \ldots) &= e + X_2 + X'_3 + \ldots \\
&\quad + [Y_2, e + X_2 + X'_3 + \ldots] \\
&\quad + \tfrac{1}{2}[Y_2, [Y_2, e + X_2 + X'_3 + \ldots]] + \ldots \\
&= e + X_2 + X'_3 + \cdots + (X_3 - X'_3) + \ldots \\
&= e + X_2 + X_3 + X''_4 + \ldots.
\end{aligned}
$$

Continuing in this way, we obtain

$$
\mathrm{Ad}(\exp Y_{m-1} \cdots \exp Y_2 \exp Y_1)e = e + X_2 + \cdots + X_m,
$$

and (7.101) follows.

Finally let $e' = \sum_{\alpha \in \Pi} x_\alpha X_\alpha + X_2 + \cdots + X_m$ be given with $x_\alpha \neq 0$ for each $\alpha \in \Pi$. We can choose $h' \in H$ so that $\mathrm{Ad}(h')e'$ is in the right side of (7.101), and then we can choose $n \in N$ to carry the result to e. Hence $\mathrm{Ad}(nh')e' = e$. Since $\mathrm{Ad}(HN)e$ is thus dense in $\mathfrak{n}$, it follows from Corollary 7.86 that $\mathrm{Ad}(G)e$ is dense in $\mathfrak{N}$.

Lemma 7.102. The ideal of all members of $S(\mathfrak{g}^*)$ vanishing on $\mathfrak{N}$ is prime.

PROOF. Let $p_1 p_2$ vanish on $\mathfrak{N}$. Define $\tilde{p}_1(g) = p_1(\mathrm{Ad}(g)e)$ and $\tilde{p}_2(g) = p_2(\mathrm{Ad}(g)e)$. Then $\tilde{p}_1$ and $\tilde{p}_2$ are holomorphic functions on G with $\tilde{p}_1 \tilde{p}_2 = 0$. Since G is a connected manifold, $\tilde{p}_1 = 0$ or $\tilde{p}_2 = 0$. Say $\tilde{p}_1 = 0$. Then p_1 vanishes on $\mathrm{Ad}(G)e$. Since Lemma 7.98 shows $\mathrm{Ad}(G)e$ to be dense in $\mathfrak{N}$, p_1 vanishes on $\mathfrak{N}$.

Lemma 7.103. Let S be the algebra of polynomials on $\mathbb{C}^n$, let I be the ideal generated by members $p_1, \ldots, p_l$ of S, let $V(I)$ be the locus of common zeros of I, and let $I(V(I))$ be the ideal of all polynomials vanishing on $V(I)$. Suppose that

(a) $I(V(I))$ is prime
(b) there is some point $y \in V(I)$ such that the differentials $(dp_i)_y$ are linearly independent for $1 \leq i \leq l$.

Then I is prime, and it coincides with $I(V(I))$.

REMARK. The proof uses more commutative algebra and algebraic geometry than do other results in this book. We shall use the two volumes of Zariski-Samuel and one section of Chevalley [1958] as references.

PROOF. Let S_y be the localization of S at y, and let M_y be the maximal ideal of S_y. Letting $x_1, \ldots, x_n$ be coordinates of $\mathbb{C}^n$ and renumbering them suitably, we can arrange by (b) that the differentials of $p_1, \ldots, p_l, x_{l+1}, \ldots, x_n$ are linearly independent at y. Write

$$y_j = \begin{cases} p_j & \text{for } 1 \leq j \leq l \\ x_j - x_j(y) & \text{for } l+1 \leq j \leq n. \end{cases}$$

For $f \in M_y$, use of the functions $x_j - x_j(y), 1 \leq j \leq n$, shows that the map $f \mapsto df_y$ carries M_y onto $\mathbb{C}^n$ and has kernel M_y^2. Hence $\dim M_y/M_y^2 = n$. Applying the same map to $y_1, \ldots, y_n$ we see that the classes of $y_1, \ldots, y_n$ generate a subspace of dimension n of M_y/M_y^2, hence all of M_y/M_y^2. By Proposition 1 of Chevalley [1958], 216, $y_1, \ldots, y_n$ generate S_y, i.e., they are "uniformizing variables"* at y in the sense of Definition 2 in Chevalley [1958], 217. By Proposition 3 of Chevalley [1958], 219, IS_y is a prime ideal in S_y and is the localization at y of the ideal $I(V')$ of a variety V' of dimension $n - l$, i.e., $IS_y = I(V')S_y$.

By the Hilbert Nullstellensatz, $I(V(I))$ is the radical of I. Therefore $I(V(I))S_y$ is the radical of IS_y. Since IS_y is prime, it equals its own radical, and we have

(7.104) $$I(V(I))S_y = IS_y.$$

Thus $I(V(I))S_y = I(V')S_y$. The ideal $I(V(I))$ is prime by (a), and $I(V')$ is prime by construction; both are contained in the ideal in S of polynomials vanishing at y. By the second Corollary 1 of Zariski-Samuel [1958], 224, $I(V(I)) = I(V')$. Therefore $V(I) = V'$, and we conclude that $\dim V(I) = n - l$. Hence $I(V(I))$ has dimension $n - l$ in the sense of Zariski-Samuel [1960], 90.

Let $I = \cap Q_i$ be the Lasker-Noether unique decomposition (see Zariski-Samuel [1958], §§IV.4–5) of I as the irredundant intersection of primary ideals, and let P_i be the radical of Q_i. Then the radical of I equals $\cap P_i$, so that $I(V(I)) = \cap P_i$. When we discard terms P_i to make this intersection irredundant, we must end up with $I(V(I)) = I(V(I))$ by uniqueness. Hence $I(V(I)) = P_{i_0}$ for some i_0 and $I(V(I)) \subsetneq P_i$ for all other i. From Zariski-Samuel [1960], 91, $\dim I(V(I)) > \dim P_i$ for $i \neq i_0$. In the sense of the definition of dimension of I given in Zariski-Samuel [1960], 196, we therefore have

(7.105) $$\dim I = \max \dim P_i = \dim I(V(I)) = n - l.$$

---

* "Uniformizing parameters" and "local parameters" are other terms that are used for the same notion.

We can now apply Macaulay's Theorem (Theorem 26, Zariski-Samuel [1960], 203), which says that an ideal I in $\mathbb{C}[X_1, \ldots, X_n]$ of dimension $n - l$ with l generators has all its primes P_i of the same dimension. Then only the i_0^{th} factor can be present in the irredundant expression $I = \cap Q_i$, and I is primary.

For a primary ideal J contained in the ideal of members of S vanishing at y, one has $J S_y \cap S = J$ by the second Corollary 1 of Zariski-Samuel [1958], 224. Using (7.104) and this identity with J equal to I and then $I(V(I))$, we obtain $I = I(V(I))$.

Proposition 7.106. $\mathcal{I}'_+ S(\mathfrak{g}^*)$ is a prime ideal in $S(\mathfrak{g}^*)$ and coincides with the ideal of all members of $S(\mathfrak{g}^*)$ vanishing on $\mathfrak{N}$.

PROOF. If X is in $\mathfrak{g}$, then

$$X \in \mathfrak{N} \iff \pi(X) \text{ is nilpotent for every finite-dimensional representation of } \mathfrak{g}, \text{ by Proposition 7.78}$$

$$\iff \operatorname{Tr}(\pi(X)^m) = 0 \text{ for every finite-dimensional representation of } \mathfrak{g} \text{ and every } m > 0, \text{ since the elementary symmetric polynomials in } k \text{ variables are in the algebra generated by all } (\lambda_1 + \cdots + \lambda_k)^m$$

$$\iff p(X) = 0 \text{ for all } p \in \mathcal{I}'_+, \text{ by Theorem 7.65.}$$

Hence $\mathfrak{N}$ is the zero locus of $\mathcal{I}'_+ S(\mathfrak{g}^*)$. In the notation of Lemma 7.103, we can now take S to be $S(\mathfrak{g}^*)$ and $p_1, \ldots, p_l$ to be as in Lemma 7.87c. Then $I = \mathcal{I}'_+ S(\mathfrak{g}^*)$, and we have just seen that $V(I) = \mathfrak{N}$. Lemma 7.102 shows that condition (a) is satisfied in Lemma 7.103, and Lemma 7.87d shows that condition (b) holds. Thus Lemma 7.103 yields the conclusions of the proposition.

Lemma 7.107. Let $h_1, \ldots, h_k$ be in $S(\mathfrak{g}^*)$, let $\{u_i\}_{i=1}^\infty$ be a basis of $U(\mathfrak{g})$, let $u_i h_j$ refer to the action of $U(\mathfrak{g})$ on $S(\mathfrak{g}^*)$ obtained from the action of $\mathrm{ad}^* \mathfrak{g}$ on $S(\mathfrak{g}^*)$, let X be in $\mathfrak{g}$, and let $O_X = \mathrm{Ad}(G)X$. Then the functions $h_1|_{O_X}, \ldots, h_k|_{O_X}$ are linearly independent if and only if the k-column matrix $D(X) = (u_i h_j(X))$ has rank k.

PROOF. For f in $S(\mathfrak{g}^*)$, let $\tilde{f}$ be the holomorphic function on G given by $\tilde{f}(g) = f(\mathrm{Ad}(g)X)$. If Z is in $\mathfrak{g}$, then Z acts on $\tilde{f}$ and f by

$$Z\tilde{f}(g) = \tfrac{d}{dz} \tilde{f}((\exp zZ)^{-1}g)|_{z=0}$$

and $\qquad\qquad Zf(\mathrm{Ad}(g)X) = \tfrac{d}{dz} f(\mathrm{Ad}((\exp zZ)^{-1}g)X)|_{z=0},$

and it is clear that the extensions of these actions to $U(\mathfrak{g})$ satisfy

$$(7.108) \qquad u\tilde{f}(g) = \widetilde{uf}(g) \qquad \text{for all } u \in U(\mathfrak{g}) \text{ and } g \in G.$$

If $h_1|_{O_X}, \ldots, h_k|_{O_X}$ are linearly dependent, choose constants $c_1, \ldots, c_k$ not all 0 such that

$$(c_1 h_1 + \ldots c_k h_k)|_{O_X} = 0.$$

Then $c_1\tilde{h}_1 + \cdots + c_k\tilde{h}_k = 0$ and so $c_1 u\tilde{h}_1 + \cdots + c_k u\tilde{h}_k = 0$ for all $u \in U(\mathfrak{g})$. Applying (7.108) with $g = 1$ and $u = u_i$, we see that

$$c_1 u_i h_1(X) + \cdots + c_k u_i h_k(X) = 0 \qquad \text{for } 1 \le i < \infty.$$

Hence the columns of $D(X)$ are dependent, and the rank is $< k$.

Conversely if $D(X)$ has rank $< k$, we can choose constants $c_1, \ldots, c_k$ not all 0 such that

$$c_1 u_i h_1(X) + \cdots + c_k u_i h_k(X) = 0 \qquad \text{for } 1 \le i < \infty.$$

Putting $f = c_1 h_1 + \cdots + c_k h_k$, we obtain $uf(X) = 0$ for all $u \in U(\mathfrak{g})$. By (7.108), $u\tilde{f}(1) = 0$ for all $u \in U(\mathfrak{g})$. Since $\tilde{f}$ is holomorphic and G is connected, $\tilde{f} = 0$. Therefore f vanishes on O_X, and $h_1|_{O_X}, \ldots, h_k|_{O_X}$ are dependent.

Theorem 7.109 (Kostant). $S(\mathfrak{g}^*) \cong \mathcal{H}' \otimes_{\mathbb{C}} \mathcal{I}'$ in the sense that the map $h \otimes p \mapsto hp$ for $h \in \mathcal{H}'$ and $p \in \mathcal{I}'$ extends to a linear isomorphism of $\mathcal{H}' \otimes_{\mathbb{C}} \mathcal{I}'$ onto $S(\mathfrak{g}^*)$.

PROOF. In view of Corollary 7.77, we are to show that the map $h \otimes p \mapsto hp$ is one-one. Thus suppose $h_1, \ldots, h_k$ are linearly independent elements in $\mathcal{H}'$ such that

$$(7.110) \qquad \sum h_j p_j = 0 \qquad \text{for some } p_1, \ldots, p_k \text{ in } \mathcal{I}'.$$

We are to show that $p_1 = \cdots = p_k = 0$.

Let us see that $h_1|_{\mathfrak{n}}, \ldots, h_k|_{\mathfrak{n}}$ are linearly independent. In fact, otherwise we can find constants $c_1, \ldots, c_k$ not all 0 with

$$(c_1 h_1 + \cdots + c_k h_k)|_{\mathfrak{n}} = 0.$$

By Proposition 7.106, $c_1 h_1 + \cdots + c_k h_k$ is in $\mathcal{I}'_+ S(\mathfrak{g}^*)$. But then Proposition 7.75 shows that $c_1 h_1 + \cdots + c_k h_k = 0$. Since $h_1, \ldots, h_k$ are assumed linearly independent, $c_1 = \cdots = c_k = 0$. We conclude that $h_1|_{\mathfrak{n}}, \ldots, h_k|_{\mathfrak{n}}$ are linearly independent.

For $X \in \mathfrak{g}$, let $O_X = \mathrm{Ad}(G)X$. In the case that $X = e$, Lemma 7.98 shows that O_e is dense in $\mathfrak{N}$, and therefore $h_1|_{O_e}, \ldots, h_k|_{O_e}$ are linearly independent. Let $D(e)$ be the matrix given for $h_1|_{O_e}, \ldots, h_k|_{O_e}$ by Lemma 7.107. Because of the linear independence, that lemma says that there exists a k-by-k block $m(e)$ of $D(e)$ with $\det m(e) \neq 0$. By continuity, $\det m(X) \neq 0$ for all X in an open neighborhood V of e in $\mathfrak{g}$. Consequently the converse statement of Lemma 7.107 shows that $h_1|_{O_X}, \ldots, h_k|_{O_X}$ are linearly independent for all X in V. For each $X \in V$, the elements p_j in (7.110) reduce to constants on O_X since p_j is in $\mathcal{I}'$. From (7.110) and the linear independence on O_X, we conclude that $p_j(X) = 0$ for $1 \leq j \leq k$ and for all $X \in V$. Since V is open and p_j is a polynomial, p_j vanishes identically for $1 \leq j \leq k$. This completes the proof.

As we noted at the start of §5, the vector-space isomorphism $\mathfrak{g} \cong \mathfrak{g}^*$ given by the Killing form induces an algebra isomorphism $S(\mathfrak{g}) \cong S(\mathfrak{g}^*)$ respecting the action by G. In particular $\mathcal{I}$ in $S(\mathfrak{g})$ corresponds to $\mathcal{I}'$ in $S(\mathfrak{g}^*)$. Let $\mathcal{H}$ be the graded subspace of $S(\mathfrak{g})$ corresponding to $\mathcal{H}'$ in $S(\mathfrak{g})$; this is invariant under the action of G since $\mathcal{H}'$ is.

Corollary 7.111. $S(\mathfrak{g}) \cong \mathcal{H} \otimes_{\mathbb{C}} \mathcal{I}$ in the sense that the map $h \otimes y \mapsto hy$ for $h \in \mathcal{H}$ and $y \in \mathcal{I}$ extends to a linear isomorphism of $\mathcal{H} \otimes_{\mathbb{C}} \mathcal{I}$ onto $S(\mathfrak{g})$.

This is immediate from Theorem 7.109. To pass from Corollary 7.111 to a result about $U(\mathfrak{g})$, we shall use the symmetrization map

$$\sigma : S(\mathfrak{g}) \to U(\mathfrak{g}).$$

Recall that σ is a vector-space isomorphism. Although σ does not respect multiplication, it does have other nice features. The adjoint actions of G on $S(\mathfrak{g})$ and $U(\mathfrak{g})$ are related by

(7.112a) $\sigma \mathrm{Ad}(g)(u) = \mathrm{Ad}(g)\sigma(u)$ for $u \in S(\mathfrak{g})$ and $g \in G$,

and the differentiated actions of $\mathfrak{g}$ on $S(\mathfrak{g})$ and $U(\mathfrak{g})$ are related by

(7.112b) $\sigma(\mathrm{ad}\, X)(u) = (\mathrm{ad}\, X)\sigma(u)$ for $u \in S(\mathfrak{g})$ and $X \in \mathfrak{g}$.

From (7.112) it follows that

(7.113) σ is a vector-space isomorphism of $\mathcal{I}$ onto $Z(\mathfrak{g})$.

Let $\mathcal{K} = \sigma(\mathcal{H})$. Because $\mathcal{H}$ is an invariant subspace for the adjoint action of G on $S(\mathfrak{g})$, (7.112a) shows that $\mathcal{K}$ is a G invariant subspace for the adjoint action of G on $U(\mathfrak{g})$.

Theorem 7.114. $U(\mathfrak{g})$ is a free $Z(\mathfrak{g})$ module, and any vector-space basis of $\mathcal{K}$ is a free $Z(\mathfrak{g})$ basis of $U(\mathfrak{g})$. Specifically, $U(\mathfrak{g}) \cong \mathcal{K} \otimes_{\mathbb{C}} Z(\mathfrak{g})$ in the sense that the map $k \otimes z \mapsto kz$ for $k \in \mathcal{K}$ and $z \in Z(\mathfrak{g})$ extends to a linear isomorphism of $\mathcal{K} \otimes_{\mathbb{C}} Z(\mathfrak{g})$ onto $U(\mathfrak{g})$.

Before coming to the proof, let us introduce a companion mapping to σ. The universal enveloping algebra $U(\mathfrak{g})$ is filtered by degree, and we let $U_d(\mathfrak{g})$ be the subspace of elements of degree $\leq d$. Then we have a vector-space direct-sum decomposition

$$U_d(\mathfrak{g}) = \sigma(S^d(\mathfrak{g})) \oplus U_{d-1}(\mathfrak{g}).$$

The composition of projection to the first summand, followed by σ^{-1}, is a linear map

$$\mathrm{gr}^d : U_d(\mathfrak{g}) \to S^d(\mathfrak{g})$$

that descends to an isomorphism

$$U_d(\mathfrak{g})/U_{d-1}(\mathfrak{g}) \xrightarrow{\sim} S^d(\mathfrak{g}).$$

It is clear from the construction that

(7.115) $\qquad\qquad \mathrm{gr}^d \circ \sigma \qquad$ is the identity on $S^d(\mathfrak{g})$,

and it is easy to see that

(7.116)
$$\mathrm{gr}^i(u_1)\mathrm{gr}^{d-i}(u_2) = \mathrm{gr}^d(u_1 u_2) \qquad \text{if } u_1 \in U_i(\mathfrak{g}) \text{ and } u_2 \in U_{d-i}(\mathfrak{g}).$$

PROOF OF THEOREM 7.114. To see that any u in $U(\mathfrak{g})$ is of the form $\sum k_m z_m$ with $k_m \in \mathcal{K}$ and $z_m \in Z(\mathfrak{g})$, we prove the result for $U_d(\mathfrak{g})$, inducting on d. For $d = 0$, we can take k and z to be scalar multiples of the identity. Assume the result for $d - 1$. If $u \in U_d(\mathfrak{g})$ is given, then $\mathrm{gr}^d u = p$ is in $S^d(\mathfrak{g})$. By Corollary 7.111, $p = \sum h_m y_m$ with $h_m \in \mathcal{H}_{i_m}$ and $y_m \in \mathcal{I}_{j_m}$, where $i_m + j_m = d$. The element $\sum \sigma(h_m)\sigma(y_m)$ has $\sigma(h_m) \in \mathcal{K}$ and $\sigma(y_m) \in Z(\mathfrak{g})$, and

$$\mathrm{gr}^d\left(u - \sum \sigma(h_m)\sigma(y_m)\right)$$
$$= p - \sum \mathrm{gr}^d(\sigma(h_m)\sigma(y_m))$$
$$= p - \sum \mathrm{gr}^{i_m}(\sigma(h_m))\mathrm{gr}^{j_m}(\sigma(y_m)) \qquad \text{by (7.116)}$$
$$= p - \sum h_m y_m \qquad\qquad\qquad\qquad \text{by (7.115)}$$
$$= 0.$$

Therefore $u - \sum \sigma(h_m)\sigma(y_m)$ is in $U_{d-1}(\mathfrak{g})$. By inductive hypothesis, this difference can be expanded in terms of $\mathcal{K}$ and $Z(\mathfrak{g})$, and then so can u.

To complete the proof, we show that the map from $\mathcal{K} \otimes_{\mathbb{C}} Z(\mathfrak{g})$ to $U(\mathfrak{g})$ is one-one. Assume the contrary. Then some nontrivial relation $\sum k_m z_m = 0$ must be valid, where $\{k_m\}$ is linearly independent in $\mathcal{K}$, k_m is in $\sigma(\mathcal{H}_{i_m})$, and z_m is in $Z(\mathfrak{g})$. Let us suppose that all $k_m z_m$ are in $U_d(\mathfrak{g})$ but at least one is not in $U_{d-1}(\mathfrak{g})$. Applying gr^d, we have

$$0 = \sum \mathrm{gr}^d(k_m z_m) = \sum \mathrm{gr}^{i_m}(k_m)\mathrm{gr}^{d-i_m}(z_m)$$

by (7.116). Since $\{\mathrm{gr}^{i_m}(k_m)\}$ is linearly independent, Corollary 7.111 shows that $\mathrm{gr}^{d-i_m}(z_m) = 0$ for all m. Hence $\mathrm{gr}^d(k_m z_m) = 0$ for all m, and $k_m z_m$ is in $U_{d-1}(\mathfrak{g})$ for all m, contradiction.

7. Dixmier-Duflo Theorem

Let $\mathfrak{g}$ be a complex Lie algebra, and let V be a $U(\mathfrak{g})$ module. The **annihilator** of V, denoted Ann V, is the set of all members u of $U(\mathfrak{g})$ such that u acts by 0 on V. This is a two-sided ideal in $U(\mathfrak{g})$.

The Dixmier-Duflo Theorem identifies the annihilators of the Verma modules that were defined in §IV.7. Fix a complex reductive Lie algebra $\mathfrak{g}$, a Cartan subalgebra $\mathfrak{h}$, and a system $\Delta^+(\mathfrak{g}, \mathfrak{h})$ of positive roots. Let $\mathfrak{n}$ be the sum of the root spaces for the positive roots, and let $\mathfrak{b} = \mathfrak{h} \oplus \mathfrak{n}$ be the corresponding Borel subalgebra. For $\lambda \in \mathfrak{h}^*$, we defined the **Verma module** $V(\lambda)$ by

$$V(\lambda) = P_{\mathfrak{b},\{1\}}^{\mathfrak{g},\{1\}}(\mathbb{C}_{\lambda-\delta}) = U(\mathfrak{g}) \otimes_{U(\mathfrak{b})} \mathbb{C}_{\lambda-\delta}.$$

By (4.99), $V(\lambda)$ has infinitesimal character λ. Therefore every member of

(7.117a) $$Z_\lambda = \{z \in Z(\mathfrak{g}) \mid \chi_\lambda(z) = 0\}$$

is in Ann $V(\lambda)$, and so is every member of the two-sided ideal that Z_λ generates:

(7.117b) $$\tilde{Z}_\lambda = Z_\lambda U(\mathfrak{g}).$$

Theorem 7.118 (Dixmier-Duflo). For $\mathfrak{g}$ a reductive Lie algebra over $\mathbb{C}$, the annihilator of the Verma module $V(\lambda)$ is exactly $\tilde{Z}_\lambda$.

The proof will occupy the remainder of this section and will use the notation and the theory of harmonics as developed in §6. That notation and theory applies to $\mathfrak{g}$ semisimple, and we shall need to pass to reductive $\mathfrak{g}$ at judicious points in the argument. In the reductive case we know that we can write

$$\mathfrak{g} = [\mathfrak{g}, \mathfrak{g}] \oplus Z_\mathfrak{g} \qquad \text{with } Z_\mathfrak{g} = \text{center}(\mathfrak{g}),$$

and we know that $[\mathfrak{g}, \mathfrak{g}]$ is semisimple. By Corollary 7.111,

$$(7.119) \qquad\qquad S([\mathfrak{g}, \mathfrak{g}]) \cong \mathcal{H} \otimes_{\mathbb{C}} \mathcal{I}_{[\mathfrak{g}, \mathfrak{g}]}.$$

Let G be a complex group with Lie algebra $\mathfrak{g}$. Since $\mathrm{Ad}(G)$ fixes $Z_\mathfrak{g}$, the adjoint action of G on $S(Z_\mathfrak{g})$ is trivial. From

$$S(\mathfrak{g}) \cong S([\mathfrak{g}, \mathfrak{g}]) \otimes_{\mathbb{C}} S(Z_\mathfrak{g}),$$

it is apparent that the invariants under $\mathrm{Ad}(G)$ are given by

$$(7.120) \qquad\qquad \mathcal{I} \cong \mathcal{I}_{[\mathfrak{g}, \mathfrak{g}]} \otimes_{\mathbb{C}} S(Z_\mathfrak{g}).$$

Tensoring (7.119) by $S(Z_\mathfrak{g})$, we thus obtain

$$(7.121) \qquad\qquad S(\mathfrak{g}) \cong \mathcal{H} \otimes_{\mathbb{C}} \mathcal{I}$$

with $\mathcal{H} \subseteq S([\mathfrak{g}, \mathfrak{g}])$ and with $\mathcal{I}$ given by (7.120).

The group G acts on $S(\mathfrak{g})$ through the semisimple part of G, with the center of G acting trivially. The group action on $\mathcal{H}$ in the semisimple case thus yields a group action on $\mathcal{H}$ in the reductive case. In the semisimple case, $\mathcal{H}$ splits as a direct sum of finite-dimensional invariant subspaces, and the same direct sum is a splitting into finite-dimensional invariant subspaces for the reductive case. Let $\mathcal{H}^{N^-}$ be the subspace of $\mathcal{H}$ fixed by N^-.

Lemma 7.122. For $\mathfrak{g}$ reductive, $\mathcal{H}^{N^-} \cap S(\mathfrak{g})\mathfrak{b} = 0$.

PROOF. Without loss of generality we may assume that $\mathfrak{g}$ is semisimple. Suppose f is in the intersection. Let p be the member of $S(\mathfrak{g}^*)$ that corresponds to f under the isomorphism $S(\mathfrak{g}) \cong S(\mathfrak{g}^*)$. Under this isomorphism, a member X of $\mathfrak{b}$ corresponds to the linear polynomial $\lambda_X(Y) = B(X, Y)$, which vanishes on $\mathfrak{n}$. Hence p vanishes on $\mathfrak{n}$. On the other hand, f in $\mathcal{H}^{N^-}$ implies p is in $(\mathcal{H}')^{N^-}$; in other words, p is fixed by N^-. Hence p vanishes on $\mathrm{Ad}(N^-)\mathfrak{n}$. Since $\mathrm{Ad}(HN)\mathfrak{n} = \mathfrak{n}$, p vanishes on $\mathrm{Ad}(N^- HN)\mathfrak{n}$. By (7.64f), $N^- HN$ is dense in G. Thus p vanishes on $\mathrm{Ad}(G)\mathfrak{n}$, which equals $\mathfrak{N}$ by Corollary 7.86. By Proposition 7.106, p is in $\mathcal{I}'_+ S(\mathfrak{g}^*)$. Since p is in $\mathcal{H}'$ also, Proposition 7.75 shows that $p = 0$. Hence $f = 0$.

We shall use this lemma to pass to a corresponding result about $U(\mathfrak{g})$. As in §6 in the semisimple case, let $\mathcal{K} = \sigma(\mathcal{H})$. The group G acts on $\mathcal{K}$ since σ respects G actions.

Lemma 7.123. Let $\mathfrak{g}$ be reductive. Fix $\lambda \in \mathfrak{h}^*$, and let $\mathcal{J}$ be the left ideal in $U(\mathfrak{g})$ generated by $\mathfrak{n}$ and all elements $h - \lambda(h)1$ with $h \in \mathfrak{h}$. Then

(a) $U(\mathfrak{g}) = \mathcal{J} \oplus U(\mathfrak{n}^-)$
(b) if Q denotes the projection to $U(\mathfrak{n}^-)$ in (a) and if u is a member of the subset $U_d(\mathfrak{g})$ of elements of degree $\leq d$, then Qu is in $U_d(\mathfrak{n}^-)$
(c) $\mathcal{K}^{N^-} \cap \mathcal{J} = 0$.

PROOF.
(a, b) Let $\{X_i\}_{i=1}^r$ be a basis of $\mathfrak{n}^-$, let $\{h_j\}_{j=1}^l$ be a basis of $\mathfrak{h}$, and let $\{Y_k\}_{k=1}^r$ be a basis of $\mathfrak{n}$. An easy induction on d shows that the elements

$$X_1^{a_1} \cdots X_r^{a_r} (h_1 - \lambda(h_1)1)^{b_1} \cdots (h_l - \lambda(h_l)1)^{b_l} Y_1^{c_1} \cdots Y_r^{c_r}$$

with $\sum a_i + \sum b_j + \sum c_k \leq d$ generate $U_d(\mathfrak{g})$. The number of these elements is the same as in the standard Poincaré-Birkhoff-Witt basis of $U_d(\mathfrak{g})$, and hence these elements are linearly independent. Then (a) follows: The elements of $\mathcal{J}$ are spanned by those basis elements with some b_j or c_k greater than 0, and the elements of $U(\mathfrak{n}^-)$ are spanned by those basis elements with all b_j and c_k equal to 0. Also (b) is immediate.

(c) Let u be in $\mathcal{K}^{N^-} \cap \mathcal{J}$. Assuming that $u \neq 0$, let u be in $U_d(\mathfrak{g})$ but not $U_{d-1}(\mathfrak{g})$. The element $\mathrm{gr}^d(u)$ is then in $\mathcal{H}^{N^-}$ and is not 0. By Lemma 7.122,

(7.124) $\mathrm{gr}^d(u)$ is not in $S(\mathfrak{g})\mathfrak{b}$.

Relative to the evident direct-sum decomposition

$$S(\mathfrak{g}) = S(\mathfrak{g})\mathfrak{b} \oplus S(\mathfrak{n}^-),$$

let P be the projection on $S(\mathfrak{n}^-)$. By (7.124), the element $P(\mathrm{gr}^d(u))$ of $S(\mathfrak{n}^-)$ is not 0, and it is homogeneous of degree d. Therefore

(7.125) $\sigma P(\mathrm{gr}^d(u))$ is not in $U_{d-1}(\mathfrak{n}^-)$.

The element $\mathrm{gr}^d(u) - P(\mathrm{gr}^d(u))$ is in $S(\mathfrak{g})\mathfrak{b}$ homogeneous of degree d, and it is thus a linear combination of monomials $X_1 \cdots X_{d-1}Y$ with Y in $\mathfrak{h}$ or $\mathfrak{n}$. When Y is in $\mathfrak{h}$, let us rewrite this monomial as

$$X_1 \cdots X_{d-1}(Y - \lambda(Y)1) + \lambda(Y)X_1 \cdots X_{d-1}.$$

For any Y, let us then apply σ. Each term is in $\mathcal{J}$ up to permutation. Since permutations introduce only lower-order terms, we see that

$$(7.126) \qquad \sigma(\mathrm{gr}^d(u)) - \sigma P(\mathrm{gr}^d(u)) \quad \text{is in } \mathcal{J} + U_{d-1}(\mathfrak{g}).$$

Now

$$\sigma P(\mathrm{gr}^d(u)) = (\sigma P(\mathrm{gr}^d(u)) - \sigma(\mathrm{gr}^d(u))) + (\sigma(\mathrm{gr}^d(u)) - u) + u.$$

The first term on the right is in $\mathcal{J} + U_{d-1}(\mathfrak{g})$ by (7.126), the second term on the right is in $U_{d-1}(\mathfrak{g})$ by (7.115) since gr^d of it is 0, and the third term on the right is in $\mathcal{J}$ by assumption. Thus

$$(7.127) \qquad \sigma P(\mathrm{gr}^d(u)) \quad \text{is in } \mathcal{J} + U_{d-1}(\mathfrak{g}).$$

But also it is in $U(\mathfrak{n}^-)$ since $P(\mathrm{gr}^d(u))$ is in $S(\mathfrak{n}^-)$. By (7.127), $\sigma P(\mathrm{gr}^d(u)) = Q(\sigma P(\mathrm{gr}^d(u)))$ is in $Q(U_{d-1}(\mathfrak{g}))$, which is contained in $U_{d-1}(\mathfrak{n}^-)$ by Lemma 7.123b. This conclusion contradicts (7.125), and (c) follows.

This completes the lemmas. But before proving the theorem, let us introduce some temporary notation. Within $S([\mathfrak{g}, \mathfrak{g}])$, we can write $\mathcal{H}$ as a direct sum of finite-dimensional irreducible G modules $\mathcal{H} = \bigoplus \mathcal{H}_i$ under the adjoint representation. Then $\mathcal{K}$ in $U([\mathfrak{g}, \mathfrak{g}])$ is a corresponding direct sum $\mathcal{K} = \bigoplus \mathcal{K}_i$. Let t_i be a nonzero lowest-weight vector of $\mathcal{K}_i$. The elements t_i form a vector-space basis of $\mathcal{K}^{N^-}$.

PROOF OF THEOREM 7.118. In the notation of (7.117), we know that $\mathrm{Ann}\, V(\lambda) \supseteq \tilde{Z}_\lambda$. Suppose strict inclusion holds. Since any two-sided ideal in $U(\mathfrak{g})$ is a module for G under the adjoint map and since $U(\mathfrak{g})$ is fully reducible, we can choose a finite-dimensional G invariant subspace F in $\mathrm{Ann}\, V(\lambda)$ disjoint from $\tilde{Z}_\lambda$. Let u be a nonzero lowest-weight vector of F.

By Theorem 7.114 we have

$$U([\mathfrak{g}, \mathfrak{g}]) \cong \mathcal{K} \otimes_{\mathbb{C}} Z([\mathfrak{g}, \mathfrak{g}]).$$

Tensoring with $U(Z_\mathfrak{g}) = Z(Z_\mathfrak{g})$, we obtain

$$(7.128) \qquad U(\mathfrak{g}) \cong \mathcal{K} \otimes_{\mathbb{C}} (Z([\mathfrak{g}, \mathfrak{g}]) \otimes_{\mathbb{C}} Z(Z_\mathfrak{g})) = \mathcal{K} \otimes_{\mathbb{C}} Z(\mathfrak{g}).$$

Let $\{z_\gamma\}_{\gamma \in \Gamma}$ be a vector-space basis of $Z(\mathfrak{g})$. By (7.128), we can write u as a finite sum

$$(7.129) \qquad u = \sum_{\gamma \in \Gamma} k_\gamma z_\gamma$$

for unique elements k_γ in $\mathcal{K}$. Applying $n^- \in N^-$, we see that

$$u = \sum_{\gamma \in \Gamma} (\mathrm{Ad}(n^-)k_\gamma)z_\gamma,$$

and we conclude that $\mathrm{Ad}(n^-)k_\gamma = k_\gamma$. Hence each k_γ is in $\mathcal{K}^{N^-}$. Each k_γ is therefore a linear combination of the elements t_i just before the proof of the theorem. Rearranging (7.129), we can therefore write

(7.130) $u = \sum_i z_i t_i$ with $z_i \in Z(\mathfrak{g})$.

Application of u to the element $1 \otimes 1$ of $V(\lambda)$ gives

$$0 = \sum_i \chi_\lambda(z_i) t_i (1 \otimes 1).$$

Thus $\sum_i \chi_\lambda(z_i) t_i$ annihilates $1 \otimes 1$. In the direct-sum decomposition of Lemma 7.123a, every member of $\mathcal{J}$ annihilates $1 \otimes 1$ and no nonzero member of $U(\mathfrak{n}^-)$ does (by (4.97)). Hence $\sum_i \chi_\lambda(z_i) t_i$ is in $\mathcal{J}$. But it is also in $\mathcal{K}^{N^-}$, and thus Lemma 7.123c shows that it is 0. Since the t_i are linearly independent, each coefficient $\chi_\lambda(z_i)$ is 0. Thus z_i is in Z_λ, and (7.130) shows that u is in $\tilde{Z}_\lambda$.

8. Translation Functors

In this section we shall define the translation functors $\psi_\lambda^{\lambda'}$. Fix a complex reductive Lie algebra $\mathfrak{g}$, and let $\mathcal{C}_{Zf}(\mathfrak{g})$ be the subcategory of all $Z(\mathfrak{g})$ finite modules of $\mathcal{C}(\mathfrak{g})$. All $U(\mathfrak{g})$ maps between two members of $\mathcal{C}_{Zf}(\mathfrak{g})$ are allowed as maps in the category. It is clear that $\mathcal{C}_{Zf}(\mathfrak{g})$ is a good category in the sense of §A.1. Corollary 7.132 and Theorem 7.133 will establish further closure properties of $\mathcal{C}_{Zf}(\mathfrak{g})$.

Proposition 7.131. If V is a $U(\mathfrak{g})$ module, then the following conditions are equivalent:
   (a) V is $Z(\mathfrak{g})$ finite
   (b) there exist an integer $N > 0$ and finitely many homomorphisms $\chi_1, \ldots, \chi_k$ of $Z(\mathfrak{g})$ into $\mathbb{C}$ such that

$$(z_1 - \chi_1(z_1)1)^N \cdots (z_k - \chi_k(z_k)1)^N$$

acts as 0 on V for all $z_1, \ldots, z_k$ in $Z(\mathfrak{g})$.
When (b) holds, the χ primary component of V can be nonzero only for $\chi = \chi_1, \ldots, \chi_k$.

REMARK. The $\chi_1, \ldots, \chi_k$ in (b) are not assumed distinct.

PROOF.

(a) $\Longrightarrow$ (b). This follows from Proposition 7.20.

(b) $\Longrightarrow$ (a). We wish to prove that the ideal $J = J(\chi_1, \ldots, \chi_k, N)$ in $Z(\mathfrak{g})$ generated by all elements of the form

$$(z_1 - \chi_1(z_1)1)^N \cdots (z_k - \chi_k(z_k)1)^N$$

has finite codimension in $Z(\mathfrak{g})$. If $\chi_1 = \chi_2$ for example, then

$$J(\chi_1, \ldots, \chi_k, N) \supseteq J(\chi_2, \ldots, \chi_k, 2N).$$

If the second of these ideals has finite codimension, then so does the first. The result is that there is no loss of generality in assuming that $\chi_1, \ldots, \chi_k$ are distinct. We prove that the ideal generated by the elements in question has finite codimension in $Z(\mathfrak{g})$.

We argue as with Corollary 7.32. In terms of the polynomial algebra $\mathbb{C}[X_1, \ldots, X_l]$ in the proof of that corollary, there are distinct points $a_1, \ldots, a_k$ such that the elements in question are all polynomials of the form

$$(P_1 - P_1(a_1))^N \cdots (P_k - P_k(a_k))^N.$$

Fixing $P_2, \ldots, P_k$, we can use (4.113c) to show that J contains all polynomials of the form

$$Q_1(P_2 - P_2(a_2))^N \cdots (P_k - P_k(a_k))^N,$$

where Q_1 is any polynomial vanishing at a_1 to order N or higher. Then we can fix such a Q_1 along with $P_3, \ldots, P_k$ to show that J contains all polynomials

$$Q_1 Q_2 (P_3 - P_3(a_3))^N \cdots (P_k - P_k(a_k))^N,$$

where Q_i vanishes at a_i to order N or higher. Continuing in this way, we see that J contains the ideal of all polynomials vanishing to order N or higher at each of $a_1, \ldots, a_k$. This is the intersection on j of the ideal of polynomials vanishing to order N or higher at a_j. Each of these ideals, as in Corollary 7.32, has finite codimension, and hence so does the intersection.

To complete the proof, let us show that the χ primary component of V can be nonzero only for $\chi = \chi_1, \ldots, \chi_k$. Indeed, if χ is different from $\chi_1, \ldots, \chi_k$, choose $z_0 \in Z(\mathfrak{g})$ by Lemma 7.14 so that $\chi(z_0) = 1$ and $\chi_j(z_0) = 0$ for $1 \leq j \leq k$. By (b), z_0 acts nilpotently on V, hence on $P_\chi(V)$. But by definition, the action of $(z_0 - 1)^n$ on $P_\chi(V)$ is 0 for some n, and hence z_0 acts invertibly. Therefore $P_\chi(V) = 0$.

Corollary 7.132. Let

$$0 \longrightarrow V_1 \longrightarrow V_2 \longrightarrow V_3 \longrightarrow 0$$

be an exact sequence in $C(\mathfrak{g})$ with V_1 and V_3 in $C_{Zf}(\mathfrak{g})$. Then V_2 is in $C_{Zf}(\mathfrak{g})$. If the χ primary component of V_2 is nonzero, then the χ primary component of V_1 or V_3 is nonzero.

PROOF. By Proposition 7.131 for V_3, we can choose $\chi_1, \ldots, \chi_k$ so that

$$(z_1 - \chi_1(z_1)1)^N \cdots (z_k - \chi_k(z_k)1)^N$$

always acts as 0 on V_3. Since $V_3 \cong V_2/V_1$, such elements send V_2 into V_1. By Proposition 7.131 for V_1, we can choose $\chi_1', \ldots, \chi_m'$ so that

$$(z_1' - \chi_1'(z_1')1)^{N'} \cdots (z_m' - \chi_m'(z_m')1)^{N'}$$

always acts as 0 on V_1. Composing the two elements, we obtain the 0 operator on V_2. By Proposition 7.131, V_2 is $Z(\mathfrak{g})$ finite. The final statement of Proposition 7.131 shows that the χ primary component of V_2 can be nonzero only for $\chi_1, \ldots, \chi_k, \chi_1', \ldots, \chi_m'$.

Theorem 7.133 (Kostant). Let $\mathfrak{g}$ be a complex reductive Lie algebra, let $\mathfrak{h}$ be a Cartan subalgebra, and let F be a finite-dimensional $U(\mathfrak{g})$ module. If V is a $Z(\mathfrak{g})$ finite $U(\mathfrak{g})$ module, then $V \otimes_{\mathbb{C}} F$ is $Z(\mathfrak{g})$ finite. Moreover, if V has generalized infinitesimal character λ, then the χ primary component of $V \otimes_{\mathbb{C}} F$ can be nonzero only when $\chi = \chi_{\lambda'}$ with $\lambda' = \lambda + \nu$ for some weight ν of F.

PROOF. We shall reduce matters to the case that V has an infinitesimal character. Taking this special case as known for the moment, let us prove the theorem by induction on the smallest integer N such that there exist $\lambda_1, \ldots, \lambda_N$ with the property that

(7.134a) $(z_1 - \chi_{\lambda_1}(z_1)1) \cdots (z_N - \chi_{\lambda_N}(z_N)1)$

always acts as 0. Such an integer N exists by Proposition 7.131. The case $N = 1$ is the case we are assuming temporarily. Let V' be the $U(\mathfrak{g})$ submodule of V of elements on which

(7.134b) $(z_1 - \chi_{\lambda_1}(z_1)1) \cdots (z_{N-1} - \chi_{\lambda_{N-1}}(z_{N-1})1)$

always acts as 0. We then have an exact sequence

(7.135) $0 \longrightarrow V' \longrightarrow V \longrightarrow V/V' \longrightarrow 0$

For $v \in V$ and $z \in Z(\mathfrak{g})$, $(z - \chi_{\lambda_N}(z)1)v$ has the property that (7.134b) of it is 0, since (7.134a) annihilates v. Therefore $(z - \chi_{\lambda_N}(z)1)v$ is in V', and $(z - \chi_{\lambda_N}(z)1)$ acts as 0 on V/V'. Thus our inductive hypothesis applies to V' and to V/V'. By Proposition 2.53a the exactness of (7.135) implies the exactness of

$$0 \longrightarrow V' \otimes_{\mathbb{C}} F \longrightarrow V \otimes_{\mathbb{C}} F \longrightarrow V/V' \otimes_{\mathbb{C}} F \longrightarrow 0.$$

By induction $V' \otimes_{\mathbb{C}} F$ and $V/V' \otimes_{\mathbb{C}} F$ are $Z(\mathfrak{g})$ finite with nonzero primary components only of the stated form. By Corollary 7.132, the same thing is true of $V \otimes_{\mathbb{C}} F$.

Thus we may assume that V has an infinitesimal character, say λ. The direct sum $\mathfrak{g} \oplus \mathfrak{g}$ has universal enveloping algebra $U(\mathfrak{g}) \otimes_{\mathbb{C}} U(\mathfrak{g})$, and this acts on $V \otimes_{\mathbb{C}} F$, the first factor acting in V and the second in F. With annihilators defined as in §7, let us prove that

$$(7.136) \quad \mathrm{Ann}_{\mathfrak{g} \oplus \mathfrak{g}}(V \otimes_{\mathbb{C}} F) = U(\mathfrak{g}) \otimes_{\mathbb{C}} \mathrm{Ann}_{\mathfrak{g}}(F) + \mathrm{Ann}_{\mathfrak{g}}(V) \otimes_{\mathbb{C}} U(\mathfrak{g}).$$

The inclusion $\supseteq$ in (7.136) is clear. For the reverse inclusion, let $\{u_i\}_{i \in I}$ be a vector-space basis of $\mathrm{Ann}_{\mathfrak{g}}(F)$, and adjoin a linearly independent set $\{u_i\}_{i \in I'}$ such that $\{u_i\}_{i \in I \cup I'}$ is a basis of $U(\mathfrak{g})$. The most general element of $U(\mathfrak{g}) \otimes_{\mathbb{C}} U(\mathfrak{g})$ can be written uniquely in the form $\sum_{i \in I \cup I'} x_i \otimes u_i$. Suppose that this element is in $\mathrm{Ann}_{\mathfrak{g} \oplus \mathfrak{g}}(V \otimes_{\mathbb{C}} F)$. The term $\sum_{i \in I} x_i \otimes u_i$ is in $U(\mathfrak{g}) \otimes_{\mathbb{C}} \mathrm{Ann}_{\mathfrak{g}}(F)$, and we shall show that $\sum_{i \in I'} x_i \otimes u_i$ is in $\mathrm{Ann}_{\mathfrak{g}}(V) \otimes_{\mathbb{C}} U(\mathfrak{g})$. If φ_V and φ_F are the homomorphisms from $U(\mathfrak{g})$ to $\mathrm{End}_{\mathbb{C}} V$ and $\mathrm{End}_{\mathbb{C}} F$, respectively, then

$$(7.137) \qquad (\varphi_V \otimes \varphi_F)\Big(\sum_{i \in I'} x_i \otimes u_i \Big) = \sum_{i \in I'} \varphi_V(x_i) \otimes \varphi_F(u_0)$$

is the corresponding member of

$$\mathrm{End}_{\mathbb{C}} V \otimes_{\mathbb{C}} \mathrm{End}_{\mathbb{C}} F \subseteq \mathrm{End}_{\mathbb{C}}(V \otimes_{\mathbb{C}} F).$$

We are assuming that (7.137) is the 0 endomorphism. Our assumptions, however, make $\{\varphi_F(u_i)\}_{i \in I'}$ linearly independent. Hence $\varphi_V(x_i) = 0$ for all $i \in I'$, and we conclude that x_i is in $\mathrm{Ann}_{\mathfrak{g}}(V)$. Thus $\sum_{i \in I'} x_i \otimes u_i$ is in $\mathrm{Ann}_{\mathfrak{g}}(V) \otimes_{\mathbb{C}} U(\mathfrak{g})$, and (7.136) is proved.

The coproduct map $h : U(\mathfrak{g}) \to U(\mathfrak{g}) \otimes_{\mathbb{C}} U(\mathfrak{g})$ given on $\mathfrak{g}$ by $X \mapsto X \otimes 1 + 1 \otimes X$ embeds $\mathrm{Ann}_{\mathfrak{g}}(V \otimes_{\mathbb{C}} F)$ into $\mathrm{Ann}_{\mathfrak{g} \oplus \mathfrak{g}}(V \otimes_{\mathbb{C}} F)$. From (7.136) we see that

$$(7.138) \quad \mathrm{Ann}_{\mathfrak{g}}(V \otimes_{\mathbb{C}} F) = h^{-1}(U(\mathfrak{g}) \otimes_{\mathbb{C}} \mathrm{Ann}_{\mathfrak{g}}(F) + \mathrm{Ann}_{\mathfrak{g}}(V) \otimes_{\mathbb{C}} U(\mathfrak{g})).$$

Let $V(\lambda)$ be the Verma module with highest weight $\lambda - \delta$. With $\tilde{Z}_\lambda$ as in (7.117b), we have

$$
\begin{aligned}
&\mathrm{Ann}_{\mathfrak{g}}(V \otimes_{\mathbb{C}} F) \\
&\quad = h^{-1}(U(\mathfrak{g}) \otimes_{\mathbb{C}} \mathrm{Ann}_{\mathfrak{g}}(F) + \mathrm{Ann}_{\mathfrak{g}}(V) \otimes_{\mathbb{C}} U(\mathfrak{g})) \qquad \text{by (7.138)} \\
&\quad \supseteq h^{-1}(U(\mathfrak{g}) \otimes_{\mathbb{C}} \mathrm{Ann}_{\mathfrak{g}}(F) + \tilde{Z}_\lambda \otimes_{\mathbb{C}} U(\mathfrak{g})) \qquad \text{since } V \text{ has} \\
&\qquad\qquad\qquad\qquad\qquad\qquad\qquad\qquad\qquad\qquad\qquad \text{infinitesimal} \\
&\qquad\qquad\qquad\qquad\qquad\qquad\qquad\qquad\qquad\qquad\qquad \text{character } \lambda \\[4pt]
&\quad = h^{-1}(U(\mathfrak{g}) \otimes_{\mathbb{C}} \mathrm{Ann}_{\mathfrak{g}}(F) + \mathrm{Ann}_{\mathfrak{g}}(V(\lambda)) \otimes_{\mathbb{C}} U(\mathfrak{g})) \qquad \text{by Theorem 7.118} \\
&\quad = \mathrm{Ann}_{\mathfrak{g}}(V(\lambda) \otimes_{\mathbb{C}} F) \qquad\qquad\qquad\qquad\qquad\qquad \text{by (7.138).}
\end{aligned}
$$

Since $\mathrm{Ann}_{Z(\mathfrak{g})}(V)$ is just the intersection of $\mathrm{Ann}_{\mathfrak{g}}(V)$ with $Z(\mathfrak{g})$, this inclusion reduces the theorem to the case that V is the Verma module $V(\lambda)$.

We apply Lie's Theorem to F regarded as a $U(\mathfrak{b})$ module, where $\mathfrak{b} = \mathfrak{h} \oplus \mathfrak{n}$. The conclusion is that there is a filtration of $\mathcal{F}^{\mathfrak{b},\{1\}}_{\mathfrak{g},\{1\}} F$ by $U(\mathfrak{b})$ submodules

$$
(7.139) \qquad\qquad \mathcal{F}^{\mathfrak{b},\{1\}}_{\mathfrak{g},\{1\}} F = S_1 \supseteq S_2 \supseteq \cdots \supseteq S_n \supseteq 0
$$

with the property that the successive quotients are one dimensional. In any one-dimensional $\mathfrak{b}$ module, the commutator subalgebra $\mathfrak{n}$ acts by 0, and $\mathfrak{h}$ acts by a member of $\mathfrak{h}^*$. That is,

$$
S_j / S_{j+1} \cong \mathbb{C}_{\nu_j} \qquad \text{as a } U(\mathfrak{b}) \text{ module.}
$$

According to Proposition 7.20 applied to the abelian Lie algebra $\mathfrak{h}$, the various ν_j are precisely the weights of F. Now

$$
V(\lambda) \otimes_{\mathbb{C}} F = P^{\mathfrak{g},\{1\}}_{\mathfrak{b},\{1\}}(\mathbb{C}_{\lambda-\delta}) \otimes_{\mathbb{C}} F \cong P^{\mathfrak{g},\{1\}}_{\mathfrak{b},\{1\}}(\mathbb{C}_{\lambda-\delta} \otimes_{\mathbb{C}} \mathcal{F}^{\mathfrak{b},\{1\}}_{\mathfrak{g},\{1\}} F)
$$

by Theorem 2.103. We apply to (7.139) the functor $\mathbb{C}_{\lambda-\delta} \otimes_{\mathbb{C}} (\cdot)$ (exact by Proposition 2.53a) and then the functor $P^{\mathfrak{g},\{1\}}_{\mathfrak{b},\{1\}}$ (exact by Proposition 2.57c), and we see that $V(\lambda) \otimes_{\mathbb{C}} F$ has a filtration by $U(\mathfrak{g})$ modules

$$
\begin{aligned}
V(\lambda) \otimes_{\mathbb{C}} F &= \mathrm{ind}^{\mathfrak{g},\{1\}}_{\mathfrak{b},\{1\}}(\mathbb{C}_{\lambda-\delta} \otimes_{\mathbb{C}} S_1) \supseteq \mathrm{ind}^{\mathfrak{g},\{1\}}_{\mathfrak{b},\{1\}}(\mathbb{C}_{\lambda-\delta} \otimes_{\mathbb{C}} S_2) \\
&\supseteq \cdots \supseteq \mathrm{ind}^{\mathfrak{g},\{1\}}_{\mathfrak{b},\{1\}}(\mathbb{C}_{\lambda-\delta} \otimes_{\mathbb{C}} S_n) \supseteq 0
\end{aligned}
$$

with successive quotients $V(\lambda + \nu_1), V(\lambda + \nu_2), \ldots, V(\lambda + \nu_n)$. The element $(z_j - \chi_{\lambda+\nu_j}(z_j)1)$ acts as 0 on $V(\lambda + \nu_j)$ for any $z_j \in Z(\mathfrak{g})$. Hence

$(z_j - \chi_{\lambda+\nu_j}(z_j)1)$ carries $\mathrm{ind}_{\mathfrak{b},\{1\}}^{\mathfrak{g},\{1\}}(\mathbb{C}_{\lambda-\delta} \otimes_{\mathbb{C}} S_j)$ into the $U(\mathfrak{g})$ submodule $\mathrm{ind}_{\mathfrak{b},\{1\}}^{\mathfrak{g},\{1\}}(\mathbb{C}_{\lambda-\delta} \otimes_{\mathbb{C}} S_{j+1})$. Hence

$$(z_1 - \chi_{\lambda+\nu_1}(z_1)1) \cdots (z_n - \chi_{\lambda+\nu_n}(z_n)1)$$

acts as 0 in $V(\lambda) \otimes_{\mathbb{C}} F$. By Proposition 7.131, $V(\lambda) \otimes_{\mathbb{C}} F$ is $Z(\mathfrak{g})$ finite and its χ primary component can be nonzero only for χ equal to one of $\chi_{\lambda+\nu_1}, \ldots, \chi_{\lambda+\nu_n}$. This completes the proof.

We are now in a position to complete some unfinished business from §V.7. The following lemma used in the vanishing theorem of that section was left unproven until now. No circularity results from this postponement because the results of §V.7 will not be applied until Chapter VIII.

Lemma 7.140 ($= 5.100$). Suppose that the nonzero member Z of $\mathcal{C}(\mathfrak{g})$ has an infinitesimal character and that the nonzero member F of $\mathcal{C}(\mathfrak{g})$ is finite dimensional. Then $F \otimes_{\mathbb{C}} Z$ has a nonzero $U(\mathfrak{g})$ submodule with an infinitesimal character.

PROOF. Theorem 7.133 says that $Z \otimes_{\mathbb{C}} F$ is $Z(\mathfrak{g})$ finite. By Proposition 7.20, $Z \otimes_{\mathbb{C}} F$ has a nonzero $U(\mathfrak{g})$ submodule with a generalized infinitesimal character. In turn, Corollary 7.27 provides a nonzero $U(\mathfrak{g})$ submodule of this with an infinitesimal character.

If F is a finite-dimensional $U(\mathfrak{g})$ module, Theorem 7.133 shows that the functor $(\cdot) \otimes_{\mathbb{C}} F$ carries $\mathcal{C}_{Zf}(\mathfrak{g})$ into itself. The functor is exact by Proposition 2.53a. We can make another functor $\mathcal{P}_\chi$ from $\mathcal{C}_{Zf}(\mathfrak{g})$ into $\mathcal{C}_{Zf}(\mathfrak{g})$ by using the projection P_χ to the χ primary component. Namely $\mathcal{P}_\chi$ carries V to $P_\chi(V)$ and carries a map f to its restriction $f|_{P_\chi(V)}$. Proposition 7.20 implies that $\mathcal{P}_\chi$ is exact. If $\chi = \chi_\lambda$, we may write $\mathcal{P}_\lambda$ for $\mathcal{P}_{\chi_\lambda}$.

Fix a Cartan subalgebra $\mathfrak{h}$ of $\mathfrak{g}$. Let λ and λ' be members of $\mathfrak{h}^*$ whose difference $\mu = \lambda' - \lambda$ is algebraically integral, and write F^μ for an irreducible finite-dimensional complex-linear representation of $\mathfrak{g}$ with μ as an extreme weight. (F^μ exists and is determined up to isomorphism by μ.) The **translation functor** $\psi_\lambda^{\lambda'}$ is defined from $\mathcal{C}_{Zf}(\mathfrak{g})$ to itself by

$$(7.141) \qquad \psi_\lambda^{\lambda'} = \mathcal{P}_{\lambda'} \circ ((\cdot) \otimes_{\mathbb{C}} F^\mu) \circ \mathcal{P}_\lambda.$$

This functor is exact, being the composition of three exact functors. Note the indices in (7.141): The subscript λ represents the source infinitesimal character, and the superscript λ' represents the target.

Let $C_\lambda(\mathfrak{g})$ be the subcategory of $C_{Zf}(\mathfrak{g})$ of $U(\mathfrak{g})$ modules with generalized infinitesimal character λ. This is a good category. We can regard $\psi_\lambda^{\lambda'}$ as an exact functor

$$(7.142) \qquad\qquad \psi_\lambda^{\lambda'} : C_\lambda(\mathfrak{g}) \to C_{\lambda'}(\mathfrak{g}),$$

and in this case we can drop the rightmost factor in the definition (7.141).

Examples obtained directly from smooth group actions are given in §1. In using translation functors with group representations, the customary technique is to pass to the underlying $(\mathfrak{g}, K)$ modules (the sets of K finite vectors) and then forget about K. The effect of bringing K back in after translation will be discussed starting in §13. Translation functors have applications also to $U(\mathfrak{g})$ modules that do not come necessarily from group representations. Verma modules provide examples. The effect on Verma modules was examined in part in the course of proving Theorem 7.133; more discussion of this case appears in the next section.

The definition of translation functor apparently involves the selection of a Cartan subalgebra and a Weyl chamber for λ (since χ_λ determines λ only up to the Weyl group $W(\mathfrak{g}, \mathfrak{h})$). But in a sense it depends on neither. First let us compare what happens in $C_{Zf}(\mathfrak{g})$ with two reference positive systems, say Δ^+ and $w\Delta^+$ for some w in the Weyl group of $\Delta(\mathfrak{g}, \mathfrak{h})$. If V is given with a generalized infinitesimal character λ relative to Δ^+ and if we want to consider $\psi_\lambda^{\lambda+\mu}(V)$, then we choose an irreducible finite-dimensional representation F with extreme weight μ and form $P_\lambda(V \otimes_\mathbb{C} F)$. If we change to $w\Delta^+$ as reference system, then $\psi_{w\lambda}^{w(\lambda+\mu)}$ leads to the same module: V has generalized infinitesimal character $w\lambda$, the same F has extreme weight $w\mu$, and $P_\lambda = P_{w\lambda}$. So ψ does not really depend on the reference system Δ^+. Moreover, the choice of Cartan subalgebra does not really affect ψ. This fact follows from the conjugacy of Cartan subalgebras noted in (7.64e).

We end this section by noting the following elementary adjoint property of $\psi_\lambda^{\lambda'}$.

Proposition 7.143. Suppose $\mu \in \mathfrak{h}^*$ is algebraically integral, and suppose F^μ and $F^{-\mu}$ are irreducible finite-dimensional complex-linear representations of $\mathfrak{g}$ with extreme weights μ and $-\mu$, respectively. Then $F^{-\mu}$ is isomorphic to the contragredient $(F^\mu)^*$ of F^μ; fix an isomorphism between these representations. For any λ in $\mathfrak{h}^*$,

$$\mathrm{Hom}_\mathfrak{g}(\psi_{\lambda+\mu}^\lambda(V_1), V_2) \cong \mathrm{Hom}_\mathfrak{g}(V_1, \psi_\lambda^{\lambda+\mu}(V_2))$$

naturally for V_1 in $C_{\lambda+\mu}(\mathfrak{g})$ and V_2 in $C_\lambda(\mathfrak{g})$.

REMARK. The displayed isomorphism depends on the choice of the isomorphism $(F^\mu)^* \cong F^{-\mu}$.

PROOF. We have

(7.144)

$\operatorname{Hom}_\mathfrak{g} (F^{-\mu} \otimes_\mathbb{C} V_1, V_2)$

$\qquad \cong \operatorname{Hom}_\mathfrak{g} (V_1, \operatorname{Hom}_\mathbb{C}(F^{-\mu}, V_2)) \qquad$ by Corollary 2.47

$\qquad \cong \operatorname{Hom}_\mathfrak{g} (V_1, (F^{-\mu})^* \otimes_\mathbb{C} V_2) \qquad$ by Proposition 2.55

$\qquad \cong \operatorname{Hom}_\mathfrak{g} (V_1, F^\mu \otimes_\mathbb{C} V_2) \qquad$ since $(F^\mu)^* \cong F^{-\mu}$

naturally in V_1 and V_2. Since, under $U(\mathfrak{g})$ maps, generalized infinitesimal characters are preserved,

$\operatorname{Hom}_\mathfrak{g} (\psi^\lambda_{\lambda+\mu}(V_1), V_2)$

$\qquad = \operatorname{Hom}_\mathfrak{g} (\mathcal{P}_\lambda(F_{-\mu} \otimes_\mathbb{C} V_1), V_2)$

$\qquad \cong \operatorname{Hom}_\mathfrak{g} (F_{-\mu} \otimes_\mathbb{C} V_1, V_2) \qquad$ by Theorem 7.133 and Proposition 7.20

$\qquad \cong \operatorname{Hom}_\mathfrak{g} (V_1, F^\mu \otimes_\mathbb{C} V_2) \qquad$ by (7.144)

$\qquad \cong \operatorname{Hom}_\mathfrak{g} (V_1, \mathcal{P}_{\lambda+\mu}(F^\mu \otimes_\mathbb{C} V_2)) \qquad$ by Theorem 7.133 and Proposition 7.20

$\qquad = \operatorname{Hom}_\mathfrak{g} (V_1, \psi^{\lambda+\mu}_\lambda(V_2))$

naturally in V_1 and V_2.

9. Integral Dominance

Although the translation functor $\psi^{\lambda'}_\lambda$ is defined as soon as $\lambda' - \lambda$ is algebraically integral, additional hypotheses are needed to get $\psi^{\lambda'}_\lambda$ to behave well. The examples in §1 give some clue what to expect. In Example 1, $\psi^{\lambda'}_\lambda$ is applied to a finite-dimensional irreducible representation, and the result has to be finite dimensional. We saw that when both parameters are dominant, the result is irreducible. But when λ' is singular, the result has to be 0 since the infinitesimal character of a finite-dimensional irreducible representation is always nonsingular. In Example 2, by contrast, $\psi^{\lambda'}_\lambda$ is never 0, and, in the case of $\pi_{\sigma,0}$, the indicated translate has a filtration with two principal-series representations as successive quotients. Closer examination of Example 2 reveals other difficulties.

If one computes $\psi_\rho^{-\rho}$ on $\pi_{1,\rho}$, the first step is to tensor $\pi_{1,\rho}$ with a three-dimensional representation. The tensor product has a filtration with successive quotients $\pi_{1,3\rho}$, $\pi_{1,\rho}$, and $\pi_{1,-\rho}$. Upon projection according to the infinitesimal character $-\rho$, the quotient $\pi_{1,3\rho}$ disappears, but the other two quotients remain. Thus $\pi_{1,\rho}$ has led to $\pi_{1,\rho}$ and $\pi_{1,-\rho}$; the quotient $\pi_{1,-\rho}$ has been created seemingly from nothing.

Introducing some hypotheses of dominance avoids such behavior. Indeed, if λ and λ' are dominant nonsingular, it turns out that $\psi_\lambda^{\lambda'}$ is an isomorphism of the category of $U(\mathfrak{g})$ modules with infinitesimal character λ onto the category of $U(\mathfrak{g})$ modules with infinitesimal character λ'. (See Theorem 7.173 in §10.) But we can get away with much less. For the principal series in Example 2 of §1, we can adopt the point of view that only the real parts of the parameters are important as a sufficient condition for good qualitative behavior. In fact, in this example, as soon as the imaginary part is not 0, there is no trouble at all; the principal-series representations in question are irreducible, and irreducibles get mapped to irreducibles. Even if the imaginary part is 0, the trouble arises only at multiples of ρ, thus only when dominance fails and some integrality is in force.

The point of this section is to quantify the hypotheses leading to good behavior. The notion that we shall introduce is that of "integral dominance." It enlarges the ordinary notion of dominance and also of dominance of the real part. Actually when dominance (or at least dominance of the real part) is in effect, much of the development in this section is unnecessary, as the remarks after Propositions 7.166 and 7.170 explain. However, "integral dominance" should be viewed as a best possible hypothesis and is therefore essential for the more advanced parts of the theory.

Let $\mathfrak{g}$ be a complex semisimple Lie algebra, and let $\mathfrak{h}$ be a Cartan subalgebra. Form the set $\Delta(\mathfrak{g}, \mathfrak{h})$ of roots, and fix a positive system $\Delta^+(\mathfrak{g}, \mathfrak{h})$. Let $\mathfrak{h}_0^*$ be the set of **real elements** of $\mathfrak{h}^*$, i.e., the $\mathbb{R}$ span of $\Delta(\mathfrak{g}, \mathfrak{h})$ in $\mathfrak{h}^*$. A member λ of $\mathfrak{h}^*$ is said to be **integrally dominant** if every $\alpha \in \Delta^+(\mathfrak{g}, \mathfrak{h})$ satisfies

$$\frac{2\langle \lambda, \alpha \rangle}{|\alpha|^2} \notin \{-1, -2, -3, \dots\}.$$

Dominant implies integrally dominant. Real-part dominant implies integrally dominant.

We can rephrase this definition. Let

$$\Delta(\lambda) = \left\{ \alpha \in \Delta(\mathfrak{g}, \mathfrak{h}) \,\middle|\, \frac{2\langle \lambda, \alpha \rangle}{|\alpha|^2} \in \mathbb{Z} \right\}.$$

This set (possibly empty) is a root system in a subspace of $\mathfrak{h}_0^*$ in the sense that it is closed under its own reflections. In fact, if α and β are in $\Delta(\lambda)$, then

$$\frac{2\langle \lambda, s_\alpha \beta \rangle}{|s_\alpha \beta|^2} = \frac{2\langle \lambda, \beta \rangle}{|\beta|^2} - \frac{2\langle \lambda, \alpha \rangle}{|\beta|^2} \frac{2\langle \beta, \alpha \rangle}{|\alpha|^2}$$

is an integer, and the assertion follows. The parameter $\lambda \in \mathfrak{h}^*$ is always algebraically integral with respect to $\Delta(\lambda)$, and λ is integrally dominant for $\Delta(\mathfrak{g}, \mathfrak{h})$ exactly if λ is dominant relative to $\Delta^+(\lambda) = \Delta(\lambda) \cap \Delta^+(\mathfrak{g}, \mathfrak{h})$.

EXAMPLE. Let $\Delta(\mathfrak{g}, \mathfrak{h})$ be a system of type B_2, namely

$$\Delta(\mathfrak{g}, \mathfrak{h}) = \{\pm e_1, \pm e_2, \pm e_1 \pm e_2\},$$

and let $\lambda = \frac{1}{2}e_1$. Then $\Delta(\lambda) = \{\pm e_1, \pm e_2\}$, which is a root system of type $A_1 + A_1$. Note for this case that $\Delta(\lambda)$ is not closed under addition within $\Delta(\mathfrak{g}, \mathfrak{h})$: e_1 and e_2 are in $\Delta(\lambda)$ and $e_1 + e_2$ is a root, but $e_1 + e_2$ is not in $\Delta(\lambda)$. Therefore $\mathfrak{h}$ and the root vectors for $\Delta(\lambda)$ do not span a Lie subalgebra of $\mathfrak{g}$.

Since $\Delta(\lambda)$ is a root system, it has a Weyl group $W(\lambda)$, which is generated by reflections s_α in the roots of $\Delta(\lambda)$. Any λ in $\mathfrak{h}^*$ is automatically integral relative to $\Delta(\lambda)$. Therefore

$$\lambda - s\lambda = (\text{sum of roots in } \Delta(\lambda))$$

if s is in $W(\lambda)$. Our first objective in this section is to prove a converse.

Proposition 7.145. If $\lambda \in \mathfrak{h}^*$ and $s \in W(\mathfrak{g}, \mathfrak{h})$ have the property that $\lambda - s\lambda$ is a sum of roots in $\Delta(\mathfrak{g}, \mathfrak{h})$, then s is in $W(\lambda)$.

The proof of this proposition is surprisingly long. We shall concentrate on the case that λ is real and derive the proposition as an easy consequence of a result known as "Chevalley's Lemma for affine Weyl groups." Recall that the ordinary version of Chevalley's Lemma (Proposition 4.146) says that the subgroup of the Weyl group fixing an element is generated by reflections in roots orthogonal to that element. We need a version of this lemma that enlarges the Weyl group to allow translations by roots.

Until further notice in this section, we shall work with an abstract situation. After we have a little terminology, we shall state Chevalley's Lemma for affine Weyl groups. We then digress to show how this result implies Proposition 7.145 and afterward return to the abstract situation to prove Chevalley's Lemma. At the end of the section, we derive a handy consequence of the proposition (Proposition 7.166) and show how it yields formulas for the effect of translation functors on Verma modules (Corollary 7.169).

The abstract situation is as follows: V is a finite-dimensional real inner product space, Δ is a reduced root system in V, and W is the Weyl group of Δ. Let $\mathbb{Z}(\Delta)$ be the lattice in V generated by the roots. The **affine Weyl group** W^{aff} is the group of transformations of V of the form

$$(7.146) \qquad w_{s,\tau} : v \mapsto sv + \tau \quad \text{with } s \in W \text{ and } \tau \in \mathbb{Z}(\Delta).$$

Because W preserves the lattice $\mathbb{Z}(\Delta)$, W^{aff} is a group. The group of translations by members of $\mathbb{Z}(\Delta)$, which we shall identify with $\mathbb{Z}(\Delta)$, is a normal subgroup, and W^{aff} is the semidirect product of W and $\mathbb{Z}(\Delta)$.

An **affine root** is a pair $\tilde{\alpha} = (\alpha, m) \in \Delta \times \mathbb{Z}$. We write Δ^{aff} for the set of affine roots. Let us abbreviate $2|\alpha|^{-2}\alpha$ as $\alpha^{\vee}$. To the affine root $\tilde{\alpha} = (\alpha, m)$, we associate the **affine function**

$$(7.147) \qquad \tilde{\alpha}(\lambda) = \langle \lambda, \alpha^{\vee} \rangle - m \qquad \text{for } \lambda \in V.$$

This function determines the pair $(\alpha^{\vee}, m)$ by means of the formulas $m = -\tilde{\alpha}(0)$ and $\langle \lambda, \alpha^{\vee} \rangle = \tilde{\alpha}(\lambda) + m$, and therefore it determines (α, m) as well (since $\alpha = 2\alpha^{\vee}/\langle \alpha^{\vee}, \alpha^{\vee} \rangle$). It will sometimes be convenient to identify Δ^{aff} with this set of affine functions, as the notation (7.147) does.

WARNING. There is one possible source of confusion in this notation. If $m = 0$, $\tilde{\alpha}$ is being identified not with the function given by the root α, but rather with the function given by the coroot $\alpha^{\vee}$. It might be more logical to write $(\alpha^{\vee})^{\sim}$ and to speak of affine coroots, but we have avoided doing so in order to keep the notation a little simpler.

Next we associate to $\tilde{\alpha}$ the half space

$$(7.148) \qquad V_{\tilde{\alpha}}^{+} = V_{\alpha,m}^{+} = \{\lambda \in V \mid \tilde{\alpha}(\lambda) \geq 0\} = \{\lambda \in V \mid \langle \lambda, \alpha^{\vee} \rangle \geq m\}$$

and its boundary hyperplane

$$(7.149) \qquad H_{\tilde{\alpha}} = H_{\alpha,m} = \{\lambda \in V \mid \tilde{\alpha}(\lambda) = 0\}\{\lambda \in V \mid \langle \lambda, \alpha^{\vee} \rangle = m\}.$$

Write $s_{\tilde{\alpha}} = s_{\alpha,m}$ for the (affine) **reflection** of V in the hyperplane $H_{\tilde{\alpha}}$:

$$(7.150)$$
$$s_{\tilde{\alpha}}(\lambda) = \lambda - \tilde{\alpha}(\lambda)\alpha = s_{\alpha,m}(\lambda) = \lambda - (\langle \lambda, \alpha^{\vee} \rangle - m)\alpha = s_{\alpha}(\lambda) + m\alpha.$$

By (7.146), $s_{\alpha,m}$ equals the transformation $w_{s_{\alpha},m\alpha}$ and hence belongs to W^{aff}. Notice that

$$(7.151) \qquad s_{\alpha,1}s_{\alpha,0}(\lambda) = s_{\alpha,1}(s_{\alpha}(\lambda)) = s_{\alpha}^{2}(\lambda) + \alpha = \lambda + \alpha.$$

Lemma 7.152. W^{aff} is generated by the affine reflections $s_{\tilde{\alpha}}$ corresponding to all affine roots.

PROOF. Write W'^{aff} for the subgroup generated by these reflections. By (7.151), W'^{aff} contains translation by α. Consequently W'^{aff} contains $\mathbb{Z}(\Delta)$. Since $s_{\alpha,0} = s_\alpha$, W'^{aff} contains W.

Proposition 7.153 (Chevalley's Lemma for affine Weyl groups). Fix λ in V and define

$$\Delta^{\mathrm{aff}}(\lambda) = \{\tilde{\alpha} \in \Delta^{\mathrm{aff}} \mid s_{\tilde{\alpha}}\lambda = \lambda\} = \{(\alpha, m) \in \Delta^{\mathrm{aff}} \mid \langle \lambda, \alpha^\vee \rangle = m\}.$$

Then the subgroup $W^{\mathrm{aff}}(\lambda)$ of W^{aff} given by

$$W^{\mathrm{aff}}(\lambda) = \{w \in W^{\mathrm{aff}} \mid w\lambda = \lambda\}$$

is generated by reflections in the members of $\Delta^{\mathrm{aff}}(\lambda)$.

PROOF THAT PROPOSITION 7.153 IMPLIES PROPOSITION 7.145. We return to the notation involving the concrete $\Delta(\mathfrak{g}, \mathfrak{h})$. Write $\lambda = \mu + i\nu$ with μ and ν in $\mathfrak{h}_0^*$. If $\lambda - s\lambda$ is a sum of roots, then $\mu - s\mu = \tau$ is a sum of roots and also $s\nu = \nu$. Let w be the member $w_{s,\tau}$ of W^{aff}, using the notation of (7.146). From $s\mu + \tau = \mu$, we have $w\mu = \mu$. Therefore Proposition 7.153 shows that

$$(7.154\mathrm{a}) \qquad\qquad w = s_{\tilde{\alpha}_1} \cdots s_{\tilde{\alpha}_n}$$

with the jth factor equal to the reflection in the member $\tilde{\alpha}_j = (\alpha_j, m_j)$ of $\Delta^{\mathrm{aff}}(\mu)$. Since W^{aff} is a semidirect product, we can factor out by the translation parts of Weyl group elements, passing from (7.154a) to

$$(7.154\mathrm{b}) \qquad\qquad s = s_{\alpha_1} \cdots s_{\alpha_n}.$$

Since $s_{\tilde{\alpha}_j}\mu = \mu$ for the member $\tilde{\alpha}_j$ of $\Delta^{\mathrm{aff}}(\mu)$, (7.150) gives $\langle \mu, \alpha_j^\vee \rangle = m_j$. Thus α_j is in $\Delta(\mu)$ and s is in $W(\mu)$. Finally the element s of $W(\mu)$ fixes ν and, by the classical version of Chevalley's Lemma (Proposition 4.146), s is the product of reflections in roots of $\Delta(\mu)$ orthogonal to ν. If s_β is such a reflection, then $\langle \mu, \beta^\vee \rangle$ is in $\mathbb{Z}$ and $\langle \nu, \beta^\vee \rangle = 0$. Therefore $\langle \lambda, \beta^\vee \rangle$ is in $\mathbb{Z}$. Hence s_β is in $W(\lambda)$, and so is s. This completes the proof.

Let us now go back to the abstract situation and prove Proposition 7.153. The group W^{aff} acts in a natural way on functions on V: $(wf)(\lambda) = f(w^{-1}\lambda)$. We are going to show that this action preserves the affine functions (7.147) and therefore defines an action of W^{aff} on Δ^{aff}.

Lemma 7.155. Suppose $\tilde{\alpha} = (\alpha, m)$ is an affine root, $w = w_{s,\tau}$ is in W^{aff}, and λ is in V.

(a) The action of W^{aff} on affine functions satisfies

$$(w\tilde{\alpha})(w\lambda) = \tilde{\alpha}(\lambda).$$

(b) W^{aff} permutes the affine roots. More precisely, $w^{-1}\tilde{\alpha}$ is equal to the affine function defined by the affine root $(s^{-1}\alpha, m - \langle \tau, \alpha^\vee \rangle)$.

(c) If $\tilde{\beta} = (\beta, n)$ is an affine root, then $\tilde{\gamma} = s_{\tilde{\beta}}\tilde{\alpha}$ is given by

$$\tilde{\gamma} = (s_\beta\alpha, \, m - n\langle \beta, \alpha^\vee \rangle).$$

(d) W^{aff} permutes the half spaces $V_{\tilde{\alpha}}^+$ and the hyperplanes $H_{\tilde{\alpha}}$:

$$w V_{\tilde{\alpha}}^+ = V_{w\tilde{\alpha}}^+, \qquad w H_{\tilde{\alpha}} = H_{w\tilde{\alpha}}.$$

(e) $s_{w\tilde{\alpha}} = w s_{\tilde{\alpha}} w^{-1}$.

PROOF. Part (a) is just a reformulation of the definition of the action of W^{aff} on functions. For (b), we have

$$
\begin{aligned}
(w^{-1}\tilde{\alpha})(\lambda) &= \tilde{\alpha}(w\lambda) \\
&= \tilde{\alpha}(s\lambda + \tau) \\
&= \langle s\lambda + \tau, \alpha^\vee \rangle - m && \text{by (7.147)} \\
&= \langle \lambda, s^{-1}\alpha^\vee \rangle - (m - \langle \tau, \alpha^\vee \rangle) \\
&= (s^{-1}\alpha, \, (m - \langle \tau, \alpha^\vee \rangle))(\lambda) && \text{by (7.147)}.
\end{aligned}
$$

Part (c) is the special case of (b) in which $w = w^{-1} = w_{s_\beta, n\beta}$. Part (d) is a formal consequence of (b) and the definitions of the half spaces and hyperplanes in terms of affine functions. For (e), the left side fixes $H_{w\tilde{\alpha}}$, and the right side fixes $w H_{\tilde{\alpha}}$; these hyperplanes are equal by (d). Since the fixed set of a reflection determines the reflection, (e) follows.

If α is in Δ, then the set $\{\mu \in V \mid \langle \mu, \alpha^\vee \rangle \notin \mathbb{Z}\}$ is open and dense in V. Hence the finite intersection

$$R(V) = \cap_{\alpha \in \Delta}\{\mu \in V \mid \langle \mu, \alpha^\vee \rangle \notin \mathbb{Z}\}$$

is open and dense in V. The members of $R(V)$ are called **regular elements**. The connected components of $R(V)$ are open and are called **cells**. If we fix a positive system Δ^+ for Δ, then one such cell is

$$C_0 = \{\mu \in V \mid 0 < \langle \mu, \alpha^\vee \rangle < 1 \text{ for all } \alpha \in \Delta^+\}.$$

Since W^{aff} maps $R(V)$ into itself, W^{aff} permutes the cells.

A **semiregular element** $\mu_0 \in V$ is an element lying on one and only one hyperplane $H_{\tilde{\alpha}}$ (hence satisfying $\tilde{\beta}(\mu_0) = 0$ for $\tilde{\beta} = \pm\tilde{\alpha}$ but for no other affine roots).

Each affine root $\tilde{\alpha}$ is nonvanishing on $R(V)$ and has constant sign on each cell (since $\mathrm{sign}\,\tilde{\alpha}$ is a continuous function from the cell to $\{\pm 1\}$.) We call an affine root **positive** if it is positive on C_0. We call an affine root $\tilde{\alpha} = (\alpha, m)$ **simple** if it is positive and if there is some semiregular element μ_0 in the closure $\overline{C}_0$ of C_0 such that $\tilde{\alpha}(\mu_0) = 0$.

Lemma 7.156. If α is simple for Δ^+, then $(\alpha, 0)$ is a simple affine root. If $\alpha^\vee$ is the highest root in some simple component of $(\Delta^\vee)^+$, then $(-\alpha, -1)$ is a simple affine root.

REMARK. $\Delta^\vee$ is the root system $\{\alpha^\vee \mid \alpha \in \Delta\}$, and $(\Delta^\vee)^+$ equals $\{\alpha^\vee \mid \alpha \in \Delta^+\}$. Its simple components correspond under $(\cdot)^\vee$ to the simple components of Δ.

PROOF. Let $\{\alpha_j\}$ be the simple roots for Δ^+, and define Λ_i by $\langle \Lambda_i, \alpha_j^\vee \rangle = \delta_{ij}$. Fix α_j and define $\mu_0 = \epsilon \sum_{i \neq j} \Lambda_i$ for $\epsilon > 0$ small enough so that $|\langle \mu_0, \alpha^\vee \rangle| < 1$ for all $\alpha \in \Delta$. Then $\mu_0 + \epsilon'\Lambda_j$ is in C_0 for small $\epsilon' > 0$, and hence μ_0 is in $\overline{C}_0$. If β is in Δ^+, then $\langle \mu_0, \beta \rangle > 0$ for $\beta \neq \alpha_j$ while $\langle \mu_0, \alpha_j \rangle = 0$. A general affine root is of the form $\tilde{\beta} = (\beta, n)$, and

$$\tilde{\beta}(\mu_0) = \langle \mu_0, \beta^\vee \rangle - n.$$

Since $|\langle \mu_0, \beta^\vee \rangle| < 1$, this can be 0 only if $n = 0$. In that case we get 0 only if $\langle \mu_0, \beta^\vee \rangle = 0$, which forces $\beta = \pm\alpha_j$. Hence μ_0 is semiregular and exhibits $(\alpha_j, 0)$ as a simple affine root.

Let $\Delta_j^\vee$ be the simple components of $\Delta^\vee$, let $\alpha_j^\vee$ be the highest root in $(\Delta_j^\vee)^+$, and let $\delta_j^\vee$ be half the sum of the positive roots of $\Delta_j^\vee$. We show that $(-\alpha_i, -1)$ is a simple affine root by using $\mu_0 = \sum c_j \delta_j$ for a suitable choice of constants $c_j > 0$. Namely we choose $c_i = \langle \delta_i, \alpha_i^\vee \rangle^{-1}$ and the other c_j so that $c_j|\langle \delta_j, \alpha^\vee \rangle| < 1$ for all $\alpha \in \Delta$. First of all, if $\alpha^\vee$ is in $(\Delta_j^\vee)^+$, then

$$\langle (1 - \epsilon)\mu_0, \alpha^\vee \rangle = (1 - \epsilon)c_j\langle \delta_j, \alpha^\vee \rangle,$$

and this is strictly between 0 and 1 if $0 < \epsilon < 1$. Hence $(1-\epsilon)\mu_0$ is in C_0. Thus μ_0 is in $\overline{C}_0$, and $\langle \mu_0, \alpha^\vee \rangle = c_j\langle \delta_j, \alpha^\vee \rangle$. Since δ_j is strictly dominant for $(\Delta_j^\vee)^+$, it follows that $\langle \mu_0, \alpha^\vee \rangle$ is an integer only if $\alpha = \pm\alpha_i$. Hence $\tilde{\alpha}(\mu_0) = 0$ only for $\tilde{\alpha} = \pm(\alpha_i, 1)$. Therefore μ_0 is a semiregular element and exhibits $(-\alpha_i, -1)$ as a simple affine root.

Lemma 7.157. If μ is a regular element in a cell other than C_0, then there exists a simple affine root $\tilde{\alpha}$ such that $\tilde{\alpha}(\mu) < 0$.

PROOF. First suppose $\langle \mu, \alpha^\vee \rangle < 0$ for some $\alpha \in \Delta^+$. In this case $\langle \mu, \alpha^\vee \rangle < 0$ for some simple root α_j in Δ^+. Thus $(\alpha_j, 0)(\mu) < 0$. By Lemma 7.156, $(\alpha_j, 0)$ is the required simple affine root.

Now suppose $\langle \mu, \alpha^\vee \rangle > 0$ for all $\alpha \in \Delta^+$. Since μ is not in C_0, $\langle \mu, \alpha^\vee \rangle > 1$ for some $\alpha \in \Delta^+$. If $\alpha_0^\vee$ denotes the largest root in the simple component of $\Delta^\vee$ to which $\alpha^\vee$ belongs, then $\langle \mu, \alpha_0^\vee \rangle > 1$ since $\alpha_0^\vee - \alpha^\vee$ is a sum of members of $(\Delta^\vee)^+$ and μ is dominant. Thus $(-\alpha_0, -1)(\mu) = -\langle \mu, \alpha_0^\vee \rangle + 1 < 0$. By Lemma 7.156, $(-\alpha_0, -1)$ is the required simple affine root.

Lemma 7.158. W^{aff} acts transitively on the set of cells, and W^{aff} is generated by the reflections in the simple affine roots.

PROOF. Let W'^{aff} be the subgroup of W^{aff} generated by the reflections in the simple affine roots. If C is a cell, we shall produce an element of W'^{aff} mapping C to C_0. In fact, fix μ in C and μ_0 in C_0, and choose $w_0 \in W'^{\text{aff}}$ so that $|w\mu - \mu_0|$ is minimized by $w = w_0$. We show that $w_0\mu$ is in C_0.

Assume the contrary. Then Lemma 7.157 provides a simple affine root $\tilde{\alpha}$ such that $\tilde{\alpha}(w_0\mu) < 0$. Then $s_{\tilde{\alpha}} w_0$ is another element of W'^{aff}, and we have

$$
\begin{aligned}
|s_{\tilde{\alpha}} w_0 \mu - \mu_0|^2 &- |w_0\mu - \mu_0|^2 \\
&= |(w_0\mu - \mu_0) - \tilde{\alpha}(w_0\mu)\alpha|^2 - |w_0\mu - \mu_0|^2 \\
&= -2\langle w_0\mu - \mu_0, \alpha \rangle \tilde{\alpha}(w_0\mu) + \tilde{\alpha}(w_0\mu)^2 |\alpha|^2 \\
&= |\alpha|^2 (-\langle w_0\mu - \mu_0, \alpha^\vee \rangle \tilde{\alpha}(w_0\mu) + \tilde{\alpha}(w_0\mu)^2) \\
&= |\alpha|^2 ((-\tilde{\alpha}(w_0\mu) + \tilde{\alpha}(\mu_0))\tilde{\alpha}(w_0\mu) + \tilde{\alpha}(w_0\mu)^2) \\
&= |\alpha|^2 \tilde{\alpha}(\mu_0)\tilde{\alpha}(w_0\mu).
\end{aligned}
$$

The right side is < 0, and we have a contradiction to the defining condition for w_0. We conclude that $w_0\mu$ is in C_0. Hence $wC = C_0$. This proves the transitivity.

Let $\tilde{\alpha} = (\alpha, m)$. To complete the proof, we show that $s_{\tilde{\alpha}} = w^{-1} s_{\tilde{\gamma}} w$ for some $w \in W'^{\text{aff}}$ and some simple affine root $\tilde{\gamma}$; this is enough by Lemma 7.152. On the hyperplane $H_{\tilde{\alpha}}$ where $\langle \mu, \alpha^\vee \rangle = m$, there are no points where $\langle \mu, \alpha^\vee \rangle = m'$ with $m' \neq m$, and the subset, for given β, where $\langle \mu, \beta^\vee \rangle$ is in $\mathbb{Z}$ is closed and nowhere dense. Hence the subset of $H_{\tilde{\alpha}}$ that lies in no hyperplane $H_{\tilde{\beta}}$ other than for $\tilde{\beta} = \pm\tilde{\alpha}$ is relatively open and dense. Let μ_1 be a point in this set, and let $\epsilon_0 > 0$ be the distance to the union of the other hyperplanes $H_{\tilde{\beta}}$. For sufficiently small $\epsilon > 0$, the element $\mu_1 + \epsilon\alpha$ is in the open ϵ_0 ball about μ_1. Hence $\tilde{\beta}(\mu_1 + \epsilon\alpha) \neq 0$ for $\tilde{\beta} \neq \pm\tilde{\alpha}$. Also

$$\tilde{\alpha}(\mu_1 + \epsilon\alpha) = \langle \mu_1 + \epsilon\alpha, \alpha^\vee \rangle - m = 2\epsilon \neq 0.$$

Therefore $\mu_1 + \epsilon\alpha$ is regular and lies in some cell C.

By what we have shown, choose $w \in W'^{\mathrm{aff}}$ with $wC = C_0$. In particular, $w(\mu_1 + \epsilon\alpha)$ is in C for ϵ as above. Letting ϵ decrease to 0, we see that $w\mu_1$ is in $\overline{C}_0$. Also Lemma 7.155a gives

$$(w\tilde{\alpha})(w\mu_1) = \tilde{\alpha}(\mu_1) = 0$$

and
$$(w\tilde{\beta})(w\mu_1) = \tilde{\beta}(\mu_1) \neq 0 \qquad \text{for } \tilde{\beta} \neq \pm\tilde{\alpha}.$$

Hence $w\mu_1$ is a semiregular element that exhibits one of $\pm w\tilde{\alpha}$ as simple. Finally $s_{w\tilde{\alpha}} = w s_{\tilde{\alpha}} w^{-1}$ by Lemma 7.155e. Hence, as required, $s_{\tilde{\alpha}}$ is of the form $w^{-1} s_{\tilde{\gamma}} w$ with $w \in W'^{\mathrm{aff}}$ and $\tilde{\gamma} = \pm w\tilde{\alpha}$ a simple affine root.

If w is in W^{aff}, then it follows that w has a decomposition as the product of reflections in simple affine roots, say as $w = s_{\tilde{\alpha}_1} \cdots s_{\tilde{\alpha}_r}$. For a given element w, if the number r is as small as possible among all such decompositions, we say that the decomposition is **reduced**. The number of factors in a reduced decomposition of w is called the length of w and is denoted $l(w)$.

Lemma 7.159. If w is in W^{aff}, then the number of positive affine roots $\tilde{\alpha}$ such that $w\tilde{\alpha}$ is negative is finite and equals the length of w.

PROOF. Let $\tilde{\alpha}$ be a simple affine root. If $\tilde{\beta} \neq \tilde{\alpha}$, then

$$(7.160) \qquad \tilde{\beta} > 0 \quad \text{implies} \quad s_{\tilde{\alpha}}\tilde{\beta} > 0.$$

In fact, if μ_0 is a semiregular element in $\overline{C}_0$ with $\tilde{\alpha}(\mu_0) = 0$, then $\mu_0 + \epsilon\tilde{\alpha}$ is in C_0 for small $\epsilon > 0$, and $\mu_0 - \epsilon\tilde{\alpha}$ is therefore in $s_{\tilde{\alpha}}C_0$. The affine root $\tilde{\beta}$ has $\tilde{\beta}(\mu_0 + t\tilde{\alpha}) > 0$ for $|t|$ sufficiently small, and thus Lemma 7.155a gives $s_{\tilde{\alpha}}\tilde{\beta}(C_0) = \tilde{\beta}(s_{\tilde{\alpha}}C_0) > 0$. So $s_{\tilde{\alpha}}\tilde{\beta} > 0$.

Still with $\tilde{\alpha}$ simple, let w in W^{aff} have $w^{-1}\tilde{\alpha} > 0$, and let $\{\tilde{\beta}_i\}$ be the positive roots with $w^{-1}\tilde{\beta}_i < 0$. Then

$$(7.161) \qquad \begin{array}{l} \text{the positive affine roots } \tilde{\gamma}_i \text{ with } w^{-1}s_{\tilde{\alpha}}\tilde{\gamma}_i < 0 \text{ are } \tilde{\alpha} \\ \text{and the } s_{\tilde{\alpha}}\tilde{\beta}_i. \end{array}$$

In fact, we are assuming that no $\tilde{\beta}_i$ equals $\tilde{\alpha}$. Hence $\tilde{\alpha}$ and the $\tilde{\gamma}_i = s_{\tilde{\alpha}}\tilde{\beta}_i$ are positive, by (7.160). Also $w^{-1}s_{\tilde{\alpha}}(\tilde{\alpha}) = -w^{-1}\tilde{\alpha}$ is < 0, and so is

$$w^{-1}s_{\tilde{\alpha}}\tilde{\gamma}_i = w^{-1}s_{\tilde{\alpha}}s_{\tilde{\alpha}}\tilde{\beta}_i < 0.$$

If $\tilde{\gamma}$ is another positive affine root with $w^{-1}s_{\tilde{\alpha}}\tilde{\gamma} < 0$, then $\tilde{\gamma} \neq \tilde{\alpha}$ and so $\tilde{\beta} = s_{\tilde{\alpha}}\tilde{\gamma}$ is > 0 by (7.160). This $\tilde{\beta}$ has $w^{-1}\tilde{\beta} = w^{-1}s_{\tilde{\alpha}}\tilde{\gamma} < 0$, and (7.161) is proved.

Now let $w_0 = s_{\tilde{\alpha}_1} \cdots s_{\tilde{\alpha}_r}$ be a reduced decomposition in W^{aff}. We prove by induction on r that the positive $\tilde{\alpha}$'s with $w_0^{-1}\tilde{\alpha} < 0$ are

$$(7.162) \qquad s_{\tilde{\alpha}_1} \cdots s_{\tilde{\alpha}_{r-1}}(\tilde{\alpha}_r), \ s_{\tilde{\alpha}_1} \cdots s_{\tilde{\alpha}_{r-2}}(\tilde{\alpha}_{r-1}), \ \ldots, \ s_{\tilde{\alpha}_1}(\tilde{\alpha}_2), \ \tilde{\alpha}_1$$

and that these roots are distinct. The base case of the induction is $r = 1$, which is handled by (7.160).

Assume the inductive statement for $r - 1$, and let $w_0 = s_{\tilde{\alpha}_1} \cdots s_{\tilde{\alpha}_r}$ be a given reduced decomposition. Then $w = s_{\tilde{\alpha}_2} \cdots s_{\tilde{\alpha}_r}$ is a reduced decomposition, and the inductive hypothesis applies. If $w^{-1}\tilde{\alpha}_1 > 0$, then (7.161) with $\tilde{\alpha} = \tilde{\alpha}_1$ shows that the positive affine roots $\tilde{\gamma}$ with $w_0^{-1}\tilde{\gamma} < 0$ are exactly those in (7.162). Moreover, by inductive hypothesis

$$(7.163) \qquad s_{\tilde{\alpha}_2} \cdots s_{\tilde{\alpha}_{r-1}}(\tilde{\alpha}_r), \ s_{\tilde{\alpha}_2} \cdots s_{\tilde{\alpha}_{r-2}}(\tilde{\alpha}_{r-1}), \ \ldots, \ \tilde{\alpha}_2.$$

are distinct. Hence $s_{\tilde{\alpha}_1}$ of these are distinct. Also $s_{\tilde{\alpha}_1}$ of none of these can equal $\tilde{\alpha}_1$ since $s_{\tilde{\alpha}_1}\tilde{\alpha}_1 < 0$. Hence the induction is complete if $w^{-1}\tilde{\alpha}_1 > 0$.

Suppose on the contrary that $w^{-1}\tilde{\alpha}_1 < 0$. Then by inductive hypothesis, $\tilde{\alpha}_1$ is one of the positive affine roots in the list (7.163), say $\tilde{\alpha}_1 = s_{\tilde{\alpha}_2} \cdots s_{\tilde{\alpha}_{j-1}}(\tilde{\alpha}_j)$. By Lemma 7.155e,

$$s_{\tilde{\alpha}_1} = s_{\tilde{\alpha}_2} \cdots s_{\tilde{\alpha}_{j-1}}s_{\tilde{\alpha}_j}(s_{\tilde{\alpha}_2} \cdots s_{\tilde{\alpha}_{j-1}})^{-1}$$

and hence

$$w_0 = s_{\tilde{\alpha}_1} \cdots s_{\tilde{\alpha}_r} = s_{\tilde{\alpha}_1} \cdots s_{\tilde{\alpha}_{j-1}}s_{\tilde{\alpha}_j}s_{\tilde{\alpha}_{j+1}} \cdots s_{\tilde{\alpha}_r} = s_{\tilde{\alpha}_2} \cdots s_{\tilde{\alpha}_j}s_{\tilde{\alpha}_j} \cdots s_{\tilde{\alpha}_r}.$$

The $s_{\tilde{\alpha}_j}^2$ drops out on the right side, and w_0 has a decomposition with $r - 2$ factors, contradiction.

PROOF OF PROPOSITION 7.153. First suppose λ is in $\overline{C}_0$. Then for any affine root $\tilde{\beta}$,

$$(7.164) \qquad \tilde{\beta}(\lambda) > 0 \quad \text{implies} \quad \tilde{\beta} > 0.$$

For w in W^{aff} fixing λ, we prove inductively on $l(w)$ that w is the product of reflections in members of $\Delta^{\mathrm{aff}}(\lambda)$. The case $l(w) = 1$ is trivial. Assume the result for length $< r$, and let $w \in W^{\mathrm{aff}}(\lambda)$ have length r. Let $w = s_{\tilde{\gamma}_r} \cdots s_{\tilde{\gamma}_1}$ be a reduced decomposition. Referring to (7.162) for the element w^{-1}, we see that the simple affine root $\tilde{\gamma}_1$ has $w\tilde{\gamma}_1 < 0$. Since $\tilde{\gamma}_1 > 0$, we must have $\tilde{\gamma}_1(\lambda) \geq 0$. If $\tilde{\gamma}_1(\lambda) > 0$, then Lemma 7.155a gives

$$w\tilde{\gamma}_1(\lambda) = w\tilde{\gamma}_1(w\lambda) = \tilde{\gamma}_1(\lambda) > 0,$$

and (7.164) shows that $w\tilde{\gamma}_1 > 0$, contradiction. So $\tilde{\gamma}_1(\lambda) = 0$, and $\tilde{\gamma}_1$ is in $\Delta^{\mathrm{aff}}(\lambda)$. Then $ws_{\tilde{\gamma}_1} = s_{\tilde{\gamma}_r} \cdots s_{\tilde{\gamma}_2}$ has shorter length than w and is in $W^{\mathrm{aff}}(\lambda)$. By inductive hypothesis it is the product of reflections in members of $\Delta^{\mathrm{aff}}(\lambda)$, and hence so is w.

In the general case, λ is in the closure $\overline{C}$ of some cell. From Lemma 7.158 we can choose $w_0 \in W^{\mathrm{aff}}$ with $w_0 C = C_0$, hence with $w_0\overline{C} = \overline{C}_0$. Then $w_0\lambda$ is in $\overline{C}_0$. If $w\lambda = \lambda$, then $w_0 w w_0^{-1}(w_0\lambda) = w_0\lambda$. From the previous paragraph, $w_0 w w_0^{-1}$ is the product of reflections in members of $\Delta^{\mathrm{aff}}(w_0\lambda)$. Say $w_0 w w_0^{-1} = s_{\tilde{\beta}_1} \cdots s_{\tilde{\beta}_n}$ with $s_{\tilde{\beta}_j}(w_0\lambda) = w_0\lambda$. Then Lemma 7.155e gives $w = s_{w_0^{-1}\tilde{\beta}_1} \cdots s_{w_0^{-1}\tilde{\beta}_n}$ and $s_{w_0^{-1}\tilde{\beta}_j}(\lambda) = w_0^{-1}s_{\tilde{\beta}_j}w_0\lambda = \lambda$. Thus each $w_0^{-1}\tilde{\beta}_j$ is in $\Delta^{\mathrm{aff}}(\lambda)$, and w is the product of reflections in these affine roots.

This completes the proof of Proposition 7.145 and the discussion of the abstract situation involving affine Weyl groups. For the remainder of the section, we return to the setting that $\mathfrak{g}$ is a complex semisimple Lie algebra, $\mathfrak{h}$ is a Cartan subalgebra, and $\Delta(\mathfrak{g}, \mathfrak{h})$ is the root system. We shall derive two consequences of Proposition 7.145 in the propositions below and use the first one to compute the ψ functor on Verma modules under hypotheses of integral dominance.

Lemma 7.165. Suppose v is a real element of $\mathfrak{h}^*$, and suppose $s \in W$ satisfies $\langle sv, v \rangle = |v|^2$. Then $sv = v$.

PROOF. We have

$$\begin{aligned}
|sv - v|^2 &= |sv|^2 - 2\langle sv, v \rangle + |v|^2 \\
&= 2(|v|^2 - \langle sv, v \rangle) && \text{since } s \text{ preserves } | \cdot |^2 \\
&= 0 && \text{since } \langle sv, v \rangle = |v|^2.
\end{aligned}$$

Therefore $sv = v$.

Proposition 7.166. Let F^μ be an irreducible finite-dimensional complex-linear representation of $\mathfrak{g}$ with μ as an extreme weight, and suppose that μ' is a weight of F^μ. Fix a positive system $\Delta^+(\mathfrak{g}, \mathfrak{h})$, and suppose that λ is a member of $\mathfrak{h}^*$ satisfying

(a) λ and $\lambda + \mu$ are integrally dominant, and
(b) $\lambda + \mu$ is at least as singular as λ, i.e., $\langle \lambda + \mu, \alpha \rangle = 0$ for every $\alpha \in \Delta(\mathfrak{g}, \mathfrak{h})$ such that $\langle \lambda, \alpha \rangle = 0$.

Then the equation

$$\lambda + \mu' = s(\lambda + \mu)$$

can hold for $s \in W$ only if $\mu' = \mu$ and s fixes $\lambda + \mu$.

PROOF. Let us extend the usual form $\langle \cdot, \cdot \rangle$ to be complex bilinear on $\mathfrak{h}^* \times \mathfrak{h}^*$. We can rewrite the given equation as

$$\lambda + \mu - s(\lambda + \mu) = \mu - \mu'.$$

Since $\mu - \mu'$ is a sum of roots, Proposition 7.145 shows that s is in $W(\lambda + \mu)$, the Weyl group of $\Delta(\lambda + \mu)$. The integral dominance of $\lambda + \mu$ given in (a) means that $\lambda + \mu$ is dominant relative to $\Delta^+(\lambda + \mu)$. Since s is in $W(\lambda + \mu)$,

$$(7.167) \qquad (\lambda + \mu) - s(\lambda + \mu) = \sum_{\alpha \in \Delta^+(\lambda+\mu)} n_\alpha \alpha \qquad \text{with } n_\alpha \geq 0.$$

Let us rewrite the given equation in a different way as

$$\mu' = s(\lambda + \mu) - \lambda.$$

Computing $\langle \mu', \mu' \rangle$ and using the Weyl group invariance of $\langle \cdot, \cdot \rangle$, we have

$$
\begin{aligned}
|\mu'|^2 &= \langle s(\lambda + \mu) - \lambda, \, s(\lambda + \mu) - \lambda \rangle \\
&= \langle \lambda + \mu, \lambda + \mu \rangle - 2\langle s(\lambda + \mu), \lambda \rangle + \langle \lambda, \lambda \rangle \\
&= \langle \lambda + \mu, \lambda + \mu \rangle - 2\langle \lambda + \mu, \lambda \rangle + \langle \lambda, \lambda \rangle + \sum_{\alpha \in \Delta^+(\lambda+\mu)} n_\alpha \langle \alpha, \lambda \rangle \\
&= \langle \lambda + \mu - \lambda, \lambda + \mu - \lambda \rangle + \sum_{\alpha \in \Delta^+(\lambda+\mu)} n_\alpha \langle \alpha, \lambda \rangle \\
&= |\mu|^2 + \sum_{\alpha \in \Delta^+(\lambda+\mu)} n_\alpha \langle \alpha, \lambda \rangle.
\end{aligned}
$$

Now $\Delta(\lambda) = \Delta(\lambda + \mu)$ since μ is algebraically integral. Then the sum on the right is over roots $\alpha \in \Delta^+(\lambda)$. Since λ is assumed integrally dominant by (a), the sum is ≥ 0. On the other hand, Theorem 4.7e says that $|\mu'|^2 \leq |\mu|^2$, and thus $\sum_{\alpha \in \Delta^+(\lambda+\mu)} n_\alpha \langle \alpha, \lambda \rangle = 0$.

We may therefore rewrite (7.167) as

$$(7.168) \qquad (\lambda + \mu) - s(\lambda + \mu) = \sum_{\substack{\alpha \in \Delta^+(\mathfrak{g},\mathfrak{h}), \\ \langle \lambda, \alpha \rangle = 0}} n_\alpha \alpha.$$

By (b), we can write this equation as

$$(\lambda + \mu) - s(\lambda + \mu) = \sum_{\substack{\alpha \in \Delta^+(\mathfrak{g},\mathfrak{h}), \\ \langle \lambda+\mu, \alpha \rangle = 0}} n_\alpha \alpha.$$

The imaginary part of this equation says that s fixes $\operatorname{Im}(\lambda + \mu)$. Taking the real part of this equation and forming its inner product with $\operatorname{Re}(\lambda + \mu)$, we obtain

$$\langle s \operatorname{Re}(\lambda + \mu), \operatorname{Re}(\lambda + \mu) \rangle = |\operatorname{Re}(\lambda + \mu)|^2.$$

By Lemma 7.165, s fixes $\operatorname{Re}(\lambda + \mu)$. Therefore s fixes $\lambda + \mu$. Returning to the given equation, we then have $\lambda + \mu' = \lambda + \mu$ and hence $\mu' = \mu$.

REMARKS. If (a) in the proposition is replaced by the condition that $\operatorname{Re}\langle \lambda, \alpha \rangle \geq 0$ and $\operatorname{Re}\langle \lambda + \mu, \alpha \rangle \geq 0$ for all $\alpha \in \Delta^+(\mathfrak{g}, \mathfrak{h})$, then Proposition 7.166 becomes much more elementary, because Proposition 7.145 is not needed. The argument begins with the observation that s fixes the imaginary part of λ. The classical version of Chevalley's Lemma (Proposition 4.146) says that s is in the Weyl group of the system Δ_I of roots orthogonal to the imaginary part of λ; as a consequence we may work within Δ_I and assume that λ is real. Then we can skip down to (7.167), which is valid at least with the α's in Δ_I^+ and the n_α's real and ≥ 0. The computation with $|\mu'|^2$ leads to (7.168), and the rest of the proof goes through with no need for change.

Corollary 7.169. Let $\mu \in \mathfrak{h}^*$ be algebraically integral. Fix a positive system $\Delta^+(\mathfrak{g}, \mathfrak{h})$, and suppose that λ is a member of $\mathfrak{h}^*$ satisfying

(a) λ and $\lambda + \mu$ are integrally dominant relative to $\Delta^+(\mathfrak{g}, \mathfrak{h})$, and
(b) $\lambda + \mu$ is at least as singular as λ, i.e., $\langle \lambda + \mu, \alpha \rangle = 0$ for every $\alpha \in \Delta(\mathfrak{g}, \mathfrak{h})$ such that $\langle \lambda, \alpha \rangle = 0$.

Then the Verma modules $V(\lambda)$ and $V(\lambda+\mu)$, defined relative to a possibly different positive system $\Delta^{+\prime}(\mathfrak{g}, \mathfrak{h})$, satisfy

$$\psi_\lambda^{\lambda+\mu}(V(\lambda)) \cong V(\lambda + \mu).$$

PROOF. Let F^μ be an irreducible finite-dimensional complex-linear representation of $\mathfrak{g}$ with μ as an extreme weight. From the proof of Theorem 7.133, we know that $V(\lambda) \otimes_{\mathbb{C}} F^\mu$ has a filtration by $U(\mathfrak{g})$ modules with successive quotients $V(\lambda + \nu_1)$, $V(\lambda + \nu_2)$, ..., $V(\lambda + \nu_n)$, where the ν_i are the weights of F^μ repeated according to their multiplicities. In passing to the $\lambda + \mu$ primary component under $Z(\mathfrak{g})$, we retain those quotients whose infinitesimal characters match $\lambda + \mu$, i.e., those for which there is an $s \in W(\mathfrak{g}, \mathfrak{h})$ with

$$\lambda + \nu_j = s(\lambda + \mu).$$

According to Proposition 7.166, the only possibility has $\nu_j = \mu$, and this weight occurs with multiplicity 1. The result follows.

Proposition 7.170. Fix a positive system $\Delta^+(\mathfrak{g}, \mathfrak{h})$, and define Verma modules relative to it. Suppose λ and η are in $\mathfrak{h}^*$, and suppose λ is integrally dominant. Then $\dim \operatorname{Hom}_{\mathfrak{g}}(V(\lambda), V(\eta))$ is 1 if $\lambda = \eta$ and is 0 otherwise.

REMARKS. As with Proposition 7.166, this result becomes much more elementary if λ is assumed to be dominant (or to have real-part dominant). In this case, Proposition 7.145 is not needed.

PROOF. By Frobenius reciprocity (Propositions 2.34 and 2.33),

$$\operatorname{Hom}_{\mathfrak{g}}(V(\lambda), V(\eta)) \cong \operatorname{Hom}_{\mathfrak{b}}(\mathbb{C}_{\lambda - \delta}, V(\eta))$$

under restriction of a map in the left Hom to the multiples of the generator $1 \otimes 1$ of $V(\lambda)$. The generator has weight $\lambda - \delta$ and must map to a vector of weight $\lambda - \delta$ in $V(\eta)$. If $\lambda = \eta$, we obtain a one-dimensional space. In general, if $\operatorname{Hom}_{\mathfrak{g}}(V(\lambda), V(\eta)) \neq 0$, we must have $\lambda = \eta - \sum_{\alpha \in \Delta^+(\mathfrak{g}, \mathfrak{h})} n_\alpha \alpha$ with the n_α's integers ≥ 0 in order for $\lambda - \delta$ to be a weight of $V(\eta)$, and we must have $\eta = s\lambda$ for some $s \in W(\mathfrak{g}, \mathfrak{h})$ in order to get a match of infinitesimal characters. Then $\lambda - s\lambda = -\sum_{\alpha \in \Delta^+(\mathfrak{g}, \mathfrak{h})} n_\alpha \alpha$. Since the right side is a sum of roots, Proposition 7.145 shows that s is in $W(\lambda)$. Since λ is integrally dominant, $\lambda - s\lambda = \sum_{\alpha \in \Delta^+(\lambda)} m_\alpha \alpha$ with the $m_\alpha \geq 0$. Comparing the sums of roots, we see that both are 0. Hence $\lambda = s\lambda = \eta$.

An argument much like the one for Proposition 7.170 shows that if $-\lambda$ is integrally dominant, then the Verma module $V(\lambda)$ is irreducible. We shall not need this result, and we leave its proof to the interested reader.

10. Overview of Preservation of Irreducibility

Starting in this section, we address the extent to which the translation functors carry irreducible $U(\mathfrak{g})$ modules to irreducible $U(\mathfrak{g})$ modules, as well as nonzero modules to nonzero modules. Let us state the three main results right away.

Theorem 7.171. Let $\mathfrak{g}$ be a complex reductive Lie algebra, let $\mathfrak{h}$ be a Cartan subalgebra, and fix a system $\Delta^+(\mathfrak{g}, \mathfrak{h})$ of positive roots. Suppose $\mu \in \mathfrak{h}^*$ is algebraically integral. Further suppose λ is a member of $\mathfrak{h}^*$ satisfying

  (i) λ and $\lambda + \mu$ are integrally dominant, and
  (ii) $\lambda + \mu$ is at least as singular as λ, i.e., $\langle \lambda + \mu, \alpha \rangle = 0$ for every $\alpha \in \Delta(\mathfrak{g}, \mathfrak{h})$ such that $\langle \lambda, \alpha \rangle = 0$.

Then $\psi_\lambda^{\lambda+\mu}$ has the following properties:

  (a) if M is an irreducible member of $\mathcal{C}_\lambda(\mathfrak{g})$, then $\psi_\lambda^{\lambda+\mu}(M)$ is irreducible in $\mathcal{C}_{\lambda+\mu}(\mathfrak{g})$ or is 0.
  (b) if M and M' are irreducible members of $\mathcal{C}_\lambda(\mathfrak{g})$ such that

$$\psi_\lambda^{\lambda+\mu}(M) \cong \psi_\lambda^{\lambda+\mu}(M') \neq 0,$$

  then $M \cong M'$.
  (c) every member N of $\mathcal{C}_{\lambda+\mu}(\mathfrak{g})$ with an infinitesimal character is of the form $N \cong \psi_\lambda^{\lambda+\mu}(M)$ for some M in $\mathcal{C}_\lambda(\mathfrak{g})$ with an infinitesimal character. If N is irreducible, then M may be taken irreducible.

Theorem 7.172. Let $\mathfrak{g}$ be a complex reductive Lie algebra, let $\mathfrak{h}$ be a Cartan subalgebra, and fix a system $\Delta^+(\mathfrak{g}, \mathfrak{h})$ of positive roots. Suppose $\mu \in \mathfrak{h}^*$ is algebraically integral. Further suppose λ is a member of $\mathfrak{h}^*$ satisfying

  (i) λ and $\lambda + \mu$ are integrally dominant, and
  (ii) λ is at least as singular as $\lambda + \mu$, i.e., $\langle \lambda, \alpha \rangle = 0$ for every $\alpha \in \Delta(\mathfrak{g}, \mathfrak{h})$ such that $\langle \lambda + \mu, \alpha \rangle = 0$.

Then $\psi_\lambda^{\lambda+\mu}$ carries nonzero members of $\mathcal{C}_\lambda(\mathfrak{g})$ to nonzero members of $\mathcal{C}_{\lambda+\mu}(\mathfrak{g})$.

Theorem 7.173. Let $\mathfrak{g}$ be a complex reductive Lie algebra, let $\mathfrak{h}$ be a Cartan subalgebra, and fix a system $\Delta^+(\mathfrak{g}, \mathfrak{h})$ of positive roots. Suppose $\mu \in \mathfrak{h}^*$ is algebraically integral. Further suppose λ is a member of $\mathfrak{h}^*$ satisfying

(i) λ and $\lambda + \mu$ are integrally dominant, and

(ii) λ and $\lambda + \mu$ are equisingular, i.e., $\langle \lambda, \alpha \rangle = 0$ and $\langle \lambda + \mu, \alpha \rangle = 0$ for the same roots $\alpha \in \Delta(\mathfrak{g}, \mathfrak{h})$.

Then $\psi_\lambda^{\lambda+\mu}$ and $\psi_{\lambda+\mu}^{\lambda}$ have the following properties:

(a) if M is an irreducible member of $\mathcal{C}_\lambda(\mathfrak{g})$, then $\psi_\lambda^{\lambda+\mu}(M)$ is irreducible in $\mathcal{C}_{\lambda+\mu}(\mathfrak{g})$.

(b) if M is an irreducible member of $\mathcal{C}_\lambda(\mathfrak{g})$, then $\psi_{\lambda+\mu}^{\lambda}\psi_\lambda^{\lambda+\mu}(M) \cong M$.

(c) if $\mathcal{C}_\lambda^{(1)}(\mathfrak{g})$ is the subcategory of modules in $\mathcal{C}_\lambda(\mathfrak{g})$ with an infinitesimal character, then $\psi_\lambda^{\lambda+\mu}$ is an isomorphism of $\mathcal{C}_\lambda^{(1)}(\mathfrak{g})$ onto $\mathcal{C}_{\lambda+\mu}^{(1)}(\mathfrak{g})$ in the sense that there is an exact functor ψ' from $\mathcal{C}_{\lambda+\mu}^{(1)}(\mathfrak{g})$ to $\mathcal{C}_\lambda^{(1)}(\mathfrak{g})$ such that $\psi'\psi_\lambda^{\lambda+\mu}$ is naturally isomorphic to the identity functor on $\mathcal{C}_\lambda^{(1)}(\mathfrak{g})$ and $\psi_\lambda^{\lambda+\mu}\psi'$ is naturally isomorphic to the identity functor on $\mathcal{C}_{\lambda+\mu}^{(1)}(\mathfrak{g})$.

The proofs of these theorems have the flavor of the proof of Theorem 7.133, where the result was reduced to the case of Verma modules using arguments with annihilators. For each conclusion in the present theorems, we shall replace our specific modules by some universal modules, show that it is enough to handle the universal modules, and make a direct computation to handle them. All of the reductions involve approximately the same tools, and we shall concentrate on just Theorem 7.171a at first. In this section we shall give the steps in the argument, omitting certain details, and in §11 we shall supply the details. In §12 we shall prove Theorem 7.171b, 7.171c, 7.172, and 7.173.

Let us now turn our attention to the proof of Theorem 7.171a. Proposition 4.87 shows that the center $Z_\mathfrak{g}$ acts by scalars in V. Hence there is no loss of generality in assuming that $\mathfrak{g}$ is semisimple in the theorem. We make that assumption from now on. To fix the ideas, let the given M be in $\mathcal{C}_\lambda(\mathfrak{g})$ until further notice; we impose no restriction on λ yet. Let F be an irreducible finite-dimensional complex-linear representation of $\mathfrak{g}$.

1) Define

$$R(\lambda) = R = U(\mathfrak{g})/\tilde{Z}_\lambda,$$

where $\tilde{Z}_\lambda$ is the two-sided ideal in (7.117b). There are two commuting actions of $U(\mathfrak{g})$ on R, both induced from actions on $U(\mathfrak{g})$ given in §I.1:

(7.174) $l_1(u)x = ux$ and $r_1(u)x = xu^t$ for $u \in U(\mathfrak{g})$, $x \in R$, where u^t is as in (1.4). Under the action by l_1, R has infinitesimal character λ since $z \in Z(\mathfrak{g})$ implies

$$l_1(z)(x + \tilde{Z}_\lambda) = (z - \chi_\lambda(z)1)(x + \tilde{Z}_\lambda) + \chi_\lambda(z)(x + \tilde{Z}_\lambda)$$

and since $x - \chi_\lambda(z)1$ is in $Z_\lambda \subseteq \tilde{Z}_\lambda$. Under the action by r_1, let us see that R has infinitesimal character $-\lambda$. In fact,

$$
\begin{aligned}
r_1(z)(x + \tilde{Z}_\lambda) &= (x + \tilde{Z}_\lambda)z^t \\
&= (z^t - \chi_\lambda(z^t)1)(x + \tilde{Z}_\lambda) + \chi_\lambda(z^t)(x + \tilde{Z}_\lambda) \\
&= \chi_\lambda(z^t)(x + \tilde{Z}_\lambda) \\
&= \chi_{-\lambda}(z)(x + \tilde{Z}_\lambda) \qquad \text{by (7.55).}
\end{aligned}
$$

We shall use the notation $R = R_\lambda^{-\lambda}$ to emphasize that R has infinitesimal character λ under the left action by $U(\mathfrak{g})$ and infinitesimal character $-\lambda$ under the right action. (The subscript is therefore the index that refers to the *left* action.)

2) Since $U(\mathfrak{g})$ acts on F, we can map $U(\mathfrak{g})$ to a subalgebra of $\mathrm{End}_{\mathbb{C}} F$. Then $\mathrm{End}_{\mathbb{C}} F$ acquires two commuting actions by $U(\mathfrak{g})$ in the manner of (7.174), l_2 from left multiplication and r_2 from right multiplication. Define

$$S = R \otimes_{\mathbb{C}} \mathrm{End}_{\mathbb{C}} F.$$

By means of (2.38), we define commuting actions l and r of $U(\mathfrak{g})$ on S by

(7.175) $l = l_1 \otimes l_2$ and $r = r_1 \otimes r_2.$

Each of these is the tensor product of a finite-dimensional representation and a module having an infinitesimal character. By Theorem 7.133 each action is $Z(\mathfrak{g})$ finite. Under l let the primary decomposition of S given in Proposition 7.20 be $S = \bigoplus_\alpha S_\alpha$, and under r let the primary decomposition be $S = \bigoplus_\beta S^\beta$. Since l and r commute, these two decompositions are compatible. That is, if we set $S_\alpha^\beta = S_\alpha \cap S^\beta$, then

(7.176) $$S = \bigoplus_{\alpha, \beta} S_\alpha^\beta.$$

Each S_α^β is a $U(\mathfrak{g})$ module under both l and r. Moreover, the algebra S satisfies the following multiplication rules:

(7.177) $$S_\alpha^\beta S_\gamma^\delta \subseteq \begin{cases} S_\alpha^\delta & \text{if } \beta = -\gamma \\ 0 & \text{otherwise.} \end{cases}$$

In fact, with obvious notation

$$l(z)(s_\alpha^\beta s_\gamma^\delta) = z s_\alpha^\beta s_\gamma^\delta = (l(z)s_\alpha^\beta)s_\gamma^\delta = \chi_\alpha(z)s_\alpha^\beta s_\gamma^\delta,$$

$$r(z)(s_\alpha^\beta s_\gamma^\delta) = s_\alpha^\beta s_\gamma^\delta z^t = s_\alpha^\beta (r(z)s_\gamma^\delta) = \chi_\delta(z)s_\alpha^\beta s_\gamma^\delta,$$

$$\chi_\beta(z)(s_\alpha^\beta s_\gamma^\delta) = (r(z)s_\alpha^\beta)s_\gamma^\delta = s_\alpha^\beta z^t s_\gamma^\delta = s_\alpha^\beta (l(z^t)s_\gamma^\delta)$$
$$= \chi_\gamma(z^t)(s_\alpha^\beta s_\gamma^\delta) = \chi_{-\gamma}(z)(s_\alpha^\beta s_\gamma^\delta),$$

and (7.177) follows. Consequently

(7.178a) $S_\gamma^{-\gamma}$ is a subalgebra of S.

The projection of S to $S_\gamma^{-\gamma}$ with kernel $\bigoplus_{\substack{\alpha \neq \gamma \\ \text{or } \beta \neq -\gamma}} S_\alpha^\beta$ is not an algebra homomorphism. However,

(7.178b)   the $S_\gamma^{-\gamma}$ component $1_\gamma^{-\gamma}$ of $1 \in S$ is the identity of $S_\gamma^{-\gamma}$.

In fact, write $1 = \sum 1_\alpha^\beta$, and let $s_\gamma^{-\gamma}$ be in $S_\gamma^{-\gamma}$. Then

$$s_\gamma^{-\gamma} = s_\gamma^{-\gamma} 1 = \sum s_\gamma^{-\gamma} 1_\alpha^\beta,$$

and (7.177) shows that all terms on the right are 0 except the one with $\alpha = \gamma$ and $\beta = -\gamma$. Thus $1_\gamma^{-\gamma}$ is a right identity, and a similar argument shows it is a left identity.

3) Returning to M, which is assumed to be irreducible and have infinitesimal character λ, let N be the $U(\mathfrak{g})$ module

$$N = M \otimes_{\mathbb{C}} F.$$

By Theorem 7.133, N has a primary decomposition, which we write

$$N = \bigoplus_\gamma N_\gamma.$$

(The subscript is consistent with thinking of the action by $U(\mathfrak{g})$ as occurring on the left.) From the irreducibility of M, M is a simple left R module. Let us deduce that

(7.179) N is a simple left S module.

In fact, let N' be a nonzero S submodule of N, and let $\sum m_i \otimes f_i$ be a nonzero element of N', written with the elements f_i independent and with $m_1 \otimes f_1 \neq 0$. Applying a suitable member of $1 \otimes \text{End}_{\mathbb{C}} F$, we can map $m_1 \otimes f_1$ to itself and the other $m_i \otimes f_i$ to 0. Thus $m_1 \otimes f_1$ is in N'. Applying all members of $1 \otimes \text{End}_{\mathbb{C}} F$ to $m_1 \otimes f_1$, we see that $m_1 \otimes F$ is in N'. Applying $R \otimes 1$, we see that $M \otimes F$ is in N'. Hence $N' = N$, and (7.179) is proved.

4) Arguing as with (7.177), we see that

$$(7.180) \qquad S_\alpha^\beta N_\gamma \subseteq \begin{cases} N_\gamma & \text{if } \beta = -\gamma \\ 0 & \text{otherwise.} \end{cases}$$

Taking (7.178) into account, we see that N_γ is a left $S_\gamma^{-\gamma}$ module. Let us deduce from (7.179) that

$$(7.181) \qquad N_\gamma \text{ is a simple left } S_\gamma^{-\gamma} \text{ module, or else } N_\gamma = 0.$$

In fact, let N' be a nonzero $S_\gamma^{-\gamma}$ submodule of N_γ, and let n' be a nonzero element of N'. By (7.179), (7.176), and (7.180), we have

$$N = Sn' = \bigoplus_{\alpha, \beta} S_\alpha^\beta n' = \bigoplus_\alpha S_\alpha^{-\gamma} n' \qquad \text{since } n' \text{ is is } N_\gamma.$$

Consequently another application of (7.180) gives $N_\gamma = S_\gamma^{-\gamma} n'$. This proves (7.181).

5) For $s \in S$ and $n \in N$, the actions of $U(\mathfrak{g})$ are related by

$$(su)n = s(un) \qquad \text{for } u \in U(\mathfrak{g}).$$

If X is in $\mathfrak{g}$, let us write $(\text{ad } X)s = Xs - sX$. Then we have

$$(l(X)s)n = (Xs)n = ((\text{ad } X)s)n + (sX)n = ((\text{ad } X)s)n + s(Xn),$$

and this relation is valid also for $s \in S_\gamma^{-\gamma}$ and $n \in N_\gamma$. For s equal to the identity 1 of S, we have $(\text{ad } X)1 = 0$ since

$$(\text{ad } X)1 \otimes 1) = X(1 \otimes 1) - (1 \otimes 1)X = (X \otimes 1 + (1 \otimes X)) - (X \otimes 1 + (1 \otimes X)) = 0.$$

Since $l(U(\mathfrak{g}))$ and $r(U(\mathfrak{g}))$ map each S_α^β into itself, it follows that

$$(\text{ad } X)1_\gamma^{-\gamma} = 0.$$

Therefore $(l(X)1_\gamma^{-\gamma})n = 1_\gamma^{-\gamma}(Xn)$, and we obtain

$$(7.182) \qquad (l(u)1_\gamma^{-\gamma})n = un \qquad \text{for } u \in U(\mathfrak{g}) \text{ and } n \in N_\gamma.$$

Combining (7.182) and (7.181), we come to this conclusion:

(7.183)
> If $l(U(\mathfrak{g}))1_\gamma^{-\gamma} = S_\gamma^{-\gamma}$, then $U(\mathfrak{g})n = S_\gamma^{-\gamma}n$ for each $n \in N_\gamma$, and hence either N_γ is an irreducible $U(\mathfrak{g})$ module or else $N_\gamma = 0$.

This completes the process of making Theorem 7.171a into a more universal statement. To prove Theorem 7.171a, we are to prove under some conditions on the original λ and this γ that $l(U(\mathfrak{g}))1_\gamma^{-\gamma} = S_\gamma^{-\gamma}$. Since the original $U(\mathfrak{g})$ module M does not enter into the definitions of S and l, M no longer needs to play a role in the proof of the theorem.

6) For x in R or S, the bilinear map

$$u' \times u \mapsto l(u')r(u)x$$

of $U(\mathfrak{g}) \times U(\mathfrak{g})$ into R or S extends to a linear map of $U(\mathfrak{g}) \otimes_{\mathbb{C}} U(\mathfrak{g})$ into R or S, and in this way we obtain actions of the algebra $U(\mathfrak{g}) \otimes_{\mathbb{C}} U(\mathfrak{g})$ on R, S, and each S_α^β. The Poincaré-Birkhoff-Witt Theorem shows that

$$U(\mathfrak{g}) \otimes_{\mathbb{C}} U(\mathfrak{g}) \cong U(\mathfrak{g} \oplus \mathfrak{g}),$$

and therefore R, S, and each S_α^β become $\mathfrak{g} \oplus \mathfrak{g}$ modules. We shall introduce a compact group K that makes $(\mathfrak{g} \oplus \mathfrak{g}, K)$ into a reductive pair and these modules into $(\mathfrak{g} \oplus \mathfrak{g}, K)$ modules. Let G be a complex semisimple Lie group with Lie algebra $\mathfrak{g}$. Let $\mathfrak{k}_0$ be a compact form of $\mathfrak{g}$ built from $\mathfrak{h}$, and write

$$\mathfrak{g} = \mathfrak{k}_0 \oplus J\mathfrak{k}_0,$$

where J denotes $\sqrt{-1}$ in $\mathfrak{g}$. Let K be the (compact) analytic subgroup of G with Lie algebra $\mathfrak{k}_0$, and let bar denote conjugation in $\mathfrak{g}$ with respect to $\mathfrak{k}_0$. Write $\mathfrak{g}_{\mathbb{C}} = \mathfrak{g} \oplus i\mathfrak{g}$, and define $\varphi : \mathfrak{g}_{\mathbb{C}} \to \mathfrak{g} \oplus \mathfrak{g}$ by

(7.184a)
$$\varphi(X + iY) = (X + JY, \bar{X} + J\bar{Y}).$$

One checks readily that φ is a *complex*-linear isomorphism, the respective values of $\sqrt{-1}$ being i for $\mathfrak{g}_{\mathbb{C}}$ and J for $\mathfrak{g} \oplus \mathfrak{g}$. Of special interest are the embeddings

(7.184b)
$$\mathfrak{g} \hookrightarrow \mathfrak{g} \oplus \mathfrak{g} \quad \text{by } X \mapsto (X, \bar{X})$$
$$\mathfrak{k}_0 \hookrightarrow \mathfrak{g} \oplus \mathfrak{g} \quad \text{by } X \mapsto (X, X).$$

The first of these exhibits $\mathfrak{g} \oplus \mathfrak{g}$ as the complexified Lie algebra of G, and the second of these gives the relationship of K to $\mathfrak{g} \oplus \mathfrak{g}$. Then $(\mathfrak{g} \oplus \mathfrak{g}, K)$

is a reductive pair. Let us write $\mathfrak{k}$ for the complexified image of $\mathfrak{k}_0$ in $\mathfrak{g} \oplus \mathfrak{g}$, i.e., $\mathfrak{k} = \{(X, X) \mid X \in \mathfrak{g}\}$.

7) We have already seen that R, S, and each S_α^β are $\mathfrak{g} \oplus \mathfrak{g}$ modules. Let us see that R, S, and each S_α^β are in fact $(\mathfrak{g} \oplus \mathfrak{g}, K)$ modules, i.e., that the action of the embedded $\mathfrak{k}_0$ lifts to K. If X is in $\mathfrak{k}_0$, X acts by $(X, X) \in \mathfrak{g} \oplus \mathfrak{g}$, hence by $X \otimes 1 + 1 \otimes X$ in $U(\mathfrak{g}) \otimes_\mathbb{C} U(\mathfrak{g})$. Thus, in the case of R, $X \cdot u$ is identified with

$$(l_1(X)r_1(1) + l_1(1)r_1(X))u = Xu - uX = (\operatorname{ad} X)u,$$

and the action of $\mathfrak{k}_0$ on R certainly lifts to K. In the case of $\operatorname{End}_\mathbb{C} F$, $X \cdot \varphi$ is identified similarly with $X\varphi - \varphi X$. If $\tilde{K}$ is a simply connected cover of K, then both l_2 and r_2 lift from representations of $\mathfrak{k}_0$ to representations of $\tilde{K}$, and $k \cdot \varphi$ is identified with $k\varphi k^{-1}$ for $k \in \tilde{K}$. The irreducibility of F forces the central elements of $\tilde{K}$ to map to scalar operators in $\operatorname{End}_\mathbb{C} F$, and thus $k\varphi k^{-1} = \varphi$ for $k \in Z_{\tilde{K}}$. It follows that the action of $\tilde{K}$ descends to an action of K, and $\operatorname{End}_\mathbb{C} F$ is a $(\mathfrak{g} \oplus \mathfrak{g}, K)$ module. Consequently S, as a tensor product, and each S_α^β are $(\mathfrak{g} \oplus \mathfrak{g}, K)$ modules. Note for X in $\mathfrak{k}_0$ and $x \in S$ that

$$(l(X)r(1) + l(1)r(X))x = Xx - xX = (\operatorname{ad} X)x.$$

The fact that this action lifts to K is a justification for the use of the notation "ad X" in this expression.

8) With $V(\lambda)$ denoting the Verma module with parameter $\lambda \in \mathfrak{h}^*$, we introduce a $(\mathfrak{g} \oplus \mathfrak{g}, K)$ module

$$(7.185) \qquad P(\lambda) = \operatorname{Hom}_\mathbb{C}(V(\lambda), V(\lambda))_K.$$

The subscript is to be read "K finite" and will be explained shortly. The action by a pure tensor $u_1 \otimes u_2$ in $U(\mathfrak{g}) \otimes_\mathbb{C} U(\mathfrak{g})$ on $\operatorname{Hom}_\mathbb{C}(V(\lambda), V(\lambda))$ is

$$(7.186a) \quad ((u_1 \otimes u_2)\Phi)(v) = u_1(\Phi(u_2^t v)) \quad \text{for } \Phi \in \operatorname{Hom}_\mathbb{C}(V(\lambda), V(\lambda)).$$

A member X of $\mathfrak{k}_0$ acts as $(X, X) \in \mathfrak{g} \oplus \mathfrak{g}$, hence by

$$(7.186b) \quad (X\Phi)(v) = ((X \otimes 1 + 1 \otimes X)\Phi)(v) = X(\Phi(v)) - \Phi(Xv).$$

The subscript K in (7.185) refers to the subspace of elements Φ in $\operatorname{Hom}_\mathbb{C}(V(\lambda), V(\lambda))$ such that $U(\mathfrak{k})\Phi$ is finite dimensional. If $\tilde{K}$ denotes a simply connected cover of K, the action of $\mathfrak{k}_0$ on $P(\lambda)$ lifts to a locally $\tilde{K}$ finite representation of $\tilde{K}$. By Proposition 2.70, we can regard $P(\lambda)$ as given by

$$P(\lambda) \cong \Gamma_{\mathfrak{g} \oplus \mathfrak{g}, \{1\}}^{\mathfrak{g} \oplus \mathfrak{g}, \tilde{K}}(\operatorname{Hom}_\mathbb{C}(V(\lambda), V(\lambda))).$$

Therefore $P(\lambda)$ is a $(\mathfrak{g} \oplus \mathfrak{g}, \tilde{K})$ module.

Let us show that $P(\lambda)$ is in fact a $(\mathfrak{g} \oplus \mathfrak{g}, K)$ module. Recall that $\mathfrak{k}_0$ was constructed from the Cartan subalgebra $\mathfrak{h}$ of $\mathfrak{g}$. Let $\mathfrak{h}_0$ be the real subalgebra of $\mathfrak{h}$ on which all roots are imaginary, and identify $\mathfrak{h}_0$ with its image $\mathfrak{t}_0 = \{(X, X) \mid X \in \mathfrak{h}_0\}$ in $\mathfrak{g} \oplus \mathfrak{g}$. Let T be the analytic subgroup of K with Lie algebra $\mathfrak{t}_0$. The argument will be based on the following consequence of Theorem 4.6c: Any finite-dimensional $(\mathfrak{k}, T)$ module is automatically a $(\mathfrak{k}, K)$ module. We have to show that the $\mathfrak{t}_0$ action on $P(\lambda)$ exponentiates to T. Thus suppose Φ is a $\mathfrak{k}$ finite member of $\mathrm{Hom}_{\mathbb{C}}(V(\lambda), \lambda)$. Without loss of generality, we may assume that Φ is a nonzero generalized weight vector corresponding to some weight $\mu \in \mathfrak{t}^*$:

$$(X - \mu(X)I)^N \Phi = 0 \qquad \text{for all } X \in \mathfrak{t}.$$

If $v \in V(\lambda)$ has $\mathfrak{h}$ weight γ, an easy computation shows that

$$(X - (\mu + \gamma)(X)1)^m (\Phi(v)) = ((X - \mu(X)1)^m \Phi)(v)$$

for all $X \in \mathfrak{h}$. Taking $m = N$ and reading from right to left, we see that $\Phi(v)$ has generalized $\mathfrak{h}$ weight $\gamma + \mu$. Since $\mathfrak{h}$ acts fully reducibly on $V(\lambda)$, we deduce that the left side is 0 for $m = 1$. Thus Φ is a weight vector for $\mathfrak{t}$. Meanwhile, $\Phi(v)$ is nonzero for some v, and thus $\gamma + \mu$ is a weight of $V(\lambda)$. Since any two weights of $V(\lambda)$ differ by a sum of roots, μ is a sum of roots. Thus the weight of the weight vector Φ is a sum of roots, and the action of $\mathfrak{t}_0$ exponentiates to T.

Therefore $P(\lambda)$ is a $(\mathfrak{g} \oplus \mathfrak{g}, K)$ module, and we can regard $P(\lambda)$ as given by

$$P(\lambda) \cong \Gamma^{\mathfrak{g} \oplus \mathfrak{g}, K}_{\mathfrak{g} \oplus \mathfrak{g}, \{1\}} (\mathrm{Hom}_{\mathbb{C}}(V(\lambda), V(\lambda))).$$

We shall use the subscript K as in (7.185) instead of the notation with Γ in order to emphasize that $P(\lambda)$ is a subspace of $\mathrm{Hom}_{\mathbb{C}}(V(\lambda), V(\lambda))$.

9) We use $\mathfrak{h} \oplus \mathfrak{h}$ as a Cartan subalgebra of $\mathfrak{g} \oplus \mathfrak{g}$. The set of roots $\Delta(\mathfrak{g} \oplus \mathfrak{g}, \mathfrak{h} \oplus \mathfrak{h})$ has the form

$$(7.187) \quad \Delta(\mathfrak{g} \oplus \mathfrak{g}, \mathfrak{h} \oplus \mathfrak{h}) = \{(\alpha, 0) \mid \alpha \in \Delta(\mathfrak{g}, \mathfrak{h})\} \cup \{(0, \beta) \mid \beta \in \Delta(\mathfrak{g}, \mathfrak{h})\},$$

because each expression $(\alpha, 0)$ and $(0, \beta)$ is a root and because their root vectors, together with $\mathfrak{h} \oplus \mathfrak{h}$, span $\mathfrak{g} \oplus \mathfrak{g}$. Since $V(\lambda)$ has infinitesimal character λ relative to $\mathfrak{g}$, it follows from (7.55) and (7.186a) that $P(\lambda)$ has infinitesimal character $(\lambda, -\lambda)$. The identity mapping is a member of $P(\lambda)$, and we denote it $1_{P(\lambda)}$. From (7.186b) we see that it is fixed by K.

10) If $\lambda \in \mathfrak{h}^*$ is integrally dominant, then there exists a $(\mathfrak{g} \oplus \mathfrak{g}, K)$ isomorphism $R(\lambda) \xrightarrow{\sim} P(\lambda)$ carrying 1 to $1_{P(\lambda)}$; moreover, the space of K fixed vectors of $P(\lambda)$ is $\mathbb{C}1_{P(\lambda)}$. This is the difficult step in the argument. It will be proved in §11.

11) Suppose that μ is algebraically integral and F^μ is an irreducible finite-dimensional complex-linear representation of $\mathfrak{g}$ with extreme weight μ. We shall see that $\operatorname{End}_\mathbb{C} F^\mu$ is irreducible for $\mathfrak{g} \oplus \mathfrak{g}$ with extreme weight $(\mu, -\mu)$. Suppose that λ and $\lambda + \mu$ are integrally dominant relative to $\Delta^+(\mathfrak{g}, \mathfrak{h})$. We shall prove that if $\lambda + \mu$ is at least as singular as λ, then the ψ functor for $\mathfrak{g} \oplus \mathfrak{g}$ satisfies $\psi_{(\lambda, -\lambda)}^{(\lambda+\mu, -\lambda-\mu)}(P(\lambda)) \cong P(\lambda + \mu)$. This step uses the comparable result for Verma modules given as Corollary 7.169. It will be carried out in §11.

12) In the context of (10), we obtain $U(\mathfrak{g}) \otimes_\mathbb{C} U(\mathfrak{g})$ isomorphisms

$$R(\lambda + \mu) \cong P(\lambda + \mu) \cong (P(\lambda) \otimes_\mathbb{C} \operatorname{End}_\mathbb{C} F^\mu)_{\lambda+\mu}^{-\lambda-\mu} \cong S_{\lambda+\mu}^{-\lambda-\mu},$$

and the K fixed vector, which is unique up to a scalar in $P(\lambda + \mu)$, must map to a K fixed vector at each end. Hence we may assume that 1 in $R(\lambda + \mu)$ maps to $1_{\lambda+\mu}^{-\lambda-\mu}$ in $S_{\lambda+\mu}^{-\lambda-\mu}$. It is obvious that 1 is cyclic under $U(\mathfrak{g}) \otimes 1$ in $R(\lambda + \mu) = U(\mathfrak{g})/\tilde{Z}_{\lambda+\mu}$, and it follows that $1_{\lambda+\mu}^{-\lambda-\mu}$ is cyclic under $l(U(\mathfrak{g}))$ in $S_{\lambda+\mu}^{-\lambda-\mu}$. Thus the hypothesis in (7.183) is satisfied, and the theorem follows.

Thus the proof of Theorem 7.171a comes down to proving the assertions in steps (10) and (11) above. All of these assertions will be proved in §11. In §12 we shall take up Theorems 7.172 and 7.173 and also the remaining parts of Theorem 7.171.

11. Details of Irreducibility

In this section we shall complete the proof of Theorem 7.171a by carrying out steps (10) and (11) of §10. Let $\mathfrak{g}$ be a complex semisimple Lie algebra, let $\mathfrak{h}$ be a Cartan subalgebra, and let λ be in $\mathfrak{h}^*$. Recall that $R(\lambda)$ is defined to be $U(\mathfrak{g})/\tilde{Z}_\lambda$, where $\tilde{Z}_\lambda$ is the two-sided ideal of $U(\mathfrak{g})$ generated by $\ker \chi_\lambda$ in $Z(\mathfrak{g})$. In (7.185) we defined $P(\lambda)$ to be $\operatorname{Hom}_\mathbb{C}(V(\lambda), V(\lambda))_K$. Both $R(\lambda)$ and $P(\lambda)$ are $(\mathfrak{g} \oplus \mathfrak{g}, K)$ modules. The following theorem restates step (10).

Theorem 7.188. If $\lambda \in \mathfrak{h}^*$ is integrally dominant, then

(a) there exists a $(\mathfrak{g} \oplus \mathfrak{g}, K)$ isomorphism $R(\lambda) \xrightarrow{\sim} P(\lambda)$ carrying 1 to $1_{P(\lambda)}$, and
(b) the space of K fixed vectors of $P(\lambda)$ is $\mathbb{C}1_{P(\lambda)}$.

Let us define a $(\mathfrak{g} \oplus \mathfrak{g}, K)$ map $F : R(\lambda) \to P(\lambda)$. First we define $\tilde{F} : U(\mathfrak{g}) \to P(\lambda)$ by

$$\tilde{F}(u) = (u \otimes 1)1_{P(\lambda)} \qquad \text{for } u \in U(\mathfrak{g}).$$

Since $P(\lambda)$ has infinitesimal character $(\lambda, -\lambda)$ under $U(\mathfrak{g}) \otimes_{\mathbb{C}} U(\mathfrak{g})$, $P(\lambda)$ has infinitesimal character λ under $U(\mathfrak{g}) \otimes 1$. Therefore $\ker \chi_\lambda$ maps to 0, and $\tilde{F}$ descends to a map $F : R(\lambda) \to P(\lambda)$ given by

$$(7.189a) \qquad F(u + \tilde{Z}_\lambda) = (u \otimes 1)1_{P(\lambda)} \qquad \text{for } u + \tilde{Z}_\lambda \in R(\lambda).$$

Clearly F carries the identity to $1_{P(\lambda)}$. The map F is set up to respect the action of $U(\mathfrak{g}) \otimes 1$. To see that it respects $1 \otimes U(\mathfrak{g})$, let X be in $\mathfrak{g}$. Then

$$\begin{aligned}
F(r(X)(u + \tilde{Z}_\lambda)) &= F(-uX + \tilde{Z}_\lambda) \\
&= (-uX \otimes 1)1_{P(\lambda)} \\
&= (u \otimes 1)(-X \otimes 1)1_{P(\lambda)} \\
&= (u \otimes 1)(1 \otimes X)1_{P(\lambda)} \qquad \text{since } \mathfrak{k} \text{ annihilates } 1_{P(\lambda)} \\
&= (1 \otimes X)(u \otimes 1)1_{P(\lambda)} \\
&= (1 \otimes X)F(u + \tilde{Z}_\lambda).
\end{aligned}$$

The equivariance follows. Consequently a more general formula for F is

$(7.189b)$

$$F(u'u + \tilde{Z}_\lambda) = (u' \otimes u^t)1_{P(\lambda)} \quad \text{for } u' + \tilde{Z}_\lambda \text{ and } u + \tilde{Z}_\lambda \text{ in } R(\lambda).$$

Since $R(\lambda)$ and $P(\lambda)$ are $(\mathfrak{g} \oplus \mathfrak{g}, K)$ modules and K is connected and F is equivariant for $\mathfrak{g} \oplus \mathfrak{g}$, F is a $(\mathfrak{g} \oplus \mathfrak{g}, K)$ map.

Let us see that F is one-one. For v in $V(\lambda)$, we have

$$F(u + \tilde{Z}_\lambda)(v) = ((u \otimes 1)1_{P(\lambda)})(v) = u(1_{P(\lambda)}(v)) = uv.$$

If $F(u + \tilde{Z}_\lambda) = 0$, the above expression is 0 for all v, and hence u is in Ann V. By the Dixmier-Duflo theorem (Theorem 7.118), u is in $\tilde{Z}_\lambda$. Thus F is one-one.

Thus Theorem 7.188a will follow if we prove that F is onto. We carry out this step by calculating the multiplicities of K types in $P(\lambda)$ and $R(\lambda)$ and seeing that they match and are finite. Along the way, the calculation for the trivial K type in $P(\lambda)$ will prove Theorem 7.188b. Since $\mathfrak{k} = \{(X, X) \mid X \in \mathfrak{g}\}$, we can take $\mathfrak{h}_K = \{(X, X) \mid X \in \mathfrak{h}\}$ as a Cartan subalgebra of $\mathfrak{k}$.

Proposition 7.190. If $\lambda \in \mathfrak{h}^*$ is integrally dominant and if F is in $\widehat{K}$, then the multiplicity of F in $P(\lambda)$ satisfies

$$[P(\lambda) : F] \leq \dim(0 \text{ weight space of } F).$$

REMARK. In fact, equality holds, as will be apparent from the last part of the proof of Theorem 7.194 below.

PROOF. The 0 weight space has the same dimension in F^* as in F, and we shall show that

(7.191) $[P(\lambda) : F^*] \leq \dim(0 \text{ weight space of } F).$

The left side is

(7.192)
$$= [\mathrm{Hom}_{\mathbb{C}}(V(\lambda), V(\lambda))_K : F^*] = [F \otimes_{\mathbb{C}} \mathrm{Hom}_{\mathbb{C}}(V(\lambda), V(\lambda))_K : 1].$$

Let us make F into a $(\mathfrak{g} \oplus \mathfrak{g}, K)$ module $\mathbb{C} \otimes_{\mathbb{C}} F$. The first summand $\mathfrak{g}$ is to act (trivially) in the factor $\mathbb{C}$, the second summand $\mathfrak{g}$ is to act on F through the complexified action from $\mathfrak{k}_0$, and K is to act by its given action on F. Now $F \otimes_{\mathbb{C}} \mathrm{Hom}_{\mathbb{C}}(V(\lambda), V(\lambda)) \cong \mathrm{Hom}_{\mathbb{C}}(V(\lambda), F \otimes_{\mathbb{C}} V(\lambda))$ as $\mathfrak{g} \oplus \mathfrak{g}$ modules. We readily check that

(7.193a) $\mathfrak{k}(F \otimes \Phi) \subseteq F \otimes U(\mathfrak{k})\Phi$ for $\Phi \in \mathrm{Hom}_{\mathbb{C}}(V(\lambda), V(\lambda))$;

hence

(7.193b) $F \otimes_{\mathbb{C}} \mathrm{Hom}_{\mathbb{C}}(V(\lambda), V(\lambda))_K \cong \mathrm{Hom}_{\mathbb{C}}(V(\lambda), F \otimes_{\mathbb{C}} V(\lambda))_K$

as $(\mathfrak{g} \oplus \mathfrak{g}, K)$ modules. Taking into account this formula and the action (7.186b), we see that (7.192) is

$$= \dim \mathrm{Hom}_{\mathfrak{g}}(V(\lambda), F \otimes_{\mathbb{C}} V(\lambda)).$$

According to the proof of Theorem 7.133, $F \otimes_{\mathbb{C}} V(\lambda)$ has a finite filtration by $U(\mathfrak{g})$ modules with successive quotients $V(\lambda + \nu_j)$, where the ν_j's are the weights of F repeated according to their multiplicities. Since $\mathrm{Hom}_{\mathfrak{g}}(V(\lambda), \cdot)$ is left exact, it follows that

$$\dim \mathrm{Hom}_{\mathfrak{g}}(V(\lambda), F \otimes_{\mathbb{C}} V(\lambda)) \leq \sum_j \dim \mathrm{Hom}_{\mathfrak{g}}(V(\lambda), V(\lambda + \nu_j)).$$

By Proposition 7.170, the terms on the right are 1 for $\nu_j = 0$ and are 0 otherwise. So the right side is the dimension of the 0 weight space of F. This proves (7.191) and hence also the proposition.

In particular, Proposition 7.190 proves Theorem 7.188b. To complete the proof of Theorem 7.188a, it is enough to prove the following formula.

Theorem 7.194 (Kostant). If λ is in $\mathfrak{h}^*$ and if F is in $\widehat{K}$, then the multiplicity of F in $R(\lambda)$ satisfies

$$[R(\lambda) : F] = \dim(0 \text{ weight space of } F).$$

The proof of $\geq$ in Theorem 7.194 will use some more of Kostant's theory of harmonics. Let us therefore revert temporarily to the notation of §6. The Lie algebra $\mathfrak{g}$ remains complex semisimple. The action that we have been regarding as from the complexification of the embedded $\mathfrak{k}_0$ on $R(\lambda)$, namely by

$$((l(X) \otimes 1) + (1 \otimes r(X))(u + \tilde{Z}_\lambda) = Xu - uX + \tilde{Z}_\lambda,$$

is nothing more than the adjoint action of $\mathfrak{g}$ on $U(\mathfrak{g})$, pushed down to $R(\lambda)$. Let $\mathcal{H}'$ and $\mathcal{K}$ be as in Theorems 7.109 and 7.114.

Lemma 7.195. The action of ad $\mathfrak{g}$ on $R(\lambda)$ is equivalent with the action of ad $\mathfrak{g}$ on $\mathcal{K} \subseteq U(\mathfrak{g})$, and this is equivalent with the coadjoint action of $\mathfrak{g}$ on $\mathcal{H}' \subseteq S(\mathfrak{g}^*)$.

PROOF. The map $\mathcal{K} \subseteq U(\mathfrak{g}) \to U(\mathfrak{g})/\tilde{Z}_\lambda = R(\lambda)$ is equivariant under ad $\mathfrak{g}$. We prove it is one-one onto. Theorem 7.114 says that any member of $U(\mathfrak{g})$ is of the form $\sum k_i z_i$ with $k_i \in \mathcal{K}$ and $z_i \in Z(\mathfrak{g})$. This element has the same image as $\sum \chi_\lambda(z_i)k_i$ in $R(\lambda)$, and hence $\mathcal{K}$ maps onto $R(\lambda)$. To see that $\mathcal{K}$ maps in one-one fashion, let $k \in \mathcal{K}$ map to 0. Then k is in $U(\mathfrak{g}) \ker \chi_\lambda$, say $k = \sum u_i z_i$ with $z_i \in \ker \chi_\lambda$. Expanding each u_i by means of Theorem 7.114, we obtain $k = \sum k_j z'_j$ with the z'_j in $\ker \chi_\lambda$. Without loss of generality, we may assume that the z'_j are linearly independent. Since 1 is not in $\ker \chi_\lambda$, the relation $k1 - \sum k_j z'_j = 0$ implies $k = 0$, by the uniqueness in Theorem 7.114. Hence $R(\lambda)$ and $\mathcal{K}$ are equivalent. By (7.112b), $\mathcal{K}$ and $\mathcal{H}'$ are equivalent. This completes the proof.

PROOF OF THEOREM 7.194. In view of Lemma 7.195, we are interested in the number of times each irreducible complex-linear finite-dimensional representation of $\mathfrak{g}$ occurs in the representation on $\mathcal{H}'$ via ad. This amounts to the same thing as the multiplicity of each K type in $(\mathcal{H}', \operatorname{Ad} K)$.

Let X be a generic element of $\mathfrak{g}$ lying in $\mathfrak{h} \cap \mathfrak{k}_0$, where $\mathfrak{h}$ is our standard Cartan subalgebra. (Any element of $\mathfrak{h} \cap \mathfrak{k}_0$ with $\alpha(X) \neq 0$ for all roots α

will do.) We shall study

the orbit O_X of X under G in $\mathfrak{g}$ and

the orbit $O_{X,K}$ of X under K in $\mathfrak{g}_0$.

In the second case the isotropy subgroup of K at X is the Cartan subgroup T of K; thus $O_{X,K} \cong K/T$. In the first case the isotropy subgroup of G at X is $H = TA$; thus $O_X \cong G/TA$. The inclusion $O_{X,K} \subseteq O_X$ corresponds to the injection $K/T \hookrightarrow G/TA$.

Each complex-valued polynomial on $\mathfrak{g}_0$ restricts to a K finite function on $O_{X,K}$, and the Stone-Weierstrass Theorem says that the space of restrictions is uniformly dense in the continuous functions on $O_{X,K}$, hence L^2 dense in $L^2(K/T)$. It follows that the restriction map carries the space of complex-valued polynomials on $\mathfrak{g}_0$ onto the space of K finite functions on $O_{X,K} \cong K/T$. Now Corollary 7.77 shows that any polynomial on $\mathfrak{g}$ is of the form $\sum h_i\, p_i$ with $h_i \in \mathcal{H}'$ and $p_i \in \mathcal{I}'$. Restricted to $O_{X,K}$, p_i is a constant. Hence restriction to $O_{X,K}$ carries $\mathcal{H}'$ onto the space of K finite functions on K/T. Thus the multiplicity of any K type τ in $\mathcal{H}'$ is $\geq$ the multiplicity of that K type in $L^2(K/T)$. Now $L^2(K/T)$ is classically induced from the trivial representation of T. Hence classical Frobenius reciprocity says that

$$[L^2(K/T) : \tau]_K = [\tau|_T : 1].$$

The right side is the dimension of the 0 weight space of τ, and consequently

(7.196) $\qquad [R(\lambda) : F] \geq \dim(0 \text{ weight space of } F).$

Let us now prove that equality holds. Since $R(w\lambda) = R(\lambda)$ for w in the Weyl group, we may assume that $\mathrm{Re}\langle \lambda, \alpha \rangle \geq 0$ for every positive root α. Then λ is integrally dominant. Since $R(\lambda)$ maps one-one into $P(\lambda)$, we have

$$
\begin{aligned}
[R(\lambda) : F] &\leq [P(\lambda) : F] \\
&\leq \dim(0 \text{ weight space of } F) \qquad \text{by Proposition 7.190} \\
&\leq [R(\lambda) : F] \qquad\qquad\qquad\quad \text{by (7.196).}
\end{aligned}
$$

Equality must hold throughout, and the proof of Theorem 7.194 is complete.

Next let us fill in the details of (11) in §9. We return to the notation of that section. Let F be an irreducible finite-dimensional complex-linear representation of $\mathfrak{g}$ with a basis $\{f_i\}$ of weight vectors and with respective weights μ_i relative to $\mathfrak{h}$. For the contragredient F^*, the dual basis $\{f_i^*\}$ is a basis of weight vectors with respective weights $-\mu_i$. Let $\{e_{ij}\}$ be the basis of $\text{End}_{\mathbb{C}} F$ consisting of the endomorphisms

$$(7.197a) \qquad\qquad f \mapsto \langle f_i^*, f \rangle f_j.$$

If h is in $\mathfrak{h}$, then

$$(7.197b) \qquad \begin{aligned} l(h)e_{ij}(f) &= he_{ij}(f) = \langle f_i^*, f \rangle h(f_j) \\ &= \mu_j(h)\langle f_i^*, f \rangle f_j = \mu_j(h)e_{ij}(f) \end{aligned}$$

and

$$(7.197c) \qquad \begin{aligned} r(h)e_{ij}(f) &= -e_{ij}hf = -\langle f_i^*, hf \rangle f_j \\ &= \langle hf_i^*, f \rangle f_j = -\mu_i(h)\langle f_i^*, f \rangle f_j = -\mu_i(h)e_{ij}(f). \end{aligned}$$

Consequently the weights of $\text{End}_{\mathbb{C}} F$ are the various $(\mu_j, -\mu_i)$. From (7.187) we see that if F has μ as an extreme weight, then $\text{End}_{\mathbb{C}} F$ has $(\mu, -\mu)$ as an extreme weight.

We can make $\text{End}_{\mathbb{C}} F \cong F \otimes_{\mathbb{C}} F^*$ into a $\mathfrak{g} \oplus \mathfrak{g}$ module by having the first $\mathfrak{g}$ act in F and the second $\mathfrak{g}$ act in F^*. With this convention

$$(7.198) \qquad\qquad \text{End}_{\mathbb{C}} F \cong F \otimes_{\mathbb{C}} F^*$$

as a $\mathfrak{g} \oplus \mathfrak{g}$ module by the isomorphism (7.197). To see that $\text{End}_{\mathbb{C}} F \cong F \otimes_{\mathbb{C}} F^*$ is irreducible, we give two proofs.

The first proof is for algebraists: The Jacobson Density Theorem (see Jacobson [1956], 28) says that irreducibility of a finite-dimensional module E for an associative algebra A over $\mathbb{C}$ is equivalent with the surjectivity of the algebra map $A \to \text{End}_{\mathbb{C}} E$. Applying this fact with E equal to the irreducible $U(\mathfrak{g})$ module F, we see that irreducibility of $F \otimes_{\mathbb{C}} F^*$ under $U(\mathfrak{g}) \otimes_{\mathbb{C}} U(\mathfrak{g})$ is equivalent with irreducibility of $F \otimes_{\mathbb{C}} F^*$ under $\text{End}_{\mathbb{C}} F \otimes_{\mathbb{C}} U(\mathfrak{g})$. In turn this irreducibility is the special case of (3) in §10 in which $M = F^*$.

The second proof is for analysts: Note first that $\text{End}_{\mathbb{C}} F$ is fully reducible since $\mathfrak{g} \oplus \mathfrak{g}$ is semisimple. If irreducibility fails, there is thus a nontrivial projection

$$Q : F \otimes_{\mathbb{C}} F^* \to F \otimes_{\mathbb{C}} F^*$$

commuting with $\mathfrak{g} \oplus \mathfrak{g}$. Write $Q = \sum_k R_k \otimes S_k$ with $R_k \in \text{End}_{\mathbb{C}} F$,

$S_k \in \mathrm{End}_{\mathbb{C}} F^*$, and the S_k independent. Since Q commutes with $(\mathfrak{g}, 0)$, we obtain

$$0 = Q(X \otimes 1) - (X \otimes 1)Q = \sum_k (R_k X - X R_k) \otimes S_k,$$

and we see that each $R_k X - X R_k$ is 0 for all $X \in \mathfrak{g}$. Hence each R_k is scalar. Thus Q is of the form $1 \otimes S$ with $S \in \mathrm{End}_{\mathbb{C}} F^*$, and the commutativity of Q with $(0, \mathfrak{g})$ shows that S is scalar. Then Q is scalar, in contradiction with its being a nontrivial projection. We conclude that $\mathrm{End}_{\mathbb{C}} F$ is irreducible as a $\mathfrak{g} \oplus \mathfrak{g}$ module.

Now we can prove the last statement of (11) in §10.

Proposition 7.199. Let $\mathfrak{g}$ be a complex semisimple Lie algebra, let $\mathfrak{h}$ be a Cartan subalgebra, and fix a system $\Delta^+(\mathfrak{g}, \mathfrak{h})$ of positive roots. Suppose $\mu \in \mathfrak{h}^*$ is algebraically integral. Further suppose λ is a member of $\mathfrak{h}^*$ satisfying

(i) λ and $\lambda + \mu$ are integrally dominant, and
(ii) $\lambda + \mu$ is at least as singular as λ, i.e., $\langle \lambda + \mu, \alpha \rangle = 0$ for every $\alpha \in \Delta(\mathfrak{g}, \mathfrak{h})$ such that $\langle \lambda, \alpha \rangle = 0$.

Then the ψ functor for $\mathfrak{g} \oplus \mathfrak{g}$ satisfies

$$\psi_{(\lambda, -\lambda)}^{(\lambda+\mu, -\lambda-\mu)}(P(\lambda)) \cong P(\lambda + \mu).$$

PROOF. Let F have μ as an extreme weight. Using (7.198) and the same kind of calculation as in (7.193), we see that

$$P(\lambda) \otimes_{\mathbb{C}} \mathrm{End}_{\mathbb{C}} F \cong \mathrm{Hom}_{\mathbb{C}}(V(\lambda) \otimes_{\mathbb{C}} F, \ V(\lambda) \otimes_{\mathbb{C}} F)_K.$$

When we project under $Z(\mathfrak{g}) \otimes 1$ according to the infinitesimal character $\lambda + \mu$, the right side becomes

$$(P_{\lambda+\mu} \otimes 1)(\mathrm{Hom}_{\mathbb{C}}(V(\lambda) \otimes_{\mathbb{C}} F, \ V(\lambda) \otimes_{\mathbb{C}} F)_K)$$
$$\cong \mathrm{Hom}_{\mathbb{C}}(V(\lambda) \otimes_{\mathbb{C}} F, \ P_{\lambda+\mu}(V(\lambda) \otimes_{\mathbb{C}} F))_K$$
$$\cong \mathrm{Hom}_{\mathbb{C}}(V(\lambda) \otimes_{\mathbb{C}} F, \ V(\lambda + \mu))_K,$$

according to Corollary 7.169. When we project under $1 \otimes Z(\mathfrak{g})$ according to the infinitesimal character $-\lambda - \mu$ and take (7.55) into account, this expression becomes

$$(1 \otimes P_{-\lambda-\mu})(\mathrm{Hom}_{\mathbb{C}}(V(\lambda) \otimes_{\mathbb{C}} F, \ V(\lambda + \mu))_K)$$
$$\cong \mathrm{Hom}_{\mathbb{C}}(P_{\lambda+\mu}(V(\lambda) \otimes_{\mathbb{C}} F), \ V(\lambda + \mu))_K$$
$$\cong \mathrm{Hom}_{\mathbb{C}}(V(\lambda + \mu), \ V(\lambda + \mu))_K$$
$$= P(\lambda + \mu),$$

as required.

12. Nonvanishing of Certain Translation Functors

In this section we shall prove the remaining parts of Theorem 7.171, as well as Theorems 7.172 and 7.173. These other proofs use approximately the same tools as the proof of Theorem 7.171a. Thus we retain the notation of §10. The Lie algebra $\mathfrak{g}$ is semisimple, and we continue to work with the $(\mathfrak{g} \oplus \mathfrak{g}, K)$ module $R(\lambda)$.

PROOF OF THEOREM 7.171b. In §10 we regarded an irreducible $U(\mathfrak{g})$ module M with infinitesimal character λ as a simple $R(\lambda)$ module M. Let us write γ for $\lambda + \mu$. Under the construction of steps (1) through (4), we passed from the simple module M for $R(\lambda)$ to a simple module N for $S = R(\lambda) \otimes_{\mathbb{C}} \text{End}_{\mathbb{C}} F$ and then to a module N_γ for $S_\gamma^{-\gamma}$ that is simple or is 0. As a result of step (12), $U(\mathfrak{g})$ maps onto $S_\gamma^{-\gamma}$. Therefore if N_γ is not 0, it is a simple module for $U(\mathfrak{g})$.

Suppose that irreducible M and M' lead to isomorphic nonzero $U(\mathfrak{g})$ modules N_γ. Then they lead to isomorphic nonzero $S_\gamma^{-\gamma}$ modules. We shall reconstruct N as an S module from N_γ and also reconstruct M as an $R(\lambda)$ module from N. It then follows that M is isomorphic with M' (as an $R(\lambda)$ module and hence as a $U(\mathfrak{g})$ module).

The reconstruction of M from N is by the formula

$$M \cong \text{Hom}_{\text{End} F}(F, N).$$

Here the isomorphism of M on the left to $\text{Hom}_{\text{End} F}(F, M \otimes_{\mathbb{C}} F)$ on the right sends m to the map $f \mapsto m \otimes f$.

To reconstruct N from N_γ, we shall relate N canonically to $S \otimes_{S_\gamma^{-\gamma}} N_\gamma$. This space is a $U(\mathfrak{g})$ module under the left action on S, and hence it is a $Z(\mathfrak{g})$ module. In the notation before (7.176), it decomposes under $Z(\mathfrak{g})$ as

$$S \otimes_{S_\gamma^{-\gamma}} N_\gamma = \bigoplus_\alpha S_\alpha \otimes_{S_\gamma^{-\gamma}} N_\gamma.$$

In particular, it is $Z(\mathfrak{g})$ finite, and its α primary component under $Z(\mathfrak{g})$ is the $U(\mathfrak{g})$ module

$$(S \otimes_{S_\gamma^{-\gamma}} N_\gamma)_\alpha = S_\alpha \otimes_{S_\gamma^{-\gamma}} N_\gamma.$$

The γ primary component is therefore the $U(\mathfrak{g})$ module $\bigoplus_\beta S_\gamma^\beta \otimes_{S_\gamma^{-\gamma}} N_\gamma$. For $\beta \neq -\gamma$, (7.177) gives

$$s_\gamma^\beta \otimes n_\gamma = s_\gamma^\beta \otimes 1_\gamma^{-\gamma} n_\gamma = s_\gamma^\beta 1_\gamma^{-\gamma} \otimes n_\gamma = 0$$

in obvious notation, and thus the γ primary component reduces to $S_\gamma^{-\gamma} \otimes_{S_\gamma^{-\gamma}} N_\gamma$, which is $U(\mathfrak{g})$ isomorphic to N_γ by (7.182).

Now let us compute

$$
\begin{aligned}
\operatorname{Hom}_S(S \otimes_{S_\gamma^{-\gamma}} N_\gamma, N) &\cong \operatorname{Hom}_{S_\gamma^{-\gamma}}(N_\gamma, N) && \text{by (C.20)} \\
&= \operatorname{Hom}_{S_\gamma^{-\gamma}}(N_\gamma, \bigoplus_\alpha N_\alpha) && \text{by step (3) of §10} \\
&= \operatorname{Hom}_{S_\gamma^{-\gamma}}(N_\gamma, N_\gamma) && \text{by (7.180)} \\
&\cong \mathbb{C} && \text{by irreducibility of } N_\gamma.
\end{aligned}
$$

In this computation the isomorphism at the first step comes by restricting maps from $S \otimes_{S_\gamma^{-\gamma}} N_\gamma$ to $S_\gamma^{-\gamma} \otimes_{S_\gamma^{-\gamma}} N_\gamma$, which is the full γ primary component. Thus N is exhibited as an irreducible quotient of $S \otimes_{S_\gamma^{-\gamma}} N_\gamma$ by an S map whose kernel has 0 intersection with the γ primary component. The kernel is $Z(\mathfrak{g})$ finite, and it follows that its γ primary component is 0. By Proposition 7.20 the kernel is contained in the sum of the α primary components of $S \otimes_{S_\gamma^{-\gamma}} N_\gamma$ for $\alpha \neq \gamma$.

To prove that N_γ determines N, it is enough to show that there is only one maximal S submodule of $S \otimes_{S_\gamma^{-\gamma}} N_\gamma$ contained in the sum of the α primary components for $\alpha \neq \gamma$. In fact, the sum of all S submodules of $S \otimes_{S_\gamma^{-\gamma}} N_\gamma$ that are contained in the sum of the α primary components for $\alpha \neq \gamma$ is again one and is clearly the unique maximal one. This completes the reconstruction of N from N_γ.

PROOF OF THEOREM 7.171c. If we are given a $U(\mathfrak{g})$ module N_γ in $\mathcal{C}_{\lambda+\mu}(\mathfrak{g}) = \mathcal{C}_\gamma(\mathfrak{g})$ having an infinitesimal character, then N_γ is an $R(\gamma)$ module. Since step (12) in §10 says that $S_\gamma^{-\gamma} \cong R(\gamma)$, we can regard N_γ as an $S_\gamma^{-\gamma}$ module. Form the S module

$$
(7.200) \qquad\qquad N_1 = S \otimes_{S_\gamma^{-\gamma}} N_\gamma.
$$

As in the proof of Theorem 7.171b, we see that the γ primary component of N_1 is isomorphic with N_γ as a $U(\mathfrak{g})$ module. Form the $R(\lambda)$ module

$$
(7.201) \qquad\qquad M_1 = \operatorname{Hom}_{\operatorname{End} F}(F, N_1).
$$

Then

$$
(7.202) \qquad\qquad M_1 \otimes_\mathbb{C} F \cong N_1
$$

as S modules under the map $\varphi \otimes f \mapsto \varphi(f)$ for $\varphi \in M_1$ and $f \in F$. Thus the $R(\lambda)$ module M_1 leads to the S module N_1 in the construction of §10, and the S module N_1 leads to the γ primary component N_γ as an $S_\gamma^{-\gamma}$ module. In other words, $\psi_\lambda^\gamma(M_1) \cong N_\gamma$.

If N_γ is irreducible, we replace N_1 in the above argument with its quotient N by the sum of all S submodules contained in the sum of the α primary components for $\alpha \neq \gamma$. Then $(N_1)_\gamma \cong N_\gamma$, and N is a simple S module. If we put $M = \operatorname{Hom}_{\operatorname{End} F}(F, N)$, then M is a simple $R(\lambda)$ module (hence an irreducible $U(\mathfrak{g})$ module), and we find that $\psi_\lambda^\gamma(M) \cong N$.

PROOF OF THEOREM 7.172. Let $N \neq 0$ be given in $\mathcal{C}_\lambda(\mathfrak{g})$, and fix a nonzero member v of N. Since $\psi_\lambda^{\lambda+\mu}$ is exact, it is enough to prove that $\psi_\lambda^{\lambda+\mu}(U(\mathfrak{g})v) \neq 0$. Since $U(\mathfrak{g})$ is Noetherian (see Jacobson [1962], 166), $U(\mathfrak{g})v$ satisfies the ascending-chain condition for its $U(\mathfrak{g})$ submodules. The set of proper $U(\mathfrak{g})$ submodules thus has a maximal element, and it follows that $U(\mathfrak{g})v$ has an irreducible quotient N_0. Again since $\psi_\lambda^{\lambda+\mu}$ is exact, it is enough to prove that $\psi_\lambda^{\lambda+\mu}(N_0) \neq 0$.

So we may assume that N_0 is a given irreducible member of $\mathcal{C}_\lambda(\mathfrak{g})$. Then N_0 has infinitesimal character λ. The hypotheses of Theorem 7.171 are satisfied for $\psi_{\lambda+\mu}^\lambda$. Thus we can choose M_0 in $\mathcal{C}_{\lambda+\mu}(\mathfrak{g})$ with $\psi_{\lambda+\mu}^\lambda(M_0) \cong N_0$, by Theorem 7.171c. By Proposition 7.143,

$$\operatorname{Hom}_{\mathfrak{g}}(\psi_\lambda^{\lambda+\mu}(N_0), M_0) \cong \operatorname{Hom}_{\mathfrak{g}}(N_0, \psi_{\lambda+\mu}^\lambda(M_0)) \cong \operatorname{Hom}_{\mathfrak{g}}(N_0, N_0).$$

The right side is not 0 because the identity map is in it. Hence the left side is not 0, and $\psi_\lambda^{\lambda+\mu}(N_0)$ cannot be 0.

PROOF OF THEOREM 7.173. The hypotheses of both Theorem 7.171 and Theorem 7.172 are satisfied. Then (a) in the present theorem follows by combining Theorems 7.171a and 7.172. For (b) we apply (a) to $\{\lambda, \lambda+\mu\}$ and then to $\{\lambda+\mu, \lambda\}$ to see that $\psi_{\lambda+\mu}^\lambda \psi_\lambda^{\lambda+\mu}(M)$ is irreducible. By Proposition 7.143,

$$\operatorname{Hom}_{\mathfrak{g}}(\psi_{\lambda+\mu}^\lambda \psi_\lambda^{\lambda+\mu}(M), M) \cong \operatorname{Hom}_{\mathfrak{g}}(\psi_\lambda^{\lambda+\mu}(M), \psi_\lambda^{\lambda+\mu}(M)),$$

and the right side contains the (nonzero) identity map since $\psi_\lambda^{\lambda+\mu}(M)$ is not 0. Hence there is a nonzero $\mathfrak{g}$ map from $\psi_{\lambda+\mu}^\lambda \psi_\lambda^{\lambda+\mu}(M)$ to M. Since domain and range are irreducible, the map is an isomorphism.

Now we consider (c), which comes from close examination of the proofs of Theorem 7.171b–c. We can identify $\mathcal{C}_\lambda^{(1)}(\mathfrak{g})$ with the category of left $R(\lambda)$ modules. The functor $M_1 \mapsto N_1$ from this category to the category of left S modules via (7.202) is invertible via (7.201), and the compositions

$$M_1 \mapsto M_1 \otimes_\mathbb{C} F \mapsto \operatorname{Hom}_{\operatorname{End} F}(F, M_1 \otimes_\mathbb{C} F)$$

and $\qquad N_1 \mapsto \operatorname{Hom}_{\operatorname{End} F}(F, N_1) \mapsto \operatorname{Hom}_{\operatorname{End} F}(F, N_1) \otimes_\mathbb{C} F$

are naturally isomorphic to the respective identity functors.

The other half of the construction of ψ is the functor $N \mapsto N_\gamma$ from left S modules to left $S_\gamma^{-\gamma}$ modules, where $\gamma = \lambda + \mu$. The "inverse" functor is to be $N_\gamma \mapsto S \otimes_{S_\gamma^{-\gamma}} N_\gamma$. We saw in the proof of Theorem 7.171c that

$$N_\gamma \mapsto S \otimes_{S_\gamma^{-\gamma}} N_\gamma \mapsto (S \otimes_{S_\gamma^{-\gamma}} N_\gamma)_\gamma$$

recovers N_γ up to natural isomorphism, and what we have to see is that

$$N \mapsto N_\gamma \mapsto S \otimes_{S_\gamma^{-\gamma}} N_\gamma$$

recovers N up to natural isomorphism. As in the proof of Theorem 7.171b (but without the assumption of any irreducibility), we have

$$\begin{aligned}
\operatorname{Hom}_S(S \otimes_{S_\gamma^{-\gamma}} N_\gamma, N) &\cong \operatorname{Hom}_{S_\gamma^{-\gamma}}(N_\gamma, N) && \text{by (C.20)} \\
&= \operatorname{Hom}_{S_\gamma^{-\gamma}}\Big(N_\gamma, \bigoplus_\alpha N_\alpha\Big) && \text{by step (3) of §10} \\
&= \operatorname{Hom}_{S_\gamma^{-\gamma}}(N_\gamma, N_\gamma) && \text{by (7.180).}
\end{aligned}$$

The identity map is a member of the right side, and we let

$$\varphi : S \otimes_{S_\gamma^{-\gamma}} N_\gamma \to N$$

be the corresponding member of the left side. The S module $\ker \varphi$ has γ primary component 0, and the corresponding $R(\lambda)$ module M therefore maps to 0 under ψ_λ^γ. By Theorem 7.172, $M = 0$. Therefore $\ker \varphi = 0$. Also since φ is an $S_\gamma^{-\gamma}$ isomorphism of $S_\gamma^{-\gamma} \otimes_{S_\gamma^{-\gamma}} N_\gamma$ onto N_γ, $N/\text{image } \varphi$ has γ primary component 0. Again the corresponding $R(\lambda)$ module M' maps to 0 under ψ_λ^γ, and Theorem 7.172 shows that $M' = 0$; consequently $N/\text{image } \varphi = 0$. We conclude that φ is an isomorphism, and the proof is complete.

13. Application to $(\mathfrak{g}, K)$ Modules with K Connected

All the development of translation functors so far has dealt with $U(\mathfrak{g})$ modules for a complex reductive Lie algebra $\mathfrak{g}$. If $(\mathfrak{g}, K)$ is a reductive pair with K connected, then the category $\mathcal{C}(\mathfrak{g}, K)$ of $(\mathfrak{g}, K)$ modules is a subcategory of $\mathcal{C}(\mathfrak{g})$, and we let $\mathcal{C}_{Zf}(\mathfrak{g}, K) = \mathcal{C}_{Zf}(\mathfrak{g}) \cap \mathcal{C}(\mathfrak{g}, K)$ be the further subcategory of $(\mathfrak{g}, K)$ modules that are $Z(\mathfrak{g})$ finite. To bring translation functors into the theory of $(\mathfrak{g}, K)$ modules, we keep with the conventions in Chapter IV. There will be no need, however, to consider only Cartan subalgebras of $\mathfrak{g}$ that are complexifications of θ stable Cartan subalgebras of $\mathfrak{g}_0$.

Proposition 7.203. Let $(\mathfrak{g}, K)$ be a reductive pair with K connected, and let $\mathfrak{h}$ be a Cartan subalgebra of $\mathfrak{g}$. Then $\mathcal{C}_{Zf}(\mathfrak{g}, K)$ is closed under tensor product with any finite-dimensional $(\mathfrak{g}, K)$ module and with passage to the $Z(\mathfrak{g})$ primary components. Hence the translation functors $\psi_\lambda^{\lambda+\mu}$ (defined relative to $\mathfrak{h}$) carry $\mathcal{C}_{Zf}(\mathfrak{g}, K)$ into itself if the finite-dimensional representation F^μ has the structure of a $(\mathfrak{g}, K)$ module.

In other words, μ is to satisfy a suitable integrality condition so that the representation F^μ on $\mathfrak{k}_0$ exponentiates to K. The proposition is clear and requires no proof.

To exploit the special properties of $\mathcal{C}_{Zf}(\mathfrak{g}, K)$, we shall use the following theorem.

Theorem 7.204 (Harish-Chandra). Let $(\mathfrak{g}, K)$ be a reductive pair with K connected, and let $\mathfrak{h}$ be a Cartan subalgebra of $\mathfrak{g}$. Every irreducible $(\mathfrak{g}, K)$ module is admissible. Moreover, for each λ in $\mathfrak{h}^*$, there exist finitely many K types $\tau_{\mu_1}, \ldots, \tau_{\mu_n}$ such that every irreducible $(\mathfrak{g}, K)$ module with infinitesimal character λ contains one of these K types.

The traditional proof of Theorem 7.204 proves much more, namely that there are, up to isomorphism, only finitely many irreducible $(\mathfrak{g}, K)$ modules with a given infinitesimal character. After one proves that admissibility follows from irreducibility, the steps are to realize each irreducible $(\mathfrak{g}, K)$ module as the underlying module of a group representation, to attach to it a global character, to prove that the global character is a locally integrable function, and to exhibit a finite-dimensional space of functions (depending on the infinitesimal character) in which this function must lie. Since the global characters of nonisomorphic irreducible admissible $(\mathfrak{g}, K)$ modules are linearly independent, the result follows.

But we need only what is stated in Theorem 7.204, and this much is accessible by the methods in this book. We shall give the argument in Chapter X.

In order to avoid repetition, let us show that Theorem 7.204 extends to (𝔤, K) modules with a fixed infinitesimal character.

Corollary 7.205. Let (𝔤, K) be a reductive pair with K connected, and let 𝔥 be a Cartan subalgebra of 𝔤. For each λ in 𝔥*, there exist finitely many K types $\tau_{\mu_1}, \ldots, \tau_{\mu_n}$ such that every nonzero (𝔤, K) module with generalized infinitesimal character λ contains one of these K types.

REMARK. The argument that we give for Theorem 7.204 in Chapter X will prove this corollary at the same time. However, we derive the corollary from the theorem now anyway.

PROOF. Let the given (𝔤, K) module be V, and let $v \neq 0$ be in V. Since $U(𝔤)v$ is finitely generated and $U(𝔤)$ is Noetherian (see Jacobson [1962], 166), $U(𝔤)v$ satisfies the ascending-chain condition for $U(𝔤)$ submodules. Letting W be a maximal proper $U(𝔤)$ submodule, we see that $U(𝔤)v/W$ is irreducible. By Theorem 7.204, $U(𝔤)v/W$ contains at least one of the K types $\tau_{\mu_1}, \ldots, \tau_{\mu_n}$. Therefore the same thing is true of $U(𝔤)v$ and also of V.

For an admissible (𝔤, K) module, we shall introduce several finiteness conditions and use Theorem 7.204 to show that they are equivalent. For background, see §A.3; the good category $\mathcal{C}$ for use in that section is the category of all admissible (𝔤, K) modules (i.e., unital left $R(𝔤, K)$ modules that are admissible under K). If V is admissible in $\mathcal{C}(𝔤, K)$, recall that a **finite filtration** for V is a finite descending chain

$$(7.206) \qquad V = V_0 \supseteq V_1 \supseteq \cdots \supseteq V_n = 0$$

of (𝔤, K) submodules. Schreier's Theorem (Theorem A.24) says that any two finite filtrations have equivalent refinements in the sense that the nonzero consecutive quotients for one refinement are isomorphic in some order to the nonzero consecutive quotients of the other. A finite filtration is a **composition series** if the consecutive quotients are irreducible (nonzero). By Schreier's Theorem any two composition series are equivalent in the sense that their sets of consecutive quotients match in some order, up to isomorphism. An admissible (𝔤, K) module

V need not have a composition series. (For example, the direct sum
of the $(\mathfrak{g}, K)$ modules of §2 that correspond to the discrete series $\mathcal{D}_n$
of $SU(1, 1)$, each taken with multiplicity 1, does not.) But if a $(\mathfrak{g}, K)$
submodule V' of V and the quotient V/V' both have composition series,
so does V. (See (A.18).) If V has a composition series with indices as
in (7.206), we say that V has **length** n.

Corollary 7.207. Let $(\mathfrak{g}, K)$ be a reductive pair with K connected. If
V is an admissible $(\mathfrak{g}, K)$ module, then the following conditions on V
are equivalent:

    (a) V is finitely generated as a $U(\mathfrak{g})$ module
    (b) V is $Z(\mathfrak{g})$ finite
    (c) V has a composition series.

PROOF.

(a) $\Longrightarrow$ (b). This is proved as Example 2 in §2.

(b) $\Longrightarrow$ (c). We may assume that V has a generalized infinitesimal
character, say λ. Let $\tau_{\mu_1}, \ldots, \tau_{\mu_n}$ be as in Corollary 7.205. We prove
that (b) implies (c) by induction on the sum of the multiplicities of
$\tau_{\mu_1}, \ldots, \tau_{\mu_n}$ in V, the trivial case for the induction being that the sum
is 0. In this case, Corollary 7.205 says that $V = 0$, and so V has a
composition series. Assume the implication known when the sum of
the multiplicities is $< N$, and let the sum of the multiplicities for V be
N. Without loss of generality we may assume V to be reducible. Let
V' be a nontrivial $(\mathfrak{g}, K)$ submodule. Corollary 7.205 shows that at least
one of $\tau_{\mu_1}, \ldots, \tau_{\mu_n}$ occurs in V' and at least one occurs in V/V'. The
inductive hypothesis is therefore applicable to both V' and V/V'. Then
each of them has a composition series, and so must V.

(c) $\Longrightarrow$ (a). Let (7.206) be a composition series for V. For $0 \le j < n$,
let v_j be an element in V_j but not V_{j-1}. Then $\{v_j \mid 0 \le j < n\}$ is a set of
generators for V as a $U(\mathfrak{g})$ module.

Still with K connected, if the admissible $(\mathfrak{g}, K)$ module V satisfies
the equivalent conditions of Corollary 7.207, we say that V has **finite
length**. By (c) this definition is consistent with the one in §A.3. Let
$\mathcal{F}(\mathfrak{g}, K)$ be the category of $(\mathfrak{g}, K)$ modules of finite length. This consists
of the admissible $(\mathfrak{g}, K)$ modules in $\mathcal{C}_{Zf}(\mathfrak{g}, K)$, by (b). The category
$\mathcal{F}(\mathfrak{g}, K)$ is closed under tensor products with finite-dimensional $(\mathfrak{g}, K)$
modules, by (a) in the corollary, and it is closed under passage to $Z(\mathfrak{g})$
primary components, by (b). This proves the following result.

Corollary 7.208. Let $(\mathfrak{g}, K)$ be a reductive pair with K connected, and let $\mathfrak{h}$ be a Cartan subalgebra of $\mathfrak{g}$. If the finite-dimensional representation F^μ of $\mathfrak{g}$ used in defining $\psi_\lambda^{\lambda+\mu}$ is a $(\mathfrak{g}, K)$ module, then $\psi_\lambda^{\lambda+\mu}$ carries $\mathcal{F}(\mathfrak{g}, K)$ into itself.

REMARK. In this section we use the hypothesis "reductive pair" rather than "pair with $\mathfrak{g}$ reductive" precisely to get the conclusion of this corollary, that the translation functors preserve finite length.

We write $\mathcal{F}_\lambda(\mathfrak{g}, K)$ for the members of $\mathcal{F}(\mathfrak{g}, K)$ with generalized infinitesimal character λ.

Corollary 7.209. Let $(\mathfrak{g}, K)$ be a reductive pair with K connected, let $\mathfrak{h}$ be a Cartan subalgebra of $\mathfrak{g}$, and fix a system $\Delta^+(\mathfrak{g}, \mathfrak{h})$ of positive roots. Suppose that λ and μ are in $\mathfrak{h}^*$, that μ is algebraically integral, that λ and $\lambda + \mu$ are integrally dominant and equisingular, and that F^μ is a finite-dimensional irreducible $(\mathfrak{g}, K)$ module with extreme weight μ to be used in defining $\psi_\lambda^{\lambda+\mu}$. Fix $F^{-\mu} = (F^\mu)^*$ as a finite-dimensional irreducible $(\mathfrak{g}, K)$ module to use in defining $\psi_{\lambda+\mu}^\lambda$. Then $\psi_{\lambda+\mu}^\lambda \psi_\lambda^{\lambda+\mu}$ is naturally isomorphic to the identity functor on $\mathcal{F}_\lambda(\mathfrak{g}, K)$, and $\psi_\lambda^{\lambda+\mu} \psi_{\lambda+\mu}^\lambda$ is naturally isomorphic to the identity functor on $\mathcal{F}_{\lambda+\mu}(\mathfrak{g}, K)$.

REMARK. Compare with Theorem 7.173c.

PROOF. Let us abbreviate $\psi = \psi_{\lambda+\mu}^\lambda$ and $\psi' = \psi_\lambda^{\lambda+\mu}$. We prove the result for $\psi'\psi$; the result for $\psi\psi'$ follows by interchanging the roles of λ and $\lambda + \mu$. Proposition 7.143 gives us an isomorphism

$$\alpha_{X,X'} : \operatorname{Hom}(\psi'X, X') \xrightarrow{\sim} \operatorname{Hom}(X, \psi X')$$

natural in X and in X'. Naturality in the X subscript means that if $\gamma_1 : X \to Y$ is given and if ε' is in $\operatorname{Hom}(\psi'Y, X')$, then

(7.210a) $$\alpha_{X,X'}(\varepsilon' \circ \psi'(\gamma_1)) = \alpha_{Y,X'}(\varepsilon') \circ \gamma_1.$$

Naturality in the X' subscript means that if $\gamma_2 : X' \to Y'$ is given and if ε is in $\operatorname{Hom}(\psi'X, X')$, then

(7.210b) $$\alpha_{X,Y'}(\gamma_2 \circ \varepsilon) = \psi(\gamma_2) \circ \alpha_{X,X'}(\varepsilon).$$

Now $1_{\psi A}$ is a member of $\operatorname{Hom}(\psi A, \psi A)$, and we define

$$T_A : \psi'\psi(A) \to A \qquad \text{by} \qquad T_A = \alpha_{\psi A, A}^{-1}(1_{\psi A}).$$

Let a map $\gamma : A \to B$ be given. Taking $\gamma_2 = \gamma$ and $\varepsilon = T_A$ in (7.210b), we have

$$\alpha_{\psi A, B}(\gamma \circ T_A) = \psi(\gamma) \circ \alpha_{\psi A, A}(T_A) = \psi(\gamma) \circ 1_{\psi A} = \psi(\gamma).$$

Taking $\gamma_1 = \psi(\gamma)$ and $\varepsilon' = T_B$ in (7.210a), we have

$$\alpha_{\psi A, B}(T_B \circ \psi' \psi(\gamma)) = \alpha_{\psi B, B}(T_B) \circ \psi(\gamma) = 1_{\psi B} \circ \psi(\gamma) = \psi(\gamma).$$

Thus $\alpha_{\psi A, B}(T_B \circ \psi' \psi(\gamma)) = \alpha_{\psi A, B}(\gamma \circ T_A)$. Since $\alpha_{\psi A, B}$ is one-one,

(7.211) $$T_B \circ \psi' \psi(\gamma) = \gamma \circ T_A.$$

Thus T is a natural transformation from $\psi' \psi$ to the identity functor in the sense of (C.16).

To prove that T is a natural isomorphism, we have only to prove that each T_A is an isomorphism. We do so by induction on the length of A. The case of length 1 is Theorem 7.173b. Let T_B be an isomorphism whenever B has length $< n$, and let A have length n. We can form an exact sequence

$$0 \longrightarrow C \longrightarrow A \longrightarrow B \longrightarrow 0$$

in which C and B have length $< n$. Application of $\psi' \psi$ gives another exact sequence since ψ' and ψ are exact, and (7.211) allows us to form a diagram

$$
\begin{array}{ccccccccc}
0 & \longrightarrow & \psi' \psi(C) & \longrightarrow & \psi' \psi(A) & \longrightarrow & \psi' \psi(B) & \longrightarrow & 0 \\
& & \downarrow{\scriptstyle T_C} & & \downarrow{\scriptstyle T_A} & & \downarrow{\scriptstyle T_B} & & \\
0 & \longrightarrow & C & \longrightarrow & A & \longrightarrow & B & \longrightarrow & 0
\end{array}
$$

in which the rows are exact and the squares commute. By inductive hypothesis, T_C and T_B are isomorphisms. A little diagram chase shows that T_A is therefore an isomorphism. This completes the induction.

14. Application to $(\mathfrak{g}, K)$ Modules with K Disconnected

If $(\mathfrak{g}, K)$ is a reductive pair with K disconnected, then $\mathcal{C}(\mathfrak{g}, K)$ is not a subcategory of $\mathcal{C}(\mathfrak{g})$, and the theory of translation functors is not immediately applicable. Let us confront some of the difficulties.

It is still meaningful to speak of a $Z(\mathfrak{g})$ finite $(\mathfrak{g}, K)$ module, and the category $\mathcal{C}_{Zf}(\mathfrak{g}, K)$ of all such is defined. However, in a $(\mathfrak{g}, K)$ module, the χ primary subspace relative to $Z(\mathfrak{g})$ need not be K invariant. In fact, although $\operatorname{Ad}(K)$ acts trivially on $Z(\mathfrak{g})$ if $(\mathfrak{g}, K)$ is of Harish-Chandra class, it does not have to act trivially on $Z(\mathfrak{g})$ otherwise. In general the $(\mathfrak{g}, K)$ submodule that plays the role of the χ primary subspace is the sum over k in K of the primary subspaces for $k\chi$, where

$$k\chi(z) = \chi(\operatorname{Ad}(k)^{-1}z) \qquad \text{for } k \in K, \ z \in Z(\mathfrak{g}).$$

That is, the appropriate primary decomposition into $(\mathfrak{g}, K)$ submodules has summands parametrized by K orbits of homomorphisms of $Z(\mathfrak{g})$ into $\mathbb{C}$. The orbits are finite, since K/K_0 is finite, and G orbits amount to the same thing as K orbits.

More specifically the **primary decomposition** of a $Z(\mathfrak{g})$ finite $(\mathfrak{g}, K)$ module V may be written

$$V = \bigoplus_{\text{orbits } K\chi} P_{K\chi}(V),$$

where $P_{K\chi}(V) = \sum_{k \in K} P_{k\chi}$ is the (finite) sum of the primary projections for the distinct members of the orbit $K\chi$. We write $\mathcal{C}_{K\chi}(\mathfrak{g}, K)$ for the category of $Z(\mathfrak{g})$ finite $(\mathfrak{g}, K)$ modules whose only primary component is the one for $K\chi$. If $K\chi_\lambda = \{\chi_\lambda\}$, we may abbreviate $\mathcal{C}_{K\chi_\lambda}(\mathfrak{g}, K)$ as $\mathcal{C}_\lambda(\mathfrak{g}, K)$.

Proposition 7.212 (Wigner's Lemma). Let $K\chi$ and $K\chi'$ be distinct K orbits of homomorphisms of $Z(\mathfrak{g})$ into $\mathbb{C}$, and let U and V be in $\mathcal{C}_{K\chi}(\mathfrak{g}, K)$ and $\mathcal{C}_{K\chi'}(\mathfrak{g}, K)$, respectively. Then $\operatorname{Ext}^n_{\mathfrak{g}, K}(U, V) = 0$ for all n.

PROOF. By Lemma 7.14, choose $z_1 \in Z(\mathfrak{g})$ with $(k\chi)(z_1) = 0$ and $(k\chi')(z_1) = 0$ for all $k \in K/K_0$. Put

$$z = \frac{1}{|K : K_0|} \sum_{k \in K/K_0} \operatorname{Ad}(k)z_1.$$

Then the action of z on any $(\mathfrak{g}, K)$ module is a $(\mathfrak{g}, K)$ map. Moreover, for l large enough, z^l acts as 0 on U and is invertible on V. Arguing as in the proof of Proposition 7.28, we see that z^l induces the 0 map on $\operatorname{Ext}^n_{\mathfrak{g}, K}(U, V)$ via the U coordinate and an invertible map on $\operatorname{Ext}^n_{\mathfrak{g}, K}(U, V)$ via the V coordinate. Therefore $\operatorname{Ext}^n_{\mathfrak{g}, K}(U, V) = 0$.

It turns out to be a nuisance to take the K orbits of homomorphisms $Z(\mathfrak{g}) \to \mathbb{C}$ into account in dealing with translation functors. The following proposition offers reassurance that in studying $(\mathfrak{g}, K)$ modules in $\mathcal{C}_{K\chi}(\mathfrak{g}, K)$, we can concentrate on the case that the orbit has one element, i.e., the module has a generalized infinitesimal character. The result is not perfect. When we address irreducibility questions in §VIII.4, the proposition will take care of one step toward handling modules in $\mathcal{C}_{K\chi}(\mathfrak{g}, K)$, but it will be necessary also to use Mackey theory as in §II.4.

Proposition 7.213. Let $(\mathfrak{g}, K)$ be a reductive pair, let $\mathfrak{h}$ be a Cartan subalgebra of $\mathfrak{g}$, let λ be in $\mathfrak{h}^*$, and let K_1 be the stablizer in K of the homomorphism $\chi_\lambda : Z(\mathfrak{g}) \to \mathbb{C}$. Form the inclusion of pairs $(\mathfrak{g}, K_1) \hookrightarrow (\mathfrak{g}, K)$. Then the exact functor $\text{induced}_{\mathfrak{g},K_1}^{\mathfrak{g},K}$ of (2.74) is an isomorphism of $\mathcal{C}_\lambda(\mathfrak{g}, K_1)$ onto $\mathcal{C}_{K\chi_\lambda}(\mathfrak{g}, K)$ in the following sense: The composition of the functor $P_\lambda \circ \mathcal{F}_{\mathfrak{g},K}^{\mathfrak{g},K_1}$ from $\mathcal{C}_{K\chi_\lambda}(\mathfrak{g}, K)$ to $\mathcal{C}_\lambda(\mathfrak{g}, K_1)$ with the functor $\text{induced}_{\mathfrak{g},K_1}^{\mathfrak{g},K}$, in either order, is naturally isomorphic with the identity functor.

REMARKS. The exactness of $\text{induced}_{\mathfrak{g},K_1}^{\mathfrak{g},K}$ is given in Proposition 2.77. That proposition shows also that $\text{induced}_{\mathfrak{g},K_1}^{\mathfrak{g},K}$ is naturally equivalent with $P_{\mathfrak{g},K_1}^{\mathfrak{g},K}$ and with $I_{\mathfrak{g},K_1}^{\mathfrak{g},K}$.

PROOF. Write $F = P_\lambda \circ \mathcal{F}_{\mathfrak{g},K}^{\mathfrak{g},K_1}$. One order for the composition is $F \circ \text{induced}_{\mathfrak{g},K_1}^{\mathfrak{g},K}$. By Frobenius reciprocity (Proposition 2.75a), we have

$$\text{Hom}_{\mathfrak{g},K_1}(P_\lambda \mathcal{F}_{\mathfrak{g},K}^{\mathfrak{g},K_1}\text{induced}_{\mathfrak{g},K_1}^{\mathfrak{g},K}(V), \, V)$$
$$\cong \text{Hom}_{\mathfrak{g},K_1}(\mathcal{F}_{\mathfrak{g},K}^{\mathfrak{g},K_1}\text{induced}_{\mathfrak{g},K_1}^{\mathfrak{g},K}(V), \, V)$$
$$\cong \text{Hom}_{\mathfrak{g},K}(\text{induced}_{\mathfrak{g},K_1}^{\mathfrak{g},K}(V), \, \text{induced}_{\mathfrak{g},K_1}^{\mathfrak{g},K}(V))$$

for V in $\mathcal{C}_\lambda(\mathfrak{g}, K_1)$. Let φ_V be the member of the left side that corresponds to the identity on the right. Then φ_V is a natural map from $F \circ \text{induced}_{\mathfrak{g},K_1}^{\mathfrak{g},K}$ to the identity functor, in the sense of (C.16). We are to prove that each φ_V is an isomorphism.

Examining the proof of Proposition 2.75a, we see that the Frobenius reciprocity isomorphism above is $e_V \circ \Phi \leftarrow \Phi$, where e_V is evaluation at 1. Consequently φ_V equals e_V. The members of

(7.214) $\mathcal{F}_{\mathfrak{g},K}^{\mathfrak{g},K_1}\text{induced}_{\mathfrak{g},K_1}^{\mathfrak{g},K}(V)$

are the functions from K to V transforming on the right under K_1, and the action by $U(\mathfrak{g})$, according to (2.74c), is given by

(7.215) $(uf)(k) = (\text{Ad}(k)^{-1}u)(f(k)).$

Meanwhile, the space (7.214) is the direct sum over the double cosets $K_1 k K_1 \in K_1 \backslash K / K_1$ of the functions supported in $K_1 k K_1$ and having the same transformation law. Applying (7.215) to the element $u = (z - \chi_\lambda(\mathrm{Ad}(k)^{-1} z))^n$ with $z \in Z(\mathfrak{g})$, we see, for n large, that this element acts as 0 on the subspace of functions in (7.214) supported in $K_1 k K_1$. In other words, this subspace of functions has generalized infinitesimal character $k \chi_\lambda$. By definition of K_1, $k \chi_\lambda = \chi_\lambda$ if and only if k is in K_1. Consequently P_λ of (7.214) equals the subspace of functions supported in $K_1 1 K_1 = K_1$. The map e_V is an isomorphism of this space onto V, its two-sided inverse being the map j_V defined in Proposition 2.75. Hence $F \circ \mathrm{induced}_{\mathfrak{g}, K_1}^{\mathfrak{g}, K}$ is naturally isomorphic to the identity functor.

The other order for the composition is $\mathrm{induced}_{\mathfrak{g}, K_1}^{\mathfrak{g}, K} \circ F$. By Frobenius reciprocity (Proposition 2.75a), we have

$$\mathrm{Hom}_{\mathfrak{g}, K}(W, \mathrm{induced}_{\mathfrak{g}, K_1}^{\mathfrak{g}, K} P_\lambda \mathcal{F}_{\mathfrak{g}, K}^{\mathfrak{g}, K_1}(W))$$
$$\cong \mathrm{Hom}_{\mathfrak{g}, K_1}(\mathcal{F}_{\mathfrak{g}, K}^{\mathfrak{g}, K_1}(W), P_\lambda \mathcal{F}_{\mathfrak{g}, K}^{\mathfrak{g}, K_1}(W))$$
$$\cong \mathrm{Hom}_{\mathfrak{g}, K_1}(P_\lambda \mathcal{F}_{\mathfrak{g}, K}^{\mathfrak{g}, K_1}(W), P_\lambda \mathcal{F}_{\mathfrak{g}, K}^{\mathfrak{g}, K_1}(W))$$

for W in $\mathcal{C}_{K \chi_\lambda}(\mathfrak{g}, K)$. Let φ_W be the member of the left side that corresponds to the identity on the right. Then φ_W is a natural map from the identity functor to $\mathrm{induced}_{\mathfrak{g}, K_1}^{\mathfrak{g}, K} \circ F$, and we are to prove that each φ_W is an isomorphism.

Let us denote by e the evaluation-at-1 mapping for

$$(7.216) \qquad \mathrm{induced}_{\mathfrak{g}, K_1}^{\mathfrak{g}, K} P_\lambda \mathcal{F}_{\mathfrak{g}, K}^{\mathfrak{g}, K_1}(W).$$

The isomorphism of the Hom's above is given by $\Phi \to e\Phi$. Taking $e\Phi = P_\lambda$ makes $\Phi = \varphi_W$. Thus $e\varphi_W = P_\lambda$, and $\varphi_W(w)(1) = P_\lambda(w)$ for $w \in W$. Since φ_W is K linear, φ_W has to be given by

$$(7.217) \qquad \varphi_W(w)(k) = P_\lambda(k^{-1} w).$$

Let us show that $\ker \varphi_W$ is 0 and image φ_W is (7.216). If w is in $\ker \varphi_W$, then (7.217) gives $P_\lambda(k^{-1} w) = 0$ for all $k \in K$. Since W is in $\mathcal{C}_{K \chi_\lambda}(\mathfrak{g}, K)$ and K_1 is the stabilizer of χ_λ, we can write

$$w = \sum_{k' K_1 \in K / K_1} w_{k'},$$

where $w_{k'}$ is in the image of $P_{k' \chi_\lambda}$. Fix attention on a term $w_{k_0'}$. We have $k_0'^{-1} w = \sum k_0'^{-1} w_{k'}$ with $k_0'^{-1} w_{k'}$ in the image of $P_{k_0'^{-1} k' \chi_\lambda}$. The only time when $P_{k_0'^{-1}k'\chi_\lambda} = P_{\chi_\lambda}$ is when $k_0' = k'K_1$. Hence the condition $P_\lambda(k_0'^{-1}w) = 0$ forces $k_0'^{-1}w_{k_0'} = 0$, i.e., $w_{k_0'} = 0$. Hence $w = 0$, and $\ker \varphi_W = 0$.

To compute image φ_W, observe for $w \in$ image P_λ that $\varphi_W(w) = j_W(w)$ with j_W as in Proposition 2.75. The proof of Proposition 2.75b shows that the K translates of the elements $j_W(w)$ generate the entire space (7.216). Since image φ_W is K invariant, it follows that image φ_W equals (7.216). Therefore φ_W is an isomorphism, and induced$_{\mathfrak{g}, K_1}^{\mathfrak{g}, K} \circ F$ is naturally isomorphic with the identity functor.

With this proposition in hand, we shall consider translation functors only on $(\mathfrak{g}, K)$ modules with an infinitesimal character, insisting that their target modules have the same property. Suppose that μ is an algebraically integral member of $\mathfrak{h}^*$ that is an extreme weight of an irreducible $(\mathfrak{g}, K)$ module F^μ that remains irreducible under $\mathfrak{g}$. (In other words, suppose that F^μ is a $(\mathfrak{g}, K)$ module, that it has just one highest weight, and that the weight in question has multiplicity 1.) Suppose that χ_λ and $\chi_{\lambda+\mu}$ are fixed by K. If V is in $\mathcal{C}_\lambda(\mathfrak{g}, K)$, then we set

$$(7.218) \qquad \psi_\lambda^{\lambda+\mu}(V) = P_{\lambda+\mu}(V \otimes_{\mathbb{C}} F^\mu).$$

A little care is required in using this functor, since the notation on the left side hides the dependence of φ on the choice of F^μ (which may not be unique for given μ).

The functor $\psi_\lambda^{\lambda+\mu}$ is exact and covariant from $\mathcal{C}_\lambda(\mathfrak{g}, K)$ to $\mathcal{C}_{\lambda+\mu}(\mathfrak{g}, K)$. Since F^μ is irreducible under $\mathfrak{g}$, the present ψ is intertwined with the ψ of the previous section by means of forgetful functors:

$$(7.219) \qquad \mathcal{F}_{\mathfrak{g}, K}^{\mathfrak{g}, K_0} \circ \psi_\lambda^{\lambda+\mu} \cong \psi_\lambda^{\lambda+\mu} \circ \mathcal{F}_{\mathfrak{g}, K}^{\mathfrak{g}, K_0} \qquad \text{on } \mathcal{C}_\lambda(\mathfrak{g}, K).$$

It follows immediately that conditions for the new $\psi_\lambda^{\lambda+\mu}$ to send nonzero modules to nonzero modules can be read off from Theorem 7.172.

The adjoint relation of Proposition 7.143 extends to the present situation provided F^μ and $F^{-\mu}$ are chosen as contragredients of each other.

Proposition 7.220. Suppose that μ is an extreme weight of an irreducible finite-dimensional $(\mathfrak{g}, K)$ module F and that F is irreducible under $\mathfrak{g}$. Let λ be a member of $\mathfrak{h}^*$ such that χ_λ and $\chi_{\lambda+\mu}$ are fixed by K, and use F and its contragredient to define ψ functors. Then

$$\operatorname{Hom}_{\mathfrak{g}, K}(\psi_{\lambda+\mu}^\lambda(V_1), V_2) \cong \operatorname{Hom}_{\mathfrak{g}, K}(V_1, \psi_\lambda^{\lambda+\mu}(V_2))$$

naturally for V_1 in $\mathcal{C}_{\lambda+\mu}(\mathfrak{g}, K)$ and V_2 in $\mathcal{C}_\lambda(\mathfrak{g}, K)$.

PROOF. The proof is the same as for Proposition 7.143.

We shall be interested in conditions under which ψ sends irreducible $(\mathfrak{g}, K)$ modules to irreducible $(\mathfrak{g}, K)$ modules. First we develop a notion of "finite length" that parallels the one in the previous section.

Proposition 7.221. Let $(\mathfrak{g}, K)$ be a reductive pair. If V is an admissible $(\mathfrak{g}, K)$ module, then V is admissible as a $(\mathfrak{g}, K_0)$ module.

PROOF. There are only finitely many irreducible representations of K whose restriction to K_0 contains a given irreducible representation π of K_0, since any such representation of K must occur in the finite-dimensional representation induced$_{K_0}^K(\pi)$. (This is immediate from classical Frobenius reciprocity.) Hence V is admissible as a $(\mathfrak{g}, K_0)$ module.

Corollary 7.222. Let $(\mathfrak{g}, K)$ be a reductive pair, and let $K\chi$ be a K orbit of homomorphisms $Z(\mathfrak{g}) \to \mathbb{C}$. Then there exist finitely many K types $\tau_{\mu_1}, \ldots, \tau_{\mu_n}$ such that every nonzero $(\mathfrak{g}, K)$ module in $\mathcal{C}_{K\chi}(\mathfrak{g}, K)$ contains one of these K types.

PROOF. Let V be a member of $\mathcal{C}_{K\chi}(\mathfrak{g}, K)$, and regard V as a $U(\mathfrak{g})$ module. Then Proposition 7.20 shows that V is the direct sum of members of the various $\mathcal{C}_{k\chi}(\mathfrak{g}, K_0)$, $k \in K/K_0$. For each $\mathcal{C}_{k\chi}(\mathfrak{g}, K_0)$, Corollary 7.205 supplies finitely many K_0 types, one of which must appear in each summand (the one depending on the summand). Only finitely many K types can contain a given K_0 type (by classical Frobenius reciprocity), and the union of these finite sets over our list of K_0 types is the required finite set of K types.

Corollary 7.223. Let $(\mathfrak{g}, K)$ be a reductive pair. If V is an admissible $(\mathfrak{g}, K)$ module, then the following conditions on V are equivalent:

  (a) V is finitely generated as a $U(\mathfrak{g})$ module
  (b) V is $Z(\mathfrak{g})$ finite
  (c) V has a composition series.

PROOF. By Proposition 7.221, V is admissible as a $(\mathfrak{g}, K_0)$ module. Then (a) and (b) are equivalent by Corollary 7.207, and (c) implies (b) by the same argument as for Corollary 7.207. Assume that (b) holds, let V be in $\mathcal{C}_{K\chi}(\mathfrak{g}, K)$ without loss of generality, and let $\tau_{\mu_1}, \ldots, \tau_{\mu_n}$ be as in Corollary 7.222. Then the same inductive argument as in the proof of Corollary 7.207 leads to a composition series.

If the admissible $(\mathfrak{g}, K)$ module satisfies the equivalent conditions of Corollary 7.223, we say that V has **finite length**. By (c), this definition is consistent with the one in §A.3 within the good category of all admissible

$(\mathfrak{g}, K)$ modules. Let $\mathcal{F}(\mathfrak{g}, K)$ be the category of $(\mathfrak{g}, K)$ modules of finite length, and let $\mathcal{F}_{K\chi}(\mathfrak{g}, K)$ be the subcategory of those in $\mathcal{C}_{K\chi}(\mathfrak{g}, K)$. When $K\chi = \{\chi_\lambda\}$, we may write $\mathcal{F}_\lambda(\mathfrak{g}, K)$ in place of $\mathcal{F}_{K\chi_\lambda}(\mathfrak{g}, K)$. When a member of $\mathcal{F}(\mathfrak{g}, K)$ is regarded as a $(\mathfrak{g}, K_0)$ module, it is in $\mathcal{F}(\mathfrak{g}, K_0)$; admissibility is by Proposition 7.221, and finite length is by preservation of $Z(\mathfrak{g})$ finiteness.

To investigate the effect of ψ on irreducibility, we shall relate $\mathcal{F}(\mathfrak{g}, K)$ to $\mathcal{F}(\mathfrak{g}, K_0)$. The tool for doing so is "complete reducibility," which is discussed in §A.2.

If M is in $\mathcal{F}(\mathfrak{g}, K)$, the **socle** of M, denoted soc(M), is the sum of all completely reducible $(\mathfrak{g}, K)$ submodules of M. It is completely reducible by Corollary A.10b and the finiteness of the length of M. We shall relate socles in $\mathcal{F}(\mathfrak{g}, K)$ and $\mathcal{F}(\mathfrak{g}, K_0)$.

Lemma 7.224. Let $(\mathfrak{g}, K)$ be a reductive pair. If M is irreducible in $\mathcal{F}(\mathfrak{g}, K)$, then M, as a $(\mathfrak{g}, K_0)$ module, is completely reducible in $\mathcal{F}(\mathfrak{g}, K_0)$.

PROOF. By Proposition 7.221, M is admissible as a $(\mathfrak{g}, K_0)$ module. It has finite length since it is $Z(\mathfrak{g})$ finite. By Corollary 7.207, M has a composition series as a $(\mathfrak{g}, K_0)$ module and in particular contains an irreducible $(\mathfrak{g}, K_0)$ submodule. By Proposition 2.84, M has the required direct-sum decomposition.

Proposition 7.225. Let $(\mathfrak{g}, K)$ be a reductive pair. If M is in $\mathcal{F}(\mathfrak{g}, K)$, then M is completely reducible in $\mathcal{F}(\mathfrak{g}, K)$ if and only if its underlying $(\mathfrak{g}, K_0)$ module is completely reducible in $\mathcal{F}(\mathfrak{g}, K_0)$.

PROOF. If M is completely reducible in $\mathcal{F}(\mathfrak{g}, K)$, then Lemma 7.224 shows it is completely reducible in $\mathcal{F}(\mathfrak{g}, K_0)$.

Conversely let M be in $\mathcal{C}(\mathfrak{g}, K)$, and suppose M is completely reducible as a $(\mathfrak{g}, K_0)$ module. Because of induction on the length, it is enough to assume that M is reducible and to prove that M decomposes nontrivially as a direct sum of $(\mathfrak{g}, K)$ modules. By Corollary 7.223, M has an irreducible $(\mathfrak{g}, K)$ submodule M'. By Proposition A.9b–c, we can choose a $(\mathfrak{g}, K_0)$ submodule M'' of M such that $M = M' \oplus M''$; the submodule M'' is not 0 since we are assuming M is reducible. Let P be the projection of M on M' along M''. This is a $(\mathfrak{g}, K_0)$ map, and, for any $k \in K, k_0 \in K_0$, and $X \in \mathfrak{g}$, we have

$$k_0(kPk^{-1}) = k(k^{-1}k_0k)Pk^{-1} = kP(k^{-1}k_0k)k^{-1} = (kPk^{-1})k_0,$$
$$X(kPk^{-1}) = k(\mathrm{Ad}(k)^{-1}X)Pk^{-1} = kP(\mathrm{Ad}(k)^{-1}X)k^{-1} = (kPk^{-1})X.$$

Thus kPk^{-1} is another $(\mathfrak{g}, K_0)$ map, and it clearly depends only on the coset kK_0. Define

$$E = \frac{1}{[K : K_0]} \sum_{K/K_0} kPk^{-1}.$$

Then E is a $(\mathfrak{g}, K)$ map, and E is 1 on M'; since $M' \neq 0$, E has 1 as an eigenvalue.

Let N' be the generalized eigenspace for E for the eigenvalue 1, and let N'' be the sum of the other generalized eigenspaces. The nature of these spaces is not clear since M is presumably infinite dimensional. The subspaces N' and N'' are $(\mathfrak{g}, K)$ submodules of M, and we shall prove that $M = N' \oplus N''$ and $N'' \neq 0$. Within M, let τ be an irreducible representation of K_0, and let us lump together as $M_{K\tau}$ all K_0 isotypic subspaces of M for irreducible representations $k\tau, k \in K$. (Here $(k\tau)(k') = \tau(k^{-1}k'k)$.) Then M is the direct sum of the spaces $M_{K\tau}$, and each $M_{K\tau}$ is K invariant. Also $M_{K\tau} = M'_{K\tau} \oplus M''_{K\tau}$, and P maps $M_{K\tau}$ to itself. By the K invariance, each kPk^{-1} maps $M_{K\tau}$ to itself, and so does E. Referring to the definition of E and recalling that $M_{K\tau}$ is finite dimensional, we see that $\operatorname{Tr} E|_{M_{K\tau}} = \operatorname{Tr} P|_{M_{K\tau}}$. Thus E cannot have all eigenvalues 1 on $M_{K\tau}$ if P is not the identity, i.e., if $M''_{K\tau} \neq 0$. Let us write $M_{K\tau} = N'_{K\tau} \oplus N''_{K\tau}$, where $N'_{K\tau}$ is the generalized eigenspace for $E|_{M_{K\tau}}$ for the eigenvalue 1 and where $N''_{K\tau}$ is the sum of the generalized eigenspaces for the other eigenvalues. What we have just shown is that $N''_{K\tau} \neq 0$ if $M''_{K\tau} \neq 0$, and certainly $M''_{K\tau} \neq 0$ for some τ since $M'' \neq 0$. Since N' and N'' are the respective sums of the $N'_{K\tau}$ and the $N''_{K\tau}$, we obtain the required $(\mathfrak{g}, K)$ decomposition $M = N' \oplus N''$ with $N'' \neq 0$.

Corollary 7.226. If M is in $\mathcal{F}(\mathfrak{g}, K)$, then the socle of M as a $(\mathfrak{g}, K)$ module is equal to the socle of M as a $(\mathfrak{g}, K_0)$ module.

PROOF. Since $\operatorname{soc}(M)$ in $\mathcal{F}(\mathfrak{g}, K)$ is completely reducible, Proposition 7.225 shows it is completely reducible in $\mathcal{F}(\mathfrak{g}, K_0)$ and hence is contained in the socle of M as a $(\mathfrak{g}, K_0)$ module.

Conversely we must show that the socle $\operatorname{soc}_0(M)$ of M as a $(\mathfrak{g}, K_0)$ module is a $(\mathfrak{g}, K)$ submodule of M. Obviously $\operatorname{soc}_0(M)$ is preserved by $U(\mathfrak{g})$; we must show it is preserved by K. Now $\operatorname{soc}_0(M)$ is the sum of the irreducible $(\mathfrak{g}, K_0)$ submodules of M. Suppose M_0 is such a submodule and k is in K. Since k normalizes K_0, kM_0 is another irreducible $(\mathfrak{g}, K_0)$ submodule. It is therefore contained in $\operatorname{soc}_0(M)$. Thus the action of K permutes a collection of subspaces spanning $\operatorname{soc}_0(M)$. Hence it preserves $\operatorname{soc}_0(M)$, as we wished to show.

Corollary 7.227. Let $(\mathfrak{g}, K)$ be a reductive pair, let K_1 be a subgroup of finite index in K, and form the inclusion of pairs $(\mathfrak{g}, K_1) \hookrightarrow (\mathfrak{g}, K)$. If V is a completely reducible $(\mathfrak{g}, K_1)$ module, then $\text{induced}_{\mathfrak{g}, K_1}^{\mathfrak{g}, K}(V)$ is a completely reducible $(\mathfrak{g}, K)$ module.

REMARK. It follows that all the reducibility of $\text{induced}_{\mathfrak{g}, K_1}^{\mathfrak{g}, K}(V)$ is detected by Mackey theory as given in Corollary 2.83.

PROOF. By Proposition 7.225, $\mathcal{F}_{\mathfrak{g}, K_1}^{\mathfrak{g}, K_0}(V)$ is completely reducible as a $(\mathfrak{g}, K_0)$ module. Proposition 2.85 thus shows that

$$\mathcal{F}_{\mathfrak{g}, K}^{\mathfrak{g}, K_0}(\text{induced}_{\mathfrak{g}, K_0}^{\mathfrak{g}, K}(\mathcal{F}_{\mathfrak{g}, K_1}^{\mathfrak{g}, K_0}(V)))$$

is completely reducible as a $(\mathfrak{g}, K_0)$ module, and it follows from Proposition 7.225 that

$$(7.228) \qquad\qquad \text{induced}_{\mathfrak{g}, K_0}^{\mathfrak{g}, K}(\mathcal{F}_{\mathfrak{g}, K_1}^{\mathfrak{g}, K_0}(V))$$

is completely reducible as a $(\mathfrak{g}, K)$ module. By Proposition 2.87b, $\text{induced}_{\mathfrak{g}, K_1}^{\mathfrak{g}, K}(V)$ is $(\mathfrak{g}, K)$ isomorphic with a submodule of (7.228), and thus the corollary follows from Proposition A.9c.

Theorem 7.229. Let $(\mathfrak{g}, K)$ be a reductive pair, let $\mathfrak{h}$ be a Cartan subalgebra of $\mathfrak{g}$, and fix a system $\Delta^+(\mathfrak{g}, \mathfrak{h})$ of positive roots. Let $\mu \in \mathfrak{h}^*$ be algebraically integral, and suppose that a finite-dimensional irreducible $(\mathfrak{g}, K)$ module F^μ is given that remains irreducible under $\mathfrak{g}$, has extreme weight μ, and is to be used in translation functors. Further suppose λ is a member of $\mathfrak{h}^*$ satisfying

   (i) λ and $\lambda + \mu$ are integrally dominant,
  (ii) K fixes χ_λ and $\chi_{\lambda+\mu}$, and
 (iii) $\lambda + \mu$ is at least as singular as λ, i.e., $\langle \lambda + \mu, \alpha \rangle = 0$ for every $\alpha \in \Delta(\mathfrak{g}, \mathfrak{h})$ such that $\langle \lambda, \alpha \rangle = 0$.

If V is an irreducible member of $\mathcal{F}_\lambda(\mathfrak{g}, K)$, then $\psi_\lambda^{\lambda+\mu}(V)$ is irreducible in $\mathcal{F}_{\lambda+\mu}(\mathfrak{g}, K)$ or is 0.

PROOF. Let us abbreviate $\psi = \psi_\lambda^{\lambda+\mu}$ and $\varphi = \psi_{\lambda+\mu}^\lambda$, with φ constructed from the contragredient of F^μ so that the adjoint formula applies. We shall work with both $\mathcal{F}(\mathfrak{g}, K)$ and $\mathcal{F}(\mathfrak{g}, K_0)$. By Corollary 7.226, there is no danger of confusion if we write soc for the socle in either category.

Let V be irreducible in $\mathcal{F}(\mathfrak{g}, K)$. We first prove

(7.230)

$$\text{multiplicity of } V \text{ in soc}(\varphi\psi V) \text{ is } \begin{cases} = 0 & \text{if } \psi V = 0 \\ = 1 & \text{if } \psi V \text{ is irreducible} \\ > 1 & \text{if } \psi V \text{ is reducible.} \end{cases}$$

In fact, Proposition 7.225 shows that V is completely reducible in $\mathcal{F}(\mathfrak{g}, K_0)$, say as $V_1 \oplus \cdots \oplus V_n$. Then $\psi(V) = \psi(V_1) \oplus \cdots \oplus \psi(V_n)$ since translation functors respect direct sums. Theorem 7.171a shows that each $\psi(V_i)$ is irreducible or is 0 in $\mathcal{F}(\mathfrak{g}, K_0)$. So $\psi(V)$ is completely reducible in $\mathcal{F}(\mathfrak{g}, K_0)$. Hence $\psi(V)$ is completely reducible in $\mathcal{F}(\mathfrak{g}, K)$, by Proposition 7.225. It follows that

$$\dim \operatorname{Hom}_{\mathfrak{g},K}(\psi(V), \psi(V)) \text{ is } \begin{cases} = 0 & \text{if } \psi V = 0 \\ = 1 & \text{if } \psi V \text{ is irreducible} \\ > 1 & \text{if } \psi V \text{ is reducible.} \end{cases}$$

Combining this formula with the computation

$$\begin{aligned} \text{multiplicity of } V &\text{ in } \operatorname{soc}(\varphi \psi V) \\ &= \dim \operatorname{Hom}_{\mathfrak{g},K}(V, \operatorname{soc}(\varphi \psi V)) \\ &= \dim \operatorname{Hom}_{\mathfrak{g},K}(V, \varphi \psi V) & \text{by Proposition A.10a} \\ &= \dim \operatorname{Hom}_{\mathfrak{g},K}(\psi V, \psi V) & \text{by Proposition 7.220,} \end{aligned}$$

we obtain (7.230).

For the moment, let us compute in $\mathcal{F}(\mathfrak{g}, K_0)$, taking V_0 as an irreducible member of $\mathcal{F}(\mathfrak{g}, K_0)$. From Theorem 7.171a we know that $\psi(V_0)$ is irreducible or is 0. If it is 0, then $\operatorname{soc}(\varphi \psi V_0) = 0$. Otherwise (7.230) shows that V_0 occurs exactly once in $\operatorname{soc}(\varphi \psi V_0)$. In this case suppose $V_0' \not\cong V_0$ is another irreducible member of $\mathcal{F}(\mathfrak{g}, K_0)$ occurring in $\operatorname{soc}(\varphi \psi V_0)$. Then we have

$$0 \neq \operatorname{Hom}_{\mathfrak{g},K_0}(V_0', \operatorname{soc}(\varphi \psi V_0)) = \operatorname{Hom}_{\mathfrak{g},K_0}(V_0', \varphi \psi V_0)$$
$$\cong \operatorname{Hom}_{\mathfrak{g},K}(\psi V_0', \psi V_0).$$

Since $\psi V_0'$ is irreducible (Theorem 7.171a), we have $\psi V_0' \cong \psi V_0$. By Theorem 7.171b, we obtain $V_0' \cong V_0$, contradiction. This proves:

(7.231)
$$\operatorname{soc}(\varphi \psi V_0) \cong \begin{cases} V_0 & \text{if } V_0 \text{ is irreducible in } \mathcal{F}(\mathfrak{g}, K_0) \text{ and } \psi V_0 \neq 0 \\ 0 & \text{if } \psi V_0 = 0. \end{cases}$$

Now let us return to our irreducible V in $\mathcal{F}(\mathfrak{g}, K)$ and its decomposition $V = V_1 \oplus \cdots \oplus V_n$ in $\mathcal{F}(\mathfrak{g}, K_0)$. From (7.231) we have

$$\operatorname{soc}(\varphi \psi V) \cong \text{submodule of } V \qquad \text{as } (\mathfrak{g}, K_0) \text{ modules.}$$

Therefore V has multiplicity at most 1 in $\operatorname{soc}(\varphi \psi V)$ as a $(\mathfrak{g}, K)$ module. By (7.230), $\psi(V)$ is irreducible or 0.

Theorem 7.232. Let $(\mathfrak{g}, K)$ be a reductive pair, let $\mathfrak{h}$ be a Cartan subalgebra of $\mathfrak{g}$, and fix a system $\Delta^+(\mathfrak{g}, \mathfrak{h})$ of positive roots. Suppose $\mu \in \mathfrak{h}^*$ is algebraically integral. Further suppose λ is a member of $\mathfrak{h}^*$ satisfying

(i) λ and $\lambda + \mu$ are integrally dominant,

(ii) K fixes χ_λ and $\chi_{\lambda+\mu}$, and

(iii) λ and $\lambda + \mu$ are equisingular, i.e., $\langle \lambda, \alpha \rangle = 0$ and $\langle \lambda + \mu, \alpha \rangle = 0$ for the same roots $\alpha \in \Delta(\mathfrak{g}, \mathfrak{h})$.

Fix a finite-dimensional irreducible $(\mathfrak{g}, K)$ module F^μ that remains irreducible under $\mathfrak{g}$, has extreme weight μ, and is to be used in defining $\psi_\lambda^{\lambda+\mu}$. Let $F^{-\mu} = (F^\mu)^*$ be used in defining $\psi_{\lambda+\mu}^\lambda$. Then

(a) $\psi_\lambda^{\lambda+\mu} : \mathcal{F}_\lambda(\mathfrak{g}, K) \to \mathcal{F}_{\lambda+\mu}(\mathfrak{g}, K)$ carries irreducible modules to irreducible modules,

(b) $\psi_{\lambda+\mu}^\lambda \psi_\lambda^{\lambda+\mu}$ is naturally isomorphic to the identity functor on $\mathcal{F}_\lambda(\mathfrak{g}, K)$, and

(c) $\psi_\lambda^{\lambda+\mu} \psi_{\lambda+\mu}^\lambda$ is naturally isomorphic to the identity functor on $\mathcal{F}_{\lambda+\mu}(\mathfrak{g}, K)$.

REMARK. Compare with Theorem 7.173 and Corollary 7.209.

PROOF. (a) For V irreducible in $\mathcal{F}(\mathfrak{g}, K)$, Theorem 7.229 shows that $\psi_\lambda^{\lambda+\mu}(V)$ is irreducible or 0, and (7.219) and Theorem 7.172 show that $\psi_\lambda^{\lambda+\mu}(V)$ is not 0.

(b) If V is irreducible in $\mathcal{F}(\mathfrak{g}, K)$, then application of (a) to $\psi_\lambda^{\lambda+\mu}$ and $\psi_{\lambda+\mu}^\lambda$ shows that $\psi_{\lambda+\mu}^\lambda \psi_\lambda^{\lambda+\mu}(V)$ is irreducible. The adjoint formula (Proposition 7.220) gives

$$\mathrm{Hom}_{\mathfrak{g},K}(\psi_{\lambda+\mu}^\lambda \psi_\lambda^{\lambda+\mu}(V), V) \cong \mathrm{Hom}_{\mathfrak{g},K}(\psi_\lambda^{\lambda+\mu}(V), \psi_\lambda^{\lambda+\mu}(V)) \neq 0,$$

and hence

(7.233) $\psi_{\lambda+\mu}^\lambda \psi_\lambda^{\lambda+\mu}(V) \cong V.$

To finish the proof, we argue as in Corollary 7.209, taking (7.233) as a starting point rather than Theorem 7.173b.

15. Application to Cohomological Induction

In this section we shall prove, under suitable conditions, that the translation functors ψ commute with cohomological induction up to shifts in parameters. We use notation as in §V.1, taking $(\mathfrak{g}, K)$ to be a reductive pair and $\mathfrak{q} = \mathfrak{l} \oplus \mathfrak{u}$ to be a θ stable parabolic subalgebra of $\mathfrak{g}$. The cohomological induction functors $\mathcal{L}_j$ and $\mathcal{R}^j$ are defined in (5.3).

In our first application, in Chapter VIII, we shall want to translate parameters only by multiples of $2\delta(\mathfrak{u})$. Later we shall want to be able to translate by other parameters, and we shall even want extensions available for the present results to parabolics that are not θ stable.

Our goal in this section is to relate translation functors for $\mathfrak{l}$ to those for $\mathfrak{g}$. In order to do so, we must first relate the data that enter the definitions of these functors—the weights and finite-dimensional representations. We start with a Cartan subalgebra $\mathfrak{h}$ of $\mathfrak{l}$; automatically $\mathfrak{h}$ is also a Cartan subalgebra of $\mathfrak{g}$. A translation functor for $(\mathfrak{g}, K)$ modules involves an algebraically integral weight $\mu \in \mathfrak{h}^*$ and a finite-dimensional $(\mathfrak{g}, K)$ module F^μ with μ as an extreme weight. The deepest results, such as Theorems 7.229 and 7.232, impose also the hypothesis

(7.234) $\qquad\qquad F^\mu$ is irreducible as a $\mathfrak{g}$ module.

We shall need to use such data to build a translation functor for $(\mathfrak{l}, L \cap K)$ modules. The weight μ is automatically algebraically integral for $\mathfrak{l}$; we need only an $(\mathfrak{l}, L \cap K)$ module with μ as extreme weight. We can find it as a composition factor of the restriction $\mathcal{F}_{\mathfrak{g}, K}^{\mathfrak{l}, L \cap K}(F^\mu)$ of F^μ to an $(\mathfrak{l}, L \cap K)$ module. Because the weight μ occurs exactly once in F^μ, there is exactly one place in a composition series of this restriction where one of the consecutive quotients contains the weight μ. Call this quotient E^μ. Since μ is extremal in F^μ with respect to $\mathfrak{g}$, it is extremal in E^μ with respect to $\mathfrak{l}$. In short,

(7.235) $\qquad$ $E^\mu = $ unique irreducible $(\mathfrak{l}, L \cap K)$ composition
$\qquad\qquad\qquad$ factor of F^μ containing the weight μ.

In this way we associate to F^μ a distinguished irreducible $(\mathfrak{l}, L \cap K)$ module E^μ of extreme weight μ. In order to have the best behavior for the corresponding translation functors, we need to impose as a final hypothesis on the pair (μ, F^μ) the condition that

(7.236) $\qquad\qquad E^\mu$ is irreducible as an $\mathfrak{l}$ module.

(Unfortunately this condition does not follow automatically from (7.234).)

Theorem 7.237. Let $(\mathfrak{g}, K)$ be a reductive pair, and let $\mathfrak{q} = \mathfrak{l} \oplus \mathfrak{u}$ be a θ stable parabolic subalgebra. Let $\mathfrak{h}$ be a Cartan subalgebra of $\mathfrak{l}$, and introduce a positive system $\Delta^+(\mathfrak{g}, \mathfrak{h})$. Suppose that $\mu \in \mathfrak{h}^*$ is algebraically integral for $\mathfrak{g}$, and that F^μ is an irreducible finite-dimensional $(\mathfrak{g}, K)$ module of extreme weight μ that remains irreducible as a $\mathfrak{g}$ module. Define an irreducible finite-dimensional $(\mathfrak{l}, L \cap K)$ module E^μ of extreme weight μ as in (7.235), and suppose (7.236) holds. Suppose that λ is a member of $\mathfrak{h}^*$ satisfying

  (i) $\lambda + \delta(\mathfrak{u})$ and $\lambda + \delta(\mathfrak{u}) + \mu$ are integrally dominant relative to $\Delta^+(\mathfrak{g}, \mathfrak{h})$,
  (ii) K fixes the $Z(\mathfrak{g})$ infinitesimal characters $\chi_{\lambda+\delta(\mathfrak{u})}$ and $\chi_{\lambda+\delta(\mathfrak{u})+\mu}$,
  (iii) $L \cap K$ fixes the $Z(\mathfrak{l})$ infinitesimal characters χ_λ and $\chi_{\lambda+\mu}$, and
  (iv) $\lambda + \delta(\mathfrak{u}) + \mu$ is at least as singular as $\lambda + \delta(\mathfrak{u})$.

Use F^μ to define a translation functor $\psi_{\lambda+\delta(\mathfrak{u})}^{\lambda+\delta(\mathfrak{u})+\mu}$ for $(\mathfrak{g}, K)$ modules, and use E^μ to define a translation functor $\psi_\lambda^{\lambda+\mu}$ for $(\mathfrak{l}, L \cap K)$ modules. If Z is an $(\mathfrak{l}, L \cap K)$ module in $\mathcal{C}_\lambda(\mathfrak{l}, L \cap K)$, then there are natural isomorphisms

$$\psi_{\lambda+\delta(\mathfrak{u})}^{\lambda+\delta(\mathfrak{u})+\mu}(\mathcal{L}_j(Z)) \cong \mathcal{L}_j(\psi_\lambda^{\lambda+\mu}(Z))$$

and

$$\psi_{\lambda+\delta(\mathfrak{u})}^{\lambda+\delta(\mathfrak{u})+\mu}(\mathcal{R}^j(Z)) \cong \mathcal{R}^j(\psi_\lambda^{\lambda+\mu}(Z))$$

for all j.

REMARKS.

1) By Corollary 5.25c, $\mathcal{L}_j(Z)$ is a direct sum of $U(\mathfrak{g})$ modules of infinitesimal character $k_j \chi_{\lambda+\delta(\mathfrak{u})}$, where k_j runs through coset representatives for $K/((L \cap K)K_0)$. Hypothesis (ii) thus implies that $\mathcal{L}_j(Z)$ has infinitesimal character $\chi_{\lambda+\delta(\mathfrak{u})}$. Similar remarks apply to $\mathcal{R}^j(Z)$. In analogous fashion, $\mathcal{L}_j(\psi_\lambda^{\lambda+\mu}(Z))$ and $\mathcal{R}^j(\psi_\lambda^{\lambda+\mu}(Z))$ have infinitesimal character $\chi_{\lambda+\delta(\mathfrak{u})+\mu}$.

2) The positive system $\Delta^+(\mathfrak{g}, \mathfrak{h})$ is used in the theorem only for defining integral dominance. It is *not* assumed that $\Delta(\mathfrak{u}) \subseteq \Delta^+(\mathfrak{g}, \mathfrak{h})$.

3) As will be apparent from the proof, the delicate part of the argument for $\mathcal{L}_j$ is the handling of ind, not the handling of Π_j. In fact, we shall see that

$$(7.238a) \qquad \psi_{\lambda+\delta(\mathfrak{u})}^{\lambda+\delta(\mathfrak{u})+\mu} \circ (\Pi_{\mathfrak{g},L\cap K}^{\mathfrak{g},K})_j \cong (\Pi_{\mathfrak{g},L\cap K}^{\mathfrak{g},K})_j \circ \psi_{\lambda+\delta(\mathfrak{u})}^{\lambda+\delta(\mathfrak{u})+\mu}$$

on $\mathcal{C}_{\lambda+\delta(\mathfrak{u})}(\mathfrak{g}, L \cap K)$, as a consequence of hypothesis (ii), with no need to assume (i) or (iv). Similarly

$$(7.238b) \qquad \psi_{\lambda+\delta(\mathfrak{u})}^{\lambda+\delta(\mathfrak{u})+\mu} \circ (\Gamma_{\mathfrak{g},L\cap K}^{\mathfrak{g},K})^j \cong (\Gamma_{\mathfrak{g},L\cap K}^{\mathfrak{g},K})^j \circ \psi_{\lambda+\delta(\mathfrak{u})}^{\lambda+\delta(\mathfrak{u})+\mu}.$$

PROOF. We prove the result for $\mathcal{L}_j$, the proof for $\mathcal{R}^j$ being completely analogous. First let us prove (7.238a). We abbreviate $\Pi_j = (\Pi_{\mathfrak{g},L\cap K}^{\mathfrak{g},K})_j$. For W in $\mathcal{C}_{\lambda+\delta(\mathfrak{u})}(\mathfrak{g}, L \cap K)$, we have

$$\psi_{\lambda+\delta(\mathfrak{u})}^{\lambda+\delta(\mathfrak{u})+\mu}\Pi_j(W)$$

$$= P_{\lambda+\delta(\mathfrak{u})+\mu}(\Pi_j(W) \otimes_\mathbb{C} F^\mu)$$

$$\cong P_{\lambda+\delta(\mathfrak{u})+\mu}(\Pi_j(W \otimes_\mathbb{C} F^\mu)) \qquad \text{by Theorem 2.103 and passage to derived functors}$$

$$= \bigoplus_\gamma P_{\lambda+\delta(\mathfrak{u})+\mu}(\Pi_j(P_{\lambda+\delta(\mathfrak{u})+\gamma}(W \otimes_\mathbb{C} F^\mu))) \qquad \text{by Theorem 7.133,}$$

the sum being over the weights γ of F^μ. By Theorem 5.21c, only the terms with $\chi_{\lambda+\delta(\mathfrak{u})+\mu}$ and $\chi_{\lambda+\delta(\mathfrak{u})+\gamma}$ in the same K orbit survive in this sum. By hypothesis (ii), $\chi_{\lambda+\delta(\mathfrak{u})+\mu} = \chi_{\lambda+\delta(\mathfrak{u})+\gamma}$ in this case. Thus the above sum reduces to

$$= P_{\lambda+\delta(\mathfrak{u})+\mu}(\Pi_j(P_{\lambda+\delta(\mathfrak{u})+\mu}(W \otimes_\mathbb{C} F^\mu)))$$

$$= P_{\lambda+\delta(\mathfrak{u})+\mu}(\Pi_j(\psi_{\lambda+\delta(\mathfrak{u})}^{\lambda+\delta(\mathfrak{u})+\mu}(W))).$$

Applying hypothesis (ii) again, we see that we can drop the outside $P_{\lambda+\delta(\mathfrak{u})+\mu}$, and then (7.238a) results.

Let

$$(7.239) \qquad 0 = F_{-1} \subsetneqq F_0 \subsetneqq F_1 \subsetneqq \cdots \subsetneqq F_N = F^\mu$$

be a filtration of F^μ by $(\bar{\mathfrak{q}}, L\cap K)$ modules in such a way that $\bar{\mathfrak{u}}$ acts trivially on each consecutive quotient $E_p = F_p/F_{p-1}$ and E_p is an irreducible $(\mathfrak{l}, L \cap K)$ module. Since this filtration is a composition series for F^μ as a $(\bar{\mathfrak{q}}, L \cap K)$ module, E^μ will be isomorphic with one of the consecutive quotients (see §A.3), and no other subquotient will contain the weight μ. Define p_0 by $E^\mu \cong E_{p_0}$.

Applying the exact functor $P_{\lambda+\delta(\mathfrak{u})+\mu}\mathrm{ind}_{\bar{\mathfrak{q}},L\cap K}^{\mathfrak{g},L\cap K}((\,\cdot\,) \otimes_\mathbb{C} Z_{\bar{\mathfrak{q}}}^\#)$ to (7.239), we obtain a filtration of $P_{\lambda+\delta(\mathfrak{u})+\mu}\mathrm{ind}_{\bar{\mathfrak{q}},L\cap K}^{\mathfrak{g},L\cap K}(F^\mu \otimes_\mathbb{C} Z_{\bar{\mathfrak{q}}}^\#)$ with consecutive quotients

$$(7.240) \qquad P_{\lambda+\delta(\mathfrak{u})+\mu}\mathrm{ind}_{\bar{\mathfrak{q}},L\cap K}^{\mathfrak{g},L\cap K}(E_p \otimes_\mathbb{C} Z_{\bar{\mathfrak{q}}}^\#).$$

According to Theorems 7.133 and 5.24a, all the generalized infinitesimal characters of $\mathrm{ind}_{\bar{\mathfrak{q}},L\cap K}^{\mathfrak{g},L\cap K}(E_p \otimes_\mathbb{C} Z_{\bar{\mathfrak{q}}}^\#)$ are of the form $\lambda + \delta(\mathfrak{u}) + \gamma$, where γ is a weight of E_p. Thus (7.240) can be nonzero only if

$$(7.241) \qquad \lambda + \delta(\mathfrak{u}) + \gamma = s(\lambda + \delta(\mathfrak{u}) + \mu)$$

for some $s \in W(\mathfrak{g}, \mathfrak{h})$. In this equation, $\lambda + \delta(\mathfrak{u}) + \mu$ is at least as singular as $\lambda + \delta(\mathfrak{u})$, by hypothesis (iv). Since γ is a weight of E_p and hence of F^μ, Proposition 7.166 shows that $\gamma = \mu$ and that s fixes $\lambda + \delta(\mathfrak{u}) + \mu$. From $\gamma = \mu$, we see that $p = p_0$. Consequently

$$P_{\lambda+\delta(\mathfrak{u})+\mu}\mathrm{ind}_{\bar{\mathfrak{q}}, L\cap K}^{\mathfrak{g}, L\cap K}(F^\mu \otimes_{\mathbb{C}} Z_{\bar{\mathfrak{q}}}^\#)$$

$$\cong P_{\lambda+\delta(\mathfrak{u})+\mu}\mathrm{ind}_{\bar{\mathfrak{q}}, L\cap K}^{\mathfrak{g}, L\cap K}(E^\mu \otimes_{\mathbb{C}} Z_{\bar{\mathfrak{q}}}^\#)$$

$$\cong P_{\lambda+\delta(\mathfrak{u})+\mu}\mathrm{ind}_{\bar{\mathfrak{q}}, L\cap K}^{\mathfrak{g}, L\cap K}(P_{\lambda+2\delta(\mathfrak{u})+\mu}(E^\mu \otimes_{\mathbb{C}} Z_{\bar{\mathfrak{q}}}^\#))$$

$$\text{since } \gamma = \mu \text{ above}$$

$$\cong \mathrm{ind}_{\bar{\mathfrak{q}}, L\cap K}^{\mathfrak{g}, L\cap K}(P_{\lambda+2\delta(\mathfrak{u})+\mu}(E^\mu \otimes_{\mathbb{C}} Z_{\bar{\mathfrak{q}}}^\#))$$

$$\cong \mathrm{ind}_{\bar{\mathfrak{q}}, L\cap K}^{\mathfrak{g}, L\cap K}((\psi_\lambda^{\lambda+\mu}(Z))_{\bar{\mathfrak{q}}}^\#).$$

Meanwhile the left side in this calculation, by Theorem 2.103, is

$$\cong P_{\lambda+\delta(\mathfrak{u})+\mu}(F^\mu \otimes_{\mathbb{C}} \mathrm{ind}_{\bar{\mathfrak{q}}, L\cap K}^{\mathfrak{g}, L\cap K}(Z_{\bar{\mathfrak{q}}}^\#))$$

$$= \psi_{\lambda+\delta(\mathfrak{u})}^{\lambda+\delta(\mathfrak{u})+\mu}(\mathrm{ind}_{\bar{\mathfrak{q}}, L\cap K}^{\mathfrak{g}, L\cap K}(Z_{\bar{\mathfrak{q}}}^\#)).$$

Thus ψ commutes with ind, apart from a shift in parameters. Combining this result with (7.238a), we obtain the conclusion of the theorem for $\mathcal{L}_j$.

16. Application to $\mathfrak{u}$ Homology and Cohomology

Finally we shall prove, under suitable conditions, that translation functors commute with the $\mathfrak{u}$ homology and cohomology functors on $(\mathfrak{g}, K)$ modules, up to shifts in parameter. In the light of the fundamental spectral sequences of Theorem 5.120, the result is a kind of "adjoint" to Theorem 7.237. We retain the notation of §15.

Theorem 7.242. Let $(\mathfrak{g}, K)$ be a reductive pair, and let $\mathfrak{q} = \mathfrak{l} \oplus \mathfrak{u}$ be a θ stable parabolic subalgebra. Let $\mathfrak{h}$ be a Cartan subalgebra of $\mathfrak{l}$, and introduce a positive system $\Delta^+(\mathfrak{g}, \mathfrak{h})$. Suppose that $\mu \in \mathfrak{h}^*$ is algebraically integral for $\mathfrak{g}$, and that F^μ is an irreducible finite-dimensional $(\mathfrak{g}, K)$ module of extreme weight μ that remains irreducible as a $\mathfrak{g}$ module. Define an irreducible finite-dimensional $(\mathfrak{l}, L\cap K)$ module E^μ of extreme weight μ as in (7.235), and suppose (7.236) holds. Suppose that λ is a member of $\mathfrak{h}^*$ satisfying

(i) $\lambda + \delta(\mathfrak{u})$ and $\lambda + \delta(\mathfrak{u}) + \mu$ are integrally dominant relative to $\Delta^+(\mathfrak{g}, \mathfrak{h})$,

(ii) K fixes the $Z(\mathfrak{g})$ infinitesimal characters $\chi_{\lambda+\delta(\mathfrak{u})}$ and $\chi_{\lambda+\delta(\mathfrak{u})+\mu}$,

(iii) $L \cap K$ fixes the $Z(\mathfrak{l})$ infinitesimal characters χ_λ and $\chi_{\lambda+\mu}$, and

(iv) $\lambda + \delta(\mathfrak{u})$ is at least as singular as $\lambda + \delta(\mathfrak{u}) + \mu$.

Use F^μ to define a translation functor $\psi_{\lambda+\delta(\mathfrak{u})}^{\lambda+\delta(\mathfrak{u})+\mu}$ for $(\mathfrak{g}, K)$ modules, and use E^μ to define a translation functor $\psi_\lambda^{\lambda+\mu}$ for $(\mathfrak{l}, L \cap K)$ modules. If X is a $(\mathfrak{g}, K)$ module in $\mathcal{C}_{\lambda+\delta(\mathfrak{u})}(\mathfrak{g}, K)$, then there are natural isomorphisms

$$P_{\lambda+\mu} H_j(\bar{\mathfrak{u}}, \psi_{\lambda+\delta(\mathfrak{u})}^{\lambda+\delta(\mathfrak{u})+\mu}(X)) \cong \psi_\lambda^{\lambda+\mu}(H_j(\bar{\mathfrak{u}}, X))$$

and
$$P_{\lambda+\mu} H^j(\mathfrak{u}, \psi_{\lambda+\delta(\mathfrak{u})}^{\lambda+\delta(\mathfrak{u})+\mu}(X)) \cong \psi_\lambda^{\lambda+\mu}(H^j(\mathfrak{u}, X))$$

as $(\mathfrak{l}, L \cap K)$ modules for all j.

REMARK. As with Theorem 7.237, $\Delta^+(\mathfrak{g}, \mathfrak{h})$ need not be correlated with $\mathfrak{u}$.

PROOF. We give the argument only for H_j. Let us abbreviate $P_j = (P_{\bar{\mathfrak{q}}, L \cap K}^{\mathfrak{l}, L \cap K})_j$ and $\mathcal{F}X = \mathcal{F}_{\mathfrak{g}, K}^{\bar{\mathfrak{q}}, L \cap K}(X)$. Recall from Proposition 3.12 that $H_j(\bar{\mathfrak{u}}, X)$ is $P_j(\mathcal{F}X)$. Under its $Z(\mathfrak{g})$ primary decomposition, $X \otimes_{\mathbb{C}} F$ is the direct sum of $U(\mathfrak{g})$ modules Y_i of generalized infinitesimal character ν_i. According to Theorem 7.56, $P_{\lambda+\mu} H_j(\bar{\mathfrak{u}}, Y_i)$ is 0 unless $\chi_{\nu_i} = \chi_{\lambda+\delta(\mathfrak{u})+\mu}$, i.e., unless Y_i equals $\psi_{\lambda+\delta(\mathfrak{u})}^{\lambda+\delta(\mathfrak{u})+\mu}(X)$. Consequently

$$(7.243) \qquad \begin{aligned} P_{\lambda+\mu} H_j(\bar{\mathfrak{u}}, \psi_{\lambda+\delta(\mathfrak{u})}^{\lambda+\delta(\mathfrak{u})+\mu}(X)) &= P_{\lambda+\mu} H_j(\bar{\mathfrak{u}}, X \otimes_{\mathbb{C}} F^\mu) \\ &= P_{\lambda+\mu} P_j(\mathcal{F}X \otimes_{\mathbb{C}} F^\mu). \end{aligned}$$

The right side here calls for a use of a Mackey isomorphism (Theorem 2.103), but F^μ is not trivial as a $\bar{\mathfrak{u}}$ module. We thus filter F^μ by $(\bar{\mathfrak{q}}, L \cap K)$ submodules F_p as in (7.239), with $\bar{\mathfrak{u}}$ acting trivially and $(\mathfrak{l}, L \cap K)$ acting irreducibly on the quotients $E_p = F_p / F_{p-1}$. Again E^μ will be isomorphic with one of the composition factors E_{p_0}, and no other composition factor will contain the weight μ. Applying Proposition D.57 to the functor $P \circ (\mathcal{F}X \otimes_{\mathbb{C}} (\cdot))$, we obtain a convergent spectral sequence with E_1 term

$$(7.244a) \qquad E_1^{p,q} = P_{p+q}(\mathcal{F}X \otimes_{\mathbb{C}} E_p),$$

with differential of bidegree $(-r, r-1)$, and with

$$(7.244b) \qquad E_r^{p,q} \Longrightarrow P_{p+q}(\mathcal{F}X \otimes_{\mathbb{C}} F).$$

Because $\bar{\mathfrak{u}}$ annihilates it, the $(\bar{\mathfrak{q}}, L \cap K)$ module E_p is obtained from the $(\mathfrak{l}, L \cap K)$ module E_p by applying the forgetful functor $\mathcal{F}_{\mathfrak{l}, L \cap K}^{\bar{\mathfrak{q}}, L \cap K}$. By Theorem 2.103, there is an isomorphism

$$P(\mathcal{F}X \otimes_{\mathbb{C}} E_p) \cong P(\mathcal{F}X) \otimes_{\mathbb{C}} E_p,$$

and (C.27a) and (C.28a1) allow us to deduce that

$$P_j(\mathcal{F}X \otimes_{\mathbb{C}} E_p) \cong P_j(\mathcal{F}X) \otimes_{\mathbb{C}} E_p$$

for all j. Consequently

(7.244c) $$E_1^{p,q} \cong P_{p+q}(\mathcal{F}X) \otimes_{\mathbb{C}} E_p.$$

Let us therefore compute $P_{\lambda+\mu}$ of (7.244c). Theorem 7.56 (with $\bar{\mathfrak{u}}$ used in place of $\mathfrak{u}$) says that the possible $Z(\mathfrak{l})$ infinitesimal characters of $P_{p+q}(\mathcal{F}X)$ are $s(\lambda + \delta(\mathfrak{u})) - \delta(\mathfrak{u})$ for s in $W^1(\mathfrak{g}, \mathfrak{h})$. Theorem 7.133 then says that the possible $Z(\mathfrak{l})$ infinitesimal characters of $P_{p+q}(\mathcal{F}X) \otimes_{\mathbb{C}} E_p$ are

$$s(\lambda + \delta(\mathfrak{u})) - \delta(\mathfrak{u}) + \gamma,$$

where γ is a weight of E_p. If $P_{\lambda+\mu}(E_1^{p,q})$ is $\neq 0$ for some p, then this expression must agree with $\lambda + \mu$ up to a member of $W(\mathfrak{l}, \mathfrak{h})$, i.e., we must have

$$s'(\lambda + \mu) = s(\lambda + \delta(\mathfrak{u})) - \delta(\mathfrak{u}) + \gamma$$

for some $s' \in W(\mathfrak{l}, \mathfrak{h})$. Since s' fixes $\delta(\mathfrak{u})$, we obtain

$$(\lambda + \delta(\mathfrak{u})) + \mu - s'^{-1}\gamma = t((\lambda + \delta(\mathfrak{u}) + \mu) - \mu)$$

with $t = s'^{-1}s$. Now we apply Proposition 7.166 to $\lambda + \delta(\mathfrak{u}) + \mu$ as a member of $\mathfrak{h}^*$, to the finite-dimensional representation $(F^\mu)^*$, and to the extreme weight $-\mu$. Since $\lambda + \delta(\mathfrak{u})$ is assumed to be at least as singular as $\lambda + \delta(\mathfrak{u}) + \mu$, the conclusion is that $-s'^{-1}\gamma = -\mu$ and that t fixes $\lambda + \delta(\mathfrak{u})$. Since the set of weights of E_p is invariant under $W(\mathfrak{l}, \mathfrak{h})$, it follows that μ is a weight of E_p and therefore that $p = p_0$. Thus $P_{\lambda+\mu} E_1^{p,q} = 0$ if $p \neq p_0$, while

(7.245) $$P_{\lambda+\mu} E_1^{p_0,q} \cong P_{\lambda+\mu}(P_{p_0+q}(\mathcal{F}X) \otimes_{\mathbb{C}} E^\mu) = \psi_\lambda^{\lambda+\mu}(P_{p_0+q}(\mathcal{F}X)).$$

The $P_{\lambda+\mu}$ part of the spectral sequence therefore converges, collapses, and degenerates, and the formula (7.244b) for the abutment gives

$$P_{\lambda+\mu}(P_{p_0+q}(\mathcal{F}X \otimes_{\mathbb{C}} F)) \cong P_{\lambda+\mu}(E_1^{p_0,q}).$$

In view of (7.243) and (7.245), this formula translates as

$$P_{\lambda+\mu} H_{p_0+q}(\bar{\mathfrak{u}}, \psi_{\lambda+\delta(\mathfrak{u})}^{\lambda+\delta(\mathfrak{u})+\mu}(X)) \cong \psi_\lambda^{\lambda+\mu}(H_{p_0+q}(\bar{\mathfrak{u}}, X)).$$

Since q is arbitrary, we obtain the required isomorphism.

CHAPTER VIII

IRREDUCIBILITY THEOREM

The Irreducibility Theorem states that the cohomologically induced modules $\mathcal{L}_S(Z)$ and $\mathcal{R}^S(Z)$ are irreducible if Z is irreducible and has an infinitesimal character that is fixed by $L \cap K$ and satisfies a certain dominance condition. The first part of the proof establishes irreducibility under a stronger assumption of dominance, by means of bottom-layer arguments and spectral sequences. The second part of the proof uses translation functors to get the full result.

Three versions of the general Irreducibility Theorem are presented. The first one assumes that L meets every component of G and that a suitable translate of the infinitesimal character of Z has real-part dominant. In the second version, the dominance assumption is weakened to integral dominance, and in the third version, the assumption that L meets every component of G is relaxed somewhat. The assumption that L meets every component of G can be dropped completely when G is in the Harish-Chandra class.

For the $(\mathfrak{g}, K)$ modules $A_\mathfrak{q}(\lambda)$, an improved Irreducibility Theorem is available, but one of its hypotheses is difficult to check. The Speh representations of $SL(2n, \mathbb{R})$ provide examples where this theorem applies, but other examples show how irreducibility can fail in general.

1. Main Theorem and Overview

This chapter contains the third of our five main theorems. The Irreducibility Theorem gives a sufficient condition for cohomological induction to carry irreducible modules to irreducible modules.

Throughout this chapter the setting will be as in Chapter V: We let $(\mathfrak{g}, K)$ be a reductive pair and $\mathfrak{q} = \mathfrak{l} \oplus \mathfrak{u}$ be a θ stable parabolic subalgebra. The corresponding Levi subgroup is denoted L, and $(\mathfrak{q}, L \cap K)$ is a θ stable parabolic subpair. For the first three sections of this chapter, we assume that

$$(8.1) \qquad L \text{ meets every component of } G, \text{ i.e., } K = (L \cap K)K_0.$$

The key hypothesis in the theorem is that the infinitesimal character of the given $(\mathfrak{l}, L \cap K)$ module Z exists and that a suitable translate of it satisfies a certain dominance condition. In §§1–2, we prove the theorem under a hypothesis of ordinary dominance, and in §3 we relax the hypothesis to one of integral dominance. When K is badly disconnected, an irreducible $(\mathfrak{l}, L \cap K)$ module need not have just one primary component to its decomposition under $Z(\mathfrak{g})$. This complication is serious, and we postpone discussion of it to §4.

Following the convention begun in §VII.13, we shall allow into the statements of theorems Cartan subalgebras that are not complexifications of θ stable Cartan subalgebras of $\mathfrak{g}_0$.

Theorem 8.2 (Irreducibility Theorem, first form). Let $(\mathfrak{g}, K)$ be a reductive pair, let $\mathfrak{q} = \mathfrak{l} \oplus \mathfrak{u}$ be a θ stable parabolic subalgebra, and suppose that L meets every component of G. Let $\mathfrak{h}$ be a Cartan subalgebra of $\mathfrak{l}$, and form the system of roots $\Delta(\mathfrak{g}, \mathfrak{h})$. Suppose Z is an irreducible admissible $(\mathfrak{l}, L \cap K)$ module with an infinitesimal character λ such that

$$\text{Re } \langle \lambda + \delta(\mathfrak{u}), \alpha \rangle \geq 0 \qquad \text{for all } \alpha \in \Delta(\mathfrak{u}).$$

Then the isomorphic $(\mathfrak{g}, K)$ modules $\mathcal{L}_S(Z)$ and $\mathcal{R}^S(Z)$ are 0 or are irreducible. If actually

$$\text{Re } \langle \lambda + \delta(\mathfrak{u}), \alpha \rangle > 0 \qquad \text{for all } \alpha \in \Delta(\mathfrak{u}),$$

then $\mathcal{L}_S(Z)$ and $\mathcal{R}^S(Z)$ are irreducible (not zero).

In this section we give an overview of the proof, postponing the details to §2. The isomorphism $\mathcal{L}_S(Z) \cong \mathcal{R}^S(Z)$ is given by Theorem 5.99.

For most of the argument, we work only with $\mathcal{R}^S(Z)$. We may assume that $\mathfrak{h}$ is the complexification of a θ stable Cartan subalgebra $\mathfrak{h}_0$ of $\mathfrak{l}_0$. The starting point is the vanishing theorem for cohomological induction, Theorem 5.99. Just as in the proof of that theorem, there is no loss of generality in introducing a positive system $\Delta^+(\mathfrak{l}, \mathfrak{h})$, defining $\Delta^+(\mathfrak{g}, \mathfrak{h}) = \Delta^+(\mathfrak{l}, \mathfrak{h}) \cup \Delta(\mathfrak{u})$, and assuming that $\lambda + \delta(\mathfrak{u})$ is $\Delta^+(\mathfrak{g}, \mathfrak{h})$ dominant. In combination with one of the fundamental spectral sequences of Corollary 5.121, Theorem 5.99 will give us a vanishing theorem for homology

$$(8.3a) \qquad P_{\lambda + 2\delta(\mathfrak{u})} H_p(\mathfrak{u}, X) = 0 \qquad \text{for } p < S$$

for all $(\mathfrak{g}, K)$ modules X of finite length, from which we shall deduce an adjoint relation

$$(8.3b) \qquad \text{Hom}_{\mathfrak{g}, K}(X, \mathcal{R}^S(Z)) \cong \text{Hom}_{\mathfrak{l}, L \cap K}(H_S(\mathfrak{u}, X), Z^{\#}).$$

By way of comparison, Proposition 5.71 gives us a K analog of the adjointness relation, namely

$$(8.4a) \quad \operatorname{Hom}_K(X', \mathcal{R}_K^S(Z')) \cong \operatorname{Hom}_{L \cap K}(H_S(\mathfrak{u} \cap \mathfrak{k}, X'), Z' \otimes_{\mathbb{C}} \textstyle\bigwedge^{\text{top}} \mathfrak{u}),$$

valid for any admissible K module X' and admissible $L \cap K$ module Z'.

Let τ_L be an irreducible representation of $L \cap K$, and let $\{\mu_L\}$ be its set of highest weights, repeated according to their multiplicities. Then the representation $\tau_L \otimes_{\mathbb{C}} \bigwedge^{\text{top}}(\mathfrak{u} \cap \mathfrak{p})$ has $\{\mu_L + 2\delta(\mathfrak{u} \cap \mathfrak{p})\}$ as its set of highest weights. *Assume* that the latter highest weights are $\Delta^+(\mathfrak{k}, \mathfrak{t})$ dominant. Since $L \cap K$ meets every component of K, Theorem 4.83 produces an irreducible representation (τ, V) of K such that the representation of $L \cap K$ on $V^{\mathfrak{u} \cap \mathfrak{k}}$ is equivalent with $\tau_L \otimes_{\mathbb{C}} \bigwedge^{\text{top}}(\mathfrak{u} \cap \mathfrak{p})$. This τ is unique up to isomorphism, and its set of highest weights is $\{\mu_L + 2\delta(\mathfrak{u} \cap \mathfrak{p})\}$ with the same multiplicities. From Corollary 5.72 we obtain

$$(8.4b) \quad \mathcal{R}_K^S(\tau_L) \cong \tau$$

and

$$(8.4c) \quad \mathcal{R}_K^S(\tau_L') \not\cong \tau \qquad \text{if } \tau_L' \not\cong \tau.$$

Taking $Z' = \tau_L$ in (8.4a) and using (8.4b), we obtain an equality of multiplicities:

$$(8.4d) \quad [X' : \tau] = [H_S(\mathfrak{u} \cap \mathfrak{k}, X') : \tau_L \otimes \textstyle\bigwedge^{\text{top}} \mathfrak{u}].$$

If τ_L occurs in the given $(\mathfrak{l}, L \cap K)$ module Z, then it follows from (8.4b) that

$$(8.4e) \quad \tau \text{ occurs in } \mathcal{R}_K^S(Z).$$

To relate (8.3) to (8.4), we use one of the Hochschild-Serre spectral sequences (Theorem 5.130a). Under some additional dominance conditions on τ_L, we shall obtain a surjective map of $L \cap K$ isotypic subspaces

$$(8.5) \quad H_S(\mathfrak{u} \cap \mathfrak{k}, X)_{\tau_L \otimes \wedge^{\text{top}} \mathfrak{u}} \twoheadrightarrow H_S(\mathfrak{u}, X)_{\tau_L \otimes \wedge^{\text{top}} \mathfrak{u}}.$$

Because of Corollary 7.223, we know from Corollary 5.25a and Theorem 5.35b that $\mathcal{R}^S(Z)$ has finite length. Under the assumption that τ_L occurs in Z and satisfies the above properties, let X be an arbitrary nonzero $(\mathfrak{g}, K)$ submodule of $\mathcal{R}^S(Z)$. We shall show that X contains the full τ isotypic subspace of $\mathcal{R}^S(Z)$. From this it follows that $\mathcal{R}^S(Z)$ has a unique irreducible $(\mathfrak{g}, K)$ submodule, namely the $(\mathfrak{g}, K)$ submodule generated by the τ isotypic subspace.

To prove this assertion about X, we use (8.3b) to convert the nonzero inclusion $X \hookrightarrow \mathcal{R}^S(Z)$ into a nonzero map

$$(8.6) \qquad\qquad \varphi : H_S(\mathfrak{u}, X) \to Z^{\#}.$$

This map must be onto since $Z^{\#}$ is irreducible. Then we can estimate multiplicities of $L \cap K$ types and K types as follows:

(8.7)

$$[Z : \tau_L]$$

$$= [Z^{\#} : \tau_L \otimes \textstyle\bigwedge^{\mathrm{top}} \mathfrak{u}]$$

$$\leq [H_S(\mathfrak{u}, X) : \tau_L \otimes \textstyle\bigwedge^{\mathrm{top}} \mathfrak{u}] \qquad \text{because (8.6) is onto}$$

$$\leq [H_S(\mathfrak{u} \cap \mathfrak{k}, X) : \tau_L \otimes \textstyle\bigwedge^{\mathrm{top}} \mathfrak{u}] \qquad \text{by (8.5)}$$

$$\leq [H_S(\mathfrak{u} \cap \mathfrak{k}, \mathcal{R}^S(Z)) : \tau_L \otimes \textstyle\bigwedge^{\mathrm{top}} \mathfrak{u}] \qquad \begin{array}{l}\text{since } \mathcal{R}^S(Z) \text{ is fully} \\ \text{reducible under } K \\ \text{and since } X \subseteq \mathcal{R}^S(Z)\end{array}$$

$$= [\mathcal{R}^S(Z) : \tau] \qquad \text{by (8.4d)}$$

$$= [\mathcal{R}_K^S(Z) : \tau] \qquad \begin{array}{l}\text{by (8.4e) and} \\ \text{Theorem 5.80}\end{array}$$

$$= [Z : \tau_L][\mathcal{R}_K^S(\tau_L) : \tau] \qquad \text{by (8.4c)}$$

$$= [Z : \tau_L] \qquad \text{by (8.4b).}$$

The left and right sides are equal, and therefore equality holds throughout. Hence

$$[H_S(\mathfrak{u} \cap \mathfrak{k}, X) : \tau_L \otimes \textstyle\bigwedge^{\mathrm{top}}(\mathfrak{u} \cap \mathfrak{k})] = [H_S(\mathfrak{u} \cap \mathfrak{k}, \mathcal{R}^S(Z)) : \tau_L \otimes \textstyle\bigwedge^{\mathrm{top}}(\mathfrak{u} \cap \mathfrak{k})].$$

Applying (8.4d) to both sides, we obtain

$$(8.8) \qquad\qquad [X : \tau] = [\mathcal{R}^S(Z) : \tau].$$

The last four lines of (8.7) show that the τ isotypic subspace of $\mathcal{R}^S(Z)$ is not 0. Meanwhile, formula (8.8) shows that X contains the entire τ isotypic subspace of $\mathcal{R}^S(Z)$, as we wished to show.

To pass from this conclusion about $\mathcal{R}^S(Z)$ to a conclusion of irreducibility, we shall invoke some duality formulas from §VI.3. The result of the argument is that $\mathcal{R}^S(Z)$ is irreducible if Z contains an $L \cap K$ type satisfying certain dominance conditions.

The problem with the above proof is that there is no guarantee that Z actually does contain an $L \cap K$ type τ_L with the necessary dominance properties. However, we can salvage matters by using the translation functors of Chapter VII. The $(\mathfrak{l}, L \cap K)$ module Z certainly contains some $L \cap K$ type τ_L, and then $Z \otimes (\bigwedge^{\text{top}}\mathfrak{u})^m$ contains the $L \cap K$ type $\tau_L \otimes (\bigwedge^{\text{top}}\mathfrak{u})^m$. For sufficiently large m, we shall see that $\tau_L \otimes (\bigwedge^{\text{top}}\mathfrak{u})^m$ satisfies the dominance conditions, and $\mathcal{R}^S(Z \otimes (\bigwedge^{\text{top}}\mathfrak{u})^m)$ will therefore be irreducible. Application of the translation functors will complete the proof.

2. Proof of Irreducibility Theorem

In this section we shall prove Theorem 8.2. Notation is as in §1. Before coming to the details of the proof, let us give some attention to the choice of Cartan subalgebras. For the purpose of describing infinitesimal characters, we start with any Cartan subpair $(\mathfrak{h}, T)$ of $(\mathfrak{l}, L \cap K)$ such that T meets every component of L. Existence of such a T is guaranteed by Proposition 4.42. Because $\Delta^+(\mathfrak{g}, \mathfrak{h}) = \Delta^+(\mathfrak{l}, \mathfrak{h}) \cup \Delta(\mathfrak{u})$, assumption (8.1) and Proposition 4.81 show that $(\mathfrak{h}, T)$ is a Cartan subpair of $(\mathfrak{g}, K)$.

Let us make some observations about the action of T. We may assume that $Z \neq 0$. Since Z has $\chi_{\lambda+\delta(\mathfrak{u})}$ as $Z(\mathfrak{l})$ infinitesimal character, it follows that $L \cap K$ fixes the homomorphism $\chi_{\lambda+\delta(\mathfrak{u})}$ of $Z(\mathfrak{l})$ into $\mathbb{C}$. Hence so does T. Thus T preserves the $W(\mathfrak{l}, \mathfrak{h})$ orbit of $\lambda + \delta(\mathfrak{u})$, and it follows that T preserves the $W(\mathfrak{l}, \mathfrak{h})$ orbit of $\lambda + n\delta(\mathfrak{u})$ for every n. Hence T fixes every homomorphism $\chi_{\lambda+n\delta(\mathfrak{u})}$ of $Z(\mathfrak{g})$. Since T meets every component of L, we see from (8.1) that T meets every component of K. Thus K fixes every homomorphism $\chi_{\lambda+n\delta(\mathfrak{u})}$ of $Z(\mathfrak{g})$.

For the purpose of describing highest weights of representations of K and $L \cap K$, we make an independent choice of Cartan subalgebra, using Theorem 4.83. Namely we start with a maximal abelian subspace $\mathfrak{t}$ of $\mathfrak{l}_0 \cap \mathfrak{k}_0$ and proceed from there. The notation $\mathfrak{t}$ will not arise in connection with infinitesimal characters and will therefore cause no confusion. We take $\Delta^+(\mathfrak{k}, \mathfrak{t}) = \Delta^+(\mathfrak{l} \cap \mathfrak{k}, \mathfrak{t}) \cup \Delta(\mathfrak{u} \cap \mathfrak{k})$. This latter construction leads to a Cartan subgroup of K; however, this subgroup will not arise explicitly until the proof of Corollary 8.28, and we introduce no notation for it at this time.

Let us now turn our attention to proving (8.3).

Lemma 8.9. For any $(\mathfrak{g}, K)$ module X of finite length,

$$P_{\lambda+2\delta(\mathfrak{u})} H_p(\mathfrak{u}, X) = 0 \qquad \text{for } p < S.$$

Here $P_{\lambda+2\delta(\mathfrak{u})}$ is the projection according to the generalized infinitesimal character $\lambda + 2\delta(\mathfrak{u})$.

REMARK. By the above argument, $L \cap K$ fixes $\chi_{\lambda+2\delta(\mathfrak{u})}$ for $Z(\mathfrak{l})$. Therefore $P_{\lambda+2\delta(\mathfrak{u})} H_p(\mathfrak{u}, X)$ is an $(\mathfrak{l}, L \cap K)$ module, not just a $U(\mathfrak{l})$ module.

PROOF. Since X has finite length, $H_p(\mathfrak{u}, X)$ is $Z(\mathfrak{l})$ finite and admissible, by Theorem 7.56 and Corollary 5.140. Then Corollary 7.223 shows that $H_p(\mathfrak{u}, X)$ has finite length as an $(\mathfrak{l}, L \cap K)$ module. Consequently if $P_{\lambda+2\delta(\mathfrak{u})} H_p(\mathfrak{u}, X)$ is $\neq 0$, then $H_p(\mathfrak{u}, X)$ has an irreducible quotient of infinitesimal character $\lambda + 2\delta(\mathfrak{u})$.

Choose $p = p_0$ as small as possible so that $P_{\lambda+2\delta(\mathfrak{u})} H_p(\mathfrak{u}, X) \neq 0$ for some $X = X_0$. Fix X_0, and let U_0 be an irreducible quotient of $H_{p_0}(\mathfrak{u}, X)$ of infinitesimal character $\lambda + 2\delta(\mathfrak{u})$. By Theorems 5.99b and 5.35b, the dominance of Re $\lambda + \delta(\mathfrak{u})$ implies that $\mathcal{R}^q(U_0) = 0$ for $q \neq S$. Thus Corollary 5.121b is applicable and produces a first-quadrant spectral sequence

$$(8.10) \qquad E_r^{pq} \Longrightarrow \operatorname{Ext}_{\mathfrak{g},K}^{p+q-S}(X_0, \mathcal{R}^S(U_0))$$

with differential of bidegree $(r, 1 - r)$ and with E_2 term

$$E_2^{pq} = \operatorname{Ext}_{\mathfrak{l}, L \cap K}^p (H_q(\mathfrak{u}, \mathcal{F}_{\mathfrak{g},K}^{q, L \cap K}(X_0)), U_0^\#).$$

Since the homology is the direct sum of its $Z(\mathfrak{g})$ primary components, Wigner's Lemma (Proposition 7.212) allows us to insert $P_{\lambda+2\delta(\mathfrak{u})}$ in this expression:

$$E_2^{pq} = \operatorname{Ext}_{\mathfrak{l}, L \cap K}^p (P_{\lambda+2\delta(\mathfrak{u})} H_q(\mathfrak{u}, \mathcal{F}_{\mathfrak{g},K}^{q, L \cap K}(X_0)), U_0^\#).$$

Then

$$E_2^{pq} = \begin{cases} \operatorname{Hom}_{\mathfrak{l}, L \cap K}(P_{\lambda+2\delta(\mathfrak{u})} H_q(\mathfrak{u}, \mathcal{F}_{\mathfrak{g},K}^{q, L \cap K}(X_0)), U_0) & \text{if } p = 0 \\ 0 & \text{if } q < p_0. \end{cases}$$

If $(p, q) = (0, p_0)$ or if $q < p_0$, we have $E_2^{pq} \cong \cdots \cong E_r^{pq} \cong \cdots \cong E_\infty^{pq}$. For $p + q = p_0$, we obtain just one nonzero term. Thus (8.10) gives

$$\operatorname{Hom}_{\mathfrak{l}, L \cap K}(P_{\lambda+2\delta(\mathfrak{u})} H_{p_0}(\mathfrak{u}, \mathcal{F}_{\mathfrak{g},K}^{q, L \cap K}(X_0)), U_0) = \bigoplus_{p+q=p_0} E_2^{pq}$$

$$\cong \bigoplus_{p+q=p_0} E_\infty^{pq} = \operatorname{Ext}_{\mathfrak{g},K}^{p_0-S}(X_0, \mathcal{R}^S(U_0)).$$

If $p_0 < S$, then the right side is 0, contradiction. So $p_0 \geq S$, and the lemma follows.

Proposition 8.11. For any $(\mathfrak{g}, K)$ module of finite length,

$$\mathrm{Hom}_{\mathfrak{g}, K}(X, \mathcal{R}^S(Z)) \cong \mathrm{Hom}_{\mathfrak{l}, L \cap K}(H_S(\mathfrak{u}, X), Z^\#).$$

PROOF. Applying Theorems 5.99b and 5.35b as in the proof of the lemma, we see from the dominance of Re $\lambda + \delta(\mathfrak{u})$ that $\mathcal{R}^q(Z) = 0$ for $q \neq S$. Thus Corollary 5.121b is applicable and produces a first-quadrant spectral sequence

$$(8.12) \qquad\qquad E_r^{pq} \Longrightarrow \mathrm{Ext}_{\mathfrak{g}, K}^{p+q-S}(X, \mathcal{R}^S(Z))$$

with differential of bidegree $(r, 1 - r)$ and with E_2 term

$$E_2^{pq} = \mathrm{Ext}_{\mathfrak{l}, L \cap K}^p(H_q(\mathfrak{u}, \mathcal{F}_{\mathfrak{g}, K}^{q, L \cap K}(X)), Z^\#).$$

Wigner's Lemma (Proposition 7.212) allows us to replace the homology $H_q(\mathfrak{u}, \mathcal{F}_{\mathfrak{g}, K}^{q, L \cap K}(X))$ by $P_{\lambda + 2\delta(\mathfrak{u})} H_q(\mathfrak{u}, \mathcal{F}_{\mathfrak{g}, K}^{q, L \cap K}(X))$ in this formula. By Lemma 8.9

$$E_2^{pq} = \begin{cases} \mathrm{Hom}_{\mathfrak{l}, L \cap K}(H_q(\mathfrak{u}, \mathcal{F}_{\mathfrak{g}, K}^{q, L \cap K}(X), Z^\#) & \text{if } p = 0 \\ 0 & \text{if } q < S. \end{cases}$$

If $(p, q) = (0, S)$ or if $q < S$, we have $E_2^{pq} \cong \cdots \cong E_r^{pq} \cong \cdots \cong E_\infty^{pq}$. For $p + q = S$, we obtain at most one nonzero term. Thus (8.12) gives the formula of the proposition.

The lemma and the proposition prove (8.3), and (8.4) was already proved in §1. The next proposition establishes (8.5).

Proposition 8.13. Let τ_L be an $L \cap K$ type whose highest weights μ_L have the property that $(\mu_L + 2\delta(\mathfrak{u} \cap \mathfrak{p})) - \sum_{\gamma \in E} \gamma + \delta_K$ is $\Delta(\mathfrak{u} \cap \mathfrak{k})$ dominant for all subsets E of distinct members of $\Delta(\mathfrak{u} \cap \mathfrak{p})$. Then there exists a surjective map of $L \cap K$ isotypic subspaces

$$H_S(\mathfrak{u} \cap \mathfrak{k}, X)_{\tau_L \otimes \wedge^{\mathrm{top}} \mathfrak{u}} \twoheadrightarrow H_S(\mathfrak{u}, X)_{\tau_L \otimes \wedge^{\mathrm{top}} \mathfrak{u}}.$$

PROOF. Theorem 5.130a produces a convergent spectral sequence

$$(8.14) \qquad\qquad E_r^{pq} \Longrightarrow H_{p+q}(\mathfrak{u}, X)$$

in $\mathcal{C}(\mathfrak{l} \cap \mathfrak{k}, L \cap K)$ with a differential of bidegree $(-r, r - 1)$ and with E_1 term

$$E_1^{pq} = H_{p+q-r(p)}(\mathfrak{u} \cap \mathfrak{k}, X) \otimes_{\mathbb{C}} V_p.$$

Here V_p is an $L \cap K$ invariant subspace of $\bigwedge (\mathfrak{u} \cap \mathfrak{p})$ defined in (5.129), and its weights are sums of $r(p)$ distinct members of $\Delta(\mathfrak{u} \cap \mathfrak{p})$.

We shall examine the $\tau_L \otimes \bigwedge^{\text{top}} \mathfrak{u}$ isotypic subspace

$$(H_n(\mathfrak{u} \cap \mathfrak{k}, X) \otimes_{\mathbb{C}} V_p)_{\tau_L \otimes \bigwedge^{\text{top}} \mathfrak{u}},$$

which by Corollary 3.8 is

$$\cong (H^{S-n}(\mathfrak{u} \cap \mathfrak{k}, X \otimes_{\mathbb{C}} \textstyle\bigwedge^{\text{top}}(\mathfrak{u} \cap \mathfrak{k})) \otimes_{\mathbb{C}} V_p)_{\tau_L \otimes \bigwedge^{\text{top}} \mathfrak{u}}$$
$$\cong (H^{S-n}(\mathfrak{u} \cap \mathfrak{k}, X) \otimes_{\mathbb{C}} \textstyle\bigwedge^{\text{top}}(\mathfrak{u} \cap \mathfrak{k}) \otimes_{\mathbb{C}} V_p)_{\tau_L \otimes \bigwedge^{\text{top}} \mathfrak{u}}$$
$$\cong (H^{S-n}(\mathfrak{u} \cap \mathfrak{k}, X) \otimes_{\mathbb{C}} V_p)_{\tau_L \otimes \bigwedge^{\text{top}}(\mathfrak{u} \cap \mathfrak{p})}.$$

This is the sum of its contributions, with multiplicities, from each K type τ' of X. Let τ' have $\{\mu'\}$ as its set of highest weights, repeated according to their multiplicities. As an $L \cap K_0$ module, $H^{S-n}(\mathfrak{u} \cap \mathfrak{k}, \tau')$ is the direct sum of the various $L \cap K_0$ modules $H^{S-n}(\mathfrak{u} \cap \mathfrak{k}, \tau_{\mu'})$, and each of them, by Kostant's Theorem (Theorem 4.139), is the direct sum of all irreducible finite-dimensional representations of $L \cap K_0$ with highest weights $w(\mu' + \delta_K) - \delta_K$ such that $w \in W^1$ and $l(w) = S - n$. For one of these terms of cohomology to be nonzero, the sum of $w(\mu' + \delta_K) - \delta_K$ and some weight of V_p must equal a highest weight of $\tau_L \otimes \bigwedge^{\text{top}}(\mathfrak{u} \cap \mathfrak{p})$. Thus there must be some set $E \subseteq \Delta(\mathfrak{u} \cap \mathfrak{p})$ with

$$(8.15) \qquad w(\mu' + \delta_K) - \delta_K + \sum_{\gamma \in E} \gamma = \mu_L + 2\delta(\mathfrak{u} \cap \mathfrak{p})$$

for some highest weight μ_L of τ_L.

Let us prove that $(\mu_L + 2\delta(\mathfrak{u} \cap \mathfrak{p})) - \sum_{\gamma \in E} \gamma + \delta_K$ is $\Delta^+(\mathfrak{k}, \mathfrak{t})$ dominant. If $\langle (\mu_L + 2\delta(\mathfrak{u} \cap \mathfrak{p})) - \sum_{\gamma \in E} \gamma + \delta_K, \alpha \rangle < 0$ for some $\alpha \in \Delta^+(\mathfrak{k}, \mathfrak{t})$, then (8.15) shows that $\langle w(\mu' + \delta_K), \alpha \rangle < 0$. Since μ' is $\Delta^+(\mathfrak{k}, \mathfrak{t})$ dominant, (4.133) shows that α is in $\Delta^+(w)$. Then by (4.137), α is in $\Delta(\mathfrak{u} \cap \mathfrak{k})$. But α cannot be in $\Delta(\mathfrak{u} \cap \mathfrak{k})$ by hypothesis in the proposition. Thus $(\mu_L + 2\delta(\mathfrak{u} \cap \mathfrak{p})) - \sum_{\gamma \in E} \gamma + \delta_K$ is $\Delta^+(\mathfrak{k}, \mathfrak{t})$ dominant.

Hence $w(\mu' + \delta_K)$ is $\Delta^+(\mathfrak{k}, \mathfrak{t})$ dominant. Since $\mu' + \delta_K$ and $w(\mu' + \delta_K)$ are both $\Delta^+(\mathfrak{k}, \mathfrak{t})$ dominant and nonsingular, $w = 1$. Since $l(1) = 0$, we must have $n = S$. Consequently

$$(8.16) \qquad (H_n(\mathfrak{u} \cap \mathfrak{k}, X) \otimes_{\mathbb{C}} V_p)_{\tau_L \otimes \bigwedge^{\text{top}} \mathfrak{u}} = 0 \qquad \text{if } n \neq S.$$

The n of interest is $n = p + q - r(p)$. For $p + q < S$, (8.16) gives

$$(8.17a) \qquad (E_1^{pq})_{\tau_L \otimes \bigwedge^{\text{top}} \mathfrak{u}} = 0.$$

Consider $p + q = S$. Here $r(0) = 0$ and $r(p) > 0$ for $p > 0$. When $p + q = S$, we see that $n = S$ for $p = 0$ and $n < S$ for $p > 0$. Thus, for $p + q = S$, (8.16) gives

$$(8.17b) \qquad (E_1^{pq})_{\tau_L \otimes \wedge^{\text{top}} \mathfrak{u}} = \begin{cases} H_S(\mathfrak{u} \cap \mathfrak{k}, X)_{\tau_L \otimes \wedge^{\text{top}} \mathfrak{u}} & \text{if } p = 0 \\ 0 & \text{if } p > 0. \end{cases}$$

From (8.17a), we see that $(E_r^{pq})_{\tau_L \otimes \wedge^{\text{top}} \mathfrak{u}}$ is annihilated by the differential for all $r \geq 1$ when $p + q = S$. Hence we obtain maps

$$(E_1^{pq})_{\tau_L \otimes \wedge^{\text{top}} \mathfrak{u}} \twoheadrightarrow (E_2^{pq})_{\tau_L \otimes \wedge^{\text{top}} \mathfrak{u}} \twoheadrightarrow \cdots \twoheadrightarrow (E_r^{pq})_{\tau_L \otimes \wedge^{\text{top}} \mathfrak{u}}$$

for $p + q = S$. Since the spectral sequence has an abutment (8.14), $E_r^{pq} \cong E_\infty^{pq}$ for large $r = r(p, q)$. Hence $(E_r^{pq})_{\tau_L \otimes \wedge^{\text{top}} \mathfrak{u}} \cong (E_\infty^{pq})_{\tau_L \otimes \wedge^{\text{top}} \mathfrak{u}}$. By (8.17b), $(E_r^{pq})_{\tau_L \otimes \wedge^{\text{top}} \mathfrak{u}}$ vanishes for $p + q = S$ except when $p = 0$. Since we have only one term contributing to E_∞^{pq} for $p + q = S$, (8.14) and (8.17b) give the required map $H_S(\mathfrak{u} \cap \mathfrak{k}, X)_{\tau_L \otimes \wedge^{\text{top}} \mathfrak{u}} \twoheadrightarrow H_S(\mathfrak{u}, X)_{\tau_L \otimes \wedge^{\text{top}} \mathfrak{u}}$. This proves the proposition.

Under the assumption that Z has an $L \cap K$ type τ_L whose highest weights are sufficiently $\Delta(\mathfrak{u} \cap \mathfrak{k})$ dominant, we can now check that each step of (8.7) is valid. As in (8.8), any nonzero $(\mathfrak{g}, K)$ submodule $\mathcal{R}^S(Z)$ therefore contains the full τ subspace of $\mathcal{R}^S(Z)$, where $\tau = \mathcal{R}_K^S(\tau_L)$ as in (8.4b). Moreover, this subspace is not 0.

Let Z^h be the Hermitian dual of Z, defined as in §VI.2. Since $W^h \cong W$ for any $L \cap K$ module, the $L \cap K$ types of Z^h are the same as those of Z. Consequently the above argument shows that any nonzero $(\mathfrak{g}, K)$ submodule contains the full τ subspace of $\mathcal{R}^S(Z^h)$.

On the other hand, (6.15) and (6.21a) show that

$$[\text{ind}_{\bar{\mathfrak{q}}, L \cap K}^{\mathfrak{g}, L \cap K}((Z^\#)_{\bar{\mathfrak{q}}})]^h \cong \text{pro}_{\mathfrak{q}, L \cap K}^{\mathfrak{g}, L \cap K}(((Z^\#)^h)_{\mathfrak{q}}).$$

Taking Γ^S of both sides and then applying (6.25) to the left side, we obtain $\mathcal{L}_S(Z)^h \cong \mathcal{R}^S(Z^h)$. Since $\mathcal{L}_S(Z) \cong \mathcal{R}^S(Z)$ by Theorem 5.99, we conclude that

$$(8.18) \qquad \mathcal{R}^S(Z)^h \cong \mathcal{R}^S(Z^h).$$

Suppose X is a nonzero $(\mathfrak{g}, K)$ submodule of $\mathcal{R}^S(Z)$. The annihilator $X^\perp$ of X, as a $(\mathfrak{g}, K)$ submodule of $\mathcal{R}^S(Z)^h$, intersects the τ subspace of $\mathcal{R}^S(Z)$ in 0, since X contains the full τ subspace of $\mathcal{R}^S(Z)$. Under (8.18), $X^\perp$ corresponds to a $(\mathfrak{g}, K)$ submodule Y of $\mathcal{R}^S(Z^h)$ that does not contain the τ subspace of $\mathcal{R}^S(Z^h)$. From above, $Y = 0$. Therefore $X^\perp = 0$ and

$X = \mathcal{R}^S(Z)$. Hence $\mathcal{R}^S(Z)$ is irreducible under the assumption that Z has an $L \cap K$ type τ_L whose highest weights are sufficiently $\Delta(\mathfrak{u} \cap \mathfrak{k})$ dominant.

Let us now drop the special assumption about some $L \cap K$ type of Z. Fix an arbitrary $L \cap K$ type τ_L of Z. Let τ_L have $\{\mu_L\}$ as set of highest weights, and form

$$\tau'_L = \tau_L \otimes_{\mathbb{C}} (\textstyle\bigwedge^{\text{top}} \mathfrak{u})^m.$$

This is an $L \cap K$ type of $Z' = Z \otimes_{\mathbb{C}} (\bigwedge^{\text{top}} \mathfrak{u})^m$, and its set of highest weights is $\{\mu'_L\} = \{\mu_L + 2m\delta(\mathfrak{u})\}$. We want to apply the above theory to Z' and τ'_L. For this purpose we need

(8.19a) $\qquad \langle \mu'_L + 2\delta(\mathfrak{u} \cap \mathfrak{p}), \alpha \rangle \geq 0 \qquad$ for all $\alpha \in \Delta^+(\mathfrak{k}, \mathfrak{t})$

in order to construct a corresponding representation τ' of K, and we need

(8.19b) $\quad \langle (\mu'_L + 2\delta(\mathfrak{u} \cap \mathfrak{p})) - \sum_{\gamma \in E} \gamma + \delta_K, \alpha \rangle \geq 0$

$$\text{for all } \alpha \in \Delta(\mathfrak{u} \cap \mathfrak{k}) \text{ and } E \subseteq \Delta(\mathfrak{u} \cap \mathfrak{p})$$

to be able to use Proposition 8.13. In (8.19a), any root $\alpha \in \Delta^+(\mathfrak{l} \cap \mathfrak{k}, \mathfrak{t})$ has $\langle \mu'_L + 2\delta(\mathfrak{u} \cap \mathfrak{p}), \alpha \rangle \geq 0$ since $\langle \mu_L, \alpha \rangle \geq 0$ and since α is orthogonal to $\delta(\mathfrak{u})$ and $\delta(\mathfrak{u} \cap \mathfrak{p})$. Thus (8.19a) and (8.19b) will be satisfied for some sufficiently large m if it is shown that

(8.20) $\qquad \langle \delta(\mathfrak{u}), \alpha \rangle > 0 \qquad$ for all $\alpha \in \Delta(\mathfrak{u} \cap \mathfrak{k})$.

However, (8.20) is just the special case of Proposition 4.68 in which $V = \mathfrak{g}$ and $\Lambda = \Delta(\mathfrak{u} \cap \mathfrak{k}) \cup \Delta(\mathfrak{u} \cap \mathfrak{p})$.

Thus for sufficiently large m, the above theory applies to Z' and τ'_L, and we see that $\mathcal{R}^S(Z \otimes_{\mathbb{C}} (\bigwedge^{\text{top}} \mathfrak{u})^m)$ is nonzero and irreducible. Within the $(\mathfrak{g}, K)$ module consisting of the m-fold tensor product of $\bigwedge^{R+S} \mathfrak{g}$ with itself, let F be the $(\mathfrak{g}, K)$ submodule generated by $\bigwedge^{R+S} \bar{\mathfrak{u}} \otimes \cdots \otimes \bigwedge^{R+S} \bar{\mathfrak{u}}$. Since the generating subspace is annihilated by $\bar{\mathfrak{u}}$ and is a one-dimensional $(\mathfrak{l}, L \cap K)$ module, F is irreducible. We form the translation functor $\psi^{\lambda + \delta(\mathfrak{u})}_{\lambda + (2m+1)\delta(\mathfrak{u})}$. Since Re $\lambda + \delta(\mathfrak{u})$ is dominant, Theorem 7.237 gives

$$\psi^{\lambda + \delta(\mathfrak{u})}_{\lambda + (2m+1)\delta(\mathfrak{u})} \mathcal{R}^S(Z \otimes_{\mathbb{C}} (\textstyle\bigwedge^{\text{top}} \mathfrak{u})^m) \cong \mathcal{R}^S(\psi^{\lambda}_{\lambda + 2m\delta(\mathfrak{u})}(Z \otimes_{\mathbb{C}} (\textstyle\bigwedge^{\text{top}} \mathfrak{u})^m)).$$

Now ψ on the right side is implemented by tensoring with $(\bigwedge^{\text{top}} \bar{\mathfrak{u}})^m$ and then projecting according to λ. The tensor product recovers just Z (cf. (5.95) and the discussion that precedes it), and thus

$$\psi^{\lambda + \delta(\mathfrak{u})}_{\lambda + (2m+1)\delta(\mathfrak{u})} \mathcal{R}^S(Z \otimes_{\mathbb{C}} (\textstyle\bigwedge^{\text{top}} \mathfrak{u})^m) \cong \mathcal{R}^S(Z).$$

The theorem about preservation of irreducibility by ψ (Theorem 7.229) and the result about nonvanishing of images under ψ (Theorem 7.232) complete the proof of Theorem 8.2.

REMARKS. Let us mention an alternate approach to some of the steps in the argument after Proposition 8.13. Under the assumption that Z has an $L \cap K$ type τ_L whose highest weights are sufficiently $\Delta^+(\mathfrak{k}, \mathfrak{t})$ dominant, we find that any nonzero $(\mathfrak{g}, K)$ submodule of $\mathcal{R}^S(Z)$ contains the full τ subspace of $\mathcal{R}^S(Z)$, which is not 0. Instead of introducing $\mathcal{R}^S(Z)^h$, we can repeat this argument with $\mathcal{L}_S(Z)$ in place of $\mathcal{R}^S(Z)$, making the expected notational changes. When the functor $\mathcal{F}^\vee$ makes an appearance, we can eliminate it by appealing to Corollary 5.141. We conclude that any nonzero $(\mathfrak{g}, K)$ quotient of $\mathcal{L}_S(Z)$ retains the full τ subspace of $\mathcal{L}_S(Z)$, which is not 0.

Since $\mathcal{L}_S(Z) \cong \mathcal{R}^S(Z)$ by Theorem 5.99, the only possibility is that $\mathcal{R}^S(Z)$ is irreducible. The argument with translation functors may then be applied to complete the proof.

3. Role of Integral Dominance

"Integral dominance" was introduced in §VII.9 as a generalization of ordinary dominance. In several places in the theory of cohomological induction, changing a hypothesis from ordinary dominance to integral dominance makes a theorem much closer to the best possible and provides for significantly wider applicability. One such place is with the Irreducibility Theorem, and we take up the generalization in this section.

Dominance of the real part of the infinitesimal character enters the proof of Theorem 8.2 in two ways—through Theorem 5.99 and through translation functors. The results invoked about translation functors are Theorems 7.237, 7.229, and 7.232, and all of these already assume only integral dominance.

Theorem 5.99 assumes ordinary dominance of the real part of an infinitesimal character and gives the isomorphism $\mathcal{L}_S(Z) \cong \mathcal{R}^S(Z)$ and the vanishing of $\mathcal{L}_j(Z)$ and $\mathcal{R}^j(Z)$ for $j \neq S$. With a small amount of work, we can weaken the hypothesis to allow a kind of integral dominance with respect to $\Delta(\mathfrak{u})$. The result is as follows.

Theorem 8.21. Suppose that the $(\mathfrak{l}, L \cap K)$ module Z is admissible and has an infinitesimal character $\lambda \in \mathfrak{h}^*$ such that no $\alpha \in \Delta(\mathfrak{u})$ has

$$\frac{2\langle \lambda + \delta(\mathfrak{u}), \alpha \rangle}{|\alpha|^2} \quad \text{in} \quad \{-1, -2, -3, \dots\}.$$

Then

(a) any nonzero $(\mathfrak{g}, L \cap K)$ submodule of $\mathrm{ind}_{\bar{\mathfrak{q}}, L \cap K}^{\mathfrak{g}, L \cap K}(Z_{\bar{\mathfrak{q}}}^\#)$ has nonzero intersection with $1 \otimes_{\mathbb{C}} Z^\#$

(b) $\mathcal{L}_j(Z)$ and $\mathcal{R}^j(Z)$ are 0 for $j < S$

(c) $\mathcal{L}_S(Z) \cong \mathcal{R}^S(Z)$.

PROOF. Going over the proof of Theorem 5.99, we see that only (a) needs a new argument. We begin as in the proof of Theorem 5.99, first arranging that $\lambda + \delta(\mathfrak{u})$ is $\Delta^+(\mathfrak{l}, \mathfrak{h})$ dominant. The hypothesis in the present theorem then makes $\lambda + \delta(\mathfrak{u})$ integrally dominant relative to $\Delta^+(\mathfrak{g}, \mathfrak{h})$. Next, we construct M, N, c, and c_0 as in the proof of Theorem 5.99. Arguing by contradiction, we again assume that N does not meet $1 \otimes_{\mathbb{C}} Z^\#$, and we are led to

$$(8.22) \qquad\qquad c > c_0$$

as in (5.101). The proof of Theorem 5.99 produces a finite-dimensional $\mathfrak{l}$ submodule F of $U(\mathfrak{u})$ and a nonzero $\mathfrak{l}$ submodule of $F \otimes_{\mathbb{C}} Z^\#$ with an infinitesimal character, which is called ν. By (5.104) we have

$$(8.23) \qquad\qquad \nu - \delta(\mathfrak{u}) = s(\lambda + \delta(\mathfrak{u}))$$

for some $s \in W(\mathfrak{g}, \mathfrak{h})$.

Since the weights of F are sums of members of $\Delta(\mathfrak{u})$, Theorem 7.133 shows that

$$\nu = \lambda + 2\delta(\mathfrak{u}) + \sum_{\beta_i \in \Delta(\mathfrak{u})} m_i \beta_i$$

for integers $m_i \geq 0$. Substituting into (8.23), we obtain

$$(8.24) \qquad \lambda + \delta(\mathfrak{u}) - s(\lambda + \delta(\mathfrak{u})) = - \sum_{\beta_i \in \Delta(\mathfrak{u})} m_i \beta_i.$$

By Proposition 7.145, s is in $W(\lambda + \delta(\mathfrak{u}))$. The integral dominance of $\lambda + \delta(\mathfrak{u})$ means that $\lambda + \delta(\mathfrak{u})$ is dominant relative to $\Delta^+(\lambda + \delta(\mathfrak{u}))$. Since s is in $W(\lambda + \delta(\mathfrak{u}))$,

$$(8.25) \qquad \lambda + \delta(\mathfrak{u}) - s(\lambda + \delta(\mathfrak{u})) = \sum_{\alpha \in \Delta^+(\lambda + \delta(\mathfrak{u}))} n_\alpha \alpha \qquad \text{with } n_\alpha \geq 0.$$

Comparing (8.24) and (8.25), we see that $\sum_{\beta_i \in \Delta(\mathfrak{u})} m_i \beta_i = 0$. Therefore $\nu = \lambda + 2\delta(\mathfrak{u})$, and

$$c + ib = \nu(h_{\delta(\mathfrak{u})}) = (\lambda + 2\delta(\mathfrak{u}))(h_{\delta(\mathfrak{u})}) = c_0 + ib.$$

Hence $c = c_0$, in contradiction to (8.22). The theorem follows.

Using Theorem 8.21 in place of Theorem 5.99 in §§1–2, we obtain our second version of the Irreducibility Theorem.

Theorem 8.26 (Irreducibility Theorem, second form). Let $(\mathfrak{g}, K)$ be a reductive pair, let $\mathfrak{q} = \mathfrak{l} \oplus \mathfrak{u}$ be a θ stable parabolic subalgebra, and suppose that L meets every component of G. Let $\mathfrak{h}$ be a Cartan subalgebra of $\mathfrak{l}$, and form the system of roots $\Delta(\mathfrak{g}, \mathfrak{h})$. Suppose Z is an irreducible admissible $(\mathfrak{l}, L \cap K)$ module with an infinitesimal character λ such that

$$\frac{2\langle \lambda + \delta(\mathfrak{u}), \alpha \rangle}{|\alpha|^2} \quad \text{is never in} \quad \{-1, -2, -3, \dots\} \qquad \text{for any } \alpha \in \Delta(\mathfrak{u}).$$

Then the isomorphic $(\mathfrak{g}, K)$ modules $\mathcal{L}_S(Z)$ and $\mathcal{R}^S(Z)$ are 0 or are irreducible. If actually

$$\frac{2\langle \lambda + \delta(\mathfrak{u}), \alpha \rangle}{|\alpha|^2} \quad \text{is never in} \quad \{0, -1, -2, -3, \dots\} \qquad \text{for any } \alpha \in \Delta(\mathfrak{u}),$$

then $\mathcal{L}_S(Z)$ and $\mathcal{R}^S(Z)$ are irreducible (not zero).

4. Irreducibility Theorem for K Badly Disconnected

When the reductive pair $(\mathfrak{l}, L \cap K)$ is in the Harish-Chandra class, the irreducible admissible $(\mathfrak{l}, L \cap K)$ module Z automatically has an infinitesimal character, since $\mathrm{Ad}(L \cap K)$ acts trivially on $Z(\mathfrak{l})$. But for the general case of $(\mathfrak{l}, L \cap K)$, Z need not have an infinitesimal character. In this situation, Theorems 8.2 and 8.26 break down. In fact, let the χ_λ primary subspace be nonzero for $\lambda \in \{\lambda_i\}$. Even if each λ_i has $\mathrm{Re}\, \lambda_i + \delta(\mathfrak{u})$ dominant relative to $\Delta^+(\mathfrak{g}, \mathfrak{h})$, the isomorphic $(\mathfrak{g}, K)$ modules $\mathcal{L}_S(Z)$ and $\mathcal{R}^S(Z)$ may be reducible. In this section we produce an example of this reducibility, and we give a theorem that tries to substitute for Theorem 8.26.

EXAMPLE. Let G be a semidirect product of $\mathbb{Z}_2$ with $G_0 = SL(3, \mathbb{C})$, where $\mathbb{Z}_2 = \{1, \gamma\}$ and

$$\gamma x \gamma^{-1} = \begin{pmatrix} 0 & 0 & 1 \\ 0 & 1 & 0 \\ 1 & 0 & 0 \end{pmatrix} (x^{-1})^t \begin{pmatrix} 0 & 0 & 1 \\ 0 & 1 & 0 \\ 1 & 0 & 0 \end{pmatrix} \qquad \text{for } x \in G_0.$$

Let $\mathfrak{h}_0 = \mathfrak{a}_0 \oplus \mathfrak{t}_0$ be the diagonal subalgebra of $\mathfrak{g}_0 = \mathfrak{sl}(3, \mathbb{C})$; here $\mathfrak{a}_0$ has real entries and $\mathfrak{t}_0$ has imaginary entries. If e_i denotes evaluation of the ith diagonal entry of $\mathfrak{h}_0$, then $e_i = \mathrm{Re}\, e_i \oplus i\, \mathrm{Im}\, e_i$, where $\mathrm{Re}\, e_i$ is in $\mathfrak{a}_0^*$ and $\mathrm{Im}\, e_i$ is in $\mathfrak{t}_0^*$. In this notation the roots are given by

$$\Delta(\mathfrak{g}, \mathfrak{h}) = \{e_i - e_j \text{ and } \bar{e}_i - \bar{e}_j \mid i \neq j\},$$

and we choose

$$\Delta^+(\mathfrak{g}, \mathfrak{h}) = \{e_i - e_j \text{ and } \bar{e}_j - \bar{e}_i \mid i < j\}$$

as a positive system. The corresponding Borel subalgebra is the θ stable parabolic with which we work. The element $\mathrm{Ad}(\gamma)$ normalizes $\mathfrak{h}$, and the action on $\mathfrak{h}^*$ effectively exchanges

$$e_1 \leftrightarrow -e_3, \qquad e_2 \leftrightarrow -e_2, \qquad \bar{e}_1 \leftrightarrow -\bar{e}_3, \qquad \bar{e}_2 \leftrightarrow -\bar{e}_2.$$

In particular, γ maps $\Delta^+(\mathfrak{g}, \mathfrak{h})$ to itself. We write $H_0 = \exp \mathfrak{h}_0$ for the diagonal subgroup consisting of all matrices

$$[h_1, h_2, h_3] = \begin{pmatrix} h_1 & 0 & 0 \\ 0 & h_2 & 0 \\ 0 & 0 & h_3 \end{pmatrix}.$$

The group L is then $\{1, \gamma\} \ltimes H_0$; it meets both components of G.
   Define a one-dimensional representation φ of H_0 by

$$\varphi[h_1, h_2, h_3] = |h_1/h_2|^i [h_3/h_1]^2,$$

where the exponent i is $\sqrt{-1}$ and where $[z] = z/|z|$ for nonzero complex z. We obtain a two-dimensional representation $(\tilde{\varphi}, Z)$ of L by the formula

$$\tilde{\varphi} = \mathrm{induced}_{H_0}^L(\varphi).$$

Since

$$\gamma\varphi[h_1, h_2, h_3] = |h_2/h_3|^i [h_3/h_1]^2,$$

$\gamma\varphi$ is inequivalent with φ; therefore $\tilde{\varphi}$ is irreducible. The representation $\tilde{\varphi}$ does not have a single infinitesimal character. Instead it has two eigenspaces under $Z(\mathfrak{h}) = U(\mathfrak{h})$, given by the differentials of φ and $\gamma\varphi$. These are

$$d\varphi = i \operatorname{Re}(e_1 - e_2) + 2i \operatorname{Im}(e_3 - e_1)$$

$$d(\gamma\varphi) = i \operatorname{Re}(e_2 - e_3) + 2i \operatorname{Im}(e_3 - e_1).$$

Also

$$\delta(\mathfrak{u}) = (e_1 - e_3) + (\bar{e}_3 - \bar{e}_1) = 2i \operatorname{Im}(e_1 - e_3).$$

Hence $d\varphi + \delta(\mathfrak{u})$ and $d(\gamma\varphi) + \delta(\mathfrak{u})$ are purely imaginary on $\mathfrak{a}_0 \oplus i\mathfrak{t}_0$; their real parts are 0 and are therefore dominant.

Despite the irreducibility of $(\tilde{\varphi}, Z)$ and the dominance just proved, $\mathcal{R}^S(Z)$ is reducible. To see this reducibility, we write

$$\tilde{\varphi}|_{L_0} = \varphi \oplus \gamma\varphi.$$

By Proposition 5.14c, $\mathcal{R}^S(Z)|_{G_0}$ is given by

$$\mathcal{R}^S(Z)|_{G_0} = \mathcal{R}^S(\varphi) \oplus \mathcal{R}^S(\gamma\varphi).$$

Now φ and $\gamma\varphi$ are one-dimensional $(\mathfrak{l}, L \cap K_0)$ modules; each has a single $L \cap K_0$ type, and the differential is just the restriction to $\mathfrak{t}_0$ of φ or $\gamma\varphi$. This restriction is $2i \operatorname{Im}(e_3 - e_1)$ in both cases. Adding $2\delta(\mathfrak{u} \cap \mathfrak{p}) = 2i \operatorname{Im}(e_1 - e_3)$, we obtain 0, which is $\Delta^+(\mathfrak{k}, \mathfrak{t})$ dominant. In other words, φ and $\gamma\varphi$ have the same unique $L \cap K_0$ type τ_L, and the assumption on τ_L in §1 between (8.4a) and (8.4b) is satisfied. The corresponding τ is trivial, and so the trivial K type is in the bottom layer of both $\mathcal{R}^S(\varphi)$ and $\mathcal{R}^S(\gamma\varphi)$.

Thus $\mathcal{R}^S(\varphi)$ and $\mathcal{R}^S(\gamma\varphi)$ both contain the trivial K type. These representations are irreducible by Theorem 8.2, and their respective infinitesimal characters are

$$\varphi + \delta(\mathfrak{u}) = i \operatorname{Re}(e_1 - e_2)$$

and
$$\gamma\varphi + \delta(\mathfrak{u}) = i \operatorname{Re}(e_2 - e_3).$$

Since the expressions on the right sides are conjugate by the Weyl group, $\mathcal{R}^S(\varphi)$ and $\mathcal{R}^S(\gamma\varphi)$ have the same infinitesimal character. In Chapter XI we shall see that an irreducible $(\mathfrak{g}, K_0)$ module containing the trivial K_0 type and having imaginary infinitesimal character in the case of complex G_0 is a spherical principal series. Moreover, two such representations with the same infinitesimal character are equivalent. Hence $\mathcal{R}^S(\varphi) \cong \mathcal{R}^S(\gamma\varphi)$ as $(\mathfrak{g}, K_0)$ modules.

The element γ acts on the direct sum $\mathcal{R}^S(\varphi) \oplus \mathcal{R}^S(\gamma\varphi)$, giving the $(\mathfrak{g}, K)$ module $\mathcal{R}^S(\tilde{\varphi})$. It is not hard to see that the equivalence of $\mathcal{R}^S(\varphi)$ and $\mathcal{R}^S(\gamma\varphi)$ forces this $(\mathfrak{g}, K)$ module to be reducible. Hence $\mathcal{R}^S(\tilde{\varphi})$ is reducible.

We could get a positive result by imposing nonsingularity of the real part of the various $\lambda_i + \delta(\mathfrak{u})$ as an extra hypothesis. Unfortunately this hypothesis eliminates many examples that one would like to understand, and the resulting theorem would be a poor substitute for Theorem 8.2.

Here is a better way of extending Theorems 8.2 and 8.26. It uses the symbols χ^L and χ^G to refer to homomorphisms $Z(\mathfrak{l}) \to \mathbb{C}$ and $Z(\mathfrak{g}) \to \mathbb{C}$, respectively.

Theorem 8.27 (Irreducibility Theorem, third form). Let $(\mathfrak{g}, K)$ be a reductive pair, and let $\mathfrak{q} = \mathfrak{l} \oplus \mathfrak{u}$ be a θ stable parabolic subalgebra. Let $\mathfrak{h}$ be a Cartan subalgebra of $\mathfrak{l}$, and form the system of roots $\Delta(\mathfrak{g}, \mathfrak{h})$. Suppose Z is an irreducible admissible $(\mathfrak{l}, L \cap K)$ module, one of whose $Z(\mathfrak{l})$ primary components is given by $\lambda \in \mathfrak{h}^*$. Define

$$L_1 = \{l \in L \mid l\chi_\lambda^L = \chi_\lambda^L\}$$

and
$$G_1 = \{g \in G \mid g\chi_{\lambda+\delta(\mathfrak{u})}^G = \chi_{\lambda+\delta(\mathfrak{u})}^G\}.$$

Assume that

(i) no $\alpha \in \Delta(\mathfrak{u})$ has $2\langle \lambda + \delta(\mathfrak{u}), \alpha \rangle/|\alpha|^2$ in $\{-1, -2, -3, \dots\}$, and

(ii) $G_1 = L_1 G_0$, i.e., L_1 meets every component of G_1.

Then the isomorphic $(\mathfrak{g}, K)$ modules $\mathcal{L}_S(Z)$ and $\mathcal{R}^S(Z)$ are irreducible or zero. If (i) is replaced by

(i') no $\alpha \in \Delta(\mathfrak{u})$ has $2\langle \lambda + \delta(\mathfrak{u}), \alpha \rangle/|\alpha|^2$ in $\{0, -1, -2, -3, \dots\}$,

then the isomorphic $(\mathfrak{g}, K)$ modules $\mathcal{L}_S(Z)$ and $\mathcal{R}^S(Z)$ are irreducible (not zero).

REMARKS.

1) Let us check that $L_1 \subseteq G_1$, and therefore that $L_1 G_0 \subseteq G_1$. For the proof, we may, without loss of generality, take $\mathfrak{h}_0$ to be a maximally split θ stable Cartan subalgebra of $\mathfrak{l}_0$; let the corresponding Cartan subgroup be $T_1 A$. By Proposition 4.42, T_1 meets every component of L_1. Since each member of L_0 fixes $\chi_{\lambda+\delta(\mathfrak{u})}^G$, it is enough to prove that each $t \in T_1$ fixes $\chi_{\lambda+\delta(\mathfrak{u})}^G$. Now $t\chi_\lambda^L = \chi_\lambda^L$, and thus t carries λ to a member of $W(\mathfrak{l}, \mathfrak{h})\lambda$. Also t fixes $\delta(\mathfrak{u})$, as does $W(\mathfrak{l}, \mathfrak{h})$. Thus $t(\lambda + \delta(\mathfrak{u}))$ is in $W(\mathfrak{l}, \mathfrak{h})(\lambda + \delta(\mathfrak{u}))$, which is contained in $W(\mathfrak{g}, \mathfrak{h})(\lambda + \delta(\mathfrak{u}))$. Consequently $t\chi_{\lambda+\delta(\mathfrak{u})}^G = \chi_{\lambda+\delta(\mathfrak{u})}^G$.

2) Put $K_1 = K \cap G_1$. Under the assumption that L_1 meets every component of G_1, $(\mathfrak{q}, L_1 \cap K)$ is a parabolic subpair of $(\mathfrak{g}, K_1)$. In fact, the inclusion $L_1 \hookrightarrow L$ forces L_1 to normalize $\mathfrak{q}$, and the meeting by L_1 of every component of G_1 means that L_1 is as large as possible.

3) Theorem 8.27 implies Theorem 8.26 (and hence also Theorem 8.2). In fact, suppose the hypotheses of Theorem 8.26 are satisfied. Then it is immediate that (i) holds in the present theorem. Moreover, the assumption in Theorem 8.26 that Z has an infinitesimal character means that $L_1 = L$. Thus the assumption in Theorem 8.26 that L meet every component of G implies (ii) in the present theorem. Theorem 8.27 allows us to conclude irreducibility, and that is the conclusion of Theorem 8.26.

4) In the example in the first part of this section, L_1 is equal to H_0, but G_1 is all of G; thus assumption (ii) fails. It is not too hard to check in any particular example whether or not (ii) is satisfied.

PROOF. We shall use Proposition 7.213 to reduce to the case of Theorem 8.26. According to that proposition, there is an irreducible $(\mathfrak{l}, L_1 \cap K)$ module Z_1 such that $Z = \text{induced}_{\mathfrak{l}, L_1 \cap K}^{\mathfrak{l}, L \cap K}(Z_1)$. Proposition 2.77 allows us to replace "induced" by P in this expression. Then we have

$$(P_{\bar{\mathfrak{q}}, L \cap K}^{\mathfrak{g}, K})_S(\mathcal{F}_{\mathfrak{l}, L \cap K}^{\bar{\mathfrak{q}}, L \cap K}(Z^{\#}))$$

$$= (P_{\bar{\mathfrak{q}}, L \cap K}^{\mathfrak{g}, K})_S(\mathcal{F}_{\mathfrak{l}, L \cap K}^{\bar{\mathfrak{q}}, L \cap K})(P_{\mathfrak{l}, L_1 \cap K}^{\mathfrak{l}, L \cap K})(Z_1^{\#}) \qquad \text{by definition of } Z_1$$

$$\cong (P_{\bar{\mathfrak{q}}, L \cap K}^{\mathfrak{g}, K})_S(P_{\bar{\mathfrak{q}}, L_1 \cap K}^{\bar{\mathfrak{q}}, L \cap K})(\mathcal{F}_{\mathfrak{l}, L_1 \cap K}^{\bar{\mathfrak{q}}, L_1 \cap K})(Z_1^{\#}) \qquad \text{by inspection of (2.74c)}$$

$$\cong (P_{\bar{\mathfrak{q}}, L_1 \cap K}^{\mathfrak{g}, K})_S(\mathcal{F}_{\mathfrak{l}, L_1 \cap K}^{\bar{\mathfrak{q}}, L_1 \cap K})(Z_1^{\#}) \qquad \text{by (C.28a1).}$$

Application of (C.27) shows that this $(\mathfrak{g}, K)$ module is

$$\cong P_{\mathfrak{g}, K_1}^{\mathfrak{g}, K}(P_{\bar{\mathfrak{q}}, L_1 \cap K}^{\mathfrak{g}, K_1})_S(\mathcal{F}_{\mathfrak{l}, L_1 \cap K}^{\bar{\mathfrak{q}}, L_1 \cap K})(Z_1^{\#}).$$

Put $X_1 = (P_{\bar{\mathfrak{q}}, L_1 \cap K}^{\mathfrak{g}, K_1})_S(\mathcal{F}_{\mathfrak{l}, L_1 \cap K}^{\bar{\mathfrak{q}}, L_1 \cap K})(Z_1^{\#})$. By (ii), L_1 meets every component of G_1. With (i) and (ii) in place, Theorem 8.26 says that X_1 is irreducible or 0, with infinitesimal character $\lambda + \delta(\mathfrak{u})$. Also X_1 is irreducible (not zero) if (i') holds. We can now apply Proposition 7.213 a second time to see that X_1 irreducible implies $P_{\mathfrak{g}, K_1}^{\mathfrak{g}, K}(X_1)$ irreducible. This completes the proof.

Hypothesis (ii) in Theorem 8.27 can be weakened. As the proof shows, if we redefine G_1 by $G_1 = L_1 G_0$, then X_1 is still irreducible. What is required is a condition for preserving irreducibility in the passage from X_1 to $P_{\mathfrak{g}, K_1}^{\mathfrak{g}, K}(X_1)$. Mackey theory gives an answer. By Corollary 7.227, $P_{\mathfrak{g}, K_1}^{\mathfrak{g}, K}(X_1)$ is fully reducible; hence it is irreducible if and only if

$$\text{Hom}_{\mathfrak{g}, K}(P_{\mathfrak{g}, K_1}^{\mathfrak{g}, K}(X_1), P_{\mathfrak{g}, K_1}^{\mathfrak{g}, K}(X_1))$$

is one dimensional. Corollary 2.83 says that

$$\text{Hom}_{\mathfrak{g}, K}(P_{\mathfrak{g}, K_1}^{\mathfrak{g}, K}(X_1), P_{\mathfrak{g}, K_1}^{\mathfrak{g}, K}(X_1)) \cong \bigoplus_{\substack{\text{double cosets} \\ K_1 k K_1}} \text{Hom}_{\mathfrak{g}, K_1 \cap kK_1 k^{-1}}(kX_1, X_1),$$

and the term corresponding to $k = 1$ on the right side is one dimensional. Thus the condition for irreducibility is that kX_1 cannot be mapped in nonzero equivariant fashion into X_1 as a $(\mathfrak{g}, K_1 \cap kK_1 k^{-1})$ module whenever $k \notin K_1$.

In the setting of Theorem 8.27, X_1 always has infinitesimal character $\chi_{\lambda+\delta(\mathfrak{u})}$, and kX_1 always has infinitesimal character $k\chi_{\lambda+\delta(\mathfrak{u})}$. The formulation of hypothesis (ii) in the theorem guarantees that these infinitesimal characters are distinct if k is not in K_1. A weaker hypothesis might allow kX_1 and X_1 to have the same infinitesimal character yet still be inequivalent.

These ideas lead to a definitive result in the case that G is in the Harish-Chandra class. Then $L_1 = L$, and our decomposition of the functor $\mathcal{L}_S$ reduces to the one in (5.8). Irreducibility is maintained at the outer step as a consequence of an analysis of K_1 types and a use of translation functors. Thus we obtain the corollary below, which is really a corollary of the above discussion and not a corollary of the theorem itself.

Corollary 8.28. Let $(\mathfrak{g}, K)$ be a reductive pair in the Harish-Chandra class, and let $\mathfrak{q} = \mathfrak{l} \oplus \mathfrak{u}$ be a θ stable parabolic subalgebra. Let $\mathfrak{h}$ be a Cartan subalgebra of $\mathfrak{l}$, and form the system of roots $\Delta(\mathfrak{g}, \mathfrak{h})$. Suppose Z is an irreducible admissible $(\mathfrak{l}, L\cap K)$ module with infinitesimal character λ such that

$$\frac{2\langle \lambda + \delta(\mathfrak{u}), \alpha \rangle}{|\alpha|^2} \quad \text{is never in} \quad \{-1, -2, -3, \dots\} \qquad \text{for any } \alpha \in \Delta(\mathfrak{u}).$$

Then the isomorphic $(\mathfrak{g}, K)$ modules $\mathcal{L}_S(Z)$ and $\mathcal{R}^S(Z)$ are 0 or are irreducible. If actually

$$\frac{2\langle \lambda + \delta(\mathfrak{u}), \alpha \rangle}{|\alpha|^2} \quad \text{is never in} \quad \{0, -1, -2, -3, \dots\} \qquad \text{for any } \alpha \in \Delta(\mathfrak{u}),$$

then $\mathcal{L}_S(Z)$ and $\mathcal{R}^S(Z)$ are irreducible (not zero).

PROOF. Assuming the first condition, we need to prove that $\mathcal{L}_S(Z)$ is irreducible or 0. Then the remaining conclusions follow from Theorem 8.27.

For $m \geq 0$, Theorem 7.237 gives

$$\mathcal{L}_S(Z) \cong \mathcal{L}_S(\psi^\lambda_{\lambda+2m\delta(\mathfrak{u})}(Z \otimes_{\mathbb{C}} (\textstyle\bigwedge^{\text{top}}\mathfrak{u})^m)$$
$$\cong \psi^{\lambda+\delta(\mathfrak{u})}_{\lambda+(2m+1)\delta(\mathfrak{u})}(\mathcal{L}_S(Z \otimes_{\mathbb{C}} (\textstyle\bigwedge^{\text{top}}\mathfrak{u})^m)).$$

If $\mathcal{L}_S(Z \otimes_{\mathbb{C}} (\bigwedge^{\text{top}}\mathfrak{u})^m)$ is irreducible or 0, then the conclusion about $\mathcal{L}_S(Z)$ follows from Theorem 7.229. Thus it is enough to prove that $\mathcal{L}_S(Z \otimes_{\mathbb{C}} (\bigwedge^{\text{top}}\mathfrak{u})^m)$ is irreducible or 0 for some $m \geq 0$.

Let $K_1 = (L \cap K)K_0$ and

$$X_{1,m} = (P^{\mathfrak{g},K_1}_{\bar{\mathfrak{q}},L\cap K})_S(\mathcal{F}^{\bar{\mathfrak{q}},L\cap K}_{\mathfrak{l},L\cap K}(Z^\# \otimes_{\mathbb{C}} (\textstyle\bigwedge^{\text{top}}\mathfrak{u})^m)).$$

In view of the discussion before the statement of the corollary, it is enough to prove that

$$\mathrm{Hom}_{\mathfrak{g},K_0}(k(\mathcal{F}^{\mathfrak{g},K_0}_{\mathfrak{g},K_1}X_{1,m}), \mathcal{F}^{\mathfrak{g},K_0}_{\mathfrak{g},K_1}X_{1,m}) = 0$$

for all elements k in K that are not in K_1. Proposition 5.14b shows that $\mathcal{F}^{\mathfrak{g},K_0}_{\mathfrak{g},K_1}X_{1,m} = X_m$, where

$$X_m = (P^{\mathfrak{g},K_0}_{\bar{\mathfrak{q}},L\cap K_0}) s (\mathcal{F}^{\bar{\mathfrak{q}},L\cap K_0}_{\mathfrak{l},L\cap K_0}(Z^{\#} \otimes_{\mathbb{C}} (\textstyle\bigwedge^{\mathrm{top}}\mathfrak{u})^m)).$$

Thus it is enough to prove that

$$\mathrm{Hom}_{\mathfrak{g},K_0}(kX_m, X_m) = 0$$

for all elements k in K that are not in K_1.

Since Z is irreducible in $C(\mathfrak{l}, L \cap K)$, Lemma 7.224 shows that Z is completely reducible as an $(\mathfrak{l}, L \cap K_0)$ module. Applying Theorem 8.2 to each irreducible constituent of $Z \otimes (\bigwedge^{\mathrm{top}}\mathfrak{u})^m$, we see that X_m is completely reducible as a $(\mathfrak{g}, K_0)$ module. Thus if we can produce finitely many K_0 types generating X_m whose transforms by each k under consideration are not in X_m, then it will follow that $\mathrm{Hom}_{\mathfrak{g},K_0}(kX_m, X_m) = 0$ for all such k, and the corollary will be proved.

Let T_0 be a maximal torus of $L \cap K_0$. Lemma 5.9 shows that T_0 is also a maximal torus of K_0. Fix a positive system $\Delta^+(\mathfrak{l} \cap \mathfrak{k}, \mathfrak{t})$, and let $\Delta^+(\mathfrak{k}, \mathfrak{t}) = \Delta^+(\mathfrak{l} \cap \mathfrak{k}, \mathfrak{t}) \cup \Delta(\mathfrak{u} \cap \mathfrak{k})$. Introduce the Cartan subgroup T of K corresponding to $\mathfrak{t}_0$ and $\Delta^+(\mathfrak{k}, \mathfrak{t})$. Let τ_L be any $L \cap K$ type occurring in Z, let $\mu_1, \ldots, \mu_r$ be its highest weights, and let $\tau_{L,1}, \ldots \tau_{L,r}$ be the irreducible representations of $L \cap K_0$ with highest weights $\mu_1, \ldots, \mu_r$. Since Z is irreducible as an $(\mathfrak{l}, L\cap K)$ module, the $L\cap K_0$ types $\tau_{L,1}, \ldots \tau_{L,r}$ generate Z as an $(\mathfrak{l}, L \cap K_0)$ module. When n is sufficiently large, (8.20) shows that the forms

$$\mu_j + 2m\delta(\mathfrak{u}) + 2\delta(\mathfrak{u}\cap\mathfrak{p}), \qquad 1 \le j \le r,$$

are dominant relative to $\Delta^+(\mathfrak{k}, \mathfrak{t})$, and Theorem 5.80a shows that the corresponding K_0 types occur in the bottom layer of X_m. Moreover these K_0 types generate X_m because a complete set of irreducible $(\mathfrak{l}, L \cap K_0)$ summands of $Z \otimes_{\mathbb{C}} (\bigwedge^{\mathrm{top}}\mathfrak{u})^m$ leads to a complete set of irreducible $(\mathfrak{g}, K_0)$ summands of X_m. Thus it is enough to prove for m sufficiently large that k times each of these K_0 types does not occur in X_m if k is in K but not K_1.

In arguing with such an element k, we can discard a factor from K_0 on the right of k. Because T meets every component of K (Proposition

4.22), we may assume that k is in T. Since T/T_0 is finite, it is enough to obtain an estimate for how large m needs to be for a single element k. Fix k.

We shall show that $k(\delta(\mathfrak{u})) \neq \delta(\mathfrak{u})$. In fact, we shall show that $k(\delta(\mathfrak{u})) = \delta(\mathfrak{u})$ forces k to be in $N_K(\mathfrak{q}) = L \cap K$, contradiction. Thus let X be in $\mathfrak{q}$. By Proposition 4.76b we can write $X = \sum X_i$ with $[h_{\delta(\mathfrak{u})}, X_i] = \nu_i X_i$ and $\nu_i \geq 0$. If $k(\delta(\mathfrak{u})) = \delta(\mathfrak{u})$, then we have

$$[h_{\delta(\mathfrak{u})}, \mathrm{Ad}(k)X_i] = [h_{k(\delta(\mathfrak{u}))}, \mathrm{Ad}(k)X_i] = [\mathrm{Ad}(k)h_{\delta(\mathfrak{u})}, \mathrm{Ad}(k)X_i]$$
$$= \mathrm{Ad}(k)[h_{\delta(\mathfrak{u})}, X_i] = \nu_i \mathrm{Ad}(k)X_i,$$

and Proposition 4.76b allows us to conclude that $\mathrm{Ad}(k)X_i$ is in $\mathfrak{q}$. Since $\mathrm{Ad}(k)X = \sum \mathrm{Ad}(k)X_i$, we see that k normalizes $\mathfrak{q}$, contradiction. Thus $k(\delta(\mathfrak{u})) \neq \delta(\mathfrak{u})$.

Let μ be the highest weight of a K_0 type τ in X_m, and let τ operate in V. By Theorem 5.35a, there exists $n = n(m)$ such that

$$\mathrm{Hom}_{L \cap K_0}(S^n(\mathfrak{u} \cap \mathfrak{p}) \otimes_{\mathbb{C}} Z^\# \otimes_{\mathbb{C}} (\textstyle\bigwedge^{\mathrm{top}}\mathfrak{u})^m, \, H^S(\bar{\mathfrak{u}} \cap \mathfrak{k}, V)) \neq 0.$$

Carrying through the same manipulations as in the proof of Corollary 5.72, we see that

$$\mathrm{Hom}_{L \cap K_0}(Z \otimes_{\mathbb{C}} (\textstyle\bigwedge^{\mathrm{top}}\mathfrak{u})^m \otimes_{\mathbb{C}} S^n(\mathfrak{u} \cap \mathfrak{p}) \otimes_{\mathbb{C}} \textstyle\bigwedge^{\mathrm{top}}(\mathfrak{u} \cap \mathfrak{p}), \, V^{\mathfrak{u} \cap \mathfrak{k}}) \neq 0.$$

Therefore μ is of the form

$$\mu = \mu_L + 2m\delta(\mathfrak{u}) + A_{n(m)} + 2\delta(\mathfrak{u} \cap \mathfrak{p})$$

with μ_L the highest weight of an $L \cap K_0$ type of Z and with $A_{n(m)}$ the sum of weights of $\mathfrak{u} \cap \mathfrak{p}$. Since $\tau_{L,1}, \ldots \tau_{L,r}$ generate Z as a $U(\mathfrak{l})$ module, we obtain

$$\mu_L = \mu_{j(m)} + B_m,$$

where B_m is a sum of weights of $\mathfrak{t}$ in $\mathfrak{l}$.

Now suppose that the K_0 type with highest weight

$$\mu = k(\mu_i + 2m\delta(\mathfrak{u}) + 2\delta(\mathfrak{u} \cap \mathfrak{p}))$$

occurs in X_m. Then we have

$$k\mu_i + 2mk(\delta(\mathfrak{u})) + 2k(\delta(\mathfrak{u} \cap \mathfrak{p})) = \mu_{j(m)} + B_m + 2m\delta(\mathfrak{u}) + A_{n(m)} + 2\delta(\mathfrak{u} \cap \mathfrak{p}).$$

We can rewrite this equation as

$$2m\delta(\mathfrak{u}) - 2mk(\delta(\mathfrak{u})) = -C_m + D_m,$$

where C_m is a sum of weights of $\mathfrak{t}$ in $\mathfrak{q}$ and D_m lies in a finite set of sums of weights of $\mathfrak{t}$ that is independent of m. Taking the inner product with $\delta(\mathfrak{u})$ and dividing by m, we obtain

$$1 - \frac{\langle k(\delta(\mathfrak{u})), \delta(\mathfrak{u}) \rangle}{|\delta(\mathfrak{u})|^2} = (\leq 0) + \tfrac{1}{m}(\text{bounded}).$$

As $m \to \infty$, the left side remains constant and is > 0 by the Schwarz inequality, while the right side has $\limsup \leq 0$. We obtain a contradiction, and the conclusion is that the K_0 type with highest weight $k(\mu_i + 2m\delta(\mathfrak{u}) + 2\delta(\mathfrak{u} \cap \mathfrak{p}))$ cannot occur in X_m if m is sufficiently large.

5. Consideration of $A_q(\lambda)$

When the reductive pair $(\mathfrak{g}, K)$ has K connected, recall from (5.6) that $A_q(\lambda)$ is defined to be $\mathcal{L}_S(\mathbb{C}_\lambda)$. The infinitesimal character of $A_q(\lambda)$ is $\lambda + \delta$. For such $(\mathfrak{g}, K)$ modules, we proved a special vanishing result in Theorem 5.109, in which the dominance of $\lambda + \delta$ was replaced as a hypothesis by the dominance of $\lambda + \delta(\mathfrak{u})$. We showed that there are cases where Theorem 5.109 applies and Theorem 5.99 does not. In this section we take up two theorems whose point of departure is that special vanishing result. The first of these calculates the multiplicities of K types in $A_q(\lambda)$ in terms of a partition function.

The second theorem deals with irreducibility. Although it may appear in §§1–2 that Theorem 5.99 is the key ingredient for irreducibility, it turns out that we do not freely get an irreducibility theorem by using Theorem 5.109 in its place. An additional hypothesis and additional work are required, and we shall settle for Theorem 8.31 below. In practice, the extra hypothesis is often difficult to check, and we shall see that it is sometimes valid and sometimes not.

After the two theorems we shall introduce four examples where Theorem 5.109 is applicable but Theorem 5.99 is not. The first example consists of the celebrated "Speh representations" of $G = SL(2n, \mathbb{R})$, whose irreducibility follows from Theorem 8.31 and a supplementary argument that is given as Proposition 8.75.

The second example is an $A_q(\lambda)$ for $G = Sp(2, \mathbb{R})$ such that Theorem 5.109 is applicable but $A_q(\lambda)$ is reducible. The reducibility is exhibited by using a different $A_{q'}(\lambda')$ for which Theorem 5.109 is not applicable. As part of this example, we shall see exactly why the extra hypothesis in Theorem 8.31 is breaking down in $\mathfrak{g} = \mathfrak{sp}(2, \mathbb{C})$.

The third example is an $A_{\mathfrak{q}}(\lambda)$ for $G = SO_0(5, 4)$. Theorem 5.109 is applicable but $A_{\mathfrak{q}}(\lambda)$ is reducible. We are interested in a nonobvious subrepresentation that plays an important role in classification questions. At this stage, we cannot carry out all the steps that prove reducibility, but we shall make a start.

The fourth and final example is a nonzero $A_{\mathfrak{q}}(\lambda)$ for $G = Sp(5, 2)$ such that $\lambda + 2\delta(\mathfrak{u} \cap \mathfrak{p})$ is not Δ_K^+ dominant. This example is then a nonzero cohomologically induced module whose bottom layer is empty.

We begin with the formula for multiplicities of K types of $A_{\mathfrak{q}}(\lambda)$. In the statement and proof, it will be convenient to identify K types with their highest weights.

Theorem 8.29. In the case of $A_{\mathfrak{q}}(\lambda)$, suppose that

$$\mathrm{Re}\,\langle \lambda + \delta(\mathfrak{u}), \alpha \rangle \geq 0 \qquad \text{for all } \alpha \in \Delta(\mathfrak{u}).$$

For $\nu \in \mathfrak{h}^*$, define $\mathcal{P}(\nu)$ to be the multiplicity of ν as a weight in $(S(\mathfrak{u} \cap \mathfrak{p}))^{\mathfrak{l} \cap \mathfrak{k} \cap \mathfrak{n}}$. Put $\Lambda = \lambda + 2\delta(\mathfrak{u} \cap \mathfrak{p})$, and let W^1 be the subset of $W(\mathfrak{k}, \mathfrak{t})$ defined in (4.137). If Λ' is Δ_K^+ dominant and integral, then the K type Λ' occurs in $A_{\mathfrak{q}}(\lambda)$ with multiplicity

$$\sum_{s \in W^1} (\det s)\mathcal{P}(s(\Lambda' + \delta_K) - (\Lambda + \delta_K)).$$

PROOF. Let V be irreducible with highest weight Λ'. We shall use the formula involving $\mathcal{R}^j(\mathbb{C}_\lambda)$ in Theorem 5.64. On the left side of that formula, only the term with $j = S$ can be nonzero, as a consequence of the vanishing theorem, Theorem 5.109. Since $A_{\mathfrak{q}}(\lambda) \cong A^{\mathfrak{q}}(\lambda)$ as a consequence of Theorem 5.109, Theorem 5.64 gives

$$(-1)^S \dim \mathrm{Hom}_K(V, A_{\mathfrak{q}}(\lambda))$$

$$= \sum_{j=0}^{S} (-1)^j \sum_{n=0}^{\infty} \dim \mathrm{Hom}_{L \cap K}(H_j(\mathfrak{u} \cap \mathfrak{k}, V), S^n(\mathfrak{u} \cap \mathfrak{p}) \otimes_{\mathbb{C}} \mathbb{C}_{\lambda + 2\delta(\mathfrak{u})}).$$

Replacing j by $S - j$ and using Corollary 3.8, we can rewrite this formula as

$$\dim \mathrm{Hom}_K(V, A_{\mathfrak{q}}(\lambda))$$

$$= \sum_{j=0}^{S} (-1)^j \sum_{n=0}^{\infty} \dim \mathrm{Hom}_{L \cap K}(H^j(\mathfrak{u} \cap \mathfrak{k}, V), S^n(\mathfrak{u} \cap \mathfrak{p}) \otimes_{\mathbb{C}} \mathbb{C}_{\lambda + 2\delta(\mathfrak{u} \cap \mathfrak{p})}).$$

Kostant's Theorem (Theorem 4.139) says that $H^j(\mathfrak{u} \cap \mathfrak{k}, V)$ is the sum over all $w \in W^1$ of length j of all $L \cap K$ types $w(\Lambda' + \delta_K) - \delta_K$, each occurring with multiplicity 1. Computing $\mathrm{Hom}_{L \cap K}$ between representations of $L \cap K$ as Hom_T between the spaces of $\mathfrak{l} \cap \mathfrak{k} \cap \mathfrak{n}$ invariants, we obtain the result of the theorem.

Now let us confront the issue of irreducibility. We introduce a complex connected Lie group $G_\mathbb{C}$ with Lie algebra $\mathfrak{g}$. Let K' be the analytic subgroup of $G_\mathbb{C}$ with Lie algebra $\mathfrak{k}_0' = \mathfrak{k}_0 \oplus i\mathfrak{p}_0$. Fixing our θ stable parabolic subalgebra $\mathfrak{q} = \mathfrak{l} \oplus \mathfrak{u}$, define L' to be the analytic subgroup of K' with Lie algebra $\mathfrak{l}_0' = \mathfrak{l} \cap \mathfrak{k}_0'$. Then $\mathfrak{k}_0'$ and $\mathfrak{l}_0'$ are compact real forms of $\mathfrak{g}$ and $\mathfrak{l}$, and $(\mathfrak{g}, K')$ and $(\mathfrak{l}, L')$ are pairs corresponding to compact connected groups.

The group $G_\mathbb{C}$ acts on $\mathfrak{g}$ via adjoint, and the quotient map $\mathfrak{g} \to \mathfrak{g}/\bar{\mathfrak{q}}$ is a $(\bar{\mathfrak{q}}, L')$ module map. Passing to symmetric algebras, we obtain a surjective $(\bar{\mathfrak{q}}, L')$ module map

$$(8.30a) \qquad \varphi_n : S^n(\mathfrak{g}) \to S^n(\mathfrak{g}/\bar{\mathfrak{q}}).$$

By Frobenius reciprocity (Proposition 2.21), we have an isomorphism

$$(8.30b) \quad \mathrm{Hom}_{\mathfrak{g},K'}(S^n(\mathfrak{g}), I_{\bar{\mathfrak{q}},L'}^{\mathfrak{g},K'}(S^n(\mathfrak{g}/\bar{\mathfrak{q}}))) \cong \mathrm{Hom}_{\bar{\mathfrak{q}},L'}(S^n(\mathfrak{g}), S^n(\mathfrak{g}/\bar{\mathfrak{q}})).$$

Let

$$(8.30c) \qquad \Phi_n : S^n(\mathfrak{g}) \to I_{\bar{\mathfrak{q}},L'}^{\mathfrak{g},K'}(S^n(\mathfrak{g}/\bar{\mathfrak{q}}))$$

be the member of the left side of (8.30b) that corresponds to φ_n on the right side.

Theorem 8.31. In the case of $A_q(\lambda)$, suppose that

$$\mathrm{Re}\,\langle \lambda + \delta(\mathfrak{u}), \alpha \rangle \geq 0 \qquad \text{for all } \alpha \in \Delta(\mathfrak{u}).$$

Let $\Phi_n : S^n(\mathfrak{g}) \to I_{\bar{\mathfrak{q}},L'}^{\mathfrak{g},K'}(S^n(\mathfrak{g}/\bar{\mathfrak{q}}))$ be the map (8.30c). If Φ_n is onto for every $n \geq 0$, then $A_q(\lambda)$ is irreducible or zero as a $(\mathfrak{g}, K)$ module.

The proof requires some preparation. From §§1–2 we know that $A_q(\lambda + 2m\delta(\mathfrak{u}))$ is irreducible for sufficiently large integers $m > 0$, and thus it is a question of showing that application of the translation functor $\psi_{\lambda+\delta+2m\delta(\mathfrak{u})}^{\lambda+\delta}$ yields the desired result. The whole point is that $\lambda+\delta$ is out of the range in which the theorems of Chapter VII assure that irreducibility is preserved and the image representation is $A_q(\lambda)$.

The result will come from the same methods as in §VII.10, but we shall replace the old $R = R(\lambda)$ by a new version adapted to the present case. For each one-dimensional $(\mathfrak{l}, L \cap K)$ module $\mathbb{C}_\lambda$, we let $\bar{\mathfrak{u}}$ act by 0 and then define

$$M(\lambda) = \mathrm{ind}_{\bar{\mathfrak{q}}}^{\mathfrak{g}} \mathbb{C}_\lambda^\# = U(\mathfrak{g}) \otimes_{U(\bar{\mathfrak{q}})} \mathbb{C}_\lambda^\#$$

as a $U(\mathfrak{g})$ module. The module $M(\lambda)$ is called a **generalized Verma module**, and its infinitesimal character is $((\lambda+\delta_L)+2\delta(\mathfrak{u}))-\delta(\mathfrak{u}) = \lambda+\delta$, by the proof of Theorem 5.24a. Before introducing the new $R(\lambda)$ and showing its relevance to $A_q(\lambda)$, let us isolate how the hypothesis on λ in the theorem enters the proof.

Lemma 8.32. Under the assumption that $\mathbb{C}_\lambda$ is a one-dimensional $(\mathfrak{l}, L \cap K)$ module such that

$$\mathrm{Re}\,\langle \lambda + \delta(\mathfrak{u}), \alpha \rangle \geq 0 \qquad \text{for all } \alpha \in \Delta(\mathfrak{u}),$$

suppose that $\mu \in \mathfrak{h}^*$

(i) is a sum of members of $\Delta(\mathfrak{u})$, and
(ii) is dominant relative to $\Delta^+(\mathfrak{l}, \mathfrak{h})$.

If $\lambda + \delta + \mu$ is conjugate to $\lambda + \delta$ by the Weyl group $W(\mathfrak{g}, \mathfrak{h})$, then $\mu = 0$.

PROOF. There is no loss of generality in passing to the real part of λ and therefore assuming that λ is real (on $i\mathfrak{t}_0 \oplus \mathfrak{a}_0$). We use notation as in the proof of Theorem 5.109, decomposing members β of $\mathfrak{h}^*$ as $\beta_1 + \beta_2$, where β_1 is carried on the center of $\mathfrak{l}$ and β_2 is carried on the semisimple part of $\mathfrak{l}$. We use corresponding indices for norms. For some $w \in W(\mathfrak{g}, \mathfrak{h})$, the conjugacy allows us to write

$$w(\lambda + \delta) = \lambda + \delta + \mu = (\lambda + \delta(\mathfrak{u}) + \mu)_1 + (\delta_L + \mu)_2.$$

Then

(8.33) $$\|\lambda + \delta\|^2 = \|w(\lambda + \delta)\|^2 = \|\lambda + \delta(\mathfrak{u}) + \mu\|_1^2 + \|\delta_L + \mu\|_2^2.$$

By (i) and the assumed dominance condition relative to $\Delta(\mathfrak{u})$,

(8.34) $$\|\lambda + \delta(\mathfrak{u}) + \mu\|_1^2 \geq \|\lambda + \delta(\mathfrak{u})\|_1^2 + \|\mu\|_1^2.$$

By (ii), $\|\delta_L + \mu\|_2^2 \geq \|\delta_L\|_2^2 + \|\mu\|_2^2$. Combining this inequality with (8.34), we see that the right side of (8.33) is

$$\geq \|\lambda + \delta(\mathfrak{u})\|_1^2 + \|\mu\|_1^2 + \|\delta_L\|_2^2 + \|\mu\|_2^2 = \|\lambda + \delta\|^2 + \|\mu\|^2.$$

Comparing the right side here with the left side of (8.33), we see that $\mu = 0$.

Lemma 8.35. Suppose that $\mathbb{C}_\lambda$ is a one-dimensional $(\mathfrak{l}, L \cap K)$ module such that
$$\mathrm{Re}\,\langle \lambda + \delta(\mathfrak{u}), \alpha \rangle \geq 0 \qquad \text{for all } \alpha \in \Delta(\mathfrak{u}),$$

and suppose that m is an integer ≥ 0. Then

$$\psi_{\lambda+\delta+2m\delta(\mathfrak{u})}^{\lambda+\delta}(M(\lambda + 2m\delta(\mathfrak{u}))) \cong M(\lambda).$$

PROOF. If E is any finite-dimensional $(\bar{\mathfrak{q}}, L \cap K)$ module, we let

$$M(E) = \operatorname{ind}_{\bar{\mathfrak{q}}}^{\mathfrak{g}}(E \otimes_{\mathbb{C}} \textstyle\bigwedge^{\text{top}}\mathfrak{u}).$$

A special case is $M(\lambda) = M(\mathbb{C}_\lambda)$.

Within the $(\mathfrak{g}, K)$ module consisting of the m-fold tensor product of $\bigwedge^{R+S}\mathfrak{g}$ with itself, let F be the irreducible $(\mathfrak{g}, K)$ submodule generated by $\bigwedge^{R+S}\bar{\mathfrak{u}} \otimes \cdots \otimes \bigwedge^{R+S}\bar{\mathfrak{u}}$. We are to consider

$$P_{\lambda+\delta}(M(\lambda + 2m\delta(\mathfrak{u})) \otimes_{\mathbb{C}} F),$$

which is isomorphic, via a Mackey isomorphism, to

(8.36) $$P_{\lambda+\delta}(M(\mathbb{C}_{\lambda+2m\delta(\mathfrak{u})} \otimes_{\mathbb{C}} F)).$$

Let

$$F = F_0 \supsetneq F_1 \supsetneq \cdots \supsetneq F_n = 0$$

be a filtration by $(\bar{\mathfrak{q}}, L \cap K)$ submodules such that each consecutive quotient is irreducible. The functor $P_{\lambda+\delta}(M(\mathbb{C}_{\lambda+2m\delta(\mathfrak{u})} \otimes_{\mathbb{C}} (\cdot)))$ is exact, and thus we obtain a filtration of (8.36) with consecutive quotients

(8.37) $$P_{\lambda+\delta}(M(\mathbb{C}_{\lambda+2m\delta(\mathfrak{u})} \otimes_{\mathbb{C}} (F_j/F_{j+1}))) \qquad \text{for } 0 \leq j \leq n - 1.$$

Each F_j/F_{j+1} is a trivial $\bar{\mathfrak{u}}$ module and a finite-dimensional irreducible $(\mathfrak{l}, L \cap K)$ module. Let its highest weight be λ_j. Then

$$M(\mathbb{C}_{\lambda+2m\delta(\mathfrak{u})} \otimes_{\mathbb{C}} (F_j/F_{j+1}))$$

has infinitesimal character $\lambda + \delta + 2m\delta(\mathfrak{u}) + \lambda_j$.

For (8.37) to be nonzero, we must have

(8.38) $$\lambda + \delta + 2m\delta(\mathfrak{u}) + \lambda_j = w(\lambda + \delta)$$

for some $w \in W(\mathfrak{g}, \mathfrak{h})$. In Lemma 8.32, put $\mu = 2m\delta(\mathfrak{u}) + \lambda_j$. Then (ii) holds since $\delta(\mathfrak{u})$ is orthogonal to $\Delta^+(\mathfrak{g}, \mathfrak{h})$. For (i), we recognize λ_j as a weight of F and $-2m\delta(\mathfrak{u})$ as the lowest weight. Therefore $\mu = \lambda_j - (-2m\delta(\mathfrak{u}))$ is a sum of positive roots. These roots may be taken in $\Delta(\mathfrak{u})$, as we see by applying $U(\mathfrak{g}) = U(\mathfrak{u})U(\mathfrak{l})U(\bar{\mathfrak{u}})$ to a lowest-weight vector and by taking into account that the lowest-weight space is an $\mathfrak{l}$ module. Thus (i) holds and the lemma shows that $\mu = 0$. In other words, (8.37) is nonzero only when F_j/F_{j+1} has highest weight $-2m\delta(\mathfrak{u})$. This weight occurs only once in F, and thus (8.37) equals $M(\mathbb{C}_{\lambda+2m\delta(\mathfrak{u})} \otimes_{\mathbb{C}} \mathbb{C}_{-2m\delta(\mathfrak{u})}) \cong M(\lambda)$.

Lemma 8.39. Suppose that $\mathbb{C}_\lambda$ is a one-dimensional $(\mathfrak{l}, L \cap K)$ module such that

$$\operatorname{Re} \langle \lambda + \delta(\mathfrak{u}), \alpha \rangle \geq 0 \qquad \text{for all } \alpha \in \Delta(\mathfrak{u}),$$

and suppose that m is an integer ≥ 0. Then

$$\psi^{\lambda+\delta}_{\lambda+\delta+2m\delta(\mathfrak{u})}(A_{\mathfrak{q}}(\lambda + 2m\delta(\mathfrak{u}))) \cong A_{\mathfrak{q}}(\lambda).$$

PROOF. We have

$$
\begin{aligned}
\psi^{\lambda+\delta}_{\lambda+\delta+2m\delta(\mathfrak{u})}&(A_{\mathfrak{q}}(\lambda + 2m\delta(\mathfrak{u}))) \\
&\cong \psi^{\lambda+\delta}_{\lambda+\delta+2m\delta(\mathfrak{u})}(\Pi_S(M(\lambda + 2m\delta(\mathfrak{u})))) \\
&\cong \Pi_S(\psi^{\lambda+\delta}_{\lambda+\delta+2m\delta(\mathfrak{u})}(M(\lambda + 2m\delta(\mathfrak{u})))) \qquad \text{by (7.238a)} \\
&\cong \Pi_S(M(\lambda)) \qquad\qquad\qquad\qquad\qquad\qquad \text{by Lemma 8.35} \\
&\cong A_{\mathfrak{q}}(\lambda),
\end{aligned}
$$

and the lemma follows.

PROOF OF THEOREM 8.31. As we mentioned earlier, we shall follow the line of argument of §VII.10, except that $R(\lambda)$ will have a new definition. We begin with the formalism. Fix $\lambda' \in \mathfrak{h}^*$. The complex associative algebra

$$Q(\lambda') = \operatorname{Hom}_{\mathbb{C}}(M(\lambda'), M(\lambda'))$$

is a $U(\mathfrak{g}) \otimes_{\mathbb{C}} U(\mathfrak{g})$ module, with a pure tensor $u_1 \otimes u_2$ acting by

$$(8.40) \qquad ((u_1 \otimes u_2)r)(v) = u_1(r(u_2^t v)) \quad \text{for } r \in Q(\lambda') \text{ and } v \in M(\lambda').$$

The fact that $M(\lambda')$ has infinitesimal character $\lambda' + \delta$ has consequences for the action on $Q(\lambda')$: $U(\mathfrak{g}) \otimes 1$ acts on $Q(\lambda')$ with infinitesimal character $\lambda' + \delta$ since $z \in Z(\mathfrak{g})$ implies

$$((z \otimes 1)r)(v) = z(r(v)) = \chi_{\lambda'}(z)r(v),$$

while $1 \otimes U(\mathfrak{g})$ acts on $Q(\lambda')$ with infinitesimal character $-(\lambda' + \delta)$, by an analogous computation and (7.55).

Let $1_{M(\lambda')}$ be the identity in $Q(\lambda')$. There is a natural map

$$(8.41) \qquad\qquad\qquad U(\mathfrak{g}) \to Q(\lambda')$$

given by $u \mapsto (u \otimes 1)1_{M(\lambda')}$, and this map respects the left $U(\mathfrak{g})$ actions on $U(\mathfrak{g})$ and $Q(\lambda')$. Formula (8.40) with $r = 1_{M(\lambda')}$ shows that the image of u is the ordinary action of u on $M(\lambda')$.

We define an adjoint action of $\mathfrak{g}$ on $Q(\lambda')$ by

$$(\text{ad } X)r = (X \otimes 1)r + (1 \otimes X)r \qquad \text{for } X \in \mathfrak{g} \text{ and } r \in Q(\lambda').$$

This is a Lie algebra action, and it consequently extends to $U(\mathfrak{g})$. Taking into account (8.40), we see that the formula on an element $v \in M(\lambda')$ is

$$(8.42) \qquad ((\text{ad } X)r)(v) = X(rv) - r(Xv) \qquad \text{for } X \in \mathfrak{g} \text{ and } r \in Q(\lambda').$$

In particular, we have

$$(8.43) \qquad\qquad\qquad (\text{ad } X)1_{M(\lambda')} = 0.$$

Also (8.42) gives

$$
\begin{aligned}
((\text{ad } X)r_1)r_2(v) &+ r_1((\text{ad } X)r_2)(v) \\
&= Xr_1r_2v - r_1Xr_2v + r_1Xr_2v - r_1r_2Xv \\
&= Xr_1r_2v - r_1r_2v \\
&= (\text{ad } X)(r_1r_2)(v),
\end{aligned}
$$

and we obtain

$$(8.44) \qquad (\text{ad } X)(r_1r_2) = ((\text{ad } X)r_1)r_2 + r_1((\text{ad } X)r_2).$$

We are interested in the members of $Q(\lambda')$ that lie in finite-dimensional ad $\mathfrak{g}$ invariant subspaces. Since $\mathfrak{g}$ is the complexification of the compact form $\mathfrak{k}_0'$, we could write the subspace of interest as

$$\text{Hom}_{\mathbb{C}}(M(\lambda'), M(\lambda'))_{\mathfrak{k}_0'}.$$

The same argument as in step (8) of §VII.10 shows that the action of ad $\mathfrak{k}_0'$ on this subspace exponentiates to an action Ad by K', and we can thus rewrite this space as

$$(8.45) \qquad R(\lambda') = \text{Hom}_{\mathbb{C}}(M(\lambda'), M(\lambda'))_{K'}.$$

As in §VII.10, we effectively have

$$R(\lambda') \cong \Gamma_{\mathfrak{g}\oplus\mathfrak{g},\{1\}}^{\mathfrak{g}\oplus\mathfrak{g},K'}(Q(\lambda')),$$

and therefore $R(\lambda')$ is a $(\mathfrak{g} \oplus \mathfrak{g}, K')$ module. Actually $R(\lambda')$ is a subalgebra of $Q(\lambda')$, as we see from (8.44) and the fact that the set of products from two finite-dimensional subspaces of $Q(\lambda')$ generates a finite-dimensional subspace of $Q(\lambda')$. Since $1_{M(\lambda')}$ is in $R(\lambda')$ by (8.43), it follows that the natural map (8.41) has image in $R(\lambda')$. Thus we have a natural map

$$(8.46) \qquad U(\mathfrak{g}) \to R(\lambda') \qquad \text{given by} \qquad u \mapsto (u \otimes 1)1_{M(\lambda')}.$$

Next let F be an irreducible finite-dimensional $U(\mathfrak{g})$ module. As in step (2) of §VII.10, $\text{End}_{\mathbb{C}} F$ becomes a $U(\mathfrak{g}) \otimes_{\mathbb{C}} U(\mathfrak{g})$ module, and so does

$$S = R(\lambda') \otimes_{\mathbb{C}} \text{End}_{\mathbb{C}} F.$$

Moreover, S has a primary decomposition under the left and right actions by $Z(\mathfrak{g}) \subseteq U(\mathfrak{g})$, and we write

$$S = \bigoplus_{\alpha, \beta} S_{\alpha}^{\beta}$$

as in (7.176). Here α is the parameter of the left action and β is the parameter of the right action. For each γ, $S_{\gamma}^{-\gamma}$ is a subalgebra of S by (7.178a), and the $S_{\gamma}^{-\gamma}$ component $1_{\gamma}^{-\gamma}$ of $1 \in S$ is the identity of $S_{\gamma}^{-\gamma}$ by (7.178b).

Let M be an irreducible $U(\mathfrak{g})$ module with infinitesimal character $\lambda' + \delta$, and suppose that M is also a left (unital) $R(\lambda')$ module such that the actions are compatible under the natural map (8.46):

$$(8.47) \qquad u(m) = ((u \otimes 1)1_{M(\lambda')})(m) \qquad \text{for } u \in U(\mathfrak{g}) \text{ and } m \in M.$$

With F as above, we let

$$N = M \otimes_{\mathbb{C}} F.$$

Then N is compatibly a $U(\mathfrak{g})$ module and a left S module, and, as in (7.179), N is simple as a left S module. The module N has a primary decomposition under the action by $Z(\mathfrak{g}) \subseteq U(\mathfrak{g})$, which we write

$$N = \bigoplus_{\gamma} N_{\gamma}.$$

As in (7.181), N_{γ} is a simple left $S_{\gamma}^{-\gamma}$ module, or else $N_{\gamma} = 0$. Tracing through the argument of step (5) of §VII.10, we conclude that

$$(8.48) \qquad \text{If } (U(\mathfrak{g}) \otimes 1)1_{\gamma}^{-\gamma} = S_{\gamma}^{-\gamma}, \text{ then } U(\mathfrak{g})n = S_{\gamma}^{-\gamma}n \text{ for}$$
$$\text{each } n \in N_{\gamma}, \text{ and hence either } N_{\gamma} \text{ is an irreducible}$$
$$U(\mathfrak{g}) \text{ module or else } N_{\gamma} = 0.$$

This completes the formalism. We take $\lambda' = \lambda + 2m\delta(\mathfrak{u})$ and $M = A_{\mathfrak{q}}(\lambda + 2m\delta(\mathfrak{u}))$, with $m \in \mathbb{Z}$ chosen large enough so that $A_{\mathfrak{q}}(\lambda + 2m\delta(\mathfrak{u}))$ is irreducible. Shortly we shall make M into an $R(\lambda + 2m\delta(\mathfrak{u}))$ module such that (8.47) holds.

The finite-dimensional irreducible representation F of $\mathfrak{g}$ is to be the representation constructed from alternating tensors in Lemma 8.35 and having lowest weight $-2m\delta(\mathfrak{u})$.

We take γ to be $\lambda + \delta$. Then $N_\gamma \cong A_{\mathfrak{q}}(\lambda)$ as a $U(\mathfrak{g})$ module, by Lemma 8.39. The argument of Proposition 7.199, in combination with two applications of Lemma 8.35, shows that

$$\psi_{(\lambda'+\delta,-\lambda'-\delta)}^{(\lambda+\delta,-\lambda-\delta)}(R(\lambda')) \cong R(\lambda)$$

as a $U(\mathfrak{g}) \oplus U(\mathfrak{g})$ module. Consequently $S_\gamma^{-\gamma} \cong R(\lambda)$ as a $U(\mathfrak{g} \oplus \mathfrak{g})$ module.

To identify $1_\gamma^{-\gamma}$ in $R(\lambda)$, we use the fact that $1_\gamma^{-\gamma}$ is annihilated by ad $\mathfrak{g}$. A member r of $R(\lambda)$ that is annihilated by ad $\mathfrak{g}$ has

$$X(rv) - r(Xv) = 0 \qquad \text{for } X \in \mathfrak{g} \text{ and } v \in V.$$

Consequently r is in $\operatorname{Hom}_{U(\mathfrak{g})}(M(\lambda), M(\lambda))$. However, r must then carry the weight space $\mathbb{C}_\lambda^\# \subseteq M(\lambda)$ to itself, and the effect of r elsewhere is determined by its effect on $\mathbb{C}_\lambda^\#$. In other words, r is scalar. Thus we may assume that $1_\gamma^{-\gamma}$ in $S_\gamma^{-\gamma}$ is identified with $1_{M(\lambda)}$. Applying (8.48) and taking into account that $N_\gamma \neq 0$, we come to this conclusion:

(8.49)   If the natural map (8.46) of $U(\mathfrak{g})$ into $R(\lambda)$ is onto, then $A_{\mathfrak{q}}(\lambda)$ is an irreducible $U(\mathfrak{g})$ module.

To prove the theorem, we thus have two tasks. One is to make $A_{\mathfrak{q}}(\lambda + 2m\delta(\mathfrak{u}))$ into an $R(\lambda + 2m\delta(\mathfrak{u}))$ module such that (8.47) holds, and the other is to prove that the natural map (8.46) carries $U(\mathfrak{g})$ onto $R(\lambda)$.

For the first task, fix a one-dimensional $(\mathfrak{l}, L \cap K)$ module $\mathbb{C}_\lambda^\#$. Then

$$A_{\mathfrak{q}}(\lambda') = (\Pi_{\mathfrak{g},L\cap K}^{\mathfrak{g},K})_S(M(\lambda')).$$

Let us abbreviate $R = R(\lambda')$, $M = M(\lambda') = \operatorname{ind}_{\bar{\mathfrak{q}},L\cap K}^{\mathfrak{g},L\cap K}(\mathbb{C}_{\lambda'}^\#)$, and $\Pi = \Pi_{\mathfrak{g},L\cap K}^{\mathfrak{g},K}$. The motivation for the definition of the action on $\Pi_S(M)$ comes from Proposition 3.77 and its proof. Because of the inclusion $\mathfrak{k}_0 \hookrightarrow \mathfrak{k}_0'$, we can identify $\operatorname{Ad}(K)$ with a subgroup of $\operatorname{Ad}(K')$. We make R into a $(\mathfrak{g}, K)$ module by using $(\operatorname{ad}, \operatorname{Ad})$. Let $m : R \otimes_{\mathbb{C}} R \to R$ be the multiplication map, and let $a : R \otimes_{\mathbb{C}} M \to M$ be the action map. Then m is a $U(\mathfrak{g})$ module map by (8.44), and a is a $U(\mathfrak{g})$ module map by (8.42). Since K and $L \cap K$ are connected, m is a $(\mathfrak{g}, K)$ map and a is a $(\mathfrak{g}, L \cap K)$ map.

We let μ refer to the inverse of any Mackey isomorphism as in Theorem 2.103 for the inclusion of pairs $(\mathfrak{g}, L \cap K) \hookrightarrow (\mathfrak{g}, K)$. One such isomorphism is

$$\mu : R \otimes \Pi(M) \to \Pi(R \otimes M).$$

Replacing M by a projective resolution M_j of M in $C(\mathfrak{g}, L \cap K)$, we obtain isomorphisms

$$\mu : R \otimes \Pi(M_j) \to \Pi(R \otimes M_j).$$

These are chain maps since μ is natural in each variable, and we thus obtain isomorphisms

$$\mu : R \otimes \Pi_j(M) \to \Pi_j(R \otimes M).$$

Following this μ with

$$\Pi_j(a) : \Pi_j(R \otimes M) \to \Pi_j(M),$$

we obtain a composition

(8.50) $\alpha = \Pi_j(a) \circ \mu$ with $\alpha : R \otimes \Pi_j(M) \to \Pi_j(M).$

We take α as the definition of an R module action on $\Pi_j(M)$.

We need to check that $\Pi_j(M)$ becomes an R module under α. The distributive laws are a consequence of the linearity of α in each variable, and what needs proof is associativity: $(r_1 r_2)v = r_1(r_2 v)$ for r_1 and r_2 in R and v in $\Pi_j(M)$. In terms of m and α, we are to prove that

(8.51) $\alpha \circ (m \otimes 1) = \alpha \circ (1 \otimes \alpha).$

We set up the diagram

$$
\begin{array}{ccccc}
R \otimes R \otimes \Pi_j(M) & \xrightarrow{\mu} & \Pi_j(R \otimes R \otimes M) & \xrightarrow{\Pi_j(1 \otimes a)} & \Pi_j(R \otimes M) \\
\downarrow{\scriptstyle m \otimes 1} & & \downarrow{\scriptstyle \Pi_j(m \otimes 1)} & & \downarrow{\scriptstyle \Pi_j(a)} \\
R \otimes \Pi_j(M) & \xrightarrow{\mu} & \Pi_j(R \otimes M) & \xrightarrow{\Pi_j(a)} & \Pi_j(M)
\end{array}
$$

The left square commutes because of naturality of μ in the first variable (with respect to the map $m : R \otimes R \to R$), and the right square commutes since M is an R module. Therefore

$$\Pi_j(a) \circ \mu \circ (m \otimes 1) = \Pi_j(a) \circ \Pi_j(1 \otimes a) \circ \mu,$$

which we rewrite as

(8.52) $$\alpha \circ (m \otimes 1) = \Pi_j(a) \circ \Pi_j(1 \otimes a) \circ \mu.$$

Next we set up the diagram

$$
\begin{array}{ccccc}
R \otimes R \otimes \Pi_j(M) & \xrightarrow{1 \otimes \mu} & R \otimes \Pi_j(R \otimes M) & \xrightarrow{1 \otimes \Pi_j(a)} & R \otimes \Pi_j(M) \\
\downarrow{\mu} & & \downarrow{\mu} & & \downarrow{\mu} \\
\Pi_j(R \otimes R \otimes M) & \xrightarrow{1} & \Pi_j(R \otimes R \otimes M) & \xrightarrow{\Pi_j(1 \otimes a)} & \Pi_j(R \otimes M)
\end{array}
$$

The left square commutes by the uniqueness of the Mackey isomorphism and the associativity of tensor product, and the right square commutes because of naturality of μ in the second variable (with respect to the map $a : R \otimes M \to M$). Therefore

$$\Pi_j(1 \otimes a) \circ \mu = \mu \circ (1 \otimes \Pi_j(a)) \circ (1 \otimes \mu),$$

which we rewrite as

$$\Pi_j(1 \otimes a) \circ \mu = \mu \circ (1 \otimes \alpha).$$

Left multiplication by $\Pi_j(a)$ and use of (8.50) gives

(8.53) $$\Pi_j(a) \circ \Pi_j(1 \otimes a) \circ \mu = \alpha \circ (1 \otimes \alpha).$$

Combining (8.52) and (8.53), we obtain (8.51). Therefore $\Pi_j(M)$ is a left R module.

We still have to check that (8.47) holds, i.e., that

(8.54) $$u(m) = ((u \otimes 1)1_{M(\lambda')})(m) \quad \text{for } u \in U(\mathfrak{g}) \text{ and } m \in \Pi_j(M).$$

Here the left side refers to the natural action of $U(\mathfrak{g})$ on $\Pi_j(M)$, and the right side refers to the action of $U(\mathfrak{g})$ via the natural map of $U(\mathfrak{g})$ into R; the equality says these actions coincide. We regard $U(\mathfrak{g})$ as a

$(\mathfrak{g}, K)$ module under $(\mathrm{ad}, \mathrm{Ad})$, and we repeat our construction, obtaining commutative diagrams

(8.55)

$$R \otimes \Pi_j(M)$$

$$\mu \downarrow \qquad \searrow \alpha_R$$

$$\Pi_J(R \otimes M) \xrightarrow{\Pi_j(a)} \Pi_j(M)$$

and

(8.56)

$$U(\mathfrak{g}) \otimes \Pi_j(M)$$

$$\mu \downarrow \qquad \searrow \alpha_{U(\mathfrak{g})}$$

$$\Pi_J(U(\mathfrak{g}) \otimes M) \xrightarrow{\Pi_j(a)} \Pi_j(M)$$

Let us imagine a three-dimensional diagram with (8.56) as top layer, (8.55) as bottom layer, and downward vertical maps induced by the natural map (8.46). The left vertical face of this three-dimensional diagram commutes because of naturality of μ in the first variable, and the vertical face at the bottom edge commutes because of (8.47) for M. Therefore the diagonal face commutes, and $\alpha_{U(\mathfrak{g})}$ is compatible with α_R. By Proposition 3.77, $\alpha_{U(\mathfrak{g})}$ is the natural action of $U(\mathfrak{g})$ on $\Pi_j(M)$, and (8.54) follows. Consequently $A_{\mathfrak{q}}(\lambda')$ is an $R(\lambda')$ module in such a way that (8.47) holds.

Our second task is to prove that the natural map (8.46) carries $U(\mathfrak{g})$ onto $R(\lambda)$. By (8.49) this step will complete the proof of the theorem. We begin by constructing an increasing filtration of the algebra $R(\lambda)$ by $(\mathfrak{g}, K')$ submodules

(8.57) $$\mathbb{C} = R_0(\lambda) \subseteq R_1(\lambda) \subseteq \cdots \qquad \text{with} \quad \bigcup_{n=1}^{\infty} R_n(\lambda) = R(\lambda).$$

To do so, first we filter $M(\lambda)$ by the subspaces

(8.58) $$M_n(\lambda) = U_n(\mathfrak{g})\mathbb{C}_\lambda^\# \cong U_n(\mathfrak{u}) \otimes_\mathbb{C} \mathbb{C}_\lambda^\#.$$

Note that $M_0(\lambda) = \mathbb{C}_\lambda^\#$. Let us write 1_λ for the member of $\mathbb{C}_\lambda^\#$ corresponding to the complex number 1. Each $M_n(\lambda)$ is a $U(\bar{\mathfrak{q}})$ submodule under the left action, since $X \in \bar{\mathfrak{q}}$ acts on $u(1_\lambda)$ by

(8.59) $$X(u(1_\lambda)) = ((\mathrm{ad}\, X)u)(1_\lambda) + u(X1_\lambda)$$

for $u \in U_n(\mathfrak{g})$. Since $U_n(\mathfrak{g})$ is stable under $\mathrm{ad}\, X$, $M_n(\lambda)$ is a $U(\bar{\mathfrak{q}})$ submodule. Now define

(8.60)
$$R_n(\lambda)$$
$$= \{r \in R(\lambda) \mid (\mathrm{Ad}(k')r)M_p(\lambda) \subseteq M_{p+n}(\lambda) \text{ for all } k' \in K' \text{ and all } p\}.$$

It is clear that the $R_n(\lambda)$ are increasing, and the presence of the k' in the definition makes each a $(\mathfrak{g}, K')$ submodule of $R(\lambda)$. Under the natural map $U(\mathfrak{g}) \to R(\lambda)$ of (8.46), we have

(8.61)
$$U_n(\mathfrak{g}) \to R_n(\lambda).$$

In fact, the natural map is carries u to its action on $M(\lambda)$, and the image of $U_n(\mathfrak{g})$ is contained in $R_n(\lambda)$ as a consequence of the inclusion $U_n(\mathfrak{g})U_p(\mathfrak{g}) \subseteq U_{n+p}(\mathfrak{g})$.

Before proving that the union of the $R_n(\lambda)$ is all of $R(\lambda)$, we establish the formula

(8.62) $\quad R_n(\lambda) = \{r \in R(\lambda) \mid (\mathrm{Ad}(k')r)\mathbb{C}_\lambda^{\#} \subseteq M_n(\lambda) \text{ for all } k' \in K'\}.$

In fact, write $R_n'(\lambda)$ for the right side. Certainly $R_n(\lambda) \subseteq R_n'(\lambda)$. Suppose that r is in $R_n'(\lambda)$. We prove that

(8.63) $\quad (\mathrm{Ad}(k')r)M_p(\lambda) \subseteq M_{p+n}(\lambda) \qquad \text{for all } k' \in K'$

by induction on p, the base case being the known case $p = 0$. Suppose inductively that p is > 0 and that (8.63) is known for $p - 1$. The space $M_p(\lambda)$ is spanned by elements Xm with $X \in \mathfrak{g}$ and $m \in M_{p-1}(\lambda)$, and we have

$$(\mathrm{Ad}(k')r)(Xm) = -((\mathrm{ad}\, X)(\mathrm{Ad}(k')r)m + X((\mathrm{Ad}(k')r)m).$$

Evidently $R_n'(\lambda)$ is $\mathrm{Ad}(K')$ stable and therefore also $\mathrm{ad}\,\mathfrak{g}$ stable. Hence the first term on the right side is in $M_{p+n-1}(\lambda)$ by induction. Similarly $(\mathrm{Ad}(k')r)m$ is in $M_{p+n-1}(\lambda)$ by induction, and hence the second term is in $M_{p+n}(\lambda)$. This completes the induction and the proof of (8.63).

Now we can prove that

(8.64)
$$\bigcup_{n=0}^{\infty} R_n(\lambda) = R(\lambda).$$

Suppose r is in $R(\lambda)$. Since $R(\lambda)$ is K' finite, r lies in some finite-dimensional $\mathrm{Ad}(K')$ stable subspace S of $R(\lambda)$. Let $\epsilon > 0$ be the smallest (necessarily positive) eigenvalue of $\operatorname{ad} h_{\delta(\mathfrak{u})}$ on $\mathfrak{u}$, and let N be the largest eigenvalue of $\operatorname{ad} h_{\delta(\mathfrak{u})}$ on S. We shall show that r is in $R_{[N/\epsilon]}(\lambda)$, where $[N/\epsilon]$ denotes the greatest integer in N/ϵ. According to (8.62), it is enough to prove that

$$(8.65) \qquad s\mathbb{C}_\lambda^\# \subseteq M_{[N/\epsilon]}(\lambda) \qquad \text{for all } s \in S.$$

We may assume that s is a weight vector under $\operatorname{ad} \mathfrak{h}$, say with weight μ. Then $s\mathbb{C}_\lambda^\#$ is contained in the $\lambda + 2\delta(\mathfrak{u}) + \mu$ weight space, which we write as $M(\lambda)_{\lambda+2\delta(\mathfrak{u})+\mu}$. Now

$$M(\lambda)_{\lambda+2\delta(\mathfrak{u})+\mu} \cong U(\mathfrak{u})_\mu \otimes_\mathbb{C} \mathbb{C}_\lambda^\#.$$

Since $\mu(h_{\delta(\mathfrak{u})}) \leq N$ and since the eigenvalues of $\operatorname{ad} h_{\delta(\mathfrak{u})}$ on $\mathfrak{u}$ are bounded below by $\epsilon > 0$,

$$U(\mathfrak{u})_\mu \subseteq U_{[N/\epsilon]}(\mathfrak{u}).$$

Hence

$$M(\lambda)_{\lambda+2\delta(\mathfrak{u})+\mu} \subseteq U_{[N/\epsilon]}(\mathfrak{u})\mathbb{C}_\lambda^\# = M_{[N/\epsilon]}(\lambda),$$

and (8.65) follows. This proves (8.64).

To complete our task, we are to prove that the natural map $U(\mathfrak{g}) \to R(\lambda)$ of (8.46) is onto. In view of (8.64) and (8.61), it is enough to prove that $U_n(\mathfrak{g}) \to R_n(\lambda)$ is onto for every n. Since $U_{n-1}(\mathfrak{g})$ maps into $R_{n-1}(\lambda)$, these maps induce maps

$$(8.66) \qquad U_n(\mathfrak{g})/U_{n-1}(\mathfrak{g}) \to R_n(\lambda)/R_{n-1}(\lambda) \qquad \text{for } n \geq 1.$$

Now $U_0(\mathfrak{g})$ maps onto $R_0(\lambda)$, and an easy induction shows that if (8.66) is onto for all $n \geq 1$, then (8.61) is onto for all $n \geq 0$. Thus it is enough to prove that (8.66) is onto for $n \geq 1$.

Let $\tau_n : U_n(\mathfrak{g})/U_{n-1}(\mathfrak{g}) \to S^n(\mathfrak{g})$ be the natural map. Using (8.58), we have vector-space isomorphisms

(8.67a)
$$M_n(\lambda)/M_{n-1}(\lambda) \longrightarrow (U_n(\mathfrak{u})/U_{n-1}(\mathfrak{u})) \otimes_\mathbb{C} \mathbb{C}_\lambda^\#$$

$$\xrightarrow{\tau_n \otimes 1} S^n(\mathfrak{u}) \otimes_\mathbb{C} \mathbb{C}_\lambda^\# \longrightarrow S^n(\mathfrak{g}/\bar{\mathfrak{q}}) \otimes_\mathbb{C} \mathbb{C}_\lambda^\#.$$

The composition is evidently a $U(\mathfrak{l})$ map, and thus we obtain a $U(\mathfrak{l})$ isomorphism

$$(8.67b) \qquad M_n(\lambda)/M_{n-1}(\lambda) \xrightarrow{\sim} S^n(\mathfrak{g}/\bar{\mathfrak{q}}) \otimes_\mathbb{C} \mathbb{C}_\lambda^\#.$$

Actually (8.67) respects the $\bar{u}$ actions also. To see this, we shall construct a $\bar{q}$ map that is a left inverse to (8.67). We begin with the action map

$$(8.68) \qquad U_n(\mathfrak{g}) \otimes_{\mathbb{C}} \mathbb{C}_\lambda^\# \to M_n(\lambda)$$

of (8.58), which respects the $\bar{q}$ module structures by (8.59). We follow this with passage to the quotient $M_n(\lambda)/M_{n-1}(\lambda)$ and observe that $U_{n-1}(\mathfrak{g}) \otimes_{\mathbb{C}} \mathbb{C}_\lambda^\#$ is in the kernel. Thus we obtain $\bar{q}$ maps

(8.69)

$$S^n(\mathfrak{g}) \otimes_{\mathbb{C}} \mathbb{C}_\lambda^\# \xrightarrow{\tau_n^{-1} \otimes 1} (U_n(\mathfrak{g})/U_{n-1}(\mathfrak{g})) \otimes_{\mathbb{C}} \mathbb{C}_\lambda^\#$$

$$\longrightarrow (U_n(\mathfrak{g}) \otimes_{\mathbb{C}} \mathbb{C}_\lambda^\#)/(U_{n-1}(\mathfrak{g}) \otimes_{\mathbb{C}} \mathbb{C}_\lambda^\#) \longrightarrow M_n(\lambda)/M_{n-1}(\lambda).$$

Under (8.68), $\bar{q}U_{n-1}(\mathfrak{g}) \otimes_{\mathbb{C}} \mathbb{C}_\lambda^\#$ maps into $M_{n-1}(\lambda)$, and therefore $\bar{q}S^{n-1}(\mathfrak{g}) \otimes_{\mathbb{C}} \mathbb{C}_\lambda^\#$ is in the kernel of (8.69). Since

$$S^n(\mathfrak{g})/(\bar{q}S^{n-1}(\mathfrak{g})) \cong S^n(\mathfrak{g}/\bar{q}) \qquad \text{as } \bar{q} \text{ modules,}$$

(8.69) descends to a $\bar{q}$ map

$$S^n(\mathfrak{g}/\bar{q}) \otimes_{\mathbb{C}} \mathbb{C}_\lambda^\# \to M_n(\lambda)/M_{n-1}(\lambda).$$

If we ignore the $\bar{q}$ actions and regard this map as defined on $S^n(\mathfrak{u}) \otimes_{\mathbb{C}} \mathbb{C}_\lambda^\#$, we see that this map is a left inverse to (8.67a). Therefore (8.67) is a $U(\bar{q})$ isomorphism.

We now define a map σ_n that plays the role of the principal symbol in partial differential equations. Fix $r \in R_n(\lambda)$ and $k' \in K'$. Let us follow the effect of $\mathrm{Ad}(k')r$ on $\mathbb{C}_\lambda^\#$ with passage to the quotient and (8.67):

$$\mathbb{C}_\lambda^\# \longrightarrow M_n(\lambda) \longrightarrow M_n(\lambda)/M_{n-1}(\lambda) \longrightarrow S^n(\mathfrak{g}/\bar{q}) \otimes_{\mathbb{C}} \mathbb{C}_\lambda^\#.$$

Thus $\mathrm{Ad}(k')r$ defines an element

$$\sigma_n(r)(k') \in \mathrm{Hom}_{\mathbb{C}}(\mathbb{C}_\lambda^\#, \, S^n(\mathfrak{g}/\bar{q}) \otimes_{\mathbb{C}} \mathbb{C}_\lambda^\#) \cong S^n(\mathfrak{g}/\bar{q}).$$

Here $\sigma_n(r)$ is a function from K' to $S^n(\mathfrak{g}/\bar{q})$. The kernel of σ_n is $R_{n-1}(\lambda)$, and σ_n is a K' equivariant map relative to the right regular representation of K' in the sense that

$$(8.70) \qquad \sigma_n(\mathrm{Ad}(k_1')r)(k_2') = \sigma_n(r)(k_2'k_1').$$

Thus σ_n descends to a one-one K' equivariant map of $R_n(\lambda)/R_{n-1}(\lambda)$ onto a space of functions from K' to $S^n(\mathfrak{g}/\bar{\mathfrak{q}})$. We shall show below that these functions $f = \sigma_n(r)$ satisfy

$$(8.71) \qquad\qquad X f(k') = (\operatorname{ad} X)(f(k'))$$

for $X \in \bar{\mathfrak{q}}$ if the action of X on f is defined to be on the right of k'.

We can map each such f to the system $\{uf(1) \mid u \in U(\mathfrak{g})\}$, thereby realizing f in

$$\operatorname{Hom}_{\mathbb{C}}(U(\mathfrak{g}), S^n(\mathfrak{g}/\bar{\mathfrak{q}})).$$

This map is one-one since f is real analytic. Because of (8.71), f is actually in

$$\operatorname{Hom}_{\bar{\mathfrak{q}}, L'}(U(\mathfrak{g}), S^n(\mathfrak{g}/\bar{\mathfrak{q}})).$$

Let Γ' be the functor of Example 1c in §II.1 that extracts the largest locally K' finite subspace. Since f is K' finite, we can regard f as in the subspace

$$\Gamma' \operatorname{Hom}_{\bar{\mathfrak{q}}, L'}(U(\mathfrak{g}), S^n(\mathfrak{g}/\bar{\mathfrak{q}})).$$

By Proposition 2.70, Γ' is naturally isomorphic to $\Gamma_{\mathfrak{g}, L'}^{\mathfrak{g}, K'}$. Therefore σ_n defines a one-one K' equivariant map

$$(8.72) \qquad\qquad \sigma_n : R_n(\lambda)/R_{n-1}(\lambda) \to \Gamma_{\bar{\mathfrak{q}}, L'}^{\mathfrak{g}, K'}(S^n(\mathfrak{g}/\bar{\mathfrak{q}})).$$

The composition of (8.66) followed by σ_n is a K' equivariant map

$$T_n' : U_n(\mathfrak{g})/U_{n-1}(\mathfrak{g}) \to \Gamma_{\bar{\mathfrak{q}}, L'}^{\mathfrak{g}, K'}(S^n(\mathfrak{g}/\bar{\mathfrak{q}})).$$

To identify T_n', we regard $\Gamma_{\bar{\mathfrak{q}}, L'}^{\mathfrak{g}, K'}$ as Γ' and evaluate at $1 \in U(\mathfrak{g})$, thereby obtaining a map

$$U_n(\mathfrak{g})/U_{n-1}(\mathfrak{g}) \to S^n(\mathfrak{g}/\bar{\mathfrak{q}}).$$

Tracking down the definitions, we see that this is the natural quotient map

$$t_n : S^n(\mathfrak{g}) \to S^n(\mathfrak{g}/\bar{\mathfrak{q}})$$

of (8.30a). By Frobenius recipocity (Proposition 2.21), T_n' is the map T_n of (8.30c). By hypothesis in the theorem, this map is onto. Therefore its second factor, which is the map σ_n of (8.72), must be onto. Since σ_n is one-one, σ_n is an isomorphism. Therefore the first factor (8.66) of T_n' must be onto, and this is what we wanted to show.

There is one final detail, the proof of (8.71). Since the space of functions in question is K' invariant, we may take $k' = 1$. Reinterpreting

X on the left side of (8.71) as a member of the complexified Lie algebra of K' and using (8.70), we are to show that

$$(8.73) \qquad \sigma_n((\mathrm{ad}\, X)r)(1) = (\mathrm{ad}\, X)(\sigma_n(r)(1)) \qquad \text{for } X \in \bar{\mathfrak{q}}.$$

To prove this identity, we regard $\sigma_n(r)(1)$ as in $S^n(\mathfrak{g}/\bar{\mathfrak{q}})$, and we choose a member $\mathfrak{u} \in U_n(\mathfrak{g})$ mapping to it under

$$(8.74) \qquad U_n(\mathfrak{g}) \to U_n(\mathfrak{g})/U_{n-1}(\mathfrak{g}) \to S^n(\mathfrak{g}) \to S^n(\mathfrak{g}/\bar{\mathfrak{q}}).$$

Then

$$r m_0 \equiv u m_0 \quad \mathrm{mod}\ M_{n-1}(\lambda) \qquad \text{for } m_0 \in \mathbb{C}_\lambda^\#.$$

We compute

$$((\mathrm{ad}\, X)r)m_0$$

$$\begin{aligned}
&= X(r(m_0)) - r(Xm_0) && \text{by (8.59)}\\
&\equiv X(u(m_0)) - r(Xm_0) \quad \mathrm{mod}\ XM_{n-1}(\lambda)\\
&\equiv X(u(m_0)) - r(Xm_0) \quad \mathrm{mod}\ M_{n-1}(\lambda) && \text{since}\\
& && \bar{\mathfrak{q}}M_{n-1}(\lambda) \subseteq M_{n-1}(\lambda)\\
&\equiv X(u(m_0)) - u(Xm_0) \quad \mathrm{mod}\ M_{n-1}(\lambda) && \text{since } Xm_0 \in \mathbb{C}_\lambda^\#\\
&= ((\mathrm{ad}\, X)u)m_0 \quad \mathrm{mod}\ M_{n-1}(\lambda) && \text{by (8.59).}
\end{aligned}$$

This result says that $(\mathrm{ad}\, X)u$ maps to $\sigma_n((\mathrm{ad}\, X)r)(1)$ under (8.74). But the maps in (8.74) respect $\mathrm{ad}\,\mathfrak{g}$. Hence $(\mathrm{ad}\, X)u$ maps also to $(\mathrm{ad}\, X)(\sigma_n(r)(1))$. This proves (8.73) and completes the proof of the theorem.

Proposition 8.75 (Hesselink). If $\mathfrak{u}$ is abelian, then the map Φ_n in (8.30c) is onto for every $n \geq 0$.

PROOF. First let us show that $[\bar{\mathfrak{u}}, \mathfrak{g}] \subseteq \bar{\mathfrak{q}}$. In fact, if this inclusion were not valid, we could find $-\beta \in \Delta(\bar{\mathfrak{u}})$ and $\alpha \in \Delta(\mathfrak{u})$ such that $-\beta + \alpha$ is in $\Delta(\mathfrak{u})$. Then $\alpha = (-\beta + \alpha) + \beta$ would exhibit a member of $\Delta(\mathfrak{u})$ as the sum of two members of $\Delta(\mathfrak{u})$ and contradict the fact that $\mathfrak{u}$ is abelian.

Consequently $\mathrm{ad}\,\bar{\mathfrak{u}}$ acts as 0 on $\mathfrak{g}/\bar{\mathfrak{q}}$ and also as 0 on $S^n(\mathfrak{g}/\bar{\mathfrak{q}})$. Therefore $S^n(\mathfrak{g}/\bar{\mathfrak{q}})$ is completely reducible as a $(\bar{\mathfrak{q}}, L')$ module. By the algebraic Borel-Weil Theorem (Theorem 4.83e), every $(\mathfrak{g}, K')$ submodule of $I_{\bar{\mathfrak{q}}, L'}^{\mathfrak{g}, K'}(S^n(\mathfrak{g}/\bar{\mathfrak{q}}))$ is of the form $I_{\bar{\mathfrak{q}}, L'}^{\mathfrak{g}, K'}(M)$ for some $(\bar{\mathfrak{q}}, L')$ submodule M of $S^n(\mathfrak{g}/\bar{\mathfrak{q}})$.

Let us apply this fact to the $(\mathfrak{g}, K')$ submodule image Φ_n, writing image $\Phi_n = I_{\bar{\mathfrak{q}}, L'}^{\mathfrak{g}, K'}(M)$ for some M. Then Φ_n may be regarded as a member Φ_n' of

$$\mathrm{Hom}_{\mathfrak{g}, K'}(S^n(\mathfrak{g}), I_{\bar{\mathfrak{q}}, L'}^{\mathfrak{g}, K'}(M)).$$

Let φ'_n be the member of $\mathrm{Hom}_{\bar{\mathfrak{q}},L'}(S^n(\mathfrak{g}), M)$ that corresponds to Φ'_n under Frobenius reciprocity. If $i : M \to S^n(\mathfrak{g}/\bar{\mathfrak{q}})$ denotes inclusion, then it follows from naturality of Frobenius reciprocity in the second variable (with respect to the map i) that $i \circ \varphi'_n = \varphi_n$. Since φ_n is onto, so is i. Hence $M = S^n(\mathfrak{g}/\bar{\mathfrak{q}})$. Therefore Φ_n in (8.30c) is onto.

EXAMPLE 1 (Speh representations). Let $G = GL_0(2n, \mathbb{R})$. We shall construct some representations of G, and the representations of interest will be their restrictions to $SL(2n, \mathbb{R})$. We begin with the standard maximally compact θ stable Cartan subalgebra $\mathfrak{h}_0 = \mathfrak{t}_0 \oplus \mathfrak{a}_0$ in which $\mathfrak{t}_0$ consists of 2-by-2 blocks $\begin{pmatrix} 0 & \theta_j \\ -\theta_j & 0 \end{pmatrix}$ down the diagonal and $\mathfrak{a}_0$ consists of 2-by-2 blocks $\begin{pmatrix} x_j & 0 \\ 0 & x_j \end{pmatrix}$ down the diagonal. According to Proposition 4.76, we can obtain a θ stable parabolic subalgebra of $\mathfrak{g}$ from any member of $i\mathfrak{t}_0$, and the member we choose is

$$(8.76) \qquad \tilde{J} = i \begin{pmatrix} J & & & \\ & J & & \\ & & \ddots & \\ & & & J \end{pmatrix}, \qquad \text{where } J = \begin{pmatrix} 0 & -1 \\ 1 & 0 \end{pmatrix}.$$

The centralizer of this element is a copy of $\mathfrak{gl}(n, \mathbb{C})$ embedded in $\mathfrak{gl}(2n, \mathbb{R})$ in the standard way, and L must be $GL(n, \mathbb{C})$ by Lemma 5.10.

Relative to the jth 2-by-2 block of $\mathfrak{h}$, define e_j and f_j in $\mathfrak{h}^*$ by

$$e_j \begin{pmatrix} x_j & -iy_j \\ iy_j & x_j \end{pmatrix} = y_j \qquad \text{and} \qquad f_j \begin{pmatrix} x_j & -iy_j \\ iy_j & x_j \end{pmatrix} = x_j.$$

Then

$$\Delta(\mathfrak{g}, \mathfrak{h}) = \{\pm e_j \pm e_k \pm (f_j - f_k) \mid j \neq k\} \cup \{\pm 2e_l \mid 1 \leq l \leq n\},$$

and

$$\Delta(\mathfrak{l}, \mathfrak{h}) = \{\pm(e_j - e_k) \pm (f_j - f_k) \mid j \neq k\}.$$

For $\tilde{J}$ as in (8.76), $e_j(\tilde{J}) = 1$ and $f_j(\tilde{J}) = 0$ for all $\tilde{J}$. Thus

$$\Delta(\mathfrak{u}) = \{e_j + e_k \pm (f_j - f_k) \mid j \neq k\} \cup \{2e_l \mid 1 \leq l \leq n\}.$$

The imaginary roots in $\Delta(\mathfrak{u})$ are the $2e_l$. Each corresponds to a root vector within the jth 2-by-2 block along the diagonal and therefore within a copy of $SL(2, \mathbb{R})$. Consequently the imaginary roots are all noncompact.

We can define

$$\Delta^+(\mathfrak{l}, \mathfrak{h}) = \{(e_j - e_k) \pm (f_j - f_k) \mid j < k\},$$

taking $\Delta^+(\mathfrak{g}, \mathfrak{h}) = \Delta(\mathfrak{u}) \cup \Delta^+(\mathfrak{l}, \mathfrak{h})$. This positive system determines a θ stable Borel subalgebra contained in $\mathfrak{q} = \mathfrak{l} \oplus \mathfrak{u}$. We readily check that

(8.77)
$$\begin{aligned}
\delta(\mathfrak{l} \cap \mathfrak{n}) &= (n-1)e_1 + (n-3)e_2 + \cdots + (-(n-1))e_n \\
\delta(\mathfrak{u} \cap \mathfrak{k}) &= \tfrac{1}{2}(n-1)(e_1 + \cdots + e_n) \\
\delta(\mathfrak{u} \cap \mathfrak{p}) &= \tfrac{1}{2}(n+1)(e_1 + \cdots + e_n) \\
\delta(\mathfrak{u}) &= n(e_1 + \cdots + e_n) \\
\delta_K &= (n-1)e_1 + (n-2)e_2 + \cdots + 0e_n \\
\delta &= (2n-1)e_1 + (2n-3)e_2 + \cdots + 1e_n.
\end{aligned}$$

We can write any member λ of $\mathfrak{h}^*$ as

$$\lambda = \sum_{j=1}^n a_j e_j + \sum_{j=1}^n b_j f_j.$$

If λ is orthogonal to $\Delta(\mathfrak{l}, \mathfrak{h})$, then all the a_j's are equal and all the b_j's are equal. The linear functional $f_1 + \cdots + f_n$ is just detecting behavior on the center of $\mathfrak{g}$, and the effect on a representation of G of changing λ by a multiple of $f_1 + \cdots + f_n$ is simply to tensor the representation by a one-dimensional representation of G. Thus we may as well take all the b_j's to be 0 and write λ as

(8.78)
$$\lambda = m(e_1 + \cdots + e_n).$$

The condition that $\mathbb{C}_\lambda$ be an $(\mathfrak{l}, L \cap K)$ module is that m be an integer.

The representations of G that we shall study are the $A_q(\lambda)$'s for which

(8.79a) $\langle \lambda + \delta(\mathfrak{u}), \alpha \rangle \geq 0$ for all $\alpha \in \Delta(\mathfrak{u})$.

However, we have already addressed irreducibility for all those representations except when

(8.79b) $\langle \lambda + \delta, \alpha \rangle \geq 0$ fails for some $\alpha \in \Delta(\mathfrak{u})$.

We have

$$\lambda + \delta(\mathfrak{u}) = (n+m)(e_1 + \cdots + e_n),$$

and (8.79a) holds for $m \geq -n$. Also

$$\lambda + \delta = (2n + m - 1)e_1 + (2n + m - 3)e_2 + \cdots + (m + 1)e_n,$$

and thus the failure in (8.79b) occurs (with $\alpha = 2e_n$) when $m < -1$. The new representations to study thus have λ given by (8.78) with

$$(8.80) \qquad\qquad m = -n, \ -n + 1, \ \ldots, \ -2.$$

There are $n - 1$ such values.

The restrictions of the $A_q(\lambda)$'s as in (8.79a) from G to $SL(2n, \mathbb{R})$ are called **Speh representations**. As we shall see shortly, they are all nonzero. A second series of Speh representations is obtained by adjusting $\tilde{J}$ in (8.76) so that the last 2-by-2 block is $\begin{pmatrix} 0 & 1 \\ -1 & 0 \end{pmatrix}$. Inspection of $\Delta(\mathfrak{u})$ shows that $\mathfrak{u}$ is abelian, and therefore the Speh representations are irreducible, by Proposition 8.75 and Theorem 8.31. Moreover they are the underlying $(\mathfrak{g}, K)$ modules of unitary representations of $SL(2n, \mathbb{R})$, as we shall see in §IX.6.

The parameter

$$\Lambda = \lambda + 2\delta(\mathfrak{u} \cap \mathfrak{p}) = (m + n + 1)(e_1 + \cdots + e_n)$$

of Theorem 8.29 is Δ_K^+ dominant and hence occurs in the bottom layer. Consequently $A_q(\lambda)$ is not zero. Moreover, Λ has the property that $\Lambda + 2\delta_K$ is $\Delta^+(\mathfrak{g}, \mathfrak{h})$ dominant. As we shall see in §X.1, it will follow that Λ is the unique "minimal K type" of $A_q(\lambda)$. The corresponding parameters Λ of the second series of Speh representations are conjugate by the Weyl group of $K = SO(2n)$ to $(m + n + 1)(e_1 + \cdots + e_{n-1} - e_n)$. Since

$$(m + n + 1)(e_1 + \cdots + e_{n-1} - e_n) \quad \text{and} \quad (m + n + 1)(e_1 + \cdots + e_{n-1} + e_n)$$

are not conjugate by the Weyl group of K, it follows that the representations of the two series of Speh representations are inequivalent.

Occasionally one speaks of the **Speh representations** of $GL(2n, \mathbb{R})$, which are induced from representations of G_0 with a Speh representation on $SL(2n, \mathbb{R})$ and a character on the positive scalar matrices. The resulting induced representations of $GL(2n, \mathbb{R})$ are irreducible, as a consequence of Mackey theory and the inequivalence proved in the previous paragraph.

EXAMPLE 2. Let $G = Sp(2, \mathbb{R})$. We shall construct an $A_{\mathfrak{q}}(\lambda)$ such that Theorem 5.109 is applicable but $A_{\mathfrak{q}}(\lambda)$ is reducible. We take the Cartan subalgebra $\mathfrak{h}_0$ to be in $\mathfrak{k}_0$. The root system $\Delta(\mathfrak{g}, \mathfrak{h})$ is of type C_2. In standard notation we have

$$
\begin{array}{ll}
e_1 - e_2 & \text{compact} \\
2e_1 & \text{noncompact} \\
e_1 + e_2 & \text{noncompact} \\
2e_2 & \text{noncompact.}
\end{array}
$$

For our $A_{\mathfrak{q}}(\lambda)$ we take

$$\Delta^+(\mathfrak{l}, \mathfrak{h}) = \{2e_2\} \qquad \text{and} \qquad \Delta(\mathfrak{u}) = \{e_1 - e_2, \, 2e_1, \, e_1 + e_2\}.$$

The relevant δ's are

$$\delta = 2e_1 + e_2, \qquad \delta(\mathfrak{u}) = 2e_1, \qquad 2\delta(\mathfrak{u} \cap \mathfrak{p}) = 3e_1 + e_2.$$

Define $\lambda = -2e_1$; then

$$\lambda + \delta = e_2, \qquad \lambda + \delta(\mathfrak{u}) = 0, \qquad \lambda + 2\delta(\mathfrak{u} \cap \mathfrak{p}) = e_1 + e_2.$$

From the nondominance of $\lambda + \delta$, we see that the usual vanishing theorem (Theorem 5.99) and the Irreducibility Theorem (Theorem 8.2) are not applicable. The dominance of $\lambda + \delta(\mathfrak{u})$ shows that the improved vanishing theorem (Theorem 5.109) does apply. We shall prove that $A_{\mathfrak{q}}(\lambda)$ is reducible.

We begin by using Theorem 8.29 to calculate the K spectrum exactly. In that theorem we have

$$\Delta(\mathfrak{u} \cap \mathfrak{p}) = \{2e_1, e_1 + e_2\} \qquad \text{and} \qquad \mathfrak{l} \cap \mathfrak{k} \cap \mathfrak{n} = 0,$$

from which we see, for integers c and d, that

$$
\mathcal{P}(ce_1 + de_2) = \begin{cases} 1 & \text{if } c \ge d \ge 0 \text{ and } c - d \text{ is even} \\ 0 & \text{otherwise.} \end{cases}
$$

Let $\Lambda' = ae_1 + be_2$. Since $W^1 = \{1, s_{e_1-e_2}\}$, Theorem 8.29 says that the multiplicity of the K type Λ' in $A_{\mathfrak{q}}(\lambda)$ is

$$\mathcal{P}((a-1)e_1 + (b-1)e_2) - \mathcal{P}((b-2)e_1 + ae_2).$$

Thus we find

$$
(8.81) \quad
\begin{array}{l}
\Lambda' = ae_1 + be_2 \\
\text{occurs in } A_{\mathfrak{q}}(\lambda) \\
\text{with multiplicity}
\end{array}
= \begin{cases} 1 & \text{if } a \ge b \ge 1 \text{ and } a - b \text{ is even} \\ 0 & \text{otherwise.} \end{cases}
$$

The highest weight $\Lambda' = ae_1 + be_2$ has inner product $a + b$ with $e_1 + e_2$. Since the roots of $\mathfrak{k}$ are orthogonal to $e_1 + e_2$, every weight of the K type Λ' has inner product $a + b$ with $e_1 + e_2$. Consequently $\Lambda = e_1 + e_2$ is a lowest weight for $A_\mathfrak{q}(\lambda)$ relative to inner products with $e_1 + e_2$.

Define a second θ stable parabolic subalgebra $\mathfrak{q}' = \mathfrak{l}' \oplus \mathfrak{u}'$ by

$$\Delta^+(\mathfrak{l}', \mathfrak{h}') = \{e_1 - e_2\} \qquad \text{and} \qquad \Delta(\mathfrak{u}') = \{2e_1, e_1 + e_2, 2e_2\}.$$

(In other words, $\mathfrak{l}' = \mathfrak{k}$.) The members of $\Delta(\mathfrak{u}')$ have positive inner product with $\Lambda = e_1 + e_2$, and it follows that $\bar{\mathfrak{u}}'$ annihilates the K type Λ of $A_\mathfrak{q}(\lambda)$.

For the subalgebra $\mathfrak{q}'$, we have

$$\delta = 2e_1 + e_2, \qquad \delta(\mathfrak{u}') = \tfrac{3}{2}e_1 + \tfrac{3}{2}e_2, \qquad 2\delta(\mathfrak{u}' \cap \mathfrak{p}) = 3e_1 + 3e_2.$$

Define $\lambda' = -2e_1 - 2e_2$, so that

$$\lambda' + \delta(\mathfrak{u}') = -\tfrac{1}{2}e_1 - \tfrac{1}{2}e_2 \qquad \text{and} \qquad \lambda' + 2\delta(\mathfrak{u}' \cap \mathfrak{p}) = e_1 + e_2.$$

Since $\lambda' + \delta(\mathfrak{u}')$ is not dominant, it appears that we do not have an applicable vanishing theorem. However, $S = 0$ here, and we do not need such a theorem. Since $S = 0$, the Euler-Poincaré formula (Theorem 5.64) for multiplicities of K types involves no cancellation. Thus the formula in Theorem 8.29 is applicable with $W^1 = \{1\}$. If $\Lambda' = a'e_1 + b'e_2$, then

(8.82) $\qquad \begin{array}{l} \Lambda' = a'e_1 + b'e_2 \\ \text{occurs in } A_{\mathfrak{q}'}(\lambda') \\ \text{with multiplicity} \end{array} = \mathcal{P}'((a' - 1)e_1 + (b' - 1)e_2)$

where $\mathcal{P}'$ is the partition function appropriate to $\mathfrak{q}'$. The relevant roots are those in

$$\Delta(\mathfrak{u}' \cap \mathfrak{p}) = \{2e_1, e_1 + e_2, 2e_2\}.$$

Now $S^1(\mathfrak{u}' \cap \mathfrak{p})$ is irreducible under K and has weights $2e_1, e_1 + e_2, 2e_2$. Since the inner product of a weight of $S^n(\mathfrak{u}' \cap \mathfrak{p})$ with $e_1 + e_2$ exceeds 2 if $n > 1$, we have

$$\mathcal{P}'(2e_1) = 1 \qquad \text{and} \qquad \mathcal{P}'(e_1 + e_2) = \mathcal{P}'(2e_2) = 0.$$

From (8.82), we see that the K type $\Lambda' = 2e_1 + 2e_2$ does not occur in $A_{\mathfrak{q}'}(\lambda')$. But (8.81) shows that it does occur in $A_\mathfrak{q}(\lambda)$.

Thus $A_\mathfrak{q}(\lambda)$ has a K type that does not occur in $A_{\mathfrak{q}'}(\lambda')$, and no $U(\mathfrak{g})$ map can carry $A_{\mathfrak{q}'}(\lambda')$ onto $A_\mathfrak{q}(\lambda)$.

The module $A_{q'}(\lambda')$ shares with $A_q(\lambda)$ the property that $\Lambda = e_1 + e_2$ is a lowest weight relative to inner products with $e_1 + e_2$. This fact follows immediately from (8.82) and the orthogonality of $e_1 + e_2$ to the roots of $\mathfrak{k}$. Consequently $\bar{\mathfrak{u}}'$ annihilates the K type Λ of $A_{q'}(\lambda')$.

Define
$$M = \operatorname{ind}_{\bar{\mathfrak{q}}'}^{\mathfrak{g}}(\mathbb{C}_{e_1+e_2}).$$

By Frobenius reciprocity (Propositions 2.34 and 2.33),

(8.83) $\operatorname{Hom}_{\mathfrak{g}}(M, V) \cong \operatorname{Hom}_{\bar{\mathfrak{q}}'}(\mathbb{C}_{e_1+e_2}, V)$

for any $U(\mathfrak{g})$ module V.

Let us observe that $M \cong A_{q'}(\lambda')$ as a $U(\mathfrak{g})$ module. In fact, $L' = K$ here, and thus
$$A_{q'}(\lambda') = \mathcal{L}_0(\mathbb{C}_{\lambda'}) = \Pi_{\mathfrak{g},K}^{\mathfrak{g},K}(\operatorname{ind}_{\bar{\mathfrak{q}}',K}^{\mathfrak{g},K}(\mathbb{C}_{\lambda'}^{\#}))$$
$$\cong \operatorname{ind}_{\bar{\mathfrak{q}}',K}^{\mathfrak{g},K}(\mathbb{C}_{\lambda'}^{\#}) \cong \operatorname{ind}_{\bar{\mathfrak{q}}',K}^{\mathfrak{g},K}(\mathbb{C}_{e_1+e_2}).$$

As a $U(\mathfrak{g})$ module, the right side is isomorphic to $U(\mathfrak{g}) \otimes_{U(\bar{\mathfrak{q}}')} \mathbb{C}_{e_1+e_2}$, which is just M.

If $V = A_q(\lambda)$ in (8.83), then the K type $\Lambda = e_1 + e_2$ provides a nonzero member of the right side of (8.83). Hence we get a nonzero $U(\mathfrak{g})$ map of $A_{q'}(\lambda')$ into $A_q(\lambda)$. We have seen that no $U(\mathfrak{g})$ map can carry $A_{q'}(\lambda')$ onto $A_q(\lambda)$, and it follows that $A_q(\lambda)$ is reducible.

It is instructive to see how the hypothesis of Theorem 8.31 breaks down for this nonabelian $\bar{\mathfrak{q}}$. In fact, it fails for $n = 1$. For each root α, let $\mathfrak{g}_\alpha$ be the root space for the root α. Then $\mathfrak{g}/\bar{\mathfrak{q}}$ has a filtration by $(\bar{\mathfrak{q}}, L')$ modules
$$\mathfrak{g}/\bar{\mathfrak{q}} \supseteq \mathfrak{g}/(\bar{\mathfrak{q}} + \mathfrak{g}_{e_1-e_2} + \mathfrak{g}_{e_1+e_2}) \supseteq 0$$

with consecutive quotients that are irreducible $(\bar{\mathfrak{q}}, L')$ modules with sets of weights
$$\{e_1 - e_2, e_1 + e_2\} \qquad \text{and} \qquad \{2e_1\},$$

respectively, and with $\bar{\mathfrak{u}}$ acting trivially. Let us write $N_{e_1+e_2}$ for the $(\bar{\mathfrak{q}}, L')$ module in which $\bar{\mathfrak{u}}$ acts trivially and $(\mathfrak{l}, L')$ acts with highest weight $e_1 + e_2$. To the exact sequence
$$0 \longrightarrow N_{e_1+e_2} \longrightarrow \mathfrak{g}/\bar{\mathfrak{q}} \longrightarrow \mathbb{C}_{2e_1} \longrightarrow 0$$

in $\mathcal{C}(\bar{\mathfrak{q}}, L')$, we apply the functor $I_{\bar{\mathfrak{q}},L'}^{\mathfrak{g},K'}$. The first part of the long exact sequence of (C.36) gives an exact sequence
$$0 \longrightarrow I_{\bar{\mathfrak{q}},L'}^{\mathfrak{g},K'}(N_{e_1+e_2}) \longrightarrow I_{\bar{\mathfrak{q}},L'}^{\mathfrak{g},K'}(\mathfrak{g}/\bar{\mathfrak{q}})$$
$$\longrightarrow I_{\bar{\mathfrak{q}},L'}^{\mathfrak{g},K'}(\mathbb{C}_{2e_1}) \longrightarrow (I_{\bar{\mathfrak{q}},L'}^{\mathfrak{g},K'})^1(N_{e_1+e_2}).$$

From the algebraic Bott-Borel-Weil Theorem (Theorem 4.155) and Kostant's Theorem (Theorem 4.139), we readily find that the rightmost term is 0. The second and fourth terms may be computed from the algebraic Borel-Weil Theorem (Theorem 4.83e) as the irreducible $(\mathfrak{g}, K')$ modules $V_{e_1+e_2}$ and V_{2e_1} with highest weights e_1+e_2 and $2e_1$, respectively. Thus our exact sequence reads

$$0 \longrightarrow V_{e_1+e_2} \longrightarrow I_{\bar{\mathfrak{q}},L'}^{\mathfrak{g},K'}(\mathfrak{g}/\bar{\mathfrak{q}}) \longrightarrow V_{2e_1} \longrightarrow 0.$$

Since short exact sequences split in $\mathcal{C}(\mathfrak{g}, K')$,

$$I_{\bar{\mathfrak{q}},L'}^{\mathfrak{g},K'}(\mathfrak{g}/\bar{\mathfrak{q}}) \cong V_{e_1+e_2} \oplus V_{2e_1}.$$

The first representation on the right side is the representation of $\mathfrak{sp}(2, \mathbb{C})$ as $\mathfrak{so}(5, \mathbb{C})$, and the second one is the adjoint. The map Φ_1 of (8.30c) is a $U(\mathfrak{g})$ map from the 10-dimensional $\mathfrak{g}$ into the 15-dimensional $I_{\bar{\mathfrak{q}},L'}^{\mathfrak{g},K'}(\mathfrak{g}/\bar{\mathfrak{q}})$ and cannot be onto.

EXAMPLE 3. Let $G = SO_0(5, 4)$. We use a maximal abelian subspace $\mathfrak{t}_0$ of $\mathfrak{k}_0$ as our θ stable Cartan subalgebra of $\mathfrak{g}_0$. The root system $\Delta(\mathfrak{g}, \mathfrak{t})$ is of type B_4. Numbering the indices suitably and otherwise using standard notation, we can arrange that the simple roots are

$$
\begin{array}{ll}
e_1 - e_2 & \text{noncompact} \\
e_2 - e_3 & \text{compact} \\
e_3 - e_4 & \text{noncompact} \\
e_4 & \text{noncompact.}
\end{array}
$$

The simple roots with which we define L are $e_1 - e_2, e_2 - e_3, e_3 - e_4$. Then we have

$$
\begin{aligned}
\Delta_K^+ &= \{e_1 \pm e_4, e_2 \pm e_3, e_3\} \\
\Delta(\mathfrak{u} \cap \mathfrak{k}) &= \{e_1 + e_4, e_2 + e_3, e_2, e_3\} \\
\Delta(\mathfrak{u} \cap \mathfrak{p}) &= \{e_1 + e_2, e_1 + e_3, e_2 + e_4, e_3 + e_4, e_1, e_4\}.
\end{aligned}
$$

Hence $S = 4$. We put

$$\lambda = -\delta(\mathfrak{u}) = -2e_1 - 2e_2 - 2e_3 - 2e_4.$$

Then $\lambda + \delta(\mathfrak{u})$ satisfies the hypothesis of Theorem 5.109.

The $(\mathfrak{l}, L \cap K)$ module $Z = \mathbb{C}_\lambda$ has the single $L \cap K$ type λ, and

$$\Lambda = \lambda + 2\delta(\mathfrak{u} \cap \mathfrak{p}) = e_1 + e_4$$

is Δ_K^+ dominant. Therefore the K type Λ is in the bottom layer and must occur in $A_q(\lambda)$ with multiplicity 1, by Theorem 5.80a. (It is possible also to see this fact directly from Theorem 8.29.)

Define $\Lambda' = 2e_1 + e_4$. Applying Theorem 8.29, we shall show that the K type Λ' occurs in $A_q(\lambda)$ with multiplicity 1. Thus suppose the sth term of the formula in the proposition is nonzero. Then we have

$$s(\Lambda' + \delta_K) - (\Lambda + \delta_K) = \sum_{\beta \in \Delta(\mathfrak{u} \cap \mathfrak{p})} n_\beta \beta \qquad \text{with } n_\beta \geq 0.$$

From Cartan-Weyl theory

$$(\Lambda' + \delta_K) - s(\Lambda' + \delta_K) = \sum_{\gamma \in \Delta_K^+} k_\gamma \gamma \qquad \text{with } k_\gamma \geq 0.$$

Adding these equations, we obtain

$$e_1 = \sum_{\gamma \in \Delta_K^+} k_\gamma \gamma + \sum_{\beta \in \Delta(\mathfrak{u} \cap \mathfrak{p})} n_\beta \beta.$$

Let us compute the inner product of both sides with $e_1 + e_2 + e_3 + e_4$. The left side gives 1, and each γ or β gives 0, 1, or 2. Only the 0's and 1's are eligible to contribute, and thus the k_γ's and n_β's are 0 for roots $e_i + e_j$. Hence we have

$$e_1 = a(e_1 - e_4) + b(e_2 - e_3) + ce_1 + de_4 + he_3$$

with a, b, c, d, h nonnegative integers and with $c + d + h = 1$. Solving, we find the two solutions

$$c = 1 \quad \text{and} \quad a = b = d = h = 0$$
$$a = d = 1 \quad \text{and} \quad b = c = h = 0.$$

These translate into

$$\sum k_\gamma \gamma = 0 \quad \text{and} \quad \sum n_\beta \beta = e_1$$
$$\sum k_\gamma \gamma = e_1 - e_4 \quad \text{and} \quad \sum n_\beta \beta = e_4.$$

The element s in the two cases must be $s = 1$ and $s = s_{e_1 - e_4}$, respectively. In the second alternative, $\sum k_\gamma \gamma = e_1 - e_4$ means that $s_{e_1 - e_4}(\Lambda' + \delta_K) = \Lambda' + \delta_K - (e_1 - e_4)$. But we readily check that this is not the case. Thus the only contribution comes from $s = 1$. For $s = 1$, we have

$s(\Lambda' + \delta_K) - (\Lambda + \delta_K) = e_1$, and e_1 occurs once as a highest weight in an $S^n(\mathfrak{u} \cap \mathfrak{p})$. We conclude that the K type Λ' occurs with multiplicity 1 in $A_\mathfrak{q}(\lambda)$.

In §X.2 we shall introduce a notion of "minimal K type," and Proposition 10.24 will show that $A_\mathfrak{q}(\lambda)$ has minimal K type Λ. From the theory of minimal K types, one can prove that any irreducible $(\mathfrak{g}, K)$ module with minimal K type Λ embeds in a certain series of parabolically induced representations. For the latter, one can compute the multiplicity of the K type Λ' and see that it is 0. Then it follows that $A_\mathfrak{q}(\lambda)$ is reducible. See the Notes for more details.

EXAMPLE 4 (Schlichtkrull). Let $G = Sp(5, 2)$. We give an example of a nonzero $A_\mathfrak{q}(\lambda)$ for which $\Lambda = \lambda + 2\delta(\mathfrak{u} \cap \mathfrak{p})$ is not Δ_K^+ dominant (and hence the bottom layer is empty).

We take the Cartan subalgebra $\mathfrak{h}_0$ to be in $\mathfrak{k}_0$. Writing $\Pi(-)$ for sets of simple roots, we have as usual

$$\Pi(\mathfrak{g}, \mathfrak{h}) = \{e_1 - e_2, \ e_2 - e_3, \ e_3 - e_4, \ e_4 - e_5, \ e_5 - e_6, \ e_6 - e_7, \ 2e_7\}.$$

The real form $\mathfrak{sp}(5, 2)$ results if we declare that $e_2 - e_3$ and $e_4 - e_5$ are noncompact and that the other simple roots are compact. We take the corresponding group to be simply connected, so that $G = Sp(5, 2)$. Then

$$\Pi(\mathfrak{k}, \mathfrak{h}) = \{e_1 - e_2, \ e_2 - e_5, \ e_5 - e_6, \ e_6 - e_7, \ 2e_7; \ \ e_3 - e_4, \ 2e_4\}$$

and $K = Sp(5) \times Sp(2)$. Define

$$\Pi(\mathfrak{l}, \mathfrak{h}) = \{e_1 - e_2; \ \ e_3 - e_4, \ e_4 - e_5, \ e_5 - e_6, \ e_6 - e_7, \ 2e_7\},$$

so that $\mathfrak{l}_0 = \mathfrak{u}(2) \oplus \mathfrak{sp}(3, 2)$. Then

$$\Pi(\mathfrak{l} \cap \mathfrak{k}, \mathfrak{h}) = \{e_1 - e_2; \ \ e_3 - e_4, \ 2e_4; \ \ e_5 - e_6, \ e_6 - e_7, \ 2e_7\}$$

and

$$\Delta(\mathfrak{u}) = \{e_1 \pm e_j \text{ with } j \geq 3\} \cup \{e_2 \pm e_j \text{ with } j \geq 3\} \cup \{2e_1, 2e_2, e_1 + e_2\}.$$

From this information we can read off

$$\delta = (7, 6; \ 5, 4; \ 3, 2, 1)$$
$$\delta_K = (5, 4; \ 2, 1; \ 3, 2, 1)$$
$$\delta_{L \cap K} = (\tfrac{1}{2}, -\tfrac{1}{2}; \ 2, 1; \ 3, 2, 1)$$
$$\delta(\mathfrak{u} \cap \mathfrak{k}) = (\tfrac{9}{2}, \tfrac{9}{2}; \ 0, 0; \ 0, 0, 0)$$
$$\delta(\mathfrak{u}) = (\tfrac{13}{2}, \tfrac{13}{2}; \ 0, 0; \ 0, 0, 0)$$
$$\delta(\mathfrak{u} \cap \mathfrak{p}) = (2, 2; \ 0, 0; \ 0, 0, 0).$$

We define
$$\lambda = (-5, -5;\ 0, 0;\ 0, 0, 0).$$

Since
$$\lambda + \delta(\mathfrak{u}) = (\tfrac{3}{2}, \tfrac{3}{2};\ 0, 0;\ 0, 0, 0),$$

$\lambda + \delta(\mathfrak{u})$ is $\Delta^+(\mathfrak{g}, \mathfrak{h})$ dominant. Therefore the hypotheses of Theorems 5.109 and 8.29 are satisfied, and we conclude (with $S = 15$) that $\mathcal{L}_j(\mathbb{C}_\lambda)$ and $\mathcal{R}^j(\mathbb{C}_\lambda)$ vanish for $j \neq S$ and that $\mathcal{L}_S(\mathbb{C}_\lambda) \cong \mathcal{R}^S(\mathbb{C}_\lambda)$. Since

$$\Lambda = \lambda + 2\delta(\mathfrak{u} \cap \mathfrak{p}) = (-1, -1;\ 0, 0;\ 0, 0, 0),$$

Λ is not Δ_K^+ dominant. Since Λ is the only candidate for the highest weight of a K type in the bottom layer, the bottom layer is empty.

We shall use Theorem 8.29 to see that the trivial representation of K occurs in $A_q(\lambda)$ with multiplicity 1. According to that theorem, we are to consider

$$(8.84) \quad s\delta_K - (\Lambda + \delta_K) = s(5, 4;\ 2, 1;\ 3, 2, 1) - (4, 3;\ 2, 1;\ 3, 2, 1)$$

with s in the subset W^1 of the Weyl group of K. The special condition for s to be in W^1 is that

$$\alpha \in \Delta^+(\mathfrak{l} \cap \mathfrak{k}) \qquad \text{implies} \qquad s^{-1}\alpha > 0.$$

We readily check that this condition implies

$$(8.85) \quad \begin{array}{l} s^{-1} \text{ fixes } 2e_3 \text{ and } 2e_4 \\ s^{-1} \text{ maps } 2e_5,\ 2e_6,\ 2e_7 \text{ to a subset of } 2e_1,\ 2e_2,\ 2e_5,\ 2e_6,\ 2e_7 \\ \qquad\qquad\qquad \text{with increasing indices} \\ s^{-1} \text{ maps } 2e_1 \text{ and } 2e_2 \text{ within } \pm 2e_1,\ \pm 2e_2,\ \pm 2e_5,\ \pm 2e_6,\ \pm 2e_7. \end{array}$$

Also for the partition function $\mathcal{P}(\nu)$ to be nonzero, ν must be a sum of members of $\Delta(\mathfrak{u} \cap \mathfrak{p})$. Since

$$(8.86) \qquad \Delta(\mathfrak{u} \cap \mathfrak{p}) = \{e_1 \pm e_3,\ e_1 \pm e_4,\ e_2 \pm e_3,\ e_2 \pm e_4\},$$

ν must be of the form

$$\nu = (a_1, a_2;\ a_3, a_4;\ 0, 0, 0)$$

with $a_1 \geq 0$ and $a_2 \geq 0$. This particular ν can occur in $S^n(\mathfrak{u} \cap \mathfrak{p})$ only for $n = a_1 + a_2$.

With v as in (8.84), the only possibilities with $a_1 \geq 0$ and $a_2 \geq 0$ are

$$s(5, 4; \ -) = \begin{cases} (5, 4; \ -) \\ (5, 3; \ -) \\ (4, 3; \ -), \end{cases}$$

and (8.85) forces

$$s(5, 4; \ 2, 1; \ 3, 2, 1) = \begin{cases} (5, 4; \ 2, 1; \ 3, 2, 1) \\ (5, 3; \ 2, 1; \ 4, 2, 1) \\ (4, 3; \ 2, 1; \ 5, 2, 1) \end{cases}$$

in the respective cases. Then v in (8.84) is

$$v = \begin{cases} (1, 1; \ 0, 0; \ 0, 0, 0) \\ (1, 0; \ 0, 0; \ 1, 0, 0) \\ (0, 0; \ 0, 0; \ 2, 0, 0). \end{cases}$$

The last two of these have something nonzero in the last three entries and are not sums from $\Delta(\mathfrak{u} \cap \mathfrak{p})$. The first of these corresponds to $s = 1$, and thus the multiplicity of the trivial K type in $A_{\mathfrak{q}}(\lambda)$ equals the multiplicity of the $L \cap K$ type

(8.87) $(1, 1; \ 0, 0; \ 0, 0, 0)$

in $S^2(\mathfrak{u} \cap \mathfrak{p})$.

From (8.86) it is apparent that the representation of

$$\mathfrak{u}(2) \oplus \mathfrak{sp}(2) \oplus \mathfrak{sp}(3)$$

on $S^1(\mathfrak{u} \cap \mathfrak{p}) = \mathfrak{u} \cap \mathfrak{p}$ is $\mathbb{C}^2 \otimes \mathbb{C}^4 \otimes \mathbb{C}$. We can drop the trivial $\mathfrak{sp}(3)$ action on the factor $\mathbb{C}$. The symmetric 2-tensors from $\mathfrak{u} \cap \mathfrak{p}$ then decompose as

(8.88) $S^2(\mathfrak{u} \cap \mathfrak{p}) \cong (S^2(\mathbb{C}^2) \otimes S^2(\mathbb{C}^4)) \oplus (\wedge^2(\mathbb{C}^2) \otimes \wedge^2(\mathbb{C}^4)).$

In the second summand, $\wedge^2(\mathbb{C}^4)$ splits as the sum of the trivial representation and a 5-dimensional irreducible representation of $\mathfrak{sp}(2)$, and thus $S^2(\mathfrak{u} \cap \mathfrak{p})$ contains $\wedge^2(\mathbb{C}^2) \otimes \mathbb{C}$. This has highest weight (8.87), and thus the trivial K type does occur in $A_{\mathfrak{q}}(\lambda)$. It is easy to check that (8.88) does not yield a second occurrence of the highest weight (8.87), and thus the trivial K type occurs exactly once in $A_{\mathfrak{q}}(\lambda)$.

CHAPTER IX

UNITARIZABILITY THEOREM

The Unitarizability Theorem gives conditions under which the existence of a positive invariant Hermitian form for a module Z implies the existence of a positive invariant Hermitian form for the cohomologically induced module $\mathcal{L}_S(Z)$. The important condition is the dominance of a certain shift of the infinitesimal character of Z. An improved version of the theorem is available when Z is one dimensional.

The proof for connected K uses calculations with a variation of a formal K character that is called the signature character. In a step-by-step fashion, one computes enough of the signature character of $\mathcal{L}_S(Z)$ relative to the Shapovalov form so that it is apparent that the signature character matches the formal character.

1. Statement of Theorem

This chapter contains the fourth of our five main theorems. The Unitarizability Theorem gives a sufficient condition for cohomological induction to carry unitary representations to unitary representations. Actually the appropriate notion of unitarity is the one for $(\mathfrak{g}, K)$ modules obtained by differentiation of the unitarity condition at the group level: Assume (as is the case for reductive pairs) that $\mathfrak{g}$ is the complexification of an $\mathrm{Ad}(K)$ invariant real form $\mathfrak{g}_0$ such that $\mathfrak{k}_0 \subseteq \mathfrak{g}_0$. A sesquilinear form $\langle \cdot, \cdot \rangle$ on a $(\mathfrak{g}, K)$ module V is **invariant** if $\langle Xv, v' \rangle = -\langle v, Xv' \rangle$ for all $X \in \mathfrak{g}_0$ and $\langle kv, kv' \rangle = \langle v, v' \rangle$ for all $k \in K$. We say that V is **infinitesimally unitary** if V possesses a positive definite Hermitian form that is invariant.

If V is an admissible infinitesimally unitary $(\mathfrak{g}, K)$ module, and if W is a $(\mathfrak{g}, K)$ submodule, then $V = W \oplus W^\perp$ and $W^\perp$ is a $(\mathfrak{g}, K)$ submodule. To show this, if P_τ denotes the projection on the K type τ, then the admissibility forces $P_\tau V = P_\tau W \oplus P_\tau W^\perp$, and then it follows that $V = W \oplus W^\perp$. An easy computation shows that $W^\perp$ is a $(\mathfrak{g}, K)$ submodule.

Throughout the chapter the setting will be as in Chapter V: We let $(\mathfrak{g}, K)$ be a reductive pair and $\mathfrak{q} = \mathfrak{l} \oplus \mathfrak{u}$ be a θ stable parabolic subalgebra. The corresponding Levi subgroup is denoted L, and $(\mathfrak{q}, L \cap K)$ is a θ stable parabolic subpair. According to the results of Chapter VI (Theorem 6.34 when $K = (L \cap K)K_0$, Theorem 6.55 in general), an invariant sesquilinear form $\langle \cdot, \cdot \rangle_L$ on Z canonically determines an invariant sesquilinear form $\langle \cdot, \cdot \rangle_G$ on $\mathcal{L}_S(Z)$, called the Shapovalov form. Moreover, if $\langle \cdot, \cdot \rangle_L$ is Hermitian, then so is $\langle \cdot, \cdot \rangle_G$.

Theorem 9.1 (Unitarizability Theorem). Let $(\mathfrak{g}, K)$ be a reductive pair, let $\mathfrak{q} = \mathfrak{l} \oplus \mathfrak{u}$ be a θ stable parabolic subalgebra, let $\mathfrak{h}$ be a Cartan subalgebra of $\mathfrak{l}$, and introduce a positive system $\Delta^+(\mathfrak{g}, \mathfrak{h})$ such that $\Delta(\mathfrak{u}) \subseteq \Delta^+(\mathfrak{g}, \mathfrak{h})$. Suppose Z is an admissible $Z(\mathfrak{l})$ finite $(\mathfrak{l}, L \cap K)$ module such that each parameter $\lambda \in \mathfrak{h}^*$ of a $Z(\mathfrak{l})$ primary component satisfies

$$\frac{2\langle \lambda + \delta(\mathfrak{u}), \alpha \rangle}{|\alpha|^2} \notin (-\infty, -1]$$

for all $\alpha \in \Delta(\mathfrak{u})$. (That is, no $\alpha \in \Delta(\mathfrak{u})$ is to have the property that $2\langle \lambda + \delta(\mathfrak{u}), \alpha \rangle / |\alpha|^2$ is a real number ≤ -1.) If $\langle \cdot, \cdot \rangle_L$ is a positive definite invariant Hermitian form on Z, then the corresponding Shapovalov form $\langle \cdot, \cdot \rangle_G$ is a positive definite invariant Hermitian form on $\mathcal{L}_S(Z)$.

REMARKS. Because $\Delta(\mathfrak{u})$ is invariant under $W(\mathfrak{l}, \mathfrak{h})$ and because $\delta(\mathfrak{u})$ is fixed by $W(\mathfrak{l}, \mathfrak{h})$, the condition on the infinitesimal characters of Z does not depend on the choice of a representative parameter λ. Similarly it does not depend on the choice of Cartan subalgebra $\mathfrak{h}$.

Chapter VI, especially §4 and (6.45b), provides formulas for the Shapovalov form. We let $\mathfrak{c} = \mathfrak{k}/(\mathfrak{l} \cap \mathfrak{k}) \cong (\mathfrak{u} \cap \mathfrak{k}) \oplus (\bar{\mathfrak{u}} \cap \mathfrak{k})$, and we fix a choice of ε_0 in $\bigwedge^{2S} \mathfrak{c}$ such that $\varepsilon_0(\xi \wedge \bar{\xi}) > 0$ if ξ is nonzero in $\bigwedge^S (\bar{\mathfrak{u}} \cap \mathfrak{k}) \subseteq \bigwedge^S \mathfrak{c}$. We may compute $\mathcal{L}_S(Z)$ from the complex

$$(9.2) \quad 0 \longrightarrow R(K) \otimes_{L \cap K} (\bigwedge^0 \mathfrak{c} \otimes_{\mathbb{C}} \text{ind}_{\bar{\mathfrak{q}}, L \cap K}^{\mathfrak{g}, L \cap K}(Z_{\bar{\mathfrak{q}}}^{\#}))$$

$$\longrightarrow R(K) \otimes_{L \cap K} (\bigwedge^1 \mathfrak{c} \otimes_{\mathbb{C}} \text{ind}_{\bar{\mathfrak{q}}, L \cap K}^{\mathfrak{g}, L \cap K}(Z_{\bar{\mathfrak{q}}}^{\#})) \longrightarrow \cdots,$$

according to (3.27). Recall from (6.44) that we are to interpret the tensor products over $L \cap K$ in this complex as submodules, not quotients, of the corresponding tensor products over $\mathbb{C}$. Anyway, a pure tensor in $\text{ind}_{\bar{\mathfrak{q}}, L \cap K}^{\mathfrak{g}, L \cap K}(\#)$ is of the form $u \otimes z$ with $u \in U(\mathfrak{g})$ and $z \in Z^{\#}$. In this notation the form $\langle \cdot, \cdot \rangle_G$ is given on the degree S term $R(K) \otimes_{L \cap K} (\bigwedge^S \mathfrak{c} \otimes_{\mathbb{C}} \text{ind}_{\bar{\mathfrak{q}}, L \cap K}^{\mathfrak{g}, L \cap K}(\#))$ of the complex by

$$(9.3) \quad \langle T \otimes \xi \otimes (u \otimes z), T' \otimes \xi' \otimes (u' \otimes z') \rangle_G$$

$$= \varepsilon_0(\xi \wedge \bar{\xi}') \langle \bar{T}', T \rangle \langle \mu'_{\bar{\mathfrak{u}}}(\bar{u}''u)z, z' \rangle_L^{\#},$$

where T and T' are in $R(K)$, ξ and ξ' are in $\bigwedge^S \mathfrak{c}$, u and u' are in $U(\mathfrak{g})$, and z and z' are in $Z^\# = Z \otimes_{\mathbb{C}} \bigwedge^{\mathrm{top}} \mathfrak{u}$. The function $\mu'_{\bar{\mathfrak{u}}} : U(\mathfrak{g}) \to U(\mathfrak{l})$ is defined after Lemma 4.123, and the expression $\langle \bar{T}', T \rangle$ is given by

$$\langle \bar{T}', T \rangle = \int_K \overline{f'(k)} f(k) \, dk \qquad \text{if } T = f \, dk \text{ and } T' = f' \, dk.$$

Theorem 9.1 easily reduces to the case that K is connected. In fact, Lemma 6.59 shows that the Shapovalov form may be computed in stages and that the second stage preserves positive definiteness. More precisely, (6.64) allows us to obtain $\mathcal{L}_S(Z)$ from a composition

$$\mathcal{L}_S(Z) \cong P^{\mathfrak{g},K}_{\mathfrak{g},(L \cap K)K_0}(\mathcal{L}_{S,1}(Z)),$$

where $\mathcal{L}_{S,1}$ is the version of $\mathcal{L}_S$ for the pair $(\mathfrak{g}, (L \cap K)K_0)$; the lemma says that $\langle \cdot, \cdot \rangle_G$ is obtained from the Shapovalov form $\langle \cdot, \cdot \rangle_{G_1}$ on $\mathcal{L}_{S,1}(Z)$ by a construction that preserves positive definiteness. Let us write $\mathcal{L}_{S,0}$ for the version of $\mathcal{L}_S$ for $(\mathfrak{g}, K_0)$. Proposition 5.14c (in combination with (C.27) and (C.28)) shows that $\mathcal{L}_{S,1}(Z)$ and $\mathcal{L}_{S,0}(Z)$ are intertwined by forgetful functors:

$$\mathcal{F}^{\mathfrak{g},K_0}_{\mathfrak{g},(L \cap K)K_0} \circ \mathcal{L}_{S,1}(Z) \cong \mathcal{L}_{S,0} \circ \mathcal{F}^{\mathfrak{l},L \cap K_0}_{\mathfrak{l},L \cap K}(Z).$$

This isomorphism respects Shapovalov forms. In fact, we have only to compare (9.2) and (9.3) for $(\mathfrak{g}, (L \cap K)K_0)$ with (9.2) and (9.3) for $(\mathfrak{g}, K_0)$, using the isomorphism (5.15a). Because the question of positive definiteness is unaffected by forgetful functors, it suffices to prove Theorem 9.1 for $\mathcal{L}_{S,0}(Z)$.

For the rest of this chapter, we may therefore assume that K is connected. In view of Lemma 5.10, L is connected also.

By Corollary 7.223, Z has finite length as an $(\mathfrak{l}, L \cap K)$ module. According to the second paragraph of the present section, if Z' is an $(\mathfrak{l}, L \cap K)$ submodule, then Z' has a complementary subspace orthogonal to Z' that is an $(\mathfrak{l}, L \cap K)$ submodule. It follows that Z is the orthogonal direct sum of finitely many irreducible $(\mathfrak{l}, L \cap K)$ submodules. The functor $\mathcal{L}_S$ respects direct sums, and (9.3) makes it clear that the Shapovalov form respects the orthogonality. Therefore it is enough to prove Theorem 9.1 when Z is irreducible.

The condition on λ in Theorem 9.1 remains in force if we change to a different θ stable Cartan subalgebra. We take $\mathfrak{h}_0$ to be maximally compact in $\mathfrak{l}_0$ and write $\mathfrak{h}_0 = \mathfrak{t}_0 \oplus \mathfrak{a}_0$ for its decomposition according to $\mathfrak{g}_0 = \mathfrak{k}_0 \oplus \mathfrak{p}_0$. Let T be the usual group defined by (4.36c) and having Lie algebra $\mathfrak{t}_0$; Lemma 4.43d says that T is connected.

2. Signature Character and Examples

The subtle point about Theorem 9.1 is that the form described by (9.3) turns out to be positive definite on the cohomology even though the factor $\langle \mu'_{\bar{u}}(\bar{u}''u)z, z'\rangle^{\#}_L$ makes it indefinite on the complex. We shall use a kind of calculus of signatures to keep track of the cancellation and to detect the final positivity.

The tool that establishes such a scheme is the "signature character" of an admissible representation of a compact group. Let B be a compact Lie group. Recall from Proposition 6.12 that if F is a finite-dimensional B module isotypic of type $\tau \in \widehat{B}$ and if F possesses a B invariant Hermitian form $\langle \cdot, \cdot \rangle$, then F has an orthogonal B invariant decomposition $F = F_+ \oplus F_0 \oplus F_-$ such that $\langle \cdot, \cdot \rangle$ is positive on F_+, 0 on F_0, and negative on F_-. Although F_+ and F_- are not canonical, the triple of multiplicities (n_+, n_0, n_-) of τ in F_+, F_0, and F_- is well defined. We continue to call it the **signature** of $\langle \cdot, \cdot \rangle$ on F with respect to τ.

Let $\mathcal{A}(B)$ be the small category of admissible locally B finite representations of B. In Proposition 1.63 we proved that the Grothendieck group $K(\mathcal{A}(B))$ is isomorphic to $\mathbb{Z}^{\widehat{B}}$. Namely if V is a member of $\mathcal{A}(B)$, we associate to V the element

$$\Theta_B(V) = \{n_\tau\}_{\tau \in \widehat{B}},$$

where n_τ is the multiplicity $[\,V : \tau\,]$. We call $\Theta_B(V)$ the **formal character** of V. If V possesses an invariant Hermitian form $\langle \cdot, \cdot \rangle$, we want to introduce a related object called the "signature character" of V.

First let us observe that if τ and τ' are distinct members of $\widehat{B}$, then the isotypic spaces V_τ and $V_{\tau'}$ are orthogonal. In fact, if W_τ and $W_{\tau'}$ are irreducible subspaces of V_τ and $V_{\tau'}$, respectively, then the restriction of $\langle \cdot, \cdot \rangle$ to $W_\tau \times W_{\tau'}$ is an invariant sesquilinear form. The formulas following (6.9) show how to convert this into a B map $E : W_\tau \to W^h_{\tau'}$. For an irreducible representation W of a compact group, $W \cong W^h$. Hence E intertwines W_τ and $W_{\tau'}$ and must be 0, by Schur's Lemma. Thus $\langle W_\tau, W_{\tau'} \rangle = 0$, and it follows that $\langle V_\tau, V_{\tau'} \rangle = 0$.

For each $\tau \in \widehat{B}$, let $(n^\tau_+, n^\tau_0, n^\tau_-)$ be the signature on V_τ. The **signature character** of V is the member of $K(\mathcal{A}(B))$ given by

$$\Theta_{B,s}(V) = \{n^\tau_+ - n^\tau_-\}_{\tau \in \widehat{B}}.$$

Since the formal character satisfies

$$\Theta_B(V) = \{n^\tau_+ + n^\tau_0 + n^\tau_-\}_{\tau \in \widehat{B}},$$

it follows that

(9.4) $\langle\,\cdot\,,\,\cdot\,\rangle$ is positive definite if and only if $\Theta_{B,s}(V) = \Theta_B(V)$.

Let us introduce the subcategory $\mathcal{F}(B)$ of finite-dimensional V in $\mathcal{A}(B)$. By the same argument as in Proposition 1.63, the Grothendieck group $K(\mathcal{F}(B))$ is isomorphic to the subgroup of members of $\mathbb{Z}^{\widehat{B}}$ that are nonzero only finitely often. The tensor-product operation on $\mathcal{F}(B)$ gives $K(\mathcal{F}(B))$ the structure of a commutative ring with identity. In fact, the linear independence of ordinary characters of inequivalent irreducible representations of B means that we can identify formal characters of members of $\mathcal{F}(B)$ with ordinary characters. The ring structure on the $\mathbb{Z}$ span of ordinary characters corresponds to decompositions of tensor products, and we carry this ring structure to $K(\mathcal{F}(B))$. Thus the ring structure is given by the linear extension of

$$\Theta_B(\tau_1)\Theta_B(\tau_2) = \Theta_B(\tau_1 \otimes_\mathbb{C} \tau_2) \qquad \text{for } \tau_1,\, \tau_2 \in \widehat{B}.$$

In particular,

(9.5) $\Theta_B(F_1)\Theta_B(F_2) = \Theta_B(F_1 \otimes_\mathbb{C} F_2) \qquad \text{for } F_1,\, F_2 \in \mathcal{F}(B).$

Suppose V_1 and V_2 are finite-dimensional representations of B carrying invariant Hermitian forms $\langle\,\cdot\,,\,\cdot\,\rangle_1$ and $\langle\,\cdot\,,\,\cdot\,\rangle_2$. Assume that each $\langle\,\cdot\,,\,\cdot\,\rangle_i$ is positive definite, zero, or negative definite. Then

(9.6) $\Theta_{B,s}(V_i) = \begin{cases} \Theta_B(V_i) & \text{if } \langle\,\cdot\,,\,\cdot\,\rangle_i \text{ is positive} \\ 0 & \text{if } \langle\,\cdot\,,\,\cdot\,\rangle_i \text{ is zero} \\ -\Theta_B(V_i) & \text{if } \langle\,\cdot\,,\,\cdot\,\rangle_i \text{ is negative.} \end{cases}$

The two forms $\langle\,\cdot\,,\,\cdot\,\rangle_1$ and $\langle\,\cdot\,,\,\cdot\,\rangle_2$ yield an invariant Hermitian form on the tensor product, and this form will be positive definite (if the $\langle\,\cdot\,,\,\cdot\,\rangle_i$ are both positive or both negative), zero (if one of the $\langle\,\cdot\,,\,\cdot\,\rangle_i$ is zero), or negative definite (if the $\langle\,\cdot\,,\,\cdot\,\rangle_i$ are definite of opposite signs). In each case we see that

$$\Theta_{B,s}(V_1 \otimes_\mathbb{C} V_2) = \Theta_{B,s}(V_1)\Theta_{B,s}(V_2).$$

Let us combine Proposition 6.12 (to write an arbitrary invariant Hermitian form as a sum of definite forms and zero), the distributive law for tensor products and direct sums, and the additivity formula

(9.7) $\Theta_{B,s}(W_1 \oplus W_2) = \Theta_{B,s}(W_1) + \Theta_{B,s}(W_2).$

From these facts we deduce

(9.8) $\Theta_{B,s}(F_1 \otimes_{\mathbb{C}} F_2) = \Theta_{B,s}(F_1)\Theta_{B,s}(F_2)$ for F_1 and F_2 in $\mathcal{F}(B)$ carrying invariant Hermitian forms.

In the same way we can make $K(\mathcal{A}(B))$ into a $K(\mathcal{F}(B))$ module. For F in $\mathcal{F}(B)$ and V in $\mathcal{A}(B)$, the definition is

(9.9) $\Theta_B(F)\Theta_B(V) = \Theta_B(F \otimes_{\mathbb{C}} V).$

To see that the right side makes sense, we write $V = \bigoplus_{\sigma \in \widehat{B}} V_\sigma$ and

$$[F \otimes_{\mathbb{C}} V : \tau] = \sum_{\sigma \in \widehat{B}} [F \otimes_{\mathbb{C}} V_\sigma : \tau].$$

By the finite-dimensional theory, $[F \otimes_{\mathbb{C}} V_\sigma : \tau]$ is nonzero only if σ contains a constituent of $F^* \otimes_{\mathbb{C}} \tau$, and this happens for only finitely many σ. It follows that $F \otimes_{\mathbb{C}} V$ is admissible and that $\Theta_B(F \otimes_{\mathbb{C}} V)$ is defined.

Lemma 9.10. If F in $\mathcal{F}(B)$ and V in $\mathcal{A}(B)$ carry invariant Hermitian forms and if $F \otimes_{\mathbb{C}} V$ is given the tensor-product form, then

(9.11) $\Theta_{B,s}(F)\Theta_{B,s}(V) = \Theta_{B,s}(F \otimes_{\mathbb{C}} V).$

PROOF. Fix $\tau \in \widehat{B}$, and let S be the set of $\sigma \in \widehat{B}$ such that σ contains a constituent of $F^* \otimes_{\mathbb{C}} \tau$. From what we saw above, the τ component of the two sides of (9.11) is the same for V as it is for the finite-dimensional representation $\bigoplus_{\sigma \in S} V_\sigma$. But (9.8) shows that (9.11) is valid for the tensor product of F and $\bigoplus_{\sigma \in S} V_\sigma$. The lemma follows.

In making calculations with formal characters and signature characters of members of $\mathcal{A}(B)$, it will be convenient to write $\Theta_B(V)$ and $\Theta_{B,s}(V)$, for V in $\mathcal{A}(B)$, as formal sums

$$\Theta_B(V) = \sum_{\tau \in \widehat{B}} n_\tau \Theta_B(\tau)$$

and

$$\Theta_{B,s}(V) = \sum_{\tau \in \widehat{B}} (n_+^\tau - n_-^\tau)\Theta_B(\tau).$$

These may be viewed as ordinary sums if V is in $\mathcal{F}(B)$. We shall also allow into this notation manipulations with tensor products by members of $\mathcal{F}(B)$.

For the next topic in this section we shall study what the signature character of (9.3) looks like in $SU(2)$ and $SU(1,1)$. By way of preliminaries, let us make a calculation in $\mathfrak{sl}(2,\mathbb{C})$ with the function $\gamma'_{\mathfrak{n}-}$ used in defining the Harish-Chandra homomorphism. Define

$$h = \begin{pmatrix} 1 & 0 \\ 0 & -1 \end{pmatrix}, \qquad e = \begin{pmatrix} 0 & 1 \\ 0 & 0 \end{pmatrix}, \qquad f = \begin{pmatrix} 0 & 0 \\ 1 & 0 \end{pmatrix},$$

let $\mathfrak{n} = \mathbb{C}e$, and let $\mathfrak{n}^- = \mathbb{C}f$. We compute $\gamma'_{\mathfrak{n}-}(f^n e^n)$. For example,

$$fffeee = ffefee - ffhee$$
$$= feffee - fhfee - ffhee$$
$$= efffee - hffee - fhfee - ffhee.$$

The first term will map to 0 under $\gamma'_{\mathfrak{n}-}$, and the h in each of the other terms can be moved to the right end, modulo terms $ffee$ whose coefficients total $-(0+2+4)$. With $f^n e^n$, the comparable result is

$$f^n e^n = ef^n e^{n-1} - nf^{n-1}e^{n-1}h - (0+2+4+\cdots+2(n-1))f^{n-1}e^{n-1}$$
$$= ef^n e^{n-1} - nf^{n-1}e^{n-1}(h + (n-1)).$$

We can therefore induct and obtain

$$(9.12) \qquad \gamma'_{\mathfrak{n}-}(f^n e^n) = (-1)^n n!(h + (n-1))(h + (n-2))\cdots(h+1)h.$$

EXAMPLE 1. $G = SU(2)$, $L =$ diagonal, $\mathfrak{q} = \mathfrak{b} = \mathfrak{l} \oplus \mathfrak{n}$, $S = 1$.

Let $Z = \mathbb{C}_N$ with the action of L given by $\begin{pmatrix} e^{i\theta} & 0 \\ 0 & e^{-i\theta} \end{pmatrix} z = e^{iN\theta}z$. Then $Z^\# = \mathbb{C}_{N+2}$, and we may take $\langle z_1, z_2 \rangle_L^\# = z_1\bar{z}_2$. The signature character of $Z^\#$ relative to L is $\Theta_{L,s}(Z^\#) = e^{i(N+2)\theta}$. In constructing $\mathcal{L}_S(Z)$, we first form $\text{ind}_{\mathfrak{b},L}^{\mathfrak{g},L}(Z_{\mathfrak{b}}^\#)$, which is an upside-down Verma module with lowest weight $N + 2$ and infinitesimal character $N + 1$. As an $\mathfrak{n}$ module, this Verma module is just $U(\mathfrak{n}) \otimes_{\mathbb{C}} Z^\#$, and it acquires an L invariant Hermitian form given by

$$\langle u \otimes z, u' \otimes z' \rangle = \langle \gamma'_{\mathfrak{n}-}(\bar{u}'^t u)z, z' \rangle_L^\#.$$

Everything is determined by knowing what happens for $u = u' = e^n$ and $z = z' = 1$. Since $e - f$ and $i(e + f)$ are in $\mathfrak{g}_0 = \mathfrak{su}(2)$, we have $\bar{e} = -f$. Thus $\bar{u}'^t u = f^n e^n$, and (9.12) shows that

$$\langle e^n \otimes 1, e^n \otimes 1 \rangle$$
$$= (-1)^n n!((N+2)+(n-1))((N+2)+(n-2))\cdots((N+2)+1)(N+2).$$

If $N + 1 > -1$ (the dominance condition in Theorem 9.1), the right side has alternating signs as a function of n, and we obtain

$$(9.13) \qquad \Theta_{L,s}(\operatorname{ind}_{\mathfrak{b},L}^{\mathfrak{g},L}(Z_{\mathfrak{b}}^{\#})) = e^{i(N+2)\theta} - e^{i(N+4)\theta} + e^{i(N+6)\theta} - \cdots$$
$$= \frac{e^{i(N+2)\theta}}{1 + e^{2i\theta}},$$

with the factor in the denominator indicating a series that is to be expanded in positive powers of $e^{i\theta}$.

There are two other factors on the right side of (9.3), corresponding to $R(K) \otimes \bigwedge^S \mathfrak{c} = R(G) \otimes \mathfrak{c}$. A summand of $R(G)$ is of the form $\tau_m \otimes \tau_m^*$, where τ_m is a representation of $SU(2)$ with highest weight $m \geq 0$. Since the shape of the Sth term of the complex is $R(G) \otimes_L (\mathfrak{c} \otimes \operatorname{ind}_{\mathfrak{b},L}^{\mathfrak{g},L} Z_{\mathfrak{b}}^{\#})$, the contribution to $\tau_m \otimes \tau_m^*$ comes from the G representation τ_m tensored with the vector space

$$\tau_m^* \otimes_L (\mathfrak{c} \otimes \operatorname{ind}_{\mathfrak{b},L}^{\mathfrak{g},L} Z_{\mathfrak{b}}^{\#}).$$

If the tensor product in the latter case were over $\mathbb{C}$, Lemma 9.10 would tell us to compute its signature character as a product. Since the tensor product is over L, we retain only the term involving the trivial representation of L. Denoting the operation of extracting the part for the trivial representation of L by a subscript 0, we obtain

$$(9.14) \quad \Theta_{G,s}(S^{\text{th}} \text{ term of complex})$$
$$= \sum_{m=0}^{\infty} \left(\Theta_{L,s}(\tau_m^*) \Theta_{L,s}(\mathfrak{c}) \Theta_{L,s}(\operatorname{ind}_{\mathfrak{b},L}^{\mathfrak{g},L} Z_{\mathfrak{b}}^{\#}) \right)_0 \Theta_G(\tau_m).$$

In this expression the factor $\Theta_{L,s}(\operatorname{ind}_{\mathfrak{b},L}^{\mathfrak{g},L} Z_{\mathfrak{b}}^{\#})$ is given by (9.13). The form on τ_m^* corresponds to a summand of the contribution $\langle \bar{T}', T \rangle$ to (9.3) and is positive. Hence the signature character for τ_m^* is just the formal character of the restriction to L:

$$\Theta_{L,s}(\tau_m^*) = e^{im\theta} + e^{i(m-2)\theta} + \cdots + e^{-im\theta} = \frac{e^{-im\theta} - e^{i(m+2)\theta}}{1 - e^{2i\theta}}.$$

As in (9.13), we are to interpret the factor $(1 - e^{2i\theta})^{-1}$ on the right side as $1 + e^{2i\theta} + e^{4i\theta} + \cdots$. The signature character of $\mathfrak{c}$ is less evident. The relevant part of (9.3) is $\varepsilon_0(\xi \wedge \bar{\xi}')$ with $\xi \in \mathfrak{c} \cong \mathfrak{n} \oplus \mathfrak{n}^-$, and the side condition on ε_0 is that $\varepsilon_0(\xi \wedge \bar{\xi}) > 0$ if ξ is nonzero in $\bigwedge^S(\bar{\mathfrak{u}} \cap \mathfrak{k}) = \mathfrak{n}^-$. Thus $\varepsilon_0(f \wedge \bar{f}) > 0$. Also $\varepsilon_0(e \wedge \bar{f}) = \varepsilon_0(f \wedge \bar{e}) = 0$, and

$$\varepsilon_0(e \wedge \bar{e}) = \varepsilon_0(e \wedge (-f)) = -\varepsilon_0(f \wedge (-e)) = -\varepsilon_0(f \wedge \bar{f}) < 0.$$

So

(9.15)
$$\Theta_{L,s}(\mathfrak{c}) = -e^{2i\theta} + e^{-2i\theta}.$$

Therefore (9.14) gives

$\Theta_{G,s}(S^{\text{th}}\text{term of complex})$

$$= \sum_{m=0}^{\infty} \left(\frac{e^{-im\theta} - e^{i(m+2)\theta}}{1 - e^{2i\theta}} \cdot (-e^{2i\theta} + e^{-2i\theta}) \cdot \frac{e^{i(N+2)\theta}}{1 + e^{2i\theta}} \right)_0 \Theta_G(\tau_m)$$

$$= \sum_{m=0}^{\infty} \left((e^{-im\theta} - e^{i(m+2)\theta})e^{iN\theta} \right)_0 \Theta_G(\tau_m)$$

$$= \Theta_G(\tau_N).$$

By itself this is not enough information to conclude that the form on $\mathcal{L}_S(Z)$ is positive definite. There are many cancelling plus and minus signs on the level of complexes, and that could still be true on the level of homology. But we shall see that this calculation implies that the signature character on homology is also $\Theta_G(\tau_N)$. Since for this example we know from Proposition 4.173 that the formal character on homology is $\Theta_G(\tau_N)$, (9.4) says that the Shapovalov form is positive definite on $\mathcal{L}_S(Z)$.

EXAMPLE 2. $G = SU(1, 1)$, $L = $ diagonal, $q = b = \mathfrak{l} \oplus \mathfrak{n}$, $S = 0$.
Let $Z = \mathbb{C}_N$ as in Example 1. Again we have $\Theta_{L,s}(Z^{\#}) = e^{i(N+2)\theta}$. Since $e + f$ and $i(e - f)$ are in $\mathfrak{g}_0 = \mathfrak{su}(1, 1)$, we find $\bar{e} = f$. With $u = u' = e^n$, we have $\bar{u}''u = (-1)^n f^n e^n$, and (9.12) shows that

$\langle e^n \otimes 1, \, e^n \otimes 1 \rangle$
$$= n!((N + 2) + (n - 1))((N + 2) + (n - 2)) \cdots ((N + 2) + 1)(N + 2).$$

If $N + 1 > -1$ (the dominance condition in Theorem 9.1), the right side is > 0 for all n, and we obtain

$$\Theta_{L,s}(\text{ind}_{b,L}^{g,L} Z_b^{\#}) = e^{i(N+2)\theta} + e^{i(N+4)\theta} + e^{i(N+6)\theta} + \cdots$$

(9.16)
$$= \frac{e^{i(N+2)\theta}}{1 - e^{2i\theta}}.$$

Again the factor in the denominator indicates a series to be expanded in positive powers of $e^{i\theta}$.

In this example, $L = K$. Since $S = 0$, the form on $\bigwedge^S \mathfrak{c}$ plays no role. The analog of (9.14) is

$$\Theta_{G,s}(S^{\text{th}} \text{ term of complex})$$

$$= \sum_{m=-\infty}^{\infty} \left(\Theta_{L,s}((e^{im\theta})^*) \Theta_{L,s}(\operatorname{ind}_{\mathfrak{b},L}^{\mathfrak{g},L} Z_{\bar{\mathfrak{b}}}^\#) \right)_0 e^{im\theta}$$

$$= \sum_{m=-\infty}^{\infty} \left(e^{-im\theta} \frac{e^{i(N+2)\theta}}{1 - e^{2i\theta}} \right)_0 e^{im\theta}$$

$$= \sum_{m=-\infty}^{\infty} (e^{i(N+2-m)\theta} + e^{i(N+4-m)\theta} + \dots)_0 \, e^{im\theta}$$

$$(9.17) \qquad\qquad = e^{i(N+2)\theta} + e^{i(N+4)\theta} + \dots.$$

Again we shall see that this calculation implies that the signature character on homology is given by (9.17). The expression (9.17) is also the formal character, and it follows from (9.4) that the Shapovalov form is positive definite.

This concludes our study of the examples of $SU(2)$ and $SU(1, 1)$. For the general case we need to be able to make similar precise calculations with a general compact connected Lie group in place of L. Reverting to our earlier notation, let B be a compact Lie group, and assume now that B is connected and has a maximal torus T. If F is any finite-dimensional representation of B, we can restrict to T and write

$$\Theta_T(F|_T) = \sum_{e^\nu \in \widehat{T}} n_\nu \Theta_T(e^\nu),$$

with n_ν equal to the multiplicity of the weight ν in F. Let us abbreviate $\Theta_T(e^\nu)$ as e^ν, writing

$$(9.18a) \qquad\qquad \Theta_T(F|_T) = \sum_{e^\nu \in \widehat{T}} n_\nu e^\nu.$$

When F carries a B invariant Hermitian form, each weight space inherits a signature $(n_+^\nu, n_0^\nu, n_-^\nu)$, and we obtain

$$(9.18b) \qquad\qquad \Theta_{T,s}(F|_T) = \sum_{e^\nu \in \widehat{T}} (n_+^\nu - n_-^\nu) e^\nu.$$

Formulas (9.18) represent computations of the formal character and signature character weight by weight. That is, from these formulas we can (in principle) recover $\Theta_B(F)$ and $\Theta_{B,s}(F)$. The reason is that the restriction to T of the ordinary character of a finite-dimensional representation of B determines the character everywhere.

More precisely, let $\Lambda \subseteq \mathfrak{t}^*$ be the lattice of differentials of characters of T. The Grothendieck group $\mathcal{A}(T)$ may be identified with the group $\mathbb{Z}^\Lambda$ of $\mathbb{Z}$ valued functions on Λ; in notation like that in (9.18), the formal character corresponding to f is $\sum_{\lambda \in \Lambda} f(\lambda)e^\lambda$. The Weyl group $W = W(B,T)$ acts on Λ and therefore on functions on Λ. Define

$$(\mathbb{Z}^\Lambda)^W = \{W \text{ invariant functions in } \mathbb{Z}^\lambda\}$$

and $\qquad (\mathbb{Z}^\Lambda)^W_{\text{finite}} = \{\text{members of } (\mathbb{Z}^\Lambda)^W \text{ of finite support}\}.$

The precise sense in which (9.18) allows for the recovery of $\Theta_B(F)$ is given in the following proposition. Because of (9.6), we can also recover $\Theta_{B,s}(F)$.

Proposition 9.19. The map $F \mapsto \Theta_T(F|_T)$ on finite-dimensional representations of B extends to an isomorphism of the Grothendieck group $K(\mathcal{F}(B))$ with $(\mathbb{Z}^\Lambda)^W_{\text{finite}}$.

PROOF. The formal character of any finite-dimensional representation of B is invariant under W, by Theorem 4.10a. The map in the proposition is onto $(\mathbb{Z}^\Lambda)^W_{\text{finite}}$, by (4.110).

In the category of admissible representations of B, the map in Proposition 9.19 is not very useful. The problem is that the restriction to T of an admissible representation need not be admissible.

To salvage matters, we shall use the Weyl character formula. We introduce a positive system $\Delta^+(\mathfrak{b}, \mathfrak{t})$ of roots in the usual way. If $\tau = \tau_\lambda$ is an irreducible representation of B with highest weight λ, then the formula says

$$(9.20a) \qquad D_B \Theta_T(\tau_\lambda|_T) = \sum_{s \in W(B,T)} (\operatorname{sgn} s)e^{s(\lambda + \delta_B)},$$

where

$$(9.20b)$$
$$D_B = e^{\delta_B} \prod_{\alpha \in \Delta^+(\mathfrak{b},\mathfrak{t})} (1 - e^{-a}) = (-1)^{|\Delta^+(\mathfrak{b},\mathfrak{t})|} e^{-\delta_B} \prod_{\alpha \in \Delta^+(\mathfrak{b},\mathfrak{t})} (1 - e^\alpha).$$

Suppose $\mu \in \mathfrak{t}^*$ is dominant and analytically integral. Then $e^{\mu + \delta_B}$ appears on the right side of (9.20a) if and only if $\mu = \lambda$, and in this case it appears with coefficient 1. To take advantage of this fact in our calculus, we shall in effect multiply formal characters of admissible representations of B by D_B and then interpret matters on T.

The step of including expressions like (9.20a) in our calculus involves going outside $\mathbb{Z}^\Lambda$, since δ_B may fail to be analytically integral. Define the slanted $\Lambda \subseteq \mathfrak{t}^*$ by

$$\Lambda = \{\lambda \in \mathfrak{t}^* \mid \lambda - \delta_B \in \Lambda\}.$$

It will often be convenient to identify a function f on Λ with the purely symbolic expression

$$\Theta(f) = \sum_{\lambda \in \Lambda} f(\lambda)e^\lambda.$$

The Weyl group $W(B, T)$ acts on Λ and therefore on functions on Λ. A $\mathbb{Z}$ valued function f on Λ is said to be $W(B, T)$ **odd** if

$$f(s\lambda) = (\operatorname{sgn} s)f(\lambda) \qquad \text{for all } \lambda \in \Lambda \text{ and } s \in W(B, T).$$

Define

$$\mathbb{Z}^\Lambda_{\text{odd}} = \{\mathbb{Z} \text{ valued odd functions on } \Lambda\}$$

and $\qquad \mathbb{Z}^\Lambda_{\text{odd,finite}} = \{\text{members of } \mathbb{Z}^\Lambda_{\text{odd}} \text{ with finite support}\}.$

The next proposition shows how we can bring the Weyl character formula into the calculus.

Proposition 9.21. The Grothendieck group $K(\mathcal{A}(B))$ may be identified with $\mathbb{Z}^\Lambda_{\text{odd}}$ by the map

$$(9.22a) \qquad \sum_{\substack{\nu \in \Lambda, \\ \nu \text{ dominant}}} n_\nu \Theta_B(\tau_\nu) \mapsto \sum_{\substack{\nu \in \Lambda, \\ \nu \text{ dominant}}} \sum_{s \in W} n_\nu (\operatorname{sgn} s)e^{s(\nu + \delta_B)}.$$

The inverse map is

$$(9.22b) \qquad f \mapsto \sum_{\substack{\nu \in \Lambda, \\ \nu \text{ dominant}}} f(\nu + \delta_B)\Theta_B(\tau_\nu).$$

Under these maps the subgroup $K(F(B))$ is identified with $\mathbb{Z}^\Lambda_{\text{odd,finite}}$.

Of course, on a finite-dimensional representation F of B, this map is

$$(9.23) \qquad \Theta_B(F) \mapsto D_B\Theta_T(F|_T)$$

by the Weyl character formula. For this reason we denote the map (9.22a) by Θ^{num}, since formally it gives the numerator of the Weyl character formula for F. But for general admissible representations (9.23) no longer makes sense, and we have to rely on the more formal definition in the proposition.

With (9.22a) the map Θ^{num} is defined on the full Grothendieck group $K(\mathcal{A}(B))$. Special cases of its values are Θ^{num} applied to the formal character and signature character of $V \in \mathcal{A}(B)$. We denote these by $\Theta_B^{\mathrm{num}}(V)$ and $\Theta_{B,s}^{\mathrm{num}}(V)$, respectively.

The group Λ acts on Λ by translation, and therefore Λ acts on $\mathbb{Z}^\Lambda$. This action gives rise to an action of the $\mathbb{Z}$ group ring $\mathbb{Z}^\Lambda_{\mathrm{finite}}$ on $\mathbb{Z}^\Lambda$. Explicitly

$$(9.24) \qquad \left(\sum_{\nu \in \Lambda} n_\nu e^\nu\right)\left(\sum_{\lambda \in \Lambda} m_\lambda e^\lambda\right) = \sum_{\gamma \in \Lambda}\left(\sum_{\substack{\nu \in \Lambda, \\ \lambda+\nu=\gamma}} n_\nu m_\lambda\right)e^\gamma.$$

Here we are assuming that the first sum on the left is finite, and then the inner sum on the right is finite as well.

Proposition 9.25. The formula (9.24) makes $\mathbb{Z}^\Lambda_{\mathrm{odd}}$ into a $(\mathbb{Z}^\Lambda)^W_{\mathrm{finite}}$ module. In the identifications of $(\mathbb{Z}^\Lambda)^W_{\mathrm{finite}}$ and $\mathbb{Z}^\Lambda_{\mathrm{odd}}$ with $K(\mathcal{F}(B))$ and $K(\mathcal{A}(B))$, this ring action corresponds to tensoring a finite-dimensional representation of B with an admissible representation. That is,

$$\Theta_B^{\mathrm{num}}(F \otimes_\mathbb{C} V) = \Theta_T(F)\Theta_B^{\mathrm{num}}(V)$$

for F finite dimensional and V admissible. If F in $\mathcal{F}(B)$ and V in $\mathcal{A}(B)$ carry invariant Hermitian forms and if $F \otimes_\mathbb{C} V$ is given the tensor product form, then

$$\Theta_{B,s}^{\mathrm{num}}(F \otimes_\mathbb{C} V) = \Theta_{T,s}(F)\Theta_{B,s}^{\mathrm{num}}(V).$$

PROOF. The first assertion is formal and easy. The second is a consequence of the Weyl character formula. In fact, consider a tensor product $F \otimes V$, with F finite dimensional and V admissible. We are to compare the multiplicities of B types of $F \otimes_\mathbb{C} V$ with certain coefficients in $\Theta_T(F)\Theta_B^{\mathrm{num}}(V)$. Each multiplicity depends only on a finite number of multiplicities of B types of V, and each coefficient depends only on a finite number of coefficients of $\Theta_B^{\mathrm{num}}(V)$. We may therefore check each of the equalities we want by replacing V with a finite-dimensional V_0. In that case the relation is immediate from (9.24).

The third assertion is proved similarly, by combining the above argument with the proof of Lemma 9.10.

Part of our calculation of signatures will use a process of continuous deformation. The representations appearing in the deformation are not representations of a compact connected group B, but only (locally finite) representations of the Lie algebra $\mathfrak{b}_0$ and the complexification $\mathfrak{b}$. We need to extend our calculus of signatures a little to allow for this generalization.

Define $\tilde{B}$ to be the simply connected cover of B. It follows from Corollary 4.4 that $\tilde{B}$ is a direct product

$$\cdot\ \tilde{B} = \tilde{B}_{ss} \times \tilde{Z}$$

of a compact connected semisimple group by a vector group isomorphic to $\mathbb{R}^n$ for some n. All of the statements below may therefore be reduced to the case of compact groups treated above and to the case of $\mathbb{R}^n$, where the theory is elementary; we omit the details. Define

$$\widehat{\mathfrak{b}} = \text{unitary dual of } \tilde{B},$$

i.e., $\widehat{\mathfrak{b}}$ is the set of equivalence classes of irreducible unitary representations of $\tilde{B}$. Any such representation is the tensor product of a (finite-dimensional) irreducible unitary representation of $\tilde{B}_{ss}$ and a (one-dimensional) unitary character of $\tilde{Z}$.

Let us review the development so far in this section, adapting definitions and results to the setting of $\widehat{\mathfrak{b}}$. Let $\mathcal{A}(\mathfrak{b})$ be the small category of admissible, locally finite, fully reducible representations of $\tilde{B}$ such that every irreducible constituent belongs to $\widehat{\mathfrak{b}}$. Just as in Proposition 1.63, we find that the Grothendieck group $K(\mathcal{A}(\mathfrak{b}))$ is isomorphic to $\mathbb{Z}^{\widehat{\mathfrak{b}}}$. Namely if V is a member of $\mathcal{A}(\mathfrak{b})$, we associate to V the **formal character**

$$\Theta_{\mathfrak{b}}(V) = \{n_{\tau}\}_{\tau \in \widehat{\mathfrak{b}}}$$

in $K(\mathcal{A}(\mathfrak{b}))$. Here n_{τ} is the multiplicity of τ in V.

Suppose F is finite dimensional and is isotypic of type $\tau \in \widehat{\mathfrak{b}}$. If F carries a $\tilde{B}$ invariant Hermitian form $\langle \cdot, \cdot \rangle$, then a generalization of Proposition 6.12 applies: V has an orthogonal $\tilde{B}$ invariant decomposition $V = V_+ \oplus V_0 \oplus V_-$ such that $\langle \cdot, \cdot \rangle$ is positive on V_+, zero on V_0, and negative on V_-; moreover, the corresponding triple of multiplicities (n_+, n_0, n_-) is well defined. We call this triple the **signature** of $\langle \cdot, \cdot \rangle$ on F with respect to τ.

Suppose next that V is a member of $\mathcal{A}(\mathfrak{b})$ carrying an invariant Hermitian form. Again the isotypic decomposition is automatically orthogonal, and each isotypic subspace V_{τ} has a signature $(n_+^{\tau}, n_0^{\tau}, n_-^{\tau})$. We define the **signature character** of V to be the member of $K(\mathcal{A}(\mathfrak{b}))$ given by

$$\Theta_{\mathfrak{b},s}(V) = \{n_+^\tau - n_-^\tau\}_{\tau \in \widehat{\mathfrak{b}}}.$$

Define $\mathcal{F}(\mathfrak{b})$ to be the subcategory of finite-dimensional representations in $\mathcal{A}(\mathfrak{b})$. The formalism involving tensor products in (9.5) through (9.11) carries over without change.

To discuss weights, let $\tilde{T}$ be the inverse image of T in $\tilde{B}$. This group is the direct product of a maximal torus in $\tilde{B}_{ss}$ with the vector group $\tilde{Z}$. Its irreducible unitary representations are parametrized by

$$\tilde{\Lambda} = \left\{ \lambda \in \mathfrak{t}^* \;\middle|\; \begin{array}{l} \lambda \text{ is algebraically integral and takes} \\ \text{purely imaginary values on } \mathfrak{t}_0 \end{array} \right\}.$$

We write $\nu \leftrightarrow e^\nu$ for the bijection between $\tilde{\Lambda}$ and $\widehat{\mathfrak{t}}$. The formalism in (9.18) of expanding characters and signature characters of finite-dimensional representations in terms of weights now extends to $\tilde{B}$.

To bring the Weyl character formula into the signature calculus for $\tilde{B}$, it is helpful to begin with the analog of Theorem 4.7 for $\tilde{B}$.

Proposition 9.26. The equivalence classes of irreducible unitary representations of $\tilde{B}$ stand in one-one correspondence with the dominant elements of $\tilde{\Lambda}$, the correspondence being that λ is the highest weight of τ_λ.

With this fact in place, we can introduce symbolic expressions like $\sum_{\lambda \in \tilde{\Lambda}} f(\lambda) e^\lambda$, and Propositions 9.21 and 9.25 both extend to $\tilde{B}$. There is no need to introduce a separate set $\tilde{\Lambda}$ denoted by a slanted letter, since δ_B already belongs to $\tilde{\Lambda}$.

3. Signature Character of ind $Z^{\#}$

Let us abbreviate $\operatorname{ind}_{\tilde{\mathfrak{q}}, L \cap K}^{\mathfrak{g}, L \cap K}(Z_{\tilde{\mathfrak{q}}}^{\#})$ as ind $Z^{\#}$. Already in the examples of §2, the computations of the signature character of ind $Z^{\#}$ required an induction and a lucky factorization. The computation in general appears much more complicated, and we cannot expect it to work out as well. But what we can expect is to be able to figure out the main term in the computation, essentially the term $(-1)^n n! h^n$ on the right side of (9.12). A key idea in the proof of Theorem 9.1 permits us to relate this main term to the signature character of ind $Z^{\#}$ that we seek. The idea is to replace Z by a one-parameter family of modules $Z_t, t \geq 0$, with $Z_0 = Z$, so that the signature character of ind $Z_t^{\#}$ for large t is determined by the main term and so that the signature does not change nontrivially for $t \geq 0$. To accomplish this miracle, we have to go outside the category of $(\mathfrak{l}, L \cap K)$ modules and consider just $U(\mathfrak{l})$ modules.

Fix $(\mathfrak{g}, K)$ with K connected, so that L and T are connected. From near the end of §1, we may assume that the $(\mathfrak{l}, L \cap K)$ module Z is irreducible and infinitesimally unitary. Let its infinitesimal character be λ. We continue to refer to the invariant positive definite Hermitian form for Z as $\langle \cdot, \cdot \rangle_L$. Let

$$Z_t = Z \otimes_{\mathbb{C}} \mathbb{C}_{t\delta(\mathfrak{u})}.$$

For $t \in 2\mathbb{Z}$ this is an $(\mathfrak{l}, L \cap K)$ module, but in general it is merely a $U(\mathfrak{l})$ module. The infinitesimal character of Z_t is $\lambda + t\delta(\mathfrak{u})$. Since by hypothesis

$$\frac{2\langle \lambda + \delta(\mathfrak{u}), \alpha \rangle}{|\alpha|^2} \notin (-\infty, -1] \qquad \text{for } \alpha \in \Delta(\mathfrak{u}),$$

the infinitesimal character satisfies

$$\frac{2\langle (\lambda + t\delta(\mathfrak{u})) + \delta(\mathfrak{u}), \alpha \rangle}{|\alpha|^2} \notin (-\infty, -1] + \frac{2t\langle \delta(\mathfrak{u}), \alpha \rangle}{|\alpha|^2}.$$

By (4.71), $\langle \delta(\mathfrak{u}), \alpha \rangle \geq 0$. Thus

$$(9.27) \qquad \frac{2\langle (\lambda + t\delta(\mathfrak{u})) + \delta(\mathfrak{u}), \alpha \rangle}{|\alpha|^2} \notin \{-1, -2, -3, \dots\}$$

as long as $t \geq 0$ and $\alpha \in \Delta(\mathfrak{u})$.

Let us form ind $Z_t^{\#}$, by which we mean

$$(9.28) \qquad \text{ind } Z_t^{\#} = \text{ind}_{\bar{\mathfrak{q}}}^{\mathfrak{g}}((Z_t)_{\bar{\mathfrak{q}}}^{\#}).$$

As a $U(\mathfrak{l})$ module, ind $Z_t^{\#}$ is just

$$(9.29) \qquad U(\mathfrak{u}) \otimes_{\mathbb{C}} Z^{\#} \otimes_{\mathbb{C}} \mathbb{C}_{t\delta(\mathfrak{u})},$$

with $\mathfrak{l}$ acting by the triple tensor product of ad and the respective actions on $Z^{\#}$ and $\mathbb{C}_{t\delta(\mathfrak{u})}$. In particular ind $Z_t^{\#}$ is fully reducible as an $\mathfrak{l} \cap \mathfrak{k}$ module and is even a member of the category $\mathcal{A}(\mathfrak{l} \cap \mathfrak{k})$ defined in §2. By the proof of Theorem 8.21a (and Theorem 5.99a), any nonzero $U(\mathfrak{g})$ submodule of ind $Z_t^{\#}$ (for $t \geq 0$) has nonzero intersection with $1 \otimes Z_t^{\#}$. Therefore the proof of Corollary 5.105 allows us to conclude that

$$(9.30) \qquad \text{ind } Z_t^{\#} \text{ is irreducible as a } U(\mathfrak{g}) \text{ module for all } t \geq 0.$$

We shall use the following observation about the $U(\mathfrak{l})$ action in (9.29): If v is an eigenvector in $U(\mathfrak{u}) \otimes_{\mathbb{C}} Z^{\#}$ under $h_{\delta(\mathfrak{u})}$ with eigenvalue c, $v \otimes 1$ is an eigenvector in ind $Z_t^{\#}$ under $h_{\delta(\mathfrak{u})}$ with eigenvalue $c + t|\delta(\mathfrak{u})|^2$.

The tensor product of $\langle \cdot, \cdot \rangle_L$ on Z with the usual inner product on $\mathbb{C}_{t\delta(\mathfrak{u})}$ is a Hermitian inner product on Z_t that is $\mathfrak{l}_0$ invariant. In fact, it is enough to see that the inner product on $\mathbb{C}_{t\delta(\mathfrak{u})}$ is $\mathfrak{l}_0$ invariant. On $\mathbb{C}_{t\delta(\mathfrak{u})}$, $[\mathfrak{l}_0, \mathfrak{l}_0]$ acts as 0, and $X \in \mathfrak{h}_0$ has

$$\langle X\zeta, \zeta' \rangle_{\mathbb{C}} = \delta(\mathfrak{u})\zeta\bar{\zeta}' = -\zeta\overline{\delta(\mathfrak{u})(X)}\bar{\zeta}' = -\langle \zeta, X\zeta' \rangle_{\mathbb{C}} \quad \text{for } \zeta, \zeta' \text{ in } \mathbb{C},$$

since $\delta(\mathfrak{u})$ vanishes on $\mathfrak{a}_0$.

Returning to (9.29), suppose $V \subseteq U(\mathfrak{u}) \otimes_{\mathbb{C}} Z^{\#}$ is an isotypic subspace for $L \cap K$, say for the $L \cap K$ type with highest weight φ. Then $V \otimes \mathbb{C}_{t\delta(\mathfrak{u})}$ is $\mathfrak{l} \cap \mathfrak{k}$ isotypic with highest weight $\varphi + t\delta(\mathfrak{u})$. In these terms the $\mathfrak{g}_0$ invariant Hermitian form on ind $Z^{\#}$ is given by

$$(9.31)$$
$$\langle u \otimes z, u' \otimes z' \rangle_0 = \langle \mu_{\bar{\mathfrak{u}}}'(\bar{u}'^t u)z, z' \rangle_L \quad \text{for } u, u' \in U(\mathfrak{u}) \text{ and } z, z' \in Z^{\#},$$

and the corresponding form on ind $Z_t^{\#}$ is given by

$$\langle u \otimes z \otimes \zeta, u' \otimes z' \otimes \zeta' \rangle_t = \langle \mu_{\bar{\mathfrak{u}}}'(\bar{u}'^t u)(z \otimes \zeta), z' \otimes \zeta' \rangle_L \quad \text{for } \zeta, \zeta' \in \mathbb{C}.$$

The way in which the right side depends on t is through the action of $\mu_{\bar{\mathfrak{u}}}'(\bar{u}'^t u)$ on the tensor product $Z^{\#} \otimes_{\mathbb{C}} \mathbb{C}_{t\delta(\mathfrak{u})}$, and we can see that the right side is a polynomial function from $\{t \in \mathbb{R}\}$ to $\mathbb{C}$.

Lemma 9.32. The Hermitian form $\langle \cdot, \cdot \rangle_t$ on ind $Z_t^{\#}$ is nondegenerate for all $t \geq 0$. If

$$V \subseteq U(\mathfrak{u}) \otimes_{\mathbb{C}} Z^{\#} \cong \text{ind } Z^{\#}$$

is the isotypic subspace under $L \cap K$ for the $L \cap K$ type with highest weight φ, then

$$V \otimes_{\mathbb{C}} \mathbb{C}_{t\delta(\mathfrak{u})} \subseteq U(\mathfrak{u}) \otimes_{\mathbb{C}} Z^{\#} \otimes_{\mathbb{C}} \mathbb{C}_{t\delta(\mathfrak{u})}$$

is the isotypic subspace under $\mathfrak{l} \cap \mathfrak{k}$ for the $\mathfrak{l} \cap \mathfrak{k}$ type with highest weight $\varphi + t\delta(\mathfrak{u})$. Consequently the restriction of $\langle \cdot, \cdot \rangle_t$ to $V \otimes_{\mathbb{C}} \mathbb{C}_{t\delta(\mathfrak{u})}$ is nondegenerate. Its signature is independent of t, for all $t \geq 0$.

REMARKS. The proof will use one detail about signatures of matrices that was not included in the remarks with Proposition 6.12: Small perturbations of a nonsingular Hermitian matrix do not change the signature. In fact, let R be nonsingular Hermitian with signature $(p, 0, q)$, let $\mathbb{C}^n = V_+ \oplus V_-$ be the decomposition of (6.14) for R, and let m be the smallest absolute value of an eigenvalue of R. If A is Hermitian with $\|A\| < m$, then any $v \neq 0$ in V_+ satisfies

(9.33) $\langle (R + A)v, v \rangle = \langle Rv, v \rangle + \langle Av, a \rangle \geq (m - \|A\|)\|v\|^2 > 0.$

Fix A, and let $V = W_+ \oplus W_-$ be the decomposition of (6.14) for $R + A$. Then (9.33) shows that $V_+ \cap W_- = 0$. Hence

$$\dim V_+ + \dim W_- \leq n,$$

i.e., $\dim V_+ \leq \dim W_+$. Similarly $\dim V_- \leq \dim W_-$. It follows that R and $R + A$ have the same signature. Since A is arbitrary with $\|A\| < m$, small perturbations of R do not affect its signature.

PROOF. The restriction of $\langle \cdot, \cdot \rangle_t$ to $Z^{\#} \otimes_{\mathbb{C}} \mathbb{C}_{t\delta(\mathfrak{u})}$ is the tensor product of the nondegenerate form $\langle \cdot, \cdot \rangle_L$ with the standard positive form on $\mathbb{C}$. It is therefore nondegenerate. Since the radical of $\langle \cdot, \cdot \rangle_t$ is a $U(\mathfrak{g})$ invariant subspace of ind $Z_t^{\#}$, it follows from (9.30) that the radical is 0. Therefore $\langle \cdot, \cdot \rangle_t$ is nondegenerate on ind $Z_t^{\#}$.

The correspondence of isotypic subspaces follows from the definition of the tensor-product action. Since the various $\mathfrak{l} \cap \mathfrak{k}$ isotypic subspaces are mutually orthogonal, the nondegeneracy of $\langle \cdot, \cdot \rangle_t$ on ind $Z_t^{\#}$ implies nondegeneracy of $\langle \cdot, \cdot \rangle_t$ on the $\mathfrak{l} \cap \mathfrak{k}$ isotypic subspace $V \otimes_{\mathbb{C}} \mathbb{C}_{t\delta(\mathfrak{u})}$.

For the assertion about the variation of the signature with t, let us identify all the spaces $V \otimes_{\mathbb{C}} \mathbb{C}_{t\delta(\mathfrak{u})}$ with V, by sending $v \otimes \zeta$ to ζv. Next let us recall from Remark 4 for Proposition 6.12 that we can use a vector-space isomorphism from V to $\mathbb{C}^n$ to convert Hermitian forms to Hermitian matrices. Let $M(t)$ be the Hermitian matrix obtained in this way from the form $\langle \cdot, \cdot \rangle_t$ carried to V. By the remarks after (9.31), the entries of $M(t)$ depend on t in polynomial fashion. The nondegeneracy statement just proved implies that $M(t)^{-1}$ is nonsingular for each $t \geq 0$. By the remarks after the statement of the lemma, the signature is a continuous function on nonsingular Hermitian matrices. Composing $t \mapsto M(t)$ with this function, we obtain a continuous function from the connected set $\{t \geq 0\}$ into a discrete set. Hence the signature is constant as a function of t. This proves the lemma.

A rational function $r(t) = p(t)/q(t)$ of one real variable will be said to have **degree** equal to the difference of the degrees of the polynomials p and q and to have **leading coefficient** equal to the quotient of the leading coefficients of p and q.

Lemma 9.34. Let $R(t)$ be a Hermitian matrix whose entries $R_{ij}(t)$ are rational functions of degree d_{ij} (with the degree of 0 being $-\infty$). Suppose that no diagonal entry is identically 0, and suppose that

$$d_{ij} \leq \tfrac{1}{2}(d_{ii} + d_{jj} - 1) \qquad \text{for } i \neq j.$$

Let p be the number of diagonal entries with positive leading coefficient and q the number with negative leading coefficient. Then there is an upper-triangular matrix $A(t)$ with 1's on the diagonal such that

$$S(t) = A(t)R(t)A(t)^*$$

is diagonal. The ith diagonal entry of $S(t)$ has the same degree and leading coefficient as the ith diagonal entry of $R(t)$. Consequently for all sufficiently large positive t, the signature of $R(t)$ is $(p, 0, q)$.

PROOF. For the proof that $S(t) = A(t)R(t)A(t)^*$ with $S(t)$ as asserted, we induct on the size of $R(t)$, the base case being size 1-by-1. Assume the result for size $n - 1$. If

$$R(t) = \begin{pmatrix} a & b \\ b^* & c \end{pmatrix}$$

with a of size $(n-1)$-by-$(n-1)$ and c of size 1-by-1, then we set

$$R'(t) = \begin{pmatrix} 1 & -c^{-1}b \\ 0 & 1 \end{pmatrix} R(t) \begin{pmatrix} 1 & -c^{-1}b \\ 0 & 1 \end{pmatrix}^* = \begin{pmatrix} a - c^{-1}bb^* & 0 \\ 0 & c \end{pmatrix}.$$

The matrix $R'(t) = \begin{pmatrix} a' & 0 \\ 0 & c \end{pmatrix}$ is Hermitian. The diagonal entries of a' have

$$a'_{ii} = a_{ii} - c^{-1}b_{in}\bar{b}_{ni}$$

and

$$\deg(c^{-1}b_{in}\bar{b}_{ni}) = -d_{nn} + d_{in} + d_{ni} \leq -d_{nn} + (d_{ii} + d_{nn} - 1) = d_{ii} - 1.$$

Hence a'_{ii} has the same degree and leading coefficient as a_{ii}. If $i \neq j$, we have

$$a'_{ij} = a_{ij} - c^{-1}b_{in}\bar{b}_{nj}$$

and

$$d'_{ij} \leq \max\{d_{ij}, -d_{nn} + d_{in} + d_{nj}\}.$$

Since

$$-d_{nn} + d_{in} + d_{nj} \leq -d_{nn} + \tfrac{1}{2}(d_{ii} + d_{nn} - 1) + \tfrac{1}{2}(d_{nn} + d_{jj} - 1)$$
$$= \tfrac{1}{2}(d_{ii} + d_{jj} - 2)$$
$$< \tfrac{1}{2}(d_{ii} + d_{jj} - 1),$$

d'_{ij} is $\leq \tfrac{1}{2}(d_{ii} + d_{jj} - 1)$. Hence the inductive hypothesis is applicable to the $(n - 1)$-by-$(n - 1)$ block $a' = a'(t)$ in $R'(t)$. Substituting from the inductive hypothesis, we obtain $S(t) = A(t)R(t)A(t)^*$ as required.

For t sufficiently large and positive, p of the diagonal entries of $S(t)$ are positive and q are negative. Thus $S(t)$ has signature $(p, 0, q)$. By Remark 3 for Proposition 6.12, $R(t)$ has the same signature.

Lemma 9.35. Let $u \in U(\mathfrak{g})$ be a product in any order of m elements of $\bar{\mathfrak{u}}, r$ elements of $\mathfrak{l}$, and n elements of $\mathfrak{u}$. Then

$$\mu'_{\bar{\mathfrak{u}}}(u) \qquad \text{is in} \qquad U_{r+\min\{m,n\}}(\mathfrak{l}) \subseteq U_{r+\frac{1}{2}(m+n)}(\mathfrak{l}).$$

PROOF. We may assume that each factor in question is an eigenvector of $\operatorname{ad} h_{\delta(\mathfrak{u})}$, so that the bracket of any two of them is in $\bar{\mathfrak{u}}, \mathfrak{l}$, or $\mathfrak{u}$ and is again an eigenvector. We induct on $N = m + n + r$, the base case being the trivial situation $N = 0$. The inductive case is $N > 0$. With N fixed, we induct on m. If $m = 0$, the result is easy, since the $\bar{\mathfrak{u}}$ factors are absent and $[\mathfrak{l}, \mathfrak{u}] \subseteq \mathfrak{u}$. For general $m > 0$, we induct on the number c of factors to the right of the last $\bar{\mathfrak{u}}$ factor. If $c = 0$, then u is in $U(\mathfrak{g})\bar{\mathfrak{u}}$ and $\mu'_{\bar{\mathfrak{u}}}(u) = 0$. If $c > 0$, we commute the rightmost $\bar{\mathfrak{u}}$ factor one position to the right, obtaining a flipped term and a commutator term. The flipped term has the same m, n, and r, and its c has been reduced by one; hence it is handled by the inductive hypothesis. The commutator term has some new m', n', and r', with

$$m' + n' + r' = m + n + r - 1.$$

Thus the inductive hypothesis on N shows that $\mu'_{\bar{\mathfrak{u}}}$ of the commutator term is in $U_{r'+\min\{m',n'\}}(\mathfrak{l})$. To complete the proof, it is enough to show that

(9.36) $$r' + \min\{m', n'\} \leq r + \min\{m, n\}.$$

The commutator arises by replacing a factor XY by $[X, Y]$, where X is in $\bar{\mathfrak{u}}$ and Y is in $\mathfrak{l}$ or $\mathfrak{u}$. If Y is in $\mathfrak{l}$, then $[X, Y]$ is in $\bar{\mathfrak{u}}$. So

$$(m', n', r') = (m, n, r - 1),$$

and (9.36) is immediate.

The hard case is that Y is in $\mathfrak{u}$. Since X and Y are eigenvectors of $\mathrm{ad}\, h_{\delta(\mathfrak{u})}$, the commutator $[X, Y]$ can be in $\bar{\mathfrak{u}}$, $\mathfrak{l}$, or $\mathfrak{u}$. In the three subcases, we have

$$(m', n', r') = (m, n - 1, r), \quad (m - 1, n - 1, r + 1), \quad \text{and} \quad (m - 1, n, r),$$

respectively. In each subcase, (9.36) is valid, and thus the induction is complete.

Lemma 9.37. In $\mathfrak{sl}(2, \mathbb{C})$, let

$$h = \begin{pmatrix} 1 & 0 \\ 0 & -1 \end{pmatrix}, \qquad e = \begin{pmatrix} 0 & 1 \\ 0 & 0 \end{pmatrix}, \qquad f = \begin{pmatrix} 0 & 0 \\ 1 & 0 \end{pmatrix}.$$

Then $[e, f^n] = nf^{n-1}(h - (n - 1))$ in $U(\mathfrak{sl}(2, \mathbb{C}))$.

PROOF. Let

$$Lf = \text{left by } f \text{ in } U(\mathfrak{sl}(2, \mathbb{C})$$
$$Rf = \text{right by } f$$
$$\mathrm{ad}\, f = Lf - Rf.$$

Then $Rf = Lf - \mathrm{ad}\, f$, and the terms on the right commute. By the binomial theorem

$$(Rf)^n e = \sum_{j=0}^{n} \binom{n}{j} (Lf)^{n-j}(-\mathrm{ad}\, f)^j e$$

$$= (Lf)^n e + n(Lf)^{n-1}(-\mathrm{ad}\, f)e + \frac{n(n-1)}{2}(Lf)^{n-2}(-\mathrm{ad}\, f)^2 e$$

since $(\mathrm{ad}\, f)^3 e = 0$, and this expression is

$$= (Lf)^n e + nf^{n-1}h + \frac{n(n-1)}{2}f^{n-2}(-2f)$$
$$= (Lf)^n e + nf^{n-1}(h - (n - 1)).$$

Thus

$$[e, f^n] = (Rf)^n e - (Lf)^n e = nf^{n-1}(h - (n - 1)).$$

We come to the main combinatorial lemma. This result will make repeated use of Lemma 9.35, and in order to keep track of the results, we need good bases for $\mathfrak{u}$ and $\bar{\mathfrak{u}}$. For this purpose we use the basis of weight vectors of $\mathfrak{t}$ constructed in (4.45). Namely for each $(\mathfrak{t}, \theta)$ weight α with $\alpha|_{\mathfrak{t}} \neq 0$, let e_α be a nonzero weight vector, with e_α and $e_{-\alpha}$ normalized so that

$$(9.38) \qquad [e_\alpha, e_{-\alpha}] = h'_\alpha \text{ in } \mathfrak{t} \text{ corresponds to } 2|\alpha|^{-2}\alpha|_{\mathfrak{t}} \text{ in } \mathfrak{t}^*$$

$$(9.39) \qquad \bar{e}_\alpha = \begin{cases} -e_{-\alpha} & \text{if } e_\alpha \text{ is in } \mathfrak{k} \\ e_{-\alpha} & \text{if } e_\alpha \text{ is in } \mathfrak{p}. \end{cases}$$

As a consequence of (4.45b), if α and α' are $(\mathfrak{t}, \theta)$ weights with $e_\alpha \neq e_{\alpha'}$, then

$$(9.40) \qquad [e_\alpha, e_{-\alpha'}] \in \mathfrak{l} \qquad \text{implies} \qquad [e_\alpha, e_{-\alpha'}] \in [\mathfrak{l}, \mathfrak{l}] \oplus \mathfrak{a}.$$

Lemma 9.41. Let $\alpha_1, \ldots, \alpha_n$ be an enumeration of the $(\mathfrak{t}, \theta)$ weights of $\Delta(\mathfrak{u} \cap \mathfrak{k})$ and of $\Delta(\mathfrak{u} \cap \mathfrak{p})$, and let e_{α_i} be a corresponding normalized weight vector (in $\mathfrak{k}$ or $\mathfrak{p}$) for α_i. Let $I = (i_1, \ldots, i_n)$ and $J = (j_1, \ldots, j_n)$, and let $|I| = \sum i_k$ and $I! = i_1! \cdots i_n!$
   (a) If $I \neq J$, then

$$\mu'_{\bar{\mathfrak{u}}}(e^{i_1}_{-\alpha_1} \cdots e^{i_n}_{-\alpha_n} e^{j_n}_{\alpha_n} \cdots e^{j_1}_{\alpha_1})$$

is in

$$U_{\frac{1}{2}(|I|+|J|-1)}(\mathfrak{l}) + U_{\frac{1}{2}(|I|+|J|)-1}(\mathfrak{l})U_1([\mathfrak{l}, \mathfrak{l}] \oplus \mathfrak{a}).$$

   (b) If $I = J$, then

$$\mu'_{\bar{\mathfrak{u}}}(e^{i_1}_{-\alpha_1} \cdots e^{i_n}_{-\alpha_n} e^{i_n}_{\alpha_n} \cdots e^{i_1}_{\alpha_1}) - (-1)^{|I|} I! \prod_{k=1}^{n} (h'_{\alpha_k})^{i_k}$$

is in

$$U_{|I|-1}(\mathfrak{l})U_1([\mathfrak{l}, \mathfrak{l}] \oplus \mathfrak{a}).$$

PROOF. We work with the notation of (a) but treat the cases $I \neq J$ and $I = J$ simultaneously. However, if $I \neq J$, we give the argument in case $i_n = j_n, \ldots, i_{r+1} = j_{r+1}, i_r < j_r$, indicating at the end how to handle $i_r > j_r$. Lemma 9.37 gives

$$(9.42)$$
$$(e^{i_1}_{-\alpha_1} \cdots e^{i_{n-1}}_{-\alpha_{n-1}}) e^{i_n}_{-\alpha_n} e_{\alpha_n} (e^{j_n-1}_{\alpha_n} e^{j_{n-1}}_{\alpha_{n-1}} \cdots e^{j_1}_{\alpha_1}) = (e^{i_1}_{-\alpha_1} \cdots e^{i_{n-1}}_{-\alpha_{n-1}})$$
$$\times (e_{\alpha_n} e^{i_n}_{-\alpha_n} - i_n e^{i_n-1}_{\alpha_n} h'_{\alpha_n} + i_n(i_n - 1)e^{i_n-1}_{-\alpha_n})(e^{j_n-1}_{\alpha_n} e^{j_{n-1}}_{\alpha_{n-1}} \cdots e^{j_1}_{\alpha_1})$$

The right side of (9.42) has three terms, corresponding to the three terms in the factor after the $\times$ sign. By Lemma 9.35, $\mu'_{\bar{u}}$ of the third term on the right side is in $U_{\frac{1}{2}(|I|+|J|-2)}(\mathfrak{l})$.

Consider the first term on the right side. We shall manipulate the part of it that corresponds to the first two factors on the right, namely

$$(e^{i_1}_{-\alpha_1} \cdots e^{i_{n-1}}_{-\alpha_{n-1}}) e_{\alpha_n} e^{i_n}_{-\alpha_n} = e_{\alpha_n} (e^{i_1}_{-\alpha_1} \cdots e^{i_{n-1}}_{-\alpha_{n-1}}) e^{i_n}_{-\alpha_n}$$

$$+ \sum_{k=1}^{n-1} \sum_{l=0}^{i_k-1} e^{i_1}_{-\alpha_1} \cdots e^{i_{k-1}}_{-\alpha_{k-1}} e^l_{-\alpha_k} [e_{\alpha_n}, e_{-\alpha_k}] e^{i_k-l-1}_{-\alpha_k} e^{i_{k+1}}_{-\alpha_{k+1}} \cdots e^{i_{n-1}}_{-\alpha_{n-1}} e^{i_n}_{-\alpha_n}.$$

Because of the factor e_{α_n}, $\mu'_{\bar{u}}$ of the term on the first line of the right side (together with the third factor from (9.42)) is 0. For the $(k, l)^{\text{th}}$ term of the sum on the second line, there are cases. We may assume $[e_{\alpha_n}, e_{-\alpha_k}]$ is not 0. Since $\alpha_n \neq \alpha_k$, (9.40) shows that $[e_{\alpha_n}, e_{-\alpha_k}]$ is in $\bar{u}$, $[\mathfrak{l}, \mathfrak{l}] \oplus \mathfrak{a}$, or $\mathfrak{u}$.

If $[e_{\alpha_n}, e_{-\alpha_k}]$ is in $\bar{u}$, the number of $\bar{u}$ factors in the total term is $|I|$, while the number of $\mathfrak{u}$ factors is $|J| - 1$. Thus Lemma 9.35 shows that $\mu'_{\bar{u}}$ of the expression in question is in $U_{\frac{1}{2}(|I|+|J|-1)}(\mathfrak{l})$.

If $[e_{\alpha_n}, e_{-\alpha_k}]$ is in $[\mathfrak{l}, \mathfrak{l}] \oplus \mathfrak{a}$, then the total term is in

$$U_{|I|-1}(\bar{u})U_1([\mathfrak{l}, \mathfrak{l}] \oplus \mathfrak{a})U_{|J|-1}(\mathfrak{u}).$$

If we take into account that the member $[e_{\alpha_n}, e_{-\alpha_k}]$ of $[\mathfrak{l}, \mathfrak{l}] \oplus \mathfrak{a}$ can be isolated from the other factors in applying Lemma 9.35, then the lemma shows that $\mu'_{\bar{u}}$ of the expression in question is in

$$U_{\frac{1}{2}(|I|+|J|)-1}(\mathfrak{l})U_1([\mathfrak{l}, \mathfrak{l}] \oplus \mathfrak{a}).$$

If $[e_{\alpha_n}, e_{-\alpha_k}]$ is in $\mathfrak{u}$, then the total term is in

$$(9.43) \qquad U_{i_1+\cdots+i_{k-1}+l}(\bar{u})U_1(\mathfrak{u})U_{i_k-l-1+i_{k+1}+\cdots+i_n}(\bar{u})U_{|J|-1}(\mathfrak{u}).$$

Thus Lemma 9.35 shows that $\mu'_{\bar{u}}$ of the expression in question is in $U_{\frac{1}{2}(|I|+|J|-1)}(\mathfrak{l})$.

Consequently $\mu'_{\bar{u}}$ of the left side of (9.42) is congruent to $\mu'_{\bar{u}}$ of the middle term on the right side, modulo

$$(9.44) \qquad U_{\frac{1}{2}(|I|+|J|-1)}(\mathfrak{l}) + U_{\frac{1}{2}(|I|+|J|)-1}(\mathfrak{l})U_1([\mathfrak{l}, \mathfrak{l}] \oplus \mathfrak{a}).$$

If we commute h'_{α_n} in (9.42) to the right end of its term, then $\mu'_{\bar{u}}$ of each error term is in $U_{\frac{1}{2}(|I|+|J|-1)}(\mathfrak{l})$, by Lemma 9.35. Now we can iterate our argument, working with one factor of $e_{\alpha_n}^{j_n}$ at a time. If $i_n = j_n$, the factor that accumulates in the surviving term is

$$(-1)^{i_n} i_n! (h'_{\alpha_n})^{i_n},$$

and we can repeat the process with $\alpha_{n-1}, \alpha_{n-2}$, etc.

If $I = J$, this process continues, and we arrive at conclusion (b). If $I \neq J$, we have set matters up so that $i_r < j_r$ for the first index r where equality fails. After i_r iterations at this stage, we are considering a multiple of

$$(9.45) \qquad \{ e_{-\alpha_1}^{i_1} \cdots e_{-\alpha_{r-1}}^{i_{r-1}} e_{\alpha_r}^{j_r - i_r} e_{\alpha_{r-1}}^{j_{r-1}} \cdots e_{\alpha_1}^{j_1} \} (h'_{\alpha_n})^{i_n} \cdots (h'_{\alpha_r})^{i_r}.$$

We treat the part of this expression in braces in the same way as the first term on the right side of (9.42). When we apply $\mu'_{\bar{u}}$ to the expression in braces, Lemma 9.35 gives us something in

$$U_{\frac{1}{2}(i_1 + j_1 + \cdots + i_{r-1} + j_{r-1} + j_r - i_r - 1)}(\mathfrak{l})$$
$$+ U_{\frac{1}{2}(i_1 + j_1 + \cdots + i_{r-1} + j_{r-1} + j_r - i_r) - 1}(\mathfrak{l}) U_1([\mathfrak{l}, \mathfrak{l}] \oplus \mathfrak{a}).$$

Taking into account the factor not in braces, we see that $\mu'_{\bar{u}}$ of (9.45) is a member of (9.44). So we arrive at conclusion (a).

Let us mention how to handle $i_r > j_r$. In this case, we combine the identity $\mu'_{\bar{u}}(x) = \mu'_u(x^t)^t$, which follows easily from the definitions, with the case of the lemma already treated for u. The result is the desired conclusion.

Proposition 9.46. Relative to the Hermitian form (9.31), the signature character of ind $Z^{\#}$ is given by

$$\Theta_{L \cap K, s}^{\text{num}}(\text{ind } Z^{\#}) = \frac{\Theta_{L \cap K}^{\text{num}}(Z^{\#})}{\prod_{\alpha \in \Delta(\mathfrak{u} \cap \mathfrak{k})} (1 + e^{\alpha}) \prod_{\alpha \in \Delta(\mathfrak{u} \cap \mathfrak{p})} (1 - e^{\alpha})}.$$

REMARKS. On the right side the mixture of information relative to $L \cap K$ with information relative to T follows the convention explained at the end of §2. To interpret it, we expand the factors in the denominator as

$$(1 + e^{\alpha})^{-1} = 1 - e^{\alpha} + e^{2\alpha} - \ldots$$

and
$$(1 - e^{\alpha})^{-1} = 1 + e^{\alpha} + e^{2\alpha} + \ldots,$$

multiply the resulting expressions, and read off $L \cap K$ signature characters by means of Proposition 9.21.

PROOF. We continue to use $(\mathfrak{t}, \theta)$ weight vectors e_α in $\mathfrak{k}$ and $\mathfrak{p}$, normalized as in the paragraph before Lemma 9.41. Fix a finite-dimensional $L \cap K$ invariant subspace $V \subseteq Z^{\#}$ and consider the signature character relative to $\langle \cdot, \cdot \rangle_t$ of the $\mathfrak{l} \cap \mathfrak{k}$ representation $U_N(\mathfrak{u}) \otimes_{\mathbb{C}} V \otimes_{\mathbb{C}} \mathbb{C}_{t\delta(\mathfrak{u})}$. Let $\{v_r\}$ be an orthonormal basis of V. In the notation of Lemma 9.41, let $\alpha_1, \ldots, \alpha_S$ have θ component $+1$ and let $\alpha_{S+1}, \ldots, \alpha_n$ have θ component -1. If $I = (i_1, \ldots, i_n)$ is given, write $I' = (i_1, \ldots, i_S)$ and $I'' = (i_{S+1}, \ldots, i_n)$. The set of elements

$$e_{\alpha_n}^{i_n} \cdots e_{\alpha_1}^{i_1} \otimes v_r \otimes 1 \qquad \text{with } |I| \leq N$$

is a basis of $U_N(\mathfrak{u}) \otimes_{\mathbb{C}} V \otimes_{\mathbb{C}} \mathbb{C}_{t\delta(\mathfrak{u})}$. For two such, we compute the form

$$
\begin{aligned}
R_{(I,r),(J,s)}(t) &= \langle e_{\alpha_n}^{j_n} \cdots e_{\alpha_1}^{j_1} \otimes v_r \otimes 1, \; e_{\alpha_n}^{i_n} \cdots e_{\alpha_1}^{i_1} \otimes v_s \otimes 1 \rangle_t \\
&= \langle (-1)^{|I|} \mu_{\bar{\mathfrak{u}}}'(\bar{e}_{\alpha_1}^{i_1} \cdots \bar{e}_{\alpha_n}^{i_n} e_{\alpha_n}^{j_n} \cdots e_{\alpha_1}^{j_1})(v_r \otimes 1), \; (v_s \otimes 1) \rangle_L \\
(9.47) \qquad &= \langle (-1)^{|I''|} \mu_{\bar{\mathfrak{u}}}'(e_{-\alpha_1}^{i_1} \cdots e_{-\alpha_n}^{i_n} e_{\alpha_n}^{j_n} \cdots e_{\alpha_1}^{j_1})(v_r \otimes 1), \; (v_s \otimes 1) \rangle_L.
\end{aligned}
$$

In this expression we consider the action $x(v_r \otimes 1)$ with $x \in U(\mathfrak{l})$, which is to be computed with $\mathbb{C}_{t\delta(\mathfrak{u})}$ in place. If $x = X$ is in $\mathfrak{l}$, then X acts as 0 on $1 \in \mathbb{C}_{t\delta(\mathfrak{u})}$ if X is in $[\mathfrak{l}, \mathfrak{l}]$, and it acts as $t\delta(\mathfrak{u})(X)$ if X is in $\mathfrak{t}$. So if x is a monomial with a factors from $[\mathfrak{l}, \mathfrak{l}]$ and b factors from $\mathfrak{t}$, then x acts by multiplying by a polynomial in t of degree $\leq b$. Referring to Lemma 9.41, we see that for $I \neq J$, $\mu_{\bar{\mathfrak{u}}}'(-)$ acts by a polynomial in t of degree $\leq \frac{1}{2}(|I| + |J| - 1)$. Hence

$$\deg R_{(I,r),(J,s)}(t) \leq \tfrac{1}{2}(|I| + |J| - 1) \qquad \text{if } I \neq J.$$

Similarly if $I = J$, then

$$
\begin{aligned}
(9.48) \qquad R_{(I,r),(I,s)}(t) &= (-1)^{|I''|} I! \prod_{k=1}^{n} \left(\frac{2\langle \delta(\mathfrak{u}), \alpha_k \rangle}{|\alpha_k|^2} \right)^{i_k} t^{|I|} \langle v_r, v_s \rangle_L \\
&\quad + (\text{polynomial in } t \text{ of degree } \leq |I| - 1),
\end{aligned}
$$

since $(-1)^{|I''|}(-1)^{|I|} = (-1)^{|I'|}$. If $r \neq s$, this is 0, and if $r = s$, it is not 0. Moreover, we see that $(I, r) \neq (J, s)$ implies that

$$\deg R_{(I,r),(J,s)} \leq \tfrac{1}{2}(\deg R_{(I,r),(I,r)} + \deg R_{(J,s),(J,s)} - 1).$$

Define $\langle I \rangle = \sum i_k \alpha_k$ and $\langle J \rangle = \sum j_k \alpha_k$. If

$$(9.49) \qquad \langle \langle I \rangle, \delta(\mathfrak{u}) \rangle \neq \langle \langle J \rangle, \delta(\mathfrak{u}) \rangle,$$

we claim that (9.47) is 0. In fact, the map $\mu'_{\bar{u}} : U(\mathfrak{g}) \to U(\mathfrak{l})$ respects ad $\mathfrak{l}$. The element

$$
(9.50) \qquad\qquad e^{i_1}_{-\alpha_1} \cdots e^{i_n}_{-\alpha_n} e^{j_n}_{\alpha_n} \cdots e^{j_1}_{\alpha_1}
$$

has weight $\langle \langle J \rangle - \langle I \rangle, \delta(\mathfrak{u}) \rangle$ under $h_{\delta(\mathfrak{u})}$, and by (9.49) we are assuming this weight is nonzero. Since ad $h_{\delta(\mathfrak{u})}$ acts by 0 on $U(\mathfrak{l})$, it follows that $\mu'_{\bar{u}}$ of (9.50) is 0. Thus (9.47) is 0.

As a consequence the matrix $R_{(I,r),(J,s)}(t)$ is block diagonal, with each block corresponding to indices I with the same weight. We apply Lemma 9.34 to the whole matrix and see that the eigenvalues of $R_{(I,r),(J,s)}(t)$, for large t, have the same signs as the leading coefficients of the diagonal entries, namely the signs of $(-1)^{|I'|} \langle v_r, v_s \rangle_L$. Therefore the signature character of $U_N(\mathfrak{u}) \otimes_{\mathbb{C}} V \otimes_{\mathbb{C}} \mathbb{C}_{t\delta(\mathfrak{u})}$ is

$$
\Theta_{\mathfrak{l} \cap \mathfrak{k}, s}(U_N(\mathfrak{u}) \otimes_{\mathbb{C}} V \otimes_{\mathbb{C}} \mathbb{C}_{t\delta(\mathfrak{u})}) = \sum_{|I| \leq N} (-1)^{|I'|} e^{\langle I \rangle} \Theta_{L \cap K}(V) e^{t\delta(\mathfrak{u})},
$$

provided t is large enough.

We multiply through by $D_{L \cap K}$, because $\Theta^{\text{num}}_{L \cap K}(V)$ counts multiplicities of highest weights, while $\Theta_{L \cap K}(V)$ counts multiplicities of all weights. Then we get

$$
(9.51) \qquad \Theta^{\text{num}}_{L \cap K, s}(U_N(\mathfrak{u}) \otimes_{\mathbb{C}} V \otimes_{\mathbb{C}} \mathbb{C}_{t\delta(\mathfrak{u})}) = \sum_{|I| \leq N} (-1)^{|I'|} e^{\langle I \rangle} \Theta^{\text{num}}_{L \cap K}(V) e^{t\delta(\mathfrak{u})}.
$$

Consider the $L \cap K$ type τ_φ with highest weight φ. Since $h_{\delta(\mathfrak{u})}$ acts as a scalar on $Z^\#$ and since the eigenvalues of $h_{\delta(\mathfrak{u})}$ on $U_{n+1}(\mathfrak{u})/U_n(\mathfrak{u})$ are bounded below by constants $c(n)$ with $c(n) \to +\infty$, all occurrences of τ_φ in $U(\mathfrak{u}) \otimes Z^\#$ are in $U_N(\mathfrak{u}) \otimes_{\mathbb{C}} Z^\#$ for some sufficiently large N. Since $U_n(\mathfrak{u})$ is finite dimensional, only finitely many $L \cap K$ types of $Z^\#$ can contribute to τ_φ. Because of the admissibility of $Z^\#$, there exists a finite-dimensional $L \cap K$ invariant subspace V of $Z^\#$ such that all occurrences of τ_φ in $U(\mathfrak{u}) \otimes_{\mathbb{C}} Z^\#$ lie in $U_n(\mathfrak{u}) \otimes_{\mathbb{C}} V$. For this N and V, we apply (9.51). Lemma 9.32 says that $\langle \cdot, \cdot \rangle_t$ is nondegenerate on the $\mathfrak{l} \cap \mathfrak{k}$ type of $U_N(\mathfrak{u}) \otimes_{\mathbb{C}} V \otimes_{\mathbb{C}} \mathbb{C}_{t\delta(\mathfrak{u})}$ with highest weight $\varphi + t\delta(\mathfrak{u})$. It says also that the signature is independent of t for $t \geq 0$. Consequently the coefficient of $e^{\varphi + \delta_{L \cap K}}$ in $\Theta^{\text{num}}_{L \cap K, s}(U_N(\mathfrak{u}) \otimes_{\mathbb{C}} V)$ is the same as the coefficient of $e^{\varphi + \delta_{L \cap K}}$ in $\sum_{|I| \leq N} (-1)^{|I'|} e^{\langle I \rangle} \Theta^{\text{num}}_{L \cap K}(V)$.

This conclusion remains valid as we let V swell to $Z^\#$, and then it remains valid when we let N tend to infinity. Since φ is arbitrary, we obtain

$$
\Theta^{\text{num}}_{L \cap K, s}(U(\mathfrak{u}) \otimes_{\mathbb{C}} Z^\#) = \left[\sum_I (-1)^{|I'|} e^{\langle I \rangle} \right] \Theta^{\text{num}}_{L \cap K}(Z^\#).
$$

The factor in brackets on the right side is

$$\Big(\prod_{\alpha \in \Delta(u \cap \mathfrak{k})} (1 - e^{\alpha} + e^{2\alpha} - \dots) \Big) \Big(\prod_{\alpha \in \Delta(u \cap \mathfrak{p})} (1 + e^{\alpha} + e^{2\alpha} + \dots) \Big),$$

and the proposition follows.

4. Signature Character of Alternating Tensors

Following the line of attack suggested by (9.3) and carried out for the examples in §2, we next compute the signature character of $\bigwedge^S \mathfrak{c}$. Recall that $\mathfrak{c} = u \cap \mathfrak{k} \oplus \bar{u} \cap \mathfrak{k}$.

Proposition 9.52.

$$\Theta_{L \cap K, s}(\textstyle\bigwedge^S \mathfrak{c}|_T) = e^{-2\delta(u \cap \mathfrak{k})} \prod_{\alpha \in \Delta(u \cap \mathfrak{k})} (1 + e^{\alpha})(1 - e^{\alpha}).$$

PROOF. As in the proof of Proposition 9.46, fix an enumeration $\alpha_1, \dots, \alpha_S$ of $\Delta(u \cap \mathfrak{k})$. Then

$$\xi_0 = e_{-\alpha_1} \wedge \cdots \wedge e_{-\alpha_S}$$

is a nonzero member of $\bigwedge^S(u \cap \mathfrak{k})$, and (9.39) shows that

$$\bar{\xi}_0 = (-1)^S e_{\alpha_1} \wedge \cdots \wedge e_{\alpha_S}.$$

The condition on ε_0 in $\bigwedge^{2S} \mathfrak{c}$ is that $\varepsilon_0(\xi_0 \wedge \bar{\xi}_0) > 0$, and the definition of the form on $\bigwedge^S \mathfrak{c}$ is

$$\langle \xi, \xi' \rangle = \varepsilon_0(\xi \wedge \bar{\xi}').$$

If P is an ordered subset of the ordered set $\{1, \dots, S\}$, we let e_P be the corresponding wedge of root vectors in $u \cap \mathfrak{k}$, and we let e_{-P} be the corresponding wedge of root vectors in $\bar{u} \cap \mathfrak{k}$. A basis of $\bigwedge^S \mathfrak{c}$ consists of all $e_{-P} \wedge e_Q$, where P and Q are subsets of $\Delta(u \cap \mathfrak{k})$ with $|P| + |Q| = S$. If $e_{-P} \wedge e_Q$ and $e_{-P'} \wedge e_{Q'}$ are given, then

$$(9.53) \qquad \langle e_{-P} \wedge e_Q, e_{-P'} \wedge e_{Q'} \rangle = (-1)^S \varepsilon_0(e_{-P} \wedge e_Q \wedge e_{P'} \wedge e_{-Q'}).$$

For this to be nonzero we must have

$$P \cup Q' = \Delta(u \cap \mathfrak{k}) \text{ disjointly} \quad \text{and} \quad Q \cup P' = \Delta(u \cap \mathfrak{k}) \text{ disjointly}.$$

So $Q' = P^c$ and $P' = Q^c$, where c indicates complement.

If $P \cap Q$ is not empty, then it follows that the span of the two vectors $e_{-P} \wedge e_Q$ and $e_{-Q^c} \wedge e_{P^c}$ is orthogonal to the other basis vectors. If we use $\langle P \rangle$ to denote the sum of the members of P, then both vectors have weight $\langle Q \rangle - \langle P \rangle$. Moreover, (6.47) gives

$$\begin{pmatrix} \langle e_{-P} \wedge e_Q, \, e_{-P} \wedge e_Q \rangle & \langle e_{-P} \wedge e_Q, \, e_{-P'} \wedge e_{Q'} \rangle \\ \langle e_{-P'} \wedge e_{Q'}, \, e_{-P} \wedge e_Q \rangle & \langle e_{-P'} \wedge e_{Q'}, \, e_{-P'} \wedge e_{Q'} \rangle \end{pmatrix} = \begin{pmatrix} 0 & c \\ \bar{c} & 0 \end{pmatrix}.$$

Hence the signature for the weight $\langle Q \rangle - \langle P \rangle$ on the two-dimensional space in question is $(1, 0, 1)$, and the contribution to the signature character is 0.

Now consider $\langle e_{-P} \wedge e_Q, \, e_{-P} \wedge e_Q \rangle$ when $Q = P^c$. If $\beta_j = \pm \alpha_j$ for each j, one easily checks that

$$e_{\beta_1} \wedge \cdots \wedge e_{\beta_s} \wedge e_{-\beta_1} \wedge \cdots \wedge e_{-\beta_s}$$

changes sign when a single β_i is replaced by its negative. Repeating this process $|Q|$ times, we see that

$$\langle e_{-P} \wedge e_Q, \, e_{-P} \wedge e_Q \rangle = (-1)^{|Q|} \langle \xi_0, \xi_0 \rangle$$

when $Q = P^c$. Therefore

$$\Theta_{L \cap K, s}\left(\textstyle\bigwedge^s \mathfrak{c}|_T\right) = \sum_{\substack{P \subseteq \Delta(\mathfrak{u} \cap \mathfrak{k}), \\ Q = P^c}} (-1)^{|Q|} e^{\langle Q \rangle - \langle P \rangle}$$

$$= \sum_{Q \subseteq \Delta(\mathfrak{u} \cap \mathfrak{k})} (-1)^{|Q|} e^{-\langle \Delta(\mathfrak{u} \cap \mathfrak{k}) \rangle} e^{2\langle Q \rangle}$$

$$= e^{-2\delta(\mathfrak{u} \cap \mathfrak{k})} \prod_{\alpha \in \Delta(\mathfrak{u} \cap \mathfrak{k})} (1 - e^{2\alpha}),$$

and the result follows.

5. Signature Character and Formal Character of $\mathcal{L}_S(Z)$

The hard work has been done, and now we can proceed quickly to the proof of Theorem 9.1. We need a few general facts about the relationship between complexes and cohomology.

Lemma 9.54. Let E be a finite-dimensional complex vector space carrying a nondegenerate Hermitian form $\langle \cdot, \cdot \rangle$ of vector-space signature $(p, 0, q)$, let S be a subspace of E such that $\langle \cdot, \cdot \rangle_{S \times S}$ is identically zero, and put

$$S^\perp = \{e \in E \mid \langle e, S \rangle = 0\} \supseteq S.$$

Then

(a) the radical of $\langle \cdot, \cdot \rangle|_{S^\perp \times S^\perp}$ is S, and thus $\langle \cdot, \cdot \rangle$ induces a nondegenerate Hermitian form $\langle \cdot, \cdot \rangle_F$ on $F = S^\perp/S$.

(b) the vector-space signature (p', q') of $\langle \cdot, \cdot \rangle_F$, satisfies

$$p' - q' = p - q.$$

PROOF. Choose an orthogonal decomposition $E = E_+ \oplus E_-$ as in Proposition 6.12, and let π_+ and π_- be the corresponding orthogonal projections. Because the restriction of $\langle \cdot, \cdot \rangle$ to any subspace of E_+ or E_- must be definite, we must have $S \cap E_+ = S \cap E_- = 0$. If S_+ and S_- denote the images of S under π_+ and π_-, then it follows that

$$\pi_+ : S \to S_+ \subseteq E_+ \qquad \text{and} \qquad \pi_- : S \to S_- \subseteq E_-$$

are isomorphisms. Write $j : S_+ \to S_-$ for the isomorphism

$$j = (\pi_-|_S) \circ (\pi_+|_S)^{-1}.$$

If $s = s_+ + s_- \in E_+ \oplus E_-$ is a member of S, then $(\pi_+|_S)^{-1}(s_+) = s$ and $(\pi_-|_S)(s) = s_-$. Hence $j(s_+) = s_-$, and we obtain

(9.55) $$S = \{s_+ + j(s_+) \in E_+ \oplus E_- \mid s_+ \in S_+\}.$$

Since $S_+ \subseteq E_+$ and $S_- \subseteq E_-$ are orthogonal, the assumption that $\langle \cdot, \cdot \rangle$ is trivial on S means that

(9.56) $$\langle s_+, s'_+ \rangle = -\langle j(s_+), j(s'_+) \rangle \qquad \text{for } s_+, s'_+ \in S_+.$$

Let F_+ be the orthogonal complement of S_+ in E_+, and let F_- be the orthogonal complement of S_- in E_-. Then

(9.57) $$E = F_+ \oplus S_+ \oplus S_- \oplus F_-.$$

Let us use (9.55) and (9.56) to check that

(9.58) $$S^\perp \cap (S_+ \oplus S_-) = S.$$

In fact, the left side clearly contains the right side. If $s_+ + s_-$ is a member of the left side, then every $s = s'_+ + j(s'_+)$ in S has

$$0 = \langle s_+ + s_-, s'_+ + j(s'_+) \rangle = \langle s_+, s'_+ \rangle + \langle s_-, j(s'_+) \rangle = \langle s_- - j(s_+), j(s'_+) \rangle.$$

Since $j(s'_+)$ is arbitrary in S_-, $s_- = j(s_+)$; therefore $s_+ + s_-$ is in S.

Combining (9.57) and (9.58), we obtain

$$S^\perp = F_+ \oplus S \oplus F_-.$$

Evidently this is a decomposition of $S^\perp$ as in Proposition 6.12. In particular, S is the radical of $\langle \cdot, \cdot \rangle_{S^\perp \times S^\perp}$. This proves (a).

We see also that the quotient $F = S^\perp/S$ is isomorphic (with its Hermitian form) to $F_+ \oplus F_-$. Thus the vector-space signature of the form on F is

$$(p', 0, q') = (\dim F_+, 0, \dim F_-).$$

By (9.57) the vector-space signature of the form on E is

$$(p, 0, q) = (\dim F_+ + \dim S_+, 0, \dim S_- + \dim F_-).$$

Since j exhibits S_+ and S_- as isomorphic vector spaces, we have $p' - q' = p - q$. This proves (b).

Lemma 9.59. Let

$$V : \quad 0 \longrightarrow V_0 \xrightarrow{\ d_0\ } V_1 \xrightarrow{\ d_1\ } V_2 \longrightarrow 0$$

be a complex of finite-dimensional complex vector spaces. Suppose that V_1 carries a nondegenerate Hermitian form $\langle \cdot, \cdot \rangle$ of vector-space signature $(p, 0, q)$ and that $V_0 \cong V_2^h$ via a nondegenerate Hermitian pairing of V_0 with V_2, also denoted $\langle \cdot, \cdot \rangle$. Suppose that the adjoint d_0 relative to this form and this pairing satisfies $d_0^* = d_1$. Write H^1 for the cohomology $\ker d_1 / \mathrm{image}\, d_0$. Then $\langle \cdot, \cdot \rangle$ descends to a Hermitian form $\langle \cdot, \cdot \rangle_{H^1}$ on H^1, and the vector-space signature $(p', 0, q')$ of $\langle \cdot, \cdot \rangle_{H^1}$ satisfies

$$p' - q' = p - q.$$

PROOF. Let $S = d_0(V_0)$ be the space of coboundaries. In the notation of Lemma 9.54, $S^\perp$ is the set of $v_1 \in V_1$ such that

$$0 = \langle v_1, d_0(V_0) \rangle = \langle d_0^*(v_1), V_0 \rangle = \langle d_1(v_1), V_0 \rangle.$$

Since the Hermitian pairing of V_0 and V_2 is nondegenerate, this equation holds if and only if $d_1(v_1) = 0$. Thus $S^\perp$ is the subspace of cocycles. Since V is a complex, the form on V_1 vanishes on S. According to Lemma 9.54a, the form on V_1 induces a nondegenerate form on $S^\perp/S$, and we have identified $S^\perp/S$ with H^1. Moreover, (b) in the lemma says that $p' - q' = p - q$. This completes the proof.

Now let us specialize to the setting of Theorem 9.1, with K connected as at the end of §1. We retain the notation of that theorem.

Lemma 9.60. $\Theta_{K,s}(\mathcal{L}_S(Z)) = \Theta_{K,s}(R(K) \otimes_{L\cap K} (\bigwedge^S \mathfrak{c} \otimes_{\mathbb{C}} \operatorname{ind} Z^\#))$.

PROOF. Fix a K type τ. We show that the coefficients of $\Theta_K(\tau)$ in the two sides of the lemma are the same. To do so, we shall apply Lemma 9.59 to the complex consisting of the degree $k = S + 1$, S, and $S - 1$ terms of the τ isotypic component of the complex

$$(9.61) \qquad R(K) \otimes_{L\cap K} (\bigwedge^k \mathfrak{c} \otimes_{\mathbb{C}} \operatorname{ind}_{\bar{\mathfrak{q}},L\cap K}^{\mathfrak{g},L\cap K}(Z^\#)).$$

Since $R(K) \cong \bigoplus_{\gamma \in \widehat{K}} V_\gamma \otimes_{\mathbb{C}} V_\gamma^*$ with K acting on V_γ and $L \cap K$ acting on V_γ^*, the isotypic component of (9.61) of interest is

$$(9.62) \qquad (\tau^*|_{L\cap K} \otimes_{\mathbb{C}} \bigwedge^k \mathfrak{c} \otimes_{\mathbb{C}} \operatorname{ind}_{\bar{\mathfrak{q}},L\cap K}^{\mathfrak{g},L\cap K}(Z^\#))_0 \tau.$$

In (9.62) the subscript 0 indicates the component corresponding to the trivial representation of $L \cap K$. The spaces are finite dimensional since $\tau^*|_{L\cap K}$ and $\bigwedge^k \mathfrak{c}$ are finite dimensional and $\operatorname{ind}_{\bar{\mathfrak{q}},L\cap K}^{\mathfrak{g},L\cap K}(Z^\#)$ is $L \cap K$ admissible (Proposition 5.96b). The maps are $d_0 = \partial$ and $d_1 = (-1)^S \partial$, the two operators ∂ being the usual ones as in (3.27). The cohomology corresponding to $k = S$ is just the τ isotypic subspace of $\mathcal{L}_S(Z)$.

The Hermitian form to be used in Lemma 9.59 is the one in (9.3), rewritten from (6.45b). For the Hermitian pairing on $V_0 \times V_2$, we can use (6.45b) also. Let v_S and v_{S+1} be members of (9.62) for $k = S$ and $S + 1$, respectively. The lemma will follow from Lemma 9.59 if we prove the adjoint formula

$$(9.63) \qquad \langle \partial v_S, v_{S+1} \rangle = (-1)^S \langle v_S, \partial v_{S+1} \rangle.$$

Starting from the left side of (9.63), we have

$$
\begin{aligned}
&\langle \partial v_S, v_{S+1} \rangle \\
&= \Psi_Z^G(\partial v_S)(\iota v_{S+1}) && \text{by (6.45a)} \\
&= \mathcal{E}_Z^G \circ \Gamma^{S+1}(\psi_Z^G) \circ \mathcal{D}_Z^G \circ \partial(v_S)(\iota v_{S+1}) && \text{by (6.33)} \\
&= (-1)^S \mathcal{E}_Z^G \circ \Gamma^{S+1}(\psi_Z^G) \circ d \circ \mathcal{D}_Z^G(v_S)(\iota v_{S+1}) && \text{by (3.35)} \\
&= (-1)^S \mathcal{E}_Z^G \circ d \circ \Gamma^S(\psi_Z^G) \circ \mathcal{D}_Z^G(v_S)(\iota v_{S+1}) && \begin{array}{l}\text{since } \Gamma^*(\psi_Z^G) \text{ is a} \\ \text{cochain map by} \\ \text{Lemma 6.42}\end{array} \\
&= (-1)^S \mathcal{E}_Z^G \circ \Gamma^S(\psi_Z^G) \circ \mathcal{D}_Z^G(v_S)(\iota \partial v_{S+1}) && \begin{array}{l}\text{since } \mathcal{E}_Z^G \text{ is a chain} \\ \text{map by Lemma 6.27}\end{array} \\
&= (-1)^S \langle v_S, \partial v_{S+1} \rangle && \text{by (6.33) and (6.45a).}
\end{aligned}
$$

This proves the lemma.

PROOF OF THEOREM 9.1. In view of (9.4), we are to prove that $\Theta_{K,s}(\mathcal{L}_S(Z))$ equals $\Theta_K(\mathcal{L}_S(Z))$. We shall compute the coefficient of $\Theta_K(\tau)$ in each of these expressions for each $\tau \in \widehat{K}$. Let us write V_τ for a concrete irreducible representation of K of type τ.

In the case of $\Theta_{K,s}(\mathcal{L}_S(Z))$, we apply Lemma 9.60, which says that we should compute the coefficient of $\Theta_K(\tau)$ in

$$(9.64) \qquad \Theta_{K,s}(R(K) \otimes_{L \cap K} (\textstyle\bigwedge^S \mathfrak{c} \otimes_{\mathbb{C}} \operatorname{ind} Z^{\#})).$$

Since $R(K) \cong \bigoplus_{\gamma \in \widehat{K}} V_\gamma \otimes_{\mathbb{C}} V_\gamma^*$ with K acting on V_γ and $L \cap K$ acting on V_γ^*, the coefficient of τ in (9.64) equals the coefficient of $\Theta_{L \cap K}(1)$ in

$$\Theta_{L \cap K,s}(V_\tau^* \otimes_{\mathbb{C}} \textstyle\bigwedge^S \mathfrak{c} \otimes_{\mathbb{C}} \operatorname{ind} Z^{\#}).$$

By Propositions 9.21 and 9.25 this is the coefficient of $e^{\delta_{L \cap K}}$ in

$$\Theta_{L \cap K,s}^{\mathrm{num}}(V_\tau^* \otimes_{\mathbb{C}} \textstyle\bigwedge^S \mathfrak{c} \otimes_{\mathbb{C}} \operatorname{ind} Z^{\#})$$
$$= \Theta_{L \cap K,s}(V_\tau^*)\Theta_{L \cap K,s}(\textstyle\bigwedge^S \mathfrak{c})\Theta_{L \cap K,s}^{\mathrm{num}}(\operatorname{ind} Z^{\#}).$$

On V_τ^* the form is the one from $R(K)$, hence positive definite and K invariant. Thus $\Theta_{L \cap K,s}(V_\tau^*) = \Theta_{L \cap K}(V_\tau^*)$. The other two signature characters are computed by Propositions 9.46 and 9.52. Substituting, we see that we seek the coefficient of $e^{\delta_{L \cap K}}$ in

$$\frac{\Theta_T(V_\tau^*)e^{-2\delta(\mathfrak{u} \cap \mathfrak{k})} \prod_{\alpha \in \Delta(\mathfrak{u} \cap \mathfrak{k})}(1 + e^\alpha)(1 - e^\alpha)\Theta_{L \cap K}^{\mathrm{num}}(Z^{\#})}{\prod_{\alpha \in \Delta(\mathfrak{u} \cap \mathfrak{k})}(1 + e^\alpha) \prod_{\alpha \in \Delta(\mathfrak{u} \cap \mathfrak{p})}(1 - e^\alpha)}.$$

Thus the coefficient of $\Theta_K(\tau)$ in $\Theta_{K,s}(\mathcal{L}_S(Z))$ is equal to the coefficient of $e^{\delta_{L \cap K}}$ in

$$(9.65) \qquad \frac{e^{-2\delta(\mathfrak{u} \cap \mathfrak{k})} \prod_{\alpha \in \Delta(\mathfrak{u} \cap \mathfrak{k})}(1 - e^\alpha)}{\prod_{\alpha \in \Delta(\mathfrak{u} \cap \mathfrak{p})}(1 - e^\alpha)} \cdot \Theta_T(V_\tau^*)\Theta_{L \cap K}^{\mathrm{num}}(Z^{\#}).$$

In the case of $\Theta_K(\mathcal{L}_S(Z))$, we apply Theorem 5.99, which says that $\mathcal{L}_j(Z) = 0$ for $j \neq S$ and that $\mathcal{L}_S(Z) \cong \mathcal{R}^S(Z)$. By Theorem 5.64 the coefficient of $\Theta_K(\tau)$ in $\Theta_K(\mathcal{L}_S(Z))$ is

$$(9.66)$$

$$= (-1)^S \sum_{j=0}^{S} (-1)^j \dim \operatorname{Hom}_K(V_\tau, \mathcal{R}^j(Z))$$

$$= (-1)^S \sum_{j=0}^{S} (-1)^j \sum_{n=0}^{\infty} \dim \operatorname{Hom}_{L \cap K}(H_j(\mathfrak{u} \cap \mathfrak{k}, V_\tau), S^n(\mathfrak{u} \cap \mathfrak{p}) \otimes_{\mathbb{C}} Z^{\#}).$$

The homology $H_j(\mathfrak{u} \cap \mathfrak{k}, V_\tau)$ is computed from the complex

$$0 \longleftarrow C_0(\mathfrak{u} \cap \mathfrak{k}, V_\tau) \longleftarrow C_1(\mathfrak{u} \cap \mathfrak{k}, V_\tau) \longleftarrow \cdots \longleftarrow C_S(\mathfrak{u} \cap \mathfrak{k}, V_\tau) \longleftarrow 0$$

in the category $\mathcal{C}(\mathfrak{l} \cap \mathfrak{k}, L \cap K)$, where $C_j(\mathfrak{u} \cap \mathfrak{k}, V_\tau) = \bigwedge^j(\mathfrak{u} \cap \mathfrak{k}) \otimes_{\mathbb{C}} V_\tau$. By Theorem C.40,

$$\sum_{j=0}^{S} (-1)^j (C_j(\mathfrak{u} \cap \mathfrak{k}, V_\tau)) = \sum_{j=0}^{S} (-1)^j (H_j(\mathfrak{u} \cap \mathfrak{k}, V_\tau))$$

in the Grothendieck group $K(\mathcal{C}(\mathfrak{l} \cap \mathfrak{k}, L \cap K))$. By Proposition 1.63a, the multiplicity of an $L \cap K$ type is a well-defined operation on this equation, and we see that (9.66) is

$$= (-1)^S \sum_{j=0}^{S} (-1)^j \sum_{n=0}^{\infty} \dim \operatorname{Hom}_{L \cap K}(C_j(\mathfrak{u} \cap \mathfrak{k}, V_\tau), S^n(\mathfrak{u} \cap \mathfrak{p}) \otimes_{\mathbb{C}} Z^\#),$$

and this equals the coefficient of $\Theta_{L \cap K}(1)$ in

$$= (-1)^S \sum_{j=0}^{S} (-1)^j \sum_{n=0}^{\infty} \Theta_{L \cap K}\left(\bigwedge^j(\mathfrak{u} \cap \mathfrak{k})^* \otimes_{\mathbb{C}} V_\tau^* \otimes_{\mathbb{C}} S^n(\mathfrak{u} \cap \mathfrak{p}) \otimes_{\mathbb{C}} Z^\#\right).$$

In turn, this is the coefficient of $e^{\delta_{L \cap K}}$ in

$$= (-1)^S \sum_{j=0}^{S} (-1)^j \sum_{n=0}^{\infty} \Theta_T(\bigwedge^j(\mathfrak{u} \cap \mathfrak{k})^*)\Theta_T(S^n(\mathfrak{u} \cap \mathfrak{p}))$$

$$\times \Theta_T(V_\tau^*)\Theta_{L \cap K}^{\mathrm{num}}(Z^\#).$$

Let P denote a subset of $\Delta(\mathfrak{u} \cap \mathfrak{k})$, and let I denote a tuple of integers indexed by the members of $\Delta(\mathfrak{u} \cap \mathfrak{p})$. As in the proofs of Propositions 9.52 and 9.46, $\langle P \rangle$ refers to the sum of the members of P, and $\langle I \rangle$ refers to the sum of the members of $\Delta(\mathfrak{u} \cap \mathfrak{p})$ weighted by the integers of I as coefficients. It is clear that

$$\Theta_T(\bigwedge^j(\mathfrak{u} \cap \mathfrak{k})^*) = \sum_{|P|=j} e^{-\langle P \rangle}$$

and

$$\Theta_T(S^n(\mathfrak{u} \cap \mathfrak{p})) = \sum_{|I|=n} e^{\langle I \rangle}.$$

Therefore

$$(-1)^S \sum_{j=0}^{S} (-1)^j \sum_{n=0}^{\infty} \Theta_T(\textstyle\bigwedge^j(\mathfrak{u} \cap \mathfrak{k})^*)\Theta_T(S^n(\mathfrak{u} \cap \mathfrak{p}))$$

$$= (-1)^S \sum_{P} \sum_{I} (-1)^{|P|} e^{-\langle P \rangle} e^{\langle I \rangle}$$

$$= \sum_{Q=P^c} \sum_{I} e^{-2\delta(\mathfrak{u} \cap \mathfrak{k})} (-1)^{|Q|} e^{\langle Q \rangle} e^{\langle I \rangle}$$

$$= \frac{e^{-2\delta(\mathfrak{u} \cap \mathfrak{k})} \prod_{\alpha \in \Delta(\mathfrak{u} \cap \mathfrak{k})} (1 - e^\alpha)}{\prod_{\alpha \in \Delta(\mathfrak{u} \cap \mathfrak{p})} (1 - e^\alpha)}.$$

We conclude that the coefficient of $\Theta_K(\tau)$ in $\Theta_K(\mathcal{L}_S(Z))$ is the coefficient of $e^{\delta_{L \cap K}}$ in

$$(9.67) \qquad \frac{e^{-2\delta(\mathfrak{u} \cap \mathfrak{k})} \prod_{\alpha \in \Delta(\mathfrak{u} \cap \mathfrak{k})} (1 - e^\alpha)}{\prod_{\alpha \in \Delta(\mathfrak{u} \cap \mathfrak{p})} (1 - e^\alpha)} \cdot \Theta_T(V_\tau^*) \Theta_{L \cap K}^{\text{num}}(Z^\#).$$

Since (9.65) and (9.67) match, (9.4) says that the Shapovalov form on $\mathcal{L}_S(Z)$ is positive definite.

6. Improved Theorem for $A_\mathfrak{q}(\lambda)$

In this section we specialize to the $(\mathfrak{g}, K)$ modules $A_\mathfrak{q}(\lambda)$ and $A^\mathfrak{q}(\lambda)$, which were defined in (5.6) as $\mathcal{L}_S(\mathbb{C}_\lambda)$ and $\mathcal{R}^S(\mathbb{C}_\lambda)$, respectively. Recall that they have infinitesimal character $\lambda + \delta$. In discussing these modules, we always assume that K is connected.

It is instructive to abstract what properties of Z were actually used in the proof of Theorem 9.1, including what the significance was of the condition on the infinitesimal character. Going over the proof, we are led to the following theorem.

Theorem 9.68. Let $(\mathfrak{g}, K)$ be a reductive pair, and let $\mathfrak{q} = \mathfrak{l} \oplus \mathfrak{u}$ be a θ stable parabolic subalgebra. Suppose that Z is a $Z(\mathfrak{l})$ finite $(\mathfrak{l}, L \cap K)$ module having the following properties:

  (i) Z carries a positive definite invariant Hermitian form $\langle \cdot, \cdot \rangle_L$.
  (ii) if Z_0 is any irreducible $U(\mathfrak{l})$ submodule of Z, then the $U(\mathfrak{g})$ module $\text{ind}_{\bar{\mathfrak{q}}}^{\mathfrak{g}}((Z_0 \otimes_{\mathbb{C}} \mathbb{C}_{t\delta(\mathfrak{u})})_{\bar{\mathfrak{q}}}^\#)$ is irreducible for all $t \geq 0$.

Then

(a) $\mathcal{L}_j(Z) \cong \mathcal{R}^j(Z) = 0$ for $j \neq S$,

(b) $\mathcal{L}_S(Z) \cong \mathcal{R}^S(Z)$, and

(c) the corresponding Shapovalov form $\langle \cdot, \cdot \rangle_G$ on $\mathcal{L}_S(Z)$ is positive definite.

PROOF. By Corollary 7.223, Z has finite length as an $(\mathfrak{l}, L \cap K)$ module. Condition (i) implies that Z is completely reducible as an $(\mathfrak{l}, L \cap K)$ module, and it follows from Lemma 7.224 that Z is completely reducible as a $U(\mathfrak{l})$ module. Condition (ii) guarantees that the maps

$$(9.69) \qquad \varphi_t : \operatorname{ind}_{\bar{\mathfrak{q}}}^{\mathfrak{g}}((Z \otimes_{\mathbb{C}} \mathbb{C}_{t\delta(\mathfrak{u})})_{\bar{\mathfrak{q}}}^{\#}) \to \operatorname{pro}_{\mathfrak{q}}^{\mathfrak{g}}((Z \otimes_{\mathbb{C}} \mathbb{C}_{t\delta(\mathfrak{u})})_{\mathfrak{q}}^{\#})$$

are one-one for all $t \geq 0$.

When $t = 0$, φ_0 reduces to φ_Z^G, and Proposition 5.96c shows that φ_0 is an isomorphism of $\operatorname{ind}_{\bar{\mathfrak{q}}, L \cap K}^{\mathfrak{g}, L \cap K}(Z_{\bar{\mathfrak{q}}}^{\#})$ onto $\operatorname{pro}_{\mathfrak{q}, L \cap K}^{\mathfrak{g}, L \cap K}(Z_{\mathfrak{q}}^{\#})$. Then (a) and (b) follow from Proposition 5.93.

Conclusion (c) is what has been proved in §§2–5. It is a formal consequence of (a), (b), and the fact that (9.69) is one-one for all $t \geq 0$.

Corollary 9.70. Let $(\mathfrak{g}, K)$ be a reductive pair with K connected, let $\mathfrak{q} = \mathfrak{l} \oplus \mathfrak{u}$ be a θ stable parabolic subalgebra, let $\mathfrak{h}_0 = \mathfrak{t}_0 \oplus \mathfrak{a}_0$ be a θ stable Cartan subalgebra of $\mathfrak{l}_0$, and introduce a positive system $\Delta^+(\mathfrak{g}, \mathfrak{h})$ such that $\Delta(\mathfrak{u}) \subseteq \Delta^+(\mathfrak{g}, \mathfrak{h})$. Suppose that $\mathbb{C}_\lambda$ is a one-dimensional $(\mathfrak{l}, L \cap K)$ module whose differential $\lambda \in \mathfrak{h}^*$ is real on $\mathfrak{t}_0$ and imaginary on $\mathfrak{a}_0$, and suppose that

$$(9.71) \qquad \operatorname{Re}\langle \lambda + \delta(\mathfrak{u}), \alpha \rangle \geq 0 \qquad \text{for all } \alpha \in \Delta(\mathfrak{u}).$$

Then the isomorphic $(\mathfrak{g}, K)$ modules $A_q(\lambda)$ and $A^q(\lambda)$ are infinitesimally unitary.

PROOF. Under the hypothesis (9.71), $Z = \mathbb{C}_\lambda$ satisfies condition (ii) in Theorem 9.68, as we see by going over the proofs of Theorem 5.109 and Corollary 5.105. Condition (i) follows from the assumption about the differential. Therefore Theorem 9.68 applies, and conclusions (b) and (c) say that $A_q(\lambda)$ and $A^q(\lambda)$ are isomorphic and infinitesimally unitary.

This corollary is applicable to most of the examples considered in §VIII.5. The details are as follows.

EXAMPLES.

1) $G = SL(2n, \mathbb{R})$ and the Speh representations. The members of the two series of Speh representations are distinct and irreducible, as was shown in §VIII.5, and they are infinitesimally unitary by Corollary 9.70.

2) $G = Sp(2, \mathbb{R})$ and reducibility. We gave an example of a reducible $A_{\mathfrak{q}}(\lambda)$, with $\mathfrak{l}$ built from the long simple root and with $\lambda = -2e_1$. For this example, $\lambda + \delta(\mathfrak{u}) = 0$ is dominant. Thus Corollary 9.70 is applicable and proves unitarity. In the course of the proof of reducibility in §VIII.5, we introduced an auxiliary $(\mathfrak{g}, K)$ module $A_{\mathfrak{q}'}(\lambda')$ with $\mathfrak{l}'$ built from the short simple root and with $\lambda' + \delta(\mathfrak{u}) = -\frac{1}{2}e_1 - \frac{1}{2}e_2$. In this case, $\lambda' + \delta(\mathfrak{u})$ is not dominant, and Corollary 9.70 is not applicable to decide unitarity. However, the relevant generalized Verma modules can be shown to be irreducible, and then Theorem 9.68 says that $A_{\mathfrak{q}'}(\lambda')$ is infinitesimally unitary.

3) $G = SO(5, 4)$ and reducibility. We gave a specific example of a reducible $A_{\mathfrak{q}}(\lambda)$, omitting some of the details. Since $\lambda + \delta(\mathfrak{u}) = 0$ is dominant, the hypotheses of Corollary 9.70 are satisfied, and $A_{\mathfrak{q}}(\lambda)$ is infinitesimally unitary. We showed that the K type with highest weight $\Lambda' = 2e_1 + e_4$ has multiplicity 1. The representation of exceptional interest here is not the $A_{\mathfrak{q}}(\lambda)$, however, but the $(\mathfrak{g}, K)$ submodule generated by the K type with highest weight Λ'. This submodule is irreducible. In fact, $A_{\mathfrak{q}}(\lambda)$ has finite length by Corollary 7.223. Since $A_{\mathfrak{q}}(\lambda)$ is infinitesimally unitary, it follows from the remarks at the beginning of this chapter that $A_{\mathfrak{q}}(\lambda)$ is fully reducible. Consequently any K type of multiplicity 1 generates an irreducible submodule. The K type with highest weight $\Lambda' = 2e_1 + e_4$ meets this condition, and therefore the $(\mathfrak{g}, K)$ submodule generated by this K type is irreducible. This irreducible representation plays a special role in using the Langlands classification to help in classifying irreducible unitary representations: Except in split F_4, this representation is the most difficult case in which to prove unitarity among all representations arising from rank-one parabolic subgroups of simple Lie groups (cf. Chapter XI). For it, the proof of unitarity using the embedding in an $A_{\mathfrak{q}}(\lambda)$ is the only known proof of unitarity. Notice the importance in Corollary 9.70 of having a conclusion that the invariant Hermitian form on $A_{\mathfrak{q}}(\lambda)$ is positive definite, not merely positive semidefinite. If the form were only semidefinite, we would not obtain the unitarity of the subrepresentation of interest.

4) $G = Sp(5, 2)$ and a nonzero $A_{\mathfrak{q}}(\lambda)$ with no bottom layer. We saw that this representation contains the trivial K type. Corollary 9.70 shows that it is infinitesimally unitary. This representation arises also as a member of the discrete series for a certain semisimple symmetric

space and is irreducible, according to the theory of those discrete series. Even if that theory is not used, for purposes of classifying irreducible unitary representations, the representation of interest is the (irreducible) subspace generated by the trivial K type.

CHAPTER X

MINIMAL K TYPES

In any finitely generated $(\mathfrak{g}, K)$ module, where $(\mathfrak{g}, K)$ is a reductive pair with K connected, every K isotypic subspace is finitely generated as a $Z(\mathfrak{g})$ module. It follows that every irreducible $(\mathfrak{g}, K)$ module is admissible.

For a reductive pair $(\mathfrak{g}, K)$, if an infinitesimal character is specified, then there exist finitely many irreducible representations of K such that every irreducible $(\mathfrak{g}, K)$ module with that infinitesimal character contains one of these K types. The notion used in the proof is that of "minimal K type" for a $(\mathfrak{g}, K)$ module. Under some hypotheses any module $A_{\mathfrak{q}}(\lambda)$ has a unique minimal K type, namely the K type with highest weight $\lambda + 2\delta(\mathfrak{u} \cap \mathfrak{p})$. In other words the minimal K type is the unique K type in the bottom layer.

More generally, under a suitable hypothesis of strict dominance, the minimal K types of a module cohomologically induced from a Levi subgroup L have highest weights exactly the sums $\lambda + 2\delta(\mathfrak{u} \cap \mathfrak{p})$, where λ is the highest weight of a minimal $L \cap K$ type of the inducing module. In particular, all the minimal K types lie in the bottom layer.

1. Admissibility of Irreducible $(\mathfrak{g}, K)$ Modules

This chapter deals with some loose ends from Chapter VII. We have left unproved Theorem 7.204 until now, and we shall prove part of it as Theorem 10.1 below and the remainder as Theorem 10.23 in §2. Along the way we shall introduce the "minimal K types" of a $(\mathfrak{g}, K)$ module. These are an important algebraic invariant of a $(\mathfrak{g}, K)$ module and especially of an irreducible $(\mathfrak{g}, K)$ module. In fact, for an irreducible $(\mathfrak{g}, K)$ module, the infinitesimal character and the minimal K types come close to characterizing the module up to equivalence.

Theorem 10.1. If $(\mathfrak{g}, K)$ is a reductive pair with K connected, then every irreducible $(\mathfrak{g}, K)$ module is admissible.

If V is an irreducible $(\mathfrak{g}, K)$ module, then $Z(\mathfrak{g})$ acts in V by scalars, according to Proposition 4.87. Hence Theorem 10.1 is a special case of the following more general theorem.

Theorem 10.2. Let $(\mathfrak{g}, K)$ be a reductive pair with K connected, let V be a finitely generated $(\mathfrak{g}, K)$ module, let $\tau \in \widehat{K}$ be a K type, and let V_τ be the K isotypic component of V of type τ. Then V_τ is finitely generated as a $Z(\mathfrak{g})$ module.

To prove Theorem 10.2, let V_0 be a finite-dimensional K invariant subspace of V such that $V = U(\mathfrak{g})V_0$. Define

$$(10.3) \qquad\qquad V_n = U_n(\mathfrak{g})V_0.$$

Here we use the convention that $U_{-1}(\mathfrak{g}) = 0$, so that $V_{-1} = 0$. The definition (10.3) in the case that $n = 0$ is consistent with the choice of V_0 already made, and we have

$$(10.4) \qquad\qquad 0 = V_{-1} \subseteq V_0 \subseteq V_1 \subseteq V_2 \subseteq \cdots$$

and

$$(10.5) \qquad\qquad U_m(\mathfrak{g})V_n \subseteq V_{m+n}.$$

By definition of $(\mathfrak{g}, K)$ module,

$$(10.6) \qquad k(uv) = (\mathrm{Ad}(k)u)(kv) \qquad \text{for } k \in K,\ u \in U(\mathfrak{g}),\ v \in V.$$

Since $U_n(\mathfrak{g})$ is preserved by Ad, it follows that V_n is a finite-dimensional K invariant subspace of V. The identity $V = U(\mathfrak{g})V_0$ implies that $V = \bigcup_{n=0}^\infty V_n$. Define

$$\mathrm{gr}\,V = \bigoplus_{n=0}^\infty (V_n/V_{n-1}).$$

This space inherits from V a locally finite representation of K; as a representation of K, $\mathrm{gr}\,V$ is equivalent with V.

Condition (10.5) allows us to make $\mathrm{gr}\,V$ into a graded module first for the graded algebra

$$(10.7) \qquad\qquad \mathrm{gr}\,U(\mathfrak{g}) = \bigoplus_{m=0}^\infty (U_m(\mathfrak{g})/U_{m-1}(\mathfrak{g}))$$

and then for the symmetric algebra $S(\mathfrak{g})$. The action is defined as follows. Suppose p is in $S^m(\mathfrak{g})$ and w is in $(\mathrm{gr}\,V)_n$. Choose representatives $u \in U_m(\mathfrak{g})$ and $v \in V_n$ for these classes. Then pw is the class in $(\mathrm{gr}\,V)_{m+n}$ represented by uv in V_{n+m}. This action is well defined as a consequence of (10.5). The right side of (10.7) is canonically isomorphic with the symmetric algebra $S(\mathfrak{g})$, as a consequence of the Poincaré-Birkhoff-Witt Theorem, and consequently $\operatorname{gr} U(\mathfrak{g})$ is an $S(\mathfrak{g})$ module. By (10.3), $\operatorname{gr} V$ is generated by V_0 as an $S(\mathfrak{g})$ module.

Since V_n is K invariant, the Lie algebra $\mathfrak{k}$ must preserve V_n. Taking (10.5) into account with $m = 1$, we see that $\mathfrak{k}$ acts by 0 on $\operatorname{gr} V$. Consequently $\operatorname{gr} V$ is a (graded) module for $S(\mathfrak{g}/\mathfrak{k}) \cong S(\mathfrak{p})$, and it is still generated by V_0. Because of (10.6), the actions of K and $\mathfrak{p}$ on $\operatorname{gr} V$ are related by

$$(10.8) \quad k(pw) = (\operatorname{Ad}(k)p)(kw) \qquad \text{for } k \in K, \ p \in S(\mathfrak{p}), \ w \in \operatorname{gr} V.$$

Let $S(\mathfrak{p})^K$ be the subalgebra of K invariants in $S(\mathfrak{p})$. Because of (10.8), the module action of any $p \in S(\mathfrak{p})^K$ on $\operatorname{gr} V$ commutes with the action of K. Consequently the action of $\mathfrak{p}$ preserves each isotypic subspace $(\operatorname{gr} V)_\tau$ of $\operatorname{gr} V$ for $\tau \in \widehat{K}$, and we conclude that $(\operatorname{gr} V)_\tau$ is an $S(\mathfrak{p})^K$ module.

Lemma 10.9. If τ is in $\widehat{K}$, then $(\operatorname{gr} V)_\tau$ is finitely generated as an $S(\mathfrak{p})^K$ module.

PROOF. Let F_τ be an irreducible K module of type τ. Then $\operatorname{Hom}_{\mathbb{C}}(F_\tau, \operatorname{gr} V)$ is a K module and an $S(\mathfrak{p})$ module under the definitions

$$(k\varphi)(x) = k(\varphi(k^{-1}x)) \qquad \text{for } k \in K, \ x \in F_\tau, \ \varphi \in \operatorname{Hom}_{\mathbb{C}}(F_\tau, \operatorname{gr} V)$$
$$(X\varphi)(x) = X(\varphi(x)) \qquad \text{for } X \in \mathfrak{p}, \ x \in F_\tau, \ \varphi \in \operatorname{Hom}_{\mathbb{C}}(F_\tau, \operatorname{gr} V).$$

As an $S(\mathfrak{p})$ module, $\operatorname{Hom}_{\mathbb{C}}(F_\tau, \operatorname{gr} V)$ is isomorphic to $F_\tau^* \otimes_{\mathbb{C}} \operatorname{gr} V$, with $\mathfrak{p}$ acting in the second factor. Since V_0 generates $\operatorname{gr} V$ as an $S(\mathfrak{p})$ module, it follows that $\operatorname{Hom}_{\mathbb{C}}(F_\tau, \operatorname{gr} V)$ is finitely generated as an $S(\mathfrak{p})$ module. The algebra $S(\mathfrak{p})$ is Noetherian, by the Hilbert Basis Theorem, and hence any $S(\mathfrak{p})$ submodule of $\operatorname{Hom}_{\mathbb{C}}(F_\tau, \operatorname{gr} V)$ satisfies the ascending-chain condition (for its submodules). We apply this fact to the $S(\mathfrak{p})$ submodule generated by the subspace $\operatorname{Hom}_K(F_\tau, \operatorname{gr} V)$ of K invariants. The conclusion is that there is a finite set $\varphi_1, \ldots, \varphi_N$ in $\operatorname{Hom}_K(F_\tau, \operatorname{gr} V)$ with the property that

$$(10.10) \qquad S(\mathfrak{p})(\operatorname{Hom}_K(F_\tau, \operatorname{gr} V)) = \sum_{j=1}^{N} S(\mathfrak{p})\varphi_j.$$

To see this conclusion, notice first that (10.10) is equivalent to the condition

$$(10.11) \qquad \operatorname{Hom}_K(F_\tau, \operatorname{gr} V) \subseteq \sum_{j=1}^{N} S(\mathfrak{p})\varphi_j.$$

We choose the φ_i as follows. Choose φ_1 arbitrarily. Once $\varphi_1, \ldots, \varphi_n$ are chosen, either (10.11) holds (and we stop) or we can choose φ_{n+1} not in $\sum_{j=1}^{n} S(\mathfrak{p})\varphi_j$. In this way we get a strictly increasing chain of submodules $\sum_{j=1}^{n} S(\mathfrak{p})\varphi_j$. By the ascending-chain condition, this chain must terminate, and then we have established (10.11).

If φ is an arbitrary member of $\mathrm{Hom}_K(F_\tau, \mathrm{gr}\, V)$, we can write

$$\varphi = \sum_{j=1}^{N} u_j \varphi_j \qquad \text{with } u_j \in S(\mathfrak{p}).$$

By (10.8), each $k \in K$ satisfies

$$\varphi = k\varphi = \sum_{j=1}^{N} (\mathrm{Ad}(k)u_j)k\varphi_j = \sum_{j=1}^{N} (\mathrm{Ad}(k)u_j)\varphi_j.$$

Integration over K therefore gives

(10.12) $$\varphi = \sum_{j=1}^{N} \left(\int_K \mathrm{Ad}(k)u_j \, dk \right)\varphi_j.$$

Since $\int_K \mathrm{Ad}(k)u_j \, dk$ is a member of $S(\mathfrak{p})^K$, this equation exhibits $\mathrm{Hom}_K(F_\tau, \mathrm{gr}\, V)$, which is the isotypic subspace of the $S(\mathfrak{p})$ module $\mathrm{Hom}_{\mathbb{C}}(F_\tau, \mathrm{gr}\, V)$ for the trivial K type, as a finitely generated $S(\mathfrak{p})^K$ module.

Let $\{x_i\}$ be a basis of F_τ. We claim that the finite set $\{\varphi_j(x_i)\}$ generates $(\mathrm{gr}\, V)_\tau$ as an $S(\mathfrak{p})^K$ module. To see this fact, consider the linear map

$$\alpha : \mathrm{Hom}_K(F_\tau, \mathrm{gr}\, V) \otimes_{\mathbb{C}} F_\tau \to (\mathrm{gr}\, V)_\tau$$

defined by $\alpha(\varphi \otimes x) = \varphi(x)$. This map is onto since $(\mathrm{gr}\, V)_\tau$ is spanned by subspaces isomorphic to F_τ. If v is given in $(\mathrm{gr}\, V)_\tau$, we can use (10.12) to write

$$v = \alpha\left(\sum_i \psi_i \otimes x_i \right) = \sum_i \psi_i(x_i) = \sum_{i,j} u_{ij}\varphi_j(x_i)$$

with $u_{ij} \in S(\mathfrak{p})^K$. The lemma follows.

As in §VII.5, we can regard $S(\mathfrak{g}^*)$ as the algebra of all (holomorphic) polynomial functions from $\mathfrak{g}$ into $\mathbb{C}$. If $\mathfrak{b}$ is any subspace of $\mathfrak{g}$, we can restrict these polynomial functions to $\mathfrak{b}$, thereby obtaining a restriction map $r_{\mathfrak{g}|\mathfrak{b}} : S(\mathfrak{g}^*) \to S(\mathfrak{b}^*)$. More generally if $\mathfrak{c}$ is a subspace of $\mathfrak{b}$, we have a restriction map $r_{\mathfrak{b}|\mathfrak{c}} : S(\mathfrak{b}^*) \to S(\mathfrak{c}^*)$.

Lemma 10.13. If $\mathfrak{a}_0$ is a maximal abelian subspace of $\mathfrak{p}_0$, then $\mathfrak{p}_0 = \bigcup_{k \in K} \mathrm{Ad}(k)\mathfrak{a}_0$. Consequently the restriction map

$$r_{\mathfrak{p}|\mathfrak{a}} : S(\mathfrak{p}^*)^K \to S(\mathfrak{a}^*)$$

is one-one.

REMARKS. Introduce the inner product $\langle \cdot, \cdot \rangle_\theta$ on $\mathfrak{g}_0$ given by

$$\langle X, Y \rangle_\theta = -\langle X, \theta Y \rangle.$$

Any member T of $\mathfrak{p}_0$ has

$$
\begin{aligned}
\langle (\mathrm{ad}\, T)X, Y \rangle_\theta &= -\langle [T, X], \theta Y \rangle = \langle X, [T, \theta Y] \rangle \\
&= -\langle X, [\theta T, \theta Y] \rangle = -\langle X, \theta [T, Y] \rangle = \langle X, (\mathrm{ad}\, T)Y \rangle_\theta,
\end{aligned}
$$

and hence the members of $\mathrm{ad}\,\mathfrak{a}_0$ form a commuting family of self-adjoint transformations. They are thus simultaneously diagonable. We call the nonzero simultaneous eigenvalues (which may be viewed as in $\mathfrak{a}_0^*$ or $\mathfrak{a}^*$) **restricted roots** of $(\mathfrak{g}_0, \mathfrak{a}_0)$.

PROOF. Let $\mathfrak{a}_0'$ be a second maximal abelian subspace of $\mathfrak{p}_0$. Choose $h \in \mathfrak{a}_0$ so that no restricted root of $(\mathfrak{g}_0, \mathfrak{a}_0)$ vanishes on h; this is possible since there are only finitely many restricted roots and their kernels are hyperplanes in $\mathfrak{a}_0$. Similarly choose $h' \in \mathfrak{a}_0'$ so that no restricted root of $(\mathfrak{g}_0, \mathfrak{a}_0')$ vanishes on h'. Then

$$(10.14) \qquad Z_{\mathfrak{p}_0}(h) = \mathfrak{a}_0 \qquad \text{and} \qquad Z_{\mathfrak{p}_0}(h') = \mathfrak{a}_0'.$$

Choose $k_0 \in K$ so that $\langle \mathrm{Ad}(k)h, h' \rangle$ has a minimum at $k = k_0$. If X is in $\mathfrak{k}_0$, then $\langle \mathrm{Ad}(\exp t X)\mathrm{Ad}(k_0)h, h' \rangle$ is a smooth function of $t \in \mathbb{R}$ that is minimized for $t = 0$. Differentiating and setting $t = 0$, we obtain

$$0 = \langle (\mathrm{ad}\, X)\mathrm{Ad}(k_0)h, h' \rangle = \langle [X, \mathrm{Ad}(k_0)h], h' \rangle = \langle X, [\mathrm{Ad}(k_0)h, h'] \rangle.$$

Since X is arbitrary in $\mathfrak{k}_0$ and since $\langle \cdot, \cdot \rangle$ is nondegenerate on $\mathfrak{k}_0$, $[\mathrm{Ad}(k_0)h, h'] = 0$. Thus $\mathrm{Ad}(k_0)h$ centralizes h', and (10.14) shows that $\mathrm{Ad}(k_0)h$ is in $\mathfrak{a}_0'$. Hence h is in $\mathrm{Ad}(k_0)^{-1}\mathfrak{a}_0'$. Since $\mathrm{Ad}(k_0)^{-1}\mathfrak{a}_0'$ is an abelian subspace of $\mathfrak{p}_0$, it is contained in the centralizer in $\mathfrak{p}_0$ of h, which is $\mathfrak{a}_0$ by (10.14). Thus $\mathrm{Ad}(k_0)^{-1}\mathfrak{a}_0' \subseteq \mathfrak{a}_0$. Since $\mathrm{Ad}(k_0)^{-1}\mathfrak{a}_0'$ is maximal abelian, equality must hold. Thus $\mathfrak{a}_0' = \mathrm{Ad}(k_0)\mathfrak{a}_0$, and we have $\mathfrak{p}_0 = \bigcup_{k \in K} \mathrm{Ad}(k)\mathfrak{a}_0$. The conclusion about the restriction map is immediate, and the lemma follows.

Lemma 10.15. With $\mathfrak{a}_0$ as in Lemma 10.13, $S(\mathfrak{a}^*)$ is finitely generated as an $r_{\mathfrak{g}|\mathfrak{a}}(S(\mathfrak{g}^*)^G)$ module.

PROOF. Without loss of generality, we may assume that $\mathfrak{g}$ is semisimple. Let $d = \dim_{\mathbb{C}} \mathfrak{g}$, and define polynomial functions q_j on $\mathfrak{g}$ by

$$(10.16) \qquad \det(tI - \operatorname{ad} X) = \sum_{j=0}^{d} t^j q_j(X) \qquad \text{for } X \in \mathfrak{g}.$$

Each q_j is in $S(\mathfrak{g}^*)^G$. Let $\alpha_1, \ldots, \alpha_d$ be the simultaneous eigenvalues of $\operatorname{ad} \mathfrak{a}_0$ (zero and restricted roots), repeated with multiplicities. Replacing X in (10.16) by a member h of $\mathfrak{a}_0$, we have

$$\prod_{j=0}^{d} (t - \alpha_j(h)) = \sum_{j=0}^{d} t^j q_j(h).$$

For $t = \alpha_i(h)$, this equation gives

$$\sum_{j=0}^{d} \alpha_i(h)^j q_j(h) = 0 \qquad \text{for } 1 \leq i \leq d.$$

Since $q_d = 1$, we can rewrite this equation as

$$(10.17) \qquad \alpha_i(h)^d = -\sum_{j=0}^{d-1} \alpha_i(h)^j q_j(h) \qquad \text{for } 1 \leq i \leq d.$$

Since $\mathfrak{g}$ is semisimple, the α_i span $\mathfrak{a}_0^*$. Thus the most general element of $S(\mathfrak{a}^*)$ is a linear combination of monomials

$$\alpha_1(h)^{n_1} \cdots \alpha_d(h)^{n_d},$$

and (10.17) allows us to introduce coefficients $q_j(h)$ and reduce each degree to less than d. Since the q_j are in $S(\mathfrak{g}^*)^G$, the monomials

$$\alpha_1(h)^{n_1} \cdots \alpha_d(h)^{n_d} \qquad \text{with all } n_i < d$$

generate $S(\mathfrak{a}^*)$ as an $r_{\mathfrak{g}|\mathfrak{a}}(S(\mathfrak{g}^*)^G)$ module.

Lemma 10.18. $S(\mathfrak{p}^*)^K$ is finitely generated as an $r_{\mathfrak{g}|\mathfrak{p}}(S(\mathfrak{g}^*)^G)$ module.

PROOF. If $\mathfrak{h}$ is a Cartan subalgebra of $\mathfrak{g}$, Theorem 7.65 exhibits $S(\mathfrak{g}^*)^G$ as isomorphic with $S(\mathfrak{h}^*)^W$, and Theorem 7.30 identifies $S(\mathfrak{h}^*)^W$ as a polynomial algebra. By the Hilbert Basis Theorem, $S(\mathfrak{g}^*)^G$ is Noetherian. Consequently its homomorphic image $r_{\mathfrak{g}|\mathfrak{a}}(S(\mathfrak{g}^*)^G)$ is Noetherian.

By Lemma 10.15, $S(\mathfrak{a}^*)$ is finitely generated as an $r_{\mathfrak{g}|\mathfrak{a}}(S(\mathfrak{g}^*)^G)$ module, and hence the submodule $r_{\mathfrak{p}|\mathfrak{a}}(S(\mathfrak{p}^*)^K)$ is finitely generated. If $r_{\mathfrak{p}|\mathfrak{a}}(p_1), \ldots, r_{\mathfrak{p}|\mathfrak{a}}(p_m)$ are generators with $p_j \in S(\mathfrak{p}^*)^K$, then any p in $S(\mathfrak{p}^*)^K$ satisfies

$$r_{\mathfrak{p}|\mathfrak{a}}(p) = \sum_{j=1}^{m} r_{\mathfrak{g}|\mathfrak{a}}(u_j)r_{\mathfrak{p}|\mathfrak{a}}(p_j)$$

for some $u_1, \ldots, u_m$ in $S(\mathfrak{g}^*)^G$. The right side is

$$= \sum_{j=1}^{m} r_{\mathfrak{p}|\mathfrak{a}}(r_{\mathfrak{g}|\mathfrak{p}}(u_j)p_j)$$

with $r_{\mathfrak{g}|\mathfrak{p}}(u_j)p_j \in S(\mathfrak{p}^*)^K$. By Lemma 10.13, we obtain

$$p = \sum_{j=1}^{m} r_{\mathfrak{g}|\mathfrak{p}}(u_j)p_j,$$

and thus $p_1, \ldots, p_m$ generate $S(\mathfrak{p}^*)^K$ as an $r_{\mathfrak{g}|\mathfrak{p}}(S(\mathfrak{g}^*)^G)$ module.

PROOF OF THEOREM 10.2. Combining Lemma 10.9, Lemma 10.18, and the isomorphism $\mathfrak{g} \cong \mathfrak{g}^*$ provided by the invariant bilinear form $\langle \cdot, \cdot \rangle$, we see that each isotypic space $(\mathrm{gr}\, V)_\tau$ is finitely generated as an $S(\mathfrak{g})^G$ module. Let $w_1, \ldots, w_r$ be homogeneous generators with w_j homogeneous of degree k_j, so that w_j is in $(V_{k_j}/V_{k_j-1})_\tau$ and

$$(10.19) \qquad (\mathrm{gr}\, V)_\tau = \sum_{j=1}^{r} S(\mathfrak{g})^G w_j.$$

For $1 \leq j \leq r$, choose a representative v_j in $(V_{k_j})_\tau$ for w_j. To prove the theorem, it is enough to show that

$$(10.20) \qquad V_\tau = \sum_{j=1}^{r} Z(\mathfrak{g})v_j.$$

It suffices to show by induction on $n \geq -1$ that

$$(10.21) \qquad (V_n)_\tau \subseteq \sum_{j=1}^{r} Z(\mathfrak{g})v_j.$$

This inclusion is obvious for $n = -1$. Suppose inductively that n is ≥ 0 and that (10.21) is known for $n - 1$. Fix v in $(V_n)_\tau$, and write w for the corresponding class in $(\operatorname{gr} V)_n$. By (10.19) we can find elements $p_j \in S(\mathfrak{g})^G$ homogeneous of degree $n - k_j$ so that

$$(10.22) \qquad\qquad w = \sum_{j=1}^{r} p_j w_j.$$

Let $\sigma : S(\mathfrak{g}) \to U(\mathfrak{g})$ be the symmetrization map. Since σ respects adjoint actions, the elements $z_j = \sigma(p_j)$ are in $Z(\mathfrak{g}) \cap U_{n-k_j}(\mathfrak{g})$ and represent p_j. By definition of the action of $S(\mathfrak{g})$ on $\operatorname{gr} V$, (10.22) means precisely that $\sum z_j v_j$ is in V_n and that

$$w - \sum z_j v_j \qquad \text{is in } V_{n-1}.$$

Since this expression evidently belongs to V_τ, the inductive hypothesis provides elements $z_j' \in Z(\mathfrak{g})$ so that

$$w - \sum z_j v_j = \sum z_j' v_j,$$

and this is (10.21). The induction is complete, and (10.20) and the theorem follow.

2. Minimal *K* Types and Infinitesimal Characters

A goal of this section is to complete the proof of Theorem 7.204, and the following theorem will achieve this goal. In the course of the argument, we shall introduce the "minimal *K* types" of a $(\mathfrak{g}, K)$ module.

Theorem 10.23. Let $(\mathfrak{g}, K)$ be a reductive pair with *K* connected, and let $\mathfrak{h}$ be a Cartan subalgebra of $\mathfrak{g}$. If $\lambda \in \mathfrak{h}^*$ is fixed, then there exist finitely many *K* types $\tau_1, \ldots, \tau_n$ such that every irreducible $(\mathfrak{g}, K)$ module with infinitesimal character λ contains one of these *K* types.

Fix a reductive pair $(\mathfrak{g}, K)$. We shall define the "minimal *K* types" of a $(\mathfrak{g}, K)$ module *V*. For the moment assume that *K* is connected. Let $\mathfrak{h}_0 = \mathfrak{t}_0 \oplus \mathfrak{a}_0$ be a maximally compact θ stable Cartan subalgebra of $\mathfrak{g}_0$. Then $\mathfrak{t}$ is a Cartan subalgebra of $\mathfrak{k}$, and we fix a positive system $\Delta^+(\mathfrak{k}, \mathfrak{t})$

for the roots of $\mathfrak{k}$. Write δ_K for half the sum of the members of $\Delta^+(\mathfrak{k}, \mathfrak{t})$. If μ is the highest weight of a K type τ_μ, define

$$\|\tau_\mu\| = |\mu + 2\delta_K| = \sqrt{\langle \mu + 2\delta_K, \mu + 2\delta_K \rangle}.$$

A **minimal K type** of the $(\mathfrak{g}, K)$ module is a K type that has minimal norm among all K types occurring in V. Since the highest weights are points in a lattice of $i\mathfrak{t}_0^*$, each bounded region contains only finitely many highest weights. Therefore

    (i) minimal K types exist for V,
    (ii) V has only finitely many minimal K types, and
    (iii) the minimal K types of V are unaffected by changing $\Delta^+(\mathfrak{k}, \mathfrak{t})$.

For examples, the principal series of $G = SL(2, \mathbb{R})$ was defined in §2 of the Introduction. For this case $K = SO(2)$, and the irreducible representations of K are parametrized by the integers. For the principal series with parameter $(+, iv)$, the K types are given by the even integers, and 0 is the unique minimal K type. For the principal series with parameter $(-, iv)$, the K types are given by the odd integers, and $+1$ and -1 are the minimal K types. The following proposition gives more examples. Recall the modules $A_\mathfrak{q}(\lambda) = \mathcal{L}_S(\mathbb{C}_\lambda)$ defined in (5.6) and studied in §§VIII.5 and IX.6.

Proposition 10.24. In the case of $A_\mathfrak{q}(\lambda)$, suppose that

    (i) $\Lambda = \lambda + 2\delta(\mathfrak{u} \cap \mathfrak{p})$ is dominant for $\Delta^+(\mathfrak{k}, \mathfrak{t})$ and
    (ii) $\Lambda + 2\delta_K$ is dominant for $\Delta(\mathfrak{u} \cap \mathfrak{p})$.

Then the K type τ_Λ with highest weight Λ is the unique minimal K type of $A_\mathfrak{q}(\lambda)$.

PROOF. If $\tau_{\Lambda'}$ is a K type, then Theorem 5.35 says that there exist weights β_j of $\mathfrak{t}$ in $\mathfrak{u} \cap \mathfrak{p}$ such that $\Lambda' = \Lambda + \sum_j \beta_j$. Adding $2\delta_K$ and computing the norm squared, we see from (ii) that

$$(10.25) \qquad |\Lambda' + 2\delta_K|^2 \geq |\Lambda + 2\delta_K|^2,$$

with equality only if $\Lambda' = \Lambda$. By (i), τ_Λ is in the bottom layer and therefore occurs in $A_\mathfrak{q}(\lambda)$. Hence (10.25) allows us to conclude that τ_Λ is the unique minimal K type.

    Minimal K types can be defined (in the same way) when K is disconnected. We need to verify that any two highest weights of an irreducible representation of K lead to the same norm, i.e., that if μ_1 and μ_2 are highest weights of τ, then $|\mu_1 + 2\delta_k|^2 = |\mu_2 + 2\delta_k|^2$. But this is a simple consequence of Proposition 4.22.

    To prove Theorem 10.23 (with K connected), we shall prove the following more precise statement.

Theorem 10.26. Let $(\mathfrak{g}, K)$ be a reductive pair with K connected, and let $\mathfrak{h}_0 = \mathfrak{t}_0 \oplus \mathfrak{a}_0$ be a maximally compact θ stable Cartan subalgebra of $\mathfrak{g}_0$. Then there exists a constant $c_\mathfrak{g}$ depending only on $\mathfrak{g}$ with the following property: Whenever V is a $(\mathfrak{g}, K)$ module with infinitesimal character λ and a minimal K type τ, then the half-sum δ of positive roots for some $\Delta^+(\mathfrak{g}, \mathfrak{h})$ satisfies

$$|\mathrm{Re}(\lambda + \delta)| \geq \|\tau\| - c_\mathfrak{g}.$$

In particular, if λ is given, then any $(\mathfrak{g}, K)$ module V with infinitesimal character λ must contain a K type of norm at most

$$\max_\delta |\mathrm{Re}(\lambda + \delta)| + c_\mathfrak{g}.$$

REMARK. Recall the definition of $\mathrm{Re}(\lambda + \delta)$: Any member ν of $\mathfrak{h}^*$ is of the form $\nu = \nu_1 + i\nu_2$, where ν_1 and ν_2 are real-valued on $i\mathfrak{t}_0 \oplus \mathfrak{a}_0$, and we put $\mathrm{Re}\,\nu = \nu_1$ and $\mathrm{Im}\,\nu = \nu_2$.

In proving Theorem 10.26, we shall use the following notation. Let τ_μ be a minimal K type of V, and fix a θ stable positive system $\Delta^+(\mathfrak{g}, \mathfrak{h})$ such that

$$\langle \mu + 2\delta_K, \alpha \rangle \geq 0 \qquad \text{for all } \alpha \in \Delta^+(\mathfrak{g}, \mathfrak{h}).$$

The restriction to $\mathfrak{t}$ of this positive system necessarily contains $\Delta^+(\mathfrak{k}, \mathfrak{t})$, since $\mu + 2\delta_K$ is dominant and nonsingular for $\Delta^+(\mathfrak{k}, \mathfrak{t})$. Fix a positive number $N = N(\mathfrak{g})$ to be defined in Lemma 10.29, and define a θ stable parabolic subalgebra $\mathfrak{q} = \mathfrak{l} \oplus \mathfrak{u}$ compatibly with $\Delta^+(\mathfrak{g}, \mathfrak{h})$ so that the set of simple roots of $\mathfrak{h}$ in $\mathfrak{l}$ is

(10.27a) $$\{\alpha \in \Delta(\mathfrak{g}, \mathfrak{h}) \mid \alpha \text{ is simple and } \langle \mu + 2\delta_K, \alpha \rangle \leq N.\}$$

If α is in $\Delta(\mathfrak{u})$, then

(10.27b) $$\langle \mu + 2\delta_K, \alpha \rangle > N.$$

Let τ_L be the representation of $L \cap K$ with highest weight $\mu + 2\delta(\mathfrak{u} \cap \mathfrak{k})$. If W is an $L \cap K$ module, write W_{τ_L} for the τ_L isotypic subspace of W under $L \cap K$. If F^μ is an irreducible representation of K of type τ_μ, then

(10.28) $$H_S(\mathfrak{u} \cap \mathfrak{k}, F^\mu) \cong \tau_L.$$

Lemma 10.29. If $N = N(\mathfrak{g})$ is large enough, then the map

$$H_S(\mathfrak{u} \cap \mathfrak{k}, V)_{\tau_L} \to H_S(\mathfrak{u}, V)_{\tau_L}$$

given by inclusion on the chain level

$$(V \otimes_{\mathbb{C}} \textstyle\bigwedge^S(\mathfrak{u} \cap \mathfrak{k}))_{\tau_L} \to (V \otimes_{\mathbb{C}} \textstyle\bigwedge^S \mathfrak{u})_{\tau_L}$$

is one-one onto.

PROOF. We shall apply the Hochschild-Serre spectral sequence (Theorem 5.130a), taking the τ_L component throughout. In the notation of that theorem, our spectral sequence has r^{th} differential of bidegree $(-r, r-1)$ and has

$$(10.30) \qquad E_1^{p,q} = (H_{p+q-r(p)}(\mathfrak{u} \cap \mathfrak{k}, V) \otimes_{\mathbb{C}} V_p)_{\tau_L},$$

where V_p and $r(p)$ are as in (5.129). Here p is nonnegative, $r(0) = 0$, and $r(p) > 0$ for $p > 0$. According to (5.129), $V_p \subseteq \bigwedge^{r(p)}(\mathfrak{u} \cap \mathfrak{p})$. In particular, $V_0 = \mathbb{C}$. The spectral sequence converges, and

$$(10.31) \qquad E_r^{p,q} \Longrightarrow H_{p+q}(\mathfrak{u}, V)_{\tau_L}.$$

Suppose that some K irreducible subspace F^γ of V with highest weight γ has the property that

$$(H_{S-i}(\mathfrak{u} \cap \mathfrak{k}, F^\gamma) \otimes_{\mathbb{C}} \bigwedge^j(\mathfrak{u} \cap \mathfrak{p}))_{\tau_L} \neq 0$$

for some i and j with $i + j > 0$. By Corollary 3.8, we have

$$(H^i(\mathfrak{u} \cap \mathfrak{k}, F^\gamma) \otimes_{\mathbb{C}} \bigwedge^{\text{top}}(\mathfrak{u} \cap \mathfrak{k}) \otimes_{\mathbb{C}} \bigwedge^j(\mathfrak{u} \cap \mathfrak{p}))_{\tau_L} \neq 0.$$

Then the sum of some highest weight of $H^i(\mathfrak{u} \cap \mathfrak{k}, F^\gamma)$ with $2\delta(\mathfrak{u} \cap \mathfrak{k})$ and a weight of $\bigwedge^j(\mathfrak{u} \cap \mathfrak{p})$ equals $\mu + 2\delta(\mathfrak{u} \cap \mathfrak{k})$. By Kostant's Theorem (Theorem 4.139), there exist an element $w \in W^1$ of length i and a set J of j weights β of $\mathfrak{t}$ in $\mathfrak{u} \cap \mathfrak{p}$ such that

$$\mu + 2\delta(\mathfrak{u} \cap \mathfrak{k}) = (w(\gamma + \delta_K) - \delta_K) + 2\delta(\mathfrak{u} \cap \mathfrak{k}) + \sum_{\beta \in J} \beta.$$

The members of J are distinct, by the discussion with (9.38). Adding $2\delta_K - 2\delta(\mathfrak{u} \cap \mathfrak{k})$ to both sides, we obtain

$$\mu + 2\delta_K = w(\gamma + 2\delta_K) + (\delta_K - w\delta_K) + \sum_{\beta \in J} \beta.$$

By Lemma 4.134, $\delta_K - w\delta_K = \sum_{\alpha \in \Delta^+(w)} \alpha$. Since w is in W^1, $\Delta^+(w)$ is contained in $\Delta(\mathfrak{u} \cap \mathfrak{k})$. Letting I be the set $\Delta^+(w)$ of i distinct weights α of $\mathfrak{u} \cap \mathfrak{k}$, we can rewrite the above identity as

$$\mu + 2\delta_K = w(\gamma + 2\delta_K) + \sum_{\alpha \in I} \alpha + \sum_{\beta \in J} \beta.$$

Hence

$$(10.32) \qquad w(\gamma + 2\delta_K) = \mu + 2\delta_K - \sum_{\alpha \in I \cup J} \alpha,$$

where $I \cup J$ is a set of $i + j$ weights of t in u. Taking the norms squared of both sides, we obtain

$$|\gamma + 2\delta_K|^2 = |\mu + 2\delta_K|^2 - 2 \sum_{\alpha \in I \cup J} \langle \mu + 2\delta_K, \alpha \rangle + | \sum_{\alpha \in I \cup J} \alpha|^2.$$

On the right side, the second term is $\leq -2(i + j)N$ by (10.27b), and the third term is bounded above by a constant depending only on $\mathfrak{g}$, say by C. Then

$$|\gamma + 2\delta_K|^2 \leq |\mu + 2\delta_K|^2 + C - 2(i + j)N.$$

Since τ_γ and τ_μ are K types occurring in V and since τ_μ is minimal, this is a contradiction if N is sufficiently large. Fix an N this large. For this N we conclude that

$$(10.33) \qquad (H_{S-i}(\mathfrak{u} \cap \mathfrak{k}, V) \otimes_{\mathbb{C}} \textstyle\bigwedge^j (\mathfrak{u} \cap \mathfrak{p}))_{\tau_L} = 0$$

for $i + j > 0$.

In terms of the spectral sequence, the relations (10.28), (10.30), and (10.33) combine to give

$$E_1^{p,q} \cong \begin{cases} H_S(\mathfrak{u} \cap \mathfrak{k}, V)_{\tau_L} & \text{if } (p, q) = (0, S) \\ 0 & \text{otherwise.} \end{cases}$$

Since the spectral sequence has bidegree $(r, 1 - r)$, it follows that it collapses, i.e,

$$E_1^{p,q} \cong E_r^{p,q}$$

for all $r \geq 1$, and this is 0 for $(p, q) \neq (0, S)$. The convergence given in Theorem 5.130a says that

$$E_r^{p,q} \cong E_\infty^{p,q}$$

for large $r = r(p, q)$. Therefore

$$(10.34) \qquad E_1^{p,q} \cong E_\infty^{p,q},$$

and this is 0 for $(p, q) \neq (0, S)$. The spectral sequence thus degenerates, and its abutment, which is given as $H^{p+q}(\mathfrak{u}, V)_{\tau_L}$ by (10.31), therefore satisfies

$$\begin{aligned} H_S(\mathfrak{u}, V)_{\tau_L} &\cong E_\infty^{0,S} && \text{from (D.28) and what follows it} \\ &\cong E_1^{0,S} && \text{by (10.34)} \\ &\cong H_S(\mathfrak{u} \cap \mathfrak{k}, V)_{\tau_L} && \text{by (10.30).} \end{aligned}$$

This completes the proof of the lemma.

PROOF OF THEOREM 10.26. The $(\mathfrak{g}, K)$ module V has infinitesimal character λ by assumption. By Theorem 7.56 the $(\mathfrak{l}, L \cap K)$ module $H_S(\mathfrak{u}, V)$ is $Z(\mathfrak{l})$ finite. According to (10.28) and Lemma 10.29, $H_S(\mathfrak{u}, V)$ contains the $L \cap K$ type τ_L. We shall show that some subquotient W_0 of $H_S(\mathfrak{u}, V)$ containing the $L \cap K$ type τ_L has an infinitesimal character.

First we choose a $Z(\mathfrak{l})$ primary component W_1 of $H_S(\mathfrak{u}, V)$ that contains τ_L. Say W_1 has generalized infinitesimal character λ_L. By Proposition 7.20, choose n so that $(z - \chi_{\lambda_L}(z)1)^n$ acts as 0 in W_1. Let W_2 be the subspace of W_1 where $(z - \chi_{\lambda_L}(z)1)$ acts as 0. Then $(z - \chi_{\lambda_L}(z)1)^{n-1}$ acts as 0 in W_1/W_2. Since τ_L occurs in W_1, τ_L must occur in W_2 or W_1/W_2. Then an easy induction produces a subquotient W_0 of $H_S(\mathfrak{u}, V)$ containing the $L \cap K$ type τ_L and having infinitesimal character λ_L.

Let us write $\mathfrak{t} = \mathfrak{t}_1 \oplus \mathfrak{t}_2$, where $\mathfrak{t}_1 = \mathfrak{t} \cap Z_{\mathfrak{l}}$ and $\mathfrak{t}_2 = \mathfrak{t} \cap [\mathfrak{l}, \mathfrak{l}]$. Then $\mathfrak{t}^* = \mathfrak{t}_1^* \oplus \mathfrak{t}_2^*$, and a member of $\mathfrak{t}^*$ decomposes correspondingly as $\gamma = \gamma_1 + \gamma_2$.

Both λ_L and τ_L tell how the subset $\mathfrak{t}_1$ of $\mathfrak{t}$ acts in the subquotient W_0, and thus

$$(\lambda_L)_1 = (\mu + 2\delta(\mathfrak{u} \cap \mathfrak{k}))_1.$$

Theorem 7.56 says that $\lambda_L = w\lambda + \delta(\mathfrak{u})$ for some $w \in W(\mathfrak{g}, \mathfrak{h})$. Hence

$$(w\lambda + \delta(\mathfrak{u}))_1 = (\mu + 2\delta(\mathfrak{u} \cap \mathfrak{k}))_1.$$

Since $(\delta_L)_1 = 0$ and $(\delta_{L \cap K})_1 = 0$, we can add $(\delta_L)_1$ to the left side and $(2\delta_{L \cap K})_1$ to the right side and obtain

(10.35) $(w\lambda + \delta)_1 = (\mu + 2\delta_K)_1.$

Then $(w\lambda + \delta)_1 = \mathrm{Re}(w\lambda + \delta)_1$, and we have

$$|\mathrm{Re}(w\lambda + \delta)|^2 \geq |(w\lambda + \delta)_1|^2$$
$$= |(\mu + 2\delta_K)_1|^2$$
(10.36) $= |\mu + 2\delta_K|^2 - |(\mu + 2\delta_K)_2|^2.$

By (10.27a), we have $0 \leq \langle \mu + 2\delta_K, \alpha \rangle \leq N$ for every simple root of $\mathfrak{l}$. For such simple roots α_j, choose $\lambda_i \in \mathfrak{t}_2^*$ with $\langle \lambda_i, \alpha_j \rangle = \delta_{ij}$. Then

$$(\mu + 2\delta_K)_2 = \sum_i \langle \mu + 2\delta_K, \alpha_i \rangle \lambda_i,$$

and

$$|(\mu + 2\delta_K)_2|^2 = \sum_{i,j} \langle \mu + 2\delta_K, \alpha_i \rangle \langle \mu + 2\delta_K, \alpha_j \rangle \langle \lambda_i, \lambda_j \rangle \leq N^2 \sum_{i,j} |\langle \lambda_i, \lambda_j \rangle|^2.$$

Defining $c_{\mathfrak{g}}$ to be the right side of this inequality and applying (10.36), we obtain

$$|\mathrm{Re}(w\lambda + \delta)|^2 \geq \|\tau_\mu\|^2 - c_{\mathfrak{g}},$$

which is equivalent with the assertion in the theorem.

The proof of Theorem 10.26 contains a great deal of information and can be pushed to give further results. Theorem 10.37 below is a sample. In §3 we shall see that this result implies irreducibility of the unitary spherical principal series for complex semisimple groups G.

Theorem 10.37. Let $(\mathfrak{g}, K)$ be a reductive pair, suppose that $\mathfrak{g}_0$ has only one conjugacy class of Cartan subalgebras, and let $\mathfrak{h}_0 = \mathfrak{t}_0 \oplus \mathfrak{a}_0$ be a θ stable Cartan subalgebra. Let V be a $(\mathfrak{g}, K)$ module with minimal K type τ_μ of highest weight μ, and suppose V has an infinitesimal character λ. Then there exists $w \in W(\mathfrak{g}, \mathfrak{h})$ such that

$$w\lambda|_\mathfrak{t} = \mu + 2\delta_K - \delta.$$

If λ is purely imaginary, then τ_μ is the trivial K type.

PROOF. We can regard $\mathfrak{h}_0$ as maximally compact. The discussion with (9.38) shows the structure of the roots. Since $\mathfrak{g}_0$ has only one conjugacy class of Cartan subalgebras, Lemma 4.41a says that there are no noncompact imaginary roots. Thus the restriction to $\mathfrak{t}$ of any member of $\Delta(\mathfrak{g}, \mathfrak{h})$ is a member of $\Delta(\mathfrak{k}, \mathfrak{t})$. Hence for a weight in $\mathfrak{t}^*$, "dominant for $\mathfrak{g}$" is the same as "dominant for $\mathfrak{k}$."

The weight $2\delta_K - \delta$ is in $\mathfrak{t}^*$. If $\alpha \in \Delta(\mathfrak{g}, \mathfrak{h})$ is simple, then $\alpha + \theta\alpha$ is not a root, since otherwise a root vector for it would be $[e_\alpha, \theta e_\alpha]$ and $\alpha + \theta\alpha$ would be noncompact imaginary. Hence $\langle \alpha, \theta\alpha \rangle \geq 0$, and $|\alpha|_\mathfrak{t}|^2 \geq |\alpha|_\mathfrak{a}|^2$. Thus $2|\alpha|_\mathfrak{t}|^2 \geq |\alpha|^2$. Since $\alpha|_\mathfrak{t}$ is in $\Delta^+(\mathfrak{k}, \mathfrak{t})$, this inequality and Proposition 4.67b give

$$\frac{2\langle 2\delta_K - \delta, \alpha \rangle}{|\alpha|^2} = \frac{2\langle 2\delta_K, \alpha|_\mathfrak{t} \rangle}{|\alpha|^2} - \frac{2\langle \delta, \alpha \rangle}{|\alpha|^2} \geq \frac{2\langle \delta_K, \alpha|_\mathfrak{t} \rangle}{|\alpha|_\mathfrak{t}|^2} - 1 \geq 0.$$

Therefore

(10.38) $2\delta_K - \delta$ is dominant.

We shall prove that the conclusion "one-one" in Lemma 10.29 holds with $N = 0$. (Note that when $N = 0$, the set (10.27a) is empty. Hence $\mathfrak{l} = \mathfrak{h}$ and $\mathfrak{q} = \mathfrak{b}$.) We begin by proving that (10.33) is 0 if $i < j$. We start from (10.32):

(10.39) $w(\gamma + 2\delta_K) = \mu + 2\delta_K - \sum_{\alpha \in I \cup J} \alpha = (\mu + 2\delta_K - \delta) + (\delta - \sum_{\alpha \in I \cup J} \alpha).$

Then we have

$$|\gamma + 2\delta_K|^2 = |\mu + 2\delta_K - \delta|^2 + |\delta - \sum_{\alpha \in I \cup J} \alpha|^2$$

(10.40)

$$+ 2\langle \mu + 2\delta_K - \delta, \delta \rangle - 2\langle \mu + 2\delta_K - \delta, \sum_{\alpha \in I \cup J} \alpha \rangle.$$

Recall in (10.39) that the weights of I are distinct, and so are the weights of J. Each $\alpha \in I \cup J$ extends to a root $\tilde{\alpha} \in \Delta(\mathfrak{g}, \mathfrak{h})$, and we may assume that these roots are distinct. (In fact, the only possible match of a weight in I with a weight in J arises from the restriction of a complex root $\tilde{\alpha}$, and we may use $\tilde{\alpha}$ and $\theta\tilde{\alpha}$ as the extensions.) Thus we have

(10.41a)
$$|\delta - \sum_{\alpha \in I \cup J} \alpha|^2 \le |\delta - \sum_{\alpha \in I \cup J} \tilde{\alpha}|^2.$$

We shall show shortly that the right side is

(10.41b) $\le |\delta|^2.$

Let us accept this fact for the moment. Observe from (10.38) that the last term on the right side of (10.40) is ≤ 0. Therefore (10.40) implies

$$|\gamma + 2\delta_K|^2 \le |\mu + 2\delta_K - \delta|^2 + |\delta|^2 + 2\langle \mu + 2\delta_K - \delta, \delta \rangle = |\mu + 2\delta_K|^2.$$

Since τ_μ is minimal, equality must hold. Thus equality holds in (10.41a), no matter how we choose the extensions $\tilde{\alpha}$. From this equality, we obtain $\sum \tilde{\alpha} = \sum \alpha$. Since no choice of the $\tilde{\alpha}$'s violates this equality, $\theta\tilde{\alpha}$ is present whenever $\tilde{\alpha}$ is present. It follows that $i \ge j$. In other words, (10.33) holds for $i < j$.

We now show that every term $E_r^{p,q}$ of the spectral sequence (10.30) is equal to 0 when $p = r$ and $q = S - r + 1$. This term is a subquotient of

(10.42) $(H_{S-i}(\mathfrak{u} \cap \mathfrak{k}, V) \otimes_{\mathbb{C}} \bigwedge^j (\mathfrak{u} \cap \mathfrak{p}))_{\tau_L}$

with

$$S - i = p + q - r(p) = r + (S - r + 1) - r(r) = S - r(r) + 1$$

and $j = r(p) = r(r)$. Hence $i = r(r) - 1$ and $j = r(r)$, and therefore $i < j$. From the vanishing of (10.42) for $i < j$, we see that $E_r^{r, S-r+1} = 0$ for $r \ge 1$. Since also $E_r^{-r, S+r-1} = 0$, we have

$$E_{r+1}^{0,S} \cong \ker(E_r^{0,S} \to E_r^{-r, S+r-1})/\text{image}(E_r^{r, S-r+1}) = E_r^{0,S}.$$

Inspecting the proof of Lemma 10.29, we see that $E_1^{0,S} \cong E_\infty^{0,S}$. Since $E_\infty^{0,S}$ is a subspace of $H_S(\mathfrak{u}, V)_{\tau_L}$, the conclusion of "one-one" in Lemma 10.29 follows with $N = 0$, apart from the proof of (10.41b).

From this conclusion in Lemma 10.29, let us prove Theorem 10.26. We have observed that $\mathfrak{l} = \mathfrak{h}$ when $N = 0$, and therefore $\mathfrak{t} \cap [\mathfrak{l}, \mathfrak{l}] = 0$. From (10.35) we know that $(w\lambda + \delta)_{\mathfrak{l}} = (\mu + 2\delta_K)_{\mathfrak{l}}$, and it follows that

$$(10.43) \qquad w\lambda|_{\mathfrak{t}} = \mu + 2\delta_K - \delta,$$

as asserted.

If λ is imaginary, so is $w\lambda$. But (10.43) shows that $w\lambda|_{\mathfrak{t}}$ is real. Therefore $\mu + 2\delta_K - \delta = 0$. Since μ is dominant and (10.38) shows $2\delta_K - \delta$ to be dominant, we conclude that $\mu = 0$. This completes the proof of the theorem except for (10.41b).

Now we prove (10.41b). Let us introduce the compact form $\mathfrak{k}_0' = (\mathfrak{k}_0 \oplus i\mathfrak{p}_0) \cap [\mathfrak{g}, \mathfrak{g}]$ of the semisimple part of $\mathfrak{g}$. By Weyl's Theorem (Theorem 4.3), let K' be a compact simply connected group with Lie algebra $\mathfrak{k}_0'$. Then $\mathfrak{t}_0' = \mathfrak{k}_0' \cap \mathfrak{h}$ is maximal abelian in $\mathfrak{k}_0'$, and we can identify $\Delta(\mathfrak{g}, \mathfrak{h})$ with $\Delta(\mathfrak{k}', \mathfrak{t}')$. If we identify $\Delta^+(\mathfrak{g}, \mathfrak{h})$ with $\Delta^+(\mathfrak{k}', \mathfrak{t}')$, then δ gets identified with $\delta_{K'}$. The member $\delta_{K'}$ of $(\mathfrak{t}')^*$ is dominant integral by Proposition 4.67b. If F is an irreducible representation of K' with highest weight $\delta_{K'}$, then the Weyl character formula says that the character of F is given in the notation of §IX.2 by

$$D_{K'}(\exp h)\Theta_{T'}(F)(\exp h) = \sum_{s \in W(\mathfrak{k}', \mathfrak{t}')} (\operatorname{sgn} s)e^{2w\delta_{K'}(h)} \quad \text{for } h \in i\mathfrak{t}_0'$$

$$= \sum_{s \in W(\mathfrak{k}', \mathfrak{t}')} (\operatorname{sgn} s)e^{w\delta_{K'}(2h)}.$$

The right side equals $D_{K'}(\exp 2h)$ and therefore is

$$= e^{2\delta_{K'}(h)} \prod_{\alpha \in \Delta^+(\mathfrak{k}', \mathfrak{t}')} (1 - e^{-2\alpha(h)}).$$

Dropping the h and dividing by

$$D_{K'} = e^{\delta_{K'}} \prod_{\alpha \in \Delta^+(\mathfrak{k}', \mathfrak{t}')} (1 - e^{-\alpha}),$$

we obtain

$$\Theta_{T'}(F) = e^{\delta_{K'}} \prod_{\alpha \in \Delta^+(\mathfrak{k}', \mathfrak{t}')} (1 + e^{-\alpha}) = e^{\delta_{K'}} \sum_{Q \subseteq \Delta^+(\mathfrak{k}', \mathfrak{t}')} e^{-\langle Q \rangle}.$$

Application of this formula with Q as the set corresponding to $I \cup J$ in $\Delta^+(\mathfrak{g}, \mathfrak{h})$ shows that $\delta - \sum_{\alpha \in I \cup J} \tilde{\alpha}$ is a weight of F. Thus (10.41b) follows from Theorem 4.10b.

3. Minimal K Types and Cohomological Induction

In this section we address the question of how the minimal K types of a cohomologically induced module $\mathcal{L}_S(W)$ are related to the minimal $L \cap K$ types of W. The result may be regarded as a generalization of Proposition 10.24, where W is assumed to be one dimensional. Hypothesis (ii) in that proposition is a $\mathfrak{u}$ dominance condition and will be replaced by a different $\mathfrak{u}$ dominance condition in the generalization.

Theorem 10.44. Let $(\mathfrak{g}, K)$ be a reductive pair in the Harish-Chandra class with underlying group G, let $(\mathfrak{q}, L \cap K)$ be a θ stable parabolic subpair with $\mathfrak{q} = \mathfrak{l} \oplus \mathfrak{u}$ and with Levi subgroup L, and suppose that L meets every component of G. If W is an irreducible $(\mathfrak{l}, L \cap K)$ module with an infinitesimal character λ that satisfies

$$(10.45a) \qquad \operatorname{Re}\langle \lambda, \alpha \rangle \geq 0 \qquad \text{for all } \alpha \in \Delta(\mathfrak{u}),$$

then the minimal K types of the isomorphic $(\mathfrak{g}, K)$ modules $\mathcal{L}_S(W)$ and $\mathcal{R}^S(W)$ are all in the bottom layer and are exactly those whose highest weights are K dominant expressions of the form

$$(10.45b) \qquad \Lambda = \mu + 2\delta(\mathfrak{u} \cap \mathfrak{p}), \qquad \begin{array}{l}\text{where } \mu \text{ is a highest weight of}\\ \text{a minimal } L \cap K \text{ type of } W.\end{array}$$

Moreover if the inequality (10.45a) is strict for all $\alpha \in \Delta(\mathfrak{u})$, then every expression (10.45b) is K dominant and is therefore a highest weight of a minimal K type of $\mathcal{L}_S(W)$.

REMARKS. For many reductive groups the infinitesimal character and the set of minimal K types form a complete set of invariants for irreducible $(\mathfrak{g}, K)$ modules. For these groups when the Irreducibility Theorem (Theorem 8.2) is applicable, Theorem 10.44 and the results on infinitesimal characters in Chapter V describe completely the effect of $\mathcal{L}_S$ on irreducible $(\mathfrak{l}, L \cap K)$ modules under the assumption (10.45a). Because of the vanishing theorem (Theorem 5.99) and the long exact sequence for the functors $\mathcal{L}_j$, the assumption (10.45a) implies that $\mathcal{L}_S$ is exact on the category $\mathcal{C}_\lambda(\mathfrak{g}, K)$ of $(\mathfrak{g}, K)$ modules of infinitesimal character λ. Therefore the effect of $\mathcal{L}_S$ is known on $(\mathfrak{l}, L \cap K)$ modules of finite length.

In Chapter XI we shall be able to use the Langlands classification to characterize the effect of $\mathcal{L}_S$ on irreducible $(\mathfrak{l}, L \cap K)$ modules using just (10.45a). We shall no longer be limited to reductive groups for which the infinitesimal character and the set of minimal K types characterize irreducible $(\mathfrak{g}, K)$ modules.

A full proof of Theorem 10.44 is beyond the scope of this book. We shall prove only some aspects of the theorem, and we postpone them until late in Chapter XI. There are three things that we shall skip:

(i) a proof that some expression (10.45b) is K dominant as long as $\mathcal{L}_S(W)$ is not 0

(ii) a proof of the last assertion of the theorem

(iii) some equivalent formulations of minimality of K types for reductive groups that are "split modulo center."

We return to these matters in §XI.11.

CHAPTER XI

TRANSFER THEOREM

The idea of generating $(\mathfrak{g}, K)$ modules by applying the composition of some Π_j and a functor ind, or of some Γ^j and a functor pro, extends from θ stable parabolic subalgebras to all germane parabolic subalgebras. This chapter examines the functors obtained in this way and relates them to one another.

The first special case to study is that of parabolic induction, which mirrors a real-analysis construction. This construction takes representations of the Levi subgroup of a real parabolic subgroup and gives representations of a reductive group. On the level of $(\mathfrak{g}, K)$ modules, it is mirrored by $\Gamma^0 \circ$ pro relative to a real parabolic subalgebra, the higher functors $\Gamma^j \circ$ pro giving 0. The corresponding functor $\Pi_0 \circ$ ind gives the same result, apart from a shift of parameters, and its higher functors $\Pi_j \circ$ ind give 0.

Relative to any germane parabolic subalgebra, the functors $\Pi_j \circ$ ind and $\Gamma^j \circ$ pro, together with the usual implicit forgetful functors, are abbreviated ${}^u\mathcal{L}_j$ and ${}^u\mathcal{R}^j$. These are the unnormalized versions of the functors of primary interest in this chapter.

The first main result about ${}^u\mathcal{L}_j$ and ${}^u\mathcal{R}^j$ is a spectral sequence for induction in stages. In practice these spectral sequences often collapse and degenerate. In this case a composition of one ${}^u\mathcal{L}_j$ or ${}^u\mathcal{R}^j$ with another functor of the same type is a single functor of the same type, with the composite degree j given by the sum of the degrees of the factors.

The second main result about ${}^u\mathcal{L}_j$ and ${}^u\mathcal{R}^j$ is the Transfer Theorem. For this result the parabolic subalgebras are Borel subalgebras containing a common Cartan subalgebra. The theorem compares the effect of ${}^u\mathcal{L}_j$ or ${}^u\mathcal{R}^j$ if two different Borel subalgebras are used that differ by a single root, which is assumed to be complex. Under the hypotheses of the theorem, the two ${}^u\mathcal{L}_j$'s or the two ${}^u\mathcal{R}^j$'s give isomorphic $(\mathfrak{g}, K)$ modules, but the degrees j differ by 1 for the two functors.

For a reductive $(\mathfrak{g}, K)$ pair in the Harish-Chandra class, the Transfer Theorem can be iterated. As a consequence one is led to study all "standard modules," those built by a functor ${}^u\mathcal{L}_j$ or ${}^u\mathcal{R}^j$ involving a Borel subalgebra and using data appropriate for the Transfer Theorem. Several different kinds of standard modules are of interest, constructed of "Π type" or "Γ type," and coming from Borel subalgebras that are maximally complex or are maximally real or produce dominance for the parameters. The main theorem relates these standard modules to one another.

The theory of standard modules becomes more symmetric if normalized versions ${}^n\mathcal{L}_j$ and ${}^n\mathcal{R}^j$ of ${}^u\mathcal{L}_j$ and ${}^u\mathcal{R}^j$ are introduced. Defining the normalized functors involves developing a theory of double covers of groups relative to one-dimensional representations.

Discrete series and "limits of discrete series" arise (for connected G) as certain modules $A_{\mathfrak{q}}(\lambda)$ when $\mathfrak{q}$ is a Borel subalgebra. They are one ingredient in stating the Langlands classification of all irreducible admissible $(\mathfrak{g}, K)$ modules. This classification constructs all irreducible admissible $(\mathfrak{g}, K)$ modules out of parabolic induction starting from a discrete series or limit, followed by passage to an irreducible quotient.

The iterated version of the Transfer Theorem exhibits other classifications as equivalent with the Langlands classification. In one such, all irreducible $(\mathfrak{g}, K)$ modules are constructed by cohomological induction applied to a principal series of a group that is "split modulo center."

In addition, the ability to follow the effect of changing Borel subalgebras allows one to write down the effect of cohomological induction on the Langlands parameters of irreducible admissible modules. The technique applies also to the effect of cohomological induction on standard modules and leads to a version of the Signature Theorem cast in the setting of the Langlands classification.

1. Parabolic Induction Globally

This chapter contains the fifth of our five main theorems. The Transfer Theorem relates

$$(11.1a) \qquad (\Pi_{\mathfrak{g},L\cap K}^{\mathfrak{g},K})_j (\mathrm{ind}_{\mathfrak{q},L\cap K}^{\mathfrak{g},L\cap K}(\mathcal{F}_{\mathfrak{l},L\cap K}^{\mathfrak{q},L\cap K}(Z)))$$

for different choices of $\mathfrak{u}$ and j, and it does the same for

$$(11.1b) \qquad (\Gamma_{\mathfrak{g},L\cap K}^{\mathfrak{g},K})^j (\mathrm{pro}_{\mathfrak{q},L\cap K}^{\mathfrak{g},L\cap K}(\mathcal{F}_{\mathfrak{l},L\cap K}^{\mathfrak{q},L\cap K}(Z))).$$

In these relationships it is not assumed that $\mathfrak{q}$ is a θ stable parabolic subalgebra, and therefore (11.1) represents a more general construction of $(\mathfrak{g}, K)$ modules than we have been using until now. As a consequence of the relationships, we shall be able to locate many cohomologically induced modules in the "Langlands classification" of all irreducible admissible $(\mathfrak{g}, K)$ modules.

For a reductive pair $(\mathfrak{g}, K)$ and its associated reductive group G, "parabolic induction" initially refers to a construction of representations of G that generalizes the construction of the principal series of $SL(2, \mathbb{R})$ in §2 of the Introduction. The upper-triangular subgroup gets replaced by a parabolic subgroup of G, and Mackey's generalization of classical induction is used for passing from representations of the parabolic subgroup to representations of G. Parabolic induction is the context in which the Langlands classification of irreducible admissible $(\mathfrak{g}, K)$ modules is stated. In this section we shall develop a little of the theory of parabolic induction on the group level. In §2 we relate this theory to algebraic constructions with $(\mathfrak{g}, K)$ modules. We shall see that the underlying $(\mathfrak{g}, K)$ module of a parabolically induced representation is given by (11.1b) with the degree j equal to 0 and with $\mathfrak{q}$ a real parabolic subalgebra of $\mathfrak{g}$. Starting in §2, we therefore expand the use of the term "parabolic induction" to cover this algebraic construction. Efforts to relate cohomological induction to parabolic induction will lead us to the other $(\mathfrak{g}, K)$ modules (11.1) and to the Transfer Theorem.

Before beginning the representation theory, we review some necessary structure theory. In this section we work with a reductive pair $(\mathfrak{g}, K)$ in the Harish-Chandra class. Let G be the associated reductive group given in Proposition 4.31, and let other notation be as in Chapter IV.

We shall introduce the ingredients of real parabolic induction first in the "minimal" case and then in the general case (using the same notation). Let $\mathfrak{a}_0$ be a maximal abelian subspace of $\mathfrak{p}_0$. By Lemma 10.13, $\mathfrak{a}_0$ is uniquely determined up to conjugacy via $\mathrm{Ad}(K)$. Let $\Sigma = \Sigma(\mathfrak{g}_0, \mathfrak{a}_0)$ be the set of restricted roots, defined as in the remarks with Lemma 10.13, and let $(\mathfrak{g}_0)_\beta$ be the eigenspace corresponding to $\beta \in \Sigma$. The eigenspace corresponding to eigenvalue 0 is of the form $\mathfrak{m}_0 \oplus \mathfrak{a}_0$, where $\mathfrak{m}_0 = Z_{\mathfrak{k}_0}(\mathfrak{a}_0)$. If we regard Σ as a subset of $\mathfrak{a}_0^*$, we may order $\mathfrak{a}_0^*$ lexicographically and thereby determine a subset $\Sigma^+ = \Sigma^+(\mathfrak{g}_0, \mathfrak{a}_0)$ of positive restricted roots. We can check that $\theta((\mathfrak{g}_0)_\beta) = (\mathfrak{g}_0)_{-\beta}$, and it follows that the negative restricted roots are exactly the members of $-\Sigma^+$. Let

$$(11.2) \qquad \mathfrak{n}_0 = \bigoplus_{\beta \in \Sigma^+} (\mathfrak{g}_0)_\beta \qquad \text{and} \qquad \mathfrak{n}_0^- = \bigoplus_{\beta \in -\Sigma^+} (\mathfrak{g}_0)_\beta.$$

Since $[(\mathfrak{g}_0)_\beta, (\mathfrak{g}_0)_\gamma] \subseteq (\mathfrak{g}_0)_{\beta+\gamma}$, $\mathfrak{n}_0$ and $\mathfrak{n}_0^-$ are Lie subalgebras of $\mathfrak{g}_0$, and $\mathfrak{m}_0$ normalizes each $(\mathfrak{g}_0)_\beta$. We have $\mathfrak{n}_0^- = \theta \mathfrak{n}_0$ and

$$(11.3) \qquad\qquad \mathfrak{g}_0 = \mathfrak{m}_0 \oplus \mathfrak{a}_0 \oplus \mathfrak{n}_0 \oplus \mathfrak{n}_0^-.$$

Let $\mathfrak{t}_0$ be a maximal abelian subspace of $\mathfrak{m}_0$. Then $\mathfrak{h}_0 = \mathfrak{t}_0 \oplus \mathfrak{a}_0$ is evidently a maximal abelian subalgebra of $\mathfrak{g}_0$ such that $\theta\mathfrak{h}_0 = \mathfrak{h}_0$ and hence is a θ stable Cartan subalgebra as defined in §IV.4. It is maximally split in the sense of Lemma 4.41 and Proposition 4.42. If $\alpha \in \Delta(\mathfrak{g}, \mathfrak{h})$ is a root, then $\mathrm{ad}\,\mathfrak{a}_0$ acts in the root space $\mathfrak{g}_\alpha$ by $\alpha|_{\mathfrak{a}_0}$, and it follows that $\alpha|_{\mathfrak{a}_0}$ is zero or a restricted root. Writing $\mathfrak{g}_\beta$ for the complexification of the restricted root space $(\mathfrak{g}_0)_\beta$, we see that

$$
\begin{aligned}
\mathfrak{g}_\beta &= \bigoplus_{\substack{\alpha \in \Delta(\mathfrak{g},\mathfrak{h}) \\ \alpha|_\mathfrak{a} = \beta}} \mathfrak{g}_\alpha \qquad \text{if } \beta \text{ is in } \Sigma \\
\mathfrak{m} &= \mathfrak{t} \oplus \bigoplus_{\substack{\alpha \in \Delta(\mathfrak{g},\mathfrak{h}) \\ \alpha|_\mathfrak{a} = 0}} \mathfrak{g}_\alpha.
\end{aligned}
$$

(11.4)

It is in this sense that the restricted roots are the restrictions of the roots.

Let us extend the above ordering of $\mathfrak{a}_0$ to an ordering of $\mathfrak{a}_0 \oplus i\mathfrak{t}_0$ that takes $\mathfrak{a}_0$ first, calling the resulting positive system $\Delta^+(\mathfrak{g}, \mathfrak{h})$. Then it is apparent that a restricted root is positive if and only if all of the roots restricting to it are positive. Since the conjugate of a positive root contributing to a restricted root is again a positive root, we obtain the important identity

$$(11.5) \qquad\qquad \delta = \rho + \delta_M,$$

where δ and δ_M are the half sums of the members of $\Delta^+(\mathfrak{g}, \mathfrak{h})$ and $\Delta^+(\mathfrak{m}, \mathfrak{t})$, and where ρ is half the sum of the members of Σ^+ with their multiplicities.

Taking (11.4) into account, we see that $\mathfrak{q} = \mathfrak{m} \oplus \mathfrak{a} \oplus \mathfrak{n}$ contains the Borel subalgebra corresponding to $\Delta^+(\mathfrak{g}, \mathfrak{h})$. It is therefore a parabolic subalgebra in the sense of §IV.6. The parabolic subalgebra $\mathfrak{q}$ is the complexification of $\mathfrak{q}_0 = \mathfrak{m}_0 \oplus \mathfrak{a}_0 \oplus \mathfrak{n}_0$ and is therefore real in the sense of that section. The Levi factor $\mathfrak{l} = \mathfrak{m} \oplus \mathfrak{a}$ is the complexification of $\mathfrak{l}_0 = \mathfrak{m}_0 \oplus \mathfrak{a}_0$, and the nilpotent radical $\mathfrak{n}$ is the complexification of $\mathfrak{n}_0$. Define A, N, and N^- to be the analytic subgroups of G with Lie algebras $\mathfrak{a}_0$, $\mathfrak{n}_0$, and $\mathfrak{n}_0^-$, respectively, and let

$$(11.6) \qquad\qquad M = Z_K(\mathfrak{a}_0).$$

The group M is a closed subgroup of K and hence is compact; it normalizes each restricted root space $(\mathfrak{g}_0)_\beta$. We shall take the following facts as known. The reductive pair $(\mathfrak{g}, K)$ continues to be in the Harish-Chandra class. (See the Notes for references.)

(11.7a) A, N, and AN are closed subgroups of G.

(11.7b) The Levi subgroup of $\mathfrak{q} = \mathfrak{m} \oplus \mathfrak{a} \oplus \mathfrak{n}$ is a direct product $L = MA$, and this decomposition is the Cartan decomposition of L as in (iii) of Definition 4.29.

(11.7c) The subgroup M centralizes A and normalizes N, and it meets every component of G. The subgroup $Q = MAN$ is closed and has Lie algebra $\mathfrak{q}_0$. Also Q is equal to its own normalizer.

(11.7d) (Iwasawa decomposition) $G = KAN$ in the sense that multiplication is a real analytic diffeomorphism from $K \times A \times N$ onto G.

The subgroup $Q = MAN$ is called a **minimal parabolic subgroup** of G.

Let us now consider a general real parabolic subalgebra of $\mathfrak{g}$ as defined after (4.75), i.e., a parabolic subalgebra that is the complexification of a subalgebra of $\mathfrak{g}_0$. We shall assign new meanings to the real subspaces $\mathfrak{a}_0$, $(\mathfrak{g}_0)_\beta$, $\mathfrak{m}_0$, $\mathfrak{n}_0$, $\mathfrak{n}_0^-$, $\mathfrak{t}_0$; to their complexifications; to the groups A, N, N^-, M; and to the symbols Σ, Σ^+, ρ, δ_M. Normally this generalized use of the notation will not invite any confusion; when there is any possibility of ambiguity, we shall insert subscripts $\mathfrak{p}$ on the symbols for the earlier special case, to emphasize that the construction began with a maximal abelian subspace of $\mathfrak{p}_0$.

Our real parabolic subalgebra $\mathfrak{q}$, being the complexification of $\mathfrak{q}_0 = \mathfrak{g}_0 \cap \mathfrak{q}$, is germane, and we let L be the corresponding Levi subgroup of G. Since $(\mathfrak{g}, K)$ is in the Harish-Chandra class, so is $(\mathfrak{l}, L \cap K)$. It is traditional to let $\mathfrak{n}$ denote the nilpotent radical of $\mathfrak{q}$. Since $\mathfrak{q}$ is real, $\mathfrak{n}$ is the complexification of a nilpotent subalgebra $\mathfrak{n}_0 \subseteq \mathfrak{g}_0$. We define

$$(11.8) \qquad \mathfrak{m}_0 = [\mathfrak{l}_0, \mathfrak{l}_0] \oplus (Z_{\mathfrak{l}_0} \cap \mathfrak{k}_0) \qquad \text{and} \qquad \mathfrak{a}_0 = Z_{\mathfrak{l}_0} \cap \mathfrak{p}_0,$$

so that $\mathfrak{l}_0 = \mathfrak{m}_0 \oplus \mathfrak{a}_0$. Since $\mathfrak{q}$ is a real parabolic, $\mathfrak{q}^- = \theta\mathfrak{q}$. Hence $\mathfrak{n}^- = \theta\mathfrak{n}$. Also (4.64) gives

$$(11.9) \qquad \mathfrak{g} = \mathfrak{m} \oplus \mathfrak{a} \oplus \mathfrak{n} \oplus \mathfrak{n}^-.$$

Intersecting this decomposition with $\mathfrak{g}_0$, we obtain

$$(11.10) \qquad \mathfrak{g}_0 = \mathfrak{m}_0 \oplus \mathfrak{a}_0 \oplus \mathfrak{n}_0 \oplus \mathfrak{n}_0^-,$$

where $\mathfrak{n}_0^- = \theta\mathfrak{n}_0$.

Let $\mathfrak{h}_0$ be a θ stable Cartan subalgebra of $\mathfrak{l}$. This contains $\mathfrak{a}_0$ since $\mathfrak{a}_0 \subseteq Z_{\mathfrak{l}_0}$, and thus we may write $\mathfrak{h}_0$ as $\mathfrak{a}_0 \oplus (\mathfrak{h}_M)_0$, where $(\mathfrak{h}_M)_0 = \mathfrak{h}_0 \cap \mathfrak{m}_0$ is a θ stable Cartan subalgebra of $\mathfrak{m}_0$. Decomposing $(\mathfrak{h}_M)_0$ according to θ, we obtain

$$(11.11) \qquad \mathfrak{h}_0 = \mathfrak{a}_0 \oplus (\mathfrak{a}_M)_0 \oplus \mathfrak{t}_0.$$

Here θ is -1 on $\mathfrak{a}_0 \oplus (\mathfrak{a}_M)_0$ and is $+1$ on $\mathfrak{t}_0$. Let $\Delta^+(\mathfrak{l}, \mathfrak{h})$ be any positive system of roots for $\mathfrak{l}$, and define $\Delta^+(\mathfrak{g}, \mathfrak{h}) = \Delta^+(\mathfrak{l}, \mathfrak{h}) \cup \Delta(\mathfrak{n})$. According to Proposition 4.76b, the member $h_{\delta(\mathfrak{n})}$ of $\mathfrak{h}$ lies in $\mathfrak{a}_0 \oplus (\mathfrak{a}_M)_0$ (i.e., has 0 component in $\mathfrak{t}$) and has the property that

$$(11.12) \qquad \alpha(h_{\delta(\mathfrak{n})}) \text{ is } \begin{cases} > 0 & \text{for } \alpha \in \Delta(\mathfrak{n}) \\ = 0 & \text{for } \alpha \in \Delta(\mathfrak{l}, \mathfrak{h}) \\ < 0 & \text{for } \alpha \in \Delta(\mathfrak{n}^-). \end{cases}$$

Lemma 11.13. The element $h_{\delta(\mathfrak{n})}$ lies in $\mathfrak{a}_0$.

PROOF. We can regard $\Delta(\mathfrak{m}, \mathfrak{h}_M)$ as the subset of $\Delta(\mathfrak{g}, \mathfrak{h})$ vanishing on $\mathfrak{a}$, and we can thus regard $W(\mathfrak{m}, \mathfrak{h}_M)$ as acting on $\mathfrak{h}$. Since $\mathfrak{m}$ normalizes $\mathfrak{n}$, $W(\mathfrak{m}, \mathfrak{h}_M)$ fixes $h_{\delta(\mathfrak{n})}$. Thus the $\mathfrak{a}_M$ component of $h_{\delta(\mathfrak{n})}$ lies in the center of $\mathfrak{m}$. But it lies in $\mathfrak{p}$ also, and it must be 0 since the center of $\mathfrak{m}$ was constructed in (11.8) to be in $\mathfrak{k}$. Hence the $\mathfrak{a}_M$ component is 0, and the lemma follows.

Now let us consider the simultaneous eigenspace decomposition of $\mathfrak{g}$ relative to ad $\mathfrak{a}$. Since $\mathfrak{a}_0$ is an abelian subalgebra of $\mathfrak{p}_0$, ad $\mathfrak{a}_0$ acts fully reducibly on $\mathfrak{g}_0$; let Σ denote the set of nonzero roots of $(\mathfrak{g}_0, \mathfrak{a}_0)$. The members of Σ will be treated as members of $\mathfrak{a}_0^*$ or $\mathfrak{a}^*$ and will be called $\mathfrak{a}$-**roots** of $\mathfrak{g}_0$ or $\mathfrak{g}$. The $\mathfrak{a}$-root space in $\mathfrak{g}_0$ for an $\mathfrak{a}$-root β will be denoted $(\mathfrak{g}_0)_\beta$, and its complexification will be denoted $\mathfrak{g}_\beta$. The set Σ is closed under $\beta \to -\beta$ because this operation corresponds to applying θ to an $\mathfrak{a}$-root vector. However, Σ need not be a root system.

Let us prove that an $\mathfrak{a}$-root space $\mathfrak{g}_\beta$ lies in $\mathfrak{n}$ or $\mathfrak{n}^-$. Since $\mathfrak{a} \subseteq \mathfrak{h}$, $\mathfrak{a}$ commutes with $\mathfrak{h}$, and it follows that $[\mathfrak{h}, \mathfrak{g}_\beta] \subseteq \mathfrak{g}_\beta$ for $\beta \in \Sigma$. Hence $\mathfrak{g}_\beta$ is a sum of root spaces relative to $\Delta(\mathfrak{g}, \mathfrak{h})$:

$$(11.14) \qquad \mathfrak{g}_\beta = \bigoplus_{\alpha|_\mathfrak{a} = \beta} \mathfrak{g}_\alpha.$$

By (11.12) and the lemma, all the α's in this sum contribute to $\mathfrak{n}$ if $\beta(h_{\delta(\mathfrak{n})}) > 0$ or to $\mathfrak{n}^-$ if $\beta(h_{\delta(\mathfrak{u})}) < 0$ or to $Z_\mathfrak{g}(\mathfrak{a}) = \mathfrak{a} \oplus \mathfrak{m}$ if $\beta(h_{\delta(\mathfrak{u})}) = 0$. Hence $\mathfrak{g}_\beta$ correspondingly is contained in $\mathfrak{n}$ or $\mathfrak{n}^-$ or $\mathfrak{a} \oplus \mathfrak{m}$.

We say that an $\mathfrak{a}$-root β is **positive** if $\mathfrak{g}_\beta \subseteq \mathfrak{n}$, or **negative** if $\mathfrak{g}_\beta \subseteq \mathfrak{n}^-$. We have just seen that $\Sigma = \Sigma^+ \cup (-\Sigma^+)$. The decomposition (11.9) respects this notion of positivity. Formula (11.14) shows that the following conditions are equivalent:

(11.15)
(a) the member β of Σ is positive
(b) some member α of $\Delta(\mathfrak{g}, \mathfrak{h})$ restricting to β is positive
(c) every member α of $\Delta(\mathfrak{g}, \mathfrak{h})$ restricting to β is positive.

Let $\rho \in \mathfrak{a}^*$ be half the sum of the members of Σ^+ repeated according to their multiplicities. If we regard ρ as a member of $\mathfrak{h}^*$ vanishing on $\mathfrak{h}_M$, then (11.14) and Lemma 11.13 give

$$(11.16) \qquad \rho = \tfrac{1}{2} \sum_{\alpha \in \Delta(\mathfrak{n})} \alpha|_\mathfrak{a} = \delta(\mathfrak{n})|_\mathfrak{a} = \delta(\mathfrak{n}).$$

The members of $\Delta(\mathfrak{g}, \mathfrak{h})$ vanishing on $\mathfrak{a}$ can be identified with the root system $\Delta(\mathfrak{m}, \mathfrak{h}_M)$ of $\mathfrak{m}$. Our choice of $\Delta^+(\mathfrak{g}, \mathfrak{h})$ determines a choice of

$\Delta^+(\mathfrak{m}, \mathfrak{h}_M)$. If δ_M denotes half the sum of the members of $\Delta^+(\mathfrak{m}, \mathfrak{h}_M)$, then it is clear from (11.16) that

$$(11.17) \qquad\qquad \delta = \rho + \delta_M.$$

Before passing to groups, let us relate the construction for a general real parabolic subalgebra $\mathfrak{q}$ to the one obtained by starting from a maximal abelian subspace $(\mathfrak{a}_p)_0$ of $\mathfrak{p}_0$. Let $(\mathfrak{a}_M)_0$ be a maximal abelian subspace of $\mathfrak{m}_0 \cap \mathfrak{p}_0$, and extend $(\mathfrak{a}_M)_0$ to a θ stable Cartan subalgebra $(\mathfrak{a}_M)_0 \oplus \mathfrak{t}_0$ of $\mathfrak{m}_0$ as in the special case. Then

$$(11.18) \qquad\qquad \mathfrak{h}_0 = \mathfrak{a}_0 \oplus (\mathfrak{a}_M)_0 \oplus \mathfrak{t}_0$$

is a θ stable Cartan subalgebra of $\mathfrak{l}$, and we can use it in place of (11.11).

Lemma 11.19. Under the definition (11.18), $\mathfrak{a}_0 \oplus (\mathfrak{a}_M)_0$ is a maximal abelian subspace of $\mathfrak{p}_0$.

PROOF. It is certainly abelian in $\mathfrak{p}_0$, and we let $(\mathfrak{a}_p)_0$ be a maximal abelian subspace containing it. If h is in $(\mathfrak{a}_p)_0$, then h centralizes $\mathfrak{a}_0$ and hence is in $\mathfrak{a}_0 \oplus \mathfrak{m}_0$. Subtracting its $\mathfrak{a}_0$ component, we may assume that h is in $\mathfrak{m}_0$. The element h is then in $\mathfrak{m}_0 \cap \mathfrak{p}_0$ and centralizes $(\mathfrak{a}_M)_0$, hence is in $(\mathfrak{a}_M)_0$.

Taking a cue from the lemma, let us put $(\mathfrak{a}_p)_0 = \mathfrak{a}_0 \oplus (\mathfrak{a}_M)_0$. Then $\mathfrak{a} \subseteq \mathfrak{a}_p \subseteq \mathfrak{h}$. We see that the nonzero restrictions of the roots in $\Delta(\mathfrak{g}, \mathfrak{h})$ to $\mathfrak{a}_p$ are the restricted roots and that the nonzero restrictions of the restricted roots to $\mathfrak{a}$ are the $\mathfrak{a}$-roots. When necessary, we can choose compatible orderings by taking $\mathfrak{a}_0$ before $(\mathfrak{a}_M)_0$ and $(\mathfrak{a}_M)_0$ before $i\mathfrak{t}_0$. If we do so, then we have

$$\mathfrak{m}_p \subseteq \mathfrak{m}, \qquad \mathfrak{a}_p \supseteq \mathfrak{a}, \qquad \mathfrak{n}_p \supseteq \mathfrak{n},$$

and also

$$\mathfrak{m}_p \oplus \mathfrak{a}_p \oplus \mathfrak{n}_p \subseteq \mathfrak{m} \oplus \mathfrak{a} \oplus \mathfrak{n}.$$

Thus $\mathfrak{m}_p \oplus \mathfrak{a}_p \oplus \mathfrak{n}_p$ is minimal (under inclusion) among real parabolic subalgebras of $\mathfrak{g}$.

Now let us discuss subgroups. We define A, N, and N^- as the analytic subgroups of G with Lie algebras $\mathfrak{a}_0$, $\mathfrak{n}_0$, and $\mathfrak{n}_0^-$, respectively. The definition of M is more complicated than in (11.6), since $\mathfrak{m}_0$ need no longer be contained in $\mathfrak{k}_0$. First we let M_0 be the analytic subgroup of G with Lie algebra $\mathfrak{m}_0$. The group $L \cap K = N_K(\mathfrak{q})$ normalizes M_0 because it normalizes $\mathfrak{m}_0$, and it thus makes sense to define

$$(11.20) \qquad\qquad M = (L \cap K) M_0.$$

We shall take the following facts as known, still assuming $(\mathfrak{g}, K)$ to be in the Harish-Chandra class. (See the Notes for references.)

(11.21a) A, N, and AN are closed subgroups of G.

(11.21b)   The Levi subgroup of $\mathfrak{q} = \mathfrak{m} \oplus \mathfrak{a} \oplus \mathfrak{n}$ is a direct product $L = MA$, and multiplication is a real analytic diffeomorphism from $M \times A$ onto L.

(11.21c)   The subgroups M_0 and M are closed. They centralize A and normalize N, and M meets every component of G. The subgroup $Q = MAN$ is a closed subgroup with Lie algebra $\mathfrak{q}_0$, and multiplication is a real analytic diffeomorphism from $M \times A \times N$ onto Q. The subgroup Q is equal to its own normalizer.

(11.21d) $G = KMAN$ in the sense that each element of G is the product of four elements in order, one from each group. The A and N components are unique, and so is the product of the K and M components.

(11.21e)   The group M normalizes each $\mathfrak{a}$-root space, and it satisfies $L \cap K = K \cap M$.

(11.21f)   The pair $(\mathfrak{m}, K \cap M)$ is a reductive pair.

(11.21g)   Left and right Haar measures on MAN can be normalized to be related by $d_l(man) = e^{-2\rho \log a} d_r(man)$.

(11.21h)   In terms of the decomposition $G = KMAN$, Haar measure on G, when suitably normalized, decomposes as $d(kman) = dk\, d_r(man)$.

The subgroup Q is called a **parabolic subgroup** of G, and the decomposition $Q = MAN$ of (11.21c) is called the **Langlands decomposition** of Q. Following tradition, we shall write

(11.22) $$g = \kappa(g)\mu(g)e^{H(g)}n(g)$$

for the decomposition of a typical element of G according to $G = KMAN$ as in (11.21d). Here $H(g)$ is uniquely defined as a member of $\mathfrak{a}_0$, and $n(g)$ is uniquely defined as a member of N. The components $\kappa(g) \in K$ and $\mu(g) \in M$ are chosen arbitrarily so that (11.22) holds. By (11.21d) the product $\kappa(g)\mu(g)$ is then uniquely defined.

To MAN, we associate the **continuous series** of representations of G parametrized by pairs (ξ, ν), where ξ is a representation of M and where

ν is in $\mathfrak{a}^*$. We shall assume that ξ is a continuous representation on a Hilbert space, in the sense of §1 of the Introduction. There is no loss of generality in assuming that $K \cap M$ acts by unitary operators (see Wallach [1988], 28–29), and we shall make this assumption. We are especially interested in ξ's that are admissible relative to $K \cap M$.

For the representation $I_{MAN}^G(\xi, \nu)$ corresponding to the pair (ξ, ν), a dense subspace of the representation space is

(11.23a)
$$X_{\text{cont}} = X_{\text{cont}}(\xi, \nu)$$
$$= \left\{ f : G \to V^\xi \text{ continuous} \,\middle|\, \begin{array}{l} f(xman) = e^{-(\nu+\rho)\log a}\xi(m)^{-1}f(x) \\ \text{for } man \in MAN \text{ and } x \in G. \end{array} \right\},$$

and the action is

(11.23b) $(I_{MAN}^G(\xi, \nu, g)f)(x) = f(g^{-1}x)$ for g and x in G.

It is clear that $I_{MAN}^G(\xi, \nu, g_1g_2) = I_{MAN}^G(\xi, \nu, g_1)I_{MAN}^G(\xi, \nu, g_2)$. The realization (11.23) of $I_{MAN}^G(\xi, \nu)$ is called the **induced picture**.

Elements of the space are determined by their restrictions to K as a consequence of the decomposition $G = KMAN$ given in (11.21d). The space of restrictions is nothing more than

(11.24a) $\left\{ f_0 : K \to V^\xi \text{ continuous} \,\middle|\, \begin{array}{l} f_0(km) = \xi(m)^{-1}f_0(k) \\ \text{for } m \in K \cap M \text{ and } k \in K \end{array} \right\}$,

and K acts by the left-regular representation.

The whole representation can be viewed in terms of restrictions to K, and this realization of $I_{MAN}^G(\xi, \nu)$ is called the **compact picture**. The action by $g \in G$ in the compact picture is given by

(11.24b) $(I_{MAN}^G(\xi, \nu, g)f_0)(k) = e^{-(\nu+\rho)H(g^{-1}k)}\xi(\mu(g^{-1}k))^{-1}f_0(\kappa(g^{-1}k))$.

Completing the representation from (11.23) or (11.24) to an action on a Hilbert space requires some argument. On the dense subspace X_{cont} given in (11.23a) or (11.24a), we impose the norm

(11.25) $$\|f\|^2 = \int_K |f(k)|_\xi^2 \, dk,$$

where $|\cdot|_\xi$ denotes the Hilbert space norm on V^ξ. The space $X_{\text{Hilb}} = X_{\text{Hilb}}(\xi, \nu)$ of our representation is to be the completion of X_{cont} in this norm. To define $I_{MAN}^G(\xi, \nu, g)$ on the completion, we need to prove that each $I_{MAN}^G(\xi, \nu, g)$ is bounded on (11.23a) or (11.24a). The first step is given in the following proposition, which uses the notation ξ^h for the continuous representation $\xi^h(m) = \xi(m^{-1})^*$ on V^ξ.

Proposition 11.26. If f_1 is in $X_{\text{cont}}(\xi, \nu)$ and f_2 is in $X_{\text{cont}}(\xi^h, -\bar{\nu})$, then

$$\langle I^G_{MAN}(\xi, \nu, g) f_1, I^G_{MAN}(\xi^h, -\bar{\nu}, g) f_2 \rangle = \langle f_1, f_2 \rangle.$$

The proof will be more understandable if we introduce integration of "densities" before giving the proof. This discussion also will explain the role of ρ in (11.23a) and the significance of integration over K in (11.25).

If G is a Lie group and H is a closed subgroup, then the quotient G/H need not carry a left-invariant Borel measure. It turns out that functions on G/H are the wrong thing to try to integrate.

Fix right Haar measures $d_r g$ and $d_r h$ on G and H, and let Δ_G and Δ_H be the modular functions defined in §I.1. For their respective groups, these functions are determined by the condition $d_r(t \cdot) = \Delta(t) d_r(\cdot)$. Then we can introduce left Haar measures $d_l g$ and $d_r h$ on G and H by means of the formulas

$$(11.27) \qquad d_l x = d_r(x^{-1}) = \Delta(x)^{-1} d_r x.$$

Fix a continuous homomorphism ω of H into the positive reals and consider continuous functions F on G that satisfy

$$(11.28) \qquad F(gh) = \omega(h)^{-1} F(g) \qquad \text{for } g \in G \text{ and } h \in H.$$

To construct such a function that is compactly supported modulo H, we can start with any $f \in C_{\text{com}}(G)$ and define

$$(11.29) \qquad F(g) = \int_H f(gh)\omega(h) \, d_l h.$$

The existence of local continuous cross-sections for G/H, in combination with an argument using a partition of unity, can be used to show that this mapping $f \mapsto F$ carries $C_{\text{com}}(G)$ onto the space $C_{\text{com}}(G/H, \omega)$ of continuous F that are compactly supported modulo H and satisfy (11.28). We omit the proof of this assertion; see the Notes for references.

For the definition of an invariant "integral" of an $F \in C_{\text{com}}(G/H, \omega)$, it is natural to try

$$(11.30) \qquad \int_{G/H} F = \int_G f(g) \, d_l g,$$

where f and F are related by (11.29). This expression is obviously linear, and it is left invariant in the sense that

$$(11.31) \qquad \int_{G/H} l(g) F = \int_{G/H} F$$

if $l(g)F$ is defined by $(l(g)F)(x) = F(g^{-1}x)$. Also if $F \geq 0$, we can arrange that f is ≥ 0 and then (11.30) is ≥ 0. But what needs proof is that (11.30) is well defined. If two different f's yield the same F, do the two f's have the same integral over G? In other words, we are to assume that

$$(11.32) \qquad\qquad \int_H f(gh)\omega(h)\,d_l h = 0$$

for all $g \in G$, and we ask whether $\int_G f(g)\,d_l g = 0$. If ψ is any member of $C_{\text{com}}(G)$, then (11.32) shows that

(11.33)
$$
\begin{aligned}
0 &= \int_G \left[\int_H \psi(g)f(gh)\omega(h)\,d_l h \right] d_l g \\
&= \int_H \left[\int_G \psi(g)f(gh)\,d_l g \right] \omega(h)\,d_l h \\
&= \int_H \left[\int_G \Delta_G(g)^{-1}\psi(g)f(gh)\,d_r g \right] \omega(h)\,d_l h && \text{by (11.27)} \\
&= \int_H \left[\int_G \Delta_G(gh^{-1})^{-1}\psi(gh^{-1})f(g)\,d_r g \right] \omega(h)\,d_l h \\
&= \int_H \left[\int_G \psi(gh^{-1})f(g)\,d_l g \right] \Delta_G(h)\omega(h)\,d_l h \\
&= \int_G f(g) \left[\int_H \psi(gh^{-1})\Delta_G(h)\omega(h)\,d_l h \right] d_l g \\
&= \int_G f(g) \left[\int_H \psi(gh)\Delta_G(h)^{-1}\omega(h)^{-1}\,d_r h \right] d_l g && \text{by (11.27)} \\
&= \int_G f(g) \left[\int_H \psi(gh)\Delta_G(h)^{-1}\Delta_H(h)\omega(h)^{-1}\,d_l h \right] d_l g && \text{by (11.27)}.
\end{aligned}
$$

If

$$(11.34) \qquad\qquad \omega(h) = \Delta_G(h)^{-1}\Delta_H(h),$$

then the inner integral at the end of (11.33) collapses to $\int_H \psi(gh)\,d_l h$. This is a function of the form (11.29) for $\omega = 1$, and the earlier partition of unity argument shows that a locally finite sum of such functions is identically one. Summing (11.33) over the corresponding ψ's, we obtain $0 = \int_g f(g)\,d_l g$ as required.

In other words, $\int_{G/H} F$ is meaningful provided (11.34) holds. We say that a continuous function F on G satisfying (11.28) for ω as in

(11.34) is a **density** for G/H. Integration as in (11.30) is meaningful for any density that is compactly supported modulo H. Dropping the "$0 =$" from the computation in (11.33), we can rewrite the result of that computation as the handy formula

$$(11.35) \qquad \int_G \psi(g) F(g) \, d_l g = \int_G f(g) \Psi(g) \, d_l g,$$

where $\Psi(g) = \int_H \psi(gh) \, d_l h$.

Let us return to the situation as in Proposition 11.26. The reductive group G has $\Delta_G = 1$, and we take $H = Q = MAN$. Comparing (11.21g) and (11.27), we see from (11.34) that $\Delta_Q(man) = e^{2\rho \log a}$. Thus densities for G/Q are continuous functions on G with

$$(11.36) \qquad F(xman) = e^{-2\rho \log a} \qquad \text{for } x \in G \text{ and } man \in Q.$$

This formula explains the role of ρ in (11.23a).

Integration of F simplifies because of the formula for Haar measure in (11.21h) corresponding to the decomposition $G = KQ$. If $f \mapsto F$, we have

$$(11.37) \qquad \begin{aligned} \int_{G/H} F &= \int_G f(g) \, dg = \int_{K \times Q} f(kq) \, dk \, d_r q \\ &= \int_K \left[\int_Q f(kq) \Delta_Q(q) \, d_l q \right] dk = \int_K F(k) \, dk \end{aligned}$$

by (11.29) since $\omega(q) = \Delta_Q(q)$. This formula explains the significance of integration over K in (11.25).

PROOF OF PROPOSITION 11.26. Let $F(x) = \langle f_1(x), f_2(x) \rangle_\xi$. This function is a density for G/Q because

$$\begin{aligned} F(xman) &= \langle e^{-(\nu+\rho) \log a} \xi(m)^{-1} f_1(x), \, e^{-\bar{\nu}+\rho) \log a} \xi(m)^* f_2(x) \rangle_\xi \\ &= e^{-2\rho \log a} \langle f_1(x), f_2(x) \rangle_\xi \\ (11.38) \qquad &= e^{-2\rho \log a} F(x). \end{aligned}$$

Here we have moved the operator $\xi(m)^*$ to the left by removing the adjoint, then cancelled it with $\xi(m)^{-1}$. The scalar factors are removed similarly, using sesquilinearity of the inner product on V^ξ. Then the left side of the desired identity is

$$= \int_K F(g^{-1}k)\, dk$$

$$= \int_{G/Q} l(g)F \qquad \text{by (11.37)}$$

$$= \int_{G/Q} F \qquad \text{by (11.31)}$$

$$= \int_K F(k)\, dk \qquad \text{by (11.37),}$$

and this is the right side.

Corollary 11.39. If ξ is unitary and ν is purely imaginary, then $I^G_{MAN}(\xi, \nu)$ is unitary.

PROOF. Under the given conditions, $\xi^h = \xi$ and $-\bar{\nu} = \nu$. When $f_1 = f_2$, Proposition 11.26 says that $\|I^G_{MAN}(\xi, \nu, g)f_1\|^2 = \|f_1\|^2$.

Corollary 11.40. If f is a continuous function on K that is right invariant under $K \cap M$ and if g is in G, then

$$\int_K e^{-2\rho H(g^{-1}k)} f(\kappa(g^{-1}k))\, dk = \int_K f(k)\, dk.$$

REMARK. Even though $\kappa(\cdot)$ is not uniquely defined, $f(\kappa(g^{-1}k))$ is uniquely defined because of the assumed right invariance under $K \cap M$.

PROOF. This follows from Corollary 11.39 with $\xi = 1$ and $\nu = 0$.

Proposition 11.41. For any ξ and ν, $I^G_{MAN}(\xi, \nu)$ extends to a continuous representation on the Hilbert-space completion X_{Hilb} of (11.23a) or (11.24a).

PROOF. Since ξ is continuous, $\|\xi(m)\|$ is bounded by a constant for m in any compact subset of M. Fix $g \in G$, and let us say that $\|\xi(\mu(g^{-1}k)^{-1})\| \le C$ for $k \in K$. If f is any member of the dense subspace (11.24a), then

$$\|I^G_{MAN}(\xi, \nu, g)f\|^2$$

$$= \int_K e^{-2\operatorname{Re}\nu H(g^{-1}k)} e^{-2\rho H(g^{-1}k)} |\xi(\mu(g^{-1}k)^{-1}) f(\kappa(g^{-1}k))|^2_\xi\, dk$$

$$\le \left\{ \sup_{k \in K} e^{-2\operatorname{Re}\nu H(g^{-1}k)} \right\} C^2 \int_K e^{-2\rho H(g^{-1}k)} |f(\kappa(g^{-1}k))|^2_\xi.$$

The integral on the right side equals $\|f\|^2$, and it follows that each $I_{MAN}^G(\xi, \nu, g)$ is bounded. An easy argument with dominated convergence shows that

$$\lim_{g \to 1} \|I_{MAN}^G(\xi, \nu, g)f - f\|^2 = 0$$

for f in the space (11.24a), and the proposition follows.

Now that $I_{MAN}^G(\xi, \nu)$ is a continuous representation in a Hilbert space, (11.24) shows that the restriction to K of $I_{MAN}^G(\xi, \nu)$ is the classical induced representation $\text{induced}_{K \cap M}^K(\xi)$. By classical Frobenius reciprocity, the multiplicity of a K type τ satisfies

$$[\text{induced}_{K \cap M}^K(\xi) : \tau] = \sum_{\omega \in (K \cap M)^\wedge} [\xi : \omega][\text{induced}_{K \cap M}^K(\omega) : \tau]$$

(11.42)
$$= \sum_{\omega \in (K \cap M)^\wedge} [\xi : \omega][\tau|_{K \cap M} : \omega].$$

The underlying $(\mathfrak{g}, K)$ module of $I_{MAN}^G(\xi, \nu)$, in the sense of §1 of the Introduction, consists of the C^∞ members of (11.23a) that are K finite. If ξ is admissible with respect to $K \cap M$, then every K type of $I_{MAN}^G(\xi, \nu)$ occurs with only finite multiplicity in $I_{MAN}^G(\xi, \nu)$ as a consequence of (11.42), and the K finite vectors are therefore automatically C^∞. In particular, the underlying $(\mathfrak{g}, K)$ module of $I_{MAN}^G(\xi, \nu)$ is admissible with respect to K, and the multiplicities are given by (11.42). In the admissible case we denote the underlying $(\mathfrak{g}, K)$ module of $I_{MAN}^G(\xi, \nu)$ by X_K or $X_K(\xi, \nu)$.

Proposition 11.43. For the continuous series $I_{MAN}^G(\xi, \nu)$, if the underlying $(\mathfrak{m}, K \cap M)$ module of ξ has infinitesimal character λ_ξ relative to $\mathfrak{h}_M$, then the underlying $(\mathfrak{g}, K)$ module $X_K(\xi, \nu)$ of $I_{MAN}^G(\xi, \nu)$ has infinitesimal character $\lambda_\xi + \nu$ relative to $\mathfrak{h} = \mathfrak{h}_M \oplus \mathfrak{a}$.

PROOF. Let f be C^∞ in the induced space, and write I_{MAN}^G for $I_{MAN}^G(\xi, \nu)$. If X is in $\mathfrak{n}_0$, then

$$I_{MAN}^G(X)I_{MAN}^G(g)f(1) = \tfrac{d}{dt} I_{MAN}^G(g)f((\exp tX)^{-1})|_{t=0} = 0$$

because $I_{MAN}^G(g)f$ is right invariant under N. As a consequence, $(I_{MAN}^G(Xu)f)(1) = 0$ for all $X \in \mathfrak{n}$ and $u \in U(\mathfrak{g})$, and Lemma 4.123b implies that

(11.44) $$I_{MAN}^G(Z)f(1) = I_{MAN}^G(\mu'_{\mathfrak{n}^-}(Z))f(1) \qquad \text{for } Z \in Z(\mathfrak{g}).$$

If $X = X_\mathfrak{a} + X_\mathfrak{m}$ is in $\mathfrak{a}_0 \oplus \mathfrak{m}_0$, then

$$
\begin{aligned}
I^G_{MAN}(X)f(1) &= \tfrac{d}{dt} f((\exp tX)^{-1})|_{t=0} \\
&= \tfrac{d}{dt}(e^{(\nu+\rho)} \otimes \xi)(\exp tX)f(1)|_{t=0} \\
&= \{(\nu + \rho)(X_\mathfrak{a}) + \xi(X_\mathfrak{m})\}f(1).
\end{aligned}
$$

Hence

(11.45)
$$
I^G_{MAN}(\mu'_{\mathfrak{n}^-}(Z))f(1) = ((\nu + \rho) \otimes \xi)(\mu'_{\mathfrak{n}^-}(Z))f(1) \quad \text{for } Z \in Z(\mathfrak{g}).
$$

If we write $\mathfrak{n}_M$ for the sum of the root spaces for $\Delta^+(\mathfrak{m}, \mathfrak{h}_M)$, we see that the coefficient on the right side of (11.45) is

$$
\begin{aligned}
&((\nu + \rho) \otimes \xi)(\mu'_{\mathfrak{n}^-}(Z)) \\
&\quad = (\nu + \rho + \lambda_\xi)(\gamma_{\mathfrak{m} \oplus \mathfrak{a}}(\mu'_{\mathfrak{n}^-}(Z))) && \text{by definition of } \lambda_\xi \\
&\quad = (\nu + \rho + \lambda_\xi)(\tau_{\mathfrak{n}_M^-} \circ \gamma'_{\mathfrak{n}_M^-} \circ \mu'_{\mathfrak{n}^-}(Z)) && \text{by (4.93)} \\
&\quad = (\nu + \rho + \lambda_\xi)(\tau_{\mathfrak{n}_M^-} \circ \gamma'_{\mathfrak{n}^- + \mathfrak{n}_M^-}(Z)) && \text{by Proposition 4.129a} \\
&\quad = (\nu + \rho + \lambda_\xi + \delta_M)(\gamma'_{\mathfrak{n}^- + \mathfrak{n}_M^-}(Z)) && \text{by (4.94)} \\
&\quad = (\nu + \lambda_\xi)(\gamma_\mathfrak{g}(Z)) && \text{by (11.17) and (4.94)} \\
&\quad = \chi_{\lambda_\xi + \nu}(Z).
\end{aligned}
$$

When we combine this calculation with (11.44) and (11.45), we obtain $I^G_{MAN}(Z)f(1) = \chi_{\lambda_\xi + \nu}(Z)f(1)$ for $Z \in Z(\mathfrak{g})$. Applying this formula to $I^G_{MAN}(g^{-1})f$ and using the fact that $\mathrm{Ad}(g)Z = Z$, we see that $I^G_{MAN}(Z)f(g) = \chi_{\lambda_\xi + \nu}(Z)f(g)$ for all $Z \in Z(\mathfrak{g})$ and $g \in G$, as asserted.

A consequence of Proposition 11.43 and the theory of §X.2 is the irreducibility of the "unitary spherical principal series" of a complex semisimple Lie group. This irreducibility and its proof played a role in the example in §VIII.4. By the **principal series** of a reductive group, we mean the continuous series induced from a minimal parabolic subgroup.

Theorem 11.46 (Parthasarathy-Rao-Varadarajan). Let G be a complex semisimple group. Then every principal-series representation $I^G_{MAN}(1, \nu)$ induced with $\xi = 1$ on $M_\mathfrak{p}$ and with ν purely imaginary on $(\mathfrak{a}_\mathfrak{p})_0$ is irreducible.

PROOF. Since $M_\mathfrak{p}$ is abelian, Proposition 11.43 shows that the underlying $(\mathfrak{g}, K)$ module V of $I_{MAN}^G(1, \nu)$ has infinitesimal character ν, which is purely imaginary. Assuming that V is reducible, let V' be a nontrivial $(\mathfrak{g}, K)$ submodule. Then each of V' and V/V' has a purely imaginary infinitesimal character, and Theorem 10.37 says that each of them contains the trivial K type. Hence V contains the trivial K type with multiplicity at least 2. But this conclusion contradicts (11.42), which shows that the trivial K type occurs in V with multiplicity exactly 1.

The conclusion of irreducibility in Theorem 11.46 extends to all the principal-series representations $I_{MAN}^G(\xi, \nu)$ of a complex semisimple group with ξ irreducible and ν purely imaginary. This extended theorem is due to Zhelobenko. It may be derived in the same way: If λ is the differential of ξ, then $I_{MAN}^G(\xi, \nu)$ has infinitesimal character $\lambda + \nu$. Since $2\delta_K = \delta$ in a complex group, the real part of the identity in Theorem 10.37 gives $|\lambda| \geq |\mu|$. However, the only K type in $I_{MAN}^G(\xi, \nu)$ whose highest weight has magnitude $\leq |\lambda|$ is the dominant Weyl group transform of λ, by Frobenius reciprocity, and this K type has multiplicity 1. Thus μ has to be this highest weight, and the above proof goes through.

2. Parabolic Induction Infinitesimally

To bring parabolic induction into the framework of the first ten chapters, we need to identify the underlying $(\mathfrak{g}, K)$ modules of parabolically induced representations. For consistency with §1, we assume temporarily that the reductive pair $(\mathfrak{g}, K)$ is in the Harish-Chandra class. Later in this section, we shall drop this assumption.

Proposition 11.47. Let $(\mathfrak{g}, K)$ be a reductive pair in the Harish-Chandra class, let G be the corresponding reductive group, and let $Q = MAN$ be the Langlands decomposition of a parabolic subgroup of G. Let ξ be a continuous representation of M in a Hilbert space, suppose that ξ is admissible under $K \cap M$, and let $V_{K \cap M}^\xi$ be the underlying $(\mathfrak{m}, K \cap M)$ module. For $\nu \in \mathfrak{a}^*$, denote by $V_{K \cap M}^{\xi, \nu+\rho} = V_{K \cap M}^\xi \otimes \mathbb{C}_{\nu+\rho} \otimes \mathbb{C}$ the $(\mathfrak{q}, K \cap M)$ module such that $(\mathfrak{m}, K \cap M)$ acts in $V_{K \cap M}^\xi$, $\mathfrak{a}$ acts in $\mathbb{C}_{\nu+\rho}$, and $\mathfrak{n}$ acts trivially. Then the underlying $(\mathfrak{g}, K)$ module of the continuous-series representation $I_{MAN}^G(\xi, \nu)$, which is admissible for K, is

$$\cong I_{\mathfrak{q}, K \cap M}^{\mathfrak{g}, K}(V_{K \cap M}^{\xi, \nu+\rho}) \cong \Gamma_{\mathfrak{g}, K \cap M}^{\mathfrak{g}, K}(\mathrm{pro}_{\mathfrak{q}, K \cap M}^{\mathfrak{g}, K \cap M}(V_{K \cap M}^{\xi, \nu+\rho})).$$

PROOF. It will be more convenient to work with an equivalent realization of the continuous-series representation, making G operate on the right instead of the left. The space, which we denote X_{cont} is

$$\left\{ f : G \to V^\xi \text{ continuous} \;\middle|\; \begin{array}{l} f(manx) = e^{(\nu+\rho)\log a}\xi(m)f(x) \\ \text{for } man \in MAN \text{ and } x \in G. \end{array} \right\},$$

and G acts by

$$I^G_{MAN}(\xi, \nu, g)f(x) = f(xg).$$

(The equivalence is given by inverting the domain variable.)

Let X_K be the underlying $(\mathfrak{g}, K)$ module of $I^G_{MAN}(\xi, \nu)$. We decompose $I^{\mathfrak{g},K}_{\mathfrak{q},K\cap M}$ as $I^{\mathfrak{g},K}_{\mathfrak{g},K\cap M}\circ\mathrm{pro}^{\mathfrak{g},K\cap M}_{\mathfrak{q},K\cap M}$ by Propositions 2.19 and 2.57b. By Frobenius reciprocity (Proposition 2.21),

$$(11.48) \quad \mathrm{Hom}_{R(\mathfrak{g},K)}(X_K, \; I^{\mathfrak{g},K}_{\mathfrak{g},K\cap M}(\mathrm{pro}^{\mathfrak{g},K\cap M}_{\mathfrak{q},K\cap M}(V^{\xi,\nu+\rho}_{K\cap M})))$$

$$\cong \mathrm{Hom}_{R(\mathfrak{g},K\cap M)}(X_K, \; \mathrm{pro}^{\mathfrak{g},K\cap M}_{\mathfrak{q},K\cap M}(V^{\xi,\nu+\rho}_{K\cap M})),$$

where we have suppressed the forgetful functor $\mathcal{F}^{\mathfrak{g},K\cap M}_{\mathfrak{g},K}$ on X_K on the right. The correspondence of Φ on the left to φ on the right is given in (2.22) as

$$\Phi(f)(r) = \varphi(rf) \qquad \text{for } r \in R(\mathfrak{g}, K), \; f \in X_K.$$

If $f \neq 0$ is in $\ker \Phi$, then $\varphi(rf) = 0$ for all r. Since $R(\mathfrak{g}, K)$ has an approximate identity, we obtain $\varphi(f) = 0$ and even $\varphi(kf) = 0$ for all $k \in K$. In other words, if we can produce a φ in the right side of (11.48) such that $\varphi(Kf) \neq 0$ when $f \neq 0$, then X_K injects into $I^{\mathfrak{g},K}_{\mathfrak{q},K\cap M}(V^{\xi,\nu+\rho}_{K\cap M})$.

Such a map φ is given initially as a map into $\mathrm{Hom}_{\mathbb{C}}(U(\mathfrak{g}), V^{\xi,\nu+\rho}_{K\cap M})$ by

$$(11.49) \quad \varphi(f)(u) = (I^G_{MAN}(\xi, \nu, u)f)(1) \qquad \text{for } u \in U(\mathfrak{g}), \; f \in X_K.$$

Here u is to act on the right of the G variable in f. We have to check that the K finiteness of f forces $\varphi(f)(u)$ to be $K \cap M$ finite, so that $\varphi(f)(u)$ is actually in $V^{\xi,\nu+\rho}_{K\cap M}$. If k is in $K \cap M$, we have

$$k(\varphi(f)(u))$$
$$= k(I^G_{MAN}(\xi, \nu, u)f)(1)$$
$$= (I^G_{MAN}(\xi, \nu, u)f)(k) \qquad\qquad \text{since } k \text{ is in } K \cap M$$
$$= (I^G_{MAN}(\xi, \nu, k)I^G_{MAN}(\xi, \nu, u)f)(1) \qquad \text{since } k \text{ is in } K$$
$$= (I^G_{MAN}(\xi, \nu, \mathrm{Ad}(k)u)I^G_{MAN}(\xi, \nu, k)f)(1).$$

As k varies, the vector-space span of $I_{MAN}^{G}(\xi, \nu, k) f$ is finite dimensional, and the operators $I_{MAN}^{G}(\xi, \nu, \mathrm{Ad}(k)u)$ lie in a finite-dimensional space. Hence $k(\varphi(f)(u))$ lies in a finite-dimensional space.

If $\varphi(Kf) = 0$, then $\varphi(kf)(1) = 0$ for all $k \in K$ and

$$0 = (I_{MAN}^{G}(\xi, \nu, k) f)(1) = f(k).$$

Hence $f = 0$. Thus $\varphi(Kf) \neq 0$ when $f \neq 0$.

Let us check that φ has image in $\mathrm{Hom}_{\mathfrak{q}}(U(\mathfrak{g}), V_{K \cap M}^{\xi, \nu+\rho})_{K \cap M}$. In fact, if x is in $\mathfrak{q}$, then $\varphi(f)(xu)$ is

$$= (I_{MAN}^{G}(\xi, \nu, xu) f)(1) = \begin{cases} \xi(x)(I_{MAN}^{G}(\xi, \nu, u) f)(1) & \text{if } x \in \mathfrak{m}_0 \\ (\nu + \rho)(x)(I_{MAN}^{G}(\xi, \nu, u) f)(1) & \text{if } x \in \mathfrak{a}_0 \\ 0 & \text{if } x \in \mathfrak{n}_0. \end{cases}$$

Consequently φ has image in $\mathrm{Hom}_{\mathfrak{q}}(U(\mathfrak{g}), V_{K \cap M}^{\xi, \nu+\rho})$. If m is in $K \cap M$, then (2.13b) gives

$$\begin{aligned} \varphi(I_{MAN}^{G}(\xi, \nu, m) f)(u) &= I_{MAN}^{G}(\xi, \nu, u) I_{MAN}^{G}(\xi, \nu, m) f(1) \\ &= (I_{MAN}^{G}(\xi, \nu, m) I_{MAN}^{G}(\xi, \nu, \mathrm{Ad}(m)^{-1}u) f)(1) \\ &= \xi(m)(I_{MAN}^{G}(\xi, \nu, \mathrm{Ad}(m)^{-1}u) f(1)) \\ &= \xi(m)(\varphi(f)(\mathrm{Ad}(m)^{-1}u)) \\ &= m(\varphi(f))(u). \end{aligned}$$

Since X_K is $K \cap M$ finite, this equivariance shows that the image of φ is $K \cap M$ finite. Consequently φ has image in $\mathrm{Hom}_{\mathfrak{q}}(U(\mathfrak{g}), V_{K \cap M}^{\xi, \nu+\rho})_{K \cap M}$. Also φ is $U(\mathfrak{g})$ equivariant since

$$\begin{aligned} \varphi(I_{MAN}^{G}(\xi, \nu, u') f)(u) &= (I_{MAN}^{G}(\xi, \nu, u) I_{MAN}^{G}(\xi, \nu, u') f)(1) \\ &= (I_{MAN}^{G}(\xi, \nu, uu') f)(1) \\ &= \varphi(f)(uu'). \end{aligned}$$

We have therefore proved that X_K injects into $I_{\mathfrak{q}, K \cap M}^{\mathfrak{g}, K}(V_{K \cap M}^{\xi, \nu+\rho})$. Since $I_{MAN}^{G}(\xi, \nu)$ is admissible by (11.42) and the admissibility of ξ, this injection will be an isomorphism if we show that each K type has the same multiplicity in $I_{MAN}^{G}(\xi, \nu)$ as in $I_{\mathfrak{q}, K \cap M}^{\mathfrak{g}, K}(V_{K \cap M}^{\xi, \nu+\rho})$. We begin with the isomorphism

(11.50)
$$\mathcal{F}_{\mathfrak{g},K}^{\mathfrak{k},K} \circ I_{\mathfrak{q},K\cap M}^{\mathfrak{g},K}$$

$$\cong \mathcal{F}_{\mathfrak{g},K}^{\mathfrak{k},K} \circ I_{\mathfrak{g},K\cap M}^{\mathfrak{g},K} \circ I_{\mathfrak{q},K\cap M}^{\mathfrak{g},K\cap M} \qquad \text{by Proposition 2.19}$$

$$\cong I_{\mathfrak{k},K\cap M}^{\mathfrak{k},K} \circ \mathcal{F}_{\mathfrak{g},K\cap M}^{\mathfrak{k},K\cap M} \circ I_{\mathfrak{q},K\cap M}^{\mathfrak{g},K\cap M} \qquad \text{by Proposition 2.69b}$$

$$\cong I_{\mathfrak{k},K\cap M}^{\mathfrak{k},K} \circ I_{\mathfrak{k}\cap\mathfrak{m},K\cap M}^{\mathfrak{k},K\cap M} \circ \mathcal{F}_{\mathfrak{q},K\cap M}^{\mathfrak{k}\cap\mathfrak{m},K\cap M} \qquad \begin{array}{l}\text{by Lemma 5.30b}\\ \text{since } \mathfrak{k}+\mathfrak{q}=\mathfrak{g}\end{array}$$

$$\cong I_{\mathfrak{k}\cap\mathfrak{m},K\cap M}^{\mathfrak{k},K} \circ \mathcal{F}_{\mathfrak{q},K\cap M}^{\mathfrak{k}\cap\mathfrak{m},K\cap M} \qquad \text{by Proposition 2.19.}$$

If F^τ is an irreducible K module of type τ, then we have

$$\mathrm{Hom}_K(F^\tau, \mathcal{F}_{\mathfrak{g},K}^{\mathfrak{k},K} \circ I_{\mathfrak{q},K\cap M}^{\mathfrak{g},K}(V_{K\cap M}^{\xi,\nu+\rho}))$$

$$\cong \mathrm{Hom}_K(F^\tau, I_{\mathfrak{k}\cap\mathfrak{m},K\cap M}^{\mathfrak{k},K}(V_{K\cap M}^{\xi,\nu+\rho})) \qquad \text{by (11.50)}$$

$$\cong \mathrm{Hom}_{K\cap M}(F^\tau, V_{K\cap M}^{\xi,\nu+\rho}) \qquad \text{by Proposition 2.21.}$$

Therefore

(11.51)
$$[I_{\mathfrak{q},K\cap M}^{\mathfrak{g},K}(V_{K\cap M}^{\xi,\nu+\rho}) : \tau] = \sum_{\omega \in (K\cap M)^\wedge} [\tau|_{K\cap M} : \omega]\,[V_{K\cap M}^{\xi,\nu+\rho} : \omega]$$

$$= \sum_{\omega \in (K\cap M)^\wedge} [\tau|_{K\cap M} : \omega]\,[V_{K\cap M}^{\xi} : \omega]$$

$$= [I_{MAN}^G(\xi,\nu) : \tau],$$

the last step holding by (11.42). This completes the proof.

Because of the notational similarity between the $(\mathfrak{g}, K)$ module in Proposition 11.47 and the cohomologically induced modules in (5.3), it is natural to ask whether use of the derived functors of $\Gamma_{\mathfrak{g},K\cap M}^{\mathfrak{g},K}$ might result in additional interesting $(\mathfrak{g}, K)$ modules. The following proposition, however, shows that all the modules from the higher derived functors of $\Gamma_{\mathfrak{g},K\cap M}^{\mathfrak{g},K}$ are 0.

Proposition 11.52. Let $(\mathfrak{g}, K)$ be a reductive pair in the Harish-Chandra class, and let $(\mathfrak{q}, K \cap M)$ be a real parabolic subpair. Then $I_{\mathfrak{q},K\cap M}^{\mathfrak{g},K}$ is exact as a functor from $\mathcal{C}(\mathfrak{q}, K \cap M)$ to $\mathcal{C}(\mathfrak{g}, K)$.

PROOF. Taking $(\cdot)^j$ of both sides of (11.50) and arguing with (C.28.a2) and (C.27a) as in the proof of Proposition 2.115, we find that

$$\mathcal{F}_{\mathfrak{g},K}^{\mathfrak{k},K} \circ (I_{\mathfrak{q},K\cap M}^{\mathfrak{g},K})^j \cong (I_{\mathfrak{k}\cap\mathfrak{m},K\cap M}^{\mathfrak{k},K})^j \circ \mathcal{F}_{\mathfrak{q},K\cap M}^{\mathfrak{k}\cap\mathfrak{m},K\cap M}.$$

Now $(I_{\mathfrak{k}\cap\mathfrak{m},K\cap M}^{\mathfrak{k},K})^j = 0$ for $j > 0$ since every module in $\mathcal{C}(\mathfrak{k}\cap\mathfrak{m}, K\cap M)$ is injective (by (C.14) and Proposition 1.18b). Hence $\mathcal{F}_{\mathfrak{g},K}^{\mathfrak{k},K} \circ (I_{\mathfrak{q},K\cap M}^{\mathfrak{g},K})^j = 0$ for $j > 0$. In particular, $(I_{\mathfrak{q},K\cap M}^{\mathfrak{g},K})^1 = 0$. By (C.36c), $I_{\mathfrak{q},K\cap M}^{\mathfrak{g},K\cap M}$ is exact.

For $(\mathfrak{g}, K)$ in the Harish-Chandra class, the Levi factor of the real parabolic pair $(\mathfrak{q}, K\cap M)$ has until now been $L = MA$, and the Levi pair has been $(\mathfrak{l}, L\cap K) = (\mathfrak{m}\oplus\mathfrak{a}, K\cap M)$. When we drop the assumption that $(\mathfrak{g}, K)$ is in the Harish-Chandra class, it will be helpful not to insist on generalizing MA precisely. Instead, when $\mathfrak{q} = \mathfrak{l} \oplus \mathfrak{n}$ is a real parabolic subalgebra of $\mathfrak{g}$, let us redefine a **Levi subgroup** L to be any subgroup of G satisfying

(11.53) $$L_0 \subseteq L \subseteq N_G(\mathfrak{q}) \cap N_G(\theta\mathfrak{q}).$$

Here the minimal choice L_0 is the analytic subgroup of G with Lie algebra $\mathfrak{l}_0$, and the maximal choice $N_G(\mathfrak{q}) \cap N_G(\theta\mathfrak{q})$ is given in (4.77a). By Proposition 4.78a, L still has Lie algebra $\mathfrak{l}_0$. The pair $(\mathfrak{l}, L\cap K)$ is still reductive and is called the corresponding **Levi subpair** of $(\mathfrak{g}, K)$. The subgroup $Q = LN$ is the corresponding **parabolic subgroup**, and $(\mathfrak{q}, L\cap K)$ is the corresponding **parabolic subpair** of $(\mathfrak{g}, K)$.

Now let $(\mathfrak{g}, K)$ be any reductive pair, and let $(\mathfrak{q}, L\cap K)$ be a real parabolic subpair with L as in the previous paragraph. We define a functor

(11.54a) $${}^u\mathcal{R}_{\mathfrak{q},L\cap K}^{\mathfrak{g},K} : \mathcal{C}(\mathfrak{l}, L\cap K) \to \mathcal{C}(\mathfrak{g}, K)$$

by

(11.54b) $${}^u\mathcal{R}_{\mathfrak{q},L\cap K}^{\mathfrak{g},K}(Z) = \Gamma_{\mathfrak{g},L\cap K}^{\mathfrak{g},K}(\mathrm{pro}_{\mathfrak{q},L\cap K}^{\mathfrak{g},L\cap K}(\mathcal{F}_{\mathfrak{l},L\cap K}^{\mathfrak{q},L\cap K}(Z))).$$

Proposition 11.47 says that ${}^u\mathcal{R}_{\mathfrak{q},L\cap K}^{\mathfrak{g},K}$ models the continuous series under some mild hypotheses, except that it does not take the factor $\mathbb{C}_\rho$ into account. The superscript "u" is an abbreviation for "unnormalized" and is to serve as a reminder of this fact. In any case ${}^u\mathcal{R}_{\mathfrak{q},L\cap K}^{\mathfrak{g},K}$ is to act on $(\mathfrak{l}, L\cap K)$ modules, taking $\mathfrak{q}$ into account; the $\mathfrak{l}$ is implicit in the notation since $\mathfrak{l} = \mathfrak{q} \cap \theta\bar{\mathfrak{q}}$.

Motivated by the form of this functor, we can introduce a generalization of the continuous series of §1. First we observe in the context of §1 that $e^{2\rho \log a} = \left| \det \mathrm{Ad}(ma)|_{\mathfrak{n}} \right|$. Thus we can define a one-dimensional $(\mathfrak{l}, L \cap K)$ module $\mathbb{C}_\rho$ to have the action $\left| \det \mathrm{Ad}(l)|_{\mathfrak{n}} \right|^{1/2}$ for $l \in L$. If ξ is a continuous representation of L in a Hilbert space V^ξ, then we can form

(11.55a)
$$\left\{ f : G \to V^\xi \text{ continuous} \;\middle|\; \begin{array}{l} f(xl) = \left| \det \mathrm{Ad}(l)|_{\mathfrak{n}} \right|^{-1/2} \xi(l)^{-1} f(x) \\ \text{for } l \in L \text{ and } x \in G. \end{array} \right\}$$

and have G act by

(11.55b) $I_{MAN}^G(\xi, g) f(x) = f(g^{-1}x).$

Proposition 11.26 and Corollaries 11.39 and 11.40 generalize to this setting, and so does (11.42). From the proposition and corollaries, we obtain the following generalization of Proposition 11.41.

Proposition 11.56. For any ξ, $I_{MAN}^G(\xi)$ extends to a continuous representation on the Hilbert-space completion of (11.55a) in the norm obtained from $L^2(K)$.

The first result of this section generalizes to this enlarged setting as follows.

Proposition 11.57. Let $(\mathfrak{g}, K)$ be a reductive pair, let G be the corresponding reductive subgroup, and let $Q = LN$ be a parabolic subgroup of G. Let ξ be a continuous representation of L in a Hilbert space, suppose that ξ is admissible under $L \cap K$, and let $V_{L \cap K}^\xi$ be the underlying $(\mathfrak{l}, L \cap K)$ module. Then the underlying $(\mathfrak{g}, K)$ module of the representation $I_{MAN}^G(\xi)$ is admissible under K and is given by $^u\mathcal{R}_{\mathfrak{q}, L \cap K}^{\mathfrak{g}, K}(V_{L \cap K}^\xi \otimes \mathbb{C}_\rho)$.

REMARKS.

1) The proof is the same as for Proposition 11.47, except for small changes.

2) In §7 we shall normalize the functor $^u\mathcal{R}_{\mathfrak{q}, L \cap K}^{\mathfrak{g}, K}$ so as to deal with $\mathbb{C}_\rho$. The definition will be

(11.58) $\mathcal{R}_{\mathfrak{q}, L \cap K}^{\mathfrak{g}, K}(V) = {}^u\mathcal{R}_{\mathfrak{q}, L \cap K}^{\mathfrak{g}, K}(V \otimes \mathbb{C}_\rho),$

and $\mathcal{R}_{\mathfrak{q}, L \cap K}^{\mathfrak{g}, K}$ will give the effect on $(\mathfrak{g}, K)$ modules corresponding to the effect of I_{MAN}^G on representations.

Corollary 11.59. Let $(\mathfrak{g}, K)$ be a reductive pair, let G be the corresponding reductive subgroup, and let $Q = LN$ be a parabolic subgroup of G. Let ξ be a continuous representation of L in a Hilbert space, suppose that ξ is admissible under $L \cap K$, and let $V_{L \cap K}^{\xi}$ be the underlying $(\mathfrak{l}, L \cap K)$ module. Then ${}^u\mathcal{R}_{\mathfrak{q}, L \cap K}^{\mathfrak{g}, K}(V_{L \cap K}^{\xi} \otimes \mathbb{C}_{\rho})^h \cong {}^u\mathcal{R}_{\mathfrak{q}, L \cap K}^{\mathfrak{g}, K}((V_{L \cap K}^{\xi})^h \otimes \mathbb{C}_{\rho})$.

PROOF. The generalized Proposition 11.26 shows that the Hilbert-space Hermitian dual of $I_{MAN}^G(\xi)$ is $I_{MAN}^G(\xi^h)$. Applying Proposition 11.57 to ξ and ξ^h, we obtain the corollary.

It is instructive to construct algebraically the pairing in Corollary 11.59. Put $V = V_{L \cap K}^{\xi}$. We have a natural $(\mathfrak{l}, L \cap K)$ bilinear map $V \times V^h \to \mathbb{C}$, and this yields an $(\mathfrak{l}, L \cap K)$ linear map

$$(11.60a) \qquad (V \otimes \mathbb{C}_{\rho}) \otimes (V^h \otimes \mathbb{C}_{\rho}) \to \mathbb{C}_{2\rho}.$$

We shall derive from this a canonical $(\mathfrak{g}, K)$ bilinear map

$$(11.60b) \qquad {}^u\mathcal{R}_{\mathfrak{q}, L \cap K}^{\mathfrak{g}, K}(V \otimes \mathbb{C}_{\rho}) \times {}^u\mathcal{R}_{\mathfrak{q}, L \cap K}^{\mathfrak{g}, K}(V^h \otimes \mathbb{C}_{\rho}) \to \mathbb{C}.$$

We do so in two steps, first obtaining from any $(\mathfrak{q}, L \cap K)$ linear map

$$(11.61a) \qquad X \otimes Y \to W$$

a canonical $(\mathfrak{g}, K)$ linear map

$$(11.61b) \qquad I(X) \otimes I(Y) \to I(W),$$

where $I = I_{\mathfrak{q}, L \cap K}^{\mathfrak{g}, K}$. Second we construct an explicit $(\mathfrak{g}, K)$ linear map

$$(11.61c) \qquad I(\mathbb{C}_{2\rho}) \to \mathbb{C}.$$

For the first step, let φ be the map (11.61a). We start from the inclusions

(11.62)
$$\operatorname{Hom}_{\mathfrak{q}, L \cap K}(I(X), X) \otimes_{\mathbb{C}} \operatorname{Hom}_{\mathfrak{q}, L \cap K}(I(Y), Y)$$
$$\subseteq \operatorname{Hom}_{\mathfrak{q} \oplus \mathfrak{q}, (L \cap K) \times (L \cap K)}(I(X) \otimes I(Y), X \otimes Y)$$
$$\subseteq \operatorname{Hom}_{\mathfrak{q}, L \cap K}(I(X) \otimes I(Y), X \otimes Y).$$

Here the first inclusion is given by $\psi_1 \otimes \psi_2 \to \psi$ with $\psi(\tilde{x} \otimes \tilde{y}) = \psi_1(\tilde{x}) \otimes \psi_2(\tilde{y})$ for $\tilde{x} \in I(X)$ and $\tilde{y} \in I(Y)$. The second inclusion is given by regarding the $(\mathfrak{q} \otimes \mathfrak{q}, (L \cap K) \times (L \cap K))$ modules as $(\mathfrak{q}, L \cap K)$ modules under the diagonal actions. On the left side of (11.62) are the maps that correspond to

$$1_{I(X)} \in \mathrm{Hom}_{\mathfrak{g},K}(I(X), I(X)) \quad \text{and} \quad 1_{I(Y)} \in \mathrm{Hom}_{\mathfrak{g},K}(I(Y), I(Y))$$

under Frobenius reciprocity (Proposition 2.21). Their tensor product is contained in the right side of (11.62) and corresponds under Frobenius reciprocity to a member of

$$\mathrm{Hom}_{\mathfrak{g},K}(I(X) \otimes I(Y), I(X \otimes Y)).$$

Thus we obtain a canonical nonzero $(\mathfrak{g}, K)$ map

$$\Phi : I(X) \otimes I(Y) \to I(X \otimes Y).$$

Following Φ with $I(\varphi)$, we obtain the required map (11.61b).

For the second step, we use Proposition 11.57 to identify $I(\mathbb{C}_{2\rho})$ as the underlying $(\mathfrak{g}, K)$ module of $I_{MAN}^G(\mathbb{C}_\rho)$. The space for this representation consists of the right $L \cap K$ invariant members of $L^2(K)$. Sending such a function f to $\int_K f(k)\, dk$ is G equivariant, by the generalization of Corollary 11.40 to the present setting. The result is the required map (11.61c).

The second proposition of this section generalizes, too. The statement is as follows, and the proof is essentially the same as for Proposition 11.52.

Proposition 11.63. Let $(\mathfrak{g}, K)$ be a reductive pair, and let $(\mathfrak{q}, L \cap K)$ be a real parabolic subpair. Then ${}^u\mathcal{R}_{\mathfrak{q},L\cap K}^{\mathfrak{g},K}$ is exact as a functor from $\mathcal{C}(\mathfrak{l}, L \cap K)$ to $\mathcal{C}(\mathfrak{g}, K)$.

Guided by the cohomological induction formulas (5.3), let us define an analog of $\mathcal{L}$ for the case of real parabolic subalgebras. If $(\mathfrak{q}, L \cap K)$ is a real parabolic pair, we define a functor

$$(11.64a) \qquad {}^u\mathcal{L}_{\mathfrak{q},L\cap K}^{\mathfrak{g},K} : \mathcal{C}(\mathfrak{l}, L \cap K) \to \mathcal{C}(\mathfrak{g}, K)$$

by

$$(11.64b) \qquad {}^u\mathcal{L}_{\mathfrak{q},L\cap K}^{\mathfrak{g},K}(Z) = \Pi_{\mathfrak{g},L\cap K}^{\mathfrak{g},K}(\mathrm{ind}_{\mathfrak{q},L\cap K}^{\mathfrak{g},L\cap K}(\mathcal{F}_{\mathfrak{l},L\cap K}^{\mathfrak{q},L\cap K}(Z))).$$

Proposition 11.65. Let $(\mathfrak{g}, K)$ be a reductive pair, let G be the corresponding reductive subgroup, and let $Q = LN$ be a parabolic subgroup of G. Let ξ be a continuous representation of L in a Hilbert space, suppose that ξ is admissible under $L \cap K$, and let $V_{L\cap K}^\xi$ be the underlying $(\mathfrak{l}, L \cap K)$ module. Then

$${}^u\mathcal{L}_{\mathfrak{q},L\cap K}^{\mathfrak{g},K}(V_{L\cap K}^\xi \otimes \mathbb{C}_{-\rho}) \cong {}^u\mathcal{R}_{\mathfrak{q},L\cap K}^{\mathfrak{g},K}(V_{L\cap K}^\xi \otimes \mathbb{C}_\rho)$$

as $(\mathfrak{g}, K)$ modules.

REMARKS.   In §7 we shall normalize the functor ${}^{u}\mathcal{L}^{\mathfrak{g},K}_{\mathfrak{q},L\cap K}$ so as to incorporate $\mathbb{C}_{-\rho}$. The definition will be

$$(11.66a) \qquad \mathcal{L}^{\mathfrak{g},K}_{\mathfrak{q},L\cap K}(V) = {}^{u}\mathcal{L}^{\mathfrak{g},K}_{\mathfrak{q},L\cap K}(V \otimes \mathbb{C}_{-\rho}).$$

Normally, however, we shall be working with the opposite parabolic subalgebra $\mathfrak{q}^{-}$, and the corresponding formula will be

$$(11.66b) \qquad \mathcal{L}^{\mathfrak{g},K}_{\mathfrak{q}^{-},L\cap K}(V') = {}^{u}\mathcal{L}^{\mathfrak{g},K}_{\mathfrak{q}^{-},L\cap K}(V' \otimes \mathbb{C}_{\rho}).$$

PROOF.   Let Z be any $(\mathfrak{l}, L \cap K)$ module. Since $\mathfrak{q}$ is real, the argument for (6.18) and (6.19) gives

$$P^{\mathfrak{g},L\cap K}_{\mathfrak{q},L\cap K}(\bar{Z}) \cong \overline{P^{\mathfrak{g},L\cap K}_{\mathfrak{q},L\cap K}(Z)} \quad \text{and} \quad I^{\mathfrak{g},L\cap K}_{\mathfrak{q},L\cap K}(\bar{Z}) \cong \overline{I^{\mathfrak{g},L\cap K}_{\mathfrak{q},L\cap K}(Z)}.$$

Similarly any $(\mathfrak{g}, L \cap K)$ module W satisfies

$$P^{\mathfrak{g},K}_{\mathfrak{g},L\cap K}(\bar{W}) \cong \overline{P^{\mathfrak{g},K}_{\mathfrak{g},L\cap K}(W)} \quad \text{and} \quad I^{\mathfrak{g},K}_{\mathfrak{g},L\cap K}(\bar{W}) \cong \overline{I^{\mathfrak{g},K}_{\mathfrak{g},L\cap K}(W)}.$$

Combining these with Easy Duality (Theorem 3.1), we obtain

$$(11.67) \qquad {}^{u}\mathcal{L}^{\mathfrak{g},K}_{\mathfrak{q},L\cap K}(Z)^{h} \cong {}^{u}\mathcal{R}^{\mathfrak{g},K}_{\mathfrak{q},L\cap K}(Z^{h})$$

for any $(\mathfrak{l}, L\cap K)$ module Z, even without the assumption of admissibility. Therefore the $(\mathfrak{l}, L \cap K)$ module $V = V^{\xi}_{L\cap K}$ satisfies

$$\begin{aligned}
{}^{u}\mathcal{L}^{\mathfrak{g},K}_{\mathfrak{q},L\cap K}(V \otimes \mathbb{C}_{-\rho})^{h} &\cong {}^{u}\mathcal{R}^{\mathfrak{g},K}_{\mathfrak{q},L\cap K}((V \otimes \mathbb{C}_{-\rho})^{h}) && \text{by (11.67)} \\
&\cong {}^{u}\mathcal{R}^{\mathfrak{g},K}_{\mathfrak{q},L\cap K}(V^{h} \otimes \mathbb{C}_{\rho}) && \\
&\cong {}^{u}\mathcal{R}^{\mathfrak{g},K}_{\mathfrak{q},L\cap K}(V \otimes \mathbb{C}_{\rho})^{h} && \text{by Corollary 11.59.}
\end{aligned}$$

Taking the Hermitian dual of both sides and using admissibility, we obtain the result of the proposition.

Proposition 11.68. Let $(\mathfrak{g}, K)$ be a reductive pair, and let $(\mathfrak{q}, L \cap K)$ be a real parabolic subpair. Then ${}^{u}\mathcal{L}^{\mathfrak{g},K}_{\mathfrak{q},L\cap K}$ is exact as a functor from $\mathcal{C}(\mathfrak{l}, L \cap K)$ to $\mathcal{C}(\mathfrak{g}, K)$.

Proof. We have

(11.69)
$$\mathcal{F}_{\mathfrak{g},K}^{\mathfrak{k},K} \circ P_{\mathfrak{q},L\cap K}^{\mathfrak{g},K}$$

$$\begin{aligned}
&\cong \mathcal{F}_{\mathfrak{g},K}^{\mathfrak{k},K} \circ P_{\mathfrak{g},L\cap K}^{\mathfrak{g},K} \circ P_{\mathfrak{q},L\cap K}^{\mathfrak{g},L\cap K} && \text{by Proposition 2.19} \\
&\cong P_{\mathfrak{k},L\cap K}^{\mathfrak{k},K} \circ \mathcal{F}_{\mathfrak{g},L\cap K}^{\mathfrak{k},L\cap K} \circ P_{\mathfrak{q},L\cap K}^{\mathfrak{g},L\cap K} && \text{by Proposition 2.69b} \\
&\cong P_{\mathfrak{k},L\cap K}^{\mathfrak{k},K} \circ P_{\mathfrak{l}\cap\mathfrak{k},L\cap K}^{\mathfrak{k},L\cap K} \circ \mathcal{F}_{\mathfrak{q},L\cap K}^{\mathfrak{l}\cap\mathfrak{k},L\cap K} && \text{by Lemma 5.26b} \\
&&& \text{since } \mathfrak{k} + \mathfrak{q} = \mathfrak{g} \\
&\cong P_{\mathfrak{l}\cap\mathfrak{k},L\cap K}^{\mathfrak{k},K} \circ \mathcal{F}_{\mathfrak{q},L\cap K}^{\mathfrak{l}\cap\mathfrak{k},L\cap K} && \text{by Proposition 2.19.}
\end{aligned}$$

Taking $(\,\cdot\,)_j$ of both sides of (11.69) and arguing with (C.28.a1) and (C.27a) as in the proof of Proposition 2.115, we find that

$$\mathcal{F}_{\mathfrak{g},K}^{\mathfrak{k},K} \circ (P_{\mathfrak{q},L\cap K}^{\mathfrak{g},K})_j \cong (P_{\mathfrak{l}\cap\mathfrak{k},L\cap K}^{\mathfrak{k},K})_j \circ \mathcal{F}_{\mathfrak{q},L\cap K}^{\mathfrak{l}\cap\mathfrak{k},L\cap K}.$$

However, $(P_{\mathfrak{l}\cap\mathfrak{k},L\cap K}^{\mathfrak{k},K})_j = 0$ for $j > 0$ since every module in $\mathcal{C}(\mathfrak{l}\cap\mathfrak{k}, L\cap K)$ is injective (by (C.14) and Proposition 1.18b). Hence $\mathcal{F}_{\mathfrak{g},K}^{\mathfrak{k},K} \circ (P_{\mathfrak{q},L\cap K}^{\mathfrak{g},K})_j = 0$ for $j > 0$. In particular, $(P_{\mathfrak{q},L\cap K}^{\mathfrak{g},K})_1 = 0$. By (C.36a), $P_{\mathfrak{q},L\cap K}^{\mathfrak{g},L\cap K}$ is exact.

Now we can define unnormalized versions of the functors with which we shall work for the remainder of this chapter. Let $(\mathfrak{g}, K)$ be a reductive pair, let G be the corresponding reductive group, and let $\mathfrak{q} = \mathfrak{l} \oplus \mathfrak{u}$ be a germane parabolic subalgebra in the sense of the equivalent conditions of Proposition 4.74. A **Levi subgroup** L is any subgroup of G satisfying

(11.70) $L_0 \subseteq L \subseteq N_G(\mathfrak{q}) \cap N_G(\theta\mathfrak{q})$.

We call $(\mathfrak{q}, L\cap K)$ a **parabolic subpair** and $(\mathfrak{l}, L\cap K)$ the corresponding **Levi subpair**. Define functors

(11.71a)
$$\begin{aligned}
&{}^{u}\mathcal{L}_{\mathfrak{q},L\cap K}^{\mathfrak{g},K} : \mathcal{C}(\mathfrak{l}, L\cap K) \to \mathcal{C}(\mathfrak{g}, K) \\
&{}^{u}\mathcal{R}_{\mathfrak{q},L\cap K}^{\mathfrak{g},K} : \mathcal{C}(\mathfrak{l}, L\cap K) \to \mathcal{C}(\mathfrak{g}, K)
\end{aligned}$$

by

(11.71b)
$$\begin{aligned}
&{}^{u}\mathcal{L}_{\mathfrak{q},L\cap K}^{\mathfrak{g},K}(Z) = \Pi_{\mathfrak{g},L\cap K}^{\mathfrak{g},K}(\mathrm{ind}_{\mathfrak{q},L\cap K}^{\mathfrak{g},L\cap K}(\mathcal{F}_{\mathfrak{l},L\cap K}^{\mathfrak{q},L\cap K}(Z))) \\
&{}^{u}\mathcal{R}_{\mathfrak{q},L\cap K}^{\mathfrak{g},K}(Z) = \Gamma_{\mathfrak{g},L\cap K}^{\mathfrak{g},K}(\mathrm{pro}_{\mathfrak{q},L\cap K}^{\mathfrak{g},L\cap K}(\mathcal{F}_{\mathfrak{l},L\cap K}^{\mathfrak{q},L\cap K}(Z))).
\end{aligned}$$

The corresponding functors involving the derived functors of Π and Γ are given by

(11.71c)
$$({}^{u}\mathcal{L}^{\mathfrak{g},K}_{\mathfrak{q},L\cap K})_{j} : \mathcal{C}(\mathfrak{l}, L\cap K) \to \mathcal{C}(\mathfrak{g}, K)$$
$$({}^{u}\mathcal{R}^{\mathfrak{g},K}_{\mathfrak{q},L\cap K})^{j} : \mathcal{C}(\mathfrak{l}, L\cap K) \to \mathcal{C}(\mathfrak{g}, K)$$

with

(11.71d)
$$({}^{u}\mathcal{L}^{\mathfrak{g},K}_{\mathfrak{q},L\cap K})_{j}(Z) = (\Pi^{\mathfrak{g},K}_{\mathfrak{g},L\cap K})_{j}(\operatorname{ind}^{\mathfrak{g},L\cap K}_{\mathfrak{q},L\cap K}(\mathcal{F}^{\mathfrak{q},L\cap K}_{\mathfrak{l},L\cap K}(Z)))$$
$$({}^{u}\mathcal{R}^{\mathfrak{g},K}_{\mathfrak{q},L\cap K})^{j}(Z) = (\Gamma^{\mathfrak{g},K}_{\mathfrak{g},L\cap K})^{j}(\operatorname{pro}^{\mathfrak{g},L\cap K}_{\mathfrak{q},L\cap K}(\mathcal{F}^{\mathfrak{q},L\cap K}_{\mathfrak{l},L\cap K}(Z))).$$

As usual, we can use $(\Pi \circ \operatorname{ind})_{j}$ and $(\Gamma \circ \operatorname{pro})^{j}$ on the right sides of (11.71d) in place of $\Pi_{j} \circ \operatorname{ind}$ and $\Gamma^{j} \circ \operatorname{pro}$. But we cannot automatically incorporate the forgetful functors into the derived functors, and therefore $({}^{u}\mathcal{L})_{j}$ and $({}^{u}\mathcal{R})^{j}$ should not be mistaken for the derived functors of ${}^{u}\mathcal{L}$ and ${}^{u}\mathcal{R}$.

It will be useful to know the effect on $({}^{u}\mathcal{L})_{j}$ and $({}^{u}\mathcal{R})^{j}$ of changing L. Suppose L_{1} and L_{2} satisfy (11.70) with $L_{1} \subseteq L_{2}$. Then

(11.72)
$$({}^{u}\mathcal{L}^{\mathfrak{g},K}_{\mathfrak{q},L_{2}\cap K})_{j} \circ P^{\mathfrak{l},L_{2}\cap K}_{\mathfrak{l},L_{1}\cap K} \cong ({}^{u}\mathcal{L}^{\mathfrak{g},K}_{\mathfrak{q},L_{1}\cap K})_{j}$$
$$({}^{u}\mathcal{R}^{\mathfrak{g},K}_{\mathfrak{q},L_{2}\cap K})^{j} \circ I^{\mathfrak{l},L_{2}\cap K}_{\mathfrak{l},L_{1}\cap K} \cong ({}^{u}\mathcal{R}^{\mathfrak{g},K}_{\mathfrak{q},L_{1}\cap K})^{j}.$$

For example, to prove the first identity, we expand the left side. Then we use the identity

$$\mathcal{F}^{\mathfrak{q},L_{2}\cap K}_{\mathfrak{l},L_{2}\cap K} \circ P^{\mathfrak{l},L_{2}\cap K}_{\mathfrak{l},L_{1}\cap K} \cong P^{\mathfrak{q},L_{2}\cap K}_{\mathfrak{q},L_{1}\cap K} \circ \mathcal{F}^{\mathfrak{q},L_{1}\cap K}_{\mathfrak{l},L_{1}\cap K},$$

which follows from Proposition 2.77. Finally we use familiar properties of the P functor to regroup the factors as on the right side of (11.72).

If $\mathfrak{q}$ is θ stable, then the relationships between the functors (11.71) and those in (5.3), apart from the absence of the forgetful functors in the definitions of $\mathcal{L}_{j}$ and $\mathcal{R}^{j}$, are

(11.73)
$$({}^{u}\mathcal{R}^{\mathfrak{g},K}_{\mathfrak{q},L\cap K})^{j}(Z) = \mathcal{R}^{j}(Z \otimes (\textstyle\bigwedge^{\mathrm{top}}\mathfrak{u})^{*})$$
$$({}^{u}\mathcal{L}^{\mathfrak{g},K}_{\bar{\mathfrak{q}},L\cap K})_{j}(Z) = \mathcal{L}_{j}(Z \otimes (\textstyle\bigwedge^{\mathrm{top}}\mathfrak{u})^{*}),$$

with one important distinction: We have assumed in Chapters V through IX that the maximal choice of L in (11.70) is used on the right side of (11.73). However, as a consequence of (11.72), a number of theorems in the θ stable case are unaffected by allowing more choices for L. For an example, let us see that the vanishing theorem (Theorem 5.99) for $\mathcal{R}$ remains valid for the wider selection of L's. In fact, let L_2 be the maximal choice of L, and let L_1 be another choice. If $Z \in \mathcal{C}(\mathfrak{l}, L_1 \cap K)$ has an infinitesimal character as in Theorem 5.99, then $I_{\mathfrak{l},L_1\cap K}^{\mathfrak{l},L_2\cap K}(Z)$ may have several infinitesimal characters, but they all satisfy the hypotheses of Theorem 5.99. Hence $({}^{\mathrm{u}}\mathcal{R}_{\mathfrak{q},L_2\cap K}^{\mathfrak{g},K})^j$ vanishes on

$$I_{\mathfrak{l},L_1\cap K}^{\mathfrak{l},L_2\cap K}(Z) \otimes (\textstyle\bigwedge^{\mathrm{top}}\mathfrak{u}) \cong I_{\mathfrak{l},L_1\cap K}^{\mathfrak{l},L_2\cap K}(Z \otimes \textstyle\bigwedge^{\mathrm{top}}\mathfrak{u})$$

for $j \neq S$, and (11.72) shows that $({}^{\mathrm{u}}\mathcal{R}_{\mathfrak{q},L_1\cap K}^{\mathfrak{g},K})^j$ vanishes on $Z \otimes \bigwedge^{\mathrm{top}}\mathfrak{u}$ for $j \neq S$.

The functors (11.71) are unnormalized in the sense that Z is not tensored with some one-dimensional module involving a sum or half sum of roots. Evidently (11.73) is asking for a shift using a sum of roots in the case of a θ stable parabolic, while Proposition 11.47 is asking for a shift using only a half sum of roots in the case of a real parabolic. This difference represents a serious problem, and we postpone the question of proper normalization of our functors until after the proof of the Transfer Theorem.

3. Preliminary Lemmas

In this section we prove two technical lemmas concerning semidirect product maps of pairs. The first says that two P functors are intertwined by forgetful functors, and the second says a similar thing about I functors. We shall apply these lemmas in several situations in this chapter.

Lemma 11.74. Let $(\mathfrak{q}, B) \rightarrow (\mathfrak{l}, B)$ be a semidirect product map of pairs with $\mathfrak{q} = \mathfrak{l} \oplus \mathfrak{u}$, let $(\mathfrak{u}, \{1\}) \hookrightarrow (\mathfrak{q}, B)$ be the associated inclusion, and suppose that $(\mathfrak{q}_1, B_1)$ is a subpair of $(\mathfrak{l}, B)$. Then

$$R(\mathfrak{q}, B) \otimes_{R(\mathfrak{q}_1+\mathfrak{u},B_1)} \mathcal{F}_{\mathfrak{q}_1,B_1}^{\mathfrak{q}_1+\mathfrak{u},B_1}(W) \cong \mathcal{F}_{\mathfrak{l},B}^{\mathfrak{q},B}(R(\mathfrak{l}, B) \otimes_{R(\mathfrak{q}_1,B_1)} W)$$

naturally for W in $\mathcal{C}(\mathfrak{q}_1, B_1)$.

PROOF. To simplify the notation, let us drop the forgetful functors, understanding that $\mathfrak{u}$ acts as 0 in W and $R(\mathfrak{l}, B) \otimes_{R(\mathfrak{q}_1,B_1)} W$ when relevant. Let $\varphi : U(\mathfrak{q}) \rightarrow U(\mathfrak{l})$ and $\iota : U(\mathfrak{l}) \rightarrow U(\mathfrak{q})$ be the homomorphisms induced by the Lie-algebra homomorphisms $\mathfrak{q} = \mathfrak{l} \oplus \mathfrak{u} \to \mathfrak{l}$ and $\mathfrak{l} \hookrightarrow \mathfrak{q}$. The map of left side to right side in the lemma is

(11.75a)
$$(u \otimes S) \otimes w \mapsto (\varphi(u) \otimes S) \otimes w \quad \text{for } u \in U(\mathfrak{q}), \ S \in R(B), \ w \in W,$$

and the map of right side to left side is

(11.75b)
$$(u' \otimes S) \otimes w \mapsto (\iota(u') \otimes S) \otimes w \quad \text{for } u' \in U(\mathfrak{l}), \ S \in R(B), \ w \in W.$$

Let us check that (11.75a) is well defined. First we check that $u \otimes S \mapsto \varphi(u) \otimes S$ is well defined. Theorem 1.119 says that it is enough to show that $uX \otimes S - u \otimes XS$ maps to 0 if X is in $\mathfrak{b}$. But

$$uX \otimes S - u \otimes XS \mapsto \varphi(u)\varphi(X) \otimes S - \varphi(u) \otimes \varphi(X)S = 0$$

since $\varphi(X)$ may be moved across the tensor-product sign in the image. Now we turn to (11.75a). We are to check that

$$(u \otimes S)X \otimes w - (u \otimes S) \otimes Xw \mapsto 0 \qquad \text{for } X \in \mathfrak{q}_1 + \mathfrak{u}$$

and $\qquad (u \otimes S)b_1 \otimes w - (u \otimes S) \otimes b_1 w \mapsto 0 \qquad \text{for } b_1 \in B_1.$

In the first case we use (1.118'a) to write

$$(u \otimes S)X \otimes w - (u \otimes S) \otimes Xw$$
$$= (u(\mathrm{Ad}(\cdot)X) \otimes S) \otimes w - (u \otimes S) \otimes Xw$$
$$\mapsto (\varphi(u)(\mathrm{Ad}(\cdot)\varphi(X)) \otimes S) \otimes w - (\varphi(u) \otimes S) \otimes \varphi(X)w$$
$$= (\varphi(u) \otimes S)\varphi(X) \otimes w - (\varphi(u) \otimes S) \otimes \varphi(X)w = 0.$$

In the second case we use (1.118'b) to write

$$(u \otimes S)b_1 \otimes w - (u \otimes S) \otimes b_1 w$$
$$= (u \otimes Sb_1) \otimes w - (u \otimes S) \otimes b_1 w$$
$$\mapsto (\varphi(u) \otimes Sb_1) \otimes w - (\varphi(u) \otimes S) \otimes b_1 w$$
$$= (\varphi(u) \otimes S)b_1 \otimes w - (\varphi(u) \otimes S) \otimes b_1 w = 0.$$

Therefore (11.75a) is well defined.

A completely analogous argument shows that (11.75b) is well defined, and it is clear that these maps are two-sided inverses of each other and respect the $(\mathfrak{q}, B)$ actions. Also the isomorphisms (11.75) are natural in W, and the identity follows.

Lemma 11.76. Let $(q, B) \to (\mathfrak{l}, B)$ be a semidirect product map of pairs with $q = \mathfrak{l} \oplus \mathfrak{u}$, let $(\mathfrak{u}, \{1\}) \hookrightarrow (q, B)$ be the associated inclusion, and suppose that (q_1, B_1) is a subpair of $(\mathfrak{l}, B)$. Then

$$\operatorname{Hom}_{q_1+\mathfrak{u},B_1}(R(q, B), \mathcal{F}_{q_1,B_1}^{q_1+\mathfrak{u},B_1}(W)) \cong \mathcal{F}_{\mathfrak{l},B}^{q,B}(\operatorname{Hom}_{q_1,B_1}(R(\mathfrak{l}, B), W))$$

naturally for W in $\mathcal{C}(q_1, B_1)$.

PROOF. The proof of this identity is similar to the one in the previous lemma. Again we drop the forgetful functors. The map from left to right in the lemma is essentially restriction. For the map from right to left, we recall the map $R(q, B) \to R(\mathfrak{l}, B)$ produced in the course of defining (11.75a). Let us call this map Φ. The map from right to left is then essentially $\operatorname{Hom}(\Phi, 1)$. It is routine to check that the maps between the two sides of the identity are well defined, are two-sided inverses of each other, respect the (q, B) actions, and are natural in W.

4. Spectral Sequences for Induction in Stages

The first of the two main tools for relating the functors $({}^{\mathfrak{u}}\mathcal{L}_{q,L\cap K}^{\mathfrak{g},K})_j$ and $({}^{\mathfrak{u}}\mathcal{R}_{q,L\cap K}^{\mathfrak{g},K})^j$ to one another is induction in stages. We formulate the result as a pair of spectral sequences.

Theorem 11.77 (spectral sequences for induction in stages). Let $(\mathfrak{g}, K)$ be a reductive pair, let $(q, L \cap K)$ be a parabolic subpair with corresponding Levi pair $(\mathfrak{l}, L \cap K)$, and let $(q_1, L_1 \cap K)$ be a parabolic subpair of $(\mathfrak{l}, L \cap K)$ with corresponding Levi pair $(\mathfrak{l}_1, L_1 \cap K)$. Put $q = \mathfrak{l} \oplus \mathfrak{u}$. Then $(q_1 + \mathfrak{u}, L_1 \cap K)$ is a parabolic subpair of $(\mathfrak{g}, K)$ with corresponding Levi pair $(\mathfrak{l}_1, L_1 \cap K)$. Suppose that Z is an $(\mathfrak{l}_1, L_1 \cap K)$ module.

(a) There exists a first-quadrant spectral sequence

$$E_r^{p,q} \implies ({}^{\mathfrak{u}}\mathcal{L}_{q_1+\mathfrak{u},L_1\cap K}^{\mathfrak{g},K})_{p+q}(Z)$$

with E_2 term

$$E_2^{p,q} = ({}^{\mathfrak{u}}\mathcal{L}_{q,L\cap K}^{\mathfrak{g},K})_p({}^{\mathfrak{u}}\mathcal{L}_{q_1,L_1\cap K}^{\mathfrak{l},L\cap K})_q(Z).$$

The differential on $E_r^{p,q}$ has bidegree $(-r, r-1)$.

(b) There exists a first-quadrant spectral sequence

$$E_r^{p,q} \implies ({}^{\mathfrak{u}}\mathcal{R}_{q_1+\mathfrak{u},L_1\cap K}^{\mathfrak{g},K})^{p+q}(Z)$$

with E_2 term

$$E_2^{p,q} = ({}^{\mathfrak{u}}\mathcal{R}_{q,L\cap K}^{\mathfrak{g},K})^p({}^{\mathfrak{u}}\mathcal{R}_{q_1,L_1\cap K}^{\mathfrak{l},L\cap K})^q(Z).$$

The differential on $E_r^{p,q}$ has bidegree $(r, 1-r)$.

REMARKS. The idea for each part is to apply a Grothendieck Spectral Sequence (Theorem D.51), but we cannot do so immediately because the derived functors of ${}^u\mathcal{L}$ and ${}^u\mathcal{R}$ are not what is at issue. We therefore begin with three lemmas that bring matters more into line with the hypotheses of Theorem D.51.

Lemma 11.78. Under the hypotheses of Theorem 11.77, there are natural isomorphisms

$$\Pi_{q,L_1\cap K}^{q,L\cap K} \circ \mathcal{F}_{\mathfrak{l},L_1\cap K}^{q,L_1\cap K} \cong \mathcal{F}_{\mathfrak{l},L\cap K}^{q,L\cap K} \circ \Pi_{\mathfrak{l},L_1\cap K}^{\mathfrak{l},L\cap K}$$

and
$$\Gamma_{q,L_1\cap K}^{q,L\cap K} \circ \mathcal{F}_{\mathfrak{l},L_1\cap K}^{q,L_1\cap K} \cong \mathcal{F}_{\mathfrak{l},L\cap K}^{q,L\cap K} \circ \Gamma_{\mathfrak{l},L_1\cap K}^{\mathfrak{l},L\cap K}.$$

PROOF. This is the special case of Lemmas 11.74 and 11.76 in which $B = L \cap K$, $B_1 = L_1 \cap K$, and $q_1 = \mathfrak{l}$.

Lemma 11.79. Under the hypotheses of Theorem 11.77,

$$\mathrm{ind}_{q_1+u,L_1\cap K}^{q,L_1\cap K} \circ \mathcal{F}_{q_1,L_1\cap K}^{q_1+u,L_1\cap K} \cong \mathcal{F}_{\mathfrak{l},L_1\cap K}^{q,L_1\cap K} \, \mathrm{ind}_{q_1,L_1\cap K}^{\mathfrak{l},L_1\cap K}$$

and
$$\mathrm{pro}_{q_1+u,L_1\cap K}^{q,L_1\cap K} \circ \mathcal{F}_{q_1,L_1\cap K}^{q_1+u,L_1\cap K} \cong \mathcal{F}_{\mathfrak{l},L_1\cap K}^{q,L_1\cap K} \, \mathrm{pro}_{q_1,L_1\cap K}^{\mathfrak{l},L_1\cap K}.$$

PROOF. This is the special case of Lemmas 11.74 and 11.76 in which $B = B_1 = L_1 \cap K$.

Lemma 11.80. Under the hypotheses of Theorem 11.77,

$$(\Pi_{\mathfrak{g},L_1\cap K}^{\mathfrak{g},L\cap K})_q \circ \mathrm{ind}_{q,L_1\cap K}^{\mathfrak{g},L_1\cap K} \circ \mathcal{F}_{\mathfrak{l},L_1\cap K}^{q,L_1\cap K} \cong \mathrm{ind}_{q,L\cap K}^{\mathfrak{g},L\cap K} \circ \mathcal{F}_{\mathfrak{l},L\cap K}^{q,L\cap K} \circ (\Pi_{\mathfrak{l},L_1\cap K}^{\mathfrak{l},L\cap K})_q$$

$$(\Gamma_{\mathfrak{g},L_1\cap K}^{\mathfrak{g},L\cap K})^q \circ \mathrm{pro}_{q,L_1\cap K}^{\mathfrak{g},L_1\cap K} \circ \mathcal{F}_{\mathfrak{l},L_1\cap K}^{q,L_1\cap K} \cong \mathrm{pro}_{q,L\cap K}^{\mathfrak{g},L\cap K} \circ \mathcal{F}_{\mathfrak{l},L\cap K}^{q,L\cap K} \circ (\Gamma_{\mathfrak{l},L_1\cap K}^{\mathfrak{l},L\cap K})^q.$$

PROOF. We prove only the first identity, the second identity being similar. For $q = 0$, we have natural isomorphisms

(11.81)
$$\Pi_{\mathfrak{g},L_1\cap K}^{\mathfrak{g},L\cap K} \circ \mathrm{ind}_{q,L_1\cap K}^{\mathfrak{g},L_1\cap K} \circ \mathcal{F}_{\mathfrak{l},L_1\cap K}^{q,L_1\cap K}$$

$$\cong \mathrm{ind}_{q,L\cap K}^{\mathfrak{g},L\cap K} \circ \Pi_{q,L_1\cap K}^{q,L\cap K} \circ \mathcal{F}_{\mathfrak{l},L_1\cap K}^{q,L_1\cap K} \qquad \text{by Proposition 2.19}$$

$$\cong \mathrm{ind}_{q,L\cap K}^{\mathfrak{g},L\cap K} \circ \mathcal{F}_{\mathfrak{l},L\cap K}^{q,L\cap K} \circ \Pi_{\mathfrak{l},L_1\cap K}^{\mathfrak{l},L\cap K} \qquad \text{by Lemma 11.78.}$$

Let W_q be a projective resolution of W in $\mathcal{C}(\mathfrak{l}, L_1 \cap K)$. Since $\mathrm{ind}_{q,L\cap K}^{\mathfrak{g},L\cap K} \circ \mathcal{F}_{\mathfrak{l},L\cap K}^{q,L\cap K}$ is exact (Proposition 2.57c), the homology of

(11.82a)
$$\mathrm{ind}_{q,L\cap K}^{\mathfrak{g},L\cap K} \circ \mathcal{F}_{\mathfrak{l},L\cap K}^{q,L\cap K} \circ \Pi_{\mathfrak{l},L_1\cap K}^{\mathfrak{l},L\cap K}(W_q)$$

is naturally isomorphic to

(11.82b) $\text{ind}_{q,L\cap K}^{g,L\cap K} \circ \mathcal{F}_{l,L\cap K}^{q,L\cap K} \circ (\Pi_{l,L_1\cap K}^{l,L\cap K})_q(W).$

In view of (11.81), it is enough to show that the homology of

$$\Pi_{g,L_1\cap K}^{g,L\cap K} \circ \text{ind}_{q,L_1\cap K}^{g,L_1\cap K} \circ \mathcal{F}_{l,L_1\cap K}^{q,L_1\cap K}(W_q)$$

is $(\Pi_{g,L_1\cap K}^{g,L\cap K})_q \circ \text{ind}_{q,L_1\cap K}^{g,L_1\cap K} \circ \mathcal{F}_{l,L_1\cap K}^{q,L_1\cap K}(W),$

and (C.26) says that this will be the case if

(11.83) $(\Pi_{g,L_1\cap K}^{g,L\cap K})_n(\text{ind}_{q,L_1\cap K}^{g,L_1\cap K} \circ \mathcal{F}_{l,L_1\cap K}^{q,L_1\cap K}(W_q)) = 0$

for all $n > 0$. By Proposition 2.115 it is enough to know that

(11.84) $\mathcal{F}_{g,L_1\cap K}^{l\cap\mathfrak{k},L_1\cap K} \circ \text{ind}_{q,L_1\cap K}^{g,L_1\cap K} \circ \mathcal{F}_{l,L_1\cap K}^{q,L_1\cap K}(W_q)$

is acyclic for $\Pi_{l\cap\mathfrak{k},L_1\cap K}^{l\cap\mathfrak{k},L\cap K}$, and a second application of Proposition 2.115 shows that it is enough to know that

(11.85) $\mathcal{F}_{g,L_1\cap K}^{l,L_1\cap K} \circ \text{ind}_{q,L_1\cap K}^{g,L_1\cap K} \circ \mathcal{F}_{l,L_1\cap K}^{q,L_1\cap K}(W_q)$

is acyclic for $\Pi_{l,L_1\cap K}^{l,L\cap K}$. But (11.85) is projective in $\mathcal{C}(l, L\cap K)$, being

$$\cong U(\mathfrak{g}) \otimes_{U(\mathfrak{q})} \mathcal{F}_{l,L_1\cap K}^{q,L_1\cap K}(W_q) \cong U(\mathfrak{u}^-) \otimes_{\mathbb{C}} W_q.$$

Hence (11.85) is acyclic in $\mathcal{C}(l, L\cap K)$ for $\Pi_{l,L_1\cap K}^{l,L\cap K}$, (11.84) is acyclic in $\mathcal{C}(l\cap\mathfrak{k}, L_1\cap K)$ for $\Pi_{l\cap\mathfrak{k},L_1\cap K}^{l\cap\mathfrak{k},L\cap K}$, and (11.83) holds. The first identity of the lemma follows.

PROOF OF THEOREM 11.77. Certainly $(q_1 + \mathfrak{u}, L_1\cap K)$ is a parabolic subpair, and $(l_1, L_1\cap K)$ is the corresponding Levi pair. To prove (a) in the theorem, we apply Variation (1) of the Grothendieck Spectral Sequence (Theorem D.51) to the composition of functors

$$\Pi_{g,L\cap K}^{g,K} \circ \Pi_{g,L_1\cap K}^{g,L\cap K} \cong \Pi_{g,L_1\cap K}^{g,K}.$$

The inner functor on the left side carries projectives to projectives by Corollary 2.35. For any $V \in \mathcal{C}(\mathfrak{g}, L_1\cap K)$, we obtain a first-quadrant spectral sequence

$$E_r^{p,q} \implies (\Pi_{g,L_1\cap K}^{g,K})_{p+q}(V)$$

with E_2 term

$$E_2^{p,q} = (\Pi_{\mathfrak{g},L\cap K}^{\mathfrak{g},K})_p (\Pi_{\mathfrak{g},L_1\cap K}^{\mathfrak{g},L\cap K})_q (V).$$

Taking

$$V = \operatorname{ind}_{q_1+u,L_1\cap K}^{\mathfrak{g},L_1\cap K} (\mathcal{F}_{\mathfrak{l}_1,L_1\cap K}^{q_1+u,L_1\cap K}(Z)),$$

we have

$$\begin{aligned}
E_2^{p,q} = \ & (\Pi_{\mathfrak{g},L\cap K}^{\mathfrak{g},K})_p (\Pi_{\mathfrak{g},L_1\cap K}^{\mathfrak{g},L\cap K})_q \\
& \circ \operatorname{ind}_{q,L_1\cap K}^{\mathfrak{g},L_1\cap K} \operatorname{ind}_{q_1+u,L_1\cap K}^{q,L_1\cap K} \mathcal{F}_{q_1,L_1\cap K}^{q_1+u,L_1\cap K} \mathcal{F}_{\mathfrak{l}_1,L_1\cap K}^{q_1,L_1\cap K} (Z) \\
\cong \ & (\Pi_{\mathfrak{g},L\cap K}^{\mathfrak{g},K})_p (\Pi_{\mathfrak{g},L_1\cap K}^{\mathfrak{g},L\cap K})_q \operatorname{ind}_{q,L_1\cap K}^{\mathfrak{g},L_1\cap K} \mathcal{F}_{\mathfrak{l},L_1\cap K}^{q,L_1\cap K} \operatorname{ind}_{q_1,L_1\cap K}^{\mathfrak{l},L_1\cap K} \mathcal{F}_{\mathfrak{l}_1,L_1\cap K}^{q_1,L_1\cap K} (Z) \\
& \text{by Lemma 11.79} \\
\cong \ & (\Pi_{\mathfrak{g},L\cap K}^{\mathfrak{g},K})_p \operatorname{ind}_{q,L\cap K}^{\mathfrak{g},L\cap K} \mathcal{F}_{\mathfrak{l},L\cap K}^{q,L\cap K} (\Pi_{\mathfrak{l},L_1\cap K}^{\mathfrak{l},L\cap K})_q \operatorname{ind}_{q_1,L_1\cap K}^{\mathfrak{l},L_1\cap K} \mathcal{F}_{\mathfrak{l}_1,L_1\cap K}^{q_1,L_1\cap K} (Z) \\
& \text{by Lemma 11.80} \\
= \ & (^{\mathfrak{u}}\mathcal{L}_{q,L\cap K}^{\mathfrak{g},K})_p (^{\mathfrak{u}}\mathcal{L}_{q_1,L_1\cap K}^{\mathfrak{l},L\cap K})_q (Z)
\end{aligned}$$

and

$$\begin{aligned}
(\Pi_{\mathfrak{g},L_1\cap K}^{\mathfrak{g},K})_{p+q}(V) &= (\Pi_{\mathfrak{g},L_1\cap K}^{\mathfrak{g},K})_{p+q} \operatorname{ind}_{q_1+u,L_1\cap K}^{\mathfrak{g},L_1\cap K} \mathcal{F}_{\mathfrak{l}_1,L_1\cap K}^{q_1+u,L_1\cap K}(Z) \\
&= (^{\mathfrak{u}}\mathcal{L}_{q_1+u,L_1\cap K}^{\mathfrak{g},K})_{p+q}(Z).
\end{aligned}$$

This proves (a). For (b) we use Variation (5) of Theorem D.51, and the argument proceeds similarly.

Corollary 11.86. Let $(\mathfrak{g}, K)$ be a reductive pair, let $(\mathfrak{q}, L \cap K)$ be a parabolic subpair with corresponding Levi pair $(\mathfrak{l}, L \cap K)$, and let $(\mathfrak{q}_1, L_1 \cap K)$ be a parabolic subpair of $(\mathfrak{l}, L \cap K)$ with corresponding Levi pair $(\mathfrak{l}_1, L_1 \cap K)$. Put $\mathfrak{q} = \mathfrak{l} \oplus \mathfrak{u}$, and suppose that Z is an $(\mathfrak{l}_1, L_1 \cap K)$ module.

(a) If $(^{\mathfrak{u}}\mathcal{L}_{q_1,L_1\cap K}^{\mathfrak{l},L\cap K})_q(Z) = 0$ except when $q = q_0$, then

$$(^{\mathfrak{u}}\mathcal{L}_{q_1+u,L_1\cap K}^{\mathfrak{g},K})_{p+q_0}(Z) \cong (^{\mathfrak{u}}\mathcal{L}_{q,L\cap K}^{\mathfrak{g},K})_p (^{\mathfrak{u}}\mathcal{L}_{q_1,L_1\cap K}^{\mathfrak{l},L\cap K})_{q_0}(Z).$$

(b) If $(^{\mathfrak{u}}\mathcal{R}_{q_1,L_1\cap K}^{\mathfrak{l},L\cap K})^q(Z) = 0$ except when $q = q_0$, then

$$(^{\mathfrak{u}}\mathcal{R}_{q_1+u,L_1\cap K}^{\mathfrak{g},K})^{p+q_0}(Z) \cong (^{\mathfrak{u}}\mathcal{R}_{q,L\cap K}^{\mathfrak{g},K})^p (^{\mathfrak{u}}\mathcal{R}_{q_1,L_1\cap K}^{\mathfrak{l},L\cap K})^{q_0}(Z).$$

REMARK. These conclusions are valid for all p, not just $p \geq 0$. In particular, $(^{\mathfrak{u}}\mathcal{L}_{q_1+u,L_1\cap K}^{\mathfrak{g},K})_j(Z)$ and $(^{\mathfrak{u}}\mathcal{R}_{q_1+u,L_1\cap K}^{\mathfrak{g},K})^j(Z)$ are both 0 if $j < q_0$.

PROOF. Under the hypotheses of vanishing except when $q = q_0$, the spectral sequences in Theorem 11.77 collapse and degenerate, and the corollary follows immediately.

5. Transfer Theorem

The second of the two main tools for relating the functors $({}^{\mathrm{u}}\mathcal{L}^{\mathfrak{g},K}_{\mathfrak{q},L\cap K})_j$ to one another and the functors $({}^{\mathrm{u}}\mathcal{R}^{\mathfrak{g},K}_{\mathfrak{q},L\cap K})^j$ to one another is the Transfer Theorem. This theorem establishes an isomorphism in the case of two functors ${}^{\mathrm{u}}\mathcal{L}$ or two functors ${}^{\mathrm{u}}\mathcal{R}$ when the parabolic subalgebras are Borel subalgebras that differ from each other by a complex root.

Theorem 11.87 (Transfer Theorem). Let $(\mathfrak{g}, K)$ be a reductive pair, and let $\mathfrak{h}_0$ be a θ stable Cartan subalgebra of $\mathfrak{g}_0$. Suppose that $\mathfrak{b}_1$ and $\mathfrak{b}_2$ are two Borel subalgebras of $\mathfrak{g}$ containing $\mathfrak{h}$ such that

(i) $(\mathfrak{b}_1, T)$ and $(\mathfrak{b}_2, T)$ are parabolic subpairs of $(\mathfrak{g}, K)$ for the same T, so that both have $(\mathfrak{h}, T)$ as the corresponding Levi pair
(ii) $\mathfrak{b}_1 \cap \mathfrak{b}_2^- = \mathfrak{h} \oplus \mathfrak{g}_\alpha$ for a complex root $\alpha \in \Delta(\mathfrak{g}, \mathfrak{h})$
(iii) $\mathfrak{g}_{\theta\alpha} \subseteq \mathfrak{b}_1$.

Let Z be an irreducible $(\mathfrak{h}, T)$ module such that $\mathfrak{h}$ acts in Z by scalars according to $\lambda \in \mathfrak{h}^*$.

(a) If

$$\frac{2\langle\lambda, \alpha\rangle}{|\alpha|^2} \quad \text{is not in } \{0, 1, 2, \dots\},$$

then $({}^{\mathrm{u}}\mathcal{L}^{\mathfrak{g},K}_{\mathfrak{b}_2,T})_j(Z \otimes \mathbb{C}_\alpha) \cong ({}^{\mathrm{u}}\mathcal{L}^{\mathfrak{g},K}_{\mathfrak{b}_1,T})_{j+1}(Z)$ for all j.
(b) If

$$\frac{2\langle\lambda, \alpha\rangle}{|\alpha|^2} \quad \text{is not in } \{0, -1, -2, \dots\},$$

then $({}^{\mathrm{u}}\mathcal{R}^{\mathfrak{g},K}_{\mathfrak{b}_2,T})^j(Z \otimes \mathbb{C}_{-\alpha}) \cong ({}^{\mathrm{u}}\mathcal{R}^{\mathfrak{g},K}_{\mathfrak{b}_1,T})^{j+1}(Z)$ for all j.

REMARKS.

1) As usual, we write $\mathfrak{h}_0 = \mathfrak{t}_0 \oplus \mathfrak{a}_0$. Following the convention for Levi pairs adopted in §2, the group T is assumed to be between T_0 (the analytic subgroup corresponding to $\mathfrak{t}_0$) and $N_K(\mathfrak{b}_1) \cap N_K(\mathfrak{b}_2)$. If $(\mathfrak{g}, K)$ is in the Harish-Chandra class, then T lies between T_0 and $Z_K(\mathfrak{h})$.

2) Even when $(\mathfrak{g}, K)$ is in the Harish-Chandra class, T can be non-abelian. To take into account arbitrary irreducible $(\mathfrak{h}, T)$ modules in which $\mathfrak{h}$ acts as a scalar, we have to allow Z to have dimension greater

than one. The reader may want to assume, however, that Z is one dimensional on a first reading of the proof.

3) The $(\mathfrak{h}, T)$ module $\mathbb{C}_\alpha$ requires some explanation. Under the hypotheses of the theorem, $\mathrm{Ad}(T)$ maps $\mathfrak{g}_\alpha$ and $\mathfrak{g}_{-\alpha}$ into themselves. The one-dimensional space $\mathfrak{g}_\alpha$ may therefore be regarded as an $(\mathfrak{h}, T)$ module, and it is this module that we denote by $\mathbb{C}_\alpha$.

4) In the presence of hypotheses (i) and (ii), hypothesis (iii) selects the more nearly real of $\mathfrak{b}_1$ and $\mathfrak{b}_2$ to associate to the lower degree of homology or cohomology. In more detail, (iii) forces $\mathfrak{b}_2$ to have the conjugate pair of roots $\{-\alpha, -\bar{\alpha}\}$, while $\mathfrak{b}_1$ has no corresponding conjugate pair. On the other hand, suppose (iii) fails. Then $-\alpha$ is a complex root, and (i) is unchanged. Hypothesis (ii) gets replaced by $\mathfrak{b}_2 \cap \mathfrak{b}_1^- = \mathfrak{g}_{-\alpha}$. Since $\mathfrak{g}_{\theta\alpha} \subseteq \mathfrak{b}_1$ is false, we must have $\mathfrak{g}_{\theta(-\alpha)} \subseteq \mathfrak{b}_1$. By (ii) it is true also that $\mathfrak{g}_{\theta(-\alpha)} \subseteq \mathfrak{b}_2$, and this condition is hypothesis (iii) for $-\alpha$. In other words, if (iii) fails for α, then it holds for $-\alpha$ when $\mathfrak{b}_1$ and $\mathfrak{b}_2$ are interchanged.

The proof of Theorem 11.87 will be preceded by several lemmas. Briefly the idea for (a) when Z is one dimensional is as follows. The isomorphism in the conclusion is to be a connecting homomorphism in a long exact sequence for the derived functors of $\Pi_{\mathfrak{g},T}^{\mathfrak{g},K}$. What is needed to obtain such an exact sequence is a $(\mathfrak{g}, T)$ module E with $\mathrm{ind}_{\mathfrak{b}_2,T}^{\mathfrak{g},T}(Z \otimes \mathbb{C}_\alpha)$ as a submodule and $\mathrm{ind}_{\mathfrak{b}_1,T}^{\mathfrak{g},T}(Z)$ as a quotient. To define E, we put $\mathfrak{q} = \mathfrak{b}_1 + \mathfrak{b}_2$. Then $\mathrm{ind}_{\mathfrak{b}_1,T}^{\mathfrak{q},T}(Z)$ is essentially a Verma module for the $\mathfrak{sl}(2, \mathbb{C})$ corresponding to α, $\mathrm{ind}_{\mathfrak{b}_2,T}^{\mathfrak{q},T}(Z \otimes \mathbb{C}_\alpha)$ is essentially an upside-down Verma module, and their weights abut. The module E is obtained by gluing these $(\mathfrak{q}, T)$ modules to form a $(\mathfrak{q}, T)$ module E_0 and then setting $E = \mathrm{ind}_{\mathfrak{q},T}^{\mathfrak{g},T}(E_0)$. For the connecting homomorphism to be an isomorphism, it is enough to do the gluing so that $(\Pi_{\mathfrak{g},T}^{\mathfrak{g},K})_j(E) = 0$ for all j.

The hypothesis on λ means that $\mathrm{ind}_{\mathfrak{b}_1,T}^{\mathfrak{q},T}(Z)$ has no finite-dimensional submodule, and the gluing can therefore be done in such a way that the sum of root vectors $e_\alpha + \theta e_\alpha$ acts invertibly on E. To prove that $(\Pi_{\mathfrak{g},T}^{\mathfrak{g},K})_j(E) = 0$, we shall prove that the multiplicity of each K type is 0. If F is an irreducible K module, the thing to prove is that

$$H^j(\mathfrak{k}, T; E^c \otimes_{\mathbb{C}} F) = 0.$$

This vanishing follows from a variant of the Hochschild-Serre spectral sequence once one shows that $e_\alpha + \theta e_\alpha$ acts invertibly in $E^c \otimes_{\mathbb{C}} F$. In turn this invertibility follows from the invertibility of $e_\alpha + \theta e_\alpha$ on E.

Lemma 11.88. Under the hypotheses on Theorem 11.87, let $q = \mathfrak{b}_1 + \mathfrak{b}_2$ and $\mathfrak{l} = \mathfrak{h} + \mathfrak{g}_\alpha + \mathfrak{g}_{-\alpha}$, so that $\mathfrak{q}$ is a germane parabolic subalgebra of $\mathfrak{g}$ with Levi factor $\mathfrak{l}$.

(a) If

$$\frac{2\langle \lambda, \alpha \rangle}{|\alpha|^2} \quad \text{is not in } \{0, 1, 2, \dots\},$$

then there exists a $(\mathfrak{q}, T)$ module E_0 such that the root vector e_α acts invertibly in E_0, θe_α acts as 0, and E_0 fits in an exact sequence

$$0 \longrightarrow \operatorname{ind}_{\mathfrak{b}_2, T}^{\mathfrak{q}, T}(Z \otimes \mathbb{C}_\alpha) \longrightarrow E_0 \longrightarrow \operatorname{ind}_{\mathfrak{b}_1, T}^{\mathfrak{q}, T}(Z) \longrightarrow 0$$

(b) If

$$\frac{2\langle \lambda, \alpha \rangle}{|\alpha|^2} \quad \text{is not in } \{0, -1, -2, \dots\},$$

then there exists a $(\mathfrak{q}, T)$ module E_0' such that the root vector e_α acts invertibly in E_0', θe_α acts as 0, and E_0' fits in an exact sequence

$$0 \longrightarrow \operatorname{pro}_{\mathfrak{b}_1, T}^{\mathfrak{q}, T}(Z) \longrightarrow E_0' \longrightarrow \operatorname{pro}_{\mathfrak{b}_2, T}^{\mathfrak{q}, T}(Z \otimes \mathbb{C}_{-\alpha}) \longrightarrow 0.$$

REMARKS. The reader is encouraged to think of Z as one dimensional in the proof. It is necessary to allow for higher-dimensional Z because T may be nonabelian, but the need to handle higher-dimensional Z is merely a technical complication.

PROOF.

(a) For the parabolic subalgebras under consideration, the roots $\pm\alpha$ and $\pm\bar{\alpha}$ distribute as

$$\mathfrak{b}_1 \leftrightarrow \{\alpha, \theta\alpha\}, \qquad \mathfrak{b}_2 \leftrightarrow \{-\alpha, \theta\alpha\} = \{-\alpha, -\bar{\alpha}\}, \qquad \mathfrak{q} \leftrightarrow \{\alpha, \theta\alpha, -\alpha\}.$$

Also

$$\mathfrak{l} \leftrightarrow \{\alpha, -\alpha\} \qquad \text{and} \qquad \mathfrak{u} \leftrightarrow \{\theta\alpha\}.$$

Each remaining root goes with all of $\mathfrak{b}_1$, $\mathfrak{b}_2$, $\mathfrak{q}$, $\mathfrak{u}$, or it goes with none of these. Thus we have $(\mathfrak{h}, T)$ isomorphisms

$$\operatorname{ind}_{\mathfrak{b}_1, T}^{\mathfrak{q}, T}(Z) = U(\mathfrak{q}) \otimes_{U(\mathfrak{b}_1)} Z \cong U(\mathfrak{g}_{-\alpha}) \otimes_{\mathbb{C}} Z$$

(11.89)

$$\operatorname{ind}_{\mathfrak{b}_2, T}^{\mathfrak{q}, T}(Z \otimes \mathbb{C}_\alpha) = U(\mathfrak{q}) \otimes_{U(\mathfrak{b}_2)} (Z \otimes \mathbb{C}_\alpha) \cong U(\mathfrak{g}_\alpha) \otimes_{\mathbb{C}} (Z \otimes \mathbb{C}_\alpha).$$

We may therefore write the weight spaces of these modules as

$$\dots, V_{\lambda-2\alpha}, V_{\lambda-\alpha}, V_\lambda \qquad \text{and} \qquad V_{\lambda+\alpha}, V_{\lambda+2\alpha}, \dots,$$

respectively.

From the definitions of $\mathrm{ind}_{\mathfrak{b}_1,T}^{\mathfrak{q},T}(Z)$ and $\mathrm{ind}_{\mathfrak{b}_2,T}^{\mathfrak{q},T}(Z \otimes \mathbb{C}_\alpha)$ in (11.89), we see directly that $\mathfrak{u}$ acts in each of these modules as 0. The isomorphism for ind in Lemma 11.79 shows that $\mathfrak{l}$ acts as in a Verma module (but with Z possibly of dimension more than 1).

As an $(\mathfrak{h}, T)$ module, the module E_0 is to be the direct sum of the two modules (11.89), and $\mathfrak{u}$ is to act by 0. (In particular, θe_α acts as 0.) We must define the action of the root vectors in $\mathfrak{l}$. Let us say that the copy of $\mathfrak{sl}(2, \mathbb{C})$ in $\mathfrak{l}$ has the standard basis h, e, f with $h \in \mathfrak{h}$, $e \in \mathfrak{g}_\alpha$, and $f \in \mathfrak{g}_{-\alpha}$; here e is a multiple of e_α. We define h and f to act in E_0 as they do in the direct sum. Also e acts on all the weight spaces of E_0 except V_λ as it does in the direct sum. To define the action of e on V_λ, we note that $V_\lambda \cong Z$ and $V_{\lambda+\alpha} \cong Z \otimes \mathbb{C}_\alpha$ as $(\mathfrak{h}, T)$ modules. Recalling from Remark 3 that $\mathbb{C}_\alpha$ is just $\mathfrak{g}_\alpha$, we take the action by e to be $z \mapsto z \otimes e$ for $z \in Z$. Unwinding the isomorphisms, we write $\varphi : V_\lambda \to V_{\lambda+\alpha}$ for the corresponding invertible linear map. We have $\varphi(tv) = \xi_{-\alpha}(t)t(\varphi v)$ for all $v \in V_\lambda$, and the action of e is given by $ev = \varphi(v)$ for $v \in V_\lambda$.

Let us check that we have defined a Lie-algebra representation of $\mathfrak{l}$ in E_0. The relation $hf - fh = -2f$ is automatic, since h and f act in E_0 the same way they act in the direct sum. For the relation $he - eh \overset{?}{=} 2e$, the question is whether equality holds on V_λ. For $v \in V_\lambda$, we have

$$hev - ehv = (\lambda + \alpha)(h)\varphi(v) - \lambda(h)\varphi(v) \qquad \text{and} \qquad 2ev = 2\varphi(v),$$

and these are equal since $\alpha(h) = 2$. For the relation $ef - fe \overset{?}{=} h$, we need to check equality on both V_λ and $V_{\lambda+\alpha}$. For $v \in V_\lambda$, we have

$$efv - fev = efv - f\varphi(v) = efv = hv$$

since $ev = 0$ in $\mathrm{ind}_{\mathfrak{b}_1,T}^{\mathfrak{q},T}(Z)$. For $v \in V_{\lambda+\alpha}$, we have

$$efv - fev = e(0) - fev = -fev = hv$$

since $fv = 0$ in $\mathrm{ind}_{\mathfrak{b}_2,T}^{\mathfrak{q},T}(Z \otimes \mathbb{C}_\alpha)$.

Thus we have defined a representation of $\mathfrak{l}$ on E_0. Since $\mathfrak{u}$ is an ideal in $\mathfrak{q}$ and we have defined $\mathfrak{u}$ to act as 0, we actually have a Lie-algebra representation of $\mathfrak{q}$. Also E_0 is a representation space for T, and the $\mathfrak{q}$ and T actions are compatible because of the definition of the action of e on V_λ in terms of φ. Thus E_0 is a $(\mathfrak{q}, T)$ module. It is clear that $\mathrm{ind}_{\mathfrak{b}_2,T}^{\mathfrak{q},T}(Z \otimes \mathbb{C}_\alpha)$ is a $(\mathfrak{q}, T)$ submodule of E_0 and that the quotient is isomorphic to $\mathrm{ind}_{\mathfrak{b}_1,T}^{\mathfrak{q},T}(Z)$.

Finally we show that e acts in invertible fashion in E_0. For each n, e maps $V_{\lambda+n\alpha}$ to $V_{\lambda+(n+1)\alpha}$. If $n \geq 1$, this map is one-one onto by (4.97). If $n = 0$, it is one-one onto by construction. Suppose $n < 0$. If e is not one-one, then the argument that proves (4.99) proves also that members z of $Z(\mathfrak{sl}(2, \mathbb{C}))$ act in $\mathrm{ind}_{\mathfrak{b}_1,T}^{\mathfrak{q},T}(Z)$ both by $\xi_{\lambda+\frac{1}{2}\alpha}(z)$ and by $\xi_{\lambda+n\alpha+\frac{1}{2}\alpha}(z)$. By Theorem 4.115

$$\lambda + n\alpha + \tfrac{1}{2}\alpha = -(\lambda + \tfrac{1}{2}\alpha).$$

Thus $2\lambda = -(1+n)\alpha$, and $2\langle\lambda, \alpha\rangle/|\alpha|^2 = -(1+n)$. Since n is an integer ≤ -1, $2\langle\lambda, \alpha\rangle/|\alpha|^2$ is an integer ≥ 0. But these values of $2\langle\lambda, \alpha\rangle/|\alpha|^2$ have been excluded by hypothesis. We conclude that $e : V_{\lambda+n\alpha} \to V_{\lambda+(n+1)\alpha}$ is one-one for all n. Since the weight spaces are finite dimensional, e is onto. This proves (a).

(b) The proof is similar to that of (a). In place of (11.89), we have

$$\mathrm{pro}_{\mathfrak{b}_1,T}^{\mathfrak{q},T}(Z) = \mathrm{Hom}_{U(\mathfrak{b}_1)}(U(\mathfrak{q}), Z)_T \cong \mathrm{Hom}_{\mathbb{C}}(U(\mathfrak{g}_{-\alpha}), Z)_T$$

$$\mathrm{pro}_{\mathfrak{b}_2,T}^{\mathfrak{q},T}(Z \otimes \mathbb{C}_{-\alpha}) = \mathrm{Hom}_{U(\mathfrak{b}_2)}(U(\mathfrak{q}), Z \otimes \mathbb{C}_{-\alpha})_T$$
$$\cong \mathrm{Hom}_{\mathbb{C}}(U(\mathfrak{g}_{\alpha}), Z \otimes \mathbb{C}_{-\alpha})_T$$

The weight spaces of these modules are

$$V_\lambda, \ V_{\lambda+\alpha}, \ V_{\lambda+2\alpha}, \ \ldots \qquad \text{and} \qquad \ldots, \ V_{\lambda-2\alpha}, \ V_{\lambda-\alpha},$$

respectively. The gluing to form E_0' is to be done in the same way as in (a) with the action of e modified so that e carries $V_{\lambda-\alpha}$ one-one onto V_λ. To prove that the action by e is one-one (and therefore invertible), we have to prove that the piece with weight spaces V_λ, $V_{\lambda+\alpha}$, $V_{\lambda+2\alpha}$, $\ldots$ has no nonzero finite-dimensional invariant subspace. This conclusion follows from the assumption that $2\langle\lambda, \alpha\rangle/|\alpha|^2$ is not in $\{0, -1, -2, \ldots\}$.

Lemma 11.90. Let $(\mathfrak{g}, K)$ be a pair, let X be in $\mathfrak{g}$, let F be a finite-dimensional $(\mathfrak{g}, K)$ module on which the action of X is nilpotent, and let E be a $(\mathfrak{g}, K)$ module on which the action of X is invertible. Then the action of X on $E \otimes_{\mathbb{C}} F$ is invertible.

PROOF. Since X acts in nilpotent fashion on F, we can choose an increasing filtration

$$0 = F_{-1} \subseteq F_0 \subseteq F_1 \subseteq \cdots \subseteq F_n = F$$

of F by subspaces such that $XF_j \subseteq F_{j-1}$ for $0 \leq j \leq n$. Then X acts as 0 on F_j/F_{j-1} and hence as $X \otimes 1$ on $E \otimes_{\mathbb{C}} (F_j/F_{j-1})$. Since X is invertible on E, $X \otimes 1$ is invertible on $E \otimes_{\mathbb{C}} (F_j/F_{j-1})$. Therefore X acts invertibly on each $(E \otimes_{\mathbb{C}} F_j)/(E \otimes_{\mathbb{C}} F_{j-1})$, and an easy induction shows that X acts invertibly on $E \otimes_{\mathbb{C}} F$.

Lemma 11.91. Let notation be as in Lemma 11.88, and let $X = e_\alpha + \theta e_\alpha$. Then

(a) X acts invertibly on $E = \operatorname{ind}_{\mathfrak{q},T}^{\mathfrak{g},T}(E_0)$
(b) X acts invertibly on $E' = \operatorname{pro}_{\mathfrak{q},T}^{\mathfrak{g},T}(E_0')$.

PROOF.

(a) Let us identify E_0 with the subspace $1 \otimes E_0$ of E. The subspace $E^{(n)} = U_n(\mathfrak{g})(E_0)$ of E is a $U(\mathfrak{q})$ submodule, and $\bigcup_{n=0}^\infty E^{(n)} = E$. Thus it is enough to prove that the element X of $\mathfrak{q}$ acts invertibly on each $E^{(n)}$. By Lemma 11.88, X acts invertibly on $E^{(0)} \cong E_0$, and it is therefore enough, by induction, to prove that X acts invertibly on each $E^{(n)}/E^{(n-1)}$.

Much of the proof is similar to an argument with (8.67). Let $\tau_n : U_n(\mathfrak{g})/U_{n-1}(\mathfrak{g}) \to S^n(\mathfrak{g})$ be the natural map. Since

$$E^{(n)} \cong U_n(\mathfrak{u}^-) \otimes_{\mathbb{C}} E_0$$

as an $(\mathfrak{l}, T)$ module, we have vector-space isomorphisms

(11.92a)
$$E^{(n)}/E^{(n-1)} \longrightarrow (U_n(\mathfrak{u}^-)/U_{n-1}(\mathfrak{u}^-)) \otimes_{\mathbb{C}} E_0$$

$$\xrightarrow{\ \tau_n \otimes 1\ } S^n(\mathfrak{u}^-) \otimes_{\mathbb{C}} E_0 \longrightarrow S^n(\mathfrak{g}/\mathfrak{q}) \otimes_{\mathbb{C}} E_0.$$

The composition is evidently a $U(\mathfrak{l})$ map, and thus we have a $U(\mathfrak{l})$ isomorphism

(11.92b)
$$E^{(n)}/E^{(n-1)} \xrightarrow{\sim} S^n(\mathfrak{g}/\mathfrak{q}) \otimes_{\mathbb{C}} E_0.$$

In fact, (11.92) respects the $\mathfrak{u}$ action also. To see this, we construct a $\mathfrak{q}$ map that is a left inverse to (11.92). We begin with the action map

(11.93)
$$U_n(\mathfrak{g}) \otimes_{\mathbb{C}} E_0 \to E^{(n)},$$

which respects the $\mathfrak{q}$ module structures. We follow this with passage to the quotient $E^{(n)}/E^{(n-1)}$ and observe that $U_{n-1}(\mathfrak{g}) \otimes_{\mathbb{C}} E_0$ is in the kernel. Thus we obtain $\mathfrak{q}$ maps

(11.94)
$$S^n(\mathfrak{g}) \otimes_{\mathbb{C}} E_0 \xrightarrow{\ \tau_n^{-1} \otimes 1\ } (U_n(\mathfrak{g})/U_{n-1}(\mathfrak{g})) \otimes_{\mathbb{C}} E_0$$

$$\longrightarrow (U_n(\mathfrak{g}) \otimes_{\mathbb{C}} E_0)/(U_{n-1}(\mathfrak{g}) \otimes_{\mathbb{C}} E_0) \longrightarrow E^{(n)}/E^{(n-1)}.$$

Under (11.93), $qU_{n-1}(\mathfrak{g}) \otimes_{\mathbb{C}} E_0$ maps into $E^{(n-1)}$, and therefore $qS^{n-1}(\mathfrak{g}) \otimes_{\mathbb{C}} E_0$ is in the kernel of (11.94). Since

$$S^n(\mathfrak{g})/(qS^{n-1}(\mathfrak{g})) \cong S^n(\mathfrak{g}/\mathfrak{q}) \qquad \text{as q modules,}$$

(11.94) descends to a q map

$$S^n(\mathfrak{g}/\mathfrak{q}) \otimes_{\mathbb{C}} E_0 \to E^{(n)}/E^{(n-1)}.$$

If we ignore the q action and regard this map as defined on $S^n(\mathfrak{u}^-) \otimes_{\mathbb{C}} E_0$, we see that this map is a left inverse to (11.92a). Therefore (11.92) is a $U(\mathfrak{q})$ isomorphism.

The action of $X \in \mathfrak{q}$ on E_0 is invertible, by Lemma 11.88, and the action of ad X on $S^n(\mathfrak{g}/\mathfrak{q})$ is nilpotent (since $e_\alpha + \theta e_\alpha$ lies in the nilpotent radical of $\mathfrak{b}_1$). Thus Lemma 11.90 shows that X acts invertibly on $E^{(n)}/E^{(n-1)}$.

(b) The argument is dual to the proof of (a). Let $E'^{(n)}$ be the space of restrictions $E'^{(n)} = \mathrm{Hom}_{U(\mathfrak{q})}(U(\mathfrak{q})U_n(\mathfrak{g}), E'_0)_T$ of members of E'. We have $E'^{(0)} \cong E'_0$ as a $(\mathfrak{q}, T)$ module, and Lemma 11.88 says that X acts invertibly on this. Arguing as in (a) leads to the isomorphism

$$\ker(E'^{(n)} \to E'^{(n-1)}) \cong \mathrm{Hom}_{\mathbb{C}}(S^n(\mathfrak{g}/\mathfrak{q}), E_0) \cong S^n(\mathfrak{g}/\mathfrak{q})^* \otimes_{\mathbb{C}} E_0$$

as a $U(\mathfrak{q})$ module, and it follows from Lemma 11.90 that X acts invertibly in $\ker(E'^{(n)} \to E'^{(n-1)})$. By induction X acts invertibly in $E'^{(n)}$ for every n.

Since X acts in one-one fashion on each $E'^{(n)}$, it follows that X acts in one-one fashion on E'. We have to show that X maps E' onto E'. Let ψ be in E'. We have just seen for each n that there is a unique φ_n in $E'^{(n)}$ with $X\varphi_n = \psi|_{U(\mathfrak{q})U_n(\mathfrak{g})}$. Because of the uniqueness, these φ_n's are compatible as n varies, and we can set $\varphi = \cup \varphi_n$. Then φ is in $\mathrm{Hom}_{U(\mathfrak{q})}(U(\mathfrak{g}), E'_0)$, and the problem is to prove that φ is T finite.

We shall use that X is $\mathrm{Ad}(T)$ finite; specifically $\mathrm{Ad}(t)X = \xi_\alpha(t)X$ for $t \in T$. If u is in $U(\mathfrak{g})$, we have $X\varphi = \psi$. Suppose that the span $\langle t\varphi \mid t \in T \rangle$ is infinite dimensional. Since X is one-one, the span $\langle X(t\varphi) \mid t \in T \rangle$ is infinite dimensional, and so is

$$\langle \xi_\alpha(t)X(t\varphi) \mid t \in T \rangle = \langle (\mathrm{Ad}(t)X)(t\varphi) \mid t \in T \rangle$$
$$= \langle t(X\varphi) \mid t \in T \rangle = \langle t\psi \mid t \in T \rangle.$$

But this is a contradiction, since ψ was assumed T finite. Thus φ is T finite, and X is invertible on E'. This proves (b).

Lemma 11.95. Let $(\mathfrak{g}, K)$ be a pair, and suppose $\mathfrak{g} = \mathbb{C}X \oplus \mathfrak{k}$ with $\mathrm{Ad}(K)X \subseteq \mathbb{C}X$. If E is a $(\mathfrak{g}, K)$ module on which the action of X is invertible, then $H^j(\mathfrak{g}, K; E) = 0$ for all $j \geq 0$.

REMARK. In our application to proving Theorem 11.87, K will be T.

PROOF. Since $\dim(\mathfrak{g}/\mathfrak{k}) = 1$, the only relevant j's are 0 and 1. For $j = 0$, (2.15) gives

$$H^0(\mathfrak{g}, K; E) = E^{\mathfrak{g}, K} = \{v \in E \mid \mathfrak{g}v = 0 \text{ and } Kv = v\} \subseteq \{v \in E \mid Xv = 0\},$$

and this is 0 since the action by X is one-one. For $j = 1$, Corollary 3.6 gives

$$H^1(\mathfrak{g}, K; E) \cong H_0(\mathfrak{g}, K; E \otimes_{\mathbb{C}} (\mathfrak{g}/\mathfrak{k})^*).$$

In the notation of (3.3), the inclusion $\mathrm{Ad}(K)X \subseteq \mathbb{C}X$ allows us to take $\mathfrak{c} = \mathbb{C}X$, and we readily check from the definition that $\mathfrak{g}$ acts by 0 on $\mathfrak{c}$. Therefore X acts by 0 on $(\mathfrak{g}/\mathfrak{k})^*$, and the action by X on $E' = E \otimes_{\mathbb{C}} (\mathfrak{g}/\mathfrak{k})^*$ is invertible. By (2.15),

$$H^1(\mathfrak{g}, K; E) \cong (E')_{\mathfrak{g}, K} \cong (E'/\mathfrak{g}E')^K \subseteq E'/\mathfrak{g}E',$$

and this is 0 since the action by X on E' is onto. This proves the lemma.

At this stage we would like to quote a variant of the Hochschild-Serre spectral sequences of Theorem 5.130. Specifically for $X = e_\alpha + \theta e_\alpha$, Lemma 11.95 gives us information about cohomology relative to the subpair $(\mathbb{C}X \oplus \mathfrak{t}, T)$ of $(\mathfrak{k}, T)$, and we would like to deduce information about cohomology relative to $(\mathfrak{k}, T)$. We do not have the Hochschild-Serre spectral sequences available in this generality, however, and we do not need such general results elsewhere. Thus we shall be content with a direct argument using exact sequences. We begin with a result from linear algebra.

Lemma 11.96. Let V be a finite-dimensional complex vector space, and let L be a one-dimensional subspace. Then the exact sequence

$$0 \longrightarrow L \longrightarrow V \longrightarrow V/L \longrightarrow 0$$

yields canonically an exact sequence

$$0 \longrightarrow L \otimes_{\mathbb{C}} \bigwedge^{p-1}(V/L) \overset{\psi}{\longrightarrow} \bigwedge^p V \overset{\varphi}{\longrightarrow} \bigwedge^p(V/L) \longrightarrow 0.$$

PROOF. Let $\varphi : \bigwedge^p V \to \bigwedge^p(V/L)$ be the map induced by the quotient map $V \to V/L$. To define ψ, let $l \otimes v$ be given with $l \in L$ and

$v \in \bigwedge^{p-1}(V/L)$. Let $\tilde{v}$ be a member of $\bigwedge^{p-1}V$ mapping onto v by the version of φ for degree $p-1$, and define $\psi(l \otimes v) = l \wedge \tilde{v}$.

To see that ψ is well defined, let $\tilde{v} + l'$ be another representative of v, with $l' \in L$. Since L is one dimensional, l and l' are dependent. Thus $l \wedge l' = 0$ and $l \wedge \tilde{v} = l \wedge (\tilde{v} + l')$. Hence ψ is well defined.

For the exactness we use bases. Let $v_1, \ldots, v_n$ be a basis of V with $v_1 \in L$. We can identify the subset

$$v_{i_1} \wedge \cdots \wedge v_{i_{p-1}} \qquad \text{for } 2 \le i_1 < \cdots < i_{p-1} \le n$$

with a basis of $\bigwedge^{p-1}(V/L)$. Under this identification,

(11.97a) $\psi(v_1 \otimes (v_{i_1} \wedge \cdots \wedge v_{i_{p-1}})) = v_1 \wedge v_{i_1} \wedge \cdots \wedge v_{i_{p-1}}$,

and ψ is certainly one-one. Meanwhile the domain of φ has basis

$$v_{i_0} \wedge \cdots \wedge v_{i_{p-1}} \qquad \text{for } 1 \le i_0 < \cdots < i_{p-1} \le n.$$

Under our identifications,

(11.97b) $\varphi(v_{i_0} \wedge \cdots \wedge v_{i_{p-1}}) = \begin{cases} v_{i_0} \wedge \cdots \wedge v_{i_{p-1}} & \text{if } i_0 \ge 2 \\ 0 & \text{if } i_0 = 1, \end{cases}$

and φ is certainly onto. Comparing (11.97a) and (11.97b), we see that $\ker \varphi = \text{image } \psi$. The lemma follows.

To apply Lemma 11.96, we return to our reductive pair $(\mathfrak{g}, K)$ and the notation as in Lemma 11.88. Let $X = e_\alpha + \theta e_\alpha$. In Lemma 11.96, put

$$L = (\mathbb{C}X \oplus \mathfrak{t})/\mathfrak{t}, \qquad V = \mathfrak{k}/\mathfrak{t}, \qquad V/L = \mathfrak{k}/(\mathbb{C}X \oplus \mathfrak{t}).$$

Then we obtain an exact sequence of T modules

$$0 \longrightarrow ((\mathbb{C}X \oplus \mathfrak{t})/\mathfrak{t}) \otimes \bigwedge^{p-1}(\mathfrak{k}/(\mathbb{C}X \oplus \mathfrak{t})) \longrightarrow \bigwedge^p(\mathfrak{k}/\mathfrak{t})$$
$$\longrightarrow \bigwedge^p(\mathfrak{k}/(\mathbb{C}X \oplus \mathfrak{t})) \longrightarrow 0.$$

If E is any $(\mathfrak{k}, T)$ module, then application of the exact contravariant functor $\text{Hom}_T(\,\cdot\,, E)$ gives the exact sequence

(11.98)

$$0 \longrightarrow \text{Hom}_T(\bigwedge^p(\mathfrak{k}/(\mathbb{C}X \oplus \mathfrak{t})), E) \xrightarrow{\ i_p\ } \text{Hom}_T(\bigwedge^p(\mathfrak{k}/\mathfrak{t}), E)$$
$$\xrightarrow{\ j_p\ } \text{Hom}_T((\mathbb{C}X \oplus \mathfrak{t})/\mathfrak{t}, \text{Hom}_{\mathbb{C}}(\bigwedge^{p-1}(\mathfrak{k}/(\mathbb{C}X \oplus \mathfrak{t})), E)) \longrightarrow 0.$$

Lemma 11.99. With notation as in (11.98), let $d_{\mathfrak{k}/\mathfrak{t}}$ be the differential

$$d_{\mathfrak{k}/\mathfrak{t}} : \mathrm{Hom}_T(\textstyle\bigwedge^p(\mathfrak{k}/\mathfrak{t}), E) \to \mathrm{Hom}_T(\textstyle\bigwedge^{p+1}(\mathfrak{k}/\mathfrak{t}), E),$$

and identify

(11.100a) $\mathrm{Hom}_T(\bigwedge^p(\mathfrak{k}/(\mathbb{C}X \oplus \mathfrak{t})), E)$

$$\cong \mathrm{Hom}_T(\textstyle\bigwedge^0((\mathbb{C}X \oplus \mathfrak{t})/\mathfrak{t}), \mathrm{Hom}_\mathbb{C}(\textstyle\bigwedge^p(\mathfrak{k}/(\mathbb{C}X \oplus \mathfrak{t})), E))$$

(11.100b) $\mathrm{Hom}_T((\mathbb{C}X \oplus \mathfrak{t})/\mathfrak{t}, \mathrm{Hom}_\mathbb{C}(\bigwedge^p(\mathfrak{k}/(\mathbb{C}X \oplus \mathfrak{t})), E))$

$$\cong \mathrm{Hom}_T(\textstyle\bigwedge^1((\mathbb{C}X \oplus \mathfrak{t})/\mathfrak{t}), \mathrm{Hom}_\mathbb{C}(\textstyle\bigwedge^p(\mathfrak{k}/(\mathbb{C}X \oplus \mathfrak{t})), E)).$$

Then $j_{p+1} \circ d_{\mathfrak{k}/\mathfrak{t}} \circ i_p$ equals the differential $d_{(\mathbb{C}X\oplus\mathfrak{t})/\mathfrak{t}}$ from the right side of (11.100a) to the right side of (11.100b).

PROOF. Let $\mathfrak{s}$ be a T invariant complement to $\mathbb{C}X \oplus \mathfrak{t}$ in $\mathfrak{k}$, and let $\mathcal{P}$ and $\mathcal{P}'$ denote the projections of $\mathfrak{k}$ on $\mathbb{C}X \oplus \mathfrak{s}$ along $\mathfrak{t}$ and on $\mathfrak{s}$ along $\mathbb{C}X \oplus \mathfrak{t}$, respectively. If ω is in the left side of (11.100a) and if $X_1, \ldots, X_p$ are in $\mathfrak{s}$, then (2.127b) gives

(11.101)
$$j_{p+1}d_{\mathfrak{k}/\mathfrak{t}}i_p\omega(X)(X_1 \wedge \cdots \wedge X_p)$$
$$= d_{\mathfrak{k}/\mathfrak{t}}i_p\omega(X \wedge X_1 \wedge \cdots \wedge X_p)$$
$$= X(i_p\omega(X_1 \wedge \cdots \wedge X_p))$$
$$+ \sum_s (-1)^s i_p\omega(\mathcal{P}[X, X_s] \wedge X_1 \wedge \cdots \wedge \widehat{X}_s \wedge \cdots \wedge X_p),$$

since all terms with an X in the argument of $i_p\omega$ vanish. On the other hand, (2.127b) gives

$$(d_{(\mathbb{C}X\oplus\mathfrak{t})/\mathfrak{t}}\omega)(X) = X\omega,$$

where $X\omega$ as refers to the operation of X on ω in

$$\mathrm{Hom}_\mathbb{C}(\textstyle\bigwedge^p(\mathfrak{k}/(\mathbb{C}X \oplus \mathfrak{t})), E).$$

This action is well defined since the formula

$$(\mathrm{ad}\,X)(X' + (\mathbb{C}X \oplus \mathfrak{t})) = [X, X'] + (\mathbb{C}X \oplus \mathfrak{t})$$

gives an action of X on $\mathfrak{k}/(\mathbb{C}X \oplus \mathfrak{t})$. For the same $X_1, \ldots, X_p$ in $\mathfrak{s}$, we have

$$(X\omega)(X_1 \wedge \cdots \wedge X_p)$$
$$= X(\omega(X_1 \wedge \cdots \wedge X_p)) + \omega((\operatorname{ad} X)(X_1 \wedge \cdots \wedge X_p))$$
$$= X(i_p\omega(X_1 \wedge \cdots \wedge X_p))$$
$$\quad + \sum_s \omega(X_1 \wedge \cdots \wedge ([X, X_s] + (\mathbb{C}X \oplus \mathfrak{t})) \wedge \cdots \wedge X_p)$$
$$= X(i_p\omega(X_1 \wedge \cdots \wedge X_p))$$
$$\quad + \sum_s (-1)^s i_p\omega(\mathcal{P}'[X, X_s] \wedge X_1 \wedge \cdots \wedge \widehat{X}_s \wedge \cdots \wedge X_p).$$

This last expression matches (11.101), and the lemma follows.

Lemma 11.102. With notation as in Lemma 11.88, suppose that $X = e_\alpha + \theta e_\alpha$ acts invertibly in a $(\mathfrak{k}, T)$ module E. Then $H^p(\mathfrak{k}, T; E) = 0$ for all $p \geq 0$.

PROOF. For any n, $\bigwedge^n(\mathfrak{k}/(\mathbb{C}X \oplus \mathfrak{t}))$ is a $(\mathbb{C}X \oplus \mathfrak{t}, T)$ module on which the action of X is nilpotent. By assumption X acts invertibly in E, and thus Lemma 11.90 shows that X acts invertibly in the $(\mathbb{C}X \oplus \mathfrak{t}, T)$ module

$$\operatorname{Hom}_{\mathbb{C}}(\textstyle\bigwedge^n(\mathfrak{k}/(\mathbb{C}X \oplus \mathfrak{t})), E).$$

Let us introduce the differentials that were defined in Lemma 11.99. By Lemma 11.95 the map

(11.103)
$$d_{(\mathbb{C}X\oplus\mathfrak{t})/\mathfrak{t}} : \operatorname{Hom}_T(\textstyle\bigwedge^0((\mathbb{C}X \oplus \mathfrak{t})/\mathfrak{t}), \operatorname{Hom}_{\mathbb{C}}(\bigwedge^n(\mathfrak{k}/(\mathbb{C}X \oplus \mathfrak{t})), E))$$
$$\to \operatorname{Hom}_T(\textstyle\bigwedge^1((\mathbb{C}X \oplus \mathfrak{t})/\mathfrak{t}), \operatorname{Hom}_{\mathbb{C}}(\bigwedge^n(\mathfrak{k}/(\mathbb{C}X \oplus \mathfrak{t})), E))$$

from the right side of (11.100a) to the right side of (11.100b) is an isomorphism.

If the element ω_p of $\operatorname{Hom}_T(\bigwedge^p(\mathfrak{k}/\mathfrak{t}), E)$ is a cocycle, i.e., if

(11.104) $d_{\mathfrak{k}/\mathfrak{t}}\omega_p = 0,$

then we shall prove that ω_p is a coboundary, i.e., ω_p is $d_{\mathfrak{k}/\mathfrak{t}}$ of something. The element $j_p\omega_p$ is in the range space of (11.103) with $n = p - 1$. Since $d_{(\mathbb{C}X\oplus\mathfrak{t})/\mathfrak{t}}$ is onto, there exists γ_{p-1} in the domain space of (11.103) with $n = p - 1$ such that $j_p\omega_p = d_{(\mathbb{C}X\oplus\mathfrak{t})/\mathfrak{t}}\gamma_{p-1}$. Applying Lemma 11.99, we can rewrite this equation as

$$j_p\omega_p = j_p d_{\mathfrak{k}/\mathfrak{t}} i_{p-1}\gamma_{p-1}.$$

Therefore $j_p(\omega_p - d_{\mathfrak{k}/\mathfrak{t}} i_{p-1} \gamma_{p-1}) = 0$, and the exactness of (11.98) implies that

$$(11.105) \qquad \omega_p - d_{\mathfrak{k}/\mathfrak{t}} i_{p-1} \gamma_{p-1} = i_p \delta_p$$

for some δ_p in the domain space of (11.103) when $n = p$. Applying $d_{\mathfrak{k}/\mathfrak{t}}$ to both sides and using (11.104), we see that

$$(11.106) \qquad d_{\mathfrak{k}/\mathfrak{t}} i_p \delta_p = 0.$$

Composition of (11.106) with j_{p+1} and a second use of Lemma 11.99 yields

$$0 = j_{p+1} d_{\mathfrak{k}/\mathfrak{t}} i_p \delta_p = d_{(\mathbb{C}X \oplus \mathfrak{t})/\mathfrak{t}} \delta_p.$$

Since $d_{(\mathbb{C}X \oplus \mathfrak{t})/\mathfrak{t}}$ is one-one, $\delta_p = 0$. Therefore (11.105) shows that $\omega_p = d_{\mathfrak{k}/\mathfrak{t}}(i_{p-1} \gamma_{p-1})$ as required.

PROOF OF THEOREM 11.87. For (a), let E_0 be as in Lemma 11.88, and put $E = \operatorname{ind}_{\mathfrak{q},T}^{\mathfrak{g},T}(E_0)$. Then Lemma 11.91 shows that $X = e_\alpha + \theta e_\alpha$ acts invertibly in E. By Lemma 11.90, X acts invertibly in any $(\mathfrak{k}, T)$ module of the form $E \otimes_{\mathbb{C}} F$, where F is a finite-dimensional $(\mathfrak{k}, T)$ module in which X acts in nilpotent fashion. Therefore Lemma 11.102 shows that

$$(11.107a) \qquad H^p(\mathfrak{k}, T; E \otimes_{\mathbb{C}} F) = 0 \qquad \text{for all } p \geq 0$$
$$(11.107b) \qquad H^p(\mathfrak{k}, T; E^c \otimes_{\mathbb{C}} F) = 0 \qquad \text{for all } p \geq 0$$

whenever F is a finite-dimensional $(\mathfrak{k}, T)$ module in which X acts in nilpotent fashion.

With q as in Lemma 11.88, Proposition 2.57c shows that $\operatorname{ind}_{\mathfrak{q},T}^{\mathfrak{g},T}$ is exact. Applying this functor to the exact sequence in Lemma 11.88a and using Proposition 2.19, we obtain an exact sequence

$$0 \longrightarrow \operatorname{ind}_{\mathfrak{b}_2,T}^{\mathfrak{g},T}(Z \otimes \mathbb{C}_\alpha) \longrightarrow E \longrightarrow \operatorname{ind}_{\mathfrak{b}_1,T}^{\mathfrak{g},T}(Z) \longrightarrow 0.$$

Since $\Pi_{\mathfrak{g},T}^{\mathfrak{g},K}$ is covariant and right exact, (C.36a) gives us a long exact sequence containing the terms

$$(11.108) \quad (\Pi_{\mathfrak{g},T}^{\mathfrak{g},K})_{j+1}(E) \longrightarrow (\Pi_{\mathfrak{g},T}^{\mathfrak{g},K})_{j+1}(\operatorname{ind}_{\mathfrak{b}_1,T}^{\mathfrak{g},T}(Z))$$
$$\longrightarrow (\Pi_{\mathfrak{g},T}^{\mathfrak{g},K})_j(\operatorname{ind}_{\mathfrak{b}_2,T}^{\mathfrak{g},T}(Z \otimes \mathbb{C}_\alpha)) \longrightarrow (\Pi_{\mathfrak{g},T}^{\mathfrak{g},K})_j(E).$$

We shall prove that the end terms of (11.108) are 0, and then (a) follows in the theorem.

In other words, it suffices to prove that $(\Pi_{\mathfrak{g},T}^{\mathfrak{g},K})_p(E) = 0$ for all $p \geq 0$. Let V be an irreducible K module. Then

$$\mathrm{Hom}_K((\Pi_{\mathfrak{g},T}^{\mathfrak{g},K})_p(E), V)$$

$$\cong \mathrm{Ext}_{\mathfrak{k},T}^p(E, (\mathcal{F}^\vee)_{\mathfrak{k},K}^{\mathfrak{k},T}(V)) \qquad \text{by (5.32a) and its proof}$$

$$\cong \mathrm{Ext}_{\mathfrak{k},T}^p(E, V) \qquad \text{by Proposition 2.33}$$

$$\cong H^p(\mathfrak{k}, T; \mathrm{Hom}_\mathbb{C}(E, V)_T) \qquad \text{by Proposition 2.117}$$

$$\cong H^p(\mathfrak{k}, T; E^c \otimes_\mathbb{C} V).$$

The right side is 0 by (11.107b), and hence V has multiplicity 0 in $(\Pi_{\mathfrak{g},T}^{\mathfrak{g},K})_p(E)$. Since V is arbitrary, $(\Pi_{\mathfrak{g},T}^{\mathfrak{g},K})_p(E) = 0$. This proves (a).

For (b), we argue similarly with E_0' as in Lemma 11.88b and E' as in Lemma 11.91b. We are led to a sequence

$$0 \longrightarrow \mathrm{pro}_{\mathfrak{b}_1,T}^{\mathfrak{g},T}(Z) \longrightarrow E' \longrightarrow \mathrm{pro}_{\mathfrak{b}_2,T}^{\mathfrak{g},T}(Z \otimes \mathbb{C}_{-\alpha}) \longrightarrow 0$$

and then, via (C.36c), to the exact sequence

$$(\Gamma_{\mathfrak{g},T}^{\mathfrak{g},K})^j(E') \longrightarrow (\Gamma_{\mathfrak{g},T}^{\mathfrak{g},K})^j(\mathrm{pro}_{\mathfrak{b}_2,T}^{\mathfrak{g},T}(Z \otimes \mathbb{C}_{-\alpha}))$$

$$\longrightarrow (\Gamma_{\mathfrak{g},T}^{\mathfrak{g},K})^{j+1}(\mathrm{pro}_{\mathfrak{b}_1,T}^{\mathfrak{g},T}(Z)) \longrightarrow (\Gamma_{\mathfrak{g},T}^{\mathfrak{g},K})^{j+1}(E').$$

To prove (b), it suffices to prove that the end terms are 0. In other words, it suffices to prove that $(\Gamma_{\mathfrak{g},T}^{\mathfrak{g},K})^p(E') = 0$ for all p. If V is an irreducible K module, we find that

$$\mathrm{Hom}_K(V, (\Gamma_{\mathfrak{g},T}^{\mathfrak{g},K})^p(E')) \cong H^p(\mathfrak{k}, T; V^c \otimes_\mathbb{C} E'),$$

and this is 0 by (11.107a).

EXAMPLE. Let $G = SU(2, 1)$ with $K = S(U(2) \times U(1))$. Using the diagonal subalgebra as a compact Cartan subalgebra, we can construct a θ stable parabolic subalgebra $\mathfrak{q} = \mathfrak{l} \oplus \mathfrak{u}$ whose Levi factor L is block diagonal with blocks of sizes 1 and 2, respectively. For this situation $S = \dim(\mathfrak{u} \cap \mathfrak{k}) = 1$. The group L is locally isomorphic to the product of a circle and $SU(1, 1)$, and we consider on L a representation V that is the tensor product of a character of the circle and a principal-series representation of $SU(1, 1)$.

The Transfer Theorem enables us to identify $U = (^\mathfrak{u}\mathcal{R}_{\mathfrak{q},L\cap K}^{\mathfrak{g},K})^1(V)$. In fact, V is of the form $^\mathfrak{u}\mathcal{R}_{\mathfrak{b},T}^{\mathfrak{l},L\cap K}(Z)$, where $\mathfrak{h} = \mathfrak{b} \cap \theta\bar{\mathfrak{b}}$ is the maximally split Cartan subalgebra

$$\mathfrak{h} = \begin{pmatrix} -2i\varphi & & \\ & i\varphi & r \\ & r & i\varphi \end{pmatrix},$$

and where Z is one dimensional. In the decomposition $\mathfrak{h} = \mathfrak{t} \oplus \mathfrak{a}$, $\mathfrak{t}$ corresponds to $r = 0$, and $\mathfrak{a}$ corresponds to $\varphi = 0$. Let the real roots be $\pm \alpha$, and let the complex roots be $\beta, \bar{\beta}, -\beta, -\bar{\beta}$. Let us say that α is positive relative to $\mathfrak{b}$ and that $\mathfrak{u}$ is built from β and $\theta \beta = -\bar{\beta}$. We must have either $\beta = \alpha + \theta \beta$ or $\theta \beta = \alpha + \beta$. Possibly by interchanging β and $\theta \beta$, we may suppose that $\beta = \alpha + \theta \beta$. Corollary 11.86 shows that

$$U \cong (^{\mathfrak{u}}\mathcal{R}^{\mathfrak{g},L}_{\mathfrak{b}_1,T})^1(Z),$$

where $\mathfrak{b}_1$ is the Borel subalgebra of $\mathfrak{g}$ corresponding to $\{\alpha, \beta, \theta\beta\}$. Let $\mathfrak{b}_2$ correspond to $\{\alpha, \beta, \bar{\beta}\}$. In Theorem 11.87 the root $\gamma = \theta\beta$ has $\mathfrak{b}_1 \cap \mathfrak{b}_2^- = \mathfrak{g}_\gamma$ and $\mathfrak{g}_{\theta\gamma} \subseteq \mathfrak{b}_1$. Under the mild condition on Z in the theorem, we obtain

$$U \cong (^{\mathfrak{u}}\mathcal{R}^{\mathfrak{g},L}_{\mathfrak{b}_2,T})^0(Z \otimes \mathbb{C}_{-\theta\beta}).$$

Proposition 11.47 therefore identifies U as a specific principal-series representation of $SU(2, 1)$.

6. Standard Modules

In this section we iterate the Transfer Theorem in order to prove equivalences among as many $(\mathfrak{g}, K)$ modules as possible of the kind in the theorem. The result will be the set of "standard modules," in terms of which various classification theorems may be stated.

Fix a reductive pair $(\mathfrak{g}, K)$. For the remainder of the chapter, we shall assume that $(\mathfrak{g}, K)$ is of the Harish-Chandra class. We introduce $(\mathfrak{g}, K)$ modules corresponding to sets of data consisting of

(11.109)

(a) a Cartan subpair $(\mathfrak{h}, T)$ with T maximal, necessarily having $T = Z_K(\mathfrak{h})$ since $(\mathfrak{g}, K)$ is in the Harish-Chandra class

(b) a parameter $\lambda \in \mathfrak{h}^*$, which will turn out to be the infinitesimal character of the $(\mathfrak{g}, K)$ modules to be constructed

(c) positive systems Δ^+_{imag} and Δ^+_{real} for the subsystems of imaginary roots and real roots within $\Delta(\mathfrak{g}, \mathfrak{h})$

(d) a class of "related" irreducible $(\mathfrak{h}, T)$ modules $Z = Z(\mathfrak{b})$ indexed by the Borel subalgebras $\mathfrak{b} = \mathfrak{h} \oplus \mathfrak{n}$, such that $\mathfrak{h}$ acts in $Z(\mathfrak{b})$ by $\lambda + \delta(\mathfrak{n})$.

Items (c) and (d) require some elaboration. In (c) it is required that
(i) and either (ii) or (ii′) hold in

$$
\begin{array}{llll}
& \text{(i)} & \langle \lambda, \alpha \rangle \geq 0 & \text{for all } \alpha \in \Delta^{+}_{\text{imag}} \\
(11.110) & \text{(ii)} & \operatorname{Re} \langle \lambda, \alpha \rangle \geq 0 & \text{for all } \alpha \in \Delta^{+}_{\text{real}} \\
& \text{(ii′)} & \operatorname{Re} \langle \lambda, \alpha \rangle \leq 0 & \text{for all } \alpha \in \Delta^{+}_{\text{real}}.
\end{array}
$$

The choice of (ii) or (ii′) will depend upon the particular setting we are
studying. We impose these conditions because the Transfer Theorem
does not allow us to change the imaginary and real roots in the Borel
subalgebra used to define a $(\mathfrak{g}, K)$ module. (Under certain conditions,
one can prove other theorems that do allow the imaginary and real roots
in the Borel subalgebra to be changed, but we shall not address this
topic.)

To make (d) precise, we use the condition that $(\mathfrak{g}, K)$ is in the Harish-
Chandra class. If G is the underlying group, then $\operatorname{Ad}(G)$ is contained
in the connected complex adjoint group $G_{\mathbb{C}}$ for $\mathfrak{g}$. The subgroup $\operatorname{Ad}(T)$
centralizes $\mathfrak{h}$ and therefore lies in the analytic subgroup $H_{\mathbb{C}}$ of $G_{\mathbb{C}}$ with
Lie algebra $\mathfrak{h}$. If γ is in the $\mathbb{Z}$ span of the roots, then γ exponentiates
to $H_{\mathbb{C}}$ and pulls back to a character ξ_{γ} of T. Let $\mathbb{C}_{\gamma}$ denote the $(\mathfrak{h}, T)$
module in which $\mathfrak{h}$ acts by γ and T acts by ξ_{γ}. This construction has the
consistency property that $\mathbb{C}_{\gamma} \otimes_{\mathbb{C}} \mathbb{C}_{\gamma'} \cong \mathbb{C}_{\gamma+\gamma'}$, and it is compatible with
the construction done in Remark 3 for Theorem 11.87 in the case that γ
is a root and $(\mathfrak{g}, K)$ is not necessarily in the Harish-Chandra class. We
shall use this new construction shortly when γ is the difference of two
half sums of roots: $\gamma = \delta(\mathfrak{n}') - \delta(\mathfrak{n}^{-})$.

In the special case that γ is in the $\mathbb{Z}$ span of the *real* roots, we shall use
a second construction. In this case γ vanishes on $\mathfrak{t}$ and ξ_{γ} is therefore
1 on the identity component T_{0}. However, ξ_{γ} need not be 1 on all of T.
(For example, take $\mathfrak{h}_{0}$ to be the diagonal subalgebra for $\mathfrak{sl}(3, \mathbb{R})$. Then
$\mathfrak{t}_{0} = 0$. If γ is the root $e_{1} - e_{2}$, then ξ_{γ} is -1 on the diagonal matrix
$\operatorname{diag}(1, -1, -1)$, which is in T.) Thus ξ_{γ} and $|\xi_{\gamma}| = 1$ have the same
differential but may be different characters. We shall make use of the
$(\mathfrak{h}, T)$ module in which $\mathfrak{h}$ acts by γ and T acts by $|\xi_{\gamma}| = 1$.

When γ is in the $\mathbb{Z}$ span of all the roots and has an understood canonical
decomposition as $\gamma = \gamma_{1} + \gamma_{2}$ with γ_{1} in the $\mathbb{Z}$ span of the nonreal roots
and γ_{2} in the $\mathbb{Z}$ span of the real roots, we can define a one-dimensional
$(\mathfrak{h}, T)$ module

$$
(11.111\text{a}) \qquad\qquad\qquad \mathbb{C}_{\gamma'}
$$

by having $\mathfrak{h}$ act by γ and by having T act by a character $\xi_{\gamma'}$ defined to
equal the character built from γ_{1} alone:

$$
(11.111\text{b}) \qquad\qquad \xi_{\gamma'} = \xi_{\gamma_{1}} = \xi_{\gamma_{1}} |\xi_{\gamma_{2}}|.
$$

Our first use of this notation will be with

$$(11.112) \qquad \gamma = \delta(\mathfrak{n}') - \delta(\mathfrak{n}^-) = \sum_{\substack{\alpha \in \Delta(\mathfrak{n}') \cap \Delta(\mathfrak{n}^-), \\ \alpha \text{ not real}}} \alpha + \sum_{\substack{\alpha \in \Delta(\mathfrak{n}') \cap \Delta(\mathfrak{n}^-), \\ \alpha \text{ real}}} \alpha.$$

In the one-dimensional $(\mathfrak{h}, T)$ module $\mathbb{C}_{(\delta(\mathfrak{n}')-\delta(\mathfrak{n}^-))'}$, the Lie algebra $\mathfrak{h}$ acts by $\delta(\mathfrak{n}') - \delta(\mathfrak{n}^-)$ and the group T acts by the character corresponding to the first term of the right side of (11.112):

$$(11.113) \qquad \xi_{(\delta(\mathfrak{n}')-\delta(\mathfrak{n}^-))'} = \prod_{\substack{\alpha \in \Delta(\mathfrak{n}') \cap \Delta(\mathfrak{n}^-), \\ \alpha \text{ not real}}} \xi_\alpha.$$

This is the $(\mathfrak{h}, T)$ module that we shall use to define the notion of "related" in (d).

In (d), let $(Z, \mathfrak{b})$ and $(Z', \mathfrak{b}')$ be pairs consisting of an irreducible $(\mathfrak{h}, T)$ module and a Borel subalgebra containing $\mathfrak{h}$. We say that $(Z, \mathfrak{b})$ and $(Z', \mathfrak{b}')$ are **related** if

$$(11.114) \qquad Z' \cong Z \otimes_{\mathbb{C}} \mathbb{C}_{(\delta(\mathfrak{n}')-\delta(\mathfrak{n}))'}.$$

"Related" is an equivalence relation on the ordered pairs $(Z, \mathfrak{b})$ as above, and two related pairs $(Z, \mathfrak{b})$ and $(Z', \mathfrak{b}')$ with $\mathfrak{b} = \mathfrak{b}'$ have $Z \cong Z'$. It is therefore meaningful (up to isomorphism) to define a function $(Z, \mathfrak{b}) \mapsto Z = Z(\mathfrak{b})$ on an equivalence class. The image of such a function on one equivalence class is a class of related modules for (d), provided $\mathfrak{h}$ acts in $Z(\mathfrak{b})$ by $\lambda + \delta(\mathfrak{n})$. If we start from a single Z and $\mathfrak{b}$ such that $\mathfrak{h}$ acts by $\lambda + \delta(\mathfrak{n})$, we can construct a class of related modules for (d) by the definition

$$Z(\mathfrak{b}') = Z \otimes_{\mathbb{C}} \mathbb{C}_{(\delta(\mathfrak{n}')-\delta(\mathfrak{n}))'},$$

but it is not necessary to distinguish a particular Z in a class in order to make sense of (d).

The need for a class of Z's in (d) is apparent from the statement of Theorem 11.87. Conclusion (a) relates a $(\mathfrak{g}, K)$ module constructed from $Z \otimes \mathbb{C}_\alpha$ to one constructed from Z, while (b) relates a $(\mathfrak{g}, K)$ module constructed from $Z \otimes \mathbb{C}_{-\alpha}$ to one constructed from Z. The significance of $\mathbb{C}_\alpha$ is that $\alpha = \delta(\mathfrak{n}_1) - \delta(\mathfrak{n}_2) = (\delta(\mathfrak{n}_1) - \delta(\mathfrak{n}_2))'$. The need for special attention to the real roots will become apparent in the proof of Theorem 11.129c.

Corresponding to a set of data (11.109) satisfying the consistency condition (11.110), we shall introduce two $(\mathfrak{g}, K)$ modules, one constructed using $({}^{u}\mathcal{L})_p$ and one constructed using $({}^{u}\mathcal{R})^p$, for each Borel subalgebra $\mathfrak{b} = \mathfrak{h} \oplus \mathfrak{n}$ of a particular kind. There will four kinds of interest, called "type BB," "type L," "type L'," and "type VZ," and each will have a more restrictive version called "strict." In (11.110) we shall impose condition (ii) in studying types BB and L, and we shall impose condition (ii') in studying types VZ and L'. In every case, the degree p of interest will be

$$(11.115) \qquad\qquad p = \dim(\mathfrak{n} \cap \mathfrak{k}),$$

and, as we shall see later, there will be a vanishing theorem in all other degrees.

For a Borel subalgebra $\mathfrak{b}$ of one of the above types, the **standard** $(\mathfrak{g}, K)$ **module** of Π type associated to the data in (11.109) and (11.110) is

$$(11.116a) \qquad\qquad ({}^{u}\mathcal{L}^{\mathfrak{g},K}_{\mathfrak{b}^{-},T})_p(Z(\mathfrak{b})),$$

and the **standard** $(\mathfrak{g}, K)$ **module** of Γ type is

$$(11.116b) \qquad\qquad ({}^{u}\mathcal{R}^{\mathfrak{g},K}_{\mathfrak{b},T})^p(Z(\mathfrak{b})).$$

Theorem 11.129 will show that the standard modules of type VZ are equivalent with those of type L', that the standard modules of type BB are equivalent with those of type L, and that the standard modules of types L and L' corresponding to dual data are related by interchanging Π and Γ. Moreover all the strict types always exist, as a consequence of Propositions 11.117, 11.126, and 11.127 below.

When we come to classification theorems, there are two settings. When (ii) is in force in (11.110), our interest will be in irreducible submodules of the standard modules of Π type and in irreducible quotients of the standard modules of Γ type. When (ii') is in force in (11.110), these roles will be reversed: Our interest will be in irreducible quotients of the standard modules of Π type and in irreducible submodules of the standard modules of Γ type.

We begin with type BB. If $\mathfrak{b} = \mathfrak{h} \oplus \mathfrak{n}$ is a Borel subalgebra such that (i) and (ii) hold in (11.110) and

(i) $\Delta^{+}_{\text{imag}} \subseteq \Delta(\mathfrak{n})$

(ii) $\Delta^{+}_{\text{real}} \subseteq \Delta(\mathfrak{n})$

(iii) each $\alpha \in \Delta(\mathfrak{n})$ has $2\langle \lambda, \alpha \rangle / |\alpha|^2 \notin \{-1, -2, \dots\}$,

we say that $\mathfrak{b}$ is of **type BB**. If (iii) is replaced by

(iii-s) $\text{Re}\langle \lambda, \alpha \rangle \geq 0$ for all $\alpha \in \Delta(\mathfrak{n})$, we say that $\mathfrak{b}$ is of **strict type BB**. Condition (iii) is that λ is integrally dominant. We can summarize (iii-s) by saying that $\mathrm{Re}\,\lambda$ is dominant; here $\mathrm{Re}\,\lambda$ denotes the complex-linear extension to $\mathfrak{h}$ of the linear functional $h \mapsto \mathrm{Re}(\lambda(h))$ on $i\mathfrak{t}_0 \oplus \mathfrak{a}_0$. Clearly condition (iii-s) implies condition (iii).

Proposition 11.117. If $\{(\mathfrak{h}, T), \lambda, \Delta^+_{\mathrm{imag}}, \Delta^+_{\mathrm{real}}, \{Z(\mathfrak{b})\}\}$ is a set of data for standard $(\mathfrak{g}, K)$ modules satisfying the consistency conditions (i) and (ii) in (11.110), then there exists a Borel subalgebra $\mathfrak{b}$ of strict type BB, and any two $\mathfrak{b}$'s of type BB lead to the same set of standard modules, up to equivalence.

REMARKS. We shall state a lemma, prove the proposition, and then prove the lemma. Note that distinct $\mathfrak{b}$'s of type BB yield their standard modules in their respective degrees p as in (11.115). These degrees may be different for different $\mathfrak{b}$'s.

Lemma 11.118. Let $\{(\mathfrak{h}, T), \lambda, \Delta^+_{\mathrm{imag}}, \Delta^+_{\mathrm{real}}, \{Z(\mathfrak{b})\}\}$ be a set of data for standard $(\mathfrak{g}, K)$ modules satisfying the consistency conditions (i) and (ii) in (11.110), and let $\mathfrak{b} = \mathfrak{h} \oplus \mathfrak{n}$ and $\mathfrak{b}' = \mathfrak{h}' \oplus \mathfrak{n}'$ be two Borel subalgebras satisfying the conditions

  (i) $\Delta^+_{\mathrm{imag}} \subseteq \Delta(\mathfrak{n}) \cap \Delta(\mathfrak{n}')$
  (ii) $\Delta^+_{\mathrm{real}} \subseteq \Delta(\mathfrak{n}) \cap \Delta(\mathfrak{n}')$
  (iii) each complex root α with $2\langle\lambda, \alpha\rangle/|\alpha|^2$ equal to a nonzero integer is in both or neither of $\Delta(\mathfrak{n})$ and $\Delta(\mathfrak{n}')$.

If $p = \dim(\mathfrak{n} \cap \mathfrak{k})$ and $p' = \dim(\mathfrak{n}' \cap \mathfrak{k})$, then

$$({}^{\mathrm{u}}\mathcal{L}^{\mathfrak{g}, K}_{\mathfrak{b}-, T})_{p+q}(Z(\mathfrak{b})) \cong ({}^{\mathrm{u}}\mathcal{L}^{\mathfrak{g}, K}_{\mathfrak{b}'-, T})_{p'+q}(Z(\mathfrak{b}'))$$

and

$$({}^{\mathrm{u}}\mathcal{R}^{\mathfrak{g}, K}_{\mathfrak{b}, T})^{p+q}(Z(\mathfrak{b})) \cong ({}^{\mathrm{u}}\mathcal{R}^{\mathfrak{g}, K}_{\mathfrak{b}', T})^{p'+q}(Z(\mathfrak{b}'))$$

for all q.

PROOF OF PROPOSITION 11.117. To define $\mathfrak{b}$, let δ_{imag} and δ_{real} be the half sums of the members of Δ^+_{imag} and Δ^+_{real}, respectively. We order $(i\mathfrak{t}_0 \oplus \mathfrak{a}_0)^*$ lexicographically by using the nonzero members of the list $\mathrm{Re}\,\lambda, \delta_{\mathrm{imag}}$, and δ_{real} as the first members of a spanning set that defines the ordering. (Positivity is to be defined lexicographically in terms of inner products with the members of the spanning set.) This lexicographic ordering determines a positive system in $\Delta(\mathfrak{g}, \mathfrak{h})$ and hence defines $\mathfrak{n}$ and $\mathfrak{b}$.

Then $\Delta(\mathfrak{n})$ has the property that $\operatorname{Re}\lambda$ is dominant since $\operatorname{Re}\lambda$ is the first member of the list. Thus (iii-s) holds in the definition of "strict type BB." If α is in Δ^+_{imag}, then (11.110) shows that $\langle\operatorname{Re}\lambda,\alpha\rangle\geq 0$. If $\langle\operatorname{Re}\lambda,\alpha\rangle > 0$, then α is in $\Delta(\mathfrak{n})$ since $\operatorname{Re}\lambda$ is the first member of the spanning set defining the ordering. If $\langle\operatorname{Re}\lambda,\alpha\rangle = 0$, then $\langle\delta_{\text{imag}},\alpha\rangle > 0$ forces α to be in $\Delta(\mathfrak{n})$. Hence $\Delta^+_{\text{imag}}\subseteq\Delta(\mathfrak{n})$, and (i) holds. Similarly $\Delta^+_{\text{real}}\subseteq\Delta(\mathfrak{n})$, and (ii) holds. Thus $\mathfrak{b} = \mathfrak{h}\oplus\mathfrak{n}$ is a Borel subalgebra of strict type BB.

If $\mathfrak{b}$ and $\mathfrak{b}'$ are two Borel subalgebras of type BB, then $\mathfrak{b}$ and $\mathfrak{b}'$ satisfy the hypotheses of Lemma 11.118. By the lemma, the standard modules arising from $\mathfrak{b}$ are equivalent as $(\mathfrak{g}, K)$ modules with the standard modules arising from $\mathfrak{b}'$.

PROOF OF LEMMA 11.118. Let us fix $\mathfrak{n}$ and consider the various possibilities for $\mathfrak{n}'$. We shall prove the lemma by induction downward on $n = \dim(\mathfrak{n}\cap\mathfrak{n}')$. The base case of the induction is that $n = \frac{1}{2}|\Delta(\mathfrak{g}, \mathfrak{h})|$. In this case we have $\mathfrak{n} = \mathfrak{n}'$, and the result is trivial. Suppose that $n \leq \frac{1}{2}|\Delta(\mathfrak{g}, \mathfrak{h})| - 1$, that the theorem has been proved for $n + 1$, and that $\mathfrak{n}'$ has $\dim(\mathfrak{n}\cap\mathfrak{n}') = n$. Among all the roots that are simple for $\mathfrak{n}'$, at least one must be negative for $\mathfrak{n}$ since $\Delta(\mathfrak{n})\neq\Delta(\mathfrak{n}')$. Let β be such a root. Then β is complex, since (i) and (ii) imply that $\Delta(\mathfrak{n})$ and $\Delta(\mathfrak{n}')$ contain the same real roots and the same imaginary roots. Moreover, (iii) implies that

$$(11.119) \qquad \frac{2\langle\lambda,\beta\rangle}{|\beta|^2} \quad \text{is 0 or is not an integer.}$$

Let s_β be the Weyl-group reflection in the root β, and define $\mathfrak{b}'' = \mathfrak{h}\oplus\mathfrak{n}''$ so that $\Delta(\mathfrak{n}'') = s_\beta\Delta(\mathfrak{n}')$. By Proposition 4.67,

$$(11.120) \qquad \Delta(\mathfrak{n})\cap\Delta(\mathfrak{n}'') = (\Delta(\mathfrak{n})\cap\Delta(\mathfrak{n}'))\cup\{-\beta\}$$

disjointly. Thus $\dim(\mathfrak{n}\cap\mathfrak{n}'') = n + 1$. Since β is complex and satisfies (11.119), it follows from (11.120) that $\mathfrak{b}''$ and $\mathfrak{b}$ satisfy hypotheses (i), (ii), and (iii) of the lemma. Consequently we can apply the inductive hypothesis to $\mathfrak{b}''$ and $\mathfrak{b}$.

Let $p'' = \dim(\mathfrak{n}''\cap\mathfrak{k})$. By inductive hypothesis

$$({}^u\mathcal{L}^{\mathfrak{g},K}_{\mathfrak{b}^-,T})_{p+q}(Z(\mathfrak{b})) \cong ({}^u\mathcal{L}^{\mathfrak{g},K}_{\mathfrak{b}''^-,T})_{p''+q}(Z(\mathfrak{b}''))$$

and
$$({}^u\mathcal{R}^{\mathfrak{g},K}_{\mathfrak{b},T})^{p+q}(Z(\mathfrak{b})) \cong ({}^u\mathcal{R}^{\mathfrak{g},K}_{\mathfrak{b}'',T})^{p''+q}(Z(\mathfrak{b}'')).$$

Thus the desired conclusion will follow if we prove that

(11.121a) $$({}^{u}\mathcal{L}_{\mathfrak{b}''-,T}^{\mathfrak{g},K})_{p''+q}(Z(\mathfrak{b}'')) \cong ({}^{u}\mathcal{L}_{\mathfrak{b}'-,T}^{\mathfrak{g},K})_{p'+q}(Z(\mathfrak{b}'))$$

and

(11.121b) $$({}^{u}\mathcal{R}_{\mathfrak{b}'',T}^{\mathfrak{g},K})^{p''+q}(Z(\mathfrak{b}'')) \cong ({}^{u}\mathcal{R}_{\mathfrak{b}',T}^{\mathfrak{g},K})^{p'+q}(Z(\mathfrak{b}')).$$

The proof now divides into two cases. For the first case, we suppose that $\theta\beta$ is in $\Delta(\mathfrak{n}')$. To handle ${}^{u}\mathcal{L}$, put $\mathfrak{n}_1 = \mathfrak{n}'^{-}$, $\mathfrak{n}_2 = \mathfrak{n}''^{-}$, and $\alpha = -\beta$. Since

$$\Delta(\mathfrak{n}_1) \cap \Delta(\mathfrak{n}_2^{-}) = \Delta(\mathfrak{n}'^{-}) \cap \Delta(\mathfrak{n}'') = \{-\beta\} = \{\alpha\}$$

and $$\{\theta\alpha\} = \{-\theta\beta\} \subseteq \Delta(\mathfrak{n}'^{-}) = \Delta(\mathfrak{n}_1),$$

hypotheses (i), (ii), and (iii) of Theorem 11.87 are satisfied. We apply (a) of the theorem to the $(\mathfrak{h}, T)$ module $Z(\mathfrak{b}')$, in which $\mathfrak{h}$ acts by scalars according to $\lambda + \delta(\mathfrak{n}')$. We have

(11.122)
$$\frac{2\langle\lambda + \delta(\mathfrak{n}'), \alpha\rangle}{|\alpha|^2} = -\frac{2\langle\lambda, \beta\rangle}{|\beta|^2} - \frac{2\langle\delta(\mathfrak{n}'), \beta\rangle}{|\beta|^2} = -\frac{2\langle\lambda, \beta\rangle}{|\beta|^2} - 1.$$

By (11.119) the right side is -1 or it is not an integer. Thus Theorem 11.87a is applicable. Since (11.114) gives

$$Z(\mathfrak{b}') \otimes_{\mathbb{C}} \mathbb{C}_{\alpha} = Z(\mathfrak{b}') \otimes_{\mathbb{C}} \mathbb{C}_{-\beta} = Z(\mathfrak{b}') \otimes_{\mathbb{C}} \mathbb{C}_{(\delta(\mathfrak{n}'') - \delta(\mathfrak{n}'))'} = Z(\mathfrak{b}''),$$

the theorem gives

$$({}^{u}\mathcal{L}_{\mathfrak{b}''-,T}^{\mathfrak{g},K})_j(Z(\mathfrak{b}'')) \cong ({}^{u}\mathcal{L}_{\mathfrak{b}'-,T}^{\mathfrak{g},K})_{j+1}(Z(\mathfrak{b}'))$$

for all j. To deduce (11.121a) from this isomorphism, we have only to show that

(11.123) $$p' = p'' + 1.$$

Applying θ to a root-space decomposition of $\mathfrak{g}$ relative to $\mathfrak{h}$, we see that

(11.124)
$$\dim(\mathfrak{n}' \cap \mathfrak{k}) = \#\{\text{compact imaginary roots in } \Delta(\mathfrak{n}')\}$$
$$+ \#\{\text{pairs } \{\gamma, \theta\gamma\} \text{ of complex roots in } \Delta(\mathfrak{n}')\}$$

and similarly for $\dim(\mathfrak{n}'' \cap \mathfrak{k})$. The only difference in these quantities for $\mathfrak{n}'$ and $\mathfrak{n}''$ is that $\{\beta, \theta\beta\}$ is present for $\mathfrak{n}'$ but not for $\mathfrak{n}''$. Then (11.123) is the consequence, and (11.121a) follows.

Still in the first case, let us now handle $^u\mathcal{R}$. Put $\mathfrak{n}_1 = \mathfrak{n}'$, $\mathfrak{n}_2 = \mathfrak{n}''$, and $\alpha = \beta$. Then

$$\Delta(\mathfrak{n}_1) \cap \Delta(\mathfrak{n}_2^-) = \{\alpha\} \qquad \text{and} \qquad \{\theta\alpha\} \subseteq \Delta(\mathfrak{n}_1),$$

and hypotheses (i), (ii), and (iii) of Theorem 11.87 are satisfied. We apply (b) of the theorem to the $(\mathfrak{h}, T)$ module $Z(\mathfrak{b}')$, in which $\mathfrak{h}$ acts by scalars according to $\lambda + \delta(\mathfrak{n}')$. We have

$$(11.125) \qquad \frac{2\langle \lambda + \delta(\mathfrak{n}'), \alpha \rangle}{|\alpha|^2} = \frac{2\langle \lambda, \beta \rangle}{|\beta|^2} + \frac{2\langle \delta(\mathfrak{n}'), \beta \rangle}{|\beta|^2} = \frac{2\langle \lambda, \beta \rangle}{|\beta|^2} + 1.$$

By (11.119) the right side is $+1$ or it is not an integer. Thus Theorem 11.87b is applicable. Since (11.114) gives

$$Z(\mathfrak{b}') \otimes_{\mathbb{C}} \mathbb{C}_{-\alpha} = Z(\mathfrak{b}') \otimes_{\mathbb{C}} \mathbb{C}_{-\beta} = Z(\mathfrak{b}') \otimes_{\mathbb{C}} \mathbb{C}_{(\delta(\mathfrak{n}'') - \delta(\mathfrak{n}'))'} = Z(\mathfrak{b}''),$$

the theorem gives

$$(^u\mathcal{R}_{\mathfrak{b}'',T}^{\mathfrak{g},K})^j(Z(\mathfrak{b}'')) \cong (^u\mathcal{R}_{\mathfrak{b}',T}^{\mathfrak{g},K})^{j+1}(Z(\mathfrak{b}'))$$

for all j. Taking (11.123) into account, we obtain (11.121b).

The second case for the proof is that $\bar{\beta} = -\theta\beta$ is in $\Delta(\mathfrak{n}')$. Then $\bar{\beta}$ is in $\Delta(\mathfrak{n}'')$ too, since $\bar{\beta} \neq \beta$. Thus we can interchange the roles of $\mathfrak{n}'$ and $\mathfrak{n}''$ and replace β by $-\beta$ to reduce matters to the first case, just as in Remark 4 with Theorem 11.87. This completes the proof.

REMARK. Although inset hypotheses (i), (ii), and (iii) of the lemma were used in the proof, conditions (i) and (ii) in (11.110) played no role. Condition (11.110) is mentioned only because of its appearance in the definition of "standard module."

Next we introduce Borel subalgebras of types L and L'. If $\mathfrak{b} = \mathfrak{h} \oplus \mathfrak{n}$ is a Borel subalgebra such that conditions (i) and (ii) hold in (11.110) and

(i) $\Delta_{\text{imag}}^+ \subseteq \Delta(\mathfrak{n})$
(ii) $\Delta_{\text{real}}^+ \subseteq \Delta(\mathfrak{n})$
(iii) a complex root α is in $\Delta(\mathfrak{n})$ if and only if $\bar{\alpha}$ is in $\Delta(\mathfrak{n})$
(iv) each complex root α in $\Delta(\mathfrak{n})$ with $2\langle \lambda, \alpha \rangle / |\alpha|^2$ in $\{-1, -2, \dots\}$ has $\text{Re}\langle \lambda, \bar{\alpha} \rangle \geq 0$,

we say that $\mathfrak{b}$ is of **type L**. If (i), (ii), (iii), and

(iv') each complex root α in $\Delta(\mathfrak{n})$ with $2\langle\lambda,\alpha\rangle/|\alpha|^2$ in $\{1, 2, \ldots\}$ has $\mathrm{Re}\langle\lambda, \bar{\alpha}\rangle \leq 0$

hold, and if conditions (i) and (ii') hold in (11.110), we say that $\mathfrak{b}$ is of **type L'**. In the definition of "type L," we say that $\mathfrak{b}$ is of **strict type L** if (iv) is changed to

(iv-s) each complex root α in $\Delta(\mathfrak{n})$ has $\mathrm{Re}\langle\lambda, \alpha + \bar{\alpha}\rangle \geq 0$.

Similarly in the definition of "type L'," we say that $\mathfrak{b}$ is of **strict type L'** if (iv') is changed to

(iv'-s) each complex root α in $\Delta(\mathfrak{n})$ has $\mathrm{Re}\langle\lambda, \alpha + \bar{\alpha}\rangle \leq 0$.

Notice that condition (iv-s) implies condition (iv) and that condition (iv'-s) implies condition (iv').

Proposition 11.126. If $\{(\mathfrak{h}, T), \lambda, \Delta_{\mathrm{imag}}^+, \Delta_{\mathrm{real}}^+, \{Z(\mathfrak{b})\}\}$ is a set of data for standard $(\mathfrak{g}, K)$ modules satisfying the consistency conditions (i) and (ii) in (11.110), then there exists a Borel subalgebra $\mathfrak{b}$ of strict type L. If the set of data satisfies the consistency conditions (i) and (ii') in (11.110), then there exists a Borel subalgebra $\mathfrak{b}$ of strict type L'.

REMARKS. It will follow from Theorem 11.129 that any two $\mathfrak{b}$'s of type L lead to the same set of standard modules, up to equivalence. A similar statement applies to type L'.

PROOF. To define $\mathfrak{b}$ of strict type L, let δ_{imag} and δ_{real} be as in the proof of Proposition 11.117, and let $\lambda = \lambda_\mathfrak{t} + \lambda_\mathfrak{a}$ be the decomposition of λ according to $\mathfrak{h} = \mathfrak{t} \oplus \mathfrak{a}$. We order $(i\mathfrak{t}_0 \oplus \mathfrak{a}_0)^*$ lexicographically by the list $\mathrm{Re}\,\lambda_\mathfrak{a}$, δ_{real}, basis of $\mathfrak{a}_0^*$, $\mathrm{Re}\,\lambda_\mathfrak{t}$, δ_{imag}, basis of $i\mathfrak{t}_0^*$. This lexicographic ordering determines a positive system in $\Delta(\mathfrak{g}, \mathfrak{h})$ and hence defines $\mathfrak{n}$ and $\mathfrak{b}$.

If α is a complex root in $\Delta(\mathfrak{n})$, then

$$0 \leq \langle \mathrm{Re}\,\lambda_\mathfrak{a}, \alpha\rangle = \tfrac{1}{2}\mathrm{Re}\langle\lambda, \alpha + \bar{\alpha}\rangle$$

since $\alpha + \bar{\alpha}$ is carried on $\mathfrak{a}$. This proves (iv-s) in the definition of "strict type L." If α is in Δ_{real}^+, then α is carried on $\mathfrak{a}$ and (11.110) shows that $\langle \mathrm{Re}\,\lambda, \alpha\rangle \geq 0$. If $\langle \mathrm{Re}\,\lambda, \alpha\rangle > 0$, then α is in $\Delta(\mathfrak{n})$ since $\langle \mathrm{Re}\,\lambda, \alpha\rangle = \langle \mathrm{Re}\,\lambda_\mathfrak{a}, \alpha\rangle$ and since $\mathrm{Re}\,\lambda_\mathfrak{a}$ is the first member of the list defining the ordering. If $\langle \mathrm{Re}\,\lambda, \alpha\rangle = 0$, then $\langle\delta_{\mathrm{real}}, \alpha\rangle > 0$ forces α to be in $\Delta(\mathfrak{n})$. Hence $\Delta_{\mathrm{real}}^+ \subseteq \Delta(\mathfrak{n})$, and (i) holds.

A similar argument, based on the fact that $\mathrm{Re}\,\lambda_\mathfrak{t}$ and δ_{imag} are the first two members of the list that can be nonzero on $\mathfrak{t}$, shows that $\Delta_{\mathrm{imag}}^+ \subseteq \Delta(\mathfrak{n})$. This proves (ii). For (iii), let α be a complex root. Since $\bar{\alpha}|_\mathfrak{a} = \alpha|_\mathfrak{a} \neq 0$

and since the first members of the list defining the ordering are a spanning set of $\mathfrak{a}_0^*$, it follows that both or neither of α and $\bar\alpha$ is in $\Delta(\mathfrak{n})$. This proves (iii). Thus $\mathfrak{b}$ is of strict type L.

To define $\mathfrak{b}$ of strict type L', we replace the list that defines the ordering by $-\operatorname{Re}\lambda_\mathfrak{a}$, $-\delta_{\mathrm{real}}$, basis of $\mathfrak{a}_0^*$, $\operatorname{Re}\lambda_\mathfrak{t}$, δ_{imag}, basis of $i\mathfrak{t}_0^*$. We readily check that $\mathfrak{b}$ is of strict type L'.

Finally we introduce Borel subalgebras of type VZ. If $\mathfrak{b} = \mathfrak{h} \oplus \mathfrak{n}$ is a Borel subalgebra such that (i) and (ii') hold in (11.110) and

  (i) $\Delta_{\mathrm{imag}}^+ \subseteq \Delta(\mathfrak{n})$
 (ii) $\Delta_{\mathrm{real}}^+ \subseteq \Delta(\mathfrak{n})$
(iii) a complex root α is in $\Delta(\mathfrak{n})$ if and only if $\theta\alpha$ is in $\Delta(\mathfrak{n})$
(iv) each complex root α in $\Delta(\mathfrak{n})$ with $2\langle\lambda, \alpha\rangle/|\alpha|^2$ in $\{-1, -2, \dots\}$ has $\operatorname{Re}\langle\lambda, \theta\alpha\rangle \geq 0$,

we say that $\mathfrak{b}$ is of **type VZ**. If (iv) is replaced by

(iv-s) each complex root α in $\Delta(\mathfrak{n})$ has $\operatorname{Re}\langle\lambda, \alpha + \theta\alpha\rangle \geq 0$,

we say that $\mathfrak{b}$ is of **strict type VZ**. Notice that condition (iv-s) implies condition (iv).

Proposition 11.127. If $\{(\mathfrak{h}, T), \lambda, \Delta_{\mathrm{imag}}^+, \Delta_{\mathrm{real}}^+, \{Z(\mathfrak{b})\}\}$ is a set of data for standard $(\mathfrak{g}, K)$ modules satisfying the consistency conditions (i) and (ii') in (11.110), then there exists a Borel subalgebra $\mathfrak{b}$ of strict type VZ.

REMARK. As in the case of types L and L', it will follow from Theorem 11.129 that any two $\mathfrak{b}$'s of type VZ lead to the same set of standard modules, up to equivalence.

PROOF. Proceeding as for Proposition 11.126, we order $(i\mathfrak{t}_0 \oplus \mathfrak{a}_0)^*$ lexicographically by using the list $\operatorname{Re}\lambda_\mathfrak{t}$, δ_{imag}, basis of $i\mathfrak{t}_0^*$, $-\operatorname{Re}\lambda_\mathfrak{a}$, $-\delta_{\mathrm{real}}$, basis of $\mathfrak{a}_0^*$. This lexicographic ordering determines a positive system in $\Delta(\mathfrak{g}, \mathfrak{h})$ and hence defines $\mathfrak{n}$ and $\mathfrak{b}$. The same kind of argument as in the proof of Proposition 11.126 shows that $\mathfrak{b}$ is of strict type VZ.

Lemma 11.128. Let $\{(\mathfrak{h}, T), \lambda, \Delta_{\mathrm{imag}}^+, \Delta_{\mathrm{real}}^+, \{Z(\mathfrak{b})\}\}$ be a set of data for standard $(\mathfrak{g}, K)$ modules satisfying (i) and either (ii) or (ii') in (11.110), and let $\mathfrak{b} = \mathfrak{h} \oplus \mathfrak{n}$ and $\mathfrak{b}' = \mathfrak{h} \oplus \mathfrak{n}'$ be two Borel subalgebras satisfying the conditions

  (i) $\Delta_{\mathrm{imag}}^+ \subseteq \Delta(\mathfrak{n}) \cap \Delta(\mathfrak{n}')$
 (ii) $\Delta_{\mathrm{real}}^+ \subseteq \Delta(\mathfrak{n}) \cap \Delta(\mathfrak{n}')$
(iii) whenever α is a complex root with $\alpha \in \Delta(\mathfrak{n}')$ but not $\Delta(\mathfrak{n})$ and is such that $2\langle\lambda, \alpha\rangle/|\alpha|^2$ is a nonzero integer, then the integer is positive, and $\theta\alpha$ is in $\Delta(\mathfrak{n}) \cap \Delta(\mathfrak{n}')$.

If $p = \dim(\mathfrak{n} \cap \mathfrak{k})$ and $p' = \dim(\mathfrak{n}' \cap \mathfrak{k})$, then

$$({}^{\mathrm{u}}\mathcal{L}_{\mathfrak{b}^-,T}^{\mathfrak{g},K})_{p+q}(Z(\mathfrak{b})) \cong ({}^{\mathrm{u}}\mathcal{L}_{\mathfrak{b}'^-,T}^{\mathfrak{g},K})_{p'+q}(Z(\mathfrak{b}'))$$

and

$$({}^{\mathrm{u}}\mathcal{R}_{\mathfrak{b},T}^{\mathfrak{g},K})^{p+q}(Z(\mathfrak{b})) \cong ({}^{\mathrm{u}}\mathcal{R}_{\mathfrak{b}',T}^{\mathfrak{g},K})^{p'+q}(Z(\mathfrak{b}'))$$

for all q.

REMARK. As with Lemma 11.118 condition (11.110) plays a role in the definition of "standard module" but no role in the proof.

PROOF. We argue as in Lemma 11.118, fixing $\mathfrak{n}$ and letting $\mathfrak{n}'$ vary subject to hypotheses (i), (ii), and (iii). The proof goes by induction downward on $n = \dim(\mathfrak{n} \cap \mathfrak{n}')$, the base case $n = \frac{1}{2}|\Delta(\mathfrak{g}, \mathfrak{h})|$ being trivial. Suppose that $n \leq \frac{1}{2}|\Delta(\mathfrak{g}, \mathfrak{h})| - 1$, that the theorem has been proved for $n+1$, and that $\mathfrak{n}'$ has $\dim(\mathfrak{n} \cap \mathfrak{n}') = n$. We choose β and define $\mathfrak{b}'' = \mathfrak{h} \oplus \mathfrak{n}''$ as in the proof of Lemma 11.118.

As in Lemma 11.118 we want to apply the inductive hypothesis to $\mathfrak{b}''$ and $\mathfrak{b}$. Because of (i) and (ii) for $\mathfrak{b}'$ and $\mathfrak{b}$, β must be complex. Therefore (11.120) shows that (i) and (ii) hold for $\mathfrak{b}''$ and $\mathfrak{b}$. To see that (iii) holds for $\mathfrak{b}''$ and $\mathfrak{b}$, let α be a complex root in $\Delta(\mathfrak{n}'')$ but not $\Delta(\mathfrak{n})$ such that $2\langle \lambda, \alpha \rangle / |\alpha|^2$ is a nonzero integer. By Proposition 4.67,

$$\Delta(\mathfrak{n}') \cap (-\Delta(\mathfrak{n})) = (\Delta(\mathfrak{n}'') \cap (-\Delta(\mathfrak{n}))) \cup \{\beta\}.$$

Therefore α is in $\Delta(\mathfrak{n}') \cap (-\Delta(\mathfrak{n}))$, and (iii) for $\mathfrak{b}'$ and $\mathfrak{b}$ shows that $2\langle \lambda, \alpha \rangle / |\alpha|^2$ is positive and that $\theta\alpha$ is in $\Delta(\mathfrak{n}) \cap \Delta(\mathfrak{n}')$. By (11.120), $\theta\alpha$ is in $\Delta(\mathfrak{n}) \cap \Delta(\mathfrak{n}'')$, and thus (iii) holds for $\mathfrak{b}''$ and $\mathfrak{b}$.

As in Lemma 11.118 the proof thus reduces to a verification of (11.121). Since (iii) holds for $\mathfrak{b}'$ and $\mathfrak{b}$, the root β satisfies one of the conditions

(a) $2\langle \lambda, \beta \rangle / |\beta|^2$ is 0 or is not an integer
(b) $2\langle \lambda, \beta \rangle / |\beta|^2$ is an integer > 0, and $\theta\beta$ is in $\Delta(\mathfrak{n}) \cap \Delta(\mathfrak{n}')$.

There are two cases. The first case is that $\theta\beta$ is in $\Delta(\mathfrak{n}')$. To handle ${}^{\mathrm{u}}\mathcal{L}$ in this case, we follow the argument of Lemma 11.118 through (11.122). By (a) and (b) the right side of (11.122) is not an integer ≥ 0. Thus Theorem 11.87a is applicable, and the rest of the verification of (11.121a) proceeds as in Lemma 11.118. To handle ${}^{\mathrm{u}}\mathcal{R}$ in this case, we continue with the argument of Lemma 11.118 through (11.125). By (a) and (b) above, the right side of (11.125) is not an integer ≤ 0. Theorem 11.87b is applicable, and the rest of the verification of (11.121b) is unchanged.

The second case is that $\theta\beta$ is not in $\Delta(\mathfrak{n}')$. Then $\theta\beta$ is not in $\Delta(\mathfrak{n}'')$ either, since $\theta\beta \neq -\beta$. Because β must satisfy (a) or (b), β must satisfy (a). For ${}^{\mathrm{u}}\mathcal{L}$, put $\mathfrak{n}_1 = \mathfrak{n}''^-$, $\mathfrak{n}_2 = \mathfrak{n}'^-$, and $\alpha = \beta$. Since

$$\Delta(\mathfrak{n}_1) \cap \Delta(\mathfrak{n}_2^-) = \Delta(\mathfrak{n}''^-) \cap \Delta(\mathfrak{n}') = \{\beta\} = \{\alpha\}$$

and
$$\{\theta\alpha\} = \{\theta\beta\} \subseteq \Delta(\mathfrak{n}''^-),$$

we can argue as above, effectively interchanging $\mathfrak{n}'$ and $\mathfrak{n}''$ and replacing β by $-\beta$. Because β satisfies (a), the analog of (11.122) (with $\mathfrak{n}''$ in place of $\mathfrak{n}'$) is still not an integer ≥ 0. Thus the argument goes through for $^u\mathcal{L}$. For $^u\mathcal{R}$, put $\mathfrak{n}_1 = \mathfrak{n}''$, $\mathfrak{n}_2 = \mathfrak{n}'$, and $\alpha = -\beta$. Again we can argue as above, and the fact that β satisfies (a) means that the analog of (11.125) is not an integer ≤ 0. Thus the argument goes through for $^u\mathcal{R}$.

Theorem 11.129. Let $D = \{(\mathfrak{h}, T), \lambda, \Delta_{\mathrm{imag}}^+, \Delta_{\mathrm{real}}^+, \{Z(\mathfrak{b})\}\}$ be a set of data for standard $(\mathfrak{g}, K)$ modules satisfying the consistency conditions (i) and (ii) in (11.110). Let $D' = \{(\mathfrak{h}, T), \lambda, \Delta_{\mathrm{imag}}^+, -\Delta_{\mathrm{real}}^+, \{Z(\mathfrak{b})\}\}$ be the "dual" set of data, which satisfies the consistency conditions (i) and (ii') in (11.110).

(a) If $\mathfrak{b}_{\mathrm{BB}}$ and $\mathfrak{b}_{\mathrm{L}}$ are Borel subalgebras of types BB and L, respectively, relative to the set of data D, then the two associated standard modules of Π type are equivalent, and the two associated standard modules of Γ type are equivalent. In fact, for each integer q,

(11.130a) $(^u\mathcal{L}_{\mathfrak{b}^-, T}^{\mathfrak{g}, K})_{p+q}(Z(\mathfrak{b}))$ with $p = p(\mathfrak{b}) = \dim(\mathfrak{n} \cap \mathfrak{k})$

is the same for $\mathfrak{b} = \mathfrak{b}_{\mathrm{BB}}$ and $\mathfrak{b}_{\mathrm{L}}$, and also

(11.130b) $(^u\mathcal{R}_{\mathfrak{b}, T}^{\mathfrak{g}, K})^{p+q}(Z(\mathfrak{b}))$ with $p = p(\mathfrak{b}) = \dim(\mathfrak{n} \cap \mathfrak{k})$

is the same for $\mathfrak{b} = \mathfrak{b}_{\mathrm{BB}}$ and $\mathfrak{b}_{\mathrm{L}}$. For $q \neq 0$, these modules are 0.

(b) If $\mathfrak{b}_{\mathrm{VZ}}$ and $\mathfrak{b}_{\mathrm{L}'}$ are Borel subalgebras of types VZ and L', respectively, relative to the set of data D', then the two associated standard modules of Π type are equivalent, and the two associated standard modules of Γ type are equivalent. In fact, for each integer q, (11.130a) is the same for $\mathfrak{b} = \mathfrak{b}_{\mathrm{VZ}}$ and $\mathfrak{b}_{\mathrm{L}'}$, and also (11.130b) is the same for $\mathfrak{b} = \mathfrak{b}_{\mathrm{VZ}}$ and $\mathfrak{b}_{\mathrm{L}'}$. For $q \neq 0$, these modules are 0.

(c) If $\mathfrak{b} = \mathfrak{h} \oplus \mathfrak{n}$ is a Borel subalgebra of type L relative to the set of data D and if

$$\Delta(\mathfrak{n}') = \{\alpha \in \Delta(\mathfrak{n}) \mid \alpha \text{ imaginary}\} \cup \{\alpha \in -\Delta(\mathfrak{n}) \mid \alpha \text{ complex or real}\},$$

then $\mathfrak{b}' = \mathfrak{h} \oplus \mathfrak{n}'$ is a Borel subalgebra of type L' relative to the set of data D' and

$$(^u\mathcal{R}_{\mathfrak{b}, T}^{\mathfrak{g}, K})^j(Z(\mathfrak{b})) \cong (^u\mathcal{L}_{\mathfrak{b}'^-, T}^{\mathfrak{g}, K})_j(Z(\mathfrak{b}'))$$

and
$$(^u\mathcal{L}_{\mathfrak{b}^-, T}^{\mathfrak{g}, K})_j(Z(\mathfrak{b})) \cong (^u\mathcal{R}_{\mathfrak{b}', T}^{\mathfrak{g}, K})^j(Z(\mathfrak{b}'))$$

for all j. For $j \neq \#\{\alpha \in \Delta(\mathfrak{n}) \mid \alpha \text{ compact imaginary}\}$, these modules are 0.

REMARK. In (c), the equality of the degrees j on the two sides of each isomorphism is consistent with the identity

(11.131)
$$\dim(\mathfrak{n} \cap \mathfrak{k}) = \dim(\mathfrak{n}' \cap \mathfrak{k}) = \#\{\alpha \in \Delta(\mathfrak{n}) \mid \alpha \text{ compact imaginary}\},$$

which follows from (11.124).

PROOF. In Lemma 11.128 let us say that the two Borel subalgebras stand in the relation $\mathfrak{b}' \to \mathfrak{b}$. For (a) in the present theorem, we shall argue as follows. If we are given $\mathfrak{b}_L$, we shall construct a strict $\mathfrak{b}_{BB}$ such that $\mathfrak{b}_{BB} \to \mathfrak{b}_L$. Then Lemma 11.128 says that $\mathfrak{b}_{BB}$ and $\mathfrak{b}_L$ lead to the same $(\mathfrak{g}, K)$ modules. Lemma 11.118, however, says that any two $\mathfrak{b}_{BB}$'s lead to the same $(\mathfrak{g}, K)$ modules. Hence all $\mathfrak{b}_{BB}$'s and all $\mathfrak{b}_L$'s lead to the same $(\mathfrak{g}, K)$ modules. This will prove (a) except for the assertion about vanishing. Conclusion (c) will show that any $\mathfrak{b}_L$ has a corresponding $\mathfrak{b}_{L'}$ leading to the same $(\mathfrak{g}, K)$ modules, but with the roles of $^u\mathcal{L}$ and $^u\mathcal{R}$ reversed. Evidently each $\mathfrak{b}_{L'}$ comes from some $\mathfrak{b}_L$, as well. Since, by (a), all $\mathfrak{b}_L$'s lead to the same $(\mathfrak{g}, K)$ modules, all $\mathfrak{b}_{L'}$'s lead to the same $(\mathfrak{g}, K)$ modules. Finally in (b) we shall show that any $\mathfrak{b}_{VZ}$ and any strict $\mathfrak{b}_{L'}$ satisfy $\mathfrak{b}_{VZ} \to \mathfrak{b}_{L'}$, and it follows that all $\mathfrak{b}_{VZ}$'s and all strict $\mathfrak{b}_{L'}$'s lead to the same $(\mathfrak{g}, K)$ modules. Since all $\mathfrak{b}_{L'}$'s lead to the same $(\mathfrak{g}, K)$ modules and since strict $\mathfrak{b}_{L'}$'s exist (Proposition 11.126), all $\mathfrak{b}_{VZ}$'s and all $\mathfrak{b}_{L'}$'s lead to the same $(\mathfrak{g}, K)$ modules. The assertions about vanishing will be proved after (c).

For the isomorphisms in (a), let $\mathfrak{b}_L = \mathfrak{h} \oplus \mathfrak{n}_L$ be a Borel subalgebra of type L relative to the set of data D. Define

(11.132) $\quad \Delta(\mathfrak{n}_{BB}) = \{\alpha \mid \operatorname{Re}\langle\lambda, \alpha\rangle > 0\} \cup \{\alpha \in \Delta(\mathfrak{n}_L) \mid \operatorname{Re}\langle\lambda, \alpha\rangle = 0\}.$

Since $\{\alpha \mid \operatorname{Re}\langle\lambda, \alpha\rangle = 0\}$ is a root system, the positive system $\{\alpha \in \Delta(\mathfrak{n}_L) \mid \operatorname{Re}\langle\lambda, \alpha\rangle = 0\}$ is defined by a basis $v_1, \ldots, v_k$ for the orthogonal complement of $\operatorname{Re}\lambda$, and then $\Delta(\mathfrak{n}_{BB})$ is defined by the basis $\operatorname{Re}\lambda, v_1, \ldots, v_k$. Hence $\Delta(\mathfrak{n}_{BB})$ is a positive system, and $\mathfrak{b}_{BB} = \mathfrak{h} \oplus \mathfrak{n}_{BB}$ is a Borel subalgebra. We see immediately that $\mathfrak{b}_{BB}$ satisfies the conditions that make it of strict type BB.

Toward proving that $\mathfrak{b}_{BB} \to \mathfrak{b}_L$, we see from (11.110) that (i) and (ii) hold in Lemma 11.128. Let $\alpha \in \Delta(\mathfrak{n}_{BB})$ be such that $\alpha \notin \Delta(\mathfrak{n}_L)$, and suppose that $2\langle\lambda, \alpha\rangle/|\alpha|^2$ is a nonzero integer. By (i) and (ii), α is necessarily complex. By (11.132), $2\langle\lambda, \alpha\rangle/|\alpha|^2$ is a positive integer. Since $-\alpha$ is in $\Delta(\mathfrak{n}_L)$, condition (iii) in the definition of "type L" forces $\theta\alpha$ to be in $\Delta(\mathfrak{n}_L)$, and condition (iv), applied to $-\alpha$, leads to

$\mathrm{Re}\langle\lambda, \theta\alpha\rangle \geq 0$. By (11.132) it follows that $\theta\alpha$ is in $\Delta(\mathfrak{n}_{BB})$, and the argument that $\mathfrak{b}_{BB} \to \mathfrak{b}_L$ is complete. By the remarks in the first paragraph of the proof, this proves the isomorphisms in (a).

For the isomorphisms in (b), suppose $\mathfrak{b}_{VZ}$ is of type VZ and $\mathfrak{b}_{L'}$ is of strict type L', relative to the set of data D'. Toward proving that $\mathfrak{b}_{VZ} \to \mathfrak{b}_{L'}$, we observe that (i) and (ii) in Lemma 11.128 are immediate. Let $\alpha \in \Delta(\mathfrak{n}_{VZ})$ be such that $\alpha \notin \Delta(\mathfrak{n}_{L'})$, and suppose that $2\langle\lambda, \alpha\rangle/|\alpha|^2$ is a nonzero integer. By (i) and (ii), α is necessarily complex. By (iii) in the definitions of "type VZ" and "type L'," $\theta\alpha$ is in $\Delta(\mathfrak{n}_{VZ}) \cap \Delta(\mathfrak{n}_{L'})$. To prove (iii) in Lemma 11.128, we need show only that the integer $2\langle\lambda, \alpha\rangle/|\alpha|^2$ is positive. Arguing by contradiction, suppose that it is negative. Applying (iv) in the definition of "type VZ," we see that $\mathrm{Re}\langle\lambda, \theta\alpha\rangle \geq 0$. On the other hand, (iv') in the definition of "type L'," applied to $-\alpha$, shows that $\mathrm{Re}\langle\lambda, \theta\alpha\rangle \leq 0$. Thus $\mathrm{Re}\langle\lambda, \theta\alpha\rangle = 0$, and we conclude that $\mathrm{Re}\langle\lambda, \theta\alpha - \alpha\rangle > 0$. This conclusion contradicts (iv'-s) for the root $-\alpha$ in the definition of strict type L', and the argument that $\mathfrak{b}_{VZ} \to \mathfrak{b}_{L'}$ is complete. By the remarks in the first paragraph of the proof, this proves the isomorphisms in (b).

For (c), we prove only the first isomorphism, the second one being similar. Starting from $\mathfrak{b}$, we define an intermediate parabolic subalgebra by using the simple roots that are imaginary to determine $\mathfrak{l}$. By (i) in the definition of "type L," the imaginary roots are generated by the imaginary roots that are simple. Therefore the Levi factor is given by

$$L = Z_G(\mathfrak{a}_0) \qquad \text{and} \qquad \mathfrak{l} = \mathfrak{h} \oplus \bigoplus_{\alpha \text{ imaginary}} \mathfrak{g}_\alpha,$$

and the parabolic subalgebra $\mathfrak{q} = \mathfrak{l} \oplus \mathfrak{u}$ has

$$\mathfrak{u} = \bigoplus_{\substack{\alpha \text{ real or} \\ \text{complex}}} \mathfrak{g}_\alpha.$$

Then $\mathfrak{b} \cap \mathfrak{l}$ is a θ stable parabolic subalgebra of $\mathfrak{l}$, and $\mathfrak{q}$ is a real parabolic subalgebra of $\mathfrak{g}$.

Under the definition of $\Delta(\mathfrak{n}')$, it is routine to check that $\mathfrak{b}'$ is a Borel subalgebra of type L' relative to the set of data D'. The corresponding intermediate parabolic subalgebra is $\mathfrak{q}^-$, and the Levi factor is the same $\mathfrak{l}$. Since $\mathfrak{q}$ and $\mathfrak{q}'$ are real, the discussion of §§1–2 is applicable. The Levi subgroup is MA, and the half sum ρ of the positive $\mathfrak{a}$-roots relative to $\mathfrak{q}$ is meaningful. According to (11.17), we have

$$2\rho = 2\delta(\mathfrak{n}) - 2\delta(\mathfrak{l} \cap \mathfrak{n}) = \delta(\mathfrak{n}) - \delta(\mathfrak{n}').$$

The action of $TA \subseteq MA$ by $e^{2\rho \log a}$ is trivial on T, and $\mathbb{C}_{2\rho}$ therefore does not necessarily match $\mathbb{C}_{\delta(\mathfrak{n})-\delta(\mathfrak{n}')}$. The correct formula is

$$(11.133) \qquad \mathbb{C}_{2\rho} = \mathbb{C}_{(\delta(\mathfrak{n})-\delta(\mathfrak{n}'))'},$$

with notation as in (11.111).

In fact, if α is a complex root and t is in T, we shall observe that

$$(11.134) \qquad \xi_{\bar{\alpha}}(t) = \overline{\xi_\alpha(t)}$$

and hence that

$$(11.135) \qquad \xi_\alpha(t)\xi_{\bar{\alpha}}(t) = \xi_\alpha(t)\overline{\xi_\alpha(t)} = 1.$$

To see (11.134), we define $(\overline{\mathrm{ad}\, Y})(X) = \overline{[Y, \bar{X}]} = [\bar{Y}, X]$ for Y and X in $\mathfrak{g}$; here the bar refers to conjugation of $\mathfrak{g}$ with respect to $\mathfrak{g}_0$. The Lie algebra $\overline{\mathrm{ad}\, \mathfrak{g}}$ coincides with ad $\mathfrak{g}$. If we define $\overline{\mathrm{Ad}(g)}(X) = \overline{\mathrm{Ad}(g)\bar{X}}$, it follows from the connectedness of $G_\mathbb{C}$ that $\overline{\mathrm{Ad}(g)}$ is in $G_\mathbb{C}$ whenever $\mathrm{Ad}(g)$ is in $G_\mathbb{C}$. In the case of a member g of G, $\mathrm{Ad}(g)$ and $\overline{\mathrm{Ad}(g)}$ restrict to the same automorphism of $\mathfrak{g}_0$, and we conclude that

$$(11.136) \qquad \mathrm{Ad}(g) = \overline{\mathrm{Ad}(g)} \qquad \text{for } g \in G.$$

Let η_α and $\eta_{\bar{\alpha}}$ be the homomorphisms of $H_\mathbb{C}$ into $\mathbb{C}^\times$ corresponding to the roots α and $\bar{\alpha}$, so that $\xi_\alpha(t) = \eta_\alpha(\mathrm{Ad}(t))$ and similarly for $\bar{\alpha}$ whenever t is in T. For any element $\mathrm{Ad}(t) \in H_\mathbb{C}$, we have

$$\eta_{\bar{\alpha}}(\mathrm{Ad}(t)) = \overline{\eta_\alpha(\overline{\mathrm{Ad}(t)})}.$$

Substituting from (11.136), we obtain (11.134) and therefore also (11.135).

From (11.135) we see that the total contribution to the action of T in $\mathbb{C}_{\delta(\mathfrak{n})-\delta(\mathfrak{n}')}$ from the complex roots is 1. Since the imaginary roots do not contribute, we obtain a match with the action of T in $\mathbb{C}_{2\rho}$ by having the total contribution from the real roots be 1. Then (11.133) follows.

We shall apply induction in stages to $({}^u\mathcal{R}^{\mathfrak{g},K}_{\mathfrak{b},T})^j$ and $({}^u\mathcal{L}^{\mathfrak{g},K}_{\mathfrak{b}',T})_j$. Since $({}^u\mathcal{R}^{\mathfrak{g},K}_{\mathfrak{q},L \cap K})^i = 0$ and $({}^u\mathcal{L}^{\mathfrak{g},K}_{\mathfrak{q},L \cap K})_i = 0$ for $i > 0$ by Propositions 11.63 and 11.68, the spectral sequences in Theorem 11.77 collapse and degenerate, and we obtain

$$(11.137a) \qquad ({}^u\mathcal{R}^{\mathfrak{g},K}_{\mathfrak{b},T})^j(Z(\mathfrak{b})) \cong {}^u\mathcal{R}^{\mathfrak{g},K}_{\mathfrak{q},L \cap K}({}^u\mathcal{R}^{\mathfrak{l},L \cap K}_{\mathfrak{b} \cap \mathfrak{l},T})^j(Z(\mathfrak{b}))$$

$$(11.137b) \qquad ({}^u\mathcal{L}^{\mathfrak{g},K}_{\mathfrak{b}'-,T})_j(Z(\mathfrak{b}')) \cong {}^u\mathcal{L}^{\mathfrak{g},K}_{\mathfrak{q},L \cap K}({}^u\mathcal{L}^{\mathfrak{l},L \cap K}_{\bar{\mathfrak{b}} \cap \mathfrak{l},T})_j(Z(\mathfrak{b}')).$$

The infinitesimal character of $Z(\mathfrak{b})$ is $\lambda + \delta(\mathfrak{n})$. Since $\mathfrak{b}$ is of type L, condition (i) of (11.110) and condition (i) of the definition of "type L" say that $\langle \lambda, \alpha \rangle \geq 0$ for all roots of $\mathfrak{b} \cap \mathfrak{k}$. Since $\delta(\mathfrak{n}) - \delta(\mathfrak{n} \cap \mathfrak{l}) = \delta(\mathfrak{u})$ is orthogonal to the roots of $\mathfrak{l}$, we have

$$(11.138) \qquad \langle \lambda + \delta(\mathfrak{n}) - \delta(\mathfrak{n} \cap \mathfrak{l}), \alpha \rangle \geq 0 \qquad \text{for } \alpha \in \Delta(\mathfrak{n} \cap \mathfrak{l}).$$

Taking (11.73) into account, we see from Theorem 5.99 that (11.138) is sufficient to ensure that

$$({}^{u}\mathcal{R}^{\mathfrak{l},L\cap K}_{\mathfrak{b}\cap\mathfrak{l},T})^{j}(Z(\mathfrak{b})) \cong ({}^{u}\mathcal{L}^{\mathfrak{l},L\cap K}_{\mathfrak{b}\cap\mathfrak{l},T})_{j}(Z(\mathfrak{b}))$$

for all j, both sides being 0 for $j \neq S = \dim(\mathfrak{n} \cap \mathfrak{l} \cap \mathfrak{k})$. Therefore the right side of (11.137a) is

$$\cong {}^{u}\mathcal{R}^{\mathfrak{g},K}_{\mathfrak{q},L\cap K}({}^{u}\mathcal{L}^{\mathfrak{l},L\cap K}_{\mathfrak{b}\cap\mathfrak{l},T})_{j}(Z(\mathfrak{b}))$$

$$\cong {}^{u}\mathcal{L}^{\mathfrak{g},K}_{\mathfrak{q},L\cap K}(({}^{u}\mathcal{L}^{\mathfrak{l},L\cap K}_{\mathfrak{b}\cap\mathfrak{l},T})_{j}(Z(\mathfrak{b})) \otimes \mathbb{C}_{-2\rho}) \qquad \text{by Proposition 11.65}$$

$$\cong {}^{u}\mathcal{L}^{\mathfrak{g},K}_{\mathfrak{q},L\cap K}({}^{u}\mathcal{L}^{\mathfrak{l},L\cap K}_{\mathfrak{b}\cap\mathfrak{l},T})_{j}(Z(\mathfrak{b}) \otimes \mathbb{C}_{-2\rho}) \qquad \begin{array}{l}\text{by Theorem 2.103 and pas-}\\ \text{sage to derived functors}\end{array}$$

$$\cong {}^{u}\mathcal{L}^{\mathfrak{g},K}_{\mathfrak{q},L\cap K}({}^{u}\mathcal{L}^{\mathfrak{l},L\cap K}_{\mathfrak{b}\cap\mathfrak{l},T})_{j}(Z(\mathfrak{b}) \otimes \mathbb{C}_{(\delta(\mathfrak{n}')-\delta(\mathfrak{n}))'}) \qquad \text{by (11.133)}$$

$$\cong {}^{u}\mathcal{L}^{\mathfrak{g},K}_{\mathfrak{q},L\cap K}({}^{u}\mathcal{L}^{\mathfrak{l},L\cap K}_{\mathfrak{b}\cap\mathfrak{l},T})_{j}(Z(\mathfrak{b}')) \qquad \text{by (11.114),}$$

and this is the right side of (11.137b). Thus the left sides of (11.137a) and (11.137b) are isomorphic, and the isomorphisms in (c) are proved.

In the course of applying Theorem 5.99, we saw that the $(\mathfrak{g}, K)$ modules in (c) are 0 for $j \neq S = \dim(\mathfrak{n} \cap \mathfrak{l} \cap \mathfrak{k})$. This number S equals the number $p = \dim(\mathfrak{n} \cap \mathfrak{k})$ by (11.131), and hence the modules (11.130a) and (11.130b) are 0 if $q \neq 0$ with $\mathfrak{b}$ equal to $\mathfrak{b}_L$ or $\mathfrak{b}_{L'}$. Because of the isomorphisms proved in (a) and (b), it follows that the modules in (11.130a) and (11.130b) are 0 if $q \neq 0$ with $\mathfrak{b}$ equal to $\mathfrak{b}_{BB}$ or $\mathfrak{b}_{VZ}$. This proves the assertions about vanishing in (a) and (b).

Corollary 11.139. In the setting of Theorem 11.129, all the standard modules are admissible and have infinitesimal character λ.

PROOF. In view of the isomorphisms in the theorem, it is enough to prove this assertion when $\mathfrak{b} = \mathfrak{b}_L$. For this $\mathfrak{b}$, we are to prove that

$$({}^{u}\mathcal{R}^{\mathfrak{g},K}_{\mathfrak{b},T})^{j}(Z(\mathfrak{b})) \qquad \text{and} \qquad ({}^{u}\mathcal{L}^{\mathfrak{g},K}_{\mathfrak{b}^{-},T})_{j}(Z(\mathfrak{b}))$$

are admissible and have infinitesimal character λ. The result about the infinitesimal character is immediate from Theorems 5.21b and 5.24. For admissibility in the case of $({}^u\mathcal{R}_{\mathfrak{b},T}^{\mathfrak{g},K})^j(Z(\mathfrak{b}))$, we compute $({}^u\mathcal{R}_{\mathfrak{b},T}^{\mathfrak{g},K})^j(Z(\mathfrak{b}))$ in stages, using (11.137a). The first stage $({}^u\mathcal{R}_{\mathfrak{b}\cap\mathfrak{l},T}^{\mathfrak{l},L\cap K})^j(Z(\mathfrak{b}))$ is a cohomological induction, and the result is admissible for $L\cap K$ by Theorem 5.35b. The second stage is a parabolic induction, and it results in a $(\mathfrak{g},K)$ module admissible for K by the steps from (11.50) to (11.51). For admissibility in the case of $({}^u\mathcal{L}_{\mathfrak{b}^-,T}^{\mathfrak{g},K})_j(Z(\mathfrak{b}))$, the admissibility results by a similar computation, using (11.137b), Theorem 5.35a, and (11.69).

7. Normalization of $\mathcal{L}$ and $\mathcal{R}$

At the end of §2 we contrasted our normalization of the functors of cohomological induction corresponding to θ stable parabolics with our proposed normalization of the parabolic induction functors corresponding to real parabolics. In the θ stable case, we defined $\mathcal{L}$ and $\mathcal{R}$ in Chapters IV and V to involve a shift by a sum of roots; this shift was implicit in our use of $Z^\# = Z \otimes_{\mathbb{C}} \bigwedge^{\text{top}}\mathfrak{u}$ and was introduced in Chapter IV to make $\mathcal{L}$ and $\mathcal{R}$ respect highest weights. But in the real case, we saw in Propositions 11.47 and 11.65, as well as in Corollary 11.39 and other places, that a shift by the *half* sum ρ of roots would make some formulas more symmetric. Indeed, in (11.58) and (11.66), we introduced notation to implement such a normalization. Let us see the ways in which a normalization involving a half sum of roots is better and the extent to which it can be implemented uniformly.

One theme of much of Harish-Chandra's work is that the irreducible representations occurring in the Plancherel formula of a reductive group G should be parametrized, to a first approximation, by the (finite-dimensional) irreducible unitary representations of a system of nonconjugate Cartan subgroups. The specific construction to implement this parametrization starts with a θ stable Cartan subalgebra $\mathfrak{h}_0 = \mathfrak{t}_0 \oplus \mathfrak{a}_0$ of $\mathfrak{g}_0$, forms any parabolic subgroup MAN whose Levi subgroup MA equals $Z_G(\mathfrak{a}_0)$, and passes to the continuous-series representations built from a "discrete-series" representation of M and a unitary character of A. Roughly, but not exactly, the discrete-series representations of M correspond to irreducible $(\mathfrak{t}, T)$ modules whose infinitesimal characters are nonsingular. The continuous-series representations so constructed exhaust the representations needed in the Plancherel formula and therefore justify Harish-Chandra's principle.

However, the parametrization of discrete series of M is not exactly by irreducible $(\mathfrak{t}, T)$ modules. There are two difficulties: that the parameter needs to be shifted in a fashion that may not preserve integrality, and that the parametrization has a certain amount of redundance to it. We shall discuss discrete series in more detail in §8. For now we mention that in the special case when $M = G$ (i.e., $\mathfrak{a}_0 = 0$) and G is connected, the exact parametrization is by members λ of $\mathfrak{t}^*$ such that

 (i) $\lambda - \delta$ is analytically integral for δ obtained from some (any) choice of a system of positive roots
 (ii) λ is nonsingular with respect to all roots,

two such representations π_λ and $\pi_{\lambda'}$ being equivalent if and only if λ is conjugate to λ' by the Weyl group of T in K (which equals the Weyl group $W(\mathfrak{k}, \mathfrak{t})$). This result is one way of formulating part of the main theorem of Harish-Chandra [1966]; the parameter λ is called the **Harish-Chandra parameter** of the discrete-series representation π_λ, and π_λ has infinitesimal character λ.

It can happen (e.g., in $SL(2, \mathbb{R})$/center) that δ is not analytically integral, and then (i) says that λ is never analytically integral. In other words, the discrete series of a connected G is parametrized by characters of T only when δ is analytically integral. Harish-Chandra dealt with this difficulty in his 1966 paper by assuming that G is "acceptable," i.e., that δ is analytically integral, and then the theory for a general connected G can be deduced from this special case.

In our situation the best argument for a normalization by a half sum of roots is the need for the definition of "related" in §6. In fact, (11.114) shows that

$$(11.140\text{a}) \qquad Z' \otimes_\mathbb{C} \mathbb{C}_{-\delta(\mathfrak{n}')'} \cong Z \otimes_\mathbb{C} \mathbb{C}_{-\delta(\mathfrak{n})'}$$

if we can write

$$(11.140\text{b}) \qquad \mathbb{C}_{(\delta(\mathfrak{n}') - \delta(\mathfrak{n}))'} \overset{?}{\cong} \mathbb{C}_{\delta(\mathfrak{n}')'}\mathbb{C}_{-\delta(\mathfrak{n})'}$$

with each factor on the right meaningful. Once (11.140) holds, we can reinterpret (11.116) in terms of functors applied to a common module, namely the common value of the two sides of (11.140a). Then Theorem 11.129 says that when we start with a module as in (11.140a), various functors yield the same result. In other words, parameter (d) in (11.109) can be simplified from a class of related modules to a single module. The result is a simpler-sounding theory with more symmetry.

We need to be able to make sense out of $\mathbb{C}_{-\delta(\mathfrak{n})'}$ in (11.140a), for all choices of $\mathfrak{n}$, in such a way that $\mathbb{C}_{-\delta(\mathfrak{n})'} \otimes \mathbb{C}_{-\delta(\mathfrak{n})'} \cong \mathbb{C}_{-2\delta(\mathfrak{n})'}$ and that

(11.140b) holds. Of course, we cannot do so directly, since in the case that $\mathfrak{a}_0 = 0$, $\delta(\mathfrak{n})$ need not be analytically integral. What we shall do instead is introduce a "double cover" of the (possibly disconnected) Cartan subgroup, recognize integrality there, and reinterpret matters in terms of our original group.

We continue to suppose that $(\mathfrak{g}, K)$ is a reductive pair in the Harish-Chandra class with underlying group G. If $\mathfrak{q} = \mathfrak{l} \oplus \mathfrak{u}$ is a germane parabolic subalgebra, we shall work with one-dimensional $(\mathfrak{l}, L \cap K)$ modules like $\bigwedge^{\mathrm{top}} \mathfrak{u}$, extracting their square roots. However, our first and most immediate applications will be to the case of a Borel subalgebra, and we shall use notation suggestive of this case.

Thus let H be a Lie group, and let ξ_μ be a one-dimensional representation of H on the space $\mathbb{C}_\mu$ with differential $\mu \in \mathfrak{h}^*$. We can regard ξ_μ as a homomorphism $\xi_\mu : H \to \mathbb{C}^\times$.

In terms of H and ξ_μ, we shall construct a Lie group $\tilde{H}$, a homomorphism π of $\tilde{H}$ onto H with two-element kernel $\{1, \varepsilon\}$, and a homomorphism $\xi_{\frac{1}{2}\mu} : \tilde{H} \to \mathbb{C}^\times$ such that

$$(11.141) \qquad \xi_{\frac{1}{2}\mu}(\tilde{h})^2 = \xi_\mu(\pi(\tilde{h})) \qquad \text{for } \tilde{h} \in \tilde{H}.$$

The definition of $\tilde{H}$ is

$$(11.142) \qquad \tilde{H} = \{(h, z) \in H \times \mathbb{C}^\times \mid \xi_\mu(h) = z^2\}.$$

This is a closed subgroup of the Lie group $H \times \mathbb{C}^\times$ and hence is a Lie group. Let $\pi : \tilde{H} \to H$ be the projection on the first component, and let $\xi_{\frac{1}{2}\mu} : \tilde{H} \to \mathbb{C}^\times$ be the projection on the second component. These are both continuous homomorphisms and hence are smooth. The kernel of π is the subset of elements $(1, z)$ in $\tilde{H}$, and evidently z has to be ± 1. Thus the kernel of π is $\{1, \varepsilon\}$, where

$$\varepsilon = (1, -1) \qquad \text{in } \tilde{H}.$$

The map π is onto H since every member of $\mathbb{C}^\times$ has a square root. It follows that the differential of π is an isomorphism and that the Lie algebra of $\tilde{H}$ may be identified with $\mathfrak{h}_0$. We call $\tilde{H}$ the **double cover** of H relative to ξ_μ, and we say that $\xi_{\frac{1}{2}\mu}$ is the **square root** of ξ_μ. The whole construction is called the **square-root construction**.

The homomorphism $\xi_{\frac{1}{2}\mu}$, given by projection to the $\mathbb{C}^\times$ component, satisfies

$$\xi_{\frac{1}{2}\mu}(\varepsilon) = -1.$$

If $\tilde{h} = (h, z)$ is in $\tilde{H}$, then

$$\xi_{\frac{1}{2}\mu}(\tilde{h})^2 = z^2 = \xi_\mu(h) = \xi_\mu(\pi(\tilde{h})),$$

and this proves (11.141). Since ξ_μ has differential μ, it follows that $\xi_{\frac{1}{2}\mu}$ has differential $\frac{1}{2}\mu$.

EXAMPLE 1. $G = SU(2)$, $H = \left\{ \begin{pmatrix} e^{i\theta} & 0 \\ 0 & e^{-i\theta} \end{pmatrix} \right\} \cong \{e^{i\theta}\}$. For $\xi_{2\delta}(e^{i\theta}) = e^{2i\theta}$, we have

$$\tilde{H} = \{(e^{i\theta}, z) \mid e^{2i\theta} = z^2\} = \{(e^{i\theta}, \pm e^{i\theta})\} \cong H \times \{1, \varepsilon\}.$$

Then $\xi_\delta(e^{i\theta}, \pm e^{i\theta}) = \pm e^{i\theta}$ as a homomorphism of $\tilde{H}$ into $\mathbb{C}^\times$. This example is generalized by the next proposition.

Proposition 11.143. Suppose that H and ξ_μ lead to $\tilde{H}$ and $\xi_{\frac{1}{2}\mu}$. If there exists a homomorphism $\tau : H \to \mathbb{C}^\times$ such that $\tau(h)^2 = \xi_\mu(h)$ for $h \in H$, then $\tilde{H} \cong H \times \{1, \varepsilon\}$ under the map φ carrying

$$(h, 1) \in H \times \{1, \varepsilon\} \mapsto (h, \tau(h)) \in \tilde{H}$$

and $$(h, \varepsilon) \in H \times \{1, \varepsilon\} \mapsto (h, -\tau(h)) \in \tilde{H}.$$

The restriction $\pi \circ \varphi : H \times \{1\} \to H$ satisfies $\pi(\varphi(h, 1)) = h$ for $h \in H$. Conversely if $\tilde{H} \cong H \times \{1, \varepsilon\}$ under an isomorphism $\varphi : H \times \{1, \varepsilon\} \to \tilde{H}$ such that $\pi(\varphi(h, 1)) = h$ for $h \in H$ and if $\iota : H \to H \times \{1\}$ denotes the map $\iota(h) = (h, 1)$, then $\tau = \xi_{\frac{1}{2}\mu} \circ \varphi \circ \iota$ is a homomorphism of H into $\mathbb{C}^\times$ whose square matches $\xi_\mu : H \to \mathbb{C}^\times$.

PROOF. If τ exists, then the given map is a homomorphism into $\tilde{H}$ and is evidently one-one and onto. Thus $\tilde{H} \cong H \times \{1, \varepsilon\}$ as asserted. Conversely if φ exists with $\pi(\varphi(h, 1)) = h$ and if $\tau = \xi_{\frac{1}{2}\mu} \circ \varphi \circ \iota$, then (11.141) gives

$$\tau(h)^2 = \xi_{\frac{1}{2}\mu}(\varphi(h, 1))^2 = \xi_\mu(\pi(\varphi(h, 1))) = \xi_\mu(h).$$

EXAMPLE 2. Let $G = GL(2, \mathbb{R})$, $H = \left\{ \begin{pmatrix} u & 0 \\ 0 & v \end{pmatrix} \mid u, v \in \mathbb{R}^\times \right\}$, and $\xi_{2\delta} \begin{pmatrix} u & 0 \\ 0 & v \end{pmatrix} = uv^{-1}$. Then

$$\tilde{H} = \left\{ \left(\begin{pmatrix} u & 0 \\ 0 & v \end{pmatrix}, z \right) \mid z^2 = uv^{-1} \right\} \cong A \times \tilde{M}$$

with $A = \left\{ \left(\begin{pmatrix} e^t & 0 \\ 0 & e^{-t} \end{pmatrix}, e^t \right) \right\}$ and

$$\tilde{M} = \left\{ \left(\begin{pmatrix} \varepsilon_1 & 0 \\ 0 & \varepsilon_2 \end{pmatrix}, z \right) \;\middle|\; z^2 = \varepsilon_1 \varepsilon_2 \text{ with } \varepsilon_1 = \pm 1 \text{ and } \varepsilon_2 = \pm 1 \right\}$$
$$= \left\{ \left(\begin{pmatrix} 1 & 0 \\ 0 & 1 \end{pmatrix}, \pm 1 \right), \left(\begin{pmatrix} -1 & 0 \\ 0 & -1 \end{pmatrix}, \pm 1 \right), \left(\begin{pmatrix} 1 & 0 \\ 0 & -1 \end{pmatrix}, \pm i \right), \left(\begin{pmatrix} -1 & 0 \\ 0 & 1 \end{pmatrix}, \pm i \right) \right\}.$$

The group $\tilde{M}$ has elements of order 4, and H does not. Proposition 11.143 shows that $\xi_{2\delta}$ has no square root within H.

EXAMPLE 3. $G = SU(2)/\text{center}$, $H = \left\{ \begin{pmatrix} e^{i\theta} & 0 \\ 0 & e^{-i\theta} \end{pmatrix} \right\} \Big/ \text{center}(G)$.
The group H is swept out by $0 \le \theta < \pi$. For $\xi_{2\delta}(e^{i\theta}) = e^{2i\theta}$, we have

$$\tilde{H} = \{ (e^{i\theta}, z) \mid 0 \le \theta < \pi \text{ and } e^{2i\theta} = z^2 \}$$
$$= \{ (e^{i\theta}, e^{i\theta}) \mid 0 \le \theta < \pi \} \cup \{ (e^{i\theta}, e^{i(\theta+\pi)}) \mid 0 \le \theta < \pi \}$$
$$\cong \{ e^{i\theta} \mid 0 \le \theta < 2\pi \}.$$

Thus $\tilde{H}$ is a circle group and is connected. It is a twofold covering group of H in the traditional sense.

In our applications, H will be the underlying group of a reductive pair $(\mathfrak{h}, T)$. According to Proposition 4.46, one-dimensional representations of H amount to the same thing as one-dimensional $(\mathfrak{h}, T)$ modules. Say the representation ξ_μ corresponds to the module $\mathbb{C}_\mu$. When we pass to the double cover $\tilde{H}$, we can let $\tilde{T} = \pi^{-1}(T)$, and then $(\mathfrak{h}, \tilde{T})$ will be a pair that is reductive in the sense of Definition 4.30. By Proposition 4.31, the corresponding group $\tilde{H}$ is reductive. The group $\tilde{T}$ is exactly the double cover of T relative to $\xi_\mu|_T$, and the square root of $\xi_\mu|_T$ on $\tilde{T}$ is nothing more than $\xi_{\frac{1}{2}\mu}|_{\tilde{T}}$.

We turn to the verification of (11.140). Let $(\mathfrak{h}, T)$ be a Cartan subpair, and let H be the corresponding underlying group. Let $\xi_{-2\delta(\mathfrak{n})'}$ be the character of H given by

$$\xi_{-2\delta(\mathfrak{n})'} = \Big(\prod_{\substack{\alpha \in \Delta(\mathfrak{n}), \\ \alpha \text{ not real}}} \xi_{-\alpha} \Big) \Big(\prod_{\substack{\alpha \in \Delta(\mathfrak{n}), \\ \alpha \text{ real}}} |\xi_{-\alpha}| \Big),$$

and let $\xi_{(\delta(\mathfrak{n}')-\delta(\mathfrak{n}))'}$ be the character of H given by

$$\xi_{(\delta(\mathfrak{n}')-\delta(\mathfrak{n}))'} = \Big(\prod_{\substack{\alpha \in \Delta(\mathfrak{n}') \cap \Delta(\mathfrak{n}^-), \\ \alpha \text{ not real}}} \xi_\alpha \Big) \Big(\prod_{\substack{\alpha \in \Delta(\mathfrak{n}') \cap \Delta(\mathfrak{n}^-), \\ \alpha \text{ real}}} |\xi_\alpha| \Big).$$

The latter definition is consistent with (11.112).

Proposition 11.144. In the case of a Cartan subpair $(\mathfrak{h}, T)$ with underlying group H, let $\mathfrak{b} = \mathfrak{h} \oplus \mathfrak{n}$ and $\mathfrak{b}' = \mathfrak{h} \oplus \mathfrak{n}'$ be Borel subalgebras obtained from two positive systems of roots. Suppose that the square-root construction leads from H and $\xi_{-2\delta(\mathfrak{n})'}$ to $\tilde{H}$ and $\xi_{-\delta(\mathfrak{n})'}$, and from H and $\xi_{-2\delta(\mathfrak{n}')'}$ to $\tilde{H}'$ and $\xi_{-\delta(\mathfrak{n}')'}$. Let $\tilde{\xi}_{(\delta(\mathfrak{n}')-\delta(\mathfrak{n}))'}$ be the lift of $\xi_{(\delta(\mathfrak{n}')-\delta(\mathfrak{n}))'}$ to $\tilde{H}'$. Then there exists a canonical isomorphism $\varphi : \tilde{H}' \to \tilde{H}$ covering the identity map on H such that

$$(11.145a) \qquad \tilde{\xi}_{(\delta(\mathfrak{n}')-\delta(\mathfrak{n}))'} \xi_{-\delta(\mathfrak{n}')'} = \xi_{-\delta(\mathfrak{n})'} \circ \varphi.$$

If Z and Z' are $(\mathfrak{h}, T)$ modules with respective actions η and η' such that $\eta' \cong \eta \otimes \xi_{(\delta(\mathfrak{n}')-\delta(\mathfrak{n}))'}$, and if $\tilde{\eta}$ and $\tilde{\eta}'$ are the lifts of Z and Z' to $(\mathfrak{h}, \tilde{T})$ and $(\mathfrak{h}, \tilde{T}')$ modules, then

$$(11.145b) \qquad \tilde{\eta}' \otimes \xi_{-\delta(\mathfrak{n}')'} \cong (\tilde{\eta} \otimes \xi_{-\delta(\mathfrak{n})'}) \circ \varphi$$

as $(\mathfrak{h}, \tilde{T}')$ modules.

PROOF. Define $\varphi : \tilde{H}' \to \tilde{H}$ by

$$\varphi(h', z) = (h', z\xi_{(\delta(\mathfrak{n}')-\delta(\mathfrak{n}))'}(h')).$$

The image is in $\tilde{H}$ because

$$\begin{aligned}
\xi_{-2\delta(\mathfrak{n})'}(h') &= \xi_{-2\delta(\mathfrak{n}')'}(h')\xi_{2\delta(\mathfrak{n}')'-2\delta(\mathfrak{n})'}(h')\\
&= \xi_{-2\delta(\mathfrak{n}')'}(h')\xi_{(\delta(\mathfrak{n}')-\delta(\mathfrak{n}))'}(h')^2\\
&= (z\xi_{(\delta(\mathfrak{n}')-\delta(\mathfrak{n}))'}(h'))^2,
\end{aligned}$$

and the computation

$$\begin{aligned}
\xi_{-\delta(\mathfrak{n})'}(\varphi(h', z)) &= \xi_{-\delta(\mathfrak{n})'}(h', z\xi_{(\delta(\mathfrak{n}')-\delta(\mathfrak{n}))'}(h'))\\
&= z\xi_{(\delta(\mathfrak{n}')-\delta(\mathfrak{n}))'}(h')\\
&= \xi_{-\delta(\mathfrak{n}')'}(h', z)\tilde{\xi}_{(\delta(\mathfrak{n}')-\delta(\mathfrak{n}))'}(h', z)
\end{aligned}$$

proves (11.145a). Passing to $(\mathfrak{h}, \tilde{T}')$ modules by differentiation, we consequently have

$$\tilde{\eta}' \otimes \xi_{-\delta(\mathfrak{n}')'} \cong \tilde{\eta} \otimes \tilde{\xi}_{(\delta(\mathfrak{n}')-\delta(\mathfrak{n}))'} \otimes \xi_{-\delta(\mathfrak{n}')'} \cong (\tilde{\eta} \otimes \xi_{-\delta(\mathfrak{n})'}) \circ \varphi,$$

and this is (11.145b). The effect of φ in the first variable tells the homomorphism of H to itself that φ covers, and this is the identity.

Now we can define normalized functors $(^n\mathcal{L})_j$ and $(^n\mathcal{R})^j$ in the case that we start from a Cartan pair $(\mathfrak{h}, T)$. Let $\mathfrak{b} = \mathfrak{h} \oplus \mathfrak{n}$ be a given Borel subalgebra, and apply the square-root construction to $(\mathfrak{h}, T)$ and $\xi_{-2\delta(\mathfrak{n})'}$, obtaining $(\mathfrak{h}, \tilde{T})$ and $\xi_{-\delta(\mathfrak{n})'}$. The functors $(^n\mathcal{L})_j$ and $(^n\mathcal{R})^j$ are to be defined on any $(\mathfrak{h}, \tilde{T})$ module $\tilde{Z}$ such that the element ε of $\tilde{T}$ acts by -1. Since $\xi_{-\delta(\mathfrak{n})'}(\varepsilon) = -1$, the $(\mathfrak{h}, \tilde{T})$ module $\tilde{Z} \otimes_{\mathbb{C}} \mathbb{C}_{\delta(\mathfrak{n})'}$ has ε act by $+1$. Thus $\tilde{Z} \otimes_{\mathbb{C}} \mathbb{C}_{\delta(\mathfrak{n})'}$ can be regarded as an $(\mathfrak{h}, T)$ module, and we can define

(11.146a) $\qquad (^n\mathcal{L}_{\mathfrak{b}^-,T}^{\mathfrak{g},K})_j(\tilde{Z}) = (^u\mathcal{L}_{\mathfrak{b}^-,T}^{\mathfrak{g},K})_j(\tilde{Z} \otimes_{\mathbb{C}} \mathbb{C}_{\delta(\mathfrak{n})'})$

(11.146b) $\qquad (^n\mathcal{R}_{\mathfrak{b},T}^{\mathfrak{g},K})^j(\tilde{Z}) = (^u\mathcal{R}_{\mathfrak{b},T}^{\mathfrak{g},K})^j(\tilde{Z} \otimes_{\mathbb{C}} \mathbb{C}_{\delta(\mathfrak{n})'})$.

The subcategory of modules $\mathcal{C}^-(\mathfrak{h}, \tilde{T})$ of $\mathcal{C}(\mathfrak{h}, \tilde{T})$ where ε acts by -1 is good in the sense of Appendix A, and $(^n\mathcal{L}_{\mathfrak{b}^-,T}^{\mathfrak{g},K})_j$ and $(^n\mathcal{R}_{\mathfrak{b},T}^{\mathfrak{g},K})^j$ carry $\mathcal{C}^-(\mathfrak{h}, \tilde{T})$ to $\mathcal{C}(\mathfrak{g}, K)$.

To unwind correspondences, we should think of $\tilde{Z}$ as $Z \otimes_{\mathbb{C}} \mathbb{C}_{-\delta(\mathfrak{n})'}$ in (11.114). If Z has infinitesimal character $\lambda + \delta(\mathfrak{n})$, then $\tilde{Z}$ has infinitesimal character λ. Corollary 11.139 thus shows that $(^n\mathcal{L}_{\mathfrak{b}^-,T}^{\mathfrak{g},K})_j$ and $(^n\mathcal{R}_{\mathfrak{b},T}^{\mathfrak{g},K})^j$ preserve infinitesimal character under some conditions on $\mathfrak{b}$, and this conclusion is in fact valid for all $\mathfrak{b}$.

Up to canonical isomorphism that covers the identity map on $T, \tilde{T}$ in (11.146) is independent of $\mathfrak{b}$, according to Proposition 11.144, and we shall work with $\tilde{T}$ without referring to $\mathfrak{b}$. With this understanding, we can now redo the development of standard modules, taking normalizations into account. A data set now is a tuple $((\mathfrak{h}, \tilde{T}), \lambda, \Delta_{\text{imag}}^+, \Delta_{\text{real}}^+, \tilde{Z})$ in which $(\mathfrak{h}, T), \lambda, \Delta_{\text{imag}}^+$, and Δ_{real}^+ are as in (11.109) and $\tilde{Z}$ is an irreducible $(\mathfrak{h}, T)$ module such that ε acts by -1 and $\mathfrak{h}$ acts by λ. We continue to assume that (i) and either (ii) or (ii') hold in (11.110). If $\mathfrak{b}$ is a Borel subalgebra of one of the types BB, L, L', or VZ, and if the appropriate one of (ii) or (ii') holds in (11.110), the associated standard modules are

(11.147)
$\qquad (^n\mathcal{L}_{\mathfrak{b}^-,T}^{\mathfrak{g},K})_p(\tilde{Z}) \quad \text{and} \quad (^n\mathcal{R}_{\mathfrak{b},T}^{\mathfrak{g},K})^p(\tilde{Z}), \qquad \text{where } p = p(\mathfrak{b}) = \dim(\mathfrak{n} \cap \mathfrak{k})$.

The statement of Theorem 11.129 simplifies notationally in the obvious way.

We can use the square-root construction also to normalize the functors $(^n\mathcal{L})_j$ and $(^n\mathcal{R})^j$ when we start from a general parabolic pair $(\mathfrak{q}, L \cap K)$. Let us write $\mathfrak{q} = \mathfrak{l} \oplus \mathfrak{u}$ as usual. The claim is that the $(\mathfrak{l}, L \cap K)$ module $\mathfrak{u}$ splits as the direct sum of irreducible $(\mathfrak{l}, L \cap K)$ modules and that the summands are irreducible and inequivalent as $U(\mathfrak{l})$ modules. In fact, under the action of a Cartan subalgebra $\mathfrak{h}$ of $\mathfrak{l}$, $\mathfrak{u}$ decomposes into inequivalent $U(\mathfrak{h})$ modules, since the roots of $\mathfrak{g}$ have multiplicity 1. Thus $\mathfrak{u}$ is the direct sum of submodules for the center $Z_{\mathfrak{l}}$ in such a way that $Z_{\mathfrak{l}}$ acts by scalars on each submodule. Each submodule is invariant under the semsimple part $[\mathfrak{l}, \mathfrak{l}]$ of $\mathfrak{l}$ and is fully reducible under $[\mathfrak{l}, \mathfrak{l}]$. In this way, $\mathfrak{u}$ is the direct sum of irreducible $U(\mathfrak{l})$ modules. These are inequivalent because the roots have multiplicity 1, and hence they have distinct infinitesimal characters. Since the pair $(\mathfrak{l}, L \cap K)$ is in the Harish-Chandra class, it follows that each of these irreducible $U(\mathfrak{l})$ modules is normalized by $L \cap K$ and hence is an irreducible $(\mathfrak{l}, L \cap K)$ module.

Let

$$(11.148) \qquad\qquad \mathfrak{u} = \mathfrak{u}^1 \oplus \cdots \oplus \mathfrak{u}^s$$

be the decomposition of $\mathfrak{u}$ into irreducible $U(\mathfrak{l})$ modules. With conjugation defined as in §VI.2, we distinguish between those $\mathfrak{u}^i$ that are **self-conjugate** (i.e., equivalent with their respective conjugates) and those that are not. We define a one-dimensional representation $\xi_{2\delta(\mathfrak{u})'}$ of L on a space $\mathbb{C}_{2\delta(\mathfrak{u})'}$ by

$$(11.149)$$
$$\xi_{2\delta(\mathfrak{u})'}(l) = \Big(\prod_{\substack{i \text{ with } \mathfrak{u}^i \\ \text{self-conjugate}}} \big| \det \mathrm{Ad}(l)|_{\mathfrak{u}^i} \big| \Big)\Big(\prod_{\substack{i \text{ with } \mathfrak{u}^i \text{ not} \\ \text{self-conjugate}}} \det \mathrm{Ad}(l)|_{\mathfrak{u}^i} \Big).$$

Then $\mathbb{C}_{2\delta(\mathfrak{u})'}$ becomes a one-dimensional $(\mathfrak{l}, L \cap K)$ module. Observe that $\xi_{2\delta(\mathfrak{u})'}(l)$ has the same absolute value as $\det \mathrm{Ad}(l)|_{\mathfrak{u}}$. Thus in checking identities involving $\xi_{2\delta(\mathfrak{u})'}(l)$, we often have no problem with the absolute value and need concentrate only on the complex-number argument.

In the special case that $\mathfrak{q}$ is a Borel subalgebra $\mathfrak{h} \oplus \mathfrak{n}$, $\xi_{2\delta(\mathfrak{u})'}$ reduces to $\xi_{2\delta(\mathfrak{n})'}$ as defined in (11.111). For general $\mathfrak{q}$, we shall show in Proposition 11.153 that (11.149) is also consistent with introducing any θ stable Cartan subalgebra $\mathfrak{h}_0$ of $\mathfrak{l}_0$, forming the set of roots $\Delta(\mathfrak{g}, \mathfrak{h})$, writing

$$2\delta(\mathfrak{u}) = \sum_{\substack{\alpha \in \Delta(\mathfrak{u}), \\ \alpha \text{ not real}}} \alpha + \sum_{\substack{\alpha \in \Delta(\mathfrak{u}), \\ \alpha \text{ real}}} \alpha,$$

and constructing $\xi_{2\delta(\mathfrak{u})'}$ as in (11.111).

With $\xi_{2\delta(\mathfrak{u})'}$ in hand, we apply the square-root construction to $(\mathfrak{l}, L \cap K)$ and $\xi_{-2\delta(\mathfrak{u})'}$, obtaining $(\mathfrak{l}, (L \cap K)^{\sim})$ and $\xi_{-\delta(\mathfrak{u})'}$. The functors $({}^n\mathcal{L})_j$ and $({}^n\mathcal{R})^j$ are to be defined on any $(\mathfrak{l}, (L \cap K)^{\sim})$ module $\tilde{W}$ such that the element ε of $(L \cap K)^{\sim}$ acts by -1. Since $\xi_{-\delta(\mathfrak{u})'}(\varepsilon) = -1$, the $(\mathfrak{l}, (L \cap K)^{\sim})$ module $\tilde{W} \otimes_{\mathbb{C}} \mathbb{C}_{\delta(\mathfrak{u})'}$ has ε act by $+1$. Thus $\tilde{W} \otimes_{\mathbb{C}} \mathbb{C}_{\delta(\mathfrak{u})'}$ can be regarded as an $(\mathfrak{l}, L \cap K)$ module, and we can define

(11.150a) $\qquad ({}^{n}\mathcal{L}^{\mathfrak{g},K}_{\mathfrak{q}^{-},L \cap K})_{j}(\tilde{W}) = ({}^{u}\mathcal{L}^{\mathfrak{g},K}_{\mathfrak{q}^{-},L \cap K})_{j}(\tilde{W} \otimes_{\mathbb{C}} \mathbb{C}_{\delta(\mathfrak{u})'})$

(11.150b) $\qquad ({}^{n}\mathcal{R}^{\mathfrak{g},K}_{\mathfrak{q},L \cap K})^{j}(\tilde{W}) = ({}^{u}\mathcal{R}^{\mathfrak{g},K}_{\mathfrak{q},L \cap K})^{j}(\tilde{W} \otimes_{\mathbb{C}} \mathbb{C}_{\delta(\mathfrak{u})'})$.

The subcategory of modules $\mathcal{C}^{-}(\mathfrak{l}, (L \cap K)^{\sim})$ of $\mathcal{C}(\mathfrak{l}, (L \cap K)^{\sim})$ where ε acts by -1 is good in the sense of Appendix A, and $({}^{n}\mathcal{L}^{\mathfrak{g},K}_{\mathfrak{q}^{-},L \cap K})_{j}$ and $({}^{n}\mathcal{R}^{\mathfrak{g},K}_{\mathfrak{q},L \cap K})^{j}$ carry $\mathcal{C}^{-}(\mathfrak{l}, (L \cap K)^{\sim})$ to $\mathcal{C}(\mathfrak{g}, K)$.

To unwind correspondences, we should think of $\tilde{W}$ as $W \otimes_{\mathbb{C}} \mathbb{C}_{-\delta(\mathfrak{u})'}$. If W has infinitesimal character $\lambda + \delta(\mathfrak{u})$, then $\tilde{W}$ has infinitesimal character λ, and

(11.151)
$\qquad ({}^{n}\mathcal{L}^{\mathfrak{g},K}_{\mathfrak{q}^{-},L \cap K})_{j}(\tilde{W}) \quad$ and $\quad ({}^{n}\mathcal{R}^{\mathfrak{g},K}_{\mathfrak{q},L \cap K})^{j}(\tilde{W}) \qquad$ have infinitesimal character λ.

This fact is a consequence of Theorems 5.24 and 5.21b.

Lemma 11.152. Let $\mathfrak{h}_{0}$ be a θ stable Cartan subalgebra of $\mathfrak{l}_{0}$. In the decomposition (11.148) of $\mathfrak{u}$ into irreducible $U(\mathfrak{l})$ modules, the following conditions on a summand $\mathfrak{u}^{i}$ are equivalent:

(a) $\mathfrak{u}^{i}$ is self-conjugate
(b) the set of roots of $\mathfrak{h}$ in $\mathfrak{u}^{i}$ is closed under conjugation
(c) there is a root of $\mathfrak{h}$ in $\mathfrak{u}^{i}$ whose conjugate is a root of $\mathfrak{h}$ in $\mathfrak{u}^{i}$.

Moreover, these conditions hold if

(d) some root of $\mathfrak{h}$ in $\mathfrak{u}^{i}$ is real.

PROOF. The weights of $\bar{\mathfrak{u}}_{i}$ are the conjugates of the weights of $\mathfrak{u}_{i}$, and it follows that (a) implies (b). Clearly (b) implies (c). Now let (c) hold. Since $\mathfrak{u} \oplus \mathfrak{u}^{-}$ is closed under conjugation, it follows that $\overline{\mathfrak{u}^{i}}$, the conjugate of $\mathfrak{u}^{i}$, occurs in $\mathfrak{u} \oplus \mathfrak{u}^{-}$. By (c), $\mathfrak{u}^{i}$ and $\overline{\mathfrak{u}^{i}}$ have a weight of $\mathfrak{h}$ in common. Since the weights of $\mathfrak{h}$ in $\mathfrak{u} \oplus \mathfrak{u}^{-}$ have multiplicity 1, $\mathfrak{u}^{i} \cap \overline{\mathfrak{u}^{i}} \neq 0$. By irreducibility, $\mathfrak{u}^{i} = \overline{\mathfrak{u}^{i}}$. Thus (a) holds.

Finally (d) implies (c) since a real root contributing to $\mathfrak{u}^{i}$ is a root in $\mathfrak{u}^{i}$ whose conjugate contributes to $\mathfrak{u}^{i}$.

Proposition 11.153. Let $(\mathfrak{q}, L \cap K)$ be a parabolic pair with $\mathfrak{q} = \mathfrak{l} \oplus \mathfrak{u}$, and let $(\mathfrak{h}, T)$ be a Cartan subpair of $(\mathfrak{l}, L \cap K)$ with underlying group H. Then for $h \in H$, the definition of $\xi_{2\delta(\mathfrak{u})'}(h)$ in (11.149) is consistent with the definition of $\xi_{2\delta(\mathfrak{u})'}(h)$ in (11.111).

Proof. The absolute values are the same, and we must show equality of the arguments. Let u decompose under l as in (11.148), fix attention on one of the summands u^i, and let $\Delta(u^i) = \{\beta_1, \ldots, \beta_m\}$. First suppose that u^i is self-conjugate. By definition the contribution of u^i to $\xi_{2\delta(u)'}(h)$ in (11.149) is the positive number $|\det \mathrm{Ad}(h)|_{u^i}|$. Meanwhile the equivalence of (a) and (b) in Lemma 11.152 says that $\Delta(u^i)$ consists of real roots and conjugate pairs. Each real root β contributes $|\xi_\beta(h)|$ to (11.111), and each conjugate pair $\{\beta, \bar{\beta}\}$ contributes $\xi_\beta(h)\xi_{\bar{\beta}}(h)$, which is positive by (11.135). Hence the total contribution from u^i is positive to both (11.111) and (11.149), and the two contributions must agree.

Next suppose that u^i is not self-conjugate. Then the contribution of u^i to $\xi_{2\delta(u)'}(h)$ in (11.149) is

$$(11.154) \qquad \det \mathrm{Ad}(h)|_{u^i} = \prod_{j=1}^{m} \xi_{\beta_j}(h).$$

On the other hand, the implication (d) $\Longrightarrow$ (a) in Lemma 11.152 shows that no β_j is real. Hence the right side of (11.154) is the contribution of u^i to (11.111). Thus again the contributions of u^i to (11.111) and (11.149) are equal. Taking the product of these contributions for $1 \leq i \leq s$, we obtain the result of the proposition.

The normalized functors $({}^n\mathcal{L}^{\mathfrak{g},K}_{q^-,L\cap K})_j$ and $({}^n\mathcal{R}^{\mathfrak{g},K}_{q,L\cap K})^j$ of (11.150) are manageable, but they become cumbersome when they are composed. We shall take up the general question of induction in stages for normalized functors shortly. Things simplify considerably, however, when the homomorphism $\xi_{2\delta(u)'} : L \to \mathbb{C}^\times$ has a square root, say $\xi^{1/2}_{2\delta(u)'} : L \to \mathbb{C}^\times$. (This is what happens when q is real; in that case we were able to take (11.58) and (11.66) as definitions of normalized functors.) Suppose that the square-root construction earlier leads from L and $\xi_{2\delta(u)}$ to $\tilde{L}$ and $\xi_{\delta(u)'}$. Proposition 11.143 says that the map $\varphi : L \times \{1, \varepsilon\} \to \tilde{L}$ given by

$$(11.155) \qquad \begin{aligned} \varphi(l, 1) &= (l, \xi^{1/2}_{2\delta(u)'}(l)) \\ \varphi(l, \varepsilon) &= (l, -\xi^{1/2}_{2\delta(u)'}(l)) \end{aligned}$$

is an isomorphism. The homomorphism $\xi_{\delta(u)'}$ is just the homomorphism to the second coordinate in $\tilde{L} \subseteq L \times \mathbb{C}^\times$. Hence we see from (11.155) that the isomorphism $L \times \{1, \varepsilon\} \cong \tilde{L}$ identifies

$$(11.156a) \qquad \xi^{1/2}_{2\delta(u)'} \otimes \mathrm{sgn} \leftrightarrow \xi_{\delta(u')};$$

here sgn is the nontrivial character of $\{1, \varepsilon\}$. In the same way, we can identify modules W in $\mathcal{C}(\mathfrak{l}, L \cap K)$ with modules $\tilde{W}$ in $\mathcal{C}^-(\mathfrak{l}, (L \cap K)^\sim)$ by the formula

$$(11.156b) \qquad\qquad W \otimes \text{sgn} \leftrightarrow \tilde{W}.$$

Under this identification we have

$$(11.156c) \qquad\qquad (W \otimes \xi_{2\delta(\mathfrak{u})'}^{1/2}) \otimes 1 \leftrightarrow \tilde{W} \otimes \xi_{\delta(\mathfrak{u})'}$$

since $\text{sgn} \otimes \text{sgn} = 1$ on $\{1, \varepsilon\}$. Let us define

$$(11.157a) \qquad (\mathcal{L}_{\mathfrak{q}^-, L \cap K}^{\mathfrak{g}, K})_j(W) = ({}^{\mathfrak{u}}\mathcal{L}_{\mathfrak{q}^-, L \cap K}^{\mathfrak{g}, K})_j(W \otimes \xi_{2\delta(\mathfrak{u})'}^{1/2})$$

$$(11.157b) \qquad (\mathcal{R}_{\mathfrak{q}, L \cap K}^{\mathfrak{g}, K})^j(W) = ({}^{\mathfrak{u}}\mathcal{R}_{\mathfrak{q}, L \cap K}^{\mathfrak{g}, K})^j(W \otimes \xi_{2\delta(\mathfrak{u})'}^{1/2}).$$

The virtue of these definitions is that no double covers are involved. If we pass to double covers, using $\tilde{W}$ as in (11.156b), then we see from (11.156c) and (11.150) that

$$(11.158a) \qquad (\mathcal{L}_{\mathfrak{q}^-, L \cap K}^{\mathfrak{g}, K})_j(W) = ({}^{\mathfrak{n}}\mathcal{L}_{\mathfrak{q}^-, L \cap K}^{\mathfrak{g}, K})_j(\tilde{W})$$

$$(11.158b) \qquad (\mathcal{R}_{\mathfrak{q}, L \cap K}^{\mathfrak{g}, K})^j(W) = ({}^{\mathfrak{n}}\mathcal{R}_{\mathfrak{q}, L \cap K}^{\mathfrak{g}, K})^j(\tilde{W}).$$

Before turning to induction in stages, we prove one preliminary result.

Proposition 11.159. Let $(\mathfrak{q}, L \cap K)$ be a parabolic pair with corresponding Levi pair $(\mathfrak{l}, L \cap K)$, and let $(\mathfrak{q}_1, L_1 \cap K)$ be a parabolic subpair of $(\mathfrak{l}, L \cap K)$ with corresponding Levi pair $(\mathfrak{l}_1, L_1 \cap K)$. Put $\mathfrak{q} = \mathfrak{l} \oplus \mathfrak{u}$, so that $(\mathfrak{q}_1 + \mathfrak{u}, L_1 \cap K)$ is a parabolic subpair of $(\mathfrak{g}, K)$ with corresponding Levi pair $(\mathfrak{l}_1, L_1 \cap K)$. Then

$$\xi_{2\delta(\mathfrak{u}+\mathfrak{u}_1)'} = (\xi_{2\delta(\mathfrak{u})'}|_{L_1})\xi_{2\delta(\mathfrak{u}_1)'}.$$

PROOF. Let $\mathfrak{u}$ decompose under L as in (11.148). For each i, let $\mathfrak{u}^i$ decompose under L_1 as

$$\mathfrak{u}^i = \mathfrak{u}^{i,1} \oplus \cdots \oplus \mathfrak{u}^{i,p_i}.$$

Also let $\mathfrak{u}_1$ decompose under L_1 as

$$\mathfrak{u}_1 = \mathfrak{u}_1^1 \oplus \cdots \oplus \mathfrak{u}_1^t.$$

For $l \in L_1$, we have

(11.160)
$$\xi_{2\delta(\mathfrak{u}+\mathfrak{u}_1)'}(l_1)$$
$$= \left(\prod_{\substack{i,j \text{ with } u^{i,j} \\ \text{self-conjugate}}} \left| \det \mathrm{Ad}(l_1)|_{u^{i,j}} \right| \right) \left(\prod_{\substack{i,j \text{ with} \\ u^{i,j} \text{ not} \\ \text{self-conjugate}}} \det \mathrm{Ad}(l_1)|_{u^{i,j}} \right)$$
$$\times \left(\prod_{\substack{k \text{ with } u_1^k \\ \text{self-conjugate}}} \left| \det \mathrm{Ad}(l_1)|_{u_1^k} \right| \right) \left(\prod_{\substack{k \text{ with } u_1^k \text{ not} \\ \text{self-conjugate}}} \det \mathrm{Ad}(l_1)|_{u_1^k} \right).$$

The second line of the right side of (11.160) equals $\xi_{2\delta(\mathfrak{u}_1)'}(l_1)$, and thus it is enough to prove that

(11.161) $\xi_{2\delta(\mathfrak{u})'}(l_1)$
$$\overset{?}{=} \left(\prod_{\substack{i,j \text{ with } u^{i,j} \\ \text{self-conjugate}}} \left| \det \mathrm{Ad}(l_1)|_{u^{i,j}} \right| \right) \left(\prod_{\substack{i,j \text{ with} \\ u^{i,j} \text{ not} \\ \text{self-conjugate}}} \det \mathrm{Ad}(l_1)|_{u^{i,j}} \right).$$

Fix i. If u^i is self-conjugate, then it follows that the $u^{i,j}$'s that are not self-conjugate occur in conjugate pairs. Hence
$$\left| \det \mathrm{Ad}(l_1)|_{u^i} \right|$$
$$= \left(\prod_{\substack{j \text{ with } u^{i,j} \\ \text{self-conjugate}}} \left| \det \mathrm{Ad}(l_1)|_{u^{i,j}} \right| \right) \left(\prod_{\substack{j \text{ with} \\ u^{i,j} \text{ not} \\ \text{self-conjugate}}} \det \mathrm{Ad}(l_1)|_{u^{i,j}} \right).$$

The left side here is the contribution of u^i to the left side of (11.161), and the right side here is the contribution of u^i to the right side of (11.161).

Now suppose that u^i is not self-conjugate. By Proposition 4.42 we can choose a Cartan subpair $(\mathfrak{h}, T)$ of $(\mathfrak{l}_1, L_1 \cap K)$ such that T meets every component of $L_1 \cap K$. To complete the verification of (11.161), it is enough to prove that

(11.162) $\det \mathrm{Ad}(l_1)|_{u^i}$
$$\overset{?}{=} \left(\prod_{\substack{j \text{ with } u^{i,j} \\ \text{self-conjugate}}} \left| \det \mathrm{Ad}(l_1)|_{u^{i,j}} \right| \right) \left(\prod_{\substack{j \text{ with} \\ u^{i,j} \text{ not} \\ \text{self-conjugate}}} \det \mathrm{Ad}(l_1)|_{u^{i,j}} \right)$$

for $t \in T$. Lemma 11.152 shows that the first product on the right side of (11.162) has no factors. Thus the right side of (11.162) reduces to
$$\left(\prod_{\text{all } j} \det \mathrm{Ad}(l_1)|_{u^{i,j}} \right),$$
and this equals the left side. The proposition follows.

Finally we are in a position to address induction in stages for the normalized functors (11.150). Although the notation is complicated, there are only two ideas and they are simple. The first idea is that the main result should be expressed by a spectral sequence analogous to the one for unnormalized functors in Theorem 11.77, with the square root of the identity in Proposition 11.159 showing that the normalizations are consistent.

The second idea is that the notation on the right side of (11.150), though convenient, is not completely precise because $\tilde{W} \otimes \mathbb{C}_{\delta(\mathfrak{u})'}$ is not exactly an $(\mathfrak{l}, L \cap K)$ module. Instead it is a module for the double cover with trivial action by the kernel. Once we incorporate a functor to make the notation rigorous, the necessary computations become almost completely formal.

The new functors in question are P, I, and the forgetful functor $\mathcal{F}$ for a map of pairs $(\mathfrak{h}, \tilde{T}) \to (\mathfrak{h}, T)$ in which i_{alg} is the identity and i_{gp} is onto. Since i_{alg} is one-one, it follows that i_{gp} has finite kernel. This is the kind of situation we encounter with covering groups. It is apparent that the forgetful functor $\mathcal{F}_{\mathfrak{h},T}^{\mathfrak{h},\tilde{T}}$ just lifts $(\mathfrak{h}, T)$ modules to $(\mathfrak{h}, \tilde{T})$ modules. Proposition 2.33 shows that the pseudoforgetful functor $(\mathcal{F}^\vee)_{\mathfrak{h},T}^{\mathfrak{h},\tilde{T}}$ is the same as $\mathcal{F}_{\mathfrak{h},T}^{\mathfrak{h},\tilde{T}}$, and Proposition 2.92 identifies P and I with the averaging functor $A_{\mathfrak{h},\tilde{T}}^{\mathfrak{h},T}$ given by

$$(11.163) \quad A_{\mathfrak{h},\tilde{T}}^{\mathfrak{h},T}(V) = \{v \in V \mid \ker(i_{\mathrm{gp}})v = v\} = \left\{ \int_{\ker(i_{\mathrm{gp}})} \tilde{t}v \, d\tilde{t} \,\middle|\, v \in V \right\}.$$

(Here it is to be understood that the Haar measure $d\tilde{t}$ on the finite group $\ker(i_{\mathrm{gp}})$ has total mass 1.) Because of this identification $\mathcal{F}_{\mathfrak{h},T}^{\mathfrak{h},\tilde{T}}$ followed by either $P_{\mathfrak{h},\tilde{T}}^{\mathfrak{h},T}$ or $I_{\mathfrak{h},\tilde{T}}^{\mathfrak{h},T}$ is naturally isomorphic with the identity functor.

Proposition 11.164. If $(\mathfrak{q}, L \cap K)$ is a parabolic pair and $(\mathfrak{l}, (L \cap K)^\sim)$ is a double cover of the Levi pair $(\mathfrak{l}, L \cap K)$, then

$$\mathcal{F}_{\mathfrak{l},L\cap K}^{\mathfrak{q},L\cap K} P_{\mathfrak{l},(L\cap K)^\sim}^{\mathfrak{l},L\cap K} \cong P_{\mathfrak{q},(L\cap K)^\sim}^{\mathfrak{q},L\cap K} \mathcal{F}_{\mathfrak{l},(L\cap K)^\sim}^{\mathfrak{q},(L\cap K)^\sim}$$

and

$$\mathcal{F}_{\mathfrak{l},L\cap K}^{\mathfrak{q},L\cap K} I_{\mathfrak{l},(L\cap K)^\sim}^{\mathfrak{l},L\cap K} \cong I_{\mathfrak{q},(L\cap K)^\sim}^{\mathfrak{q},L\cap K} \mathcal{F}_{\mathfrak{l},(L\cap K)^\sim}^{\mathfrak{q},(L\cap K)^\sim}.$$

PROOF. This kind of identity with P or I replaced by the averaging functor A is clear from the definitions, and hence the result follows from Proposition 2.92.

In the definition (11.150a), we can make the right sides precise by replacing $\tilde{W} \otimes_{\mathbb{C}} \mathbb{C}_{\delta(\mathfrak{u})'}$ by $P_{\mathfrak{l},(L\cap K)^\sim}^{\mathfrak{l},L\cap K}(\tilde{W} \otimes_{\mathbb{C}} \mathbb{C}_{\delta(\mathfrak{u})'})$. Then the right side of (11.150a) is

$$\cong ({}^{\mathfrak{u}}\mathcal{L}_{\mathfrak{q}^-,L\cap K}^{\mathfrak{g},K})_j \, P_{\mathfrak{l},(L\cap K)^\sim}^{\mathfrak{l},L\cap K}(\tilde{W} \otimes \mathbb{C}_{\delta(\mathfrak{u})'})$$

$$\cong (P_{\mathfrak{q}^-,L\cap K}^{\mathfrak{g},K})_j \, \mathcal{F}_{\mathfrak{l},L\cap K}^{\mathfrak{q}^-,L\cap K} \, P_{\mathfrak{l},(L\cap K)^\sim}^{\mathfrak{l},L\cap K}(\tilde{W} \otimes \mathbb{C}_{\delta(\mathfrak{u})'})$$

$$\cong (P_{\mathfrak{q}^-,L\cap K}^{\mathfrak{g},K})_j \, P_{\mathfrak{q}^-,(L\cap K)^\sim}^{\mathfrak{q}^-,L\cap K} \, \mathcal{F}_{\mathfrak{l},(L\cap K)^\sim}^{\mathfrak{q}^-,(L\cap K)^\sim}(\tilde{W} \otimes \mathbb{C}_{\delta(\mathfrak{u})'}) \quad \text{by Proposition} \quad 11.164$$

$$\cong (P_{\mathfrak{q}^-,(L\cap K)^\sim}^{\mathfrak{g},K})_j \, \mathcal{F}_{\mathfrak{l},(L\cap K)^\sim}^{\mathfrak{q}^-,(L\cap K)^\sim}(\tilde{W} \otimes \mathbb{C}_{\delta(\mathfrak{u})'}) \quad \text{by (C.28a1).}$$

A similar computation applies to (11.150b), and the more precise formulations of (11.150) are therefore

(11.165a) $({}^{\mathfrak{n}}\mathcal{L}_{\mathfrak{q}^-,L\cap K}^{\mathfrak{g},K})_j(\tilde{W}) = (P_{\mathfrak{q}^-,(L\cap K)^\sim}^{\mathfrak{g},K})_j \mathcal{F}_{\mathfrak{l},(L\cap K)^\sim}^{\mathfrak{q}^-,(L\cap K)^\sim}(\tilde{W} \otimes_{\mathbb{C}} \mathbb{C}_{\delta(\mathfrak{u})'})$

(11.165b) $({}^{\mathfrak{n}}\mathcal{R}_{\mathfrak{q},L\cap K}^{\mathfrak{g},K})^j(\tilde{W}) = (I_{\mathfrak{q},(L\cap K)^\sim}^{\mathfrak{g},K})^j \mathcal{F}_{\mathfrak{l},(L\cap K)^\sim}^{\mathfrak{q},(L\cap K)^\sim}(\tilde{W} \otimes_{\mathbb{C}} \mathbb{C}_{\delta(\mathfrak{u})'}).$

The notation for induction in stages will be as in Proposition 11.159. Namely the parabolic pairs for proceeding in two steps will be $(\mathfrak{q}, L \cap K)$ on the outside and $(\mathfrak{q}_1, L_1 \cap K)$ on the inside, and the parabolic pair for proceeding in one step will be $(\mathfrak{q}_1 + \mathfrak{u}, L_1 \cap K)$. Let

(11.166)
$$\xi_{\delta(\mathfrak{u}_1)'}, \, (L_1)_1^\sim, \, \text{and } (\mathfrak{l}_1, (L_1 \cap K)_1^\sim) \longleftrightarrow \text{square root of} \quad \xi_{2\delta(\mathfrak{u}_1)'}$$
$$\xi_{\delta(\mathfrak{u})'}, \, \quad L_2^\sim \,, \, \text{and } (\mathfrak{l}, (L \cap K)_2^\sim) \longleftrightarrow \text{square root of} \quad \xi_{2\delta(\mathfrak{u})'}$$
$$\xi_{\delta(\mathfrak{u})'}|_{(L_1)_2^\sim}, \, (L_1)_2^\sim, \, \text{and } (\mathfrak{l}_1, (L_1 \cap K)_2^\sim) \longleftrightarrow \text{square root of} \quad \xi_{2\delta(\mathfrak{u})'}|_{L_1}$$
$$\xi_{\delta(\mathfrak{u}+\mathfrak{u}_1)'}, \, (L_1)_3^\sim, \, \text{and } (\mathfrak{l}_1, (L_1 \cap K)_3^\sim) \longleftrightarrow \text{square root of} \quad \xi_{2\delta(\mathfrak{u}+\mathfrak{u}_1)'}.$$

We introduce also a pair $(\mathfrak{l}_1, (L_1 \cap K)_{12}^\sim)$ with corresponding group $(L_1)_{12}^\sim$ given by

(11.167)
$$(L_1)_{12}^\sim = \{(l_1, z_1, z_2) \in L_1 \times \mathbb{C}^\times \times \mathbb{C}^\times \mid \xi_{2\delta(\mathfrak{u}_1)'} = z_1^2 \text{ and } \xi_{2\delta(\mathfrak{u})'} = z_2^2\}.$$

Let φ be the homomorphism

$$\varphi : (L_1)_{12}^\sim \to (L_1)_3^\sim \qquad \text{given by} \qquad (l_1, z_1, z_2) \mapsto (l_1, z_1 z_2).$$

This map has image in $(L_1)_3^\sim$ by Proposition 11.159, it is onto, and its kernel is $\{(1, 1, 1), (1, -1, -1)\}$. The corresponding map of pairs

$(\mathfrak{l}_1, (L_1 \cap K)_{\widetilde{12}}) \to (\mathfrak{l}_1, (L_1 \cap K)_{\widetilde{3}})$ is of the kind considered in Proposition 2.92.

The projections

$$(l_1, z_1, z_2) \mapsto (l_1, z_1) \qquad \text{and} \qquad (l_1, z_1, z_2) \mapsto (l_1, z_2)$$

each have two-element kernels, and they carry $(L_1)_{\widetilde{12}}$ onto $(L_1)_{\widetilde{1}}$ and $(L_1)_{\widetilde{2}}$, respectively. Lifting the characters $\xi_{\delta(u_1)'}$ and $\xi_{\delta(u)'}$ back to $(L_1)_{\widetilde{12}}$, we can regard them both as defined on $(L_1)_{\widetilde{12}}$. Because of Proposition 11.159, these lifts then satisfy

$$(11.168) \qquad (\xi_{\delta(u)'}|_{(L_1)_{\widetilde{12}}}) \xi_{\delta(u_1)'} = \xi_{\delta(u+u_1)'} \circ \varphi.$$

Let $\tilde{Z}$ be in $\mathcal{C}^-(\mathfrak{l}_1, (L_1 \cap K)_{\widetilde{3}})$, and define

$$(11.169) \qquad \tilde{\tilde{Z}} = \mathcal{F}_{\mathfrak{l}_1,(L_1 \cap K)_{\widetilde{3}}}^{\mathfrak{l}_1,(L_1 \cap K)_{\widetilde{12}}}(\tilde{Z}).$$

The elements $(1, -1, 1)$ and $(1, 1, -1)$ in $(L_1)_{\widetilde{12}}$ act as -1 in $\tilde{\tilde{Z}}$, and (11.168) allows us to write

$$(11.170) \qquad \tilde{Z} \otimes \mathbb{C}_{\delta(u+u_1)'} \cong A_{\mathfrak{l}_1,(L_1 \cap K)_{\widetilde{12}}}^{\mathfrak{l}_1,(L_1 \cap K)_{\widetilde{3}}} (\tilde{\tilde{Z}} \otimes \mathbb{C}_{\delta(u_1)'} \otimes \mathbb{C}_{\delta(u)'}),$$

where A is the averaging functor. In view of Proposition 2.92, we are allowed to replace A by P or I in this identity.

We shall calculate the expresion that was called somewhat imprecisely

$$(11.171) \qquad ({}^u\mathcal{R}_{\mathfrak{q},L \cap K}^{\mathfrak{g},K})^i [({}^u\mathcal{R}_{\mathfrak{q}_1,L_1 \cap K}^{\mathfrak{l},L \cap K})^j (\tilde{Z} \otimes \mathbb{C}_{\delta(u+u_1)'})]$$

in the notation of (11.150). A more precise form of the expression in brackets is

$$({}^u\mathcal{R}_{\mathfrak{q}_1,L_1 \cap K}^{\mathfrak{l},L \cap K})^j I_{\mathfrak{l}_1,(L_1 \cap K)_{\widetilde{3}}}^{\mathfrak{l}_1,L_1 \cap K} (\tilde{Z} \otimes \mathbb{C}_{\delta(u+u_1)'})$$

$$\cong ({}^u\mathcal{R}_{\mathfrak{q}_1,L_1 \cap K}^{\mathfrak{l},L \cap K})^j I_{\mathfrak{l}_1,(L_1 \cap K)_{\widetilde{12}}}^{\mathfrak{l}_1,L_1 \cap K} (\tilde{\tilde{Z}} \otimes \mathbb{C}_{\delta(u_1)'} \otimes \mathbb{C}_{\delta(u)'}) \quad \begin{array}{l} \text{by (11.170) and} \\ \text{Proposition 2.19} \end{array}$$

$$\cong (I_{\mathfrak{q}_1,L_1 \cap K}^{\mathfrak{l},L \cap K})^j \mathcal{F}_{\mathfrak{l}_1,L_1 \cap K}^{\mathfrak{q}_1,L_1 \cap K} I_{\mathfrak{l}_1,(L_1 \cap K)_{\widetilde{12}}}^{\mathfrak{l}_1,L_1 \cap K} (\tilde{\tilde{Z}} \otimes \mathbb{C}_{\delta(u_1)'} \otimes \mathbb{C}_{\delta(u)'})$$

$$\cong (I_{\mathfrak{q}_1,L_1 \cap K}^{\mathfrak{l},L \cap K})^j I_{\mathfrak{q}_1,(L_1 \cap K)_{\widetilde{12}}}^{\mathfrak{q}_1,L_1 \cap K} \mathcal{F}_{\mathfrak{l}_1,(L_1 \cap K)_{\widetilde{12}}}^{\mathfrak{q}_1,(L_1 \cap K)_{\widetilde{12}}} (\tilde{\tilde{Z}} \otimes \mathbb{C}_{\delta(u_1)'} \otimes \mathbb{C}_{\delta(u)'})$$
$$\text{by Proposition 11.164}$$

$$\cong (I_{\mathfrak{q}_1,(L_1 \cap K)_{\widetilde{12}}}^{\mathfrak{l},L \cap K})^j \mathcal{F}_{\mathfrak{l}_1,(L_1 \cap K)_{\widetilde{12}}}^{\mathfrak{q}_1,(L_1 \cap K)_{\widetilde{12}}} (\tilde{\tilde{Z}} \otimes \mathbb{C}_{\delta(u_1)'} \otimes \mathbb{C}_{\delta(u)'})$$

$$\cong (I_{\mathfrak{l},(L\cap K)_{\widetilde{2}}}^{\mathfrak{l},L\cap K} I_{\mathfrak{q}_1,(L_1\cap K)_{\widetilde{12}}}^{\mathfrak{l},(L\cap K)_{\widetilde{2}}})^j \mathcal{F}_{\mathfrak{l}_1,(L_1\cap K)_{\widetilde{12}}}^{\mathfrak{q}_1,(L_1\cap K)_{\widetilde{12}}}(\widetilde{\widetilde{Z}} \otimes \mathbb{C}_{\delta(\mathfrak{u}_1)'} \otimes \mathbb{C}_{\delta(\mathfrak{u})'})$$

$$\cong I_{\mathfrak{l},(L\cap K)_{\widetilde{2}}}^{\mathfrak{l},L\cap K}(I_{\mathfrak{q}_1,(L_1\cap K)_{\widetilde{12}}}^{\mathfrak{l},(L\cap K)_{\widetilde{2}}})^j[\mathcal{F}_{\mathfrak{l}_1,(L_1\cap K)_{\widetilde{12}}}^{\mathfrak{q}_1,(L_1\cap K)_{\widetilde{12}}}(\widetilde{\widetilde{Z}} \otimes \mathbb{C}_{\delta(\mathfrak{u}_1)'}) \otimes \mathcal{F}_{\mathfrak{l}_1,(L_1\cap K)_{\widetilde{12}}}^{\mathfrak{q}_1,(L_1\cap K)_{\widetilde{12}}}(\mathbb{C}_{\delta(\mathfrak{u})'})]$$

$$\cong I_{\mathfrak{l},(L\cap K)_{\widetilde{2}}}^{\mathfrak{l},L\cap K}[(I_{\mathfrak{q}_1,(L_1\cap K)_{\widetilde{12}}}^{\mathfrak{l},(L\cap K)_{\widetilde{2}}})^j(\mathcal{F}_{\mathfrak{l}_1,(L_1\cap K)_{\widetilde{12}}}^{\mathfrak{q}_1,(L_1\cap K)_{\widetilde{12}}}(\widetilde{\widetilde{Z}} \otimes \mathbb{C}_{\delta(\mathfrak{u}_1)'})) \otimes \mathbb{C}_{\delta(\mathfrak{u})'}],$$

the last isomorphism following from Corollary 2.97. The $(\mathfrak{l}_1, (L_1 \cap K)_{\widetilde{12}})$ module $\widetilde{\widetilde{Z}} \otimes \mathbb{C}_{\delta(\mathfrak{u}_1)'}$ has $(1, -1, 1)$ act by 1 and $(1, 1, -1)$ act by -1. In the imprecise (but usually convenient) notation of (11.150), we regard $\widetilde{\widetilde{Z}} \otimes \mathbb{C}_{\delta(\mathfrak{u}_1)'}$ as a module for $(\mathfrak{l}_1, (L_1 \cap K)_{\widetilde{12}})$ pushed down by a homomorphism with kernel $\{(1, 1, 1), (1, -1, 1)\}$. We may use the projection $(l_1, z_1, z_2) \mapsto (l_1, z_2)$ for this purpose, the image being a module for $(\mathfrak{l}_1, (L_1 \cap K)_{\widetilde{2}})$. In this notation the last line above is

$$= I_{\mathfrak{l},(L\cap K)_{\widetilde{2}}}^{\mathfrak{l},L\cap K}[(\,^{\mathrm{u}}\mathcal{R}_{\mathfrak{q}_1,(L_1\cap K)_{\widetilde{2}}}^{\mathfrak{l},(L\cap K)_{\widetilde{2}}})^j(\widetilde{\widetilde{Z}} \otimes \mathbb{C}_{\delta(\mathfrak{u}_1)'}) \otimes \mathbb{C}_{\delta(\mathfrak{u})'}]$$

$$= I_{\mathfrak{l},(L\cap K)_{\widetilde{2}}}^{\mathfrak{l},L\cap K}[(\,^{\mathrm{n}}\mathcal{R}_{\mathfrak{q}_1,(L_1\cap K)_{\widetilde{2}}}^{\mathfrak{l},(L\cap K)_{\widetilde{2}}})^j(\widetilde{\widetilde{Z}}) \otimes \mathbb{C}_{\delta(\mathfrak{u})'}].$$

Applying $(\,^{\mathrm{u}}\mathcal{R}_{\mathfrak{q},L\cap K}^{\mathfrak{g},K})^i$ to both sides of the above computation, we may thus interpret all of (11.171) as

$$(\,^{\mathrm{u}}\mathcal{R}_{\mathfrak{q},L\cap K}^{\mathfrak{g},K})^i(\,^{\mathrm{u}}\mathcal{R}_{\mathfrak{q}_1,L_1\cap K}^{\mathfrak{l},L\cap K})^j I_{\mathfrak{l}_1,(L_1\cap K)_{\widetilde{3}}}^{\mathfrak{l}_1,L_1\cap K}(\widetilde{Z} \otimes \mathbb{C}_{\delta(\mathfrak{u}+\mathfrak{u}_1)'})$$

$$\cong (\,^{\mathrm{u}}\mathcal{R}_{\mathfrak{q},L\cap K}^{\mathfrak{g},K})^i I_{\mathfrak{l},(L\cap K)_{\widetilde{2}}}^{\mathfrak{l},L\cap K}[(\,^{\mathrm{n}}\mathcal{R}_{\mathfrak{q}_1,(L_1\cap K)_{\widetilde{2}}}^{\mathfrak{l},(L\cap K)_{\widetilde{2}}})^j(\widetilde{\widetilde{Z}}) \otimes \mathbb{C}_{\delta(\mathfrak{u})'}]$$

$$= (I_{\mathfrak{q},L\cap K}^{\mathfrak{g},K})^i \mathcal{F}_{\mathfrak{l},L\cap K}^{\mathfrak{q},L\cap K} I_{\mathfrak{l},(L\cap K)_{\widetilde{2}}}^{\mathfrak{l},L\cap K}[(\,^{\mathrm{n}}\mathcal{R}_{\mathfrak{q}_1,(L_1\cap K)_{\widetilde{2}}}^{\mathfrak{l},(L\cap K)_{\widetilde{2}}})^j(\widetilde{\widetilde{Z}}) \otimes \mathbb{C}_{\delta(\mathfrak{u})'}]$$

$$\cong (I_{\mathfrak{q},L\cap K}^{\mathfrak{g},K})^i I_{\mathfrak{q},(L\cap K)_{\widetilde{2}}}^{\mathfrak{q},L\cap K} \mathcal{F}_{\mathfrak{l},(L\cap K)_{\widetilde{2}}}^{\mathfrak{q},(L\cap K)_{\widetilde{2}}}[(\,^{\mathrm{n}}\mathcal{R}_{\mathfrak{q}_1,(L_1\cap K)_{\widetilde{2}}}^{\mathfrak{l},(L\cap K)_{\widetilde{2}}})^j(\widetilde{\widetilde{Z}}) \otimes \mathbb{C}_{\delta(\mathfrak{u})'}]$$

$$\text{by Proposition 11.164}$$

$$\cong (I_{\mathfrak{q},(L\cap K)_{\widetilde{2}}}^{\mathfrak{g},K})^i \mathcal{F}_{\mathfrak{l},(L\cap K)_{\widetilde{2}}}^{\mathfrak{q},(L\cap K)_{\widetilde{2}}}[(\,^{\mathrm{n}}\mathcal{R}_{\mathfrak{q}_1,(L_1\cap K)_{\widetilde{2}}}^{\mathfrak{l},(L\cap K)_{\widetilde{2}}})^j(\widetilde{\widetilde{Z}}) \otimes \mathbb{C}_{\delta(\mathfrak{u})'}]$$

$$= (\,^{\mathrm{n}}\mathcal{R}_{\mathfrak{q},L\cap K}^{\mathfrak{g},K})^i(\,^{\mathrm{n}}\mathcal{R}_{\mathfrak{q}_1,(L_1\cap K)_{\widetilde{2}}}^{\mathfrak{l},(L\cap K)_{\widetilde{2}}})^j(\widetilde{\widetilde{Z}}) \quad \text{by (11.165b)}.$$

A similar computation works for the $\mathcal{L}$ functor. Combining the results of these computations with Theorem 11.77, we obtain the following theorem.

Theorem 11.172 (spectral sequences for induction in stages). Let $(\mathfrak{g}, K)$ be a reductive pair, let $(\mathfrak{q}, L \cap K)$ be a parabolic subpair with corresponding Levi pair $(\mathfrak{l}, L \cap K)$, and let $(\mathfrak{q}_1, L_1 \cap K)$ be a parabolic subpair of $(\mathfrak{l}, L \cap K)$ with corresponding Levi pair $(\mathfrak{l}_1, L_1 \cap K)$. Put $\mathfrak{q} = \mathfrak{l} \oplus \mathfrak{u}$, so that $(\mathfrak{q}_1 + \mathfrak{u}, L_1 \cap K)$ is a parabolic subpair of $(\mathfrak{g}, K)$ with corresponding Levi pair $(\mathfrak{l}_1, L_1 \cap K)$. Let $(\mathfrak{l}_1, (L_1 \cap K)_1^{\sim})$, $(\mathfrak{l}, (L \cap K)_2^{\sim})$, $(\mathfrak{l}_1, (L_1 \cap K)_2^{\sim})$, and $(\mathfrak{l}_1, (L_1 \cap K)_3^{\sim})$ be defined as in (11.166) Suppose that $\tilde{Z}$ is in $\mathcal{C}^-(\mathfrak{l}_1, (L_1 \cap K)_3^{\sim})$, and let $\tilde{\tilde{Z}}$ be defined as in (11.169).

(a) The module $({}^n\mathcal{L}_{q_1^-,(L_1 \cap K)_2^{\sim}}^{\mathfrak{l},(L \cap K)_2^{\sim}})_q(\tilde{\tilde{Z}})$ is in $\mathcal{C}^-(\mathfrak{l}, (L \cap K)_2^{\sim})$, and there exists a first-quadrant spectral sequence

$$E_r^{p,q} \implies ({}^n\mathcal{L}_{q_1^-+u^-,L_1 \cap K}^{\mathfrak{g},K})_{p+q}(\tilde{Z})$$

with E_2 term

$$E_2^{p,q} = ({}^n\mathcal{L}_{q^-,L \cap K}^{\mathfrak{g},K})_p({}^n\mathcal{L}_{q_1^-,(L_1 \cap K)_2^{\sim}}^{\mathfrak{l},(L \cap K)_2^{\sim}})_q(\tilde{\tilde{Z}}).$$

The differential on $E_r^{p,q}$ has bidegree $(-r, r-1)$.

(b) The module $({}^n\mathcal{R}_{q_1,(L_1 \cap K)_2^{\sim}}^{\mathfrak{l},(L \cap K)_2^{\sim}})^q(\tilde{\tilde{Z}})$ is in $\mathcal{C}^-(\mathfrak{l}, (L \cap K)_2^{\sim})$, and there exists a first-quadrant spectral sequence

$$E_r^{p,q} \implies ({}^n\mathcal{R}_{q_1+u,L_1 \cap K}^{\mathfrak{g},K})^{p+q}(\tilde{Z})$$

with E_2 term

$$E_2^{p,q} = ({}^n\mathcal{R}_{q,L \cap K}^{\mathfrak{g},K})^p({}^n\mathcal{R}_{q_1,(L_1 \cap K)_2^{\sim}}^{\mathfrak{l},(L \cap K)_2^{\sim}})^q(\tilde{\tilde{Z}}).$$

The differential on $E_r^{p,q}$ has bidegree $(r, 1-r)$.

Corollary 11.173. Let $(\mathfrak{g}, K)$ be a reductive pair, let $(\mathfrak{q}, L \cap K)$ be a parabolic subpair with corresponding Levi pair $(\mathfrak{l}, L \cap K)$, and let $(\mathfrak{q}_1, L_1 \cap K)$ be a parabolic subpair of $(\mathfrak{l}, L \cap K)$ with corresponding Levi pair $(\mathfrak{l}_1, L_1 \cap K)$. Put $\mathfrak{q} = \mathfrak{l} \oplus \mathfrak{u}$, so that $(\mathfrak{q}_1 + \mathfrak{u}, L_1 \cap K)$ is a parabolic subpair of $(\mathfrak{g}, K)$ with corresponding Levi pair $(\mathfrak{l}_1, L_1 \cap K)$. Let $(\mathfrak{l}_1, (L_1 \cap K)_1^{\sim})$, $(\mathfrak{l}, (L \cap K)_2^{\sim})$, $(\mathfrak{l}_1, (L_1 \cap K)_2^{\sim})$, and $(\mathfrak{l}_1, (L_1 \cap K)_3^{\sim})$ be defined as in (11.166) Suppose that $\tilde{Z}$ is in $\mathcal{C}^-(\mathfrak{l}_1, (L_1 \cap K)_3^{\sim})$, and let $\tilde{\tilde{Z}}$ be defined as in (11.169).

(a) If $({}^n\mathcal{L}_{q_1^-,(L_1 \cap K)_2^{\sim}}^{\mathfrak{l},(L \cap K)_2^{\sim}})_q(\tilde{\tilde{Z}}) = 0$ except when $q = q_0$, then

$$({}^n\mathcal{L}_{q_1^-+u^-,L_1 \cap K}^{\mathfrak{g},K})_{p+q_0}(\tilde{Z}) \cong ({}^n\mathcal{L}_{q^-,L \cap K}^{\mathfrak{g},K})_p({}^n\mathcal{L}_{q_1^-,(L_1 \cap K)_2^{\sim}}^{\mathfrak{l},(L \cap K)_2^{\sim}})_{q_0}(\tilde{\tilde{Z}}).$$

(b) If $({}^n\mathcal{R}_{q_1,(L_1 \cap K)_2^{\sim}}^{\mathfrak{l},(L \cap K)_2^{\sim}})^q(\tilde{\tilde{Z}}) = 0$ except when $q = q_0$, then

$$({}^n\mathcal{R}_{q_1+u,L_1 \cap K}^{\mathfrak{g},K})^{p+q_0}(\tilde{Z}) \cong ({}^n\mathcal{R}_{q,L \cap K}^{\mathfrak{g},K})^p({}^n\mathcal{R}_{q_1,(L_1 \cap K)_2^{\sim}}^{\mathfrak{l},(L \cap K)_2^{\sim}})^{q_0}(\tilde{\tilde{Z}}).$$

PROOF. Under the hypotheses of vanishing except when $q = q_0$, the spectral sequences in Theorem 11.172 collapse and degenerate, and the corollary follows immediately.

As we mentioned earlier, Theorem 11.172 and Corollary 11.173 simplify when one or more of $\xi_{2\delta(\mathfrak{u}+\mathfrak{u}_1)'}$, $\xi_{2\delta(\mathfrak{u})'}$, and $\xi_{2\delta(\mathfrak{u}_1)'}$ have square roots, since we can then use the functors (11.158). Rather than formulate results in each of these cases, we shall treat each situation individually as it arises.

8. Discrete Series and Limits

For a reductive pair $(\mathfrak{g}, K)$ in the Harish-Chandra class, the Langlands classification identifies all the irreducible admissible $(\mathfrak{g}, K)$ modules. We shall state the Langlands classification in the next section and then translate it into the language of standard $(\mathfrak{g}, K)$ modules built from a Borel subalgebra of strict type L. Afterward we shall combine this classification with Theorem 11.129 and induction in stages to obtain the following results:

(i) a table of eight settings for classifications of irreducible admissible $(\mathfrak{g}, K)$ modules

(ii) an indication of two classifications besides the one due to Langlands

(iii) Langlands parameters for cohomologically induced modules when the Langlands parameters are known for the inducing module and a dominance condition holds for the infinitesimal character

(iv) versions of the Irreducibility Theorem and the Signature Theorem stated in terms of Langlands parameters.

The Langlands classification involves "discrete series" and "limits of discrete series," and we shall discuss these representations in this section. For orientation, first assume that the underlying group G is connected. Assume further that rank G = rank K, i.e., that $\mathfrak{g}_0$ has a Cartan subalgebra $\mathfrak{t}_0$ lying in $\mathfrak{k}_0$. By Lemma 4.43d the corresponding Cartan subgroup T is connected, hence abelian. Introduce any positive system $\Delta^+(\mathfrak{g}, \mathfrak{t})$, and let $\mathfrak{b}$ be the corresponding Borel subalgebra of $\mathfrak{g}$. Since θ fixes each member of $\Delta^+(\mathfrak{g}, \mathfrak{t})$, $\mathfrak{b}$ is a θ stable parabolic subalgebra. With notation as in (5.6), we shall be interested in certain $(\mathfrak{g}, K)$ modules $A_\mathfrak{q}(\mu)$ for $\mathfrak{q} = \mathfrak{b}$. It is implicit in the notation that $\mathbb{C}_\mu$ is a $(\mathfrak{t}, T)$ module. Therefore μ is analytically integral and must be real on $i\mathfrak{t}_0$.

For the **discrete series** the assumption is that

$$(11.174) \qquad \langle \mu + \delta(\mathfrak{n}), \alpha \rangle > 0 \qquad \text{for all } \alpha \in \Delta^+(\mathfrak{g}, \mathfrak{t}).$$

In this case Theorem 8.2 shows that

(11.175a) $\qquad A_{\mathfrak{b}}(\mu)$ is nonzero irreducible,

and Theorem 9.1 shows that

(11.175b) $\qquad A_{\mathfrak{b}}(\mu)$ is infinitesimally unitary.

Referring to Proposition 10.24, let us show that

(11.175c) $\quad \Lambda = \mu + 2\delta(\mathfrak{n} \cap \mathfrak{p})$ is the unique minimal K type of $A_{\mathfrak{b}}(\mu)$.

In fact, if α is simple in $\Delta^+(\mathfrak{k}, \mathfrak{t})$, then we have

$$
\begin{aligned}
\frac{2\langle \Lambda, \alpha \rangle}{|\alpha|^2} &= \frac{2\langle \mu + 2\delta(\mathfrak{n} \cap \mathfrak{p}), \alpha \rangle}{|\alpha|^2} \\
&= \frac{2\langle \mu + \delta(\mathfrak{n}), \alpha \rangle}{|\alpha|^2} + \frac{2\langle \delta(\mathfrak{n}), \alpha \rangle}{|\alpha|^2} - 2\frac{2\langle \delta(\mathfrak{n} \cap \mathfrak{k}), \alpha \rangle}{|\alpha|^2}.
\end{aligned}
$$

The first term on the right is an integer > 0, the second term is an integer ≥ 1, and the third term is -2. Thus the left side is ≥ 0, and (i) holds in Proposition 10.24. Also

$$
\Lambda + 2\delta(\mathfrak{n} \cap \mathfrak{k}) = \mu + 2\delta(\mathfrak{n} \cap \mathfrak{p}) + 2\delta(\mathfrak{n} \cap \mathfrak{k}) = (\mu + \delta(\mathfrak{n})) + \delta(\mathfrak{n})
$$

exhibits $\Lambda + 2\delta(\mathfrak{n} \cap \mathfrak{k})$ as the sum of two strictly dominant forms. Hence it is dominant, and (ii) holds in Proposition 10.24. This proves (11.175c).

We can reinterpret these facts in the notation of §7 by introducing the **Harish-Chandra parameter** $\lambda = \mu + \delta(\mathfrak{n})$. For a module $\mathbb{C}_\lambda$ in $C^-(\mathfrak{t}, \tilde{T})$ with

(11.176) $\qquad \langle \lambda, \alpha \rangle > 0 \qquad$ for all $\alpha \in \Delta^+(\mathfrak{g}, \mathfrak{t})$,

we have

$$
\begin{aligned}
({}^{\mathfrak{n}}\mathcal{R}_{\mathfrak{b},T}^{\mathfrak{g},K})^S(\mathbb{C}_\lambda) &= ({}^{\mathfrak{u}}\mathcal{R}_{\mathfrak{b},T}^{\mathfrak{g},K})^S(\mathbb{C}_\lambda \otimes_{\mathbb{C}} \mathbb{C}_{\delta(\mathfrak{n})}) \\
&\cong \mathcal{R}^S(\mathbb{C}_{\lambda+\delta(\mathfrak{n})} \otimes_{\mathbb{C}} (\textstyle\bigwedge^{\mathrm{top}}\mathfrak{b})^*) \cong \mathcal{R}^S(\mathbb{C}_{\lambda-\delta(\mathfrak{n})}) \cong A_{\mathfrak{b}}(\mu)
\end{aligned}
$$

by (11.150), (11.73), and Theorem 5.99c. Under the assumption (11.176), let V_K^λ denote the $(\mathfrak{g}, K)$ module $({}^{\mathfrak{n}}\mathcal{R}_{\mathfrak{b},T}^{\mathfrak{g},K})^S(\mathbb{C}_\lambda)$. (Because of

(11.176), the parameter λ uniquely determines $\mathfrak{b}$; thus we are justified in dropping $\mathfrak{b}$ from the notation.) Then (11.175) says that

(11.177a) V_K^λ is nonzero irreducible

(11.177b) V_K^λ is infinitesimally unitary

(11.177c)

$\Lambda = \lambda + \delta(\mathfrak{n}) - 2\delta(\mathfrak{n} \cap \mathfrak{k})$ is the unique minimal K type of V_K^λ.

Also we know from Corollary 5.25 that

(11.177d) V_K^λ has infinitesimal character λ.

In the presence of (11.177a) and (11.177b), Theorem 0.6 says that V_K^λ is the underlying $(\mathfrak{g}, K)$ module of an irreducible unitary representation of G. Following Harish-Chandra, we write π_λ for this representation. We have earlier used the notation V_K^ξ for the underlying $(\mathfrak{g}, K)$ module of a representation ξ of G. For the case of $\xi = \pi_\lambda$, $V_K^{\pi_\lambda}$ is just V_K^λ, and we shall stick with the simpler notation V_K^λ.

As defined in §3 of the Introduction, "discrete series" are irreducible unitary representations with some (or equivalently any) nonzero matrix coefficient in $L^2(G)$. The relationship of the π_λ's to this notion is as follows.

Theorem 11.178. Let the reductive group G be connected. If $\mathfrak{g}_0$ has rank $G \neq \operatorname{rank} K$, then G has no discrete series. If $\mathfrak{g}_0$ has rank $G = \operatorname{rank} K$, fix a Cartan subalgebra $\mathfrak{t}_0$ of $\mathfrak{g}_0$ lying in $\mathfrak{k}_0$. Then

   (a) each representation π_λ is in the discrete series of G
   (b) the representations π_λ, as λ (and $\mathfrak{b}$) vary, exhaust the discrete series of G
   (c) π_λ is equivalent with $\pi_{\lambda'}$ if and only if λ and λ' are conjugate via $W(\mathfrak{k}, \mathfrak{t})$.

We omit the proof, referring the reader to the Notes. All the conclusions in the theorem are difficult except (c). For (c), suppose λ and λ' are conjugate by an element of $W(\mathfrak{k}, \mathfrak{t})$. Then the conjugacy is implemented by a member of K, and this member of K can be shown to carry π_λ to $\pi_{\lambda'}$. Conversely suppose π_λ is equivalent with $\pi_{\lambda'}$. Then π_λ and $\pi_{\lambda'}$ have minimal K type parameters as in (11.177c) that are conjugate via $W(\mathfrak{k}, \mathfrak{t})$, and we can argue as follows that λ and λ' are conjugate via $W(\mathfrak{k}, \mathfrak{t})$. Fix a positive system $\Delta^+(\mathfrak{k}, \mathfrak{t})$. By the direct part of (c), we may assume that λ and λ' are dominant for $\Delta^+(\mathfrak{k}, \mathfrak{t})$. Hence their minimal K type parameters are now equal. Let $\mathfrak{b} = \mathfrak{t} \oplus \mathfrak{n}$ and $\mathfrak{b}' = \mathfrak{t} \oplus \mathfrak{n}'$ be the Borel subalgebras of $\mathfrak{g}$ relative to which λ and λ' are dominant. Because of the equality of the minimal K type parameters, formula (11.177c) and the equality $2\delta(\mathfrak{n} \cap \mathfrak{k}) = 2\delta(\mathfrak{n}' \cap \mathfrak{k})$ imply

$$\lambda + \delta(\mathfrak{n}) = \lambda' + \delta(\mathfrak{n}').$$

The left side is strictly dominant for $\mathfrak{n}$, and the right side is strictly dominant for $\mathfrak{n}'$. Hence $\mathfrak{n} = \mathfrak{n}'$ and therefore also $\lambda = \lambda'$.

It will be convenient to allow the term **discrete series** to refer also to the underlying $(\mathfrak{g}, K)$ module of a discrete-series representation.

Still with G connected, let us suppose that the module $\mathbb{C}_\lambda$ in $C^-(\mathfrak{t}, \tilde{T})$ has

(11.179) $\qquad \langle \lambda, \alpha \rangle \geq 0 \qquad$ for all $\alpha \in \Delta^+(\mathfrak{g}, \mathfrak{t})$,

with equality for some α. In this case, $\mathfrak{b}$ is not determined by λ, and we need to retain it in the notation. Let $V_K^{\lambda, \mathfrak{b}} = ({}^n \mathcal{R}_{\mathfrak{b}, T}^{\mathfrak{g}, K})^S(\mathbb{C}_\lambda)$.

Proposition 11.180. With G connected and with λ satisfying (11.179), the $(\mathfrak{g}, K)$ module $V_K^{\lambda, \mathfrak{b}}$ is 0 if and only if $\langle \lambda, \alpha \rangle = 0$ for some simple root α in $\Delta^+(\mathfrak{g}, \mathfrak{t})$ that is compact.

PROOF. Suppose that $\langle \lambda, \alpha \rangle = 0$ for some simple root α in $\Delta^+(\mathfrak{g}, \mathfrak{t})$ that is compact. Put $\Lambda = \lambda + \delta(\mathfrak{n}) - 2\delta(\mathfrak{n} \cap \mathfrak{k})$. If Λ' is $\Delta^+(\mathfrak{k}, \mathfrak{t})$ dominant and integral, then, according to Theorem 8.29, the K type with highest weight Λ' occurs in $V_K^{\lambda, \mathfrak{b}}$ with multiplicity

(11.181) $\qquad \sum_{s \in W(\mathfrak{k}, \mathfrak{t})} (\det s) \mathcal{P}(s(\Lambda' + \delta(\mathfrak{n} \cap \mathfrak{k})) - (\Lambda + \delta(\mathfrak{n} \cap \mathfrak{k}))),$

where $\mathcal{P}(\nu)$ is the multiplicity of ν as a weight in $S(\mathfrak{n} \cap \mathfrak{p})$. Since α is compact and is simple for $\Delta^+(\mathfrak{g}, \mathfrak{t})$, the root reflection s_α carries $\Delta(\mathfrak{n} \cap \mathfrak{p})$ to itself. Thus

(11.182) $\qquad \mathcal{P}(s_\alpha \nu) = \mathcal{P}(\nu).$

Also

$$\frac{2\langle \Lambda + \delta(\mathfrak{n} \cap \mathfrak{k}), \alpha \rangle}{|\alpha|^2} = \frac{2\langle \lambda, \alpha \rangle}{|\alpha|^2} + \frac{2\langle \delta(\mathfrak{n}) - \delta(\mathfrak{n} \cap \mathfrak{k}), \alpha \rangle}{|\alpha|^2} = 0$$

since both terms in the expansion are 0, and hence

(11.183) $s_\alpha(\Lambda + \delta(\mathfrak{n} \cap \mathfrak{k})) = \Lambda + \delta(\mathfrak{n} \cap \mathfrak{k}).$

Thus

$$\mathcal{P}(s_\alpha s(\Lambda' + \delta(\mathfrak{n} \cap \mathfrak{k})) - (\Lambda + \delta(\mathfrak{n} \cap \mathfrak{k})))$$
$$= \mathcal{P}(s(\Lambda' + \delta(\mathfrak{n} \cap \mathfrak{k})) - s_\alpha(\Lambda + \delta(\mathfrak{n} \cap \mathfrak{k}))) \qquad \text{by (11.182)}$$
$$= \mathcal{P}(s(\Lambda' + \delta(\mathfrak{n} \cap \mathfrak{k})) - (\Lambda + \delta(\mathfrak{n} \cap \mathfrak{k}))) \qquad \text{by (11.183).}$$

Consequently the terms of (11.181) cancel in pairs, and the multiplicity is 0.

Conversely suppose $\langle \lambda, \alpha \rangle \neq 0$ for every simple root of $\Delta^+(\mathfrak{g}, \mathfrak{t})$ that is compact. With $\Lambda = \lambda + \delta(\mathfrak{n}) - 2\delta(\mathfrak{n} \cap \mathfrak{k})$, we show that $\langle \Lambda, \alpha \rangle \geq 0$ for α simple in $\Delta^+(\mathfrak{k}, \mathfrak{t})$. If $\langle \lambda, \alpha \rangle > 0$, then the argument is as in the proof of (11.175c). Otherwise we have $\langle \lambda, \alpha \rangle = 0$ and $2\langle \delta(\mathfrak{n}), \alpha \rangle / |\alpha|^2 \geq 2$. Thus

$$\frac{2\langle \Lambda, \alpha \rangle}{|\alpha|^2} = \frac{2\langle \lambda, \alpha \rangle}{|\alpha|^2} + \frac{2\langle \delta(\mathfrak{n}), \alpha \rangle}{|\alpha|^2} - 2\frac{2\langle \delta(\mathfrak{n} \cap \mathfrak{k}), \alpha \rangle}{|\alpha|^2} \geq 2 - 2 = 0.$$

Then Λ is dominant for K, and the K type τ_Λ is in the bottom layer of $V_K^{\lambda, \mathfrak{b}}$. In particular, $V_K^{\lambda, \mathfrak{b}}$ is not 0.

When (11.179) holds and $V_K^{\lambda, \mathfrak{b}}$ is not 0, $V_K^{\lambda, \mathfrak{b}}$ has the following properties:

(11.184a) $V_K^{\lambda, \mathfrak{b}}$ is irreducible

(11.184b) $V_K^{\lambda, \mathfrak{b}}$ is infinitesimally unitary

(11.184c)
        $\Lambda = \lambda + \delta(\mathfrak{n}) - 2\delta(\mathfrak{n} \cap \mathfrak{k})$ is the unique minimal K type of $V_K^{\lambda, \mathfrak{b}}$

(11.184d) $V_K^{\lambda, \mathfrak{b}}$ has infinitesimal character λ.

The verification that Λ in (c) is K dominant was carried out in the proof of Proposition 11.180, and the proofs of the other statements are the same as for (11.177). By Theorem 0.6, $V_K^{\lambda, \mathfrak{b}}$ is the underlying $(\mathfrak{g}, K)$ module of an irreducible unitary representation $\pi_{\lambda, \mathfrak{b}}$ of G. The module $V_K^{\lambda, \mathfrak{b}}$ or the representation $\pi_{\lambda, \mathfrak{b}}$ is called a **limit of discrete series**. Arguing as for Theorem 11.178 (but now taking $\mathfrak{b}$ and $\mathfrak{b}'$ as specified with λ and λ'), we see in the nonzero case that

(11.184e)
                $V_K^{\lambda, \mathfrak{b}}$ is equivalent with $V_K^{\lambda', \mathfrak{b}'}$ if and only if
                $(\lambda, \mathfrak{b})$ is conjugate to $(\lambda', \mathfrak{b}')$ via $W(\mathfrak{k}, \mathfrak{t})$.

Now let us drop the assumption that G is connected. By Corollary 7.223, the restriction to G_0 of a discrete-series representation of G is a finite direct sum of discrete series of G_0. By Theorem 11.178, there are no discrete series unless $\operatorname{rank} G = \operatorname{rank} K$. Thus we may assume that $\mathfrak{g}_0$ has a Cartan subalgebra $\mathfrak{t}_0$ contained in $\mathfrak{k}_0$. Let $T = Z_K(\mathfrak{t}_0)$ be the corresponding Cartan subgroup of G.

First we suppose that $G = TG_0$. Let $\mathfrak{b}$ be a Borel subalgebra of $\mathfrak{g}$, and let $\tilde{Z}$ be an irreducible module in $\mathcal{C}^-(\mathfrak{t}, \tilde{T})$ with infinitesimal character λ satisfying

$$(11.185) \qquad \langle \lambda, \alpha \rangle \geq 0 \qquad \text{for all } \alpha \in \Delta^+(\mathfrak{g}, \mathfrak{t}).$$

Anticipating the introduction shortly of a group representation $\xi = \xi(\tilde{Z}, \mathfrak{b})$, put

$$V_K^\xi = V_K^{\xi(\tilde{Z}, \mathfrak{b})} = ({}^n\mathcal{R}_{\mathfrak{b},T}^{\mathfrak{g},K})^S(\tilde{Z}).$$

By Theorems 8.2 and 9.1,

$$(11.186a) \qquad V_K^{\xi(\tilde{Z},\mathfrak{b})} \text{ is zero or is irreducible}$$

$$(11.186b) \qquad V_K^{\xi(\tilde{Z},\mathfrak{b})} \text{ is infinitesimally unitary.}$$

Hence $V_K^{\xi(\tilde{Z},\mathfrak{b})}$, if nonzero, is the underlying $(\mathfrak{g}, K)$ module of an irreducible unitary representation $\xi = \xi(\tilde{Z}, \mathfrak{b})$ of G. By Proposition 5.14 and passage to derived functors,

$$(11.187) \qquad \mathcal{F}_{\mathfrak{g},K}^{\mathfrak{g},K_0}(V_K^{\xi(\tilde{Z},\mathfrak{b})}) \cong ({}^n\mathcal{R}_{\mathfrak{b},T}^{\mathfrak{g},K})^S(\mathcal{F}_{\mathfrak{t},T^\sim}^{\mathfrak{t},(T_0)^\sim}(\tilde{Z})).$$

The forgetful functor applied to $\tilde{Z}$ is a direct sum of copies of $\mathbb{C}_\lambda$, and hence the theory in the connected case implies that

$(11.188a)$

$V_K^{\xi(\tilde{Z},\mathfrak{b})}$ is zero if and only if

$\quad \langle \lambda, \alpha \rangle = 0$ for some simple $\alpha \in \Delta^+(\mathfrak{g}, \mathfrak{t})$ that is compact

$(11.188b)$

$V_K^{\xi(\tilde{Z},\mathfrak{b})}$ is in the discrete series if and only if λ is nonsingular.

In the nonzero case, (11.188a) and the proof of Proposition 11.180 show that $\Lambda = \lambda + \delta(\mathfrak{n}) - 2\delta(\mathfrak{n} \cap \mathfrak{k})$ is dominant for K. Consequently, by Theorem 4.83, there exists a unique K type such that T acts by

$$\tilde{Z} \otimes_{\mathbb{C}} \mathbb{C}_{\delta(\mathfrak{n})} \otimes_{\mathbb{C}} \mathbb{C}_{-2\delta(\mathfrak{n} \cap \mathfrak{k})}$$

in the highest-weight space. From (11.184c), (11.187), and the defini-
tions it is clear that this K type is the unique minimal K type of $V_K^{\xi(\tilde{Z},\mathfrak{b})}$.
Just as in the proofs of (11.184e) and Theorem 11.178c, it follows in the
nonzero case that

(11.188c) $\qquad$ $V_K^{\xi(\tilde{Z},\mathfrak{b})}$ is equivalent with $V_K^{\xi(\tilde{Z}',\mathfrak{b}')}$ if and only if

$\qquad\qquad\qquad$ $(\tilde{Z},\mathfrak{b})$ and $(\tilde{Z}',\mathfrak{b}')$ are conjugate via $W(K,\tilde{T})$.

Finally every discrete series is of the form (11.188c) as a consequence
of the completeness in Theorem 11.178b.

For general G, let $G_1 = TG_0$. Again let

(11.189) $\qquad\qquad\qquad$ $V_K^{\xi(\tilde{Z},\mathfrak{b})} = ({}^n\mathcal{R}_{\mathfrak{b},T}^{\mathfrak{g},K})^S(\tilde{Z})$.

By Corollary 8.28,

(11.190a) $\qquad\qquad$ $V_K^{\xi(\tilde{Z},\mathfrak{b})}$ is zero or is irreducible.

Theorem 9.1 still shows that

(11.190b) $\qquad\qquad$ $V_K^{\xi(\tilde{Z},\mathfrak{b})}$ is infinitesimally unitary.

Since $P_{\mathfrak{g},TK_0}^{\mathfrak{g},K}$ carries nonzero modules to nonzero modules, it follows
from (11.188a) that

(11.190c)

$\qquad$ $V_K^{\xi(\tilde{Z},\mathfrak{b})}$ is zero if and only if

$\qquad\qquad$ $\langle \lambda, \alpha \rangle = 0$ for some simple $\alpha \in \Delta^+(\mathfrak{g},\mathfrak{t})$ that is compact

(11.190d)

$\qquad$ $V_K^{\xi(\tilde{Z},\mathfrak{b})}$ is in the discrete series if and only if λ is nonsingular.

The nonzero $V_K^{\xi(\tilde{Z},\mathfrak{b})}$'s for which (11.190d) fails are called **limits of
discrete series**. Every discrete-series module is of the form (11.190d)
as a consequence of the same conclusion for $G_1 = TG_0$.

In order to use (11.189) to work with discrete series and limits, it is
helpful to understand the structure of T. This is given in the following
proposition.

Proposition 11.191. Let the reductive pair $(\mathfrak{g}, K)$ be in the Harish-Chandra class, and suppose that rank G = rank K. Fix a Cartan subalgebra $\mathfrak{t}_0$ of $\mathfrak{g}_0$ lying in $\mathfrak{k}_0$. Then the Cartan subgroup $T = Z_K(\mathfrak{t}_0)$ of G is a commuting product $T = Z_K(\mathfrak{g}_0)T_0$.

PROOF. Let $G_\mathbb{C}$ be the complex adjoint group, and let $T_\mathbb{C}$ be the analytic subgroup with Lie algebra $\mathfrak{t}$. Fix a positive system $\Delta^+(\mathfrak{g}, \mathfrak{t})$, let $\alpha_1, \ldots, \alpha_l$ be the simple roots, and let $e_{\alpha_1}, \ldots, e_{\alpha_l}$ be corresponding nonzero root vectors. If t is in T, then $\mathrm{Ad}(t)$ is in $T_\mathbb{C}$ since $(\mathfrak{g}, K)$ is in the Harish-Chandra class. The group $\mathrm{Ad}(T)$ is compact, and it follows that there exist complex numbers $e^{i\theta_1}, \ldots, e^{i\theta_l}$ of absolute value 1 such that $\mathrm{Ad}(t)e_{\alpha_j} = e^{i\theta_j}e_{\alpha_j}$ for $1 \leq j \leq l$. If we define $h \in \mathfrak{t}_0$ so that $\alpha_j(h) = i\theta_j$ for all j and put $t_0 = \exp h$, then $\mathrm{Ad}(t_0)e_{\alpha_j} = e^{i\theta_j}e_{\alpha_j}$ for all j. Consequently tt_0^{-1} is in T, and $\mathrm{Ad}(tt_0^{-1})$ centralizes $\mathfrak{t}$ and the e_{α_j}. Since $Z_\mathfrak{g}(tt_0^{-1})$ is conjugate closed and since the smallest conjugate-closed subalgebra of $\mathfrak{g}$ containing $\mathfrak{t}$ and the e_{α_j} is $\mathfrak{g}$ itself, we deduce that tt_0^{-1} is in $Z_K(\mathfrak{g}_0)$. Hence $t = (tt_0^{-1})t_0$ is the required decomposition.

Proposition 11.191 implies that every irreducible representation of T is of the form $\eta = \eta'\xi_\mu$. Here η' is an irreducible representation of $Z_K(\mathfrak{g}_0) = Z_T(\mathfrak{g}_0)$ and ξ_μ is a one-dimensional character of T_0 with differential $\mu \in i\mathfrak{t}_0^*$. The notation means that η acts on the space of the representation η', with $z \in Z_K(\mathfrak{g}_0)$ acting by $\eta'(z)$ and with $t \in T_0$ acting by $\xi_\mu(t)$. Conversely if we are given η' and ξ_μ, then $\eta = \eta'\xi_\mu$ is a well-defined representation of T if and only if η' and ξ_μ act by the same scalar on $T_0 \cap Z_K(\mathfrak{g}_0) = Z_{K_0}(\mathfrak{g}_0)$. With this notation we can describe the K types of a discrete series or limit when T meets every component of K.

Proposition 11.192. Let the reductive pair $(\mathfrak{g}, K)$ be in the Harish-Chandra class, and suppose that rank G = rank K. Fix a Cartan subalgebra $\mathfrak{t}_0$ of $\mathfrak{g}_0$ lying in $\mathfrak{k}_0$, and suppose that T meets every component of K. Then the K types of a (nonzero) discrete series or limit $\xi(\tilde{Z}, \mathfrak{b})$ may be described as follows: Write $\mathfrak{b} = \mathfrak{t} \oplus \mathfrak{n}$, let η' be the restriction to $Z_T(\mathfrak{g}_0)$ of $\tilde{Z} \otimes \mathbb{C}_{\delta(\eta)}$, and let λ be the infinitesimal character of $\tilde{Z}$. Define $\Lambda = \lambda + \delta(\mathfrak{n}) - 2\delta(\mathfrak{n} \cap \mathfrak{k})$. Then Λ is the unique weight of a dominant irreducible representation $\eta_\Lambda = \eta'\xi_\Lambda$ of T. Let τ_Λ be the unique irreducible representation of K whose highest weight in the sense of Theorem 4.83 is η_Λ. Then

(a) τ_Λ occurs in $\xi(\tilde{Z}, \mathfrak{b})$ with multiplicity 1
(b) the most general K type in $\xi(\tilde{Z}, \mathfrak{b})$ corresponds to a dominant irreducible representation $\eta_{\Lambda'}$ of T with $\eta_{\Lambda'} = \eta'\xi_{\Lambda'}$, where $\xi_{\Lambda'}$ is

the character of T_0 with differential Λ' and Λ' is of the form

$$\Lambda' = \Lambda + \sum_{\beta_i \in \Delta(\mathfrak{n} \cap \mathfrak{p})} n_i \beta_i \qquad \text{with } n_i \geq 0$$

(c) τ_Λ is the unique minimal K type of $\xi(\tilde{Z}, \mathfrak{b})$.

REMARK. It will be clear from the proof that the K type in (b) occurs in ξ with the same multiplicity that the representation of K_0 with highest weight Λ' occurs in the $(\mathfrak{g}, K_0)$ module $A_\mathfrak{b}(\lambda - \delta)$. This multiplicity is given by Theorem 8.29.

PROOF. We have

$$V_K^{\xi(\tilde{Z},\mathfrak{b})} = ({}^{\mathfrak{n}}\mathcal{R}_{\mathfrak{b},T}^{\mathfrak{g},K})^S(\tilde{Z}) = ({}^{\mathfrak{u}}\mathcal{R}_{\mathfrak{b},T}^{\mathfrak{g},K})^S(\tilde{Z} \otimes \mathbb{C}_{\delta(\mathfrak{n})}) = \mathcal{R}^S(\tilde{Z} \otimes \mathbb{C}_{-\delta(\mathfrak{n})}).$$

Here $\tilde{Z} \otimes \mathbb{C}_{-\delta(\mathfrak{n})}$ may be regarded as a $(\mathfrak{t}, T)$ module with unique weight $\lambda - \delta$. The second half of the proof of Proposition 11.180 shows that

$$\Lambda = \lambda + \delta(\mathfrak{n}) - 2\delta(\mathfrak{n} \cap \mathfrak{k}) = (\lambda - \delta(\mathfrak{n})) + 2\delta(\mathfrak{n} \cap \mathfrak{p})$$

is dominant for $\Delta^+(\mathfrak{k}, \mathfrak{t})$, and thus $\tilde{Z} \otimes \mathbb{C}_{-\delta(\mathfrak{n})} \otimes \mathbb{C}_{2\delta(\mathfrak{n} \cap \mathfrak{p})}$ is a dominant representation of T. If τ_λ denotes the corresponding irreducible representation of K (given by Theorem 4.83), then τ_Λ is in the bottom layer for ξ and hence has multiplicity 1. This proves (a).

Theorem 5.35 shows that the most general K type of ξ has the property that its corresponding dominant representation $\eta_{\Lambda'}$ of T occurs in

$$\eta_\Lambda \otimes_\mathbb{C} S^n(\mathfrak{n} \cap \mathfrak{p}) \qquad \text{for some } n.$$

The members of $Z_T(\mathfrak{g}_0)$ act as 1 on $S^n(\mathfrak{n} \cap \mathfrak{p})$, and therefore $\eta_{\Lambda'}$ occurs in

$$(\eta' \otimes \xi_\Lambda) \otimes_\mathbb{C} (1 \otimes S^n(\mathfrak{n} \cap \mathfrak{p})) = \eta' \otimes_\mathbb{C} (\xi_\Lambda \otimes S^n(\mathfrak{n} \cap \mathfrak{p})).$$

Conclusion (b) follows.

For (c) we have

$$\Lambda + 2\delta(\mathfrak{n} \cap \mathfrak{k}) = \lambda + \delta(\mathfrak{n})$$

and
$$\Lambda' + 2\delta(\mathfrak{n} \cap \mathfrak{k}) = \lambda + \delta(\mathfrak{n}) + \sum n_i \beta_i.$$

Since $\lambda + \delta(\mathfrak{n})$ is dominant and the β_i are in $\Delta(\mathfrak{n})$, it follows that $|\Lambda' + 2\delta(\mathfrak{n} \cap \mathfrak{k})|^2 \geq |\Lambda + 2\delta(\mathfrak{n} \cap \mathfrak{k})|^2$ with equality only if $\Lambda' = \Lambda$.

9. Langlands Parameters

Let $(\mathfrak{g}, K)$ be a reductive pair in the Harish-Chandra class, and let G be the underlying reductive group. The "Langlands classification" relates all irreducible admissible $(\mathfrak{g}, K)$ modules to "standard continuous-series representations" of G. Continuous-series representations were defined in (11.23). We say that a continuous-series representation $I_{MAN}^G(\xi, \nu)$ is **standard** if

(i) ξ is a discrete-series or limit of discrete-series representation of M

(ii) ν satisfies $\langle \operatorname{Re}\nu, \beta \rangle \geq 0$ for all $\mathfrak{a}$-roots β that are positive relative to $\mathfrak{n}$.

If (i) holds and if (ii) is replaced by the stronger condition

(ii') ν has $\operatorname{Re}\nu = 0$,

then $I_{MAN}^G(\xi, \nu)$ is called a **standard tempered representation**.

To any standard continuous-series representation

$$(11.193) \qquad\qquad I_{MAN}^G(\xi, \nu)$$

of G, we can associate canonically a standard tempered representation of a certain subgroup M' as follows: By the discussion following Lemma 11.19, we can choose a minimal parabolic subgroup $M_\mathfrak{p} A_\mathfrak{p} N_\mathfrak{p} \subseteq MAN$. That construction uses a Cartan subalgebra $\mathfrak{h}_0 = \mathfrak{a}_0 \oplus (\mathfrak{a}_M)_0 \oplus \mathfrak{t}_0$ as in (11.18), as well as a positive system $\Delta^+(\mathfrak{g}, \mathfrak{h})$ that takes $\mathfrak{a}_0$ as larger than $(\mathfrak{a}_M)_0$ and takes $(\mathfrak{a}_M)_0$ as larger than $i\mathfrak{t}_0$. Let

$$\alpha_1, \ldots, \alpha_i, \alpha_{i+1}, \ldots \alpha_j, \alpha_{j+1}, \ldots \alpha_l$$

be the simple roots, with $\alpha_{j+1}, \ldots, \alpha_l$ carried on $\mathfrak{t}$ and $\alpha_{i+1}, \ldots, \alpha_l$ carried on $\mathfrak{a}_M \oplus \mathfrak{t}$. The subset of simple roots $\alpha_{i+1}, \ldots, \alpha_l$ defines the parabolic subalgebra $\mathfrak{m} \oplus \mathfrak{a} \oplus \mathfrak{n}$ of $\mathfrak{g}$ by the procedure of Proposition 4.57, and we define another parabolic subalgebra $\mathfrak{m}' \oplus \mathfrak{a}' \oplus \mathfrak{n}'$ to correspond to the subset

$$\{\alpha_k \text{ simple} \mid \langle \operatorname{Re}\nu, \alpha_k \rangle = 0\}.$$

Since each α_k with $k \geq i+1$ is in this set,

$$\mathfrak{m}' \oplus \mathfrak{a}' \oplus \mathfrak{n}' \supseteq \mathfrak{m} \oplus \mathfrak{a} \oplus \mathfrak{n}.$$

The roots α contributing to $\mathfrak{n}'$ are those with $\langle \operatorname{Re}\nu, \alpha_k \rangle > 0$, i.e.,

$$\mathfrak{n}' = \bigoplus_{\substack{\mathfrak{a} \in \Delta^+(\mathfrak{g}, \mathfrak{h}), \\ \langle \operatorname{Re}\nu, \alpha_k \rangle > 0}} \mathfrak{g}_\alpha,$$

and these are closed under conjugation since $\mathrm{Re}\,\nu$ is real. Hence $\mathfrak{m}' \oplus \mathfrak{a}' \oplus \mathfrak{n}'$ is real, and we can form the corresponding parabolic subgroup $M'A'N'$. The definition of $\mathfrak{a}'$ is

$$\mathfrak{a}' = \{h' \in \mathfrak{a} \mid \alpha_k(h') = 0 \text{ for all } \alpha_k \text{ with } \langle \mathrm{Re}\,\nu, \alpha_k \rangle = 0\}.$$

We let

$$\overset{*}{\mathfrak{n}''} = \bigoplus_{\substack{\mathfrak{a} \in \Delta^+(\mathfrak{g}, \mathfrak{h}), \\ \alpha|_{\mathfrak{a}_p} \neq 0, \\ \langle \mathrm{Re}\,\nu, \alpha_k \rangle = 0}} \mathfrak{g}_\alpha$$

and

$$\mathfrak{a}_0'' = \text{orthogonal complement of } \mathfrak{a}_0' \text{ in } \mathfrak{a}_0, \text{ so that } \mathfrak{a}_0 = \mathfrak{a}_0'' \oplus \mathfrak{a}_0'.$$

Then M' is a reductive group in the Harish-Chandra class, and $MA''N''$ is a parabolic subgroup of M'. The member $h_{\mathrm{Re}\,\nu}$ of $\mathfrak{a}_0$ corresponding to $\mathrm{Re}\,\nu$ in $\mathfrak{a}_0^*$ is in $\mathfrak{a}_0'$ by definition, and hence it is orthogonal to $\mathfrak{a}_0''$. In other words,

$$(11.194) \qquad \mathrm{Re}\,\nu(h'') = 0 \qquad \text{for } h'' \in \mathfrak{a}''.$$

The **associated standard tempered representation** is

$$(11.195) \qquad I_{MA''N''}^{M'}(\xi, \nu|_{\mathfrak{a}''}).$$

According to (11.194), $\nu|_{\mathfrak{a}''}$ is purely imaginary. By Corollary 11.39, the representation (11.195) is unitary. Since (11.195) has finite length as an $(\mathfrak{m}', K \cap M')$ module, it is the direct sum of finitely many irreducible unitary representations of M'.

We can recover the given standard continuous-series representation (11.193) from (11.195) by the **induction-in-stages** formula

$$(11.196) \qquad I_{MAN}^{G}(\xi, \nu) = I_{M'A'N'}^{G}(I_{MA''N''}^{M'}(\xi, \nu|_{\mathfrak{a}''}), \nu|_{\mathfrak{a}'}).$$

This formula is a global version of the corresponding formula in Proposition 2.19 for the I functor, in view of Proposition 11.47. To verify (11.196), observe that a map from right to left in (11.196) is $F \mapsto F(\cdot)(1)$, where F is in the space of continuous members for $I_{M'A'N'}^{G}(-)$, $F(x)$ is in the space of continuous members for $I_{MA''N''}^{M'}(-)$, and $F(x)(1)$ is in the space of ξ. It is easy to see that this map is an isomorphism.

The special feature of the decomposition (11.196) is that the inner representation (of M') is unitary, as we have already observed, while the outer representation has

(11.197) $\langle \operatorname{Re} \nu|_{\mathfrak{a}'}, \beta \rangle > 0$ for every positive $\mathfrak{a}'$-root β.

In fact, $\operatorname{Re} \nu$ is carried on $\mathfrak{a}'$ by (11.194), and this inner product is therefore positive by the definition of $\mathfrak{n}'$.

Ostensibly the construction of the associated standard tempered representation (11.195) depends on the choice of $\mathfrak{h}$ and certain features of a positive system $\Delta^+(\mathfrak{g}, \mathfrak{h})$, but we can readily check that this is not so. For example, we could work with the "simple" $\mathfrak{a}$-roots $\alpha_1|_{\mathfrak{a}}, \ldots, \alpha_i|_{\mathfrak{a}}$ throughout, and then no choices are involved. Thus the associated standard tempered representation is canonical.

Theorem 11.198 (Langlands classification).

(a) Every standard tempered representation is the direct sum of finitely many irreducible standard tempered representations.

(b) A standard tempered representation $I_{MAN}^G(\xi, \nu)$ is irreducible if $\langle \nu, \beta \rangle \neq 0$ for every $\mathfrak{a}$-root β.

(c) If the associated standard tempered representation of a standard continuous-series representation $I_{MAN}^G(\xi, \nu)$ is irreducible, then $I_{MAN}^G(\xi, \nu)$ has a unique irreducible quotient and this quotient is infinitesimally equivalent with an irreducible subrepresentation of $I_{MAN^-}^G(\xi, \nu)$.

(d) Every irreducible admissible $(\mathfrak{g}, K)$ module is equivalent with the underlying $(\mathfrak{g}, K)$ module of the unique irreducible quotient of a standard continuous-series representation whose associated standard tempered representation is irreducible.

REMARKS.

1) We omit the proof, giving references in the Notes.

2) A unique irreducible quotient as in (c) is called a **Langlands quotient**. Briefly the main conclusion is (d): Every irreducible admissible $(\mathfrak{g}, K)$ module is infinitesimally equivalent with a Langlands quotient. Except in the situation of (b), it will normally be inconvenient for us to decide when standard tempered representations are irreducible, and we shall consequently allow the term "Langlands quotient" to refer to any irreducible quotient of *any* standard continuous-series representation. When an irreducible admissible $(\mathfrak{g}, K)$ module V is exhibited as a Langlands quotient, we shall say that V has **Langlands quotient parameters** (MAN, ξ, ν).

3) The way we have stated the classification, the Langlands quotient parameters of an irreducible admissible $(\mathfrak{g}, K)$ module are not unique,

even if one insists that the associated standard tempered representation is irreducible. All the nontrivial ingredients in the nonuniqueness come from the tempered case. In this case with M, A, ξ, and ν fixed, any N is allowable, by (ii) in the definition of standard continuous-series representation at the beginning of the section. In addition, when an irreducible standard tempered representation is realized as $I^G_{MAN}(\xi, \nu)$, it is sometimes possible to enlarge or shrink MAN, obtaining the same representation with new parameters. A simple example occurs for $G = SL^{\pm}(2, \mathbb{R})$, in which there is a unique limit of discrete-series representation; this representation can be regarded as a limit of discrete series (with $MAN = G$) or as a full principal-series representation (with MAN equal to a minimal parabolic subgroup). A qualitatively different example occurs in $G = SU(2, 1)$ with the limit of discrete series built from $\lambda = 0$ when both simple roots are noncompact; here λ is orthogonal to a compact root. This representation similarly can be regarded as a limit of discrete series (with $MAN = G$) or as a full principal-series representation (with MAN equal to a minimal parabolic subgroup). There are two standard ways of formulating a theorem about the tempered case that asserts uniqueness up to conjugacy of the parameters M, A, ξ, and ν: One is to insist that MAN is as small as possible (in which case the representation is said to be given with "nondegenerate data"), and the other is to insist that MAN is as large as possible (in which case the representation is said to be realized as "final"). Since we shall not be investigating irreducibility of standard tempered representations, these distinctions will not concern us. Thus we shall live with the possibility that the Langlands quotient parameters are not unique.

4) Despite the nonuniqueness, the magnitude $|\mathrm{Re}\, \nu|$ can be shown to be the same for all sets of Langlands quotient parameters of a given irreducible admissible $(\mathfrak{g}, K)$ module. A useful fact that comes out of the proof of the Langlands classification is that, in the context of (c) of the theorem, if V has Langlands quotient parameters (MAN, ξ, ν) and if V' is another subquotient of $I^G_{MAN}(\xi, \nu)$, then the Langlands quotient parameters $(M'A'N', \xi', \nu')$ of V' have $|\mathrm{Re}\, \nu'| < |\mathrm{Re}\, \nu|$.

5) If condition (ii) in the definition of standard continuous-series representation is changed to say that $\langle \mathrm{Re}\, \nu, \beta \rangle \leq 0$ for all $\mathfrak{a}$-roots β that are positive relative to $\mathfrak{n}$, then there is a variant of Theorem 11.198 in which the Langlands quotients in (c) and (d) are replaced by **Langlands subrepresentations**. This variant of the Langlands classification follows from Theorem 11.198 by using Hermitian duals. In fact, Proposition 11.26 and admissibility show that

$$(11.199) \qquad I^G_{MAN}(\xi, \nu)^h \cong I^G_{MAN}(\xi, -\bar\nu)$$

when ξ is a (necessarily unitary) discrete series or limit of discrete series. When $\langle \text{Re}\, \nu, \beta \rangle \geq 0$ for β positive, we have $\langle \text{Re}\,(-\bar{\nu}), \beta \rangle \leq 0$. If V is infinitesimally equivalent with a Langlands quotient of $I_{MAN}^G(\xi, \nu)$, then V^h is infinitesimally equivalent with an irreducible subrepresentation of $I_{MAN}^G(\xi, \nu)^h$, since $(\cdot)^h$ is contravariant and exact, and (11.199) shows that V^h is infinitesimally equivalent with a Langlands subrepresentation of $I_{MAN}^G(\xi, -\bar{\nu})$. This argument can be reversed, and it follows that (c) and (d) of Theorem 11.198 are equivalent with (c) and (d) of the variant of Theorem 11.198. Then it is meaningful to speak of **Langlands subrepresentation parameters** or **Langlands submodule parameters**.

6) If an irreducible admissible $(\mathfrak{g}, K)$ module V has Langlands quotient parameters (MAN, ξ, ν), then V has Langlands submodule parameters (MAN^-, ξ, ν). This fact is immediate from the infinitesimal equivalence asserted in (c) of the theorem.

With $I_{MAN}^G(\xi, \nu)$ equal to a standard continuous-series representation, let $V_{K \cap M}^{\xi}$ denote the underlying $(\mathfrak{m}, K \cap M)$ of ξ. According to Proposition 11.47, the underlying $(\mathfrak{g}, K)$ module of $I_{MAN}^G(\xi, \nu)$ is

$$(11.200) \qquad X_K(\xi, \nu) \cong {}^{\mathrm{u}}\mathcal{R}_{\mathfrak{q}, K \cap M}^{\mathfrak{g}, K}(V_{K \cap M}^{\xi} \otimes \mathbb{C}_{\nu + \rho}),$$

where $\mathfrak{q} = \mathfrak{m} \oplus \mathfrak{a} \oplus \mathfrak{n}$, $\mathfrak{m}$ acts in $V_{K \cap M}^{\xi}$, and $\mathfrak{a}$ acts in $\mathbb{C}_{\nu + \rho}$. Since ξ is a discrete series or limit, $\mathfrak{m}_0$ must have a Cartan subalgebra $\mathfrak{t}_0$ lying in $\mathfrak{k}_0 \cap \mathfrak{m}_0$. If $T = Z_K(\mathfrak{t}_0)$ is the corresponding Cartan subgroup of M, then $V_{K \cap M}^{\xi}$ is given as in (11.189) by

$$(11.201) \qquad V_{K \cap M}^{\xi} = ({}^{\mathrm{n}}\mathcal{R}_{\mathfrak{b}, T}^{\mathfrak{m}, K \cap M})^S(\tilde{Z})$$

for some Borel subalgebra $\mathfrak{b}$ and some irreducible $(\mathfrak{t}, \tilde{T})$ module $\tilde{Z}$ such that ε acts by -1. Since all roots in $\Delta(\mathfrak{m}, \mathfrak{t})$ are imaginary, the number S is the number of positive compact roots, namely $\frac{1}{2} \dim_{\mathbb{R}}((K \cap M)/T)$.

Lemma 11.202. For any $Z \in \mathcal{C}(\mathfrak{t}, T)$, any Borel subalgebra $\mathfrak{b}$ of $\mathfrak{m}$ containing $\mathfrak{t}$, and any $\nu \in \mathfrak{a}^*$,

$$({}^{\mathrm{u}}\mathcal{L}_{\mathfrak{b}^-, T}^{\mathfrak{m}, K \cap M})_j(Z) \otimes \mathbb{C}_{\nu} \cong ({}^{\mathrm{u}}\mathcal{L}_{\mathfrak{b}^- + \mathfrak{a}, T}^{\mathfrak{m} + \mathfrak{a}, K \cap M})_j(Z \otimes \mathbb{C}_{\nu})$$

and

$$({}^{\mathrm{u}}\mathcal{R}_{\mathfrak{b}, T}^{\mathfrak{m}, K \cap M})^j(Z) \otimes \mathbb{C}_{\nu} \cong ({}^{\mathrm{u}}\mathcal{R}_{\mathfrak{b} + \mathfrak{a}, T}^{\mathfrak{m} + \mathfrak{a}, K \cap M})^j(Z \otimes \mathbb{C}_{\nu})$$

for all $j \geq 0$.

PROOF. We shall first treat the special case $\nu = 0$. For the first identity we have

(11.203a)
$$({}^{u}\mathcal{L}_{\mathfrak{b}^{-},T}^{\mathfrak{m},K\cap M})_{j}(Z) \otimes \mathbb{C}_{0} = \mathcal{F}_{\mathfrak{m},K\cap M}^{\mathfrak{m}+\mathfrak{a},K\cap M}(\Pi_{\mathfrak{m},T}^{\mathfrak{m},K\cap M})_{j}\operatorname{ind}_{\mathfrak{b}^{-},T}^{\mathfrak{m},T}\mathcal{F}_{\mathfrak{t},T}^{\mathfrak{b}^{-},T}(Z)$$

and

(11.203b)
$$({}^{u}\mathcal{L}_{\mathfrak{b}^{-}+\mathfrak{a},T}^{\mathfrak{m}+\mathfrak{a},K\cap M})_{j}(Z \otimes \mathbb{C}_{0}) = (\Pi_{\mathfrak{m}+\mathfrak{a},T}^{\mathfrak{m}+\mathfrak{a},K\cap M})_{j}\operatorname{ind}_{\mathfrak{b}^{-}+\mathfrak{a},T}^{\mathfrak{m}+\mathfrak{a},T}\mathcal{F}_{\mathfrak{b}^{-},T}^{\mathfrak{b}^{-}+\mathfrak{a},T}\mathcal{F}_{\mathfrak{t},T}^{\mathfrak{b}^{-},T}(Z).$$

Now Lemma 11.74 gives

$$\operatorname{ind}_{\mathfrak{b}^{-}+\mathfrak{a},T}^{\mathfrak{m}+\mathfrak{a},T}\mathcal{F}_{\mathfrak{b}^{-},T}^{\mathfrak{b}^{-}+\mathfrak{a},T}(\mathcal{F}_{\mathfrak{t},T}^{\mathfrak{b}^{-},T}(Z)) \cong \mathcal{F}_{\mathfrak{m},T}^{\mathfrak{m}+\mathfrak{a},T}\operatorname{ind}_{\mathfrak{b}^{-},T}^{\mathfrak{m},T}(\mathcal{F}_{\mathfrak{t},T}^{\mathfrak{b}^{-},T}(Z))$$

if we take $\mathfrak{q} = \mathfrak{l} \oplus \mathfrak{u} = \mathfrak{m} \oplus \mathfrak{a}$, $\mathfrak{q}_{1} = \mathfrak{b}^{-}$, and $B = B_{1} = T$. Substitution into (11.203b) gives

$$({}^{u}\mathcal{L}_{\mathfrak{b}^{-}+\mathfrak{a},T}^{\mathfrak{m}+\mathfrak{a},K\cap M})_{j}(Z \otimes \mathbb{C}_{0}) = (\Pi_{\mathfrak{m}+\mathfrak{a},T}^{\mathfrak{m}+\mathfrak{a},K\cap M})_{j}\mathcal{F}_{\mathfrak{m},T}^{\mathfrak{m}+\mathfrak{a},T}\operatorname{ind}_{\mathfrak{b}^{-},T}^{\mathfrak{m},T}\mathcal{F}_{\mathfrak{t},T}^{\mathfrak{b}^{-},T}(Z).$$

Comparing this formula with (11.203a), we see that the desired formula for ${}^{u}\mathcal{L}$ will follow if we show that

(11.204) $$\mathcal{F}_{\mathfrak{m},K\cap M}^{\mathfrak{m}+\mathfrak{a},K\cap M}(\Pi_{\mathfrak{m},T}^{\mathfrak{m},K\cap M})_{j}(W) \overset{?}{\cong} (\Pi_{\mathfrak{m}+\mathfrak{a},T}^{\mathfrak{m}+\mathfrak{a},K\cap M})_{j}\mathcal{F}_{\mathfrak{m},T}^{\mathfrak{m}+\mathfrak{a},T}(W)$$

naturally for $W \in \mathcal{C}(\mathfrak{m}, T)$.

Lemma 11.74 gives us a natural isomorphism

(11.205) $$\Pi_{\mathfrak{m}+\mathfrak{a},T}^{\mathfrak{m}+\mathfrak{a},K\cap M}\mathcal{F}_{\mathfrak{m},T}^{\mathfrak{m}+\mathfrak{a},T} \cong \mathcal{F}_{\mathfrak{m},K\cap M}^{\mathfrak{m}+\mathfrak{a},K\cap M}\Pi_{\mathfrak{m},T}^{\mathfrak{m},K\cap M}$$

if we take $\mathfrak{q} = \mathfrak{l} \oplus \mathfrak{u} = \mathfrak{m} \oplus \mathfrak{a}$, $\mathfrak{q}_{1} = \mathfrak{m}$, $B = K \cap M$, and $B_{1} = T$. Let W_{j} be a projective resolution of W in $\mathcal{C}(\mathfrak{m}, T)$. Since $\mathcal{F}_{\mathfrak{m},K\cap M}^{\mathfrak{m}+\mathfrak{a},K\cap M}$ is exact, the homology of

$$\mathcal{F}_{\mathfrak{m},K\cap M}^{\mathfrak{m}+\mathfrak{a},K\cap M}\Pi_{\mathfrak{m},T}^{\mathfrak{m},K\cap M}(W_{j})$$

is naturally isomorphic to

$$\mathcal{F}_{\mathfrak{m},K\cap M}^{\mathfrak{m}+\mathfrak{a},K\cap M}(\Pi_{\mathfrak{m},T}^{\mathfrak{m},K\cap M})_{j}(W).$$

In view of (11.205), formula (11.204) will follow if we show that the homology of

$$\Pi_{\mathfrak{m}+\mathfrak{a},T}^{\mathfrak{m}+\mathfrak{a},K\cap M}\mathcal{F}_{\mathfrak{m},T}^{\mathfrak{m}+\mathfrak{a},T}(W_{j})$$

is

$$(\Pi_{\mathfrak{m}+\mathfrak{a},T}^{\mathfrak{m}+\mathfrak{a},K\cap M})_{j}\mathcal{F}_{\mathfrak{m},T}^{\mathfrak{m}+\mathfrak{a},T}(W),$$

and (C.26) says that this will be the case if

$$(11.206) \qquad (\Pi_{\mathfrak{m}+\mathfrak{a},T}^{\mathfrak{m}+\mathfrak{a},K\cap M})_n(\mathcal{F}_{\mathfrak{m},T}^{\mathfrak{m}+\mathfrak{a},T}(W_j)) = 0$$

for all $n > 0$. By Proposition 2.115 it is enough to know that

$$(11.207) \qquad \mathcal{F}_{\mathfrak{m}+\mathfrak{a},T}^{\mathfrak{k}\cap\mathfrak{m},T}(\mathcal{F}_{\mathfrak{m},T}^{\mathfrak{m}+\mathfrak{a},T}(W_j))$$

is acyclic for $\Pi_{\mathfrak{k}\cap\mathfrak{m},T}^{\mathfrak{k}\cap\mathfrak{m},K\cap M}$, and a second application of Proposition 2.115 shows that it is enough to know that

$$(11.208) \qquad \mathcal{F}_{\mathfrak{m}+\mathfrak{a},T}^{\mathfrak{m},T}(\mathcal{F}_{\mathfrak{m},T}^{\mathfrak{m}+\mathfrak{a},T}(W_j))$$

is acyclic for $\Pi_{\mathfrak{m},T}^{\mathfrak{m},K\cap M}$. But (11.208) is isomorphic with W_j, which is projective in $C(\mathfrak{m}, T)$ and hence is acyclic for $\Pi_{\mathfrak{m},T}^{\mathfrak{m},K\cap M}$. Thus (11.207) is acyclic in $C(\mathfrak{k}\cap\mathfrak{m}, T)$ for $\Pi_{\mathfrak{k}\cap\mathfrak{m},T}^{\mathfrak{k}\cap\mathfrak{m},K\cap M}$, and (11.206) holds. This proves (11.204).

This completes the proof of the special case $\nu = 0$ for the $^u\mathcal{L}$ functors. To pass to general ν from the case $\nu = 0$, we regard $\mathbb{C}_\nu$ as a $(\mathfrak{t} + \mathfrak{a}, T)$ module with trivial action by $(\mathfrak{t}, T)$. Since the Mackey isomorphism Theorem 2.103 also works for the derived functors of P (by Proposition 2.53a, (C.27a), and (C.28a1)), we have

$$(^u\mathcal{L}_{\mathfrak{b}^-+\mathfrak{a},T}^{\mathfrak{m}+\mathfrak{a},K\cap M})_j(Z \otimes \mathbb{C}_\nu)$$

$$= (P_{\mathfrak{b}^-+\mathfrak{a},T}^{\mathfrak{m}+\mathfrak{a},K\cap M})_j \mathcal{F}_{\mathfrak{t}+\mathfrak{a},T}^{\mathfrak{b}^-+\mathfrak{a},T}(Z \otimes \mathbb{C}_\nu)$$

$$= (P_{\mathfrak{b}^-+\mathfrak{a},T}^{\mathfrak{m}+\mathfrak{a},K\cap M})_j(\mathcal{F}_{\mathfrak{t}+\mathfrak{a},T}^{\mathfrak{b}^-+\mathfrak{a},T}(Z \otimes \mathbb{C}_0) \otimes \mathcal{F}_{\mathfrak{t}+\mathfrak{a},T}^{\mathfrak{b}^-+\mathfrak{a},T}(\mathbb{C}_\nu))$$

$$\cong ((P_{\mathfrak{b}^-+\mathfrak{a},T}^{\mathfrak{m}+\mathfrak{a},K\cap M})_j\mathcal{F}_{\mathfrak{t}+\mathfrak{a},T}^{\mathfrak{b}^-+\mathfrak{a},T}(Z \otimes \mathbb{C}_0)) \otimes \mathbb{C}_\nu$$

$$\cong (^u\mathcal{L}_{\mathfrak{b}^-+\mathfrak{a},T}^{\mathfrak{m}+\mathfrak{a},K\cap M})_j(Z \otimes \mathbb{C}_0) \otimes \mathbb{C}_\nu.$$

This last expression is

$$\cong (^u\mathcal{L}_{\mathfrak{b}^-,T}^{\mathfrak{m},K\cap M})_j(Z) \otimes \mathbb{C}_\nu$$

by the special case $\nu = 0$, and the desired identity for the functors $^u\mathcal{L}$ is proved.

The argument for the $^u\mathcal{R}$ functors is completely analogous, with Lemma 11.76 used in place of Lemma 11.74 and with Corollary 2.97 used in place of Theorem 2.103.

For the real parabolic subalgebra q, the representation $\xi_{2\rho}$ of $L = MA$ is positive and has a positive square root, which we shall denote ξ_ρ. Specifically $\xi_\rho(ma) = e^{\rho \log a}$ for $m \in M$ and $a \in A$. Consequently (11.157) provides definitions of functors $\mathcal{L}^{\mathfrak{g},K}_{\mathfrak{q}^-,L \cap K}$ and $\mathcal{R}^{\mathfrak{g},K}_{\mathfrak{q},L \cap K}$ that are normalized but do not involve double covers.

In addition, if $\tilde{Z}$ is an irreducible $(\mathfrak{t}, \tilde{T})$ module that defines an $(\mathfrak{m}, K \cap M)$ module $V^\xi_{K \cap M}$ as in (11.201), then $\tilde{T}$, which is given by

$$(11.209a) \qquad \tilde{T} = \{(t, z) \in T \times \mathbb{C}^\times \mid \xi_{2\delta(\mathfrak{b})}(t) = z^2\},$$

is the same group as

$$(11.209b) \qquad \{(t, z) \in T \times \mathbb{C}^\times \mid \xi_{2\delta}(t) = z^2\}.$$

(In fact, we have only to use Proposition 11.159 and note that $\xi_{2\rho}(t) = 1$ for all $t \in T$.) Hence $\tilde{Z} \otimes \mathbb{C}_\nu$ is an irreducible $(\mathfrak{t} \oplus \mathfrak{a}, \tilde{T})$ module that can be used as data for passing to a $(\mathfrak{g}, K)$ module.

Proposition 11.210. For the standard continuous-series representation $I^G_{MAN}(\xi, \nu)$ with ξ such that $V^\xi_{K \cap M} = ({}^n \mathcal{R}^{\mathfrak{m}, K \cap M}_{\mathfrak{b}, T})^S(\tilde{Z})$, the underlying $(\mathfrak{g}, K)$ module is

$$X_K(\xi, \nu) \cong (\mathcal{R}^{\mathfrak{g},K}_{\mathfrak{q},K \cap M})({}^n \mathcal{R}^{\mathfrak{m}+\mathfrak{a}, K \cap M}_{\mathfrak{b}+\mathfrak{a}, T})^S(\tilde{Z} \otimes \mathbb{C}_\nu) \cong ({}^n \mathcal{R}^{\mathfrak{g},K}_{\mathfrak{b}+\mathfrak{a}+\mathfrak{n}, T})^S(\tilde{Z} \otimes \mathbb{C}_\nu).$$

PROOF. For the first conclusion we have

$X_K(\xi, \nu)$

$\cong {}^u \mathcal{R}^{\mathfrak{g},K}_{\mathfrak{q},K \cap M}(V^\xi_{K \cap M} \otimes \mathbb{C}_{\nu+\rho})$ by (11.200)

$= {}^u \mathcal{R}^{\mathfrak{g},K}_{\mathfrak{q},K \cap M}(({}^n \mathcal{R}^{\mathfrak{m},K \cap M}_{\mathfrak{b},T})^S(\tilde{Z}) \otimes \mathbb{C}_{\nu+\rho})$ by (11.201)

$= {}^u \mathcal{R}^{\mathfrak{g},K}_{\mathfrak{q},K \cap M}(({}^u \mathcal{R}^{\mathfrak{m},K \cap M}_{\mathfrak{b},T})^S(\tilde{Z} \otimes \mathbb{C}_{\delta(\mathfrak{b})'}) \otimes \mathbb{C}_{\nu+\rho})$ by (11.146)

$\cong {}^u \mathcal{R}^{\mathfrak{g},K}_{\mathfrak{q},K \cap M}(({}^u \mathcal{R}^{\mathfrak{m}+\mathfrak{a}, K \cap M}_{\mathfrak{b}+\mathfrak{a}, T})^S(\tilde{Z} \otimes \mathbb{C}_{\delta(\mathfrak{b})'} \otimes \mathbb{C}_\nu) \otimes \mathbb{C}_\rho)$ by Lemma 11.202

$= {}^u \mathcal{R}^{\mathfrak{g},K}_{\mathfrak{q},K \cap M}(({}^n \mathcal{R}^{\mathfrak{m}+\mathfrak{a}, K \cap M}_{\mathfrak{b}+\mathfrak{a}, T})^S(\tilde{Z} \otimes \mathbb{C}_\nu) \otimes \mathbb{C}_\rho)$ by (11.146)

$= \mathcal{R}^{\mathfrak{g},K}_{\mathfrak{q},K \cap M}({}^n \mathcal{R}^{\mathfrak{m}+\mathfrak{a}, K \cap M}_{\mathfrak{b}+\mathfrak{a}, T})^S(\tilde{Z} \otimes \mathbb{C}_\nu).$

For the second conclusion we begin with the fourth line from the top in the above calculation. Lemma 11.202 shows that it is

$$(11.211) \qquad \cong {}^u \mathcal{R}^{\mathfrak{g},K}_{\mathfrak{q},K \cap M}({}^u \mathcal{R}^{\mathfrak{m}+\mathfrak{a}, K \cap M}_{\mathfrak{b}+\mathfrak{a}, T})^S(\tilde{Z} \otimes \mathbb{C}_{\delta(\mathfrak{b})'} \otimes \mathbb{C}_{\nu+\rho}).$$

Since Proposition 11.52 gives $({}^{u}\mathcal{R}^{\mathfrak{g},K}_{\mathfrak{q},K\cap M})^{j} = 0$ for $j > 0$, the spectral sequence of Theorem 11.77b for this situation collapses and degenerates and shows that (11.211) is

$$\cong ({}^{u}\mathcal{R}^{\mathfrak{g},K}_{\mathfrak{b}+\mathfrak{a}+\mathfrak{n},T})^{S}(\tilde{Z} \otimes \mathbb{C}_{\nu} \otimes \mathbb{C}_{\delta(\mathfrak{b})'} \otimes \mathbb{C}_{\rho}).$$

Now $\mathbb{C}_{\delta(\mathfrak{b})'} \otimes \mathbb{C}_{\rho}$ is isomorphic to $\mathbb{C}_{\delta'}$ as a $\tilde{T}$ module and as a $\mathfrak{t} \oplus \mathfrak{a}$ module, hence as a $(\mathfrak{t} \oplus \mathfrak{a}, \tilde{T})$ module. Therefore the above expression is

$$\cong ({}^{u}\mathcal{R}^{\mathfrak{g},K}_{\mathfrak{b}+\mathfrak{a}+\mathfrak{n},T})^{S}(\tilde{Z} \otimes \mathbb{C}_{\nu} \otimes \mathbb{C}_{\delta'})$$
$$= ({}^{n}\mathcal{R}^{\mathfrak{g},K}_{\mathfrak{b}+\mathfrak{a}+\mathfrak{n},T})^{S}(\tilde{Z} \otimes \mathbb{C}_{\nu}).$$

This completes the proof.

The first conclusion of Proposition 11.210 tells how the underlying $(\mathfrak{g}, K)$ module of any standard continuous-series representation is given by cohomological induction followed by parabolic induction. In the terminology of §6, the second conclusion is that these $(\mathfrak{g}, K)$ modules coincide with the standard modules of Γ type built from the Cartan subalgebra $\mathfrak{t}_0 \oplus \mathfrak{a}_0$ and a Borel subalgebra of strict type L. If we bring Theorem 11.198 into the discussion, we obtain a reformulation of the Langlands classification, the main conclusion being that every irreducible admissible $(\mathfrak{g}, K)$ module is an irreducible quotient of a standard module of Γ type built from a Borel subalgebra of strict type L. We continue to call such a quotient a **Langlands quotient**.

If we replace ν by its negative, then the proof of Proposition 11.210 is still valid. The result is a corresponding statement about the variant of standard continuous-series representations in which it is assumed that ν satisfies $\langle \text{Re}\, \nu, \beta \rangle \leq 0$ for all $\mathfrak{a}$-roots β that are positive relative to $\mathfrak{n}$: These $(\mathfrak{g}, K)$ modules coincide with the standard modules of Γ type built from the Cartan subalgebra $\mathfrak{t}_0 \oplus \mathfrak{a}_0$ and a Borel subalgebra of strict type L'. Taking Remarks 5 and 6 for Theorem 11.198 into account, we see that every irreducible admissible $(\mathfrak{g}, K)$ module is an irreducible submodule of a standard module of Γ type built from a Borel subalgebra of strict type L'. We call such a submodule a **Langlands submodule**.

Theorem 11.129, reformulated in the notation of (11.147) to take normalized functors into account, related all the standard $(\mathfrak{g}, K)$ modules to those of Γ type for Borel subalgebras of type L or L'. In view of the above completeness results for these two special kinds, we see that every irreducible admissible $(\mathfrak{g}, K)$ module arises in each of the following ways:

Module type	Borel type	Occurrence as a Langlands
Γ	L	quotient
Γ	L$'$	submodule
Γ	BB	quotient
Γ	VZ	submodule
Π	L	submodule
Π	L$'$	quotient
Π	BB	submodule
Π	VZ	quotient

Any of the above eight settings can thus be used to state a classification of the irreducible admissible $(\mathfrak{g}, K)$ modules, and we have an
alternate classification result if "Borel type" is replaced by "strict Borel
type." Whether "strict" is imposed as a condition or not, it follows from
Theorem 11.129 that these eight classifications are equivalent. The two
classifications involving Borel subalgebras of type BB are closely related
to the classification due to Beilinson and Bernstein of irreducible admissible $(\mathfrak{g}, K)$ modules in terms of "$\mathcal{D}$ modules." Another classification of
irreducible admissible $(\mathfrak{g}, K)$ modules is in terms of cohomological induction and is due to Vogan and Zuckerman. Effectively the foundation
for this classification is Borel subalgebras of type VZ or of strict type
VZ.

Let us understand in more detail the relationship between standard
modules built from Borel subalgebras of strict type L$'$ and those built
from Borel subalgebras of strict type VZ. To fix the ideas, let us work
with $(\mathfrak{g}, K)$ modules of type Γ. The construction in both cases begins
with a θ stable Cartan subalgebra $\mathfrak{h}_0 = \mathfrak{t}_0 \oplus \mathfrak{a}_0$ and with $T = Z_K(\mathfrak{h}_0)$.
To the Cartan pair $(\mathfrak{h}_0, T)$, we can associate a double cover $(\mathfrak{h}_0, \tilde{T})$. The
actual definition of $\tilde{T}$ uses a homomorphism $\xi_{2\delta'} : T \to \mathbb{C}^\times$ that depends
on the choice of a positive system of roots, but any two choices lead to
$\tilde{T}$'s that are canonically isomorphic by an isomorphism that covers the
identity map of T to itself. Also we suppose we are given a tuple of data
$(\Delta_{\mathrm{imag}}^+, \Delta_{\mathrm{real}}^+, \lambda, \nu, \tilde{Z})$, where

$\Delta_{\mathrm{imag}}^+ =$ positive system for the imaginary roots of $\Delta(\mathfrak{g}, \mathfrak{h})$

$\Delta_{\mathrm{real}}^+ =$ positive system for the real roots of $\Delta(\mathfrak{g}, \mathfrak{h})$

$\quad\lambda =$ member of $\mathfrak{t}^*$ such that $\langle \lambda, \alpha \rangle \geq 0$ for all $\alpha \in \Delta_{\mathrm{imag}}^+$

$\quad\nu =$ member of $\mathfrak{a}^*$ such that $\mathrm{Re}\langle \nu, \alpha \rangle \leq 0$ for all $\alpha \in \Delta_{\mathrm{real}}^+$

$\quad\tilde{Z} =$ irreducible $(\mathfrak{t}, \tilde{T})$ module of infinitesimal character λ.

In terms of these data, we know from §6 that we can always associate two Borel subalgebras of $\mathfrak{g}$ relative to $\mathfrak{h}$: $\mathfrak{b}_{L'}$ of strict type L' and $\mathfrak{b}_{VZ}$ of strict type VZ. Then $\tilde{Z} \otimes \mathbb{C}_\nu$ is an $(\mathfrak{h}, \tilde{T})$ module, and the standard modules of interest are

$$({}^n\mathcal{R}^{\mathfrak{g},K}_{\mathfrak{b}_{L'},T})^{p_{L'}}(\tilde{Z} \otimes \mathbb{C}_\nu) \qquad \text{and} \qquad ({}^n\mathcal{R}^{\mathfrak{g},K}_{\mathfrak{b}_{VZ},T})^{p_{VZ}}(\tilde{Z} \otimes \mathbb{C}_\nu),$$

where $p_{L'} = \dim(\mathfrak{n}_{L'} \cap \mathfrak{k})$ and $p_{VZ} = \dim(\mathfrak{n}_{VZ} \cap \mathfrak{k})$. There may be some choice in defining $\mathfrak{b}_{L'}$ and $\mathfrak{b}_{VZ}$, but Theorem 11.129 assures us that we can use any choice, obtaining isomorphic $(\mathfrak{g}, K)$ modules in the end.

In practice, however, one works with these $(\mathfrak{g}, K)$ modules in a different form, written in terms of induction in stages. For strict type L' the relevant intermediate Levi subgroup is

$$MA = Z_G(\mathfrak{a}_0),$$

while for strict type VZ the relevant intermediate Levi subgroup is

$$L = Z_G(\mathfrak{t}_0).$$

Evidently

$$\mathfrak{m} \oplus \mathfrak{a} = \mathfrak{h} \oplus \bigoplus_{\substack{\alpha \in \Delta(\mathfrak{g},\mathfrak{h}), \\ \alpha=0 \text{ on } \mathfrak{a}_0}} \mathfrak{g}_\alpha$$

and

$$\mathfrak{l} = \mathfrak{h} \oplus \bigoplus_{\substack{\alpha \in \Delta(\mathfrak{g},\mathfrak{h}), \\ \alpha=0 \text{ on } \mathfrak{t}_0}} \mathfrak{g}_\alpha.$$

Therefore

$$\Delta(\mathfrak{m} \oplus \mathfrak{a}, \mathfrak{h}) \quad \text{consists entirely of imaginary roots}$$

and

$$\Delta(\mathfrak{l}, \mathfrak{h}) \quad \text{consists entirely of real roots.}$$

A reductive group having a θ stable Cartan subalgebra with only real roots is said to be **split modulo center**; the group L is of this kind.

As we saw in Proposition 11.210 and Lemma 11.202, we can write the standard $(\mathfrak{g}, K)$ module built from $\mathfrak{b}_{L'}$ as

(11.212a)
$$({}^n\mathcal{R}^{\mathfrak{g},K}_{\mathfrak{b}_{L'},T})^{p_{L'}}(\tilde{Z} \otimes \mathbb{C}_\nu) \cong \mathcal{R}^{\mathfrak{g},K}_{\mathfrak{m}+\mathfrak{a}+\mathfrak{n},K\cap M}({}^n\mathcal{R}^{\mathfrak{m}+\mathfrak{a},K\cap M}_{\mathfrak{h}+(\mathfrak{m}\cap\mathfrak{n}_{L'}),T})^{p_{L'}}(\tilde{Z} \otimes \mathbb{C}_\nu)$$

$$\cong \mathcal{R}^{\mathfrak{g},K}_{\mathfrak{m}+\mathfrak{a}+\mathfrak{n},K\cap M}(({}^n\mathcal{R}^{\mathfrak{m},K\cap M}_{\mathfrak{t}+(\mathfrak{m}\cap\mathfrak{n}_{L'}),T})^{p_{L'}}(\tilde{Z}) \otimes \mathbb{C}_\nu),$$

where $\mathfrak{n}$ denotes the sum of the root spaces for the nonimaginary roots that are positive for $\mathfrak{b}_{L'}$. Here the number $p_{L'}$ is

$$(11.212\text{b}) \qquad\qquad p_{L'} = \#\{\text{positive compact imaginary roots}\},$$

$({}^{n}\mathcal{R}^{\mathfrak{m}, K \cap M}_{\mathfrak{t}+(\mathfrak{m}\cap\mathfrak{n}_{L'}), T})^{p_{L'}}(\tilde{Z})$ is the underlying $(\mathfrak{m}, K \cap M)$ module of a discrete-series or limit representation of M whose Harish-Chandra parameter is dominant for Δ^{+}_{imag}, and the parameter $\nu \in \mathfrak{a}^{*}$ satisfies $\langle \operatorname{Re}\nu, \beta \rangle \le 0$ for every positive $\mathfrak{a}$-root β (including the restrictions to $\mathfrak{a}$ of the members of Δ^{+}_{real}).

To write the standard $(\mathfrak{g}, K)$ module built from $\mathfrak{b}_{\text{VZ}}$ in tidy fashion using induction in stages, we need an analog of the equality of (11.209a) with (11.209b). We introduce the linear functional 2ρ for the split-modulo-center group L. The homomorphism $\xi_{2\rho}$ is everywhere positive, equal to 1 on T. Hence it has a square root, which we denote ξ_{ρ}. If $\mathfrak{u}$ denotes the sum of the root spaces for the nonreal roots that are positive for $\mathfrak{b}_{\text{VZ}}$, then Proposition 11.159 gives us the identity

$$(11.213) \qquad\qquad \xi_{2\delta'} = (\xi_{2\delta(\mathfrak{u})}|_{T})\xi_{2\rho}.$$

(No superscript "prime" is needed for $2\delta(\mathfrak{u})$ since $\Delta(\mathfrak{u})$ contains no real roots.) Recall that the double cover $\tilde{T}$ is built from T and $\xi_{2\delta'}$ and is therefore given by

$$(11.214\text{a}) \qquad \tilde{T} = \{(t, z) \in T \times \mathbb{C}^{\times} \mid \xi_{2\delta'}(t) = z^{2}\}.$$

With

$$(L \cap K)^{\sim} = \{(l, z) \in (L \cap K) \times \mathbb{C}^{\times} \mid \xi_{2\delta(\mathfrak{u})}(l) = z^{2}\},$$

we see from (11.213) that the subgroup

$$(11.214\text{b}) \qquad \tilde{T}_{2} = \{(t, z) \in T \times \mathbb{C}^{\times} \mid \xi_{2\delta(\mathfrak{u})}(t) = z^{2}\}.$$

is the same group as $\tilde{T}$. Hence the irreducible $(\mathfrak{h}, \tilde{T})$ module $\tilde{Z} \otimes \mathbb{C}_{\nu}$ is also an irreducible $(\mathfrak{h}, \tilde{T}_{2})$ module and can be used for passing via the parabolic induction functor $\mathcal{R}^{\mathfrak{l}, (L\cap K)^{\sim}}_{\mathfrak{h}+(\mathfrak{l}\cap\mathfrak{n}_{\text{VZ}}), \tilde{T}}(\tilde{Z} \otimes \mathbb{C}_{\nu})$ to an $(\mathfrak{l}, (L \cap K)^{\sim})$ module. The analog of (11.212) for the Borel subalgebra $\mathfrak{b}_{\text{VZ}}$ is then

(11.215a)
$$({}^{n}\mathcal{R}^{\mathfrak{g}, K}_{\mathfrak{b}_{\text{VZ}}, T})^{p_{\text{VZ}}}(\tilde{Z} \otimes \mathbb{C}_{\nu}) \cong ({}^{n}\mathcal{R}^{\mathfrak{g}, K}_{\mathfrak{l}+\mathfrak{u}, L\cap K})^{p_{\text{VZ}}} \mathcal{R}^{\mathfrak{l}, (L\cap K)^{\sim}}_{\mathfrak{h}+(\mathfrak{l}\cap\mathfrak{n}_{\text{VZ}}), \tilde{T}}(\tilde{Z} \otimes \mathbb{C}_{\nu})$$

with

(11.215b)
$$p_{VZ} = \#\{\text{positive compact imaginary roots}\} + \tfrac{1}{2}\#\{\text{positive complex roots}\}.$$

The motivation for this formula comes from Corollary 11.173, but, as with Proposition 11.210, the proof is a little easier if one unwinds the definitions and appeals directly to Theorem 11.77b or its corollary. Here the intermediate $(\mathfrak{l}, (L \cap K)^{\sim})$ module is a principal series for the group L, which is split modulo center, and the standard $(\mathfrak{g}, K)$ module is obtained by cohomological induction from that.

In applications, sometimes we are given a $(\mathfrak{g}, K)$ module that is written as parabolically induced from a discrete series or limit on M and a one-dimensional representation of A, and we want to convert it to a $(\mathfrak{g}, K)$ module that is cohomologically induced from a principal series of a group split modulo center. Or we might want to go in the reverse direction. The above discussion shows how to proceed: We extract the data $(\Delta_{imag}^+, \Delta_{real}^+, \lambda, \nu, \tilde{Z})$, form $\mathfrak{b}_{VZ}$ or $\mathfrak{b}_{L'}$ as appropriate, and rewrite the $(\mathfrak{g}, K)$ module using (11.215) or (11.212).

The first main consequence of this discussion is that we can give the effect of cohomological induction on Langlands parameters when a suitable dominance condition is in force.

Theorem 11.216. Let $(\mathfrak{g}, K)$ be a reductive pair in the Harish-Chandra class with underlying group G, let $\mathfrak{q} = \mathfrak{l} \oplus \mathfrak{u}$ be a θ stable parabolic subalgebra of $\mathfrak{g}$, and let $L = N_G(\mathfrak{q}) \cap N_G(\theta\mathfrak{q})$ be the corresponding Levi subgroup. Let $(\mathfrak{h}, T)$ be a Cartan subpair of $(\mathfrak{l}, L \cap K)$ with $\mathfrak{h}_0 = \mathfrak{t}_0 \oplus \mathfrak{a}_0$ and $T = Z_{L \cap K}(\mathfrak{h})$, so that T equals $Z_K(\mathfrak{h})$ and $(\mathfrak{h}, T)$ is a Cartan subpair of $(\mathfrak{g}, K)$. Let W be an irreducible admissible $(\mathfrak{l}, L \cap K)$ module with $(M_L A_L N_L, \xi_L, \nu_L)$ as Langlands quotient parameters, where

$$M_L A_L = Z_L(\mathfrak{a}_0),$$

$(\mathfrak{t}, \tilde{T}_L) = \text{double cover of } (\mathfrak{t}, T) \text{ relative to } \mathfrak{m}_L,$

$V_{K \cap M_L}^{\xi_L} = ({}^n\mathcal{R}_{\mathfrak{b},T}^{\mathfrak{m}_L, K \cap M_L})^p(\tilde{Z}_L) \quad$ for an irreducible $(\mathfrak{t}, \tilde{T})$ module $\tilde{Z}_L$ with infinitesimal character λ dominant for a Borel subalgebra $\mathfrak{b}$ of $\mathfrak{m}_L$,

$p = \#\{\mathfrak{b} \text{ positive compact imaginary roots in } \Delta(\mathfrak{m}_L, \mathfrak{t})\},$

and $\qquad \langle \operatorname{Re} \nu_L, \beta \rangle \geq 0 \quad$ for every positive $\mathfrak{a}$-root β of $\mathfrak{l}$.

Suppose that the member $\lambda + \mathrm{Re}\, \nu_L + \delta(\mathfrak{u})$ of $\mathfrak{h}^*$ is dominant for $\Delta(\mathfrak{u})$. Then the isomorphic cohomologically induced modules $\mathcal{R}^S(W)$ and $\mathcal{L}_S(W)$, defined as in (5.3) with S as in (5.5), either are both 0 or else are irreducible and have Langlands quotient parameters (MAN, ξ, ν) described as follows: Define $\Delta^+(\mathfrak{l}, \mathfrak{h})$ to consist of the roots contributing to $\mathfrak{n}_L \oplus \mathfrak{b}$, and let $\Delta^+(\mathfrak{g}, \mathfrak{h}) = \Delta^+(\mathfrak{l}, \mathfrak{h}) \cup \Delta(\mathfrak{u})$. Then

$$A = A_L \quad \text{and} \quad \nu = \nu_L,$$

$$MA = Z_G(\mathfrak{a}_0),$$

$$\Delta(\mathfrak{n}) \text{ is such that } \begin{cases} \Delta(\mathfrak{n}) \text{ closed under conjugation,} \\ \Delta(\mathfrak{n}) \supseteq \Delta^+_{\mathrm{real}}(\mathfrak{l}, \mathfrak{h}), \\ \langle \mathrm{Re}\, \nu_L, \alpha \rangle \geq 0 \text{ for } \alpha \in \Delta(\mathfrak{n}), \end{cases}$$

$$(\mathfrak{t}, \tilde{T}_G) = \text{double cover of } (\mathfrak{t}, T) \text{ relative to } \mathfrak{m},$$

$$\tilde{Z} = \tilde{Z}_L \otimes \mathbb{C}_{\delta(\mathfrak{u})} \quad \text{as an irreducible } (\mathfrak{t}, \tilde{T}_G) \text{ module,}$$

$$\mathfrak{b}' = \text{Borel subalgebra of } \mathfrak{m} \oplus \mathfrak{a} \text{ built from } \Delta^+_{\mathrm{imag}}(\mathfrak{g}, \mathfrak{h}),$$

$$p' = \#\{\text{compact imaginary roots in } \Delta^+(\mathfrak{g}, \mathfrak{h})\},$$

and

$$V^{\xi}_{K \cap M} = (^{\mathfrak{n}}\mathcal{R}^{\mathfrak{m}, K \cap M}_{\mathfrak{b}', T})^{p'}(\tilde{Z}).$$

REMARKS.

1) The $(\mathfrak{l}, L \cap K)$ module W has infinitesimal character $\lambda + \nu$ by Proposition 11.43. Theorem 5.99c shows that the $(\mathfrak{g}, K)$ modules $\mathcal{R}^S(W)$ and $\mathcal{L}_S(W)$ are isomorphic, and Corollary 8.28 shows that they are 0 or irreducible.

2) Roughly the conclusion of Theorem 11.216 is that MA is $Z_G(\mathfrak{a}_0)$, N is arbitrary subject to the conditions we know it must satisfy, ν equals ν_L, and ξ has Harish-Chandra parameter $\lambda + \delta(\mathfrak{u})$. However, $\lambda + \delta(\mathfrak{u})$ does not completely determine ξ, since M is often disconnected, and more information needs to be given. Capturing this additional information is what makes the general statement of Theorem 11.216 cumbersome.

3) Some explanation is appropriate for how $\tilde{Z}$ becomes a $(\mathfrak{t}, \tilde{T}_G)$ module. The $\mathfrak{t}$ action is is given just by the tensor product, and the question is really about $\tilde{T}_G$. The group $\tilde{T}_L$ comes from the double-cover construction applied to T and $\xi_{2\delta(\Delta^+_{\mathrm{imag}}(\mathfrak{l}, \mathfrak{h}))}$, while $\tilde{T}_G$ comes from the construction applied to T and $\xi_{2\delta(\Delta^+_{\mathrm{imag}}(\mathfrak{g}, \mathfrak{h}))}$. However,

$$\xi_{2\delta(\Delta^+_{\mathrm{imag}}(\mathfrak{l}, \mathfrak{h}))}(t) = \xi_{2\delta(\Delta^+(\mathfrak{l}, \mathfrak{h}))'}(t)$$

for $t \in T$ since the complex roots in $\Delta^+(\mathfrak{l}, \mathfrak{h})$ are closed under conjugation. Also since $\Delta(\mathfrak{u})$ contains no real roots, we have

$$\xi_{2\delta(\Delta^+_{\mathrm{imag}}(\mathfrak{g},\mathfrak{h}))}(t) = \xi_{2\delta(\Delta^+(\mathfrak{g},\mathfrak{h}))'}(t)\xi_{-2\delta_c(\mathfrak{u})}(t),$$

where $2\delta_c(\mathfrak{u})$ is the sum of the complex roots in $\Delta(\mathfrak{u})$. The complex roots contributing to $2\delta_c(\mathfrak{u})$ come in pairs β and $\theta\beta$ whose associated characters are equal on T. Thus $\xi_{-2\delta_c(\mathfrak{u})}$ is the square of a character $\xi_{-\delta_c(\mathfrak{u})}$ of T. Let us abbreviate $2\delta'_L = 2\delta(\Delta^+(\mathfrak{l}, \mathfrak{h}))'$ and $2\delta'_G = 2\delta(\Delta^+(\mathfrak{g}, \mathfrak{h}))'$. Then

$$\tilde{T}_L = \{(t, z_1) \mid \xi_{2\delta'_L}(t) = z_1^2\}$$

and
$$\tilde{T}_G = \{(t, z) \mid \xi_{2\delta'_G}(t) = (\xi_{\delta_c(\mathfrak{u})}(t)z)^2\}.$$

Let ε_L and ε_G be the respective elements $(1, -1)$. We are given that $\tilde{Z}_L$ is a $(\mathfrak{t}, \tilde{T}_L)$ module in which ε_L acts as -1, and we want $\tilde{Z}$ to be a $(\mathfrak{t}, \tilde{T}_G)$ module in which ε_G acts as -1. Define

$$\tilde{T}_{\mathfrak{u}} = \{(t, z_2) \mid \xi_{2\delta(\mathfrak{u})}(t) = z_2^2\},$$

put $\varepsilon_{\mathfrak{u}} = (1, -1)$, and regard $\mathbb{C}_{\delta(\mathfrak{u})}$ as a $(\mathfrak{t}, \tilde{T}_{\mathfrak{u}})$ module in which $\varepsilon_{\mathfrak{u}}$ acts as -1. Let $\tilde{T}_L \circ \tilde{T}_{\mathfrak{u}}$ be the group

$$\tilde{T}_L \circ \tilde{T}_{\mathfrak{u}} = \{(t, z_1, z_2) \mid \xi_{2\delta'_L}(t) = z_1^2 \text{ and } \xi_{2\delta(\mathfrak{u})}(t) = z_2^2\}.$$

This group acts on $\tilde{Z} = \tilde{Z}_L \otimes \mathbb{C}_{\delta(\mathfrak{u})}$ by

(11.217) $\qquad (t, z_1, z_2)(z_L \otimes c) = (t, z_1)z_L \otimes (t, z_2)c.$

Meanwhile, the map $(t, z_1, z_2) \mapsto (t, z_1 z_2)$ is a homomorphism of $\tilde{T}_L \circ \tilde{T}_{\mathfrak{u}}$ onto $\tilde{T}_G$ with a two-element kernel. The nontrivial element of the kernel is $(1, -1, -1)$, and

$$(1, -1, -1)(z_L \otimes c) = \varepsilon_L z_L \otimes \varepsilon_{\mathfrak{u}} c = z_L \otimes c.$$

Hence the action (11.217) descends to $\tilde{T}_G$. A preimage of ε_G in $\tilde{T}_L \circ \tilde{T}_{\mathfrak{u}}$ is $(1, -1, 1)$, and thus

$$\varepsilon_G(z_L \otimes c) = (1, -1, 1)(z_L \otimes c) = \varepsilon_L z_L \otimes c = -z_L \otimes c.$$

In other words, ε_G acts in $\tilde{Z}$ by -1. In this way, $\tilde{Z}$ becomes a member of $\mathbb{C}^-(\mathfrak{t}, \tilde{T}_G)$.

PROOF. Since $(\mathfrak{g}, K)$ is in the Harish-Chandra class, Proposition 4.38d shows that Cartan subgroups are given by centralizers. Thus the Cartan subpair associated to $\mathfrak{h}_0$ and L is $(\mathfrak{h}, Z_{L \cap K}(\mathfrak{h}))$, and the Cartan subpair associated to $\mathfrak{h}_0$ and G is $(\mathfrak{h}, Z_K(\mathfrak{h}))$. The elements of $Z_K(\mathfrak{h})$ normalize each root space, and hence they normalize also $\mathfrak{q}$ and $\theta\mathfrak{q}$. Consequently these elements are in $L = N_G(\mathfrak{q}) \cap N_G(\theta\mathfrak{q})$, and the two centralizers are equal. We denote the common centralizer by T.

We know that $(M_L A_L N_L^-, \xi_L, \nu_L)$ is a set of Langlands submodule parameters for W, and Proposition 11.210 shows that W is an irreducible submodule of

$$X = ({}^n\mathcal{R}^{\mathfrak{l}, L \cap K}_{\mathfrak{h} + \mathfrak{n}_L^- + \mathfrak{n}_1, T})^p (\tilde{Z}_L \otimes \mathbb{C}_\nu),$$

where $\mathfrak{n}_1$ is the nilradical of the Borel subalgebra $\mathfrak{b}$ of $\mathfrak{m}_L$ and where

$$p = \#\{\text{compact imaginary roots in } \Delta^+(\mathfrak{l}, \mathfrak{h})\}.$$

The number S, in view of (5.5), is

$$S = \#\{\text{compact imaginary roots in } \Delta(\mathfrak{u})\}$$
$$+ \tfrac{1}{2}\#\{\text{complex roots in } \Delta(\mathfrak{u})\}.$$

We may assume that $\mathcal{R}^S(W)$ is nonzero, hence irreducible by Corollary 8.28. On the category $\mathcal{C}_{\lambda + \nu}(\mathfrak{l}, L \cap K)$ of $(\mathfrak{l}, L \cap K)$ modules of infinitesimal character $\lambda + \nu$, $\mathcal{R}^S$ is an exact functor, as a consequence of Theorems 5.35b and 5.99b. Therefore $\mathcal{R}^S(W)$ is an irreducible submodule of $\mathcal{R}^S(X)$. If we show that

$$\mathcal{R}^S(X) \cong \mathcal{R}^{\mathfrak{g}, K}_{\mathfrak{m} + \mathfrak{a} + \mathfrak{n}^-, K \cap M}(V^\xi_{K \cap M} \otimes \mathbb{C}_\nu),$$

then it follows that $\mathcal{R}^S(W)$ has the indicated Langlands quotient parameters.

The algebra $\mathfrak{h} + \mathfrak{n}_L^- + \mathfrak{n}_1$ is a Borel subalgebra of $\mathfrak{l}$ of strict type L', and Theorem 11.129 allows us to replace it by the Borel subalgebra $\mathfrak{h} + \mathfrak{n}_{VZ}$ built from $\Delta(\mathfrak{u}_1) \cup (-\Delta^+_{\text{real}}(\mathfrak{l}, \mathfrak{h}))$. Thus

$$X \cong ({}^n\mathcal{R}^{\mathfrak{l}, L \cap K}_{\mathfrak{h} + \mathfrak{n}_{VZ}, T})^{p''} (\tilde{Z}_L \otimes \mathbb{C}_\nu)$$

with

$$p'' = \#\{\text{compact imaginary roots in } \Delta^+(\mathfrak{l}, \mathfrak{h})\}$$
$$+ \tfrac{1}{2}\#\{\text{complex roots in } \Delta^+(\mathfrak{l}, \mathfrak{h})\}.$$

Using (11.73), (11.150), and a Mackey isomorphism, we therefore have

$$
\begin{aligned}
\mathcal{R}^S(X) &= ({}^u\mathcal{R}^{\mathfrak{g},K}_{\mathfrak{q},L\cap K})^S (X \otimes \textstyle\bigwedge^{\text{top}}\mathfrak{u}) \\
&\cong ({}^u\mathcal{R}^{\mathfrak{g},K}_{\mathfrak{q},L\cap K})^S (({}^u\mathcal{R}^{\mathfrak{l},L\cap K}_{\mathfrak{h}+\mathfrak{n}_{\mathrm{VZ}},T})^{p''}(\tilde{Z}_L \otimes \mathbb{C}_{\delta(\mathfrak{n}_{\mathrm{VZ}})'} \otimes \mathbb{C}_\nu) \otimes \mathbb{C}_{2\delta(\mathfrak{u})}) \\
&\cong ({}^u\mathcal{R}^{\mathfrak{g},K}_{\mathfrak{q},L\cap K})^S ({}^u\mathcal{R}^{\mathfrak{l},L\cap K}_{\mathfrak{h}+\mathfrak{n}_{\mathrm{VZ}},T})^{p''}(((\tilde{Z}_L \otimes \mathbb{C}_{\delta(\mathfrak{n}_{\mathrm{VZ}})'}) \otimes \mathbb{C}_{2\delta(\mathfrak{u})}) \otimes \mathbb{C}_\nu).
\end{aligned}
$$

Let us use induction in stages (Theorem 11.77b) to combine the two ${}^u\mathcal{R}$ functors. If S is changed to any other index, then the right side is 0 for any index in place of p''. Therefore the spectral sequence in the theorem collapses and degenerates, and the right side above is

$$
\begin{aligned}
&\cong ({}^u\mathcal{R}^{\mathfrak{g},K}_{\mathfrak{h}+\mathfrak{n}_{\mathrm{VZ}}+\mathfrak{u},T})^{S+p''}(((\tilde{Z}_L \otimes \mathbb{C}_{\delta(\mathfrak{n}_{\mathrm{VZ}})'}) \otimes \mathbb{C}_{2\delta(\mathfrak{u})}) \otimes \mathbb{C}_\nu) \\
&= ({}^n\mathcal{R}^{\mathfrak{g},K}_{\mathfrak{h}+\mathfrak{n}_{\mathrm{VZ}}+\mathfrak{u},T})^{S+p''}((\tilde{Z}_L \otimes \mathbb{C}_{\delta(\mathfrak{n}_{\mathrm{VZ}})'} \otimes \mathbb{C}_{2\delta(\mathfrak{u})} \otimes \mathbb{C}_\nu) \otimes \mathbb{C}_{-\delta(\mathfrak{n}_{\mathrm{VZ}}+\mathfrak{u})'}).
\end{aligned}
$$

Here $\mathfrak{h} + \mathfrak{n}_{\mathrm{VZ}} + \mathfrak{u}$ is a Borel subalgebra of $\mathfrak{g}$ of strict type VZ, and

$$
\begin{aligned}
S + p'' = \,&\#\{\text{compact imaginary roots in } \Delta^+(\mathfrak{g}, \mathfrak{h})\} \\
&+ \tfrac{1}{2}\#\{\text{complex roots in } \Delta^+(\mathfrak{g}, \mathfrak{h})\}.
\end{aligned}
$$

Let $\mathfrak{b}' = \mathfrak{h} + \mathfrak{n}'$. The subalgebra $\mathfrak{h} + \mathfrak{n}' + \mathfrak{n}^-$ is a Borel subalgebra of $\mathfrak{g}$ of strict type L', and the isomorphism of (11.215) with (11.212) allows us to read off the desired Langlands submodule parameters. The group MAN and the parameter ν are as asserted, and the $(\mathfrak{m}, K \cap M)$ module is

$$
V^\xi_{K\cap M} = ({}^n\mathcal{R}^{\mathfrak{m},K\cap M}_{\mathfrak{t}+\mathfrak{n}',T})^{p'}((\tilde{Z}_L \otimes \mathbb{C}_{\delta(\mathfrak{n}_{\mathrm{VZ}})'} \otimes \mathbb{C}_{2\delta(\mathfrak{u})}) \otimes \mathbb{C}_{-\delta(\mathfrak{n}_{\mathrm{VZ}}+\mathfrak{u})'}).
$$

To complete the proof we have only to identify

(11.218) $\qquad (\tilde{Z}_L \otimes \mathbb{C}_{\delta(\mathfrak{n}_{\mathrm{VZ}})'} \otimes \mathbb{C}_{2\delta(\mathfrak{u})}) \otimes \mathbb{C}_{-\delta(\mathfrak{n}_{\mathrm{VZ}}+\mathfrak{u})'}$

with $\tilde{Z}$. Proposition 11.144 tells us that we can replace

$$
\mathbb{C}_{\delta(\mathfrak{n}_{\mathrm{VZ}})'} \qquad \text{by} \qquad \mathbb{C}_{\delta'_L} \otimes \mathbb{C}_{(\delta(\mathfrak{n}_{\mathrm{VZ}})-\delta_L)'}
$$

if we apply a certain canonical isomorphism to $\tilde{T}_L$ that covers the identity map of T. We can also replace

$$
\mathbb{C}_{-\delta(\mathfrak{n}_{\mathrm{VZ}}+\mathfrak{u})'} \qquad \text{by} \qquad \mathbb{C}_{-\delta'_G} \otimes \mathbb{C}_{(\delta_G-\delta(\mathfrak{n}_{\mathrm{VZ}}+\mathfrak{u}))'}
$$

if we apply a certain canonical isomorphism to $\tilde{T}_G$ that covers the identity map of T. Thus we can regard (11.218) as

$$\cong (\tilde{Z}_L \otimes \mathbb{C}_{\delta'_L} \otimes (\mathbb{C}_{(\delta(\mathfrak{n}_{VZ}) - \delta_L)'} \otimes \mathbb{C}_{(\delta_G - \delta(\mathfrak{n}_{VZ} + \mathfrak{u}))'}) \otimes \mathbb{C}_{2\delta(\mathfrak{u})}) \otimes \mathbb{C}_{-\delta'_G}$$

$$\cong (\tilde{Z}_L \otimes \mathbb{C}_{\delta'_L} \otimes \mathbb{C}_{2\delta(\mathfrak{u})}) \otimes \mathbb{C}_{-\delta'_G}.$$

Remark 3 shows how to regard this module as $\tilde{Z}_L \otimes \mathbb{C}_{\delta(\mathfrak{u})}$, and the theorem follows.

Corollary 11.219. Let $(\mathfrak{g}, K)$ be a reductive pair whose underlying group G is connected, let $\mathfrak{q} = \mathfrak{l} \oplus \mathfrak{u}$ be a θ stable parabolic subalgebra of $\mathfrak{g}$, and let L be the corresponding Levi subgroup. Fix a Cartan subalgebra of $\mathfrak{l}$, and suppose that $\mathbb{C}_\lambda$ is a one-dimensional $(\mathfrak{l}, L \cap K)$ module for which

$$\operatorname{Re}\langle \lambda + \delta, \alpha \rangle \geq 0 \qquad \text{for all } \alpha \in \Delta(\mathfrak{u}).$$

Then $A_{\mathfrak{q}}(\lambda)$ either is 0 or irreducible and has Langlands quotient parameters (MAN, ξ, ν) described as follows: Let $\mathfrak{h}_0 = \mathfrak{t}_0 \oplus \mathfrak{a}_0$ be a maximally split θ stable Cartan subalgebra of $\mathfrak{l}$, let $\Delta^+(\mathfrak{l}, \mathfrak{h})$ be a positive system for the roots of $\mathfrak{l}$ taking $\mathfrak{a}_0$ before $i\mathfrak{t}_0$, let $\Delta^+(\mathfrak{g}, \mathfrak{h}) = \Delta^+(\mathfrak{l}, \mathfrak{h}) \cup \Delta(\mathfrak{u})$, and let ρ_L be half the sum of the positive restricted roots of $\mathfrak{l}$, counting multiplicities. Then

$$A = \exp \mathfrak{a}_0 \quad \text{and} \quad \nu = \rho_L,$$

$$MA = Z_G(\mathfrak{a}_0),$$

$$\Delta(\mathfrak{n}) \text{ is such that} \begin{cases} \Delta(\mathfrak{n}) \text{ closed under conjugation,} \\ \Delta(\mathfrak{n}) \supseteq \Delta^+_{\mathrm{real}}(\mathfrak{l}, \mathfrak{h}), \\ \langle \rho_L, \alpha \rangle \geq 0 \text{ for } \alpha \in \Delta(\mathfrak{n}), \end{cases}$$

$$(\mathfrak{t}, \tilde{T}) = \text{double cover of } (\mathfrak{t}, T) \text{ relative to } \mathfrak{m},$$

$$\mathbb{C}_{\lambda+\delta} = \text{one-dimensional } (\mathfrak{t}, \tilde{T}) \text{ module,}$$

$$\mathfrak{b}' = \text{Borel subalgebra of } \mathfrak{m} \oplus \mathfrak{a} \text{ built from } \Delta^+_{\mathrm{imag}}(\mathfrak{g}, \mathfrak{h}),$$

$$p' = \#\{\text{compact imaginary roots in } \Delta^+(\mathfrak{g}, \mathfrak{h})\},$$

and
$$V^\xi_{K \cap M} = ({}^{\mathrm{n}}\mathcal{R}^{\mathfrak{m}, K \cap M}_{\mathfrak{b}', T})^{p'}(\mathbb{C}_{\lambda+\delta}).$$

REMARKS. The group L is connected by Lemma 5.10, and T, although possibly disconnected, is contained in L, as we saw in the first paragraph of the proof of Theorem 11.216. To define the action of $\tilde{T}$ on $\mathbb{C}_{\lambda+\delta}$, we let the action of T on $\mathbb{C}_\lambda$ be that of a subgroup of L, and we let $\tilde{T}$ act on $\mathbb{C}_\delta$ in the usual way, with ε acting by -1. Then $\tilde{T}$ acts on the tensor product, which we write as $\mathbb{C}_{\lambda+\delta}$.

PROOF. This is the special case of Theorem 11.216 in which $W = \mathbb{C}_\lambda$. Note that the present proof, invoking Theorem 11.216 as it does, is using only the vanishing theorem in the form of Theorem 5.99, not in the improved form for $A_\mathfrak{q}(\lambda)$ that is given in Theorem 5.109. In any event, the Langlands parameters for W as an $(\mathfrak{l}, L \cap K)$ module consist of a minimal parabolic subgroup $M_\mathfrak{p} A_\mathfrak{p} N_\mathfrak{p}$ of L with $A_\mathfrak{p} = A$, ξ_L equal to the restriction from L to $M_\mathfrak{p}$ of the action of L on $\mathbb{C}_\lambda$, and $\nu_L = \rho_L$.

That these are the correct Langlands quotient parameters follows from the existence of the map (11.61c). Use of that map says that we should arrange for ξ_L to be trivial on the intersection of $M_\mathfrak{p}$ with the semisimple part of L. Indeed, ξ_L has this property because the semisimple part of L has to act trivially on $\mathbb{C}_\lambda$.

10. Cohomological Induction and Standard Modules

Let $(\mathfrak{g}, K)$ be a reductive pair in the Harish-Chandra class with underlying group G. In this section we shall study the effect of cohomological induction on the underlying $(\mathfrak{g}, K)$ modules of standard continuous-series representations.

The setting will be approximately that in Theorem 11.216, but we shall sometimes be able to drop one annoying hypothesis (see (11.224) below). We start with

(i) a θ stable parabolic subalgebra $\mathfrak{q} = \mathfrak{l} \oplus \mathfrak{u}$ of $\mathfrak{g}$,
(ii) a Levi subgroup L for $\mathfrak{l}$ chosen as large as possible, namely $L = N_G(\mathfrak{q}) \cap N_G(\theta\mathfrak{q})$,
(iii) a Cartan pair $(\mathfrak{h}, T)$ for both $(\mathfrak{l}, L \cap K)$ and $(\mathfrak{g}, K)$, with $\mathfrak{h}_0 = \mathfrak{t}_0 \oplus \mathfrak{a}_0$ and $T = Z_{L \cap K}(\mathfrak{h}) = Z_K(\mathfrak{h})$, and
(iv) a member λ of $i\mathfrak{t}_0^*$ such that

$$(11.220) \qquad \langle \lambda + \delta(\mathfrak{u}), \alpha \rangle \geq 0 \qquad \text{for all } \alpha \in \Delta(\mathfrak{u}).$$

Let $A = \exp \mathfrak{a}_0$, and let ν be a parameter varying through $\mathfrak{a}^*$. We consider the two standard continuous-series representations

$$(11.221) \qquad I_{MAN}^G(\xi, \nu) \qquad \text{and} \qquad I_{M_L A N_L}^L(\xi_L, \nu)$$

and their underlying modules, subject to the following conditions:

$$MA = Z_G(\mathfrak{a}_0) \quad \text{and} \quad M_L A = Z_L(\mathfrak{a}_0),$$

$$N \supseteq N_L,$$

$$\langle \mathrm{Re}\, \nu, \beta \rangle \geq 0 \quad \text{for every positive } \mathfrak{a}\text{-root } \beta \text{ of } \mathfrak{g},$$

$$\mathfrak{b} = \text{Borel subalgebra of } \mathfrak{m}_L,$$

$$\mathfrak{b}' = \text{Borel subalgebra of } \mathfrak{m},$$

$$\mathfrak{b}' \supseteq \mathfrak{b} \quad \text{and} \quad \mathfrak{b}' \supseteq \mathfrak{m}_\beta \text{ if } \mathfrak{m}_\beta \subseteq \mathfrak{u},$$

(11.222)

$$\tilde{Z}_L = \text{irreducible } (\mathfrak{t}, \tilde{T}_L) \text{ module}$$
$$\text{of infinitesimal character } \lambda \text{ dominant for } \mathfrak{b},$$

$$\tilde{Z} = \text{irreducible } (\mathfrak{t}, \tilde{T}_G) \text{ module}$$
$$\text{of infinitesimal character } \lambda + \delta(\mathfrak{u}) \text{ dominant for } \mathfrak{b}',$$

$$\tilde{Z} = \tilde{Z}_L \otimes \mathbb{C}_{\delta(\mathfrak{u})},$$

$$\xi_L = \xi(\tilde{Z}_L, \mathfrak{b}) \text{ as a representation of } M_L, \text{ and}$$

$$\xi = \xi(\tilde{Z}, \mathfrak{b}') \text{ as a representation of } M.$$

To make a match with the conditions in Theorem 11.216, we identify the positive imaginary roots and the positive real roots as

(11.223)

$$\Delta^+_{\mathrm{imag}}(\mathfrak{g}, \mathfrak{h}) = \text{members of } \Delta(\mathfrak{m}, \mathfrak{t}) \text{ contributing to } \mathfrak{b}'$$

$$\Delta^+_{\mathrm{imag}}(\mathfrak{l}, \mathfrak{h}) = \text{members of } \Delta(\mathfrak{m}_L, \mathfrak{t}) \text{ contributing to } \mathfrak{b}$$

$$\Delta^+_{\mathrm{real}}(\mathfrak{g}, \mathfrak{h}) = \text{real roots contributing to } \mathfrak{n}$$

$$\Delta^+_{\mathrm{real}}(\mathfrak{l}, \mathfrak{h}) = \text{real roots contributing to } \mathfrak{n}_L.$$

For both the imaginary roots and the real roots, the positive system for $\mathfrak{g}$ contains the positive system for $\mathfrak{l}$. We readily check that the conditions (11.222) and the definitions (11.223) imply all the conditions in Theorem 11.216 except

(11.224) $$\langle \lambda + \mathrm{Re}\, \nu + \delta(\mathfrak{u}), \alpha \rangle \overset{?}{\geq} 0 \qquad \text{for all } \alpha \in \Delta(\mathfrak{u}).$$

Actually we do not want to assume (11.224) except when necessary. Since $\mathfrak{u}$ is θ stable, assuming (11.224) would force also

$$\langle \lambda + \mathrm{Re}\, \nu + \delta(\mathfrak{u}), \theta\alpha \rangle \geq 0.$$

The two inequalities together are equivalent with

$$|\langle \mathrm{Re}\, \nu, \alpha \rangle| \leq \langle \lambda + \delta(\mathfrak{u}), \alpha \rangle$$

and have the effect of bounding or partially bounding $\mathrm{Re}\, \nu$. In order to handle all standard continuous-series representations, we want to avoid any such bound, and therefore we do not want to assume (11.224).

Theorem 11.225. Let $(\mathfrak{g}, K)$ be a reductive pair in the Harish-Chandra class with underlying group G, let $\mathfrak{q} = \mathfrak{l} \oplus \mathfrak{u}$ be a θ stable parabolic subalgebra of $\mathfrak{g}$, let $L = N_G(\mathfrak{q}) \cap N_G(\theta\mathfrak{q})$ be the corresponding Levi subgroup, and let $(\mathfrak{h}, T)$ be a Cartan subpair of both $(\mathfrak{l}, L \cap K)$ and $(\mathfrak{g}, K)$ with $\mathfrak{h}_0 = \mathfrak{t}_0 \oplus \mathfrak{a}_0$ and $T = Z_{L \cap K}(\mathfrak{h}) = Z_K(\mathfrak{h})$. Let $I_{MAN}^G(\xi, \nu)$ and $I_{M_L AN_L}^L(\xi_L, \nu)$ be standard continuous-series representations whose respective underlying modules $X_K(\xi, \nu)$ and $X_{L \cap K}(\xi_L, \nu)$ are related as in (11.222). Then

$$X_K(\xi, \nu) \cong \mathcal{L}_S(X_{L \cap K}(\xi_L, \nu)),$$

where $S = \dim(\mathfrak{u} \cap \mathfrak{k})$. Moreover $\mathcal{L}_j(X_{L \cap K}(\xi_L, \nu)) = 0$ for $j \neq S$.

REMARK. This theorem does not assume the bound (11.224) on ν, and thus the infinitesimal character can be outside the range where a vanishing theorem holds and where $\mathcal{L}_S$ coincides with $\mathcal{R}^S$.

PROOF. Let $X_K'(\xi, \nu)$ and $X_{L \cap K}'(\xi_L, \nu)$ be the underlying modules for $I_{MAN^-}^G(\xi, \nu)$ and $I_{M_L AN_L^-}^L(\xi_L, \nu)$, respectively. By an argument similar to the one in Theorem 11.216, we shall prove that

$$(11.226) \qquad \mathcal{R}^j(X_{L \cap K}'(\xi_L, \nu)) \cong \begin{cases} X_K'(\xi, \nu) & \text{for } j = S \\ 0 & \text{for } j \neq S. \end{cases}$$

Assuming (11.226) for the moment, we write

$$\begin{aligned} \mathcal{L}_j(X_{L \cap K}(\xi_L, \nu))^h &\cong \mathcal{R}^j(X_{L \cap K}(\xi_L, \nu)^h) && \text{by (6.24) and (6.21a)} \\ &\cong \mathcal{R}^j(X_{L \cap K}(\xi_L^h, \nu^h)) && \text{by Corollary 11.59} \\ &\cong \mathcal{R}^j(X_{L \cap K}(\xi_L, -\bar{\nu})) && \text{since } \xi_L \text{ is unitary.} \end{aligned}$$

Now $-\bar{\nu}$ is negative for N, and hence (11.226) shows that this expression is

$$\cong \begin{cases} X_K(\xi, -\bar{\nu}) & \text{for } j = S \\ 0 & \text{for } j \neq S. \end{cases}$$

But another application of Corollary 11.59 gives

$$X_K(\xi, -\bar{\nu}) \cong X_K(\xi^h, \nu^h) \cong (X_K(\xi, \nu))^h,$$

and hence

$$\mathcal{L}_j(X_{L \cap K}(\xi_L, \nu))^h \cong \begin{cases} (X_K(\xi, \nu))^h & \text{if } j = S \\ 0 & \text{if } j \neq S. \end{cases}$$

Taking the Hermitian dual of both sides and using admissibility, we obtain the theorem.

Thus we are to prove (11.226). To do so, we let $\mathfrak{n}_1$ be the nilradical of the Borel subalgebra $\mathfrak{b}$ of $\mathfrak{m}_L$, and we write

$$X'_K(\xi, \nu) = ({}^{\mathfrak{n}}\mathcal{R}^{\mathfrak{l}, L \cap K}_{\mathfrak{h} + \mathfrak{n}_L^- + \mathfrak{n}_1, T})^p (\tilde{Z}_L \otimes \mathbb{C}_\nu),$$

with p equal to the number of compact roots in $\Delta^+_{\text{imag}}(\mathfrak{l}, \mathfrak{h})$. If p is changed to any other index, the resulting module is 0. Therefore a Mackey isomorphism and Corollary 11.173 give

$$
\begin{aligned}
&\mathcal{R}^S(X'_K(\xi, \nu)) \\
&= ({}^{\mathfrak{u}}\mathcal{R}^{\mathfrak{g}, K}_{\mathfrak{q}, L \cap K})^S (({}^{\mathfrak{u}}\mathcal{R}^{\mathfrak{l}, L \cap K}_{\mathfrak{h} + \mathfrak{n}'_L + \mathfrak{n}_1, T})^p (\tilde{Z}_L \otimes \mathbb{C}_{\delta(\mathfrak{n}'_L + \mathfrak{n}_1)'} \otimes \mathbb{C}_\nu) \otimes \mathbb{C}_{2\delta(\mathfrak{u})}) \\
&= ({}^{\mathfrak{u}}\mathcal{R}^{\mathfrak{g}, K}_{\mathfrak{q}, L \cap K})^S ({}^{\mathfrak{u}}\mathcal{R}^{\mathfrak{l}, L \cap K}_{\mathfrak{h} + \mathfrak{n}'_L + \mathfrak{n}_1, T})^p (\tilde{Z}_L \otimes \mathbb{C}_{\delta(\mathfrak{n}'_L + \mathfrak{n}_1)'} \otimes \mathbb{C}_\nu \otimes \mathbb{C}_{2\delta(\mathfrak{u})}) \\
&= ({}^{\mathfrak{u}}\mathcal{R}^{\mathfrak{g}, K}_{\mathfrak{h} + \mathfrak{u} + \mathfrak{n}'_L + \mathfrak{n}_1, T})^{S+p} (\tilde{Z}_L \otimes \mathbb{C}_{\delta(\mathfrak{n}'_L + \mathfrak{n}_1)'} \otimes \mathbb{C}_\nu \otimes \mathbb{C}_{2\delta(\mathfrak{u})}) \\
&= ({}^{\mathfrak{n}}\mathcal{R}^{\mathfrak{g}, K}_{\mathfrak{h} + \mathfrak{u} + \mathfrak{n}'_L + \mathfrak{n}_1, T})^{S+p} (\tilde{Z}_L \otimes \mathbb{C}_{\delta(\mathfrak{n}'_L + \mathfrak{n}_1)'} \\
&\qquad\qquad \otimes \mathbb{C}_\nu \otimes \mathbb{C}_{2\delta(\mathfrak{u})} \otimes \mathbb{C}_{-\delta(\mathfrak{u} + \mathfrak{n}_L^- + \mathfrak{n}_1)'}).
\end{aligned}
$$

In this last expression, we want to change the Borel subalgebra $\mathfrak{h} + \mathfrak{u} + \mathfrak{n}_L^- + \mathfrak{n}_1$ to the Borel subalgebra $\mathfrak{h} + \mathfrak{n}^- + \mathfrak{n}'$, which is of strict type L', and then the arguments given with Theorem 11.216 will complete the proof for $\mathcal{R}^S$. We cannot use Theorem 11.129 directly, since $\mathfrak{h} + \mathfrak{u} + \mathfrak{n}_L^- + \mathfrak{n}_1$ is not of one of the usual types.

Thus we shall go back to Lemma 11.128. In the terminology of the proof of Theorem 11.129, we shall prove that

$$(\mathfrak{h} + \mathfrak{u} + \mathfrak{n}_L^- + \mathfrak{n}_1) \to (\mathfrak{h} + \mathfrak{n}^- + \mathfrak{n}').$$

Conditions (i) and (ii) in the lemma are immediate from (11.223). For (iii), let α be a complex root in $\Delta(\mathfrak{u} + \mathfrak{n}_L^- + \mathfrak{n}_1)$ with α not in $\Delta(\mathfrak{n}^- + \mathfrak{n}')$, and suppose that $2\langle \lambda + \nu + \delta(\mathfrak{u}), \alpha \rangle / |\alpha|^2$ is a nonzero integer. Since $\mathfrak{n}_1 \subseteq \mathfrak{n}'$ and $\mathfrak{n}_L^- \subseteq \mathfrak{n}^-$, it follows that α is in $\Delta(\mathfrak{u})$. Since $\mathfrak{u}$ is θ stable, $\theta\alpha$ is in $\Delta(\mathfrak{u})$ and therefore also in $\Delta(\mathfrak{u} + \mathfrak{n}_L^- + \mathfrak{n}_1)$. Meanwhile $-\alpha$ is in $\Delta(\mathfrak{n}^- + \mathfrak{n}')$, and $\Delta(\mathfrak{n}^- + \mathfrak{n}')$ is closed under conjugation. Therefore $-\bar{\alpha} = \theta\alpha$ is in $\Delta(\mathfrak{n}^- + \mathfrak{n}')$. Thus $\theta\alpha$ is in $\Delta(\mathfrak{n}^- + \mathfrak{n}') \cap \Delta(\mathfrak{u} + \mathfrak{n}_L^- + \mathfrak{n}_1)$, as required by (iii) in the lemma. To complete the verification of (iii), we need to prove that the integer $2\langle \lambda + \nu + \delta(\mathfrak{u}), \alpha \rangle / |\alpha|^2$ is positive. Since α is in $\Delta(\mathfrak{u})$, (11.220) shows that $\langle \lambda + \delta(\mathfrak{u}), \alpha \rangle \geq 0$ and hence that

$$\langle \lambda + \operatorname{Re} \nu + \delta(\mathfrak{u}), \alpha + \theta\alpha \rangle \geq 0.$$

Meanwhile α, being complex, has $\theta\alpha \in \Delta(\mathfrak{n}^-)$ since $\theta\alpha$ is in $\Delta(\mathfrak{n}^- + \mathfrak{n}')$ and $\Delta(\mathfrak{n}')$ contains only imaginary roots. By the positivity condition on $\mathrm{Re}\,\nu$, $\langle \mathrm{Re}\,\nu, \theta\alpha\rangle \leq 0$. Therefore

$$\langle \lambda + \mathrm{Re}\,\nu + \delta(\mathfrak{u}), \theta\alpha - \alpha\rangle \leq 0.$$

Subtracting these two inequalities, we obtain $\langle \lambda + \mathrm{Re}\,\nu + \delta(\mathfrak{u}), \alpha\rangle \geq 0$, and thus (iii) holds in the lemma.

The lemma allows us to change the Borel subalgebra from $\mathfrak{h} + \mathfrak{u} + \mathfrak{n}_L^- + \mathfrak{n}_1$ to $\mathfrak{h} + \mathfrak{n}^- + \mathfrak{n}'$, and the theorem follows for $\mathcal{R}^S$. For $\mathcal{R}^j$ with $j \neq S$, we trace through the above argument and are led to

$$\mathcal{R}^j(X'_{L \cap K}(\xi_L, \nu)) \cong \mathcal{R}^{\mathfrak{g}, K}_{\mathfrak{m} + \mathfrak{a} + \mathfrak{n}^-, K \cap M}(((^{\mathfrak{n}}\mathcal{R}^{\mathfrak{m}, K \cap M}_{\mathfrak{t} + \mathfrak{n}', T})^l(\tilde{Z}_L) \otimes \mathbb{C}_\nu) \otimes \mathbb{C}_{\delta(\mathfrak{u})})$$

with l not equal to the index p' in Theorem 11.216. The right side is 0, and the proof is complete.

We turn to corollaries of Theorem 11.225. The first one rephrases a special case of the Irreducibility Theorem. One approach to classifying irreducible unitary representations is to regard MAN and a discrete series or limit ξ as fixed and to search for all parameters ν with real part in the positive Weyl chamber such that the underlying $(\mathfrak{g}, K)$ module of $I^G_{MAN}(\xi, \nu)$ admits a nonzero positive semidefinite invariant Hermitian form. For such ν, if there is a unique Langlands quotient, then that quotient will be infinitesimally unitary. Typically one works with a nonzero invariant Hermitian form for a family of ν's, and one knows positivity at some point. Knowledge of irreducibility of $I^G_{MAN}(\xi, \nu)$ on a connected set of ν's containing that point allows one to conclude that the form remains positive for the whole connected set of ν's.

Corollary 11.227. Let $(\mathfrak{g}, K)$ be a reductive pair in the Harish-Chandra class with underlying group G, let $\mathfrak{q} = \mathfrak{l} \oplus \mathfrak{u}$ be a θ stable parabolic subalgebra of $\mathfrak{g}$, let $L = N_G(\mathfrak{q}) \cap N_G(\theta\mathfrak{q})$ be the corresponding Levi subgroup, and let $(\mathfrak{h}, T)$ be a Cartan subpair of both $(\mathfrak{l}, L \cap K)$ and $(\mathfrak{g}, K)$ with $\mathfrak{h}_0 = \mathfrak{t}_0 \oplus \mathfrak{a}_0$ and $T = Z_{L \cap K}(\mathfrak{h}) = Z_K(\mathfrak{h})$. Let $I^G_{MAN}(\xi, \nu)$ and $I^L_{M_L A N_L}(\xi_L, \nu)$ be standard continuous-series representations whose respective underlying modules $X_K(\xi, \nu)$ and $X_{L \cap K}(\xi_L, \nu)$ are related as in (11.222). If $X_{L \cap K}(\xi_L, \nu)$ is irreducible and if

$$|\langle \mathrm{Re}\,\nu, \alpha\rangle| \leq \langle \lambda + \delta(\mathfrak{u}), \alpha\rangle \qquad \text{for all } \alpha \in \Delta(\mathfrak{u}),$$

then $X_K(\xi, \nu)$ is zero or is irreducible.

REMARK. The only way that $X_K(\xi, \nu)$ can be zero is for ξ to be zero.

PROOF. The assumed inequality is equivalent with the assumption that $\langle \lambda + \mathrm{Re}\,\nu + \delta(\mathfrak{u}), \alpha\rangle \geq 0$ for all $\alpha \in \Delta(\mathfrak{u})$. Hence the corollary follows immediately from Theorem 11.225 and Corollary 8.28.

The second corollary is a version of the Signature Theorem appropriate for ruling out Langlands quotients as infinitesimally unitary.

Corollary 11.228. Let $(\mathfrak{g}, K)$ be a reductive pair in the Harish-Chandra class with underlying group G, let $\mathfrak{q} = \mathfrak{l} \oplus \mathfrak{u}$ be a θ stable parabolic subalgebra of $\mathfrak{g}$, let $L = N_G(\mathfrak{q}) \cap N_G(\theta\mathfrak{q})$ be the corresponding Levi subgroup, and let $(\mathfrak{h}, T)$ be a Cartan subpair of both $(\mathfrak{l}, L \cap K)$ and $(\mathfrak{g}, K)$ with $\mathfrak{h}_0 = \mathfrak{t}_0 \oplus \mathfrak{a}_0$ and $T = Z_{L\cap K}(\mathfrak{h}) = Z_K(\mathfrak{h})$. Assume also that L meets every component of G. Let $I^G_{MAN}(\xi, \nu)$ and $I^L_{M_L AN_L}(\xi_L, \nu)$ be standard continuous-series representations whose respective underlying modules $X_K(\xi, \nu)$ and $X_{L\cap K}(\xi_L, \nu)$ are related as in (11.222). Suppose that $X_K(\xi, \nu)$ and $X_{L\cap K}(\xi_L, \nu)$ have

  (i) irreducible associated standard tempered representations
 (ii) a K type τ_Λ and an $L \cap K$ type τ_{Λ_L}, respectively, having highest weights Λ and Λ_L related by

$$\Lambda = \Lambda_L + 2\delta(\mathfrak{u} \cap \mathfrak{p})$$

 (iii) nonzero invariant Hermitian forms $\langle \cdot, \cdot \rangle_G$ and $\langle \cdot, \cdot \rangle_L$ such that the signature of $\langle \cdot, \cdot \rangle_G$ on the K type τ_Λ is not zero and equals the signature of $\langle \cdot, \cdot \rangle_L$ on the $L \cap K$ type τ_{Λ_L}.

Whenever $\tau_{\Lambda'}$ is a K type with Λ' as a highest weight and $\tau_{\Lambda'_L}$ is an $L \cap K$ type with Λ'_L as a highest weight such that

$$\Lambda' = \Lambda'_L + 2\delta(\mathfrak{u} \cap \mathfrak{p}),$$

then the signature of $\langle \cdot, \cdot \rangle_G$ on the K type $\tau_{\Lambda'}$ equals the signature of $\langle \cdot, \cdot \rangle_L$ on the $L \cap K$ type $\tau_{\Lambda'_L}$.

REMARKS.

1) In applications to classification questions, G may be assumed connected, at least initially, and then L automatically meets every component of G. In this case there is no loss of generality in assuming (i) in order to work with all Langlands quotients. Condition (ii) will be met by some minimal K type τ_Λ of $X_K(\xi, \nu)$, according to Theorem 10.44, and τ_{Λ_L} will be a minimal K type of $X_{L\cap K}(\xi_L, \nu)$. For a connected group the minimal K types of a standard continuous-series representation are known to have multiplicity 1, and an invariant form will have to be nonzero on each minimal K type if (i) holds. Thus (iii) becomes a question of adjusting a single sign appropriately.

2) The power of this corollary comes in its applicability without a bound on ν. Specifically the corollary applies even if ν does not satisfy the condition that $|\langle \operatorname{Re} \nu, \alpha \rangle| \leq \langle \lambda + \delta(\mathfrak{u}), \alpha \rangle$ for all $\alpha \in \Delta(\mathfrak{u})$.

PROOF. First we show that $\langle \cdot, \cdot \rangle_G$ is the only invariant sesquilinear form on $X_K(\xi, \nu)$, up to a scalar factor. Let Y be the Langlands quotient of $X_K(\xi, \nu)$, which is unique by (i) and by Theorem 11.198c, and let $\pi : X_K(\xi, \nu) \rightarrow Y$ be the quotient map. To show the asserted uniqueness, we shall show that any invariant sesquilinear form on $X_K(\xi, \nu)$ vanishes on $\ker \pi$.

Such a form yields a $(\mathfrak{g}, K)$ map $\varphi : X_K(\xi, \nu) \rightarrow X_K(\xi, \nu)^h$, and we know that

$$X_K(\xi, \nu)^h \cong X_K(\xi^h, \nu^h) \cong X_K(\xi, -\bar{\nu})$$

by Corollary 11.59. Since $-\bar{\nu}$ is negative relative to N, $X_K(\xi, -\bar{\nu})$ has a unique irreducible submodule Y', by Remark 5 with Theorem 11.198. Since φ is not 0, the image of φ must contain Y'. Now Y' has a Langlands quotient $\mathfrak{a}^*$ parameter ν' with $|\mathrm{Re}\, \nu'| = |\mathrm{Re}\, \nu|$ by Remarks 4 and 6 with Theorem 11.198. Since Remark 4 shows that the irreducible subquotients of $X_K(\xi, \nu)$ other than Y have Langlands quotient $\mathfrak{a}^*$ parameters ν'' with $|\mathrm{Re}\, \nu''| < |\mathrm{Re}\, \nu|$, it follows that $Y \cong Y'$. From this isomorphism we shall deduce that φ maps $X_K(\xi, \nu)$ into Y'.

In fact, to conclude that $\varphi(X_K(\xi, \nu)) \subseteq Y'$, recall from Corollary 7.223 that $X_K(\xi, \nu)$ has finite length as a $(\mathfrak{g}, K)$ module. Therefore it is enough to prove that $\varphi(V) = 0$ for every proper $(\mathfrak{g}, K)$ submodule of $X_K(\xi, \nu)$. We do so by induction on the length of V, the case of length 0 being the base case for the induction. Given V, choose a maximal proper $(\mathfrak{g}, K)$ submodule V_0 of V. By inductive hypothesis, $\varphi(V_0) = 0$. Therefore $\varphi(V)$ is irreducible or 0. If it is 0, the induction is complete. Otherwise $\varphi(V) = Y'$, and φ exhibits V/V_0 as isomorphic to Y', hence to Y. This conclusion is a contradiction, since Y occurs only once as a subquotient of $X_K(\xi, \nu)$ and already occurs in $X_K(\xi, \nu)/V$. Therefore φ maps $X_K(\xi, \nu)$ into Y'. Consequently $\varphi(\ker \pi) = 0$, and the form vanishes on $\ker \pi$.

Thus $\langle \cdot, \cdot \rangle_G$ is the unique invariant Hermitian form on $X_K(\xi, \nu)$, up to a scalar, and similarly $\langle \cdot, \cdot \rangle_L$ is the unique invariant Hermitian form on $X_{L \cap K}(\xi_L, \nu)$, up to a scalar.

By Theorem 11.225, we have $X_K(\xi, \nu) \cong \mathcal{L}_S(X_{L \cap L}(\xi_L, \nu))$. In the notation of (ii), the K type τ_Λ is then in the bottom layer. According to (iii), $\langle \cdot, \cdot \rangle_L$ is not 0 on the $L \cap K$ type τ_{Λ_L}, and it follows from the Signature Theorem (Theorem 6.34) that the Shapovalov form of $\langle \cdot, \cdot \rangle_L$ on $X_K(\xi, \nu)$ is not 0. By the uniqueness in the previous paragraph, $\langle \cdot, \cdot \rangle_G$ is a multiple of the Shapovalov form, and comparison of (iii) with the statement of Theorem 6.34 shows that the multiple is positive. Thus Theorem 6.34 is applicable in comparing the signatures of $\langle \cdot, \cdot \rangle_G$ and $\langle \cdot, \cdot \rangle_L$ on the bottom layer, and the corollary follows.

The third corollary extends an earlier result (Corollary 11.219) that identifies the Langlands quotient parameters of $A_q(\lambda)$'s.

Corollary 11.229. Let $(\mathfrak{g}, K)$ be a reductive pair whose underlying group G is connected, let $\mathfrak{q} = \mathfrak{l} \oplus \mathfrak{u}$ be a θ stable parabolic subalgebra of $\mathfrak{g}$, and let L be the corresponding Levi subgroup. In terms of a maximally compact θ stable Cartan subalgebra of $\mathfrak{l}$, suppose that $\mathbb{C}_\lambda$ is a one-dimensional $(\mathfrak{l}, L \cap K)$ module for which λ is real, λ is carried on the $\mathfrak{k}$ part of the Cartan subalgebra,

$$\langle \lambda + \delta(\mathfrak{u}), \alpha \rangle \geq 0 \qquad \text{for all } \alpha \in \Delta(\mathfrak{u}),$$

and $\qquad \langle \lambda + 2\delta(\mathfrak{u} \cap \mathfrak{p}), \alpha \rangle \geq 0 \qquad \text{for all } \alpha \in \Delta(\mathfrak{u} \cap \mathfrak{k}).$

Let $\mathfrak{h}_0 = \mathfrak{t}_0 \oplus \mathfrak{a}_0$ be a maximally split θ stable Cartan subalgebra of $\mathfrak{l}$, let $\Delta^+(\mathfrak{l}, \mathfrak{h})$ be a positive system for the roots of $\mathfrak{l}$ taking $\mathfrak{a}_0$ before $i\mathfrak{t}_0$, let $\Delta^+(\mathfrak{g}, \mathfrak{h}) = \Delta^+(\mathfrak{l}, \mathfrak{h}) \cup \Delta(\mathfrak{u})$, and let ρ_L be half the sum of the positive restricted roots of $\mathfrak{l}$, counting multiplicities. Suppose that the projection of $\lambda + \delta$ on $\mathfrak{t}$ is dominant with respect to $\Delta^+_{\text{imag}}(\mathfrak{g}, \mathfrak{h})$. Then the K type of $A_q(\lambda)$ with highest weight $\lambda + 2\delta(\mathfrak{u} \cap \mathfrak{p})$ generates an irreducible $(\mathfrak{g}, K)$ submodule V with Langlands quotient parameters (MAN, ξ, ν) described as follows:

$$A = \exp \mathfrak{a}_0 \quad \text{and} \quad \nu = \rho_L,$$

$$MA = Z_G(\mathfrak{a}_0),$$

$$\Delta(\mathfrak{n}) \text{ is such that} \begin{cases} \Delta(\mathfrak{n}) \text{ closed under conjugation,} \\ \Delta(\mathfrak{n}) \supseteq \Delta^+_{\text{real}}(\mathfrak{l}, \mathfrak{h}), \\ \langle \rho_L, \alpha \rangle \geq 0 \text{ for } \alpha \in \Delta(\mathfrak{n}), \end{cases}$$

$$(\mathfrak{t}, \tilde{T}) = \text{double cover of } (\mathfrak{t}, T) \text{ relative to } \mathfrak{m},$$

$$\mathbb{C}_{\lambda+\delta} = \text{one-dimensional } (\mathfrak{t}, \tilde{T}) \text{ module,}$$

$$\mathfrak{b}' = \text{Borel subalgebra of } \mathfrak{m} \oplus \mathfrak{a} \text{ built from } \Delta^+_{\text{imag}}(\mathfrak{g}, \mathfrak{h}),$$

$$p' = \#\{\text{compact imaginary roots in } \Delta^+(\mathfrak{g}, \mathfrak{h})\},$$

and $\qquad V^\xi_{K \cap M} = ({}^n\mathcal{R}^{\mathfrak{m}, K \cap M}_{\mathfrak{b}', T})^{p'}(\mathbb{C}_{\lambda+\delta}).$

REMARKS.

1) As in Corollary 11.219, we know that L is connected, that T is contained in L, and that the double cover $\tilde{T}$ acts naturally on $\mathbb{C}_{\lambda+\delta}$.

2) It is understood that δ is defined relative to $\Delta^+(\mathfrak{g}, \mathfrak{h})$. The dominance of $\lambda + \delta$ with respect to $\Delta^+_{\text{imag}}(\mathfrak{g}, \mathfrak{h})$ makes $V^\xi_{K \cap M}$ a discrete series or limit of

discrete series. In the presence of the other hypotheses, this hypothesis is automatic if $\Delta(\mathfrak{l}, \mathfrak{h})$ has no imaginary roots.

3) In view of the previous remark, the corollary gives the Langlands quotient parameters of the Speh representations of $SL(2n, \mathbb{R})$, which are defined in Example 1 of §VIII.5. It gives the parameters also of one of the irreducible constituents of each of the reducible $A_{\mathfrak{q}}(\lambda)$'s in Examples 2 and 3 of that same section.

PROOF. If we abbreviate $M_L = M \cap L$ and $N_L = N \cap L$, then $M_L A N_L$ is a minimal parabolic subgroup of L. The group M_L is the product of a central torus in L and the M group $M_{L_{ss}}$ for the semisimple part of L, and we let ξ_L be the one-dimensional representation of M_L that matches the action of the central torus on $\mathbb{C}_\lambda$ and is trivial on $M_{L_{ss}}$. Then $\mathbb{C}_\lambda$ is the Langlands quotient of $X_{L \cap K}(\xi_L, \rho_L)$, and we let $\varphi : X_{L \cap K}(\xi_L, \rho_L) \to \mathbb{C}_\lambda$ be the quotient mapping.

We form the diagram

$$
\begin{array}{ccc}
U(\mathfrak{k}) \otimes_{U(\bar{\mathfrak{q}} \cap \mathfrak{k})} X_{L \cap K}(\xi_L, \rho_L) & \xrightarrow{\operatorname{ind}_{\bar{\mathfrak{q}} \cap \mathfrak{k}, L \cap K}^{\mathfrak{k}, L \cap K}(\varphi)} & U(\mathfrak{k}) \otimes_{U(\bar{\mathfrak{q}} \cap \mathfrak{k})} \mathbb{C}_\lambda \\
\Big\downarrow{\beta_X} & & \Big\downarrow{\beta_A} \\
U(\mathfrak{g}) \otimes_{U(\bar{\mathfrak{q}})} X_{L \cap K}(\xi_L, \rho_L)) & \xrightarrow{\operatorname{ind}_{\bar{\mathfrak{q}}, L \cap K}^{\mathfrak{g}, L \cap K}(\varphi)} & U(\mathfrak{g}) \otimes_{U(\bar{\mathfrak{q}})} \mathbb{C}_\lambda
\end{array}
$$

in which β_X and β_A are the one-one $(\mathfrak{k}, L \cap K)$ maps given by (5.74). This diagram commutes since effectively β_X and β_A act in the respective first factors while $\operatorname{ind}_{\bar{\mathfrak{q}} \cap \mathfrak{k}, L \cap K}^{\mathfrak{k}, L \cap K}(\varphi)$ and $\operatorname{ind}_{\bar{\mathfrak{q}}, L \cap K}^{\mathfrak{g}, L \cap K}(\varphi)$ act in the respective second factors. Applying Π_S^K to this diagram, we obtain the commutative diagram

$$
\begin{array}{ccc}
\mathcal{L}_S^K(X_{L \cap K}(\xi_L, \rho_L)) & \xrightarrow{\mathcal{L}_S^K(\varphi)} & \mathcal{L}_S^K(\mathbb{C}_\lambda) \\
\Big\downarrow{\mathcal{B}_X} & & \Big\downarrow{\mathcal{B}_A} \\
\mathcal{L}_S(X_{L \cap K}(\xi_L, \rho_L)) & \xrightarrow{\mathcal{L}_S(\varphi)} & \mathcal{L}_S(\mathbb{C}_\lambda)
\end{array}
$$

in which $\mathcal{B}_X$ and $\mathcal{B}_A$ are bottom-layer maps as in (5.76).

The $L \cap K$ type with highest weight λ occurs with multiplicity 1 in $X_{L \cap K}(\xi_L, \rho_L)$, and φ is one-one on that $L \cap K$ type and is 0 on the other $L \cap K$ types. Thus $\mathcal{L}_S^K(\varphi)$ is one-one on the multiplicity-one K type with highest weight $\Lambda = \lambda + 2\delta(\mathfrak{u} \cap \mathfrak{p})$ and is 0 on the other K types. By Theorem 5.80a, $\mathcal{B}_A$ is one-one onto for the K type with highest weight Λ. Consequently $\mathcal{B}_A \circ \mathcal{L}_S^K(\varphi)$ maps onto the multiplicity-one K type of $\mathcal{L}_S(\mathbb{C}_\lambda)$ with highest weight Λ.

Thus the same thing must be true of $\mathcal{L}_S(\varphi) \circ \mathcal{B}_X$, and we conclude that the image of $\mathcal{L}_S(\varphi)$ contains the multiplicity-one K type with highest weight Λ. Since $\mathcal{L}_S(\mathbb{C}_\lambda) \cong A_{\mathfrak{q}}(\lambda)$ and since $\mathcal{L}_S(\varphi)$ is a $(\mathfrak{g}, K)$ map, the image of $\mathcal{L}_S(\varphi)$ contains the $(\mathfrak{g}, K)$ module V in the statement of the corollary. The module V is irreducible since $A_{\mathfrak{q}}(\lambda)$ is infinitesimally unitary (Theorem 9.1) and since $A_{\mathfrak{q}}(\lambda)$ is therefore completely reducible.

By Theorem 11.225 we may regard $\mathcal{L}_S(\varphi)$ as a $(\mathfrak{g}, K)$ map of $X_K(\xi, \rho_L)$ into $A_{\mathfrak{q}}(\lambda)$ with image containing V. Since V is a direct summand of $A_{\mathfrak{q}}(\lambda)$, we can compose $\mathcal{L}_S(\varphi)$ with projection on V, thereby obtaining a $(\mathfrak{g}, K)$ map of $X_K(\xi, \rho_L)$ onto V. Since $X_K(\xi, \rho_L)$ is the underlying $(\mathfrak{g}, K)$ module of a standard continuous-series representation, the corollary follows.

11. Minimal K Type Formula

Our objective in this section and the next is to prove aspects of Theorem 10.44, which tells how the minimal K types of a cohomologically induced $(\mathfrak{g}, K)$ module $\mathcal{L}_S(W)$ are related to the minimal $L \cap K$ types of the $(\mathfrak{l}, L \cap K)$ module W. We shall assume the dominance condition (10.45a) on the infinitesimal character of W, and we shall assume that some expression (10.45b) is K dominant whenever $\mathcal{L}_S(W)$ is not 0.

The idea is as follows. For a standard module with a unique Langlands quotient, we show that the minimal K types of the Langlands quotient coincide with the minimal K types of the standard module. This step requires knowing an alternate characterization of the minimal K types in groups split modulo center, and we shall state such a characterization as Theorem 11.254 below without proof. Since $\mathcal{L}_S$ is exact for the range of infinitesimal characters under consideration, it follows that it is enough to handle the case that W and $\mathcal{L}_S(W)$ are standard modules as in Theorem 11.225. We can reduce further to the case that W is a principal series for a group that is split modulo center, and then a direct calculation handles matters.

In many situations in the argument, when we seek to identify the minimal K types of $\mathcal{L}_S(W)$ in terms of the minimal $L \cap K$ types of W, there is no loss of generality in assuming that G is connected. In fact, one of the hypotheses of Theorem 10.44 is that L meets every component of G. According to Proposition 5.14 (when combined with a passage to derived functors), $\mathcal{L}_S$ for $(\mathfrak{g}, (L \cap K)K_0)$ is intertwined by forgetful functors with $\mathcal{L}_S$ for $(\mathfrak{g}, K_0)$, and thus the norms of $(L \cap K)K_0$ types are the same as those in the connected case. In this way matters reduce to the case that G is connected.

Now let us come to the proof. We begin with the special case where a direct calculation will ultimately handle matters. Thus we assume that we are considering standard modules $W = X_{L \cap K}(\xi_L, \nu)$ and $\mathcal{L}_S(W) = X_K(\xi, \nu)$ related as in (11.222) and Theorem 11.225, where L is split modulo center and W is a principal series. Using Proposition 11.43, we write $\lambda + \nu$ for the infinitesimal character of W (instead of just λ as in the statement of Theorem 10.44); here λ is the infinitesimal character of ξ_L. By the argument in the previous paragraph, we may assume that the underlying group G is connected, and then Lemma 5.10 shows that L is connected.

Another point of view about this situation is that $X_K(\xi, \nu)$ is given and we regard $X_K(\xi, -\nu)$ as a standard module built from a Borel subalgebra of strict type L'. We convert to an isomorphic standard module built from a Borel subalgebra of strict type VZ, and we write this standard module as cohomologically induced from a principal series of a group L split modulo center. If we put ν back in place of $-\nu$, we arrive at the situation (11.222) under study. From this point of view, the formula for the minimal K types of $X_K(\xi, \nu)$ in terms of the minimal K types of $X_{L \cap K}(\xi_L, \nu)$ may be regarded as an absolute formula for the minimal K types of $X_K(\xi, \nu)$.

Theorem 11.230 (Minimal K type formula). Let $(\mathfrak{g}, K)$ be a reductive pair with underlying group G connected, and let $X_K(\xi, \nu)$ be the underlying module of a standard continuous-series representation constructed from a θ stable Cartan subalgebra $\mathfrak{h}_0 = \mathfrak{t}_0 \oplus \mathfrak{a}_0$. Suppose that $X_K(\xi, \nu)$ has infinitesimal character $\lambda + \nu$, that the θ stable parabolic subalgebra $\mathfrak{q} = \mathfrak{l} \oplus \mathfrak{u}$, the Levi group L split modulo center, and the principal series $X_{L \cap K}(\xi_L, \nu)$ of L are constructed from $X_K(\xi, \nu)$ as in (11.222), and that other notation is as in (11.222). Extend $\mathfrak{t}_0$ to a maximally compact θ stable Cartan subalgebra $(\mathfrak{t}')_0 = \mathfrak{t}_0 \oplus (\mathfrak{t}_r)_0$ of $\mathfrak{l}_0$, and suppose that there exists $\mu \in \mathfrak{t}_r^*$ such that $\lambda + \mu$ is the highest weight of a minimal $L \cap K$ type of $X_{L \cap K}(\xi_L, \nu)$ and $\lambda + \mu + 2\delta(\mathfrak{u} \cap \mathfrak{p})$ is dominant for K. Then every K dominant expression $\lambda + \mu' + 2\delta(\mathfrak{u} \cap \mathfrak{p})$ with $\mu' \in \mathfrak{t}_r^*$ and $\lambda + \mu'$ the highest weight of a minimal $L \cap K$ type of $X_{L \cap K}(\xi_L, \nu)$ is the highest weight of a minimal K type of $X_K(\xi, \nu)$, and every minimal K type of $X_K(\xi, \nu)$ has its highest weight of this form.

This theorem will establish the special case of Theorem 10.44 that we mentioned, and its proof will occupy the remainder of this section. We begin by introducing further notation. Our given θ stable Cartan subalgebra $\mathfrak{h}_0$ of $\mathfrak{l}_0$ has $\mathfrak{h} = \mathfrak{t} \oplus \mathfrak{a}$. Since W is a principal series and L is split modulo center, $\Delta(\mathfrak{l}, \mathfrak{h})$ has only real roots. As we saw in §9, L

equals $Z_G(\mathfrak{t}_0)$. We shall work with a maximally compact θ stable Cartan subalgebra $(\mathfrak{h}_{mc})_0$ of $\mathfrak{l}_0$, and a construction that we describe toward the end of this section shows that we can take $\mathfrak{h}_{mc} = \mathfrak{t}' \oplus \mathfrak{a}'$ such that

$$\mathfrak{t}' = \mathfrak{t} \oplus \mathfrak{t}_r \qquad \text{and} \qquad \mathfrak{a} = \mathfrak{a}' \oplus \mathfrak{a}_r,$$

with $i\,(\mathfrak{t}_r)_0$ conjugate to $(\mathfrak{a}_r)_0$ via the complex adjoint group $\mathrm{Ad}(G_{\mathbb{C}})$. Here $G_{\mathbb{C}}$ is a connected complex group with Lie algebra $\mathfrak{g}$. The specific kind of conjugating map that we produce will be called a **Cayley transform** and will in particular be a conjugation by a member of the analytic subgroup $L_{\mathbb{C}}$ of $G_{\mathbb{C}}$ with Lie algebra $\mathfrak{l}$. We assume that a particular Cayley transform has been fixed so that we can identify $\Delta(\mathfrak{g}, \mathfrak{h})$ and $\Delta(\mathfrak{g}, \mathfrak{h}_{mc})$. By Lemma 5.9, $(\mathfrak{h}_{mc})_0$ is maximally compact in $\mathfrak{g}_0$, and thus $(\mathfrak{t}')_0$ is a Cartan subalgebra of both $\mathfrak{k}_0$ and $\mathfrak{l}_0 \cap \mathfrak{k}_0$.

We introduce and fix any positive system $\Delta^+(\mathfrak{l}, \mathfrak{h})$ that takes $\mathfrak{a}_r$ before $\mathfrak{a}'$; it will not matter whether such a positive system is consistent with the condition on real roots in (11.223). Then we define $\Delta^+(\mathfrak{g}, \mathfrak{h}) = \Delta^+(\mathfrak{l}, \mathfrak{h}) \cup \Delta(\mathfrak{u})$. This notion of positivity transfers to $\Delta(\mathfrak{g}, \mathfrak{h}_{mc})$ in such a way that any positive root $\alpha = \alpha_{\mathfrak{t}'} + \alpha_{\mathfrak{a}'}$ with $\alpha_{\mathfrak{t}'} \neq 0$ has $\alpha_{\mathfrak{t}'} - \alpha_{\mathfrak{a}'}$ positive. By restriction we therefore obtain positive systems $\Delta^+(\mathfrak{k}, \mathfrak{t}')$ and $\Delta^+(\mathfrak{l} \cap \mathfrak{k}, \mathfrak{t}')$. Let

$$\delta = \text{half the sum of the members of } \Delta^+(\mathfrak{g}, \mathfrak{h}_{mc})$$

$$\delta_K = \text{half the sum of the members of } \Delta^+(\mathfrak{k}, \mathfrak{t}')$$

$$\delta_L = \text{half the sum of the members of } \Delta^+(\mathfrak{l}, \mathfrak{h}_{mc})$$
$$\text{(all supported of } \mathfrak{t}_r \oplus \mathfrak{a}' \text{ since } L \text{ is split}$$
$$\text{modulo center)}$$

$$\delta_{L \cap K} = \text{half the sum of the members of } \Delta^+(\mathfrak{l} \cap \mathfrak{k}, \mathfrak{t}')$$
$$\text{(all supported on } \mathfrak{t}_r)$$

$$\delta(\mathfrak{u}) = \text{half the sum of the members of } \Delta(\mathfrak{u})$$

$$\delta_M = \text{half the sum of the members of } \Delta^+(\mathfrak{g}, \mathfrak{h}_{mc})$$
$$\text{vanishing on } \mathfrak{t}_r \oplus \mathfrak{a}'$$

$$\delta_{K \cap M} = \text{half the sum of the counterparts in } \Delta^+(\mathfrak{g}, \mathfrak{h}_{mc})$$
$$\text{of the compact imaginary roots in } \Delta^+(\mathfrak{g}, \mathfrak{h}).$$

In the case of $\delta_{K \cap M}$, the imaginary roots in $\Delta(\mathfrak{g}, \mathfrak{h})$ certainly remain imaginary when considered in $\Delta(\mathfrak{g}, \mathfrak{h}_{mc})$, but the compact imaginary roots do not necessarily remain compact imaginary: The Cayley transform need not preserve $\mathfrak{k}$ and $\mathfrak{p}$.

Let E be the projection of $\mathfrak{t}'^*$ on $\mathfrak{t}^*$ along $\mathfrak{t}_r^*$. We shall make use of the following identity, whose proof is deferred until the end of this section.

Lemma 11.231. $2(\delta_K - \delta_{K \cap M}) = \delta - \delta_L - \delta_M + E(2\delta_K)$.

In our situation, G and L are connected. Because L is split modulo center, it is true also that $M_L = T$, and thus the underlying $(\mathfrak{m}_L, K \cap M_L)$ module of ξ_L is nothing more than the $(\mathfrak{t}, \tilde{T})$ module $\tilde{Z}_L$, which we can regard as a $(\mathfrak{t}, T)$ module. In accordance with (11.222), $\tilde{Z}_L$ has infinitesimal character λ. The identity component T_0 of T is central in L, and it acts with differential λ in $X_{L \cap K}(\xi_L, \nu)$. Consequently the highest weight of any minimal $L \cap K$ type of $X_{L \cap K}(\xi_L, \nu)$ equals λ on $\mathfrak{t}$. Following the existence assumption that was stated in the first paragraph of this section and again in Theorem 11.230, let $\mu \in \mathfrak{t}_r^*$ be such that $\lambda + \mu$ is the highest weight of a minimal $L \cap K$ type of $X_{L \cap K}(\xi_L, \nu)$ for which

$$(11.232a) \qquad \Lambda = \lambda + \mu + 2\delta(\mathfrak{u} \cap \mathfrak{p}) \qquad \text{is } \Delta_K^+ \text{ dominant.}$$

(Here $\Delta_K^+ = \Delta^+(\mathfrak{k}, \mathfrak{t}')$.) Lemma 11.231 shows that another formula for Λ is

$$(11.232b) \qquad \Lambda = (\lambda + \delta(\mathfrak{u}) + \delta_M - 2\delta_{K \cap M}) - E(2\delta_K) + 2\delta_{L \cap K} + \mu.$$

Let us abbreviate

$$(11.232c) \qquad \lambda_{\min} = \lambda + \delta(\mathfrak{u}) + \delta_M - 2\delta_{K \cap M},$$

so that (11.232b) can be rewritten as

$$(11.232d) \qquad \Lambda = \lambda_{\min} - E(2\delta_K) + 2\delta_{L \cap K} + \mu.$$

The expression (11.232c) has an interpretation in terms of the representation $\xi = \xi(\tilde{Z}, \mathfrak{b}')$ of M. Let $M^\#$ be the subgroup of M that corresponds to the pair $(\mathfrak{m}, T(K \cap M)_0)$. By (5.8) we can form ξ in stages, first constructing the representation $\xi^\#$ of $M^\#$ with

$$V_{K \cap M^\#}^{\xi^\#} = ({}^n \mathcal{R}_{\mathfrak{b}', T}^{\mathfrak{m}, K \cap M^\#})^S(\tilde{Z})$$

and then obtaining ξ by a finite-step induction

$$V_{K \cap M}^{\xi} = I_{\mathfrak{m}, K \cap M^\#}^{\mathfrak{m}, K \cap M}(V_{K \cap M^\#}^{\xi^\#}).$$

Proposition 11.192 is applicable to $\xi^\#$ since T meets every component of $K \cap M^\#$. The $(\mathfrak{t}, \tilde{T})$ module $\tilde{Z} = \tilde{Z}_L \otimes \mathbb{C}_{\delta(\mathfrak{u})}$ has infinitesimal character $\lambda + \delta(\mathfrak{u})$, and (11.232c) shows that $\lambda_{\min}$ is the parameter for $\xi^\#$ that is called Λ in Proposition 11.192. The $(\mathfrak{t}, T)$ module $\tilde{Z} \otimes \mathbb{C}_{-\delta(\mathfrak{b}')} \otimes \mathbb{C}_{2\delta(\mathfrak{b}' \cap \mathfrak{p})}$ is the space of a finite-dimensional irreducible representation $\eta_{\lambda_{\min}}$ of T that is dominant in the sense of Theorem 4.83. If $\xi_{\lambda_{\min}}$ denotes the character of T_0 with differential $\lambda_{\min}$, then $\eta_{\lambda_{\min}}$ decomposes according to $T = Z_T(\mathfrak{m}_0)T_0$ as

$$(11.233) \qquad\qquad \eta_{\lambda_{\min}} = \eta' \xi_{\lambda_{\min}}$$

for some η' in $Z_T(\mathfrak{m}_0)\widehat{}$. According to Proposition 11.192, the corresponding representation $\sigma_{\lambda_{\min}}$ of $K \cap M^{\#}$ has multiplicity 1 in $\xi^{\#}_{K \cap M^{\#}}$, and the other $K \cap M^{\#}$ types of $\xi^{\#}$ are of the form $\sigma_{\lambda'}$, where $\sigma_{\lambda'}$ corresponds to the dominant representation $\eta_{\lambda'} = \eta' \xi_{\lambda'}$ of $T = Z_T(\mathfrak{m}_0)T_0$ and where

$$(11.234) \qquad \lambda' = \lambda_{\min} + \sum_{\substack{\alpha_i \in \Delta^+(\mathfrak{g}, \mathfrak{h}_{mc}), \\ \alpha_i = 0 \text{ on } \mathfrak{t}_r + \mathfrak{a}'}} m_i \alpha_i \qquad \text{with } m_i \geq 0.$$

We shall need to relate the representations $\sigma_{\lambda'}$ of $K \cap M^{\#}$ to the given representation ξ_L of $K \cap M_L = T$. Formulas (11.232c) and (11.234) are sufficient to handle the part that can be expressed in terms of weights, and what is needed is some information about the factor $Z_T(\mathfrak{m}_0)$ of T, which captures any disconnectedness of T and $M^{\#}$. Thus the important thing is to relate the representation η' in (11.233) and the restriction of ξ_L to this subgroup of T. Our technique is to take advantage of the inclusion

$$(11.235) \qquad\qquad T \subseteq L \cap K,$$

which follows since $T = Z_K(\mathfrak{t}_0 \oplus \mathfrak{a}_0)$ and $L \cap K = Z_K(\mathfrak{t}_0)$.

Lemma 11.236. If $\alpha \in \Delta(\mathfrak{g}, \mathfrak{h}_{mc})$ vanishes on $\mathfrak{t}_r \oplus \mathfrak{a}'$, then there exists a one-dimensional representation τ_α of L with unique weight α. This representation has the property that $\tau_\alpha(t) = 1$ for $t \in Z_T(\mathfrak{m}_0)$.

PROOF. The first conclusion is immediate from Theorem 4.86. For the second conclusion, let H_{mc} be the Cartan subgroup of L with Lie algebra $(\mathfrak{h}_{mc})_0$. This is connected by Lemma 4.43d, and τ_α restricts to the exponentiated root ξ_α on it, because the differentials match. Since the center Z_G is in H_{mc} and since ξ_α is trivial on Z_G, τ_α descends to a one-dimensional representation of $L' = L/\ker \operatorname{Ad}_G$, which is a linear group. Write $L' = L''V$ as the direct product of a group L'' with compact center and a vector group V. A suitable finite covering group $\tilde{L}''$ of L'' has a complexification $\tilde{L}''_{\mathbb{C}}$ such that τ_α lifts to $\tilde{L}''$ and extends holomorphically to a one-dimensional representation τ_α of $\tilde{L}''_{\mathbb{C}}$. The group $\tilde{L}'' \times V$ is a finite covering of L'. If $V_{\mathbb{C}}$ denotes a complexification of V, then $\tilde{L}'_{\mathbb{C}} = \tilde{L}''_{\mathbb{C}} \times V_{\mathbb{C}}$ is a complexification of $\tilde{L}'$ and τ_α extends to a holomorphic one-dimensional representation of $\tilde{L}_{\mathbb{C}}$ if we define τ_α to be 1 on $V_{\mathbb{C}}$.

The Cayley transform is a member of the complex adjoint group of L and lifts to a member g of $\tilde{L}'_{\mathbb{C}}$. Since τ_α is one-dimensional, its differential has

$$d\tau_\alpha(gX) = g \circ d\tau_\alpha(X) \circ g^{-1} = d\tau_\alpha(X) \qquad \text{for } X \in \mathfrak{l}.$$

Since the restriction of $d\tau_\alpha$ to $\mathfrak{h}_{mc}$ is the root $\alpha \in \Delta(\mathfrak{g}, \mathfrak{h}_{mc})$, it follows that the restriction of $d\tau_\alpha$ to $\mathfrak{h}$ is the corresponding root, say $\tilde{\alpha} \in \Delta(\mathfrak{g}, \mathfrak{h})$. The root $\tilde{\alpha}$ vanishes on $\mathfrak{a}$, and hence $d\tau_\alpha(i\mathfrak{a}_0) = 0$. Consequently τ_α is trivial on the subgroup $\exp(i\mathfrak{a}_0)$ of $\tilde{L}'_{\mathbb{C}}$ and on the subgroup $\tilde{L}' \cap \exp i\mathfrak{a}_0$ of $\tilde{L}'$. In other words, if $t_1 \in L$ has $\mathrm{Ad}_G(t_1)$ in the exponential of $i\mathfrak{a}_0$, then $\tau_\alpha(t_1) = 1$.

Now suppose that t is in $Z_T(\mathfrak{m}_0)$, and let $T_{\mathbb{C}}$ and $A_{\mathbb{C}}$ be the analytic subgroups of $G_{\mathbb{C}}$ with Lie algebras $\mathfrak{t}$ and $\mathfrak{a}$, respectively. Since $\mathrm{Ad}_G(t) = 1$ on $\mathfrak{h} = \mathfrak{t} \oplus \mathfrak{a}$, $\mathrm{Ad}_G(t)$ is a member of $\mathrm{Ad}_G(T_{\mathbb{C}} A_{\mathbb{C}})$, say $\mathrm{Ad}_G(t) = t_{\mathbb{C}} a_{\mathbb{C}}$ with $t_{\mathbb{C}} \in \mathrm{Ad}_G(T_{\mathbb{C}})$ and $a_{\mathbb{C}} \in \mathrm{Ad}_G(A_{\mathbb{C}})$. Since $\mathrm{Ad}_G(t)$ can be regarded as unitary on $\mathfrak{g}$, $t_{\mathbb{C}}$ and $a_{\mathbb{C}}$ are separately unitary. Therefore $t_{\mathbb{C}} = \mathrm{Ad}_G(t_0)$ for some $t_0 \in T_0$. Both t and $a_{\mathbb{C}}$ centralize $\mathfrak{m}$ and hence so does t_0. Write $t_0 = \exp h$ with $h \in \mathfrak{t}_0$. The root $\tilde{\alpha}$ vanishes on $\mathfrak{a}$, and its root space is therefore in $\mathfrak{m}$. If X is nonzero in this root space, we have

$$X = \mathrm{Ad}(t_0)X = \mathrm{Ad}(\exp h)X = e^{\tilde{\alpha}(h)}X = e^{\alpha(h)}X.$$

Since τ_α has differential α, we have $d\tau_\alpha(h) = \alpha(h)$ and therefore

$$\tau_\alpha(t_0) = \tau_\alpha(\exp h) = e^{\alpha(h)} = 1.$$

Now let us consider $a_{\mathbb{C}}$. Put $t_1 = t t_0^{-1}$. This element is in T, and $\mathrm{Ad}_G(t_1) = a_{\mathbb{C}}$ is in the exponential of $i\mathfrak{a}_0$. By the result of the previous paragraph, $\tau_\alpha(t_1) = 1$. Therefore $\tau_\alpha(t) = \tau_\alpha(t_1 t_0) = 1$.

Lemma 11.237. The representations $\eta_{\lambda_{\min}}$ and ξ_L of T are related by

$$\eta_{\lambda_{\min}} = (\xi_L)(\tau_{2\delta(\mathfrak{u}\cap\mathfrak{p})}|_T),$$

where $\tau_{2\delta(\mathfrak{u}\cap\mathfrak{p})}$ is the one-dimensional representation of $L \cap K$ on $\bigwedge^{\text{top}}(\mathfrak{u} \cap \mathfrak{p})$.

REMARK. Before the proof we give an example to illustrate how some ostensibly trivial representations of T can be nontrivial.

EXAMPLE. Let $G = SU(2, 1)$ with

$$\mathfrak{a}_0 = \mathbb{R} \begin{pmatrix} 0 & 0 & 0 \\ 0 & 0 & 1 \\ 0 & 1 & 0 \end{pmatrix}, \quad \mathfrak{t}_0 = \mathbb{R} \begin{pmatrix} -2i & 0 & 0 \\ 0 & i & 0 \\ 0 & 0 & i \end{pmatrix}, \quad (\mathfrak{t}_r)_0 = \mathbb{R} \begin{pmatrix} 0 & 0 & 0 \\ 0 & i & 0 \\ 0 & 0 & -i \end{pmatrix}.$$

Here $M = M^{\#} = T = T_0 = Z_T(\mathfrak{m}_0) = \exp \mathfrak{t}_0$. We use the usual notation for roots relative to $\mathfrak{t}' = \mathfrak{t} \oplus \mathfrak{t}_r$. The system $\Delta(\mathfrak{l}, \mathfrak{t}')$ is just $\{\pm(e_2 - e_3)\}$, and we can take $\Delta(\mathfrak{u}) = \{e_1 - e_2, e_1 - e_3\}$ with $\Delta(\mathfrak{u} \cap \mathfrak{p}) = \{e_1 - e_3\}$. Suppose ξ_L is trivial, so that $\lambda = 0$ and $\tilde{Z}_L = \mathbb{C}$. Then $\tilde{Z} = \mathbb{C}_{\delta(\mathfrak{u})} = \mathbb{C}_{e_1 - \frac{1}{2}(e_2 + e_3)}$, and the element $t_\theta = \mathrm{diag}(e^{-2i\theta}, e^{i\theta}, e^{i\theta})$ of T acts on $\tilde{Z}$ by $e^{-3i\theta}$. Thus $\xi(t_\theta) = e^{-3i\theta} = \sigma_{\lambda_{\min}}(t_\theta) = \eta_{\lambda_{\min}}(t_\theta)$. The representation $\tau_{2\delta(\mathfrak{u} \cap \mathfrak{p})} = \tau_{e_1 - e_3}$ is defined on the full diagonal subgroup and has $\tau_{e_1 - e_3}(t_\theta) = e^{-3i\theta}$. Thus $\eta_{\lambda_{\min}} = (\xi_L)(\tau_{2\delta(\mathfrak{u} \cap \mathfrak{p})}|_T)$, in agreement with the statement of the lemma.

PROOF. We have

$$V_{K \cap M^{\#}}^{\xi^{\#}} = ({}^{\mathrm{n}}\mathcal{R}_{\mathfrak{b}', T}^{\mathfrak{m}, K \cap M^{\#}})^S(\tilde{Z})$$
$$= ({}^{\mathrm{u}}\mathcal{R}_{\mathfrak{b}', T}^{\mathfrak{m}, K \cap M^{\#}})^S(\tilde{Z} \otimes \mathbb{C}_{\delta(\mathfrak{b}')}) = \mathcal{R}^S(\tilde{Z} \otimes \mathbb{C}_{-\delta(\mathfrak{b}')}).$$

From Corollary 5.72 and Theorem 5.80, the action of T in the highest-weight space of $\sigma_{\lambda_{\min}}$, namely $\eta_{\lambda_{\min}}$, matches the tensor product of the action in $\tilde{Z} \otimes \mathbb{C}_{-\delta(\mathfrak{b}')}$ by the action in $\mathbb{C}_{2\delta(\mathfrak{b}' \cap \mathfrak{p})}$. That is, T acts in

$$\tilde{Z} \otimes \mathbb{C}_{-\delta(\mathfrak{b}')} \otimes \mathbb{C}_{2\delta(\mathfrak{b}' \cap \mathfrak{p})}$$

by $\eta_{\lambda_{\min}}$. Since $\tilde{Z} = \tilde{Z}_L \otimes \mathbb{C}_{\delta(\mathfrak{u})}$, the representation $(\eta_{\lambda_{\min}})(\tau_{2\delta(\mathfrak{u} \cap \mathfrak{p})}|_T)^{-1}$ of T occurs in

$$\tilde{Z}_L \otimes \mathbb{C}_{\delta(\mathfrak{u})} \otimes \mathbb{C}_{-\delta(\mathfrak{b}')} \otimes \mathbb{C}_{2\delta(\mathfrak{b}' \cap \mathfrak{p})} \otimes \mathbb{C}_{-2\delta(\mathfrak{u} \cap \mathfrak{p})}.$$

The representation of T in $\tilde{Z}_L$ is ξ_L, and thus we want to see that the representation of T in

$$\mathbb{C}_{\delta(\mathfrak{u})} \otimes \mathbb{C}_{-\delta(\mathfrak{b}')} \otimes \mathbb{C}_{2\delta(\mathfrak{b}' \cap \mathfrak{p})} \otimes \mathbb{C}_{-2\delta(\mathfrak{u} \cap \mathfrak{p})}$$

is trivial.

If we take into account Remark 3 with Theorem 11.216, we see that this representation of T is to be interpreted as

$$(11.238) \qquad \mathbb{C}_{\delta(\mathfrak{u}) - \delta(\mathfrak{b}')} \otimes \mathbb{C}_{-2(\delta(\mathfrak{u} \cap \mathfrak{p}) - 2\delta(\mathfrak{b}' \cap \mathfrak{p}))}.$$

Roots are to be interpreted relative to $\mathfrak{t} \oplus \mathfrak{a}$. The roots of $\mathfrak{u}$ that are not roots of $\mathfrak{b}'$ are exactly the complex positive roots, which come in pairs

$\alpha \pm \beta$ built with $\alpha \in \mathfrak{t}^*$ and $\beta \in \mathfrak{a}^*$ and yielding the same character of T: $\xi_{\alpha+\beta}|T = \xi_{\alpha-\beta}|T$. If we write this restriction as ξ_α, the representation of T on $\mathbb{C}_{\delta(\mathfrak{u})-\delta(\mathfrak{b}')}$ counts one factor of ξ_α for each pair, or

$$(11.239) \qquad \prod_{\substack{\text{complex pairs} \\ \alpha \pm \beta \text{ in } \Delta(\mathfrak{u})}} \xi_\alpha$$

in all. (Note that a factor ξ_α may occur more than once, since complex pairs $\alpha \pm \beta$ and $\alpha \pm \gamma$ with $\beta \neq \pm\gamma$ may be possible when $\mathfrak{t}_0$ is not a Cartan subalgebra of $\mathfrak{k}_0$.)

For the other factor of (11.238), the imaginary roots of $\Delta(\mathfrak{g}, \mathfrak{h})$ contributing to $\mathfrak{u}$ have root vectors in $\mathfrak{k}$ or $\mathfrak{p}$; of these imaginary roots, the compact roots do not contribute to $2\delta(\mathfrak{u} \cap \mathfrak{p})$ or $2\delta(\mathfrak{b}' \cap \mathfrak{p})$, while the noncompact roots contribute to both $2\delta(\mathfrak{u} \cap \mathfrak{p})$ and $2\delta(\mathfrak{b}' \cap \mathfrak{p})$, for a net total of 0. There are no real roots in $\mathfrak{u}$, and a complex pair $\alpha \pm \beta$ has root vectors $e_{\alpha+\beta}$ and $\theta e_{\alpha+\beta}$ that contribute

$$e_{\alpha+\beta} + \theta e_{\alpha+\beta} \quad \text{to } \mathfrak{k} \qquad \text{and} \qquad e_{\alpha+\beta} - \theta e_{\alpha+\beta} \quad \text{to } \mathfrak{p}.$$

The group T acts on the two-dimensional span of these root vectors by the character ξ_α defined in the previous paragraph, and the contribution of the complex pair $\alpha \pm \beta$ to $2\delta(\mathfrak{u} \cap \mathfrak{p}) - 2\delta(\mathfrak{b}' \cap \mathfrak{p})$ comes from the factor ξ_α on the $\mathfrak{p}$ part of this two-dimensional space. The contribution from all complex pairs to $2\delta(\mathfrak{u} \cap \mathfrak{p}) - 2\delta(\mathfrak{b}' \cap \mathfrak{p})$ is therefore equal to (11.239), and it follows that T acts trivially on (11.238).

Lemma 11.240. If Λ_0 is the highest weight of a minimal K type τ_{Λ_0} of $X_K(\xi, \nu)$, then τ_{Λ_0} has a weight of the form $\lambda' + \omega$ in $\mathfrak{t}'^*$ such that

    (a) λ' is in $\mathfrak{t}^*$ and ω is in $\mathfrak{t}_r^*$

    (b) $\sigma_{\lambda'}$ occurs in $\tau_{\Lambda_0}|_{K \cap M^\#}$

    (c) $\sigma_{\lambda'}$ corresponds to an irreducible representation $\eta_{\lambda'}$ of T under Theorem 4.83

    (d) $\lambda' + \omega$ is dominant for $\Delta^+(\mathfrak{l} \cap \mathfrak{k}, \mathfrak{t}')$, and the corresponding irreducible representation $\tau_{\lambda'+\omega}$ of $L \cap K$ contains $\eta_{\lambda'}$ in its restriction to T.

PROOF. We know that $\mathcal{R}^{\mathfrak{g},K}_{m+\mathfrak{a}+n, K \cap M}$ models parabolic induction from MAN to G and that $I^{\mathfrak{m}, K \cap M}_{\mathfrak{m}, K \cap M^\#}$ models induction from $M^\#$ to M. Since

$$\begin{aligned} X_K(\xi, \nu) &\cong \mathcal{R}^{\mathfrak{g},K}_{m+\mathfrak{a}+n, K \cap M}(V^\xi_{K \cap M}) \\ &\cong \mathcal{R}^{\mathfrak{g},K}_{m+\mathfrak{a}+n, K \cap M^\#}((I^{\mathfrak{m}, K \cap M}_{\mathfrak{m}, K \cap M^\#}(V^{\xi^\#}_{K \cap M^\#})) \otimes \mathbb{C}_\nu) \\ &\cong \mathcal{R}^{\mathfrak{g},K}_{m+\mathfrak{a}+n, K \cap M^\#}(V^{\xi^\#}_{K \cap M^\#} \otimes \mathbb{C}_\nu), \end{aligned}$$

Frobenius reciprocity (cf. (11.42)) shows that

$$[X_K(\xi, \nu) : \tau_{\Lambda_0}] = \sum_{\gamma \in (K \cap M^\#)^\widehat{}} [\xi^\# : \gamma][\tau_{\Lambda_0}|_{K \cap M^\#} : \gamma].$$

Since τ_{Λ_0} occurs in $X_K(\xi, \nu)$ by assumption, there exists some $K \cap M^\#$ type $\sigma_{\lambda'}$ of $\xi^\#$ such that $\sigma_{\lambda'}$ occurs in $\tau_{\Lambda_0}|_{K \cap M^\#}$. Choose v_0 in the space of τ_{Λ_0} such that $\tau_{\Lambda_0}(K \cap M^\#)v_0$ is this copy of $\sigma_{\lambda'}$ and such that v_0 is a λ' weight vector under $\mathfrak{t}$, i.e.,

$$\tau_{\Lambda_0}(h)v_0 = \lambda'(h)v_0 \qquad \text{for } h \in \mathfrak{t}_0.$$

Then $\tau_{\Lambda_0}(T)v_0$ is the representation space for a copy of the representation $\eta_{\lambda'}$ of T defined in (c) of the lemma. Let

$$V_0 = \{v \text{ in space of } \tau_{\Lambda_0} \mid \tau_{\Lambda_0}(h)v = \lambda'(h)v \text{ for all } h \in \mathfrak{t}_0\}.$$

Since $L = Z_G(\mathfrak{t}_0)$, $\tau_{\Lambda_0}|_{L \cap K}$ acts in V_0. Since $T \subseteq L \cap K$ and the representation $\eta_{\lambda'}$ of T occurs in V_0, some irreducible constituent of $\tau_{\Lambda_0}|_{L \cap K}$ in V_0 contains $\eta_{\lambda'}$ when restricted to T. The highest weight of this representation of $L \cap K$ is necessarily of the form $\lambda' + \omega$ with $\omega \in \mathfrak{t}_r^*$ and with $\lambda' + \omega$ dominant for $\Delta^+(\mathfrak{l} \cap \mathfrak{k}, \mathfrak{t}')$, and the lemma follows.

Now we can proceed with the direct calculation that will handle the special case. Let Λ_0 be the highest weight of a minimal K type τ_{Λ_0} of $X_K(\xi, \nu)$. The element Λ of $\mathfrak{t}^*$ is by assumption Δ_K^+ dominant, and the corresponding K type τ_Λ is therefore in the bottom layer of $X_K(\xi, \nu)$. Since τ_Λ thus occurs in $X_K(\xi, \nu)$, the minimality of Λ_0 gives

(11.241) $$|\Lambda_0 + 2\delta_K|^2 \leq |\Lambda + 2\delta_K|^2.$$

Let $\lambda' + \omega$ be as in Lemma 11.240. Since $\lambda' + \omega$ is a weight of τ_{Λ_0}, Theorem 4.10b gives

(11.242a) $$|\lambda' + \omega|^2 \leq |\Lambda_0|^2,$$

while Theorem 4.7d gives

(11.242b) $$\lambda' + \omega = \Lambda_0 - \sum n_j \beta_j \qquad \text{with } n_j \geq 0 \text{ and } \beta_j \in \Delta_K^+.$$

Therefore

(11.243)
$$|\lambda + 2\delta(\mathfrak{u})|^2 + |\omega + E(2\delta_K)|^2$$

$\leq |\lambda + 2\delta(\mathfrak{u}) + \sum m_i \alpha_i|^2 + |\omega + E(2\delta_K)|^2$ with α_i as in (11.234) since $\lambda + 2\delta(\mathfrak{u})$ is dominant

$= |\lambda + 2\delta(\mathfrak{u}) + E(2\delta_K) + \sum m_i \alpha_i + \omega|^2$

$= |\lambda + \delta(\mathfrak{u}) + \delta_M - 2\delta_{K \cap M} + \sum m_i \alpha_i + \omega + 2\delta_K|^2$ by Lemma 11.231

$= |(\lambda_{\min} + \sum m_i \alpha_i) + \omega + 2\delta_K|^2$ by (11.232c)

$= |\lambda' + \omega + 2\delta_K|^2$ by (11.234)

$= |\lambda' + \omega|^2 + 2\langle \lambda' + \omega, 2\delta_K \rangle + |2\delta_K|^2$

$\leq |\Lambda_0|^2 + 2\langle \Lambda_0 - \sum n_j \beta_j, 2\delta_K \rangle + |2\delta_K|^2$ by (11.242)

$\leq |\Lambda_0|^2 + 2\langle \Lambda_0, 2\delta_K \rangle + |2\delta_K|^2$ since β_j is in Δ_K^+

$= |\Lambda_0 + 2\delta_K|^2$

$\leq |\Lambda + 2\delta_K|^2$ by (11.241)

$= |\lambda + \mu + 2\delta(\mathfrak{u} \cap \mathfrak{p}) + 2\delta(\mathfrak{u} \cap \mathfrak{k}) + 2\delta_{L \cap K}|^2$ by (11.232a)

$= |\lambda + 2\delta(\mathfrak{u})|^2 + |\mu + 2\delta_{L \cap K}|^2.$

Going over the computation (11.243), we note for later reference that equality throughout would force

(11.244a) $\lambda_{\min} = \lambda'$ since all m_i equal 0

(11.244b) $\lambda' + \omega = \Lambda_0$ since all n_j equal 0

(11.244c) τ_Λ is a minimal K type   since $|\Lambda_0 + 2\delta_K|^2 = |\Lambda + 2\delta_K|^2.$

Using (11.244a) and Lemma 11.240, we can rewrite (11.244b) as

(11.244d) $\Lambda_0 = \lambda_{\min} + \omega,$ where the restriction of the $L \cap K$ type $\tau_{\lambda_{\min} + \omega}$ to T contains $\eta_{\lambda_{\min}}$.

Subtracting $2\langle \lambda, 2\delta(\mathfrak{u}) \rangle + |2\delta(\mathfrak{u})|^2$ from both sides of (11.243), we obtain

$$|\lambda + \omega + E(2\delta_K)|^2 \leq |\lambda + \mu + 2\delta_{L \cap K}|^2$$

or

(11.245) $|(\lambda + \omega + E(2\delta_K) - 2\delta_{L \cap K}) + 2\delta_{L \cap K}|^2 \leq |(\lambda + \mu) + 2\delta_{L \cap K}|^2.$

Equality in (11.245) will force (11.244). To prove such an equality, we shall show that

$$(11.246) \qquad \lambda + \omega + E(2\delta_K) - 2\delta_{L \cap K}$$

is the highest weight of an $L \cap K$ type τ in $X_{L \cap K}(\xi_L, \nu)$. Since $\lambda + \mu$ is the highest weight of an $L \cap K$ type that is minimal, equality will follow.

We start by using Lemma 11.231, (11.232c), and (11.234) to see that

$$(11.247) \qquad \begin{aligned} \lambda + \omega + E(2\delta_K) - 2\delta_{L \cap K} &= \lambda_{\min} + \omega - 2\delta(\mathfrak{u} \cap \mathfrak{p}) \\ &= (\lambda' + \omega) - \sum m_i \alpha_i - 2\delta(\mathfrak{u} \cap \mathfrak{p}). \end{aligned}$$

In view of Lemmas 11.240d, 11.236, and 11.235, the expression (11.246) is therefore the highest weight of the irreducible representation

$$\tau = \tau_{\lambda' + \omega} \otimes \bigotimes_i (\tau_{\alpha_i})^{-m_i} \otimes (\tau_{2\delta(\mathfrak{u} \cap \mathfrak{p})})^{-1}$$

of $L \cap K$. For the argument that τ occurs in $X_{L \cap K}(\xi_L, \nu)$, Frobenius reciprocity says that it is enough to show that $\tau|_T$ contains ξ_L. The differential of τ, restricted to $\mathfrak{t}_0$, is the scalar

$$\lambda' - \sum m_i \alpha_i - 2\delta(\mathfrak{u} \cap \mathfrak{p})|_{\mathfrak{t}} = \lambda$$

by (11.247). Hence we have agreement on T_0. For the factor $Z_T(\mathfrak{m}_0)$ of T, we appeal to Lemma 11.240d, which says that $\tau_{\lambda' + \omega}|_T$ contains $\eta_{\lambda'} = \eta' \xi_{\lambda'}$. Therefore $\tau_{\lambda' + \omega}|_{Z_T(\mathfrak{m}_0)}$ contains η'. Bringing in Lemma 11.236, we see that $\tau|_{Z_T(\mathfrak{m}_0)}$ contains $\eta' \otimes (\tau_{2\delta(\mathfrak{u} \cap \mathfrak{p})})^{-1}|_{Z_T(\mathfrak{m}_0)}$, and this equals $\xi_L|_{Z_T(\mathfrak{m}_0)}$ by Lemma 11.237. Thus (11.246) is the highest weight of an $L \cap K$ type in $X_{L \cap K}(\xi_L, \nu)$, and equality holds in (11.245).

Consequently all the conclusions of (11.244) are valid. Conclusion (11.244c) shows that τ_Λ is a minimal K type. Conclusion (11.244d), in combination with the equality of (11.232d) with (11.232a), shows that

$$\begin{aligned} \Lambda_0 &= \lambda_{\min} - E(2\delta_K) + 2\delta_{L \cap K} + (\omega + E(2\delta_K) - 2\delta_{L \cap K}) \\ &= \lambda + (\omega + E(2\delta_K) - 2\delta_{L \cap K}) + 2\delta(\mathfrak{u} \cap \mathfrak{p}), \end{aligned}$$

where $\lambda + (\omega + E(2\delta_K) - 2\delta_{L \cap K})$ is the highest weight of a minimal $L \cap K$ type in $X_{L \cap K}(\xi_L, \nu)$, according to (11.246) and the equality in (11.245). Therefore every Δ_K^+ dominant expression (11.232a) is the highest weight of a minimal K type of $X_K(\xi, \nu)$, and every minimal K type has highest weight Λ_0 of this form.

To finish the proof of Theorem 11.230 and thereby complete our treatment of the special case, we need to define the Cayley transform and to prove Lemma 11.231.

We begin with the construction of the Cayley transform. Let us say that two orthogonal roots α and β are **strongly orthogonal** if $\alpha + \beta$ and $\alpha - \beta$ are not roots. (Since α and β are orthogonal, it cannot be true that just one of $\alpha + \beta$ and $\alpha - \beta$ is a root.) We start with a maximal orthogonal set of real roots in $\Delta(\mathfrak{l}, \mathfrak{h})$. If two of the roots α and β in this set fail to be strongly orthogonal, then $\alpha + \beta$ and $\alpha - \beta$ are strongly orthogonal, and we can replace the subset $\{\alpha, \beta\}$ by the subset $\{\alpha + \beta, \alpha - \beta\}$. Iterating this process, we may assume that the members of our maximal orthogonal real roots are mutually strongly orthogonal. Let us write $\alpha_1, \dots, \alpha_l$ for these roots (although notation $\tilde{\alpha}_1, \dots \tilde{\alpha}_l$ would be more in keeping with the proof of Lemma 11.236). To each α_j, we shall associate a conjugation $\mathbf{c}_j$ within the $\mathfrak{sl}(2, \mathbb{C})$ subalgebra of $\mathfrak{g}$ that corresponds to α_j, and $\mathbf{c}_j$ will have the effect that $\begin{pmatrix} z & 0 \\ 0 & -z \end{pmatrix}$ maps to $\begin{pmatrix} 0 & -iz \\ iz & 0 \end{pmatrix}$. Within $\mathfrak{g}_0$, a part of the θ stable Cartan subalgebra $\mathfrak{h}_0$ that lies in $\mathfrak{p}_0$, namely $\mathbb{R}\begin{pmatrix} 1 & 0 \\ 0 & -1 \end{pmatrix}$, thus gets replaced by something that lies in $\mathfrak{k}_0$, namely $\mathbb{R}\begin{pmatrix} 0 & -1 \\ 1 & 0 \end{pmatrix}$. The image of the real root α_j under this transformation is noncompact imaginary. After all the $\mathbf{c}_j$'s have been applied (in any order), there will be no real roots of $\mathfrak{l}$ left (since $\{\alpha_1, \dots, \alpha_l\}$ was maximal orthogonal among the real roots of $\mathfrak{l}$), and Lemma 4.43 allows us to conclude that the resulting Cartan subalgebra $\mathfrak{h}_{\mathrm{mc}}$, when intersected with $\mathfrak{l}_0$, is maximally compact in $\mathfrak{l}_0$. The composition $\mathbf{c} = \prod_{j=1}^{l} \mathbf{c}_j$ is the **Cayley transform**; we use the term "Cayley transform" also to refer to $\mathbf{c}^{-1}$.

Let us define $\mathbf{c}_j$. Since α_j is real, the root space $\mathfrak{g}_{\alpha_j}$ is closed under conjugation and therefore contains a member e_j of $\mathfrak{g}_0$. Put $f_j = -\theta e_j$, so that f_j is in $\mathfrak{g}_{-\alpha_j}$. Since $-\langle \cdot, \theta(\cdot) \rangle$ is positive definite on $\mathfrak{g}_0$,

$$-\langle e_j, \theta e_j \rangle = \langle e_j, f_j \rangle$$

is positive. Normalizing e_j (and therefore also $f_j = -\theta e_j$) by a positive scalar, we may assume that $\langle e_j, f_j \rangle = 2/|\alpha_j|^2$. Now define $h_j = 2|\alpha_j|^{-2} h_{\alpha_j}$ in $\mathfrak{a}_0$, so that $\alpha_j(h_j) = 2$. If h' is in $\mathfrak{h}$, then the element $[e_j, f_j]$ of $\mathfrak{h}$ satisfies

$$\langle [e_j, f_j], h' \rangle = \langle e_j, [f_j, h'] \rangle = -\langle e_j, [h', f_j] \rangle$$
$$= \alpha_j(h') \langle e_j, f_j \rangle = 2|\alpha_j|^{-2} \alpha_j(h') = \langle h_j, h' \rangle,$$

and it follows that $[e_j, f_j] = h_j$. If we identify

$$h_j \leftrightarrow \begin{pmatrix} 1 & 0 \\ 0 & -1 \end{pmatrix}, \quad e_j \leftrightarrow \begin{pmatrix} 0 & 1 \\ 0 & 0 \end{pmatrix}, \quad f_j \leftrightarrow \begin{pmatrix} 0 & 0 \\ 1 & 0 \end{pmatrix},$$

then we see that $\{h_j, e_j, f_j\}$ spans a copy of $\mathfrak{sl}(2, \mathbb{R})$ in $\mathfrak{g}_0$. The mapping $\mathbf{c}_j$ is

$$(11.248) \quad \mathbf{c}_j = \mathrm{Ad}(\exp \tfrac{\pi i}{4}(e_j + f_j)) \quad \text{in the complex adjoint group,}$$

which corresponds to conjugation by $\frac{\sqrt{2}}{2} \begin{pmatrix} 1 & i \\ i & 1 \end{pmatrix}$. (This matrix conjugates $\begin{pmatrix} 1 & 0 \\ 0 & -1 \end{pmatrix}$ to $\begin{pmatrix} 0 & -i \\ i & 0 \end{pmatrix}$, as it is supposed to.) From (11.248) we readily check that

(i) $\mathbf{c}_j(h_j) = -i(e_j - f_j)$
(ii) $\mathbf{c}_j$ fixes the kernel of α_j in $\mathfrak{h}$
(iii) $\mathbf{c}_j$ carries $\mathfrak{l}$ to itself.

If $j \neq k$, then the strong orthogonality of α_j and α_k implies that e_j and f_j commute with e_k and f_k. Consequently

(iv) $\mathbf{c}_j$ commutes with $\mathbf{c}_k$ if $j \neq k$, and hence $\mathbf{c} = \prod_{j=1}^{l} \mathbf{c}_j$ is independent of the order of the factors.

We now put

$$(\mathfrak{a}_r)_0 = \sum_{j=1}^{l} \mathbb{R}h_j, \qquad (\mathfrak{t}_r)_0 = \sum_{j=1}^{l} \mathbb{R}(e_j - f_j),$$

$$\mathfrak{a}_0' = \text{orthogonal complement of } (\mathfrak{a}_r)_0 \text{ in } \mathfrak{a}_0,$$

and $$\mathfrak{t}_0' = (\mathfrak{t}_r)_0 \oplus \mathfrak{t}_0.$$

Then $\mathbf{c}$ carries $\mathfrak{a}_r$ to $\mathfrak{t}_r$ and $\mathfrak{h} = \mathfrak{t} \oplus \mathfrak{a}_r \oplus \mathfrak{a}'$ to $\mathfrak{h}_{\mathrm{mc}} = \mathfrak{t} \oplus \mathfrak{t}_r \oplus \mathfrak{a}'$, while fixing $\mathfrak{t}$ and $\mathfrak{a}'$.

The transpose map of $\mathbf{c}^{-1}$, which we denote $\mathbf{c}$ also, sends $\mathfrak{h}^*$ to $\mathfrak{h}_{\mathrm{mc}}^*$, and sends roots to roots. The real roots α_j in $\Delta(\mathfrak{g}, \mathfrak{h})$ used to define $\mathbf{c}$ become noncompact imaginary roots in $\Delta(\mathfrak{g}, \mathfrak{h}_{\mathrm{mc}})$. Under $\mathbf{c}$, the roots carried on $\mathfrak{t}$ in $\mathfrak{h}$, which are the roots of $\mathfrak{m}$, are sent to roots carried on $\mathfrak{t}$ in $\mathfrak{h}_{\mathrm{mc}}$. However, compactness and noncompactness of these imaginary roots is not necessarily preserved by $\mathbf{c}$. The correct result is as follows.

Proposition 11.249. If β in $\Delta(\mathfrak{g}, \mathfrak{h})$ is carried on $\mathfrak{t}$, then β is orthogonal to all of $\alpha_1, \ldots, \alpha_l$ and can fail to be strongly orthogonal to at most one of $\alpha_1, \ldots, \alpha_l$. Also

(a) if β is strongly orthogonal to all of $\alpha_1, \ldots, \alpha_l$, then a root vector for β is a root vector for $\mathbf{c}(\beta)$, and $\mathbf{c}$ therefore preserves compactness/noncompactness of β, while
(b) if β fails to be strongly orthogonal to α_k, then a root vector for $\mathbf{c}(\beta)$ is a linear combination of a root vector for $\beta + \alpha_k$ and a root vector for $\beta - \alpha_k$, and $\mathbf{c}$ therefore reverses compactness/noncompactness of β.

PROOF. Suppose β is carried on $\mathfrak{t}$. Certainly β is orthogonal to $\alpha_1, \ldots, \alpha_l$, which are carried on $\mathfrak{a}$. If β fails to be strongly orthogonal to α_i and α_j with $i \neq j$, then $\beta \pm \alpha_i$ and $\beta \pm \alpha_j$ are roots with positive inner product. Hence the differences $\pm \alpha_i \mp \alpha_j$ are roots, in contradiction to the strong orthogonality of α_i and α_j.

Now let e_β be a root vector for β. We are to consider

$$\prod_{j=1}^{l} \text{Ad}(\exp \tfrac{\pi i}{4}(e_j + f_j))e_\beta.$$

If α_j is strongly orthogonal to β, then each term of the exponential series for $\text{Ad}(\exp \tfrac{\pi i}{4}(e_j + f_j))e_\beta$ after the 0th is 0, and $\text{Ad}(\exp \tfrac{\pi i}{4}(e_j + f_j))e_\beta = e_\beta$. This proves (a) and reduces (b) to a calculation of

$$(11.250) \qquad \text{Ad}(\exp \tfrac{\pi i}{4}(e_j + f_j))e_\beta = \sum_{n=0}^{\infty} \tfrac{1}{n!}(\tfrac{\pi i}{4}\text{ad}(e_k + f_k))^n e_\beta.$$

Since α_k is orthogonal but not strongly orthogonal to β, the α_k root string through β is $\beta - \alpha_k, \beta, \beta + \alpha_k$. Thus the α_k subalgebra $\mathfrak{sl}(2, \mathbb{C})$ within $\mathfrak{g}$ acts by an irreducible three-dimensional representation on the span of the root vectors for the root string, and

$$(\text{ad } e_k)(\text{ad } f_k)e_\beta = (\text{ad } f_k)(\text{ad } e_k)e_\beta = 2e_\beta.$$

Hence

$$(\text{ad}(e_k + f_k))^2 e_\beta = 4e_\beta.$$

The even-numbered terms of (11.250) are therefore

$$e_\beta - \tfrac{1}{2!}\tfrac{\pi^2}{4^2}(4e_\beta) + \tfrac{1}{4!}\tfrac{\pi^4}{4^4}(4^2 e_\beta) - \tfrac{1}{6!}\tfrac{\pi^6}{4^6}(4^3 e_\beta) + \cdots = (\cos \tfrac{\pi}{2})e_\beta = 0,$$

while the odd-numbered terms are

$$\tfrac{\pi i}{4}\text{ad}(e_k + f_k)\{e_\beta - \tfrac{1}{3!}\tfrac{\pi^2}{4^2}(4e_\beta) + \tfrac{1}{5!}\tfrac{\pi^4}{4^4}(4^2 e_\beta) - \ldots\} = \tfrac{i}{2}(\sin \tfrac{\pi}{2})\text{ad}(e_k + f_k)e_\beta.$$

Thus (b) follows.

PROOF OF LEMMA 11.231. Let us decompose a typical root α in $\Delta(\mathfrak{g}, \mathfrak{h}_{\text{mc}})$ according to $\mathfrak{h}_{\text{mc}} = \mathfrak{t} \oplus \mathfrak{t}_r \oplus \mathfrak{a}'$ as $\alpha = \alpha_{\mathfrak{t}} + \alpha_{\mathfrak{t}_r} + \alpha_{\mathfrak{a}'}$, and let us write the prospective identity as

$$(11.251) \qquad [-\delta(\mathfrak{u})] + [2\delta_K - E(2\delta_K)] + [\delta_M] + [-2\delta_{K \cap M}] \overset{?}{=} 0.$$

For each root the contribution to all bracketed terms but the first on the left side of (11.251) is carried on $\mathfrak{t}$. We shall decompose the set of positive roots into small sets of related elements and see that the sum of the contributions to the left side of (11.251) from all members of each such small set is 0.

Case 1. $\alpha_{\mathfrak{t}} = 0$. Then α is in $\Delta(\mathfrak{l}, \mathfrak{h}_{mc})$ and the singleton set $\{\alpha\}$ contributes 0 to each term of (11.251).

Case 2. $\alpha_{\mathfrak{t}} \neq 0, \alpha_{\mathfrak{a}'} \neq 0$. Here α is complex.

Subcase 2a. $\alpha_{\mathfrak{t}_r} \neq 0$. The small set consists of the four roots $\alpha_{\mathfrak{t}} \pm \alpha_{\mathfrak{t}_r} \pm \alpha_{\mathfrak{a}'}$. These roots together contribute $-2\alpha_{\mathfrak{t}}, 2\alpha_{\mathfrak{t}}, 0, 0$ to the respective bracketed terms, for a total of 0.

Subcase 2b. $\alpha_{\mathfrak{t}_r} = 0$. The small set consists of the two roots $\alpha_{\mathfrak{t}} \pm \alpha_{\mathfrak{a}'}$. These roots together contribute $-\alpha_{\mathfrak{t}}, \alpha_{\mathfrak{t}}, \alpha_{\mathfrak{t}}, -\alpha_{\mathfrak{t}}$ to the respective bracketed terms, for a total of 0.

Case 3. $\alpha = \alpha_{\mathfrak{t}}$ strongly orthogonal to all α_i used in building the Cayley transform from $\Delta(\mathfrak{g}, \mathfrak{h}_{mc})$ to $\Delta(\mathfrak{g}, \mathfrak{h})$. Here α is imaginary.

Subcase 3a. α compact. Then α remains compact after the Cayley transform. The singleton set $\{\alpha_{\mathfrak{t}}\}$ contributes $-\frac{1}{2}\alpha_{\mathfrak{t}}, \alpha_{\mathfrak{t}}, \frac{1}{2}\alpha_{\mathfrak{t}}, -\alpha_{\mathfrak{t}}$ to the bracketed terms, for a total of 0.

Subcase 3b. α noncompact. Then α remains noncompact after the Cayley transform. The singleton set $\{\alpha_{\mathfrak{t}}\}$ contributes $-\frac{1}{2}\alpha_{\mathfrak{t}}, 0, \frac{1}{2}\alpha_{\mathfrak{t}}, 0$ to the bracketed terms, for a total of 0.

Case 4. $\alpha = \alpha_{\mathfrak{t}}$ not strongly orthogonal to α_i. The index i is unique by Proposition 11.249. The small set is $\{\alpha_{\mathfrak{t}}, \alpha_{\mathfrak{t}} \pm \alpha_i\}$.

Subcase 4a. $\alpha_{\mathfrak{t}}$ compact. Then $\alpha_{\mathfrak{t}}$ is noncompact after Cayley transform, according to Proposition 11.249. The set contributes $-\frac{3}{2}\alpha_{\mathfrak{t}}, \alpha_{\mathfrak{t}}, \frac{1}{2}\alpha_{\mathfrak{t}}, 0$, for a total of 0.

Subcase 4b. $\alpha_{\mathfrak{t}}$ noncompact. Then $\alpha_{\mathfrak{t}}$ is compact after Cayley transform, according to Proposition 11.249. The set contributes $-\frac{3}{2}\alpha_{\mathfrak{t}}, 2\alpha_{\mathfrak{t}}, \frac{1}{2}\alpha_{\mathfrak{t}}, -\alpha_{\mathfrak{t}}$, for a total of 0.

Case 5. $\alpha_{\mathfrak{t}} \neq 0, \alpha_{\mathfrak{a}'} = 0$, and $\alpha_{\mathfrak{t}_r}$ is neither 0 nor $\pm\alpha_i$. These are the only roots that are left. The α_n's form an orthogonal basis of $\mathfrak{t}_r^*$, and we can therefore write

$$\alpha = \alpha_{\mathfrak{t}} + \frac{1}{2}\sum_{j=1}^{l} \frac{2\langle \alpha, \alpha_j \rangle}{|\alpha_j|^2}\alpha_j.$$

Parseval's equality gives

$$4 = \frac{4|\alpha_{\mathfrak{t}}|^2}{|\alpha|^2} + \sum_{j=1}^{l} \frac{4\langle \alpha, \alpha_j \rangle^2}{|\alpha_j|^2|\alpha|^2}.$$

The terms in the grouped sum are integers, and at most three can be nonzero. If $|\frac{2\langle\alpha,\alpha_j\rangle}{|\alpha_j|^2}| = 2$ for some j, at most the jth term and one other term in the grouped sum can be nonzero. Apart from signs, the allowable expressions are therefore

$$\alpha = \begin{cases} \alpha_t + \alpha_i & \text{(a)} \\ \alpha_t + \alpha_i + \frac{1}{2}\alpha_j & \text{(b)} \\ \alpha_t + \frac{3}{2}\alpha_i & \text{(c)} \\ \alpha_t + \frac{1}{2}\alpha_i & \text{(d)} \\ \alpha_t + \frac{1}{2}\alpha_i + \frac{1}{2}\alpha_j & \text{(e)} \\ \alpha_t + \frac{1}{2}\alpha_i + \frac{1}{2}\alpha_j + \frac{1}{2}\alpha_k. & \text{(f)} \end{cases}$$

In Subcase (a), we consider the small set $\{\alpha_t, \alpha_t \pm \alpha_i\}$. This was treated as Case 4. In Subcase (b), we consider the small set

$$\{\alpha_t \pm \tfrac{1}{2}\alpha_j, \alpha_t \pm \alpha_i \pm \tfrac{1}{2}\alpha_j\}.$$

When $\alpha_t + \frac{1}{2}\alpha_j$ is compact, these six roots contribute a total of $-3\alpha_t, 3\alpha_t, 0, 0$ to the respective bracketed terms, for a total of 0. When $\alpha_t + \frac{1}{2}\alpha_j$ is noncompact, we get the same thing. This handles Subcase (b). In view of (a) and (b), we may assume for Subcases (c) through (f) that α is strongly orthogonal to all α_n's that do not appear explicitly. A small set consists of α with all possible sign changes allowed in the t_r part. Half the terms are compact, and half are noncompact. There is no contribution to M. If there are s roots in a small set, the s roots contribute a total of $-\frac{1}{2}s\alpha_t, \frac{1}{2}s\alpha_t, 0, 0$, for a total of 0. This completes the proof.

12. Cohomological Induction and Minimal K Types

In this section we shall conclude our discussion of Theorem 10.44. Our goal is to relate the minimal K types of $\mathcal{L}_S(W)$ to the minimal $L \cap K$ types of W. As was mentioned at the start of §11, three stages of generality are involved:

(a) the case that $W = X_{L\cap K}(\xi_L, \nu)$ and $\mathcal{L}_S(W) = X_K(\xi, \nu)$ are standard modules related as in (11.222) and Theorem 11.225, where L is split modulo center and W is a principal series,

(b) the case that $W = X_{L\cap K}(\xi_L, \nu)$ and $\mathcal{L}_S(W) = X_K(\xi, \nu)$ are standard modules related as in (11.222) and Theorem 11.225, but with no special assumption on L and W, and

(c) the general case of W and $\mathcal{L}_S(W)$, the only restrictions being that W is irreducible and that the infinitesimal character satisfies (10.45a).

Theorem 11.230 in §11 has already dealt with (a), and we treat (b) and (c) now.

Let us begin with (b). As with (a), there is no loss in generality in assuming that G and L are connected. We continue to write $\mathfrak{q} = \mathfrak{l} \oplus \mathfrak{u}$ for the θ stable parabolic subalgebra occurring in the cohomological induction, and we denote by $\mathfrak{h}_0 = \mathfrak{t}_0 \oplus \mathfrak{a}_0$ the θ stable Cartan subalgebra of $\mathfrak{l}_0$ used in defining the modules in question. Following Theorem 11.230, we can construct a θ parabolic subalgebra $\mathfrak{q}_1 = \mathfrak{l}_1 \oplus \mathfrak{u}_1$ of $\mathfrak{l}$, a corresponding Levi subgroup L_1 of L that is split modulo center, and a principal series $X_{L_1 \cap K}(\xi_{L_1}, \nu)$ of L_1 so that the highest weights of the minimal $L \cap K$ types of $X_{L \cap K}(\xi_L, \nu)$ are all the $L \cap K$ dominant expressions

$$(11.252a) \qquad \Lambda_L = \lambda + \mu_1 + 2\delta(\mathfrak{u}_1 \cap \mathfrak{p}),$$

where λ is the infinitesimal character of ξ_{L_1} and $\lambda + \mu_1$ is the highest weight of a minimal $L_1 \cap K$ type of $X_{L_1 \cap K}(\xi_{L_1}, \nu)$. When we apply the construction of Theorem 11.230 to $X_K(\xi, \nu)$, we are led to the θ stable parabolic subalgebra $\mathfrak{q} \oplus \mathfrak{u}_1$ of $\mathfrak{g}$, the same Levi subgroup L_1, and the same principal series $X_{L_1 \cap K}(\xi_L, \nu)$. The highest weights of the minimal K types of $X_K(\xi, \nu)$ are all the K dominant expressions

$$(11.252b) \qquad \Lambda = \lambda + \mu_1 + 2\delta((\mathfrak{u} + \mathfrak{u}_1) \cap \mathfrak{p})$$

for which $\lambda + \mu_1$ is the highest weight of a minimal $L_1 \cap K$ type of $X_{L_1 \cap K}(\xi_{L_1}, \nu)$.

Any K dominant expression (11.252b) leads to a relation

$$(11.252c) \qquad \Lambda = \Lambda_1 + 2\delta(\mathfrak{u} \cap \mathfrak{p})$$

with Λ_1 as in (11.252a), and Λ_1 has to be $L \cap K$ dominant since $2\delta(\mathfrak{u} \cap \mathfrak{p})$ is orthogonal to the roots of $L \cap K$. Conversely any K dominant expression (11.252c) yields a K dominant expression (11.252b) and hence a minimal K type of $X_K(\xi, \nu)$. This completes our discussion of Theorem 10.44 for the second stage of generality, called (b) at the start of the section.

To pass to stage (c), we again may assume that G and L are connected. Let λ be the infinitesimal character of W, and assume that λ satisfies the dominance condition (10.45a). We may assume that $\mathcal{L}_S(W)$ is not 0. Because of the vanishing theorem (Theorem 5.99), $\mathcal{L}_S$ is exact on the category $C_\lambda(\mathfrak{l}, L \cap K)$ of $(\mathfrak{l}, L \cap K)$ modules of infinitesimal character λ. Since W is by asssumption irreducible, the Langlands classification (Theorem 11.198) allows us to choose a standard module $X_{L \cap K}(\xi_L, \nu)$ such that W is the unique Langlands quotient of $X_{L \cap K}(\xi_L, \nu)$. Since $\mathcal{L}_S$ is exact, $\mathcal{L}_S(W)$ is a quotient of $\mathcal{L}_S(X_{L \cap K}(\xi_L, \nu))$, which Theorem 11.225 identifies as a certain standard module $X_K(\xi, \nu)$. We shall prove the following theorem.

Theorem 11.253. Let $(\mathfrak{g}, K)$ be a reductive pair with underlying group G connected, let $X_K(\xi, \nu)$ be the underlying $(\mathfrak{g}, K)$ module of a standard continuous series as in Theorem 11.198, and suppose that $X_K(\xi, \nu)$ has a unique Langlands quotient. Then every minimal K type of $X_K(\xi, \nu)$ lies in the Langlands quotient.

Once the theorem has been proved, we apply it to L and $X_{L \cap K}(\xi, \nu)$ to see that the minimal $L \cap K$ types of W coincide with the minimal $L \cap K$ types of $X_{L \cap K}(\xi_L, \nu)$. From the previous stages of our argument, we know that the minimal K types of $X_K(\xi, \nu)$ are precisely the K types of the bottom layer corresponding to these $L \cap K$ types. These K types all occur in $\mathcal{L}_S(W)$, because they are in the bottom layer of $\mathcal{L}_S(W)$, and therefore all the K types of $\mathcal{L}_S(W)$ have the required form.

Thus we have reduced the third stage of the argument to Theorem 11.253. To prove Theorem 11.253, we need to know how to recover the infinitesimal character of ξ, at least up to some conjugacy, from a minimal K type of $X_K(\xi, \nu)$. For this purpose we introduce an equivalent formulation of minimality for K types of principal series in a group G split modulo center.

Let G be a connected group split modulo center, and let $(\mathfrak{h}_{mc})_0 = \mathfrak{t}'_0 \oplus \mathfrak{a}'_0$ be a maximally compact θ stable Cartan subalgebra of $\mathfrak{g}_0$. With μ in $\mathfrak{t}'^*$, let τ_μ be a K type occurring in the ξ principal series of G, i.e., having $\tau_{\lambda+\mu}|_{M_\mathfrak{p}} \supseteq \xi$. We say that τ_μ is **small** if, in some positive system $\Delta^+(\mathfrak{g}, \mathfrak{h}_{mc})$ such that $\mu + 2\delta_K$ is $\Delta^+(\mathfrak{g}, \mathfrak{h}_{mc})$ dominant, $\mu + 2\delta_K - \delta$ is of the form $\gamma - \sum c_i \alpha_i$, where each c_i is ≥ 0, each α_i is in $\Delta^+(\mathfrak{g}, \mathfrak{h}_{mc})$, and γ is a member of $\mathfrak{h}_{mc}^*$ orthogonal to all roots. Here is a formal statement of the fact we need about this definition.

Theorem 11.254. Let $(\mathfrak{g}, K)$ be a reductive pair whose underlying group G is connected, suppose that G is split modulo center, and let τ_μ be a K type occurring in the ξ principal series of G. Then τ_μ is minimal if and only if it is small.

REMARKS. We omit the proof, which may be found in §7 of Vogan [1979]. That section introduces also a notion of "fine" for the K type

τ_μ in the context of Theorem 11.254, and the theorem is that minimal, small, and fine are equivalent. The line of argument shows that minimal implies small by a refinement of the arguments of §X.2 of this book, small implies fine by an elementary argument, and all fine K types have the same norm. Consequently all fine K types are minimal, and the three notions are equivalent.

We shall apply Theorem 11.254 to the group L. Write the minimal K type formula of Theorem 11.230 as

$$(11.255) \qquad \Lambda + 2\delta_K = (\delta - \delta_L) + \lambda + \mu + 2\delta_{L \cap K},$$

where $\lambda + \mu$ is the highest weight of a minimal $L \cap K$ type $\tau_{\lambda+\mu}$ for a principal series of a group L that is split modulo center. Let us abbreviate

$$\Delta^+ = \Delta^+(\mathfrak{g}, \mathfrak{h}_{mc})$$
$$\Delta_L^+ = \Delta^+(\mathfrak{l}, \mathfrak{h}_{mc})$$
$$\Delta_K^+ = \Delta^+(\mathfrak{k}, \mathfrak{t}')$$
$$\Delta_{L \cap K}^+ = \Delta^+(\mathfrak{l} \cap \mathfrak{k}, \mathfrak{t}').$$

We introduce a new positive system $(\Delta^+)'$ by changing the notion of positivity on $\Delta_L = \Delta(\mathfrak{l}, \mathfrak{h}_{mc})$ (and only there) so that $\tau_{\lambda+\mu}$ is exhibited as small (since $\tau_{\lambda+\mu}$ minimal is equivalent with $\tau_{\lambda+\mu}$ small, by Theorem 11.254 applied to the group L). This means in particular that $\mu + 2\delta_{L \cap K}$ is to be $(\Delta_L^+)'$ dominant. With "primes" referring to objects in the new ordering, we have

$$\begin{array}{ccc}
\Delta_{L \cap K}^+ = (\Delta_{L \cap K}^+)' & \text{and} & \Delta_K^+ = (\Delta_K^+)', \\
\delta_{L \cap K} = \delta'_{L \cap K} & \text{and} & \delta_K = \delta'_K, \\
\multicolumn{3}{c}{\delta - \delta_L = \delta' - \delta'_L.}
\end{array}$$

$$(11.256)$$

Since $\tau_{\lambda+\mu}$ is small, we can write

$$(11.257) \qquad \mu + 2\delta_{L \cap K} - \delta'_L = - \sum_{\beta_i \in (\Delta_L^+)'} c_i \beta_i \qquad \text{with } c_i \geq 0.$$

Substituting from (11.256) and (11.257) into (11.255), we obtain

$$(11.258) \qquad \Lambda + 2\delta_K - \delta' = \lambda - \sum_{\beta_i \in (\Delta_L^+)'} c_i \beta_i \qquad \text{with } c_i \geq 0.$$

We shall use (11.258) to recover λ from Λ. But we need one more fact, namely that

(11.259) $\Lambda + 2\delta_K$ is $(\Delta^+)'$ dominant.

To see this, we let β be $(\Delta^+)'$ simple and we compute $\langle \Lambda + 2\delta_K, \beta \rangle$, distinguishing two cases:

(i) $\beta|_{\mathfrak{t}} \neq 0$. Then β is in Δ^+ and thus has $\langle \lambda, \beta \rangle \geq 0$. Also each β_i is the sum of simple roots other than β, and so $\langle -\sum c_i \beta_i, \beta \rangle \geq 0$. Since also $\langle \delta', \beta \rangle \geq 0$, we conclude that $\langle \Lambda + 2\delta_K, \beta \rangle \geq 0$ from (11.258).

(ii) $\beta|_{\mathfrak{t}} = 0$. Then β is in Δ_L. Hence we have $\langle \lambda, \beta \rangle = 0 = \langle \delta - \delta_L, \beta \rangle$. Since $\mu + 2\delta_{L \cap K}$ is $(\Delta_L^+)'$ dominant, it follows from (11.255) that $\langle \Lambda + 2\delta_K, \beta \rangle \geq 0$.

Lemma 11.260. Under the assumption that $(\mathfrak{g}, K)$ is a reductive pair with underlying group G connected, Λ determines λ in formula (11.258) in the following sense. Suppose Δ_K^+ is fixed and Λ is any Δ_K^+ dominant form. Let $(\Delta^+)'$ and $(\Delta^+)''$ be positive systems such that $\Lambda + 2\delta_K$ is dominant for each and such that

(i) $\Lambda + 2\delta_K - \delta' = \lambda' - \sum c_i' \beta_i'$ with all $c_i' \geq 0$ and
$\Lambda + 2\delta_K - \delta'' = \lambda'' - \sum c_i'' \beta_i''$ with all $c_i'' \geq 0$
(ii) λ' and λ'' are dominant for $(\Delta^+)'$ and $(\Delta^+)''$, respectively
(iii) β_i' and β_i'' are positive for $(\Delta^+)'$ and $(\Delta^+)''$, respectively, for all i
(iv) $\langle \lambda', \beta_i' \rangle = 0$ and $\langle \lambda'', \beta_i'' \rangle = 0$ for all i.

Then $\lambda' = \lambda''$.

PROOF THAT LEMMA 11.260 IMPLIES THEOREM 11.253. Write $S = MAN$. If a standard continuous-series representation $I_S^G(\xi, \nu)$ has a unique Langlands quotient, we denote that quotient by $J(S, \xi, \nu)$. Assume the theorem is false. Among all counterexamples $(J(S, \xi, \nu), \tau_\Lambda)$, choose one with $|\Lambda + 2\delta_K|^2$ as small as possible; we can do so since the range of $|\Lambda' + 2\delta_K|^2$ is a discrete set of positive numbers.

For our chosen counterexample, τ_Λ lies in $I_S^G(\xi, \nu)$ but not in $J(S, \xi, \nu)$. Therefore it lies in some other subquotient of $I_S^G(\xi, \nu)$. By the Langlands classsification (Theorem 11.198d), this other subquotient is necessarily of the form $J(S', \xi', \nu')$. By Remark 4 with Theorem 11.198,

(11.261) $|\mathrm{Re}\,\nu'| < |\mathrm{Re}\,\nu|$.

However, $J(S', \xi', \nu')$ has the same infinitesimal character as $I_S^G(\xi, \nu)$ and $J(S, \xi, \nu)$, and this infinitesimal character is given by Proposition

11.43. Thus if we let λ and λ' be the infinitesimal characters of ξ and ξ', respectively, then $\lambda' + \nu'$ is conjugate to $\lambda + \nu$ and necessarily $\lambda' + \mathrm{Re}\,\nu'$ is conjugate to $\lambda + \mathrm{Re}\,\nu$. From (11.261) we conclude that

$$(11.262) \qquad\qquad\qquad |\lambda'| > |\lambda|.$$

Now τ_Λ is a minimal K type for $I_S^G(\xi, \nu)$, and Lemma 11.260 shows that we can recover λ (possibly up to some conjugacy) from τ_Λ. Then it follows from (11.262) that τ_Λ is not a minimal K type of $I_{S'}^G(\xi', \nu')$. Since τ_Λ does occur in $I_{S'}^G(\xi, \nu')$, it follows that a minimal K type $\tau_{\Lambda'}$ of $I_{S'}^G(\xi', \nu')$ has $|\Lambda' + 2\delta_K|^2 < |\Lambda + 2\delta_K|^2$. Since τ_Λ is minimal for $J(S', \xi', \nu')$, $\tau_{\Lambda'}$ cannot occur in $J(S', \xi', \nu')$. Thus $(J(S', \xi', \nu'), \tau_{\Lambda'})$ provides a smaller counterexample, contradiction.

We are left with proving Lemma 11.260. This is essentially a result about abstract root systems, and we shall frame some preparatory lemmas in that context. Thus we suppose that V is a real inner-product space in which there is an abstract reduced root system, i.e., a finite set of nonzero elements of V that span V, are closed under their own reflections, and are such that twice an element in Δ is never in Δ. Let $W(\Delta)$ be the Weyl group. For the first two lemmas below, fix a positive system Δ^+ for Δ, and let $\overline{C}$ be the closed positive Weyl chamber.

Lemma 11.263. Let v be in V, and let v_0 be the unique point in $\overline{C}$ closest to v. Then

    (a) $\langle v - v_0, w \rangle \leq 0$ for every w in $\overline{C}$, and
    (b) $\langle v - v_0, v_0 \rangle = 0$.

PROOF. Existence and uniqueness of v_0 are well-known properties of closed convex sets in Hilbert space. Let w be in $\overline{C}$. For $0 \leq \epsilon \leq 1$, $v_0 + \epsilon(w - v_0)$ is in $\overline{C}$ by convexity. For such ϵ,

$$|v_0 + \epsilon(w - v_0) - v|^2 \geq |v_0 - v|^2$$

and hence $\qquad -2\epsilon \langle w - v_0, v - v_0 \rangle + \epsilon^2 |w - v_0|^2 \geq 0.$

Dividing by ϵ and letting ϵ tend to 0, we obtain

$$(11.264) \qquad\qquad\qquad \langle w - v_0, v - v_0 \rangle \leq 0.$$

Since v_0 is in $\overline{C}$ and $\overline{C}$ is a cone, $2v_0$ and $\frac{1}{2}v_0$ are in $\overline{C}$. Putting $w = 2v_0$ and then $w = \frac{1}{2}v_0$ in (11.264), we obtain (b). Then (a) follows immediately from (11.264) and (b).

Relative to the positive system Δ^+ that we have fixed, let $\alpha_1, \ldots, \alpha_l$ be the simple roots. The **fundamental weights** $\omega_1, \ldots, \omega_l$ in V are defined by the condition $2\langle \omega_i, \alpha_j \rangle / |\alpha_j|^2 = \delta_{ij}$. The members of $\overline{C}$ are exactly those elements of V for which the expansion in the basis $\{\omega_j\}$ has all coefficients ≥ 0.

For each subset $\mathcal{F}$ of $\{1, \ldots, l\}$, the set

$$(11.265) \qquad \{\alpha_j\}_{j \in \mathcal{F}} \cup \{\omega_j\}_{j \notin \mathcal{F}}$$

is a basis of V. In fact, each subset in braces is linearly independent, and the two subsets are orthogonal to each other. Thus (11.265) is a linearly independent set of l elements and hence is a basis.

Lemma 11.266. For each v in V, there exists a unique subset $\mathcal{F}$ of $\{1, \ldots, l\}$ such that the decomposition

$$(11.267a) \qquad v = \sum_{j \notin \mathcal{F}} b_j \omega_j - \sum_{j \in \mathcal{F}} a_j \alpha_j$$

of v relative to the basis (11.265) has

$$(11.267b) \qquad \begin{array}{ll} b_j > 0 & \text{for } j \notin \mathcal{F} \\ a_j \geq 0 & \text{for } j \in \mathcal{F}. \end{array}$$

For this $\mathcal{F}$, the vector $\sum_{j \notin \mathcal{F}} b_j \omega_j$ is the vector v_0 of Lemma 11.263.

REMARK. We write $\mathcal{F}(v)$ for the subset $\mathcal{F}$ of $\{1, \ldots, l\}$ given by the lemma.

PROOF. For existence let v_0 be as in Lemma 11.263, and define $\mathcal{F}$ to consist of those j's not needed for the expansion of v_0 in terms of the ω_j's:

$$(11.268) \qquad v_0 = \sum_{j \notin \mathcal{F}} b_j \omega_j \qquad \text{with all } b_j > 0.$$

Define a_j for $1 \leq j \leq l$ by

$$(11.269) \qquad v_0 - v = \sum_{\text{all } j} a_j \alpha_j.$$

Then Lemma 11.263a applied to $w = \omega_j$ gives $a_j \geq 0$ for all j. Using this fact, taking the inner product of (11.268) and (11.269), and applying Lemma 11.263b, we obtain

$$0 = \langle v_0 - v, v_0 \rangle = \sum_{j \notin \mathcal{F}} a_j b_j$$

with each term on the right ≥ 0. From this relation we conclude $a_j = 0$ for $j \notin \mathcal{F}$, and then (11.267) follows from (11.268) and (11.269).

For uniqueness suppose $\mathcal{F}$ is a subset of $\{1, \ldots, l\}$ such that

$$v = \sum_{j \in \mathcal{F}} b_j \omega_j - \sum_{j \in \mathcal{F}} a_j \alpha_j$$

with all $b_j > 0$ and all $a_j \geq 0$. Put $v_0' = \sum_{j \notin \mathcal{F}} b_j \omega_j$. Then v_0' is in $\overline{C}$, and we see directly that

$$(11.270) \qquad\qquad \langle v - v_0', v_0' \rangle = 0.$$

If w is in $\overline{C}$, we find that

$$
\begin{aligned}
|w - v|^2 &= |(w - v_0') + (v_0' - v)|^2 \\
&= |w - v_0'|^2 + 2\langle w - v_0', v_0' - v \rangle + |v_0' - v|^2 \\
&= |w - v_0'|^2 + 2\langle w, v_0' - v \rangle + |v_0' - v|^2
\end{aligned}
$$

by (11.270). Here $v_0' - v = \sum_{j \in \mathcal{F}} a_j \alpha_j$ clearly has inner product ≥ 0 with any $w = \sum c_j \omega_j$ having all $c_j \geq 0$, and so $|w - v|^2 \geq |v_0' - v|^2$ with equality only if $w = v_0'$. Taking $w = v_0$, we see that $v_0' = v_0$. Then linear independence of the ω_j's forces $\mathcal{F}$ to be as in the existence half of the proof.

Lemma 11.271. Fix a positive system Δ^+ for Δ in V. Suppose v and v' in V are dominant, α is simple in Δ^+, and w in $W(\Delta)$ satisfies $l(s_\alpha w) = l(w) + 1$, where l is the length function on $W(\Delta)$ relative to Δ^+. Then $\langle s_\alpha w v, v' \rangle \leq \langle w v, v' \rangle$.

PROOF. The condition $l(s_\alpha w) = l(w) + 1$ means that $w^{-1}\alpha > 0$. (Otherwise $-w^{-1}\alpha$ is a positive root whose image under w is negative and is sent to a positive root by s_α.) Now

$$wv - s_\alpha wv = \frac{2\langle wv, \alpha \rangle}{|\alpha|^2}\alpha = \frac{2\langle v, w^{-1}\alpha \rangle}{|\alpha|^2}\alpha$$

and hence $\qquad \langle wv, v' \rangle - \langle s_\alpha wv, v' \rangle = \dfrac{2\langle v, w^{-1}\alpha \rangle \langle \alpha, v' \rangle}{|\alpha|^2},$

from which Lemma 11.271 follows.

Lemma 11.272. In V, let $(\Delta^+)'$ and $(\Delta^+)''$ be two positive systems for Δ, and let s be the member of the Weyl group $W(\Delta)$ with $s(\Delta^+)' = (\Delta^+)''$. If v is a vector that is dominant for both $(\Delta^+)'$ and $(\Delta^+)''$, then $sv = v$.

PROOF. Let $v' = sv$, and write s as a minimal product $s_{\gamma_n} \cdots s_{\gamma_1}$ of simple reflections in $(\Delta^+)'$. Applying Lemma 11.271 recursively, we see that

$$\langle v', v' \rangle = \langle s_{\gamma_n} \cdots s_{\gamma_1} v, v' \rangle \leq \cdots \leq \langle s_{\gamma_2} s_{\gamma_1} v, v' \rangle \leq \langle s_{\gamma_1} v, v' \rangle \leq \langle v, v' \rangle.$$

Since $|v| = |v'|$, we have equality in the Schwarz inequality and must have $v' = v$.

PROOF OF LEMMA 11.260. There is no loss of generality in assuming that $\mathfrak{g}$ is semisimple, and then our results on abstract root systems can be applied with $V = i\mathfrak{t}_0' \oplus \mathfrak{a}_0'$ and $\Delta = \Delta(\mathfrak{g}, \mathfrak{h}_{\mathrm{mc}})$. By Lemma 11.266 the properties in the $(\Delta^+)'$ system characterize λ' as the unique nearest $(\Delta^+)'$ dominant form to $\Lambda + 2\delta_K - \delta'$. In more detail the lemma says that if $\alpha_1, \ldots, \alpha_l$ are the simple roots relative to $(\Delta^+)'$, then the set $\mathcal{F} = \mathcal{F}(\Lambda + 2\delta_K - \delta')$ of indices has

(11.273a)
$$\Lambda + 2\delta_K - \delta' = \sum_{j \notin \mathcal{F}} b_j \omega_j - \sum_{j \in \mathcal{F}} a_j \alpha_j$$

(11.273b)
$$\lambda' = \sum_{j \notin \mathcal{F}} b_j \omega_j$$

(11.273c) $b_j > 0$ for $j \notin \mathcal{F}$ and $a_j \geq 0$ for $j \in \mathcal{F}$.

Let $\mathcal{F}' = \{j \mid \langle \Lambda + 2\delta_K, \alpha_j \rangle = 0\}$, and let W' be the subgroup of the Weyl group $W(\Delta)$ fixing $\Lambda + 2\delta_K$. Then $\mathcal{F}' \subseteq \mathcal{F}$. (In fact, if j is in $\mathcal{F}'$ and j is not in $\mathcal{F}$, then (11.273a) gives

$$0 = \frac{2\langle \Lambda + 2\delta_K, \alpha_j \rangle}{|\alpha_j|^2} = b_j + 1,$$

in contradiction to (11.273c).) Also Chevalley's Lemma (Proposition 4.146) and the $(\Delta^+)'$ dominance of $\Lambda + 2\delta_K$ imply that W' is generated by the reflections s_{α_j} with j in $\mathcal{F}'$. Then $s \in W'$ implies $s\lambda' = \lambda'$ because each $i \notin \mathcal{F}$ has

$$s_{\alpha_i}\left(\sum b_j \omega_j\right) = \sum b_j \omega_j - \frac{2\sum b_j \langle \omega_j, \alpha_i \rangle}{|\alpha_i|^2} \alpha_i = \sum b_j \omega_j,$$

by (11.273b).

Let s be the unique element of $W(\Delta)$ such that $s(\Delta^+)' = (\Delta^+)''$. Since $\Lambda + 2\delta_K$ is by assumption dominant for both $(\Delta^+)'$ and $(\Delta^+)''$, Lemma 11.272 shows that $s(\Lambda + 2\delta_K) = \Lambda + 2\delta_K$. Then s is in W', and it follows that $s\lambda' = \lambda'$. Moreover $s\delta' = \delta''$. Hence application of s to the identity

$$\Lambda + 2\delta_K - \delta' = \lambda' - \sum c_i' \beta_i'$$

gives
$$\Lambda + 2\delta_K - \delta'' = \lambda' - \sum c_i' s\beta_i'.$$

Thus Lemma 11.266 characterizes both λ' and λ'' as the unique nearest $(\Delta^+)''$ dominant form to $\Lambda + 2\delta_K - \delta''$, and we conclude that $\lambda' = \lambda''$, as required.

CHAPTER XII

EPILOG: WEAKLY UNIPOTENT
REPRESENTATIONS

Consideration of irreducible infinitesimally unitary $(\mathfrak{g}, K)$ modules that are not properly unitarily induced by either of the two standard constructions in this book leads one to a definition of "weakly unipotent" $(\mathfrak{g}, K)$ modules. One-dimensional modules are the prototypes. The facts about vanishing and unitarizability for cohomological induction from weakly unipotent modules are similar to those for the case of the prototypes, where the induced modules are the $A_{\mathfrak{q}}(\lambda)$'s. The theorems use a weaker dominance condition than the one used for ordinary $(\mathfrak{g}, K)$ modules.

Questions of irreducibility of the cohomologically induced modules in this situation lead to the introduction of "Dixmier algebras." The inducing module is assumed given as an irreducible module for a generalized pair involving a Dixmier algebra. Under suitable hypotheses the cohomologically induced module is an irreducible module for a different generalized pair that involves an "induced" Dixmier algebra.

This book is about building unitary representations, particularly irreducible ones, for reductive Lie groups. We have used two constructions:

(i) parabolic induction with a unitary representation on the Levi factor of a real parabolic subgroup, and

(ii) cohomological induction with a unitary representation on the Levi factor of a θ stable parabolic subalgebra and with the dominance condition of Theorem 5.99 in place.

It is time to take stock of how much we have. In doing so, we shall state most results without proof, and we shall indicate some directions for research. References appear in the Notes.

Fix a reductive pair $(\mathfrak{g}, K)$. Since we are interested only in having an idea of where things stand, we shall not insist on maximum generality. We shall always assume that $(\mathfrak{g}, K)$ is in the Harish-Chandra class. When it is convenient, we shall assume also that K is connected.

The Langlands classification (Theorem 11.198) gives a starting point. The set of irreducible infinitesimally unitary $(\mathfrak{g}, K)$ modules is a subset of the set of irreducible admissible $(\mathfrak{g}, K)$ modules (cf. Theorems 0.3 and 10.1), and the latter set is parametrized by the Langlands classification. One can write down a checkable necessary and sufficient condition in terms of Langlands quotient parameters for the Langlands quotient to carry a nonzero invariant Hermitian form, and one can give the form concretely in terms of an integral operator (sometimes a singular integral operator). The question is whether this operator is definite. Occasionally this question can be answered, but usually it is unmanageable. This difficulty of deciding the unitarity goes hand-in-hand with the difficulty of understanding the representation itself. Passage to the Langlands quotient is simply a mysterious operation. In the cases where unitarity of a Langlands quotient can be settled, the success often occurs because the Langlands quotient has been realized in some other way.

Thus a direct attack on unitarity via the Langlands classification is not a viable option. First we need to develop some understanding of the representations whose unitarity we are investigating. For a frame of reference, let us think of an irreducible infinitesimal unitary $(\mathfrak{g}, K)$ module V as given. If we can exhibit this module as obtained by one of our unitarity-preserving constructions (i) or (ii) starting from a proper Levi pair $(\mathfrak{l}, L \cap K)$, we shall say that V is **properly unitarily induced**. In this case we should regard V as understood (and therefore of no further interest in this chapter).

Theorem 12.1. Let V be an irreducible infinitesimally unitary $(\mathfrak{g}, K)$ module, let $\mathfrak{h}$ be a Cartan subalgebra of $\mathfrak{g}$, let $\mathfrak{h}_{ss} = \mathfrak{h} \cap [\mathfrak{g}, \mathfrak{g}]$, and let λ be the infinitesimal character of V. Unless the restriction of λ to $\mathfrak{h}_{ss}$ is in the $\mathbb{R}$ linear span of the roots, V is parabolically induced from an irreducible infinitesimally unitary $(\mathfrak{l}, L \cap K)$ module for some proper real parabolic subpair.

So we may restrict attention to $(\mathfrak{g}, K)$ modules with real infinitesimal character. To see what it means for V not to be cohomologically induced, we shall make use of the terms "good" and "weakly good" from Definition 0.49 for the infinitesimal character of an $(\mathfrak{l}, L \cap K)$ module relative to a θ stable parabolic subalgebra $\mathfrak{q} = \mathfrak{l} \oplus \mathfrak{u}$.

Theorem 12.2. Let V be an irreducible infinitesimally unitary $(\mathfrak{g}, K)$ module, let $\mathfrak{h}$ be a Cartan subalgebra of $\mathfrak{g}$, let $\mathfrak{h}_{ss} = \mathfrak{h} \cap [\mathfrak{g}, \mathfrak{g}]$, and let λ be the infinitesimal character of V. Suppose that $\lambda|_{\mathfrak{h}_{ss}}$ is in the $\mathbb{R}$ linear span of the roots.Then there exists a constant C depending only on the reductive pair $(\mathfrak{g}, K)$ with the following property: If $\left| \lambda|_{\mathfrak{h}_{ss}} \right| \geq C$, then there exist a proper θ stable parabolic pair $(\mathfrak{q}, L \cap K)$ and an infinitesimally unitary module Z for the Levi pair $(\mathfrak{l}, L \cap K)$ with infinitesimal character in the weakly good range such that V is cohomologically induced from Z.

In other words we may assume that λ is small or at worst moderate in size, as well as real. These conditions are not conclusive, and it is in fact not known how to tell definitely whether the given infinitesimally unitary $(\mathfrak{g}, K)$ module V is properly unitarily induced. But at least these conditions give us a clue as to a class of irreducible $(\mathfrak{g}, K)$ modules that might be reasonable to study.

DEFINITION 12.3. Let $(\mathfrak{g}, K)$ be a reductive pair, let $\mathfrak{h}$ be a Cartan subalgebra, and let $\mathfrak{h}_{ss} = \mathfrak{h} \cap [\mathfrak{g}, \mathfrak{g}]$. A finite-length $(\mathfrak{g}, K)$ module V with an infinitesimal character λ is **weakly unipotent** if
  (i) the restriction of λ to $\mathfrak{h}_{ss}$ is in the $\mathbb{R}$ linear span of the roots
  (ii) whenever F is a finite-dimensional $U(\mathfrak{g})$ module and μ is an
       infinitesimal character appearing in $V \otimes_{\mathbb{C}} F$, then $\left| \lambda|_{\mathfrak{h}_{ss}} \right| \le \left| \mu|_{\mathfrak{h}_{ss}} \right|$.

The significance of this definition will be given in a moment. One-dimensional $(\mathfrak{g}, K)$ modules satisfy the condition, and they are probably not properly unitarily induced in the case that $\mathfrak{g}_0$ has no compact simple factor. But even for infinitesimally unitary modules, the definition is not perfect as a substitute for "not properly unitarily induced." Here is how the notions of "weakly unipotent" and "not properly unitarily induced" compare for $SL(2, \mathbb{R})$.

EXAMPLE. $G = SL(2, \mathbb{R})$ or a finite covering group. Let us identify $\mathfrak{h}^*$ with $\mathbb{C}$ by identifying a root with 2. If a $(\mathfrak{g}, K)$ module V has infinitesimal character $\lambda \in \mathbb{C}$, condition (i) in Definition 12.3 says we may restrict attention to λ real. The weights of the finite-dimensional representations of G are integers. According to Theorem 7.133, the candidates for μ in (ii) are the members of $\lambda + \mathbb{Z}$. Thus (ii) is satisfied if $|\lambda| \le \frac{1}{2}$. When $|\lambda| > \frac{1}{2}$, (ii) is satisfied only for the trivial representation.

The irreducible unitary representations of $SL(2, \mathbb{R})$ consist of the discrete series and limits, which are properly unitarily induced via cohomological induction and have integral infinitesimal character; the unitary principal series, which is properly unitarily induced via parabolic induction and has purely imaginary infinitesimal character; the trivial representation, which has infinitesimal character 1; and the complementary series, which we discuss shortly. Apart from the complementary series, the trivial representation is the only irreducible unitary representation that is not properly unitarily induced, while the trivial representation, the two limits of discrete series, and the spherical principal series with parameter 0 are weakly unipotent.

Now let us come to the complementary series. The principal series comes in two parts, corresponding to the two characters of the subgroup

$M_\mathfrak{p} = \left\{ \pm \begin{pmatrix} 1 & 0 \\ 0 & 1 \end{pmatrix} \right\}$. The trivial character of $M_\mathfrak{p}$ yields the spherical principal series, and those members of the spherical principal series with λ real and $|\lambda| < 1$ are infinitesimally unitary. When made unitary, these representations with $\lambda \neq 0$ are called the **complementary series** of $SL(2, \mathbb{R})$. The complementary-series representation with parameter λ is equivalent to the one with parameter $-\lambda$, and the complementary series are not properly unitarily induced. But only those complementary series with $|\lambda| \leq \frac{1}{2}$ are weakly unipotent.

Finally we mention that the nonspherical principal series is never infinitesimally unitary for real $\lambda \neq 0$. However, those representations in this series with $|\lambda| \leq \frac{1}{2}$ are still weakly unipotent. Thus weakly unipotent does not imply infinitesimally unitary.

In the above example, we see for infinitesimally unitary representations how close the condition "not properly unitarily induced" is to the condition "weakly unipotent." In general, evidence suggests that every infinitesimally unitary irreducible $(\mathfrak{g}, K)$ module that is not properly unitarily induced either is weakly unipotent or is an interior point of a unitary series generalizing the complementary series of $SL(2, \mathbb{R})$. Discussion of the latter kind of representation is beyond the scope of this book, and we restrict attention to weakly unipotent $(\mathfrak{g}, K)$ modules.

Let us return to the significance of "weakly unipotent." The point is that the $A_\mathfrak{q}(\lambda)$'s are cohomologically induced from one-dimensional weakly unipotent modules and the corresponding modules cohomologically induced from other weakly unipotent modules share many of the properties of the $A_\mathfrak{q}(\lambda)$'s. The conditions in Definition 12.3 are chosen so that the proofs for the $A_\mathfrak{q}(\lambda)$'s adapt to work for the generalization without substantial change. The result uses the notion "weakly fair" from Definition 0.52 and is as follows.

Theorem 12.4. Let $\mathfrak{q} = \mathfrak{l} \oplus \mathfrak{u}$ be a θ stable parabolic subalgebra of $\mathfrak{g}$, let $(\mathfrak{q}, L \cap K)$ be the corresponding parabolic pair, and let $(\mathfrak{l}, L \cap K)$ be the Levi pair. If Z is a weakly unipotent $(\mathfrak{l}, L \cap K)$ module having an infinitesimal character that is in the weakly fair range, then

(a) $\mathcal{L}_j(Z) = \mathcal{R}^j(Z) = 0$ for $j \neq S = \dim(\mathfrak{u} \cap \mathfrak{k})$
(b) $\mathcal{L}_S(Z) = \mathcal{R}^S(Z)$.

Furthermore if Z is infinitesimally unitary with invariant form $\langle \cdot, \cdot \rangle_L$, then the corresponding Shapovalov form $\langle \cdot, \cdot \rangle_G$ on $\mathcal{L}_S(Z)$ is positive definite, and consequently $\mathcal{L}_S(Z)$ is infinitesimally unitary.

The fact that the dominance hypothesis in Theorem 12.4 is "weakly fair" rather than "weakly good" is a decided advantage. Under a mild condition on $\mathfrak{l}$, Propositon 12.5 below says for a given infinitesimally unitary $(\mathfrak{l}, L \cap K)$ module Z with a real infinitesimal character we can always find a θ stable parabolic subalgebra $\mathfrak{q} = \mathfrak{l} \oplus \mathfrak{u}$ for which the infinitesimal character is in the weakly fair range. If Z is weakly unipotent, then Theorem 12.4 applies for this choice of $\mathfrak{q}$ and yields an infinitesimally unitary $(\mathfrak{g}, K)$ module. Actually Proposition 12.5 will be phrased so as to give a better conclusion, and Example 1 after the proposition will show the meaning of the improved version in terms of normalized cohomological induction. Example 2 after the proposition, by contrast, shows that we cannot always choose $\mathfrak{q}$ so that the infinitesimal character is in the weakly good range; without the weak unipotence of Z, we then have no sure way of using Z to construct an infinitesimally unitary $(\mathfrak{g}, K)$ module.

Proposition 12.5. Let $(\mathfrak{l}, L \cap K)$ be a Levi pair for $(\mathfrak{g}, K)$, and let $\mathfrak{z}_L$ be the center of $\mathfrak{l}$. Suppose that the centralizer $Z_{\mathfrak{g}}(\mathfrak{z}_L \cap \mathfrak{k})$ equals $\mathfrak{l}$. If a real infinitesimal character λ of $\mathfrak{l}$ is given that vanishes on $\mathfrak{z}_L \cap \mathfrak{p}$, then there exists a θ stable parabolic subalgebra $\mathfrak{q} = \mathfrak{l} \oplus \mathfrak{u}$ of $\mathfrak{g}$ such that $\lambda - \delta(\mathfrak{u})$ is in the weakly fair range.

SKETCH OF PROOF. Fix any θ stable Cartan subalgebra $\mathfrak{h}_0 = \mathfrak{t}_0 \oplus \mathfrak{a}_0$ of $\mathfrak{l}$, and introduce a positive system for $\Delta(\mathfrak{g}, \mathfrak{h})$ such that $\lambda|_{\mathfrak{z}_L}$ is dominant. We shall adjust the notion of positivity on the root subsystem Δ' of roots orthogonal to $\lambda|_{\mathfrak{z}_L}$. Since $\lambda|_{\mathfrak{z}_L \cap \mathfrak{p}} = 0$ by hypothesis, it follows that Δ' is θ stable and contains $\Delta(\mathfrak{l}, \mathfrak{h})$. A spanning set for $i\mathfrak{t}_0 \oplus \mathfrak{a}_0$ consisting of a basis for $\mathfrak{z}_L \cap \mathfrak{t}_0$ followed by a basis of $i\mathfrak{t}_0 \oplus \mathfrak{a}_0$ determines a positive system for Δ' and therefore for $\Delta(\mathfrak{g}, \mathfrak{h})$. We take $\mathfrak{u}$ to be the sum of the root spaces for the positive roots not in $\Delta(\mathfrak{l}, \mathfrak{h})$, and then we check that $\langle \lambda, \alpha|_{\mathfrak{z}_L} \rangle \geq 0$, as required.

EXAMPLES.

1) Let $G = SL(2, \mathbb{R})$ and $L = T$. For $\lambda = 0$, there are two choices of Borel subalgebra $\mathfrak{b} = \mathfrak{t} \oplus \mathfrak{n}$, and both lead to $\lambda - \delta(\mathfrak{n})$ in the weakly fair range. To interpret this conclusion in terms of modules, we use normalized cohomological induction. If $\tilde{Z}$ is an $(\mathfrak{l}, (L \cap K)^{\sim})$ module with infinitesimal character λ, then (11.146a) and (11.73) give

$$(^{\mathfrak{n}}\mathcal{L}^{\mathfrak{g},K}_{\mathfrak{b},T})_j(\tilde{Z}) = (^{\mathfrak{u}}\mathcal{L}^{\mathfrak{g},K}_{\mathfrak{b},T})_j(\tilde{Z} \otimes \mathbb{C}_{\delta(\mathfrak{n})}) = \mathcal{L}_j(\tilde{Z} \otimes \mathbb{C}_{-\delta(\mathfrak{n})}).$$

The infinitesimal character of the $(\mathfrak{t}, T)$ module that is being cohomologically induced is $\lambda - \delta(\mathfrak{n})$, and the conditions "weakly good" and "weakly fair" (which are equivalent in this case) ask that

$$\langle (\lambda - \delta(\mathfrak{n})) + \delta(\mathfrak{n}), \alpha|_{\mathfrak{z}_L} \rangle \geq 0 \qquad \text{for } \alpha \in \Delta(\mathfrak{u}).$$

The extreme case is $\lambda = 0$, and the cohomological induction yields the two limits of discrete series.

2) Let $G = Sp(2, \mathbb{R})$, so that a Cartan subalgebra $\mathfrak{h}_0$ of $\mathfrak{k}_0$ is a Cartan subalgebra of $\mathfrak{g}_0$. Regard the root system $\Delta(\mathfrak{g}, \mathfrak{h})$ as of type C_2 with usual notation, and let $\mathfrak{l}$ be built from $\pm e_2$. The group L is of the form $L = U(1) \times Sp(1, \mathbb{R})$, and $Sp(1, \mathbb{R}) = SL(2, \mathbb{R})$. Let π be an irreducible representation of L that is a character on $U(1)$ with differential me_1 and is a discrete series on $SL(2, \mathbb{R})$ with extreme weight $(N + 1)e_2$ and infinitesimal character Ne_2. Say $m > 2$ and $N > m+2$. The corresponding $(\mathfrak{l}, L \cap K)$ module Z has infinitesimal character $\lambda = me_1 + Ne_2$. There are only two possibilities for $\Delta(\mathfrak{u})$. One is $\{2e_1, e_1 \pm e_2\}$, and the other is the set of negatives of this. If λ is to be in the weakly good range, we want $\lambda + \delta(\mathfrak{u})$ to be dominant for $\Delta(\mathfrak{u})$. Since $m > 2$, $2e_1$ has to be in $\Delta(\mathfrak{u})$, and $\Delta(\mathfrak{u})$ must be $\{2e_1, e_1 \pm e_2\}$. Then $\lambda + \delta(\mathfrak{u}) = (m + 2)e_1 + Ne_2$, and the condition $N > m + 2$ forces the inner product with $e_1 - e_2$ to be negative. Hence neither choice of $\mathfrak{u}$ makes λ be in the weakly good range. However, $\lambda|_{\mathfrak{z}_L} = me_1$, and λ is in the weakly fair range when $\Delta(\mathfrak{u}) = \{2e_1, e_1 \pm e_2\}$.

It is of great interest to know how to construct all weakly unipotent $(\mathfrak{g}, K)$ modules. This is a topic of widespread current research, partly through examples and partly through attempts at a general theory. Let us give a longer list of examples.

EXAMPLES.

1) One-dimensional $(\mathfrak{g}, K)$ modules, as we have already noted, are weakly unipotent.

2) Any $(\mathfrak{g}, K)$ module with infinitesimal character 0 is weakly unipotent. For example, principal-series modules with $\mathfrak{a}_p$ parameter 0 have this property if $\mathfrak{g}_0$ is split modulo center.

3) $A_\mathfrak{q}(\lambda)$ is weakly unipotent if $\lambda + \delta(\mathfrak{u}) = 0$. The proof of this assertion uses the techniques of §VII.15. Some of the examples in §VIII.5 have this property.

   (a) Among the Speh representations of $G = SL(2n, \mathbb{R})$, the one with $\lambda = -n(e_1 + \cdots + e_n)$ in (8.78) has $\lambda + \delta(\mathfrak{u}) = 0$.
   (b) For $G = Sp(2, \mathbb{R})$ with $\mathfrak{l}$ built from the long simple root, we constructed an $A_\mathfrak{q}(\lambda)$ such that Theorem 5.109 is applicable but $A_\mathfrak{q}(\lambda)$ is reducible. It has $\lambda + \delta(\mathfrak{u}) = 0$.
   (c) For $G = SO_0(5, 4)$, we constructed an $A_\mathfrak{q}(\lambda)$ that plays an exceptional role in classification questions. It has $\lambda + \delta(\mathfrak{u}) = 0$.

4) Arthur has defined some "special unipotent" representations, and they can be redefined in terms of their annihilators in the universal enveloping algebra. These are weakly unipotent.

5) The "metaplectic representation" of the double cover of $SL(2, \mathbb{R})$ is weakly unipotent, and it is not "special unipotent" in the sense of Example 4. The metaplectic representation was discussed at the end of Chapter I. The underlying $(\mathfrak{g}, \tilde{K})$ module was constructed as an irreducible module

$$V = \{P(x)e^{-x^2/2} \mid P \in \mathbb{C}[x]\}$$

for the generalized pair $(\mathcal{A}, \tilde{K})$, where $\mathcal{A}$ is the Weyl algebra $\mathbb{C}[x, \frac{d}{dx}]$. To obtain a $(\mathfrak{g}, \tilde{K})$ module, we used a nonobvious homomorphism of algebras $\varphi : U(\mathfrak{g}) \to \mathcal{A}$ and pulled back the action. Then V split into even and odd parts as $V = V^+ \oplus V^-$, and we found each of V^+ and V^- to be an irreducible $(\mathfrak{g}, \tilde{K})$ module. In fact, V^- is in the discrete series and hence is properly unitarily cohomologically induced. But V^+ is not; it is out of range for the discrete series and limits. Each of V^+ and V^- has infinitesimal character $|\lambda| = \frac{1}{2}$, in the notation of the $SL(2, \mathbb{R})$ example above, and is therefore weakly unipotent.

Example 5 is especially illuminating. Possibly after combining some weakly unipotent $(\mathfrak{g}, K)$ modules (V^+ and V^- in the above example), we were able to get the sum to be a simple module for a nice algebra into which $U(\mathfrak{g})$ maps. Actually in the above example, one checks that the Casimir operator $\Omega = \frac{1}{2}h^2 + ef + fe$ maps into the scalar $-\frac{3}{8}$, and thus $\Omega + \frac{3}{8}$ maps to 0. In other words if V has infinitesimal character λ and if $\tilde{Z}_\lambda$ denotes the smallest two-sided ideal in $U(\mathfrak{g})$ containing χ_λ, then the algebra homomorphism $U(\mathfrak{g}) \to \mathcal{A}$ factors through $U(\mathfrak{g})/\tilde{Z}_\lambda$.

The algebra $\mathcal{A}$ thus has something to do with annihilators. Actually if V is any irreducible $(\mathfrak{g}, K)$ module, then the algebra $\mathcal{A} = U(\mathfrak{g})/\text{Ann}_{U(\mathfrak{g})}(V)$ will act on V, and V will be a simple module for $\mathcal{A}$. But the point is that $\mathcal{A}$ in Example 5 is a concrete and relatively uncomplicated algebra.

We have seen two situations before of the same kind. The first one was in §VII.10 in the proof of the first result about preservation of irreducibility by translation functors (Theorem 7.171). In this case we initially defined the algebra abstractly as $\mathcal{A} = R(\lambda) = U(\mathfrak{g})/\tilde{Z}_\lambda$. As must always be the case when $U(\mathfrak{g})$ maps by a homomorphism of algebras to an associative algebra with identity, $U(\mathfrak{g} \oplus \mathfrak{g})$ acts on $\mathcal{A}$ through left and right multiplication as in (7.174). To realize $\mathcal{A}$ more concretely, we let $\mathfrak{k}_0$ be a compact form of $\mathfrak{g}$ and embedded it in $\mathfrak{g} \oplus \mathfrak{g}$ diagonally: $X \in \mathfrak{k}_0$ maps to (X, X). If K' is a compact connected group whose Lie algebra is the image of this map, then $(\mathfrak{g} \oplus \mathfrak{g}, K')$ is a reductive pair and $\mathcal{A}$ is a $(\mathfrak{g} \oplus \mathfrak{g}, K')$ module. Our concrete realization of $\mathcal{A}$ was that $R(\lambda)$ is isomorphic to $P(\lambda) = \operatorname{Hom}_{\mathbb{C}}(V(\lambda), V(\lambda))_{K'}$, where $V(\lambda)$ is a Verma module.

The second situation was in §VIII.5 in the discussion of the irreducibility of $A_{\mathfrak{q}}(\lambda)$. In this case we defined the algebra to be of the form

$$\mathcal{A} = \operatorname{Hom}_{\mathbb{C}}(M(\lambda'), M(\lambda'))_{K'},$$

where $M(\lambda')$ is a generalized Verma module for $\mathfrak{g}$ and the parameter $\lambda' = \lambda + 2m\delta(\mathfrak{u})$. Again $\mathcal{A}$ is a $(\mathfrak{g} \oplus \mathfrak{g}, K')$ module, and the map of $U(\mathfrak{g})$ into $\mathcal{A}$ factors through $U(\mathfrak{g})/\tilde{Z}_{\lambda'+\delta}$.

These algebras $\mathcal{A}$ are the algebras of "twisted differential operators" mentioned in Theorem 0.53. (They can be interpreted as algebras of operators on sections over a complex manifold $G_{\mathbb{C}}/Q$.) In both cases the $(\mathfrak{g}, K)$ module under study was irreducible under $\mathcal{A}$. To address irreducibility under $U(\mathfrak{g})$, we asked whether $U(\mathfrak{g})$ mapped onto $\mathcal{A}$. In the first case it did, because the initial definition of $\mathcal{A}$ was as a quotient of $U(\mathfrak{g})$. In the second case we found a sufficient condition for $U(\mathfrak{g})$ to map onto $\mathcal{A}$ (Theorem 8.31); if $U(\mathfrak{g}) \rightarrow \mathcal{A}$ is reinterpreted as a mapping between algebras of differential operators, the condition is that the corresponding symbol mapping is onto.

One way of interpreting the second example is that quotients of $U(\mathfrak{g})$ are sometimes a little smaller than they "ought" to be, as when not all twisted differential operators on some $G_{\mathbb{C}}/Q$ come from $U(\mathfrak{g})$. For that reason, $U(\mathfrak{g})$ may fail to act irreducibly when we would like it to. In this case we can sometimes add the missing elements to make a Dixmier algebra that *does* act irreducibly. To have this setting available as an option for investigating irreducibility more generally, we make the following definition. We confine ourselves to the connected case.

DEFINITION 12.6. Let $(\mathfrak{g}, K)$ be a reductive pair. A **Dixmier algebra** $\mathcal{A}$ relative to $\mathfrak{g}$ is actually a 4-tuple $(\mathcal{A}, \varphi, K', \tau)$ in which $\mathcal{A}$ is an associative algebra with identity, $\varphi : U(\mathfrak{g}) \rightarrow \mathcal{A}$ is an algebra homomorphism, K' is a compact connected Lie group whose Lie algebra is $\mathfrak{k}_0 \oplus i\mathfrak{p}_0$ and thus has $(\mathfrak{k}_0')_{\mathbb{C}} \cong \mathfrak{g}$, and τ is a locally K' finite representation of K' on $\mathcal{A}$ by algebra automorphisms such that

(i) the complexified differential of τ satisfies

$$d\tau(X)(a) = \varphi(X)a - a\varphi(X) \quad \text{for } X \in \mathfrak{g} \text{ and } a \in \mathcal{A}$$

(ii) the representation τ of K' and the $\mathfrak{g} \oplus \mathfrak{g}$ module structure on $\mathcal{A}$ given by $(X, Y)a = \varphi(X)a - a\varphi(Y)$ for X and Y in $\mathfrak{g}$ make $\mathcal{A}$ into a $(\mathfrak{g} \oplus \mathfrak{g}, K')$ module

(iii) the $(\mathfrak{g} \oplus \mathfrak{g}, K')$ module $\mathcal{A}$ has finite length.

The Dixmier algebra has a **generalized pair structure** if the inclusion $\mathfrak{k}_0 \hookrightarrow \mathfrak{k}_0 \oplus i\mathfrak{p}_0$ is the differential of a group homomorphism $\pi : K \to K'$ such that

(iv) $\tau(\pi(k))\varphi(X) = \varphi(\mathrm{Ad}(k)X)$ for $X \in \mathfrak{k}$ and $k \in K$.

When the Dixmier algebra has a generalized pair structure, $(\mathcal{A}, K)$ becomes a generalized pair under the following definitions. The "adjoint action" of K on $\mathcal{A}$ that is required by (1.87) is given by composing $\pi : K \to K'$ and the representation τ of K' on $\mathcal{A}$. The homomorphism $\iota : U(\mathfrak{k}) \to \mathcal{A}$ that is required for (1.127) is given by composing the inclusion $U(\mathfrak{k}) \hookrightarrow U(\mathfrak{g})$ and the homomorphism $\varphi : U(\mathfrak{g}) \to \mathcal{A}$. The compatibility condition (1.127) just amounts to (iv) in Definition 12.6.

EXAMPLE.   Metaplectic representation for the double cover of $SL(2, \mathbb{R})$. In the basis $\{x, \frac{d}{dx}\}$ of the subspace of $\mathcal{A}$ of elements of total degree 1, the matrix of the map $a \mapsto \varphi(X)a - a\varphi(X)$ turns out to be

$$\begin{bmatrix} 1 & 0 \\ 0 & -1 \end{bmatrix} \text{ for } X = h, \qquad \begin{bmatrix} 0 & -i \\ 0 & 0 \end{bmatrix} \text{ for } X = e, \qquad \begin{bmatrix} 0 & 0 \\ i & 0 \end{bmatrix} \text{ for } X = f.$$

For the general element of $\mathfrak{su}(2)$, (i) therefore says that

$$d\tau \begin{pmatrix} i\theta & \beta \\ -\bar\beta & -i\theta \end{pmatrix} \quad \text{has matrix} \quad \begin{bmatrix} i\theta & -i\beta \\ -i\bar\beta & -i\theta \end{bmatrix}.$$

Thus the representation τ of $SU(2)$ is such that

$$\tau(k) \quad \text{has matrix} \quad \begin{bmatrix} -i & 0 \\ 0 & 1 \end{bmatrix} k \begin{bmatrix} -i & 0 \\ 0 & 1 \end{bmatrix}^{-1}.$$

Let us verify (iv). The group K is the double cover of $SO(2)$, and a general element of K is $\exp t \begin{pmatrix} 0 & 1 \\ -1 & 0 \end{pmatrix}$. The map π is the composition of descent into $SO(2)$, followed by inclusion into $SU(2)$. Thus

$$\pi(\exp t \begin{pmatrix} 0 & 1 \\ -1 & 0 \end{pmatrix}) = \begin{pmatrix} \cos t & \sin t \\ -\sin t & \cos t \end{pmatrix}.$$

So $\tau(\pi(\exp t \begin{pmatrix} 0 & 1 \\ -1 & 0 \end{pmatrix}))$ has matrix $\begin{bmatrix} \cos t & -i\sin t \\ -i\sin t & \cos t \end{bmatrix}$. In (iv) it is enough to take $X = e - f$. The right side of (iv) is just $\varphi(X)$ since $\mathrm{Ad}(\exp t \begin{pmatrix} 0 & 1 \\ -1 & 0 \end{pmatrix})$ fixes $e - f$. For the left side we are thus to check that $\tau(\pi(\exp t \begin{pmatrix} 0 & 1 \\ -1 & 0 \end{pmatrix}))$ fixes $\varphi(e - f) = \frac{1}{2}i(x^2 - (\frac{d}{dx})^2)$, and direct computation verifies this.

When a Dixmier algebra $\mathcal{A}$ has a generalized pair structure, we can set up what might be called a "generalized map of pairs" $(U(\mathfrak{g}), K) \to (\mathcal{A}, K)$. Namely the map of algebras is φ, and the map of groups is the identity. If V is a compatible $(\mathcal{A}, K)$ module, then application of the forgetful functor $\mathcal{F}_{\mathcal{A},K}^{U(\mathfrak{g}),K}$ to V yields a $(\mathfrak{g}, K)$ module. In the example this was the way that the metaplectic representation was constructed as a $(\mathfrak{g}, K)$ module from the compatible $(\mathcal{A}, K)$ module $\{P(x)e^{-x^2/2}\}$.

In the passage from a one-dimensional $(\mathfrak{l}, L \cap K)$ module $\mathbb{C}_\lambda$ to a $(\mathfrak{g}, K)$ module $A_\mathfrak{q}(\lambda)$, we arranged that $A_\mathfrak{q}(\lambda)$ was a compatible $(\mathcal{A}_G, K)$ module for a certain Dixmier algebra $\mathcal{A}_G$ for G. In fact, we can interpret $\mathbb{C}_\lambda$ as a compatible $(\mathcal{A}_L, L \cap K)$ module for a one-dimensional Dixmier algebra $\mathcal{A}_L$ for L. Then our construction of $\mathcal{A}_G$ may be regarded as a relative construction: We formed $\mathcal{A}_G$ from $\mathcal{A}_L$. We shall now generalize this construction, obtaining "induced Dixmier algebras." The construction will work whenever we start with an $(\mathfrak{l}, L \cap K)$ module Z and use any Dixmier algebra with it that has a generalized pair structure (e.g., $\mathcal{A}_L = U(\mathfrak{l})/\mathrm{Ann}_{U(\mathfrak{l})}(Z)$). But the interest is in the case that Z is irreducible, is weakly unipotent, and has infinitesimal character in the weakly fair range. If Z is irreducible as an $(\mathcal{A}_L, L \cap K)$ module, then $\mathcal{L}_S(Z)$ will be irreducible as an $(\mathcal{A}_G, K)$ module. This result generalizes what was shown for the $A_\mathfrak{q}(\lambda)$'s in §VIII.5 and may be regarded as a step toward understanding whether $\mathcal{L}_S(Z)$ is irreducible as a $(\mathfrak{g}, K)$ module.

Thus suppose $(\mathfrak{g}, K)$ is a reductive pair with K connected, and let $\mathfrak{q} = \mathfrak{l} \oplus \mathfrak{u}$ be a θ stable parabolic subalgebra of $\mathfrak{g}$ with corresponding Levi pair $(\mathfrak{l}, L \cap K)$. Suppose we are given a Dixmier algebra $(\mathcal{A}_L, \varphi_L, K'_L, \tau_L)$ for $\mathfrak{l}$ with a generalized pair structure defined by $\pi_L : L \cap K \to K'_L$, and suppose V_L is a compatible $(\mathcal{A}_L, L \cap K)$ module. We shall assume that there exists a connected group K' with Lie algebra $\mathfrak{k}_0 \oplus i\mathfrak{p}_0$, as well as maps π_G and i', so that the diagram

(12.7)

$$
\begin{array}{ccc}
K & \xrightarrow{\;\pi_G\;} & K' \\[4pt]
{\scriptstyle i}\big\uparrow & & \big\uparrow{\scriptstyle i'} \\[4pt]
L \cap K & \xrightarrow{\;\pi_L\;} & K'_L
\end{array}
$$

commutes. Here $i : L \cap K \to K$ is the inclusion.

Dropping a forgetful functor, we may regard V_L also as an $(\mathfrak{l}, L \cap K)$ module. Put $V_\mathfrak{g} = \mathrm{ind}_{\bar{\mathfrak{q}}, L \cap K}^{\mathfrak{g}, K}(V_L^\#) \cong U(\mathfrak{u}) \otimes_\mathbb{C} V_L^\#$, the isomorphism holding on the level of $(\mathfrak{l}, L \cap K)$ modules. A member T of $\mathrm{End}_\mathbb{C}(V_\mathfrak{g})$ is said to be of **type** $\mathcal{A}_L$ if, for each $u \in U(\mathfrak{u})$, there are finitely many elements $u_1, \ldots, u_n$ in $U(\mathfrak{u})$ and members $E_1, \ldots, E_n$ of $\mathcal{A}_L$ so that for each v in

$V_L^\#$,

$$T(u \otimes v) = \sum_i u_i \otimes E_i v.$$

Using the defining relations for $V_{\mathfrak{g}}$ and an inductive argument, one checks that it is equivalent to require this condition for every u in $U(\mathfrak{g})$, or to require that the u_i belong only to $U(\mathfrak{g})$. Define

$$\mathcal{A}_{\mathfrak{g}} = \{T \in \operatorname{End}_{\mathbb{C}}(V_{\mathfrak{g}}) \mid T \text{ is of type } \mathcal{A}_L\}.$$

It is easy to check that $\mathcal{A}_{\mathfrak{g}}$ is an algebra. Because of our equivalent definitions of "type $\mathcal{A}_L$", the members of $U(\mathfrak{g})$, in their operation on $V_{\mathfrak{g}}$, yield elements of $\mathcal{A}_{\mathfrak{g}}$. Consequently we get an algebra homomorphism $\varphi_{\mathfrak{g}} : U(\mathfrak{g}) \to \mathcal{A}_{\mathfrak{g}}$.

When V_L is one dimensional, $V_{\mathfrak{g}}$ is a generalized Verma module and the condition "type $\mathcal{A}_L$" is empty. Thus $\mathcal{A}_{\mathfrak{g}}$ is the algebra of all linear maps from the generalized Verma module to itself.

There is an equivalent definition of $\mathcal{A}_{\mathfrak{g}}$ that allows us to impose an "adjoint" action by K_L'. We write

$$V_{\mathfrak{g}} \cong U(\mathfrak{u}) \otimes_{\mathbb{C}} V_L^\# \cong \sum_{\substack{\nu = \text{weight of} \\ \text{center of } \mathfrak{l}}} U^\nu(\mathfrak{u}) \otimes_{\mathbb{C}} V_L^\#,$$

where $U^\nu(\mathfrak{u})$ is the finite-dimensional subspace of $U(\mathfrak{u})$ where the center of $\mathfrak{l}$ acts with weight ν. For $T \in \operatorname{End}_{\mathbb{C}}(V_{\mathfrak{g}})$, we can restrict T to $U^\nu(\mathfrak{u}) \otimes_{\mathbb{C}} V_L^\#$ and look at the projection of the result to $U^\mu(\mathfrak{u}) \otimes_{\mathbb{C}} V_L^\#$. Then we get a linear map

$$T_{\mu\nu} : U^\nu(\mathfrak{u}) \otimes_{\mathbb{C}} V_L^\# \to U^\mu(\mathfrak{u}) \otimes_{\mathbb{C}} V_L^\#.$$

Because of the finite-dimensionality of $U^\nu(\mathfrak{u})$ and $U^\mu(\mathfrak{u})$, we may regard $T_{\mu\nu}$ as a member of

$$\operatorname{Hom}_{\mathbb{C}}(U^\nu(\mathfrak{u}) \otimes \textstyle\bigwedge^{\text{top}}\mathfrak{u}, \; U^\mu(\mathfrak{u}) \otimes \textstyle\bigwedge^{\text{top}}\mathfrak{u}) \otimes_{\mathbb{C}} \operatorname{End}_{\mathbb{C}}(V_L).$$

One can show that T is in $\mathcal{A}_{\mathfrak{g}}$ if and only if all the $T_{\mu\nu}$ are in

$$(12.8) \qquad \operatorname{Hom}_{\mathbb{C}}(U^\nu(\mathfrak{u}) \otimes \textstyle\bigwedge^{\text{top}}\mathfrak{u}, \; U^\mu(\mathfrak{u}) \otimes \textstyle\bigwedge^{\text{top}}\mathfrak{u}) \otimes_{\mathbb{C}} \mathcal{A}_L.$$

To introduce the action of K_L', we observe from the commutativity of (12.7) that K_L' has a well-defined action on $\mathfrak{u}$. This action yields an action of K_L' in the Hom factor of (12.8). Combining this with the action by τ_L in the $\mathcal{A}_L$ factor, we obtain an action of K'_L on (12.8), hence on $\mathcal{A}_{\mathfrak{g}}$. Then $\mathcal{A}_{\mathfrak{g}}$ is a $(\mathfrak{g} \oplus \mathfrak{g}, K'_L)$ module.

Finally we let $\mathcal{A}_G$ be the K' finite members of $\mathcal{A}_{\mathfrak{g}}$, which we can describe more precisely as

$$\mathcal{A}_G = I^{\mathfrak{g} \oplus \mathfrak{g}, K'}_{\mathfrak{g} \oplus \mathfrak{g}, K'_L}(\mathcal{A}_{\mathfrak{g}}).$$

We take τ_G to be the resulting representation of K' on $\mathcal{A}_G$. If the map $i : K'_L \to K'$ is one-one, the I functor here reduces to a Γ functor. In any case, as with Proposition 2.70, we can regard $\mathcal{A}_G$ as a subspace of $\mathcal{A}_{\mathfrak{g}}$, and it is easy to see that $\varphi_{\mathfrak{g}} : U(\mathfrak{g}) \to \mathcal{A}_{\mathfrak{g}}$ has its image in this subspace. Thus we can define φ_G to be the map $\varphi_{\mathfrak{g}}$ viewed with range in $\mathcal{A}_G$. The 4-tuple $(\mathcal{A}_G, \varphi_G, K', \tau_G)$ is then a Dixmier algebra (called the **induced Dixmier algebra**), and the commutativity of (12.7) yields a generalized pair structure for it. When V_L is one dimensional, $\mathcal{A}_G$ coincides with the algebra constructed for the proof of Theorem 8.31 concerning $A_{\mathfrak{q}}(\lambda)$'s.

The $(\mathfrak{g}, K)$ module $\mathcal{L}_S(V_L)$, where $S = \dim(\mathfrak{u} \cap \mathfrak{k})$, is a compatible module for the generalized pair $(\mathcal{A}_G, K)$. The proof of this assertion is rather similar to the "first task" after (8.49) in the proof of Theorem 8.31. Then the theorem is as follows.

Theorem 12.9. With notation as above, suppose that the $(\mathfrak{l}, L \cap K)$ module V_L has an infinitesimal character in the weakly fair range and is weakly unipotent. If V_L is irreducible as an $(\mathcal{A}_L, L \cap K)$ module, then $\mathcal{L}_S(V_L)$ is irreducible or zero as an $(\mathcal{A}_G, K)$ module.

APPENDIX A

MISCELLANEOUS ALGEBRA

This appendix collects some miscellaneous results about rings, associative algebras, and modules. Included are discussions of tensor products, complete reducibility, modules of finite length, and Grothendieck groups.

1. Good Categories

Let R be a ring, with or without identity. A **good category** C of R modules consists of

(i) some class of left R modules closed under passage to submodules, quotients, and finite direct sums (the **modules** of the category), and

(ii) the sets $\text{Hom}_R(A, B)$ of *all* R linear maps from A to B, for each A and B in (i) (the **maps** of the category).

An example is the category of all left R modules. If R has an identity, then another good category is the subcategory of all **unital** left R modules, i.e., those in which the identity of R acts as the identity in the module.

Let k be a field. An **associative algebra** R over k is a ring with the structure of a k vector space such that the ring multiplication is bilinear from $R \times R$ to R. In other words,

$$(A.1) \qquad c(rs) = (cr)s = r(cs) \qquad \text{for } c \in k,\ r \in R,\ s \in R.$$

If R has an identity and if M is a unital left R module, then M automatically inherits the structure of a k vector space in such a way that the R multiplication and the scalar multiplication satisfy

$$(A.2a) \qquad c(rm) = (cr)m = r(cm) \qquad \text{for } c \in k,\ r \in R,\ m \in M.$$

To prove (A.2a) we define $cm = (c1)m$ for $c \in k$ and $m \in M$. Then the axioms for a vector space follow from the fact that M is a unital module for the subalgebra $k1$. Using (A.1), we have

$$c(rm) = (c1)(rm) = ((c1)r)m = (cr)m$$

803

and also

$$r(cm) = r((c1)m) = (r(c1))m = ((c1)r)m = (cr)m.$$

Thus (A.2a) holds. The analog of (A.2a) for right R modules is

(A.2b) $\qquad c(mr) = m(cr) = (cm)r \qquad$ for $c \in k$, $r \in R$, $m \in M$.

Many of the associative algebras encountered in this book do not have an identity. Instead they have something called an "approximate identity." Motivation for this notion appears in Chapter I, and the definition is as follows.

DEFINITION A.3. Suppose R is a ring and I is a directed set. An **approximate identity** for R indexed by I is a collection $\{\chi_i \mid i \in I\}$ of elements of R having the following properties: If $i \leq i'$, then $\chi_i \chi_{i'} = \chi_{i'} \chi_i = \chi_i$; and if $r \in R$, there is an $i \in I$ such that $\chi_i r = r\chi_i = r$. A left R module M is called **approximately unital** if for any $m \in M$ there is an $i \in I$ such that $\chi_i m = m$.

We regard the approximate identity in R as fixed, and we consider all approximately unital left R modules relative to that one approximate identity. Any such module has the following property: For any finite set $m_1, \ldots, m_n$ in M, there is an $i \in I$ such that $\chi_i m_j = m_j$ for $1 \leq j \leq n$. In fact, if $\chi_i m_1 = m_1$ and $\chi_{i'} m_2 = m_2$, then the directedness of I gives us an i'' with $i \leq i''$ and $i' \leq i''$; for the index i'' we have $\chi_{i''} m_1 = m_1$ and $\chi_{i''} m_2 = m_2$. The validity of the property in question extends by induction from two elements of M to n elements.

Proposition A.4. Let R be an associative algebra over k with an approximate identity, and let M be an approximately unital left R module. Then M has canonically the structure of a vector space over k such that the compatibility condition (A.2a) holds.

PROOF. Let the approximate identity be $\{\chi_i \mid i \in I\}$. If $m \in M$ is given, choose $i \in I$ with $\chi_i m = m$, and define $cm = (c\chi_i)m$ for $c \in k$. Let us prove that cm is well defined. If also $\chi_{i'} \in I$ has $\chi_{i'} m = m$, then the directedness of I implies that we may assume $i \leq i'$. Then

$$(c\chi_i)m = (c\chi_i)(\chi_{i'} m) = ((c\chi_i)\chi_{i'})m = (c\chi_i)m$$
$$= ((c\chi_{i'})\chi_i)m = (c\chi_{i'})(\chi_i m) = (c\chi_{i'})m.$$

Thus cm is well defined.

It is a routine matter to check that M is a vector space. To verify (A.2a), let $i \in I$ be such that $\chi_i m = m$ and $\chi_i rm = rm$ and $\chi_i((cr)m) = (cr)m$. Then

$$c(rm) = (c\chi_i)(rm) = ((c\chi_i)r)m = (\chi_i(cr))m = \chi_i((cr)m) = (cr)m,$$

and

$$r(cm) = r((c\chi_i)m) = (r(c\chi_i))m = ((cr)\chi_i)m = (cr)(\chi_i m) = (cr)m.$$

This proves (A.2a).

In this book it is always assumed that modules for an associative algebra satisfy (A.2a) *or* (A.2b). Proposition A.4 gives circumstances under which we do not have to verify this additional assumption.

Now let R be an associative algebra over k, and let X and Y be left R modules. Because of (A.2) the abelian group $\operatorname{Hom}_R(X, Y)$ can acquire the structure of a k vector space from X, and it can acquire the structure of a k vector space from Y. But there is no reason for these two vector-space structures to coincide. However, if R has an approximate identity and X and Y are approximately unital, then every member of $\operatorname{Hom}_R(X, Y)$ is k linear. In fact, in obvious notation we have

$$\varphi(cx) = \varphi(c(\chi_i x)) = \varphi((c\chi_i)x) = (c\chi_i)(\varphi(x)) = c(\chi_i(\varphi(x))) = c(\varphi(x)),$$

provided χ_i fixes x and $\varphi(x)$. Thus φ is k linear and we get the same definition for $c\varphi$ from $(c\varphi)(x) = \varphi(cx)$ as from $(c\varphi)(x) = c(\varphi(x))$.

The book makes considerable use of the fact that the approximately unital left R modules form a good category. The above paragraph says: If R is an associative algebra over k, then $\operatorname{Hom}_R(X, Y)$ is canonically a k vector space and, as such, is a vector subspace of $\operatorname{Hom}_k(X, Y)$.

Corresponding remarks apply to tensor products over an associative algebra R. If X is a right R module and Y is a left R module, then $X \otimes_R Y$ is an abelian group with two possibly different k vector-space structures, one from X and one from Y. However, if R has an approximate identity and if X and Y are approximately unital, then the two vector-space structures coincide. Moreover, $X \otimes_R Y$ is the quotient vector space of $X \otimes_k Y$ by the linear span

$$\langle xr \otimes y - x \otimes ry \mid r \in R, \ x \in X, \ y \in Y \rangle.$$

For an associative algebra R without an approximate identity, it will be inconvenient to allow cases where $\operatorname{Hom}_R(X, Y)$ and $X \otimes_R Y$ are not vector spaces. To get around this problem, we can enlarge R to contain an identity by defining $\tilde{R} = k1 + R$ with R as a two-sided ideal. Then $\operatorname{Hom}_{\tilde{R}}(X, Y)$ consists of the R linear maps from X to Y that are also k linear, and $X \otimes_R Y$ is the vector space $X \otimes_k Y$ with the R action identified in X and Y. On the other hand, if R has an identity or an approximate identity, then $\operatorname{Hom}_R(X, Y) = \operatorname{Hom}_{\tilde{R}}(X, Y)$ and $X \otimes_R Y = X \otimes_{\tilde{R}} Y$. *In this book we adopt the convention when R is an associative algebra over k that* $\operatorname{Hom}_R(X, Y)$ *means* $\operatorname{Hom}_{\tilde{R}}(X, Y)$ *and* $X \otimes_R Y$ *means* $X \otimes_{\tilde{R}} Y$.

With this convention in place, let R and S be associative algebras over k, and let X, Y, and Z be vector spaces over k with module structures to be specified. We shall state some associativity formulas for the use of $\otimes$ and Hom with these modules. Let $(X^R, {}^R Y^S, Z^S)$ refer to the situation in which X is a right R module, Y is a left R module and right S module such that the R action commutes with the S action, and Z is a right S module. Then there is a canonical k vector-space isomorphism

$$(A.5) \qquad \operatorname{Hom}_S(X \otimes_R Y, Z) \cong \operatorname{Hom}_R(X, \operatorname{Hom}_S(Y, Z)).$$

Similarly in the situation $({}^R X, {}^S Y^R, {}^S Z)$, there is a canonical k vector-space isomorphism

$$(A.6) \qquad \operatorname{Hom}_S(Y \otimes_R X, Z) \cong \operatorname{Hom}_R(X, \operatorname{Hom}_S(Y, Z))$$

Finally in the situation $(X^R, {}^R Y^S, {}^S Z)$, there is a canonical k vector-space isomorphism

$$(A.7) \qquad (X \otimes_R Y) \otimes_S Z = X \otimes_R (Y \otimes_S Z).$$

For proofs of these formulas, see §II.8 of Knapp [1988]. Sharper statements appear below in Appendix C as (C.19), (C.20), and (C.21).

2. Completely Reducible Modules

Let R be a ring, and let $\mathcal{C}$ be a good category of left R modules. A module in $\mathcal{C}$ is said to be **completely reducible** or **fully reducible** if it is the direct sum of (possibly infinitely many) irreducible modules in $\mathcal{C}$. The first three propositions show that the completely reducible modules in $\mathcal{C}$ form a good category. More precisely they show that decompositions into irreducible modules behave rather like bases of vector spaces.

One difficulty with the use of bases for vector spaces is that they are not canonical; sometimes it is more convenient to do linear algebra without them. Proposition A.13 shows that the study of completely reducible modules may to a large extent be reduced to linear algebra, in a way that allows us to use "coordinate-free" methods.

Proposition A.8. Suppose that M is in C and that $\{M_s \mid s \in S\}$ is a family of irreducible R submodules of M. Assume that the sum of these submodules is all of M. Then M is completely reducible. More precisely there is a subset T of S with the property that

$$M = \bigoplus_{t \in T} M_t.$$

PROOF. Call a subset U of S independent if the sum $\sum_{u \in U} M_u$ is direct. This condition means that for every finite subset $\{u_1, \ldots, u_n\}$ of U and every set of elements $m_i \in M_{u_i}$, the equation

$$m_1 + \cdots + m_n = 0$$

implies that each m_i is 0. From this formulation it follows that the union of any increasing chain of independent subsets of S is itself independent. By Zorn's Lemma there is a maximal independent subset T of S. By definition the sum $M_0 = \sum_{t \in T} M_t$ is direct. Consequently the problem is to show that M_0 is all of M. By the hypothesis on S, it suffices to show that each M_s is contained in M_0. If s is in T, this conclusion is obvious. Thus suppose s is not in T. By the maximality of T, $T \cup \{s\}$ is not independent. Consequently the sum $M_s + M_0$ is not direct, and we see that $M_s \cap M_0 \neq 0$. But this intersection is an R submodule of M_s. Since M_s is irreducible, a nonzero R submodule must be all of M_s. Thus M_s is contained in M_0, as we wished to show.

Proposition A.9. Suppose that M is a completely reducible member of C, say $M = \bigoplus_{s \in S} M_s$ with all M_s irreducible. Let N be any R submodule of M. Then

- (a) the quotient module M/N is completely reducible. In more detail there is a subset T of S with the property that the submodule $M_T = \bigoplus_{t \in T} M_t$ of M maps isomorphically onto M/N.
- (b) N is a direct summand of M. In more detail, $M = N \oplus M_T$, where M_T is as in (a).
- (c) N is completely reducible. In more detail choose T as in (a), and write T' for the complement of T in S. Then N maps isomorphically onto the quotient M/M_T of M, and this quotient is isomorphic to $M_{T'}$.

PROOF. Each irreducible R submodule M_s of M maps to an R submodule $\overline{M}_s$ of M/N. This image either is irreducible (and is isomorphic to M_s) or is zero. We let U be the subset of S for which it is irreducible. Then M/N is evidently the sum of the irreducible R submodules $\{\overline{M}_s \mid s \in U\}$. By Proposition A.8 there is a subset T of U so that

$$M/N = \bigoplus_{t \in T} \overline{M}_t.$$

This proves (a).

For (b) we use the following elementary observation: If N and N' are R submodules of M, then $M = N \oplus N'$ if and only if N' maps isomorphically onto the quotient M/N. For (c) the same observation implies that N maps isomorphically onto the quotient M/M_T, and then $N \cong M_{T'}$ since $M = M_T \oplus M_{T'}$.

Corollary A.10. The completely reducible modules in $\mathcal{C}$ form a good category. More precisely

(a) if M is a completely reducible module in $\mathcal{C}$, then any R submodule or quotient of M is completely reducible
(b) arbitrary sums (not necessarily direct) of completely reducible modules are completely reducible
(c) if M is an module in $\mathcal{C}$, then there is a unique maximal completely reducible submodule of M.

REMARK. The submodule in (c) is called the **socle** of M and is written $\mathrm{soc}(M)$.

PROOF. Part (a) follows from Proposition A.9c, and part (b) follows from Proposition A.8. For (c), form the sum of all the irreducible R submodules of M. This module is completely reducible by (b), and it obviously contains every completely reducible R submodule of M.

We now investigate the good category of all completely reducible modules in $\mathcal{C}$. Our interest is in the nature of the submodules and the maps among them.

Lemma A.11. Suppose that E is an irreducible module in $\mathcal{C}$ and that $M = \bigoplus_{a \in A} M_a$ is a direct-sum decomposition of another module in $\mathcal{C}$. Then

$$\mathrm{Hom}_R(E, M) \cong \bigoplus_{a \in A} \mathrm{Hom}_R(E, M_a).$$

REMARK. The hypothesis that E be irreducible is critical here. Without it a map into a direct sum might have nonzero projections into infinitely many of the summands, and then it could not be represented as a finite sum of maps into summands.

PROOF. Suppose φ is in $\mathrm{Hom}_R(E, M)$. Write φ_a for the composition of φ with the projection $M \to M_a$. Since M is a direct sum, we have

$$\varphi(e) = \sum_a \varphi_a(e) \qquad \text{for all } e \in E,$$

with only finitely many terms $\varphi_a(e)$ nonzero for each $e \in E$. We must show that only finitely many of the maps φ_a are nonzero. If $\varphi = 0$, then all φ_a are 0; so assume φ is not 0. Then φ carries E isomorphically onto an irreducible R submodule of M. Let $e \in E$ be any nonzero element, and write

$$\varphi(e) = m_1 + \cdots + m_n \qquad \text{with } m_i \in M_{a_i}.$$

Because $\varphi(e)$ is a nonzero element of $\varphi(E)$, we conclude that $\varphi(E)$ has nonzero intersection with $M' = M_{a_1} \oplus \cdots \oplus M_{a_n}$. Since $\varphi(E)$ is irreducible, it follows that $\varphi(E)$ is contained in M'. That is, φ_a is nonzero only if a is one of the a_i.

Lemma A.11 will enable us to study maps between completely reducible modules in terms of maps between irreducible modules. The latter are described by the next result. Part (c) is due to Dixmier.

Proposition A.12 (Schur's Lemma). Suppose E and F are irreducible modules in a good category $\mathcal{C}$ of R modules.

(a) If E and F are not isomorphic, then $\mathrm{Hom}_R(E, F) = 0$.

(b) $\mathrm{Hom}_R(E, E)$ is a division algebra D_E, i.e., a ring with identity in which every nonzero element is invertible.

(c) If R is an associative algebra over an algebraically closed field k and if the vector-space dimension of E over k is less than the cardinality of k, then $D_E = k$.

REMARK. Following the convention in §1, we assume in (c) that the members of $\mathrm{Hom}_R(E, E)$ are k linear. Then it follows that $k \subseteq D_E$ and that D_E is an associative algebra over k with identity.

PROOF. Suppose that φ is nonzero in $\mathrm{Hom}_R(E, F)$. Then $\ker \varphi$ is a proper R submodule of E, and we must have $\ker \varphi = 0$. Similarly image φ is a nonzero R submodule of F, and we have image $\varphi = F$. Therefore φ is an isomorphism of E onto F. This proves (a) and (b).

For (c) let e be a nonzero element of E. The map $\varphi \mapsto \varphi(e)$ is k linear and one-one from D_E into E. Thus D_E is a division algebra over k of dimension less than the cardinality of k. Arguing by contradiction, let us assume that D_E is not equal to k; say D_E contains an element φ not in k.

The smallest division subalgebra of D_E containing k and φ is a field isomorphic to $k(\varphi)$. The element φ cannot be algebraic, since k is algebraically closed and φ is not in k. Therefore φ is transcendental. In the transcendental extension $k(X)$, the set of elements $\{(X - c)^{-1} \mid c \in k\}$ is linearly independent over k, and hence $\dim_k k(X)$ is $\geq$ the cardinality of k. Consequently the dimension D_E over k is $\geq$ the cardinality of k. This conclusion contradicts our earlier observation that the dimension of D_E is strictly less than the cardinality of k. So the assumption that D_E contains an element not in k must be false. This completes the proof.

Modules over a division algebra have many of the properties of vector spaces over fields, such as existence of bases and the notions of dimension and rank. We shall reinterpret the submodule structure of each "isotypic submodule" of M in this setting.

Proposition A.13. Let E be an irreducible module in a good category $\mathcal{C}$ of left R modules, and let M be any R module in $\mathcal{C}$. Define

$$M^E = \operatorname{Hom}_R(E, M).$$

Then

(a) M has a unique largest R submodule M_E that is a direct sum of copies of E. The image of every mapping in M^E belongs to M_E, and therefore

$$M^E \cong \operatorname{Hom}_R(E, M_E).$$

(b) M^E is a unital right D_E module, where $D_E = \operatorname{Hom}_R(E, E)$, and E is a unital left D_E module.
(c) with R acting on $M^E \otimes_{D_E} E$ by acting on the second factor only, the map

$$\Phi : M^E \otimes_{D_E} E \to M \qquad \text{given by} \qquad \Phi(\psi \otimes e) = \psi(e)$$

of left R modules is an isomorphism onto its image, which is precisely the R submodule M_E described in (a).

(d) the left R submodules of M_E are in one-one correspondence with the right D_E submodules (i.e., right vector subspaces) of M^E by the maps

$$N \mapsto \operatorname{Hom}_R(E, N) \subseteq \operatorname{Hom}_R(E, M) = M^E \quad \text{if } N \subseteq M_E$$

and

$$W \mapsto W \otimes_{D_E} E \subseteq M^E \otimes_{D_E} E \cong M_E \quad \text{if } W \subseteq M^E.$$

(e) for any left R module N, there is a canonical isomorphism

$$\operatorname{Hom}_R(M_E, N_E) \cong \operatorname{Hom}_{D_E}(M^E, N^E)$$

defined as follows. Suppose φ is in $\operatorname{Hom}_R(M_E, N_E)$. Composition with φ carries $\operatorname{Hom}_R(E, M)$ into $\operatorname{Hom}_R(E, N)$; this map respects the right action of D_E and hence induces a map

$$\varphi^E \in \operatorname{Hom}_{D_E}(M^E, N^E).$$

The isomorphism is given in terms of the isomorphisms Φ_M for M and Φ_N for N in (c) by

$$\varphi(\Phi_M(\psi \otimes e)) = \Phi_N(\varphi^E(\psi) \otimes e) \quad \text{for } \psi \in M^E.$$

REMARK. The submodule M_E is called the **isotypic submodule** of M of type E. If R is an associative algebra over a field k, then M_E is called an **isotypic subspace**.

PROOF.
(a) Let $\{M_s \mid s \in S\}$ be the set of all R submodules of M isomorphic to E, and let M_E be their sum. Then Proposition A.8 shows that $M_E = \bigoplus_{s \in T} M_s$ for a subset T of S, and certainly M_E has the maximality property claimed. If ψ is a nonzero map in M^E, then $\psi(E)$ is a submodule of M isomorphic to E. Hence $\psi(E) \subseteq M_E$ by construction, and part (a) follows.

(b) With $\varphi \in D_E = \operatorname{Hom}_R(E, E)$, we can form $\psi\varphi = \psi \circ \varphi$ if ψ is in $\operatorname{Hom}_R(E, M_E)$, and we can form $\varphi e = \varphi(e)$ if e is in E. These definitions give the required unital D_E module structures.

(c) We begin with the bi-additive map $\Phi : M^E \times E \to M$ given by $\Phi(\psi, e) = \psi(e)$. If φ is in D_E, then

$$\Phi(\psi \circ \varphi, e) = (\psi \circ \varphi)(e) = \psi(\varphi(e)) = \Phi(\psi, \varphi(e)).$$

By the universal mapping property of tensor products, Φ induces a map (also denoted Φ) satisfying

$$\Phi : M^E \otimes_{D_E} E \to M.$$

The induced map Φ is R linear since

$$\Phi(r(\psi \otimes e)) = \Phi(\psi \otimes re) = \psi(re) = r(\psi(e)) = r(\Phi(\psi \otimes e)).$$

Since ψ is in M^E, $\psi(e)$ is in M_E; thus Φ has image in M_E.

To see that Φ is onto M_E, fix an isomorphism $\alpha_s \in \mathrm{Hom}_R(E, M_s)$ for each $s \in T$. For any element $m \in M_E$, we can find a finite subset T' of T such that $m = \sum_{s \in T'} m_s$ with $m_s \in M_s$. If we let $e_s = \alpha_s^{-1}(m_s)$, then $\Phi(\sum_{s \in T'} \alpha_s \otimes e_s) = m$.

To see that Φ is one-one, we use (a) and Lemma A.11 to write

$$M^E = \mathrm{Hom}_R(E, M) = \mathrm{Hom}_R(E, M_E)$$
$$= \mathrm{Hom}_R\left(E, \bigoplus_{s \in T} M_s\right) = \bigoplus_{s \in T} \mathrm{Hom}_R(E, M_s).$$

Each summand on the right side is isomorphic to D_E. That is, the collection of isomorphisms $\{\alpha_s\}_{s \in T}$ from the previous paragraph is a basis of M^E as a right D_E vector space. Consequently every element of $M^E \otimes_{D_E} E$ may be written as a finite sum $\sum \alpha_s \otimes e_s$ with $e_s \in E$. The image of the element $\sum \alpha_s \otimes e_s$ is $\sum \alpha_s(e_s)$. If this is 0, then each $\alpha_s(e_s)$ is 0 because of the independence of the M_s's. Since α_s is an isomorphism, it follows that $e_s = 0$ for each s. Therefore $\sum \alpha_s \otimes e_s = 0$. This proves (c).

(d) The composition in one order is

(A.14) $\qquad N \mapsto \mathrm{Hom}_R(E, N) \mapsto \mathrm{Hom}_R(E, N) \otimes_{D_E} E.$

For $N = M_E$, the map Φ, when applied to the composition, recovers M_E, since (c) says that Φ is onto. For general N, we can write $M_E = N \oplus N'$. When we apply Φ to (A.14) for N and N' separately, we get R submodules of N and N', respectively. To have a match for all of M_E, we must get all of N and N'.

The composition in the other order is

(A.15) $\qquad W \mapsto W \otimes_{D_E} E \mapsto \mathrm{Hom}_R(E, W \otimes_{D_E} E).$

For $W = M^E$, the image corresponds under the map $\mathrm{Hom}(1, \Phi)$ to $\mathrm{Hom}_R(E, M_E) = M^E$. For general W, we can write $M^E = W \oplus W'$.

When we apply $\mathrm{Hom}(1, \Phi)$ to (A.15) for W, we get an R submodule of M^E that contains W. In fact, for any $w \in W$, $\mathrm{Hom}_R(E, E \otimes_{D_E} E)$ contains the map $e \mapsto w \otimes e$. Composing with Φ gives $e \mapsto w(e)$. Thus the members of W are in the image. Similarly the members of W' are in the image for W'. The direct sum of the images must be M^E, and thus the images must be exactly W and W'.

(e) The computation

$$\varphi(\Phi_M(\psi \otimes e)) = \varphi(\psi(e)) = (\varphi \circ \psi)(e)) = \Phi_N(\varphi \circ \psi \otimes e) = \Phi_N(\varphi^E(\psi) \otimes e)$$

proves the asserted formula. For the inverse, suppose we are given a map $\tau \in \mathrm{Hom}_{D_E}(M^E, N^E)$. Then τ induces an R linear map

$$\tau'_E : M^E \otimes_{D_E} E \rightarrow N^E \otimes_{D_E} E$$

defined by

$$\tau'_E(\psi \otimes e) = \tau(\psi) \otimes e.$$

Composition with the isomorphism of (c) gives an R linear map

$$\tau_E = \Phi_N \circ \tau'_E \circ \Phi_M^{-1} : M_E \rightarrow N_E.$$

We show that $\varphi \mapsto \varphi^E$ and $\tau \mapsto \tau_E$ are inverses. If a map φ in $\mathrm{Hom}_R(M_E, N_E)$ is given, we are to calculate $(\varphi^E)_E \in \mathrm{Hom}_R(M_E, N_E)$. It is enough to find the effect of $(\varphi^E)_E$ on elements $\Phi_M(\psi \otimes e)$ with $\psi \in M^E$ and $e \in E$. For such an element,

$$(\varphi^E)_E(\Phi_M(\psi \otimes e)) = \Phi_N((\varphi^E)'(\psi \otimes e)) = \Phi_N(\varphi^E(\psi) \otimes e)$$
$$= \varphi^E(\psi)(e) = \varphi(\psi(e)) = \varphi(\Phi_M(\psi \otimes e)).$$

Thus $(\varphi^E)_E = \varphi$. Similarly, for $\tau \in \mathrm{Hom}_{D_E}(M^E, N^E)$, we find that $(\tau_E)^E = \tau$. This proves (e).

Corollary A.16. Let M and N be modules in a good category $\mathcal{C}$ of left R modules, and suppose that M is completely reducible. Write $\mathcal{E}$ for the set of isomorphism classes of irreducible R modules in $\mathcal{C}$. Then

(a) with notation as in Proposition A.13,

$$M = \bigoplus_{E \in \mathcal{E}} M_E \cong \bigoplus_{E \in \mathcal{E}} (M^E \otimes_{D_E} E).$$

(b) the left R submodules of M are in one-one correspondence with families $\{W^E \mid E \in \mathcal{E}\}$ of right D_E submodules (i.e, right vector subspaces) of M^E.

(c) there is a canonical isomorphism

$$\operatorname{Hom}_R(M, N) = \operatorname{Hom}_R(M, \operatorname{soc}(N)) \cong \prod_{E \in \mathcal{E}} \operatorname{Hom}_{D_E}(M^E, N^E).$$

More precisely an R module map from M to N is specified by giving, for each class E of irreducible R modules in $\mathcal{C}$, an arbitrary right vector-space map from M^E to N^E.

PROOF.

(a) Let us write $M = \bigoplus_{s \in S} M_s$. Each M_s is contained in some M_E, and hence $M = \sum M_E$. If M_E has nonzero intersection with $M_{E_1} + \cdots + M_{E_n}$, then there is a nonzero R linear map from E into $M_{E_1} + \cdots + M_{E_n}$. We can write each M_{E_j} as a sum of irreducible subspaces isomorphic to E_j, and Proposition A.8 shows that

$$M_{E_1} + \cdots + M_{E_n} = \bigoplus M_s'$$

with each M_s' isomorphic to one of $E_1, \ldots, E_n$. If $E_1, \ldots, E_n$ are inequivalent with E, then Lemma A.11 and Proposition A.12a show that $\operatorname{Hom}_R(E, M_{E_1} + \cdots + M_{E_n}) = 0$, contradiction. We conclude that the sum $M = \sum M_E$ is direct. This proves the equality in the formula, and the subsequent isomorphism follows from Proposition A.13c.

(b) If N is a left R submodule of M, then $N_E \subseteq M_E$ for every E. Conversely (a) shows that a system of N_E's defines a R submodule N. Thus (b) is a restatement of Proposition A.13d.

(c) The first isomorphism is trivial. We have

$$\operatorname{Hom}_R(M, N) \cong \prod_{E \in \mathcal{E}} \operatorname{Hom}_R(M_E, N) = \prod_{E \in \mathcal{E}} \operatorname{Hom}_R(M_E, N_E),$$

and the rest follows from Proposition A.13e.

3. Modules of Finite Length

Let $\mathcal{C}$ be a good category of R modules. In this section we shall introduce the subcategory $\mathcal{F}$ of modules in $\mathcal{C}$ of "finite length," and we shall investigate complete reducibilty for members of $\mathcal{F}$. On this level the topics of this section are not really part of homological algebra, but their applications intersect with it.

A **finite filtration** of a module M in $\mathcal{C}$ is a finite descending chain

$$(A.17) \qquad M = M_0 \supseteq M_1 \supseteq \cdots \supseteq M_n = 0$$

of R submodules (necessarily in $\mathcal{C}$). We do not insist on this particular indexing, and, with the obvious adjustments, we allow also a finite increasing chain to be called a finite filtration. Relative to (A.17), the modules $M_i/M_{i+1}, 0 \le i \le n - 1$, are called the **consecutive quotients** of the filtration. The finite filtration is called a **composition series** if the consecutive quotients are all irreducible (i.e., simple) R modules; in particular, they are to be nonzero. The consecutive quotients in this case are called **composition factors**.

If M has a composition series, then we say that M has **finite length**. This notion is closed under passage to submodules and quotients. In fact, if (A.17) is a composition series of M and if M' is a submodule of M, then

$$M' = M_0 \cap M' \supseteq M_1 \cap M' \supseteq \cdots \supseteq M_n \cap M' = 0$$

is a finite filtration of M' in which each consecutive quotient is irreducible or 0. Discarding redundant terms (which lead to 0 as a consecutive quotient), we obtain a composition series for M'. A similar argument works for M/M'.

Let us show that

$$(A.18) \qquad \text{If } M' \text{ and } M/M' \text{ have finite length, then so does } M.$$

To do so, we take a composition series for M/M', pull it back to M, and concatenate it to a composition series for M'. The result is a composition series for M, and (A.18) follows. In particular, the direct sum of two members in $\mathcal{C}$ of finite length has finite length. Consequently the modules in $\mathcal{C}$ of finite length form a good subcategory, which we denote by $\mathcal{F}$.

Let

$$(A.19a) \qquad M = M_0 \supseteq M_1 \supseteq \cdots \supseteq M_m = 0$$

$$(A.19b) \qquad M = N_0 \supseteq N_1 \supseteq \cdots \supseteq N_n = 0$$

be two finite filtrations of M. We say that the second is a **refinement** of the first if there is a one-one function $f : \{0, \dots, m\} \to \{0, \dots, n\}$ with $M_i = N_{f(i)}$ for $0 \le i \le m$.

The two finite filtrations (A.19) of M are said to be **equivalent** if $m = n$ and if there is an invertible function g on $\{0, \dots, m\}$ such that

$M_i/M_{i+1} \cong N_{g(i)}/N_{g(i)+1}$ for $0 \le i \le m$. (Here by convention we take $M_{m+1} = N_{m+1} = 0$.)

In proving results, we take the **second isomorphism theorem** as our starting point: If M_1 and M_2 are submodules of M, then

(A.20)
$$(M_1 + M_2)/M_2 \cong M_1/(M_1 \cap M_2) \quad \text{under the correspondence}$$
$$m_1 + m_2 + M_2 \mapsto m_1 + (M_1 \cap M_2).$$

Lemma A.21 (Zassenhaus). Let M_1, M_2, M_1', and M_2' be submodules of a module M in $\mathcal{C}$ with $M_1' \subseteq M_1$ and $M_2' \subseteq M_2$. Then

$$((M_1 \cap M_2) + M_1')/((M_1 \cap M_2') + M_1')$$
$$\cong ((M_1 \cap M_2) + M_2')/((M_1' \cap M_2) + M_2').$$

PROOF. By the second isomorphism theorem (A.20)

(A.22)
$$(M_1 \cap M_2)/(((M_1 \cap M_2') + M_1') \cap (M_1 \cap M_2))$$
$$\cong ((M_1 \cap M_2) + (M_1 \cap M_2') + M_1')/((M_1 \cap M_2') + M_1')$$
$$= ((M_1 \cap M_2) + M_1')/((M_1 \cap M_2') + M_1').$$

Next, we have

$$((M_1 \cap M_2') + M_1') \cap (M_1 \cap M_2) = ((M_1 \cap M_2') + M_1') \cap M_2$$
$$= (M_1 \cap M_2') + (M_1' \cap M_2).$$

Hence we can rewrite (A.22) as

(A.23) $$(M_1 \cap M_2)/((M_1 \cap M_2') + (M_1' \cap M_2))$$
$$\cong ((M_1 \cap M_2) + M_1')/((M_1 \cap M_2') + M_1').$$

The left side of (A.23) is symmetric under interchange of the indices 1 and 2. Hence so is the right side, and the lemma follows.

Theorem A.24 (Schreier). Any two finite filtrations of a module M in $\mathcal{C}$ have equivalent refinements.

PROOF. Let the two finite filtrations be given as in (A.19), and define

(A.25)
$$M_{ij} = (M_i \cap N_j) + M_{i+1} \quad \text{for } 0 \le j \le n$$
$$N_{ji} = (M_i \cap N_j) + N_{j+1} \quad \text{for } 0 \le i \le m.$$

Then we obtain respective refinements of (A.19a) and (A.19b) given by

$$M = M_{00} \supseteq M_{01} \supseteq \cdots \supseteq M_{0n}$$
$$\supseteq M_{10} \supseteq M_{11} \supseteq \cdots \supseteq M_{1n} \cdots \supseteq M_{m-1,n} = 0$$

(A.26)

$$M = N_{00} \supseteq N_{01} \supseteq \cdots \supseteq N_{0m}$$
$$\supseteq N_{10} \supseteq N_{11} \supseteq \cdots \supseteq N_{1m} \cdots \supseteq N_{n-1,m} = 0.$$

The containments $M_{in} \supseteq M_{i+1,0}$ and $N_{jm} \supseteq N_{j+1,0}$ are equalities in (A.26), and the only nonzero consecutive quotients are therefore of the form $M_{ij}/M_{i,j+1}$ and $N_{ji}/N_{j,i+1}$. For these we have

$$M_{ij}/M_{i,j+1}$$
$$= ((M_i \cap N_j) + M_{i+1})/((M_i \cap N_{j+1} + M_{i+1}) \qquad \text{by (A.25)}$$
$$\cong ((M_i \cap N_j) + N_{j+1})/((M_{i+1} \cap N_j) + N_{j+1}) \qquad \text{by Lemma A.21}$$
$$= N_{ji}/N_{j,i+1} \qquad \text{by (A.25),}$$

and thus the refinements (A.26) are equivalent.

Corollary A.27. If M is a module of finite length in $\mathcal{C}$, then

(a) any finite filtration of M in which all consecutive quotients are nonzero can be refined to a composition series, and

(b) any two composition series of M are equivalent.

PROOF. We apply Theorem A.24 to a given filtration and a known composition series. After discarding redundant terms from each refinement (those that lead to 0 as a consecutive quotient), we arrive at a refinement of our given finite filtration that is equivalent with the known composition series. Hence the refinement is a composition series. This proves (a). If we specialize this argument to the case that the given filtration is a composition series, then we obtain (b).

If M has a composition series of the form (A.17), we say that M has **length** n. According to Corollary A.27, this notion of length is independent of the particular composition series that we use. The proof of (A.18) shows that if M' is a submodule of M, then the length of M is the sum of the lengths of M' and M/M'. In particular, if M' is a length n submodule of a length n module M, then $M' = M$.

Corollary A.27 implies that the composition factors for a given composition series depend only on M. Moreover, if $M' \supseteq M''$ are submodules of M such that M'/M'' is irreducible, then M'/M'' is a composition factor of M. This fact follows by eliminating redundant terms from the finite filtration

$$M \supseteq M' \supseteq M'' \supseteq 0$$

and applying Corollary A.27a to the result.

4. Grothendieck Group

Let C be a good category of left R modules. We say that C is **small** if the class of modules in C is a set. In this case the **Grothendieck group** $K(C)$ is the abelian group obtained by the following construction as a quotient of the free abelian group $\mathcal{F}(C)$ on the modules of C: Each exact sequence

$$(A.28a) \qquad\qquad 0 \longrightarrow A \longrightarrow B \longrightarrow C \longrightarrow 0$$

in C yields a relation

$$(A.28b) \qquad\qquad (B) = (A) + (C)$$

in $\mathcal{F}(C)$, and $K(C)$ is the quotient of $\mathcal{F}(C)$ by the group generated by all such relations.

As a practical matter, it may be necessary to go through some contortions to create small good categories. For example, the category of all finite-dimensional complex vector spaces is not small. But there are only countably many nonisomorphic such vector spaces, and it is possible to replace this category for many purposes by one that *is* small. Namely a construction in set theory that we shall not reproduce allows one to form a subcategory of finite-dimensional complex vector spaces that is good in the sense of §1, contains at least one finite-dimensional vector space from each isomorphism class, and is closed under a number of desirable operations; the construction involves the formation of a very large set, and each vector space in the category is to take its elements from this set. Abusing terminology, we then speak of the "small good category of all finite-dimensional complex vector spaces."

More generally there is an analog of this construction for any good category with the property that there is a set of modules containing a representative from each isomorphism class in the category; the construction produces a small good category among whose modules is at least one from each isomorphism class. For many purposes (and the construction of the Grothendieck group is such a purpose), it does not matter too much how we restrict a category to be small, as long as the result is still a *good* category.

Anyway, fix a small good category C. As in (A.28), we denote modules by A, B, ... and the corresponding members of $\mathcal{F}(C)$ and $K(C)$ by (A), (B), ... Applying (A.28) to the exact sequence of all 0, we see that

$$(A.29) \qquad\qquad (0) \text{ is the additive identity } 0 \text{ in } K(C).$$

The special case $B = A \oplus C$ in (A.28) yields

(A.30) $\qquad (A \oplus C) = (A) + (C) \qquad$ in $K(\mathcal{C})$.

Proposition A.31. In the small good category $\mathcal{C}$ of all finite-dimensional vector spaces over a field k, the map $(A) \mapsto \dim A$ extends to a well-defined isomorphism of $K(\mathcal{C})$ onto $\mathbb{Z}$.

REMARK. The extended map will be called dim.

PROOF. The map extends to a homomorphism of $\mathcal{F}(\mathcal{C})$ and descends to $K(\mathcal{C})$ as a consequence of the identity

$$\dim(\text{domain}) = \dim(\text{kernel}) + \dim(\text{image})$$

familiar from linear algebra. Since 1 is in the image and the image is a group, the map is onto $\mathbb{Z}$. To see that the map of $K(\mathcal{C})$ is one-one, let $\sum n_i (A_i)$ have $\sum n_i \dim A_i = 0$. We are to prove that $\sum n_i (A_i)$ is in the subgroup of $\mathcal{F}(\mathcal{C})$ generated by the relations.

Let us separate indices i with $n_i > 0$ from those with $n_i < 0$, writing

(A.32) $\qquad \sum n_i (A_i) = \sum k_j (A_j) - \sum l_{j'} (A_{j'})$

in $\mathcal{F}(\mathcal{C})$, with all k_j and $l_{j'} > 0$. The given condition is that

(A.33) $\qquad \sum k_j \dim A_j = \sum l_{j'} \dim A_{j'}$,

and we are to prove that the right side of (A.32) is in the relation subgroup. By (A.30), the right side of (A.32) is congruent, modulo the relation subgroup, to an element $(V) - (W)$, where $\dim V = \dim W$ by (A.33). Since $\dim V = \dim W$, V and W are isomorphic. Thus there is an exact sequence

$$0 \longrightarrow V \longrightarrow W \longrightarrow 0 \longrightarrow 0,$$

and $(V) - (W)$ is a relation. Consequently the homomorphism dim is one-one on $K(\mathcal{C})$.

APPENDIX B

DISTRIBUTIONS ON MANIFOLDS

For a smooth manifold X with a countable base, $C^\infty(X)$ gets a topology defined by suprema of derivatives on compact sets within coordinate patches. This topology is metrizable.

The continuous dual $\mathcal{E}'(X)$ of $C^\infty(X)$ is defined to be the space of distributions of compact support on X. The support of a member T of $\mathcal{E}'(X)$ is the complement of the union of all open subsets U_0 of X such that T vanishes on all members of $C^\infty(X)$ compactly supported in U_0. Indeed the support is compact, and it behaves qualitatively like the support of a function.

For two smooth manifolds X and Y, members S of $\mathcal{E}'(X)$ and T of $\mathcal{E}'(Y)$ determine a member $S \times T$ of $\mathcal{E}'(X \times Y)$, and an analog of Fubini's Theorem is valid for $S \times T$ on all of $C^\infty(X \times Y)$.

A member of $\mathcal{E}'(\mathbb{R}^m)$ can be regarded as a distribution on $\mathbb{R}^m \times \mathbb{R}^n$ supported on $\mathbb{R}^m \times \{0\}$. Conversely any member of $\mathcal{E}'(\mathbb{R}^m \times \mathbb{R}^n)$ supported on $\mathbb{R}^m \times \{0\}$ is the sum of products $T_j \times \partial^\beta$ of members of $\mathcal{E}'(\mathbb{R}^m)$ by tranverse derivatives at 0. This result can be transferred to a local result about distributions on manifolds that are supported on immersed submanifolds.

1. Topology on $C^\infty(X)$

The foundations of Hecke algebras in Chapter I involve distributions of compact support on Lie groups. This appendix develops the necessary analytic background concerning distributions on manifolds. See the Notes for references.

The word "manifold" will refer to a C^∞ manifold with a countable base for its topology. A manifold is not assumed to be connected but *is* assumed to be equidimensional. For a manifold X, we let $C^\infty(X)$ be the space of complex-valued smooth functions on X, and we let $C_{\text{com}}^\infty(X)$ be the subspace of such functions with compact support.

The countability assumption makes X paracompact: Every open cover has a locally finite subcover $\{U_k\}$, necessarily at most countable. Moreover, there exists a **smooth partition of unity** $\{p_k\}$ subordinate to the subcover: $p_k \in C^\infty(X)$, $p_k \geq 0$, support$(p_k) \subseteq U_k$, $\sum p_k = 1$.

Another consequence of the countability assumption is the existence of an **exhausting sequence**, an increasing sequence of compact sets with union X such that each set in the sequence is contained in the interior of the next set in the sequence. If $\{K_j\}$ is an exhausting sequence and K is compact, then $K \subseteq K_{j_0}$ for some j_0.

To topologize $C^\infty(X)$, we first introduce notation for the case of an open subset of Euclidean space. Let V be open in $\mathbb{R}^n$. If $\alpha = (\alpha_1, \ldots, \alpha_n)$ is a multi-index, we use the abbreviations

$$\partial^\alpha = \left(\frac{\partial}{\partial x_1}\right)^{\alpha_1} \cdots \left(\frac{\partial}{\partial x_n}\right)^{\alpha_n} \quad \text{and} \quad |\alpha| = \alpha_1 + \cdots + \alpha_n.$$

If we want to emphasize the variables, we shall use ∂_x^α instead of ∂^α. For two multi-indices β and α, we write $\beta \leq \alpha$ if $\beta_j \leq \alpha_j$ for all j. Addition and subtraction of multi-indices are similarly defined entry by entry.

The **Leibniz rule** gives a formula for applying ∂^α to a product in $C^\infty(V)$:

$$(\mathrm{B}.1) \qquad \partial^\alpha(hp) = \sum_{\beta \leq \alpha} c_{\beta,\alpha}(\partial^\beta h)(\partial^{\alpha-\beta} p)$$

for suitable integers $c_{\beta,\alpha}$. For a composition $h \circ \varphi$, where φ is smooth from an open set in one Euclidean space into an open set in a possibly different Euclidean space, there is a similar formula:

$$(\mathrm{B}.2) \qquad \partial^\alpha(h \circ \varphi) = \sum_{\beta \leq \alpha} ((\partial^\beta h) \circ \varphi)\varphi_{\beta,\alpha}$$

for suitable scalar-valued functions $\varphi_{\beta,\alpha}$ independent of h. For any compact subset C of V and any multi-index α, we define a seminorm by

$$(\mathrm{B}.3\mathrm{a}) \qquad \|h\|_{V,C,\alpha} = \sup_{x \in C} |\partial^\alpha h(x)| \qquad \text{for } h \in C^\infty(V).$$

Returning to $C^\infty(X)$, let (φ, U) be a compatible chart, let $L \subseteq U$ be a compact set, and let α be a multi-index. If f is in $C^\infty(X)$, then $h = f \circ \varphi^{-1}|_{\varphi(U)}$ is in $C^\infty(\varphi(U))$, and it makes sense to define a seminorm by

$$(\mathrm{B}.3\mathrm{b}) \qquad \|f\|_{\varphi,U,L,\alpha} = \|f \circ \varphi^{-1}|_{\varphi(U)}\|_{\varphi(U),\varphi(L),\alpha}.$$

This collection of seminorms, as the tuple (φ, U, L, α) varies, defines a topology on $C^\infty(X)$. Namely a basic open set about 0 is determined by a finite set F of tuples (φ, U, L, α) and a positive number ϵ; the open set is

$$(\mathrm{B}.4) \qquad \{f \in C^\infty(X) \mid \|f\|_{\varphi,U,L,\alpha} < \epsilon \text{ for } (\varphi, U, L, \alpha) \in F\}.$$

Basic open sets about other members of $C^\infty(X)$ are given by translation. The resulting topology makes $C^\infty(X)$ into a locally convex (Hausdorff) linear topological space.

Using (B.1) and the definition of the topology, one sees easily that

(B.5)
$$\begin{pmatrix} \text{pointwise} \\ \text{multiplication} \end{pmatrix} : C^\infty(X) \times C^\infty(X) \to C^\infty(X) \qquad \text{is continuous.}$$

We shall be interested in detecting continuity of linear maps A from or to $C^\infty(X)$. In every case the domain and range of A will both be locally convex linear topological spaces whose topologies are defined by seminorms. Thus let $\mathcal{X}$ and $\mathcal{Y}$ be two such spaces with seminorms $\|\cdot\|_{\mathcal{X},\beta}$ and $\|\cdot\|_{\mathcal{Y},\gamma}$. A linear map $A : \mathcal{X} \to \mathcal{Y}$ is continuous if and only if, for each γ, there exist a finite subset F of β's and a constant $C_{\gamma,F}$ such that

(B.6)
$$\|A(f)\|_{\mathcal{Y},\gamma} \le C_{\gamma,F} \sum_{\beta \in F} \|f\|_{\mathcal{X},\beta} \qquad \text{for all } f \in \mathcal{X}.$$

Proposition B.7. For the manifold X, the topology on $C^\infty(X)$ is metrizable.

PROOF. Starting from an atlas of compatible charts (φ, U) for which $\overline{U}$ is compact, we form a locally finite refinement (φ_k, U_k). Let $\{p_k\}$ be a smooth partition of unity subordinate to $\{U_k\}$, and let $L_k =\text{support}(p_k)$. The subcollection of seminorms $\|\cdot\|_{\varphi_k, U_k, L_k, \alpha}$ from (B.3), as k and α vary, is countable, and the claim is that it determines the same topology on $C^\infty(X)$ as the full collection. Since a countable collection of seminorms determines a topology defined by a pseudometric (i.e., a function d that is a metric except that $d(x, y) = 0$ need not imply $x = y$), this claim will prove the proposition.

Let $\mathcal{X}$ be $C^\infty(X)$ with the topology from the subcollection, and let $\mathcal{Y}$ be $C^\infty(X)$ with the topology from the full collection. We are to prove that the identity map $I : \mathcal{X} \to \mathcal{Y}$ is continuous. In view of (B.6), it is enough to prove an inequality of the form

(B.8)
$$\|f\|_{\varphi, U, L, \alpha} \le C \sum_{(k,\beta) \in F} \|f\|_{\varphi_k, U_k, L_k, \beta}.$$

Choose N so that $\sum_{k \le N} p_k = 1$ on L, and let F consist of all pairs (k, β) with $k \le N$ and $|\beta| \le |\alpha|$. We shall prove (B.8) for this F.

We have $f = \sum_{k \leq N} f p_k$ on L and therefore

$$\| f \|_{\varphi, U, L, \alpha} \leq \sum_{k \leq N} \| f p_k \|_{\varphi, U, L, \alpha}.$$

Hence it is enough to prove that

(B.9) $$\| f p_k \|_{\varphi, U, L, \alpha} \leq C \sum_{|\beta| \leq |\alpha|} \| f \|_{\varphi_k, U_k, L_k, \beta}.$$

The left side of (B.9) is

$$= \sup_{x \in \varphi(L)} |\partial^\alpha (f \circ \varphi^{-1}(x) \, p_k \circ \varphi^{-1}(x))|$$

$$= \sup_{x \in \varphi(L \cap L_k)} |\partial^\alpha (f \circ \varphi^{-1}(x) \, p_k \circ \varphi^{-1}(x))|$$

since support$(p_k) \subseteq L_k$. Applying the Leibniz rule (B.1), we see that this expression is

$$\leq \sum_{|\beta| \leq |\alpha|} C_\beta \sup_{x \in \varphi(L \cap L_k)} |\partial^\beta (f \circ \varphi^{-1})(x)|.$$

In turn the βth sup here is

$$= \sup_{x \in \varphi(L \cap L_k)} |\partial^\beta (f \circ \varphi_k^{-1} \circ (\varphi_k \circ \varphi^{-1}))(x)|.$$

Expanding this expression by (B.2), we see that it is

$$\leq \sum_{|\gamma| \leq |\beta|} C_\gamma \| f \|_{\varphi_k, U_k, L_k, \gamma}.$$

Thus (B.9) follows, and the proof is complete.

As a consequence of Proposition B.7, the topology on $C^\infty(X)$ may be described in terms of sequences. If f_n and f are in $C^\infty(X)$, then $f_n \to f$ if and only if $\| f_n - f \|_{\varphi, U, L, \alpha} \to 0$ for all tuples (φ, U, L, α), i.e., if and only if $f_n - f$, when referred to any local coordinate system, converges to 0 uniformly on compact sets, along with each of its derivatives of arbitrary order.

2. Distributions and Support

On a manifold X a **distribution with compact support** is a continuous linear functional on $C^\infty(X)$. Following tradition, we denote the vector space of such distributions by $\mathcal{E}'(X)$. The reason for the phrase "with compact support" will be given in Proposition B.15 below.

In view of (B.6), a linear functional T on $C^\infty(X)$ is continuous if and only if there exists a finite set F of tuples (φ, U, L, α) depending on T such that

$$(\text{B}.10) \qquad\qquad |T(f)| \leq C \sum_F \|f\|_{\varphi, U, L, \alpha}$$

for some C and for all f in $C^\infty(X)$.

EXAMPLES. (1) On any X, evaluation at a point p is a distribution δ_p called the **Dirac distribution** at p.

(2) In an open set V of Euclidean space, let L be a compact subset, let $d\mu$ be a Borel measure on V supported on L, and let α be a multi-index. Then

$$T(f) = \int_V \partial^\alpha f(x) \, d\mu(x)$$

is an example of a member of $\mathcal{E}'(V)$. It is easy to verify (B.10) for this example.

There is a simple operation on distributions that is often handy. Let T be in $\mathcal{E}'(X)$ and h be in $C^\infty(X)$. Define hT by

$$(\text{B}.11) \qquad\qquad (hT)(f) = T(hf).$$

Since $f \mapsto hf$ is continuous on $C^\infty(X)$ by (B.5), it follows that hT is in $\mathcal{E}'(X)$. Thus $\mathcal{E}'(X)$ becomes a $C^\infty(X)$ module.

The **support** of a member T of $\mathcal{E}'(X)$ is the complement of the union of all open $U_0 \subseteq X$ such that T vanishes on the subspace $C^\infty_{\text{com}}(U_0)$ of $C^\infty(X)$. The support of T is a closed set.

The first proposition and corollary below show that support behaves reasonably. For example, T vanishes on $C^\infty_{\text{com}}((\text{support } T)^c)$, $T = 0$ is the only member of $\mathcal{E}'(X)$ with empty support, and the support of $T_1 + T_2$ is contained in the union of the supports of T_1 and T_2. (By contrast the support of a member of the dual of l^∞ does not behave so reasonably.)

Proposition B.12. Let U_α be open sets in X. If a member T of $\mathcal{E}'(X)$ vanishes on $C^\infty_{\text{com}}(U_\alpha)$ for all α, then it vanishes on $C^\infty_{\text{com}}(\cup_\alpha U_\alpha)$.

PROOF. If f is in $C^\infty_{\text{com}}(\cup_\alpha U_\alpha)$, then the compactness of support(f) means that f is in C^∞_{com} of a finite union of U_α's. Hence it is enough to prove the lemma for two open sets, U_1 and U_2, and their union.

Thus suppose T vanishes on $C^\infty_{\text{com}}(U_1)$ and $C^\infty_{\text{com}}(U_2)$, and let f be in $C^\infty_{\text{com}}(U_1 \cup U_2)$. Choose an open set V in X with compact closure such that

$$\text{support}(f) \subseteq V \subseteq \overline{V} \subseteq U_1 \cup U_2.$$

Put $V_1 = U_1 \cap V$ and $V_2 = U_2 \cap V$, and let $\{p_1, p_2, p_0\}$ be a smooth partition of unity subordinate to the cover $\{V_1, V_2, X - \text{support}(f)\}$ of X. Then $f = p_1 f + p_2 f$ with $p_1 f \in C^\infty_{\text{com}}(V_1)$ and $p_2 f \in C^\infty_{\text{com}}(V_2)$. Since $T(p_1 f) = T(p_2 f) = 0$, we obtain $T(f) = 0$.

Corollary B.13. If T is in $\mathcal{E}'(X)$, then T vanishes on the functions in $C^\infty_{\text{com}}((\text{support } T)^c)$.

PROOF. By definition of support, there are open sets U_α with union $(\text{support } T)^c$ such that T vanishes on $C^\infty_{\text{com}}(U_\alpha)$ for all α. Thus we can apply Proposition B.12.

Corollary B.14. If $T \in \mathcal{E}'(X)$ has support K and if $f \in C^\infty(X)$ vanishes in a neighborhood of K, then $T(f) = 0$.

PROOF. Let $\{K_j\}$ be an exhausting sequence, and let p_j be a member of $C^\infty_{\text{com}}(X)$ that is 1 on K_j. Then fp_j is in $C^\infty_{\text{com}}((\text{support } T)^c)$, and $fp_j \to f$ in $C^\infty(X)$. By Corollary B.13, $T(fp_j) = 0$. Since T is continuous, $T(f) = 0$.

Proposition B.15. Each member T of $\mathcal{E}'(X)$ has compact support.

PROOF. Let $\{K_j\}$ be an exhausting sequence. We argue by contradiction. If support(T) is not compact, certainly support(T) is not contained in any K_j. Thus choose f_j in $C^\infty_{\text{com}}(K_j^c)$ with $T(f_j) \neq 0$. Put $h_j = T(f_j)^{-1} f_j$, so that $T(h_j) = 1$. If $K \subseteq X$ is compact, then $K \subseteq K_j$ for j sufficiently large, and $h_j|_K = 0$ for such j. Hence h_j tends to 0 in $C^\infty(X)$, while $T(h_j)$ is identically 1 and does not tend to $T(0) = 0$. Thus T is not continuous, contradiction.

The relationship between the support K of a distribution T and the compact sets L that appear in (B.10) is subtle. In general, it is not enough in (B.10) to use only sets L that are contained in K. The next proposition gives a positive result. Further discussion of this point will occur in §4.

Proposition B.16. Let $T \in \mathcal{E}'(X)$ have compact support K, and let K' be any compact neighborhood of K. If T satisfies an inequality

$$|T(f)| \leq C \sum_{F} \|f\|_{\varphi,U,L,\alpha},$$

then T satisfies also

$$|T(f)| \leq C' \sum_{F'} \|f\|_{\varphi,U,L\cap K',\alpha'},$$

where the (φ, U, L)'s for F' are the same as those for F but additional lower-order α''s are allowed.

PROOF. Choose $\chi \in C_{\text{com}}^{\infty}(X)$ that is 1 on a neighborhood of K and is 0 off K'. Then $(1 - \chi)f$ vanishes on a neighborhood of K, and Corollary B.14 gives $T((1 - \chi)f) = 0$. Therefore

$$T(f) = T(\chi f) + T((1 - \chi)f) = T(\chi f),$$

and we have $$|T(f)| = |T(\chi f)| \leq C \sum_{F} \|\chi f\|_{\varphi,U,L,\alpha}.$$

Using the Leibniz rule (B.1) and the fact that χ vanishes on K'^{c}, we obtain the desired inequality.

It is often useful to extend distributions to vector-valued functions. The following version of this extension will suffice for our purposes. Suppose V is a complex vector space — possibly infinite dimensional, but with no topology. If X is a manifold, define $C^{\infty}(X, V)_f$ to be the vector space of smooth functions f on X taking values in a finite-dimensional subspace of V, the subspace possibly depending on f.
Then

(B.17a) $$C^{\infty}(X, V)_f \cong C^{\infty}(X) \otimes_{\mathbb{C}} V.$$

The map from right to left is $h(x) \otimes v \mapsto h(x)v$. To see that it is onto, let $\{v_{\alpha}\}$ be a basis of V, let $f \in C^{\infty}(X, V)_f$ be given, and let $\{u_1, \ldots, u_N\}$ be a basis for a subspace of V containing the image of f. Put

$$f(x) = \sum_{j=1}^{N} f_j(x)u_j.$$

If we write $u_j = \sum_{\alpha \in S} a_{j\alpha} v_\alpha$ for a finite set S, then

$$f(x) = \sum_{\alpha \in S} \left(\sum_{j=1}^{N} f_j(x) a_{j\alpha} \right) v_\alpha,$$

exhibits f as in the image of $C^\infty(X) \otimes_{\mathbb{C}} V$.

The extension of $T \in \mathcal{E}'(X)$ to $C^\infty(X, V)_f$, also denoted T, is the map corresponding to $T \otimes 1_V$ under (B.17a). This map satisfies

(B.17b) $\qquad\qquad\qquad T : C^\infty(X, V)_f \to V.$

If $A : V \to W$ is linear, then A acts on the values of $C^\infty(X, V)_f$ and carries $C^\infty(X, V)_f$ into $C^\infty(X, W)_f$. The extensions of T to $C^\infty(X, V)_f$ and $C^\infty(X, W)_f$ are then related by

(B.17c) $\qquad\qquad\qquad T(Af) = A(Tf).$

In fact, under (B.17a), A corresponds to $1_V \otimes A$, and (B.17c) is nothing more than the equality

$$(T \otimes 1_W)(1_{C^\infty(X)} \otimes A) = T \otimes A = (1_{\mathbb{C}} \otimes A)(T \otimes 1_V).$$

With V and X as above, it is natural to ask for an interpretation of $\mathcal{E}'(X) \otimes_{\mathbb{C}} V$ analogous to the one for $C^\infty(X) \otimes_{\mathbb{C}} V$ in (B.17a). If we define $\mathcal{E}'(X, V)_f$ as the vector space of linear functions $T : C^\infty(X) \to V$ such that $T(C^\infty(X))$ is contained in a finite-dimensional subspace and T is continuous into that subspace, then the same kind of argument as for (B.17a) shows that

$$\mathcal{E}'(X) \otimes_{\mathbb{C}} V \cong \mathcal{E}'(X, V)_f.$$

It turns out that this is not such a helpful interpretation of $\mathcal{E}'(X) \otimes_{\mathbb{C}} V$. More useful is the isomorphism

(B.18) $\qquad \mathcal{E}'(X) \otimes_{\mathbb{C}} V \cong \mathcal{E}'(X) \otimes_{C^\infty(X)} C^\infty(X, V)_f,$

where the commutative algebra $C^\infty(X)$ acts on $\mathcal{E}'(X)$ by (B.11) and on $C^\infty(X, V)_f$ by pointwise multiplication. To prove (B.18), we simply combine (A.7) and (B.17a), writing

$$\begin{aligned} \mathcal{E}'(X) \otimes_{\mathbb{C}} V &\cong (\mathcal{E}'(X) \otimes_{C^\infty(X)} C^\infty(X)) \otimes_{\mathbb{C}} V \\ &\cong \mathcal{E}'(X) \otimes_{C^\infty(X)} (C^\infty(X)) \otimes_{\mathbb{C}} V) \\ &\cong \mathcal{E}'(X) \otimes_{C^\infty(X)} C^\infty(X, V)_f. \end{aligned}$$

In (B.18) the isomorphism from left to right is implemented by $T \otimes v \mapsto T \otimes v$, the v on the right being a constant function in $C^\infty(X, V)_f$. In other words, the map from left to right is an obvious inclusion. The point is that the inclusion is onto. The inverse is given as follows. If $v(\cdot)$ is in $C^\infty(X, V)_f$, let $\{v_i\}$ be a basis for a finite-dimensional subspace of V containing the image of $v(\cdot)$, and let $\{v_i^*\}$ be the dual basis for the dual space, so that

(B.19a) $$v(x) = \sum_i \langle v(x), v_i^* \rangle v_i.$$

Then

$$T \otimes v(\cdot) = \sum_i T \otimes \langle v(\cdot), v_i^* \rangle v_i = \sum_i \langle v(\cdot), v_i^* \rangle T \otimes v_i$$

within $\mathcal{E}'(X) \otimes_{C^\infty(X)} C^\infty(X, V)_f$, where $\langle v(\cdot), v_i^* \rangle T$ has the meaning in (B.11). Hence the inverse map is

(B.19b) $$T \otimes v(\cdot) \mapsto \sum_i \langle v(\cdot), v_i^* \rangle T \otimes v_i.$$

3. Fubini's Theorem

Fubini's Theorem for distributions addresses the evaluation of product distributions on the product $X \times Y$ of two manifolds. If S and T are given on X and Y, respectively, the theorem constructs $S \times T$ on $X \times Y$ and says that $S \times T$ can be evaluated in either order by computing the effect of one factor and then the effect of the other.

This result is, of course, an analog of the corresponding result in measure theory. It is often helpful to keep in mind the parallel between measures and distributions. In notation that emphasizes this parallel, we often write

$$Sf(x) = \int_X f(x) \, dS(x)$$

if f is in $C^\infty(X)$ and S is in $\mathcal{E}'(X)$. The right side here means nothing more than the left side, but the integral notation does help keep track of changes of variables and of the ingredients of Fubini's Theorem.

If h_X is in $C^\infty(X)$ and h_Y is in $C^\infty(Y)$, we can build a **product function** in $C^\infty(X \times Y)$ by the definition

$$(h_X \times h_Y)(x, y) = h_X(x) h_Y(y).$$

The condition on our product distribution $S \times T$ is that its effect on $h_X \times h_Y$ is to yield $S(h_X) T(h_Y)$.

Theorem B.20 (Fubini's Theorem). Let S and T be distributions of compact support on X and Y, respectively. Then

(a) the operators

$$\mathcal{S}f(y) = S(f(\cdot, y)) = \int_X f(x, y)\, dS(x)$$

and

$$\mathcal{T}f(x) = T(f(x, \cdot)) = \int_Y f(x, y)\, dT(y)$$

carry $C^\infty(X \times Y)$ continuously to $C^\infty(Y)$ and $C^\infty(X)$, respectively.

(b) there exists a unique distribution $S \times T$ of compact support on $X \times Y$ such that

$$(S \times T)(h_X \times h_Y) = S(h_X)T(h_Y)$$

for all $h_X \in C^\infty(X)$ and $h_Y \in C^\infty(Y)$, and $S \times T$ has the property that

$$\int_{X \times Y} f(x, y)\, d(S \times T)(x, y) = \int_X \left[\int_Y f(x, y)\, dT(y) \right] dS(x)$$

$$= \int_Y \left[\int_X f(x, y)\, dS(x) \right] dT(y)$$

for all $f \in C^\infty(X \times Y)$.

The proof will be achieved through the two lemmas below. The first says that sums of product functions, on which $S \times T$ has a known value, are dense in $C^\infty(X \times Y)$. Consequently the first lemma establishes the uniqueness.

The second lemma gives conclusion (a) of the theorem for $\mathcal{S}$, and then the conclusion for $\mathcal{T}$ follows just by change of notation. As a consequence,

(B.21) $\qquad f \to T(\mathcal{S}f) \qquad$ and $\qquad f \to S(\mathcal{T}f)$

are in $\mathcal{E}'(X \times Y)$. Since they have the required form for $S \times T$ on product functions, each one establishes the existence of $S \times T$. Because of the uniqueness, they must be equal. This equality is the asserted property of $S \times T$ in conclusion (b).

Lemma B.22. The subspace of all sums of functions $h_X \times h_Y$, with $h_X \in C^\infty(X)$ and $h_Y \in C^\infty(Y)$, is dense in $C^\infty(X \times Y)$.

PROOF. We begin with two reductions. The subspace $C^\infty_{\text{com}}(X \times Y)$ is dense in $C^\infty(X \times Y)$. (In fact, if K_j is an exhausting sequence in $X \times Y$ and if $p_j \in C^\infty_{\text{com}}(X \times Y)$ is 1 on K_j, then $fp_j \to f$ for all $f \in C^\infty(X \times Y)$.) Consequently it is enough to approximate members of $C^\infty_{\text{com}}(X \times Y)$ by sums of product functions.

Next the availability of smooth partitions of unity reduces the lemma to a local result in Euclidean space, provided we insist that the approximating functions have compact support, so that they extend to global C^∞ functions.

Thus we may assume that X and Y are cubes in Euclidean spaces and that f is in $C^\infty_{\text{com}}(X \times Y)$. Without loss of generality, we may take $X = (-\pi, \pi)^m$ and $Y = (-\pi, \pi)^n$. Choose $p_1 \in C^\infty_{\text{com}}(X)$ and $p_2 \in C^\infty_{\text{com}}(Y)$ such that

$$f(x, y) = p_1(x)p_2(y)f(x, y).$$

Let us write the multiple Fourier series of f as

$$f(x, y) \sim \sum_{\mathbf{j},\mathbf{k}} c_{\mathbf{j},\mathbf{k}} e^{i\mathbf{j}\cdot x} e^{i\mathbf{k}\cdot y}.$$

If we put

$$s_N(x, y) = \sum_{|\mathbf{j}| \leq N, \, |\mathbf{k}| \leq N} c_{\mathbf{j},\mathbf{k}} e^{i\mathbf{j}\cdot x} e^{i\mathbf{k}\cdot y},$$

then s_N converges to f uniformly on $X \times Y$ and so do its partial derivatives of all orders. However, s_N does not meet our requirement of having compact support. Thus we put

$$f_N(x, y) = p_1(x)p_2(y)s_N(x, y).$$

Then f_N is a sum of product functions, and use of the Leibniz rule shows that $f_N \to f$ in $C^\infty(X \times Y)$.

Lemma B.23. If S is a distribution of compact support on X, then

$$\mathcal{S}f(y) = S(f(\cdot, y))$$

carries $C^\infty(X \times Y)$ continuously into $C^\infty(Y)$.

PROOF. Let $\{(\varphi_m, U_m)\}$ be a locally finite cover of X by compatible charts such that each $\overline{U}_m$ is compact, and let $\{p_m\}$ be a smooth partition of unity subordinate to this cover. Choose N so that $\sum_{m \leq N} p_m = 1$ on support(S). Using the Leibniz rule, we see that $h \to S(p_m h)$ is a distribution on X whose support is a compact subset of U_m. Evidently $S = \sum_{m \leq N} S(p_m \cdot)$, and thus it is enough to consider a single $S(p_m \cdot)$.

Changing notation, we may assume that (φ, U) is a compatible chart for X with $\overline{U}$ compact, that $p \in C^\infty(X)$ is compactly supported in U, and that we are to consider $S_p = S(p \cdot)$ in place of S. Let K' be a compact neighborhood of support(p) contained in U, so that Proposition B.16 applies to S_p and K'. For $h \in C^\infty(\varphi(U))$, let $\overline{S}_p(h) = S(p(h \circ \varphi^{-1}))$; this is the local version of S_p in the chart (φ, U).

Fix $y \in Y$, and let (ψ, V) be a compatible chart for Y with $y \in V$. Put $\bar{y} = \psi(y)$, and let the vector v be so small that the closed ball of radius $|v|$ centered at $\bar{y}$ lies in $\psi(V)$. Fox $x \in U$, let $\bar{x} = \varphi(x)$, and put $\bar{f} = f \circ (\varphi^{-1} \times \psi^{-1})$. Applying Taylor's formula in one variable to $\bar{f}(\bar{x}, \bar{y} + tv)$ and then taking $t = 1$, we have

$$\bar{f}(\bar{x}, \bar{y} + v) = \bar{f}(\bar{x}, \bar{y}) + \sum_j v_j \frac{\partial \bar{f}}{\partial \bar{y}_j}(\bar{x}, \bar{y})$$

$$+ \int_0^1 \sum_{j,k} v_j v_k \frac{\partial^2 \bar{f}}{\partial \bar{y}_j \partial \bar{y}_k}(\bar{x}, \bar{y} + sv)(1 - s)\, ds.$$

Writing $r(\bar{x}, \bar{y}, v)$ for the remainder term, we see that the X seminorms of r satisfy

$$\|r(\cdot, \bar{y}, v)\|_{\varphi(U), \varphi(K'), \alpha} \leq C(\bar{f}, \bar{y}, \alpha)|v|^2$$

for constants $C(\bar{f}, \bar{y}, \alpha)$. Thus we have

$$(\text{B.24}) \quad \overline{S}_p(\bar{f}(\cdot, \bar{y} + v)) = \overline{S}_p(\bar{f}(\cdot, \bar{y})) + \sum_j v_j \overline{S}_p\left(\frac{\partial \bar{f}}{\partial \bar{y}_j}(\cdot, \bar{y})\right) + O(|v|^2),$$

with the error term depending on $\bar{f}$ and $\bar{y}$.

Formula (B.24) proves that $\overline{S}_p(\bar{f}(\cdot, \bar{y}))$ is differentiable at $\bar{y}$ and that

$$\frac{\partial}{\partial \bar{y}_j} \overline{S}_p(\bar{f}(\cdot, \bar{y})) = \overline{S}_p\left(\frac{\partial \bar{f}}{\partial \bar{y}_j}(\cdot, \bar{y})\right).$$

Allowing $\bar{y}$ to move and then iterating, we obtain

$$(\text{B.25}) \quad \partial_{\bar{y}}^\beta \overline{S}_p(\bar{f}(\cdot, \bar{y})) = \overline{S}_p(\partial_{\bar{y}}^\beta f(\cdot, \bar{y}))$$

for all β. In particular, $\bar{y} \rightarrow \overline{S}_p(\bar{f}(\,\cdot\,, \bar{y}))$ is in $C^\infty(\psi(V))$. Hence the composition

$$y \longrightarrow \psi(y) = \bar{y} \longrightarrow \overline{S}_p(\bar{f}(\,\cdot\,, \bar{y})) = S_p(f(\,\cdot\,, y))$$

is in $C^\infty(V)$. Since the chart (ψ, V) is arbitrary, $y \rightarrow S_p(f(\,\cdot\,, y))$ is in $C^\infty(Y)$.

To prove that the resulting map of $C^\infty(X \times Y)$ to $C^\infty(Y)$ is continuous, it is enough to estimate

$$\|S_p(f(\,\cdot\,, y))\|_{\psi, V, M, \beta}$$

for a chart (ψ, V) with $\overline{V}$ compact in Y and with M compact in V. We have

$$\begin{aligned}
\|S_p(f(\,\cdot\,, y))\|_{\psi, V, M, \beta} &= \|S_p(f(\,\cdot\,, \psi^{-1}\bar{y}))\|_{\psi(V), \psi(M), \beta} \\
&= \|\overline{S}_p(\bar{f}(\,\cdot\,, \bar{y}))\|_{\psi(V), \psi(M), \beta} \\
&= \sup_{\bar{y} \in \psi(M)} |\partial_{\bar{y}}^\beta \overline{S}_p(\bar{f}(\,\cdot\,, \bar{y}))| \\
&= \sup_{\bar{y} \in \psi(M)} |\overline{S}_p(\partial_{\bar{y}}^\beta \bar{f}(\,\cdot\,, \bar{y}))| \qquad \text{by (B.25)} \\
&\leq C \sum_\alpha \sup_{(\bar{x}, \bar{y}) \in \varphi(K') \times \psi(M)} |\partial_{\bar{x}}^\alpha \partial_{\bar{y}}^\beta \bar{f}(\bar{x}, \bar{y})|,
\end{aligned}$$

the last step holding since $\overline{S}_p$ is in $\mathcal{E}'(\varphi(U))$ with support contained in $\varphi(\text{support } p)$. The right side is

$$= C \sum_\alpha \|f\|_{\varphi \times \psi, U \times V, K' \times M, \alpha + \beta},$$

and the desired continuity follows from (B.6).

4. Distributions Supported on Submanifolds

If $\varphi : W \rightarrow X$ is a smooth map between manifolds, then φ induces a **push-forward** map $T \mapsto \varphi_*(T)$ of $\mathcal{E}'(W)$ into $\mathcal{E}'(X)$, given by

(B.26) $$\varphi_*(T)(f) = T(f \circ \varphi).$$

Continuity of $\varphi_*(T)$ is an easy consequence of (B.2). If $\psi : X \to Y$ is a second smooth map, then it is clear that

(B.27)
$$(\psi \circ \varphi)_* = \psi_* \circ \varphi_*.$$

In case φ is one-one and is regarded as an inclusion, we can ask about a reverse construction, namely the extent to which members of $\mathcal{E}'(X)$ that are supported on W can be constructed in a simple way from $\mathcal{E}'(W)$. This question is intimately connected with the question whether the compact sets L that appear in (B.10) can be taken to be subsets of the support.

In this section we shall give a positive answer to this question in the case that X is the product of W and another manifold, hence a positive answer locally whenever W is an immersed submanifold of X. Except for notation, the result is really a statement about distributions in Euclidean space, and we shall concentrate on that situation first. Note that if $\varphi : W \to W \times W'$ is the inclusion $w \mapsto (w, p)$, then $\varphi_*(T) = T \times \delta_p$, where δ_p is the evaluation-at-p distribution.

It will be convenient to refer to the "order" of a distribution of compact support. If T is such a distribution, the maximum of $|\alpha|$ for α appearing in the finitely many tuples (φ, U, L, α) of (B.10) is a candidate for the order. The **order** of T is the least candidate among all valid inequalities (B.10).

We shall work first with distributions in Euclidean space, and then the only (φ, U) required in (B.10) is the identity chart. Moreover, L can be taken to be a fixed compact neighborhood of the support, by Proposition B.16.

Theorem B.28 (Schwartz). Let T be a member of $\mathcal{E}'(\mathbb{R}^m \times \mathbb{R}^n)$ of order k that is supported in $\mathbb{R}^m \times \{0\}$. Then T has a decomposition

$$T = \sum_{|\beta| \le k} T_\beta \times \partial^\beta,$$

where T_β is in $\mathcal{E}'(\mathbb{R}^m)$ and ∂^β is the βth derivative at 0 in $\mathbb{R}^n$.

REMARK. One can prove also that T_β has order $\le k - |\beta|$, but we shall not need this more refined result.

PROOF. The main step will be to show for a function $r(x, y)$ that

(B.29) $\partial_x^\alpha \partial_y^\beta r(\cdot, 0) = 0$ whenever $|\alpha + \beta| \le k$ $\implies$ $T(r) = 0.$

Assuming (B.29), let f be given and form the Taylor expansion in y about 0:

(B.30)
$$f(x, y) = \sum_{\alpha=0,\, |\beta| \le k} \partial_y^\beta f(x, 0) y^\beta / \beta! + r(x, y),$$

where $y^\beta/\beta! = (y_1^{\beta_1}/\beta_1!) \cdots (y_n^{\beta_n}/\beta_n!)$. We apply $\partial_x^\alpha \partial_y^\beta$ to both sides. At $y = 0$, $\partial_x^\alpha \partial_y^\beta$ picks off the αth derivative of the βth term in the sum, and we see as a consequence that

$$\partial_x^\alpha \partial_y^\beta r(x, 0) = 0 \qquad \text{for } x \in \mathbb{R}^m \text{ and for } |\alpha + \beta| \le k.$$

By (B.29), $T(r) = 0$. If we apply T to (B.30), we therefore obtain

$$T(f) = \sum_{|\beta| \le k} T(\partial_y^\beta f(x, 0) y^\beta/\beta!).$$

Define T_β on $C^\infty(\mathbb{R}^m)$ by $T_\beta(h) = T(h(x)y^\beta/\beta!)$. Then Fubini's Theorem gives

$$(T_\beta \times \partial^\beta)(f) = \int_{\mathbb{R}^n} \left[\int_{\mathbb{R}^m} f(x, y) \, d\partial^\beta(y) \right] dT_\beta(x)$$
$$= T_\beta(\partial_y^\beta f(x, 0)) = T(\partial_y^\beta f(x, 0) y^\beta/\beta!),$$

and T has the form required by the theorem.

Thus it is enough to prove (B.29). Suppose T is compactly supported in $\{(x, 0) \mid |x| < N\}$. Let $\chi \in C^\infty(\mathbb{R}^n)$ depend only on $|y|$ and be such that $\chi(y) = 1$ for y near 0 and $\chi(y) = 0$ for $|y| \ge 1$. For $\epsilon > 0$, define $\chi_\epsilon(y) = \chi(\epsilon^{-1}y)$. Then

(B.31)
$$\sup_{y \in \mathbb{R}^n} |\partial^\beta \chi_\epsilon(y)| \le C_\beta \epsilon^{-|\beta|}$$

with C_β independent of y and ϵ.

By Corollary B.14, we have $T(r) = T(r\chi_\epsilon)$. Inequality (B.10) and Proposition B.16 give

$$|T(r)| = |T(r\chi_\epsilon)| \le C \sum_{|\alpha+\beta| \le k} \sup_{|x| \le N, |y| \le 1} |\partial_x^\alpha \partial_y^\beta (r\chi_\epsilon)|$$
$$\le C' \sum_{|\alpha+\beta'+\beta''| \le k} \sup_{|x| \le N, |y| \le \epsilon} |\partial_x^\alpha \partial_y^{\beta'} r| \sup_y |\partial_y^{\beta''} \chi_\epsilon|$$
$$\le C'' \sum_{|\alpha+\beta'| \le k} \epsilon^{|\alpha+\beta'|-k} \sup_{|x| \le N, |y| \le \epsilon} |\partial_x^\alpha \partial_y^{\beta'} r|,$$

by (B.31). If we can prove that

(B.32)
$$\lim_{\epsilon \downarrow 0} \left(\epsilon^{|\alpha+\beta|-k} \sup_{|x| \le N, |y| \le \epsilon} |\partial_x^\alpha \partial_y^{\beta'} r| \right) = 0$$

for $|\alpha + \beta| \leq k$, then we obtain $T(r) = 0$, which is the conclusion of (B.29).

Fix x, apply Taylor's formula with integral remainder in the variable t about $t = 0$ to $\partial_y^\beta r(x, ty)$, and put $t = 1$. The result is

$$\partial_y^\beta r(x, y) =$$

$$\partial_y^\beta r(x, 0) + \sum y_j \left(\frac{\partial}{\partial y_j}\right)(\partial_y^\beta r)(x, 0) + \cdots$$

$$+ \frac{1}{l!} \sum y_{j_1} \cdots y_{j_l} \left(\frac{\partial^l}{\partial y_{j_1} \cdots \partial y_{j_l}}\right)(\partial_y^\beta r)(x, 0)$$

$$+ \frac{1}{l!} \int_0^1 \sum_{j_1, \ldots, j_{l+1}} y_{j_1} \cdots y_{j_{l+1}} \left(\frac{\partial^{l+1}}{\partial y_{j_1} \cdots \partial y_{j_{l+1}}}\right)(\partial_y^\beta r)(x, sy) (1 - s)^l \, ds.$$

Taking $l = k - |\alpha| - |\beta|$ and applying ∂_x^α, we see that all terms on the right are 0 except for ∂_x^α of the integral, as a consequence of the assumption in (B.29). For $|x| \leq N$ and $|y| \leq \epsilon$, the integrand is bounded by a multiple of $\epsilon^{l+1} = \epsilon^{k+1-|\alpha|-|\beta|}$. Hence

$$\sup_{|x| \leq N, \, |y| \leq \epsilon} |\partial_x^\alpha \partial_y^\beta r(x, y)| \leq C''' \epsilon^{k+1-|\alpha|-|\beta|},$$

and (B.32) follows. This proves (B.29) and hence also the theorem.

Corollary B.33. Let T be a distribution of order k on $X \times Y$ with compact support in $X \times \{p\}$. Let (φ, U) be a chart about p, and let D^β be the distribution $(\varphi^{-1})_*(h \mapsto \partial^\beta h(\varphi(p)))$ on Y corresponding to ∂^β at $\varphi(p)$. Then T has a decomposition

$$T = \sum_{|\beta| \leq k} T_\beta \times D^\beta,$$

where T_β is in $\mathcal{E}'(X)$.

APPENDIX C

ELEMENTARY HOMOLOGICAL ALGEBRA

This appendix collects a number of definitions and results in elementary homological algebra. Proofs are omitted in the first four sections. Detailed references appear in the Notes, and a general reference is Knapp [1988]. The topics are good categories, complexes and chain maps, projectives and injectives, functors, exactness and one-sided exactness, natural isomorphisms, derived functors, connecting homomorphisms, long exact sequences, long exact sequences of derived functors, Grothendieck groups, the Euler-Poincaré Principle, and finite length.

1. Projectives and Injectives

In this section after (C.1), it is always assumed that C is a good category of R modules in the sense of (C.1).

(C.1) Let R be a ring, with or without identity. Recall from §A.1 that a **good category** C of R modules consists of

(i) some class of left R modules closed under passage to submodules, quotients, and finite direct sums (the **modules** of the category), and

(ii) the sets $\mathrm{Hom}_R(A, B)$ of *all* R linear maps from A to B, for each A and B in (i) (the **maps** of the category).

When R is an associative algebra over a field k, the members of $\mathrm{Hom}_R(A, B)$ are assumed to be k linear.

(C.2) A **complex** in C is a finite or infinite sequence of modules and maps in C of one of the forms

$$X: \quad \cdots \longleftarrow X_{n-1} \xleftarrow{\partial_{n-1}} X_n \xleftarrow{\partial_n} X_{n+1} \longleftarrow \cdots$$

$$Y: \quad \cdots \longrightarrow Y_{n-1} \xrightarrow{d_{n-1}} Y_n \xrightarrow{d_n} Y_{n+1} \longrightarrow \cdots$$

with $\partial_{n-1}\partial_n = 0$ or $d_n d_{n-1} = 0$, respectively, for all n. The **homology** of a complex like X with decreasing indices is

$$H_n(X) = \ker \partial_{n-1}/\mathrm{image}\ \partial_n.$$

and the **cohomology** of a complex like Y with increasing indices is

$$H^n(Y) = \ker d_n / \text{image } d_{n-1}.$$

The two kinds of complexes come to the same thing by replacing n by $-n$. The complex X is **exact at** X_n if image $\partial_n = \ker \partial_{n-1}$. It is an **exact sequence** or **exact complex** if it is exact at every X_n.

(C.3) In a diagram in $\mathcal{C}$ of the form

$$X: \quad \cdots \longleftarrow X_{n-1} \xleftarrow{\partial_{n-1}} X_n \xleftarrow{\partial_n} X_{n+1} \longleftarrow \cdots$$
$$\qquad \qquad F_{n-1} \downarrow \qquad F_n \downarrow \qquad F_{n+1} \downarrow$$
$$X': \quad \cdots \longleftarrow X'_{n-1} \xleftarrow{\partial'_{n-1}} X'_n \xleftarrow{\partial'_n} X'_{n+1} \longleftarrow \cdots$$

in which X and X' are complexes and the squares commute, we say that $F = \{F_n\}$ is a **chain map** (or **cochain map** if the indices are increasing). A chain map $F = \{F_n\}$ yields maps on homology also denoted by F, namely $F_n : H_n(X) \to H_n(X')$. In the above diagram if $X_{-2} = X'_{-2} = 0$, $X_{-1} = M$, $X'_{-1} = M'$, and $F_{-1} = f$, we refer to a **chain map** over f. In the corresponding situation with increasing indices, we refer to a **cochain map** over f.

(C.4) Let $F : X \to X'$ and $G : X \to X'$ be chain maps between complexes with decreasing indices. A system $s = \{s_n\}$ of maps in $\mathcal{C}$ with $s_n : X_n \to X'_{n+1}$ is a **homotopy** between F and G if

$$\partial'_n s_n + s_{n-1} \partial_{n-1} = F_n - G_n$$

for all n. For the case of cochain maps between complexes with increasing indices, a homotopy is to have $s_n : X_n \to X'_{n-1}$ with

$$d_{n-1} s_n + d_n s_{n+1} = F_n - G_n$$

for all n. Homotopic chain or cochain maps induce the same maps on homology or cohomology.

(C.5) A module P in $\mathcal{C}$ is **projective** for $\mathcal{C}$ is whenever a diagram

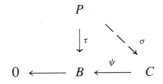

$$0 \longleftarrow B \xleftarrow{\psi} C$$

in C has ψ mapping onto B, then there exists $\sigma : P \to C$ such that the diagram commutes. If P is projective and if the diagram

$$P$$

$$A' \xleftarrow{\varphi} A \xleftarrow{\psi} A''$$

in C is exact at A (i.e., has $\ker \varphi = \text{image } \psi$) and if $\varphi\tau = 0$, then there exists $\sigma : P \to A''$ such that the diagram commutes.

(C.6) Within C an exact sequence

$$0 \longleftarrow M \longleftarrow X_0 \longleftarrow X_1 \longleftarrow X_2 \longleftarrow \ \ldots$$

is called a **projective resolution** of M if $X_0, X_1, X_2, \ldots$ are all projective. Suppose that

$$X: \quad 0 \longleftarrow M \longleftarrow X_0 \longleftarrow X_1 \longleftarrow X_2 \longleftarrow \ \ldots$$

$$\downarrow f \qquad\quad \downarrow F_0 \qquad \downarrow F_1 \qquad \downarrow F_2$$

$$X': \quad 0 \longleftarrow M' \longleftarrow X'_0 \longleftarrow X'_1 \longleftarrow X'_2 \longleftarrow \ \ldots$$

is a diagram in C in which X and X' are complexes, X' is exact, and all X_n are projective for $n \geq 0$. Then there exists a chain map $F : X \to X'$ over f. If $F : X \to X'$ and $G : X \to X'$ are two chain maps over f, then F and G are homotopic.

(C.7) A module I in C is **injective** for C if whenever a diagram

$$I$$

$$0 \longrightarrow B \xrightarrow{\varphi} C$$

in C has φ mapping one-one into C, then there exists $\sigma : C \to I$ such that the diagram commutes. If I is injective and if the diagram

$$I$$

$$A' \xrightarrow{\psi} A \xrightarrow{\varphi} A''$$

in $\mathcal{C}$ is exact at A (i.e., has $\ker \varphi = \text{image } \psi$), and if $\tau \psi = 0$, then there exists $\sigma : A'' \to I$ such that the diagram commutes.

(C.8) Within $\mathcal{C}$ an exact sequence

$$0 \longrightarrow M \longrightarrow X_0 \longrightarrow X_1 \longrightarrow X_2 \longrightarrow \cdots$$

is called an **injective resolution** of M if $X_0, X_1, X_2, \ldots$ are all injective. Suppose that

$$
\begin{array}{ccccccccc}
X: & 0 \longrightarrow & M & \longrightarrow & X_0 & \longrightarrow & X_1 & \longrightarrow & X_2 & \longrightarrow \cdots \\
 & & \downarrow{\scriptstyle f} & & \downarrow{\scriptstyle F_0} & & \downarrow{\scriptstyle F_1} & & \downarrow{\scriptstyle F_2} & \\
X': & 0 \longrightarrow & M' & \longrightarrow & X'_0 & \longrightarrow & X'_1 & \longrightarrow & X'_2 & \longrightarrow \cdots
\end{array}
$$

is a diagram in $\mathcal{C}$ in which X and X' are complexes, X is exact, and all X'_n are injective for $n \geq 0$. Then there exists a cochain map $F : X \to X'$ over f. If $F : X \to X'$ and $G : X \to X'$ are two cochain maps over f, then F and G are homotopic.

2. Functors

In this section we generally work with two good categories. Typically $\mathcal{C}$ is a good category of R modules, and $\mathcal{C}'$ is a good category of S modules.

(C.9) A **covariant functor** $F : \mathcal{C} \to \mathcal{C}'$ is an assignment of a module $F(A)$ of $\mathcal{C}'$ to each module A of $\mathcal{C}$ and of a map $F(\varphi)$ in $\text{Hom}_S(F(A), F(B))$ to each map φ in $\text{Hom}_R(A, B)$ in such a way that

(i) the identity map of $\text{Hom}_R(A, A)$ is carried to the identity map of $\text{Hom}_S(F(A), F(A))$, and

(ii) $F(\varphi_2 \varphi_1) = F(\varphi_2) F(\varphi_1)$ whenever φ_1 is in $\text{Hom}_R(A, B)$ and φ_2 is in $\text{Hom}_R(B, C)$.

A **contravariant functor** $F : \mathcal{C} \to \mathcal{C}'$ is an assignment of a module $F(A)$ of $\mathcal{C}'$ to each module A of $\mathcal{C}$ and of a map $F(\varphi)$ in $\text{Hom}_S(F(B), F(A))$ to each map φ in $\text{Hom}_R(A, B)$ in such a way that (i) holds above and

(ii') $F(\varphi_2 \varphi_1) = F(\varphi_1) F(\varphi_2)$ whenever φ_1 is in $\text{Hom}_R(A, B)$ and φ_2 is in $\text{Hom}_R(B, C)$.

(C.10) A functor $F : \mathcal{C} \to \mathcal{C}'$ is **additive** if

$$F(\varphi_1 + \varphi_2) = F(\varphi_1) + F(\varphi_2)$$

whenever φ_1 and φ_2 are in the same $\mathrm{Hom}_R(A, B)$. Any functor in a hypothesis below is tacitly assumed to be additive. Such functors send the 0 map to the 0 map, complexes to complexes, 0 modules to 0 modules, and finite direct sums to finite direct sums.

(C.11) A functor $F : \mathcal{C} \to \mathcal{C}'$ is **exact** if it transforms exact sequences into exact sequences. A functor is exact if and only if it transforms **short exact sequences**, i.e., exact sequences of the form

$$0 \longrightarrow A \longrightarrow B \longrightarrow C \longrightarrow 0$$

into exact sequences. If X is a complex and F is an exact functor, then F carries the homology or cohomology of X to the homology or cohomology of $F(X)$.

(C.12) A functor F is **left exact** if whenever

$$0 \longrightarrow A \longrightarrow B \longrightarrow C \longrightarrow 0 \qquad (F \text{ covariant})$$

$$0 \longleftarrow A \longleftarrow B \longleftarrow C \longleftarrow 0 \qquad (F \text{ contravariant})$$

is short exact, then

$$0 \longrightarrow F(A) \longrightarrow F(B) \longrightarrow F(C)$$

is exact. It is **right exact** if whenever the above five-term sequence is short exact, then

$$F(A) \longrightarrow F(B) \longrightarrow F(C) \longrightarrow 0$$

is exact. If F is left exact, then the exactness of the four-term sequence from $0, A, B, C$ implies the exactness of the above four-term sequence from $0, F(A), F(B), F(C)$. An analogous conclusion is valid for right exact F.

(C.13) With $\mathcal{C}$ denoting a good category of R modules and with $\mathcal{C}'$ denoting the good category of all $\mathbb{Z}$ modules (abelian groups),

  (i) if V is in $\mathcal{C}$, then $\mathrm{Hom}_R(\cdot, V)$ is a functor from $\mathcal{C}$ to $\mathcal{C}'$, and it is contravariant and left exact.
  (ii) if U is in $\mathcal{C}$, then $\mathrm{Hom}_R(U, \cdot)$ is a functor from $\mathcal{C}$ to $\mathcal{C}'$, and it is covariant and left exact.
  (iii) if U is a right R module, then $U \otimes_R (\cdot)$ is a functor from $\mathcal{C}$ to $\mathcal{C}'$, and it is covariant and right exact.

If R is an associative algebra over a field k, then the target category for the three functors $\mathrm{Hom}_R(\cdot, V)$, $\mathrm{Hom}_R(U, \cdot)$, and $U \otimes_R (\cdot)$ may be taken to be the good category of all k modules (vector spaces over k), according to the conventions at the end of §A.1.

(C.14) A module P in $\mathcal{C}$ is projective if and only if $\mathrm{Hom}_R(P, \cdot)$ is an exact functor. A module I in $\mathcal{C}$ is injective if and only if $\mathrm{Hom}_R(\cdot, I)$ is an exact functor.

(C.15) In the good category of all vector spaces over a field k, every module is projective and injective. If V is a vector space over k, then the three functors $\mathrm{Hom}_k(\cdot, V)$, $\mathrm{Hom}_k(V, \cdot)$, and $V \otimes_k (\cdot)$ are all exact.

(C.16) Let $F : \mathcal{C} \to \mathcal{C}'$ and $G : \mathcal{C} \to \mathcal{C}'$ be functors, either both covariant or both contravariant. A **natural map** $T = \{T_A\}$ from F to G is a system of maps in $\mathcal{C}'$, namely $T_A : F(A) \to G(A)$ for each module A in $\mathcal{C}$, such that whenever $\varphi : A \to A'$ is a map in $\mathcal{C}$, then the appropriate one of the following diagrams commutes:

$$
\begin{array}{ccc}
F(A) & \xrightarrow{\ T_A\ } & G(A) \\
{\scriptstyle F(\varphi)}\big\downarrow & & \big\downarrow{\scriptstyle G(\varphi)} \\
F(A') & \xrightarrow{\ T_{A'}\ } & G(A')
\end{array}
\qquad (F \text{ covariant})
$$

$$
\begin{array}{ccc}
F(A) & \xrightarrow{\ T_A\ } & G(A) \\
{\scriptstyle F(\varphi)}\big\uparrow & & \big\uparrow{\scriptstyle G(\varphi)} \\
F(A') & \xrightarrow{\ T_{A'}\ } & G(A')
\end{array}
\qquad (F \text{ contravariant})
$$

The above natural map is a **natural isomorphism** if each T_A is an isomorphism; in this case $\{T_A^{-1}\}$ is a natural map. Functors that are naturally isomorphic have the same exactness properties (exact, left exact, right exact).

(C.17) Let k be a field, and let A, B, and C be vector spaces over k. Then there is a unique k linear isomorphism Φ of left side to right side in

$$\mathrm{Hom}_k(A \otimes_k B, \, C) \cong \mathrm{Hom}_k(A, \, \mathrm{Hom}_k(B, C))$$

such that

$$\Phi(\varphi)(a)(b) = \varphi(a \otimes b)$$

for all $\varphi \in \mathrm{Hom}_k(A \otimes_k B, \, C)$, $a \in A$, and $b \in B$. Moreover, Φ is **natural** in each variable in the following sense: If any two of A, B, C are fixed and the other is allowed to vary, then Φ is a natural isomorphism of the functors $\mathrm{Hom}_k(A \otimes B, \, C)$ and $\mathrm{Hom}_k(A, \, \mathrm{Hom}_k(B, C))$.

(C.18) Let k be a field, and let A, B, and C be vector spaces over k. Then there is a unique k linear isomorphism Φ of left side to right side in

$$(A \otimes_k B) \otimes_k C = A \otimes_k (B \otimes_k C)$$

such that $\Phi((a \otimes b) \otimes c) = a \otimes (b \otimes c)$ for all $a \in A$, $b \in B$, and $c \in C$. Moreover, Φ is natural in each of the three variables.

(C.19) Let R and S be algebras over a field k. Let $(A^R, {}^R B^S, C^S)$ refer to the situation in which A is a right R module, B is a left R module and right S module such that the R action commutes with the S action, and C is a right S module. Then the isomorphism

$$\operatorname{Hom}_k(A \otimes_k B, C) \cong \operatorname{Hom}_k(A, \operatorname{Hom}_k(B, C))$$

of (C.17) induces a k vector-space isomorphism

$$\operatorname{Hom}_S(A \otimes_R B, C) \cong \operatorname{Hom}_R(A, \operatorname{Hom}_S(B, C))$$

that is natural in each variable. (It is understood that the definition of good category in (C.1) has to be enlarged to allow right modules and two-sided modules for this statement to make sense.)

(C.20) Let R and S be algebras over a field k. In the situation $({}^R A, {}^S B^R, {}^S C)$ for modules A, B, C (see (C.19)), the isomorphism

$$\operatorname{Hom}_k(B \otimes_k A, C) \cong \operatorname{Hom}_k(A \otimes_k B, C) \cong \operatorname{Hom}_k(A, \operatorname{Hom}_k(B, C))$$

built from (C.17) induces a k vector-space isomorphism

$$\operatorname{Hom}_S(B \otimes_R A, C) \cong \operatorname{Hom}_R(A, \operatorname{Hom}_S(B, C))$$

that is natural in each variable.

(C.21) Let R and S be algebras over a field k. In the situation $(A^R, {}^R B^S, {}^S C)$ for modules A, B, C (see (C.19)), the isomorphism

$$(A \otimes_k B) \otimes_k C = A \otimes_k (B \otimes_k C)$$

of (C.18) induces a k vector-space isomorphism

$$(A \otimes_R B) \otimes_S C = A \otimes_R (B \otimes_S C)$$

that is natural in each variable.

(C.22) Suppose that U and U' are modules in a good category $\mathcal{C}'$ of S modules such that

$$\operatorname{Hom}_S(U, V) \cong \operatorname{Hom}_S(U', V)$$

naturally in V. Then $U \cong U'$, the isomorphism from left to right being given by the member of $\mathrm{Hom}_S(U, U')$ that corresponds to 1 in $\mathrm{Hom}_S(U', U')$. Moreover, if $F : \mathcal{C} \to \mathcal{C}'$ and $G : \mathcal{C} \to \mathcal{C}'$ are functors and there are isomorphisms

$$\mathrm{Hom}_S(F(A), V) \cong \mathrm{Hom}_S(G(A), V)$$

natural for A in $\mathcal{C}$ and V in $\mathcal{C}'$, then $F(A) \cong G(A)$ naturally in A. Corresponding results are valid for both statements if the roles of the two variables in Hom are interchanged.

3. Derived Functors

In this section until (C.28), it is assumed that $F : \mathcal{C} \to \mathcal{C}'$ is a functor from a good category $\mathcal{C}$ of R modules to a good category $\mathcal{C}'$ of S modules. After (C.23) it is always assumed that $\mathcal{C}$ has enough projectives and enough injectives in the sense of (C.23).

(C.23)

  (a) The good category $\mathcal{C}$ has **enough projectives** if every module in $\mathcal{C}$ is a quotient of a projective in $\mathcal{C}$. In this case every module M in $\mathcal{C}$ has a projective resolution in $\mathcal{C}$.
  (b) The good category $\mathcal{C}$ has **enough injectives** if every module in $\mathcal{C}$ is a submodule of an injective in $\mathcal{C}$. In this case every module M in $\mathcal{C}$ has an injective resolution in $\mathcal{C}$.

(C.24) The **derived functors** F_n or F^n are defined on modules as follows. If M is a module, form a projective or injective resolution of M according to the first three columns of the following table:

Exactness	Covariant or Contravariant	Resolution	F_n or F^n
right	covariant	projective	F_n
right	contravariant	injective	F_n
left	covariant	injective	F^n
left	contravariant	projective	F^n

Apply F to the resolution, drop the $F(M)$ term, and define the homology or cohomology of the resulting complex to be $F_n(M)$ or $F^n(M)$. (Which is which is indicated in the fourth column of the table.) Then $F_n(M)$ or $F^n(M)$ is well defined up to canonical isomorphism.

(C.25) The derived functors F_n or F^n are defined on maps as follows. If $\varphi : M \to M'$ is a map in $\mathcal{C}$, form resolutions X of M and X' of M' as in (C.24), connect them by φ, and form a chain or cochain map $\Phi : X \to X'$ over φ. Then

$$F(\Phi) : F(X) \to F(X') \qquad (F \text{ covariant})$$

$$F(\Phi) : F(X') \to F(X) \qquad (F \text{ contravariant})$$

is a chain or cochain map and induces maps on homology or cohomology that are denoted $F_n(\varphi)$ or $F^n(\varphi)$. Then $F_n(\varphi)$ or $F^n(\varphi)$ is well defined up to the canonical isomorphisms of (C.24).

(C.26) Under the definitions (C.24) and (C.25), the F_n's or F^n's are functors. The 0th derived functor of F is naturally isomorphic with F. Naturally isomorphic functors have naturally isomorphic derived functors. Sometimes it is convenient to compute derived functors using resolutions more general than those in (C.6) and (C.8). Suppose for definiteness in (C.24) that F is a right-exact covariant functor. A module X in $\mathcal{C}$ is called **acyclic for** F if $F_n(X) = 0$ for all $n > 0$. A projective module P is always acyclic because $F_n(P)$ can be computed from the projective resolution

$$0 \longleftarrow P \longleftarrow P \longleftarrow 0.$$

An **acyclic resolution** of M is an exact complex

$$0 \longleftarrow M \longleftarrow X_0 \longleftarrow X_1 \longleftarrow X_2 \longleftarrow \cdots$$

with X_i acyclic for F for all $i \geq 0$. In this case the nth derived functor may be calculated as the nth homology of the complex

$$0 \longleftarrow F(X_0) \longleftarrow F(X_1) \longleftarrow F(X_2) \longleftarrow \cdots$$

A parallel result is valid in each of the other cases of (C.24).

(C.27) When $G : \mathcal{C} \to \mathcal{C}'$ is a one-sided exact functor with derived functors G_n or G^n and when $F : \mathcal{C}' \to \mathcal{C}''$ is an exact functor,

(a) if F is covariant, then $F \circ G$ is one-sided exact, and its derived functors satisfy $(F \circ G)_n \cong F \circ G_n$ or $(F \circ G)^n \cong F \circ G^n$.

(b) if F is contravariant, then $F \circ G$ is one-sided exact, and its derived functors satisfy $(F \circ G)^n \cong F \circ G_n$ or $(F \circ G)_n \cong F \circ G^n$.

(C.28) Let $\tilde{\mathcal{C}}, \mathcal{C}$, and $\mathcal{C}'$ be good categories, and suppose $\tilde{\mathcal{C}}$ and $\mathcal{C}$ have enough projectives and enough injectives. Let $F : \tilde{\mathcal{C}} \to \mathcal{C}$ be an exact functor, and let $G : \mathcal{C} \to \mathcal{C}'$ be a one-sided exact functor with derived functors G_n or G^n.

(a) Suppose that F is covariant.

(a1) Suppose G_n or G^n is defined from projective resolutions and F carries projectives to projectives. Then $G \circ F$ is one-sided exact, and its derived functors satisfy $(G \circ F)_n \cong G_n \circ F$ or $(G \circ F)^n \cong G^n \circ F$.

(a2) Suppose G_n or G^n is defined from injective resolutions and F carries injectives to injectives. Then $G \circ F$ is one-sided exact, and its derived functors satisfy $(G \circ F)_n \cong G_n \circ F$ or $(G \circ F)^n \cong G^n \circ F$.

(b) Suppose that F is contravariant.

(b1) Suppose G_n or G^n is defined from projective resolutions and F carries injectives to projectives. Then $G \circ F$ is one-sided exact, and its derived functors satisfy $(G \circ F)^n \cong G^n \circ F$ or $(G \circ F)_n \cong G_n \circ F$.

(b2) Suppose G_n or G^n is defined from injective resolutions and F carries projectives to injectives. Then $G \circ F$ is one-sided exact, and its derived functors satisfy $(G \circ F)^n \cong G^n \circ F$ or $(G \circ F)_n \cong G_n \circ F$.

4. Long Exact Sequences

This section continues with the notation of §3. Let $F : \mathcal{C} \to \mathcal{C}'$ be a functor from a good category $\mathcal{C}$ of R modules to a good category $\mathcal{C}'$ of S modules, and assume that $\mathcal{C}$ has enough projectives and enough injectives in the sense of (C.23).

(C.29) In a good category, a diagram

$$
\begin{array}{ccccccccc}
0 & \longrightarrow & A & \stackrel{\psi}{\longrightarrow} & B & \stackrel{\varphi}{\longrightarrow} & C & \longrightarrow & 0 \\
 & & \downarrow{\scriptstyle \varphi_A} & & \downarrow{\scriptstyle \varphi_B} & & \downarrow{\scriptstyle \varphi_C} & & \\
0 & \longrightarrow & A' & \stackrel{\psi'}{\longrightarrow} & B' & \stackrel{\varphi'}{\longrightarrow} & C' & \longrightarrow & 0
\end{array}
$$

with exact rows and commuting squares induces a map

$$\rho : \ker \varphi_C \to A'/\text{image } \varphi_A$$

with $\ker \rho = \varphi(\ker \varphi_B)$ and with

$$\text{image } \rho = \psi'^{-1}(\text{image } \varphi_B)/\text{image } \varphi_A.$$

The map ρ is called a **connecting homomorphism**.

(C.30) In a good category, a diagram

$$
\begin{array}{ccc}
B_{n+1} & \xrightarrow{\varphi_{n+1}} & C_{n+1} \\
\downarrow{\scriptstyle\psi_B} & & \downarrow{\scriptstyle\psi_C} \\
B_n & \xrightarrow{\varphi_n} & C_n \\
\downarrow{\scriptstyle\varphi_B} & & \downarrow{\scriptstyle\varphi_C} \\
B_{n-1} & \xrightarrow{\varphi_{n-1}} & C_{n-1}
\end{array}
$$

in which the columns B and C are complexes and the squares commute has $\varphi_n|_{\ker \varphi_B}$ mapping to $\ker \varphi_C / \text{image } \psi_C$ with image given by $\varphi_n(\ker \varphi_B)/\text{image } \psi_C$ and with kernel given by $\ker \varphi_B \cap \varphi_n^{-1}(\text{image } \psi_C)$. The kernel of the descended map contains image ψ_B.

(C.31) In a good category, an infinite diagram

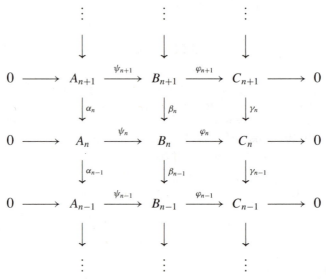

(in which the columns A, B, and C are complexes, the rows are exact, and the squares commute), induces an exact sequence (**long exact sequence**) on homology

$$
\cdots \longrightarrow H_{n+1}(B) \xrightarrow{\varphi_{n+1}} H_{n+1}(C) \xrightarrow{\rho_n} H_n(A) \xrightarrow{\psi_n}
$$

$$
H_n(B) \xrightarrow{\varphi_n} H_n(C) \xrightarrow{\rho_{n-1}} H_{n-1}(A) \longrightarrow \cdots
$$

Here ψ_n and φ_n refer to the maps on homology induced by the chain maps ψ and φ, and ρ_n is induced by the connecting homomorphism for the top two indicated rows.

(C.32) In a good category, the following conditions on a short exact sequence

$$0 \longrightarrow R \xrightarrow{\psi} S \xrightarrow{\varphi} T \longrightarrow 0$$

are equivalent:

(i)   there exists a map $\bar{\varphi} : T \to S$ with $\varphi\bar{\varphi} = 1_T$
(ii) there exists a map $\bar{\psi} : S \to R$ with $\bar{\psi}\psi = 1_R$
(iii) $S \cong R \oplus T$ with ψ corresponding to the inclusion of R and φ corresponding to the projection onto T.

In this case we say that the short exact sequence **splits**.

(C.33) The short exact sequence in (C.32) splits if R is injective or T is projective.

(C.34) In the good category $\mathcal{C}$, let the short exact sequence in (C.32) be given, and let X and Z be projective resolutions of R and T, respectively. Then there exists a projective resolution Y of S such that

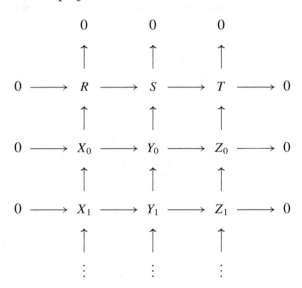

has commuting squares and exact rows. Moreover Y_n can be taken as $X_n \oplus Z_n$, and the horizontal maps can be taken as the inclusion of X_n into $X_n \oplus Z_n$ and the projection of $X_n \oplus Z_n$ on Z_n.

(C.35) In the good category $\mathcal{C}$, let the short exact sequence in (C.32) be given, and let X and Z be injective resolutions of R and T, respectively.

Then there exists an injective resolution Y of S such that

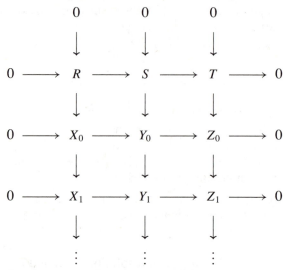

has commuting squares and exact rows. Moreover Y_n can be taken as $X_n \oplus Z_n$, and the horizontal maps can be taken as the inclusion of X_n into $X_n \oplus Z_n$ and the projection of $X_n \oplus Z_n$ on Z_n.

(C.36) In the good category $\mathcal{C}$, let the short exact sequence (C.32) be given, and let $F : \mathcal{C} \to \mathcal{C}'$ be one-sided exact. Then the derived functors of F on R, S, and T form a long exact sequence as follows:

(a) if F is covariant and right exact, the long exact sequence is

$$0 \longleftarrow F(T) \longleftarrow F(S) \longleftarrow F(R) \longleftarrow F_1(T) \longleftarrow F_1(S) \longleftarrow F_1(R)$$

$$\longleftarrow F_2(T) \longleftarrow F_2(S) \longleftarrow F_2(R) \longleftarrow F_3(T) \longleftarrow \cdots .$$

(b) f F is contravariant and right exact, the long exact sequence is

$$0 \longleftarrow F(R) \longleftarrow F(S) \longleftarrow F(T) \longleftarrow F_1(R) \longleftarrow F_1(S) \longleftarrow F_1(T)$$

$$\longleftarrow F_2(R) \longleftarrow F_2(S) \longleftarrow F_2(T) \longleftarrow F_3(R) \longleftarrow \cdots .$$

(c) if F is covariant and left exact, the long exact sequence is

$$0 \longrightarrow F(R) \longrightarrow F(S) \longrightarrow F(T) \longrightarrow F^1(R) \longrightarrow F^1(S) \longrightarrow F^1(T)$$

$$\longrightarrow F^2(R) \longrightarrow F^2(S) \longrightarrow F^2(T) \longrightarrow F^3(R) \longrightarrow \cdots .$$

(d) if F is contravariant and left exact, the long exact sequence is

$$0 \longrightarrow F(T) \longrightarrow F(S) \longrightarrow F(R) \longrightarrow F^1(T) \longrightarrow F^1(S) \longrightarrow F^1(R)$$

$$\longrightarrow F^2(T) \longrightarrow F^2(S) \longrightarrow F^2(R) \longrightarrow F^3(T) \longrightarrow \cdots .$$

5. Euler-Poincaré Principle

Let C be a small good category of left R modules. The Grothendieck group $K(C)$ was defined in §A.4.

Proposition C.37. If C and C' are small good categories and if $F : C \to C'$ is an exact functor, then the map $(A) \mapsto (F(A))$ extends to a homomorphism of $K(C)$ into $K(C')$.

PROOF. The map extends to a homomorphism of $\mathcal{F}(C)$ into $K(C')$. If

$$0 \longrightarrow A \longrightarrow B \longrightarrow C \longrightarrow 0$$

is exact in C, then the exactness of F implies that

$$0 \longrightarrow F(A) \longrightarrow F(B) \longrightarrow F(C) \longrightarrow 0 \qquad (F \text{ covariant})$$

$$0 \longrightarrow F(C) \longrightarrow F(B) \longrightarrow F(A) \longrightarrow 0 \qquad (F \text{ contravariant})$$

is exact in C'. Hence $(F(B)) = (F(A)) + (F(C))$ in $\mathcal{F}(C')$. Then the relations of $\mathcal{F}(C)$ map to 0 in $K(C')$, and the homomorphism descends to $K(C)$.

Corollary C.38. Let C be a small good category whose objects are finite-dimensional k vector spaces with additional structure and whose maps are k linear (among other things). Then the map $(A) \mapsto \dim A$ extends to a well-defined homomorphism of $K(C)$ into $\mathbb{Z}$.

PROOF. Let C' be the small good category of all finite-dimensional k vector spaces. The assumption is that there is a well-defined exact forgetful functor $F : C \to C'$. By Proposition C.37, F yields a well-defined homomorphism of $K(C)$ into $K(C')$. Composing this homomorphism with the homomorphism dim of Proposition A.31, we obtain the corollary.

The Euler-Poincaré Principle arose originally in connection with the combinatorics of triangulated surfaces. If a finite triangulation of a surface S yields γ_0 vertices, γ_1 edges, and γ_2 faces, and if a finite refinement yields the corresponding integers $\gamma_0', \gamma_1', \gamma_2'$, then

(C.39a) $$\gamma_0 - \gamma_1 + \gamma_2 = \gamma_0' - \gamma_1' + \gamma_2'.$$

Once simplicial homology is introduced, we can recognize γ_i as the dimension of the space C_i of simplicial i-chains on S. If β_i denotes the dimension of the ith homology, then the identity that "explains" (C.39a) is

(C.39b) $$\gamma_0 - \gamma_1 + \gamma_2 = \beta_0 - \beta_1 + \beta_2.$$

In turn, (C.39b) is "explained" by applying the theorem below in the small good category $\mathcal{C}$ of all finite-dimensional real vector spaces: For the complex

$$0 \longrightarrow C_0 \longrightarrow C_1 \longrightarrow C_2 \longrightarrow 0,$$

the theorem asserts the equality of two members of $K(\mathcal{C})$; then dim of these two members must be the same, and the result is (C.39b).

Theorem C.40 (Euler-Poincaré Principle). Let

$$X: \qquad 0 \xrightarrow{d_{-1}} X_0 \xrightarrow{d_0} X_1 \xrightarrow{d_1} \cdots \xrightarrow{d_{n-1}} X_n \xrightarrow{d_n} 0$$

be a finite complex in a small good category $\mathcal{C}$. Then

$$\sum_{i=0}^{n} (-1)^i (X_i) = \sum_{i=0}^{n} (-1)^i (H^i(X))$$

in $K(\mathcal{C})$.

PROOF. The short exact sequences

$$0 \longrightarrow \ker d_i \longrightarrow X_i \longrightarrow \text{image } d_i \longrightarrow 0,$$

$$0 \longrightarrow \text{image } d_{i-1} \longrightarrow \ker d_i \longrightarrow H^i(X) \longrightarrow 0,$$

in $\mathcal{C}$ yield identities

$$(X_i) = (\ker d_i) + (\text{image } d_i)$$

$$(H^i(X)) = (\ker d_i) - (\text{image } d_{i-1})$$

in $K(\mathcal{C})$ for $0 \le i \le n$. The alternating sums of these identities are

$$\sum_{i=0}^{n} (-1)^i (X_i) = \sum_{i=0}^{n} (-1)^i (\ker d_i) + \sum_{i=0}^{n} (-1)^i (\text{image } d_i)$$

and

$$\sum_{i=0}^{n} (-1)^i (H^i(X)) = \sum_{i=0}^{n} (-1)^i (\ker d_i) - \sum_{i=0}^{n} (-1)^i (\text{image } d_{i-1})$$

$$= \sum_{i=0}^{n} (-1)^i (\ker d_i) + \sum_{i=0}^{n} (-1)^{i-1} (\text{image } d_{i-1}),$$

and the right sides are equal since $d_{-1} = 0$ and $d_n = 0$.

Corollary C.41. If

$$0 \longrightarrow X_0 \longrightarrow X_1 \longrightarrow \cdots \longrightarrow X_n \longrightarrow 0$$

is a finite exact complex in a small good category $\mathcal{C}$, then $\sum_{i=0}^{n} (-1)^i (X_i) = 0$ in $K(\mathcal{C})$.

PROOF. This is the special case of Theorem C.40 in which $H^i(X) = 0$ for $0 \le i \le n$.

Corollary C.42. Let $\mathcal{C}$ and $\mathcal{C}'$ be small good categories, and let $F_0, F_1, F_2, \ldots$ be functors from $\mathcal{C}$ to $\mathcal{C}'$ with the property, for each C in $\mathcal{C}$, that $F_i(C) = 0$ for i sufficiently large. Suppose that the exactness of

$$0 \longrightarrow R \longrightarrow S \longrightarrow T \longrightarrow 0$$

in $\mathcal{C}$ always implies that one of the four long sequences in (C.36) is exact. Then

$$\sum_{i \ge 0} (-1)^i \{(F_i(S)) - (F_i(R)) - (F_i(T))\} = 0$$

in $K(\mathcal{C}')$. Consequently the map $(S) \mapsto \sum_{i \ge 0} (-1)^i (F_i(S))$ extends to a homomorphism from $K(\mathcal{C})$ into $K(\mathcal{C}')$. In particular, if

$$M_N \supseteq M_{N-1} \supseteq \cdots \supseteq M_0 \supseteq M_{-1} = 0$$

is a sequence of modules in $\mathcal{C}$, then

$$(\text{C.43}) \qquad \sum_{i \ge 0} (-1)^i (F_i(M_N)) = \sum_{i \ge 0} \sum_{n=0}^{N} (-1)^i (F_i(M_n/M_{n-1}))$$

in $K(\mathcal{C}')$.

PROOF. The first conclusion follows immediately by applying Corollary C.41 to the long exact sequence. (The exactness of the short sequence involving R, S, T plays no role.) Consequently we obtain the indicated homomorphism. To prove (C.43), we proceed by induction on N, the case $N = 0$ being trivial. Assume inductively that (C.43) holds for $N - 1$. Because of the homomorphism property, we deduce from $(M_N) = (M_{N-1}) + (M_N/M_{N-1})$ that

$$\sum_{i \ge 0} (-1)^i (F_i(M_N)) = \sum_{i \ge 0} (-1)^i (F_i(M_{N-1})) + \sum_{i \ge 0} (-1)^i (F_i(M_N/M_{N-1})).$$

Adding this identity to (C.43) for $N - 1$, we obtain (C.43) for N, and the induction is complete.

The corollary suggests that the F_i might be derived functors of a single functor, yet it is important not to insist that they be of this form. It turns out that a small good category may be too large for $K(\mathcal{C})$ to be interesting. For example, if A and B are modules with $A \oplus B = A$, then $(A)+(B) = (A)$ and hence $(B) = 0$ in $K(\mathcal{C})$. This happens in a category of vector spaces if A is infinite dimensional and B is finite dimensional. Thus the presence of one large module in the category can eliminate much interesting structure from $K(\mathcal{C})$. For this reason it may be necessary to pass to a good subcategory of the small good category to get an interesting $K(\mathcal{C})$, and in doing so we may have to abandon having enough projectives and enough injectives. It is therefore important to construct all necessary derived functors before passing to the subcategory.

EXAMPLE 1 IN THE CONTEXT OF CHAPTER II. Suppose $(\mathfrak{g}, K)$ is a pair, $\mathcal{C}_0$ is the category of $(\mathfrak{g}, K)$ modules, $\mathcal{C}_0'$ is the category of complex vector spaces, and $F_i(M) = H_i(\mathfrak{g}, K; M)$ is the ith relative homology functor. To construct F_i as a derived functor, we need to use the projectives in $\mathcal{C}_0$, and these are always infinite dimensional or 0 if $\mathfrak{g}$ is not the complexified Lie algebra of K. Unfortunately as soon as we allow any infinite-dimensional vector spaces in $\mathcal{C}_0'$, we give up the possibility of counting dimensions as in Corollary C.38. One way around this difficulty is to pass to the subcategory $\mathcal{C}$ of finite-dimensional $(\mathfrak{g}, K)$ modules. By using standard complexes, one can see that the functors F_i carry $\mathcal{C}$ into the category $\mathcal{C}'$ of finite-dimensional vector spaces. Combining Corollaries C.42 and C.38, we find that the map $(E) \mapsto \sum_{i \geq 0} (-1)^i \dim H_i(\mathfrak{g}, K; E)$ is a well-defined homomorphism from the Grothendieck group of finite dimensional $(\mathfrak{g}, K)$ modules into $\mathbb{Z}$.

In applying (C.43) we shall encounter situations where we have good control of $F_i(M)$ only for very special modules M in a collection that is too small to form a good category. If we attempt to enlarge the class of modules to make it a good category, we may lose control of $F_i(M)$.

EXAMPLE 2 IN THE CONTEXT OF CHAPTER II. Suppose $\mathfrak{g}$ is a complex Lie algebra. Within the category $\mathcal{C} = \mathcal{C}(\mathfrak{g})$ of all $U(\mathfrak{g})$ modules, let $\mathcal{S}$ be the subclass of $U(\mathfrak{g})$ modules that are free of finite rank. For the functors F_i, we take the Lie algebra homology functors $H_i(\mathfrak{g}, \cdot)$. Initially these carry modules in $\mathcal{C}$ to complex vector spaces, sometimes infinite dimensional. Formula (C.43) is uninteresting in this setting, since all terms are 0.

If M is a module in $\mathcal{S}$, then $F_0(M)$ is easily seen to be a complex vector space of dimension equal to the rank of M as a $U(\mathfrak{g})$ module, and $F_i(M) = 0$ for $i > 0$. We would like (C.43) to be valid for $\mathcal{S}$ with terms in K of the category of finite-dimensional complex vector spaces. The trouble is that the class $\mathcal{S}$ is not closed under passage to arbitrary quotients and therefore does not constitute a good category. Nevertheless, a useful version of (C.43) is true: Suppose that

$$M_N \supseteq M_{N-1} \supseteq \cdots \supseteq M_0 \supseteq M_{-1} = 0$$

is a finite descending chain of $U(\mathfrak{g})$ submodules of a $U(\mathfrak{g})$ module M_N and that each quotient M_n/M_{n-1} belongs to $\mathcal{S}$. Then Proposition C.44 below will show that

$$\sum_{i \geq 0} (-1)^i \dim F_i(M_N) = \sum_{n=0}^{N} \dim F_0(M_n/M_{n-1}).$$

Meanwhile, the long exact sequence for homology that corresponds to

$$0 \longrightarrow M_{n-1} \longrightarrow M_n \longrightarrow M_n/M_{n-1} \longrightarrow 0$$

shows that $F_i(M_{n-1}) \cong F_i(M_n)$ for $i > 0$. Hence $F_i(M_N) = 0$ for $i > 0$, and we conclude that

$$\dim F_0(M_N) = \sum_{n=0}^{N} \dim F_0(M_n/M_{n-1}).$$

Returning to the general case, we give a result saying that (C.43) applies under weaker hypotheses than those in Corollary C.42.

Proposition C.44. Let $\mathcal{C}$ and $\mathcal{C}'$ be small good categories, and let $F_0, F_1, F_2, \ldots$ be functors from $\mathcal{C}$ to $\mathcal{C}'$. Suppose that the exactness of

$$0 \longrightarrow R \longrightarrow S \longrightarrow T \longrightarrow 0$$

in $\mathcal{C}$ always implies that one of the four long sequences in (C.36) is exact. Suppose in addition that $\mathcal{C}''$ is a small good subcategory of $\mathcal{C}'$ that is closed under the formation of extensions in $\mathcal{C}'$. That is, if

$$0 \longrightarrow A \longrightarrow B \longrightarrow C \longrightarrow 0$$

is a short exact sequence in $\mathcal{C}'$ and if A and C belong to $\mathcal{C}''$, then B belongs to $\mathcal{C}''$. Next suppose that $\mathcal{S}$ is a collection of modules S in $\mathcal{C}$ having the following two properties:

(ii) $F_i(S)$ belongs to $\mathcal{C}''$ for every i, and
(ii) $F_i(S) = 0$ for i sufficiently large.

Suppose finally that
$$M_N \supseteq M_{N-1} \supseteq \cdots \supseteq M_0 \supseteq M_{-1} = 0$$
is a finite descending chain in $\mathcal{C}$ and that each M_n/M_{n-1} belongs to the collection $\mathcal{S}$. Then

(a) $F_i(M_N)$ belongs to $\mathcal{C}''$ for all i,
(b) $F_i(M_N) = 0$ for i sufficiently large, and
(c) the formula
$$\sum_{i \geq 0} (-1)^i (F_i(M_N)) = \sum_{i \geq 0} \sum_{n=0}^{N} (-1)^i (F_i(M_n/M_{n-1}))$$
is valid in $K(\mathcal{C}'')$.

PROOF. We proceed by induction on N. If $N = 0$, then the hypothesis says simply that M_0 belongs to $\mathcal{S}$. Thus (a) and (b) amount to hypotheses (i) and (ii) on $\mathcal{S}$. The two sides of the formula in (c) are the same in this case.

Suppose therefore that $N > 0$ and that the result is known for $N - 1$. In particular, the result is known for M_{N-1} and the descending chain with M_N omitted. The short exact sequence
$$0 \longrightarrow M_{N-1} \longrightarrow M_N \longrightarrow M_N/M_{N-1} \longrightarrow 0$$
gives rise to a long exact sequence in $\mathcal{C}'$, either

(C.45) $\cdots \longrightarrow F_i(M_{N-1}) \longrightarrow F_i(M_N) \longrightarrow F_i(M_N/M_{N-1}) \longrightarrow \cdots$

or a variant. We work with (C.45), the variants being handled similarly. Let ρ be either of the maps at the ends of (C.45). By inductive hypothesis the first term in (C.45) belongs to $\mathcal{C}''$; since $\mathcal{C}''$ is a good category, the quotient $F_i(M_{N-1})/\text{image}(\rho)$ belongs to $\mathcal{C}''$. Similarly the third term in (C.45) belongs to $\mathcal{C}''$ by hypothesis (i); since $\mathcal{C}''$ is a good category, the submodule $\ker(\rho)$ belongs to $\mathcal{C}''$. Consequently
$$0 \longrightarrow F_i(M_{N-1})/\text{image}(\rho) \longrightarrow F_i(M_N) \longrightarrow \ker(\rho) \longrightarrow 0$$
is a short exact sequence in $\mathcal{C}'$ with the second and fourth terms in $\mathcal{C}''$. Since by hypothesis $\mathcal{C}''$ is closed under extensions, we see that $F_i(M_N)$ is in $\mathcal{C}''$. This proves (a).

A similar but easier argument proves (b), that $F_i(M) = 0$ for i large enough. Thus (C.45) is a finite exact complex in the small good catgeory $\mathcal{C}''$. By Corollary C.41,
$$\sum_{i \geq 0} (-1)^i (F_i(M_N)) = \sum_{i \geq 0} (-1)^i (F_i(M_{N-1})) + \sum_{i \geq 0} (-1)^i ((F_i(M_N/M_{N-1}))$$
in $K(\mathcal{C}'')$. Replacing the first sum on the right side by the expression given by inductive hypothesis, we obtain (c).

APPENDIX D

SPECTRAL SEQUENCES

A complex with a compatible filtration yields a spectral sequence that gives relationships among the various homology or cohomology groups that are involved. The prototype is the case of a short exact sequence of complexes, where these relationships are captured in the long exact sequence in homology or cohomology.

A double complex has two associated filtered complexes, thus two associated spectral sequences. These two spectral sequences are related. Because of this relationship, one is able to connect the derived functors of a composition with the composition of derived functors.

1. Spectral Sequence of a Filtered Complex

Spectral sequences arise in representation theory chiefly as a tool to analyze derived functors of a composition. When one of the two functors is exact, the situation is quite simple and can be handled by the elementary results (C.27) and (C.28). But the situation quickly escalates in complexity when neither functor is exact. Spectral sequences provide a uniform method for analyzing the situation without special tricks.

In the case of derived functors of a composition, the analysis in question takes place in a double complex. In turn the analysis of a double complex reduces to that of a filtered complex, which is where we begin.

Let $\mathcal{C}$ be a good category of R modules, in the sense of (C.1), and suppose $\mathcal{C}$ is closed under countable direct sums. All constructions in this section will take place in this category $\mathcal{C}$. A **differential** on a module A is a map $d : A \to A$ such that $d^2 = 0$. (If

$$\cdots \longrightarrow {}^{n-1}A \xrightarrow{d} {}^{n}A \xrightarrow{d} {}^{n+1}A \longrightarrow \cdots$$

is a complex, then an example is obtained by taking d as the differential on the direct sum $A = \bigoplus_{n=-\infty}^{\infty} {}^{n}A$. In the end we are interested almost exclusively in differentials of this form, but it is notationally and conceptually convenient not to impose this extra structure at first.)

The **cohomology** of A is $H(A) = (\ker d)/(\text{image } d)$; we may also call this the **homology** of A. The abstract goal is to compute $H(A)$.

A **spectral sequence** is a family $(E_r, d_r)_{r \geq 0}$ of modules with differentials such that $E_{r+1} = H(E_r)$. Although in this book we are interested in only one special kind of spectral sequence (the spectral sequence of a filtered complex), it is helpful to begin with a brief discussion on a formal level. Typically the modules E_r should be thought of as approximations to some module E_∞ related to $H(A)$ that we would like to compute, but that is difficult to understand directly. The modules E_0 and E_1, and often E_2, will be much easier to compute. Once we know the existence of a spectral sequence connected with $H(A)$, we are faced with two problems. The first problem is "convergence"; we want E_∞ to be a limit of E_r in some precise sense as r tends to infinity. Typically this step is addressed by appealing to a general sufficient condition for convergence (such as Proposition D.26 or D.26′ below). The second problem is "collapsing"; we have to relate E_∞ to computable terms like E_2. This step is much more delicate, and usually requires strong hypotheses in each particular application.

If M is a module in $\mathcal{C}$, a **decreasing filtration** of M is a system $\{M^p\}_{p \in \mathbb{Z}}$ of submodules of M such that

(D.1a) $$\cdots \supseteq M^{p-1} \supseteq M^p \supseteq M^{p+1} \supseteq \cdots$$

with $\bigcup_p M^p = M$. If in addition we have $\bigcap_p M^p = 0$, the filtration is said to be **separated**. (The terminology arises from the toplogy on M defined by the filtration, in which $\{m + M^p\}$ is a neighborhood base at M. From a filtration as in (D.1a) we can always form the successive quotients

(D.1b) $$\mathrm{gr}^p(M) = M^p / M^{p+1}$$

and the **associated graded module**

(D.1c) $$\mathrm{gr}(M) = \bigoplus_{p=-\infty}^{\infty} (M^p / M^{p+1}).$$

For separated filtrations, $\mathrm{gr}(M)$ retains some of the structure of M. For example, if $\mathcal{C}$ is a category of vector spaces over a field and M is finite dimensional with a separated filtration, then $\dim M = \dim \mathrm{gr}(M)$. More generally if M has a separated filtration and if M and $\mathrm{gr}(M)$ lie in a small good subcategory $\mathcal{C}'$ of $\mathcal{C}$, then $(M) = (\mathrm{gr}(M))$ in the Grothendieck group $K(\mathcal{C}')$ defined in §A.4; in the context of finite-dimensional vector spaces, equality in $K(\mathcal{C}')$ implies equality of dimensions, by Corollary C.38.

Spectral sequences of the kind in this book compute not the module of direct interest but rather its associated graded module with respect to some filtration. Here is the underlying situation when the filtration is decreasing: Suppose A is a module with a differential d, and suppose that A is endowed with a decreasing filtration $\{A^p\}_{p \in \mathbb{Z}}$ that satisfies

$$d(A^p) \subseteq A^p \qquad \text{for all } p \in \mathbb{Z}.$$

The module of interest is the cohomology

$$H(A) = (\ker d)/(\text{image } d).$$

Each submodule A^p is a module with a differential and so has its own cohomology $H(A^p)$. The inclusion of A^p in A induces a map

$$H(A^p) \to H(A).$$

We write D^p for the image of this map:

(D.2)
$$D^p = ((A^p \cap \ker d) + \text{image } d)/(\text{image } d).$$
$$\cong (A^p \cap \ker d)/(A^p \cap \text{image } d).$$

Evidently D^p is a submodule of $H(A)$. The first formula of (D.2) shows that $D^p \supseteq D^{p+1}$ for all p. If $z + (\text{image } d)$ is in $H(A)$, so that z is in $\ker d$, then the equality $\bigcup_p A^p = A$ forces z to be in A^{p_0} for some p_0; hence $z + (\text{image } d)$ is in D^{p_0} and $\bigcup_p D^p = H(A)$. Consequently $\{D^p\}_{p \in \mathbb{Z}}$ is a decreasing filtration of $H(A)$.

Even if $\{A^p\}_{p \in \mathbb{Z}}$ is separated as a filtration of A, $\{D^p\}_{p \in \mathbb{Z}}$ need not be separated as a filtration of $H(A)$. Parts (c) of Propositions D.26 and D.26′ below give a sufficient condition for $\{D^p\}_{p \in \mathbb{Z}}$ to be separated. In (D.15) we shall define E_∞^p and E_∞ in such a way that

$$E_\infty^p \cong D^p/D^{p+1} \qquad \text{and} \qquad E_\infty = \bigoplus_{p=-\infty}^{\infty} E_\infty^p \cong \text{gr}(H(A)).$$

The spectral sequence of the filtered complex $\{A^p\}_{p \in \mathbb{Z}}$, which we have yet to define, will provide information about $\text{gr}(H(A))$ rather than $H(A)$. The idea is that we are willing to settle for this kind of information in cases where we cannot compute $H(A)$ itself.

Although the above situation is the important one for representation theory, there is a second situation of interest. We define an **increasing**

filtration of a module M in C to be a system $\{M^p\}_{p\in\mathbb{Z}}$ of submodules such that

(D.3a) $\cdots \subseteq M^{p-1} \subseteq M^p \subseteq M^{p+1} \subseteq \cdots$

with $\bigcup_p M^p = M$. The filtration again is **separated** if $\bigcap_p M^p = 0$. In any case the successive quotients are denoted

(D.3b) $\mathrm{gr}^p(M) = M^p/M^{p-1}$

and the **associated graded module** is

(D.3c) $$\mathrm{gr}(M) = \bigoplus_{p=-\infty}^{\infty} (M^{p+1}/M^p).$$

In this context, our presumed underlying complex has decreasing indices

$$\cdots \longleftarrow {}^{n-1}A \xleftarrow{\ \partial\ } {}^nA \xleftarrow{\ \partial\ } {}^{n+1}A \longleftarrow \cdots .$$

Thus we suppose that A comes with a differential ∂, and we suppose that there is an increasing filtration $\{A^p\}_{p\in\mathbb{Z}}$ with $\partial(A^p) \subseteq A^p$. Of course, this situation reduces to the first one by replacing p by $-p$ (and n by $-n$), but we shall later want p and n to be ≥ 0 and shall have to distinguish the two situations for this purpose.

Back in the first situation (with a decreasing filtration) we introduce modules Z_r^p of approximate cocycles

(D.4)
$$Z_r^p = \{z \in A^p \mid dz \in A^{p+r}\} \qquad \text{for } r \geq 0$$
$$Z_{-1}^p = \{z \in A^p \mid dz \in A^p\} = A^p.$$

(Recall that the A^p's decrease to 0 as p tends to $+\infty$.) Also we shall use modules that approximate the full module of coboundaries in A^p:

(D.5) $dZ_r^{p-r} \subseteq Z_s^p \subseteq A^p \qquad$ for $r \geq -1$ and all s.

For $r \geq 0$, the modules Z_r^p satisfy two basic inclusions

(D.6a) $Z_{r+1}^p \subseteq Z_r^p \qquad$ and $\qquad Z_{r-1}^{p+1} \subseteq Z_r^p$

and the identity

(D.6b) $Z_{r+1}^p \cap Z_{r-1}^{p+1} = Z_r^{p+1}.$

In terms of the Z_r^p's, let

(D.7) $$E_r^p = Z_r^p/(dZ_{r-1}^{p-r+1} + Z_{r-1}^{p+1}) \qquad \text{for } r \geq 0.$$

This quotient is meaningful by (D.5) for $r - 1$ and by the second inclusion of (D.6a). We shall see that in nice cases E_r^p is an approximation to the module E_∞^p discussed after (D.2).

An alternate expression for E_r^p is

(D.8) $$E_r^p \cong (Z_r^p + A^{p+1})/(dZ_{r-1}^{p-r+1} + A^{p+1}) \qquad \text{for } r \geq 0.$$

In fact, we map the numerator of (D.7) to (D.8) by

$$z_r^p \mapsto z_r^p + dZ_{r-1}^{p-r+1} + A^{p+1}.$$

The denominator of (D.7) maps to the 0 coset, and thus we get a map of (D.7) onto (D.8). To see that the result is one-one, let z_r^p map to the 0 coset. Then z_r^p is in $dZ_{r-1}^{p-r+1} + A^{p+1}$, say $z_r^p = dz_{r-1}^{p-r+1} + a^{p+1}$. From

$$da^{p+1} = dz_r^p \in A^{p+r},$$

we see that a^{p+1} is in Z_{r-1}^{p+1}. Hence z_r^p represents 0 in E_r^p. This proves (D.8).

Formula (D.8) allows us easily to compute E_r^p for $r = 0$ and $r = 1$. For $r = 0$, we have $Z_0^p = A^p$ and $Z_{-1}^{p+1} = A^{p+1}$, by (D.4). Thus (D.7) gives

(D.9a) $$E_0^p = A^p/A^{p+1}.$$

In other words, $\bigoplus_p E_0^p = \mathrm{gr}(A)$. For $r = 1$, (D.8) gives

$$E_1^p \cong (Z_1^p + A^{p+1})/(dA^p + A^{p+1}),$$

which we can write as

(D.9b) $$E_1^p \cong H(A^p/A^{p+1}).$$

(Notice that d is meaningful on A^p/A^{p+1} since $d(A^p) \subseteq A^p$ and $d(A^{p+1}) \subseteq A^{p+1}$.)

The transition from E_0^p to E_1^p is thus accomplished by taking cohomology $H(\cdot)$. The proposition below shows that the same thing is true from E_r^p to E_{r+1}^p; the argument uses (D.7) and not (D.8). In formula (D.7) we have

$$E_r^p = Z_r^p/(dZ_{r-1}^{p-r+1} + Z_{r-1}^{p+1})$$

and

$$E_r^{p+r} = Z_r^{p+r}/(dZ_{r-1}^{p+1} + Z_{r-1}^{p+r+1}).$$

The numerator of E_r^p has $d(Z_r^p) \subseteq Z_r^{p+r}$, while the denominator has $d(dZ_{r-1}^{p-r+1}) = 0 \subseteq dZ_{r-1}^{p+1}$ and $d(Z_{r-1}^{p+1}) = dZ_{r-1}^{p+1}$. Thus d induces a map

(D.10) $$d_r^p : E_r^p \to E_r^{p+r}.$$

Proposition D.11. The maps d_r^p satisfy

$$\ker d_r^p = (Z_{r+1}^p + Z_{r-1}^{p+1})/(dZ_{r-1}^{p-r+1} + Z_{r-1}^{p+1})$$

$$\text{image } d_r^{p-r} = (dZ_r^{p-r} + Z_{r-1}^{p+1})/(dZ_{r-1}^{p-r+1} + Z_{r-1}^{p+1})$$

$$(\ker d_r^p)/(\text{image } d_r^{p-r}) \cong E_{r+1}^p.$$

PROOF. The numerator of the kernel is the subset of $z_r^p \in Z_r^p$ with

(D.12) $$dz_r^p \in dZ_{r-1}^{p+1} + Z_{r-1}^{p+r+1}.$$

If $z_r^p \in Z_{r-1}^{p+1}$, then dz_r^p satisfies (D.12). If $z_r^p \in Z_{r+1}^p$, then $dz_r^p \in A^{p+r+1}$ with $d(dz_r^p) = 0$; hence dz_r^p is in Z_{r-1}^{p+r+1} and satisfies (D.12). Consequently all the members z_r^p of $Z_{r+1}^p + Z_{r-1}^{p+1}$ satisfy (D.12). In the reverse direction, let $z_r^p \in Z_r^p$ have dz_r^p as in (D.12). We are to show that z_r^p is in $Z_{r+1}^p + Z_{r-1}^{p+1}$. Say $dz_r^p = dz_{r-1}^{p+1} + z_{r-1}^{p+r+1}$. Write

$$z_r^p = (z_r^p - z_{r-1}^{p+1}) + z_{r-1}^{p+1}.$$

The second term on the right side is in Z_{r-1}^{p+1}, and we need the first term to be in Z_{r+1}^p. It is in A^p since $A^{p+1} \subseteq A^p$, and d of it is

$$d(z_r^p - z_{r-1}^{p+1}) = z_{r-1}^{p+r+1} \in A^{p+r+1};$$

hence $z_r^p - z_{r-1}^{p+1}$ is in Z_{r+1}^p. This completes the identification of the kernel.

The numerator of the image is dZ_r^{p-r} plus the denominator of E_r^p, and the formula for the image follows since $dZ_{r-1}^{p-r+1} \subseteq dZ_r^{p-r}$ by (D.6).

Finally we have

$$(\ker d_r^p)/(\text{image } d_r^{p-r})$$

$$\cong (Z_{r+1}^p + Z_{r-1}^{p+1})/(dZ_r^{p-r} + Z_{r-1}^{p+1})$$

$$\cong Z_{r+1}^p/(Z_{r+1}^p \cap (dZ_r^{p-r} + Z_{r-1}^{p+1}))$$

since $R \supseteq S$ implies $(R + T)/(S + T) \cong R/(R \cap (S + T))$

$$= Z_{r+1}^p/(dZ_r^{p-r} + (Z_{r+1}^p \cap Z_{r-1}^{p+1}))$$

since $R \supseteq S$ implies $R \cap (S + T) = S + (R \cap T)$

$$= Z_{r+1}^p/(dZ_r^{p-r} + Z_r^{p+1}) \qquad \text{by (D.6b)}$$

$$= E_{r+1}^p,$$

and the proof is complete.

We can interpret Proposition D.11 as follows. Put $E_r = \bigoplus_p E_r^p$. The elements of E_r^p are said to be of **filtration degree** p. The maps d_r^p allow us to view E_r as a complex, but with a coboundary operator d_r of **degree** r rather than the usual degree 1 (or -1 in the case of an ordinary boundary operator ∂ that decreases indices). The sequence E_r of graded modules with coboundary operators d_r of degree r is called the **spectral sequence** attached to the module A with a decreasing filtration and with coboundary operator d.

We saw that $E_0 = \text{gr}(A)$, and the operator d_0 is what d induces on each A^p/A^{p+1}. Proposition D.11 concludes that E_1 is isomorphic to the cohomology of (E_0, d_0), i.e., $E_1 \cong \bigoplus_p H(A^p/A^{p+1})$; this conclusion is in agreement with (D.9b). The hope is that E_0, E_1, E_2, ... will tend toward a module E_∞ that we shall now define. Let

(D.13) $Z_\infty^p = A^p \cap \ker d$ and $B_\infty^p = A^p \cap \text{image } d.$

In analogy with (D.7), put

(D.14) $E_\infty^p = Z_\infty^p/(Z_\infty^{p+1} + B_\infty^p).$

An alternate expression for E_∞^p is

(D.15) $E_\infty^p = (Z_\infty^p + \text{image } d)/(Z_\infty^{p+1} + \text{image } d) \cong D^p/D^{p+1},$

where D^p, as in (D.2), is the image of $H(A^p)$ in $H(A)$. (The derivation of (D.15) from (D.14) is similar to the derivation of (D.8) from (D.7), and we omit it.) Following what we did with E_r^p, we put $E_\infty = \bigoplus_p E_\infty^p$.

As we noted after (D.2), $\{D^p\}_{p\in\mathbb{Z}}$ is a decreasing filtration of $H(A)$ with

$$(D.16) \qquad\qquad E_\infty \cong \operatorname{gr}(H(A)).$$

We shall introduce hypotheses that force this filtration to be separated. The same hypotheses will make E_0, E_1, E_2, ... "converge" to E_∞, and the whole process that started with E_0 will give us a handle on $\operatorname{gr}(H(A))$.

We are ready to introduce the hypothesis that A comes from a complex. We suppose that

$$(D.17a) \qquad\qquad A = \bigoplus_{n=-\infty}^{\infty} {}^nA$$

with $d({}^nA) \subseteq {}^{n+1}A$. The elements of ${}^nA$ are said to have **total degree** n. Let

$$(D.17b) \qquad\qquad A^{p,q} = {}^{p+q}A \cap A^p.$$

The integer q is called the **complementary degree**. Notice that

$$(D.17c) \qquad\qquad d : A^{p,q} \to A^{p,q+1}.$$

Extraction of elements of total degree n involves the following additional definitions:

$$(D.18) \quad
\begin{aligned}
Z_r^{p,q} &= {}^{p+q}A \cap Z_r^p = \{z \in A^{p,q} \mid dz \in A^{p+r,q+1-r}\} \quad \text{for } r \geq 0 \\
Z_{-1}^{p,q} &= {}^{p+q}A \cap Z_{-1}^p = A^{p,q} \\
E_r^{p,q} &= Z_r^{p,q}/(dZ_{r-1}^{p-r+1,q+r-2} + Z_{r-1}^{p+1,q-1}) \quad \text{for } r \geq 0 \\
E_0^{p,q} &= A^{p,q}/A^{p+1,q-1} \\
Z_\infty^{p,q} &= {}^{p+q}A \cap Z_\infty^p = A^{p,q} \cap \ker d \\
B_\infty^{p,q} &= {}^{p+q}A \cap B_\infty^p = A^{p,q} \cap \operatorname{image} d \\
D^{p,q} &= (Z_\infty^{p,q} + \operatorname{image} d)/(\operatorname{image} d) \\
E_\infty^{p,q} &= Z_\infty^{p,q}/(Z_\infty^{p+1,q-1} + B_\infty^{p,q}) \cong D^{p,q}/D^{p+1,q-1}.
\end{aligned}$$

The isomorphism in the last line follows from (D.15). Property (D.17) makes

$$A^p = \bigoplus_q A^{p,q} \quad \text{and} \quad Z_r^p = \bigoplus_q Z_r^{p,q},$$

$$(D.19)$$

$$Z_\infty^p = \bigoplus_q Z_\infty^{p,q} \quad \text{and} \quad B_\infty^p = \bigoplus_q B_\infty^{p,q}, \quad \text{and} \quad D^p = \bigoplus_q D^{p,q}.$$

Using the isomorphism

(D.20a) $$\bigoplus_i (X_i / Y_i) \cong \left(\bigoplus_i X_i\right) / \left(\bigoplus_i Y_i\right),$$

we obtain

(D.20b) $$E_r^p \cong \bigoplus_q E_r^{p,q} \qquad \text{and} \qquad E_\infty^p = \bigoplus_q E_\infty^{p,q}.$$

Since d increases total degree by 1, so does d_r. From $d_r(E_r^p) \subseteq E_r^{p+r}$, we deduce

$$d_r(E_r^{p,q}) \subseteq E_r^{p+r,q+1-r}.$$

We say that d_r has **bidegree** $(r, 1-r)$.

EXAMPLE. Suppose we have a complex

$$S: \qquad \cdots \longrightarrow S_{n-1} \xrightarrow{d} S_n \xrightarrow{d} S_{n+1} \longrightarrow \cdots$$

and a subcomplex

$$R: \qquad \cdots \longrightarrow R_{n-1} \longrightarrow R_n \longrightarrow R_{n+1} \longrightarrow \cdots .$$

The subcomplex R is to have $R_n \subseteq S_n$ for all n, and its maps are to be restrictions of the maps of S. Put $T_n = S_n/R_n$, so that the diagram is

$$
\begin{array}{ccccccccc}
 & & \vdots & & \vdots & & \vdots & & \\
 & & \downarrow & & \downarrow & & \downarrow & & \\
0 & \longrightarrow & R_{n-1} & \longrightarrow & S_{n-1} & \longrightarrow & T_{n-1} & \longrightarrow & 0 \\
 & & \downarrow & & \downarrow & & \downarrow & & \\
0 & \longrightarrow & R_n & \longrightarrow & S_n & \longrightarrow & T_n & \longrightarrow & 0 \\
 & & \downarrow & & \downarrow & & \downarrow & & \\
0 & \longrightarrow & R_{n+1} & \longrightarrow & S_{n+1} & \longrightarrow & T_{n+1} & \longrightarrow & 0 \\
 & & \downarrow & & \downarrow & & \downarrow & & \\
 & & \vdots & & \vdots & & \vdots & &
\end{array}
$$

The rows are exact, and the verticals are complexes. The squares commute because d is consistently defined on R_n and S_n and because d descends to T_n.

Define

$$A = A^0 = \bigoplus_{n=-\infty}^{\infty} S_n, \qquad A^1 = \bigoplus_{n=-\infty}^{\infty} R_n, \qquad A^2 = 0.$$

We are forced to take $A^p = A$ for $p < 0$ and $A^p = 0$ for $p > 2$. The total degree is the index n, and thus

$$A^{0,q} = S_q \qquad \text{and} \qquad A^{1,q} = R_{q+1}.$$

From (D.9a) we have

(D.21a) $\qquad\qquad E_0^{p,q} = 0 \quad$ unless $p = 0$ or $p = 1$,

(D.21b) $\qquad\qquad E_0^{0,q} = T_q, \qquad E_0^{1,q} = R_{q+1}.$

From (D.21a) and Proposition D.11, we obtain

(D.22a) $\qquad E_r^{pq} = 0 \qquad$ for $r \geq 0$ unless $p = 0$ or $p = 1$.

Since d_r has bidegree $(r, \, 1 - r)$, (D.21a) implies also

(D.22b) $\qquad\qquad E_2^{p,q} = E_3^{p,q} = E_4^{p,q} = \dots .$

Elementary calculation with (D.7) shows that the actual values of $E_2^{p,q}$ are given by (D.22a) and

(D.22c)
$$E_2^{0,q} = (S_q \cap \ker d)/((R_q \cap \ker d) + (S_q \cap \text{image } d)) = E_\infty^{0,q}$$
$$E_2^{1,q} = (R_q \cap \ker d)/(R_q \cap \text{image } d) = E_\infty^{1,q}$$

Formulas (D.9b) and (D.22c), together with Proposition D.11 and the last line of (D.18), are approximately equivalent with the exactness of the long exact sequence in (C.31). In fact, (D.9b) gives

$$E_1^{0,q} \cong H^q(T) \qquad \text{and} \qquad E_1^{1,q} \cong H^{q+1}(R).$$

Unwinding the isomorphisms shows that $d_1 : E_1^{0,q} \to E_1^{1,q}$ corresponds to the connecting homomorphism

$$d_1 : H^q(T) \to H^{q+1}(R)$$

of (C.29). This instance of d_1, together with the maps induced on cohomology from inclusion $R \to S$ and passage to the quotient $S \to T$, gives us a complex

(D.23)
$$\cdots \longrightarrow H^q(R) \longrightarrow H^q(S) \longrightarrow H^q(T) \xrightarrow{d_1}$$
$$H^{q+1}(R) \longrightarrow H^{q+1}(S) \longrightarrow \cdots .$$

Proposition D.11 allows us to compute E_2 as the cohomology of (E_1, d_1). Hence

(D.24a)
$$E_2^{0,q} \cong \ker(H^q(T) \to H^{q+1}(R))$$
$$E_2^{1,q} \cong H^{q+1}(R)/\mathrm{image}(H^q(T) \text{ in } H^{q+1}(R)).$$

Meanwhile the definitions of D^p and $D^{p,q}$ give us

$$D^{0,q} = H^q(S)$$
$$D^{1,q} = \mathrm{image}(H^{q+1}(R) \text{ in } H^{q+1}(S))$$
$$D^{2,q} = 0.$$

Hence the last line of (D.18) yields

(D.24b)
$$E_\infty^{0,q} \cong D^{0,q}/D^{1,q-1} = H^q(S)/\mathrm{image}(H^q(R) \text{ in } H^q(S))$$
$$E_\infty^{1,q} \cong D^{1,q}/D^{2,q-1} = \mathrm{image}(H^{q+1}(R) \text{ in } H^{q+1}(S)).$$

Now we can put everything together. We have

$E_\infty^{0,q} \cong H^q(S)/\mathrm{image}(H^q(R) \text{ in } H^q(S))$	by (D.24b)
$\quad$ maps onto $H^q(S)/\ker(H^q(S) \to H^q(T))$	since (D.23) is a complex
$\cong \mathrm{image}(H^q(S) \text{ in } H^q(T))$	
$\subseteq \ker(H^q(T) \to H^{q+1}(R))$	since (D.23) is a complex
$\cong E_2^{0,q}$	by (D.24a).

Since (D.22c) shows that the composition is an isomorphism $E_\infty^{0,q} \cong E_2^{0,q}$, we conclude that (D.23) is exact at S and T. Similarly

$E_2^{1,q} \cong H^{q+1}(R)/\mathrm{image}(H^q(T) \text{ in } H^{q+1}(R)))$	by (D.24a)
$\quad$ maps onto $H^{q+1}(R)/\ker(H^{q+1}(R) \to H^{q+1}(S))$	since (D.23) is a complex
$\cong \mathrm{image}(H^{q+1}(R) \text{ in } H^{q+1}(S))$	
$\cong E_\infty^{1,q}$	by (D.24b),

and the isomorphism $E_2^{1,q} \cong E_\infty^{1,q}$ of (D.22c) allows us to conclude that (D.23) is exact at R.

Note that this example has

$$H(A) = D^0 \supseteq D^1 \supseteq D^2 = 0.$$

Thus $\bigcap_p D^p = 0$, and the filtration of $H(A)$ is separated. We can read off the same conclusion from (D.24b) and the fact that $E_\infty^{2,q} = 0$.

Let us return to the general case. In practice, filtrations are not usually finite. Thus we seek a more widely applicable test for the nice behavior in the example.

Proposition D.25. The following conditions on a spectral sequence $\{E_r\}$ constructed from a decreasing filtration are equivalent:

(a) $A^0 = A$ and $A^{p,q} = 0$ whenever $q < 0$.

(b) the filtration of A is separated, and $E_0^{p,q} = 0$ whenever $p < 0$ or $q < 0$.

(c) the filtration of A is separated, and $\mathrm{gr}^p({}^nA) = 0$ for p not in $\{0, 1, \ldots, n\}$.

PROOF.

(a) $\Longrightarrow$ (b). By (D.18),

$$E_0^{p,q} = A^{p,q}/A^{p+1,q-1}.$$

If $q < 0$, then $A^{p,q} = 0$ and hence $E_0^{p,q} = 0$. If $p < 0$, then

$$A^{p,q} = {}^{p+q}A \cap A^p = {}^{p+q}A \cap A = {}^{p+q}A$$

and $\qquad A^{p+1,q-1} = {}^{p+q}A \cap A^{p+1} = {}^{p+q}A \cap A = {}^{p+q}A,$

so that $E_0^{p,q} = {}^{p+q}A/{}^{p+q}A = 0$.

To see that the filtration is separated, we use that $A^{p,q} = 0$ for $q < 0$. Then

$$A^p = \bigoplus_n ({}^nA \cap A^p) = \bigoplus_n A^{p,n-p} = \bigoplus_{n \geq p} A^{p,n-p} \subseteq \bigoplus_{n \geq p} {}^nA.$$

The intersection of these on p is 0, and thus the filtration is separated.

(b) $\Longrightarrow$ (a). Suppose $A^0 \neq A$. Choose x not in A^0. Since $\bigcup_p A^p = A$, there is some $p < 0$ with $x \in A^p$ and $x \notin A^{p+1}$. Then $E_0^p = A^p/A^{p+1}$ is nonzero. Since $E_0^p \cong \bigoplus_q E_0^{p,q}$, some $E_0^{p,q}$ is nonzero, contradiction.

Consider $A^{p,q}$ with $q < 0$. From (a) we obtain

$$0 = E_0^{p,q} = E_0^{p+1,q-1} = E_0^{p+2,q-2} = \ldots.$$

From (D.18), $E_0^{p,q} = A^{p,q}/A^{p+1,q-1}$. Therefore

$$0 = A^{p,q}/A^{p+1,q-1} = A^{p+1,q-1}/A^{p+2,q-2} = \ldots$$

and $\qquad\qquad A^{p,q} = A^{p+1,q-1} = A^{p+2,q-2} = \ldots.$

If x is in $A^{p,q}$, this chain of equalities shows that

$$x \in A^p \cap A^{p+1} \cap A^{p+2} \cap \ldots,$$

and this intersection is 0 since the filtration is separated. Thus $A^{p,q} = 0$.
  (b) $\Longleftrightarrow$ (c). Since $E_0^{p,q} = A^{p,q}/A^{p+1,n-p-1}$ and $\mathrm{gr}^p(A) = A^p/A^{p+1}$, we have

$$\mathrm{gr}^p({}^nA) = ({}^nA \cap A^p)/({}^nA \cap A^{p+1}) = A^{p,n-p}/A^{p+1,n-p-1} = E_0^{p,n-p}.$$

Thus the condition that $E_0^{p,q} = 0$ whenever $p < 0$ or $q < 0$ implies that $\mathrm{gr}^p({}^nA) = 0$ whenever $p \notin \{0, 1, \ldots, n\}$, and conversely. This completes the proof.

Motivated by (b) in Proposition D.25, we say that the spectral sequence $\{E_r\}$ is **first quadrant** if it satisfies the equivalent conditions of the proposition. First-quadrant spectral sequences are the kind that arise most often in representation theory. If $p < 0$ or $q < 0$, then $E_r^{p,q} = 0$ for all $r \geq 0$, as a consequence of (b) and Proposition D.11.

Proposition D.26. A first-quadrant spectral sequence has
  (a) $E_r^{p,q} = E_{r+1}^{p,q}$ for $r > \max\{p, q+1\}$
  (b) $E_r^{p,q} = E_\infty^{p,q}$ for $r > \max\{p, q+1\}$
  (c) $\bigcap_p D^p = 0$
  (d) ${}^nA = 0$ for $n < 0$
  (e) $H^n(A) = D^{0,n} \supseteq D^{1,n-1} \supseteq \cdots \supseteq D^{n,0} \supseteq D^{n+1,-1} = 0$ for $n \geq 0$.

PROOF. For (a) let $p \geq 0$ and $q \geq 0$, and suppose $r > \max\{p, q+1\}$. We have

$$d_r(E_r^{p,q}) \subseteq E_r^{p+r,q+1-r} = 0 \quad \text{since } r > q+1$$

and $\qquad\qquad d_r(E_r^{p-r,q-1+r}) = d_r(0) = 0 \quad \text{since } r > p.$

Thus

$$E_{r+1}^{p,q} = (\ker d_r|_{E_r^{p,q}})/d_r(E_r^{p-r,q-1+r}) = E_r^{p,q}/0 = E_r^{p,q}.$$

In (b), when $r > \max\{p, q+1\}$, the point is that $E_r^{p,q} = E_\infty^{p,q}$ factor by factor:

$$Z_r^{p,q} = \{z \in A^{p,q} \mid dz \in A^{p+r,q-r}\} = \{z \in A^{p,q} \mid dz = 0\} = Z_\infty^{p,q},$$

$$Z_{r-1}^{p+1,q-1} = \{z \in A^{p+1,q-1} \mid dz \in A^{p+r,q-r}\}$$
$$= \{z \in A^{p+1,q-1} \mid dz = 0\} = Z_\infty^{p+1,q-1},$$

$$dZ_{r-1}^{p-r+1,q+r-2} = d\{z \in {}^{p+q-1}A \cap A^{p-r+1} \mid dz \in A^{p,q}\}$$
$$= d\{z \in {}^{p+q-1}A \mid dz \in A^{p,q}\} \qquad \text{since } A^0 = A \text{ by}$$
$$\text{Proposition D.25a}$$
$$= A^{p,q} \cap \text{image } d$$
$$= B_\infty^{p,q}.$$

Skipping (c) for the moment, consider (d). By Proposition D.25a,

$${}^nA = {}^nA \cap A = {}^nA \cap A^0 = A^{0,n},$$

and this is 0 if $n < 0$, again by Proposition D.25a.

In (e) we have $E_\infty^{p,q} \cong D^{p,q}/D^{p+1,q-1}$ and thus

$$D^{0,n} \supseteq D^{1,n-1} \supseteq \cdots \supseteq D^{n,0} \supseteq D^{n+1,-1}.$$

But

$$D^{n+1,-1} = (Z_\infty^{n+1,-1} + \text{image } d)/(\text{image } d)$$
$$\subseteq (A^{n+1,-1} + \text{image } d)/(\text{image } d) = 0$$

and

$$D^{0,n} = (Z_\infty^{0,n} + \text{image } d)/(\text{image } d)$$
$$= ((A^{0,n} \cap \ker d) + \text{image } d)/(\text{image } d)$$
$$= (({}^nA \cap \ker d) + \text{image } d)/(\text{image } d)$$
$$= (({}^nA \cap \ker d) + d({}^{n-1}A))/d({}^{n-1}A) \qquad \text{by (D.20a)}$$
$$= H^n(A).$$

Finally (c) follows by using (e) to write

$$D^p = \bigoplus_{q \geq 0} D^{p,q} \subseteq \bigoplus_{q \geq 0} D^{0,p+q} = \bigoplus_{q \geq 0} H^{p+q}(A) = \bigoplus_{k \geq p} H^k(A).$$

The intersection of these over all p is 0.

The spectral sequence $\{E_r\}$ is said to **converge** if, for each (p, q),

$$E_r^{p,q} = E_{r+1}^{p,q} = \cdots = E_\infty^{p,q}$$

for r sufficiently large. It is said to **abut** on $H(A)$ if $\bigcap_p D^p = 0$, and we write

(D.27) $$E_r^{p,q} \Longrightarrow H^{p+q}(A)$$

to indicate this relationship. Here $H(A)$ is called the **abutment** of $\{E_r\}$. In the cases of the example and of Proposition D.26, $\{E_r\}$ was convergent and abutted on $H(A)$.

The language with which spectral sequences are applied requires some explanation. Typically some complex A is mentioned, and then a statement of the following kind is made as a proposition: "There exists a first-quadrant spectral sequence

$$E_r^{p,q} \Longrightarrow H^{p+q}(A)$$

with E_2 term (or sometimes just the E_1 term)

$$E_2^{p,q} = (\text{some expression in } p \text{ and } q).$$

The differential on $E_r^{p,q}$ has bidegree $(r, 1 - r)$."

Let us sort out the meaning of this statement. In the first place, A implicitly has a filtration, but we are not told what it is. The fact that the index $p + q$ on $H(A)$ is a superscript ought to tell us that the differential on A increases the total degree and therefore that the filtration is decreasing. But this deduction is not absolutely reliable, since some authors use other conventions about indices. However, we can make this deduction positively from the fact that the bidegree is $(r, 1 - r)$. (We shall note shortly that the second kind of spectral sequence, coming from an increasing filtration with a differential that decreases the total degree, has bidegree $(-r, r - 1)$.) The spectral sequence is given as first quadrant; therefore it converges and abuts on $H(A)$, by Proposition D.26. The notation (D.27) allows us to match the total degrees in the formula

$$E_\infty \cong \mathrm{gr}(H(A)).$$

Namely it tells us that

(D.28) $$\bigoplus_{p+q=n} E_\infty^{p,q} \cong \mathrm{gr}(H^n(A)).$$

Implicitly we are to understand that

$$E_\infty^{p,q} \cong H^{p+q}(A)^p / H^{p+q}(A)^{p+1}$$

because the filtration on $H^n(A)$ is the one in Proposition D.26e. Finally knowledge of $E_2^{p,q}$ and the bidegree might give us information about $E_r^{p,q}$ for large r and therefore a second formula for $E_\infty^{p,q}$.

A proposition announcing a spectral sequence is most helpful when we can see that $E_2^{p,q} = E_\infty^{p,q}$. In this case we say that the spectral sequence **collapses**.

A spectral sequence **degenerates** if there exists an integer r such that, for every n, one has

$$E_r^{n-q,q} = 0 \qquad \text{for } q \neq q(n).$$

If the spectral sequence degenerates and also converges, then $E_\infty^{n-q,q} = 0$ for $q \neq q(n)$. Hence for each n, the sum implicit on the right side of (D.28) has just one term, and we have

$$H^{p_0+q_0}(A) = E_\infty^{p_0,q_0}.$$

In a spectral sequence that collapses and degenerates, we end up with a formula for the cohomology of A: $H^{p_0+q_0}(A) = E_2^{p_0,q_0}$ with $E_2^{p_0,q_0}$ given to us explicitly.

Even if a first-quadrant spectral sequence does not collapse or degenerate, it may be possible to deduce helpful information by using the Grothendieck group defined in §A.4. Suppose that all modules of interest lie in a small good subcategory C' of C. From (D.28) we obtain

$$(D.29) \qquad (H^n(A)) = \sum_{p+q=n} (E_\infty^{p,q})$$

in the Grothendieck group $K(C')$. If in addition only finitely many $E_2^{p,q}$ are nonzero, we obtain, inductively for $r \geq 2$, a second identity in $K(C')$:

$$\sum_{p,q} (-1)^{p+q} (E_2^{p,q}) = \sum_{p,q} (-1)^{p+q} (E_r^{p,q}) \qquad \text{by Theorem C.40}$$

$$= \sum_{p,q} (-1)^{p+q} (E_\infty^{p,q}) \qquad \text{by Proposition D.26b}$$

$$(D.30) \qquad = \sum_n (-1)^n (H^n(A)) \qquad \text{by (D.29).}$$

We have concentrated so far on the spectral sequence corresponding to a decreasing filtration with a differential that increases the total degree. Let us indicate the modifications in the theory necessary to handle an increasing filtration with a differential that decreases the total degree. For everything up to the example, the theory for an increasing filtration follows from that for a decreasing filtration by replacing filtration degrees, total degrees, and complementary degrees (including those of the differential) by their negatives. The differential d_r has bidegree $(-r, r - 1)$. The example can be reworked as corresponding to an increasing filtration, say with

$$A^{-1} = 0, \qquad A^0 = \bigoplus_{n=-\infty}^{\infty} R_n, \qquad A = A^1 = \bigoplus_{n=-\infty}^{\infty} S_n,$$

and with differential that decreases n by 1.

The new part begins with Proposition D.25, which has the following counterpart.

Proposition D.25′. The following conditions on a spectral sequence $\{E_r\}$ constructed from an increasing filtration are equivalent:

(a) $A^{-1} = 0$ and $^nA \subseteq A^n$ for all n.
(b) the filtration of A is separated, and $E_0^{p,q} = 0$ whenever $p < 0$ or $q < 0$.
(c) the filtration of A is separated, and $\mathrm{gr}^p(^nA) = 0$ for p not in $\{0, 1, \ldots, n\}$.

PROOF.

(a) $\Longrightarrow$ (b). From $A^{-1} = 0$ in (a), we have $A^{p,q} = 0$ for $p < 0$ and hence $E_0^{p,q} = A^{p,q}/A^{p-1,q+1} = 0$ for those p. Now let $q < 0$. Then (a) gives $^{p+q}A \subseteq A^{p+q} \subseteq A^{p-1} \subseteq A^p$. Hence

$$A^{p,q} = {}^{p+q}A \cap A^p = {}^{p+q}A = {}^{p+q}A \cap A^{p+1} = A^{p-1,q+1},$$

and $E_0^{p,q} = A^{p,q}/A^{p-1,q+1}$ is 0. The condition $A^{-1} = 0$ forces the filtration to be separated.

(b) $\Longrightarrow$ (a). Suppose $A^{-1} \neq 0$. Choose $x \neq 0$ in A^{-1}. Since the filtration is separated, $\bigcap_p A^p = 0$; thus there is some $p < 0$ with $x \in A^p$

and $x \notin A^{p-1}$. Then $E_0^p = A^p/A^{p-1}$ is not 0. Hence $E_0^{p,q}$ is not 0 for some q, contradiction.

To see that $^nA \subseteq A^n$, assume the contrary and let $x \in {}^nA$ with $x \notin A^n$. Since $\bigcup_p A^p = A$, there is some $p > n$ with $x \in A^p$ and $x \notin A^{p-1}$. Since x is in ^nA, x is in $A^{p,n-p}$ but not $A^{p-1,n-p+1}$. Then $E_0^{p,n-p} = A^{p,n-p}/A^{p-1,n-p+1}$ is not 0. Since $n - p < 0$, this contradicts (b).

(b) $\Longleftrightarrow$ (c). Since $E_0^{p,q} = A^{p,q}/A^{p-1,n-p+1}$ and $\mathrm{gr}^p(A) = A^p/A^{p-1}$, we have

$$\mathrm{gr}^p({}^nA) = ({}^nA \cap A^p)/({}^nA \cap A^{p-1}) = A^{p,n-p}/A^{p-1,n-p+1} = E_0^{p,n-p}.$$

Thus the condition that $E_0^{p,q} = 0$ whenever $p < 0$ or $q < 0$ implies that $\mathrm{gr}^p({}^nA) = 0$ whenever $p \notin \{0, 1, \ldots, n\}$, and conversely. This completes the proof.

We say that the spectral sequence $\{E_r\}$ constructed from an increasing filtration is **first quadrant** if the equivalent conditions of Proposition D.25′ are satisfied. For such a spectral sequence, $E_r^{p,q} = 0$ for all $r \geq 0$ when $p < 0$ or $q < 0$.

Proposition D.26′. A first-quadrant spectral sequence constructed from an increasing filtration has

(a) $E_r^{p,q} = E_{r+1}^{p,q}$ for $r > \max\{p, q+1\}$
(b) $E_r^{p,q} = E_\infty^{p,q}$ for $r > \max\{p, q+1\}$
(c) $\bigcap_p D^p = 0$
(d) $^nA = 0$ for $n < 0$
(e) $0 = D^{-1,n+1} \subseteq D^{0,n} \subseteq \cdots \subseteq D^{n-1,1} \subseteq D^{n,0} = H_n(A)$ for $n \geq 0$.

With minor modifications, the proof is the same as for Proposition D.26.

2. Spectral Sequences of a Double Complex

As in §1, let $\mathcal{C}$ be a good category of R modules closed under countable direct sums. All constructions in this section will take place in this category $\mathcal{C}$.

A **double complex** is a system of modules $C^{p,q}$ and maps

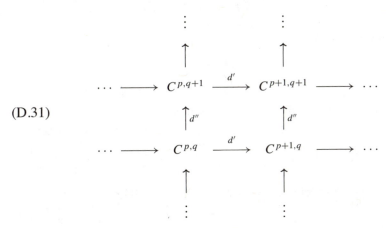

(D.31)

such that each row is a complex, each column is a complex, and each square commutes. (Alternatively we allow all arrows to go in the reverse direction, down or to the left. We occasionally refer to this variant as a **double complex of the second kind.**)

We introduce a modified system $d''^{\#}$ of vertical maps by defining $d''^{\#}$ to be $(-1)^p d''$ on $C^{p,q}$. With the maps $d''^{\#}$ replacing d'', each column of (D.31) is still a complex and has the same cohomology as before. The squares now anticommute:

(D.32) $$d'd''^{\#} = -d''^{\#}d'.$$

Define $A = \bigoplus_{p,q} C^{p,q}$, and define $d : A \to A$ by

(D.33) $$d = d' + d''^{\#}.$$

Since $d'^2 = 0$ and $(d''^{\#})^2 = 0$, (D.32) shows that $d^2 = 0$.

We observe that d carries $C^{p,q}$ into $C^{p+1,q} \oplus C^{p,q+1}$. If we put

(D.34a) $$^nA = \bigoplus_{p+q=n} C^{p,q},$$

then

(D.34b) $$\cdots \longrightarrow {}^{n-1}A \xrightarrow{d} {}^nA \xrightarrow{d} {}^{n+1}A \longrightarrow \cdots$$

is a complex with

(D.34c) $$A = \bigoplus_{n=-\infty}^{\infty} {}^nA.$$

We are in a position to apply the theory of §1 as soon as we impose a decreasing filtration $\{A^p\}_{p \in \mathbb{Z}}$ with $d(A^p) \subseteq A^p$ for all p.

There are two such filtrations of interest, both of them separated. The **first filtration** of the double complex has

(D.35a) $$A^p = \bigoplus_{\substack{h \geq p, \\ \text{all } q}} C^{h,q},$$

and it is clear that this filtration is separated. Then

(D.35b) $$A^{p,q} = {}^{p+q}A \cap A^p = \bigoplus_{\substack{h \geq p, \\ \text{all } q', \\ h+q'=p+q}} C^{h,q'} = \bigoplus_{h \geq 0} C^{p+h,q-h}.$$

Let ${}^{I}E_r^{p,q}$ be the associated spectral sequence. By (D.9b) we have

$$ {}^{I}E_1^p \cong H(A^p/A^{p+1}, d). $$

But d gives the same map on A^p/A^{p+1} as $d''^\#$. Hence

$$ {}^{I}E_1^p \cong H(A^p/A^{p+1}, d'') \cong H\left(\bigoplus_{\text{all } q} C^{p,q}, d'' \right), $$

and we obtain

(D.36) $$ {}^{I}E_1^{p,q} \cong H^q(C^{p,*}, d''). $$

To compute ${}^{I}E_2^{p,q}$, we rewrite

$$ {}^{I}E_1^p \cong (Z_1^p + A^{p+1})/(dA^p + A^{p+1}). $$

Here d_1 is induced by applying d to an element $z_1^p \in Z_1^p$. Such an element has $dz_1^p \in A^{p+1}$, by definition, and we know that $z_1^p \in Z_1^p \subseteq A^p$. If we decompose z_1^p according to (D.35a) as

$$ z_1^p = \sum_{\text{all } q} c^{pq} + \sum_{\substack{h \geq p+1, \\ \text{all } q}} c^{hq}, $$

the second term is in A^{p+1}, and thus there is no loss of generality in assuming $z_1^p = \sum_{\text{all } q} c^{pq}$. Since z_1^p is in Z_1^p, we have $dz_1^p \in A^{p+1}$. We know that $d'(A^p) \subseteq A^{p+1}$, and thus $d''z_1^p$ is in A^{p+1}. But $d''(C^{p,q}) \subseteq C^{p,q+1}$. From the form of z_1^p, we then see that $d''z_1^p = 0$. Hence $dz_1^p = d'z_1^p$, and we can regard the differential d_1 as what is induced by d'. By Proposition D.11 we obtain

(D.37) $$ {}^{I}E_2^{p,q} \cong H^p(H^q(C^{*,*}, d''), d'). $$

The double complex is said to be **first quadrant** if $C^{p,q} = 0$ for all (p, q) with $p < 0$ or $q < 0$. Proposition D.26 yields the following conclusion.

Proposition D.38. Let $\{C^{p,q}\}$ be a first-quadrant double complex of the form (D.31), and let (A, d) be the associated complex given by (D.33) and (D.34). Then there exists a first-quadrant spectral sequence

$$E_r^{p,q} \implies H^{p+q}(A),$$

namely the spectral sequence of the first filtration, with E_1 and E_2 terms

$$^{\mathrm{I}}E_1^{p,q} \cong H^q(C^{p,*}, d'')$$

and $\qquad\qquad ^{\mathrm{I}}E_2^{p,q} \cong H^p(H^q(C^{*,*}, d''), d').$

The differential has bidegree $(r, 1 - r)$.

The **second filtration** of the double complex $\{C^{p,q}\}$ in (D.31) has

$$A^{p'} = \bigoplus_{\substack{\text{all } p, \\ k \geq p'}} C^{pk},$$

and this filtration is clearly separated. Let $^{\mathrm{II}}E_r^{p,q}$ be the associated spectral sequence. Define $\tilde{C}^{p,q} = C^{qp}$, and transplant d' and d'' to $\tilde{C}^{p,q}$. Calculating as for (D.36), we obtain

(D.39) $\qquad\qquad ^{\mathrm{II}}E_1^{p,q} \cong H^q(\tilde{C}^{p,*}, d').$

Then the same kind of argument as for (D.37) yields

(D.40) $\qquad\qquad ^{\mathrm{II}}E_2^{p,q} \cong H^p(H^q(\tilde{C}^{*,*}, d'), d'').$

The analog of Proposition D.38 is as follows.

Proposition D.41. Let $\{C^{p,q}\}$ be a first-quadrant double complex of the form (D.31), and let (A, d) be the associated complex given by (D.33) and (D.34). Then there exists a first-quadrant spectral sequence

$$E_r^{p,q} \implies H^{p+q}(A),$$

namely the spectral sequence of the second filtration, with E_1 and E_2 terms

$$^{\mathrm{II}}E_1^{p,q} \cong H^q(\tilde{C}^{p,*}, d')$$

and $\qquad\qquad ^{\mathrm{II}}E_2^{p,q} \cong H^p(H^q(\tilde{C}^{*,*}, d'), d'').$

The differential has bidegree $(r, 1 - r)$.

The two propositions both give us conclusions about $H^{p+q}(A)$, but the aspects of $H^{p+q}(A)$ that are addressed are normally different. In each case the conclusions are really about $\mathrm{gr}(H(A))$, but the filtrations are different. The nice case is if, for one of the filtrations, the spectral sequence degenerates, so that the contribution to $H^n(A)$ changes at only one level of the filtration. Below we give one computable generic example and one example that is nice in this sense. In the next section the main theorem will be proved by reducing matters to a different nice example of a double complex.

EXAMPLE 1. Let X be the complex torus $\mathbb{C}/(\mathbb{Z}+i\mathbb{Z})$, and let $\mathcal{C}$ be the category of complex vector spaces. Let $C^{p,q}(X)$ be the space of smooth (p, q) forms on X, where p refers to the number (0 or 1) of dz's and q refers to the number (0 or 1) of $d\bar{z}$'s. We take $C^{p,q}(X) = 0$ if p or q is not in $\{0, 1\}$. Then we have operators $\bar{\partial}$ and ∂ defined as follows.:

$$\bar{\partial}f = \frac{\partial f}{\partial \bar{z}} \, d\bar{z} \qquad\qquad \text{for } f \in C^{0,0}(X)$$

$$\bar{\partial}(f\, dz) = -\frac{\partial f}{\partial \bar{z}} \, dz \wedge d\bar{z} \qquad \text{for } f\, dz \in C^{1,0}(X)$$

$$\partial f = \frac{\partial f}{\partial z} \, dz \qquad\qquad \text{for } f \in C^{0,0}(X)$$

$$\partial(f\, d\bar{z}) = \frac{\partial f}{\partial z} \, dz \wedge d\bar{z} \qquad \text{for } f\, d\bar{z} \in C^{0,1}(X)$$

with $\bar{\partial}$ and ∂ equal to 0 otherwise. If d is the deRham operator, given on complex-valued forms by

$$df = \frac{\partial f}{\partial x} \, dx + \frac{\partial f}{\partial y} \, dy$$

$$d(p\, dx + q\, dy) = \left(\frac{\partial q}{\partial x} - \frac{\partial p}{\partial y}\right) dx \wedge dy,$$

then we readily check that $d = \partial + \bar{\partial}$. Hence

$$0 = d^2 = (\partial + \bar{\partial})^2 = \partial^2 + \partial\bar{\partial} + \bar{\partial}\partial + \bar{\partial}^2 = \partial\bar{\partial} + \bar{\partial}\partial,$$

and ∂ anticommutes with $\bar{\partial}$. Consequently if we replace ∂ by $\partial^{\#}$ in the diagram

$$C^{0,1}(X) \xrightarrow{\ \partial\ } C^{1,1}(X)$$

$$\Big\uparrow{\scriptstyle \bar{\partial}} \qquad\qquad\qquad \Big\uparrow{\scriptstyle \bar{\partial}}$$

$$C^{0,0}(X) \xrightarrow{\ \partial\ } C^{1,0}(X)$$

we obtain a commuting square, hence a double complex. The cohomology with respect to d is **deRham cohomology** $H^n(X)$, and the cohomology $H^{p,q}(X)$ with respect to $\bar{\partial}$ is **Dolbeault cohomology**. We can compute each of these explicitly, using Fourier series in two variables. The results are

(D.42a) $H^0(X) = \mathbb{C}, \qquad H^1(X) = \mathbb{C}^2, \qquad H^2(X) = \mathbb{C},$

(D.42b)
$$H^{0,0}(X) = \mathbb{C}, \qquad H^{0,1}(X) = \mathbb{C}$$
$$H^{1,0}(X) = \mathbb{C}, \qquad H^{1,1}(X) = \mathbb{C}.$$

Cocycles representing independent cohomology elements in (D.42a) and (D.42b) are

(D.42c)
$$1 \text{ for } H^0(X), \qquad dx \text{ and } dy \text{ for } H^1(X), \qquad dx \wedge dy \text{ for } H^2(X),$$
$$1 \text{ for } H^{0,0}(X), \qquad d\bar{z} \text{ for } H^{0,1}(X)$$
$$dz \text{ for } H^{1,0}(X), \qquad dz \wedge d\bar{z} \text{ for } H^{1,1}(X).$$

Let us see what Proposition D.38 says about this situation. The associated complex has

$$^0A = C^{0,0}(X), \qquad {}^1A = C^{0,1}(X) \oplus C^{1,0}(X), \qquad {}^2A = C^{1,1}(X),$$

and the differential is d. The cohomology of (A, d) is therefore deRham cohomology. The space $^I E_2^{p,q}$ can be nonzero only for p and q in $\{0, 1\}$, and the bidegree shows all d_r are 0 for $r \geq 2$. Thus the spectral sequence collapses, and we have

$$^I E_\infty^{p,q} = H^p(H^q(C^{*,*}(X), \bar{\partial}), \partial).$$

The inside term $H^q(C^{*,*}(X), \bar{\partial})$ is given in (D.42b) and is always $\mathbb{C}$ for p and q in $\{0, 1\}$. After one checks that ∂ is always the 0 operator on E_1, it follows that $^I E_\infty^{p,q}$ is $\mathbb{C}$ for p and q in $\{0, 1\}$. The interesting case is total degree 1, where the proposition gives

$$\mathrm{gr}(H^1(X)) \cong {}^I E_\infty^{1,0} \oplus {}^I E_\infty^{0,1} \cong \mathbb{C}^2.$$

This result is consistent with (D.42a). However, the spectral sequence has not actually calculated $H^1(X)$, only a composition series of it.

We can examine the role in this context of the Grothendieck group of the category of all finite-dimensional vector spaces. Let us apply

the homomorphism dim, defined in Proposition A.31, to both sides of
(D.29) and (D.30). From (D.29) we obtain, for example,

$$\dim H^1(X) = \dim H^{0,1}(X) + \dim H^{1,0}(X),$$

which is consistent with (D.42). In (D.30), the first two equalities
are trivial because the spectral sequence $^IE_r^{p,q}$ collapses, and the third
equality is a consequence of (D.29). Thus the equality (D.30) tells us
nothing new. But the common value of the two sides is of interest. By
deRham's Theorem, it is the Euler characteristic of the torus X, and the
known value of 0 is consistent with (D.42a).

EXAMPLE 2. In our category C, let $\{C^{p,q}\}$ be a first-quadrant double
complex (D.31) with the property that each row and column is exact
except for the 0th row and column. (To make the assumption complete,
we should have the $(-1)^{\text{st}}$ row and column consist completely of 0's.)
The claim is that

(D.43) $$H^n(C^{*,0}, d') \cong H^n(C^{0,*}, d'')$$

for all n. In fact, we apply Proposition D.38. The assumed exactness
makes

$$H^q(C^{p,*}, d'') = \begin{cases} H^q(C^{0,*}, d'') & \text{for } p = 0 \\ 0 & \text{for } p > 0. \end{cases}$$

Consideration of the bidegree shows that the induced d' has to be the 0
operator, and

$$^IE_2^{p,q} = \begin{cases} H^q(C^{0,*}, d'') & \text{for } p = 0 \\ 0 & \text{for } p > 0. \end{cases}$$

The spectral sequence collapses and degenerates, and we have

(D.44a) $$H^n(A) \cong H^n(C^{0,*}, d'').$$

We can argue similarly with the second filtration and Proposition D.41,
and we obtain

(D.44b) $$H^n(A) \cong H^n(C^{*,0}, d').$$

Then (D.43) follows by combining (D.44a) and (D.44b).

Finally let us discuss double complexes $\{C^{p,q}\}$ of the second kind. The arrows now point to the left or down. We define ^nA and A as in (D.34). The **first filtration** has

$$A^p = \bigoplus_{\substack{h \leq p \\ \text{all } q}} C^{hq},$$

and the **second filtration** has

$$A^{p'} = \bigoplus_{\substack{\text{all } p, \\ k \leq p'}} C^{pk}.$$

Both of these filtrations are separated. The double complex again is **first quadrant** if $C^{p,q} = 0$ for all (p, q) with $p < 0$ or $q < 0$. We obtain analogs of Propositions D.38 and D.41 for this complex, with H_* replacing H^* everywhere. The bidegree is $(-r, r-1)$ for each filtration.

3. Derived Functors of a Composition

In this section we suppose that $\mathcal{C}$, $\mathcal{C}'$, and $\mathcal{C}''$ are good categories of modules (possibly with different rings), that they are closed under countable direct sums, and that $\mathcal{C}$ and $\mathcal{C}'$ have enough projectives and enough injectives. Shortly we shall suppose that $F : \mathcal{C} \to \mathcal{C}'$ and $G : \mathcal{C}' \to \mathcal{C}''$ are (one-sided exact) functors, and we shall investigate the derived functors of $G \circ F$.

Before treating derived functors, we need a particular construction of a double complex. Let X be a complex in $\mathcal{C}'$:

(D.45a) $X :$ $0 \longrightarrow C_0 \xrightarrow{d_0} C_1 \xrightarrow{d_1} C_2 \longrightarrow \cdots.$

Let $H^p(X)$ be the pth cohomology of X, and let

(D.45b) $0 \longrightarrow H^p(X) \longrightarrow M_{p,0} \longrightarrow M_{p,1} \longrightarrow \cdots$

be an injective resolution of $H^p(X)$. We shall construct a double complex

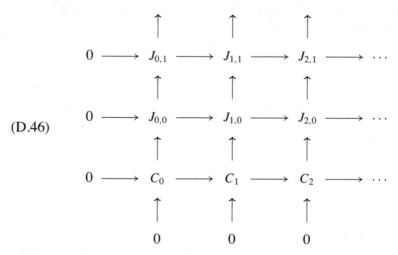

(D.46)

such that each column is an injective resolution and such that

(D.47) $\qquad H^p(J_{*,q}) = M_{p,q} \qquad$ for all (p, q).

To do so, we let $Z_i = \ker d_i$ and $B_i = \operatorname{image} d_{i-1}$, so that $H^i(X) = Z_i/B_i$. Let

$$0 \longrightarrow B_p \longrightarrow K_{p,0} \longrightarrow K_{p,1} \longrightarrow \cdots$$

be an injective resolution of B_p. From (C.35) and the exact sequence

$$0 \longrightarrow B_p \longrightarrow Z_p \longrightarrow H^p(X) \longrightarrow 0,$$

we obtain a diagram

$$
\begin{array}{ccccccccc}
 & & \vdots & & \vdots & & \vdots & & \\
 & & \uparrow & & \uparrow & & \uparrow & & \\
0 & \longrightarrow & K_{p,0} & \longrightarrow & I_{p,0} & \longrightarrow & M_{p,0} & \longrightarrow & 0 \\
 & & \uparrow & & \uparrow & & \uparrow & & \\
0 & \longrightarrow & B_p & \longrightarrow & Z_p & \longrightarrow & H^p(X) & \longrightarrow & 0 \\
 & & \uparrow & & \uparrow & & \uparrow & & \\
 & & 0 & & 0 & & 0 & &
\end{array}
$$

(D.48)

such that $I_{p,q} = K_{p,q} \oplus M_{p,q}$, the horizontal maps of $K_{p,q}$ and $I_{p,q}$ are respectively the standard inclusion and the standard projection, the middle column is an injective resolution, and the squares commute. Similarly (C.35) and the exact sequence

$$0 \longrightarrow Z_p \longrightarrow C_p \xrightarrow{d_p} B_{p+1}(X) \longrightarrow 0,$$

yield a diagram

(D.49)

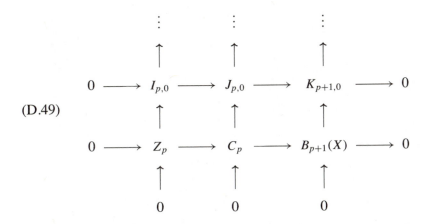

such that $J_{p,q} = I_{p,q} \oplus K_{p+1,q}$, the horizontal maps of $I_{p,q}$ and $J_{p,q}$ are respectively the standard inclusion and the standard projection, the middle column is an injective resolution, and the squares commute.

Combining (D.48) and (D.49) gives a diagram

with commuting squares. The diagram (D.46) results by taking compositions of horizontal maps above and collapsing the piece

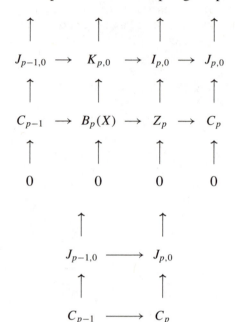

to

$$
\begin{array}{ccc}
\uparrow & & \uparrow \\
J_{p-1,0} & \longrightarrow & J_{p,0} \\
\uparrow & & \uparrow \\
C_{p-1} & \longrightarrow & C_p \\
\uparrow & & \uparrow \\
0 & & 0
\end{array}
$$

In the resulting diagram (D.46), the columns are injective resolutions and the squares commute. Writing $\hookrightarrow$ for a one-one map and $\twoheadrightarrow$ for an onto map, we see that the kernel of $J_{p,q} \to J_{p+1,q}$ comes from

$$
J_{p,q} \ \twoheadrightarrow \ K_{p+1,q} \ \hookrightarrow \ I_{p+1,q} \ \hookrightarrow \ J_{p+1,q}
$$

and is $\ker(J_{p,q} \twoheadrightarrow K_{p+1,q}) = I_{p,q}$ since the second and third maps are one-one. The image of $J_{p-1,q} \to J_{p,q}$ comes from

$$
J_{p-1,q} \ \twoheadrightarrow \ K_{p,q} \ \hookrightarrow \ I_{p,q} \ \hookrightarrow \ J_{p,q}
$$

and is the image of $K_{p,q}$ under the two inclusions. Hence the pth cohomology of the row $J_{*,q}$ is $I_{p,q}/K_{p,q} \cong M_{p,q}$. This completes the construction.

If we had started with a complex

$$
\text{(D.50)} \qquad X': \qquad 0 \longleftarrow C_0 \longleftarrow C_1 \longleftarrow C_2 \longleftarrow \cdots \, ,
$$

we could have arranged to have all the arrows in (D.46) point to the left or down and for the $J_{p,q}$ to be projectives. The above argument requires only notational changes to handle this case.

Recall the table in (C.24) that associates a choice of projective or injective resolution to be used with the various kinds of one-sided exact functors. We now suppose that $F : \mathcal{C} \to \mathcal{C}'$ and $G : \mathcal{C}' \to \mathcal{C}''$ are one-sided exact functors. Under additional hypotheses we shall construct a spectral sequence relating the composition of derived functors with the derived functors of the composition $G \circ F$. One hypothesis will be that the target four-term exact sequences of F are the source four-term exact sequences whose exactness is preserved by G. (See the end of (C.12).) The other hypothesis will be that F carries projectives or injectives, whichever are used for F, to projectives or injectives, whichever are used for G.

The two hypotheses together yield eight variations. We prove the result for one of the eight variations in the following theorem and state all eight variations afterward.

Theorem D.51 (Grothendieck Spectral Sequence). Let $F : \mathcal{C} \to \mathcal{C}'$ and $G : \mathcal{C}' \to \mathcal{C}''$ be covariant left-exact functors such that F carries injectives to injectives. If V is a module in $\mathcal{C}$, then there exists a first-quadrant spectral sequence

$$E_r^{p,q} \implies (G \circ F)^{p+q}(V)$$

with E_2 term $$E_2^{p,q} \cong G^p(F^q(V)).$$

The differential has bidegree $(r, 1-r)$.

PROOF. Let

$$0 \longrightarrow V \longrightarrow I_0 \longrightarrow I_1 \longrightarrow \cdots$$

be an injective resolution of V in $\mathcal{C}$. In (D.45a) we take X to be the complex

$$X : \qquad 0 \longrightarrow F(I_0) \longrightarrow F(I_1) \longrightarrow \cdots .$$

This has

(D.52) $$H^p(X) = F^p(V).$$

Define $M_{p,q}$ as in (D.45b) to be the members of an injective resolution of $H^p(X)$. We form (D.46) with $C_p = F(I_p)$ and apply G to obtain the first-quadrant double complex

$$
\begin{array}{ccccccc}
\vdots & & \vdots & & \vdots & \\
\uparrow & & \uparrow & & \uparrow & \\
G(J_{0,1}) & \longrightarrow & G(J_{1,1}) & \longrightarrow & G(J_{2,1}) & \longrightarrow & \cdots \\
\uparrow & & \uparrow & & \uparrow & \\
G(J_{0,0}) & \longrightarrow & G(J_{1,0}) & \longrightarrow & G(J_{2,0}) & \longrightarrow & \cdots
\end{array}
$$

(D.53)

Thus $C^{p,q} = G(J_{p,q})$ for $p \geq 0$ and $q \geq 0$. (We take $C^{p,q} = 0$ if $p < 0$ or $q < 0$.)

We shall calculate the spectral sequence of the first filtration of (D.53) by applying Proposition D.38. Since F carries injectives to injectives, $F(I_p)$ is injective. Thus

$$0 \longrightarrow 0 \longrightarrow F(I_p) \longrightarrow J_{p,0} \longrightarrow J_{p,1} \longrightarrow \cdots$$

is an injective resolution of 0, and

$$0 \longrightarrow (G \circ F)(I_p) \longrightarrow G(J_{p,0}) \longrightarrow G(J_{p,1}) \longrightarrow \cdots$$

has all cohomology 0. Consequently

(D.54) $$0 \longrightarrow G(J_{p,0}) \longrightarrow G(J_{p,1}) \longrightarrow \cdots$$

has all cohomology 0 in degree $q > 0$ and has cohomology $(G \circ F)(I_p)$ in degree 0. Therefore

$$
{}^{I}E_1^{p,q} \cong H^q(C^{p,*}, d'') = \begin{cases} (G \circ F)(I_p) & \text{for } q = 0 \\ 0 & \text{for } q > 0. \end{cases}
$$

Since $G \circ F$ is defined by injective resolutions,

$$
{}^{I}E_2^{p,q} \cong H^p(H^q(C^{*,*}, d''), d') = \begin{cases} (G \circ F)^p(V) & \text{for } q = 0 \\ 0 & \text{for } q > 0. \end{cases}
$$

This spectral sequence collapses and degenerates, and we obtain

(D.55) $$H^n(\{C^{*,*}\}) \cong (G \circ F)^n(V).$$

Next we shall calculate the spectral sequence of the second filtration of (D.53) by applying Proposition D.41. The E_1 term is

(D.56) $\qquad {}^{\mathrm{II}}E_1^{p,q} \cong H^q(\tilde{C}^{p,*}, d') = H^q(C^{*,p}, d') = H^q(G(J_{*,p}), d').$

Consider the complexes

$$J_{q-1,p} \xrightarrow{\alpha} J_{q,p} \xrightarrow{\beta} J_{q+1,p}$$

and

$$G(J_{q-1,p}) \xrightarrow{G(\alpha)} G(J_{q,p}) \xrightarrow{G(\beta)} G(J_{q+1,p}).$$

Relative to the direct-sum decomposition $J_{q,p} = I_{q,p} \oplus K_{q+1,p}$, β is 0 on $I_{q,p}$ and is the inclusion of $K_{q+1,p}$ in $I_{q+1,p}$ on $K_{q+1,p}$. Since G is left exact and G respects direct sums, $G(\beta)$ is 0 on $G(I_{q,p})$ and $G(\beta)$ is one-one on $K_{q+1,p}$. Therefore

$$\ker G(\beta) = G(I_{q,p}).$$

Similarly relative to $J_{q-1,p} = I_{q-1,p} \oplus K_{q,p}$, α is 0 on $I_{q-1,p}$ and $G(\alpha)$ is 0 on $G(I_{q-1,p})$. Since G respects direct sums,

$$\text{image } G(\alpha) = G(K_{q,p}).$$

Now

$$I_{q,p} = K_{q,p} \oplus M_{q,p}$$

implies

$$G(I_{q,p}) = G(K_{q,p}) \oplus G(M_{q,p}).$$

Consequently

$$H^q(G(J_{*,p}), d') = \ker G(\beta)/\text{image } G(\alpha) = G(I_{q,p})/G(K_{q,p}) \cong G(M_{q,p}),$$

and (D.56) shows that the E_1 term is

$$^{\mathrm{II}}E_1^{p,q} \cong G(M_{q,p}).$$

So the E_2 term is

$$^{\mathrm{II}}E_2^{p,q} \cong H^p(G(M_{q,p}), d'').$$

Recalling from (D.45b) that $M_{q,*}$ defines an injective resolution of $H^q(X)$, we see that

$$^{\mathrm{II}}E_2^{p,q} \cong G^p(H^q(X)).$$

Substitution from (D.52) yields

$$^{\mathrm{II}}E_2^{p,q} \cong G^p(F^q(V)).$$

Finally Proposition D.41 gives

$$^{\mathrm{II}}E_2^{p,q} \Longrightarrow H^{p+q}(\{C^{*,*}\}),$$

and the right side has been identified by (D.55) as $(G \circ F)^{p+q}(V)$. This completes the proof.

As we mentioned earlier, Theorem D.51 comes in eight variations. We now list all eight. Variations 2 and 5 are the most important for representation theory, and Variation 5 is the one explicitly stated in Theorem D.51.

VARIATIONS OF THEOREM D.51.

(1) F covariant right exact, G covariant right exact
                F assumed to send projectives to projectives,

$$E_2^{p,q} \cong G_p(F_q(V)), \quad E_r^{p,q} \Longrightarrow (G \circ F)_{p+q}(V), \quad \text{bidegree } (-r, r-1).$$

(2) F covariant right exact, G contravariant left exact
                F assumed to send projectives to projectives,

$$E_2^{p,q} \cong G^p(F_q(V)), \quad E_r^{p,q} \Longrightarrow (G \circ F)^{p+q}(V), \quad \text{bidegree } (r, 1-r).$$

(3) F contravariant left exact, G covariant left exact
                F assumed to send projectives to injectives,

$$E_2^{p,q} \cong G^p(F^q(V)), \quad E_r^{p,q} \Longrightarrow (G \circ F)^{p+q}(V), \quad \text{bidegree } (r, 1-r).$$

(4) F contravariant left exact, G contravariant right exact
                F assumed to send projectives to injectives,

$$E_2^{p,q} \cong G_p(F^q(V)), \quad E_r^{p,q} \Longrightarrow (G \circ F)_{p+q}(V), \quad \text{bidegree } (-r, r-1).$$

(5) F covariant left exact, G covariant left exact
                F assumed to send injectives to injectives,

$$E_2^{p,q} \cong G^p(F^q(V)), \quad E_r^{p,q} \Longrightarrow (G \circ F)^{p+q}(V), \quad \text{bidegree } (r, 1-r).$$

(6) F covariant left exact, G contravariant right exact
                F assumed to send injectives to injectives,

$$E_2^{p,q} \cong G_p(F^q(V)), \quad E_r^{p,q} \Longrightarrow (G \circ F)_{p+q}(V), \quad \text{bidegree } (-r, r-1).$$

(7) F contravariant right exact, G covariant right exact
                F assumed to send injectives to projectives,

$$E_2^{p,q} \cong G_p(F_q(V)), \quad E_r^{p,q} \Longrightarrow (G \circ F)_{p+q}(V), \quad \text{bidegree } (-r, r-1).$$

(8) F contravariant right exact, G contravariant left exact
                F assumed to send injectives to projectives,

$$E_2^{p,q} \cong G^p(F_q(V)), \quad E_r^{p,q} \Longrightarrow (G \circ F)^{p+q}(V), \quad \text{bidegree } (r, 1-r).$$

4. Derived Functors of a Filtered Module

Although derived functors of compositions are the chief source of spectral sequences in representation theory, there are other ways in which spectral sequences of interest arise. We consider one such way in this section.

We suppose that C and C' are good categories of modules, that they are closed under countable direct sums, and that C has enough projectives and enough injectives. We work with a finite filtration in C and with a one-sided exact functor $F : C \to C'$, adapting the numbering of the filtration to the exactness of F. The main result is as follows.

Proposition D.57.

(a) Let F be covariant and left exact, and let

$$X = X_0 \supseteq X_1 \supseteq \cdots \supseteq X_N \supseteq X_{N+1} = 0$$

be a filtration in C of a module X in C. Then there exists a convergent spectral sequence

$$E_r^{p,q} \Longrightarrow F^{p+q}(X)$$

with differential of bidegree $(r, 1 - r)$ and with E_1 term

$$E_1^{p,q} = F^{p+q}(X_p / X_{p+1}).$$

(b) Let F be covariant and right exact, and let

$$0 = X_{-1} \subseteq X_0 \subseteq X_1 \subseteq \cdots \subseteq X_N = X$$

be a filtration in C of a module X in C. Then there exists a convergent spectral sequence

$$E_r^{p,q} \Longrightarrow F_{p+q}(X)$$

with differential of bidegree $(-r, r - 1)$ and with E_1 term

$$E_1^{p,q} = F_{p+q}(X_p / X_{p-1}).$$

REMARKS. The spectral sequences are not quite first quadrant, but they still converge and abut on $H(A)$. In (a), for example, it will be evident from the proof that $A^0 = A$ and that $A^{p,q} = 0$ whenever $q < -N$. The proof of (a) and (b) in Proposition D.26 is valid for

$$r > N + \max\{p, q + 1\}.$$

There are versions of the proposition for F contravariant, but we omit their statements.

PROOF. We prove (a), the argument for (b) being similar. Let

$$0 \longrightarrow X_N \longrightarrow I_{0,N} \longrightarrow I_{1,N} \longrightarrow \cdots$$

be an injective resolution of X_N. From an injective resolution

$$0 \longrightarrow X_{N-1}/X_N \longrightarrow I_{0,N-1} \longrightarrow I_{1,N-1} \longrightarrow \cdots$$

of X_{N-1}/X_N, (C.35) allows us to make $I_{*,N-1} \oplus I_{*,N}$ into an injective resolution of X_{N-1} in such a way that

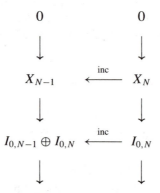

commutes. Continuing in this way inductively by using (C.35) with an injective resolution of X_p/X_{p+1}, we arrive at a commutative diagram

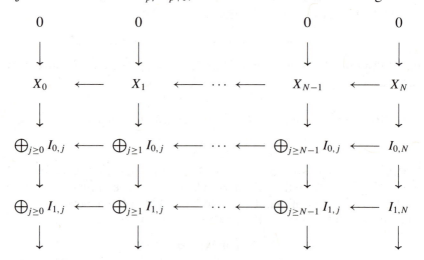

in which the horizontal maps are inclusions and the columns are injective resolutions. Then we apply F to this diagram, dropping the terms $F(X_p)$

and obtaining

$$
\begin{array}{ccccccc}
0 & & 0 & & & & 0 \\
\downarrow & & \downarrow & & & & \downarrow \\
\bigoplus_{j\geq 0} F(I_{0,j}) & \longleftarrow & \bigoplus_{j\geq 1} F(I_{0,j}) & \longleftarrow & \cdots & \longleftarrow & F(I_{0,N}) \\
\downarrow & & \downarrow & & & & \downarrow \\
\bigoplus_{j\geq 0} F(I_{1,j}) & \longleftarrow & \bigoplus_{j\geq 1} F(I_{1,j}) & \longleftarrow & \cdots & \longleftarrow & F(I_{1,N}) \\
\downarrow & & \downarrow & & & & \downarrow
\end{array}
$$

(D.58)

Define

$$
A^p = \bigoplus_{\substack{i\geq 0,\\ j\geq p}} F(I_{i,j}),
$$

so that

$$
A^0 \supseteq A^1 \supseteq \cdots \supseteq A^N \supseteq 0
$$

is a decreasing filtration of $A = A^0$. The vertical maps in (D.58) define a differential. We introduce a grading into A by setting

$$
{}^n A = \bigoplus_{j\geq 0} F(I_{n,j}).
$$

The desired spectral sequence is the one constructed from these data by the method of §1. From (D.9a) we have

$$
E_0^p = A^p / A^{p+1} = \bigoplus_{i\geq 0} F(I_{i,p}),
$$

and from (D.9b) we have

$$
E_1^p = H(A^p / A^{p+1}) \cong \bigoplus_{i\geq 0} F^i(X_p / X_{p+1}).
$$

Thus

$$
E_1^{p,q} \cong F^{p+q}(X_p / X_{p+1}).
$$

From (D.16) we have

$$
E_\infty^0 \oplus \cdots \oplus E_\infty^N \cong \operatorname{gr} H(A) = \operatorname{gr} H(A^0) = \operatorname{gr} \bigoplus_{i\geq 0} F^i(X_0).
$$

Hence

$$
E_\infty^{0,p+q} \oplus \cdots \oplus E_\infty^{p,q} \oplus \cdots \oplus E_\infty^{N,N-p-q} \cong \operatorname{gr} F^{p+q}(X_0),
$$

and the formula for the abutment follows.

NOTES

Introduction

Some books about the representation theory of reductive groups with emphasis on the analytic side are Warner [1972], Knapp [1986], and Wallach [1988]. Two books about the theory with emphasis on the algebraic side are Vogan [1981a] and Knapp [1988]. The first half of Vogan [1987a] is another treatment of the theory, discussing both analysis and algebra, but giving only some of the proofs. Many of the results cited below have expositions in one or another of these books.

Harish-Chandra [1951] in a single paper introduced Verma modules (defined in §IV.7 of the text), gave a construction of all the finite-dimensional irreducible representations of semisimple Lie algebras (as quotients of Verma modules), proved the Harish-Chandra isomorphism characterizing the center of the universal enveloping algebra (see Theorem 4.95), classified infinitesimal characters (see §IV.8), and proved that each K type has finite multiplicity in an irreducible infinite-dimensional representation of a complex semisimple Lie group. The later paper Harish-Chandra [1953] extended this last result from complex semisimple Lie groups to all real semisimple Lie groups and proved many more basic facts about the representation theory of real semisimple Lie groups.

Some of the results in Harish-Chandra's 1951 paper were obtained years before their publication. Chevalley [1948a] had given an algebraic construction of all the finite-dimensional irreducible representations of semisimple Lie algebras, and Chevalley [1948b] pointed out that Harish-Chandra had simultaneously and independently obtained a different proof of the same theorem.

Bargmann [1947] had earlier classified the irreducible unitary representations of $SL(2, \mathbb{R})$. His classification had depended on an assumption about existence of suitably smooth vectors in each representation, and Gårding [1947] proved this assumption by showing that the C^∞ vectors are always dense. Harish-Chandra needed the corresponding fact Theorem 0.1 that the analytic vectors are dense, and this he proved for semisimple groups as Theorem 3 in his 1953 paper. Later Nelson [1959] extended this result to all Lie groups. Another fact that Harish-Chandra needed was that the center of the universal enveloping algebra acts by scalars on the analytic vectors in an irreducible unitary representation. This is the result we state as Theorem 0.2. The proof combines Gårding's construction, the results of Segal [1952], and a supplementary argument by Mautner that Harish-Chandra [1953] quotes on pp. 226–227. Armed with these preliminary results, Harish-Chandra [1953] proved the present Theorems 0.3, 0.4, and 0.6b (with no assumption of linearity for G) as Lemma 33, Theorem 5, and Theorem 8, respectively. Theorem 0.6a for any particular G follows from Theorem 9 of Harish-Chandra [1953] once one knows that any irreducible admissible $(\mathfrak{g}, K)$ module is the underlying $(\mathfrak{g}, K)$ module of an irreducible Banach-space representation of the group. Such a result follows from the Subquotient Theorem, which is Theorem 4 of Harish-Chandra [1954a] for linear G and is in Lepowsky [1973] and Rader [1976] for general G.

Bargmann [1947], as we mentioned earlier, classified the irreducible unitary representations of $SL(2, \mathbb{R})$. More precisely he worked with the isomorphic group $SU(1, 1)$, which has the advantage of making the action of the standard maximal compact subgroup more evident. His approach effectively was to find all possible irreducible $(\mathfrak{g}, K)$ modules, discard those that cannot be infinitesimally unitary, and give global realizations of the remaining ones. At the end of the paper, Bargmann mentioned a plan to publish similar results for $SL(2, \mathbb{C})$ in a sequel, but this second paper became unnecessary after Gelfand-Naimark [1947b] independently obtained the same results.

Meanwhile, in about 1946 Mackey began a systematic study of the representation theory of locally compact groups. He was heavily influenced by Weil [1940], which juxtaposed representation theories for compact groups and for locally compact abelian groups. Although Weil's book had generalized induced representations to compact groups from Frobenius's theory for finite groups, induced representations did not play a significant role in Weil's theory and did not appear a fruitful topic for further study. Instead Mackey came to the discovery of induced representations for locally compact groups by a roundabout route. Mackey began by looking for locally compact groups for which he could classify the irreducible unitary representations. Together with Gleason he worked out the representation theory of the semidirect product of $\mathbb{Z}$ and a two-element group, and subsequently he generalized this construction to some other semidirect products. Gelfand-Naimark [1947a] handled the $ax + b$ group for the line, which is another semidirect product. In all cases the point of view was one of classification, and the idea was to work backwards from a given irreducible representation to see what properties it might have. (See Mackey [1992], 36, for a discussion of this point.)

While working on these matters, Mackey paused to study the Heisenberg commutation relations more closely. In the one-dimensional case these can be written $PQ - QP = cI$ for unknown unbounded self-adjoint operators P and Q on a common Hilbert space. With hindsight it is known that this equation reflects the bracket relations in the Lie algebra of the Heisenberg group (the semidirect product group given as three-by-three upper triangular matrices with 1's on the diagonal), but the Heisenberg group was not in vogue at this time. Weyl had observed that the n-dimensional version of the Heisenberg equation is formally equivalent with the equation $U(x)V(y) = e^{ix \cdot y}V(y)U(x)$ for unknown unitary representations U and V of $\mathbb{R}^n$ on a common Hilbert space H, and Stone [1930] had proved that the only jointly irreducible solutions of the n-dimensional equation, up to unitary equivalence, are $U(x)f(t) = f(x + t)$ and $V(y)f(t) = e^{iy \cdot t}f(t)$ for $f \in L^1(\mathbb{R}^n)$. Mackey saw that Stone's uniqueness theorem could be extended to any (separable) locally compact group G and its dual $\widehat{G}$. In the extended theorem, U, V, and $e^{ix \cdot y}$ get replaced by a unitary representation of G on H, a unitary representation of $\widehat{G}$ on H, and the natural map $G \times \widehat{G} \to S^1$. A second generalization of Stone's uniqueness theorem results when one regards the representation of $\widehat{G}$ on H as a mapping of a certain class of functions on G to bounded linear operators on H and then interprets the representation of $\widehat{G}$ as a projection-valued measure on G. Proofs of these two generalizations of Stone's uniqueness theorem appear in Mackey [1949a], the first generalization following easily from the second. Mackey comments about the proof of the second generalization that the argument is inspired by the techniques in Gelfand-Naimark [1947a], and he observes that G no longer needs to be abelian for the result to be valid.

From the second generalization it is only a small step to ask whether uniqueness still holds if the projection-valued measure is defined on a G-space S. The answer is negative, and the compensating positive result is the Imprimitivity Theorem, which appeared in Mackey [1949b]. Mackey realized at this stage that his work had solved the classification problem that he had set aside: He had obtained a theory for addressing the irreducible unitary representations of any semidirect product. Examining this theory more closely, he was led to the idea that there should be a construction for passing from representations of a closed subgroup to representations of the whole group. Such representations were first called "imprimitive" but were soon called "induced." They are the subject of Mackey [1950a] and [1950b].

During this period multiplier representations, which were a type of representation generalizing the concrete principal series in Bargmann's theory for $SL(2, \mathbb{R})$, were the realization of choice for writing down new representations. Gelfand-Naimark [1950] used multiplier representations heavily in their treatment of the representation theory of the complex classical groups.

As Mackey developed his theory of induced representations, he became aware that multiplier representations were always induced representations. Perhaps other authors came to the same realization at this time, but the statement does not appear in the literature until later. Knowledge of the relationship allows one to go beyond special examples in treating representations, and only Mackey's theory had this feature in 1950.

Harish-Chandra [1955], [1956a], and [1956b] introduced holomorphic discrete series as a generalization of the discrete-series representations of $SL(2, \mathbb{R})$. An exposition appears in Chapter VI of Knapp [1986]. The special case of this theory when the underlying group is compact was discovered independently at about the same time by Borel and Weil (see Serre [1954]) and by Tits [1955], 112–113. An exposition appears in §V.7 of Knapp [1986].

Sections 2, 3, and 4 of this text, starting after Theorem 0.9, are an abbreviated version of the exposition in Knapp [1993]. The Bott-Borel-Weil Theorem appears in Bott [1957], and an exposition of the proof is in Baston-Eastwood [1989]. This theorem, which concerns representations of G and sheaf cohomology, is more or less equivalent with an algebraic theorem of Kostant's (see Kostant [1961] and the present Theorem 4.139) concerning representations of a maximal torus and Lie algebra cohomology relative to a maximal nilpotent subalgebra of the Lie algebra of G. This equivalence is discussed on an analytic level in Cartier [1961]. Section IV.11 of the present text addresses the equivalence of Kostant's Theorem with an algebraic analog of the Bott-Borel-Weil Theorem. For more detail about representation theory in locally convex complete linear topological spaces, see the beginning of Harish-Chandra [1966].

For background on the bundle theory and complex geometry in §3, see Steenrod [1951] and Wells [1973], respectively. For the improved theorem about a holomorphic structure on $G \times_L V$ that is mentioned after (0.19), see Tirao-Wolf [1970]. The question of closed range for $\bar{\partial}$ when G is compact is addressed in Wells [1973]. For Cartan's Theorem B, mentioned before (0.32), see Gunning-Rossi [1965], 243.

Harish-Chandra classified the discrete series for semisimple groups in a series of three monumental papers ending with Harish-Chandra [1966]. The nature of the parametrization is to describe discrete series in terms of their global characters. Harish-Chandra did embed each discrete-series representation into $L^2(G)$, but the embedding is not so explicit as to reveal properties of the representation. See Chapters IX and XII of Knapp [1986] and §XI.8 of the present text for discussion of the discrete series.

Soon after Harish-Chandra's work, Griffiths proved a vanishing theorem for some cohomology in a setting generalizing that in the Bott-Borel-Weil Theorem: G is allowed to be noncompact, Γ is a discrete subgroup of G with G/Γ compact, the sections are required to be Γ invariant, and the parameter has a suitable regularity property. Langlands [1966] observed that a formal computation of global characters suggested that the cohomology in the remaining degree realizes an appropriate discrete-series representation. Langlands made the leap of taking $\Gamma = 1$ and then made a specific conjecture about how to realize all discrete-series representations in an L^2 version of $\bar{\partial}$ cohomology of line bundles over G/B, where B is a compact Cartan subgroup. Written at about the same time, the one-page paper Kostant [1966] makes a reference to a conjectural realization of all discrete series in terms of sheaf or Dolbeault cohomology sections over orbits of G in the dual of its Lie algebra.

The proof of a version of the Langlands conjecture analogous to Theorem 0.33, with the parameter assumed very regular, is largely in Schmid [1967] and [1970], with some details of proof postponed to Schmid [1971]. Aguilar-Rodriguez [1987] removed the restrictive hypothesis on the parameter. The first success in proving the Langlands conjecture in terms of L^2 cohomology was in Narasimhan-Okamoto [1970], which handles the case that the bundles are over G/K and G/K is complex. Afterward Schmid completely proved the Langlands conjecture in a series of papers ending with Schmid [1976]. For some intermediate history of this development, including discussion of work by other authors, see Knapp [1986], 736.

For $H^{0,j}(G/L, V)$ when L need no longer be compact, Wong [1992] proved that $\bar{\partial}$ has closed range if V is finite dimensional. Rawnsley-Schmid-Wolf [1983], under various complex-analysis assumptions on G/L, produced some harmonic representatives and showed their square integrability. Barchini-Knapp-Zierau [1992] and Barchini [1993] exhibited harmonic representatives in the generality of Wong's work, using an intertwining operator, and Zierau [1993] proved proved the square integrability of these harmonic representatives in some cases.

Zuckerman [1978] gave a series of lectures introducing an algebraic analog of the Dolbeault-cohomology section construction and establishing a number of properties of this analog. His construction of $(\mathfrak{g}, K)$ modules was somewhat more general than for connected reductive groups with compact center, where it reduces to (0.42). In any event the theory hinged on the Duality Theorem, for which Zuckerman gave a precise statement. P. Trauber and Zuckerman together gave a number of ideas toward a proof of this theorem, but they did not succeed in giving a complete argument and did not publish their work. Zuckerman [1978] was able to obtain the effect on infinitesimal characters, bounds on multiplicities of K types, and, apart from the gap in the Duality Theorem, a vanishing theorem and some form of the Irreducibility Theorem. Detailed citations are in the Notes for the individual chapters below. For a proof that (0.41) is an injective resolution under the conditions stated, see Knapp [1993]. Zuckerman did not publish his work. The book Vogan [1981a] incorporates Zuckerman's theory, while avoiding the Duality Theorem, and it extends the theory and puts it in a broader context.

Sections 5, 7, and 8 are an abbreviated version of the exposition Vogan [1993]. In §5 it is noted that $\mathcal{R}^j$ is not the appropriate functor to model unitary representations. In fact, Wong [1992] showed for Z finite dimensional that $H^{0,j}(G/L, Z^{\#})$ is the "maximal globalization" of $\mathcal{R}^j(Z)$ in the sense of Schmid [1985], and the maximal globalization is too large to admit an invariant inner product unless it is finite dimensional. By way of illustration, consider $H^{0,j}(G/L, Z^{\#})$ when $j = 0$, $G = SU(1, 1)$, L is the diagonal subgroup (maximal torus), and Z is one dimensional with a positive integral parameter. This representation reduces via (0.20) and the process described in §VI.1 of Knapp [1986] to a multiplier representation in the space of all analytic functions in the unit disc. To impose an inner product, one needs a subspace of functions with controlled behavior near the boundary; the subspace of functions that extend to be analytic in a neighborhood of the closed unit disc is an example.

The functor Π cannot be defined using "largest K finite quotients." In fact, let $K = \{k_\theta\}$ be the circle group, put $X = d/d\theta$, regard the complexified Lie algebra as $\mathbb{C} \cdot X$, and consider the universal enveloping algebra $\mathbb{C}[X]$ as a left $\mathbb{C}[X]$ module. The finite-dimensional quotients

$$\mathbb{C}[X]/(X)(X \pm i)(X \pm 2i) \cdots (X \pm ni)$$

are K finite, and their dimensions are unbounded. Every quotient of $\mathbb{C}[X]$ is of the form $\mathbb{C}[X]/(P(X))$ for some polynomial P. If $P = 0$, the result is not K finite. If $P \neq 0$, the result is finite dimensional. Thus there can be no largest K finite quotient.

In the early 1980s Bernstein studied the relationship between the Vogan-Zuckerman classification of irreducible $(\mathfrak{g}, K)$ modules (given in terms of cohomological induction) and the Beilinson-Bernstein classification (given in terms of $\mathcal{D}$ modules). He observed that the $\mathcal{D}$ module construction commutes with (infinite) direct sums while the cohomological induction construction by means of Γ commutes with direct products. This fact suggested that there might be some difficulty in making the classification correspond if $\Gamma^S \circ \text{pro}$ were used to describe cohomological induction. He therefore sought a functor defined by means of projective resolutions. In our notation the functor Π' used in Bernstein [1983] for an inclusion of pairs $(\mathfrak{g}, L) \hookrightarrow (\mathfrak{g}, K)$ is given on the level of $(\mathfrak{k}, K)$ modules by

$$\mathcal{F}^{\mathfrak{k},K}_{\mathfrak{g},K} \circ \Pi'(V) = R(K) \otimes_{(\mathfrak{k},L)} V.$$

The $\mathfrak{g}$ action is imposed by taking advantage of the isomorphism

$$\mathfrak{g} \otimes_{\mathbb{C}} \Pi'(V) \cong \Pi'(\mathfrak{g} \otimes_{\mathbb{C}} V)$$

given in (3.75) and the anticipated (3.78a), and the result is a functor naturally isomorphic with our Π.

Some treatments of unitarizability have effectively defined Π as $\Pi = \Gamma^{2S}$. This is the case with Vogan [1984], Wallach [1984], and Wallach [1988]. This approach is aesthetically unsatisfactory and leads to technical complications. For example, it is not so obvious with this definition of Π that Π is adjoint to an exact functor, and then Frobenius reciprocity (Chapter II) is not readily available.

Detailed citations for the results mentioned in §5 appear in the Notes below for the individual chapters. For citations in connection with the Hecke algebra and the Π functor as in §6, see the Notes for Chapter I. More detail about the results of §§7–8 may be found in Vogan [1993], and citations for the individual results appear in the Notes below for the individual chapters. Citations in connection with the Transfer Theorem appear in the Notes for Chapter XI.

Chapter I

For background about Haar measure and the modular function, see Chapter VI of Loomis [1953]. Other sources are Chapter Three of Hewitt-Ross [1963] and Chapter II of Nachbin [1965].

Distributions were introduced into Lie theory in Bruhat [1956], which uses them as a tool for addressing reducibility questions concerning Mackey induction. The Hecke algebra $R(K)$ for the compact case does not explicitly appear in Bruhat's paper, but it may be regarded as implicit in it. The properties of $R(K)$ given in §2 are translations of classical results into this terminology.

In early 1977 Flath, in studying automorphic forms in the context of representation theory, was looking for a completely algebraic approach to decomposing representations of $G_1 \times G_2$. In the course of his studies, he introduced the $(\mathfrak{g}, K)$ module $U(\mathfrak{g}) \otimes_{U(\mathfrak{k})} R(K)$, realized it was an algebra, and noted the validity of Theorem 1.117. Flath called the algebra a "Hecke algebra" and, at the suggestion of Deligne, identified it with the convolution algebra of bi-K-finite distributions on G supported on K. He presented this material in his 1977 Corvallis lecture and published the contents of his lecture in Flath [1979]. Borel and Trauber proposed the Hecke algebra to Zuckerman's seminar (Zuckerman [1978]) as a potential tool for proving the Duality Theorem.

As mentioned in the Preface, in 1985 the authors became aware of various gaps and difficulties in a proposed proof of the Signature Theorem (Theorem 6.34) of the text) and in the literature on which the proof was based, and the preprint Knapp-Vogan [1986] sought to repair the foundations. This paper took up the study of the Hecke algebra $R(\mathfrak{g}, K)$ where the Zuckerman seminar had left it. For applications one is led to consider other Hecke algebras $R(\mathfrak{q}, B)$, where $\mathfrak{q}$ is a complex subalgebra of $\mathfrak{g}$ and B is a closed subgroup of K. In these applications, $\mathfrak{q}$ need not be the complexification of a subalgebra of $\mathfrak{g}_0$, and a definition of $R(\mathfrak{q}, B)$ as distributions on a group Q is therefore not immediately available. One can define $R(\mathfrak{q}, B)$ as an algebra of distributions on G, but such a definition is at best clumsy.

For this reason the paper Knapp-Vogan [1986] observed that $R(\mathfrak{g}, K)$ has an equivalent abstract definition that is equally applicable to $R(\mathfrak{q}, B)$. In retrospect this abstract definition is the same as Flath's original definition. The paper stated a number of properties of the abstract $R(\mathfrak{g}, K)$ that were left unproved until Knapp [1988]. The proofs in Knapp [1988] of these properties are computational and lengthy. The idea of using vector-valued distributions, introduced in Duflo-Vergne [1987], made a simplification possible, and the present treatment may be regarded as a more conceptual development of $R(\mathfrak{g}, K)$ that takes advantage of vector-valued distributions.

See Sweedler [1969], 153–156, for smash products in the context of Hopf algebras. Hermite polynomials are mentioned in the example at the end of the chapter, and one may consult Howe-Tan [1992], 104–106, for their properties.

Chapter II

The examples in §§1–2 have different historical origins. The first is Lie-algebra cohomology, coming from the explicit complex (2.127) with $K = \{1\}$ or from the derived functors of the invariants functor when $K = \{1\}$. This example, which is an instance of our Example 2, was introduced in Chevalley-Eilenberg [1948] and is treated thoroughly in Cartan-Eilenberg [1956].

Hochschild [1956] introduced relative homological algebra with respect to a pair $(\mathfrak{g}, \mathfrak{k})$ satisfying some properties. Roughly speaking, this theory may be regarded as the full generality of our Example 2. With this notion, $H^*(\mathfrak{g}, K; \mathbb{C})$ gives the cohomology of G/K if G is compact. A more general application is to the cohomology of compact locally symmetric spaces $\Gamma \backslash G/K$. This cohomology plays a role in the theory of automorphic forms, and the cohomology spaces of interest are $H^*(\mathfrak{g}, K; C^\infty(\Gamma \backslash G))$. See Borel-Wallach [1980] for further discussion of this topic.

Kostant [1961] studied representations of compact connected Lie groups cohomologically. In the terminology of Chapter IV of the present text, a representation of G on a vector space V leads by restriction to an action of any parabolic subalgebra q on V. If one takes the Lie-algebra cohomology of V with respect to the nilpotent radical of q, then the action of the Levi factor of q survives, and one gets a representation of the corresponding Levi subgroup. Kostant's Theorem, given in the text in §IV.9, determines this representation. Schmid [1976] used a version of this theory for noncompact groups in his proof of the Langlands conjecture. Vogan [1979] made the theory a starting point for investigating infinite-dimensional representation theory; some of this material is reproduced in §X.2 of the text. What Example 3 of Chapter II does (in the company of the general semidirect product map of pairs introduced in §III.2) is to deal with these situations abstractly and give a new functorial treatment of them.

Example 1 and most of its subexamples come from Zuckerman [1978]. Zuckerman had ind and pro as in Example 1a, and he used Γ' as in Example 1c in place of the functor Γ that we introduce in Example 1b. He was also aware of Examples 1d and 1e and consequently could handle the effect of disconnectedness of the compact groups in some situations. The material involving these functors in §§2–4 is part of Zuckerman [1978]. Chapter 1 of Borel-Wallach [1980], Enright-Wallach [1980], and §§6.1-6.2 of Vogan [1981a] all treat this material.

Bernstein [1983] gave a preliminary definition of the functor Π of Example 1b, as was mentioned in the Notes for the Introduction. Also he introduced the adjoint $\mathcal{F}^\vee$ as a functor from $(\mathfrak{k}, K)$ modules to $(\mathfrak{k}, L)$

modules. His definition in our notation was

$$\mathcal{F}^{\vee}(X) = \Big(\prod_{\gamma \in \widehat{K}} X_{\gamma} \Big)_L.$$

Although Zuckerman [1978] was aware of the Hecke algebra $R(\mathfrak{g}, K)$ and looked for a functor Π, the lectures did not put these two notions together. Bernstein [1983] recognized Γ' as isomorphic with a (K finite) change-of-rings functor of Hom type built from Hecke algebras, and he defined Π to be the corresponding tensor-product change-of-rings functor. All of this mathematics, with some refinements added, appears in Knapp-Vogan [1986] and in Bien [1990]. The treatment in Knapp [1988] maintains this context, supplying additional details.

The present book recognizes all these examples as part of the same theory, involving general maps of pairs and functors P and I for each such map of pairs.

Example 1e in §4 is not very different from the theory in Mackey [1951] for induced representations in the context of open subgroups. In Mackey's setting, Proposition 2.80 becomes what is known as the Mackey Subgroup Theorem; before Mackey's work it was unknown even for finite groups.

The classical Mackey isomorphism says that induction commutes with tensoring with a finite-dimensional representation. This is a special case of the identity

$$\text{induced}_{H_1 \times H_2}^{G_1 \times G_2}(\pi_1 \otimes \pi_2) \cong \text{induced}_{H_1}^{G_1}(\pi_1) \otimes \text{induced}_{H_2}^{G_2}(\pi_2),$$

which is proved as Theorem 5.2 of Mackey [1952] under very general hypotheses; the special case has $H_1 = G_1 = G_2$ and then restricts both sides to the diagonal subgroup of $G_1 \times G_1$. Versions of Theorem 2.95 for pro and for the Zuckerman functor appear in Enright-Wallach [1980], and the general cases of Theorems 2.95 and 2.103 for inclusions of pairs are in Knapp-Vogan [1986].

Theorem 2.122 concerning the Koszul resolution is due to Koszul [1950] when $K = \{1\}$ and to Hochschild [1956] in the general case. The standard resolutions (2.123) and (2.124) played an important role in Zuckerman [1978] and are reproduced in Enright-Wallach [1980].

Chapter III

Easy Duality (Theorem 3.1) is an elementary result, and historically it has generally been recognized for a particular setting as soon as the P and I functors have been defined in that setting.

Zuckerman [1978] introduced Hard Duality in the form of Case (ii) of Theorem 3.5c for the I functor as a conjecture. (Recall that Case (ii) is the case of an inclusion of pairs $(\mathfrak{g}, L) \hookrightarrow (\mathfrak{g}, K)$.) With Trauber he gave several ideas toward proofs. Among other things Trauber and Zuckerman showed how to write down the pairing in the Duality Theorem and thereby to establish the $(\mathfrak{k}, K)$ isomorphism. What was missing was a proof that the pairing is $\mathfrak{g}$ equivariant.

Enright-Wallach [1980] gave a proof of this $\mathfrak{g}$ equivariance, and therefore of Case (ii) of Theorem 3.5c for the I functor. The main idea is that the explicit pairing satisfies some naturality properties that force it to be $\mathfrak{g}$ invariant. Their argument involved two subsidiary ideas that play an important role in this chapter. One is the device (3.75) that converts a Lie-algebra action into a map in the category, and the other is the fact given as Proposition 3.77 that the Lie-algebra action on $\Pi_j(V)$ is related to Π_j of the Lie-algebra action on V by a Mackey isomorphism.

In giving a naturality argument such as that in Enright-Wallach [1980], it is difficult to keep track of what requires proof, and the paper is not explicit on one important point, the form of the isomorphism in the paper's (4.7). If the wrong isomorphism is used, various diagrams will not commute. An expanded, corrected account appears in §§6.2–6.3 of Wallach [1988].

Another account, giving additional information, is in Knapp-Vogan [1986]. In order to introduce invariant Hermitian forms into cohomologically induced modules (as is done in Vogan [1984], Wallach [1984], and the present Chapter VI), it is necessary to have a version of the functor Π available in place of Γ. Although Vogan [1984] and Wallach [1984] effectively define $\Pi = \Gamma^m$ when $\bigwedge^m \mathfrak{c}$ is a trivial L module and avoid this problem, the paper Knapp-Vogan [1986] recognized the desirability of a more functorial approach. In addition to defining Π by a change of rings, it cast Hard Duality as a relationship between Π_j and Γ^{m-j}, as in Case (ii) of Theorem 3.5b.

Bernstein [1983] mentions Hard Duality of the kind in Case (ii) of Theorem 3.5b, and some sketchy details attributed to Bernstein are given in Proposition 4.1 of Bien [1990]. In an argument of this kind, the commutativity of diagrams on the level of homology/cohomology does not necessarily lift to commutativity on the level of cycles/cocycles, and

Bien's sketch does not give enough detail so that one can tell how this matter has been addressed.

Other authors sought to shorten the proof of Hard Duality given in Knapp-Vogan [1986], and they did so at the expense of dropping Π from consideration. Benoist [1986] and Wigner [1987] gave proofs of the result in the case that $\mathfrak{k}$ is semisimple. Duflo-Vergne [1987] gave a proof applicable to a general inclusion of pairs $(\mathfrak{g}, L) \hookrightarrow (\mathfrak{g}, K)$, adding two new ideas to the theory. One is the use of vector-valued distributions, as has already been mentioned in the Notes for Chapter I. The other is that one can write down explicit formulas (using vector-valued distributions) on the level of cochains for the action of $\mathfrak{g}$ on $\Gamma^j(V)$. Proposition 3.80 expresses these ideas here.

The extension of Hard Duality to handle all maps of pairs $(\mathfrak{h}, L) \to (\mathfrak{g}, K)$ with $i_{\mathrm{alg}}(\mathfrak{h}) = \mathfrak{g}$ is new. Some comments are in order about instances of this theorem other than for an inclusion of pairs. Poincaré duality, in the form of the third identity in Corollary 3.6, appears in Borel-Wallach [1980] under an admissibility assumption. More generally the case of a general semidirect product map of pairs relates $\mathfrak{u}$ homology and cohomology with an $\mathfrak{l}$ action in place. Normally this kind of homology and cohomology is defined directly from an explicit complex (as, for example, in §3 of Hochschild-Serre [1953]), and then duality is proved by a direct calculation similar to that in §4. Our definition, however, is by means of P and I functors involving the semidirect product Lie algebra. The hard step is to relate the two definitions. The relationship is stated as Proposition 3.12 and is proved as a special case of Proposition 3.41. Proposition 3.12 is a sufficiently complicated special case that its proof can serve as a prototype for the proof of Proposition 3.41. Proposition 3.41 proves Case (i) of Hard Duality, and the brief supplementary argument in §8 combines Cases (i) and (ii) to prove Hard Duality in complete generality.

Chapter IV

For treatments of Cartan-Weyl theory, see Chapter IV of Knapp [1986] and the various books mentioned in the Notes for that chapter. The issue of a theory for disconnected groups was raised in Kostant [1961]. Vogan [1987a] treats this topic, working with Cartan subgroups that range from "large" to "small." In this sense the Cartan subgroups of the present text are always "large."

The material in §§3–6 is based in broad outline on Vogan [1981a], 1–4, 26–35. The term "Harish-Chandra class" is in common usage for the class of reductive groups defined in §3 of Harish-Chandra [1975]. Proofs of some of the results left unproved in §4 may be found in Knapp [1986]: §IV.4 for the present Theorem 4.35, Proposition 11.16b and §V.3 for Lemma 4.41, Proposition 11.16a and §V.4 for Lemma 4.43, §IV.3 for Proposition 4.55.

Proposition 4.87 appears in a strong form in §2.6 of Dixmier [1974]. The Harish-Chandra isomorphism (Theorem 4.95) appeared originally in Harish-Chandra [1951], and there are by now several expositions in books. See, for example, Humphreys [1972], Knapp [1986], Varadarajan [1974], Wallach [1988], and Warner [1972]. Verma modules were introduced in Harish-Chandra [1951]; they are so named because of later work by Verma [1968] in this area. Theorem 4.115 is due to Harish-Chandra [1951]. For the Hilbert Nullstellensatz, see Zariski-Samuel [1960], 164–167.

Kostant's Theorem is the main result of Kostant [1961]. Kostant's proof made special use of the Casimir operator. The argument in our §§9–10, which uses the full center of the universal enveloping algebra, is taken from §3.2 of Vogan [1981a]. Chevalley's Lemma (Proposition 4.146) is attributed to Chevalley by Harish-Chandra [1956c]. Theorem 4.149 is a combination of an observation of Allan Cooper (that $H^k(\mathfrak{u}, V)$ becomes a $Z(\mathfrak{g})$ module in a natural way) and a result of Casselman-Osborne [1975]; the treatment is based on §3.1 of Vogan [1981a].

Statements of the analytic Bott-Borel-Weil Theorem and references concerning it appear in the Introduction and the Notes for the Introduction. For the algebraic Bott-Borel-Weil Theorem, see Enright-Wallach [1980]. As an introduction to cohomological induction, Knapp [1988] is organized around having a proof of the algebraic Bott-Borel-Weil Theorem as the first main result and an important piece of motivation.

Chapter V

The cohomological induction functor $\mathcal{R}^j$ is due to Zuckerman [1978], except that Zuckerman defined Γ in the connected case as in Example 1c of §II.4 and by various ad hoc formulas in general. This several-step definition of Γ is reproduced in §6.2 of Vogan [1981a]. The present Lemma 5.10 shows that the treatment of the disconnected case is a relatively minor matter, and these Notes will tend to ignore it.

The effect on infinitesimal character (Corollary 5.25), the effect on admissibility (Theorem 5.35b), the Euler-Poincaré Principle (Theorem 5.64), the vanishing theorem (Theorem 5.99), and the fundamental spectral sequences (Theorem 5.120b) are all due to Zuckerman in the case of $\mathcal{R}^j$. Zuckerman proved that the vanishing theorem is a consequence of Hard Duality, but Hard Duality was not fully proved until later. Although Zuckerman made general definitions, he tended to emphasize $A_q(\lambda)$.

Vogan [1981a] elaborated on Zuckerman's results, beginning in §6.2. The Vogan book avoids Hard Duality, and the treatment of the vanishing theorem is therefore more indirect. (See Propositions 6.3.34 and 8.2.15 for part of the indirect argument.) With Hard Duality proved in Chapter III of the present book, we have reverted to Zuckerman's original program of deriving the vanishing theorem from Hard Duality. The details follow the lectures Wallach [1982]. The proof of Theorem 5.35b is taken from Theorem 6.3.12 of Vogan [1981a], which in turn is a translation into algebra of an analytic argument using an "order-of-vanishing filtration" in Schmid [1967].

The results about $\mathcal{L}_j$ follow once one has the definition of Π_j. See Bernstein [1983], Knapp-Vogan [1986], and the Notes for the Introduction. The argument for Theorem 5.35a appears in Knapp [1988].

The paper Speh-Vogan [1980], written in 1977, independently gave a construction for passing from irreducible representations of L to irreducible representations of G, and Vogan [1984] matched this construction with cohomological induction in degree S. The point of Speh-Vogan [1980] was to investigate reducibility of standard representations (defined here in Chapter XI), and this paper may be regarded as progress toward the Irreducibility Theorem in Chapter VIII. The weakness of the construction is that it was not done by a functor; in effect it therefore amounts only to a match of parameters of representations of L with parameters of representations of G.

The side benefit of Speh-Vogan [1980] is that it introduced the notion of bottom layer. A K type τ_μ of a $(\mathfrak{g}, K)$ module V is defined in that paper to be on the "bottom layer" relative to a Borel subalgebra $\mathfrak{b}$ if every K type of V has highest weight of the form $\mu + Q$ with Q a sum of roots of $\mathfrak{b}$. Speh and Vogan in Theorem 4.17 proved the match of multiplicities as in Corollary 5.85, provided a strict dominance condition holds. But there was no bottom-layer map at this stage. That was introduced in Definition 6.9 of Vogan [1984], and the validity of the present Theorem 5.80 was observed at that time. Proposition 6.15 of that paper shows under a dominance condition that the Speh-Vogan definition of bottom layer of K types coincides with the K types in the image of the bottom-layer map.

The improved vanishing result in Theorem 5.109 is in Vogan [1984]. The Hochschild-Serre spectral sequence (Theorem 5.130) is in Vogan [1979], where it was introduced in order to obtain Corollary 5.140 and results like those in §X.2 of the present text. The result is so named because of its similarity with the spectral sequence in Hochschild-Serre [1953] for relating Lie-algebra cohomology for a Lie algebra $\mathfrak{g}$, an ideal $\mathfrak{h}$, and the quotient $\mathfrak{g}/\mathfrak{h}$.

Chapter VI

The construction of an invariant sesquilinear form for $\operatorname{ind}_{\mathfrak{q},L\cap K}^{\mathfrak{g},L\cap K}(Z_{\bar{\mathfrak{q}}}^{\#})$ as in (6.31) appears in Jantzen [1979]. The form at this stage is already normally called the "Shapolvalov form," as a result of earlier work of Shapovalov [1972]. See also Kostant [1969] and [1975a]. Zuckerman [1978] saw that Hard Duality was the tool needed to carry the form from $(\mathfrak{g}, L\cap K)$ modules to $(\mathfrak{g}, K)$ modules. The form was in place in Wallach [1982], except that Γ^S was used in place of Π_S.

The idea of using signatures as in Remark 1 with Proposition 6.12 appears for the first time in Vogan [1984], and a calculus is developed for working with signatures for a family of representations indexed by a parameter.

The Signature Theorem itself (Theorem 6.34) is an unpublished theorem of Vogan, obtained in late 1984 after the introduction of the bottom-layer map. A version of the theorem appears in Proposition 10.9 of Vogan [1986]. The part of the proof given in Proposition 6.50 was inspired by Lemma 8.7 of Enright [1983].

Chapter VII

The translation functors evolved from the work of Bernstein-Gelfand-Gelfand [1975] and Jantzen [1974] for categories of $U(\mathfrak{g})$ modules. A direct avenue for using translation functors for group representations was opened by Zuckerman [1977], who worked with $(\mathfrak{g}, K)$ modules and their global characters. For Zuckerman the motivating example was discrete series, and his approach immediately allows one to move the parameter of such a representation. Some of the theory of realizing discrete series (see the Notes for the Introduction) had previously been proved only for generic discrete-series representations, and the translation functors made the theory applicable to all discrete series.

We have chosen a more algebraic approach based on Kostant [1975b], for four reasons. One is that such an approach is more in the spirit of this book, a second is that it is applicable to Verma modules and other $U(\mathfrak{g})$ modules on which K does not operate, a third is that it gives more definitive results about irreducibility, and a fourth is that it eventually leaves us in a better position to handle $(\mathfrak{g}, K)$ modules with K disconnected.

Generalized infinitesimal characters and the primary decomposition as in §2 are in Kostant [1975b]. Wigner's Lemma (Proposition 7.28) appears in Borel-Wallach [1980], 26, where it is called an analog of a result of Wigner's about the continuous cohomology of real Lie groups.

The theory of invariants of finite reflection groups, as reported in §3, is due to Chevalley [1955]. Our presentation follows the exposition in Varadarajan [1974]. The effect in §4 of $Z(\mathfrak{l})$ on $\mathfrak{u}$ homology and cohomology is implicit in Corollary 3.10 of Vogan [1979] and explicit in Corollary 3.1.6 of Vogan [1981a].

For (7.64a–e) see §1.9 of Dixmier [1974]. For (7.64f) see (5.19) and Theorem 5.12 of Knapp [1986]. Theorems 7.65 and 7.71 are unpublished theorems of Chevalley quoted in Harish-Chandra [1953], 200; our exposition is taken from §7.3 of Dixmier [1974].

The results of §6 are due to Kostant [1959] and [1963]. We have largely followed §§8.1-8.3 of Dixmier [1974], who in turn made use of a simplification in Kostant's proofs due to Varadarajan [1968]. The proof of Lemma 7.103 is taken directly from Kostant. It uses results in Chevalley [1958] and Zariski-Samuel [1958] and [1960].

Annihilators were introduced by Dixmier [1970] and [1971] into the representation theory of reductive groups as a possible aid in studying irreducible unitary representations. For the Dixmier-Duflo theorem (Theorem 7.118), see Dixmier [1970] and [1971] and Duflo [1977]. Our exposition follows §8.4 of Dixmier [1974].

Theorem 7.133 is the main result of Kostant [1975b]. Kostant notes the possibility of proving his theorem from the Dixmier-Duflo theorem, but he builds the paper around a different proof. The adjoint relation Proposition 7.143 is due to Zuckerman [1977].

Chevalley's Lemma for affine Weyl groups (Proposition 7.153) may be found in Bourbaki [1968]. Our proof largely follows §VII.7 of Helgason [1978]. The fact that this result implies Proposition 7.145 is noted in Exercise 1 of Bourbaki [1968], 227.

Zuckerman [1977] proved under a dominance hypothesis that an irreducible admissible $(\mathfrak{g}, K)$ module is carried by translation functors into a "primary " $(\mathfrak{g}, K)$ module (one whose character is a multiple of an irreducible character). Vogan [1981a], making use of a classification, sharpened this conclusion to show that each irreducible admissible $(\mathfrak{g}, K)$ module is sent to an irreducible or to 0. The theory here, based on the theorem of Kostant's in §8, applies to $U(\mathfrak{g})$ modules, not just $(\mathfrak{g}, K)$ modules. Classification results are of no help, and a new approach is needed. The argument for Theorems 7.171-7.173, given in §§10–12, is based on the approach in Vogan [1990]. Part of the argument in §11 is in the spirit of Kostant [1963], and Theorem 7.194 is from that paper.

References for the traditional proof of Theorem 7.204, quoted in §13, are as follows: Each irreducible admissible $(\mathfrak{g}, K)$ module is the underlying module of a group representation by the Subquotient Theorem (cf. Harish-Chandra [1954a], Lepowsky [1973], and Rader [1976]). Global characters are introduced in Harish-Chandra [1954b], and they are proved to be given by locally integrable functions in Harish-Chandra [1965]. That the linear span of the irreducible characters for a given infinitesimal character is finite dimensional follows from Harish-Chandra [1956c]. Characters of infinitesimally inequivalent irreducible representations are linearly independent as a consequence of Harish-Chandra [1954b].

The proof of Corollary 7.209 is from Zuckerman [1977]. As with Proposition 7.28, Wigner's Lemma in the form of Proposition 7.212 appears in Borel-Wallach [1980], 26, where it is called an analog of a result of Wigner's about the continuous cohomology of real Lie groups. Wallach [1988], 114, offers a different approach to the relationship between $(\mathfrak{g}, K)$ modules and $(\mathfrak{g}, K_0)$ modules. Theorems 7.229 and 7.232 are new. The results of §§15–16 have Vogan [1981a], 488-494, as their point of departure.

Chapter VIII

Zuckerman [1978] deduced a form of the Irreducibility Theorem for $A_{\mathfrak{q}}(\lambda)$ from the vanishing theorem and Hard Duality. For such a proof in published form, see §6.6 of Wallach [1988].

Proposition 4.18 of Vogan [1984] establishes the Irreducibility Theorem at the level of generality in the present Theorem 8.2. The argument uses ideas from Speh-Vogan [1980] and in particular relies on a classification of irreducible admissible $(\mathfrak{g}, K)$ modules.

In the present book, we have reverted to Zuckerman's original idea, giving a proof that uses the vanishing theorem and Hard Duality. The other two tools are the bottom-layer map and the Hochschild-Serre spectral sequence (Theorem 5.130).

The third form of the Irreducibility Theorem is new, as is Corollary 8.28. Corollary 8.28 is applied several times in Chapter XI.

Theorem 8.29 is due to Zuckerman [1978]. Earlier such a formula had been a conjecture attributed to Blattner about the multiplicities of K types in each discrete-series representation. Hecht-Schmid [1975] proved this conjecture. In introducing cohomological induction, Zuckerman [1978] was producing an algebraic analog of an analytic construction. When $\mathfrak{q} = \mathfrak{b}$ was a Borel subalgebra and the Levi subalgebra $\mathfrak{l}$ was contained in $\mathfrak{k}$, the module $A_{\mathfrak{b}}(\lambda)$ in the presence of a strict dominance condition was intended as the algebraic analog of a discrete-series representation. In fact, Zuckerman knew enough about the K type structure of $A_{\mathfrak{b}}(\lambda)$ so that $A_{\mathfrak{b}}(\lambda)$ had to be the underlying $(\mathfrak{g}, K)$ module of a discrete-series representation as soon as it was irreducible. Thus Theorem 8.29 represented an algebraic proof and generalization of the Blattner formula.

The first appearance of Speh representations was in Speh [1981]. Vogan [1986] worked with these representations as cohomologically induced. Sahi-Stein [1990] gave an analytic realization. The Vogan paper sketched two proofs of irreducibility. One proof began with a result along the lines of Theorem 8.31 and was heavily influenced by Borho-Brylinski [1982]. In the literature the relevant topic for dealing with this kind of question is "normality of orbit closures." The first result in this direction was the one of Kostant [1963] that we have given as Lemma 7.103. Further papers on this topic are Borho-Kraft [1979], and Kraft-Procesi [1979] and [1981] and [1982]. For Proposition 8.75, see Hesselink [1979].

The other proof of irreducibility sketched in Vogan [1986] uses Conze-Berline–Duflo [1977]. See also Joseph [1980] and [1983].

In connection with Example 3, Corollary 9.70 shows that $A_{\mathfrak{q}}(\lambda)$ is infinitesimally unitary. Since the K type Λ' has multiplicity 1, it follows (as is shown in §IX.6) that the K type Λ' generates an irreducible $(\mathfrak{g}, K)$ submodule V of $A_{\mathfrak{q}}(\lambda)$. It is the $(\mathfrak{g}, K)$ module V that is of interest. Baldoni-Silva–Knapp [1986] wanted to classify irreducible unitary representations that arise in the Langlands classification from a maximal parabolic subgroup, and there was only one $(\mathfrak{g}, K)$ module V' for which unitarity could not be settled by known methods. Starting from the Langlands parameters of V', Vogan constructed the $A_{\mathfrak{q}}(\lambda)$ in this example and suggested to Baldoni-Silva and Knapp that V' might arise within it. Indeed, V' coincides with V, and therefore V' is infinitesimally unitary.

Some assertions about Example 3 are made without proof in the last paragraph of the discussion. One is that any irreducible $(\mathfrak{g}, K)$ module with minimal K type Λ appears as a quotient in a certain series of parabolically induced representations. This is proved in Vogan [1979]. For an argument that is within the scope of this book, one can take the answer given from Vogan [1979] and use the minimal K type formula to obtain the K type Λ as the minimal K type for that series. Application of the Langlands classification (Theorem 11.198) and the uniqueness result (Lemma 11.260) then shows that the given $(\mathfrak{g}, K)$ module appears as a quotient in that series. Once the particular series of parabolically induced representations is known, one can compute the multiplicity of the K type Λ' from classical Frobenius reciprocity (11.42) and see that it is 0. Therefore $A_{\mathfrak{q}}(\lambda)$ is reducible. See Baldoni-Silva–Knapp [1986] for further discussion.

Example 4 was first discovered as a unitary point within the Langlands classification for $G = Sp(5, 2)$, and this result is reported in Baldoni-Silva–Knapp [1989]. Schlichtkrull [1987] identified this representation as occurring discretely in L^2 of a certain semisimple symmetric space G/H, he recognized the $(\mathfrak{g}, K)$ module as an $A_{\mathfrak{q}}(\lambda)$, and he showed how the Blattner multiplicity formula led nontrivially to the occurrence of the trivial K type in the representation.

Chapter IX

Zuckerman [1978] conjectured that $A_{\mathfrak{q}}(\lambda)$ is infinitesimally unitary if $\langle \lambda + \delta, \alpha \rangle \geq 0$ for all α in $\Delta(\mathfrak{u})$, and Vogan [1981a] extended this conjecture to say that $\mathcal{R}^j(Z)$ is infinitesimally unitary if Z is infinitesimally unitary and its infinitesimal character λ has $\operatorname{Re}\langle \lambda + \delta(\mathfrak{u}), \alpha \rangle \geq 0$ for all α in $\Delta(\mathfrak{u})$.

Vogan [1984] proved the extended conjecture and effectively Theorem 9.1. The method involved tracking the signature of an invariant form on a Langlands quotient while moving the parameter of the representation from the tempered case to the representation in question. In other words, the proof used classification results in an intricate fashion.

Hearing that Vogan had proved this result by keeping track of signatures, Wallach [1984] invented signature characters and gave a relatively elementary proof of the extended conjecture. This proof is reproduced for $A_{\mathfrak{q}}(\lambda)$ in §6.7 of Wallach [1988], but Wallach has informed the authors that the top displayed formula on p. 199 is incorrect and needs to be adjusted as in our (9.48) and the lines after it.

Although the Vogan proof is longer, it gives information that is apparently not readily available otherwise. Namely Theorem 9.1 admits a converse. Say G is connected. If Z is irreducible and has infinitesimal character λ such that $\mathrm{Re}\langle\lambda+\delta(\mathfrak{u}), \alpha\rangle > 0$ for all α in $\Delta(\mathfrak{u})$, then $\mathcal{R}^S(Z)$ infinitesimally unitary implies Z infinitesimally unitary. This is Theorem 1.3b of Vogan [1984], and it is needed for a full proof of our Theorem 12.2. (See the Notes for Chapter XII.)

The wording of Theorem 9.1 comes from the proof by Wallach [1984]. Theorem 9.68 is in §7 of Vogan [1984].

Chapter X

Theorem 10.2 is from Theorem 1 of Harish-Chandra [1953]. Our exposition follows §3.4 of Wallach [1988]. Lemma 10.13 is due to Cartan, and the proof is a simple variant of the one in Hunt [1956].

The material in §2 is based on Vogan [1979]. But the goal of the section is less ambitious than the goal of that paper, and thus the treatment has been simplified considerably. Inequality (10.41b) is in Kostant [1961], and we have used an argument in Wallach [1988], 398, in its proof.

For commentary on Theorem 10.44, see the Notes for the last two sections of Chapter XI.

Chapter XI

Parabolic induction dates to Mackey's realization in 1949 or 1950 that multiplier representations, including the principal series for $SL(2, \mathbb{R})$, are an instance of his theory of induction of representations for locally compact groups. Gelfand-Naimark [1950] had introduced an extensive system of multiplier representations for complex groups, and these could all then be viewed as induced. A systematic role for parabolic induction appears in the announcement Harish-Chandra [1954c]. Many of the structure-theoretic results about groups in (11.7) and (11.21), as well as about Lie algebras later in §1, appear in Chapters III and VI of Helgason [1962], and historical references appear at the ends of those chapters. The measure-theoretic results in (11.21) are proved in Chapter X of Helgason [1962].

The induced and compact pictures appear at various times in early work of Harish-Chandra, and the realizations of representations in Gelfand-Naimark [1950] may be regarded in retrospect as a noncompact picture obtained by restrictions of the induced picture to $N^- =$

θN. Kunze-Stein [1960], [1961], and [1967] exploited the relationships among the three pictures.

In Mackey's original theory of induction, some work was involved in introducing a Hilbert-space structure so that representations induced from unitary representations are unitary. Mackey [1952] invented "quasi-invariant measures" for this purpose. For Lie groups Bruhat [1956] showed how the use of modular functions made the theory of quasi-invariant measures unnecessary. Densities as in §1 are a variation on Bruhat's theory. The argument for Lie groups that $f \mapsto F$ with F as in (11.29) carries $C_{\text{com}}(G)$ onto $C_{\text{com}}(G/H, \omega)$ is in Bruhat [1956]. Actually the existence of local continuous cross-sections from G/H into G can be avoided in Bruhat's argument; the map is onto whenever G is a locally compact group and H is a closed subgroup.

The irreducibility of the unitary spherical principal series of a complex semisimple group (Theorem 11.46) is due to Parthasarathy-Rao-Varadarajan [1967], and the generalization to the full unitary principal series is due to Želobenko [1968]. Kostant [1969] and [1975a] proved that the unitary spherical principal series of any real semisimple group is irreducible. The nature of the reducibility of the full unitary principal series is now completely understood; see Knapp-Zuckerman [1982] and the references in that paper.

Although the action of the Lie algebra in parabolically induced representations was considered by several authors, it was not until the expansion in Vogan [1981a] of the work of Zuckerman [1978] that the space of the representation was addressed abstractly. See Proposition 6.3.5 of Vogan's book for the present Propositions 11.47 and 11.52. Proposition 11.65 is new.

The functor $({}^{\text{u}}\mathcal{R}^{\mathfrak{g}, K}_{\mathfrak{q}, L \cap K})^j$ for an arbitrary germane parabolic subpair $(\mathfrak{q}, L \cap K)$, as in (11.73), is in Chapter 6 of Vogan [1981a]. The results of the present §§3–4 are on pp. 351–355 of that book.

The idea of the need for the Transfer Theorem arose historically from the desire to compare classification theorems for irreducible admissible $(\mathfrak{g}, K)$ modules. There were three such theorems: (a) the Langlands classification (see Langlands [1973] and Knapp-Zuckerman [1977]), (b) the Vogan-Zuckerman classification (see Vogan [1979] and [1981a]), and the Beilinson-Bernstein classification (see Beilinson-Bernstein [1981]). The paper Speh-Vogan [1980] had worked with the Langlands classification, even though it was eventually seen to be a paper about cohomological induction. (See the Notes for Chapter V.) This influence of the Langlands classification on proofs about cohomological induction carried over to Vogan [1981a] and [1984]. So it was natural to try to relate the Langlands and Vogan-Zuckerman classifications, and it was consequently natural also to relate these to the Beilinson-Bernstein classification.

The Transfer Theorem (Theorem 11.87) appears in a first form in Vogan [1983], where a result like Lemma 11.128 is proved. The statement and proof of the Transfer Theorem are essentially the inductive step of that argument. The actual proof of the Transfer Theorem (i.e., the inductive argument in Vogan [1983]) proceeds along lines different from the ones in our §5, obtaining an adjoint statement about cohomology and deducing the result from that. The more direct argument in the present proof depends critically upon an argument for $\mathfrak{sl}(2, \mathbb{C})$ given in Lemma 11.88. The construction in this lemma should be compared with Khoroshkin [1981]; published work on this subject was brought to the authors' attention by P.-Y. Gaillard. For further work on the Transfer Theorem, see Hecht-Miličić-Schmid-Wolf [1987].

The Transfer Theorem addresses the effect of changing one root from positive to negative only in the case that the root is complex. Further theories result from considering imaginary or real roots. When the root is imaginary and compact, the theory is closely related to the theory of the "Penrose transform." See Demazure [1968] and [1976], Murray-Rice [1989], and the proof of the Bott-Borel-Weil Theorem in Baston-Eastwood [1989]. When the root is real, the theory is closely related to the theory of intertwining operators of Knapp-Stein [1980]; the actual analog of the Transfer Theorem is the special case of $\mathfrak{sl}(2, \mathbb{R})$, which is addressed in Kunze-Stein [1960]. When the root is imaginary and noncompact, the theory concerns character identities as in Schmid [1975]. See §§XII.9 and XIV.13 of Knapp [1986].

The idea of "standard modules," arises in connection with each of the classification theorems, and our definitions are arranged so that Theorem 11.129 results. Some of the conclusions of Theorem 11.129 appear in other language (and with other proofs) in §§8.1-8.2 of Vogan [1981a].

The question of normalization, as in §7, does not need to be addressed until arbitrary germane parabolic subalgebras are used in the construction of the functors ${}^u\mathcal{L}_j$ and ${}^u\mathcal{R}^j$ of (11.71), and we have chosen to delay the question of normalization as long as possible. The suggestion of using double covers comes from the work of Duflo [1982] in the context of algebraic (not necessarily reductive) groups, and it is in place already in Vogan [1987a]. Adams-Barbasch-Vogan [1992] makes critical use of normalizations. It does not, however, handle the real roots in a special way, as we have done; our special treatment of real roots is motivated by the way in which Harish-Chandra [1965] handles real roots in normalizing orbital integrals.

As soon as more than one normalized functor is in use, it is necessary to address the relationships among the normalizations. In the later results of §7, we have chosen a direct approach to this problem, which is simple but messy. A more complicated but tidier approach is to do all the normalizations simultaneously by means of an inverse limit construction; this is the approach used in Adams-Barbasch-Vogan [1992].

The discrete series were classified by Harish-Chandra [1966], and the Notes for the Introduction describe their history. For an exposition of the mathematics, see Theorem 9.20 and the first five sections of Chapter XII of Knapp [1986]. Zuckerman [1978] knew because of K type analysis that the $A_{\mathfrak{b}}(\mu)$'s of (11.74) are discrete series and that they account for all the discrete series when G is connected. Wallach [1988] takes the approach of constructing the discrete series for connected groups as in (11.74). He proves their irreducibility, their unitarity, and finally their square integrability. For completeness he has to fall back upon Harish-Chandra's analytic arguments using global character theory. For discrete series of disconnected groups, see §27 of Harish-Chandra [1975].

Harish-Chandra [1956b] mentions some representations that turn out to be limits of discrete series. Further such representations came up in papers by several other authors, and a complete treatment of them is in Knapp-Zuckerman [1982].

Let G be a reductive group in the Harish-Chandra class, and suppose G has compact center. An irreducible admissible representation of G is called **tempered** if its K finite matrix coefficients are in $L^{2+\epsilon}(G)$ for all $\epsilon > 0$. Langlands [1973] classified the irreducible admissible representations of G in terms of irreducible tempered representations of the M's of parabolic subgroups, and he showed that the irreducible tempered representations are always irreducible constituents of the standard continuous-series representations $I_{MAN}^{G}(\xi, \nu)$ with ξ is in discrete series and ν imaginary. See also Trombi [1977]. One exposition of this work is in Chapter VIII of Knapp [1986]. Another is in Borel-Wallach [1980].

The classification of the irreducible tempered representations was carried out by Knapp-Zuckerman [1977] for enough G's to complement the Langlands classification for connected linear groups. See Chapter XIV of Knapp [1986] for an exposition. Mirković [1986] extended the Knapp-Zuckerman work to general G. Theorem 11.198 combines the Langlands results with aspects of the classification of irreducible tempered representations and with a result about irreducible quotients due to Miličić [1977].

The nonuniqueness in associating data to an irreducible tempered representation, as in Remark 3 with Theorem 11.198, is addressed in detail for connected G in Knapp-Zuckerman [1982]. A notion of "nondegenerate data" allows for the formulation of a simply-stated uniqueness result. When the Langlands classification is combined with the classification of irreducible tempered representations, the combination displays some odd nonuniqueness that is difficult to classify directly. See Knapp-Zuckerman [1977] for an example. For proofs of the assertions about $|\text{Re } \nu|$ in Remark 4 with Theorem 11.198, see Proposition 8.61 of Knapp [1986]; the result first appears in Borel-Wallach [1980].

Section 9 gives a table with eight settings for classifications. Chapter 8 of Vogan [1981a] compares the classifications corresponding to the two settings $(\Gamma, L', \text{submodule})$ and $(\Gamma, VZ, \text{submodule})$.

Theorem 11.225 is stated for the functor $\mathcal{R}^j$ in Proposition 8.2.15 of Vogan [1981a], but with an extra restriction on the parameter. If ν is assumed to have real part in the negative of the closed positive Weyl chamber, then no further restriction on ν is necessary.

The second form of the Signature Theorem (Corollary 11.228) was obtained by Vogan in 1984, but its proof has not been published before this. Baldoni-Silva–Knapp [1986] make extensive use of this result, quoting it as Theorem 3.4. That paper shows that the power of the theorem comes from its applicability to all ν with real part in the closed positive Weyl chamber.

Corollary 11.229, giving Langlands parameters for certain $A_q(\lambda)$'s whose data are outside the weakly good range, is new.

The minimal K type formula is stated and proved in Theorem 7.17 of Vogan [1979] for the case that the infinitesimal character λ of the representation ξ is nonsingular, and the result is stated in the equivalent form (11.232d) without a restriction on λ in Knapp [1983a]. An unpublished sequel to Vogan [1979] leads to the formula in general, as well. The easy steps of the proof in general are in §XV.1 of Knapp [1986]; the hard step is the proof of K dominance, especially when the M parameter is singular. The sequel to the Vogan paper visualizes an indirect argument that takes into account the results of Speh-Vogan [1980], while the Knapp paper visualizes a direct but lengthy computation reducing matters to the case of nondegenerate data and using a detailed analysis of root types. Lemma 11.231 is from Knapp [1983a], and Proposition 11.249 is from Knapp-Wallach [1976].

Theorem 11.253 is one of the main results of Vogan [1979]. This theorem depends of the equivalence of "minimal" and "small" in Theorem 11.254, which is proved in §7 of the Vogan paper. The uniqueness result Lemma 11.260 is a thinly disguised form of the uniqueness in Proposition 4.1 of Vogan [1979]; the relatively short proof that we give is due to Carmona [1983].

Chapter XII

For discussion and early history of efforts toward classifying irreducible unitary representations, see the Notes of Knapp [1986] for Chapter XVI. The checkable necessary and sufficient condition for a Langlands quotient to carry an invariant Hermitian form evolved in Knapp-Zuckerman [1977] and Knapp-Speh [1982]; it appears as Theorem 16.6 in Knapp [1986]. The condition involves an integral or singular integral operator. For the Knapp-Zuckerman result, it is enough to use the intertwining operator introduced by Langlands [1973], but the Knapp-Speh result requires the broader theory of intertwining operators in Knapp-Stein [1980].

One technique for establishing positivity of the invariant form is continuity arguments. Sally [1967] and Kostant [1969] and [1975a] were the first to use such arguments; see Knapp [1986], 741–742, for more of the early history of this topic. Some representative later papers that show the success and the limitations of this method are Duflo [1979], Baldoni-Silva–Barbasch [1983], Baldoni-Silva–Knapp [1986] and [1989], Chen [1993], and Vogan [1994]. All these papers use additional methods to handle particular unitary representations, and all these papers deal only with situations where the Langlands quotients are attached to parabolic subgroups with A of dimension ≤ 2. Papers handling situations with higher-dimensional A include Vogan [1986], which classifies the irreducible unitary representations for $GL(n, \mathbb{R})$ and $GL(n, \mathbb{C})$ and $GL(n, \mathbb{H})$, and Barbasch [1989], which classifies the irreducible unitary representations for the remaining complex classical groups. Clozel [1987] gives an exposition of some of the work by Vogan and Barbasch.

Theorem 12.1 is due to Vogan and appears as Theorem 16.10 of Knapp [1986]. Theorem 12.2 is alluded to in the introduction of Vogan [1984], and, except for one point, the proof is conversational in the context of the present Chapter XI. Let us give the argument when rank $G = $ rank K. We view V as the unique Langlands quotient of some standard $X_K(\xi, \nu)$ and transform $X_K(\xi, \nu)$ to the form in the paragraph before Theorem 11.230. If we assume that the infinitesimal character has large real part on $\mathfrak{h}_{ss}$, then its inner product is large with some simple root α. The idea is to take $\mathfrak{l}$ to be built from the remaining simple roots and $\mathfrak{u}$ to be built from the positive roots not used for $\mathfrak{l}$. Since V is infinitesimally unitary, the contribution of ν to the infinitesimal character cannot be large (see Helgason-Johnson [1969] or Problem 7 in Knapp [1988], 666), and consequently α cannot be real. Since rank $G = $ rank K, the simple roots that are real span $\mathfrak{a}^*$, and $\theta\alpha - \alpha$ is therefore a sum of real roots. Then $\theta\alpha - \alpha$ is the sum of roots of $\mathfrak{l}$, and it follows that $\mathfrak{u}$ is θ stable. Thus we end up in the situation described at the beginning of §XI.10. The dominance condition is in place for the roots of $\mathfrak{u}$. Consequently $\mathcal{L}_S$ is an exact functor, and most of the theorem follows. This argument proves everything except that Z is infinitesimally unitary. To prove this part, we appeal to a converse of Theorem 9.1, which is discussed in the Notes for Chapter IX.

An overview of the remainder of the chapter, starting with Definition 12.3, is in §4 of Vogan [1993]. The definition and the unitarizability asserted in Theorem 12.4 are in Vogan [1984]; see Proposition 8.5. Dixmier algebras and induced Dixmier algebras are defined in Vogan [1990]. That paper develops some of the background needed for the proof of Theorem 12.9, which is announced but not proved in Vogan [1993]. A prototype for the proof appears as the formal part of the present Theorem 8.31 (through (8.48)).

Arthur's "special unipotent" representations are introduced in Arthur [1984], 10–11. Barbasch-Vogan [1985] redefines them in terms of annihilators. (Actually the redefinition is given only for complex G, but the real case is treated in the same way.) Proposition 5.18 of that paper verifies that "special unipotent" implies "weakly unipotent."

Other examples of irreducible unitary representations constructed in concert with algebras appear in Barbasch [1990], Binegar-Zierau [1991], Brylinski-Kostant [1994], and Gross-Wallach [1994]. See also Kazhdan-Savin [1990] and McGovern [1990] and [1994].

Appendix A

Homological algebra is developed in Cartan-Eilenberg [1956] in the context of all unital modules for a given ring with identity. All the elementary results work for a good category, but some care is required when it is no longer assumed that the ring has an identity. For example, the identity $R \otimes_R M \cong M$ may no longer be valid. More commentary about good categories appears with the Notes for Appendix C. For proofs of the associativity formulas (A.5), (A.6), and (A.7) for general rings (not just algebras), see Cartan-Eilenberg [1956], 26-27.

For Proposition A.9 and Corollary A.10 concerning complete reducibility, see Bourbaki [1958], 32–34. Those pages discuss isotypic submodules also. In Bourbaki, the term "semisimplicity" is used in place of "complete reducibility." Other authors use the term "full reducibility."

For Dixmier's result, see §2.6 of Dixmier [1974], for example; actually that book proves something more in the case that the algebra is a universal enveloping algebra.

The subject of finite length is discussed in Jacobson [1951]. The second isomorphism theorem and the result of Zassenhaus (Lemma A.21) are on p. 136, and Schreier's Theorem (Theorem A.24) is on p. 138.

Appendix B

Three book-length treatments of distributions are Schwartz [1966], Gelfand-Shilov [1964], and Hörmander [1983]. Treves [1967] contains a substantial amount of material on distributions and gives a bibliography.

A more thorough Euclidean version of the material in §§1–2 may be found in Hörmander [1983]. For smooth partitions of unity, see Hörmander's §1.4. For a discussion of distributions and support, see his Chapter II. Appropriate background material on linear topological spaces is in Chapter I of Treves [1967]. For Fubini's Theorem, see Theorem 40.4 of Treves [1967]. The proof of Theorem B.28 is taken from Theorem 2.3.5 of Hörmander [1983].

Appendix C

Two books about homological algebra are Cartan-Eilenberg [1956] and Hilton-Stammbach [1971]. The setting of a "good category" may be replaced by an "abelian category" in Appendix C. The change would offer a slight simplification in a few situations in the text where an artificial ring-and-module setting has to be introduced, but it would impose a layer of unnecessary structure in almost all other situations.

Most of the results in (C.1) through (C.36) are proved in Knapp [1988]. Detailed references are in the table at the end of these Notes.

The statement in (C.26) about computing derived functors by means of acyclic resolutions is not proved in Knapp [1988]. The idea is to set up a two-dimensional commutative diagram in the first quadrant with exact rows and columns such that the first nonzero row is the acyclic resolution of M, the first nonzero column is a projective resolution of

M, and the other columns are all projective resolutions. This is done by applying (C.34) inductively to short exact sequences extracted from the bottom nonzero row. Then we apply F to the whole diagram, obtaining a commutative diagram all of whose rows and columns but the first are exact. A complicated diagram chase or a simple application of (D.44) then establishes the desired isomorphisms of the homology of the first row with the homology of the first column.

Appendix D

Spectral sequences are due to Leray [1946] and were developed by Cartan [1951], Koszul [1950], and Serre [1951]. The development of §1 is based on Serre [1951] as amplified by Livesay [1985]. The material in §§2–3 is taken from Wallach [1982].

Item	Pages in Knapp [1988]	Item	Pages in Knapp [1988]
(C.1)	199	(C.19)	96–97
(C.2)	200–201	(C.20)	97
(C.3)	201	(C.21)	98
(C.4)	202–203	(C.22)	227–230
(C.5)	203–205	(C.23)	231–232
(C.6)	205–206	(C.24)	232–234
(C.7)	206–208	(C.25)	234
(C.8)	208–210	(C.26)	235–236
(C.9)	210–212	(C.27)	236–237
(C.10)	212–214	(C.28)	237–238
(C.11)	214–217, 221–223	(C.29)	239–240
(C.12)	224–225	(C.30)	240–241
(C.13)	218–219	(C.31)	241–243
(C.14)	219–220	(C.32)	246
(C.15)	220–221	(C.33)	246–247
(C.16)	225–226	(C.34)	249–253
(C.17)	51–53	(C.35)	254–257
(C.18)	57–59	(C.36)	257–258

REFERENCES

In the text of this book, a name followed by a bracketed year is an allusion to the list that follows. The date is followed by a letter in case of ambiguity.

Adams, J., and D. A. Vogan, L-groups, projective representations, and the Langlands classification, *Amer. J. Math.* 114 (1992), 45–138.

Adams, J., D. Barbasch, and D. A. Vogan, *The Langlands Classification and Irreducible Characters for Real Reductive Groups*, Birkhäuser, Boston, 1992.

Aguilar-Rodriguez, R., Connections between representations of Lie groups and sheaf cohomology, Ph.D. Diss., Harvard University, 1987.

Arthur, J., On some problems suggested by the trace formula, *Lie Group Representations II*, Lecture Notes in Math. 1041, Springer-Verlag, New York, 1984, pp. 1–49.

Baldoni Silva, M. W., and D. Barbasch, The unitary spectrum for real rank one groups, *Invent. Math.* 72 (1983), 27–55.

Baldoni-Silva, M. W., and A. W. Knapp, Unitary representations induced from maximal parabolic subgroups, *J. Func. Anal.* 69 (1986), 21–120.

Baldoni-Silva, M. W., and A. W. Knapp, A construction of unitary representations in parabolic rank two, *Annali della Scuola Normale Superiore di Pisa* 16 (1989), 579–601.

Barbasch, D., Unipotent representations and unitarity, *Non-Commutative Harmonic Analysis and Lie Groups*, Lecture Notes in Math. 1243, Springer-Verlag, New York, 1987, pp. 73–85.

Barbasch, D., The unitary dual for complex classical Lie groups, *Invent. Math.* 96 (1989), 103–176.

Barbasch, D., Representations with maximal primitive ideal, *Operator Algebras, Unitary Representations, Enveloping Algebras, and Invariant Theory*, A. Connes, M. Duflo, A. Joseph, and R. Rentschler (eds.), Birkhäuser, Boston, 1990, pp. 317–331.

Barbasch, D., Unipotent representations and derived functor modules, *The Penrose Transform and Analytic Cohomology in Representation Theory*, M. Eastwood, J. Wolf, and R. Zierau (eds.), Contemporary Mathematics 154, American Mathematical Society, Providence, 1993, pp. 225–238.

Barbasch, D., and D. A. Vogan, Unipotent representations of complex semisimple groups, *Ann. of Math.* 121 (1985), 41–110.

Barchini, L., Szegö mappings, harmonic forms, and Dolbeault cohomology, *J. Func. Anal.* 118 (1993), 351–406.

Barchini, L., A. W. Knapp, and R. Zierau, Intertwining operators into Dolbeault cohomology representations, *J. Func. Anal.* 107 (1992), 302–341.

Bargmann, V., Irreducible unitary representations of the Lorentz group, *Ann. of Math.* 48 (1947), 568–640.

Baston, R. J., and M. G. Eastwood, *The Penrose Transform: Its Interaction with Representation Theory*, Oxford University Press, Oxford, 1989.

Beilinson, A., and J. Bernstein, Localisation de $\mathfrak{g}$-modules, *C. R. Acad. Sci. Paris* 292 (1981), Série I, 15–18.

Benoist, Y., Un plan pour une demonstration de la dualité de Zuckerman, manuscript, December 1986.

Bernstein, I. N., I. M. Gelfand, and S. I. Gelfand, Differential operators on the base affine space and a study of $\mathfrak{g}$-modules, *Lie Groups and Their Representations (Summer School of the Bolyai János Mathematical Society)*, I. M. Gelfand (ed.), Halsted Press, New York, 1975, pp. 21–64.

Bernstein, J., Lecture "Zuckerman's construction and $\mathcal{D}$ modules," August 1, 1983.

Bien, F. V., $\mathcal{D}$-*modules and Spherical Representations*, Princeton University Press, Princeton, 1990.

Binegar, B., and R. Zierau, Unitarization of a singular representation of $SO(p, q)$, *Commun. Math. Phys.* 138 (1991), 245–258.

Borel, A., and N. Wallach, *Continuous Cohomology, Discrete Subgroups, and Representations of Reductive Groups*, Princeton University Press, Princeton, 1980.

Borho, W., and J.-L. Brylinski, Differential operators on homogeneous spaces I, *Invent. Math.* 69 (1982), 437–476.

Borho, W., and H. Kraft, Über Bahnen und deren Deformationen bei linearen Aktionen reductiver Gruppen, *Comment. Math. Helv.* 54 (1979), 61–104.

Bott, R., Homogeneous vector bundles, *Ann. of Math.* 66 (1957), 203–248.

Bourbaki, N., *Eléments de Mathématique, II - Algèbre, Chapitre 8, Modules et Anneaux Semi-Simples*, Actualités scientifiques et industrielles 1261, Hermann, Paris, 1958.

Bourbaki, N., *Eléments de Mathématique, Groupes et Algèbres de Lie, Chapitres 4, 5 et 6*, Actualités scientifiques et industrielles 1337,

Hermann, Paris, 1968.

Bruhat, F., Sur les représentations induites des groupes de Lie, *Bull. Soc. Math. France* 84 (1956), 97–205.

Brylinski, R., and B. Kostant, Minimal representations, geometric quantization, and unitarity, *Proc. Nat. Acad. Sci. USA* 91 (1994), 6026–6029.

Carmona, J., Sur la classification des modules admissibles irréductibles, *Non Commutative Harmonic Analysis and Lie Groups*, Lecture Notes in Math. 1020, Springer-Verlag, New York, 1983, pp. 11–34.

Cartan, H., Notions d'algèbre différentielle; application aux groupes de Lie et aux variétés où opère un groupe de Lie, *Colloque de topologie (espaces fibrés), Bruxelles, 1950*, Georges Thone, Liège, Masson et Cie., Paris, 1951, pp. 15–27.

Cartan, H., and S. Eilenberg, *Homological Algebra*, Princeton University Press, Princeton, 1956.

Cartier, P., Remarks on "Lie algebra cohomology and the generalized Borel-Weil theorem," by B. Kostant, *Ann. of Math.* 74 (1961), 388–390.

Casselman, W., and M. S. Osborne, The n-cohomology of representations with an infinitesimal character, *Compositio Math.* 31 (1975), 219–227.

Chen, Z., Unitary representation induced from maximal parabolic subgroups for split F_4, preprint, 1993.

Chevalley, C., Sur la classification des algèbres de Lie simples et de leurs représentations, *C. R. Acad. Sci. Paris* 227 (1948a), 1136–1138.

Chevalley, C., Sur les représentations des algèbres de Lie simples, *C. R. Acad. Sci. Paris* 227 (1948b), 1197.

Chevalley, C., Invariants of finite groups generated by reflections, *Amer. J. Math.* 77 (1955), 778–782.

Chevalley, C., *Fondements de la Géométrie Algébrique*, duplicated notes, Secrétariat Mathématique, L'Institut Henri Poincaré, Paris, 1958.

Chevalley, C., and S. Eilenberg, Cohomology theory of Lie groups and Lie algebras, *Trans. Amer. Math. Soc.* 63 (1948), 85–124.

Clozel, L., Progrès récents vers la classification du dual unitaire des groupes réductifs réels, Exposé 681, *Séminaire Bourbaki, 1986/87*, Astérisque 152–153 (1987), 229–252.

Conze-Berline, N., and M. Duflo, Sur les représentations induites des groupes semi-simples complexes, *Compositio Math.* 34 (1977), 307–336.

Demazure, M., Une démonstration algébrique d'un théorème de Bott, *Invent. Math.* 5 (1968), 349–356.

Demazure, M., A very simple proof of Bott's theorem, *Invent. Math.* 33 (1976), 271–272.

Dixmier, J., Idéaux primitifs dans l'algèbre enveloppante d'une algèbre de Lie semi-simple complexe, *C. R. Acad. Sci. Paris* 271 (1970), Série A, 134–136.

Dixmier, J., Idéaux primitifs dans l'algèbre enveloppante d'une algèbre de Lie semi-simple complexe, *C. R. Acad. Sci. Paris* 272 (1971), Série A, 1628–1630.

Dixmier, J., *Algèbres Enveloppantes*, Gauthier-Villars Editeur, Paris, 1974. English translation: *Enveloping Algebras*, North-Holland Publishing Co., Amsterdam, 1977.

Duflo, M., Sur la classification des idéaux primitifs dans l'algèbre enveloppante d'une algèbre de Lie semi-simple, *Ann. of Math.* 105 (1977), 107–120.

Duflo, M., Représentations unitaires irréductibles des groupes simples complexes de rang deux, *Bull. Soc. Math. France* 107 (1979), 55–96.

Duflo, M., Théorie de Mackey pour les groupes de Lie algébriques, *Acta Math.* 149 (1982), 153–213.

Duflo, M., and M. Vergne, Sur le foncteur de Zuckerman, *C. R. Acad. Sci. Paris* 304 (1987), Série I, 467–469.

Enright, T. J., Unitary representations for two real forms of a semisimple Lie algebra: A theory of comparison, *Lie Group Representations I*, Lecture Notes in Math. 1024, Springer-Verlag, New York, 1983, pp. 1–29.

Enright, T. J., and N. R. Wallach, Notes on homological algebra and representations of Lie algebras, *Duke Math. J.* 47 (1980), 1–15.

Flath, D., Decomposition of representations into tensor products, *Automorphic Forms, Representations and L-Functions*, Proc. Symp. Pure Math. 33, Part 1, American Mathematical Society, Providence, 1979, pp. 179–183.

Gårding, L., Note on continuous representations of Lie groups, *Proc. Nat. Acad. Sci. USA* 33 (1947), 331–332.

Gelfand, I. M., and M. A. Naimark, Unitary representations of the group of linear transformations of the straight line (Russian), *Doklady Akad. Nauk SSSR,* 55 (1947a), 571–574.

Gelfand, I. M., and M. A. Naimark, Unitary representations of the Lorentz group (Russian), *Izvestiia Akad. Nauk SSSR, Ser. Mat.* 11 (1947b), 411–504.

Gelfand, I. M., and M. A. Naimark, *Unitary Representations of the Classical Groups* (Russian), Trudy Mat. Inst. Steklov 36, Moskow-Leningrad, 1950. German translation: Akademie-Verlag, Berlin, 1957.

Gelfand, I. M., and G. E. Shilov, *Generalized Functions*, Vol. 1, Academic Press, New York, 1964.

Griffiths, P., and W. Schmid, Locally homogeneous complex manifolds, *Acta Math*. 123 (1969), 253–302.

Gross, B. H., and N. R. Wallach, A distinguished family of unitary representations for the exceptional groups of real rank = 4, *Lie Theory and Geometry: in Honor of Bertram Kostant*, J.-L. Brylinski, R. Brylinski, V. Guillemin, V. Kac (eds.), Birkhäuser, Boston, 1994, in press.

Gunning, R. C., and H. Rossi, *Analytic Functions of Several Complex Variables*, Prentice-Hall, Englewood Cliffs, N.J., 1965.

Harish-Chandra, On some applications of the universal enveloping algebra of a semisimple Lie algebra, *Trans. Amer. Math. Soc*. 70 (1951), 28–96.

Harish-Chandra, Representations of a semisimple Lie group on a Banach space I, *Trans. Amer. Math. Soc*. 75 (1953), 185–243.

Harish-Chandra, Representations of semisimple Lie groups II, *Trans. Amer. Math. Soc*. 76 (1954a), 26–65.

Harish-Chandra, Representations of semisimple Lie groups III, *Trans. Amer. Math. Soc*. 76 (1954b), 234–253.

Harish-Chandra, Representations of semisimple Lie groups V, *Proc. Nat. Acad. Sci. USA* 40 (1954c), 1076–1077.

Harish-Chandra, Representations of semisimple Lie groups IV, *Amer. J. Math*. 77 (1955), 743–777.

Harish-Chandra, Representations of semisimple Lie groups V, *Amer. J. Math*. 78 (1956a), 1–41.

Harish-Chandra, Representations of semisimple Lie groups VI, *Amer. J. Math*. 78 (1956b), 564–628.

Harish-Chandra, The characters of semisimple Lie groups, *Trans. Amer. Math. Soc*. 83 (1956c), 98–163.

Harish-Chandra, Invariant eigendistributions on a semisimple Lie group, *Trans. Amer. Math. Soc*. 119 (1965), 457–508.

Harish-Chandra, Discrete series for semisimple Lie groups II, *Acta Math*. 116 (1966), 1–111.

Harish-Chandra, Harmonic analysis on real reductive groups I, *J. Func. Anal*. 19 (1975), 104–204.

Hecht, H., D. Miličić, W. Schmid, and J. A. Wolf, Localization and standard modules for real semisimple groups I, *Invent. Math*. 90

(1987), 297–332.

Hecht, H., and W. Schmid, A proof of Blattner's conjecture, *Invent. Math.* 31 (1975), 129–154.

Helgason, S., *Differential Geometry and Symmetric Spaces*, Academic Press, New York, 1962.

Helgason, S., *Differential Geometry, Lie Groups, and Symmetric Spaces*, Academic Press, New York, 1978.

Helgason, S., and K. Johnson, The bounded spherical functions on symmetric spaces, *Advances in Math.* 3 (1969), 586–593.

Hesselink, W., The normality of closures of orbits in a Lie algebra, *Comment. Math. Helv.* 54 (1979), 105–110.

Hewitt, E., and K. A. Ross, *Abstract Harmonic Analysis*, Vol. I, Springer-Verlag, Berlin, 1963. Second edition: Springer-Verlag, New York, 1979.

Hilton, P. J., and U. Stammbach, *A Course in Homological Algebra*, Springer-Verlag, New York, 1971.

Hochschild, G., Relative homological algebra, *Trans. Amer. Math. Soc.* 82 (1956), 246–269.

Hochschild, G., and J.-P. Serre, Cohomology of Lie algebras, *Ann. of Math.* 57 (1953), 591–603.

Hörmander, L., *The Analysis of Linear Partial Differential Operators I*, Springer-Verlag, Berlin, 1983.

Howe, R., and E. C. Tan, *Non-Abelian Harmonic Analysis, Applications of $SL(2, \mathbb{R})$*, Springer-Verlag, New York, 1992.

Humphreys, J. E., *Introduction to Lie Algebras and Representation Theory*, Springer-Verlag, New York, 1972.

Hunt, G. A., A theorem of Élie Cartan, *Proc. Amer. Math. Soc.* 7 (1956), 307–308.

Jacobson, N., *Lectures in Abstract Algebra*, Vol. I, D. Van Nostrand, Princeton, 1951.

Jacobson, N., *Structure of Rings*, American Mathematical Society, Providence, 1956.

Jacobson, N., *Lie Algebras*, Interscience Publishers, New York, 1962.

Jantzen, J. C., Zur Charakterformel gewisser Darstellungen halbeinfacher Gruppen und Lie-Algebren, *Math. Zeitschrift* 140 (1974), 127–149.

Jantzen, J. C., *Moduln mit einem höchsten Gewicht*, Lecture Notes in Math. 750, Springer-Verlag, New York, 1979.

Joseph, A., Kostant's problem, Goldie rank and the Gelfand-Kirillov conjecture, *Invent. Math.* 56 (1980), 191–213.

Joseph, A., On the classification of primitive ideals in the enveloping algebra of a semisimple Lie algebra, *Lie Group Representations I*,

Lecture Notes in Math. 1024, Springer-Verlag, New York, 1983, pp. 30–76.

Kazhdan, D., and G. Savin, The smallest representations of simply laced groups, *Festschrift in Honor of I. I. Piatetski-Shapiro on the Occasion of His Sixtieth Birthday*, Part I, S. Gelbart, R. Howe, and P. Sarnak (eds.), Israel Math. Conf. Proc. 2, Weizmann Science Press of Israel, Jerusalem, 1990, pp. 209–223.

Khoroshkin, S. M., Irreducible representations of Lorentz groups (Russian), *Func. Anal. and Its Applications* 15 (1981), No. 2, 50–60. English translation: 15 (1981), 114–122.

Knapp, A. W., Minimal K-type formula, *Non Commutative Harmonic Analysis and Lie Groups*, Lecture Notes in Math. 1020, Springer-Verlag, New York, 1983a, pp. 107–118.

Knapp, A. W., Unitary representations and basic cases, *Lie Group Representations I*, Lecture Notes in Math. 1024, Springer-Verlag, New York, 1983b, pp. 77–98.

Knapp, A. W., *Representation Theory of Semisimple Groups: An Overview Based on Examples*, Princeton University Press, Princeton, 1986.

Knapp, A. W., *Lie Groups, Lie Algebras, and Cohomology*, Princeton University Press, Princeton, 1988.

Knapp, A. W., Introduction to representations in analytic cohomology, *The Penrose Transform and Analytic Cohomology in Representation Theory*, M. Eastwood, J. Wolf, and R. Zierau (eds.), Contemporary Mathematics 154, American Mathematical Society, Providence, 1993, pp. 1–19.

Knapp, A. W., and B. Speh, Status of classification of irreducible unitary representations, *Harmonic Analysis*, Lecture Notes in Math. 908, Springer-Verlag, New York, 1982, pp. 1–38.

Knapp, A. W., and E. M. Stein, Intertwining operators for semisimple groups II, *Invent. Math.* 60 (1980), 9–84.

Knapp, A. W., and D. A. Vogan, Duality theorems in relative Lie algebra cohomology, duplicated notes, SUNY Stony Brook and Massachusetts Institute of Technology, 1986.

Knapp, A. W., and N. R. Wallach, Szegö kernels associated with discrete series, *Invent. Math.* 34 (1976), 163–200; 62 (1980), 341–346.

Knapp, A. W., and G. Zuckerman, Classification theorems for representations of semisimple Lie Groups, *Non-Commutative Harmonic Analysis*, Lecture Notes in Math. 587, Springer-Verlag, New York, 1977, pp. 138–159.

Knapp, A. W., and G. J. Zuckerman, Classification of irreducible tempered representations of semisimple groups, *Ann. of Math.*

116 (1982), 389–501. (See also 119 (1984), 639.)

Kostant, B., The principal three-dimensional subgroup and the Betti numbers of a complex simple Lie group, *Amer. J. Math.* 81 (1959), 973–1032.

Kostant, B., Lie algebra cohomology and the generalized Borel-Weil theorem, *Ann. of Math.* 74 (1961), 329–387.

Kostant, B., Lie group representations on polynomial rings, *Amer. J. Math.* 85 (1963), 327–404.

Kostant, B., Orbits, symplectic structures and representation theory, *Proceedings of the United States–Japan Seminar in Differential Geometry, Kyoto, Japan, 1965*, Nippon Hyoronsha, Co., Tokyo, 1966, p. 71.

Kostant, B., On the existence and irreducibility of certain series of representations, *Bull. Amer. Math. Soc.* 75 (1969), 627–642.

Kostant, B., On the existence and irreducibility of certain series of representations, *Lie Groups and Their Representations (Summer School of the Bolyai János Mathematical Society)*, I. M. Gelfand, (ed.), Halsted Press, New York, 1975a, pp. 231–329.

Kostant, B., On the tensor product of a finite and an infinite dimensional representation, *J. Func. Anal.* 20 (1975b), 257–285.

Koszul, J.-L., Homologie et cohomologie des algèbres de Lie, *Bull. Soc. Math. France* 78 (1950), 65–127.

Kraft, H., and C. Procesi, Closures of conjugacy classes of matrices are normal, *Invent. Math.* 53 (1979), 227–247.

Kraft, H., and C. Procesi, Minimal singularities in GL_n, *Invent. Math.* 62 (1981), 503–515.

Kraft, H., and C. Procesi, On the geometry of conjugacy classes in classical groups, *Comment. Math. Helv.* 57 (1982), 539–602.

Kunze, R. A., and E. M. Stein, Uniformly bounded representations and harmonic analysis of the 2×2 real unimodular group, *Amer. J. Math.* 82 (1960), 1–62.

Kunze, R. A., and E. M. Stein, Uniformly bounded representations II, *Amer. J. Math.* 83 (1961), 723–786.

Kunze, R. A., and E. M. Stein, Uniformly bounded representations III, *Amer. J. Math.* 89 (1967), 385–442.

Langlands, R. P., Dimension of spaces of automorphic forms, *Algebraic Groups and Discontinuous Subgroups*, Proc. Symp. Pure Math. 9, American Mathematical Society, Providence, 1966, pp. 253–257.

Langlands, R. P., On the classification of irreducible representations of real algebraic groups, mimeographed notes, Institute for Advanced Study, Princeton, N.J., 1973. See *Representation Theory and Harmonic Analysis on Semisimple Lie Groups*, P. J.

Sally and D. A. Vogan (eds.), Math. Surveys and Monographs 31, American Mathematical Society, Providence, 1989, pp. 101–170.

Lepowsky, J., Algebraic results on representations of semisimple Lie groups, *Trans. Amer. Math. Soc.* 176 (1973), 1–44.

Lepowsky, J., and G. W. McCollum, On the determination of irreducible modules by restriction to a subalgebra, *Trans. Amer. Math. Soc.* 176 (1973), 45–57.

Leray, J., Structure de l'anneau d'homologie d'une représentation, *C. R. Acad. Sci. Paris* 222 (1946), 1419–1422.

Lipsman, R. L., On the characters and equivalence of continuous series representations, *J. Math. Soc. Japan* 23 (1971), 452–480.

Livesay, G. R., Course of Lectures "Topology," Cornell University, Ithaca, N.Y., Fall 1985.

Loomis, L. H., *An Introduction to Abstract Harmonic Analysis*, D. Van Nostrand, New York, 1953.

Mackey, G. W., A theorem of Stone and von Neumann, *Duke Math. J.* 16 (1949a), 313–326.

Mackey, G. W., Imprimitivity for representations of locally compact groups I, *Proc. Nat. Acad. Sci. USA* 35 (1949b), 537–545.

Mackey, G. W., Imprimitivité pour les représentations des groupes localement compacts II, *C. R. Acad. Sci. Paris* 230 (1950a), 808–809.

Mackey, G. W., Imprimitivité pour les représentations des groupes localement compacts III, *C. R. Acad. Sci. Paris* 230 (1950b), 908–909.

Mackey, G. W., On induced representations of groups, *Amer. J. Math.* 73 (1951), 576–592.

Mackey, G. W., Induced representations of locally compact groups I, *Ann. of Math.* 55 (1952), 101–139.

Mackey, G. W., Harmonic analysis and unitary group representations: the development from 1927 to 1950, *L'emergence de l'analyse harmonique abstraite (1930–1950)*, Cahiers des Séminaire d'Histoire des Mathématiques, 2e - Volume 2, Université Pierre et Marie Curie, Paris, 1992, pp. 13–42.

Mautner, F. I., Unitary representations of locally compact groups II, *Ann. of Math.* 52 (1950), 528–556.

McGovern, W. M., Dixmier algebras and the orbit method, *Operator Algebras, Unitary Representations, Enveloping Algebras, and Invariant Theory*, A. Connes, M. Duflo, A. Joseph, and R. Rentschler (eds.), Birkhäuser, Boston, 1990, pp. 397–416.

McGovern, W. M., Rings of regular functions on nilpotent orbits II, *Commun. in Algebra* 22 (1994), 765–772.

Miličić, D., Asymptotic behaviour of matrix coefficients of the discrete series, *Duke Math. J.* 44 (1977), 59–88.

Mirković, I., Classification of irreducible tempered representations of semisimple groups, Ph.D. Diss., University of Utah, Salt Lake City, 1986.

Murray, M. K., and J. W. Rice, The generalised Borel-Weil theorem and integral geometry, *J. London Math. Soc.* 39 (1989), 121–128.

Nachbin, L., *The Haar Integral*, D. Van Nostrand, Princeton, 1965.

Narasimhan, M. S., and K. Okamoto, An analogue of the Borel-Weil-Bott theorem for hermitian symmetric pairs of non-compact type, *Ann. of Math.* 91 (1970), 486–511.

Nelson, E., Analytic vectors, *Ann. of Math.* 70 (1959), 572–615.

Parthasarathy, K. R., R. Ranga Rao, and V. S. Varadarajan, Representations of complex semi-simple Lie groups and Lie algebras, *Ann. of Math.* 85 (1967), 383–429.

Rader, C., *Spherical functions on a semi-simple Lie group*, Lecture Notes in Representation Theory, Department of Mathematics, University of Chicago, Chicago, 1976.

Rawnsley, J., W. Schmid, J. A. Wolf, Singular unitary representations and indefinite harmonic theory, *J. Func. Anal.* 51 (1983), 1–114.

Sahi, S., and E. M. Stein, Analysis in matrix space and Speh's representations, *Invent. Math.* 101 (1990), 379-393.

Sally, P. J., Analytic continuation of the irreducible unitary representations of the universal covering group of SL(2,$\mathbb{R}$), *Memoirs Amer. Math. Soc.* 69 (1967).

Schlichtkrull, H., Eigenspaces of the Laplacian on hyperbolic spaces: composition series and integral transforms, *J. Func. Anal.* 70 (1987), 194–219.

Schmid, W., Homogeneous complex manifolds and representations of semisimple Lie groups, Ph.D. Diss., University of California, Berkeley, 1967. See *Representation Theory and Harmonic Analysis on Semisimple Lie Groups*, P. J. Sally and D. A. Vogan (eds.), Math. Surveys and Monographs 31, American Mathematical Society, Providence, 1989, pp. 223–286.

Schmid, W., On the realization of the discrete series of a semisimple Lie group, *Complex Analysis, 1969*, H. L. Resnikoff and R. O. Wells (eds.), Rice University Studies, Vol. 56, No. 2, 1970, pp. 99–108.

Schmid, W., On a conjecture of Langlands, *Ann. of Math.* 93 (1971), 1–42.

Schmid, W., On the characters of the discrete series: the Hermitian symmetric case, *Invent. Math.* 30 (1975), 47–144.

Schmid, W., L^2-cohomology and the discrete series, *Ann. of Math.* 103 (1976), 375–394.

Schmid, W., Boundary value problems for group invariant differential equations, *Elie Cartan et les mathématiques d'aujourd'hui (Lyon, 25–29 juin 1984)*, Astérisque, Numéro hors série (1985), 311–321.

Schwartz, L., *Théorie des distributions*, 2 volumes, Actualités scientifiques et industrielles 1091 and 1122, Hermann et Cie, Editeurs, Paris, 1950–51. Second edition: Actualités scientifiques et industrielles 1245 and 1122, 1957–59. New edition: combined volume, 1966.

Segal, I. E., Hypermaximality of certain operators on Lie groups, *Proc. Amer. Math. Soc.* 3 (1952), 13–15.

Serre, J.-P., Homologie singulière des espaces fibrés, *Ann. of Math.* 54 (1951), 425–505.

Serre, J.-P., Représentations linéaires et espaces homogènes Kählérians des groupes de Lie compacts, Exposé 100, *Séminaire Bourbaki, 6ᵉ année, 1953/54*, Inst. Henri Poincaré, Paris, 1954. Reprinted with corrections: 1965.

Shapovalov, N. N., On a bilinear form on the universal enveloping algebra of a complex semisimple Lie algebra (Russian), *Func. Anal. and Its Applications* 6 (1972), No. 4, 65–70. English translation: 6 (1972), 307–312.

Speh, B., Unitary representations of SL(n,ℝ) and the cohomology of congruence subgroups, *Non Commutative Harmonic Analysis and Lie Groups*, Lecture Notes in Math. 880, Springer-Verlag, New York, 1981, pp. 483–505.

Speh, B., and D. A. Vogan, Reducibility of generalized principal series representations, *Acta Math.* 145 (1980), 227–299.

Steenrod, N., *The Topology of Fibre Bundles*, Princeton University Press, Princeton, 1951.

Stone, M. H., Linear transformations in Hilbert space III, *Proc. Nat. Acad. Sci. USA* 16 (1930), 172-175.

Sweedler, M. E., *Hopf Algebras*, W. A. Benjamin, New York, 1969.

Tirao, J. A., and J. A. Wolf, Homogeneous holomorphic vector bundles, *Indiana U. Math. J.* 20 (1970), 15–31.

Tits, J., Sur certaines classes d'espaces homogènes de groupes de Lie, Acad. Roy. Belg. Cl. Sci. Mém. Coll. 29 (1955), No. 3.

Treves, F., *Topological Vector Spaces, Distributions and Kernels*, Academic Press, New York, 1967.

Trombi, P. C., The tempered spectrum of a real semisimple Lie group, *Amer. J. Math.* 99 (1977), 57–75.

Varadarajan, V. S., On the ring of invariant polynomials on a semisimple Lie algebra, *Amer. J. Math.* 90 (1968), 308–317.

Varadarajan, V. S., *Lie Groups, Lie Algebras, and Their Representations*, Prentice-Hall, Englewood Cliffs, N.J., 1974. Second edition: Springer-Verlag, New York, 1984.

Verma, D.-N., Structure of certain induced representations of complex semisimple Lie algebras, *Bull. Amer. Math. Soc.* 74 (1968), 160–166, 628.

Vogan, D. A., The algebraic structure of the representation of semisimple Lie groups I, *Ann. of Math.* 109 (1979), 1–60.

Vogan, D. A., *Representations of Real Reductive Lie Groups*, Birkhäuser, Boston, 1981a.

Vogan, D. A., Singular unitary representations, *Non Commutative Harmonic Analysis and Lie Groups*, Lecture Notes in Math. 880, Springer-Verlag, New York, 1981b, pp. 506–535.

Vogan, D. A., Irreducible characters of semisimple Lie groups III, *Invent. Math.* 71 (1983), 381–417.

Vogan, D. A., Unitarizability of certain series of representations, *Ann. of Math.* 120 (1984), 141–187.

Vogan, D. A., The unitary dual of GL(n) over an archimedean field, *Invent. Math.* 83 (1986), 449–505.

Vogan, D. A., *Unitary Representations of Reductive Lie Groups*, Princeton University Press, Princeton, 1987a.

Vogan, D. A., Representations of reductive Lie groups, *Proceedings of the International Congress of Mathematicians 1986*, Vol. I, American Mathematical Society, Providence, 1987b, pp. 245–266.

Vogan, D. A., Irreducibility of discrete series representations for semisimple symmetric spaces, *Representations of Lie Groups, Kyoto, Hiroshima, 1986*, Advanced Studies in Pure Math. 14, Kinokuniya Company, Ltd., Tokyo, 1988a, pp. 191–221.

Vogan, D. A., Noncommutative algebras and unitary representations, *The Mathematical Heritage of Hermann Weyl*, Proc. Symp. Pure Math. 48, American Mathematical Society, Providence, 1988b, pp. 35–60.

Vogan, D. A., Dixmier algebras, sheets, and representation theory, *Operator Algebras, Unitary Representations, Enveloping Algebras, and Invariant Theory*, A. Connes, M. Duflo, A. Joseph, and R. Rentschler (eds.), Birkhäuser, Boston, 1990, pp. 333–395.

Vogan, D. A., Associated varieties and unipotent representations, *Harmonic Analysis on Reductive Groups*, W. Barker and P. Sally (eds.), Birkhäuser, Boston, 1991, pp. 315–388.

Vogan, D. A., Unipotent representations and cohomological induction, *The Penrose Transform and Analytic Cohomology in Representation Theory*, M. Eastwood, J. Wolf, and R. Zierau (eds.), Contemporary Mathematics 154, American Mathematical Society, Providence, 1993, pp. 47–70.

Vogan, D. A., The unitary dual of G_2, *Invent. Math.* 116 (1994), 677–791.

Vogan, D. A., and G. J. Zuckerman, Unitary representations with non-zero cohomology, *Compositio Math.* 53 (1984), 51–90.

Wallach, N. R., Lecture Series "Representations constructed by derived functors," Institute for Advanced Study, Princeton, N. J., Oct.–Nov. 1982.

Wallach, N. R., On the unitarizability of derived functor modules, *Invent. Math.* 78 (1984), 131–141.

Wallach, N., *Real Reductive Groups I*, Academic Press, San Diego, 1988.

Wallach, N., *Real Reductive Groups II*, Academic Press, San Diego, 1992.

Warner, G., *Harmonic Analysis on Semi-Simple Lie Groups*, 2 vol., Springer-Verlag, New York, 1972.

Weil, A., *L'intégration dans les groupes topologiques et ses applications*, Actualités scientifiques et industrielles 869, Hermann et Cie, Editeurs, Paris, 1940. Second edition: Actualités scientifiques et industrielles 869–1145, 1951.

Wells, R. O., *Differential Analysis on Complex Manifolds*, Prentice-Hall, Englewood Cliffs, N.J., 1973. Second edition: Springer-Verlag, New York, 1980.

Wigner, D., Sur l'homologie relative des algèbres de Lie et une conjecture de Zuckerman, *C. R. Acad. Sci. Paris* 305 (1987), Série I, 59–62.

Wong, H. W., Dolbeault cohomologies and Zuckerman modules associated with finite rank representations, Ph.D. Diss., Harvard University, 1992.

Zariski, O., and P. Samuel, *Commutative Algebra*, Vol. I, D. Van Nostrand, Princeton, 1958.

Zariski, O., and P. Samuel, *Commutative Algebra*, Vol. II, D. Van Nostrand, Princeton, 1960.

Želobenko, D. P., The analysis of irreducibility in a class of elementary representations of a complex semisimple Lie group (Russian), *Izvestiia Akad. Nauk SSSR, Ser. Mat.* 32 (1968), 108–133.

Zierau, R., Unitarity of certain Dolbeault cohomology representations, *The Penrose Transform and Analytic Cohomology in Representation Theory*, M. Eastwood, J. Wolf, and R. Zierau (eds.), Contemporary Mathematics 154, American Mathematical Society, Providence, 1993, pp. 239–259.

Zuckerman, G., Tensor products of finite and infinite dimensional representations of semisimple Lie groups, *Ann. of Math.* 106 (1977), 295–308.

Zuckerman, G. J., Lecture Series "Construction of representations via derived functors," Institute for Advanced Study, Princeton, N. J., Jan.–Mar. 1978.

INDEX OF NOTATION

See also the list of Standard Notation on page xvii. In the list below, Latin, German, and script letters appear together and are followed by Greek symbols, special subscripts and superscripts, and non-letters.

933

INDEX